Human Biology

Concepts and Current Issues

Michael D. Johnson
West Virginia University

An Imprint of Addison Wesley Longman, Inc.
San Francisco Boston New York
Capetown Hong Kong London Madrid Mexico City
Montreal Munich Paris Singapore Sydney Tokyo Toronto

Publisher: Daryl Fox
Senior Project Editor: Carol Pritchard-Martinez
Development Editor: Alice Fugate

Market Development Manager: Deirdre McGill
Senior Marketing Manager: Lauren Harp

Managing Editor: Wendy Earl
Production Supervisor: Sharon Montooth
Art and Design Director: Bradley Burch
Text and Cover Designer: Kathleen Cunningham
Art Coordinator: Kelly Murphy
Photo Researcher: Diane Austin
Artists: Steve Oh, Laura Southworth, and Imagineering Scientific and Technical Artworks, Inc.
Compositor: GTS Graphics, Inc.

Cover photograph: Professor Dr. Gunther von Hagens, Institute für Plastination, Heidelberg, Germany.

The credits section of this text begins on page C1 and is considered an extension of the copyright page.

Library of Congress Cataloging-in-Publication Data

Johnson, Michael D.
Human biology: concepts and current issues / Michael D. Johnson.
p. cm.
Includes indes.
ISBN 0-8053-5071-3
1. Human biology. I. Title

QP34.5.J645 2000
612—dc21 00-064467

ISBN 0-0853-5071-3
10 9 8 7 6 5 4 3 2 1–QWD–04 03 02 01 00

1301 Sansome Street
San Francisco, California 94111

ABOUT THE AUTHOR

Michael D. Johnson teaches physiology and human biology at West Virginia University. He has been honored with several teaching awards during his more than twenty years on the faculty, including the West Virginia University Foundation Outstanding Teacher award and the Distinguished Teacher Award of the School of Medicine. An accomplished researcher, Dr. Johnson earned his Ph.D. in Physiology at the University of Michigan, completed a Postdoctoral Research Fellowship at Harvard Medical School, and has published over 25 papers on renal, cardiovascular, and endocrine physiology.

Whether teaching undergraduates or medical students, Michael Johnson has a keen interest in instilling in students an appreciation for the dynamic nature of science and an understanding of how new scientific knowledge raises ethical, political, economic, and social issues. He is a member of the teaching section of the American Physiological Society, the Human Anatomy and Physiology Society, the National Association of Biology Teachers, and the American Association for the Advancement of Science.

ABOUT THE EDITOR

Alice E. Fugate has authored more than 40 books and articles on science, health, and business. Her titles include *Don't Snore Anymore* (Crown Publishers), *The Physician's Guide to Treating Hypertension and Other Cardiovascular Conditions* (Berkley Books), *The Complete Idiot's Guide to Beautiful Skin* (MacMillan) and *As We Age: The Total Guide to Heath and Wellness* (Reader's Digest Books). Currently a contributing editor to Intercom Magazine, Alice has taught at Washington University and St. Louis Community College. Her publications have won awards from the National Quality Assurance Institute, The Women's Commerce Association, and the Society for Technical Communication.

BRIEF CONTENTS

DETAILED CONTENTS

Additional resources included with your textbook purchase.

As the owner of this book, you can get the most from your financial investment and study time by using a variety of study aids that accompany the book at no additional charge.

Each new copy of the text includes the following items:

1. A 12-month, prepaid subscription to **The Human Biology Place** web site*
2. Interactive Physiology for ***Human Biology*** **CD-ROM***

How to activate your prepaid subscription to the Human Biology Place web site:

1. Point your web browser to **www.humanbiology.com**
2. Click "Register Here."
3. Enter your pre-assigned access code exactly as it appears below:

ACCESS CODE WSHB-WACKO-CREPY-WHELM-STAIR-OBESE

4. Click "Submit."
5. Complete the online registration form to establish your personal user ID and password.
6. Once your personal ID and password are confirmed, go back to **www.humanbiology.com** to enter the site with your new ID and password.

Your access code can be used only once to establish your subscription, which is not transferable.

If you have purchased a used copy of this book, your access code may not be valid. If your instructor is recommending or requiring use of the **Human Biology Place**, or you would like to have access to the additional study tools available on the **Human Biology Place** web site, you can find more information on purchasing a subscription directly online at **www.humanbiology.com**. You also have the option of purchasing a subscription at your local college bookstore by placing an order for ISBN 0-8053-5089-6.

The **Human Biology Place** provides an interactive environment for understanding key concepts as well as preparing for exams. For each chapter Concept Review activities and electronic self-tests are provided. In the Issues Forum you can pursue different perspectives on issues raised in the text and can post your own commentary on them. You'll find lots of Web-based research projects (WebQuests), collaborative learning activities, news articles, and links to other great web sites as well as a glossary with audio pronunciation. A feature called Syllabus Manager allows instructors to create their own online syllabi.

Turn the page to learn more . . .

* These items are guaranteed only with the purchase of a **new** copy of ***Human Biology***. Please ask your textbook retailer for more information.

How to contact Customer Service and Technical Support:

Instructors can obtain examination copies by contacting their local Benjamin Cummings - Addison Wesley sales representative. To purchase additional study aids, including the Student Study Guide, visit our web site (**www.aw.com/bc**), or call Addison Wesley Customer Service at 800-282-0693.

For technical support for the **Human Biology Place** web site or **Interactive Physiology for *Human Biology* CD-ROM**, please visit the technical support web site at **http://www.awlonline.com/techsupport**. For further technical support on the CD-ROM, please send an email to **media.support@pearsoned.com**, and for web site assistance, please send an email to **online.support@pearsoned.com**. In your email message, please provide a detailed description of your computer system and the technical problem.

If you would like to receive regular news updates on special promotions for ***Human Biology*** please subscribe to our email service by sending a request to **question@awl.com.**

How to contact us:

We invite your comments and suggestions via email at **question@awl.com**. As you complete your human biology course, please take a few minutes to fill out the postage-paid questionnaire at the end of the text.

In the meantime, we wish you great success in your human biology course!

Benjamin Cummings Science
1301 Sansome Street
San Francisco, CA 94111

Chapter 10 continues

Chapter 19 continues

Chapter 21 continues

PREFACE

In 1997 an adult animal was cloned (Dolly the sheep), raising the distinct possibility that humans might be cloned in our lifetime. It was a feat that, until that time, many biologists had considered impossible. In 2000 the first draft of the entire human genome was completed—all 3 billion base pairs, representing the set of instructions for creating a human being. How will these events affect us? What will come next, and how will we cope with the speed of change in the decades to come?

Those of us who find these issues fascinating—and yes, even exciting!—must do our very best to help our students feel comfortable with, and interested in, science. Science is too much fun and far too important in our lives to be left to scientists. With that in mind, I wrote this book.

The Focus Is on the Student

This book was written expressly for students who do not yet have a strong background in science, so that they, too, might share in the joy and wonder of science and of human biology. Every effort was made to make the book accurate and up-to-date while keeping it inviting, accessible, and easy to read. Given the wide range of student background and ability in the typical science course for non-science majors, it can be challenging to engage students and take them to the next level. There are, however, many creative ways to reach out to students who are intimidated by science or who are struggling to overcome language or learning barriers. For example, this text makes liberal use of familiar analogies to connect concepts with everyday experiences. The look and feel of the text is intentionally like that of a news magazine, peppered with short features likely to be of interest to the student and with a strong visual appeal.

With human biology as its subject, the text focus naturally is on human structure and function, health and wellness. These subjects provide a great avenue for introducing science, as students are naturally interested in how their bodies function and often are curious about disorders and diseases. We capitalize on this curiosity with Health Watch boxes highlighting timely health topics. In addition, organ system chapters generally conclude with a section covering the more common human diseases and disorders.

The didactic structure of the text is designed both to draw students in and to make learning easy and rewarding. We begin by asking interesting and relevant questions—questions about which students themselves may have wondered—on the chapter-opening pages. These not only motivate students to read on, but also reinforce the idea that science always begins with a question. To help students stay on track, concept-style headings, recaps within the text, and extensive end-of-chapter material ensure that key lessons stand out and underscore the relevance of the information students are working to master.

A specific goal of this book is to help students become sufficiently comfortable with and interested in science that they begin to use science in their daily lives. The book shows by example that science is both a body of knowledge *and* a process for acquiring new knowledge that *all* persons can learn to use and enjoy. Directions in Science boxes reveal how scientific knowledge changes over time. Uncertainty in science is embraced and explained. The very simple Try It Yourself features bring an element of active learning to the course by encouraging students to test facts or concepts themselves.

Current Issues essays are the centerpiece of the text. Highlighting recent ethical, social, political, and economic issues raised by new findings in science, the Current Issues essays raise awareness of important scientific controversies facing society. Readers are encouraged to continue the dialog in the Issues Forum area of the text's website, The Human Biology Place.

The Organization Fits the Course

A fairly standard format for courses in human biology has been established across the country, which this book was designed to accommodate. Chapters 1 and 2 introduce science and the scientific method and provide basic coverage of the principles of chemistry to students for whom these subjects are new. Chapters 3 and 4 introduce the idea of levels of organization in living organisms. Chapter 3 describes cells, and Chapter 4 builds on knowledge of how cells function to introduce tissues, organs, and organ systems. Next, Chapters 5–16 review the structures and functions of the human organ systems, using a typical organ systems approach. Chapter 17 describes how cells normally reproduce and differentiate, setting the stage for Chapter 18, which is devoted exclusively to cancer. Genetics and inheritance are discussed in Chapter 19, followed by Chapter 20 on human development and aging. The last two chapters look at humans at the level of the biosphere. Chapter 21 discusses the origins of life and the evolution of humans, and Chapter 22 describes human populations and how humans fit into (and influence) ecosystems.

It is always a challenge to decide what to include and when to introduce topics. Surveys and reviews consistently showed that chapters should present the material in discrete, digestible, carefully paced units. Accordingly, we break the typically lengthy subject of the cardiovascular system into two chapters (Chapter 7 covers blood and Chapter 8 covers the heart and blood vessels), and offer separate chapters on the skeleton and on muscle. The desire to present topics in the most logical context led to the decision to cover HIV and AIDS in Chapter 9 (The Immune System), where its danger as a threat to the immune system is best understood. Coverage of sexually transmitted diseases (with mention of HIV) is incorporated into Chapter 16 (Reproductive Systems), right after the subjects of birth control and infertility. Cancer is afforded a chapter of its own (Chapter 18) because of its importance and because cell reproduction and differentiation (Chapter 17) are weighty subjects by themselves. DNA technology and genetic engineering quite naturally fell into Chapter 19 following a discussion of genetics and inheritance. Finally, evolution and ecosystems are each covered in single chapters so that the book is manageable within one semester.

Unifying Themes Tie the Subjects Together

Recurrent themes build the connections between the various subjects. **Homeostasis**, the state of dynamic equilibrium in which the internal environment of an organism is maintained fairly constant, is one of those recurrent themes. Indeed, the concept of homeostasis is *the* unifying theme in any course that seeks to understand how an organism functions. The concept of homeostasis ties in with another recurrent theme: that **structure and function are intimately related**. Structure/function relationships are the very core of the study of anatomy and physiology, and both of these fields in turn rely on the most unifying concept in all of biology, **evolution**. Only in the context of evolution can anatomy and physiology be fully understood; without the concept of evolution, very little in biology makes sense.

Several themes related to science run throughout the book. The theme that **science is a process** for gaining new information is explained in Chapter 1, but is not thereafter forgotten. Directions in Science boxes, Try It Yourself elements, and numerous examples within the text show the scientific method in action, highlight examples of scientific progress over time, and demonstrate the process of formulating and testing hypotheses. With practice and reinforcement, students become comfortable with science and are able to use the scientific method themselves.

This book also explores the **role of science in society.** The Current Issues essays in particular consider how new scientific knowledge can impact society. I hope students will formulate their own positions on these timely and often controversial topics, and will see that their personal opinions (as long as they are well-informed) do matter. Thus the theme of **personal and societal choices** appears throughout the book. Students will learn that by either omission or commission, we (both as individuals and as a society) are constantly making choices regarding the use of science and scientific information.

Features That Promote Learning

As mentioned, this book is designed with the student in mind. Some features facilitate learning; others reinforce themes or highlight relevance.

Chapter-Opening Questions Many texts begin with a daunting list of factually oriented "learning objectives." Although the mastery of a certain body of knowledge is certainly important, we may question whether this is the best way to motivate students. We would rather have students get the message that science (and human biology) addresses subjects that are inherently interesting and useful to them so that they are willing to become lifelong learners. Accordingly, in this book each chapter opens with a series of questions of a type that students might ask. They are repeated in the text where the question is discussed.

Concept-Style Headings and Section Recaps A wise teacher once told me, "First you tell them what you're going to tell them; then you tell them; then you tell them what you told them." With this axiom in mind, the headings in this book express concepts whenever possible, and short summaries called **Recap** are inserted periodically to reiterate the main lesson. These two elements can be used to help students preview and review the text as well.

Spotlights on Structure and Function Selected passages are marked to highlight examples of the interrelationship of structure and function. These frequent reminders help students to appreciate this recurring theme in biology and to contemplate the role evolution plays in the development of every living structure.

Try It Yourself Boxes Nothing reinforces a lesson like being able to prove it to yourself. Most chapters include active-learning exercises called **Try It Yourself,** which encourage students to verify facts for themselves rather than take them on faith. For example, it's easy to quickly demonstrate the existence of the blind spot in the eye (p. 308) or that ligaments and tendons stabilize moveable joints (p. 119).

Boxes on Science and Health As mentioned, it's important that students see the relevance of what they are learning to circumstances in their own lives. **Current Issues** boxes cover controversial subjects of ethical, social, political, and economic importance. We ask, for example: Should human stem cells derived from human embryos or fetuses be used for research or transplantation? What are some of the ethical, social, and legal consequences of the Human Genome Project? These boxes help develop students' critical thinking skills as well as encourage participation in societal decisions. Boxes called **Directions in Science** reveal that science is an ongoing process. They may describe how certain scientific discoveries were made, highlight controversies raised by new discoveries, or demonstrate how our knowledge has changed or is still changing. They cover such issues as how our understanding of motion sickness changed over time, and how advances in the treatment of ulcers were made.

Health Watch boxes tap into students' natural interest in such questions as "Should I breast-feed my baby?" and "Are cell phones safe?" Some provide more in-depth discussion of topics broached in the text, such as artherosclerosis and resistance to antibiotics.

End-of-Chapter Pedagogy Upon concluding each chapter students can review the key lessons of each section in the **Chapter Summary** and test their recall of the **Terms You Should Know.** This is followed by an extensive list of basic objective questions in the **Test Yourself** section, which will give students a chance to see what they know and what they need to review. Answers to Test Yourself questions can be found at the back of the book. **Concept Review** questions call for recall of basic facts and concepts, and **Apply What You Know** questions test students' ability to use what they have learned in a new context. Some instructors will choose to use the more open-ended Apply What You Know questions to stimulate class discussion. A unique feature called **Case In Point** follows, which asks students to apply what they've learned to an actual situation related to human biology reported in news or research. Chapters conclude with a brief list of **References and Additional Reading**. The list highlights highly readable popular lay-science books, history, and even fiction related to human biology.

If You Can't Cover It All

There is never enough time to teach all that is interesting, exciting, and relevant about human biology in one semester. Even the best instructors have to make choices. Fortunately, this book allows for a certain degree of flexibility. Instructors wishing to emphasize the basics of human anatomy and physiology or focus on the medical aspects of human biology could omit or de-emphasize the last two chapters. Instructors should feel free to present the organ system chapters in a different order if they feel more comfortable doing so. More environmentally oriented instructors may want to allow sufficient time for the last two chapters, even if it means that they must move quickly or selectively through the organ system chapters or skip the chapter on cancer. Within chapters, Diseases and Disorders sections could be omitted or considered optional since they are not necessary for a full understanding of the material. Those interested in a more cellular or molecular approach might want to give greater emphasis to chapters 4 and 17–20 and move quickly through the organ systems chapters. All of these approaches are equally valid.

However much you cover, dig in and enjoy your course!

Michael D. Johnson

ACKNOWLEDGMENTS

This book is a product of the hard work and dedication of dozens of people at Benjamin Cummings Science, all of whom shared a common goal: to produce the very best Human Biology book possible. It is also a product of hundreds of insightful faculty and student reviewers around the country. I am honored to have had their help.

Three individuals in particular stand out as having had a major hand in shaping the book. Benjamin Cummings' Publisher **Daryl Fox**, an incredibly unique individual in his own right, shared my vision for the book and had the foresight to allow me to proceed with this project. Along the way he patiently guided me toward a book that would fit the market. Senior Project Editor **Carol Pritchard-Martinez** managed all aspects of the development of the book, as well as its media and print supplements. She quickly learned how to get the most out of me while making me feel good about it. If anyone could juggle ten balls at once, while helping those around her juggle theirs as well, it would be Carol. As a result of her efforts, the book and all of its supplements were finished on time. Development Editor **Alice Fugate** is a writer without parallel. Her sense of organization of the written word is superb. Alice reworked and reorganized my first-draft chapters and provided insightful analyses of the reviews. She also developed some outstanding original material for this book, including many of the Case in Point features. The result is a seamless flow of science and prose that seems to come from one voice but in reality comes from both Alice and me. Thank you, Alice.

Backing up these key individuals was a whole team of hardworking and talented people. **Lauren Fogel** had a strong hand in the early stages of the project, shaping the chapter organization and providing the necessary push to get me going initially. Now Executive Producer of multimedia for Benjamin Cummings, she has provided vision and support for the development of the media for this text. **Claire Brassert**, **Susan Teahan**, and **Aura Mayo** commissioned the many reviews for this text and kept all of us organized.

The marketing team, led by **Lauren Harp** and **Stacy Treco**, participated throughout the creative process to ensure that the book would be responsive to the needs of teachers and students alike. The Market Development Manager, **Deirdre McGill,** held focus groups, met with instructors, and worked tirelessly to set up class tests to ensure the book's development would be influenced by extensive student input. Deirdre and Lauren have been my cheerleaders, giving me enthusiastic encouragement just when it seemed that I needed it most.

The production team at **Wendy Earl Productions** has a stellar reputation, and lived up to it. **Wendy** set the tone for the production process, ably supported by **Bradley Burch.** One of Brad's greatest contributions was commissioning the brilliant text designer, **Kathleen Cunningham**, to design the text and its cover. The exceptional chapter openers also were developed by Kathleen. Much hard work, organizational skill and patience was required of the production editors, **David Rich** and **Sharon Montooth**, who were instrumental in coordinating the copyediting (kudos to copyeditor **Richard Reser**), typesetting and layout of the text. While many artists were involved, the first person I'd like to thank for the wonderful art program in this text is Art Coordinator **Kelly Murphy**. She examined every piece of art and every photograph carefully for both style and content. Her organization was so good that she could put her finger on any piece of art within seconds, no matter when I called. The majority of the artwork in this book was produced by **Imagineering;** it has been a great pleasure to work with such a professional studio. I'd also like to thank artists **Laura Southworth** and **Steve Oh** for developing several of the more complex illustrations in this text. Photoresearcher **Diane Austin** worked tirelessly to find just the right photos, including the stunning chapter-opening photos.

Every chapter was reviewed at least three times by instructors, all of whom have a strong commitment to teaching and to the learning process. I am grateful for the insight of the many reviewers and class testers who put so much effort into helping improve the text. Special thanks go to **Robert Sullivan** of Marist College not only for endless reviewing but also for writing the *Instructor's Guide* and developing the *Lecture Slides*; to **Debra Chapman,** whose early reviews of the text had a decided influence on its development and whose *Test Bank* is probably the best around; to **Judy Stewart**, whose cheerful insight not only urged my work forward but will also touch those students who use her wonderful *Study Guide***;** and to **Bert Atsma and Bev Clendening** whose *Laboratory Manual* will complement the best human biology courses.

Last, but not least, I'd like to thank my wife, Pamela, for her whole-hearted support and understanding over the past three years.

SUPPLEMENTS

For the Instructor

Instructor's Guide and Test Bank (0–8053–5079–9)
By Robert Sullivan, Marist College
and Debra Chapman, Wilkes University

The Instructor's Guide and Test Bank that accompany Johnson's *Human Biology* were written by experienced teachers who have been instrumental in the development of the text. The Instructor's Guide by Robert Sullivan contains lecture outlines and insightful lecture hints, creative suggestions for student assignments and class demonstrations, correlations to the text's CD-ROM and website, as well as a wonderful array of additional resources. Answers to the text's end-of-chapter questions are also provided. The Test Bank, by Debra Chapman, contains more than 2000 test questions covering all the major concepts covered in each chapter. Questions and answers have been thoroughly checked for appropriateness and accuracy. All test questions are cross-referenced to the text.

Computerized Test Bank (0–8053–5080–2)

The printed test bank is available in a cross-platform CD-ROM using the latest version of Test Gen EQ. Functions of the software include the ability to add, edit, and print selected test questions as well as the random creation of tests.

Transparency Acetates (0–8053–5080–7)

Full color acetates of all drawings in the book are available to instructors at no cost.

Digital Library: Instructors' Presentation CD-ROM (0–8053–5072–1)

This presentation CD-ROM contains all illustrations in the book, which can be projected in lecture or imported into PowerPoint, tests, and Web sites. Illustrations can be selected in three formats: with labels and leaders, with leaders only, and without labels and leaders.

Lecture Slides in PowerPoint (0–0853–5064–0)
By Robert Sullivan, Marist College

PowerPoint slide shows for 30 lectures are included in this cross-platform CD-ROM. Selected text illustrations are utilized in each lecture. Instructors may build on and/or modify lectures.

Course Management Systems, including WebCT and Blackboard, for *Human Biology*. (Available Summer 2001)

Useful for on-line course management and distance learning courses. Content from the text's Test Bank as well as its Web site (The Human Biology Place) can be pre-loaded into WebCT and Blackboard, saving instructors time in preparing on-line course materials. Both the WebCT and Blackboard systems reside on your university's servers, providing you with complete control over your on-line environment. Visit http://cms.awlonline.com for more information.

For the Student

Interactive Physiology for Human Biology CD-ROM
Nick Kapp, Skyline College, Consultant

This CD-ROM is included with the text. Includes 23 narrated tutorials of key physiological processes from the award-winning, animated, Interactive Physiology series developed by adam.com and Benjamin Cummings. See "How to Use This Book and Its Media" for a more complete description. Not available separately.

The Human Biology Place Web site

12-month subscriptions are included with purchase of a new text. Stand-alone subscriptions are also available. See "How to Use This Book and Its Media" for a more complete description. Visit http://www.humanbiology.com.

Study Guide to accompany Johnson's *Human Biology* (0–8053–5087-X)
By Judith Stewart, Southern Nevada Community College

Written by an experienced instructor who was one of the textbook's key reviewers, this Study Guide offers useful reviews of key concepts and a wide variety of practice questions to ensure students' success in class and on exams.

Laboratory Manual for *Human Biology* (Available Spring 2001) (0–8053–5067–5)
By Bert Atsma, Union County Community College
and Beverly Clendening, Hofstra University

Two experienced instructors and lab manual authors have teamed up to provide a lab manual for the one-semester human biology course. Written with Johnson's text in mind, this manual will provide a variety of lab activities exploring fundamental concepts for approximately 30 lab sessions.

REVIEWERS

Cindy Beck
The Evergreen State College

Peter Biesemeyer
North Country Community College

Joanna Borucinska
University of Hartford

Hessel Bouma III
Calvin College

Brian Bradley
University of Maryland—Baltimore County

Michael Bucher
College of San Mateo

Timothy H. Carter
St. John's University

Debra Chapman
Wilkes University

Beverly Clendening
Hofstra University

Mary C. Colavito
Santa Monica College

Tom Denton
Auburn University—Montgomery

Charles Dick
Pasco-Hernando Community College

Diane Dudzinski
Washington State Community College

Constance Eames
Dutchess Community College

Isaac Elegbe
The College of New Rochelle

Michael Emsley
George Mason University

David Enock
Community College of Allegheny County

Gina Erickson
Highline Community College

Jeffrey W. Flosi
University of Houston—Downtown

Mary Fox
University of Cincinnati

David E. Fulford
Edinboro University of Pennsylvania

Claudette Giscombe
University of Southern Indiana

Sheldon R. Gordon
Oakland University

James Grass
City College of San Francisco

Kenneth Gregg
Winthrop University

T. Daniel Griffiths
Northern Illinois University

N. Gail Hall
Boston College

Joseph F. Hawkins
College of Southern Idaho

Clare Hays
Metropolitan State College of Denver

Mark Hoover
Penn State—Altoona

Charles Hummel
New York Institute of Technology

Susan Hutchins
Itasca Community College

Alison Jassen
Northwestern Connecticut Community College

Ronald Jenkins
Samford University

Robyn Jordan
Northeast Louisiana University

Nick Kapp
Skyline College

Gwendolyn M. Kinebrew
John Carroll University

Vasily Kolchenko
Pace University

Kenneth D. Laser
Edison Community College

Roger Lloyd
Florida Community College—Jacksonville

John Lovell
Kent State University—Tuscarawas

Ardath Lunbeck
Sheridan College

Jacqueline Maly
Windward Community College

Barbara Mania-Farnell
Purdue University—Calumet

Patricia Matthews
Grand Valley State University

Rodger J. McCormick
West Virginia Wesleyan College

Michael McDarby
Fulton-Montgomery Community College

James L. McDonel
Penn State University

Jacqueline S. McLaughlin
Penn State University—Lehigh Valley

Diane A. Merlos
Grossmont College

Rhonda Meyers
Lower Columbia College

Javanika Mody
Anne Arundel Community College

Pam Monaco
Molloy College

Scott Moody
Ohio University

Michael S. Mooring
Point Loma Nazarene University

David Mork
St. Cloud State University

Bernard F. Murphy
Hagerstown Community College

Don Naber
University of Maine

John J. Natalini
Quincy University

Richard Packauskas
Fort Hays State University

Lee D. Peachey
University of Pennsylvania

Delois M. Powell
University of Maryland—Eastern Shore

Elsa C. Price
Wallace Community College

S. Rana
St. John's University

Dennis Rich
Naugatuck Valley Community—Technical College

Laura H. Ritt
Burlington County College

Barbra Roller
Florida International University

Marc M. Roy
Beloit College

Veena Sallan
Owensboro Community College

John W. Sherman
Erie Community College—North

Richard H. Shippee
Vincennes University

Walt Sinnamon
Southern Wesleyan University

Pat Steubing
University of Nevada—Las Vegas

Gregory J. Stewart
State University of West Georgia

Judy Stewart
Community College of Southern Nevada

Sandra K. Stewart
Vincennes University

Mark Sturtevant
Northern Arizona University

Robert J. Sullivan
Marist College

Jill Turchany
Rock Valley College

Karen Van Meter
Des Moines Area Community College

Samuel E. Wages
South Plains College

Wayne Weber
University of Wisconsin—Platteville

David R. Weisbrot
William Paterson University

Susan Whittemore
Keene State College

Roberta Williams
University of Nevada—Las Vegas

Douglas B. Woodmansee
Wilmington College

Lamond Woods
San Joaquin Delta College

Marcus Young Owl
California State University—Long Beach

FOCUS GROUP PARTICIPANTS

Beverly Clendening
Hofstra University

Charles Dick
Pasco-Hernando Community College

William Dopirak, Jr.
Three Rivers Community-Technical College

Joan Felice
Minneapolis Community and Technical College

Joseph Glass
Camden County College

Patricia Hauslein
St. Cloud State University

Allison Jassen
Northwestern Connecticut Community-Technical College

Martia Kapper
Central Connecticut State University

Vasily Kolchenko
Pace University

Michele K. H. Malotky
University of Wisconsin—River Falls

Maria Menna-Peeper
St. Francis College

Pam Monaco
Molloy College

Bernie Murphy
Hagerstown Community College

Frank Pearce
West Valley College

Ron Quinta
Ohlone College

Richard Roth
University of Hartford

Ron Salyk
Bloomfield College

J. Small
Rollins College

Ron Stephens
Ferrum College

Tom Terry
University of Connecticut

Carlene Tonini
College of San Mateo

Pat Werronen
Minneapolis Community and Technical College

CLASS TESTERS

Don Alford
Metropolitan State College

Bert Atsma
Union County College

Ray Bailey
University of Alaska—Anchorage

Claudia Barreto
University of Wisconsin—Milwaukee

Cindy Beck
The Evergreen State College

Kim Blake
Mitchell College

Hessel Bouma III
Calvin College

William Brett
Indiana State University

Beverly Clendening
Hofstra University

Walter Coombs
Western New England College

John Cowlishaw
Oakland University

Charles Dick
Pasco-Hernando Community College

Felidia Ellis
Gulf Coast Community College

David Enock
Community College of Allegheny County

David Ferris
University of South Carolina—Spartanburg

Tom Freeland
Walsh University

April Gardner
University of Northern Colorado

Guadalupe Garza
Texas A&M International University

Donald Glassman
Des Moines Area Community College

Kenneth Gregg
Winthrop University

Martin Hahn
William Paterson University

Madeline Hall
Cleveland State University

Lisa Harvey
Victor Valley College

Scott Helgeson
Des Moines Area Community College

Carl Heuther
University of Cincinnati

Barbara Hunnicutt
Seminole Community College

Janice Ito
Leeward Community College

Paul James
Miami University of Ohio

Robyn Jordan
Northeast Louisiana University

Paul Jones
Miami University of Ohio

Nick Kapp
Skyline College

Mary Jo Kenyon
Minnesota State University, Moorhead

Dr. Krebs
Cleveland State University

Peter Lardner
Flagler College

Deborah Leonard
SUNY College at Fredonia

Martin Levin
Eastern Connecticut State University

Ken Lindberg
Bethune-Cookman College

Bob Ling
Kankakee Community College

Bonnie Lustigman
Montclair State University

Bill Mackay
Edinboro University

Debra Martin
St. Mary's University of Minnesota

Patricia Matthews
Grand Valley State University

Ann Maxwell
Angelo State University

Deanna McCullough
University of Houston—Downtown

Carey Miller
Brookdale Community College

Aaron Moe
Concordia University

Doris Morgan
Middlesex County College

R. Ward Rhees
Brigham Young University

Barbra Roller
Florida International University

Andrew Ross
Prince George's Community College

Harry Schutte
Ohio University—Athens

Roy Shake
Abilene Christian University

John Sherman
Erie Community College

Doris Shoemaker
Dalton State College

Jeff Simpson
Metropolitan State College

Pat Stranahan
Metropolitan State University

Mary Talbot
San Jose City College

Betty Taylor
York College of Pennsylvania

Deb Temperly
Delta College

Sara Sybesma Tolsma
Northwestern College

Carlene Tonini
College of San Mateo

Michael Troyan
Penn State University

Alexander Varkey
Liberty University

Arnold Weisshaar
Prince George's Community College

Tim Zehnder
Gardner-Webb University

How to Get the Most Out of This Book and Its Media

Your textbook includes several built-in tools that can help you study more effectively. To get in the right frame of mind for learning, the first question to ask is not "What do I *have* to know?" but "What do I *want* to know?" We invite you to begin each chapter by contemplating the **chapter-opening questions,** which raise questions that you may have wondered about before. You'll find answers to these questions within the chapter; each question is repeated in the margin adjacent to the text passage that addresses it.

Reminders of key themes and applications in human biology are found throughout the book. **Health Watch** boxes explain many common diseases and their treatments and offer advice on health and wellness. **Directions in Science** boxes reveal how advances in science pose new questions for science to answer. **Try It Yourself** exercises are easy and fun, and remind you not to accept facts on faith but always to look for evidence. **Current Issue** features ask you to contemplate ethical issues that have emerged along with great advances in biology, posing such questions as "Should we clone humans?" and "How should scarce organs be allocated?" **Focus on Structure and Function** passages highlight a key theme in biology: On all levels of organization, the structure and function of living things are intimately related.

To help you home in on key ideas and prepare for tests, **Concept Heads** and **Recaps** allow you to preview and review the chapter. Chapter-ending **Summaries** encourage you to review and master the chapter one section at a time. The list of **Terms You Should Know** thoughtfully includes just the most important terms in the chapter. If a term does not seem familiar, use the page reference to return to the chapter for review. The **Test Yourself** section provides a wide array of objective questions to help you identify what you know and what you need to review. **Concept Review** questions will help you see if you have a handle on the 'big picture,' and **Apply What You Know** questions give you the opportunity to contemplate how the concepts you've mastered apply to the real world. The **Case in Point** stories that end each chapter feature actual news and research that you will now be prepared to understand and evaluate. Annotated **References** describe books and articles that you may find useful and even entertaining.

The Interactive Physiology for *Human Biology* CD-ROM

Packaged with this textbook is a very special CD-ROM developed by adam.com and Benjamin Cummings that will make it much easier for you to understand and review important topics in systems physiology. It contains award-winning narrated animations of 23 key physiological processes such as the cardiac cycle, the membrane potential, and gas exchange. Arranged by system, these animations can be reviewed before, during, or after you read the corresponding material in the text.

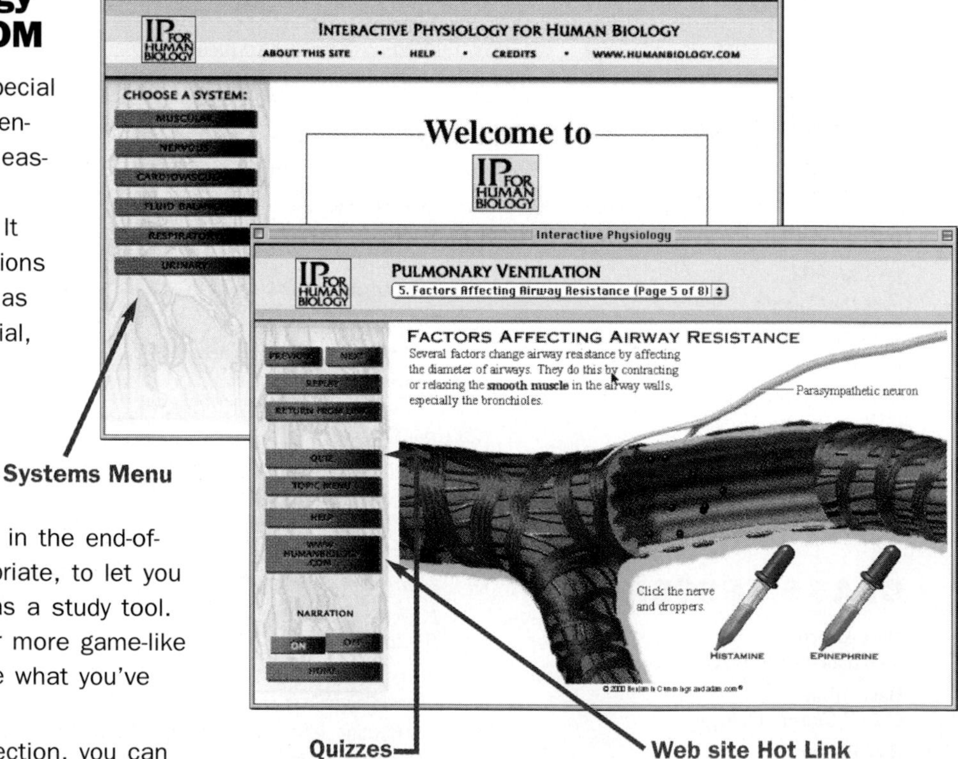

Each topic is cross-referenced in the end-of-chapter summaries where appropriate, to let you know when to use the CD-ROM as a study tool. Each topic concludes with one or more game-like quizzes and activities to reinforce what you've learned.

If your computer has an internet connection, you can move directly from the tutorial CD-ROM to the Human Biology Place web site. http://www.humanbiology.com

The Human Biology Place Web site (*www.humanbiology.com*)

If you have purchased a new copy of this book, the first page will contain an access code and password for a 12-month subscription to the online Human Biology Place. The site provides several tools to help you prepare for tests, including practice tests, tutorial animations, links to other great websites, cooperative learning activities, and news articles.

A special area called the **Issues Forum** invites subscribers to post commentary on issues raised in the text.

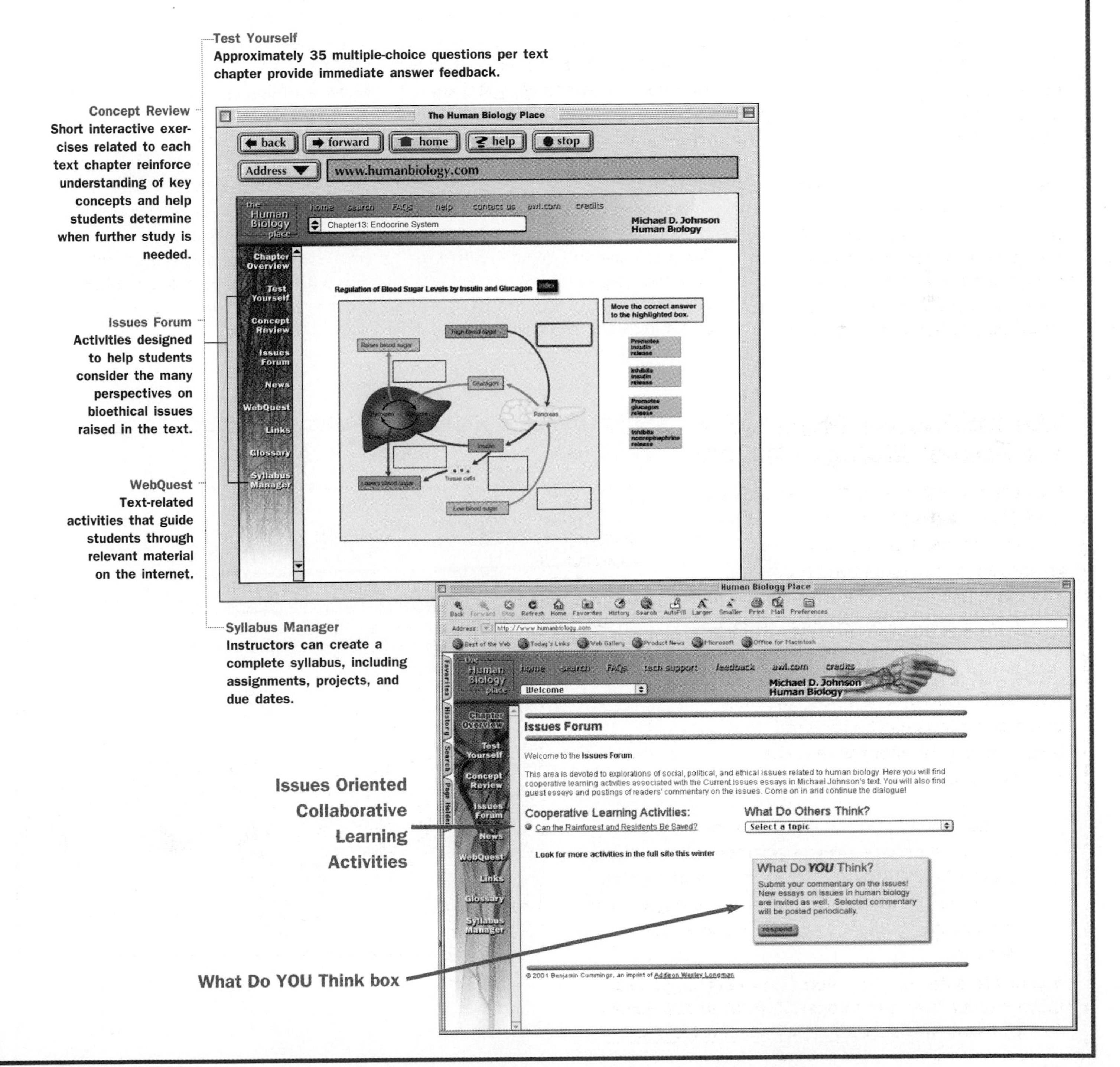

CHAPTER 1

HUMAN BIOLOGY, SCIENCE, AND SOCIETY

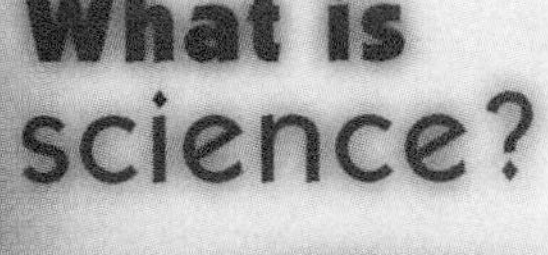

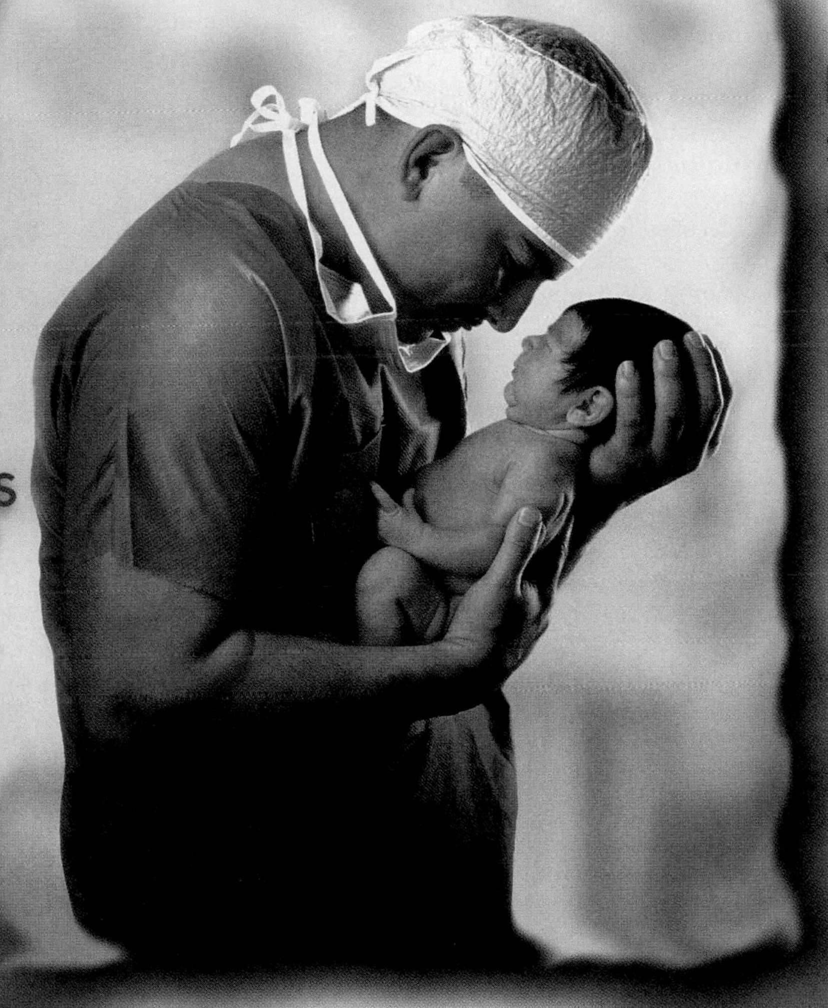

How does science affect **your life?**

What do all living things **have in common?**

How are humans different **from other animals?**

YOU WERE BORN INTO EXCITING TIMES, WHEN SCIENTIFIC DIScoveries are happening more rapidly than at any other time in human history. Like the industrial revolution of the nineteenth century and the discovery of DNA in the twentieth century, today's scientific innovations will change the human condition forever.

In your lifetime it may be possible to select or modify your children's features before they are born. You may even be able to have a clone (a copy) made of yourself. At the very least, certain diseases that threaten us now will become curable. Perhaps your grandchildren will not even know what AIDS is because the disease will have disappeared.

What you are witnessing is the power of science. **Science** is the study of the natural world. The **natural world** includes all matter (material) and all energy, and therefore it includes all living organisms because organisms are also made of matter and energy (Figure 1.1). Biology is just one of many branches of science. More specifically, **biology** (from the Greek words *bios*, life, and *logos*, science) is the study of living organisms and life's processes, or more simply the study of life. Other branches of science are chemistry, physics, geology, astronomy, and related fields such as medicine.

This text is specifically about *human* biology. We will explore what it means to be alive. We will see how the molecules that make up our bodies are created from molecules in the air and in our food and drink. We will learn how our cells grow and divide, and how we evolved from single-celled organisms that arose from nonliving chemical elements nearly 3.5 billion years ago. We will explore the functioning of our bodies, why we get the diseases we do, and how we manage to survive them. We will look at how we develop into adults, reproduce, and influence the destinies of other organisms on Earth.

With the power of science comes an awesome responsibility. All of us, individually and collectively, must choose how to use the knowledge that science will give us. Will human cloning be acceptable? How will we prevent global warming? Should your insurance company be able to reject you for coverage because genetic testing shows you *may* get cancer 40 years from now?

This text is written with the conviction that each of us must be prepared to make responsible decisions concerning not only our own immediate health and well-being, but the long-term well-being of our species as well. As is fitting for a course in human biology, we will consider all aspects of human interaction with the natural world. We'll contemplate human functioning within the environment, and human impact upon the environment. Along the way we'll confront a variety of social and personal issues and how we might make choices concerning them. Because biology is the study of life, we begin by defining life itself.

Recap **Science is the study of the natural world, which consists of all matter and energy. Biology is the study of living organisms.**

1.1 The characteristics of life

What is life? On the one hand, this question seems so easy, and on the other hand so abstract that it is more like a riddle than a serious question. We all think we can recognize life even if we can't define it easily. Children learn early to distinguish between living and nonliving things. Remember that game you played as a child that asked, "animal, vegetable, or mineral?" That game distinguishes what is alive (animals and plants) from what is not (minerals).

Most biologists accept the following criteria as signs of life:

What do all living things have in common?

- *Living things have a different molecular composition than nonliving things.* All components of the natural world, both living and nonliving, are comprised of the same set of approximately 100 different chemical *elements*. However, only a few of the elements are present in any abundance in living organisms. In addition, living organisms are able to combine elements in unique ways, creating certain *molecules* (combinations of elements) that nonliving things cannot create. These molecules of life (proteins, carbohydrates, lipids, and nucleic acids) are found in all living organisms, and are often still present in the remains of dead organisms. Variations in these molecules in different life forms account for the diversity of life as we know it.
- *Living things require energy and raw materials.* The creation of the molecules of life doesn't happen by accident, at least under the present conditions on Earth. The transformation of molecules from one form to another requires energy. The term **metabolism** refers to the physical and chemical processes involved in transforming energy and molecules so that life can be maintained. All living things take in raw materials and energy from the environment and metabolize them into the molecules and energy they need to survive. Plants use the energy of sunlight and chemicals obtained from soil, water, and air. Animals and all other forms of life ultimately obtain their energy and raw materials from water, air, plants, or other animals.

(a) Photo taken from the Hubble Space Telescope showing a tiny portion of the universe. Studies by astronomers have shown that all matter on Earth originated inside stars or with the Big Bang.

(b) The natural world comprises all matter and energy. An erupting volcano spewing liquid rock and heat is the result of energy that still remains from the creation of Earth nearly 4.6 billion years ago.

(c) Studying this unusual fish from deep waters off California allows biologists to understand the processes by which a species successfully survives. Many different environments exist in the world, but the same physical and chemical laws govern them all.

(d) Jane Goodall has dedicated her life to studying the needs and behaviors of chimpanzees. The DNA of humans and chimps is almost the same, yet important physical and behavioral differences are obvious. Evolution examines how these differences arose.

Figure 1.1 Studies of the natural world.

- *Living things are composed of cells.* A **cell** is the smallest unit that exhibits all the characteristics of life (Figure 1.2). All cells come only from existing cells. There is always at least one cell in any living thing, and some organisms (called *unicellular* organisms) are *only* one cell. *Multicellular* organisms are composed of many cells and even many different types of cells.
- *Living things maintain homeostasis.* The internal environment of all living organisms must be maintained within a narrow range of chemical composition and physical conditions compatible with life. The maintenance of a relatively constant internal environment is called **homeostasis.** Living things have developed remarkable ways of regulating their internal environment despite sometimes dramatic changes in the external environment. Single cells and unicellular organisms are surrounded by a membrane that allows the cell (or organism) to maintain internal homeostasis by providing a selective barrier to the entry and exit of various substances. In multicellular organisms the tissues, organs, and organ systems all work together to maintain homeostasis of the fluid that surrounds all cells.

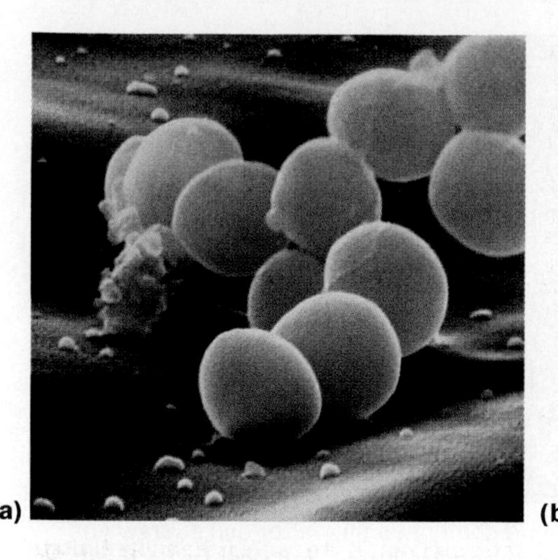

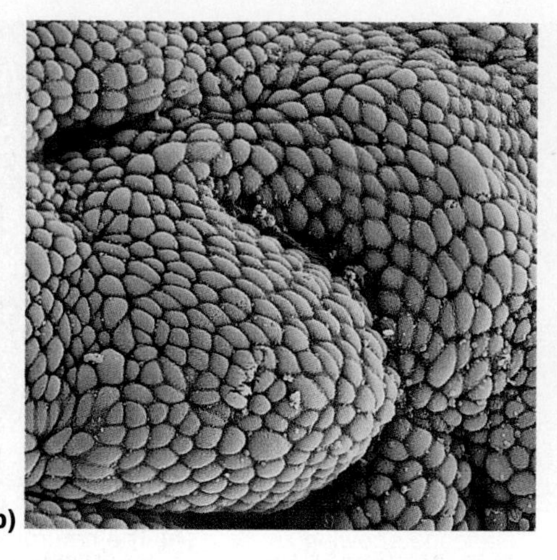

Figure 1.2 Cells are the smallest units of life. Some organisms consist of just one cell (unicellular), whereas others contain many cells (multicellular). **(a)** *Staphylococcus aureus,* the bacterium that causes food poisoning (SEM × 50,000). **(b)** Some of the many cells that line the inner surface of the human stomach (SEM × 500).

- *Living things respond to their external environment.* Stay out in the cold too long and you are likely to respond by moving to a warm room. Plants respond to their environment by turning their leaves toward light or by growing roots toward sources of nutrients and water. Even bacteria respond to their environment, by moving toward nutrients (and away from noxious stimuli) and by increasing their growth rate.
- *Living things grow and reproduce.* Living organisms have the capacity to grow and ultimately produce more living organisms like themselves (Figure 1.3). The ability to grow and reproduce is determined by the genetic material in cells, called DNA. Some nonliving things can get larger, of course; examples are glaciers and volcanic mountains. However, they cannot create copies of themselves.
- *Populations of living things evolve.* The various forms of life may change over many generations in a process known as evolution. Evolution explains why there are so many different forms of life on Earth today.

Although all these characteristics are necessary to describe life fully, not all of them apply to every living thing all the time. Individual organisms do not evolve, nor do they necessarily reproduce or always respond to their surroundings. However, populations of similar organisms have the *capacity* to perform these functions.

> ***Recap*** **All living things are composed of cells. Living things require energy and raw materials, maintain homeostasis, respond to their external environments, grow and reproduce, and evolve over many generations.**

1.2 How humans fit into the natural world

Living things are grouped according to their characteristics

To find order in the diversity of life, biologists have long sought ways to categorize living things. In 1969 a classification system of **five kingdoms** was proposed. In this system the fundamental criteria for classifying organisms are the presence or absence of a nucleus, the number of cells, and the type of metabolism. Organisms lacking a nucleus (the prokaryotes) comprise one kingdom, and organisms with cell nuclei (the eukaryotes) comprise the remaining four kingdoms.

The five-kingdom classification system places all single-celled prokaryotic organisms in the kingdom **Monera,** which comprises the bacteria, some of the oldest, smallest, simplest, and most successful organisms on Earth. Of the four kingdoms of eukaryotes, three of them **(Animalia, Plantae,** and **Fungi)** consist of multicellular organisms. The criteria for classifying animals, plants, and fungi are based largely on the organism's life cycle, structure, and mode of nutrition. Plants, for example, contain a green pigment called chlorophyll that allows them to capture the energy of sunlight, which they convert for their own use in a process called **photosynthesis.** Animals get the energy they need by eating plants or other animals, which requires structures specialized to digest and absorb flood. Most animals are able to move about to obtain food. Fungi (yeasts, molds, and mushrooms) are *decomposers,* meaning that they obtain their energy from decaying material. The fourth group of eukaryotes, the kingdom **Protista,** comprises those eukaryotes that do not fit the animal, plant, or fungi kingdom definitions. The kingdom protista includes protozoa, algae, and slime molds. All animals, plants, and fungi are thought to have evolved from single-celled protistan-type organisms.

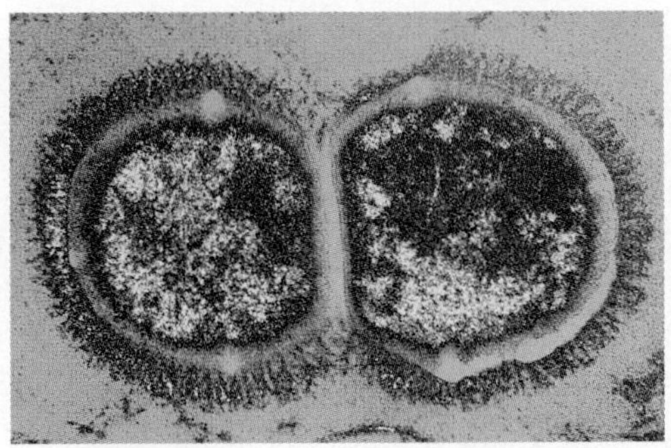

Figure 1.3 Living things grow and reproduce.

In recent years new techniques in molecular biology and biochemistry have distinguished two fundamentally different types of procaryotes. For this reason, many biologists now advocate a classification system that begins with **domains,** a higher classification level that encompasses kingdoms. In the three-domain system the organisms of the kingdom monera are distributed across two domains **(Eubacteria** and **Archaea),** each comprised of a kingdom, while all organisms whose cells have nuclei fall into the third domain, the **Eucarya.** The domain Eucarya thus comprises all four eukaryotic kingdoms.

Other classification systems have been proposed, including systems with five, seven, and eight kingdoms. All of these classification systems are subject to change as new information is discovered. There is no debate, however, concerning the classification of humans within the animal kingdom.

Humans belong to a subgroup of the animal kingdom called *vertebrates,* defined as animals with a nerve cord and a backbone. Within the vertebrates, humans are subclassified as *mammals,* defined as vertebrates with mammary glands who nurse their young. Among the mammals, we, along with apes and monkeys, are further classified as *primates.*

The smallest unit of any classification system is the species. A **species** is defined as one or more populations of organisms with similar physical and functional characteristics that interbreed and produce fertile offspring under natural conditions. All living humans belong to the same **genus** (the second smallest unit of classification) and species, called *homo sapiens.* We share common features that make us different from any other species on earth, and we can interbreed.

No one knows how many species of living organisms exist on Earth. Estimates range from about 3 million to 30 million, but only about 2 million species have been identified so far.

Recap **Classification systems place living things into increasingly inclusive groups. The most inclusive is a domain and the smallest is a species. Humans belong to the kingdom Animalia within the domain Eucarya. Our genus and species is *homo sapiens.***

The defining features of humans

Humans are not the largest animal, nor the fastest or strongest. Our eyesight and hearing are not the best. We can't fly, we swim poorly, and we don't dig holes in the ground very well with our hands. One might wonder how we have managed to survive at all. Nevertheless, we possess several features that, taken together, define how we are different from other organisms:

How are humans different from other animals?

- *Bipedalism.* Humans are the only mammal to stand upright and walk on two legs. *Bipedalism* (from the Latin words *bis*, twice, and *pes*, foot) frees the hands and forearms for carrying items ranging from weapons to infants.
- *Opposable thumbs.* Humans and several other primates have thumbs that can be moved into position to oppose the tips of the fingers. However, only humans have the well-developed muscles that enable us to exert a certain type of precise control over the thumb and fingers. For instance, we tend to pick up small objects between the tip of the thumb and the tip of either the index or second finger (Figure 1.5). In contrast, chimpanzees more naturally grasp objects between the thumb and the side of the index finger.
- *Large brain.* Humans have a large brain mass relative to body size. The evolution of a large brain seems to have coincided with the advent of stone tools, leading some scientists to suggest that a large brain was required for the complex motions associated with tool use. Others dispute this claim. An alternate hypothesis is that a large brain was necessary for language and that language developed as a consequence of the

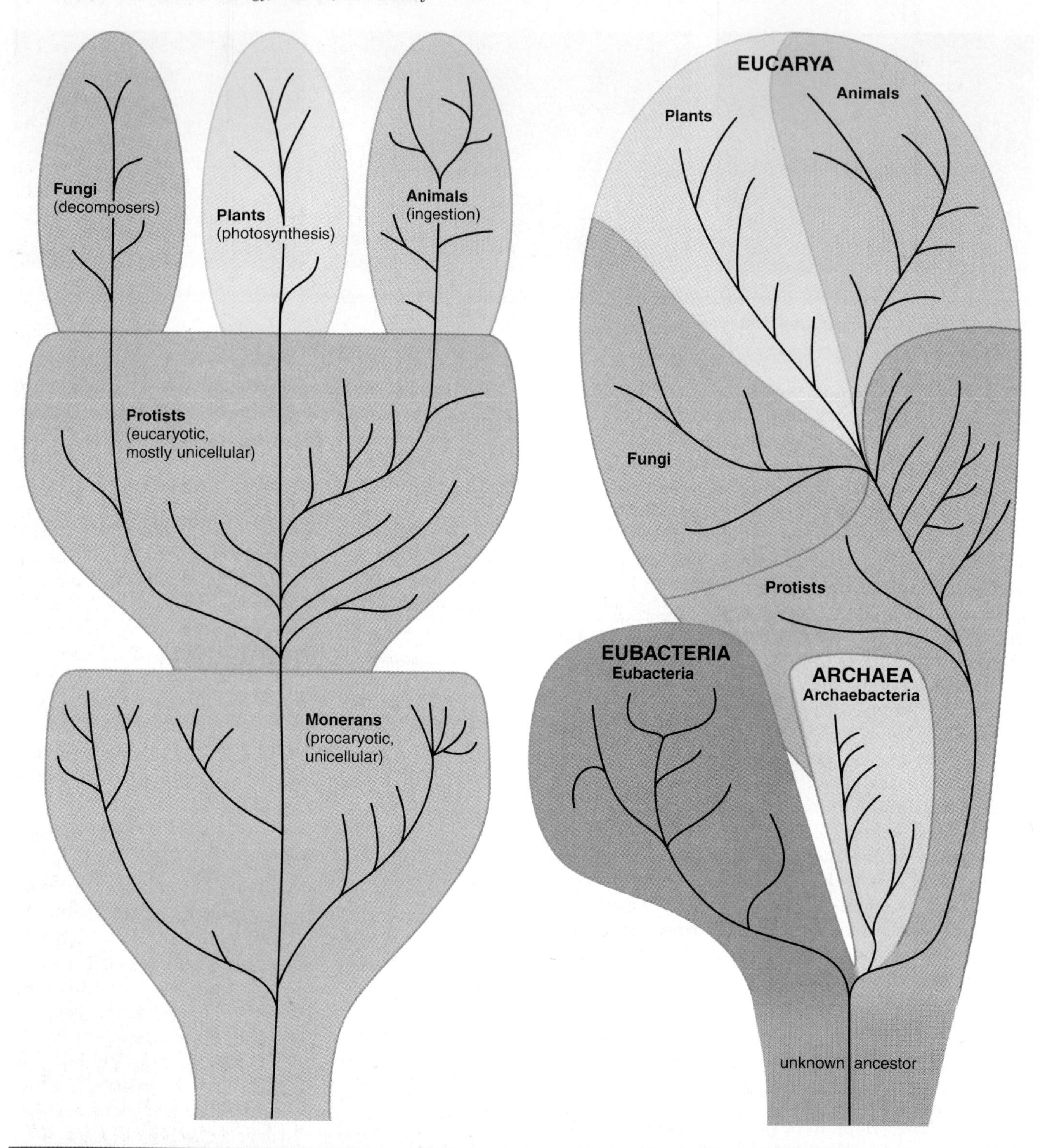

Figure 1.4 Categorizing organisms. (a) The five-kingdom system classifies organisms based on the presence or absence of a cell nucleus, whether or not the organisms are multicellular or unicellular, and certain characteristics of structure, life cycle, and mode of nutrition. **(b)** The three-domain system places prokaryotes in one of two domains based upon their biochemical and molecular characteristics, and puts all eukaryotes in a single domain.

increased importance of social interactions among humans.

- *Capacity for complex language.* Many animals vocalize (produce sounds) to warn, threaten, or identify other members of their species, and a few (such as dolphins) have developed fairly complex forms of communication. However, humans have developed both complex vocal lan-

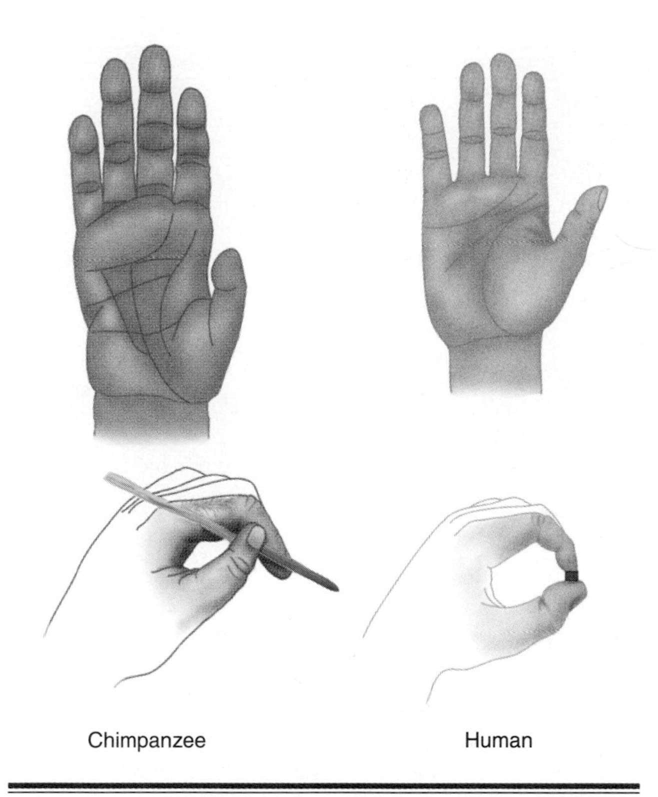

Figure 1.5 How humans and chimpanzees hold small objects. Although the hands of humans and chimpanzees seem similar, only humans tend to hold objects between the tip of a thumb and the tips of the fingers.

guage and a system of signs, symbols, and gestures that communicate concepts and emotions. Throughout the world, every group of human beings has developed a complex spoken language. Humans have also placed their languages into written form, permitting communication over great distances and spans of time (Figure 1.6).

It should be stressed that none of these features necessarily make us any better than any other species, only different.

Recap **Humans walk on two legs (are bipedal) and can grasp small objects between the tips of the thumb and first finger. Humans also have large brains relative to body mass, and the capacity for complex spoken and written languages.**

Human biology can be studied on any level of biological organization

Figure 1.7 shows how humans fit into the grand scheme of things in the natural world and provides a preview of some of the issues we will consider in this book. The figure also shows how humans—or any living thing, for that matter—can be studied on any level of biological organization, from the level of the atom to the level of the biosphere. This text examines human biology on progressively larger scales.

Our study of human biology begins with the smallest units of life. Like rocks, water, and air, humans are composed of small units of natural elements called *atoms* and *molecules.*We'll introduce the basic chemistry of living things in Chapter 2. In Chapter 3 we'll see how the atoms and molecules of living things are arranged into the smallest of living units called *cells*. Next, we'll learn how, in multicellular organisms, groups of similar cells become *tissues*, how groups of tissues that carry out a specific function constitute an *organ*, and that in an *organ system* organs may work together to carry out a more general function (Chapter 4). The structures and functions of specific human organs and organ systems are the subjects of Chapters 5–16. For example, we'll learn why—and how—more blood flows through the lungs than through any other organ and what happens to your dinner as it makes its way through your digestive system. Chapters 17–20 will consider humans as a complete *organism*, including how cells reproduce, how we inherit traits from our parents, and how we develop, age, and die. Finally, we'll discuss how *communities* of living organisms evolved and how life began (Chapter 21), and how humans fit into and alter the *ecosystems* in which we live and the entire *biosphere* of the natural world (Chapter 22).

The bottom section of Figure 1.7 lists some current issues, controversies, and "hot topics" that relate

Figure 1.6 Language. Human societies throughout the world have developed complex written languages.

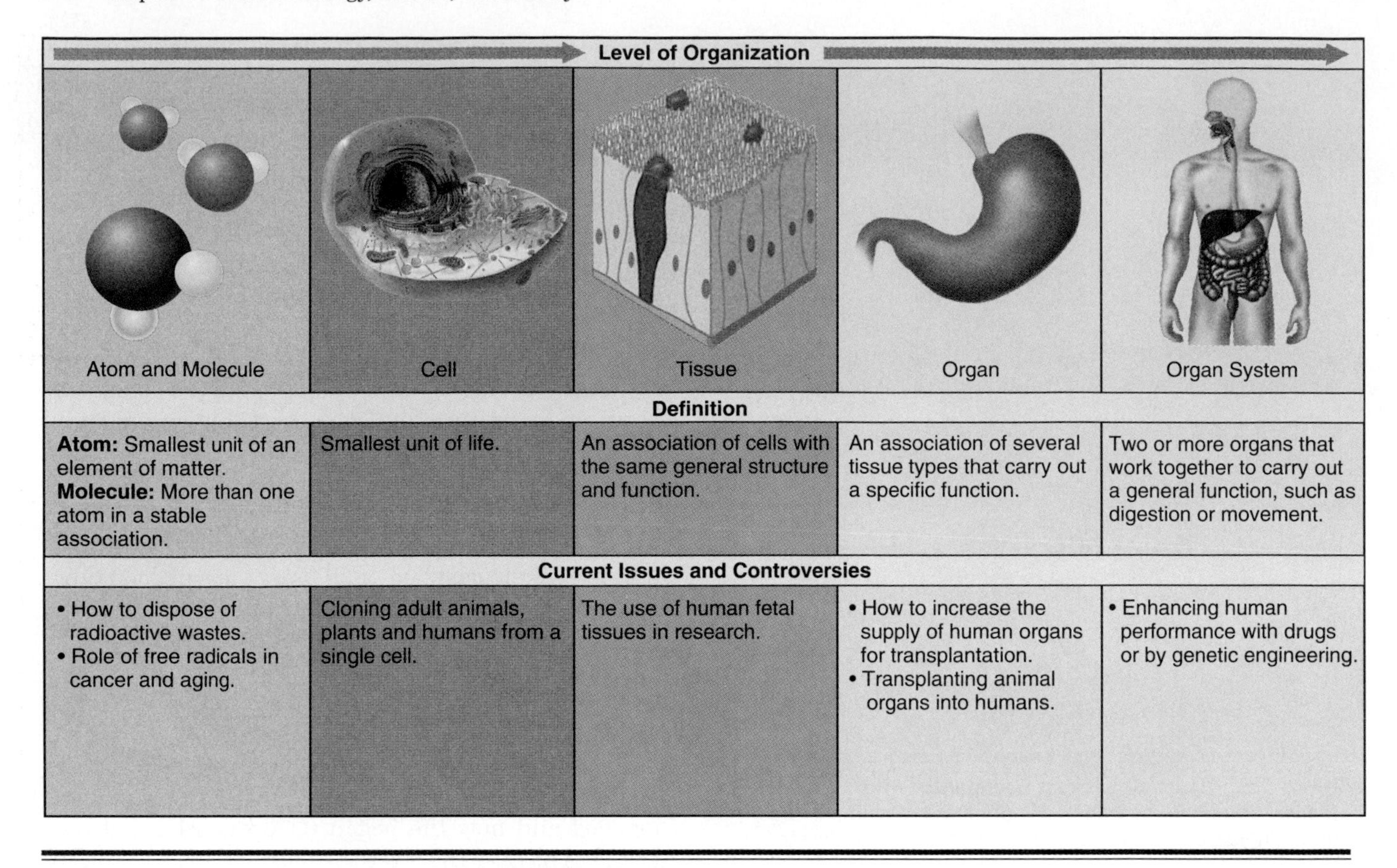

Figure 1.7 Levels of organization in human biology. Human biology affects a number of current issues in our society, as indicated by the section on page 9.

to human biology. Many of these issues and controversies are influenced by fields outside the sphere of science, such as economics, law, politics, and ethics. What you learn in this course will help you make informed decisions about these and future issues that will come up in your lifetime.

Your ability to make good judgments in the future and to feel comfortable with your decisions will depend on your critical thinking skills. We turn now to a discussion of what science is, how we can use the methodology of science to improve our critical thinking skills, and how science influences our lives.

Recap **Humans, along with all other organisms, live in a world that is highly organized. From atoms to cells, from organs to organism, and from community to biosphere, each level of biological organization is increasingly inclusive.**

1.3 Science is both a body of knowledge and a process

We've already said that science is the study of the natural world. More explicitly, science is two things: *knowledge* (organized, reliable information) about the natural world, and the *process* that we use to get that knowledge. **Scientific knowledge** refers to information about the natural world. The process of science, or the way scientific knowledge is acquired, is generally called the **scientific method,** although in practice this term encompasses a variety of methods. Throughout this book you will be presented with scientific knowledge, but it's good to remember that this information was obtained slowly over time by the scientific method.

What is science?

Scientific knowledge enables us to describe and predict the natural world. Scientific knowledge is also empirical, meaning that it relies on observation and experimentation. Through the scientific method, scientists strive to accumulate information that is as free as possible of bias, embellishment, or interpretation.

The scientific method is a process for testing ideas

Although there is more than one way to gather information about the natural world, the scientific method is a systematic process for developing and testing predictions (Figure 1.8). You probably already use the scientific method, or at least elements of it, in your own everyday problem solving. Sometimes we go through the steps informally without

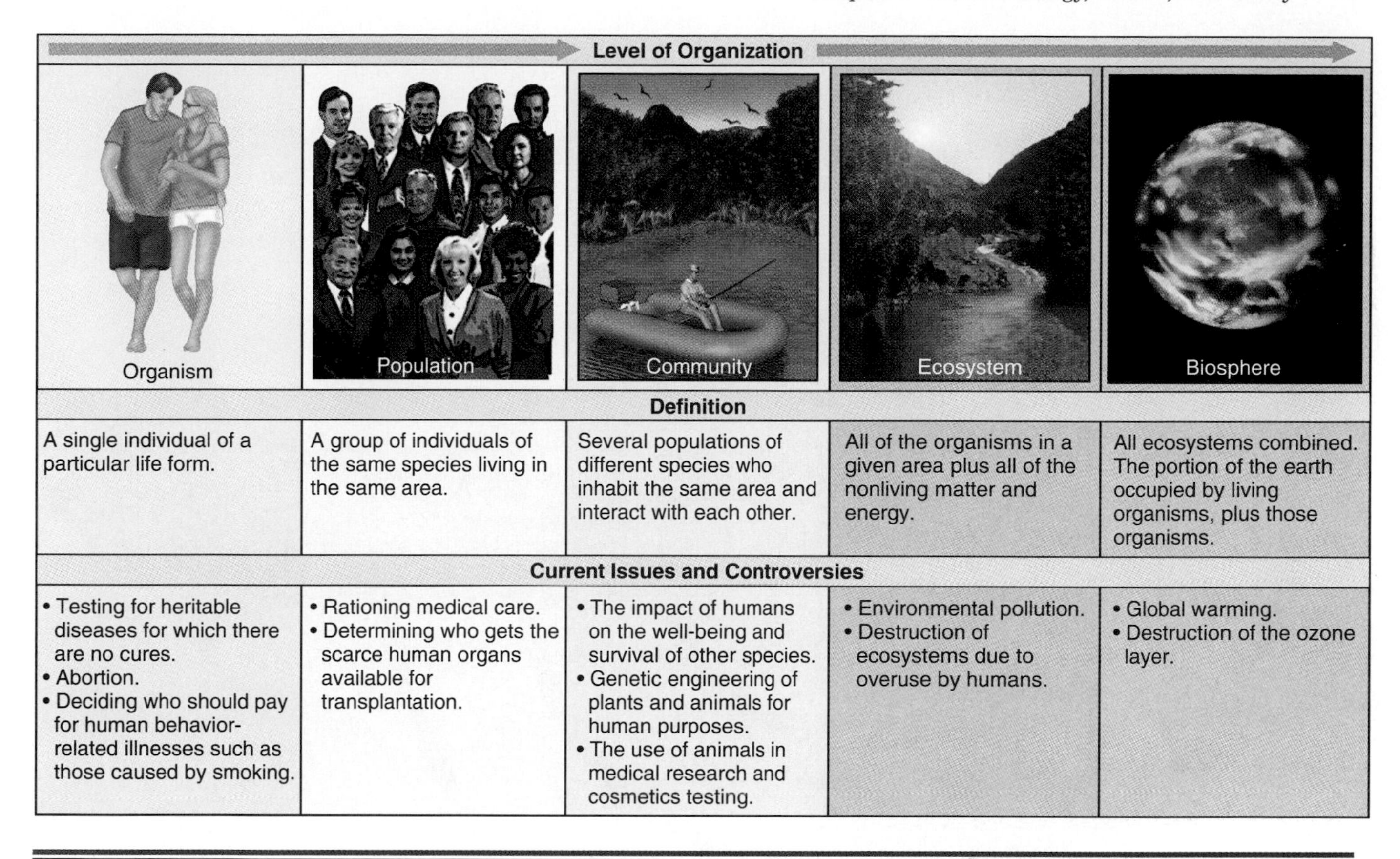

Level of Organization	Organism	Population	Community	Ecosystem	Biosphere
Definition	A single individual of a particular life form.	A group of individuals of the same species living in the same area.	Several populations of different species who inhabit the same area and interact with each other.	All of the organisms in a given area plus all of the nonliving matter and energy.	All ecosystems combined. The portion of the earth occupied by living organisms, plus those organisms.
Current Issues and Controversies	• Testing for heritable diseases for which there are no cures. • Abortion. • Deciding who should pay for human behavior-related illnesses such as those caused by smoking.	• Rationing medical care. • Determining who gets the scarce human organs available for transplantation.	• The impact of humans on the well-being and survival of other species. • Genetic engineering of plants and animals for human purposes. • The use of animals in medical research and cosmetics testing.	• Environmental pollution. • Destruction of ecosystems due to overuse by humans.	• Global warming. • Destruction of the ozone layer.

Figure 1.7 continued

thinking about them consciously, but they are there nevertheless.

Step 1: Observe and generalize When we observe the world around us and make generalizations from what we learn, we are employing *inductive reasoning* (extrapolating from the specific to the general case). Usually we don't even think about it and don't bother to put our observations and generalizations into any kind of formal language, but we do it just the same. For example, you are probably convinced that it will *always* be colder in winter than in summer (a generalization) because you have observed that every winter in the past was colder than the preceding summer (specific observation). The difference between common experience and good science is that science uses generalization to make a prediction that can be tested. Taking an example from biological research, let's start with two observations and a generalization:

> *Observation 1:* Rats given a particular drug (call it Drug X) have lower blood pressures than rats not fed the drug.
> *Observation 2:* Independently, researchers in Canada showed that Drug X lowers blood pressures in dogs and cats.
>
> *Generalization*: Drug X lowers blood pressure in all mammals.

Step 2: Formulate a hypothesis Observations and generalizations are used to develop a *hypothesis.* A **hypothesis** is a tentative statement about the natural world. Importantly, it is a statement that can lead to testable deductions.

> *Hypothesis*: Drug X would be a safe and effective treatment for high blood pressure in humans.

Step 3: Make a testable prediction Hypotheses that cannot be tested are idle speculation, so much hot air. But many hypotheses are so sweeping and comprehensive that ways must be found to test them under a variety of conditions. For example, you probably would not be convinced that Drug X is safe and effective for all people under all conditions until you had at least tested it in quite a few people under many different conditions. In order to have confidence in your hypothesis, you must make testable predictions (also called *working hypotheses*) based on the hypothesis and then test them one at a time. Predictions employ *deductive reasoning* (applying the general case to the specific). Often they are put in the form of an "if . . . then" statement, in which the "if" part of the statement is the hypothesis. For example:

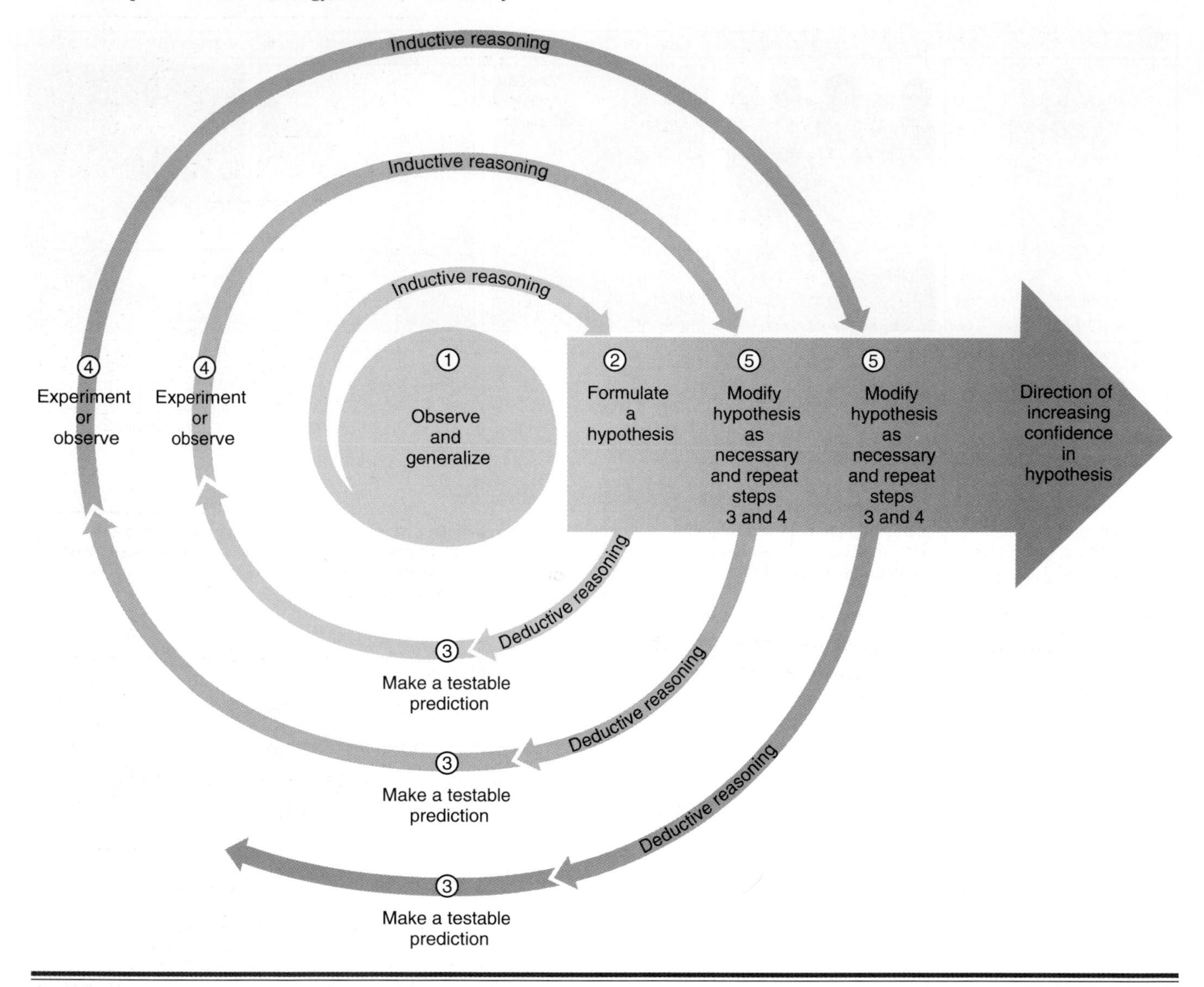

Figure 1.8 The scientific method. Observations and generalizations lead to the formulation of a hypothesis. From the hypothesis, specific predictions are made that can be tested by experimentation or observation. The results either support the hypothesis or require that it be modified to fit the new facts. The cycle is repeated. Ultimately the scientific method moves in the direction of increased confidence in the modified hypothesis.

Prediction: *If* Drug X is a safe and effective treatment for high blood pressure in humans, *then* 10 mg/day of Drug X will lower blood pressure in people with high blood pressure within one month.

Notice that the prediction is very specific. In this example the prediction specifies the dose of drug, the medical condition of the persons on whom it will be tested, what the drug is expected to do if the prediction is correct, and a specified time period for the test. Its specificity makes it testable—yes or no, true or false.

Step 4: Experiment or observe The truth or falsehood of your prediction is determined by observation or by experimentation. An **experiment** is a carefully planned and executed manipulation of the natural world that has been designed to test your prediction. The experiment that you conduct (or the observations you make) will depend on the specific nature of the prediction.

When testing a prediction, scientists try to design experiments that can be conducted under strictly controlled conditions. Experiments conducted under specific, controlled conditions are called *controlled experiments* because they have the distinct advantage of accounting for all of the possible **variables** (factors that might vary during the course of the experiment) except for the one variable of interest, called the *controlled variable*. In this case the controlled variable is blood pressure. Therefore, in the example of Drug X, the steps of your controlled experiment might be as follows (Figure 1.9):

- Select a large group of human subjects. In this case you would specifically choose people with high blood pressure rather than people with normal blood pressure.
- *Randomly* divide the larger pool of subjects into two groups. Designate one the **experimental group** and the other the **control group.** The importance of random assignment to the two groups is that all other factors that might affect the outcome (such as age or gender differences in responsiveness to the drug, or other previous health problems) are automatically equalized between the two groups. In effect, the control group will account for all unknown factors. If you treat the experimental and control groups identically at this point, the probability is that the average blood pressures of the groups will be equal, as long as the groups were large enough.
- Treat subjects in the two groups *exactly* the same except that only the experimental group gets the drug. Treat the experimental group with 10 mg/day of Drug X for one month and the control group with a **placebo,** or "false treatment." If Drug X is delivered in a pill, the placebo would be an identical-looking pill with no drug in it. If Drug X is administered as an injection in a saline solution, the placebo would be an injection of the same volume of saline.

  When working with human subjects, scientists must deal with the power of suggestion. It is important that no subjects in either group know which group has received the drug. If members of the experimental group were told that they were getting a drug to lower their blood pressure, this might accidentally influence the outcome of the experiment. To eliminate the power of suggestion as a variable, experiments are done "blind," meaning that the subjects are not told whether they will get the placebo or the drug. Sometimes experiments are done "double-blind," so that even the person administering the drugs and placebos does not know which is which until the experiment is over.
- Measure blood pressures in both groups at the end of one month and compare them using the appropriate statistical (mathematical) tests. If the group's blood pressures are found to be statistically different, then your prediction is verified and your hypothesis receives support.

Step 5: Modify hypothesis as necessary and repeat steps 3 and 4 If your prediction turns out to be false, you will have to modify your hypothesis to fit the new findings and repeat steps 3 and 4. For example, perhaps the drug would lower blood pressure if you increased the dose or gave it for a longer period of time.

Even if your prediction turns out to be true, you're still not done. This is because you've tested

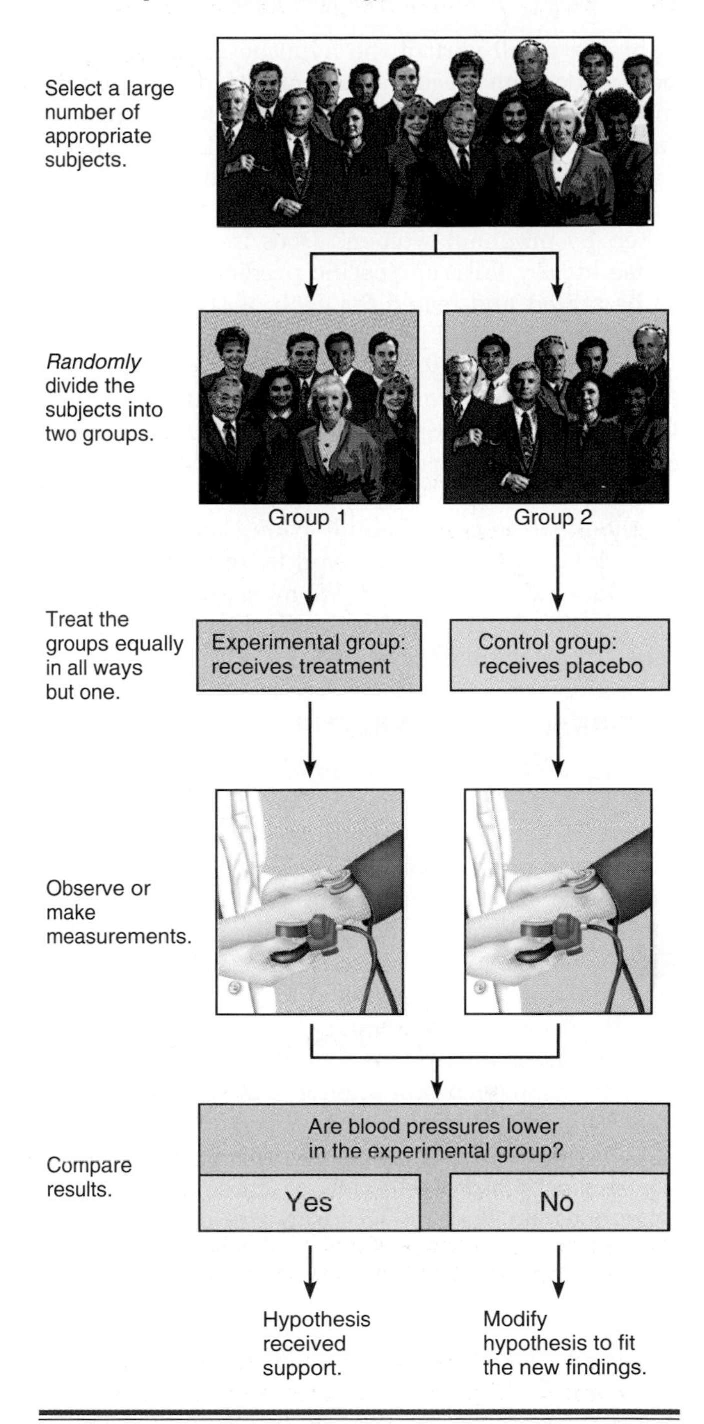

Figure 1.9 The steps in a controlled experiment. In this example the experimental variable is blood pressure. Because of the way the experiment is designed, the only difference between the two groups is the presence or absence of the experimental treatment itself.

only one small part of the hypothesis (its effectiveness under one specific set of conditions), not all the infinite possibilities. How do we know the drug is safe, and how safe is "safe," anyway? Does it cause a dangerous drop in blood pressure in people with normal blood pressures? Does it cause birth defects if taken by pregnant women? Does its long-term use cause kidney failure? Specific predictions may have to be stated and tested for each of these questions before the drug will be allowed on the market.

Only after many scientists have tried repeatedly (and failed) to disprove a hypothesis do they begin to have relatively more confidence in it. A hypothesis cannot be proved true, it can only be supported or disproved. As this example shows, the scientific method is a process of elimination that may be limited by our approach and even by our preconceived notions about what to test. We move toward the best explanation for the moment, with the understanding that it may change in the future.

Making the findings known

New information is not of much use if hardly anybody knows of it. For that reason, scientists need to let others know of their findings. Often they publish the details of their work in scientific journals as a way of announcing their findings to the world. Articles in *peer-reviewed* journals are subjected to the scrutiny of several experts (the scientists' peers) who must approve the article before it can be published. Peer-reviewed journals often contain the most accurate scientific information.

An unspoken assumption in any conclusion is always that the results are valid only for the conditions under which the experiment was done. This is why scientific articles go into such detail about exactly how the experiment was performed. Complete documentation allows other scientists to repeat the experiments themselves or to develop and test their own predictions based on the findings of others.

Try to apply the scientific method to a hypothesis dealing with some aspect of evolution or to a global problem such as cancer or AIDS, and you begin to appreciate how scientists can spend a lifetime of discovery in science (and enjoy every minute!). At times the process seems like three steps forward and two steps back. You can bet that at least some of what you learn in this book will not be considered accurate even 10 years from now. Nevertheless, even through our mistakes we make important new observations, some of which may lead to rapid advances in science and technology. Just in the last 100 years we've developed antibiotics, sent people to the moon, and put computers on every desk.

Many people associate science with certainty, whereas in reality scientists are constantly dealing with uncertainty. That is why we find scientists who don't agree or who change their minds. Building and testing hypotheses is slow, messy work, requiring that scientists constantly question and verify each other.

A well-tested hypothesis becomes a theory

Many people think that a theory is a form of idle speculation or guess. Scientists use the word quite differently. To scientists, a **theory** is a broad hypothesis that has been extensively tested and supported over time and that explains a broad range of scientific facts with a high degree of reliability.

A theory is the highest status that any hypothesis can achieve. Even theories, however, may be modified over time as new and better information emerges. Only a few hypotheses have been elevated to the status of theories in biology. Among them are the theory of evolution and the cell theory of life.

Recap **The scientific method is a systematic process of observation, hypothesis building, and hypothesis testing. A theory is a broad hypothesis that has withstood numerous tests.**

1.4 Sources of scientific information vary in style and quality

We are constantly bombarded by scientific information, some of it accurate, some not. What can you believe when the facts seem to change so quickly? All of us need to know how to find good information and evaluate it critically. Different sources of scientific information may have very different goals, so look for those that can best inform you at your own level of understanding and interest.

Some scientific knowledge is highly technical. As a result, scientists have a tendency to speak on a technical level and primarily to each other. As we've already mentioned, scientists often communicate by means of articles in specialized peer-reviewed journals such as *Nature* and *Science*. Articles in peer-reviewed journals are concise, accurate, and documented so thoroughly that another scientist ought to be able to duplicate the work just by reading the article. Generally they refer extensively to previous literature on the subject. Articles in peer-reviewed journals make for laborious reading, and they are usually as dry as toast, but bear in mind that their purpose is primarily to inform other experts.

Other helpful print sources are the increasing number of science magazines and nonfiction books designed for the relatively well-educated public. The goal is to inform the interested reader who may have only a limited background in science. The authors are usually science writers or experts who are able to translate the finer scientific points into language that we can all understand. The information is generally accurate and readable, although some of the details may be difficult at times. Generally they are sufficiently referenced that you can go more deeply into the subject if you wish.

General interest news magazines and daily newspapers also report on selected hot topics in science. Their main goal is to get the information out as quickly as possible to a wide audience. Coverage is timely, but less in-depth than in science magazines, and may not include the details you need to check the validity of the statements. A decided plus is that magazines and newspapers often discuss social, political, economic, legal, or ethical ramifications of the scientific findings, something generally lacking from the previous sources. Although the scientific information is usually accurate, it may be taken out of context and perhaps even slightly misunderstood by the reporter, who may not be particularly well versed in science. The best articles will give you enough information to enable you find the original source of scientific information. Television channels such as Discovery also present science-related topics to the public.

Since the 1980s, scientists and researchers have used the Internet to communicate and share ideas. The recent expansion of the World Wide Web has made the Internet more accessible to the general public, opening exciting new sources of scientific information. Nearly all universities now have Web pages; these end in ".edu" (for "educational") rather than ".com" (for "commercial"). A number of scientific and professional organizations have created Web sites that offer helpful information for both scientists and consumers. Examples of organizations with Web sites include the National Institutes of Health, American Cancer Society, and American Heart Association. The Web sites of government agencies and nonprofit organizations generally end in ".gov" and ".org," respectively.

Be aware that the Internet can also be a source of misinformation. At present the Internet is less closely regulated than print and broadcast media, so it can be difficult to tell the difference between objective reports and advertisements. In addition, participants in online chat rooms and special interest groups may promote their own opinions as proven truths. It pays to be skeptical.

Recap **The best sources of scientific information translate difficult or complex information accurately into understandable terms and have enough references that you can check the information if you wish.**

1.5 Learning to be a critical thinker

Scientists are people. Many are motivated by a strong curiosity or a sense of wonder and awe about how the natural world works. Exploring the frontiers of knowledge requires a great deal of creativity and imagination. Like many people, however, scientists may at times leap to conclusions or resist new ideas. A few may be driven by self-interest. To combat these natural human tendencies, good scientists consistently try to use certain tools of critical thinking.

Try It Yourself

Evaluating a Scientific Claim

Throughout your life you will have to evaluate the validity of scientific claims in the popular media. In most cases you'll make your judgment based solely on the information given.

As practice, choose an article—a newspaper or magazine article, or information posted on the Internet—on a topic related to human biology and do the following:

- State the article's primary claim. It may help to write the claim down.
- List the "evidence" provided to support the claim.
- Examine each piece of evidence to see if it meets a basic standard of scientific acceptability.

Three criteria will help you decide if the evidence is acceptable or not:

1. *Scientific evidence is observation-based and quantifiable* (example: "From 1970 to 2000 the number of spotted owls seen per year by hikers in Oregon's national forests declined from 5,000 to 2,000"). In contrast, inferential evidence is merely descriptive (example: "The population of spotted owls declined from 1970 to 2000").
2. *The source and quality of the evidence should be apparent.* Cited, verifiable references to previous work by others is a big plus. At the very least, the experimental design should be sufficiently described that you can understand how the data were collected, whether or not there is a control group, the sample size, and so on. You should be convinced that the data were collected and presented properly.
3. *The evidence should be free of obvious bias.* Who presented the data, and what is their hidden agenda, if any?

If most of the evidence presented does not meet these three simple tests, there is no reason that you should accept the claim. In such cases you would be wise to reserve judgment until a later time. Meanwhile, pay attention to what other scientists say about the claim—and be sure to apply the same criteria to their evidence as well.

You too can learn to use these tools, regardless of whether you choose a career in science.

Become a skeptic

Good scientists combine creativity and imagination with **skepticism,** a questioning attitude. If you've ever bought something based on claims about how well it works and then been disappointed, you know the value of skepticism. Question everything and dig a little deeper before believing something you read and hear. Questions you might ask yourself include:

- Who says that a particular statement is true?
- What evidence is presented?
- Are the persons speaking on a subject qualified by training or skill to speak authoritatively about it?
- Are they being paid, and if so how might that affect what they have to say?
- Where's the evidence to back up a claim?

Skepticism is particularly important for claims that are new, startling, and not yet verified by other scientists. Listen carefully to the debate between scientists in the public arena. A new scientific claim may take several years to be checked out adequately.

Appreciate the value of statistics

Statistics is the mathematics of organizing and interpreting numerical information, or **data.** Scientists use statistics to determine how much confidence they should place in the information. Most scientists would be willing to accept experimental results with confidence if (according to statistical tests) they would get the same outcome 19 out of every 20 times they repeat the experiment, or 95% of the time. When you see numerical averages followed by a smaller "+/−" number, the smaller number represents the range in which they have a certain confidence, called the "standard error." In graphs, the standard errors are represented as small lines that extend above and below the average number.

Statistics are used in many disciplines. During elections and public polls, for example, we may hear that "52% of the respondents indicated that they will vote for the President. The poll has a margin for error of +/−3%." This should tell you the pollsters are relatively certain that the actual percentage who will vote for the President is somewhere between 49 and 55%, still too close to call.

Learn how to read graphs

Just as a picture is worth a thousand words, a graph is an efficient way to display data obtained from observations and experimental results. Graphs can also be used to clarify the meaning of experimental results.

Most graphs are plotted on two lines, or axes (sing. axis). The horizontal axis at the bottom is called the *abscissa* (from math you may know this as the x-axis), and the vertical axis is called the *ordinate* (y-axis). By convention the *independent variable,* such as time, distance, age, or another category that defines groups, is generally plotted on the abscissa. The *dependent variable,* so called because its variation may depend on the independent variable, is plotted on the ordinate.

Graphs can take a variety of forms, from plots of individual data points to lines or bars of average values (Figure 1.10). When reading a graph, first check the scales and the legends on the abscissa and the ordinate to determine what the graph is about. Be careful to look for a "split axis," in which the scale changes. An example is shown in Figure 1.11. A split axis is sometimes a convenient way of representing data that cover a wide range on one axis, but it can also be used to deliberately mislead people unfamiliar with reading graphs.

Distinguish anecdotes from scientific evidence

Anecdotal evidence takes the form of a testimonial or short unverified report. Although an anecdote may be true as stated, it in no way implies scientific or statistical certainty. Advertising agencies sometimes use anecdotes to influence you. The actor on television who looks sincerely into the camera and says "Drug X worked for me" may be telling the truth, but this does not prove the drug will work for everyone.

Nonscientists (and even scientists) often say things like "my grandmother swears by this remedy." Again, it may be true, but it is not scientific evidence—perhaps the remedy only works for grandmothers. Listen carefully to how the evidence for a statement is presented.

Separate facts from conclusions

A *fact* is a verifiable piece of information, whereas a *conclusion* is a judgment based on the facts. The news media often mix facts with conclusions without indicating which is which. Almost every evening on the business news we hear statements like "The Dow Jones Industrial Average declined 50 points today on renewed concern over the consumer price index." The first half of the sentence (about the decline) is a verifiable fact. The second half is conjecture on the part of the reporter.

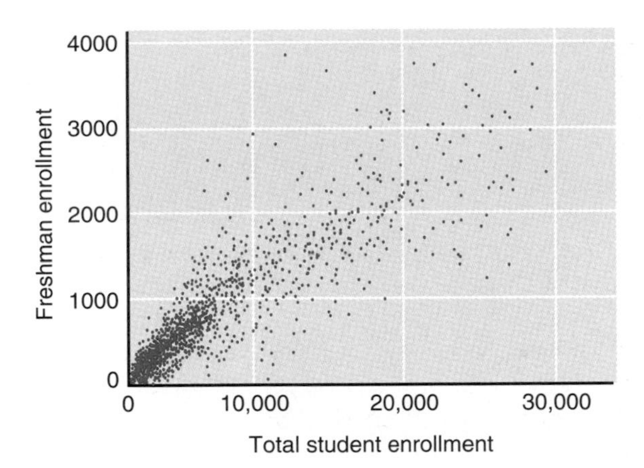

(a)

(b)

(c)

Figure 1.10 Types of graphs. Each of these graphs reports the relationship between freshman enrollment and total student enrollment at approximately 1,500 U.S. colleges and universities. **(a)** A scatter plot showing enrollments at each individual college. Each data point is known as an observation. **(b)** A line graph drawn representing the best straight line fit of the data in (a). **(c)** A bar graph in which university enrollments are lumped together in three class sizes and the freshman enrollments are then averaged. This graph also shows typical standard error bars, indicating that the data have been analyzed statistically.

Or consider the following: "The average global temperature was 0.1°C higher this year than last year, proving that global warming is upon us." Again, fact is followed by conclusion. Is the conclusion justified? We would need a lot more information to tell. For example, the conclusion might not be warranted if fluctuations up and down in the range of 0.3°C can occur in different years just by random chance.

Understand the difference between correlation and causation

Just because there seems to be a pattern or relationship (a correlation) between two variables does not prove that one caused the other. For example, suppose you read that the percentage of homosexuals is higher among college graduates than among those with a high school diploma. Does that mean that a college education *causes* homosexuality?

Part of the confusion regarding correlation and causation arises because the two *are* often associated. For example, college graduates earn higher average salaries than high school graduates, a well-known correlation for which there is good evidence of a true causation. But correlation alone is not proof of causation. Be skeptical of statements about cause that are based only on correlation.

Recap **Healthy skepticism, a basic understanding of statistics, and an ability to read graphs are important tools for critical thinking. Know anecdotal evidence when you see it, and appreciate the differences between fact and conclusion and between correlation and causation.**

1.6 The role of science in society

How do we place science in its proper perspective in our society? Why do we bother spending billions of dollars on scientific research when there are people starving in the streets? These are vital questions for all of us, so perhaps we need to take a hard look at why we study the natural world in the first place.

Science improves technology and the human condition

Science gives us information about the natural world upon which we can base our societal decisions.

How does science affect your life?

Throughout history some of the greatest benefits of science have been derived from the *application* of science, called **technology,** for the betterment of humankind (Figure 1.12). Time and time again, scientific knowledge has led to technological advances that have increased the productivity and hence prosperity of both industries and nations.

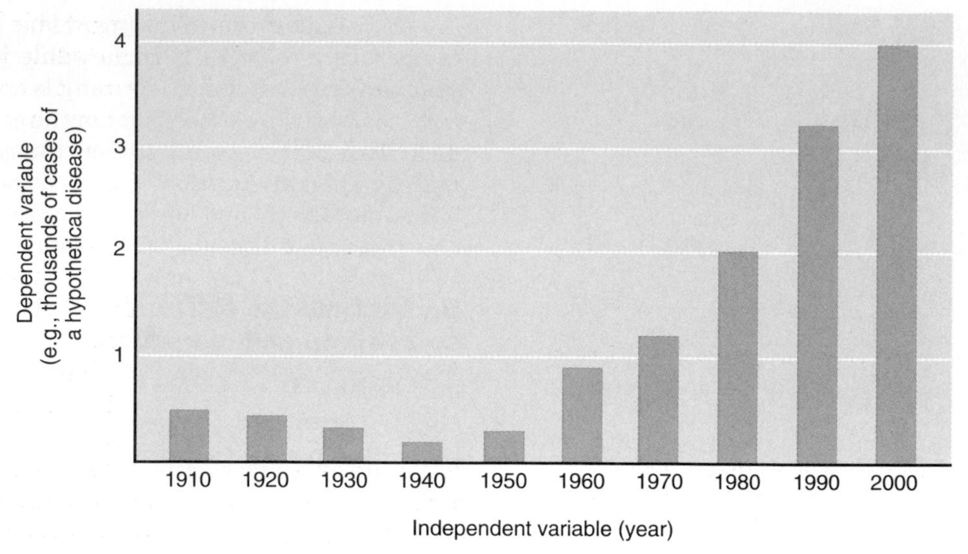

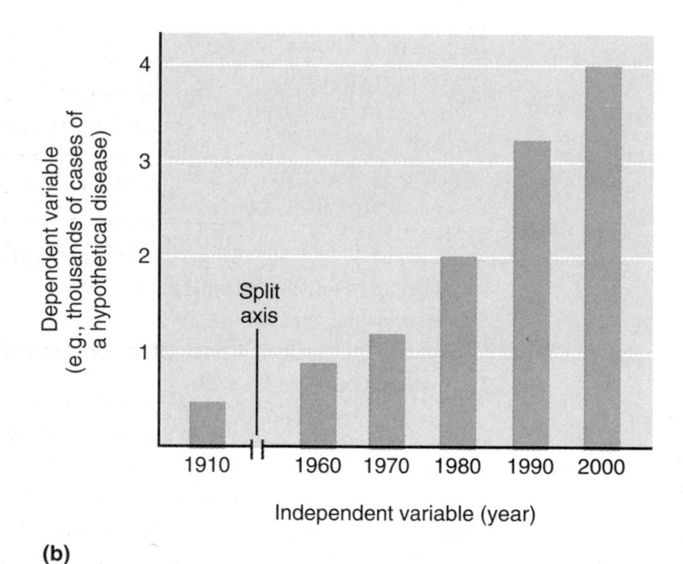

Figure 1.11 How a split axis affects a graph. The graph in **(b)** is redrawn from the data in **(a)** by splitting the abscissa and omitting the data for the years 1920–1950. The effect is a consolidated graph that fits in less space, but it might mislead you into thinking that the number of cases of the disease has been rising steadily since 1910, instead of only since 1960.

Science has given us larger crop yields, more consistent weather predictions, improved construction materials, better health care, and more efficient and cleaner sources of power, to name just a few benefits. It has made possible transportation and communication around the Earth.

Many people are concerned that overuse of our technological capabilities may lead to environmental abuse. Because this is possible, science can be deployed to detect problems early on and give us a chance to make decisions that will positively impact the future. For example, we can see the early warning role of science as scientists seek to determine whether global warming is occurring, and, if so, what to do about it. Only by understanding a problem can we learn how to solve it. Science helps us correct our mistakes.

Science has limits

Scientific knowledge is limited to physical explanations for observable events in the natural world. It cannot prove or disprove the existence of, or importance to us, of nonphysical events that fall outside the natural world, such as faith or spiritual experiences. Many scientists have a strong faith or belief that cannot be tested by science, because it does not depend on logical proof or material evidence. They believe that the search for meaning and the search for knowledge are complementary, not contradictory.

In addition, science alone cannot provide us with the "right" answers to political, economic, social, legal, or ethical dilemmas. Humans have minds, a moral sense, and a sense of history and future. How to use scientific knowledge is up to all of us, not just scientists. For example, given the current state of

(a) Modern farmers grow food crops much more efficiently than in the past, thanks to advances in such diverse fields as genetics, chemistry, and even the aerospace industry. Global positioning satellites and computers allow farmers to administer fertilizers precisely where needed, thereby eliminating waste, reducing environmental degradation, and improving yields.

(b) These scientists are studying the tree-top ecosystem of a rain forest in Panama. Studies such as this improve our understanding of the interrelationships of organisms in an ecosystem and have yielded rare natural chemical compounds that may prove useful in human and animal medicine.

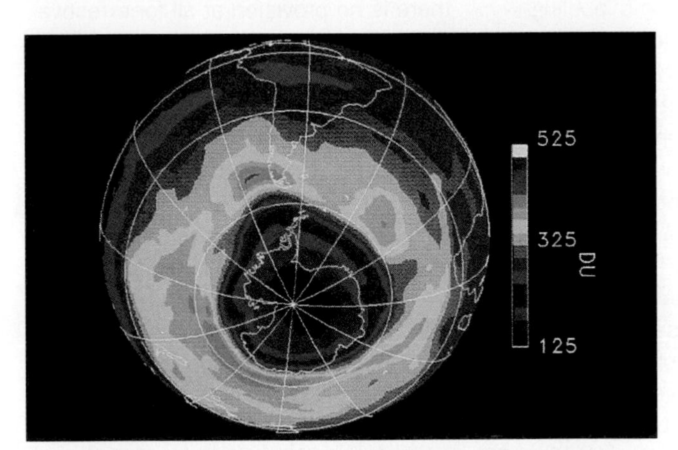

(c) A satellite map documenting depletion of the ozone layer over Antarctica in 1990. The DU color scale measures Dobson units, a measure of atmospheric ozone. The area of greatest depletion appears black. Studies such as this one allow scientists to document when and where ozone depletion occurs so that they can better understand its causes and cures.

knowledge about how cells grow and divide, scientists could probably clone an adult human being. Whether or not we should permit cloning, and if so under what circumstances, is now an important topic of public debate. It's not for scientists to decide alone.

This does not mean scientists are without moral obligation. As experts in their fields, scientists are in a unique position to advise us about the application of scientific knowledge, even if the choices ultimately rest with all of us. Not to advise us might be an abdication of social responsibility.

A practical limitation of science is that some information, including data that might be useful in improving human health, cannot be obtained by observation or experimentation. Our society places very high value on human life, and therefore we don't experiment on humans unless the experiment is likely to be of direct benefit to the subject (the use of experimental cancer drugs falls into this category). This is why it is hard to investigate the danger of street drugs like cocaine or anabolic steroids. No good scientist would ever deliberately give healthy humans a drug likely to cause injury or death, even if the resulting information could save countless lives in the future. As a substitute for human subjects, our society currently permits experimentation on animals, provided that federal guidelines are strictly followed (Figure 1.13).

The importance of making informed choices

You live in a science-oriented society. Throughout this book we present a common theme: Every day you make decisions about how you and society choose to use the knowledge that science gives us. Whether or not you are conscious of it, whether or not you deliberately take action, you make these choices daily.

Should Olympic athletes be allowed to use body-building drugs? Do you think the use of pesticides is justified in order to feed more people? How do you feel about the cloning of human beings? Are you willing to eat a proper diet in order to stay healthy, and, by the way, what *is* a proper diet? Who should pay for health care for the poor?

Text continues on page 20

Figure 1.12 (at left) Benefits of science. These photos show typical scenes of scientists and the application of science (technology) for the betterment of the human condition.

CurrentIssue

Are Dietary Supplements Safe and Effective?

In 1998, baseball slugger Mark McGwire hit a record 70 home runs. He was also taking a legal but controversial dietary supplement called androstenedione, or "Andro," a hormone-like compound that supposedly helps weightlifters increase muscle mass.

The group of products known as *dietary supplements* includes not only vitamins and minerals, but also a wide array of hormone-based compounds (androstenedione and melatonin), amino acids, and herbs (echinacea, ginseng, and gingko biloba). They are supposed to be taken only "as a supplement to the diet," rather than for the treatment of specific diseases. According to some surveys, more than half of the adult U.S. population uses dietary supplements. Mark McGwire is just one of many who believe that the particular products they are using are safe and that they work as advertised.

But how safe and effective are they? Although many dietary supplements *may* be both safe and effective, most have never been *tested* for safety and effectiveness. Dietary supplements are exempt from the strict regulations governing prescription and over-the-counter drugs. Why are the laws governing drugs and dietary supplements different, and what implications does that have for you, the consumer?

Testing and Approving New Drugs: Science in Action

Most new drugs are discovered and tested by pharmaceutical companies in a time-consuming and expensive process that is a model for the scientific method in action. By law, the pharmaceutical company must *prove beyond a reasonable doubt that a new drug is both safe and effective* before it can be sold to the public. The U.S. Food and Drug Administration (FDA) is responsible for seeing that the law is upheld. The process of new drug testing and approval includes several steps. First, studies are conducted in at least two species of animals (usually rats and mice) to determine the drug's safety (it does no harm) and effectiveness (it has a positive effect on a specific disease or condition). Next, three stages of tests are conducted on humans under carefully controlled conditions. In *Phase 1*, experiments are closely monitored and are usually conducted on only a small group (perhaps 20–80) healthy volunteers. The primary issue is safety. If the drug appears to be safe, *Phase II* clinical trials are conducted on larger groups of humans (perhaps several hundred). Safety is still closely monitored, but now an additional goal is to determine whether the drug is effective in treating a specific human disease or condition. If after Phase II the drug has (so far) proven to be both safe and effective, *Phase III* clinical trials are performed on 1,000–4,000 people in order to be as sure as possible that the consequences of the widespread use of the drug are known and understood.

The FDA closely monitors all of the studies. If it approves the drug for sale it then oversees the drug's labeling—the information on the drug's packaging and package inserts. The labels must describe the drug's ingredients, explain how the drug should be used, and warn of possible adverse reactions and interactions with other drugs. Even after the drug is available for prescription use, the company must send periodic reports to the FDA on its manufacturing processes and must report all adverse reactions to the drug.

All this is expensive. On average it takes 12–15 years to bring one new drug to the market, at a cost to the company of about $500 million. The companies can afford to go through this process only because they can patent the drug, giving them exclusive ownership and marketing rights for a certain number of years. Although patients sometimes complain about the high cost of prescription medications, pharmaceutical companies respond that the price reflects their development costs.

Dietary Supplements are Exempt from the Drug Approval Process

Dietary supplements occur naturally in herbs and minerals. Since many of them have been in use for hundreds and even thousands of years, they can't be "discovered" and they can't be patented. Without patent protection, few, if any, manufacturers could afford $500 million for product research and testing. Recognizing this, in the early 1990s representatives of the dietary supplements industry asked Congress to exempt dietary supplements from the FDA drug approval process. Congress passed the Dietary Supplements Health and Education Act (DSHEA) in 1994. The DSHEA holds dietary supplements to a much looser standard called Generally Recognized As Safe (GRAS), meaning that they can be sold *until such time that the product is proven unsafe.* Note that the burden of scientific proof (and the expense) for determining safety is shifted from the manufacturer to the FDA, and that there is no provision at all for effectiveness. As a result, dietary supplements are virtually unregulated, since the FDA does not have the budget for the necessary scientific studies.

Several other provisions of the DSHEA favor the dietary supplements industry. First, quality control of the dietary supplements manufacturing process need not be reported to the FDA. There is no required assurance, then, that a dietary supplement contains the stated amount of active ingredient and is free of impurities or contaminants. Second, the manufacturer is not required to report adverse side effects to the FDA, even though the FDA is responsible for ruling on the safety of the product. The FDA must rely on voluntary information supplied by consumers and health professionals. Third, manufacturers have considerable latitude in advertising their products as long as they do not refer to the treatment of specific conditions or diseases. They can even make "structure-function" claims based on the scientific literature, such as "calcium builds strong bones," but not statements about specific diseases like "This product reduces the risk of osteoporosis."

Enactment result of the DSHEA resulted in rapid growth of the entire dietary supplements industry. Since the 1994 law was passed U.S. sales of dietary supplements have soared by nearly 80%, to nearly $16 billion in 2000.

Questions of Safety and Effectiveness Remain

Opinions differ over how safe and effective the dietary supplements are and what should be done about it. Opponents of extensive (and expensive) regulation and testing point out that many dietary supplements have been in use for a long time. They argue that any potential adverse effects should have emerged by now, so regulation and additional testing isn't needed. Proponents of regulation and testing argue that the availability of more purified extracts increases the chances of abuse. They also point out that interactions between dietary supplements and modern pharmaceutical drugs are largely unknown and could potentially be dangerous. A few adverse interactions are suspected but not yet well documented. One that has been documented is an adverse interaction between the herbal antidepressant St. John's wort and the heart drug digoxin.

What little data is available regarding safety is somewhat contradictory. For example, from 1993 to 1998 the FDA reported only about 500 "adverse events" and fewer than 50 deaths per year associated with dietary supplements. Remember, however, that there is no requirement that the FDA be notified of adverse events. In contrast to the FDA's numbers, the American Association of Poison Control Centers received nearly 7,000 reports involving dietary supplements in 1998 alone.

Faced with the "anything-goes" mentality of a few manufacturers and a current lack of federal oversight, even industry leaders now admit that a better regulatory system may be needed. In the meantime, a few independent research and testing laboratories now routinely test the products of various manufacturers. The results have not always been positive for all manufacturers. For example, in a recent random test of the potency of a saw palmetto product commonly used for prostate enlargement, an independent laboratory reported that four of 21 products tested did not meet a defined potency standard. Individual states may also test dietary supplements. In 1998, California investigators found that nearly a third of all imported Asian herbal remedies they tested contained lead, arsenic, mercury, or drugs not mentioned on the label. Although the results of independent tests may be painful for some manufacturers, they encourage self-regulation of the industry, improve consumer confidence, and relieve the need for strict federal regulation.

Testing the actual content of these products is just the first step. In the long run consumers will want to have some assurance that the products are safe and that the product claims are truthful. How to achieve that goal is an issue we will need to confront. In the meantime, it's buyer beware.

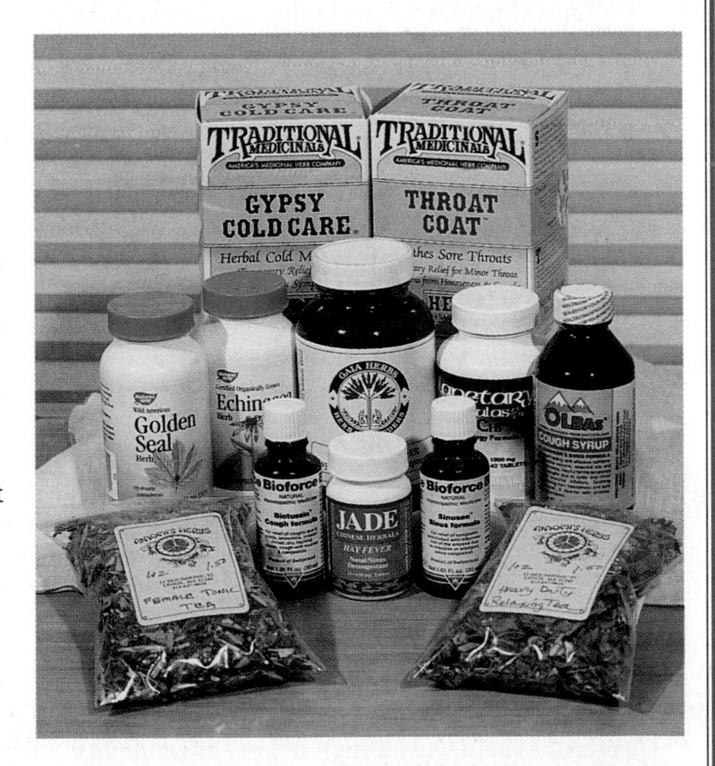

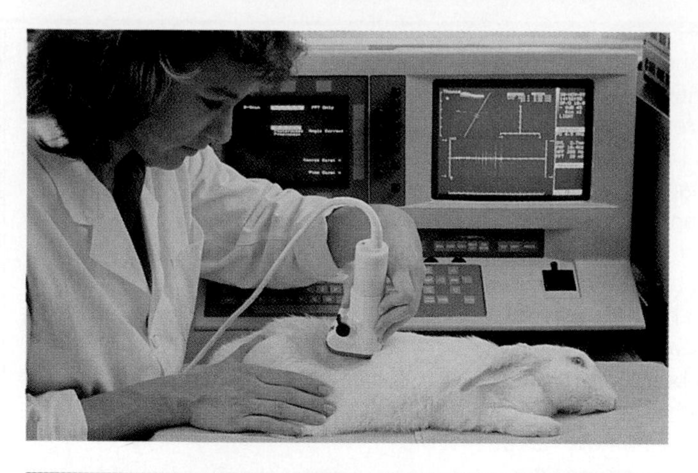

Figure 1.13 Animals in research. In this society we allow the use of animals for research in certain circumstances. This researcher is using a non-invasive technique for measuring blood flow in the skin as part of a study of vascular diseases.

Our knowledge has advanced so rapidly that by the end of this book you will know more about genetics and evolution than the scientists who originally developed the theories about them. With knowledge comes the responsibility for making choices. From global warming to genetic engineering to personal health, each of us must deal with issues that concern our well-being and the future of the biological world in which we live. We owe it to ourselves, as individuals and as a society, to acquire the knowledge and skill needed to make intelligent decisions. Your choices can make a difference.

Recap **Science and technology have improved the human condition. Science cannot, however, resolve moral dilemmas. Scientists can advise us on issues of science, but how scientific knowledge should be put to use is our choice.**

Chapter Summary

The characteristics of life

- All living things acquire both matter and energy from their environment, transforming them for their own purposes.
- The basic unit of life is a single cell.
- Living things maintain homeostasis, respond to their external environments, and reproduce.

How humans fit into the natural world

- The biological world can be organized into kingdoms. A common five-kingdom scheme contains Monera, Protista, Fungi, Plantae, and Animalia.
- Features that define humans are bipedalism, well-developed opposable thumbs, a large brain, and the capacity for complex language.
- Humans are part of communities of different organisms living together in various ecosystems.

Science is both a body of knowledge and a process

- Scientific knowledge allows us to describe and predict the natural world.
- The *scientific method* is a way of thinking, a way of testing statements about the natural world (*hypotheses*) by trying to prove them false.
- A *theory* is a hypothesis that has been extensively tested and that explains a broad range of scientific facts with a high degree of reliability.

Sources of scientific information vary in style and quality

- Articles in peer-reviewed scientific journals are generally are written for other scientists. They are difficult to read but very accurate.
- Popular science journals, books, and television shows present scientific knowledge efficiently to the general public.
- Web sites on the Internet range widely in terms of quality and accuracy of information.

Learning to be a critical thinker

- Skepticism is a questioning attitude ("prove it to me"). Critical thinking requires skepticism.
- Knowing how to read graphs and understand basic statistics can help you evaluate numerical data.
- Being able to recognize an anecdote, tell fact from conclusion, and distinguish between correlation and causation can help you evaluate the truth of a claim.

The role of science in society

- The application of science is called technology.
- Science is limited to physical explanations of observable events.
- How to use science is up to us.

Terms you should know

biology, 2
control group, 11
data, 14
experiment, 10
experimental group, 11
homeostasis, 3
hypothesis, 9
science, 2
scientific method, 9
theory, 12
variable, 10

Test Yourself

Matching Exercise

____ 1. The kingdom of life to which humans belong
____ 2. The kingdom of life to which the bacteria belong
____ 3. The process of science
____ 4. An experiment designed and conducted under strictly controlled conditions
____ 5. A main hypothesis that has withstood the test of repeated experimentation
____ 6. A questioning attitude
____ 7. The application of science

a. scientific method
b. controlled experiment
c. theory
d. Animalia
e. technology
f. Monera
g. skepticism

Fill in the Blank

8. The study of living organisms and life's processes is called ____________.
9. The natural world consists of all ____________ and all ____________.
10. The ability to walk on two legs is called ____________.
11. The first step in the scientific method is to observe and ____________.
12. Scientists design and conduct ____________ in order to test their predictions and hypotheses.
13. The ____________ group in a controlled experiment, receives the treatment and the ____________ group receives a false treatment called a placebo.
14. The field of mathematics that deals with organizing and interpreting numbers (probability) is called ____________.
15. In graphs, the independent variable is usually plotted on the ____________ and the dependent variable is usually plotted on the ordinate.

True or False?

16. All living things are comprised of cells (or are a single cell).
17. A hypothesis must be testable if it is to be of any use.
18. Conclusions drawn from an experiment are only valid for the conditions under which the experiments were done.
19. Some of the very best scientific evidence is anecdotal evidence.
20. A good correlation between two variables always means that one causes the other.

Concept Review

1. Living things have a different molecular composition than nonliving things. What *other* characteristic of living things permits this?
2. Explain what is meant by the term *homeostasis*.
3. Name four features that together contribute to our uniqueness and define us as human.
4. Describe the difference between a hypothesis and a prediction (or working hypothesis).
5. Discuss the role of scientists in helping us solve economic, social, and ethical dilemmas.

Apply What You Know

1. A magician has a coin that he says (hypothesizes) has heads on both sides, but he's unwilling to show you both sides. To convince you, he flips it three times and gets heads each time. Do you believe that the coin has two heads? What if he gets heads 10 times in a row? 100 times? What would it take (by coin flip) to prove that the coin does *not* have two heads? With this example, explain the difference between having relative confidence in the truth of a hypothesis, proving it to be true, and proving it to be false.
2. From the example used in the text that "Drug X would be a safe and effective treatment for high blood pressure in humans," create a specific predictive statement designed to test any aspect of its safety you wish. Compare your hypothesis with those of other class members.
3. Your roommate is writing a paper on the subject of cocaine and birth defects in humans and wonders why there don't seem to be published reports of controlled experiments in humans on the subject; all the studies are on rats! Describe to her how such a controlled experiment would have to be designed and conducted, and convince her that it would never be permitted by any responsible regulatory agency.

Case in Point

Child Denied Entrance to School

School officials in a small unnamed community in Pennsylvania denied a first-grader entrance to school recently because he had not had his required vaccinations for polio, mumps, and measles. The parents had refused all vaccinations for the child on the grounds that they carry a small risk. School officials explained that if vaccinations were not required of all school children, enough children would fail to get them that the risk of an epidemic of these diseases could return.

QUESTIONS:

1. Under what circumstances should government be able to compel people to do what is good for others?
2. How should the decision be made in this particular case, and who do you think is best qualified to make the decision? Discuss your opinions with other members of the class.

INVESTIGATE:

Kalb, C., and D. Foote. Necessary Shots? *Newsweek* Sept. 13, 1999, pp. 73–74.

References and Additional Reading

Brennan, Richard P. *Dictionary of Scientific Literacy*. New York: Wiley, 334 pp. (A handy quick reference guide to the vocabulary and core knowledge of science.)

Carey, Stephen S. *A Beginner's Guide to Scientific Method*. 2nd ed. Belmont, CA: Wadsworth, 1997. 200 pp. (A description of the methods underlying scientific research. Designed for nonscience students.)

May, Robert M. How Many Species Inhabit the Earth? *Scientific American* 267(4):42, Oct. 1992. (Describes efforts to catalog the number of species that inhabit Earth.)

Peirce, Andrea. *The American Pharmaceutical Association Practical Guide to Natural Medicines*. New York: Morrow, 1999. (An encyclopedia of dietary supplements and natural medicines.)

Sagan, Carl. *The Demon-Haunted World: Science as a Candle in the Dark*. New York: Ballantine, 1996. 457 pp. (A book that expresses the curiosity, amazement, and skepticism of science by a leading science author.)

CHAPTER 2

THE CHEMISTRY OF LIVING THINGS

Why is water essential **to life?**

How do organisms create **so many different** proteins?

What binds atoms **to each other?**

Why is carbon **a common** structural element?

How are organisms chemically different **from each other?**

AS WE MENTIONED IN CHAPTER 1, THE NATURAL WORLD CONsists of matter and energy. *Chemistry* is the study of matter and the energy that causes matter to combine, break apart, and recombine into all the substances, both living and nonliving, that exist in the natural world. Every substance is made from the same basic units of matter and the same types of energy (Figure 2.1). Everything in the natural world is governed by the laws of chemistry, including you and me.

Yet there is something very special about the chemistry of living things. Living organisms have the ability to grow and reproduce, unlike inanimate objects. Living things have evolved to take advantage of the rules of chemistry that govern the natural world. Living organisms create special combinations of matter not generally found in nonliving things. They also have developed the ability to store energy so that they may later turn it to their own purposes.

In this chapter we consider how the laws of chemistry serve life. We begin with an introduction to basic chemistry.

2.1 All matter consists of elements

Matter is anything that has mass and occupies space. All matter is comprised of elements. An **element** is a fundamental (pure) form of matter that cannot be broken down to a simpler form. Aluminum and iron are elements, and so are oxygen and hydrogen. There are just over 100 known elements, and together they account for all matter on Earth.

Atoms are the smallest functional units of an element

Elements are made up of particles called atoms. An **atom** is the smallest unit of any element that still retains the physical and chemical properties of that element. Although we now know that atoms can be split apart under unusual circumstances (such as a nuclear reaction), atoms are the smallest units of matter that can take part in chemical reactions. So, for all practical purposes, atoms are the smallest functional units of matter.

Even the largest atoms are so small that we can see them only with specialized microscopes. Chemists can also infer what they look like from studying their physical properties (Figure 2.2).

The central core of an atom is called the *nucleus*. The nucleus is comprised of positively charged particles called **protons** and a nearly equal number of neutral particles called **neutrons,** all tightly bound together. An exception is the smallest atom, hydrogen, whose nucleus consists of only a single proton. Smaller negatively charged particles called **electrons** orbit the nucleus. Because electrons are constantly moving, their precise position at any one time is unknown. You may think of electrons as occupying one or more spherical clouds of negative charge around the nucleus called *shells*. Each shell can accommodate only a certain number of electrons. The first shell, the one closest to the nucleus, can hold two electrons, the second can accommodate eight, and the third shell (if there is one) can contain even more. Each type of atom has a unique number of electrons. Under most circumstances the number of electrons equals the number of protons, and, as a result, the entire atom is electrically neutral.

Protons and neutrons have about the same mass and both have much more mass than electrons. (*Mass* is measured chemically and is not dependent on gravity. For the purpose of this text, however, mass and *weight* are about the same. Over 99.9% of an atom's mass is due to the protons and neutrons in its nucleus.)

In chemical equations and sometimes in writing, atoms are designated by one- or two-letter symbols taken from English or Latin. For example, oxygen is designated by the letter O, nitrogen by N, sodium by Na (from the Latin word for sodium, *natrium*), and

(a)

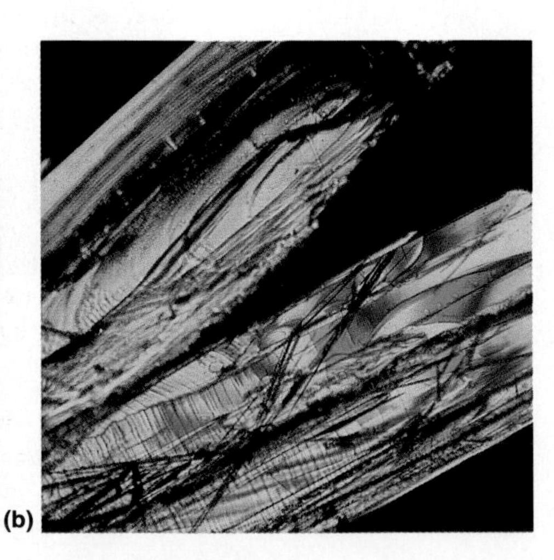

(b)

Figure 2.1 All matter is made of atoms. (a) A magnified view (×15,000) of a portion of a skeletal muscle cell. **(b)** A magnified view (×30) of a crystal of aspirin. The three most common atoms in both muscle and aspirin are carbon, hydrogen, and oxygen.

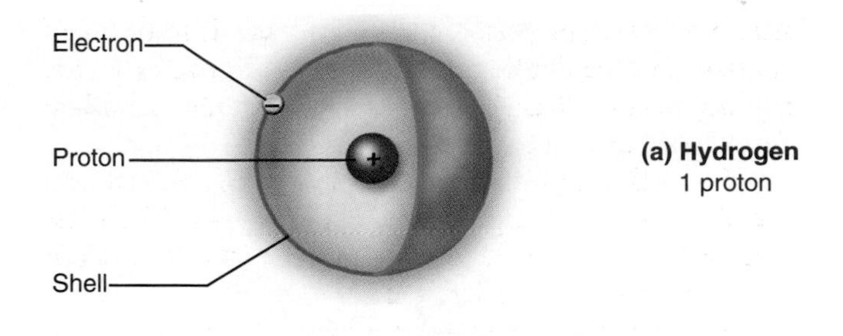

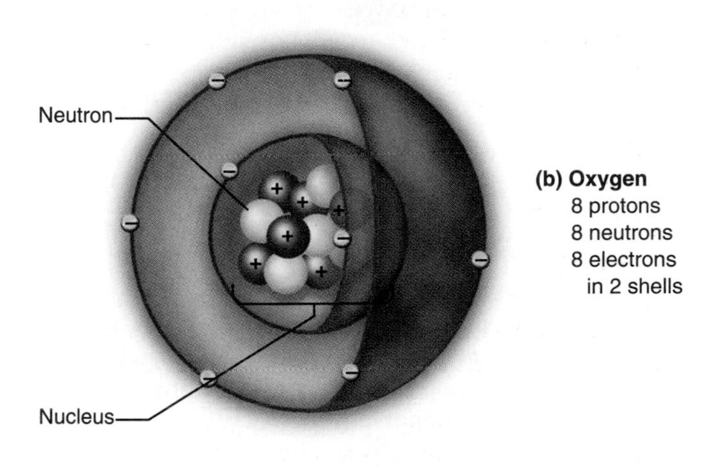

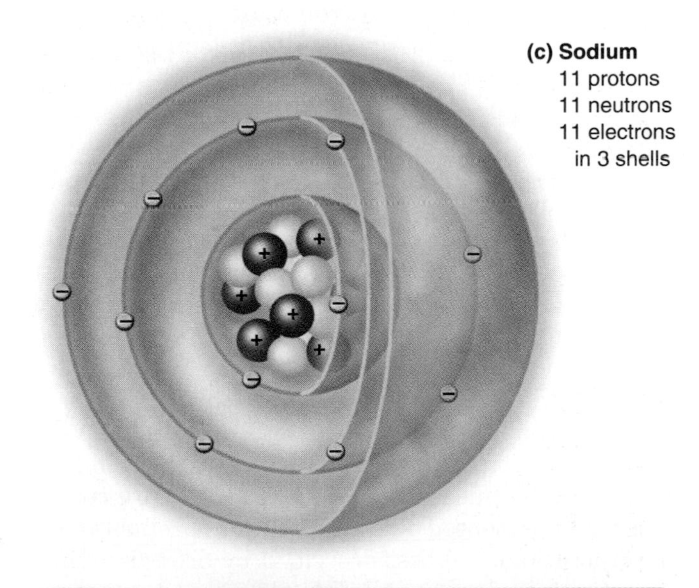

Figure 2.2 The structure of atoms. Atoms consist of a nucleus comprising positively charged protons and neutral neutrons surrounded by spherical shells of negatively charged electrons.

potassium by K (Latin *kalium*). A subscript numeral following the symbol indicates the numbers of atoms of that element. For example, the chemical formula O_2 represents two atoms of oxygen linked together, the most stable form of elemental oxygen.

In addition to a symbol, atoms have an *atomic number* representing the characteristic number of protons in the nucleus and an *atomic mass* (or mass number), which is generally fairly close to the total number of neutrons and protons.

Isotopes have a different number of neutrons

Although all the atoms of a particular element have the same number of protons, the number of neutrons can vary slightly. Atoms with either more or fewer neutrons than the usual number for that element are called **isotopes.** Isotopes of an element have the same atomic number as the more common atoms but a different atomic mass. For example, elemental carbon typically consists of atoms with six protons and six neutrons, for an atomic mass of 12. The isotope of carbon known as "carbon-14" has an atomic mass of 14 because it has two extra neutrons.

Isotopes are always identified by a superscript mass number preceding the symbol. For instance, the carbon-14 isotope is designated ^{14}C. The superscript mass number of the more common elemental form of carbon is generally omitted because it is understood to be twelve.

Many isotopes are unstable. Such isotopes are called *radioisotopes* because they tend to give off energy (in the form of radiation) and particles until they reach a more stable state. The radiation emitted by radioisotopes can be dangerous to living organisms because the energy can damage tissues.

Certain radioisotopes have a number of important scientific and medical uses. Because the rate of decay to more stable energy states is known for each radioisotope, scientists can determine when rocks and fossils were formed by measuring the amount of radioisotope still present. ^{14}C is commonly used for this purpose. In medicine, radioisotopes are used to "tag" molecules so their location within the body can be tracked by radiation sensors. For example, physicians use radioisotopes to locate areas of damaged tissue in a patient's heart after a heart attack. Radioisotopes are also used to target and kill certain kinds of cancer. Certain radioisotopes that emit energy for long periods of time are used as a power supply in heart pacemakers.

Recap **Atoms are made up of protons, neutrons, and electrons. Radioisotopes are unstable atoms with an unusual number of neutrons that give off energy and particles as they decay to a more stable state.**

2.2 Atoms combine to form molecules

A **molecule** consists of a stable association between two or more atoms. For example, a molecule of water is two atoms of hydrogen plus one atom of oxygen (written H_2O). A molecule of ordinary table salt (written NaCl) is one atom of sodium (Na) plus one atom of chlorine (Cl). A molecule of hydrogen gas (written H_2) is two atoms of hydrogen. In order to understand *why* atoms join together to form molecules, we need to know more about energy.

Energy fuels life's activities

Energy is the capacity to do work; the capacity to cause some change in matter. Joining atoms is one type of work, and breaking up molecules is another—and both require energy. Stored energy that is not actually performing any work at the moment is called **potential energy** because it has the *potential* to make things happen. Energy that is actually *doing* work—that is, energy in motion—is called **kinetic energy.**

You can visualize the difference between potential energy and kinetic energy in the water held behind a dam: there is tremendous potential energy in the water held in reserve. When the water is released, potential energy is converted into kinetic energy: rushing water that can be put to work turning turbines. Similarly, the spark of a match converts the potential energy in firewood to kinetic energy in the form of heat and light.

Potential energy is stored in the bonds that hold atoms together in all matter, both living and nonliving. Living organisms take advantage of this general principle of chemistry by using certain molecules to store energy for their own use. When the chemical bonds of these energy-storage molecules are broken, potential energy becomes kinetic energy (Figure 2.3). We rely on this energy to power "work" such as breathing, moving, digesting food, and many other tasks.

Matter is the most stable when it is at the *lowest possible energy level,* that is, when it contains the least potential energy. This has important implications for the formation of molecules, because even single atoms contain energy.

Electrons have potential energy

Recall that electrons carry a negative charge, whereas protons within the nucleus have a positive charge. Electrons are attracted to the positively charged nucleus and repelled by each other. As a result of these opposing attractive and repulsive forces, each electron occupies a specific shell around the nucleus. Each shell corresponds to a specific level of electron potential energy, and each shell farther out represents a higher potential energy level than the preceding one closer to the nucleus. When an electron moves to a shell closer to the nucleus it loses energy. In order to move to a shell that is farther from the nucleus, the electron must absorb energy.

(a)

(b)

Figure 2.3 Energy. (a) Potential energy is locked up in the chemical bonds of the fats, carbohydrates, and other energy-storage molecules in this person's tissues. **(b)** Kinetic energy is energy in motion.

Although the energy levels of electrons can be described as a series of concentric shells, the actual location of an electron within the volume of that shell is hard to pin down. Within any given shell, one or two electrons are most likely to be located in certain volumes of space called *orbitals* (Figure 2.4). An orbital describes the area in which the electron is likely to be found 90% of the time. The innermost electron energy shell is represented by one spherical orbital containing one or two electrons. The second shell, which can contain up to eight electrons, consists of one spherical orbital and three dumbbell-shaped orbitals, each oriented on a different plane.

Atomic structures are depicted in several different ways, because electron energy levels are conveniently represented by shells, whereas probable electron locations are best described by orbitals. While our discussions of chemical bonds will utilize images of electrons occupying concentric circles corresponding to energy shells, bear in mind that this simple model does not accurately describe an electron's actual location.

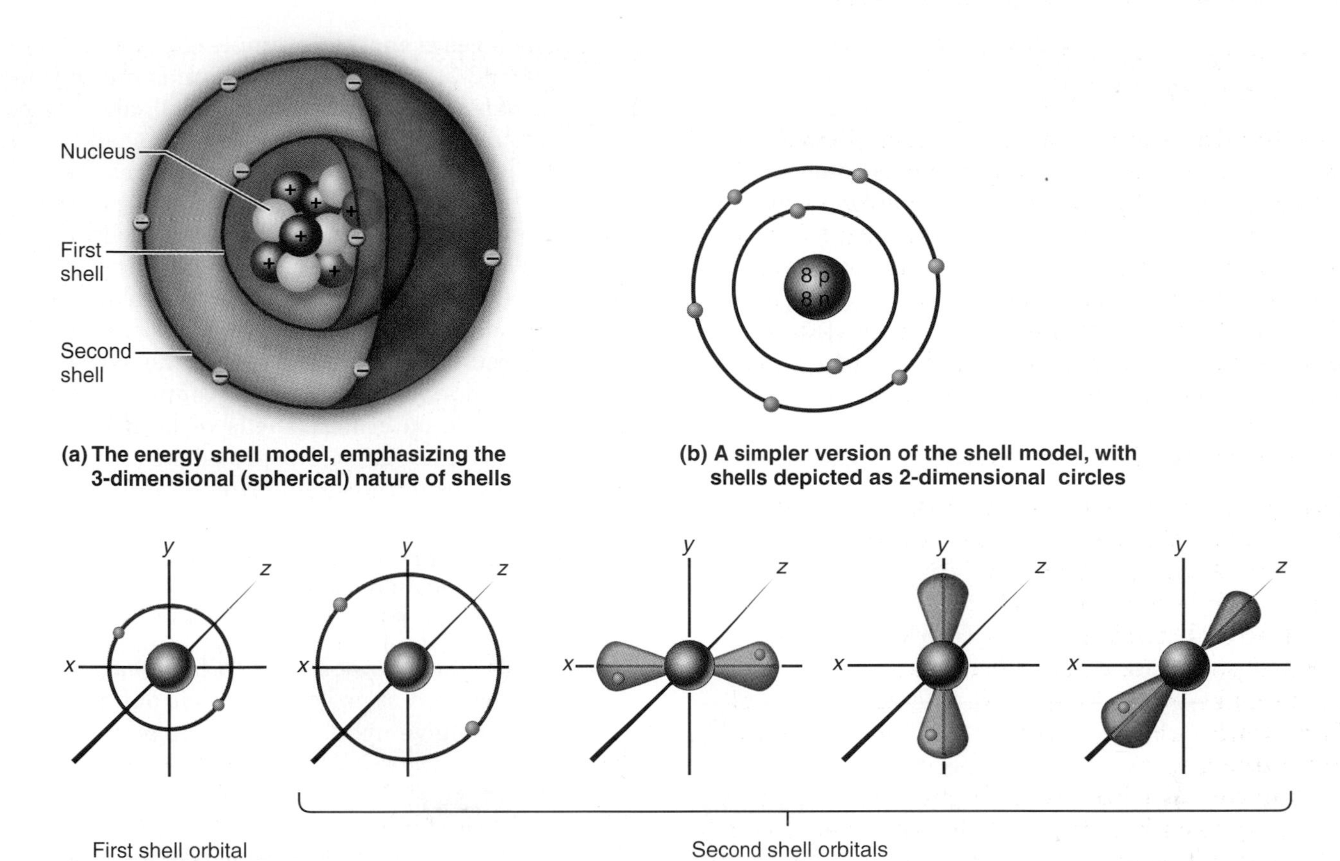

Figure 2.4 Three ways to depict atomic structure. The atom depicted here is oxygen.

Recap **The energy levels of electrons are called shells. Electrons in the shell closest to the nucleus have the least potential energy, and those in the outermost shell have the most potential energy. The probable location of an electron is described by its orbital.**

Chemical bonds link atoms to form molecules

A key concept in chemistry is that *atoms are most stable when their outermost occupied electron shell is completely filled.* An atom whose outermost electron shell is not normally filled tends to interact with one or more other atoms in a way that fills its outermost shell. Such interactions generally cause the atoms to be bound to each other by attractive forces called **chemical bonds.** The three principal types of chemical bonds are called covalent, ionic, and hydrogen bonds (Table 2.1).

What binds atoms to each other?

Table 2.1 Summary of the three types of chemical bonds

Type	Strength	Description	Examples
Covalent bond	Strong	A bond in which the sharing of electrons between atoms results in each atom having a maximally filled outermost shell of electrons.	The bonds between hydrogen and oxygen in a molecule of water.
Ionic bond	Moderate	The bond between two oppositely charged atoms or molecules that were formed by the permanent transfer of one or more electrons.	The bond between Na^+ and Cl^- in salt.
Hydrogen bond	Weak	The bond between oppositely charged regions of molecules that contain covalently bonded hydrogen atoms.	The bonds between molecules of water.

Covalent bonds involve sharing electrons One way that an atom can fill its outermost shell is by sharing a pair of electrons with another atom. An electron-sharing bond between atoms is called a **covalent bond.** Covalent bonds between atoms are among the strongest chemical bonds in nature, so strong that they rarely break apart. In structural formulas, a covalent bond is depicted as a line drawn between two atoms.

Spotlight on Structure & Function

Hydrogen gas offers an example of how a covalent (electron-sharing) bond fills the outermost shells of two atoms (Figure 2.5a). Each of the two hydrogen atoms have just one electron in the first shell, which could accommodate two electrons. When joined together by a covalent bond (forming H_2, a gas) each atom has, in effect, a "full" first shell of two electrons. As a result, H_2 gas is more stable than the same two hydrogen atoms by themselves. The sharing of one pair of electrons, as in H_2, is called a *single bond.*

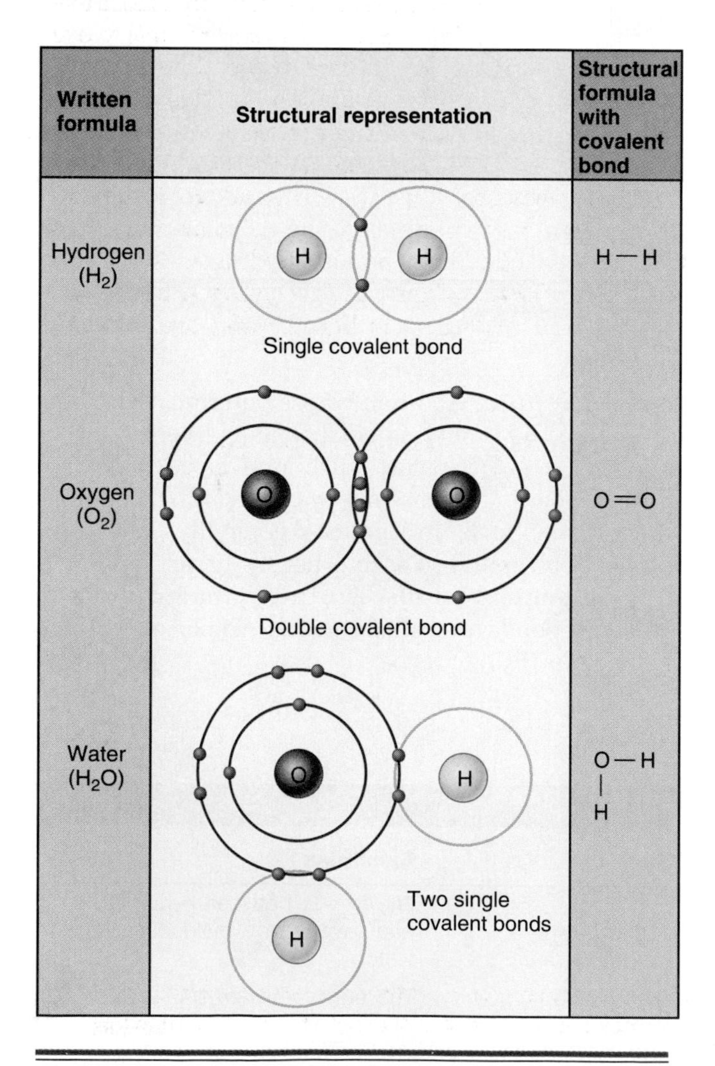

Figure 2.5 Covalent bonds. Sharing pairs of electrons is a way for atoms to fill their outermost shell.

Oxygen gas is another example of covalent bonding (Figure 2.5b). An oxygen atom has 8 electrons: two of these fill the first electron shell, and the remaining six occupy the second electron shell (which can accommodate eight). Two oxygen atoms may join to form a molecule of oxygen gas by sharing two pairs of electrons, thus completing the outer shells of both atoms. When two pairs of electrons are shared, the bond is called a *double bond.* In structural formulas, double bonds are indicated by two parallel lines.

A molecule of water forms from one oxygen and two hydrogen atoms because this combination completely fills the outermost shells of both hydrogen and oxygen (Figure 2.5c). The prevalence of water on earth follows from the simple rule described above: matter is most stable when it contains the least potential energy. That is, both hydrogen and oxygen are more stable when together (as H_2O) than as independent atoms. ■

Ionic bonds occur between oppositely charged ions A second way that atoms can fill their outer shell of electrons is to give up electrons completely (if they have only one or two electrons in their outermost shell) or to take electrons from other atoms (if they need one or two to fill their outermost shell). Such a loss or gain of electrons gives the atom a net charge, because now there are fewer (or more) electrons than protons in the nucleus. The net charge is positive (+) for each electron lost and negative (−) for each electron gained.

An electrically charged atom or molecule is called an **ion.** Examples of ions are sodium (Na^+), chloride (Cl^-), calcium (Ca^{2+}), and hydrogen phosphate (HPO_4^-). Notice that ions can have a shortage or surplus of more than one electron (Ca^{2+} has lost two electrons).

Ever heard the expression "opposites attract"? It should come as no surprise that oppositely charged ions are attracted to each other. This attractive force is called an **ionic bond** (Figure 2.6). In aqueous (watery) solutions, where ionic bonds are not as strong as covalent bonds, ions tend to dissociate (break away) from each other relatively easily. In the human body, for example, almost all of the sodium is in the form of Na^+ and most of the chlorine is in its ionized form, called *chloride* (Cl^-).

Ions in aqueous solutions are sometimes called *electrolytes* because solutions of water containing ions are good conductors of electricity. As you will see, cells can control the movement of certain ions, creating electrical forces essential to the functioning of nerves, muscles, and other living tissues.

Weak hydrogen bonds form between polar molecules A third type of attraction occurs between molecules that do not have a net charge. Glance back at the wa-

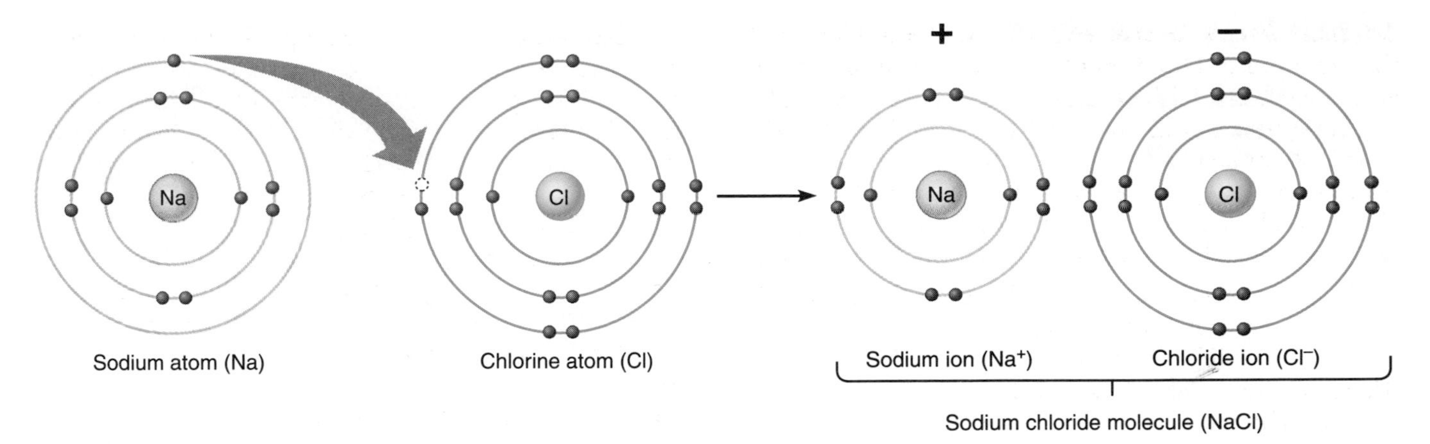

Figure 2.6 Ionic bonds. Electrically charged ions form when atoms give up or gain electrons. The oppositely charged ions are attracted to each other, forming an ionic bond.

ter molecule in Figure 2.5 and note that the two hydrogen atoms are not found at opposite ends of the water molecule, but are fairly close together. Although the oxygen and the two hydrogen atoms share electrons, the sharing is unequal. The shared electrons in a water molecule actually spend slightly more of their time near the oxygen atom than near the hydrogen atoms because the oxygen atom attracts electrons more strongly than do the hydrogen atoms. Although the water molecule is neutral overall, the uneven sharing gives the oxygen end a partial negative charge and the hydrogen end a partial positive charge.

Spotlight on Structure & Function

Molecules such as water that are electrically neutral overall but still have partially charged ends, or *poles,* are called *polar* molecules. According to the principle that opposites attract, polar molecules arrange themselves so that the negative pole of one molecule is oriented toward (attracted by) the positive pole of another molecule. The weak attractive force between oppositely charged regions of polar molecules that contain covalently bonded hydrogen is called a **hydrogen bond** (Figure 2.7). Hydrogen bonds between water molecules in liquid water are so weak

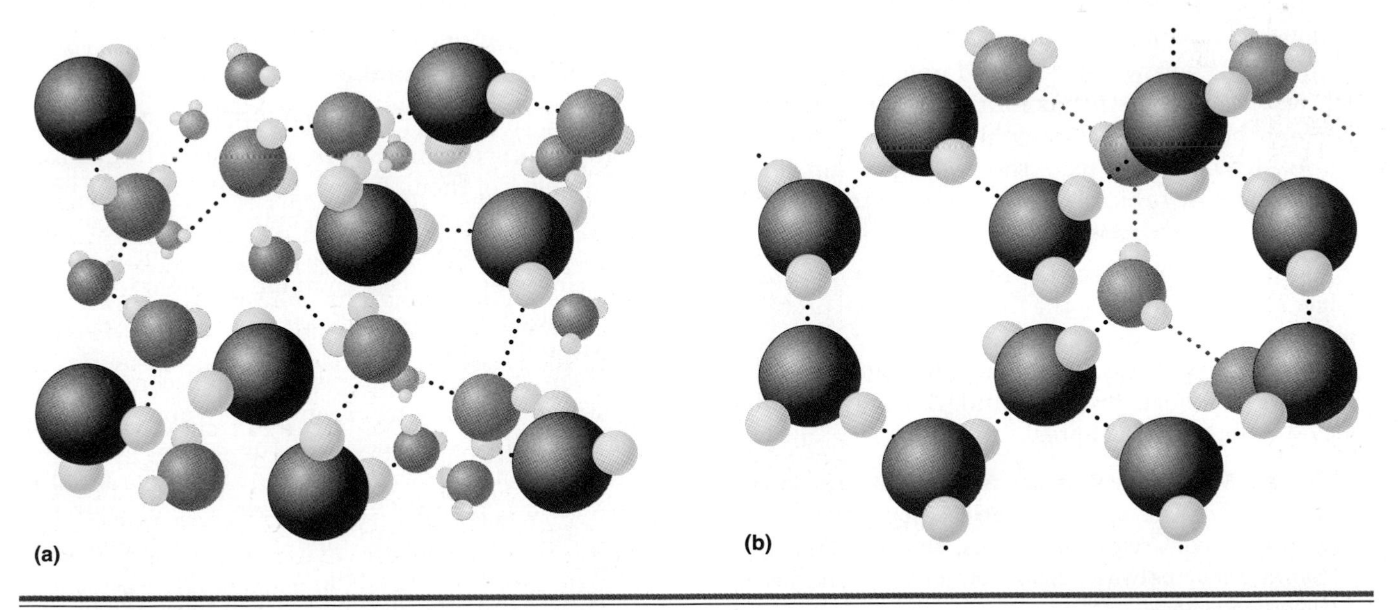

Figure 2.7 Hydrogen bonds. (a) In water, weak hydrogen bonds continually form, break, and re-form between hydrogen and oxygen atoms of adjacent water molecules. **(b)** Ice is a solid because stable hydrogen bonds form between each water molecule and four of its neighbors.

Health Watch

Free Radicals and Antioxidants

An especially unstable class of molecules are oxygen free radicals, sometimes just called **free radicals.** Free radicals are oxygen-containing molecules that have an unpaired electron in their outer shell. They are exceptionally unstable because any unpaired electron has a very high potential energy. Consequently free radicals have a strong tendency to *oxidize* (remove electrons from) another molecule. They set in motion a destructive cascade of events in which electrons are removed from stable compounds, producing still more unstable compounds. Free radicals damage body tissues, and many scientists now believe that they contribute to the aging process.

One of the most destructive free radical molecules is molecular oxygen with an extra electron (O_2^-) called *superoxide.* Other important free radicals include *peroxide* (H_2O_2) and *hydroxyl* (OH). The latter is a hydroxide ion (OH^-) that is missing an electron.

Some free radicals are accidentally produced in small amounts during the normal process of energy transfer within living cells. Exposure to chemicals, radiation, ultraviolet light, cigarette smoke, alcohol, and air pollution may also create free radicals.

We now know that certain enzymes and nutrients called **antioxidants** are the body's natural defense against oxygen free radicals. Antioxidants prevent oxidation either by preventing the formation of free radicals in the first place or by inactivating them quickly before they can damage other molecules. Important antioxidants include vitamin E, vitamin C, beta-carotene, and an enzyme called superoxide dismutase.

Many health experts believe that the antioxidant vitamins reduce the chance of certain cancers and heart disease. For example, several studies show that people who eat lots of fresh tomatoes and tomato products experience lower rates of cancer. Other studies report that high dietary intakes of beta-carotene and vitamin C are associated with a reduced risk of cardiovascular death.

Some scientists hypothesize that maintaining a high level of antioxidants in the body may even slow the aging process. This hypothesis has received support from experiments in fruit flies. In one experiment it was reported that fruit flies bred for long life have higher than normal amounts of superoxide dismutase. In another, an experimental manipulation that increased the production of superoxide dismutase also prolonged the flies' average life span.

Although the evidence is not yet convincing that antioxidants can slow the aging process in humans, it is clear that antioxidants are a healthy addition to your diet. Good sources of antioxidant vitamins include fruit, yellow/orange vegetables, green vegetables, and whole-grain breads and cereals.

---

that they continually break and reform, allowing water to flow. When water gets cold enough to freeze, each water molecule forms four stable, unchanging hydrogen bonds with its neighbors. When water is vaporized (becomes a gas), the hydrogen bonds are broken and stay broken as long as the water is in the gas phase. ■

Hydrogen bonds are very important in biological molecules. They provide the force that gives proteins their three-dimensional shape, and they keep the two strands of the DNA molecule together. The structures of both proteins and DNA will be described later in this chapter.

> ***Recap*** **Strong covalent bonds form between atoms when they share pairs of electrons, ionic bonds form between oppositely charged ions, and weak hydrogen bonds occur between oppositely charged regions of polar molecules.**

Living organisms contain only certain elements

Although there are nearly 100 different elements in nature, living organisms are constructed from a limited number of them. In fact, about 99% of your body weight consists of just six elements: oxygen, carbon, hydrogen, nitrogen, calcium, and phosphorus (Table 2.2). However, the less common elements are still important, and life as we know it would not be possible without them.

Even the largest atoms are small compared to the structures in living organisms. To appreciate the vast size differences between the atoms, cells, and organs in your body, imagine that a sodium atom is the size of a penny. On this scale, one of your red blood cells would be 1/2 mile in diameter, and your heart would be larger than the entire earth!

Next, let's look at some of the most important matter of living systems: water, hydrogen ions, and a host of molecules that contain a backbone of carbon atoms.

Table 2.2 The most common and important elements in living organisms*

Element	Atomic Symbol	Atomic Number	Atomic Mass	% of Humans by Weight	Functions in Life
Oxygen	O	8	16.0	65	Part of water and most organic molecules. Also molecular oxygen
Carbon	C	6	12.0	18	The backbone of all organic molecules
Hydrogen	H	1	1.0	10	Part of all organic molecules and of water
Nitrogen	N	7	14.0	3	Component of proteins and nucleic acids
Calcium	Ca	20	40.1	2	Constituent of bone. Also essential for the action of nerves and muscles
Phosphorus	P	15	31.0	1	Part of cell membranes and of energy storage molecules. Also a constituent of bone
Potassium	K	19	39.1	0.3	Important in nerve action
Sulfur	S	16	32.1	0.2	Structural component of most proteins
Sodium	Na	11	23.0	0.1	The primary ion in body fluids. Also important for nerve action
Chlorine	Cl	17	35.5	0.1	Component of digestive acid. Also a major ion in body fluids
Magnesium	Mg	12	24.3	Trace	Important for the action of certain enzymes and for muscle contraction
Iron	Fe	26	55.8	Trace	A constituent of hemoglobin, the oxygen-carrying molecule

*The elements are listed in descending order of their contribution to total body weight. Atomic number represents the number of protons in the nucleus. Atomic mass is roughly equivalent to the total number of protons and neutrons because electrons have very little mass. Note that 99% of your body weight is accounted for by just six elements.

2.3 Life depends on water

No molecule is more essential to life than water. Indeed, it accounts for 60% of your body weight. The following properties of water are especially important to living organisms:

- Water molecules are polar.
- Water is a liquid at body temperature.
- Water can absorb and hold heat energy.

These properties make water an ideal solvent and an important factor in temperature regulation.

Water is the biological solvent

A **solvent** is a liquid in which other substances dissolve. Water is the ideal solvent in living organisms specifically because it is a polar liquid at body temperature. As the solvent of life, water is the substance in which the many chemical reactions of living organisms take place. No other liquid could support life as we know it. Let's look at a simple example of a substance dissolving in water to better understand how the polar nature of water facilitates the reaction.

Why is water essential to life?

Consider the common and important solid, crystals of sodium chloride (NaCl), or table salt. Crystals of table salt consist of a regular, repeating pattern of sodium and chloride ions held together by ionic bonds (Figure 2.8). When salt is placed in water, individual ions at the surface of the crystal are pulled away from the crystal and are immediately surrounded by the polar water molecules. The water molecules form such a tight cluster around each ion that the ions are prevented from reassociating back into the crystalline form. In other words, water keeps the ions dissolved. Note that the water molecules are oriented around ions according to the principle that opposite charges attract.

Because water is a liquid at body temperature it can flow freely. This makes it an excellent medium for transporting dissolved substances from one place to another. Indeed, the blood in our cardiovascular system is over 90% water. As a liquid, water also occupies space. It fills our cells (the intracellular space) and the spaces between cells (the intercellular space).

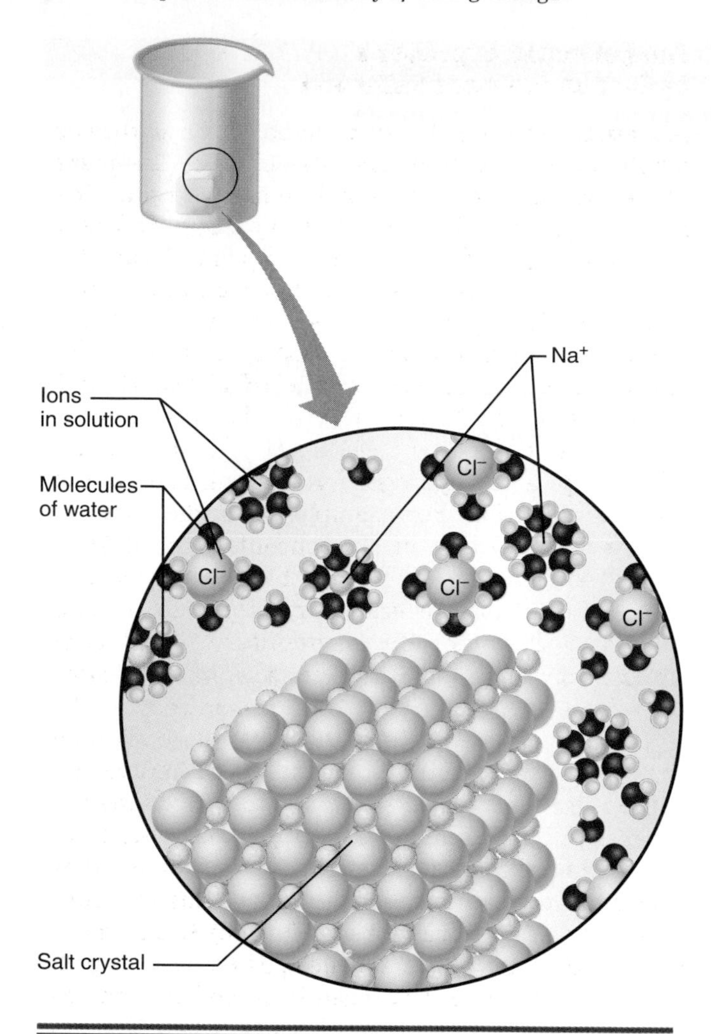

Figure 2.8 How water keeps ions in solution. The slightly negative ends of polar water molecules are attracted to positive ions, whereas the slightly positive ends of water molecules are attracted to negative ions. The water molecules pull the ions away from the crystal and prevent them from reassociating with each other.

Water helps regulate body temperature

An important property of water is that it can absorb and hold a large amount of heat energy with only a modest increase in temperature. In fact, it absorbs heat better than most other liquids. Water thus may prevent large increases in body temperature when excess heat is produced. Water also holds heat well when there is a danger of too much heat loss (such as when standing outside in shorts on a cool day). The ability of water to absorb and hold heat helps prevent rapid changes in body temperature with changes in metabolism or in the environment.

Our bodies generate heat during metabolism. We usually generate more heat than we need in order to maintain a constant body temperature of 98.6° Fahrenheit (37° Celsius), so losing heat is generally more of a priority than conserving it. One way we can lose heat quickly is by evaporation of water (Figure 2.9). When water is in contact with air, hydrogen bonds between some of the water molecules at the surface of the water are broken, and water molecules escape into the air as water vapor. It takes energy to break all those hydrogen bonds, and that energy comes from heat generated by the body and transported to the skin by the blood. Evaporation of sweat is just one of the mechanisms for the removal of heat from the body. How the body regulates body temperature will be discussed in more detail in Chapter 4.

You can demonstrate the cooling power of evaporation for yourself. The next time you perspire heavily, notice that your exposed skin may actually feel cool to the touch.

Recap **Most biological molecules dissolve readily in water because water is a polar molecule. The liquid nature of water facilitates the transport of biological molecules. Water absorbs and holds heat and can lower body temperature via evaporation.**

2.4 The importance of hydrogen ions

One of the most important ions in the body is the hydrogen ion (a single proton without an electron). In this section we will see how hydrogen ions are created and why it is so important to maintain an appropriate concentration of them.

Figure 2.9 Water contributes to the regulation of body temperature.

Acids donate hydrogen ions, bases accept them

Although the covalent bonds between hydrogen and oxygen in water are strong and thus rarely broken, it can happen. When it does, the electron from one hydrogen atom is transferred to the oxygen atom completely and the water molecule breaks into two ions—a *hydrogen ion* (H^+) and a *hydroxide ion* (OH^-).

In pure water, only a very few molecules of water are dissociated (broken apart) into H^+ and OH^- at any one time. However, there are other sources of hydrogen ions in aqueous solutions. An **acid** is any molecule that can donate (give up) an H^+. When added to pure water, acids produce an *acidic* solution, one with a higher H^+ concentration than that of pure water. (By definition, an aqueous solution with the same concentration of H^+ as that of pure water is a *neutral* solution). Common acidic solutions are vinegar, carbonated beverages, and orange juice. Conversely, a **base** is any molecule that can accept (combine with) an H^+. When added to pure water, bases produce a basic or *alkaline* solution, one with a lower H^+ concentration than that of pure water. Common alkaline solutions include baking soda in water, detergents, and drain cleaner.

Because acids and bases have opposite effects on the H^+ concentration of solutions, they are said to neutralize each other. You have probably heard that a spoonful of baking soda in water is a time-honored way to counteract an "acid stomach." Now you know that this home remedy is based on sound chemical principles.

The pH scale expresses hydrogen ion concentration

Scientists use the pH scale to indicate the acidity or alkalinity of a solution. The **pH scale** is a measure of the hydrogen ion concentration of a solution. The scale ranges from 0 to 14, with the pH of pure water defined as a pH of 7.0, the neutral point. A pH of 7 corresponds to a hydrogen ion concentration of 10^{-7} moles/liter (a *mole* is a term used by chemists to indicate a certain number of atoms, ions, or molecules). An *acidic* solution has a pH of *less* than 7 whereas a *basic* solution has a pH of *greater* than 7. Each whole number change in pH represents a 10-fold change in the hydrogen ion concentration in the opposite direction. For example, an acidic solution with a pH of 6 has an H^+ concentration of 10^{-6} moles/liter (10 times greater than pure water), whereas an alkaline solution with a pH of 8 has an H^+ concentration of 10^{-8} moles/liter (1/10 that of water). Figure 2.10 shows the pH scale and indicates the pHs of some common substances and body fluids.

The pH of blood is 7.4, just slightly more alkaline than neutral water. The hydrogen ion concentration of blood plasma is low relative to other ions (the hydrogen ion concentration of blood plasma is less than 1 *millionth* that of sodium ions, for example). It is important to maintain homeostasis of this low concentration of hydrogen ions in the body because hydrogen ions are small, mobile, positively charged, and highly reactive. Hydrogen ions tend to displace other positive ions in molecules, and when they do they alter molecular structures and change the ability of the molecule to function properly.

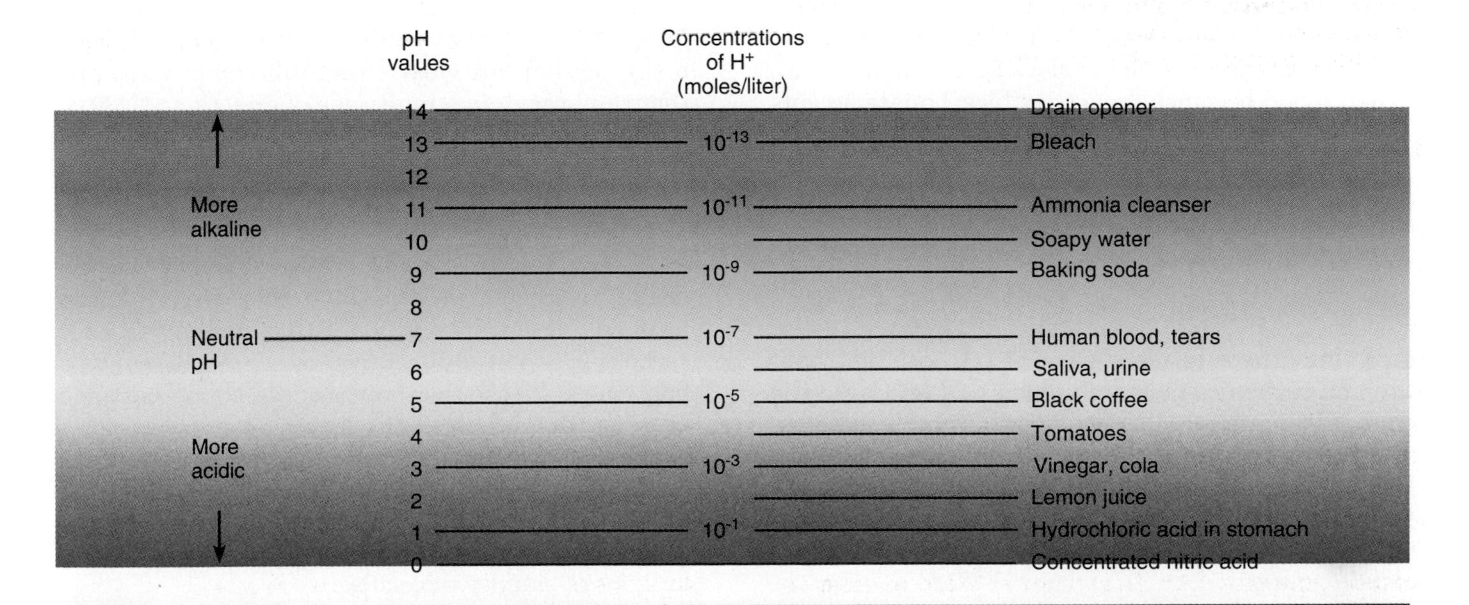

Figure 2.10 The pH scale. The pH scale is an indication of the H^+ concentration of a solution.

Changes in the pH of body fluids can affect how molecules are transported across the cell membrane and how rapidly certain chemical reactions occur. They may even alter the shapes of proteins that are structural elements of the cell. In other words, a change in the hydrogen ion concentration can be dangerous because it threatens homeostasis.

Recap **Acids can donate a hydrogen ion to a solution, whereas bases can accept hydrogen ions from a solution. The pH scale indicates the hydrogen ion concentration of a solution. The normal pH of blood is 7.4.**

Buffers minimize changes in pH

A **buffer** is any substance that tends to minimize the changes in pH that might otherwise occur when an acid or base is added to a solution. Buffers are essential to our ability to maintain homeostasis of pH in body fluids.

In biological solutions such as blood or urine, buffers are present as *pairs* of related molecules that have opposite effects. One of the pair is the acid form of the molecule (capable of donating an H^+) and the other is the base form (capable of accepting an H^+). When an acid is added and the number of H^+ ions increases, the base form of the buffer pair will accept some of the H^+, minimizing the fall in pH that might otherwise occur. Conversely, when a base is added that might take up too many H^+ ions, the acid form of the buffer pair will release additional H^+ and thus minimize the rise in pH. Buffer pairs are like absorbent sponges that can pick up excess water and then can be wrung out to release water when necessary.

One of the most important buffer pairs in body fluids such as blood is bicarbonate (HCO_3^-; the base form) and carbonic acid (H_2CO_3; the acid form). When blood becomes too acidic, bicarbonate accepts excess H^+ according to the following reaction:

$$HCO_3^- + H^+ \longrightarrow H_2CO_3$$

When blood becomes too alkaline, carbonic acid donates H^+ by the reverse reaction:

$$HCO_3^- + H^+ \longleftarrow H_2CO_3$$

In a biological solution such as blood, both bicarbonate and carbonic acid take up and release H^+ all the time. Ultimately a chemical *equilibrium* is reached in which the rates of the two chemical reactions are the same, as represented by the following combined equation:

$$HCO_3^- + H^+ \longleftrightarrow H_2CO_3$$

When excess acid is produced, the combined equation shifts to the right as the bicarbonate combines with the H^+. The reverse is true for alkalinity.

There are many other buffers in the body as well. The more buffers that are present in a body fluid, the more stable the pH will be.

Recap **Buffers tend to minimize changes in pH in a solution. They help us maintain a stable pH in body fluids.**

2.5 The organic molecules of living organisms

Organic molecules are molecules that contain carbon and other elements held together by covalent bonds. The name "organic" came about at a time when scientists believed that all organic molecules were created only by living organisms and all "inorganic" molecules came from nonliving matter. Today we know that organic molecules can be synthesized in the laboratory under the right conditions and that they probably existed on Earth before there was life.

Carbon is the common building block of organic molecules

Carbon (Figure 2.11) is relatively rare in the natural world, representing less than 0.03% of the earth's crust. However, living organisms actively accumulate it. Carbon accounts for about 18% of body weight in humans.

Carbon is the common building block of all organic molecules because of the many ways it can form strong covalent bonds with other atoms. Carbon has six electrons, two in the first shell and four in the second. Because carbon is most stable when its second shell is filled with eight electrons, *its natural tendency is to form four covalent bonds with other molecules.* This makes carbon an ideal structural component, one that can branch in a multitude of directions.

Why is carbon a common structural element?

Using the chemist's convention that a line between the chemical symbols of atoms represents a pair of shared electrons in a covalent bond, Figure 2.12 shows some of the many structural possibilities for carbon. Carbon can form covalent bonds with hydrogen, nitrogen, oxygen, or another carbon. It can form double covalent bonds with oxygen or another carbon. It can even form five- or six-membered carbon rings, with or without double bonds between carbons.

In addition to their complexity, there is almost no limit to the size of organic molecules derived from carbon. Some, called *macromolecules* (from the Greek *makros,* "long"), consist of thousands or even millions of smaller molecules.

(a) Diamonds are formed only under conditions of extreme temperature and pressure. The structure of diamond resembles the steel framework of a large building, where each atom is covalently bonded to four neighboring carbon atoms. This structure explains the hardness of diamonds.

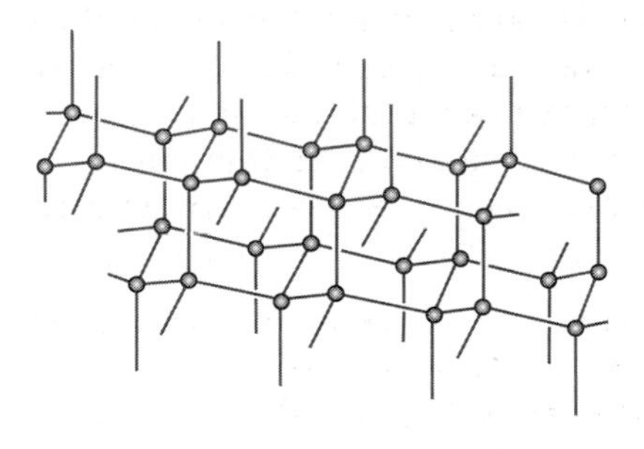

(b) Graphite is produced as a result of decay of older carbon-based substances. Its structure consists of layers of hexagonal rings of carbon atoms. Graphite is fairly soft (hence its use in pencils) because these layers of carbon atoms can slide past one another.

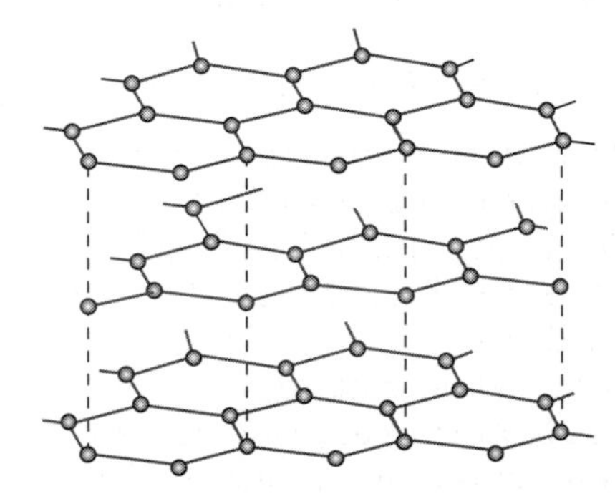

Figure 2.11 Carbon. Graphite and diamond are elemental forms of carbon. Although both solids are made solely of carbon atoms, their structures are very different owing to the ways their carbon atoms bond to one another.

Macromolecules are synthesized and broken down within the cell

Macromolecules are built (synthesized) within the cell itself. In a process called **dehydration synthesis** (also called the condensation reaction), smaller molecules called subunits are joined together by covalent bonds, like pearls on a string. The name of the process accurately describes what is happening, for each time a subunit is added, the equivalent of a water molecule is removed ("dehydration") (Figure 2.13). The subunits needed to synthesize macromolecules come from the foods you eat and from the biochemical reactions in your body that break other large molecules down to smaller ones.

The synthesis of macromolecules from smaller molecules requires energy. That is one reason we need energy to survive and grow. It is no accident that children seem to eat enormous amounts of food. Growing children require energy to make the macromolecules necessary to create new cell membranes, muscle fibers, and other body tissues.

Some macromolecules are synthesized specifically for the purpose of storing energy within our cells. The ability to store energy internally allows organisms to

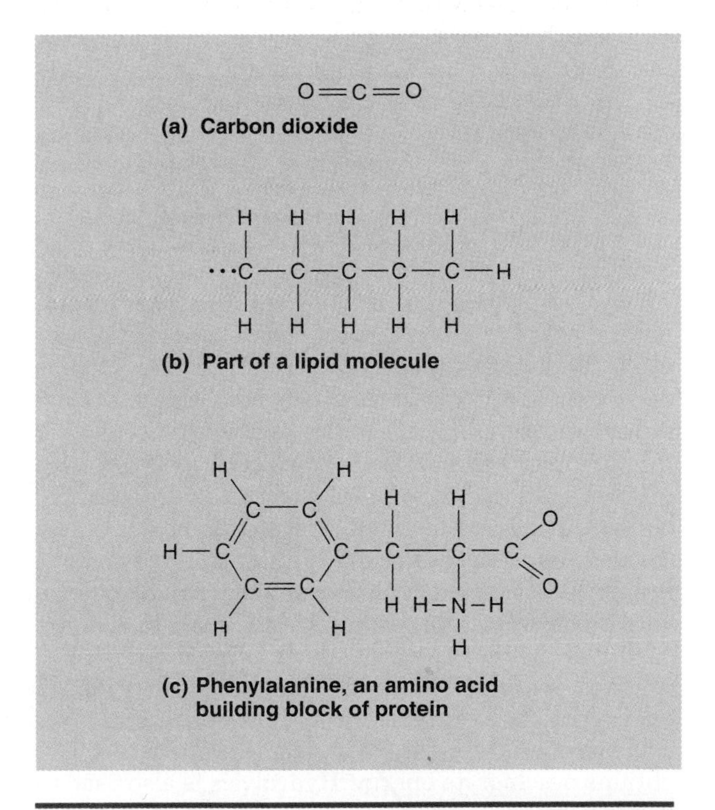

Figure 2.12 Examples of the structural diversity of carbon. **(a)** In carbon dioxide, carbon forms two covalent bonds with each oxygen. **(b)** Lipid molecules contain long chains of carbon atoms covalently bound to hydrogen. **(c)** Carbon is the backbone of the amino acid phenylalanine.

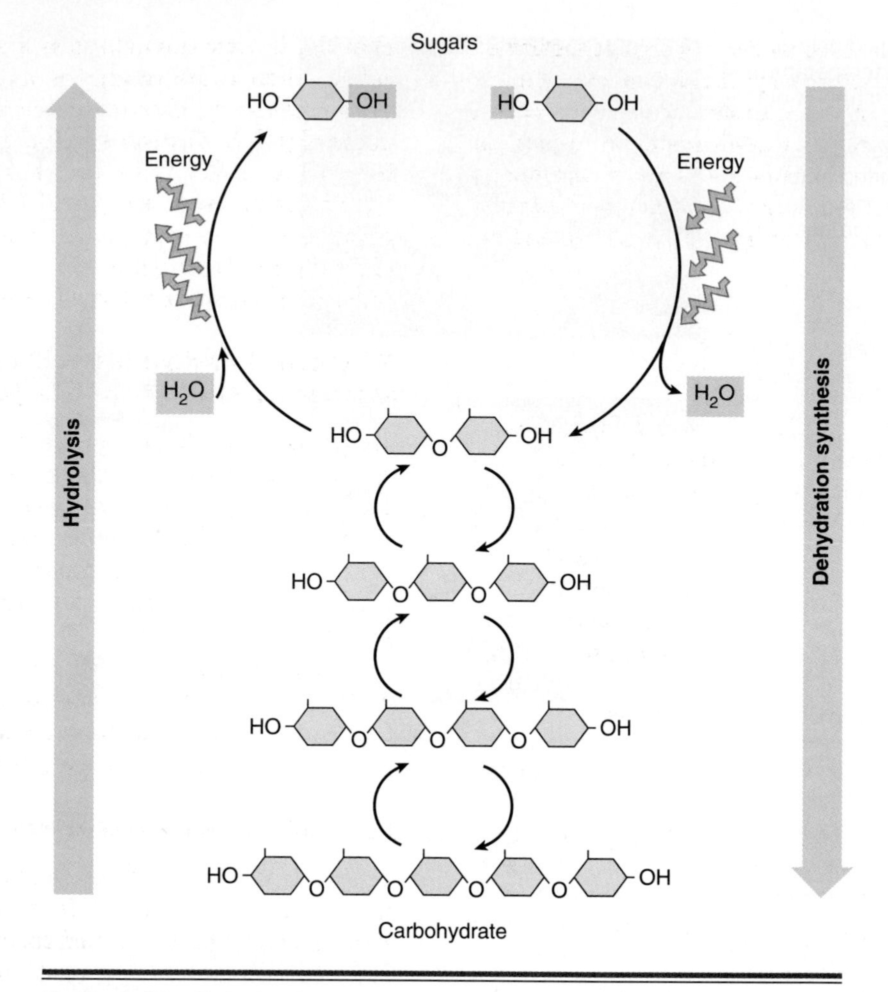

Figure 2.13 Dehydration synthesis and hydrolysis. The synthesis of larger molecules by dehydration synthesis requires energy, whereas the breakdown of molecules into smaller units by hydrolysis liberates stored energy. In this example the smallest units are simple sugars and the macromolecule is a carbohydrate.

survive even when food is not plentiful. Other macromolecules serve as structural components of cells or of extracellular (outside the cell) structures such as bone. Still others direct the many activities of the cell or serve as signaling molecules between cells.

Organic macromolecules are broken down by a process called **hydrolysis.** During hydrolysis the equivalent of a water molecule is added each time a covalent bond between single subunits in the chain is broken. Notice that hydrolysis is essentially the reverse of dehydration synthesis, and thus it should not surprise you that the breakdown of macromolecules releases energy. The energy was stored as potential energy in the covalent bonds between atoms. Hydrolysis of energy-storage molecules is how the body obtains much of its energy. Hydrolysis is also used to break down molecules of food during digestion, to recycle materials so that they can be used again, and to get rid of substances that are no longer needed by the body.

Living organisms synthesize four classes of organic molecules, known as *carbohydrates, lipids, proteins,* and *nucleic acids.* The many different molecules within each class are constructed of the same handful of chemical elements. However, there is essentially no limit to the number of different molecules that could be created. No one knows for sure how many different organic molecules there are in humans. On a chemical level, the tremendous diversity among the many species of organisms on Earth is due to differences in their organic molecules, especially their proteins and nucleic acids.

How are organisms chemically different from each other?

Recap **Carbon is a key element of organic molecules because of the multiple ways it can form strong covalent bonds with other molecules. Organic molecules are synthesized by dehydration synthesis, a process that requires energy, and broken down by hydrolysis, which liberates energy. The four classes of organic molecules are carbohydrates, lipids, proteins, and nucleic acids.**

2.6 Carbohydrates: Used for energy and structural support

A clue to the basic structure of carbohydrates is found in their name. Carbohydrates have a backbone of carbon atoms with hydrogen and oxygen attached in the same proportion as they appear in water (2-to-1), hence the carbon is "hydrated," or combined with water. Most living organisms use carbohydrates for energy, and plants use at least one carbohydrate (cellulose) as structural support.

Monosaccharides are simple sugars

The simplest kind of carbohydrate is called a **monosaccharide** (meaning "one sugar"). Monosaccharides have relatively simple structures consisting of carbon, hydrogen, and oxygen in a 1-2-1 ratio. The most common monosaccharides contain 5 or 6 carbon atoms arranged in either a 5-membered or 6-membered ring.

Glucose, fructose, ribose, and *deoxyribose* are four of the most important monosaccharides in humans (Figure 2.14). Glucose, a 6-carbon monosaccharide, is very important as a source of energy for cells. When more energy is available than can be used right away, glucose molecules can be linked together by dehydration synthesis to form larger carbohydrate molecules. Ribose and deoxyribose are both 5-carbon monosaccharides that are components of nucleotide molecules, discussed below. The only difference between the two is that deoxyribose has one less oxygen atom than ribose.

Oligosaccharides: More than one monosaccharide linked together

Oligosaccharides are short strings of monosaccharides linked together (the prefix "oligo" means "a few") by dehydration synthesis. One common oligosaccharide is table sugar, or sucrose. Sucrose is also called a *disaccharide* because it consists of just two monosaccharides (glucose + fructose). Another is maltose (glucose + glucose).

Some oligosaccharides are covalently bonded to certain cell-membrane proteins (called glycoproteins). Glycoproteins participate in linking adjacent cells together and in cell-cell recognition and communication.

Polysaccharides store energy

Complex carbohydrates called **polysaccharides** ("poly" means "many") form when thousands of monosaccharides are joined together into straight or branched chains by dehydration synthesis. Polysaccharides are a convenient way for cells to stockpile extra energy by locking it in the bonds of the polysaccharide molecule.

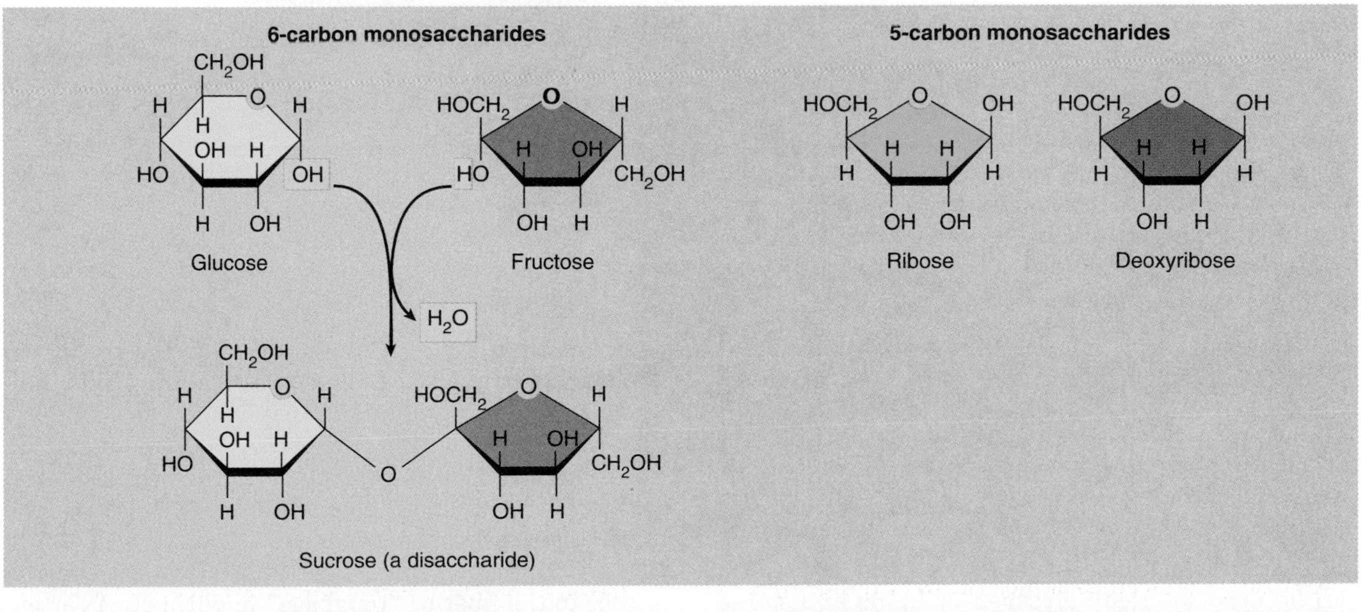

Figure 2.14 Monosaccharides. By convention, in a ringed structure the symbol C for carbon is often omitted because its presence is inferred by the union of two bond lines at an angle.

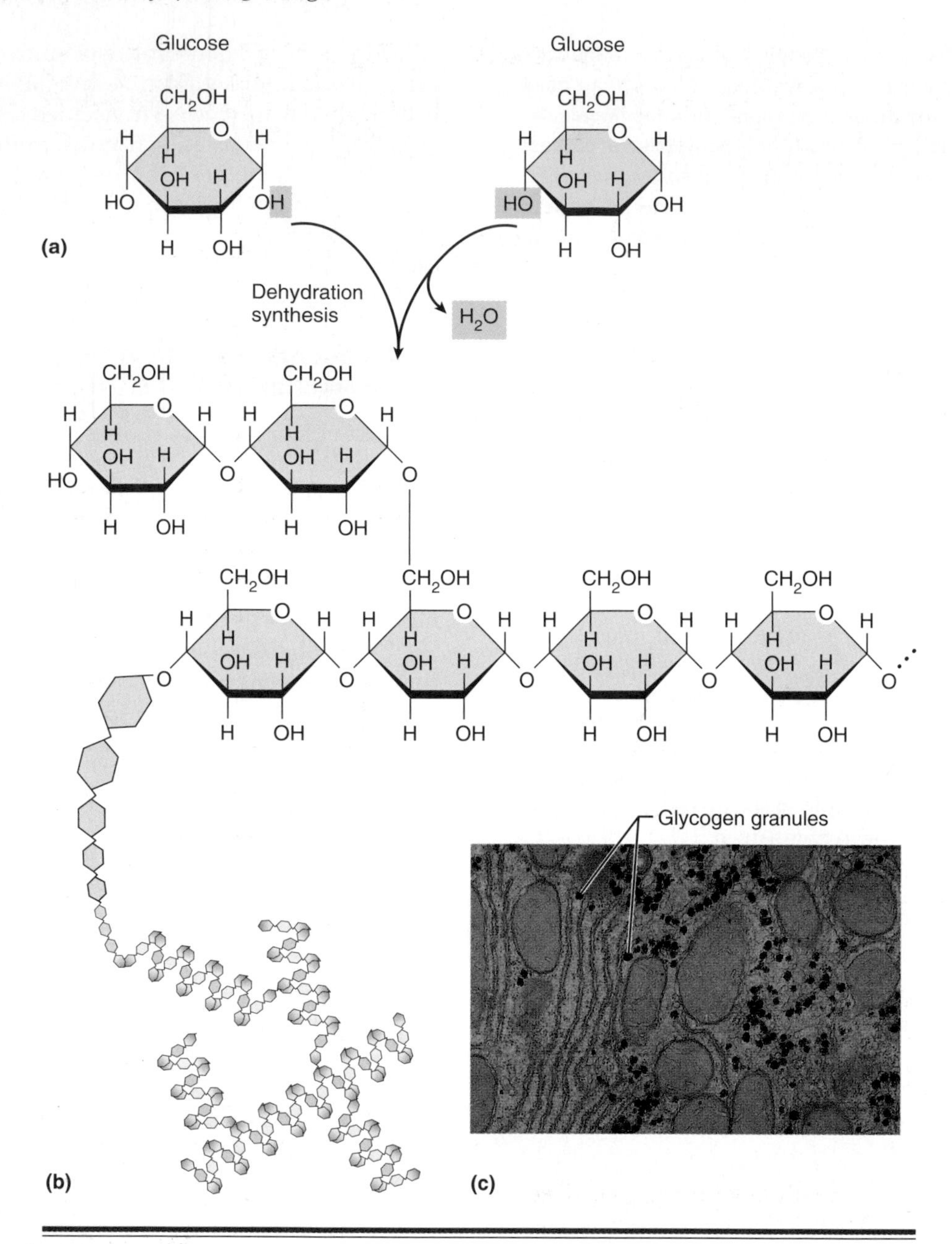

Figure 2.15 Glycogen is the storage carbohydrate in animals. (a) Glycogen is formed by dehydration synthesis from glucose subunits. **(b)** A representation of the highly branched nature of glycogen. **(c)** An animal cell with granules of stored glycogen.

The most important polysaccharides in living organisms consist of long chains of glucose monosaccharides. In animals the storage polysaccharide is **glycogen** (Figure 2.15), whereas in plants it is **starch.** The flour we obtain by grinding plant grains is high in starch, which we then utilize for our own energy needs by breaking it down to glucose. Any glucose not consumed for energy in the short term can be used to create glycogen or lipids and stored within our cells for later use.

Cellulose is a slightly different form of glucose polysaccharide. Plants use it for structural support rather than for energy storage. The nature of the chemical bonds in cellulose are such that most animals, including humans, cannot break cellulose down to glucose units (which is why we cannot digest wood). But there's plenty of energy locked in the chemical bonds of cellulose, as witnessed by the heat generated by a wood fire.

Undigested cellulose in the food we eat contributes to the fiber or "roughage" in our diet. A certain amount of fiber is thought to be beneficial because it increases the movement of wastes through the digestive tract. The more rapid excretion of wastes de-

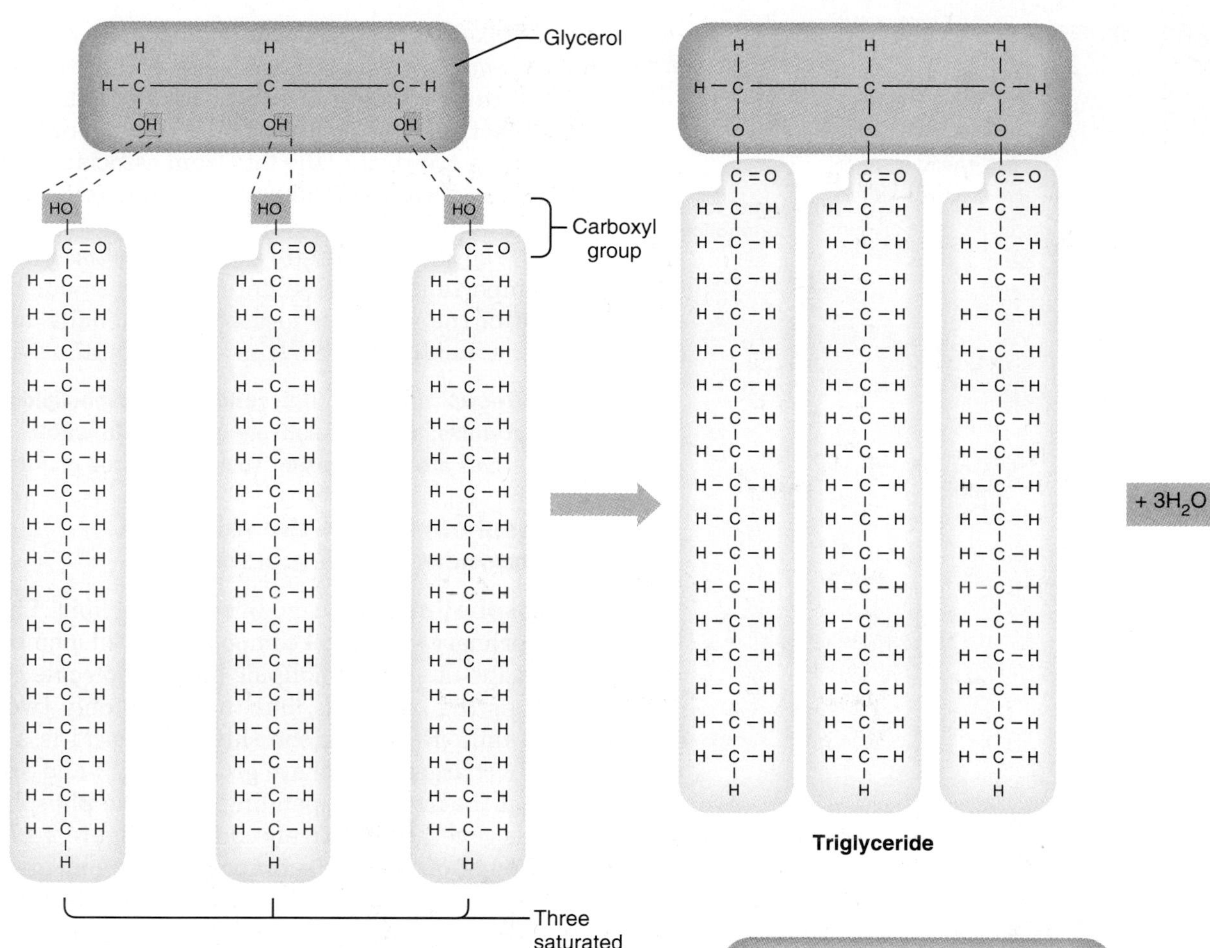

Figure 2.16 Triglycerides. (a) Triglycerides (neutral fats) are synthesized from glycerol and three fatty acids. This particular triglyceride is a saturated fat. **(b) (at right)** An unsaturated fat.

(b) Triglyceride with unsaturated fatty acid tails

creases the time of exposure to any carcinogens (cancer-causing agents) that may be in the waste material.

> ***Recap*** **Carbohydrates contain carbon, hydrogen, and oxygen in a 1-2-1 ratio. Simple sugars such as glucose provide immediate energy for cells. Complex carbohydrates called polysaccharides store energy (in animals and plants) and provide structural support (in plants).**

2.7 Lipids: Insoluble in water

The most important physical characteristic of **lipids** for biology is that they are relatively insoluble in water. The most important subclasses of lipids in your body are *triglycerides, phospholipids,* and *steroids.*

Triglycerides are energy storage molecules

Triglycerides, also called neutral fats or just fats, are synthesized from a molecule of glycerol and three fatty acids (Figure 2.16). **Fatty acids** are chains of

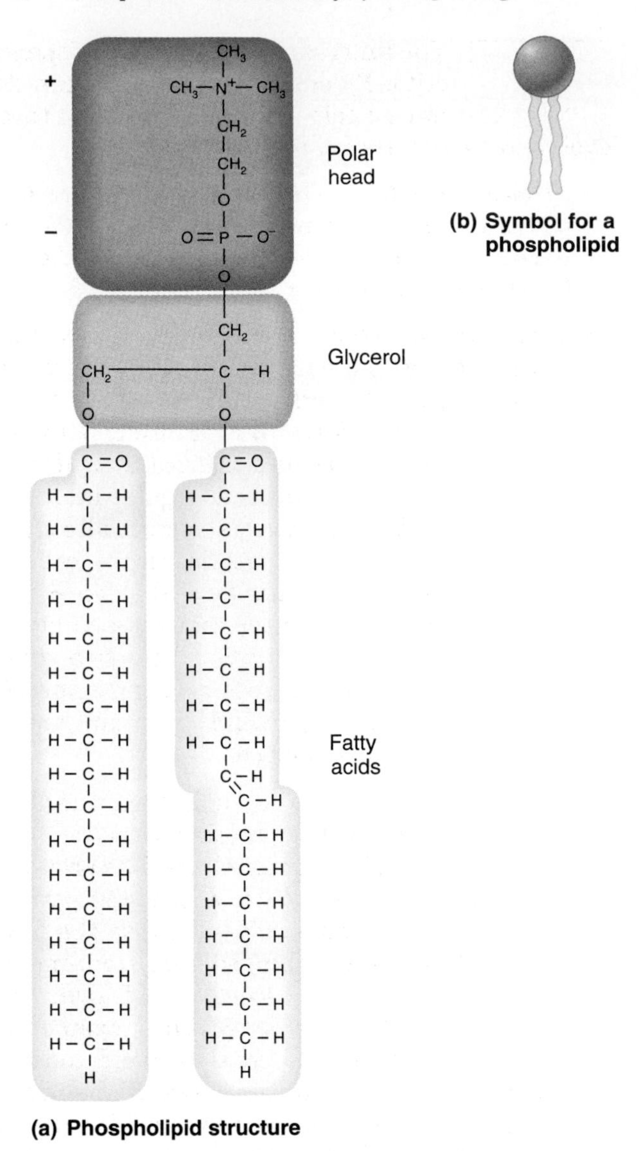

Figure 2.17 Phospholipids. Phospholipids are the primary constituent of animal cell membranes.

hydrocarbons (usually about 16–18 carbons long) that end in a group of atoms known as a carboxyl group (COOH). Fats vary by the length of their fatty acid tails and the ratio of hydrogen atoms to carbon atoms in the tails.

Saturated fats have a full complement of two hydrogen atoms for each carbon in their tails. In saturated fats, the tails are fairly straight, allowing them to pack closely together. As a result, saturated fats are generally solid at room temperature. Animal fats, most notably butter and bacon grease, are saturated fats. A diet rich in saturated fats is thought to contribute to the development of cardiovascular disease.

Unsaturated fats, also called *oils,* have fewer than two hydrogen atoms on one or more of the carbon atoms in the tails. As a result, double bonds form between adjacent carbons, putting kinks in the tails and preventing the fats from associating closely together. As a result, unsaturated fats (oils) are generally liquid at room temperature.

Triglycerides are stored in adipose tissue and are an important source of stored energy in our bodies. Most of the energy is located in the bonds between carbon and hydrogen in the fatty acid tails.

Recap **Lipids (triglycerides, phospholipids, and steroids) are all relatively insoluble in water. Triglycerides are an important source of stored energy.**

Phospholipids are the primary component of cell membranes

Phospholipids are a modified form of lipid. They are the primary structural component of cell membranes.

Like fats, phospholipids have a molecule of glycerol as the backbone, but they have only two fatty acid tails. Replacing the third fatty acid is a negatively charged phosphate group (PO_4^-) and another group that varies depending upon the phospholipid but is generally positively charged (Figure 2.17). The presence of charged groups on one end gives the phospholipid the special property that one end of the molecule is polar and thus soluble in water, whereas the other end (represented by the two fatty acid tails) is neutral and therefore relatively insoluble in water.

Steroids are composed of four rings

Steroids do not look at all like the lipids described above but are classified as lipids because they are relatively insoluble in water. Steroids consist of a backbone of three 6-membered carbon rings and one 5-membered carbon ring to which any number of different groups may be attached.

One steroid you may be familiar with is **cholesterol.** High levels of cholesterol in our blood are associated with cardiovascular disease. However, we all need a certain amount of cholesterol. It is a normal and essential structural component of animal cell membranes and is also the source of several important hormones, including the sex hormones estrogen and testosterone (Figure 2.18). Our bodies manufacture cholesterol even though we generally get more than we need in our diets.

Recap **Phospholipids, an important component of cell membranes, have a polar (water-soluble) head and two fatty acid (water-insoluble) tails. Steroids, such as cholesterol, have a four-ring structure.**

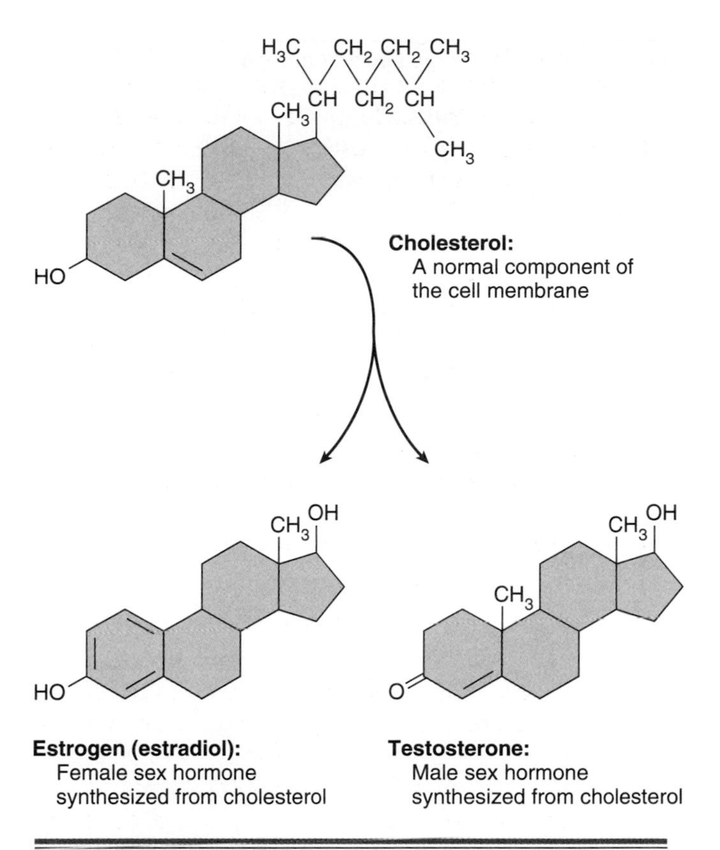

Figure 2.18 Steroids.

2.8 Proteins: Complex structures constructed of amino acids

Proteins are macromolecules constructed from long strings of single units called **amino acids.** All human proteins are constructed from only 20 different amino acids (Figure 2.19). Each amino acid has an *amino* group (NH_3) on one end, a carboxyl acid group on the other, a C-H group in the middle, and an additional group (designated "R") that represents everything else. Some of the R groups are completely neutral, others are neutral but polar, and a few carry a net charge (either positive or negative). Differences in the charge and structure of the amino acids affect the shape and functions of the proteins constructed from them. Our bodies can synthesize (make) 11 of the amino acids if necessary. However, most of them, including the 9 we cannot synthesize, are taken in as part of the food we eat.

How do organisms create so many different proteins?

Like complex carbohydrates and fats, proteins are formed by dehydration synthesis. A single string of 3 to 100 amino acids is called a **polypeptide.** A polypeptide is generally referred to as a **protein** when it is longer than 100 amino acids and has a complex structure and a function. Some proteins consist of several polypeptides linked together.

Spotlight on Structure & Function

The function of every protein depends critically on its structure. We can define protein structure on at least three levels and sometimes four (Figure 2.20):

- *Primary structure.* The primary structure of a protein is represented by its amino acid sequence. In writing, each amino acid is indicated by a three-letter code (review Figure 2.19).
- *Secondary structure.* The secondary structure describes how the chain of amino acids is oriented in space. A common secondary structure of proteins is an *alpha (α) helix.* An α helix is a right-hand spiral that is stabilized by hydrogen bonds between amino acids at regular intervals. Another common secondary structure also stabilized by hydrogen bonds is a flat ribbon called a *beta (β) sheet.* A β sheet is formed when hydrogen bonds join two primary sequences of amino acids side by side. Aside from these two structures, proteins can coil into an almost infinite variety of shapes depending on which amino acids make up the sequence.
- *Tertiary structure.* Tertiary structure, the third level, refers to how the protein twists and folds to form a three-dimensional shape. The protein's three-dimensional structure depends in part on its sequence of amino acids, because the locations of the polar and charged groups within the chain determines the locations of hydrogen bonds that hold the whole sequence together. In addition, occasionally a covalent bond called a disulfide (S-S) bond forms between the sulfur molecules of two cysteine amino acids. Finally, proteins tend to fold in such a way that neutral amino acids are more likely to end up in the interior, whereas charged and polar amino acids are more likely to face the outside (aqueous environment). Proteins acquire their characteristic tertiary structure by a folding process that occurs either during synthesis or shortly thereafter.
- *Quaternary structure.* The quaternary (fourth) structure of some proteins refers to the number of polypeptide chains comprising the protein (if there is more than one) and how they are associated with each other.

The human body has thousands of different proteins, each serving a different function. Some proteins are primarily for structural support. Others are involved in muscle contraction. Others form part of the cell membrane, where they help transmit information and materials into and out of cells. Still others, called *enzymes,* regulate the rates of biochemical reactions within cells (see page 44). ■

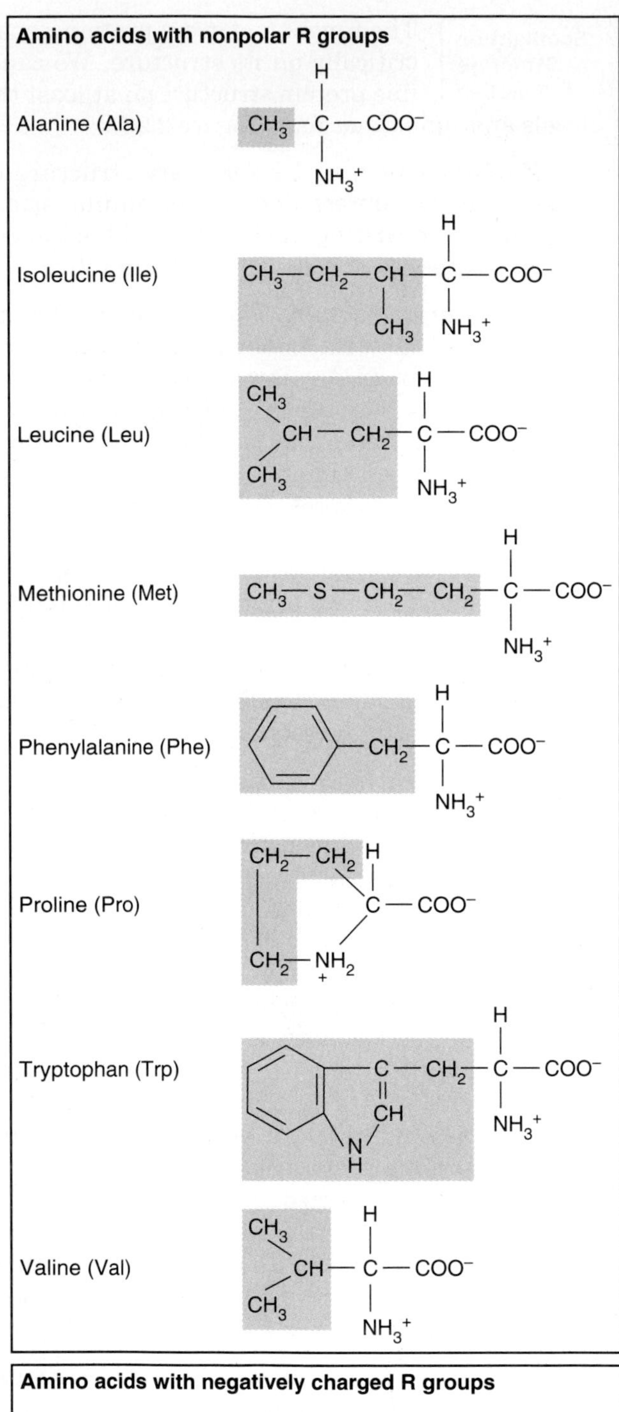

Amino acids with negatively charged R groups

Aspartic acid (Asp)

Glutamic acid (Glu)

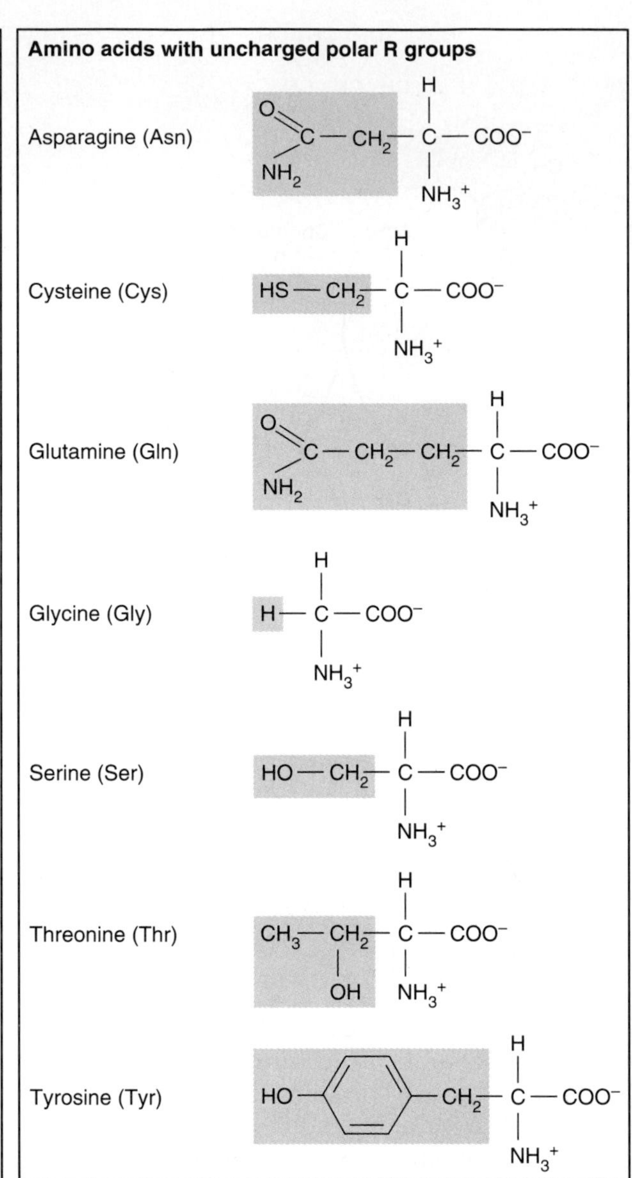

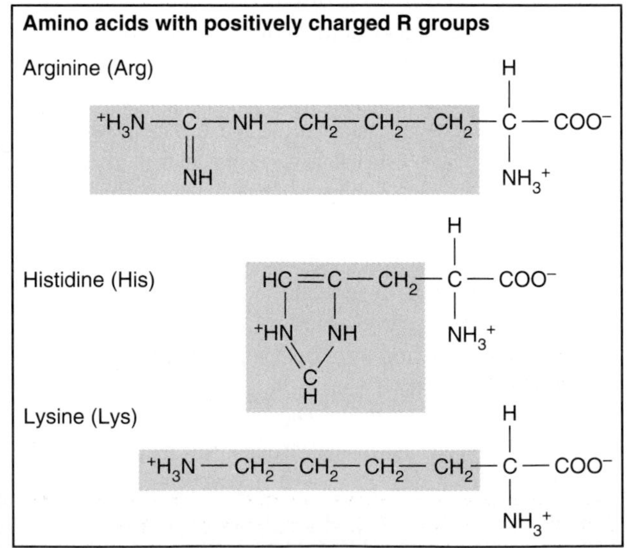

Figure 2.19 The twenty amino acid building blocks of proteins. The portions of amino acids that make them different from each other, called R groups, are colored. The three-letter codes in parentheses designate the amino acids in written formulas.

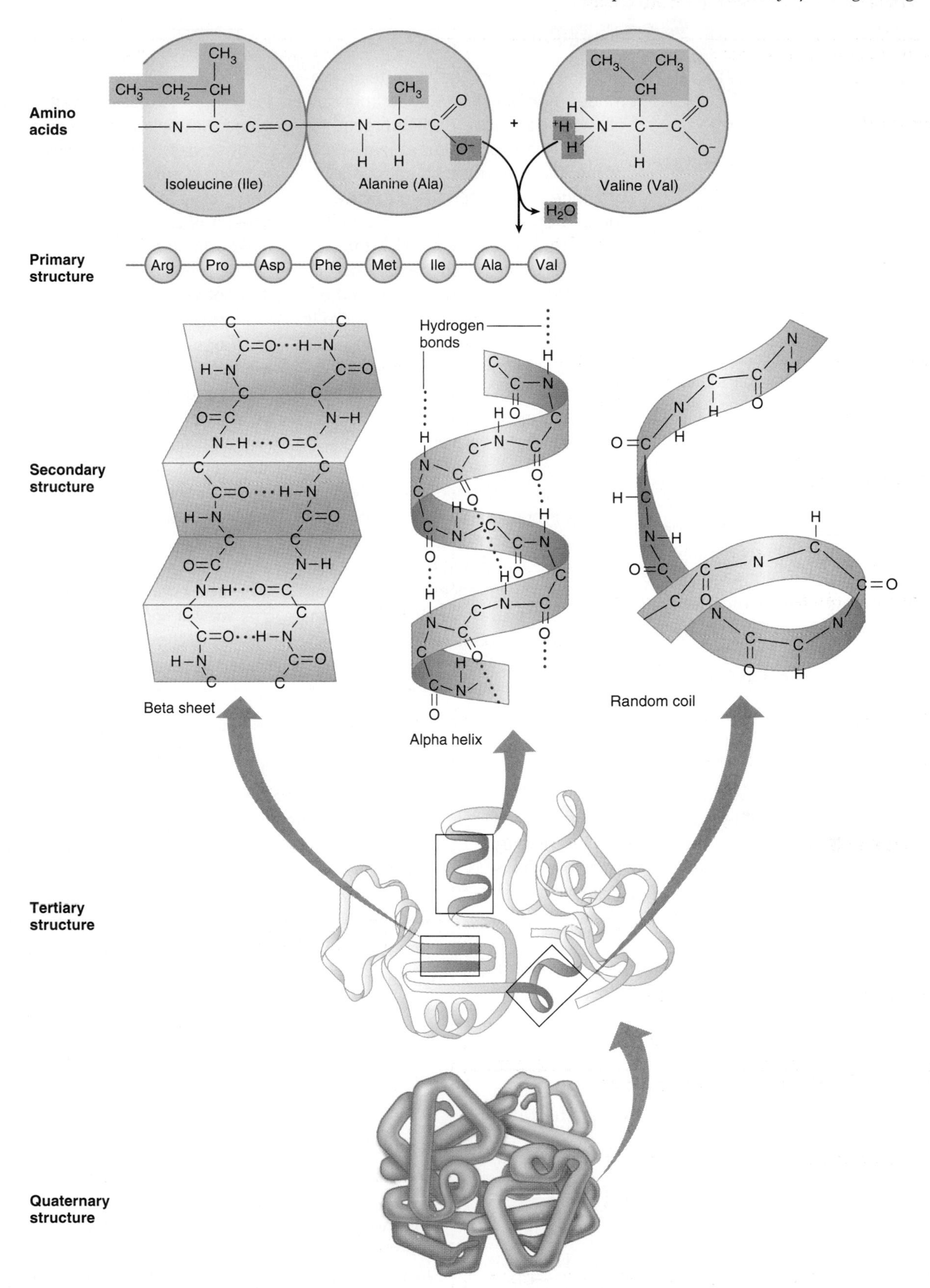

Figure 2.20 The synthesis and structure of proteins. Proteins are created from amino acids by dehydration synthesis. In the diagrams of secondary structure the R-groups have been omitted so that the basic backbone can be seen more easily.

Directions in Science

Why Protein Folding Is Important

A misfolded protein is a useless piece of junk, because protein function is dependent on protein shape. The shape of a protein, in turn, is determined by how the amino acid chain twists and folds back on itself. Some misfolded proteins cause human diseases, including cystic fibrosis and quite possibly Alzheimer's disease.

In the 1950s, researchers discovered that the various twists and folds of proteins are caused by weak hydrogen bonds that form between amino acids at different points in the amino acid chain. It soon became apparent that the final shape of a protein is determined by the precise locations of these hydrogen bonds. But what causes the protein to fold into precisely the right shape in the first place so that these bonds can form?

A breakthrough came in 1960 with the discovery that if proteins are made to unfold slightly, most (but not all) will twist back into their proper shape. This implies that at least part of the folding process is automatic, dependent only upon the attractive forces between the protein's amino acids. However, we now know that not all proteins can refold themselves or even fold themselves properly in the first place. For some proteins, there is a critical phase during the folding process when they either twist correctly or become hopelessly tangled. Proteins that cannot fold themselves are made to fold properly by special "chaperone proteins," whose sole function is to shepherd the protein into its proper shape at just the right time, like little tugboats pushing a large ship to turn in the right direction.

The consequences of protein misfolding can be catastrophic. Consider *Alzheimer's disease,* a degenerative nerve disorder found in 10% of all people over age 65 that kills 100,000 Americans every year. Dr. Alzheimer, for whom the disease was named, noted that patients who died of the disease had unusual tangles of protein fibers surrounding degenerating nerve cells in their brains. The fibers consist of an abnormal protein that no longer functions properly and that cannot be removed in the usual manner. It is not yet known whether this abnormal protein *causes* Alzheimer's disease or is a result of it. Look for answers soon.

One disease known to be caused by protein misfolding is *cystic fibrosis.* Patients with cystic fibrosis lack a critical protein that normally transports chloride ions across the cell membrane. Apparently the normal chloride-transporting protein is missing because the folding process goes awry at a critical step. It's a safe bet that we will find still other diseases in the future that are due to misfolded proteins.

Because the links that determine the secondary and tertiary structures of protein are relatively weak hydrogen bonds, they may be broken by nearby charged molecules. This means that *the shape of proteins can change* in the presence of charged or polar molecules. The ability to change shape is essential to the functions of certain proteins.

Protein structure can also be damaged, sometimes permanently, by high temperatures or changes in pH. **Denaturation** refers to permanent disruption of protein structure, leading to a loss of biological function. An egg becomes hard when it is exposed to high temperatures because the soluble proteins in the egg become denatured and clump together as a solid mass.

Most proteins are water soluble (meaning they dissolve in water). There are exceptions, however. Many of the proteins that are part of our cell membranes either are water insoluble or have water insoluble regions. You will learn more about why this is important in Chapter 3.

Recap **Proteins are complex molecules consisting of strings of amino acids. The function of a protein relates to its shape, which is determined by its amino acid sequence and how the chain of amino acids twists and folds. A denatured protein loses its shape and function.**

Enzymes facilitate biochemical reactions

An **enzyme** is a protein that functions as a biological catalyst. A **catalyst** is a substance that speeds up the rate of a chemical reaction without itself being altered or consumed by the reaction. Enzymes help biochemical reactions to occur, but they do not change the final result of the reaction. That is, they can only speed reactions that would have happened anyway, although much more slowly. A chemical reaction that could take hours by itself might reach the same point in minutes or seconds in the presence of an enzyme.

Without help from thousands of enzymes, most biochemical reactions in our cells would occur too slowly to sustain life. Each enzyme facilitates a particular chemical reaction or group of reactions. Some enzymes break molecules apart; others join molecules together. In general, the enzyme takes one or more

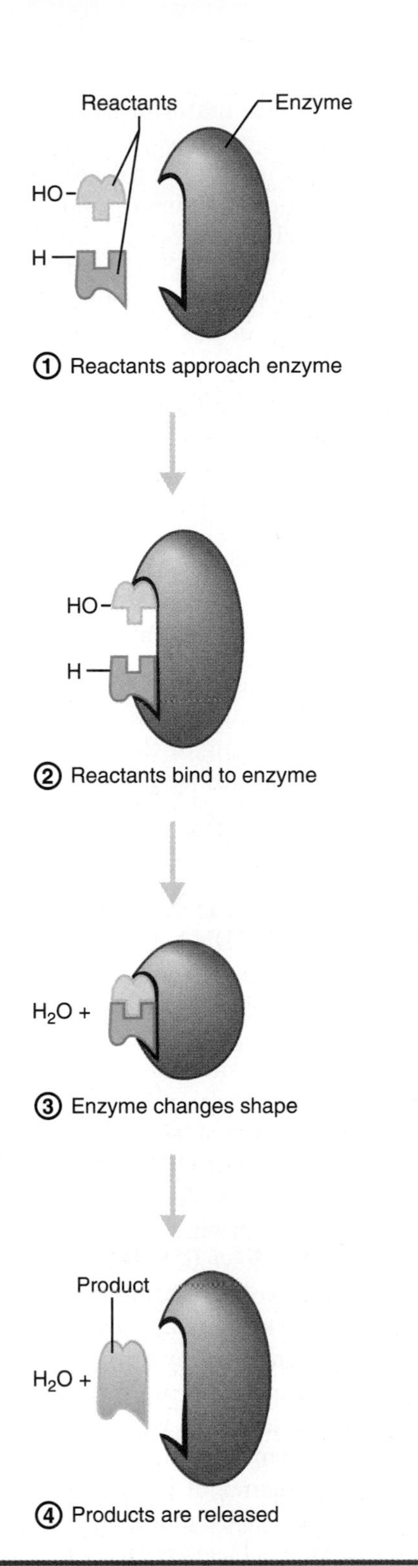

Figure 2.21 Enzymes facilitate chemical reactions. This particular enzyme facilitates a dehydration synthesis reaction in which two reactants join to create one larger product plus a molecule of water. Note that the enzyme is not used up during the reaction.

reactants (also called substrates) and turns them into one or more *products*. Enzymes serve as catalysts because, as proteins, they can change shape. The ability to change shape allows them to bind to other molecules and orient them so that they may interact. Figure 2.21 depicts how a typical enzyme works.

Just how important are enzymes? As one example, the reason we can digest glycogen and starch is that we possess specific enzymes that break the chemical bonds between the glucose monosaccharides in these molecules. In contrast, we cannot digest cellulose because we lack the right enzyme to break it apart. Termites can utilize cellulose only because their digestive systems harbor bacteria that have a cellulose-digesting enzyme.

The changeable shape of an enzyme shows why homeostasis within our cells is so important. Protein shape is in part determined by the chemical and physical environment inside a cell, including temperature, pH, and the concentrations of certain ions. Any deviation from homeostasis can affect the shapes and biological activities of dozens of different enzymes and thus alter the course of biochemical reactions within the cell.

Recap **Enzymes are proteins that facilitate biochemical reactions in the body. Without enzymes, many biochemical reactions would occur too slowly to sustain life.**

Try It Yourself

Demonstrating an Enzyme in Your Saliva

You can investigate the presence of an enzyme in one of your own body fluids, saliva, with the following demonstration.

Place a small soda cracker in your mouth and mix it with saliva. Don't swallow it, but move the mixture around in your mouth. After a minute you may experience a sweet taste that was not present initially. This is because your saliva contains an enzyme that can break the bonds in the starch of the cracker, producing the sweet-tasting disaccharide maltose. What you have just demonstrated is that the hydrolysis (digestion) of starch actually begins in your mouth, even before food reaches your stomach.

How can you be sure that the breakdown of starch wasn't caused by the *watery* component of your saliva, instead of an enzyme? You could test that question as an alternative hypothesis. Wet another soda cracker with tap water for the same length of time that you moistened the first one with saliva, and then put it in your mouth. It probably won't taste sweet.

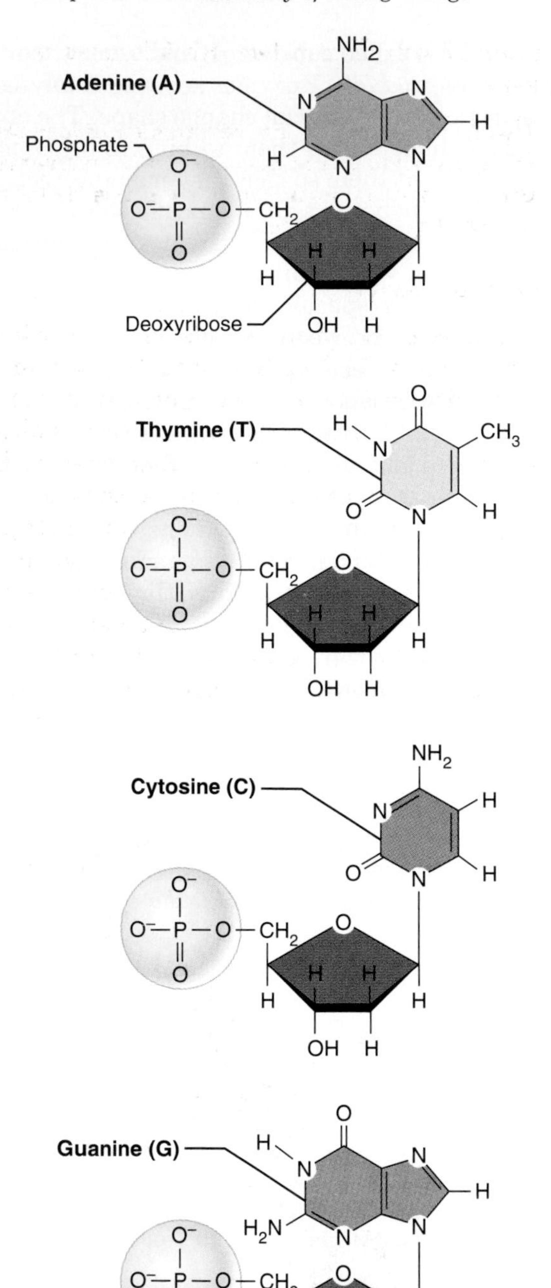

Figure 2.22 The four nucleotides that comprise DNA. The phosphate and sugar groups are identical in all four nucleotides.

2.9 Nucleic acids: Storing genetic information and making proteins

A very important class of organic molecules are the *nucleic acids,* **DNA (deoxyribonucleic acid)** and **RNA (ribonucleic acid).** You have probably heard of such subjects as cloning, genetic engineering, and DNA "fingerprinting." These subjects relate to the nucleic acids, DNA and RNA.

DNA, the genetic material in living things, directs everything the cell does. It is both the organizational plan and the set of instructions for carrying the plan out. Because it directs and controls all of life's processes including growth, development, and reproduction, DNA is key to life itself. RNA, a closely related macromolecule, is responsible for carrying out the instructions of DNA. In some viruses RNA (rather than DNA) serves as the genetic material.

To fully appreciate the importance of DNA and RNA, consider that

- DNA contains the instructions for producing RNA.
- RNA contains the instructions for producing proteins.
- Proteins direct most of life's processes.

Both DNA and RNA are composed of smaller molecular subunits called **nucleotides.** Nucleotides consist of 1) a 5-carbon sugar, 2) a single- or double-ringed structure containing nitrogen called a *base,* and 3) one or more phosphate groups. There are only eight different nucleotides, four in DNA and four in RNA.

Figure 2.22 shows the structures of the four nucleotides that make up DNA. Each nucleotide is composed of a 5-carbon sugar molecule called *deoxyribose* (like the 5-carbon sugar ribose but missing one oxygen atom), a phosphate group, and one of four different nitrogen-containing base molecules (*adenine, thymine, cytosine,* or *guanine*). To form a single strand of DNA, these nucleotides hook together by covalent bonds between the phosphate and sugar groups. The complete molecule of DNA is actually composed of two intertwined strands of nucleotides that are connected by weak hydrogen bonds (Figure 2.23). The sequence of one strand determines the sequence of the other (they are complementary strands), for adenine can form hydrogen bonds only with thymine (A with T) and cytosine can form bonds only with guanine (C with G).

The "code" for making a specific protein resides in the specific sequence of base pairs in one of the two strands of the DNA molecule. Notice that the entire genetic code is based entirely on the sequence of only four different molecular units (the four nucleotides). You will learn more about DNA, the genetic code, and inheritance in Chapters 17 and 19.

A single molecule of DNA carries the code for making a lot of different proteins. It is like an entire bookshelf of information, too big to be read all at once. In order to carry out their function, portions of the DNA molecule are transcribed into smaller fragments of RNA. RNA is structurally very similar to DNA with a few exceptions (Figure 2.24):

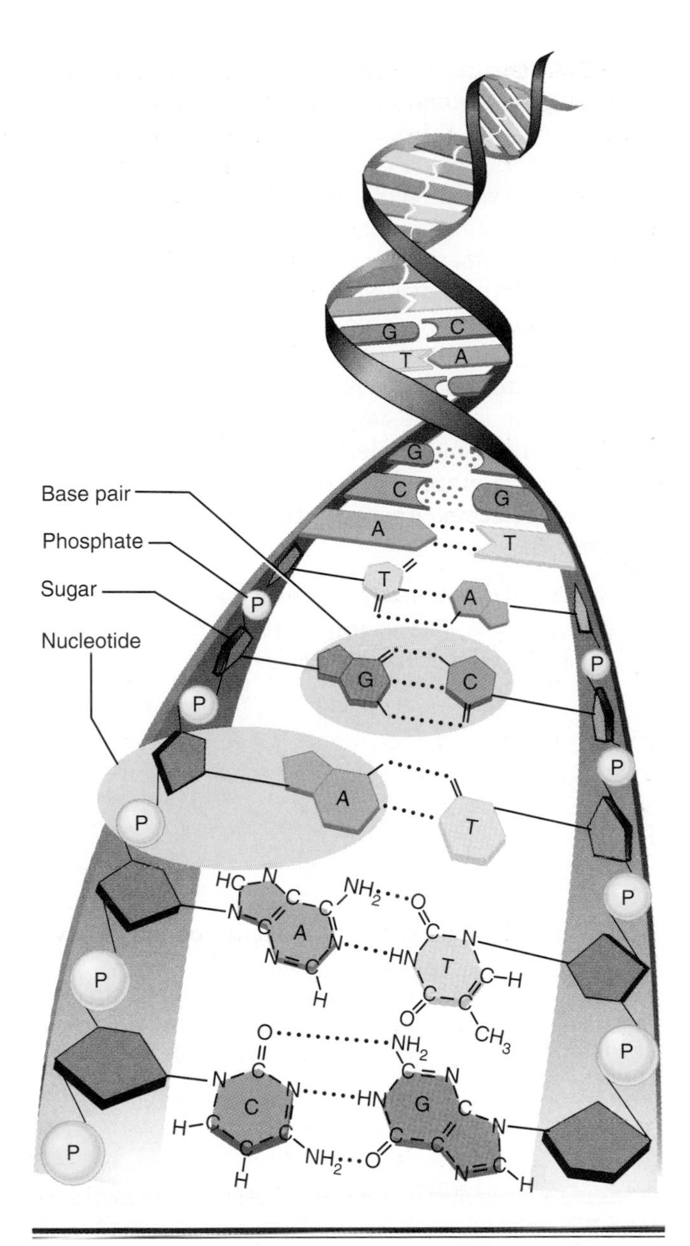

Figure 2.23 The double helical structure of DNA. The single strands of DNA are formed by dehydration synthesis. The two strands are held together by two hydrogen bonds between adenine and thymine and three hydrogen bonds between cytosine and guanine.

- The sugar unit in all four of the nucleotides in RNA is *ribose* rather than deoxyribose (hence the name ribonucleic acid).
- One of the four nitrogen-containing base molecules is different (uracil is substituted for thymine).
- RNA is a single-stranded molecule, representing a complementary copy of a portion of only *one* strand of DNA.
- RNA is shorter, representing only the segment of DNA that codes for a single protein.

Chapter 17 will discuss how RNA is used to make proteins.

> ***Recap*** **DNA and RNA are constructed of long strings of nucleotides. Double-stranded DNA represents the genetic code for life, and RNA is responsible for carrying out those instructions.**

ATP carries energy

One additional related nucleotide with a very important function is **ATP (adenosine triphosphate).** ATP is identical to the adenine-containing nucleotide in RNA except that it has two additional phosphate groups. ATP consists of an adenine base, the five-carbon sugar ribose (together they are called *adenosine*), and three phosphate groups *(triphosphate)* (Figure 2.25). ATP is a universal energy source for cells because the bonds between the phosphate groups contain a great deal of potential energy. Any time a cell needs energy for virtually any function, it can break the bond between the last two

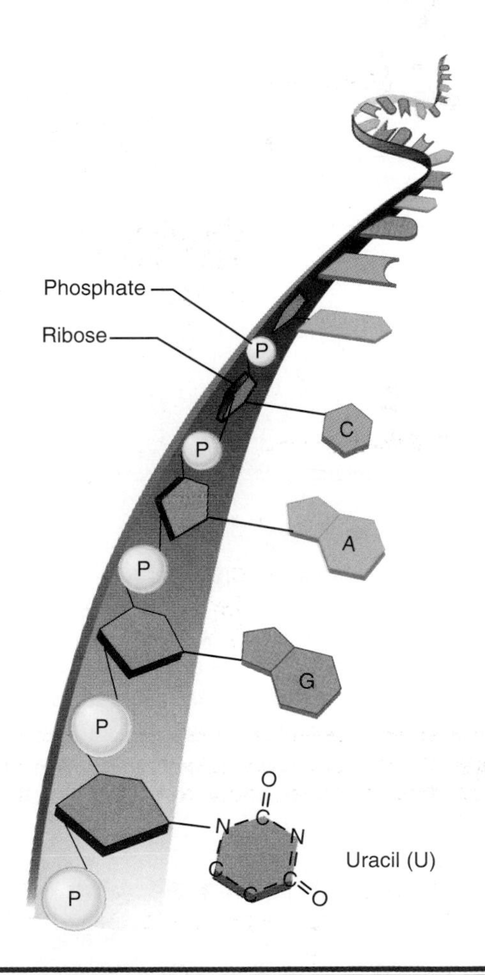

Figure 2.24 The structure of RNA. RNA is a single strand of nucleotides in which the base uracil substitutes for thymine. The sugar is ribose.

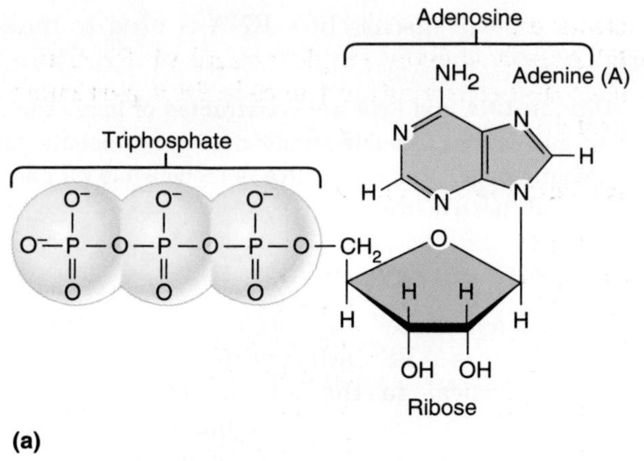

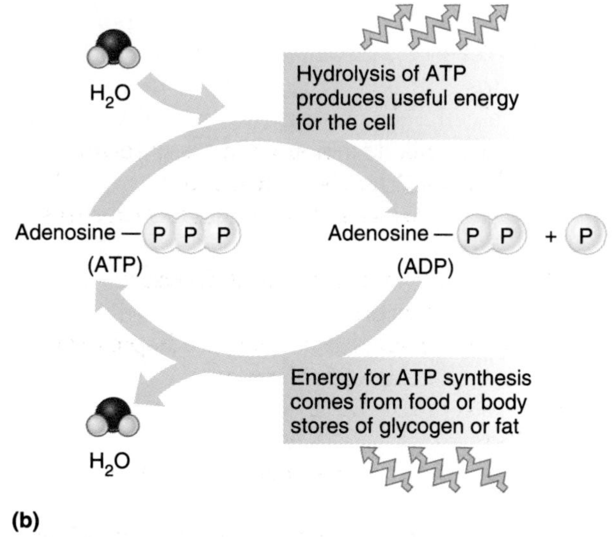

phosphate groups of ATP and energy will be released according to the following equation:

$$ATP \longrightarrow ADP + P_i + energy$$

The breakdown of ATP produces *ADP (adenosine diphosphate)* plus an inorganic phosphate group (P_i), which is not attached to an organic molecule, plus energy that is now available to do work. The reaction is reversible, meaning that ATP is replenished by using another source of energy to reattach P_i to ADP. The energy to replenish ATP may come from stored energy in the food we eat, or from the breakdown of energy storage molecules such as glycogen or fat. You will learn more about ATP as an energy source when we discuss energy utilization by muscles (Chapter 6).

Recap **ATP is a nearly universal source of quick energy for cells. The energy is stored in the chemical bonds between phosphate groups.**

Figure 2.25 Adenosine triphosphate (ATP). (a) The structure of ATP. **(b)** The breakdown (hydrolysis) of ATP yields energy for the cell. The reaction is reversible, meaning that ATP may be resynthesized using energy from other sources.

Chapter Summary

All matter consists of elements

- Atoms, the smallest functional unit of any element, comprise a nucleus and a cloud of electrons.
- The protons and neutrons in an atom's nucleus account for most of its mass.
- Radioisotopes are unstable isotopes with more or fewer neutrons than the usual number for that atom.

Atoms combine to form molecules

- Energy exists as either kinetic energy or potential energy.
- Three types of chemical bonds account for the structures of molecules: covalent, ionic, and hydrogen bonds. Covalent bonds are the strongest; hydrogen bonds are the weakest.
- Over 99% of your body weight consists of just six elements: oxygen, carbon, hydrogen, nitrogen, calcium, and phosphorus.

Life depends on water

CD Fluid Balance/Introduction to Body Fluids

- The polar nature of the water molecule accounts for its physical properties and for why it is such a good solvent for most other molecules and ions.
- Water is important in human temperature regulation.

The importance of hydrogen ions

- Molecules that can donate a hydrogen ion (H^+) are called acids. Molecules that can accept H^+ are called bases.
- The hydrogen ion concentration of a solution is expressed as pH.

- Buffers are pairs of molecules that tend to minimize changes in pH when an acid or base is added to a solution.

The organic molecules of living organisms

- The backbone of all organic molecules is carbon.
- Organic molecules are formed by a process called dehydration synthesis (requiring energy) and broken down by a process called hydrolysis (which releases energy).

Carbohydrates: Used for energy and structural support

- Monosaccharides, or simple sugars, are a source of quick energy for cells.
- Complex carbohydrates (polysaccharides) are formed by linking simple sugars (monosaccharides) together by dehydration synthesis.
- Carbohydrates are primarily energy-storage molecules. Plants use them for structural supports as well.
- In animals the storage molecule is glycogen; in plants it is starch.

Lipids: Insoluble in water

- Lipids include fats and oils, phospholipids, and steroids. Lipids are insoluble in water.
- Fats store energy. Phospholipids and cholesterol are important structural components of the cell membrane. The sex hormones are steroids synthesized from cholesterol.

Proteins: Complex structures constructed of amino acids

CD Fluid Balance/Introduction to Body Fluids

- Proteins have unique three-dimensional structures that depend on their primary structure (their amino acid sequences). Living organisms construct a tremendous number of different proteins using just 20 different amino acids.
- The human body contains thousands of proteins, each with a different function.
- Enzymes are proteins that facilitate the rates of chemical reactions.

Nucleic acids: Storing genetic information and making proteins

- ATP is a nucleotide with three phosphate groups that serves as an energy transfer molecule for living cells.
- DNA is composed of two long strands of nucleotides intertwined into a double helix. DNA is constructed from just four different DNA nucleotides.
- RNA is a shorter single strand of RNA nucleotides, representing the code for a particular protein.

Terms You Should Know

acid, 33	hydrogen bond, 29
atom, 24	ionic bond, 28
ATP, 47	lipid, 39
base, 33	molecule, 26
catalyst, 44	neutron, 24
covalent bond, 28	pH, 33
DNA, 46	protein, 41
electron, 24	proton, 24
enzyme, 44	RNA, 46

Test Yourself

Matching Exercise

_____ 1. An unstable atom with more or fewer neutrons than its typical number
_____ 2. A union between the electron structures of atoms
_____ 3. Consists of neutrons and protons
_____ 4. A simple sugar
_____ 5. The lower-energy form of an important energy-transfer molecule
_____ 6. The sugar in DNA
_____ 7. A steroid in our diet
_____ 8. A common polar molecule in living organisms
_____ 9. An atom that has permanently gained or lost electrons
_____10. Atom that serves as the backbone of organic molecules

a. nucleus
b. monosaccharide
c. cholesterol
d. chemical bond
e. water
f. deoxyribose
g. radioisotope
h. ion
i. carbon
j. ADP

Fill in the Blank

11. An _____________ solution has a pH greater than 7.
12. A string of 10 amino acids would be called a _____________.
13. Electrons position themselves in distinct _____________ around the nucleus of an atom.

14. The three-letter abbreviation for adenosine triphosphate is ______.
15. The complex carbohydrate that is the storage polysaccharide in animals is ______________.
16. ______________ are proteins that serve as biological catalysts.
17. An electron-sharing bond between atoms is called a ______________ bond.
18. The energy of motion is called ______________ energy.
19. Triglycerides, the storage form of fat, are composed of one molecule of glycerol and three molecules of ______________.
20. Fats that are liquids at room temperature are called ________.

True or False?

21. Lipids are generally very soluble in water.
22. DNA is restricted to the nucleus of the cell.
23. Dehydration synthesis requires energy.
24. The most common atom (by weight) in the human body is oxygen.
25. The prefix "poly" means "many."

Concept Review

1. Describe the electrical charges and relative masses of protons, neutrons, and electrons.
2. Explain why two atoms of hydrogen tend to combine into a molecule of hydrogen gas (H_2).
3. Explain why polar and charged molecules tend to be soluble in water.
4. How is a covalent bond different from an ionic bond?
5. Name the one element found in all organic molecules.
6. Distinguish between *saturated* and *unsaturated* fats.
7. Explain why proteins come in an almost unlimited variety of shapes.
8. Discuss the importance of enzymes in living organisms.
9. Describe the role of ATP in energy transfer within a cell.
10. How is the structure of RNA different from DNA?

Apply What You Know

1. Athletes are sometimes advised to eat large amounts of foods containing complex carbohydrates (such as pasta) for a day or two before a competitive event. Explain the reasoning behind this advice.
2. Physicians become very concerned about the potential for irreversible brain damage when body temperatures approach 105°F. Which of the four classes of macromolecules do you think is most likely affected by high temperatures? Explain.
3. Many detergents for clothes contain enzymes. How might enzymes improve the ability of a detergent to remove food and grass stains (organic material)?

Case in Point

The FDA Considers Extending the Use of Food Irradiation

The Food Irradiation Coalition, a consortium of food industry trade organizations, health organizations, and academic and consumer groups, has petitioned the Food and Drug Administration (FDA) to extend the permitted use of food irradiation to a variety of ready-to-eat foods. The list of foods that currently are irradiated to help eliminate microorganisms includes some strawberries, mushrooms, onions, tomatoes, tropical fruits, and spices. The categories of food covered by the petition include meat (most importantly, ground beef), poultry products, and fruit and vegetable products. The coalition argues that food irradiation is a safe and effective method for reducing incidents of the food-borne illnesses that kill 5,000 persons in the U.S. each year. Food irradiation also can increase the shelf life of certain foods.

Food is irradiated by placing it near radioisotopes such as cobalt 60 (^{60}Co) or cesium 137 (^{137}Cs) that have been packed into stainless steel rods. The radioisotopes emit high-energy, short-wave–length gamma rays that pass through the food in much the same way that microwaves pass through food. The high energy of the gamma rays disrupts DNA, thereby killing most of the bacteria, mites, parasites, and insects the food may contain. The radioisotopes themselves do not touch the food and no radioactive residue is left behind.

Still, some people are concerned. Food irradiation does destroy some vitamins and nutrients as well as living cells. In addition, low levels of irradiation do not necessarily kill all microorganisms, and so there is a legitimate concern that food irradiation might give consumers a false sense of security that the food is completely sterile. Some people worry that food irradiation might be used inappropriately to salvage foods that had begun to spoil. In addition, there is the problem of disposal of the radioactive waste produced as a result of the manufacture and later disposal of the stainless steel rods of radioactive materials. Finally, although food irradiation should (in theory) be harmless, there are few controlled studies to show that it is indeed completely safe.

QUESTIONS:

1. How might you test the hypothesis that a lifetime of food irradiation does not affect the longevity (length of life) of rats?
2. Why do you think that very few controlled experiments have been performed to demonstrate the safety of food irradiation in humans?

INVESTIGATE:

Sugarman, Carol. An End to Food Scares? Reconsidering Irradiation, With All Its Pros and Cons. *The Washington Post* Nov. 12, 1997, p. E01.

References and Additional Reading

Chapman, Carolyn. *Basic Chemistry for Biology.* Boston: McGraw-Hill, 1998. (An introduction to biological chemistry for beginners.)

The Merck Index: An Encyclopedia of Chemicals, Drugs, and Biologicals. Edited by Susan Budavari. Chapman and Hall, 1996; 12th edition. (A standard reference text to over 10,000 chemicals, drugs, and biological molecules.)

Morrison, Philip, Phylis Morrison, and the Office of Charles and Ray Eames. *Powers of Ten.* New York: Scientific American Books, 1982. (Stunning photographs and artists' depictions of objects of different sizes, from galaxies to subatomic particles.)

Waites, Gillian and Percy Harrison. *The Cassell Dictionary of Chemistry.* Cassell Academic, May 1999. (A reference source.)

CHAPTER 3

STRUCTURE AND FUNCTION OF CELLS

Why are cells almost always microscopic?

What are the functions of a cell's organelles?

Where do cells get the energy they need?

How do cells obtain nutrients?

SCIENTISTS FIRST OBSERVED LIVING CELLS UNDER A MICROscope in 1674. Since then, countless observations and experiments have confirmed the **cell doctrine,** which consists of three basic principles:

1. All living things are composed of cells and cell products.
2. A single cell is the smallest unit that exhibits all the characteristics of life.
3. All cells come only from preexisting cells.

If we examine any part of the human body under a microscope, we find living cells and/or cell products. "Cell products" include materials composed of cells that are no longer living (such as the outer layer of your skin) and substances resulting from cellular activity. We see no living units smaller than cells, and we find that all of our cells—tens of trillions of them—derived from earlier cells going all the way back to our first cell, a fertilized egg. And even that original cell came from preexisting cells, the sperm and egg from our parents.

3.1 Cells are classified according to their internal organization

All cells are surrounded by an outer membrane called the **plasma membrane.** The plasma membrane encloses the material inside the cell, which is mostly water but also contains ions, enzymes, and other structures the cell requires to maintain life.

All living cells are classified as either eukaryotes or prokaryotes, depending on their internal organization.

Eukaryotes have a nucleus, cytoplasm, and organelles

Human cells, like those of most species, are **eukaryotes** (*eu-* means "true" and *karyote* means "nucleus"). Nearly every eukaryotic cell has three basic structural components:

1. A *plasma membrane.* The plasma membrane forms the outer covering of the cell.
2. The *nucleus.* Nucleus is a general term for "core." In the last chapter, we saw that chemists define a nucleus as the core of an atom. In biology, the nucleus is a membrane-bound compartment that houses the cell's genetic material and functions as its "information center." Most eukaryotic cells have one nucleus. There are a few exceptions, to be discussed in later chapters.
3. The **cytoplasm** ("cell material"). The cytoplasm includes everything inside the cell except the nucleus. It is composed of a soft, gel-like fluid called the *cytosol* ("cell solution"). The cytosol contains a variety of microscopic structures called **organelles** ("little organs") that carry out specialized functions, such as digesting nutrients or packaging cellular products.

Prokaryotes lack a nucleus and organelles

Prokaryotes (*pro-* means "before" and *karyote* means "nucleus") are the bacteria (kingdom Monera). Prokaryotes have a plasma membrane but do not contain a nucleus. They also lack most of the organelles found in eukaryotes. Nevertheless they are living organisms that fit the definition of a cell according to the cell doctrine.

In the rest of this chapter and throughout the book, we concentrate on the structure and function of eukaryotic cells. However, we discuss bacteria again in terms of how they can make us ill (Chapter 9) and in the context of evolution (Chapter 21).

3.2 Cell structure reflects cell function

Spotlight on Structure & Function

Eukaryotic cells are remarkably alike in their structural features regardless of which organism they come from. This is because all cells carry out certain activities to maintain life, and there is a strong link between structure and function.

All cells must gather raw materials, excrete wastes, make macromolecules (the molecules of life), and grow and reproduce. These are not easy tasks. The specific activities carried out by a living cell (and the structures required to perform them) would rival those of any large city or even a country! There is an outer structure that defines its border; an infrastructure for support; an information center; manufacturing facilities; refining, packaging and shipping centers; transportation systems for supplying raw materials and energy; stockpiles of energy; and mechanisms for recycling or removing toxic waste. Cells even possess sophisticated defense mechanisms to combat invaders.

Most of the structural differences between cells reflect differences in function (Figure 3.1). Muscle cells contain numerous organelles to produce the necessary energy for muscle contraction. Nerve cells are long and thin, reaching from your toes to your spinal cord to carry nerve impulses. The cells that line the kidney tubules are cube shaped. Essentially every cell has a specialized function of some sort or it would be of little use to the organism.

Cells that serve the same function are often remarkably similar between species. For example, a human nerve cell has more in common (structurally and functionally) with a nerve cell in a cockroach than it does with a human liver cell. Furthermore,

(a) A portion of a muscle cell of the heart (×2,500)

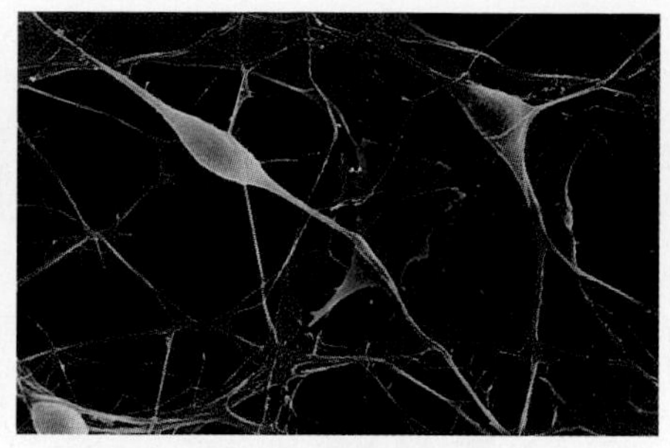

(b) Nerve cells of the central nervous system (×520)

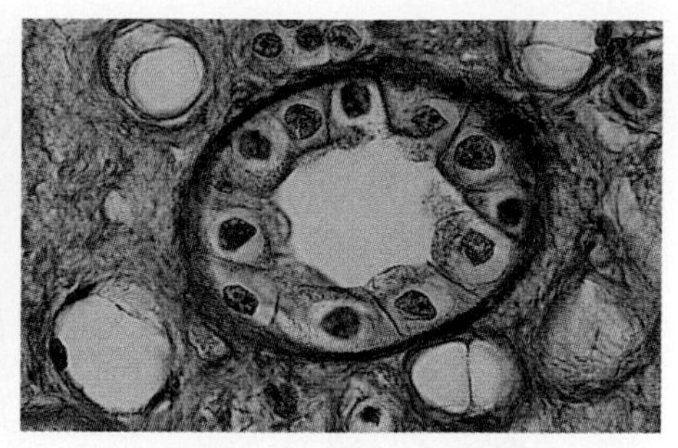

(c) Cells lining a tubule of a kidney (×250)

Figure 3.1 Human cells vary in shape.

cells in a mouse are not that much different in size from those in an elephant; it's just that an elephant has more of them. ■

Recap **Common features of nearly all eukaryotic cells are a plasma membrane, a nucleus, organelles, and the cytoplasm. Differences in cell shape and internal organization reflect differences in function.**

Cells remain small to stay efficient

Despite their structural differences, all cells have at least one feature in common: they are small in one or more dimensions, requiring considerable magnification to be seen at all (Figure 3.2). Despite the incredible complexity of human cells, not one can be seen with the naked eye. Even nerve cells that are over three feet long are so thin that we can't see them. Given the variety of life forms on Earth, why aren't there any giant cells? (An exception is the egg of some species.)

Why are cells almost always microscopic?

The answer shows that nature obeys certain simple and understandable principles:

- The total metabolic activities of a cell are proportional to its volume of cytoplasm, which is in effect its size. To support its activities, every cell needs raw materials in proportion to its size. Every cell also needs a way to get rid of its wastes.
- All raw materials, energy, and waste can enter or leave the cell only by crossing the plasma membrane.
- As objects get larger, their volume increases more than their surface area. For both spheres and cubes, for example, an eightfold increase in volume is accompanied by only a fourfold increase in surface area.

The larger a cell gets, then, the more likely that its growth and metabolism will be limited by its ability to supply itself across the plasma membrane (Figure 3.3). Put another way, the smaller a cell is the more effectively it can obtain raw materials and get rid of wastes.

Some cells have numerous microscopic projections of the plasma membrane called *microvilli*. Microvilli are an effective way to increase surface area relative to volume. Plasma membranes with microvilli are especially common in cells that transport substances into and out of the body, such as the cells that line the digestive tract and the tubules of the kidneys.

Recap **Cells exchange materials with their environment across their plasma membrane. Cells divide to remain small, because this makes them more efficient at obtaining nutrients and expelling wastes.**

3.3 A plasma membrane surrounds the cell

Consider a house. Its walls and roof are composed of special materials that prevent rain and wind from entering. They also form a barrier that allows the temperature inside the house to stay warmer or cooler than the temperature outside. At the same time, the house interacts with its environment. Windows allow light in; doors open and close to allow entry and exit. Water and power lines permit the regulated entry of water and energy, and sewer lines remove wastes. The house exchanges information with the outside world via mail slots, telephone lines, and computer cables.

Spotlight on Structure & Function

The exterior structure of a living cell is its plasma membrane. Like the roof and walls of a house, the plasma membrane must permit the movement of some substances into and out of the cell, yet restrict the movement of others. It must also allow information to be transferred across the membrane.

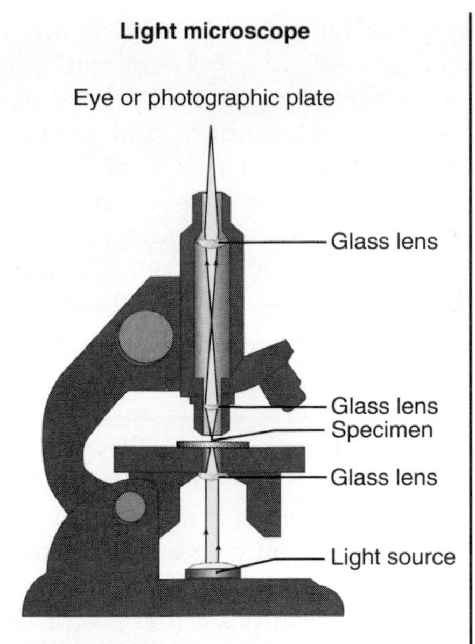

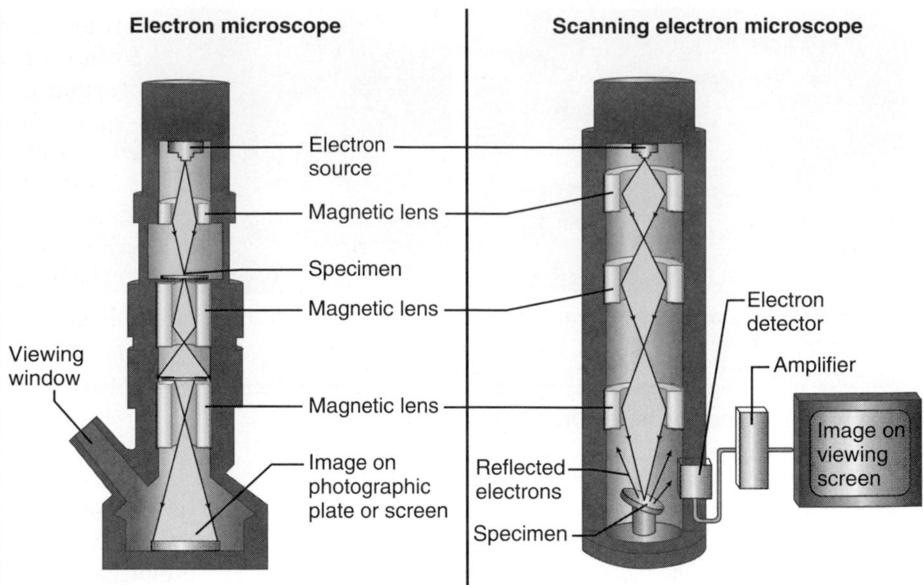

(a) The light microscope. Most of us have seen or used a light microscope. The light microscope uses visible light and a series of glass lenses to magnify a small sample as much as 1000-fold. Focus can be difficult to maintain if objects are at different depths in a sample. Light does not transmit well through very dense or thick samples. Light microscopes have been in use over 300 years.

(b) The transmission electron microscope. A transmission electron microscope (TEM) bombards the sample with a beam of electrons, some of which pass through the sample. Electrons behave like light waves but since the wavelengths of electrons are shorter than visible light, the image has greater clarity at any magnification. A good electron microscope can magnify up to about 100,000 times, a hundred times greater than the light microscope. The images are two-dimensional (flat) because the sample must be very thin, but the magnification is sufficiently high that one can see the structural details of organelles within single cells.

(c) The scanning electron microscope. The scanning electron microscope (SEM) also focuses beams of electrons on the object. It produces what appears to be a three-dimensional view of the *surface* of an object. A narrow beam of electrons is scanned over the surface of an object which has been coated with a thin coat of metal so that the electron beam cannot penetrate. As the beam passes, the metal gives off secondary electrons which can be recorded to produce a visual picture of the object's surface. Scanning electron microscopes have revealed stunning images of the relationships between cells and of the outer surfaces of cells. The scanning electron microscope can also magnify up to about 100,000 times.

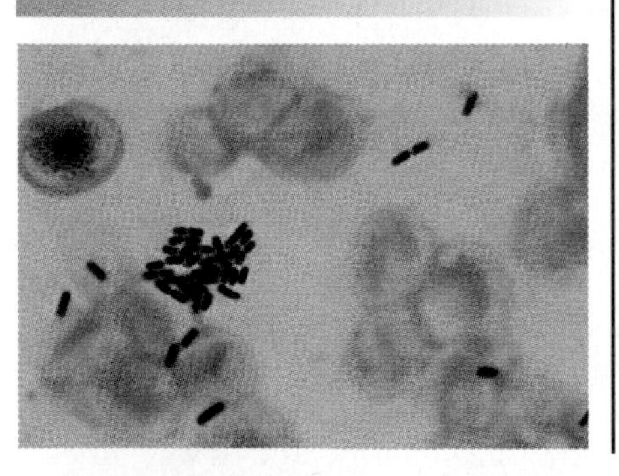

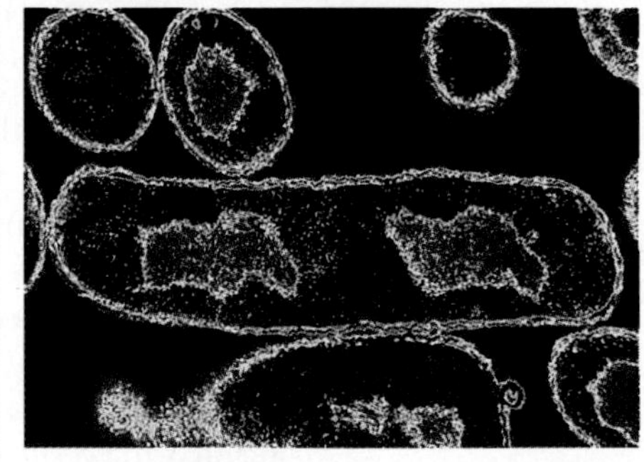

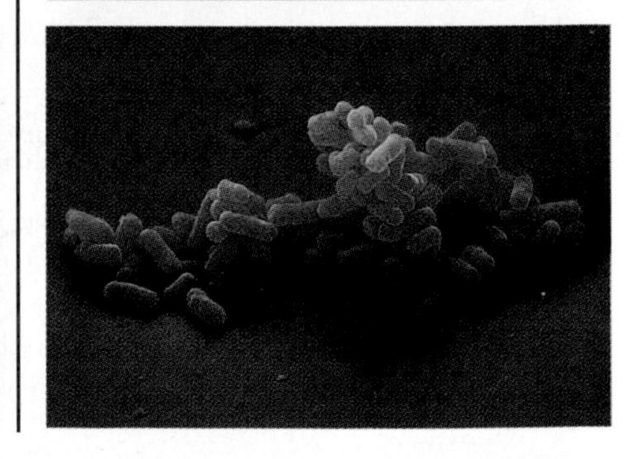

Figure 3.2 Visualizing cells with microscopes. Photographs taken by the various methods of microscopy are called *photomicrographs.* All three of the photomicrographs shown here are of *Escherichia coli,* a normally harmless bacterium found in the digestive tract.

The plasma membrane is a lipid bilayer

The plasma membrane is constructed of two layers of phospholipids, called a *lipid bilayer,* plus some cholesterol and various proteins (Figure 3.4). Each of the three components contributes to the membrane's overall structural and functional properties:

- *Phospholipids.* Recall that phospholipids are a particular type of lipid with a polar head and neutral nonpolar tails. In the plasma membrane the two layers of phospholipids are arranged so that the nonpolar tails meet in the center of the membrane. One layer of polar (water-soluble)

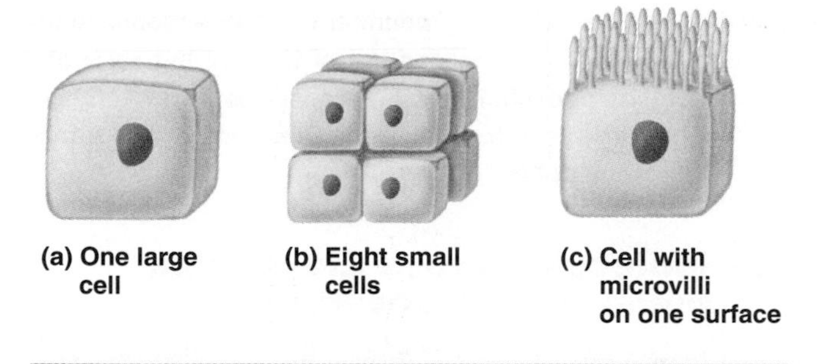

Figure 3.3 Cell size and plasma membrane shape affect surface area and volume. The volume is the same in these three groups of cells, but the ratio of volume to surface area is different in **(a), (b),** and **(c).** Compared to the cell in **(a),** the eight small cells in **(b)** have twice the surface area. **(c)** The surface area of any cell can be increased by the presence of microvilli.

heads faces the watery solution on the outside of the cell, and the other layer of polar heads faces the watery solution of the cell's cytoplasm (see Try It Yourself, page 59).

- *Cholesterol.* By itself, a bilayer of phospholipids would be soft because it would be basically a double layer of oil. Cholesterol molecules make the plasma membrane more rigid.
- *Proteins.* Various proteins are embedded in the phospholipid bilayer of the plasma membrane. Like the doors, windows, and wires of a house, they provide the means for transporting molecules and information across the plasma membrane. Some proteins span the entire membrane; others protrude from only one surface. Plasma membrane proteins generally have one region that is electrically neutral and another that is electrically charged (either $^+$ or $^-$). The charged regions tend to extend out of the membrane and thus are in contact with water, whereas the neutral portions are often embedded within the phospholipid bilayer. ■

The phospholipid bilayer of the plasma membrane is only about 3.5 nanometers thick (a nanometer, abbreviated nm, is a billionth of a meter), too small to be seen in detail even with microscopes. To appreciate relative sizes, imagine as we did in Chapter 2 that a single sodium ion is the size of a penny. At this scale the phospholipid bilayer of a typical cell would be about 13 inches thick. It is no wonder that many substances are restricted from passing through the membrane unless there is some sort of channel or transport mechanism available.

Although we have likened the plasma membrane to the exterior of a house, there are two key differences. First, the plasma membrane of animal cells is not rigid. If you could touch a plasma membrane, it would probably feel like a wet sponge, giving way under your touch and springing back when you remove your hand. Most cells do maintain a certain shape, but it is mainly due to a supporting network of fibers inside the cell, the fluid it contains, and the limitations imposed by surrounding cells, and not to the stiffness of the plasma membrane itself.

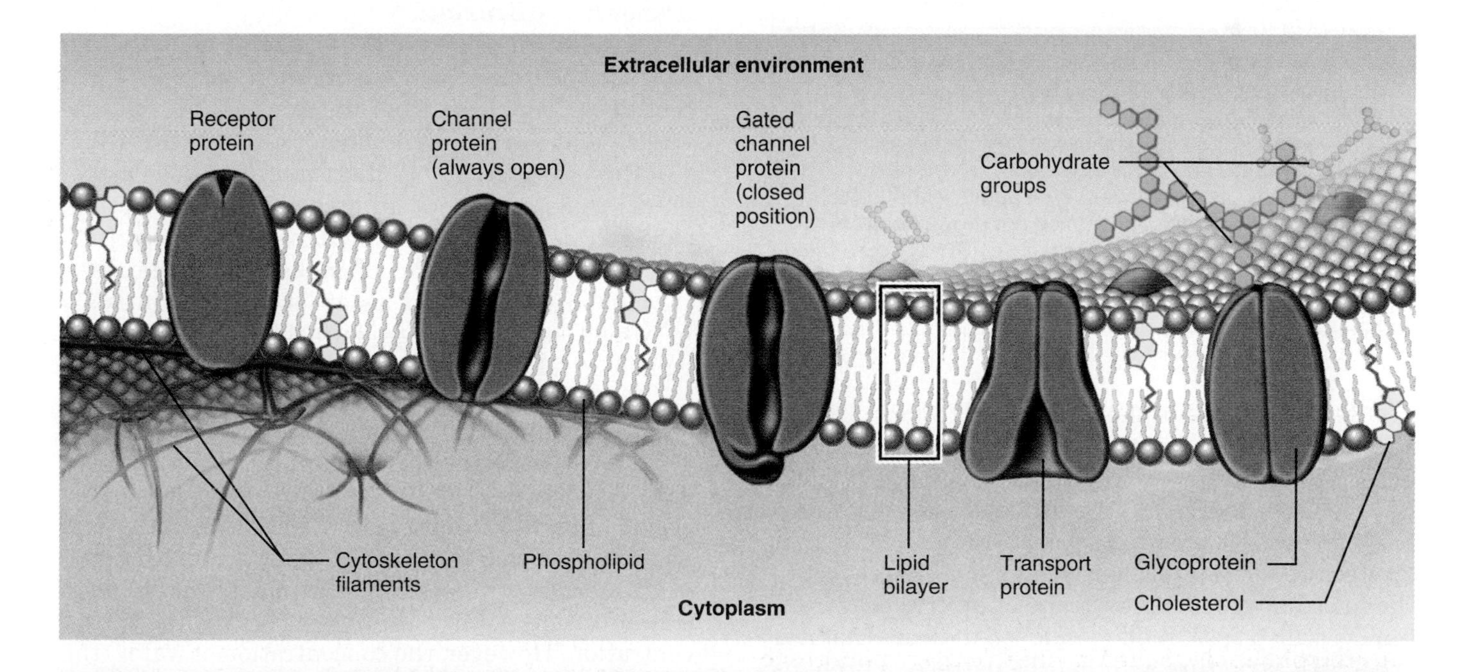

Figure 3.4 The plasma membrane.

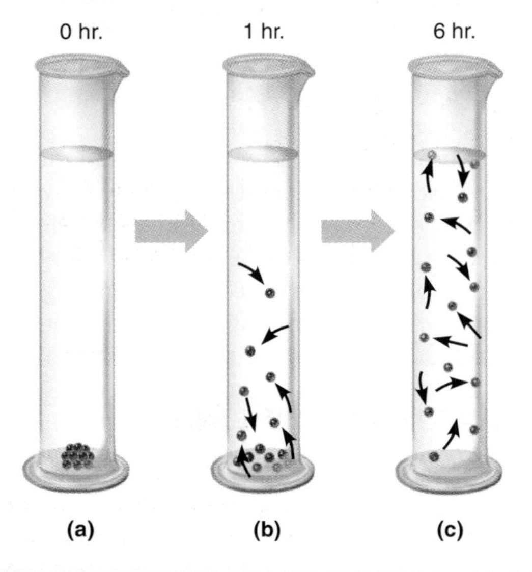

Figure 3.5 Diffusion. A high concentration of a water-soluble substance will eventually become equally distributed throughout a solution. **(a)** Crystals of a blue dye were placed in a graduated cylinder of water at zero hour. **(b)** At one hour, net diffusion of molecules of the dye has proceeded partway up the cylinder. Meanwhile, although it is not visible, water is also undergoing net diffusion, from the top of the cylinder toward the bottom. **(c)** By six hours the dye molecules and water molecules are evenly distributed. Diffusion of both water and dye molecules continues, but at equal rates in all directions.

Second, the phospholipids and proteins are not anchored to specific positions in the plasma membrane. Many proteins drift about in the lipid bilayer like icebergs floating on the surface of the sea. Imagine if you were to get up in the morning and find that the front door to your house had moved three feet to the left! The plasma membrane of animal cells is often described as a "fluid mosaic" to indicate that it is not a rigid structure and that the pattern of proteins within it constantly changes.

***Recap* The plasma membrane allows some substances to move into and out of the cell but restricts others. Its phospholipid bilayer contains cholesterol for support and proteins for information transfer and transport of molecules.**

3.4 Molecules cross the plasma membrane in several ways

How do cells obtain nutrients?

The plasma membrane creates a barrier between the cell's external environment and the processes of life going on within. Life was impossible until these functions could be enclosed and concentrated in one place, keeping what was needed for growth and reproduction inside and limiting the entry of other materials. Molecules (and ions) cross the plasma membrane in three major ways: (1) passive transport (diffusion), (2) active transport, and (3) endocytosis or exocytosis. These three processes are discussed below.

Passive transport: principles of diffusion and osmosis

Passive transport is "passive" because it transports a molecule without requiring the cell to expend any energy. Passive transport relies on the mechanism of diffusion.

Molecules in a gas or a liquid move about randomly, colliding with other molecules and changing direction. The movement of molecules from one region to another as the result of this random motion is known as **diffusion.**

If there are more molecules in one region than in another, then strictly by chance more molecules will tend to diffuse away from the area of high concentration and toward the region of low concentration. In other words, the *net* diffusion of molecules requires that there be a difference in concentration, called a *concentration gradient,* between two points. Once the concentration of molecules is the same throughout the solution, a state of equilibrium exists in which molecules are diffusing randomly but equally in all directions.

Figure 3.5 illustrates diffusion by showing what happens when crystals that dissolve in water are placed at the bottom of a container of water. Over time, the dissolved molecules diffuse away from their region of highest concentration until the entire fluid has the same concentration of dissolved molecules.

Water also diffuses from the region of its highest concentration toward the region of its lowest concentration. However, the concentration of water (the liquid, or solvent) in a solution is *opposite* to that of the molecules other than water (the solutes). The

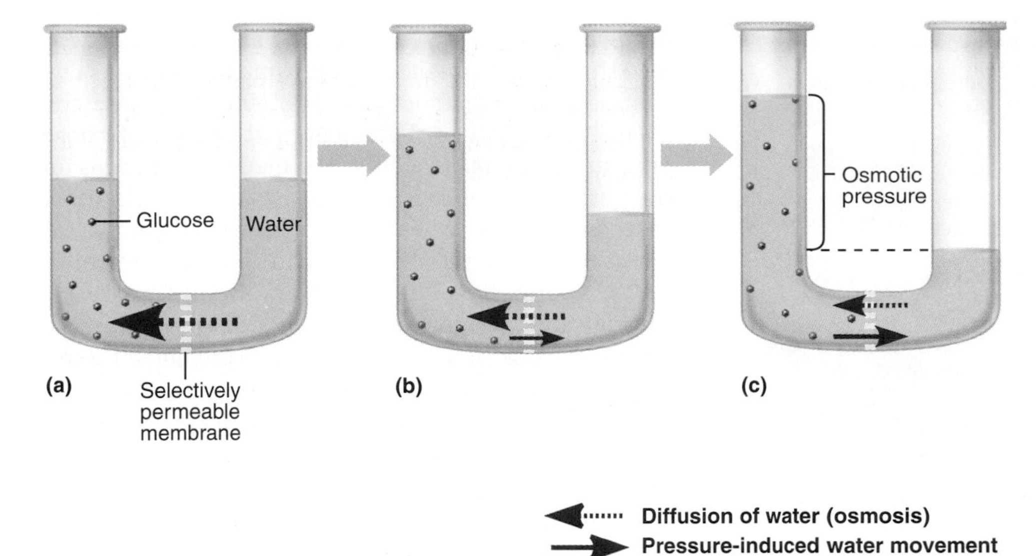

Figure 3.6 Generation of osmotic pressure by osmosis. Starting in **(a),** there is a net movement of water from right to left until the diffusion of water (osmosis) is opposed by movement of water due to osmotic pressure **(c).**

higher the concentration of solutes, the lower the concentration of water. Pure water is the solution with the highest possible concentration of water. Net diffusion of water is always toward the solution with the higher concentration of solutes and away from the solution with a higher concentration of water.

Not all substances diffuse readily into and out of living cells. The plasma membrane is **selectively permeable,** meaning that it allows some substances to cross by diffusion but not others. It is highly permeable to water, but not to all ions or molecules. The net diffusion of water across a selectively permeable membrane is called **osmosis.**

Figure 3.6 demonstrates the process of osmosis. In Figure 3.6a, a selectively permeable membrane—in this case permeable only to water—separates pure water from a solution of glucose in water. Although the glucose cannot diffuse, the water will diffuse toward its region of lower concentration, from right to left. As osmosis occurs, the volume in the left chamber rises, creating a fluid pressure that begins to oppose the continued osmosis of water. Eventually the movement of water from left to right (because of differences in fluid pressure) equals the movement from right to left (by osmosis), and there is no further net change in the volumes of water on each side of the membrane.

The fluid pressure required to exactly oppose osmosis is called *osmotic pressure.* In Figure 3.6c osmotic pressure is represented by the extra weight of the higher column of water on the left than on the right.

Recap **Molecules diffuse from a region of higher concentration to a region of lower concentration. The net diffusion of water across a selectively permeable membrane is called osmosis.**

Passive transport moves with the concentration gradient

Most substances cross cell membranes by passive transport. Passive transport always proceeds "downhill" with respect to the concentration gradient,

Try It Yourself

Oil and Water Don't Mix

To demonstrate that oil and water don't mix, place water and cooking oil in a jar and shake it. Notice that the oil tends to separate from the water, first as small droplets, then as larger drops that join together, and finally as a single layer of oil at the top of the jar.

The reason for the eventual separation of the water and oil is that water is a polar molecule and oils are uncharged and nonpolar molecules. When water and oil are mixed together, the water molecules are attracted to each other. Thus they tend to join together, excluding oil from the regions they occupy. Over time the oil is forced into ever-larger droplets until it is completely separated from the water. It rises to the top because oil is less dense than water.

Although you probably don't have any phospholipids available, you would get a different result from your demonstration if you could repeat it with phospholipids. Phospholipids do not separate from water. On microscopic examination you would see numerous small spheres, each comprised of a bilayer of phospholipids enclosing a small volume of water. Such spheres are essentially nonliving versions of the cell's plasma membrane and cytoplasm. The arrangement is a completely natural one based on the chemistry of the phospholipids and water.

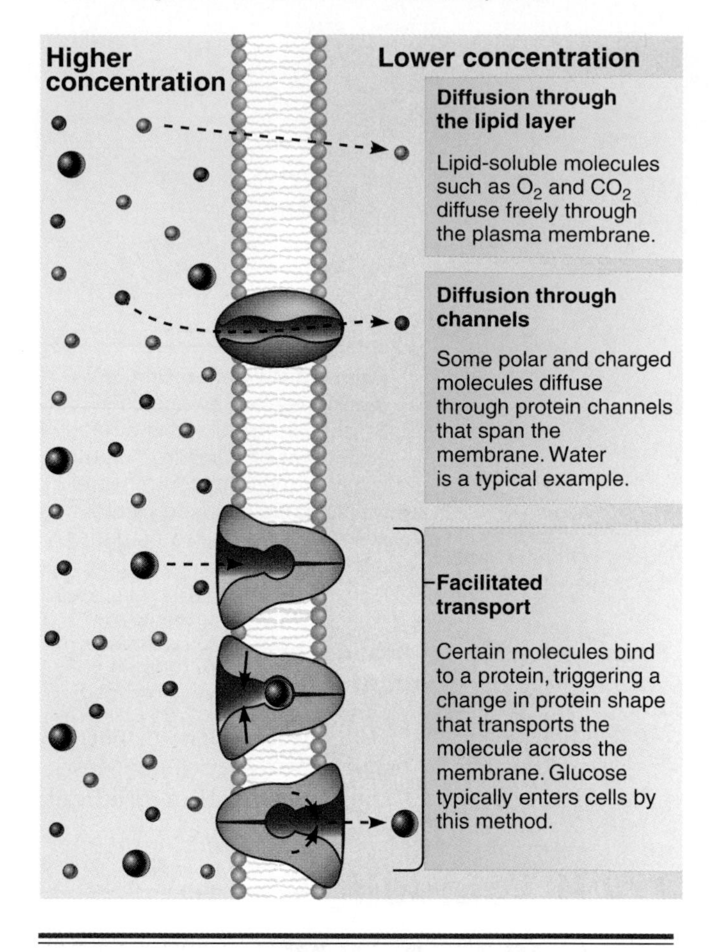

Figure 3.7 The three forms of passive transport. All involve transport down a concentration gradient without the expenditure of additional energy.

meaning that it relies on diffusion in some way. Three forms of passive transport across the cell membrane are: (1) diffusion through the lipid bilayer, (2) diffusion through channels, and (3) facilitated transport (Figure 3.7).

Diffusion through the lipid bilayer The lipid bilayer structure of the plasma membrane allows the free passage of some molecules while restricting others. For instance, small uncharged nonpolar molecules can diffuse right through the lipid bilayer as if it did not exist. Such molecules simply dissolve in the lipid bilayer, passing through it as one might imagine a ghost walking through a wall. Polar or electrically charged molecules, on the other hand, cannot cross the lipid bilayer because they are not soluble in lipids.

Two important lipid-soluble molecules are oxygen (O_2), which diffuses into cells and is used up in the process of metabolism, and carbon dioxide (CO_2), a waste product of metabolism, which diffuses out of cells and is removed from the body by the lungs. Another substance that crosses the lipid bilayer by diffusion is urea, a neutral waste product removed from the body by the kidneys.

Diffusion through channels Water and many ions diffuse through channels in the plasma membrane. The channels are constructed of proteins that span the entire lipid bilayer. The sizes and shapes of these protein channels, as well as the electrical charges on the various amino acid groups that line the channel, determine which molecules can pass through.

Some channels are open all the time (typical of water channels). The diffusion of any molecule through the membrane is largely determined by the number of channels through which the molecule can fit. Other channels are "gated," meaning that they can open and close under certain conditions. Gated channels are particularly important in regulating the transport of ions (sodium, potassium, and calcium) in cells that are electrically excitable, such as nerves (Chapter 11). Look back at Figure 3.4, in which are represented a number of the proteins that are instrumental in transport.

Facilitated transport In **facilitated transport,** also called *facilitated diffusion,* the molecule does not pass through a channel at all. Instead it attaches to a membrane protein, triggering a change in the protein's shape or orientation that transfers the molecule to the other side of the membrane and releases it there. Once the molecule is released, the protein returns to its original form. A protein that carries a molecule across the plasma membrane in this manner, rather than opening a channel through it, is called a **transport** (carrier) **protein.**

Facilitated transport is highly selective for particular substances. The direction of movement is always from a region of high concentration to one of lower concentration, and thus it does not require the cell to expend energy. The normal process of diffusion is simply being "facilitated" by the transport protein. Glucose and other simple sugars enter most cells by this method.

> ***Recap*** **Diffusion and facilitated transport are both forms of passive transport. They do not require the expenditure of energy.**

Active transport requires energy

All methods of passive transport allow substances to move only down their concentration gradients, in the direction they would normally diffuse if there were no barrier. However, **active transport** can move substances through the plasma membrane *against* their concentration gradient. Active transport allows a cell to accumulate essential molecules even when their concentration outside the cell is relatively low, and to get rid of molecules that it does not need. Active transport requires the expenditure of energy.

Like facilitated transport, active transport is accomplished by proteins that span the plasma membrane. The difference is that active transport proteins must have some source of energy in order to transport certain molecules. Some active transport proteins use the high-energy molecule ATP for this purpose (Figure 3.8a). They break ATP down to ADP and a phosphate group (P_i) and use the released energy to transport one or more molecules across the plasma membrane against their concentration gradient. Imagine that the active transport protein is a conveyor belt moving objects uphill, powered by a gasoline engine. In this analogy, ATP would be the gasoline and ADP the exhaust (with one important modification: the ADP "exhaust" within the cell can be recycled to ATP).

Proteins that actively transport molecules across the plasma membrane are sometimes called "pumps." Some pumps can transport several different molecules at once and even in both directions at the same time. One of the most important plasma membrane pumps is the **sodium-potassium pump,** which uses energy derived from breaking down ATP to transport sodium out of the cell and potassium into the cell.

Not all active transport pumps use ATP as the energy source. Some derive energy from the "downhill" facilitated transport of one molecule and use it to transport another molecule "uphill," against its concentration gradient (Figure 3.8b). This type of transport is sometimes called "co-transport" because two different molecules are being transported at once (one actively, one passively). It is analogous to an old-fashioned mill that grinds grain into flour by using energy derived from the downhill movement of water.

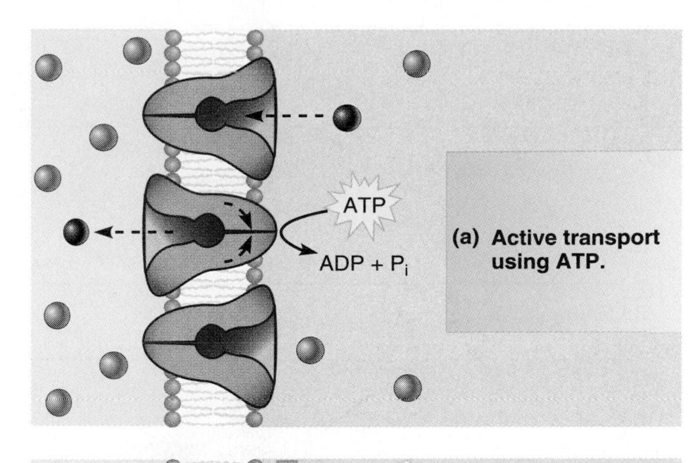

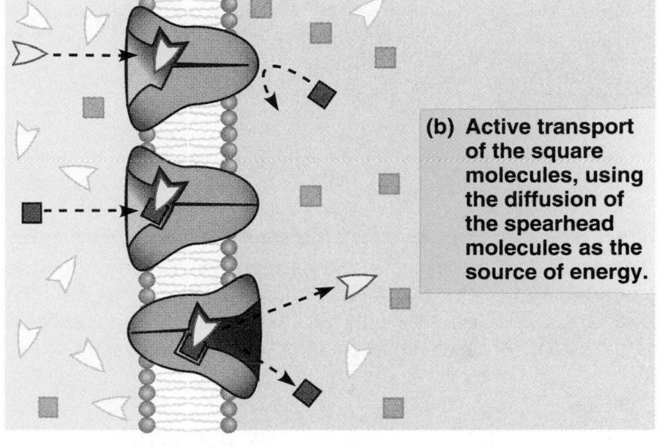

Figure 3.8 Active transport. A cell can employ active transport to move a molecule against a concentration gradient. Because this is an "uphill" effort, energy is required. **(a)** ATP offers the energy needed to change the shape of the carrier protein. ATP is converted to ADP + P_i as it liberates energy. **(b)** Some carrier proteins transport two molecules simultaneously. The downhill transport of one molecule provides the energy for the uphill transport of the other.

Endocytosis and exocytosis move materials in bulk

Most ions and small molecules move across the cell membrane by one or more of the passive and active transport mechanisms just described. However, some molecules are too big to be transported by these methods. To move large molecules or transport several kinds of molecules in bulk, some cells resort to endocytosis and/or exocytosis. The principle is the same, but the direction of movement is different. **Endocytosis** moves materials into the cell and **exocytosis** moves materials out of the cell.

Figure 3.9 depicts the processes of endocytosis and exocytosis. In endocytosis, groups of molecules or even entire cells (such as bacteria) are surrounded by a pocket of plasma membrane. Eventually the pocket pinches off, forming a membrane-bound *vesicle* within the cell. What happens to the vesicle depends on what it contains. Some white blood cells engulf and destroy bacteria by endocytosis (Chapter 9).

In exocytosis, a vesicle already present within the cell fuses with the plasma membrane and releases its contents into the fluid surrounding the cell. This is how certain cells release toxic waste products, get rid of indigestible material, or secrete their special products.

Information can be transferred across the plasma membrane

Receptor proteins that span the plasma membrane can receive and transmit information across the membrane. The information received by receptor proteins generally causes something to happen within the cell even though no molecules cross the membrane.

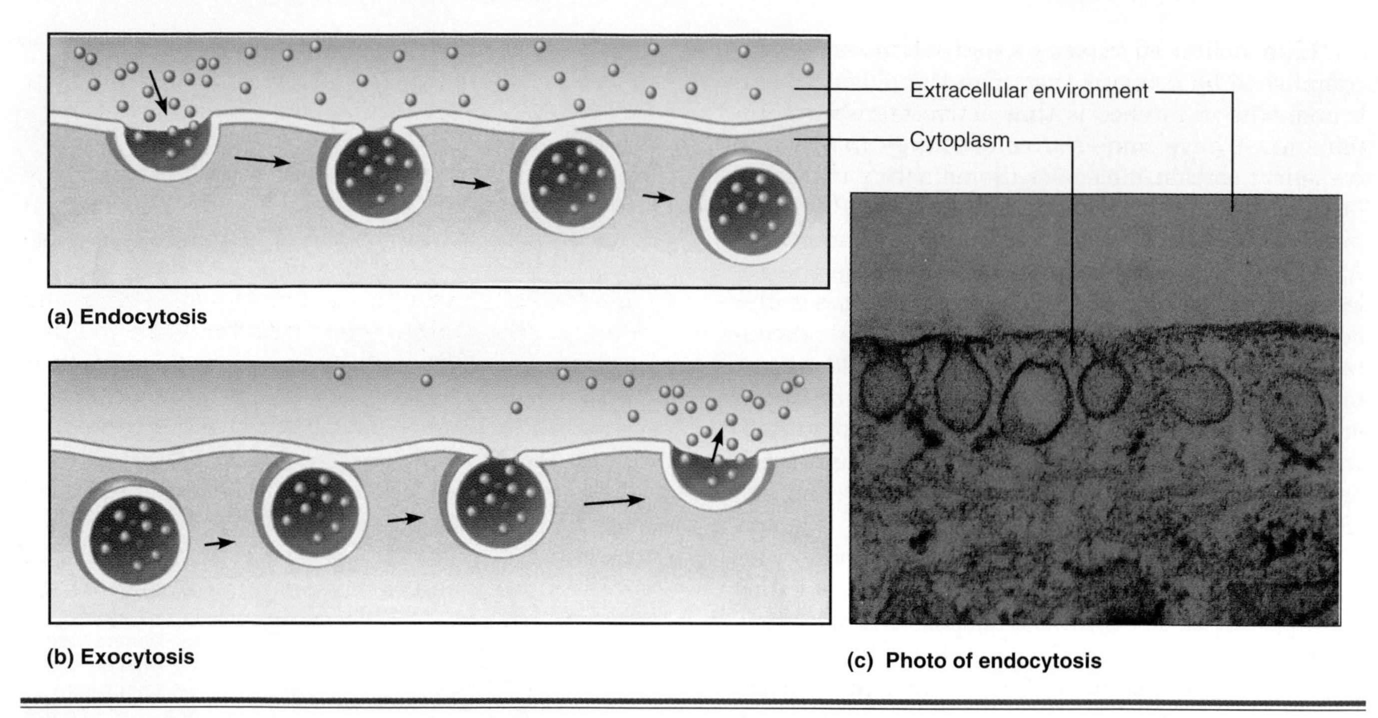

Figure 3.9 Endocytosis and exocytosis. (a) In endocytosis, material is surrounded by the cell membrane and brought into the cell. **(b)** In exocytosis, a membraneous vesicle fuses with the plasma membrane, expelling its contents outside the cell. **(c)** Photomicrograph showing various stages of endocytosis.

Figure 3.10 illustrates how a receptor protein works. A molecule approaches the membrane and binds to a specific *receptor site* in a lock-and-key fashion. This triggers a series of biochemical events that ultimately cause changes within the cell. Receptor proteins are highly specific for a particular molecule or group of similar molecules. For example, the receptor protein for the hormone insulin responds only to insulin and not to any other hormone. Furthermore, different cells have different sets of receptor proteins, which explains why some cells and tissues respond to a particular hormone and others do not.

We discuss receptor proteins and hormones in more detail in Chapter 13. For now, remember that certain molecules can influence what happens inside a cell merely by coming in contact with the cell's outer surface.

The sodium-potassium pump helps maintain cell volume

Probably the most critical task facing a cell is how to maintain its volume. Why? Recall that the plasma membrane is soft and flexible. It cannot withstand much stretching or high fluid pressures. Furthermore, cells tend to accumulate certain materials depending on what is available in their *extracellular* environment (the area outside the cell, beyond the plasma membrane). Cells already contain a nucleus and organelles. In addition, they produce or stockpile many macromolecules, including amino acids, sugars, lipids, ions, and many others. These molecules are necessary for the cell to function normally, but they add a lot of solute molecules within the cell.

Because water can diffuse across the plasma membrane rather easily, you might expect that water would diffuse into the cell, toward the high cytoplasmic solute concentration. This inward diffusion

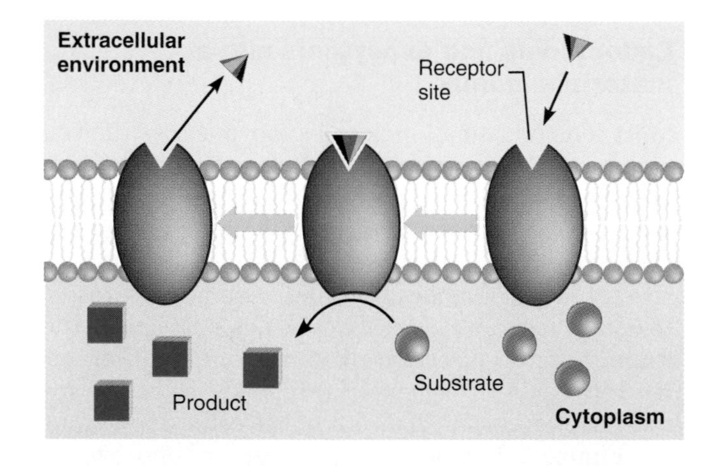

Figure 3.10 Receptor protein action. A specific molecule approaches a receptor site on a receptor protein and binds to it. The binding of the molecule to the receptor protein causes a series of chemical reactions within the cell. In this example, a particular cellular product (the squares) is produced from substrate molecules (the circles).

would increase cell volume, eventually causing the cell to swell and even rupture.

The only way to avoid this is for the cell to keep the solute concentration in its cytoplasm identical to the solute concentration of the extracellular fluid. Then there is no net driving force for the diffusion of water. What the cell actually does is get rid of ions it *doesn't* need in large quantities (primarily sodium) in exchange for those it must stockpile. This is the primary function of the sodium-potassium pump.

Using energy derived from the breakdown of ATP, the sodium-potassium pump transports three sodium ions out of the cell and brings in two potassium ions (Figure 3.11). This three-for-two exchange lowers the number of ions within the cell slightly. More importantly, the plasma membrane is much more permeable to potassium than to sodium because it contains many potassium channels but very few sodium channels. Effectively, the cell keeps the number of sodium ions in the cytoplasm low by pumping sodium ions out at the same rate that they leak in. As Figure 3.11b shows, this effectively controls cell volume.

To reduce cell volume, the cell can increase the activity of the sodium-potassium pumps, getting rid of more sodium than usual. Water will follow to maintain osmotic equilibrium. To expand cell volume, the cell lowers the activity of the pumps and water will be retained along with the extra sodium. Because potassium can diffuse so quickly no matter how much is pumped, the rate of potassium transport by the pumps is not relevant to the control of cell volume.

A single red blood cell may have over a hundred sodium-potassium pumps in its plasma membrane. In addition, the pump is essential to the ability of nerve cells to generate an electric current, as we see in later chapters.

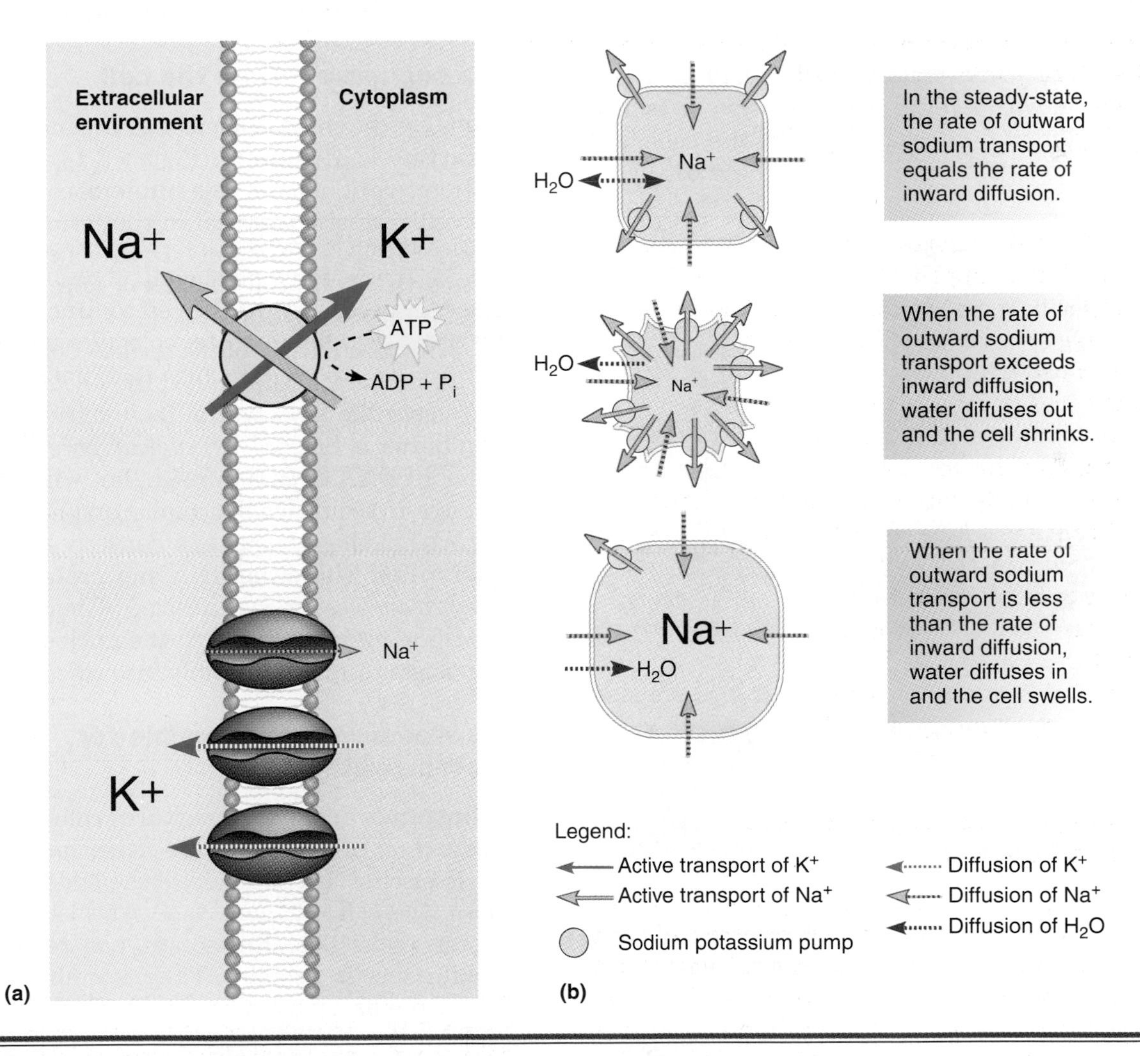

Figure 3.11 Control of cell volume by the sodium-potassium pump (Na^+–K^+ pump). (a) The cell membrane contains Na^+–K^+ pumps, and also channels that permit the rapid outward diffusion of K^+ but only a slow inward diffusion of Na^+. **(b)** The rate of transport by the Na^+–K^+ pumps determines cell volume.

Isotonic extracellular fluid also maintains cell volume

Because water can diffuse across the cell membrane so easily, the ability of a human cell to control its volume also depends on the solute concentration in the extracellular fluid (Figure 3.12). An **isotonic** extracellular fluid has the same solute concentration as the *intracellular* fluid (*iso-* means "same" and *-tonic* means "strength"). Cells maintain a normal volume in isotonic extracellular solutions because the concentration of water is the same inside and out. In humans, isotonic extracellular fluid is equivalent to about nine grams of salt dissolved in a liter of solution. Regulatory mechanisms in the body ensure that the extracellular fluid solute concentration is kept relatively constant at that level.

When cells are placed in a **hypertonic** solution, one with a concentration of solutes *higher* than the intracellular fluid, water diffuses out of the cells and the cells shrink. Eventually this impairs normal function and the cells die.

Conversely, when cells are placed in a **hypotonic** solution with a *lower* concentration of solutes than intracellular fluid, water enters the cells and causes them to swell. Pure water is the most hypotonic solution possible. Most human cells will quickly swell, burst, and die when placed in pure water.

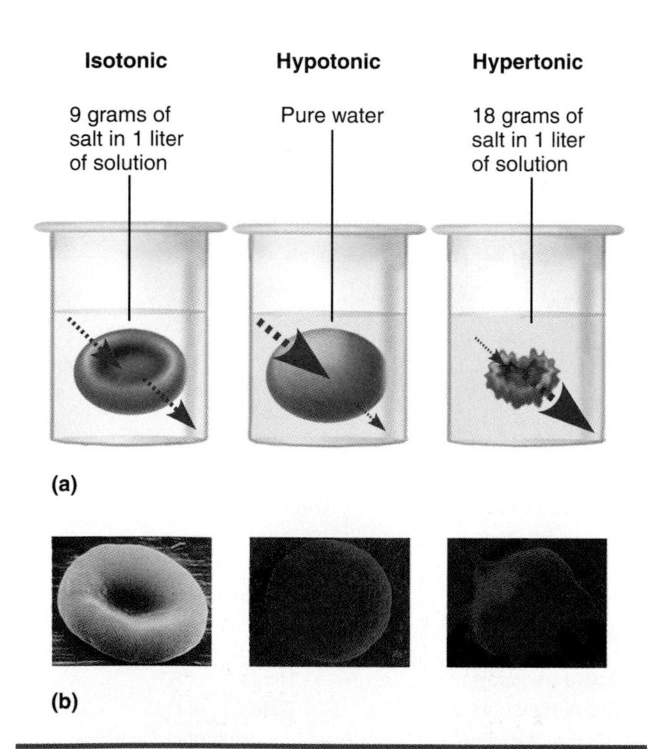

Figure 3.12 The effect of extracellular fluid tonicity on cell volume. (a) Water movement into and out of human red blood cells placed in isotonic, hypotonic, or hypertonic solutions. The amount of water movement is indicated by the sizes of the arrows. **(b)** Scanning electron micrographs of red blood cells placed in similar solutions.

Recap **The sodium-potassium pump is an essential component of the regulation of cell volume. In addition, homeostatic regulatory mechanisms keep the extracellular fluid solute concentration relatively constant.**

3.5 Internal structures carry out specific functions

So far we have discussed the plasma membrane that surrounds the cell. Now we move inside the cell, where we find a number of different membrane-bound and nonmembrane-bound structures. The membrane-bound structures are called organelles, because they are like an organ in that they have a specific function to perform, but are smaller. Figure 3.13 presents a cutaway view of an animal cell with its nucleus and organelles.

What are the functions of a cell's organelles?

The nucleus controls the cell

Arguably the most conspicuous organelle of a living eukaryote is its *nucleus* (Figure 3.14). As the information center of a cell, the **nucleus** contains most of the cell's genetic material in the form of long molecules of DNA.[1] Ultimately, DNA controls nearly all the activities of a cell. Details of how DNA controls cellular function are discussed in Chapters 17 and 19.

The outer surface of the nucleus consists of a double-layered membrane, called the *nuclear membrane,* that keeps the DNA within the nucleus. The nuclear membrane is bridged by *nuclear pores* that are too small for DNA to pass through, but which permit the passage of certain small proteins and RNA molecules.

Within the nucleus is a dense region called the **nucleolus,** where the RNA and proteins that comprise *ribosomes* are synthesized. The components of the ribosomes pass through the nuclear pores to the cytoplasm for final assembly into ribosomes.

Ribosomes are responsible for protein synthesis

Ribosomes are small structures comprised of RNA and certain proteins that are either floating freely in the cytosol or are attached to the endoplasmic reticulum, the cell organelle that synthesizes most biological molecules. Ribosomes are responsible for making specific proteins. They assemble amino acids into proteins by connecting the appropriate amino acids in the correct sequence according to an RNA template. We describe this process in more detail in Chapter 17.

[1]As you will learn in Chapter 17, mitochondria have their own DNA.

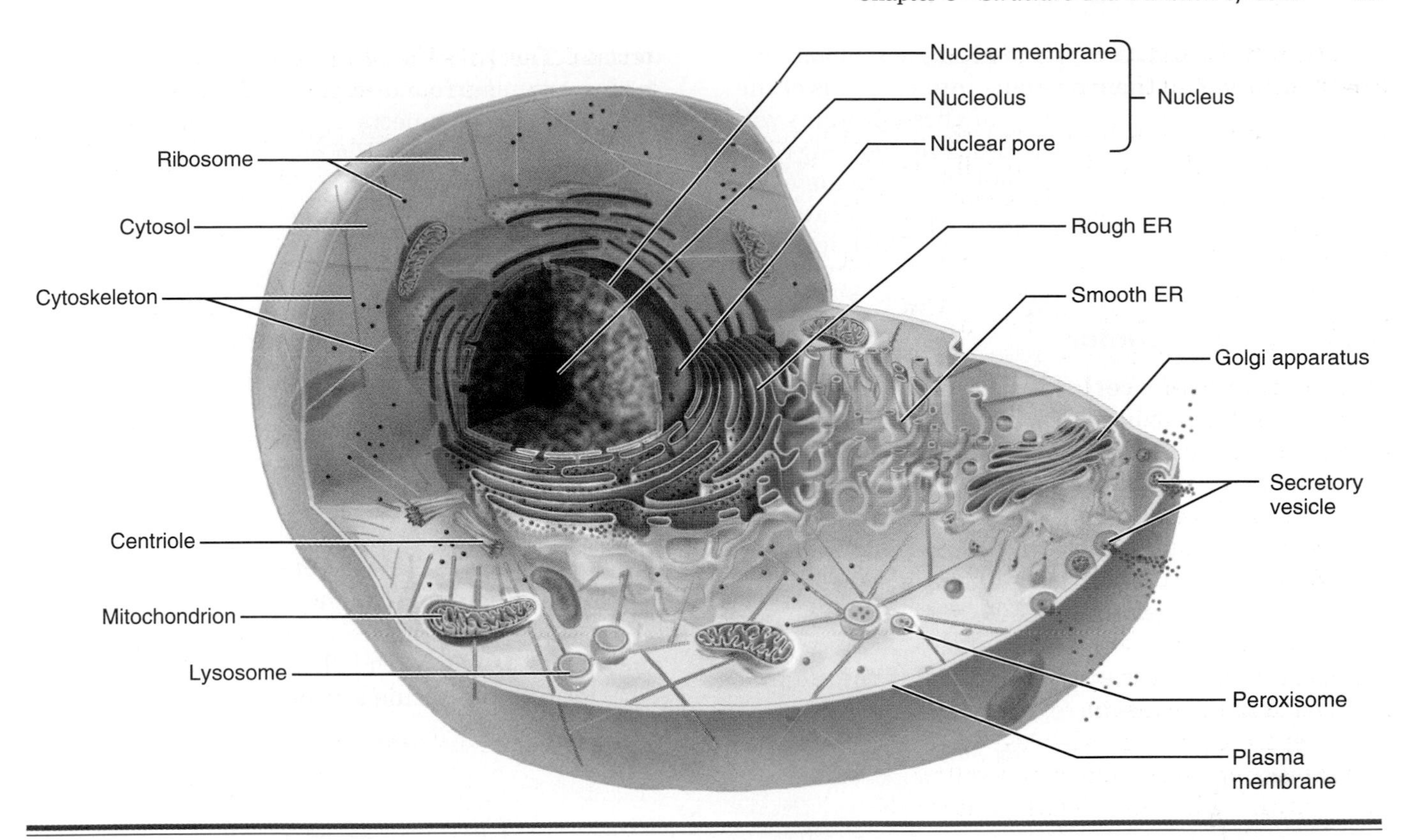

Figure 3.13 A typical animal cell. Not shown in this cell are cilia, a flagellum, glycogen granules, or fat deposits, as only certain cells have these features.

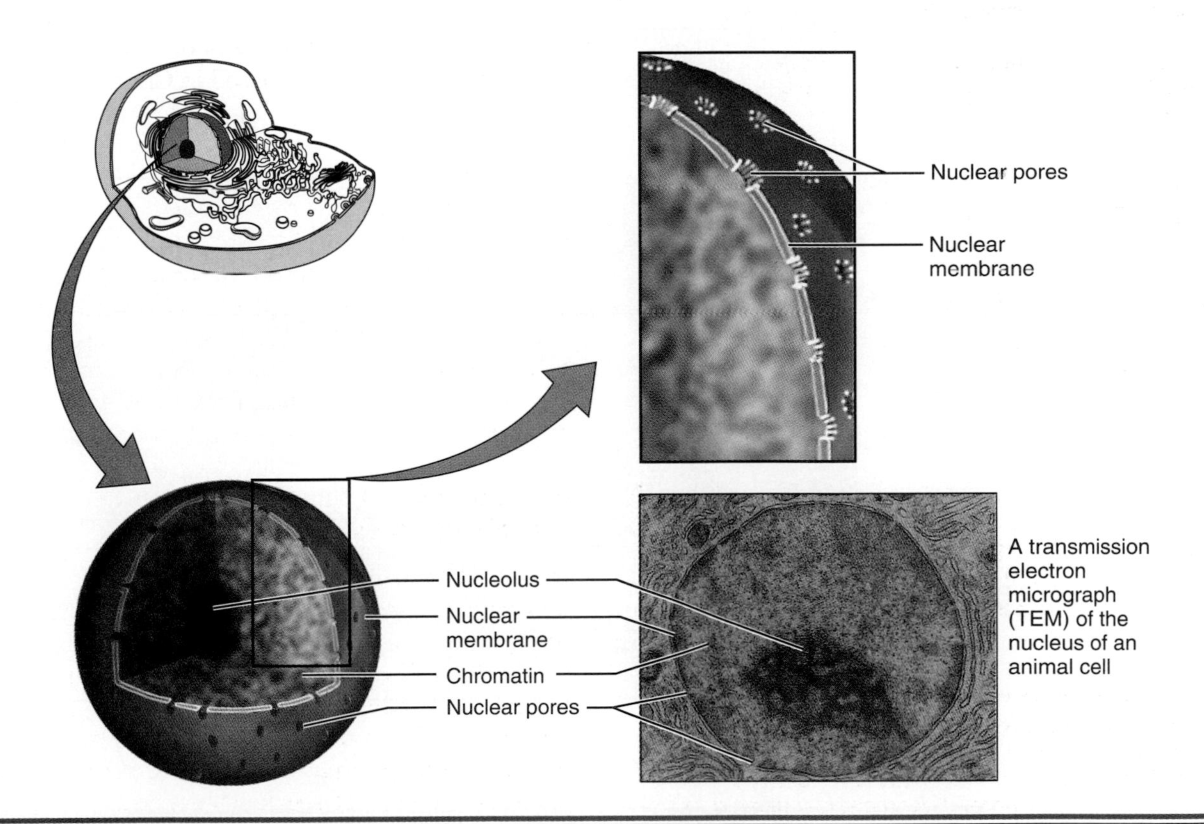

Figure 3.14 The nucleus. The nucleus contains the cell's genetic material. The nucleolus produces the protein and RNA components of ribosomes. These components exit the nucleus via nuclear pores.

Ribosomes that are attached to the endoplasmic reticulum release their proteins into the folds of the endoplasmic reticulum. Many of these proteins will eventually be secreted by the cell. Free-floating ribosomes generally produce proteins for immediate use by the cell, such as enzymes that serve as catalysts for chemical reactions within the cytoplasmic fluid.

The endoplasmic reticulum is the manufacturing center

The **endoplasmic reticulum** (ER), in conjunction with its attached ribosomes, synthesizes most of the chemical compounds made by the cell. If a cell were an industrial city, then the ER would be the city's steel mills, sawmills, and chemical plants. Like the output of a steel mill, the materials manufactured by the ER are often not in their final form. They are refined and packaged by the Golgi apparatus, discussed later.

Figure 3.15 shows the structure of the ER and its role in the manufacture of proteins and other materials. The ER is an extensively folded, membranous system surrounding a fluid-filled space. A portion of the ER connects to the nuclear membrane. Some regions of the ER's outer surface are dotted with ribosomes, giving them a granular appearance and earning them the name *rough ER*. Regions without ribosomes are called *smooth ER*.

The rough ER is involved in the synthesis of proteins, as you might guess from the presence of ribosomes. Most of the proteins synthesized by the attached ribosomes are released into the fluid-filled space of the ER. Eventually they enter the smooth ER, where they are packaged for transfer to the Golgi apparatus. A few proteins are released directly into the cytosol for immediate use by the cell.

The smooth ER synthesizes macromolecules other than protein. Most notable among these are the lipids, including some hormones. Numerous enzymes embedded in the inner surface of the ER membrane facilitate the chemical reactions necessary for macromolecule synthesis.

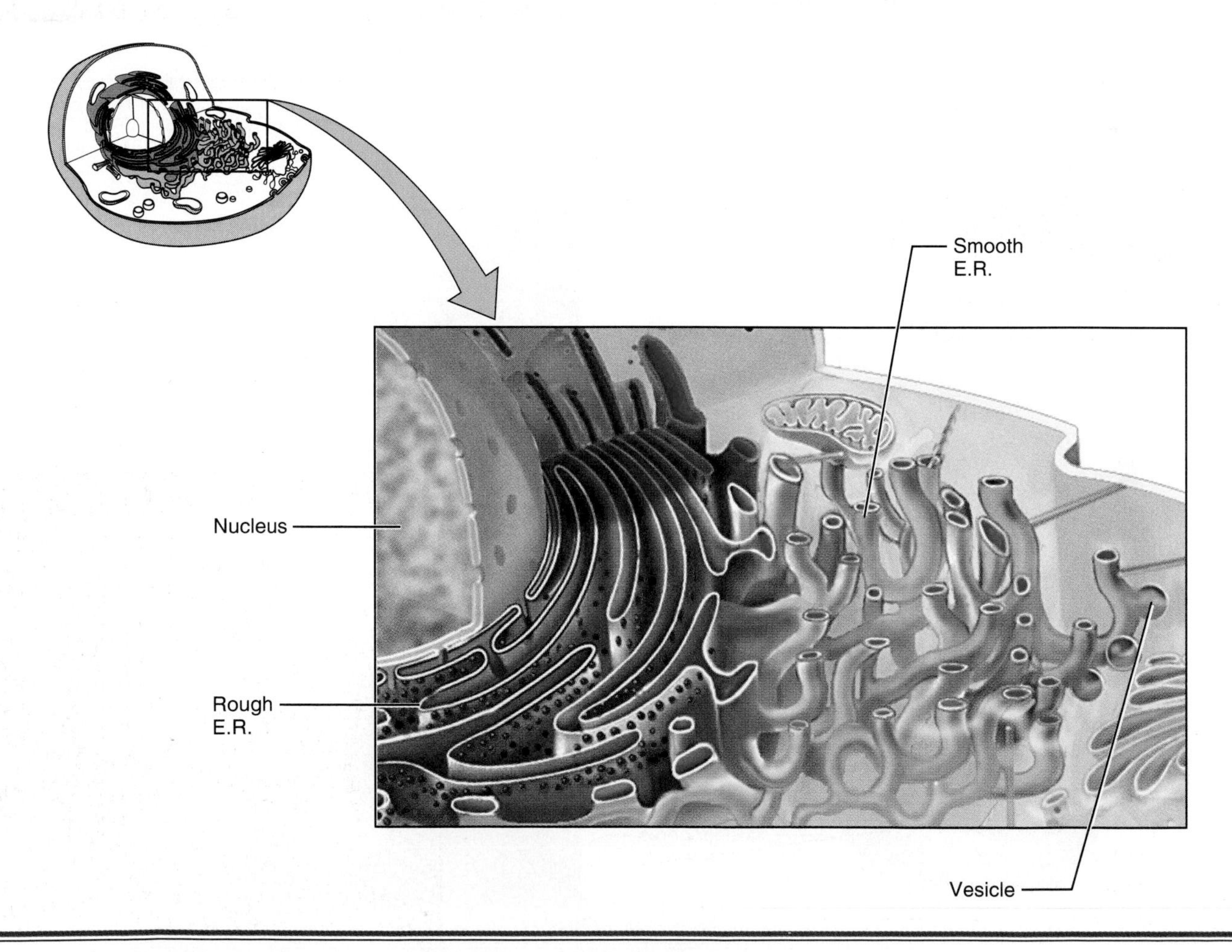

Figure 3.15 The endoplasmic reticulum (ER). The rough ER is studded with ribosomes, where proteins are made. The smooth ER packages the proteins and other products of the ER and prepares them for shipment to the Golgi apparatus in vesicles.

The smooth ER is also responsible for packaging the proteins and lipids for delivery to the Golgi apparatus. Newly synthesized proteins and lipids collect in the outermost layers of smooth ER. There, small portions of the fluid-filled space are surrounded by ER membrane and pinched off, forming vesicles that contain fluid, proteins, and lipids. The vesicles migrate to the Golgi apparatus, fuse with the Golgi apparatus membrane, and release their contents into the Golgi apparatus for further processing.

Recap **The nucleus contains most of the cell's genetic material. Ribosomes comprised of RNA and ribosomal proteins are responsible for protein assembly. The endoplasmic reticulum modifies some of these proteins, manufactures most other macromolecules in rough form, and transfers them to the Golgi apparatus.**

The Golgi apparatus refines, packages, and ships

The **Golgi apparatus** is the cell's refining, packaging, and shipping center. To continue our analogy of the cell as an industrial city, here is where steel bars are shaped into nails and screws, raw lumber assembled into doors and windows, and grain turned into bread.

Figure 3.16 diagrams the structure of the Golgi apparatus and the processes that occur there. In cross section the Golgi apparatus appears as a series of interconnected fluid-filled spaces surrounded by membrane, much like a stack of plates. Like the ER, it contains enzymes that further refine the products of the ER into final form.

The contents of the Golgi apparatus move outward by a slow but continuous process. At the outermost layer of the Golgi apparatus, the products are finally ready to be packaged into vesicles and shipped to their final destinations.

Vesicles: membrane-bound storage and shipping containers

Vesicles are membrane-bound spheres that enclose something within the cell. Sometimes they contain it inside the cell; other times they move it to another

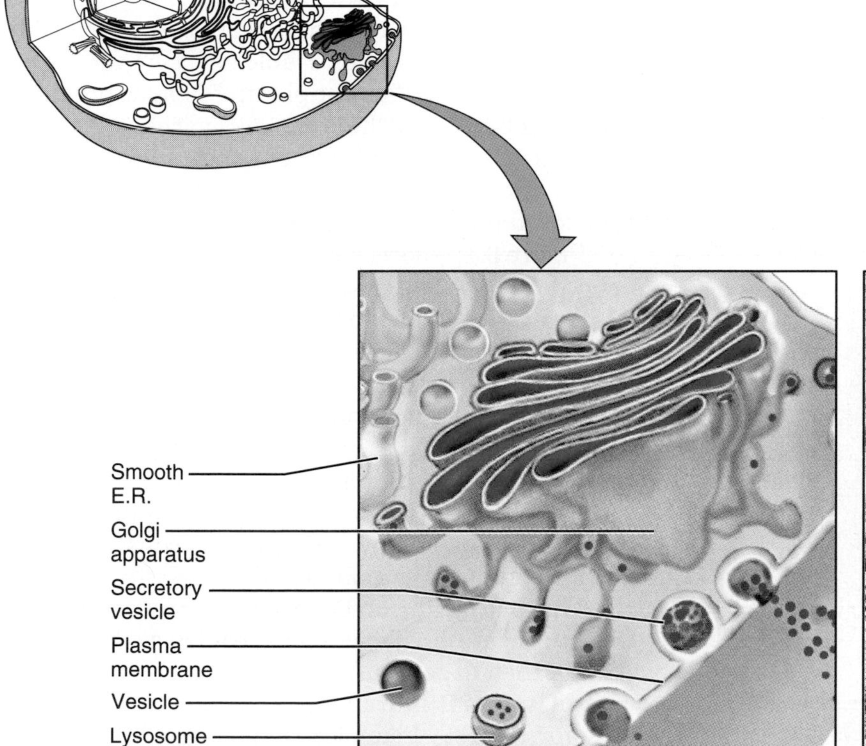

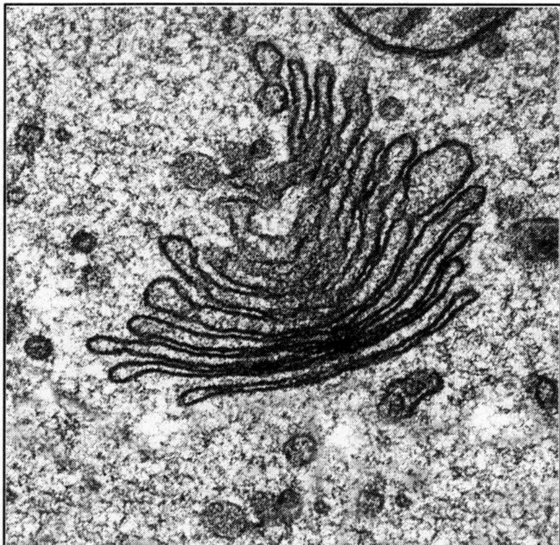

Figure 3.16 The Golgi apparatus. The Golgi apparatus receives substances from the ER, refines them into final products, and packages them into vesicles for their final destinations.

location. There are several types of vesicles, each with a different origin and purpose.

Vesicles to ship and store cellular products These vesicles enclose and transport the products of the ER and Golgi apparatus. Each vesicle contains only one product out of the thousands of substances made by the Golgi apparatus.

The contents of each vesicle depend on certain proteins in the vesicle membrane that act as "shipping labels." They determine which product is to be put into the vesicle and where the vesicle will be sent. If a vesicle's products are not immediately needed it remains in the cell cytoplasm, like a box stored in a warehouse awaiting shipment.

Secretory vesicles Secretory vesicles contain products destined for export from the cell. They migrate to the plasma membrane and release their contents out of the cell by exocytosis. Because most secretory products are made in the Golgi apparatus, secretory vesicles generally derive from Golgi apparatus membrane.

Endocytotic vesicles These structures enclose bacteria and raw materials from the extracellular environment. They bring them into the cell by endocytosis.

Peroxisomes and lysosomes These vesicles contain enzymes so powerful they must be kept within the vesicle to avoid damaging the rest of the cell. Both are produced by the Golgi apparatus. Figure 3.17 shows their functions.

The enzymes in **peroxisomes** destroy various toxic wastes produced within the cell, including hydrogen peroxide (H_2O_2). They also destroy compounds that have entered the cell from outside, such as alcohol. The detoxification process occurs entirely within the peroxisome.

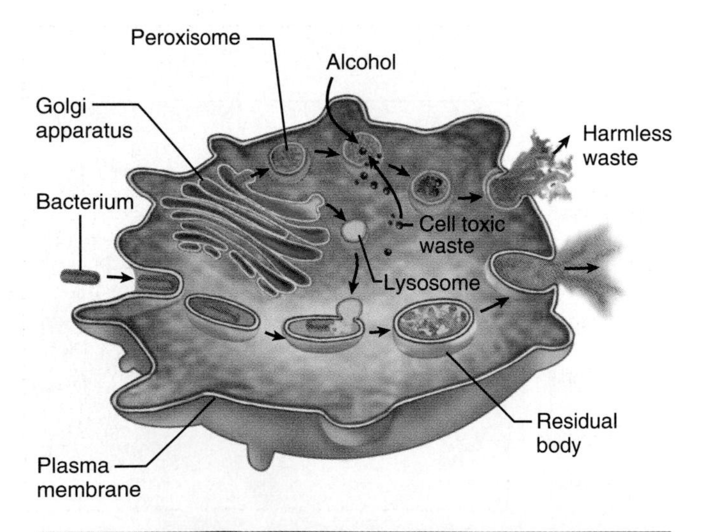

Figure 3.17 Lysosomes and peroxisomes. Lysosomes formed by the Golgi apparatus fuse with endocytotic vesicles containing a bacterium. Following digestion of the bacterium the residual waste is excreted by exocytosis. Peroxisomes take up toxic wastes (including alcohol) and degrade them to harmless waste, which is also excreted.

Lysosomes (from Gr. *lysis,* meaning "dissolution," and *soma,* or "body") contain powerful digestive enzymes. Lysosomes fuse with endocytotic vesicles within the cell, digesting bacteria and other large objects. Lysosomes also perform certain housekeeping tasks, such as dissolving and removing damaged mitochondria and other cellular debris. When their digestive task is complete they become "residual bodies," analogous to small bags of compacted waste. Residual bodies can be stored within the cell, but usually their contents are eliminated from the cell by exocytosis.

Recap **The Golgi apparatus refines cellular products to their final form and packages them in membrane-bound vesicles. Some vesicles store, ship, and/or secrete cellular products; others digest and remove toxic waste and cellular debris.**

Mitochondria provide energy

Nearly all of a cell's functions require energy. Energy is available in the chemical bonds of the food you eat, but cells cannot use it directly. Most energy in ingested nutrients must be converted to a more usable form before it can power the chemical and physical activities of living cells.

Mitochondria are the organelles responsible for providing most of this usable energy; they are often called the cells' "power plants." Not surprisingly, their number within different cells varies widely as a function of energy requirements. A cell with a high rate of energy consumption, such as a muscle cell, may contain over 1,000 mitochondria.

Figure 3.18 shows a photograph of a single mitochondrion and diagrams its structure and function. A smooth outer membrane similar to the plasma membrane covers the entire surface. Contained within the outer membrane is an inner membrane with numerous folds that increase its surface area. The inner membrane and the fluid in its folds contain hundreds of protein enzymes, which serve as catalysts to break down chemical bonds in our food and release the energy. This process consumes oxygen and produces carbon dioxide.

The energy liberated within the mitochondria is used to create high-energy molecules such as ATP. ATP is then exported from the mitochondria to the cytosol, where it is available as a quick source of energy for the cell. Like electric power, ATP is useful for a variety of purposes. We have already seen one application for ATP, and that is to provide the energy to transport sodium and potassium across the plasma membrane.

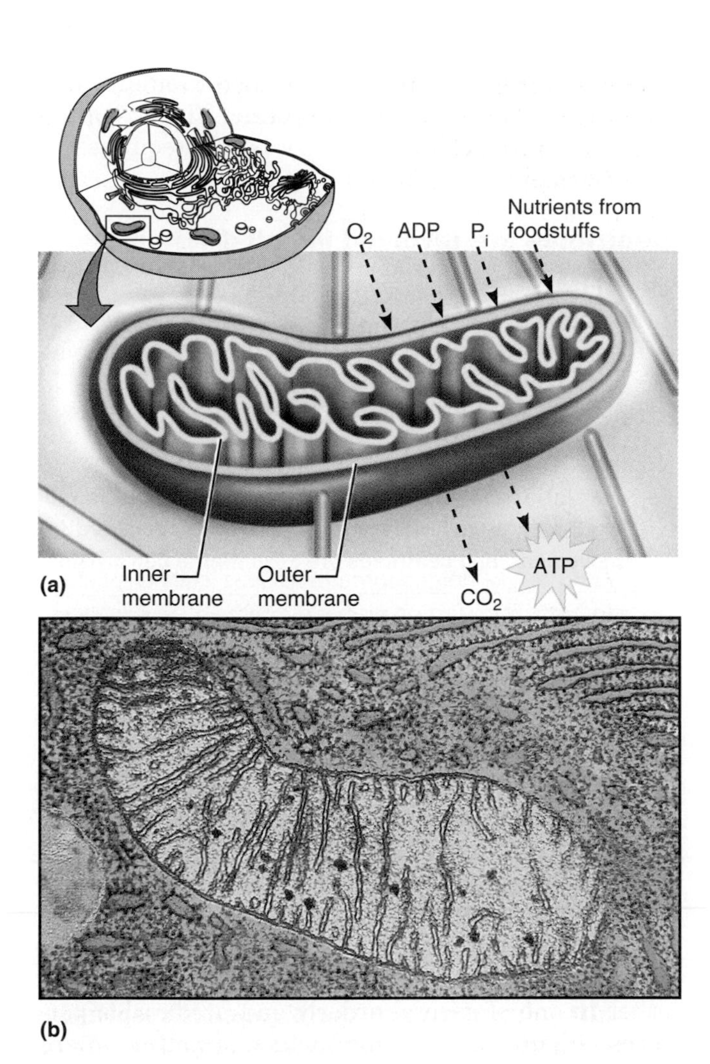

Figure 3.18 Mitochondria. **(a)** The structure and overall function of a mitochondrion. **(b)** A photomicrograph of a mitochondrion.

Fat and glycogen: Sources of energy

The mitochondria generally manufacture ATP as it is needed. To avoid the possibility that the cell might run out of fuel, energy is stored in some cells in raw form. These energy stores are not enclosed in any membrane-bound container. They are more like large piles of coal on the ground, awaiting delivery to the power plants (mitochondria) for conversion to electricity (ATP).

Some cells store raw energy as lipids (fat). Our so-called "fat cells" are so specialized for this purpose that most of their volume consists of large droplets of stored lipids. Dieting and exercise tend to reduce the amount of stored fat—that is, it makes the fat cells leaner. However, it does not reduce their number. They are available to store fat again, which is why it is so hard to *keep* lost weight off.

Other cells store energy as glycogen granules (review Figure 2.15). Muscle cells rely on glycogen granules rather than on fat deposits because the energy stored in the chemical bonds of glycogen can be used to produce ATP more quickly than the energy derived from fat.

Recap **Mitochondria manufacture ATP for the cell. Some cells maintain stockpiles of fats or glycogen that can be used to make more ATP as needed.**

3.6 Cells have structures for support and movement

The soft plasma membrane is supported by an internal scaffolding that helps the cell maintain its shape. In addition, some cells have specialized structures to help them move around, and all cells contain structures that are involved in moving cellular components during cell division. Structural elements for support and movement include the cytoskeleton, cilia and flagella, and centrioles.

The cytoskeleton supports the cell

The **cytoskeleton** (Figure 3.19) consists of a loosely structured network of fibers called microtubules and microfilaments. As their names imply, microtubules

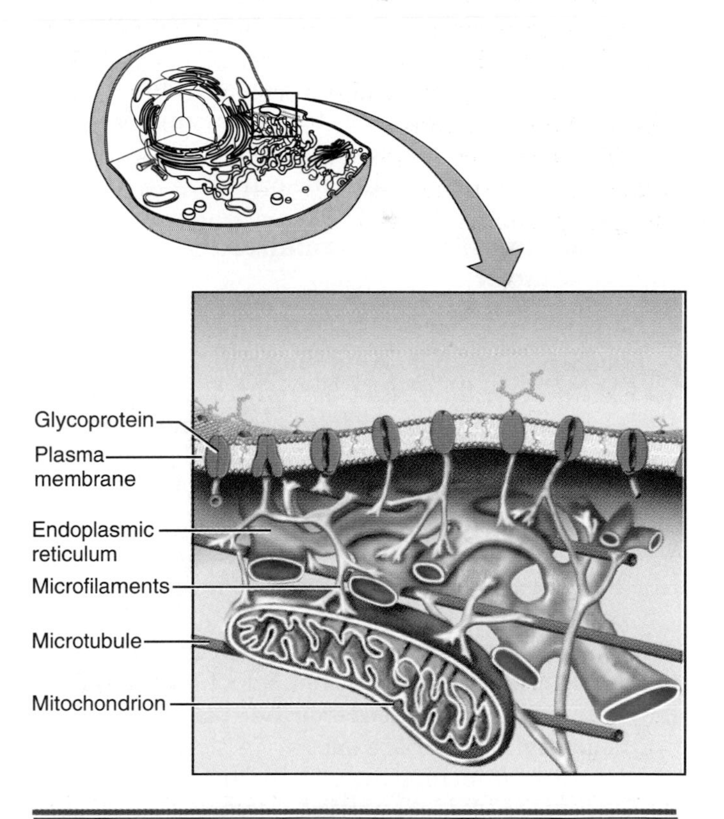

Figure 3.19 The cytoskeleton. The cytoskeleton consists of microtubules and microfilaments that attach to and support the cell's organelles and plasma membrane.

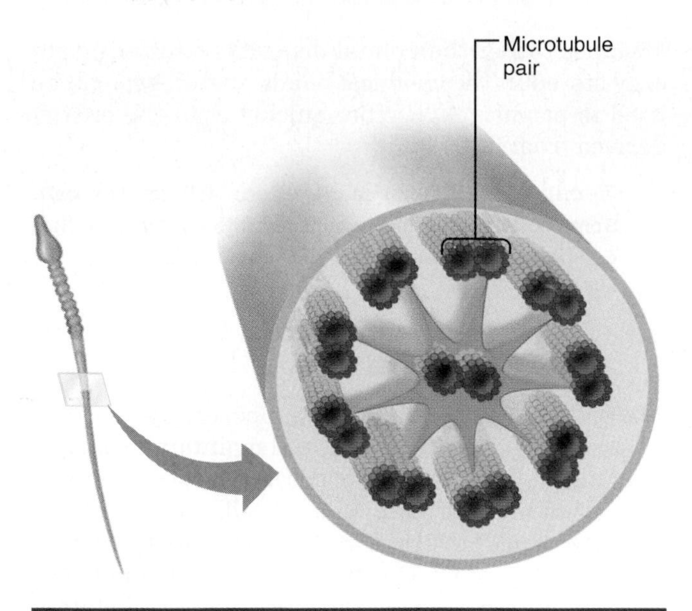

Figure 3.20 **Flagella.** A cross-sectional view of a flagellum showing that it is composed of nine pairs of microtubules surrounding a single central pair.

are tiny hollow tubes and microfilaments are thin solid fibers, both composed of protein. They attach to each other and to proteins in the plasma membrane, called *glycoproteins,* which typically have carbohydrate group components.

The cytoskeleton forms a framework for the soft plasma membrane, much as tent poles support a nylon tent. The cytoskeleton also supports and anchors the other structures within the cell.

Cilia and flagella are specialized for movement

Spotlight on Structure & Function

A few cells have hair-like **cilia** (singular: cilium) or **flagella** (singular: flagellum) that extend from the surface. Cilia are generally only 2–10 microns long (1 micron equals one millionth of a meter). In cells that have them, they are numerous. Cilia move materials along the surface of a cell with a brushing motion. They are common on the surfaces of cells that line the airways and in certain ducts within the body.

In humans, the longer flagella (approximately 200 microns long) are found only on sperm cells (Figure 3.20). The whiplike movement of the flagellum moves the entire sperm cell from one place to another.

Cilia and flagella are similar in structure. They are comprised primarily of protein microtubules held together by connecting elements and surrounded by a plasma membrane. Nine pairs of microtubules surround a single pair in the center. The entire structure bends when temporary linkages form between adjacent pairs of microtubules, causing the pairs to slide past each other. The formation and release of these temporary bonds requires energy in the form of ATP. ■

Centrioles are involved in cell division

Centrioles are short, rodlike microtubular structures located near the nucleus. Centrioles are essential to the process of cell division because they participate in aligning and dividing the genetic material of the cell. We discuss them in Chapter 17 when we describe how a cell divides.

Recap **The cytoskeleton forms a supportive framework for the cell. Cilia and flagella are specialized for movement, and centrioles are essential to cell division.**

3.7 Cells use and transform matter and energy

Living cells are able to release the energy stored in the chemical bonds of molecules and use it to build, store, and break down still other molecules as required in order to maintain life. **Metabolism** is the sum of all of the chemical reactions in the organism.

In a single cell, thousands of different chemical reactions are possible at any one time. Some of these chemical reactions are organized as **metabolic pathways** in which one reaction follows after another in one of several orderly and predictable patterns (Figure 3.21). Some metabolic pathways are linear, in which the **product** (or end material) from one chemical reaction becomes the **substrate** (starting material) for the next. Other metabolic pathways form a cycle in which substrate molecules enter and product molecules exit, but the basic chemical cycle repeats over and over again.

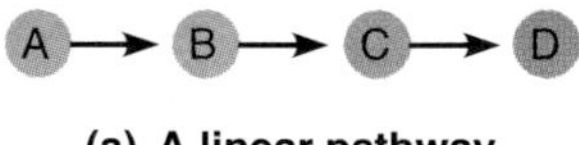

(a) A linear pathway

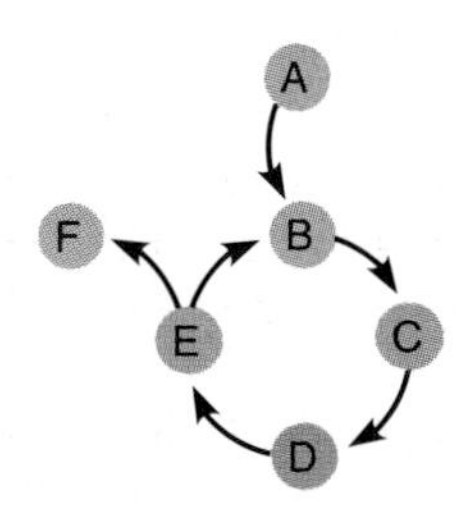

(b) A cyclic pathway in which B through E repeat over and over again

Figure 3.21 **Types of metabolic pathways.**

Anabolism assembles molecules; catabolism breaks down molecules

There are two basic types of metabolic pathways:

1. **Anabolism** (from the Greek *anabole,* a throwing up): molecules are assembled into larger molecules that contain more energy, a process that requires energy. The assembly of a protein from many amino acids is an example of an anabolic pathway.
2. **Catabolism** (Greek *katabole,* a throwing down): larger molecules are broken down, a process that releases energy. The breakdown of glucose into water, carbon dioxide, and energy is an example of a catabolic pathway.

Two facts are important about metabolic pathways. First, nearly every chemical reaction requires a specific enzyme. The specificity and availability of key enzymes is how the cell regulates and controls the rates of chemical reactions. Second, the metabolic activities of a living cell require a lot of energy. Energy is required for the creation of larger molecules by anabolic pathways. Energy is also used to power cellular activities such as active transport and muscle contraction.

Cells get their energy by catabolism of molecules that serve as chemical stores of energy. The most immediately useful source of energy, a sort of "energy cash" if you will, is ATP (adenosine triphosphate). The energy in ATP is locked in the chemical bond between the second and third phosphate group. Every time the third phosphate group is removed from ATP, energy is released that the cell can use to do work. The reaction is reversible, meaning that the application of energy to ADP and phosphate can re-create ATP, storing the energy for later use. The equation is written as:

Where do cells get the energy they need?

$$ATP \Leftrightarrow ADP + P_i + energy$$

In this equation, P_i is used as the abbreviation for the inorganic phosphate (PO_4^{2-}) to distinguish it from the chemical symbol for pure phosphorus (P).

Recap **Metabolism refers to all of a cell's chemical processes. Metabolic pathways either create molecules and use energy (anabolism) or break them down and liberate energy (catabolism).**

Cellular respiration: cells use oxygen to produce ATP

The chemical reactions that produce ATP occur primarily in the mitochondria and involve the catabolism of carbohydrates (especially glucose), lipids (fats), and proteins. Let's look at the catabolism of glucose as an example.

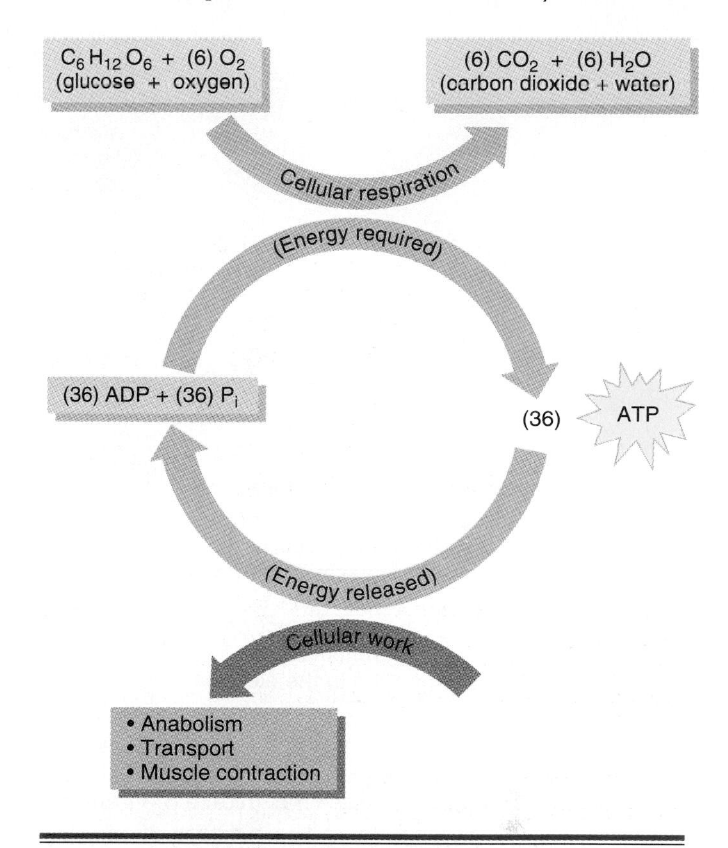

Figure 3.22 Glucose provides energy for the cell. The complete catabolism of glucose uses oxygen, produces carbon dioxide and water, and generates 36 molecules of ATP that can be used to do cellular work.

Recall that glucose is a six-carbon sugar molecule with the chemical formula $C_6H_{12}O_6$. Glucose is an excellent energy source. The complete catabolism of glucose to water and carbon dioxide requires 6 molecules of oxygen and produces 36 molecules of ATP (Figure 3.22). The production of ATP from glucose is a multistep process that can be divided into three main stages:

1. *Glycolysis:* the glucose molecule is split into two three-carbon molecules. Oxygen is not required for this process.
2. *The Krebs cycle:* the three-carbon molecules are disassembled. Most of the energy is transferred to electron transport molecules in the form of high-energy electrons and hydrogen ions. Oxygen is used and carbon dioxide is produced.
3. *The electron transport system:* the high-energy electrons and hydrogen ions are stripped of their energy, and the energy is used to create ATP. Oxygen is used and water is produced.

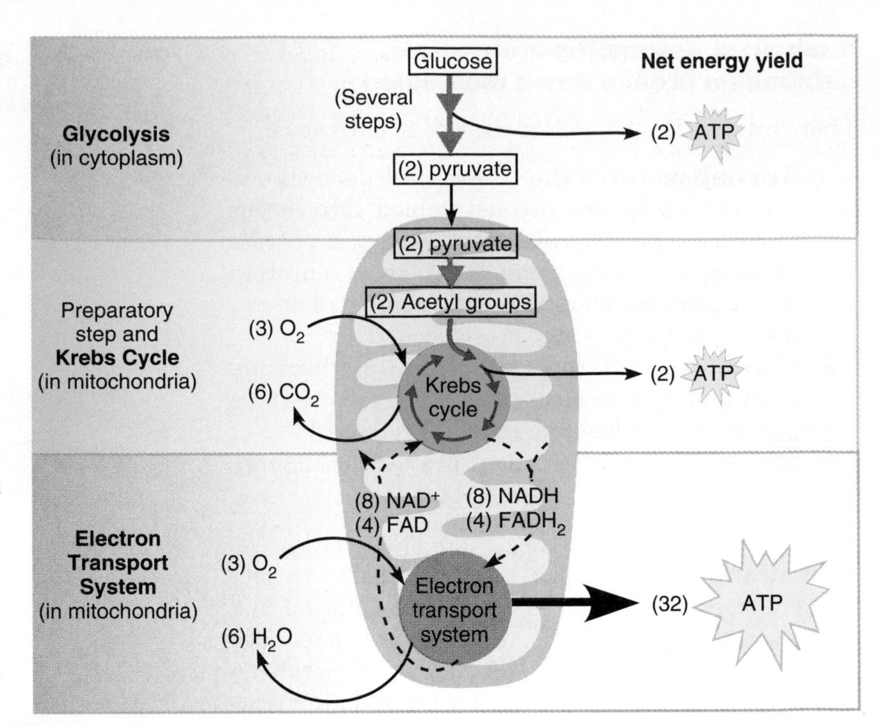

Figure 3.23 (at right) The synthesis of ATP using glucose as the energy source. Glycolysis occurs in the cytoplasm in the absence of oxygen. Cellular respiration (represented by the Krebs cycle and the electron transport system) occurs in mitochondria. Most of the ATP is synthesized in the electron transport system.

Cellular metabolic processes that utilize oxygen and produce carbon monoxide in the process of making ATP are collectively called **cellular respiration** (also called aerobic metabolism). The last two steps of the complete breakdown of glucose (the Krebs cycle and the electron transport system) represent cellular respiration, and both take place within mitochondria. The whole point of cellular respiration is to release the energy locked in the chemical bonds of energy-storage molecules gradually and then use the energy to make ATP. The three stages of the complete metabolism of glucose are described next and summarized in Figure 3.23.

Glycolysis: glucose is split into two pyruvate molecules The first step in the complete breakdown of glucose is called **glycolysis** (*glyco-* means "sweet," referring to sugar, and *-lysis* means "to break"). During glycolysis the 6-carbon glucose molecule is broken into two 3-carbon molecules called *phosphoglyceraldehyde* (PGAL). The process requires two molecules of ATP to provide the necessary energy to get the process started and is catalyzed by several enzymes.

Then, each of the two PGAL molecules is converted in a series of steps to *pyruvate.* This process requires several enzymes and generates four molecules of ATP. The net ATP yield from glycolysis, then, is only two molecules of ATP. The entire process of glycolysis takes place in the cell cytoplasm. Because it does not require oxygen, glycolysis is the only way for cells to produce ATP when oxygen is not available.

The Krebs cycle generates high-energy electron transport molecules At this point the pyruvate molecules enter the mitochondria, where all the rest of the ATP-generating reactions occur. In a preparatory step, a carbon-containing molecule (CO_2) is removed to form an *acetyl group.* The acetyl group joins with an enzyme called *Coenzyme A* to form *acetyl-CoA,* which then delivers the acetyl group to the **Krebs cycle.** The Krebs cycle is a series of sequential steps in which each acetyl group is completely disassembled into CO_2, H^+, and negatively charged electrons.

The yield of ATP from the preparatory step and the Krebs cycle together is only two more molecules of ATP (one for each pyruvate). Most of the energy is still contained in the liberated hydrogen ions and electrons, which are transferred to low-energy *electron transport molecules* called **NAD^+** and **FAD,** forming higher-energy **NADH** and **$FADH_2$.** The complete metabolism of glucose generates eight molecules of NADH and four of $FADH_2$. The Krebs cycle also generates six molecules of CO_2 waste, accounting for the original six carbons in glucose, and uses the equivalent of three molecules of O_2.

The electron transport system produces most of the ATP So far, the glucose molecule has been completely dismantled and CO_2 has been generated as a waste product. But only four new ATP molecules have been created. The rest of the energy is still in the electrons and hydrogen ions bound to the electron transport molecules.

At this point the electron transport molecules move to the inner membrane of the mitochondria and undergo another series of reactions in the **electron transport system.** Here, the energy-rich electrons are transferred sequentially from one molecule to another. The sequential transfer is important because it allows the energy in the electrons to be released in manageable quantities. The sequence of events is as follows.

NADH and $FADH_2$ release the H^+ and high-energy electrons they acquired in the Krebs cycle. The electrons pass from one molecule to the next in the electron transport chain. Each time an electron is transferred, the transport molecule acquires some of its energy. The energy is used to transport H^+ from the inner compartment of the mitochondria to the outer compartment.

Because the concentration of H^+ is now higher in the outer compartment than in the inner compartment, H^+ has a tendency to diffuse back to the inner compartment. However, it can only diffuse through channels. These channels are actually enzymes that use the energy derived from the diffusion of H^+ to catalyze the synthesis of ATP from ADP and P_i.

Figure 3.24 illustrates the events of the electron transport system. The entire sequence is so efficient that the 8 molecules of NADH and four molecules of $FADH_2$ that enter the electron transport system are able to generate 32 molecules of high-energy ATP. As the hydrogen ions are released and the electrons transfer their energy, they both combine with oxygen to form stable water. The low-energy NAD^+ and FAD molecules, now lacking hydrogen ions and electrons, are recycled and used again.

> ***Recap*** **Cellular respiration uses oxygen to produce ATP. Within mitochondria, energy acquired from glycolysis is transferred to electron transport molecules in the Krebs cycle and released in a controlled fashion in the electron transport system, driving the formation of ATP.**

Fats and proteins are additional energy sources

So far we have concentrated on the cellular catabolism of glucose. Normally the blood glucose concentration remains fairly constant even between meals because glycogen (the storage form of glucose in humans) is constantly being catabolized to replenish the glucose that is used by cells.

However, most of the body's energy reserves do not take the form of glycogen. In fact, the body stores only about 1% of its total energy reserves as

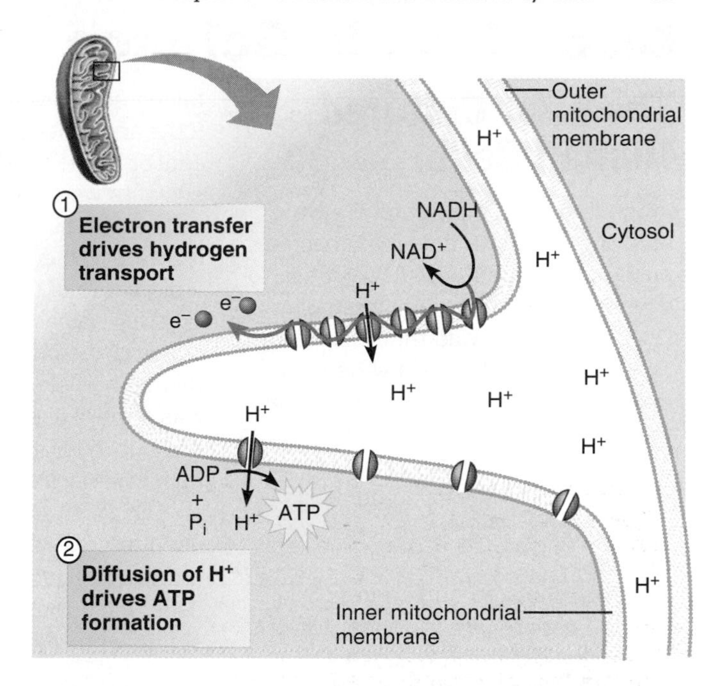

Figure 3.24 Steps in the electron transport system. **(a)** During electron transfer, the high-energy molecules of NADH (and $FADH_2$) give up electrons and hydrogen ions. As the electrons are transferred from one molecule in the system to the next, their energy is released and used to transport hydrogen ions into the space between the two mitochondrial membranes. **(b)** The energy provided by the diffusion of hydrogen ions through channels in the inner mitochondrial membrane is used to drive the synthesis of ATP from ADP and P_i.

glycogen; about 78% are stored as fats and 21% as protein. After glycogen, our bodies may utilize fats and proteins. Figure 3.25 illustrates the metabolic pathways for fat, glycogen, and protein.

Energy is constantly being transferred into and out of the body. Immediately after a meal, when plenty of glucose, lipids, and amino acids are readily available in the blood, we tend to use primarily glucose as an energy source. When we eat more calories than we can use immediately, the excess energy goes to replenish the body's stores of glycogen and the rest is converted to fat and stored in fat tissue. After we have not eaten for hours, the body uses glycogen and eventually fats and even proteins for energy.

Pound for pound, fats carry over twice the energy of carbohydrates such as glycogen. During fat catabolism, triglycerides (the storage form of fat) are first broken down to glycerol and fatty acids. The glycerol can be converted to glucose in the liver or it can be converted to pyruvic acid, which then enters the Krebs cycle. Enzymes break down the fatty acid tails to two-carbon acetyl groups, which also enter the Krebs cycle. Each molecule of triglyceride yields

Directions in Science

The First Cultured Human Cells

In the late 1940s researchers were studying a virus that was as much a public concern as the AIDS virus is today: the microbe that causes polio. One impediment was that there was no way to grow the polio virus in human cells in the laboratory because there was no way to keep human cells alive outside the body for prolonged periods of time. Most human cells failed to grow properly in laboratory dishes.

That changed on February 8, 1951. Researchers at Johns Hopkins University, led by Dr. G. O. Gey, obtained some living cancer cells from 31-year-old Henrietta Lacks, who had cancer of the cervix. Gey and his associates were able to keep these cells and their offspring alive in the laboratory, thereby establishing the first culture of human cells.

Gey named them *HeLa* cells in honor of the patient who donated them. The pseudonym Helen Lane was initially used to protect her identity, but today it seems appropriate to honor her by using her true name.

Although *HeLa* cells were first used to grow the polio virus in the laboratory, they proved to have other uses, including the study of cancer itself. Eventually Gey established a stable stock culture of HeLa cells and made it available to other researchers around the world. HeLa cells have proven invaluable for our understanding not only of the mechanisms underlying the development of cancer, but also our understanding of cell nutrition, viral growth, protein synthesis, and drug effects on cells.

Henrietta Lacks died a few years after the discovery of her cancer, but her cells live on. Indeed, HeLa cells have become so widely used in human cell research that there are over 1,000 references to HeLa cells in the scientific literature each year.

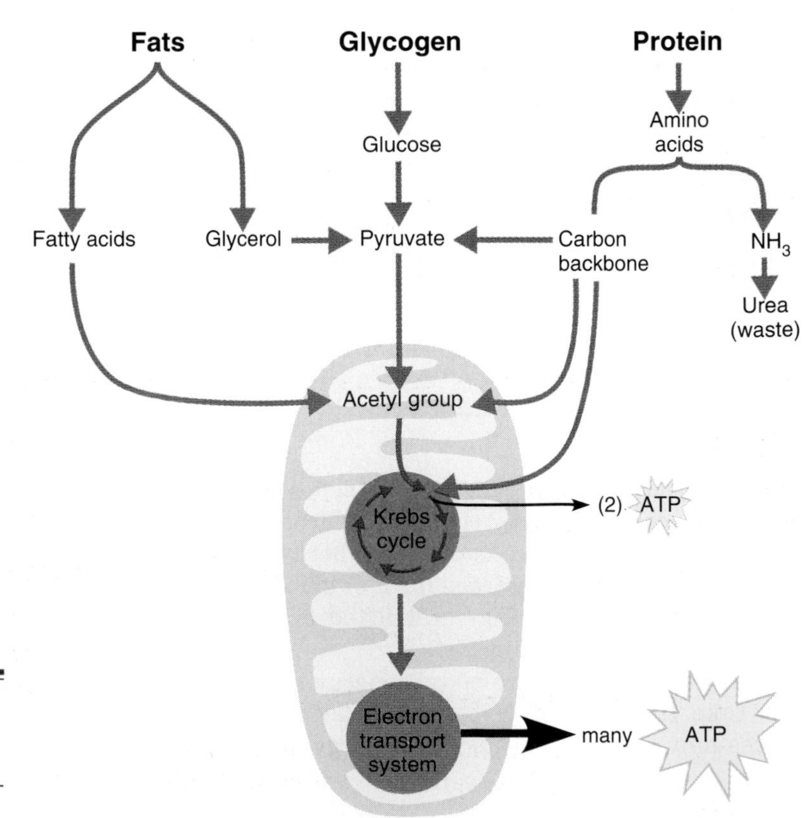

Figure 3.25 (at right) Metabolic pathways for fats, glycogen, and proteins as sources of cellular energy. All three sources produce pyruvate and acetyl groups. In addition, a few components of protein can enter the Krebs cycle directly.

a great deal of ATP because the fatty acid tails are generally 16–18 carbons long and so generate many acetyl groups.

Proteins carry about the same amount of energy per pound as glycogen. Proteins are first broken down to amino acids, and the amine group ($-NH_2$) of each amino acid is removed. The carbon backbones then enter the Krebs cycle at various points, depending on the specific amino acid. The amine groups are converted to urea by the liver and then excreted in urine as waste.

Proteins serve primarily as enzymes and structural components of the body, not as a stored form of energy. Nevertheless, during starvation protein catabolism increases significantly. Prolonged loss of protein can cause muscle wasting, but at least it keeps the individual alive.

Anaerobic pathways make energy available without oxygen

The mitochondria require oxygen to complete the chemical reactions of the Krebs cycle and the electron transport chain. However, a small amount of ATP *can* be made in humans by anaerobic pathways (without oxygen) for at least brief periods of time. Glycolysis, for example, is an anaerobic metabolic pathway. In the absence of oxygen glucose is broken down to pyruvate (glycolysis occurs), but then the pyruvate cannot proceed through the Krebs cycle and electron transport chain (Figure 3.26). Instead, the pyruvate is converted to *lactic acid*. The buildup of lactic acid is what causes the burning sensation and perhaps cramps associated with muscle fatigue when not enough oxygen is available to muscle tissue. When oxygen is again available, the lactic acid is metabolized by aerobic pathways.

Because glycolysis is the only step that can occur without oxygen, glucose (and glucose derived from glycogen) is the only fuel that can be used. Fat metabolism always requires oxygen. Note also that glycolysis can produce only two molecules of ATP per molecule of glucose instead of the usual 36.

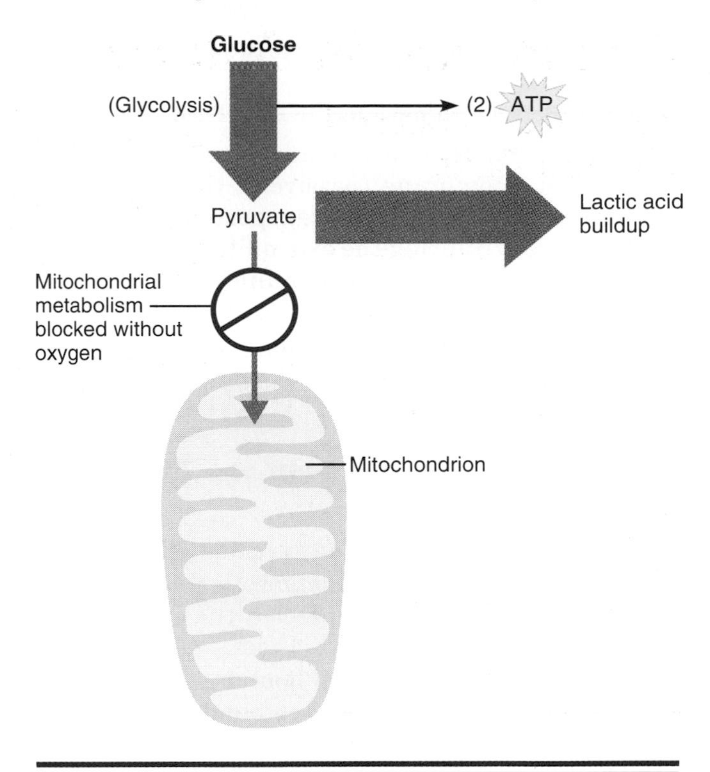

Figure 3.26 Anaerobic metabolism. In the absence of oxygen, glycolysis is the only ATP-producing step available. Glycolysis without oxygen results in lactic acid buildup.

Recap **Carbohydrates and fats are the main sources of stored energy for making ATP, although proteins can be used if necessary. In the absence of oxygen, only glucose can be utilized.**

Chapter Summary

Cells are classified according to their internal organization

- All cells have a plasma membrane that surrounds and encloses the cytoplasm.
- Eukaryotic cells have a nucleus.

Cell structure reflects cell function

- Limits to cell size are imposed by the mathematical relationship between cell volume and cell surface area.
- Various types of microscopes with magnifications up to 100,000-fold enable us to visualize cells and their structures.

A plasma membrane surrounds the cell

- The plasma membrane is a bilayer of phospholipids that also contains cholesterol and various proteins.

Molecules cross the plasma membrane in several ways

CD Fluid Balance/Introduction to Body Fluids

- Some molecules are transported across the plasma membrane passively (by diffusion), whereas others are transported by active processes requiring the expenditure of energy.
- Receptor proteins transfer information across the plasma membrane.
- The sodium-potassium pump is a plasma membrane protein with a critical role in the maintenance of cell volume.

Internal structures carry out specific functions

- The nucleus directs all of the cell's activities.
- Ribosomes, the endoplasmic reticulum, and the Golgi apparatus participate in the synthesis of life's molecules.
- Vesicles are membrane-bound spheres that transport, store, and ship cellular products and toxic or dangerous materials.
- Mitochondria make energy available for the cell in the form of the high-energy molecule ATP.

Cells have structures for support and movement

- A cytoskeleton of microtubules and microfilaments serves as structural support and anchors the various organelles.
- Cilia and flagella provide for movement in certain types of cells. Both cilia and flagella are made of pairs of protein microtubules.

Cells use and transform matter and energy

CD Muscular System/Muscle Metabolism

- The creation and destruction of molecules either requires energy or liberates energy.
- The most readily useful form of energy for cells is ATP.
- The production of ATP from glucose requires three consecutive stages: glycolysis, the Krebs cycle, and the electron transport system.
- Cells can utilize glycogen, fats, or proteins for energy.
- Only a small amount of ATP can be made in the absence of oxygen.

Terms You Should Know

active transport, 60
ATP, 61
cytoplasm, 54
cytoskeleton, 69
diffusion, 58
electron transport system, 73
endocytosis/exocytosis, 61
endoplasmic reticulum, 66
glycolysis, 72
Golgi apparatus, 67
Krebs cycle, 72
metabolism, 70
mitochondria, 68
NAD^+, 72
nucleus, 64
osmosis, 59
passive transport, 58
plasma membrane, 54
ribosomes, 64
sodium-potassium pump, 61
vesicles, 67

Test Yourself

Matching Exercise

____ 1. Microscopic structures within the cell other than the nucleus
____ 2. The outer covering of a cell
____ 3. Photographs taken through microscopes
____ 4. The net diffusion of water across a selectively permeable membrane
____ 5. Movement of molecules from one region to another as a result of random motion
____ 6. The information center of a living cell
____ 7. Cell organelle that refines, packages, and ships cellular products
____ 8. The cell's power plants
____ 9. The process whereby glucose is split into two 3-carbon molecules, yielding two molecules of ATP
____10. The cell's internal structural support

a. micrographs
b. diffusion
c. Golgi apparatus
d. glycolysis
e. organelles
f. mitochondria
g. nucleus
h. plasma membrane
i. cytoskeleton
j. osmosis

Fill in the Blank

11. Single-celled organisms without nuclei are called ________________.
12. The plasma membrane consists of phospholipids, ________________, and proteins.
13. Water diffuses through protein ________________ in the plasma membrane.
14. Active transport requires ________________.
15. Cells shrink when they are placed in a ________________ solution.
16. The ________________ ________________ begins the synthesis of most of the chemical compounds made by a cell.
17. Vesicles in the cell called ________________ contain digestive enzymes that fuse with other vesicles containing bacteria.

18. The starting material for a chemical reaction is called the ________________.
19. The ____________ ____________ is a repetitive cycle in which 2-carbon acetyl groups are broken down to CO_2, H^+, and electrons.
20. In the absence of oxygen, the catabolism of glucose produces ____________________, which can cause a burning sensation and may contribute to muscle cramps.

True or False?

21. Large animals have correspondingly larger cells.
22. The most immediately useful form of energy for most cells is ATP.
23. Pound for pound, more energy is stored in fat than in glycogen.
24. Most of the ATP produced by mitochondria is produced in the electron transport system, not in the Krebs cycle.
25. The sodium-potassium pump actively transports both sodium and potassium out of the cell.

Concept Review

1. Explain why being small is advantageous to a cell.
2. Compare/contrast a light microscope, a transmission electron microscope, and a scanning electron microscope.
3. Describe how phospholipids are oriented in the plasma membrane and why they orient naturally that way.
4. Define *passive transport* and name the three passive transport methods that are used to transport different molecules across the plasma membrane.
5. Compare/contrast endocytosis and exocytosis.
6. Describe the activity of the sodium-potassium pump and indicate its importance to the cell.
7. Define *vesicles* and name at least two different types of vesicles.
8. Describe the internal structure and the function of cilia.
9. Explain the importance of the electron transport molecules (NAD^+ and FAD) to the ability of the mitochondria to produce ATP.
10. Describe what happens to a cell's ability to produce ATP when oxygen is not available.

Apply What You Know

1. Imagine that you are shown two cells under the microscope; one is small, has lots of mitochondria, and contains numerous glycogen granules. The other is somewhat larger and has only a few mitochondria and no glycogen granules. Which cell do you think is more metabolically active? Explain your reasoning.
2. The sodium-potassium pump is a large protein molecule. Where do you think the sodium-potassium pumps are made in the cell, and how do you think they become inserted into the lipid bilayer of the plasma membrane?
3. What do you suppose would happen to a cell's volume if it were to run completely out of energy (ATP)? Explain.

Case in Point

Supplemental Energy from NADH?

Chronic fatigue syndrome is a disease characterized by severe prolonged fatigue with flulike symptoms, including weakness, fever, sore throat, headache, and sleep disturbances. The causes of the disease are unknown. Researchers at Georgetown University Medical Center reported recently that the symptoms of chronic fatigue syndrome may be improved by the administration of the electron transport molecule NADH. In the study, 26 chronic fatigue sufferers were treated with either a placebo or an over-the-counter NADH supplement for four weeks. 31% of the patients responded favorably to NADH, in contrast to only an 8% favorable response to the placebo. Results of the study suggest that NADH might be a useful tool in the management of this perplexing disease.

QUESTIONS:

1. How might NADH be working to improve a patient's energy?

2. What other molecule might you hypothesize would also improve the symptoms of chronic fatigue syndrome, based on this finding?

INVESTIGATE:

Forsyth, L.M., H.G. Preuss, A.L. MacDowell, L. Chiazze, Jr., G.D. Birkmayer, and J.A. Bellanti. Therapeutic effects of oral NADH on the symptoms of patients with chronic fatigue syndrome. *Annals of Allergy Asthma and Immunology* 82(2):185–191, 1999.

References and Additional Reading

Becker, Wayne M., Jane B. Reece, and Martin F. Phoenie. *The World of the Cell*. Menlo Park, CA: Benjamin Cummings, 1996. (A comprehensive textbook of cell biology for the student particularly interested in this subject.)

Margulis, Lynn and Dorian Sagan. *What is Life?* New York: Simon & Schuster, 1995. (An evolutionary, comparative and sometimes philosophical look at this age-old question. Well illustrated.)

Sherer, W.F., J.T. Syverton, and G.O. Gey. Studies on the Propagation in Vitro of Poliomyelitis Viruses. IV. Viral Multiplication in a Stable Strain of Human Malignant Epithelial Cells (Strain HeLa) Derived from an Epidermoid Carcinoma of the Cervix. *Journal of Experimental Medicine* 97:695–709, 1953. (The original research paper describing HeLa cells; of historical significance for the especially interested student only.)

Thomas, Lewis. *The Lives of a Cell: Notes of a Biology Watcher*. New York: Viking Press, 1994. (A classic, easy-to-read series of essays.)

CHAPTER 4

FROM CELLS TO ORGAN SYSTEMS

Why do cells assemble **into groups?**

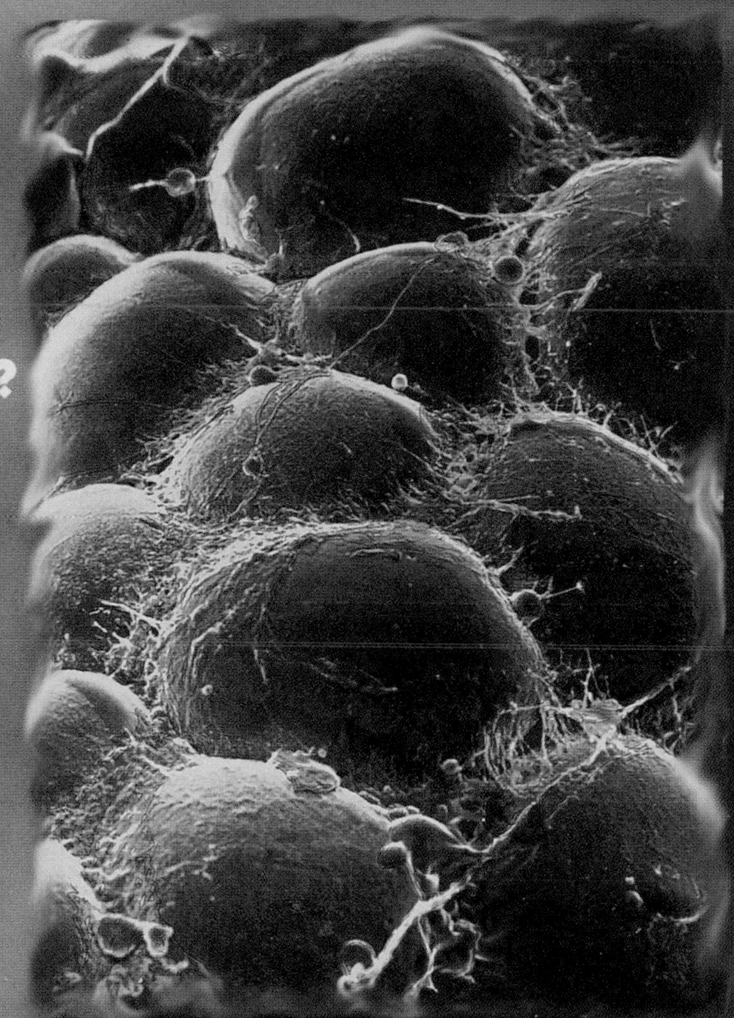

What are the functions of your skin?

How does your body maintain homeostasis?

Why does survival depend on a stable internal environment?

FOR NEARLY TWO-THIRDS OF THE HISTORY OF LIFE, ALL organisms consisted of one cell. There are still plenty of single-celled organisms today; in fact they far outnumber multicellular organisms. Theirs is a simple, uncomplicated life. They get their raw materials and energy from the fluid in which they are bathed, they dump their wastes into that same fluid, and they reproduce by dividing in two.

There are, however, disadvantages to being a single cell. The single-celled organism is completely at the mercy of its immediate external environment for every requirement of life. If the pond in which it lives dries up, it will die. If salt levels in the water rise, if the temperature gets too hot, or if its food runs out, it dies. There must be another way!

Why do cells assemble into groups?

There is another way, and that is for cells to join together. In this chapter we look at how cells are organized into the tissues, organs, and organ systems that make up your body. We consider the structure and function of your skin as an example of an organ system. And we discuss how your cells, tissues, and organs work together to maintain the health and stability of your body.

4.1 Tissues are groups of cells with a common function

A *multicellular organism* consists of many cells that collectively share the functions of life. Advantages to multicellularity include greater size (the better to eat, rather than be eaten) and the ability to seek out or maintain an environment conducive to life.

All cells in a multicellular organism have a specialized function that benefits the organism in some way. However, specialization is not enough. The specialized functions must be organized and integrated if they are to be useful.

As an example, the activity of a single cell in your heart is insignificant because the cell is so small. The beating of your heart requires that hundreds of thousands of such cells be arranged end to end, so that their functions are coordinated to produce a single heartbeat.

Tissues are groups of specialized cells that are similar in structure and that perform common functions. There are four major types of tissues: epithelial, connective, muscle, and nervous tissues.

4.2 Epithelial tissues cover body surfaces and cavities

Spotlight on Structure & Function

Most **epithelial tissues** consist of sheets of cells that line or cover various surfaces and body cavities. Two epithelial tissues you know about are your skin and the lining of your mouth. Other epithelial tissues line the inner surfaces of your digestive tract, lungs, bladder, blood vessels, and the tubules of your kidneys.

Epithelial tissues are more than just linings. They protect underlying tissues. Often they are smooth to reduce friction; the smooth epithelial tissue lining your blood vessels helps blood flow more easily through your body, for instance. Some are highly specialized for transporting materials. Epithelial tissues (and cells) absorb water and nutrients across your intestines into your blood. They also secrete waste products across the tubules of your kidneys so you can eliminate them in urine.

A few epithelial tissues are *glandular epithelia* that form the body's glands. **Glands** are epithelial tissues that are specialized to synthesize and secrete a product. *Exocrine glands* (*exo-* means "outside" or "outward") secrete their products into a hollow organ or duct. Examples of exocrine glands are the glands in your mouth that secrete saliva, sweat glands in your skin, and glands in your stomach that produce digestive acid. *Endocrine glands* (*endo-* means "within") secrete substances called *hormones* into the bloodstream. One endocrine gland is the thyroid gland, which secretes several hormones that help regulate your body's growth and metabolism. We describe various glands throughout the book where appropriate. ■

Epithelial tissues are classified according to cell shape

Biologists classify epithelial tissues into three types according to the shapes of the cells (Figure 4.1):

- *Squamous epithelium* consists of one or more layers of flattened cells. (*Squama* means "plate-like." Think of squamous epithelium as "squashed flat.") Squamous epithelium forms the outer surface of the skin and lines the inner surfaces of the blood vessels, lungs, mouth and throat, and vagina.
- *Cuboidal epithelium* is comprised of cube-shaped cells. Cuboidal epithelium forms the kidney tubules and also covers the surfaces of the ovaries.
- *Columnar epithelium* is comprised of tall rectangular (column-shaped) cells. Columnar epithelium lines parts of the digestive tract, certain reproductive organs, and the larynx. Certain cells within columnar epithelium, called *goblet cells,* secrete mucus, a thick fluid that lubricates the tissues and traps bacteria, viruses, and irritating particles.

In addition to classification by cell shape, epithelial tissues are further described by the number of cell

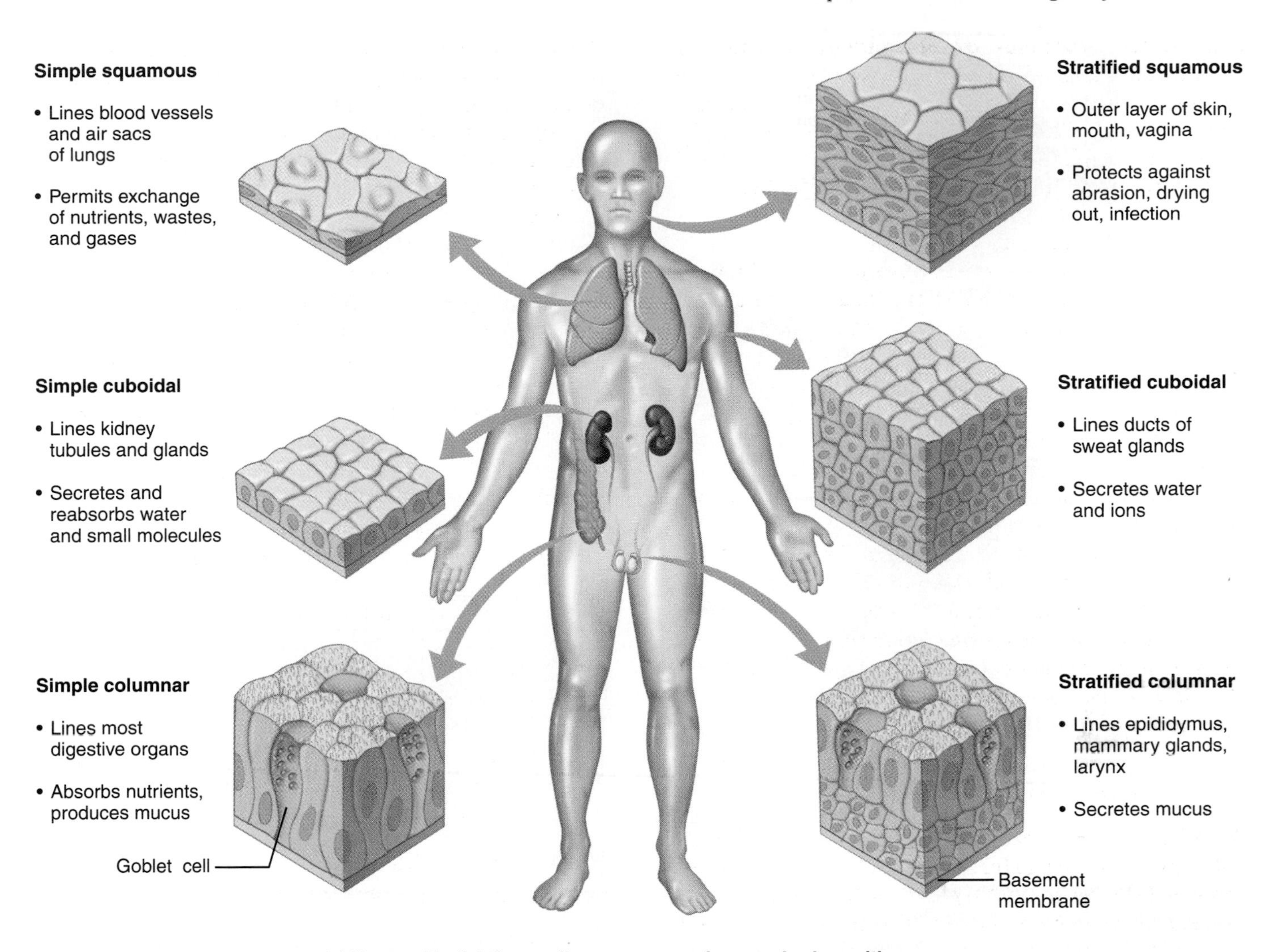

(a) Most epithelial tissues line or cover surfaces or body cavities

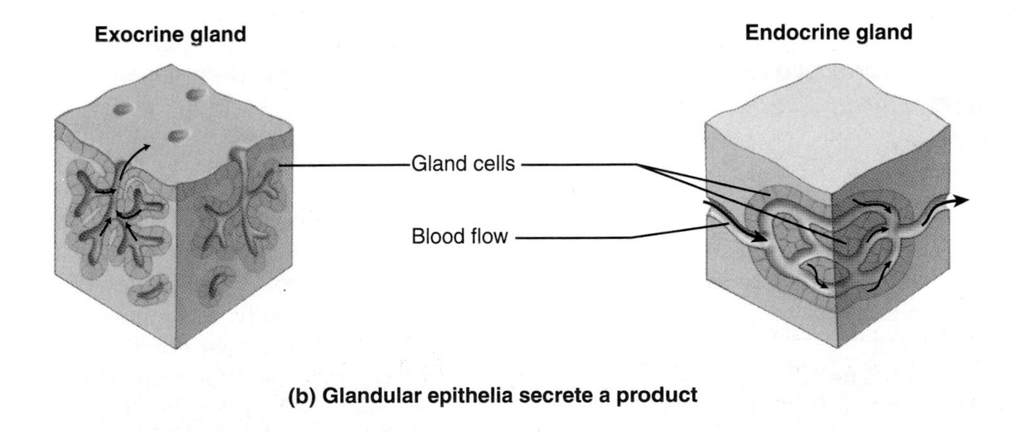

(b) Glandular epithelia secrete a product

Figure 4.1 Types of epithelial tissues.

layers comprising the tissue. A *simple epithelium* is a single layer of cells, whereas a *stratified epithelium* consists of multiple layers (or strata). Simple epithelium is thin so molecules can pass through it easily. Stratified epithelium is thicker and provides protection for underlying cells.

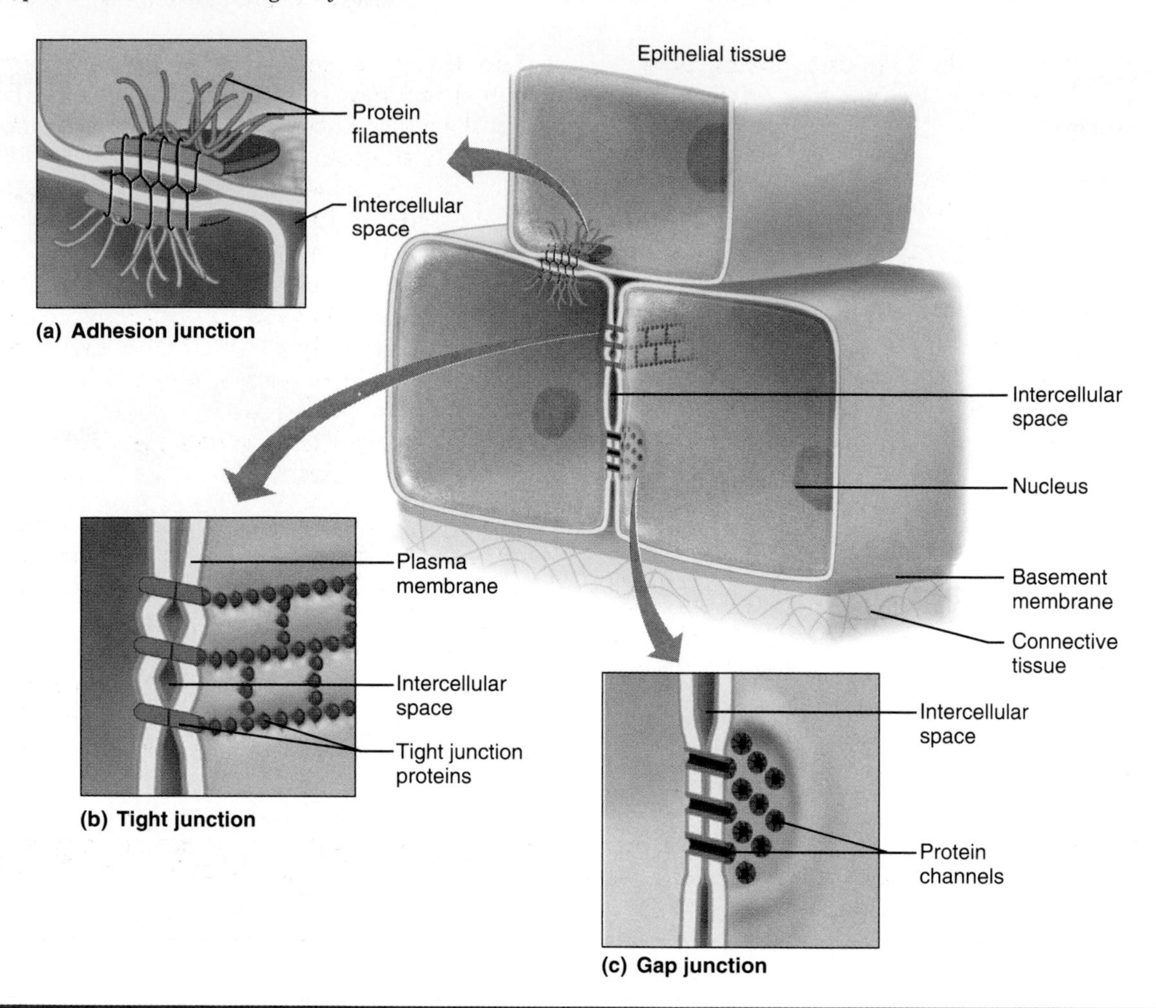

Figure 4.2 Examples of junctions between cells. Although this figure represents three types of cell junctions, only one type generally is present in any given tissue. **(a)** Tight junctions form leak-proof seals between cells. **(b)** Adhesion junctions anchor two cells together, yet allow flexibility of movement. **(c)** Gap junctions provide for the direct transfer of water and ions between adjacent cells.

The basement membrane provides structural support

Directly beneath the cells of an epithelial tissue is a supporting noncellular layer called the **basement membrane** (seen in Figure 4.1), and beneath that is generally a layer of connective tissue (described later). The basement membrane functions as a sort of mortar that anchors the cells to the stronger connective tissue underneath.

The basement membrane is comprised primarily of protein secreted by the epithelial cells. It is a cellular product and should not be confused with the plasma membrane that is a part of every living cell.

In addition to being attached to a basement membrane, epithelial cells may be connected to each other by several types of junctions comprised of various proteins. Three different types of junctions may hold the cells together, depending on the type of epithelial tissue (Figure 4.2):

- *Tight junctions* seal the plasma membranes of adjacent cells so tightly together that nothing can pass between the cells. Tight junctions are particularly important in epithelial layers that must control the movement of substances into or out of the body. Examples include the cells that line the digestive tract (which bring in nutrients) and the cells that form the tubules of the kidneys (which remove waste products from the body).
- *Adhesion junctions,* sometimes called "spot desmosomes," are looser in structure. The protein filaments of adhesion junctions allow for some movement between cells so the tissues can

stretch and bend. Adhesion junctions in the epithelium of your skin, for instance, allow you to move freely.

- *Gap junctions* represent connecting channels made of proteins that permit the movement of ions or water between two adjacent cells. They are commonly found in the epithelial cells in the liver, heart, and some muscle tissues.

Recap **Epithelial tissues line body surfaces and cavities, and form glands. They are classified according to cell shape (squamous, cuboidal, or columnar) and the number of cell layers (simple or stratified).**

4.3 Connective tissue supports and connects body parts

Connective tissue supports the softer organs of the body against the forces of gravity and connects the various parts of the body together. It also stores fat and produces the cells of the blood.

Unlike epithelial tissue, most connective tissues have few living cells. Most of their structure consists of nonliving extracellular material, the *matrix,* that is synthesized by connective tissue cells and released into the space between them. The strength of connective tissue comes from the matrix, not from the living cells themselves. The few living cells rarely make contact with each other, and so direct cell-to-cell junctions are not present.

Connective tissues are so diverse that any classification system is really a matter of convenience (Table 4.1). Broadly, we can divide them into fibrous and specialized connective tissues.

Fibrous connective tissues provide strength and elasticity

Fibrous connective tissues connect various body parts, providing strength, support, and flexibility. Figure 4.3 shows the structural elements of fibrous connective tissue.

As indicated by their name, fibrous connective tissues consist of several types of fibers and cells embedded in a gel-like ground substance. **Collagen fibers,** made of protein, confer strength and are

Table 4.1 Types of connective tissues

Type	Structure	Attributes	Locations
Fibrous Connective Tissue			
Loose	Mostly collagen and elastin fibers in no particular pattern; more ground substance	Flexible but only moderately strong	Surrounds internal organs, muscles, blood vessels
Dense	Mostly collagen in a parallel arrangement of fibers; less ground substance	Strong	In tendons, ligaments, and the lower layers of skin
Elastic	High proportion of elastin fibers	Stretches and recoils easily	Surrounds hollow organs that change shape or size regularly
Reticular (lymphoid)	Mostly thin, interconnecting reticular fibers of collagen	Serves as a flexible internal framework	In soft organs such as liver, spleen, tonsils, and lymph glands
Special Connective Tissues			
Cartilage	Primarily collagen fibers in a ground substance containing a lot of water	Maintains shape and resists compression	Embryonic tissue that becomes bone. Also the nose, vertebral disks, and the lining of joint cavities
Bone	Primarily hard mineral deposits of calcium and phosphate	Very strong	Forms the skeleton
Blood	Blood cells, platelets, and blood fluid called plasma	Transports materials and assists in defense mechanisms	Within cardiovascular system
Adipose tissue	Primarily cells called adipocytes filled with fat deposits	Stores energy in the form of fat	Under the skin, around some internal organs

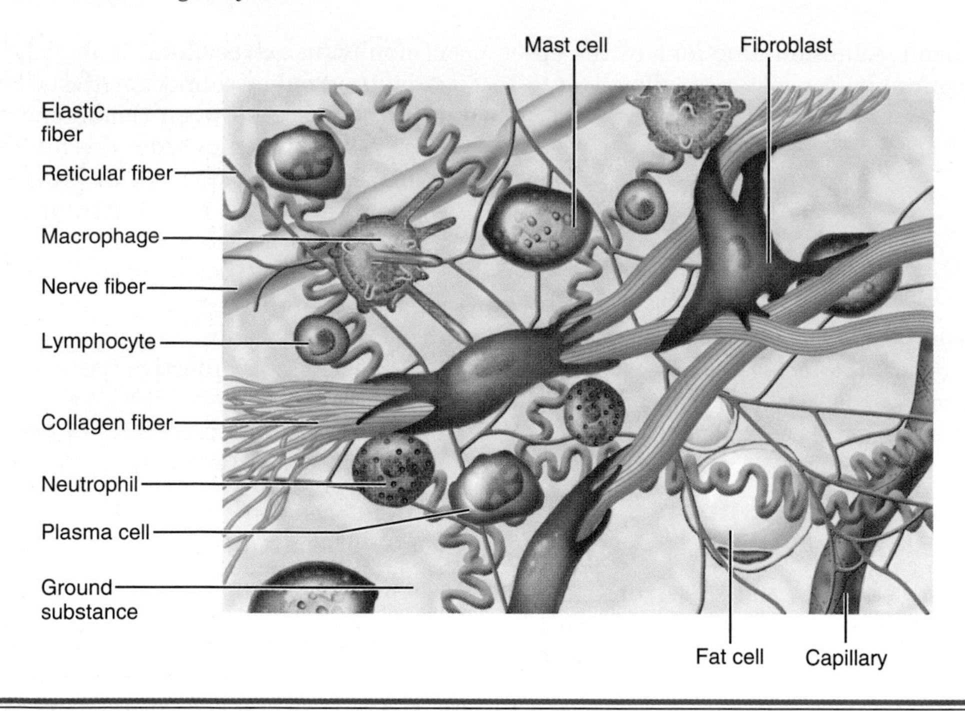

Figure 4.3 Fibrous connective tissue. The main elements are three types of fibers (collagen, elastic, and reticular) and a variety of cells (fibroblasts, fat cells, mast cells, and several types of white blood cells) in a ground substance of polysaccharides, proteins, and water. Blood vessels and nerves pass through or are associated with connective tissue. Fibrous connective tissues vary in terms of their relative proportions of cells and fibers, and also fiber orientation.

slightly flexible. Most fibrous connective tissues also contain thinner coiled **elastic fibers,** made primarily of the protein *elastin,* which can stretch without breaking. Some fibrous connective tissue also contains thinner fibers of collagen, called **reticular fibers,** that interconnect with each other. The reticular fibers often serve as an internal structural framework for some of the "soft" organs such as the liver, spleen, and lymph nodes.

The various fibers are embedded in a *ground substance* consisting of water, polysaccharides, and proteins that ranges in consistency from a soft gel to almost rubbery. The ground substance contains several types of cells, among them fat cells, mast cells, various white blood cells (macrophages, neutrophils, lymphocytes, and plasma cells), and, most importantly, **fibroblasts.** The fibroblasts are the cells responsible for producing and secreting the proteins that comprise the collagen, elastic, and reticular fibers. The fat cells, of course, store fat, and both the mast cells and white blood cells are involved in the body's immune system (Chapter 9).

Fibrous connective tissues are subclassified according to the density and arrangement of their fiber types:

- *Loose connective tissue,* also called *areolar* connective tissue, is the most common type. It surrounds many internal organs, muscles, and blood vessels. Loose connective tissue contains a few collagen fibers and elastic fibers in no particular pattern, giving it a great deal of flexibility but only a modest amount of strength.
- *Dense connective tissue,* found in tendons, ligaments, and in lower layers of skin, has more collagen fibers. The fibers are oriented primarily in one direction, especially in the tendons and ligaments in and around our joints. Dense connective tissue is the strongest connective tissue when pulled in the same direction as the orientation of the fibers, but it can tear if the stress comes from the side. There are very few blood vessels in dense connective tissue to supply the few living cells. This is why, if you strain a tendon or ligament, it can take a long time to heal. Figure 4.4 compares loose and dense connective tissue.
- *Elastic connective tissue* surrounds organs that have to change shape or size regularly. Examples include the stomach, which must stretch to accommodate food; the bladder, which stretches to store urine; and the vocal cords, which vibrate to produce sounds. Elastic connective tissue contains a high proportion of elastic fibers, which stretch and recoil easily.

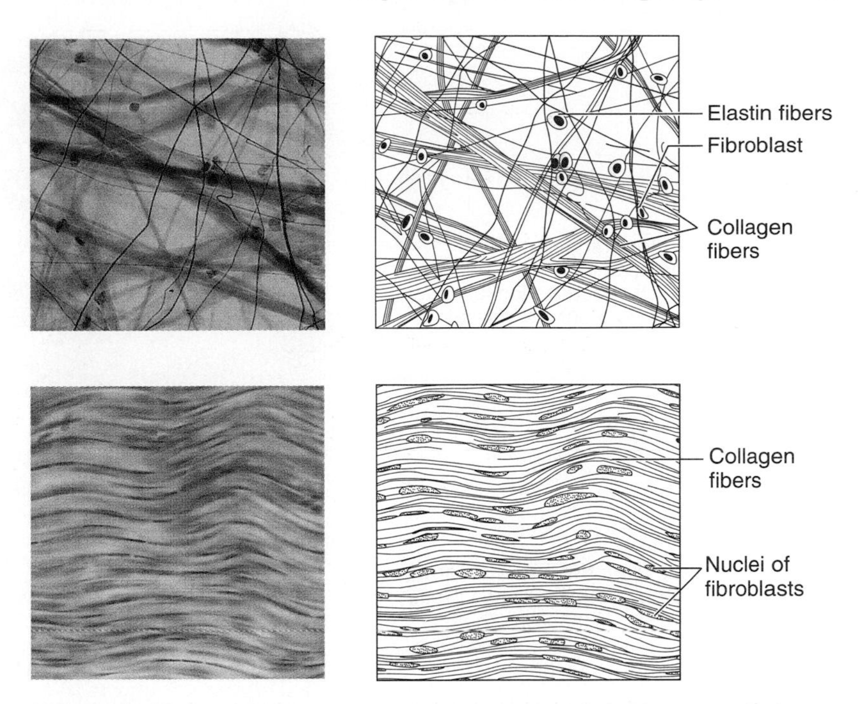

(a) Loose areolar connective tissue (×160). In loose connective tissue the collagen and elastin fibers are arrayed in a random pattern.

(b) Dense connective tissue (×160). In dense connective tissue the fibers are primarily collagen fibers. In tendons and ligaments the fibers are oriented all in the same direction, with fibroblasts occupying narrow spaces between adjacent fibers.

Figure 4.4 Examples of fibrous connective tissues.

- *Reticular connective tissue* (also called lymphoid tissue) serves as the internal framework of soft organs such as the liver and the tissues of the lymphatic system (spleen, tonsils, and lymph nodes). It consists of thin, branched reticular fibers (comprised of collagen) that form an interconnected network.

***Recap* Fibrous connective tissues hold other body parts together. Fibrous connective tissues contain extracellular fibers of strong but flexible proteins, a few living cells, and a ground substance of polysaccharides, proteins, and water.**

Specialized connective tissues serve special functions

The so-called *specialized connective tissues* are a diverse group that includes cartilage, bone, blood, and adipose tissue. Each is specialized to provide particular functions in the body.

Cartilage **Cartilage** is the transition tissue from which bone develops (Chapter 5). It also maintains the shape of certain body parts (such as the soft tip of your nose) and protects and cushions joints. Disks of cartilage separate and cushion the vertebrae in your backbone, for instance, and cartilage forms the tough, smooth surfaces that reduce friction in some body joints.

Like dense connective tissue, cartilage consists primarily of collagen fibers. The difference is that the ground substance of cartilage, which is produced by cells called *chondroblasts,* contains a great deal more water. This is why cartilage functions so well as a cushion. As cartilage develops the cells become enclosed in small chambers called *lacunae* (Figure 4.5a). There are no blood vessels in cartilage, so the mature cells (called chondrocytes) obtain their nutrients only by diffusion through the ground substance from blood vessels located outside the cartilage itself. Consequently, cartilage is slow to heal when injured.

Bone Like cartilage, **bone** is a specialized connective tissue that contains only a few living cells. Most of the matrix of bone consists of hard mineral deposits of calcium and phosphate. However, unlike cartilage, bone contains numerous blood vessels, which is why it can heal within four to six weeks after being injured. We discuss bone in more detail in Chapter 5 when we discuss the skeletal system.

Blood **Blood** consists of cells suspended in a fluid matrix called plasma. It is considered a connective tissue because all blood cells derive from earlier cells (called *stem cells*) located within bone. Red blood cells transport oxygen and nutrients to body cells and carry away the waste products of the cells' metabolism. White blood cells function in the immune system that defends the body, and platelets participate in the mechanisms that cause blood to clot

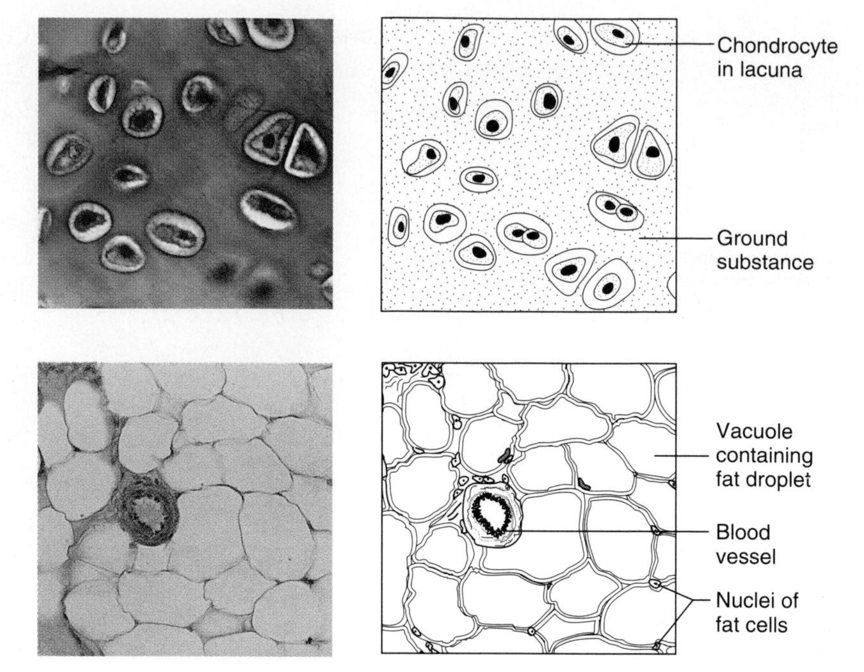

(a) Cartilage from the trachea (×300). Mature cartilage cells, called chondrocytes, become trapped in chambers called lacunae within the hard, rubbery ground substance.

(b) Adipose tissue from the subcutaneous layer under the skin (×140). Adipose tissue consists almost entirely of fat cells. The fat deposit within a fat cell can become so large that the nucleus is pushed to the side.

Figure 4.5 Examples of special connective tissues.

following an injury. You learn more about the functions of blood in Chapter 7.

Adipose tissue Adipose tissue is highly specialized for fat storage (Figure 4.5b). It has few connective tissue fibers and almost no ground substance. Most of its volume is occupied by **adipocytes** (fat cells). Adipose tissue is located primarily under the skin, where it serves as a layer of insulation. It also forms a protective layer around internal organs such as the kidneys.

The number of adipocytes you have is partly determined by your genetic inheritance. When you eat more food than your body can use, some of the excess energy is stored in your adipocytes as fat (the fat cells get "fatter"). When you lose weight the fat cells slim down too. However, weight loss reduces the volume of fat cells but not their number. *Liposuction* (literally, "fat suction") is a surgical technique that has been used to remove fat cells from the body permanently.

> ***Recap*** **Among the specialized connective tissues, cartilage and bone provide support, blood transports materials throughout the body, and adipose tissue stores energy in the form of fat.**

4.4 Muscle tissues contract to produce movement

Muscle tissue consists of cells that are specialized to shorten, or contract, resulting in movement of some kind. Muscle tissue is composed of tightly packed cells called muscle *fibers*. The fibers are generally long and thin and parallel in alignment to each other (Figure 4.6). The cytoplasm of a muscle fiber contains proteins, which interact to make the cell contract.

There are three types of muscle tissue: *skeletal, cardiac,* and *smooth.* They vary somewhat in terms of body location, structure, and function, but they all do essentially the same thing—when stimulated, they contract. An entire chapter (Chapter 6) is devoted to muscles as an organ system. For now, we will focus on differences between the three types of muscle tissue.

Skeletal muscles move body parts

Skeletal muscle tissue connects to tendons, which attach to bones. When skeletal muscles contract they cause body parts to move. The individual fibers are thin cylinders too small to be seen with the naked eye, but they may be as long as the entire muscle (Figure 4.6a). Each muscle fiber has many nuclei, a phenomenon that comes about because many cells fuse end to end during development, producing one long fiber.

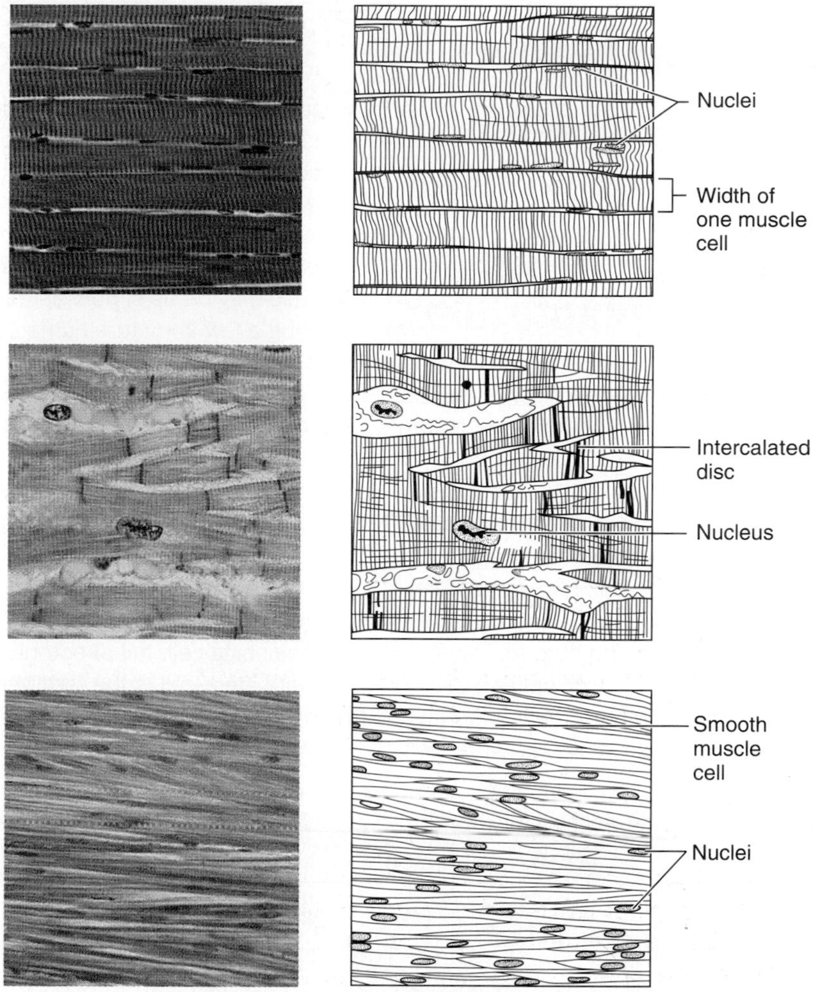

(a) Skeletal muscle (approx. ×100). Skeletal muscle cells are very long and have many nuclei.

(b) Cardiac muscle (×225). Cardiac muscle cells interconnect with each other.

(c) Sheet of smooth muscle (×250). Smooth muscle cells are thin and tapered.

Figure 4.6 Muscle tissue.

A skeletal muscle may contain thousands of individual fibers, all aligned parallel to each other. This parallel arrangement enables them to all pull together, shortening the muscle between its two points of attachment. Skeletal muscle is called *voluntary* muscle because we can exert conscious control over its activity.

Cardiac muscle cells activate each other

Cardiac muscle tissue (Greek *kardia,* the heart) is found only in the heart. The individual cells are much shorter than skeletal muscle fibers, and they have only one nucleus (Figure 4.6b). Like skeletal muscle, the cells are arranged parallel to each other. Cardiac muscle cells are short and blunt-ended, with gap junctions between the ends of adjoining cells. The gap junctions represent direct electrical connections between adjoining cells, so when one cell is activated it activates its neighbors down the line. Because of these gap junctions the entire heart contracts in a coordinated fashion.

Cardiac muscle is considered involuntary because the heart can contract rhythmically entirely on its own, without any conscious thought on our part and without any stimulation by nerves.

Smooth muscle surrounds hollow structures

Smooth muscle tissue surrounds hollow organs and tubes, including blood vessels, digestive tract, uterus, and bladder. These slim cells are much smaller than skeletal muscle cells and have only one nucleus, like cardiac muscle (Figure 4.6c). The cells are aligned roughly parallel to each other. In blood vessels they are generally aligned in a circular fashion

CurrentIssue

Should Human Stem Cells be Used for Research or Transplantation?

Michael J. Fox has Parkinson's disease, a debilitating neurological disorder marked by progressive tremor of the hands and legs. Parkinson's disease is caused by the death and degeneration of certain nerve cells in an area of the brain (called the midbrain) that coordinates muscle movements. Right now there is no cure for Parkinson's disease, but Michael and the nearly a million other Americans with the disease—including U.S. Attorney General Janet Reno, Muhammad Ali, and Billy Graham—hope that progress can be made soon. And maybe it will, if researchers working with human *stem cells* have their way. According to these researchers, the cure for Parkinson's disease may be as simple as transplanting a few immature human nerve cells into the patient's brain.

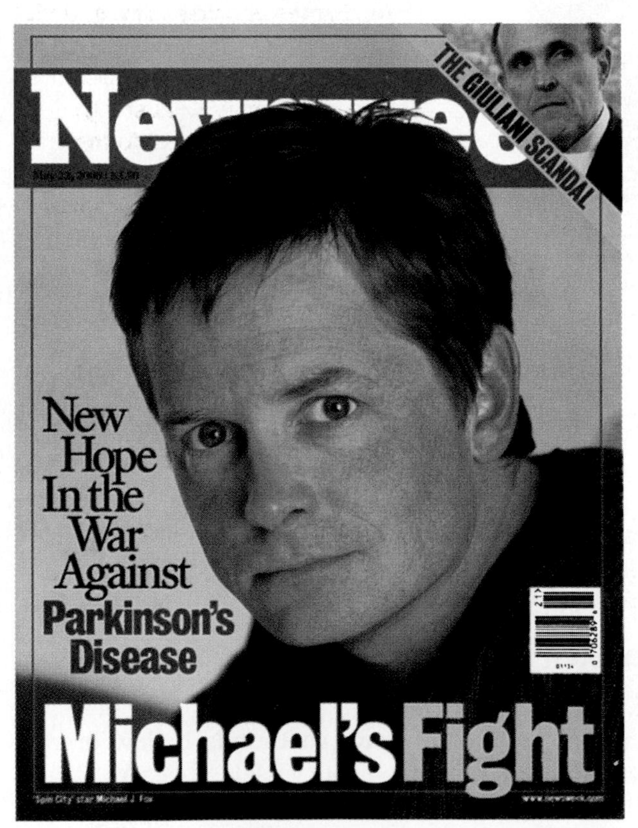

Actor Michael J. Fox has Parkinson's disease, one of the first diseases that might be cured as a result of stem cell research.

Human Stem Cells Hold the Key

The key to curing not only Parkinson's disease but many other diseases as well may be a type of cell called a **stem cell,** so-called because all of the various types of cells of the body originate (stem) from it. A stem cell is the universal beginning cell, full of potential. The ultimate stem cell is the fertilized human egg. Every human being starts as a single fertilized egg and ends up as a collection of tens of trillions of specialized cells. In other words, all specialized cells originate (stem) from that one cell, but along the way they go through structural and functional changes. They *differentiate,* meaning that they become different from the cells from which they originated. Some become muscle; others become skin; still others become nerve cells in the brain.

Stem cells have several properties that may make them (or their progeny) ideal candidates for the treatment of certain diseases:

1. Because they are still undifferentiated, stem cells represent the ancestors of all specialized cells. They can become anything if we can just learn how to make them differentiate into the type of cell we want. Indeed, stem cells are sometimes called "pluripotent" meaning that they "possess many possibilities."
2. Stem cells are easier to separate from each other than are adult cells because they don't adhere as tightly to each other.
3. Stem cells grow in culture better than do most adult cells.
4. Stem cells are easier to transplant into a patient. Instead of having to transplant a solid piece of adult tissue, stem cells or only partially differentiated cells derived from them could be injected and allowed to migrate to their target site, where hopefully they would develop into healthy adult cells.
5. A tremendous obstacle to most adult transplants is tissue rejection, which occurs when the recipient's immune system attacks the transplanted (foreign) tissue. Less differentiated cells seem to provoke a less vigorous immune response than do adult cells.

The Source of Human Stem Cells

The ideal source of human stem cells would be the very youngest-stage human embryo, in the first several days after fertilization when there are only eight or fewer cells. At this stage all the cells are still undifferentiated and alike (all eight cells are still true stem cells). Currently, the only embryos of this age that might be available are those created "in excess of clinical need" by *in vitro* fertilization techniques at private fertility clinics. These are so rare that they are effectively only available to a few researchers working in private fertility clinics.

For the treatment of specific human diseases, the cells do not need to be as completely undifferentiated as the first eight cells of the human embryo. If it's nerve cells that are needed, a good source of young nerve cells is the very first neural tissue in the human fetus. Non-living human fetuses are much more common than "extra" embryos as a consequence of the more than one million legal abortions performed in the U.S.A. each year.

Patients with Parkinson's disease are the most likely to benefit first from cells taken from human fetuses. Already, over a 100 human patients have received fetal nerve cell transplants worldwide, some with encouraging results. Researchers are also investigating the use of fetal cells in treating Alzheimer's disease (a degenerative brain disorder), spinal cord injuries, diabetes mellitus (a disease of blood sugar regulation) leukemia (a blood cancer), and radiation accidents.

Not surprisingly, the harvesting and use of either embryonic cells or fetal tissues (we'll refer to them collectively as stem cells) is highly controversial. On the one hand, patient advocacy groups recognize the potential benefits of stem cells and promote efforts to harvest and use them. On the other, some human rights groups object strongly to harvesting or using human stem cells under any circumstances. The federal government is caught in the middle.

Objections to the Use of Human Stem Cells

Objections to the use of human stem cells generally concern human rights—specifically the status of the embryo or fetus from which stem cells are obtained. Many people argue that an embryo or fetus at any age is a human being, with the same legal rights as any other living person. Since an embryo or an aborted fetus cannot give informed consent, opponents assert that stem cells derived from human embryos or fetuses should not be used under any circumstances, not even to help another person.

Others argue that an aborted fetus has the same standing as any other dead person—that is, as a cadaver. Cadaveric tissues and organs are routinely and legally used for research, teaching, and transplantation. According to the law governing the procurement of tissues or organs from a cadaver, called the Uniform Anatomical Gift Act, consent may be given by the person before death (not possible for an embryo or a fetus) or by family members after death. Although the parents are logically the people to make this decision, one could question whether parents who choose abortion are acting in the best interest of the fetus in the first place. The status of a frozen embryo in terms of the Uniform Anatomical Gift Act is not clear.

The Federal Government Redefines Its Position

The use of human embryonic cells or fetal tissues for either research or patient treatment is not against the law. The current statutes that govern the National Institutes of Health, however; do not allow them to provide funding for "research in which a human embryo or embryos are destroyed, discarded, or knowingly subjected to risk of injury or death greater than that allowed under" other laws. For years, NIH has interpreted this to mean that they cannot fund any research involving human stem cells. In practice what this has meant is that research with human stem cells can be conducted with private money (for example, in a fertility clinic) but not with federal money (which often funds university researchers).

University-based researchers and patient groups have long chafed at the seemingly contradictory nature of the regulations and have been lobbying the federal government for a change. As a result the Department of Health and Human Services (DHHS), which oversees NIH, took a closer look at the wording of the law in 1999. At that time the DHHS ruled that the law did not actually forbid research on later generations of stem cells *derived* from human embryonic or fetal cells (called cell lines) because the cells cannot develop into a person once they are removed from the embryo or the fetus.

In August of 2000 the NIH approved new guidelines under which human stem cell research could be funded:

- Researchers may *use* stem cell lines that were derived from embryos as long as they do not create the stem cell line themselves. The cell lines must be derived privately from frozen "excess" embryos created by fertility treatments at private clinics. Under no circumstances may embryos be created expressly for research. Donors of embryos must sign strictly worded consent forms.
- Federal researchers could *derive and use* new stem cell lines from fetal tissues as long as they follow the established standards for other fetal tissue research.

The new guidelines expressly prohibit certain activities, including using stem cells to try to create a human embryo, and combining stem cells with animal embryos.

The ruling that federally funded researchers can use human stem cells as long as they get them from somebody in the private sector is an odd compromise at best. Look for this issue to be revisited in the future.

around the vessel. When smooth muscle cells shorten, the diameter of the blood vessel is reduced.

Smooth muscle cells taper at both ends, and there are gap junctions between adjacent cells so that when one contracts, nearby cells also contract. Like cardiac muscle, smooth muscle is also involuntary in that we cannot exert control over its contractions.

4.5 Nervous tissue transmits impulses

Nervous tissue consists primarily of cells that are specialized for generating and transmitting electrical impulses throughout the body. It forms a rapid communication network for the body. Nervous tissue is located in the brain, the spinal cord, and the nerves that transmit information to and from various organs. Chapter 11 is devoted to the nervous system, so we describe nervous tissue only briefly here.

Nervous tissue cells that generate and transmit electrical impulses are called **neurons** (Figure 4.7). Neurons can be as long as the distance from your spinal cord to the tip of your toe. Neurons typically have three basic parts: (1) the *cell body* where the nucleus is located; (2) *dendrites,* numerous cytoplasmic extensions that extend from the cell body and receive signals from other neurons; and (3) a long extension called an *axon* that transmits electrical impulses over long distances.

Nervous tissue also includes another type of cell called a **glial cell** that does not transmit electrical impulses. Glial cells play a supporting role by surrounding and protecting neurons and supplying them with nutrients.

> ***Recap*** **The common feature of all muscle tissues (skeletal, cardiac, and smooth) is that they contract, producing movement. Nervous tissues serve as a communication network by generating and transmitting electrical impulses.**

4.6 Organs and organ systems perform complex functions

Many of the more complex functions of multicellular organisms (such as pumping blood or digesting food) cannot be carried out by one tissue type alone. **Organs** are structures composed of two or more tissue types joined together that perform a specific function or functions.

Spotlight on Structure & Function Your heart is an organ. Most of it consists of cardiac muscle, but there is also smooth muscle in the blood vessels that supply the cardiac muscle. The heart also contains nervous tissue that affects the rate at which the heart beats. It contains some connective tissue, primarily in the heart valves that open and close to control blood flow within the heart, and even a thin layer of epithelial tissue that lines the heart chambers. These tissues function together to pump blood, so together they constitute the organ known as the heart. Some organs have several functions. For example, the kidneys remove wastes and help control blood pressure. ■

The human body is organized by organ systems

Organ systems are groups of organs that together serve a broad function that is important to the survival of the whole organism, such as protection, movement, or excretion of wastes. A good example is the organ system responsible for the digestion of food. Your *digestive system* includes your mouth, throat, esophagus, stomach, intestines, and even your liver, pancreas, and gall bladder. All of these organs must interact and be controlled and coordinated to accomplish their overall function.

The box on pages 92 and 93 depicts the eleven organ systems of the human body. Some organ systems perform several functions and so will be discussed in several chapters in this book. For example,

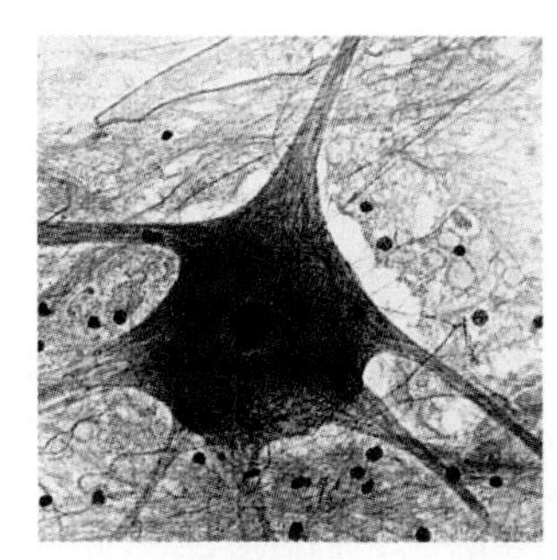

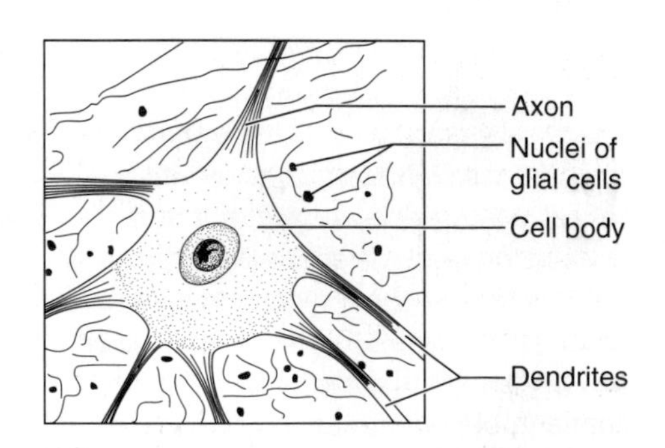

A neuron (×170). The neuron is the functional unit of nervous tissue. The single neuron shown here is surrounded by numerous supporting cells called glial cells. The cell bodies of the glial cells do not stain well, but their nuclei are clearly visible.

Figure 4.7 Nervous tissue.

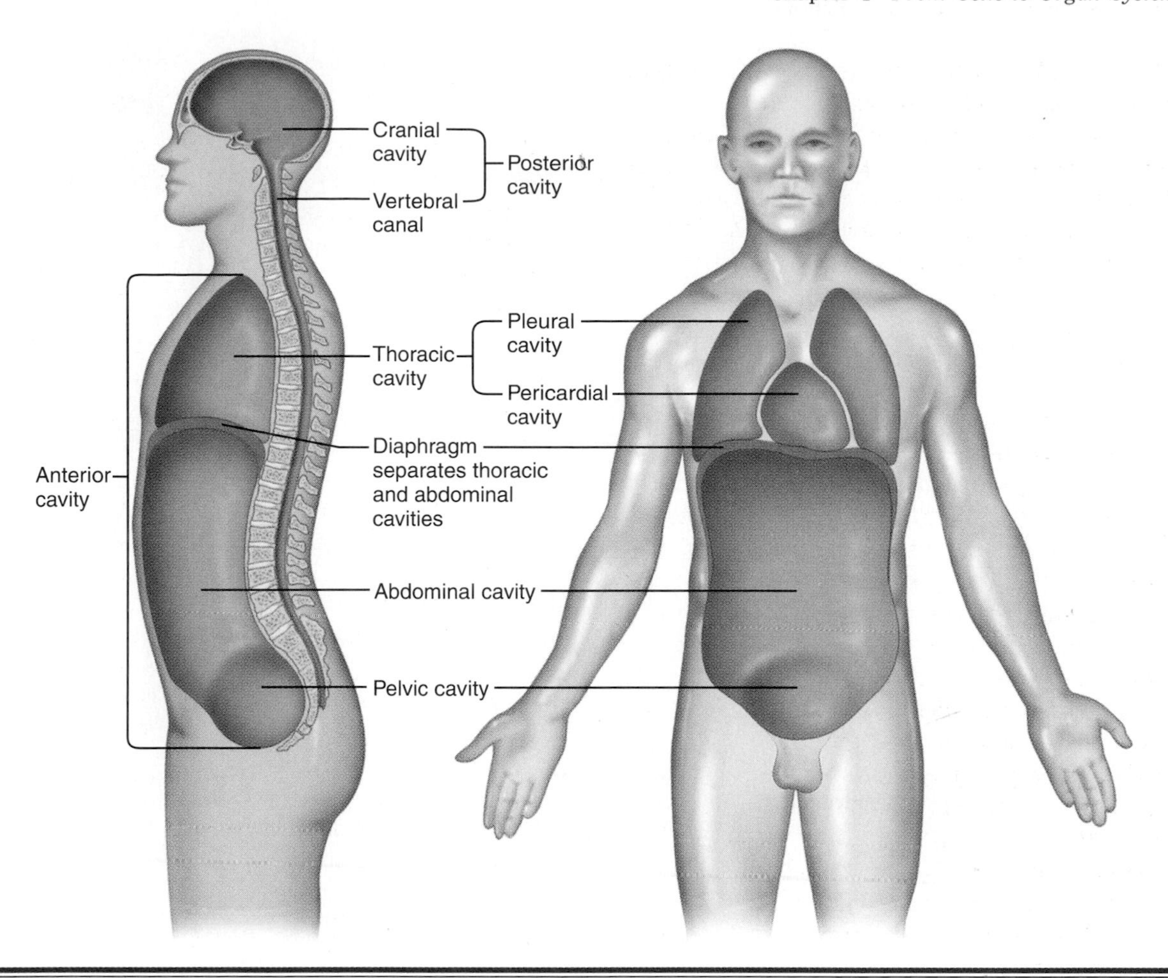

Figure 4.8 The main body cavities. The pelvic cavity and the abdominal cavities are continuous (not separated by a membrane).

the lymphatic system has important functions related to defense, the return of certain body fluids to the circulation, and digestion. Other organ systems will be covered in a single chapter.

> ***Recap*** **An organ consists of several tissue types that join together to perform a specific function. An organ system is a group of organs that shares a broad function important for survival.**

Tissue membranes line body cavities

Some of the organs and organ systems are located in hollow cavities within the body (Figure 4.8). The large *anterior cavity* is divided into the **thoracic cavity** and **abdominal cavity** by the diaphragm between them. The thoracic cavity is in turn divided into two *pleural cavities,* each containing a lung, and the *pericardial cavity,* which encloses the heart. The lower part of the abdominal cavity is sometimes called the *pelvic cavity.* The smaller *posterior cavity* consists of the **cranial cavity** and the **spinal cavity** (vertebral canal). There are many smaller cavities as well, such as the synovial cavities in moveable joints.

Tissue membranes consisting of a layer of epithelial tissue and a layer of connective tissue line each body cavity and form our skin. There are four major types of tissue membranes:

- *Serous membranes:* Line and lubricate body cavities to reduce friction between internal organs.
- *Mucous membranes:* Line the airways, digestive tract, and reproductive passages. Goblet cells within the epithelial layer secrete *mucus,* which lubricates the membrane's surface and entraps foreign particles.
- *Synovial membranes:* Line the very thin cavities between bones in moveable joints. These membranes secrete a watery fluid that lubricates the joint.
- *Cutaneous membrane:* Our outer covering. You know it as skin, and it serves several functions to be discussed later in this chapter.

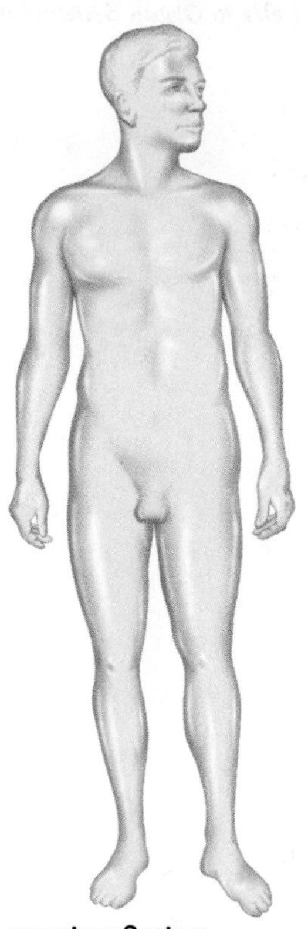

Integumentary System
- Protects us from injury, infection, and dehydration
- Participates in temperature control
- Receives sensory input from the external environment

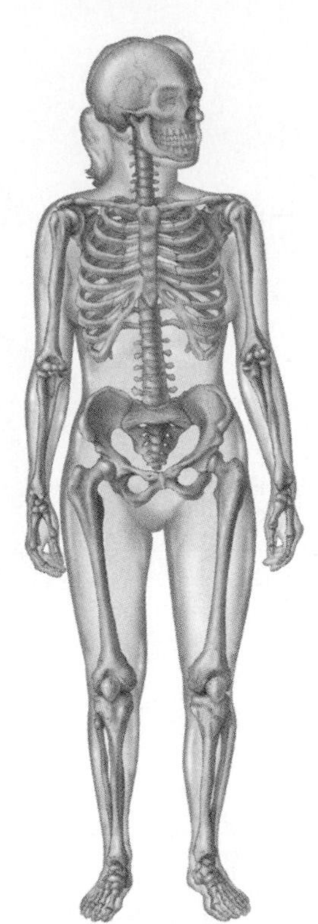

Skeletal System
- Protects, supports and anchors body parts
- Provides the structural framework for movement
- Produces blood cells
- Stores minerals

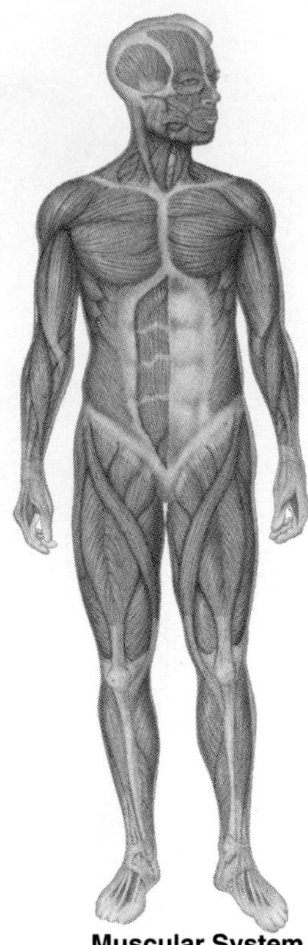

Muscular System
- Produces movement or resists movement
- Generates heat

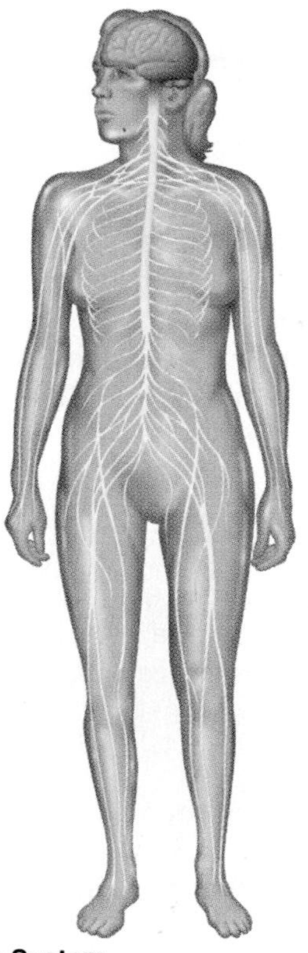

Nervous System
- Detects both external and internal stimuli
- Controls and coordinates rapid responses to these stimuli
- Integrates the activities of other organ systems

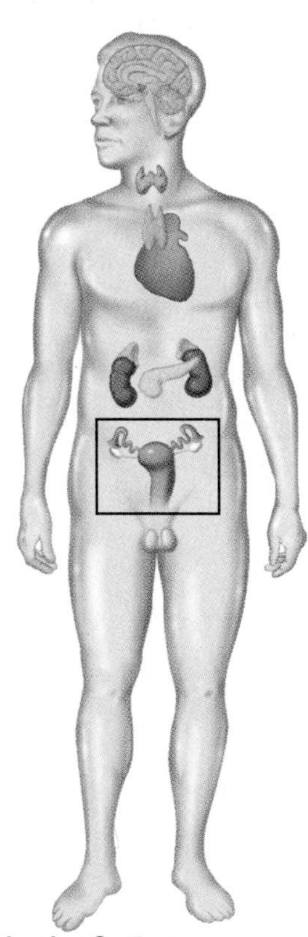

Endocrine System
- Produces hormones that regulate many body functions
- Participates with the nervous system in integrative functions

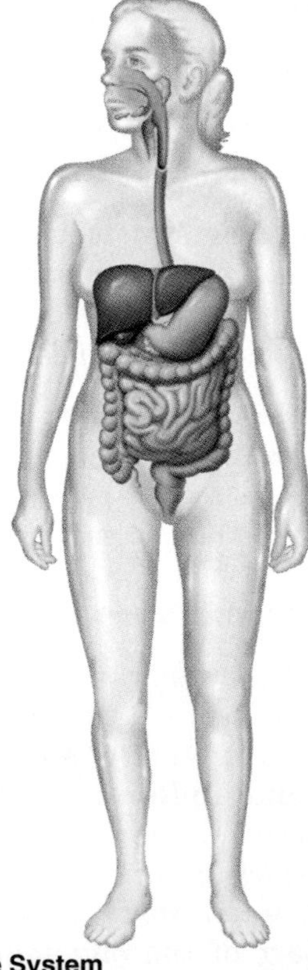

Digestive System
- Provides the body with water and nutrients
- (The liver) synthesizes certain proteins and lipids for the body
- (The liver) inactivates many chemicals, including hormones, drugs and poisons

The eleven organ systems of the human body.

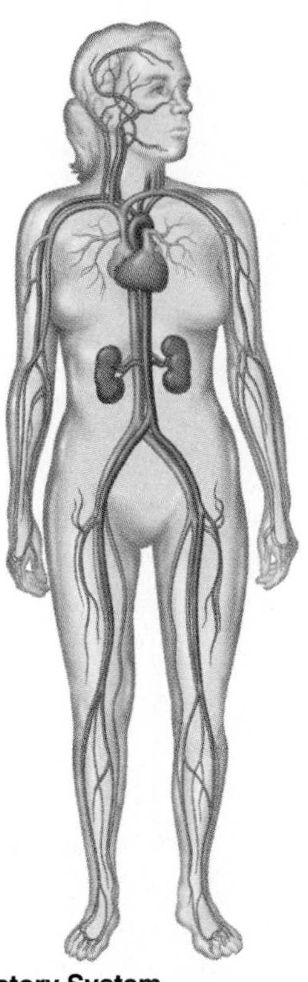

Circulatory System
- Transports materials to and from all cells
- Participates in the maintenance of body temperature
- Participates in mechanisms of defense against disease and injury

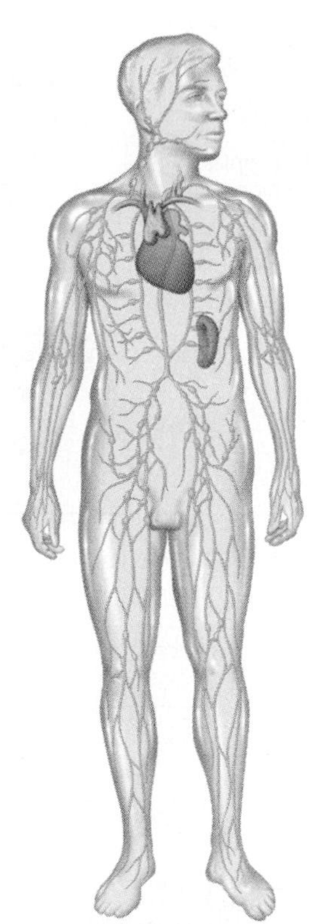

Lymphatic System
- Returns excess tissue fluid to the circulatory system
- Participates in both general and specific (immune) defense responses

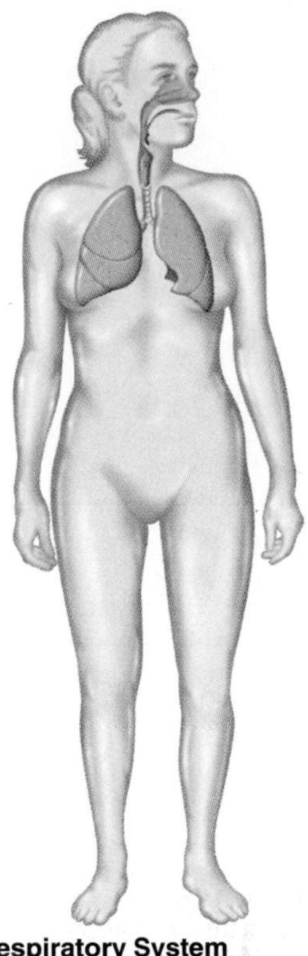

Respiratory System
- Exchanges gases (oxygen and carbon dioxide) between air and blood
- Participates in the production of sound (vocalization)

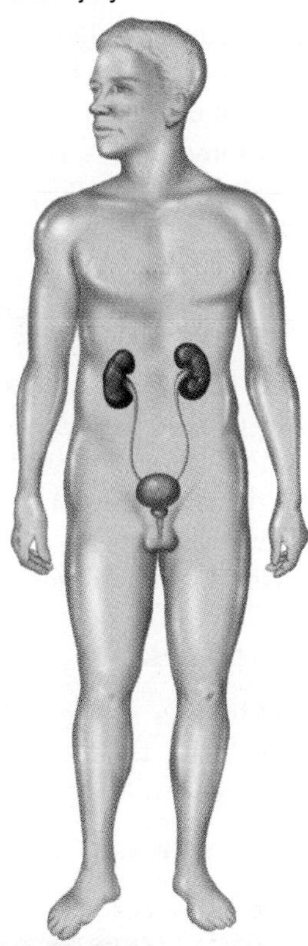

Urinary System
- Maintains the volume and composition of body fluids
- Excretes some waste products

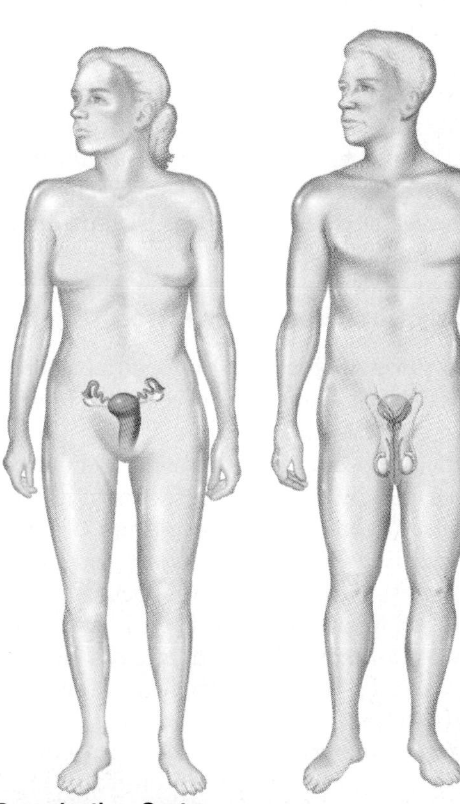

Reproductive System
- Female: Produces eggs
- Female: Nurtures the fertilized egg, developing embryo and fetus until birth
- Male: Produces sperm
- Male: Participates in the delivery of sperm to the female

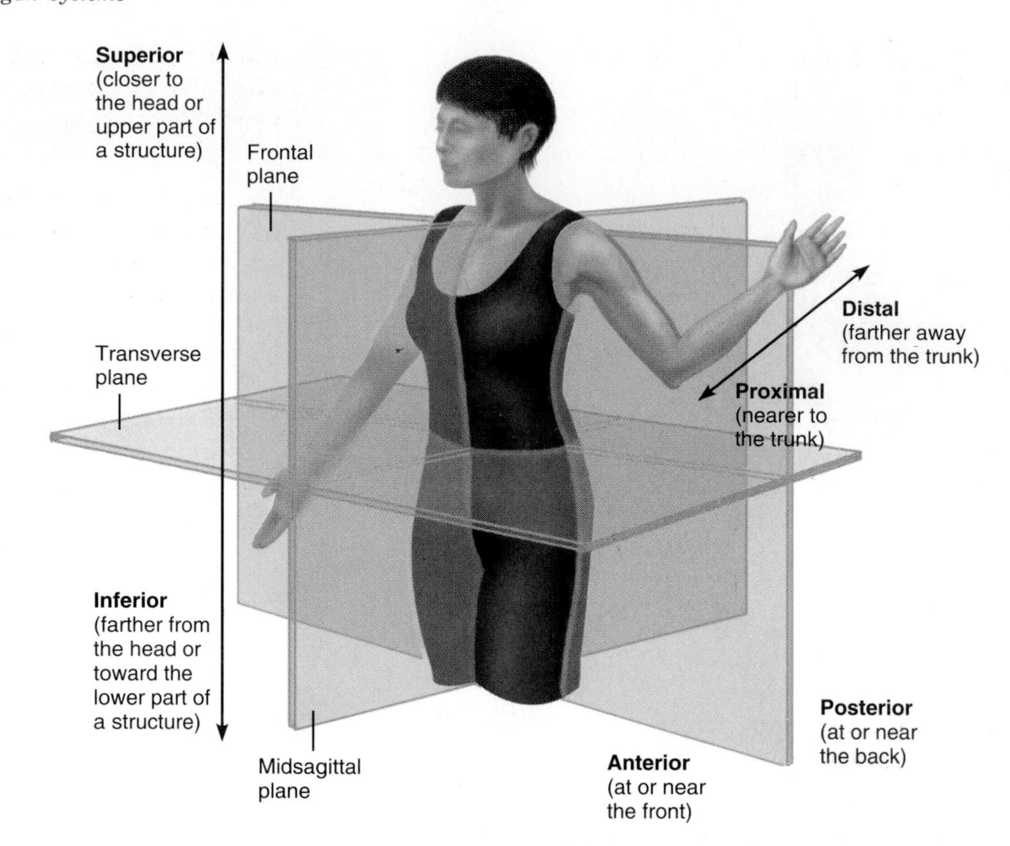

Figure 4.9 Planes of symmetry and terms used to describe position or direction in the human body. The frontal plane divides the body into anterior and posterior sections, the midsagittal plane divides it into left and right, and the transverse plane divides it into superior and inferior sections. Proximal and distal refer to points closer to or farther away from a point of reference, usually the trunk.

By now you may have noticed that "membrane" is a general term for a thin layer that covers or surrounds something. You have been introduced to three different membranes so far: the plasma membrane of phospholipids surrounding every cell, the basement membrane of extracellular material on which epithelial tissue rests, and tissue membranes consisting of several layers of tissue sandwiched together that cover or surround cavities, organs, and entire organ systems.

Describing body position or direction

When describing parts of the body, biologists use precise terms to define position and direction. It will help you to learn a few of these terms.

Generally speaking, an organ or even the entire body can be described by three planes known as the *midsagittal, frontal,* and *transverse* planes (Figure 4.9). These planes divide the body into left and right, front and back, and top and bottom, respectively. *Anterior* means at or near the front and *posterior* means at or near the back. *Proximal* means nearer to (in closer proximity to) any point of reference, usually the body trunk, and *distal* means farther away. For example, your wrist is distal to your elbow. *Superior* means situated above or directed upward, and *inferior* means situated below or directed downward. There are dozens of such terms, each with a precise meaning, and we will introduce their meanings when they first arise.

Recap **The body's hollow cavities are lined by tissue membranes that support, protect, and lubricate cavity surfaces. Locations of cavities and body parts are described relative to three planes (midsagittal, frontal, and transverse), using pairs of directional terms such as "anterior" and "posterior."**

4.7 The skin as an organ system

The proper name for the organ system represented by skin is the **integumentary system** (from the Latin *integere,* meaning "to cover"). We describe the skin here as a representative organ system; other organ systems are covered later in the book.

What are the functions of your skin?

Skin has many functions

Spotlight on Structure & Function

The skin has several different functions related to its role as the outer covering of our body:

- Protection from dehydration (helps prevent our bodies from drying out)
- Protection from injury (such as abrasion)
- Defense against invasion by bacteria and viruses

- Regulation of body temperature
- Synthesis of an inactive form of vitamin D
- Sensation: Provides information about the external world via receptors for touch, vibration, pain, and temperature

Skin consists of epidermis and dermis

Recall that skin is a tissue membrane, and that tissue membranes contain layers of epithelial and connective tissue. The outer layer of the skin's epithelial tissue is the **epidermis** and the inner layer of connective tissue is the **dermis** (Figure 4.10).

The skin rests on a supportive layer called the *hypodermis* (*hypo-* means "under"), consisting of loose connective tissue containing fat cells. The hypodermis is flexible enough to allow the skin to move and bend. The fat cells in the hypodermis insulate against excessive heat loss and cushion against injury. ■

Epidermal cells are replaced constantly The epidermis consists of multiple layers of squamous epithelial cells. A key feature of the epidermis is that it is constantly being replaced as cells near the base of the epidermis divide repeatedly, pushing older cells toward the surface.

The more numerous of two cell types are called **keratinocytes,** which produce a tough, waterproof protein called *keratin*. Actively dividing keratinocytes located near the base of the epidermis are sometimes called *basal cells*. As keratinocytes derived from the

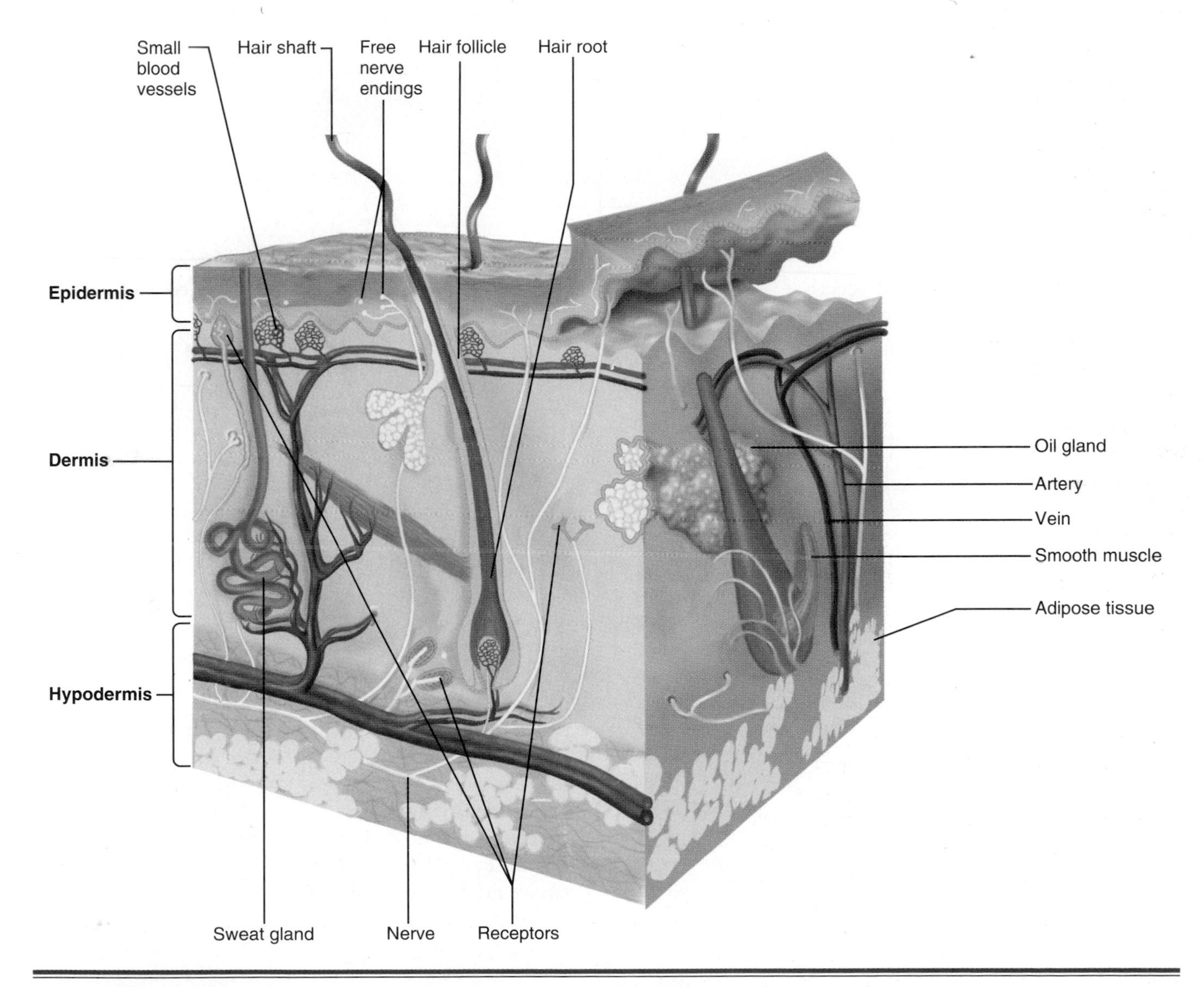

Figure 4.10 The skin. The two layers of skin (epidermis and dermis) rest on a supportive layer (the hypodermis) comprised of loose connective tissue and adipose tissue.

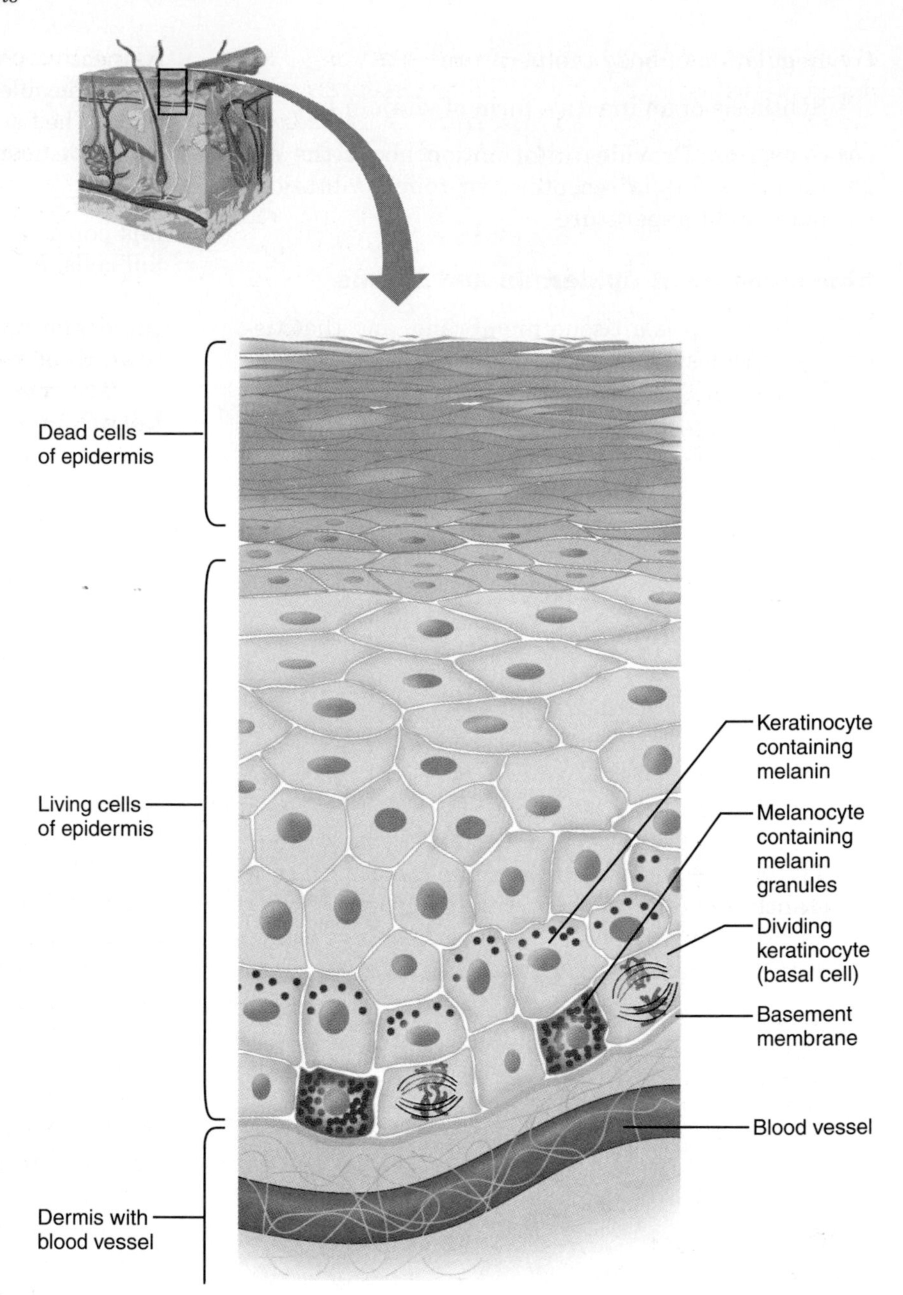

Figure 4.11 The epidermis. Living cells near the base of the epidermis divide, pushing more mature cells toward the surface. As cells migrate toward the surface they die and dry out, forming a tough, waterproof barrier. The cells of the epidermis are supplied only by blood vessels located in the dermis.

basal cells move toward the skin surface they flatten and become squamous. Eventually they die and dry out, creating a nearly waterproof barrier that covers and protects the living cells below (Figure 4.11). The rapid replacement of keratinocytes allows the skin to heal quickly after injury.

One reason the outer layers of epidermal cells die is that the epidermis lacks blood vessels, so as mature cells are pushed farther from the dermis they can no longer obtain nutrients. The dead cells of the outer layers are shed over time, accounting for the white flakes you sometimes find on your skin or on dark clothing, especially when your skin is dry.

Less numerous cells called **melanocytes** located near the base of the epidermis produce a dark-brown pigment called *melanin*. Melanin accumulates inside keratinocytes and protects us against the sun's ultraviolet radiation. Exposure to sunlight increases the activity of melanocytes, accounting for the ability of some people to develop a suntan. Because all humans have about the same number of melanocytes, racial differences in skin color also reflect either differences in melanocyte activity or differences in the rate of breakdown of melanin once it is produced.

Fibers in dermis provide strength and elasticity The dermis is primarily dense connective tissue, consisting of collagen, elastic, and reticular fibers embedded in a ground substance of water, polysaccharides, and proteins. The fibers allow the skin to stretch when we move and give it strength to resist abrasion and tearing. Our skin becomes less flexible and more wrinkled as we age because the number of fibers in the dermis decreases.

The surface of the dermis has many small projections called *papillae* that contain sensory nerve endings and small blood vessels. When the skin is rubbed excessively—such as from wearing tight shoes—the epidermis and dermis separate and a fluid-filled blister develops between them.

The most abundant living cells in the dermis are the fibroblasts that produce the various fibers, but there are also mast cells, white blood cells, and occasional fat cells. Other structures in the dermis include:

- *Hair:* Each hair has a *shaft* above the skin's surface and a *root* below the surface. Hair is actually comprised of several layers of cells enclosed in an outer layer of overlapping, dead, flattened keratinocytes. The root of a hair is surrounded by a sheath of several layers of cells called the *follicle.* The cells at the very base of the follicle are constantly dividing to form the hair root. As new hair cells are formed at the base, the hair root is pushed upward toward the skin's surface.
- *Smooth muscle:* Attached to the base of the hair follicle, it contracts when you are frightened or cold, causing your hair to become more erect.
- *Oil glands:* Also known as sebaceous glands, these secrete an oily fluid that moistens and softens hair and skin.
- *Sweat glands:* These produce sweat, a watery fluid containing dissolved ions and small amounts of metabolic wastes. Sweat glands help regulate body temperature.
- *Blood vessels:* These supply the cells of the dermis and epidermis with nutrients and remove their wastes. The blood vessels also help regulate body temperature. They dilate to facilitate heat loss when we are too hot and constrict to prevent heat loss when we are too cool. The dermis also contains lymph vessels, which drain fluids and play a role in the immune system.
- *Sensory nerve endings:* These provide information about the outside environment. Separate receptors exist to detect heat, cold, light touch, deep pressure, and vibration. You learn more about these nerve endings in Chapter 12.

As mentioned earlier, the skin synthesizes an inactive form of vitamin D. It is not known which cell type in the skin is responsible. What we do know is that a cholesterol-like molecule in the skin becomes an inactive form of vitamin D when it is exposed to the ultraviolet rays of sunlight. The inactive form must then be modified in the liver and kidneys before it becomes active.

Recap **The skin is an organ because it consists of different tissues serving common functions. Functions of skin include protection, temperature regulation, vitamin D synthesis, and sensory reception.**

4.8 Multicellular organisms must maintain homeostasis

Although multicellularity offers many advantages to organisms, it presents certain disadvantages that must be overcome. For example, cells that are surrounded entirely by other cells will not be able to obtain their nutrients directly from the organism's external environment and will be constantly exposed to the waste products of surrounding cells.

Why does survival depend on a stable internal environment?

The environment that surrounds the cells of a multicellular organism (their external environment) is the **internal environment** of the organism. The internal environment is a clear fluid called the **interstitial fluid** (the Latin noun *interstitium* means "the space between," in this case the space between cells). Every cell gets nutrients from the interstitial fluid around it and dumps wastes into it. In a multicellular organism, the interstitial fluid is the equivalent of the ocean, lake, or tiniest drop of fluid that surrounds and nourishes single-celled organisms.

Because every cell must receive all its requirements for life from the surrounding interstitial fluid, it follows that the composition of this fluid must be kept fairly constant to sustain life. In the long run, nutrients consumed by the cell must be replaced and wastes must be removed or the cell will die.

Relative constancy of the conditions within the internal environment is called **homeostasis** (*homeo-* means "unchanging" or "the same" and *-stasis* means "standing"). The maintenance of homeostasis is so important for life that multicellular organisms, including humn beings, devote a significant portion of their total metabolic activities to it. While small changes in the internal environment do occur from time to time, the activities of cells, tissues, organs, and organ systems are carefully integrated and regulated to keep these changes within acceptable limits.

Health Watch

Suntans, Smoking, and Your Skin

Many of us think suntans are attractive and would like to have one. A suntan indicates we have leisure time to spend basking in the sun, and it gives the skin a healthy-looking "glow." But are suntans really healthy?

The answer is clearly no. Sunlight, including the amount needed to produce a consistent tan, ages skin prematurely and is a risk factor for cancer of the skin. The rays of sunlight penetrate to the dermis and damage its collagen and elastin fibers. Elastin fibers begin to clump together, leading at first to fine wrinkles and later to a wrinkled, leathery skin texture.

Age spots are more common with advancing age, especially in Caucasians who have had excessive exposure to the sun.

Prolonged exposure to sunlight can also damage small blood vessels. Sometimes the vessels remain permanently dilated, leading to a condition called *telangiectases,* or spider veins. Sunlight also damages the keratinocytes and melanocytes in the epidermis. The keratinocytes become rough and thickened and no longer fit together as a smooth interlocking layer. The melanocytes begin to produce melanin unevenly, leading to patches of darker pigmentation known as freckles, age spots, or liver spots.

Perhaps you have heard that ultraviolet (UV) rays combine "good" UVA rays that tan your skin and "bad" UVB rays that cause sunburn. Tanning lotions with high "SPF" (sun protection factor) numbers are designed to block the UVB rays, which at least prevents the acute damage (and pain) of a sunburn. However, even the UVA rays aren't good for you. They penetrate more deeply than the UVB rays and are responsible for most of the long-term effects that age skin prematurely.

What about smoking? Heavy smokers are nearly five times more likely to develop premature wrinkles. Smoking damages and thickens the elastin fibers in the dermis. It also dehydrates keratinocytes in the epidermis, causing the epidermis to develop a rough texture. Finally, smoking narrows blood vessels, reducing the flow of blood to the skin. As a result the skin of smokers heals more slowly from injury than the skin of nonsmokers.

Smoking causes changes in the elastin fibers in the dermis and a loss of water from the cells of the epidermis. The result is a dehydrated, wrinkled skin.

Homeostasis is maintained by negative feedback

In living organisms, homeostasis is maintained by **negative feedback** control systems (Figure 4.12). Negative feedback control systems operate in such a way that deviations from the desired condition are automatically detected and counteracted. A negative feedback control system has the following components:

How does your body maintain homeostasis?

- A **controlled variable.** The focal point of any negative feedback control loop is the *controlled variable.* A controlled variable is any physical or chemical property that might vary from time to time and that must be controlled in order to maintain homeostasis. Examples of controlled variables are blood pressure, body temperature, and the concentration of glucose in blood.
- A **sensor** (or *receptor*). The sensor monitors the current value of the controlled variable and sends the information (via either nerves or hormones) to the control center.
- A **control center.** The control center receives input from the sensor and compares it to the cor-

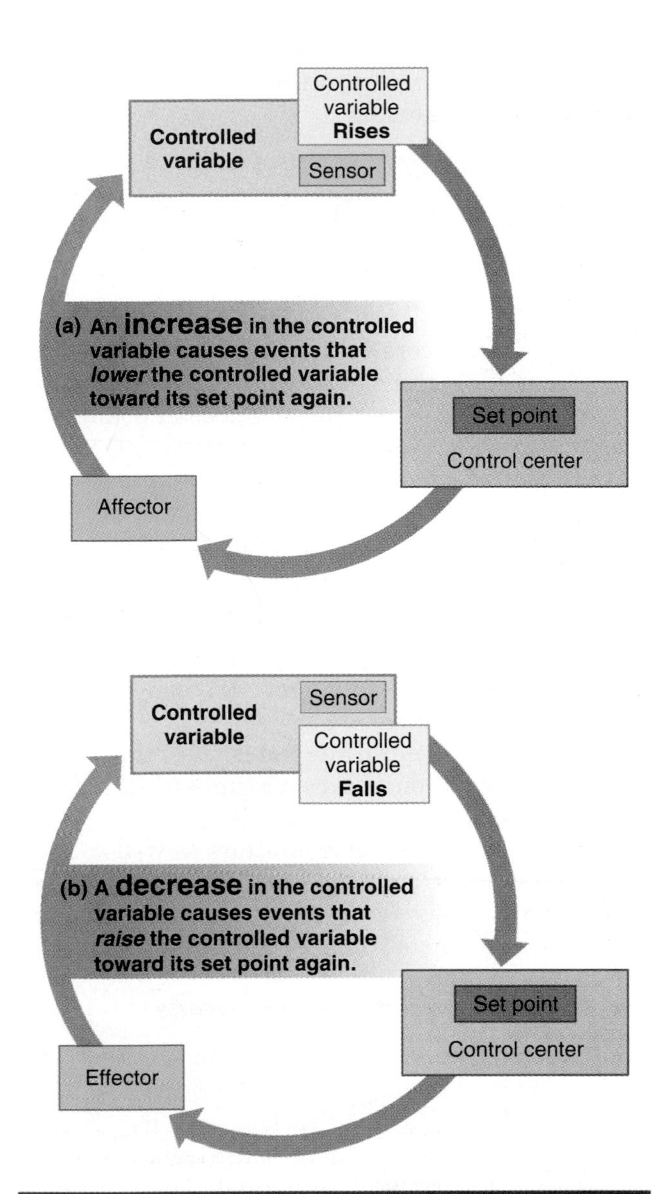

Figure 4.12 Components of a negative feedback control system. The focal point of the control system is the controlled variable. A sensor monitors the controlled variable and sends signals to a control center, which compares the current value of the controlled variable with its set point. If the controlled variable and set point do not match, the control center sends signals to effectors that take action to reverse the difference between the controlled variable and its set point.

rect, internally set value of the controlled variable, sometimes called the **"set point."** When the current value and the set point are not in agreement, the control center sends signals (again, via either nerves or hormones) to an effector.

- An **effector.** The effector takes the necessary action to correct the imbalance, in accordance with the signals it receives from the control center.

The cycle is called *negative* feedback because any change in the controlled variable triggers a series of events that ultimately opposes ("negates") the initial change, returning the variable to its set point. In other words, homeostasis is maintained.

Negative feedback helps maintain core body temperature

A prime example of negative feedback is offered by homeostasis of your body temperature. In this case, multiple organ systems participate in maintaining homeostasis (Figure 4.13).

The controlled variable is your core temperature, meaning the temperature near the center of your body. Temperature sensors in your skin and internal organs monitor core temperature. These sensors transmit signals via nerves to the control center, located in a region of your brain called the hypothalamus. The control center uses different combinations of effector mechanisms to raise or lower core temperature as needed.

When your core temperature falls *below* its set point, the hypothalamus

- Sends more nerve impulses to blood vessels in the skin that cause the vessels to narrow. This restricts blood flow to your skin, reducing heat loss.
- Stimulates your skeletal muscles, causing brief bursts of muscular contraction known as shivering. Shivering generates heat.

Try It Yourself

Negative Feedback Systems

Negative feedback systems generally keep controlled variables within their acceptable range most of the time. Any variations are so small that we usually do not notice them.

You can demonstrate the small, ongoing changes that occur in controlled variables with the following exercise. In this case, the controlled variables are your balance and your ability to control voluntarily the position of your limbs.

First, stand still and look at a distant point. Everything will look stable, as if your body is not moving at all. Now look at the same point through powerful binoculars, and your field of vision appears unsteady.

The reason is that your body is constantly making small muscular adjustments to keep you upright and hold your hands and arms steady. You become aware of these tiny motions only because the binoculars have magnified their effect on your vision.

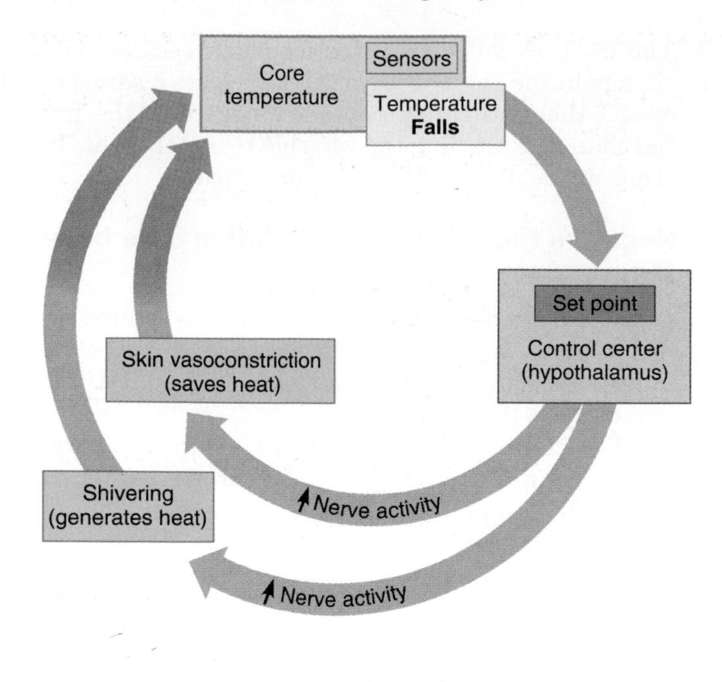

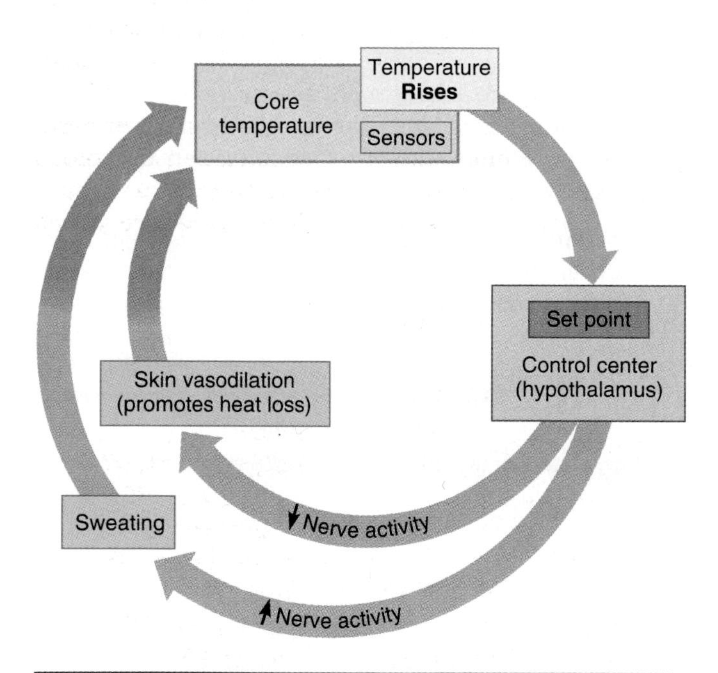

Figure 4.13 Negative feedback control of core temperature. Note that different combinations of effector mechanisms may be activated, depending on the direction of the initial change in core temperature.

When your core temperature rises *above* its set point, the hypothalamus

- Sends fewer nerve impulses to blood vessels in the skin, causing them to dilate. This increases blood flow to your skin and promotes heat loss.
- Activates your sweat glands. As perspiration evaporates from your skin, you lose heat.

Even when your core temperature is normal, your hypothalamus is transmitting some nerve impulses to the blood vessels in your skin. Small changes in temperature, then, can be handled effectively just by increasing or decreasing the normal number of signals. Only when the variations in temperature are large are sweating or shivering called into play.

From this example, we can make the following points about negative feedback control:

1. Many sensors may be active at once. In this case, sensors throughout the body monitor body temperature.
2. The control center integrates all of this incoming information and comes up with an appropriate response.
3. There can be multiple effectors as well as multiple sensors, and they may belong to different organ systems. Both skin and muscles function in returning body temperature to its set point.

Positive feedback amplifies events

Positive feedback control systems are relatively uncommon in living organisms. In positive feedback, a change in the controlled variable sets in motion a series of events that *amplify* the original change, rather than returning it to normal. The process of childbirth once labor has started is governed by positive feedback mechanisms.

Obviously something must terminate positive feedback events. The contractions of childbirth end when the child is born. The important point is that positive feedback is not a mechanism for maintaining homeostasis.

Recap **All multicellular organisms must maintain homeostasis of their internal environment. In a negative feedback control system, any change in a controlled variable sets in motion a series of events that reverse the change, maintaining homeostasis.**

Chapter Summary

Tissues are groups of cells with a common function

- The four types of tissues are epithelial tissue, connective tissue, muscle tissue, and neural tissue.

Epithelial tissues cover body surfaces and cavities

- Epithelial tissues are sheets of cells that cover or line body surfaces and form the glands.
- Epithelial tissues are supported by a noncellular layer called the basement membrane.

Connective tissue supports and connects body parts

- Fibrous connective tissues contain several types of extracellular fibers and only a few living cells. They support and connect body parts.
- Cartilage, blood, bone, and adipose tissue are classified as special connective tissues.

Muscle tissues contract to produce movement

- Muscle tissue is composed of either skeletal, cardiac, or smooth muscle cells.
- Skeletal muscles are attached to bones by tendons.

Nervous tissue transmits impulses

- Neurons are specialized for conduction of electrical impulses.
- Glial cells surround and protect neurons and supply them with nutrients.

Organs and organ systems perform complex functions

- The human body is comprised of eleven organ systems, each of which has at least one broad function.
- Membranes consisting of layers of epithelial and connective tissues line the body cavities.
- Positions of body parts are described on three planes: midsagittal, frontal, and transverse.

The skin as an organ system

- The skin functions as a protective barrier, participates in the maintenance of homeostasis, and provides us with sensory information about the external environment.
- Skin has two layers: an outer epithelial layer called the epidermis and an inner connective tissue layer called the dermis.
- Skin also contains nerves, blood vessels, glands, hair follicles, and smooth muscle.

Multicellular organisms must maintain homeostasis

- In a multicellular organism, the external environment of every cell is the internal environment of the organism.
- Relative constancy of the internal environment is called homeostasis.
- Homeostasis is maintained by negative feedback control systems.
- In a negative feedback control system, a change in the controlled variable sets in motion a sequence of events that tends to reverse (or negate) the initial change.
- In the regulation of body temperature, sensors located throughout the body send information about temperature to the control center, located in the hypothalamus of the brain.
- Possible responses to a change in body temperature include dilating or constricting the blood vessels to the skin, shivering (if temperature is too low), and sweating (if temperature is too high).

Terms You Should Know

basement membrane, 82
cell junctions, 82
connective tissue, 83
controlled variable, 98
dermis, 95
endocrine gland, 80
epidermis, 95
epithelial tissue, 80
exocrine gland, 80
homeostasis, 97
internal environment, 97
muscle tissue, 86
negative feedback, 98
nervous tissue, 90
neuron, 90
organ system, 90
set point, 99

Test Yourself

Matching Exercise

_____ 1. Relative constancy of the internal environment

_____ 2. Epithelial tissue composed of flattened cells

_____ 3. Epithelial tissue composed of more than one layer of cells

_____ 4. Layer of extracellular protein immediately beneath an epithelial tissue

_____ 5. Watertight junction between two epithelial cells
_____ 6. Junction between two epithelial cells that permits the cytoplasmic exchange of ions and water
_____ 7. Specialized connective tissue that forms the shape of the nose
_____ 8. Type of muscle found in the walls of the bladder and blood vessels
_____ 9. Pigment-producing cells in the skin
_____10. A method for generating heat by the body

a. stratified epithelium
b. tight junction
c. cartilage
d. homeostasis
e. shivering
f. squamous epithelium
g. melanocytes
h. smooth
i. gap junction
j. basement membrane

Fill in the Blank

11. The external environment of every cell is the ______________________ of the body.
12. Organized groups of cells with similar types of functions are called ____________.
13. A gland that secretes its product into the bloodstream is called an ______________ gland.
14. Connective tissue cells called ______________________ produce the fibrous protein that comprises collagen fibers.
15. The ______________ is the outer layer of the skin.
16. The cavity of the body that contains the brain is called the ____________ cavity.
17. The ______________ plane divides a structure into left and right segments.
18. Cells of the epidermis that produce the tough waterproof protein called keratin are called ______________.
19. In a feedback loop, the ______________________ integrates incoming information from sensors and sends the appropriate signals to the effectors.
20. The desired value of a controlled variable is called its _____ __________.

True or False?

21. Blood cells are considered to be a specialized connective tissue because they originate from cells in bone.
22. When you diet, fat cells die and disappear.
23. Cardiac muscle is found only in the heart.
24. Both the epidermis and the dermis of the skin have blood vessels.
25. The controlled variable in terms of temperature is core temperature, not skin temperature.

Concept Review

1. Describe some advantages and disadvantages of multicellularity.
2. List the four types of tissues in the human body.
3. Compare/contrast endocrine and exocrine glands.
4. Distinguish between an organ and an organ system.
5. List the 11 organ systems of the body and give at least one function of each.
6. Define interstitial fluid.
7. Name the two cavities of the anterior body cavity that are separated from each other by the diaphragm.
8. Compare/contrast positive and negative feedback.
9. Define set point.
10. Describe the function of a control center in a negative feedback control system.

Apply What You Know

1. Your roommate says that the concept of homeostasis is being violated when the rate of respiration goes up during exercise, because the rate of respiration clearly is not being held constant. Explain to him where his thinking is faulty.
2. What do you think would be some of the problems associated with severe third-degree burns, in which both the epidermis and the dermis of the skin are severely damaged or destroyed?

Case in Point

Producing Skin Substitutes

Despite intensive research effort we still do not have a readily available substitute for human skin. Nevertheless, some progress is being made. Several small commercial companies now grow layers of skinlike tissue from human keratinocytes (the primary cell type in the epidermis), using keratinocytes taken from the patients themselves. Skin substitutes grown from a patient's own cells eliminate potential problems of tissue rejection, but it takes 16–21 days to grow the tissue. In the meantime, the damaged area can be covered with several layers of synthetic materials that protect the patient from dehydration and infection while encouraging the regeneration of the patient's own skin. However, the synthetic materials are not biodegradable and so they must eventually be removed.

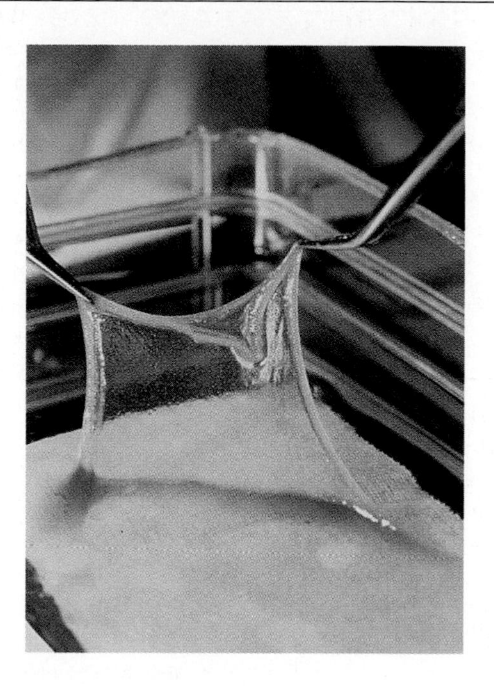

An ideal skin substitute would be available when the patient needed it, would not need to be removed once put in place, would mimic the structure and function of natural human skin, and would be permanent. Several skin substitute products (Apligraf and Dermagraft) that fulfill the first two of these criteria are now commercially available for certain uses, including treatment of skin ulcers. These products use cultured human cells derived from the foreskins of newborn infants. Although the current generation of skin-substitute products do not need to be removed, they still tend to be absorbed or rejected by the patient because they are not derived from the patient's own tissue. Look for continued improvements in these and newer generations of skin substitutes to come.

QUESTIONS:

1. From what you know of the structure and function of skin, what do you think might be some of the problems associated with the development of artificial skin products?
2. In the case study, the statement was made that problems of tissue rejection are eliminated when the patient's own cells are used to grow the tissue. What is meant by "tissue rejection," and which organ system of the body do you think is responsible for it?

INVESTIGATE:

Morgan, Jeffrey R. and Martin L. Yarmush. Bioengineered Skin Substitutes. *Science and Medicine* 4(4):6, 1997.

References and Additional Reading

Diamond, Jared. Best Size and Number of Human Body Parts. *Natural History* 6:78, 1994. (Why do we have the size and number of organs that we do?)

Gordon, Marsha and Alice E. Fugate. *The Complete Idiot's Guide to Beautiful Skin*. New York: Alpha Books, 1998. (A practical guide to skin care.)

Maynard-Moody, Steven. *The Dilemma of the Fetus: Fetal Research, Medical Progress, and Moral Politics.* New York: St. Martin's Press, 1995. (A look at the moral politics of fetal tissue research with a focus on the importance of social and political conflict over science.)

The Promise of Tissue Engineering. *Scientific American* 280(4):59, 1999. (A series of five articles on the subject of tissue engineering.)

CHAPTER 5

THE SKELETAL SYSTEM

What are bones made of?

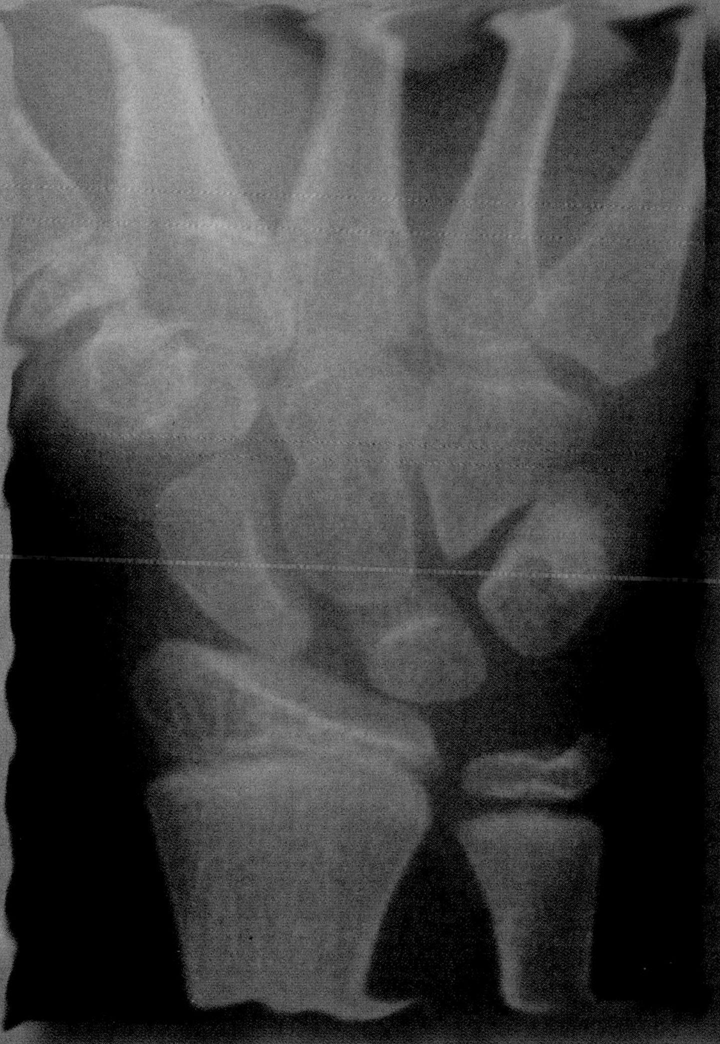

What are the functions of the skeleton?

Does regular exercise affect bones?

How do joints resist wear and tear?

THE HUMAN BODY IS CAPABLE OF AN AWESOME ARRAY OF physical activities. With training, some individuals can run a mile in less than four minutes or lift more than their own weight. Exquisitely sensitive motor skills allow us to thread a needle, turn our head to focus on a single star, and throw a baseball into the strike zone. Considered individually, any one of these activities might not seem very amazing, but for a single structure (the human body) to be capable of all of them is remarkable indeed. From an engineering standpoint it would be like designing a bulldozer that is strong enough to flatten a building, yet delicate enough to pick up a dime.

This chapter describes the skeletal system, the organ system for support, protection, and movement. We examine the structure of bone, how bones develop, and how they remodel and repair themselves. We review how the bones fit together to comprise the skeleton. We take a look at how joints enable bones and muscles to work together. Finally, we consider what can go wrong with the skeletal system.

5.1 The skeletal system consists of connective tissue

The skeletal system comprises three types of connective tissue:

- *Bones,* comprising the hard elements of the skeleton
- *Ligaments,* consisting of dense fibrous connective tissue that binds the bones to each other
- *Cartilage,* a specialized connective tissue consisting primarily of fibers of collagen and elastin in a gel-like fluid called ground substance

Bones are the hard elements of the skeleton

Spotlight on Structure & Function

Although most of the mass of **bones** consists of nonliving extracellular crystals of calcium minerals, bones also contain several types of living bone cells, nerves, and blood vessels. Indeed, bones bleed when cut during orthopedic surgery or when they break.

Bones perform five important functions. The first three are the same as the functions of the skeleton, which consists primarily of bone:

- Support: Bones form the structure (the skeleton) to which skeletal muscles attach. The skeleton also supports the soft organs.
- Movement: Bones support and interact with muscles, making it possible for our bodies to move.
- Protection: As the hard elements of the skeleton, bones surround and protect many of our delicate internal organs.
- Formation of blood cells: Some bones contain cells that are responsible for producing different types of blood cells.
- Mineral storage: Bones store minerals, including calcium and phosphate, that are important to body metabolism and function. ■

Recap **Bones contribute to support, movement, and protection. They also produce the blood cells and store minerals.**

Bone contains living cells

A typical long bone, so called because it is longer than it is wide, consists of a cylindrical shaft (called the *diaphysis*), with an enlarged knob called an *epiphysis* at each end (Figure 5.1). Dense **compact bone** forms the shaft and covers each end. A central cavity in the shaft is filled with *yellow bone marrow*. Yellow bone marrow is primarily fat that can be utilized for energy. Inside each epiphysis is **spongy bone.** Spongy bone is less dense than compact bone, allowing the bones to be light but strong. Spongy bone is a latticework of hard, relatively strong *trabeculae* (from Latin, meaning "little beams") composed of calcium minerals and living cells. In certain long bones, most notably the long bones of the upper arms and legs (humerus and femur, respectively), the spaces between the trabeculae are filled with *red bone marrow*. Special cells called *stem cells* in the red bone marrow are responsible for the production of red and white blood cells and platelets.

The outer surface of the bone is covered by a tough layer of connective tissue, the *periosteum,* which contains specialized bone-forming cells. If the end of a long bone forms a moveable joint with another bone, the joint surface is covered by a smooth layer of cartilage that reduces friction.

Taking a closer look at bone, we see that compact bone is comprised largely of extracellular deposits of calcium phosphate enclosing and surrounding living cells called **osteocytes** (from the Greek words for "bone" + "cells"). Osteocytes are arranged in rings in cylindrical structures called **osteons** (sometimes called *Haversian systems*). As bone develops and becomes hard, the osteocytes become trapped in hollow chambers called *lacunae*. However, the osteocytes remain in contact with each other via thin canals called *canaliculi*. Within the canaliculi, extensions of the cell cytoplasm and the presence of gap junctions between adjacent cells keeps the osteocytes in direct contact with each

What are bones made of?

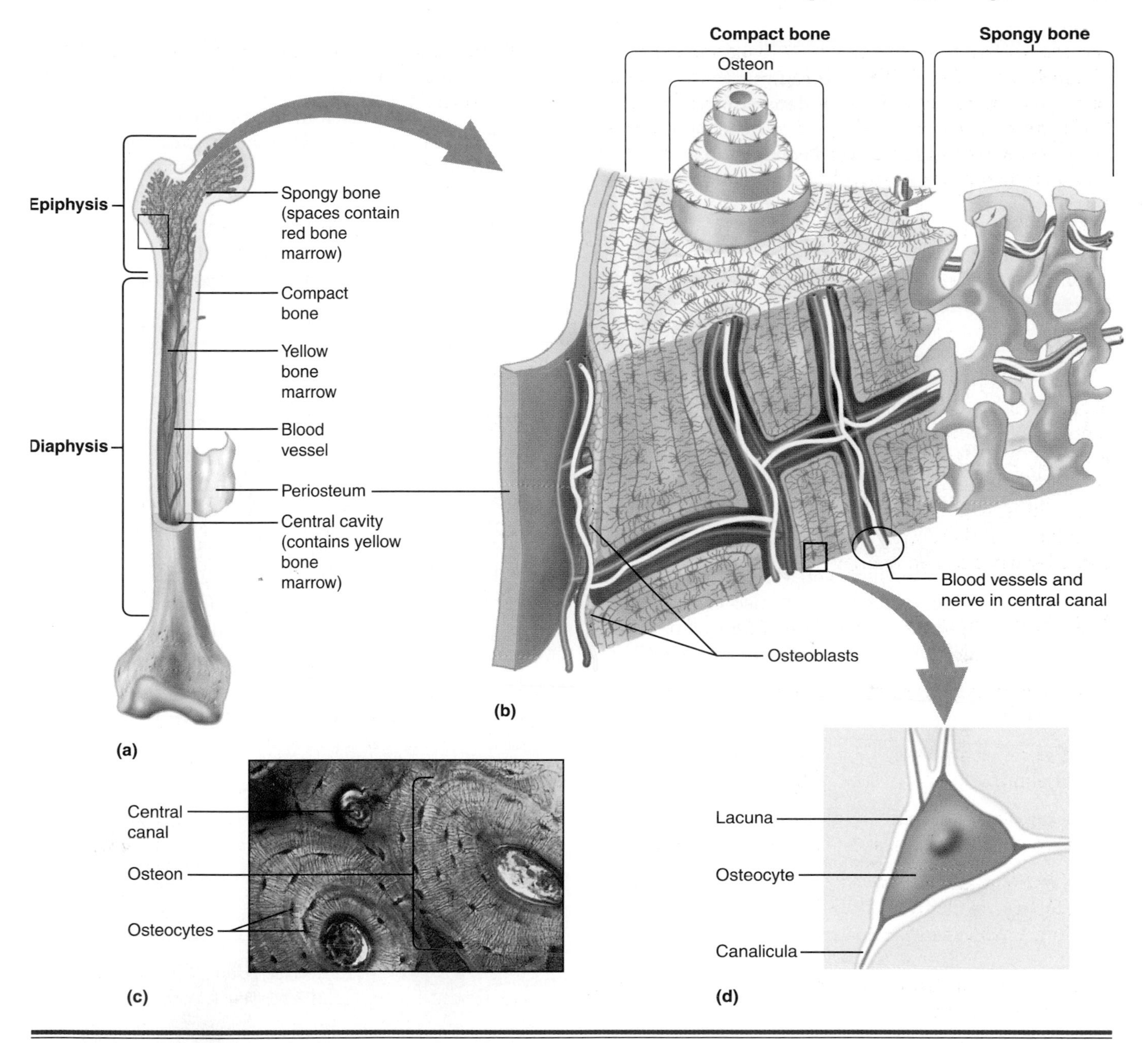

Figure 5.1 Structure of bone. (a) A partial cut through a long bone. **(b)** A closer view of a section of bone shows that compact bone is a nearly solid structure with central canals for the blood vessels and nerves. **(c)** A photograph of an osteon of compact bone showing osteocytes embedded within the solid structure. **(d)** A single osteocyte in a lacuna. Osteocytes remain in contact with each other by cytoplasmic extensions that extend into the canaliculi between cells.

other. Osteocytes nearest the center of an osteon receive nutrients by diffusion from blood vessels that pass through a **central** (Haversian) **canal.** These cells then pass nutrients on to adjacent cells via the gap junctions. In this way, all the osteocytes can be supplied with nutrients even though most osteocytes are not located near a blood vessel. Waste products produced by the osteocytes diffuse in the opposite direction and are removed from the bone by the blood vessels.

In spongy bone, osteocytes do not need to rely on central canals for nutrients and waste removal. The slender trabecular structure of spongy bone gives each osteocyte access to nearby blood vessels in red bone marrow.

Recap **Bone may be compact or spongy in appearance. Long bones have a hollow shaft of compact bone filled with yellow marrow; spongy bone with red bone marrow is found in the epiphyses.**

Ligaments hold bones together

Spotlight on Structure & Function

Ligaments attach bone to bone. Ligaments consist of dense fibrous connective tissue, meaning that they are a regular array of closely packed collagen fibers all oriented in the same direction with a few fibroblasts in between. Ligaments confer strength to certain joints while still permitting movement of the bones in relation to each other. ■

Cartilage lends support

Cartilage, as you already know, contains fibers of collagen and/or elastin in a ground substance of water and other substances. Cartilage is smoother and more flexible than bone. Cartilage is found where support under pressure is important and where some movement is necessary.

There are three types of cartilage in the human skeleton. *Fibrocartilage* consists primarily of collagen fibers arranged in thick bundles. It withstands both pressure and tension well. The intervertebral disks between the vertebrae, and also certain disk-like supportive structures in the knee joint called menisci, are made of fibrocartilage. *Hyaline cartilage* is a smooth, almost glassy cartilage of thin collagen fibers. Hyaline cartilage forms the embryonic structures that later become the bones. It also covers the ends of mature bones in joints, creating a smooth, low-friction surface. *Elastic cartilage* is mostly elastin fibers, and so it is highly flexible. It lends structure to the outer ear and to the epiglottis, a flap of tissue that covers the larynx during swallowing.

5.2 Bone development begins in the embryo

In the earliest stages of fetal development, even before organs develop, the rudimentary models of the future bones are created out of hyaline cartilage by cartilage-forming cells called **chondroblasts.** At about two to three months of fetal development the cartilage models begin to dissolve and are replaced by bone. This process is called *ossification*. Although ossification is slightly different for flat bones and long bones, we will concentrate on the process for long bones.

Figure 5.2 illustrates how ossification occurs in a long bone. First, chondroblasts die and the matrix they produced gradually breaks down inside the future shaft of the bone, making room for blood vessels

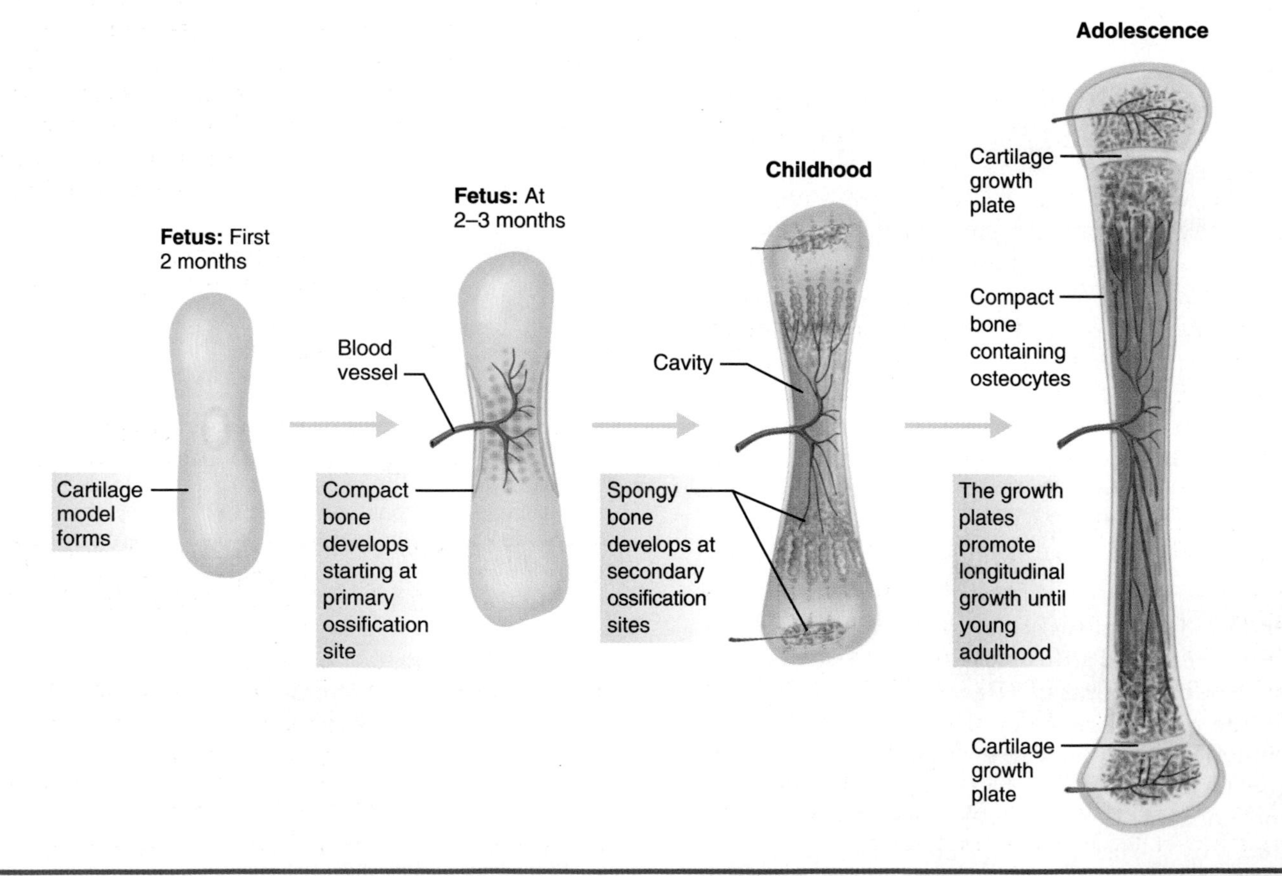

Figure 5.2 How bone develops. The first two phases occur in the fetus. Bones continue to grow longer throughout childhood and adolescence because of growth at the growth plates.

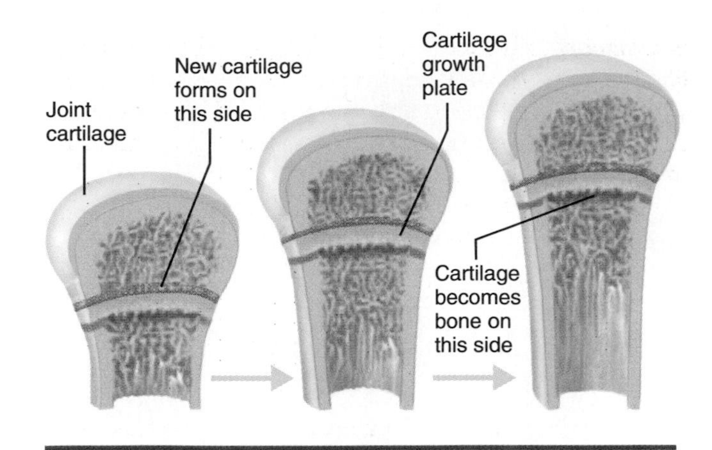

Figure 5.3 How long bones increase in length. Growth is concentrated at the growth plate. Chondrocytes produce new cartilage at the outer surface and cartilage is being converted to bone at the inner surface.

to develop. The blood vessels carry bone-forming cells called **osteoblasts** (from the Greek words for "bone" and "to build") into the area from the developing periosteum. The osteoblasts secrete a mixture of proteins (including collagen) called *osteoid,* which forms internal structure and provides strength to bone. They also secrete enzymes that facilitate the crystallization of hard mineral salts of calcium phosphate, known as *hydroxyapatite,* around and between the osteoid matrix. As more and more hydroxyapatite is deposited, the osteoblasts become embedded in the hardening bone tissue. In mature compact bone, approximately one-third of the matrix is osteoid and two-thirds is crystals of hydroxyapatite.

Eventually the rate at which osteoblasts produce the osteoid matrix and stimulate the mineral deposits declines, and they become the mature osteocytes embedded in their individual lacunae. Mature osteocytes continue to maintain the bone matrix, however. Without them the matrix would slowly disintegrate.

Bones continue to lengthen throughout childhood and adolescence. This is because a narrow strip of cartilage called the **growth plate** (or *epiphyseal plate*) remains in each epiphysis. Chondroblast activity (and hence the development of new cartilage as a model for the lengthening bone) is concentrated on the outside of the plate, whereas the conversion of the cartilage model to bone by osteoblasts is concentrated on the inside of the plate (Figure 5.3). In effect, the bone lengthens as the two growth plates migrate farther and farther apart. Bones also grow in width as osteoblasts lay down more bone on the outer surface just below the periosteum.

The bone development process is controlled by hormones, chemicals secreted by the endocrine glands. The most important hormone in preadolescents is growth hormone, which stimulates the bone-lengthening activity of the growth plate. During puberty the sex hormones (testosterone and estrogen) also stimulate the growth plate, at least initially. But at about age 18 in females and 21 in males these same sex hormones signal the growth plates to stop growing, and the cartilage is replaced by bone tissue. At this point the bones can no longer lengthen, though they can continue to grow in width.

Recap **Bone-forming cells called osteoblasts produce a protein mixture (including collagen) that becomes bone's structural framework. They also secrete an enzyme that facilitates mineral deposition within the protein matrix.**

5.3 Mature bone undergoes remodeling and repair

Even though bones stop growing longer, this does not mean they remain the same throughout life. Bone is a dynamic tissue that undergoes constant replacement, remodeling, and repair. Remodeling may be so extensive that there is a noticeable change in bone shape over time, even in adults.

Bone remodeling and repair is in part due to a third type of bone cell called an **osteoclast** (from the Greek words for "bone" + "to break"). Osteoclasts cut through mature bone tissue, dissolving the hydroxyapatite and digesting the osteoid matrix in their path. The released calcium and phosphate ions enter the blood. The areas from which bone has been removed attract new osteoblasts, which lay down new osteoid matrixes and stimulate the deposition of new hydroxyapatite crystals.

Table 5.1 summarizes the four types of cells that contribute to bone development and maintenance.

Bones can change in shape, size, and strength

Over time, constant remodeling can actually change the shape of a bone. The key is that compression

Table 5.1 Cells involved in bone development and maintenance

Type of Cell	Function
Chondroblasts	Cartilage-forming cells that build a model of the future bone
Osteoblasts	Young bone-forming cells that cause the hard extracellular matrix of bone to develop
Osteocytes	Mature bone cells that maintain the structure of bone
Osteoclasts	Bone-dissolving cells

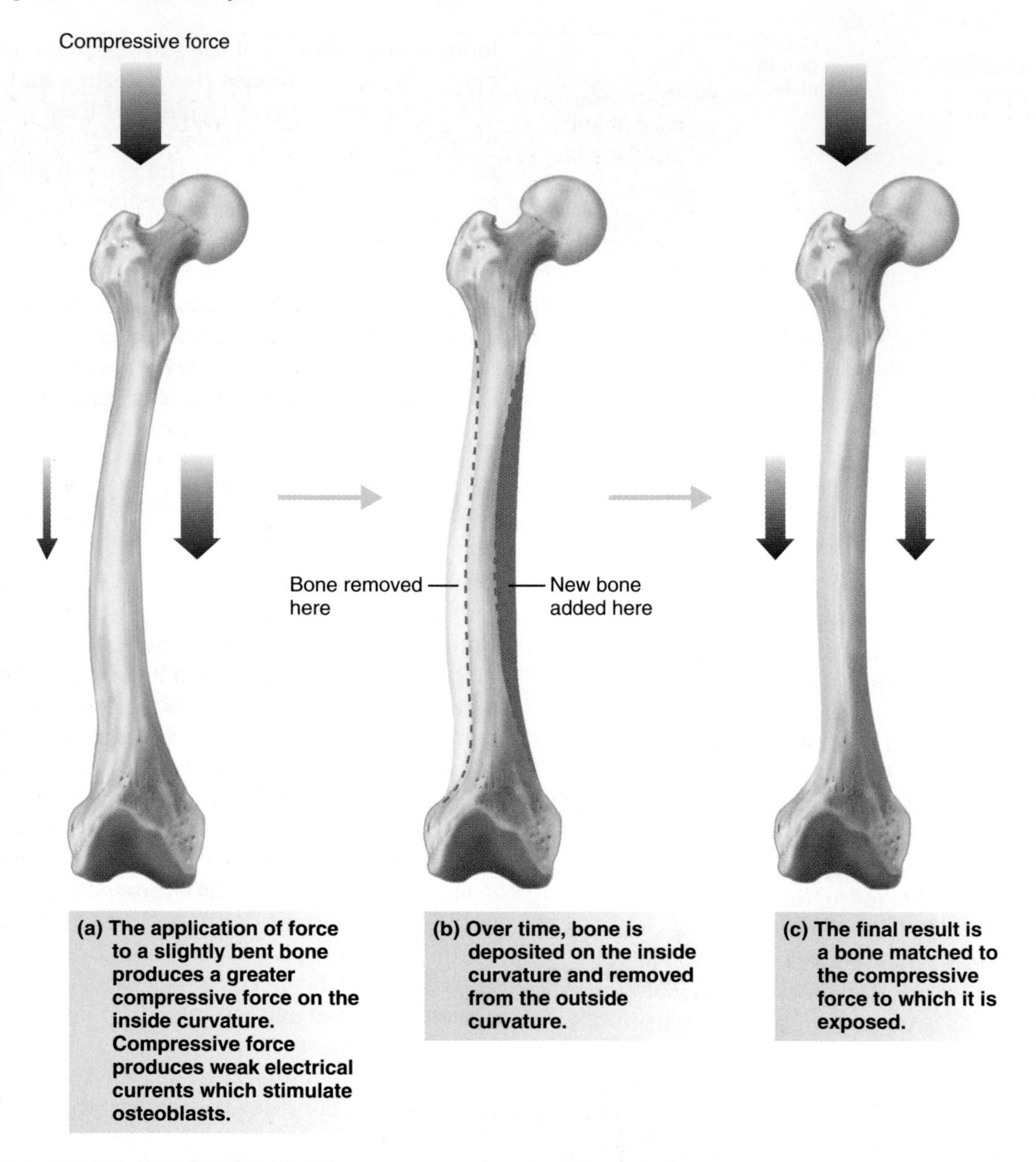

Figure 5.4 Bone remodeling.

stress on a bone, such as the force of repeated jogging on the legs, causes tiny electrical currents within the bone. These electrical currents stimulate the bone-forming activity of osteoblasts. The compressive forces and the electric currents are greatest at the inside curvature of the long bone undergoing stress (Figure 5.4). Thus, in the normal course of bone turnover, new bone is laid down in regions under high compressive stress and bone is reabsorbed in areas of low compressive stress. The final shape of a bone, then, tends to match the compressive forces to which it is exposed.

Weight-bearing exercise increases overall bone mass and strength. The effect is pronounced enough that the bones of trained athletes may be visibly thicker and heavier than those of nonathletes. You don't have to be a professional athlete to notice this benefit, however. If you begin a regular program of any weight-bearing exercise, such as jogging or weight lifting, your bones will become denser and stronger as your osteoblasts produce more bone tissue.

Does regular exercise affect bones?

The maintenance of homeostasis of bone structure depends on the precise balance of the activities of osteoclasts and osteoblasts. **Osteoporosis** is a common bone condition in which bones lose a great deal of mass (seemingly becoming "porous") due to an imbalance over many years in the rates of activities of these two types of bone cells.

Health Watch

Osteoporosis

Osteoporosis is a condition caused by excessive bone loss over time, leading to brittle, easily broken bones. Symptoms include hunched posture, difficulty walking, and an increased likelihood of bone fractures, especially of the vertebral column, hip, and forearm. Osteoporosis is a major health problem in the U.S., accounting for approximately 300,000 debilitating hip fractures every year.

Progressive bone loss can occur if even a slight imbalance develops between the rates of bone breakdown by osteoclasts and bone formation by osteoblasts. Over decades, this can cause a significant reduction in bone density and mass. For example, if you lost only 0.5% of your bone mass per year, in thirty years this would add up to about 15%. This is roughly what occurs during normal aging and would not be enough to produce osteoporosis. However, if you were to lose double that amount per year, you would lose almost a third of your bone mass over the same time. This is enough to cause bones to fracture easily, and the outward signs of osteoporosis would become apparent. This illustrates why it is so important that homeostatic regulatory systems stay in nearly exact balance throughout life.

Older women have an especially high risk of developing osteoporosis due to the normal decline in estrogen after menopause. Although the condition is most common in older women, it can affect older men as well. Other risk factors include a sedentary lifestyle, low calcium intake, and being underweight.

What causes osteoporosis? It's not just a decline in the formation of new bone by osteoblasts. Recent studies show that people who lose the most bone mass actually have an increase in the activity of both osteoblasts and osteoclasts and an increased overall rate of bone turnover. This finding has led to a promising new test for a person's risk of the condition. The test takes advantage of the fact that certain collagen molecules in bone can be detected in blood once the bone tissue has been broken down by osteoclasts. The only other way to diagnose osteoporosis is by methods that determine bone density, but these tests are only useful after bone loss has already occurred.

The good news is that osteoporosis can be prevented. There are two important strategies: get enough calcium and vitamin D, and maintain a consistent exercise program throughout your life. Calcium is crucial for the formation of new bone tissue. Current recommendations call for a daily intake of about 1,000–1,500 mg per day for adults. Women who have gone through menopause may benefit from additional calcium beyond the normal daily intake. Both men and women can benefit from weight-bearing exercise such as walking because regular weight-bearing exercise increases bone mass.

In addition, some physicians recommend estrogen hormones for postmenopausal women. Estrogen replacement therapy may carry other health risks, but many doctors believe its benefits outweigh the drawbacks. Newer treatments, such as the drug *alendronate* to inhibit the activity of osteoclasts or the hormone *calcitonin* administered as a nasal spray to stimulate the activity of osteoblasts, may be able to increase bone density with fewer risks.

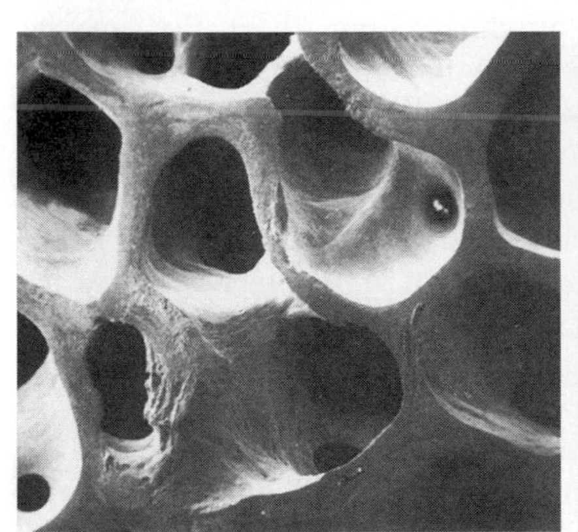

(a) SEM of spongy bone of a normal 44-year-old man.

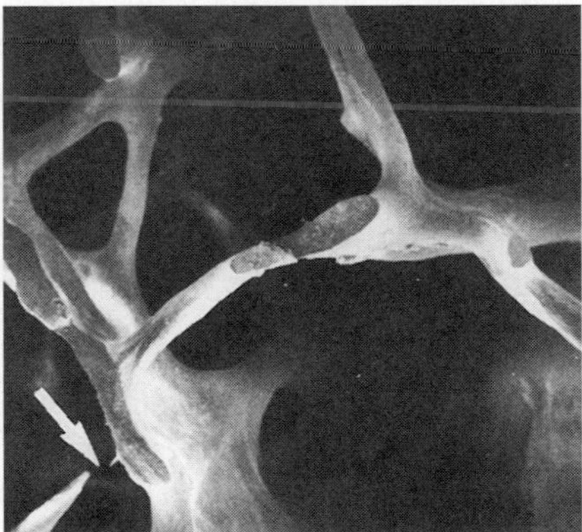

(b) SEM of spongy bone of a 47-year-old woman with osteoporosis.

(c) Osteoporosis can lead to repeated compression fractures of the spine and a permanent change in spine curvature.

Osteoporosis.

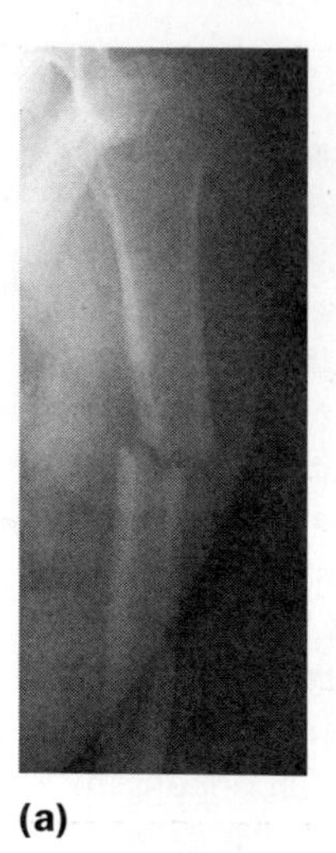

(a)

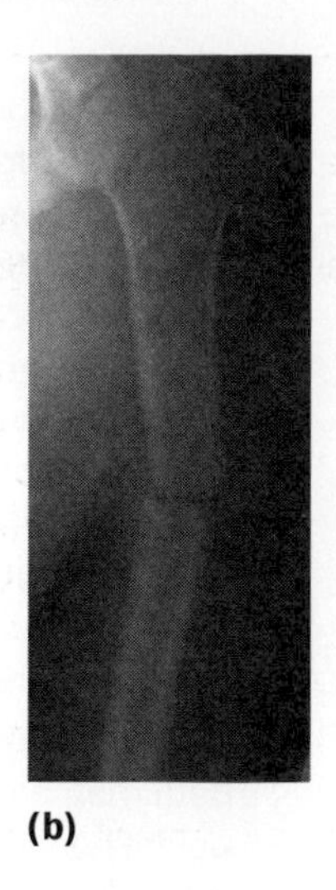

(b)

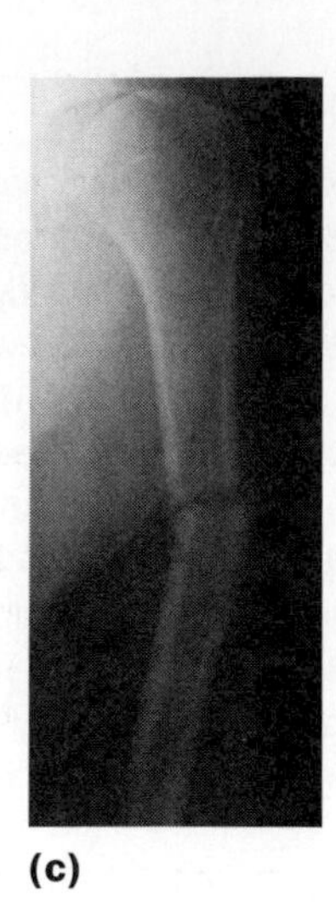

(c)

Figure 5.5 Stages of bone repair. **(a)** An x-ray of the broken left humerus of a 23-year-old male. **(b)** The same bone at 7 weeks. A prominant callus surrounds the break. **(c)** Thirteen weeks. Although healing is not yet complete, the callus is regressing and healing is underway.

Bone cells are regulated by hormones

Like bone growth, the rates of activities of osteoblasts and osteoclasts in adulthood are regulated by hormones that function to maintain calcium homeostasis. When blood levels of calcium fall below a given point, *parathyroid hormone (PTH)* stimulates the osteoclasts to secrete more bone-dissolving enzymes. The increased activity of osteoclasts causes more bone to be dissolved, releasing calcium and phosphate into the bloodstream. If calcium levels rise, then another hormone called *calcitonin* stimulates osteoblast activity, causing calcium and phosphate to be removed from blood and deposited in bone. Although the total bone mass of young adults doesn't change much, it's estimated that almost 10% of their bones may be remodeled and replaced each year. We discuss more about these and other types of hormonal regulation in Chapter 12 (The Endocrine System).

Bones undergo repair

When you break (fracture) a bone, the blood vessels supplying the bone bleed into the area, producing a mass of clotted blood called a *hematoma*. Inflammation, swelling, and pain generally accompany the hematoma in the days immediately after a fracture. The repair process begins within days as fibroblasts migrate to the area. Some of the fibroblasts become chondroblasts, and together they produce a tough fibrocartilage bond between the two broken ends of the bone called a *callus* (Figure 5.5). A callus can be felt as a hard raised ring at the point of the break. Then osteoclasts arrive and begin to remove dead fragments of the original bone and the blood cells of the hematoma. Finally, osteoblasts arrive to deposit osteoid matrix and encourage the crystallization of calcium phosphate minerals, converting the callus into bone. Eventually the temporary union becomes dense and hard again. Bones rarely break in the same place twice because the repaired union remains slightly thicker than the original bone.

The repair process can take weeks to months, depending on your age and the bone involved. In general, the repair process slows with age.

Recap **Healthy bone replacement and remodeling depend on the balance of activities of bone-resorbing osteoclasts and bone-forming osteoblasts. When a bone breaks, a fibrocartilage callus forms between the broken ends of a bone that is later replaced with bone.**

5.4 The skeleton protects, supports, and permits movement

Now that we have reviewed the dynamic nature of bone tissue, we turn to how all of those bones are classified and organized. Bones can be classified into four types based on shape: long, short, flat, and irregular. So far we have discussed *long bones,* which include the bones of the limbs and fingers. *Short bones* (the bones of the wrists), are approximately as wide as they are long. *Flat bones* (including the cranial bones and the sternum), are thin and flattened, with only a small amount of spongy bone sandwiched between two layers of compact bone. *Irregular bones* such as the coxal (hip) bones and the vertebrae include a variety of shapes that don't fit into the other categories. A few flat and irregular bones, including the sternum and the hip bones, contain red bone marrow that produces blood cells.

What are the functions of the skeleton?

The 206 bones of the human body and the various connective tissues that hold them together comprise the **skeleton** (Figure 5.6). The skeleton has three important functions. First, it serves as a structural framework for support of the soft organs. Second, it protects certain organs from physical injury. The brain, for example, is enclosed within the bones of the skull, and the heart and lungs are

Directions in Science

Helping Bones Heal Faster

Bones heal slowly, at least compared to some tissues. Worse yet, sometimes they fail to heal at all on their own. Is there anything that can be done to initiate the process or speed it up a bit?

As discussed, we know two facts: (1) weight-bearing exercise results in bigger bones, and (2) bone compression produces weak electrical currents in bone. Based on these facts, scientists hypothesized that weak electrical currents generated during exercise might stimulate osteoblasts and osteoclasts, which in turn might *cause* the increase in bone mass observed after long-term exercise training. Of course, just because two events appear together does not prove that one causes the other. Nevertheless, scientists developed a simple experiment to test a related hypothesis: "If weak electrical currents stimulate osteoblasts and osteoclasts, then the application of electrical current to a healing bone might speed the healing process."

Experiments have shown that weak electrical currents do indeed speed the rate of bone healing. As a result, weak electrical current is now sometimes used clinically as a bone-healing treatment.

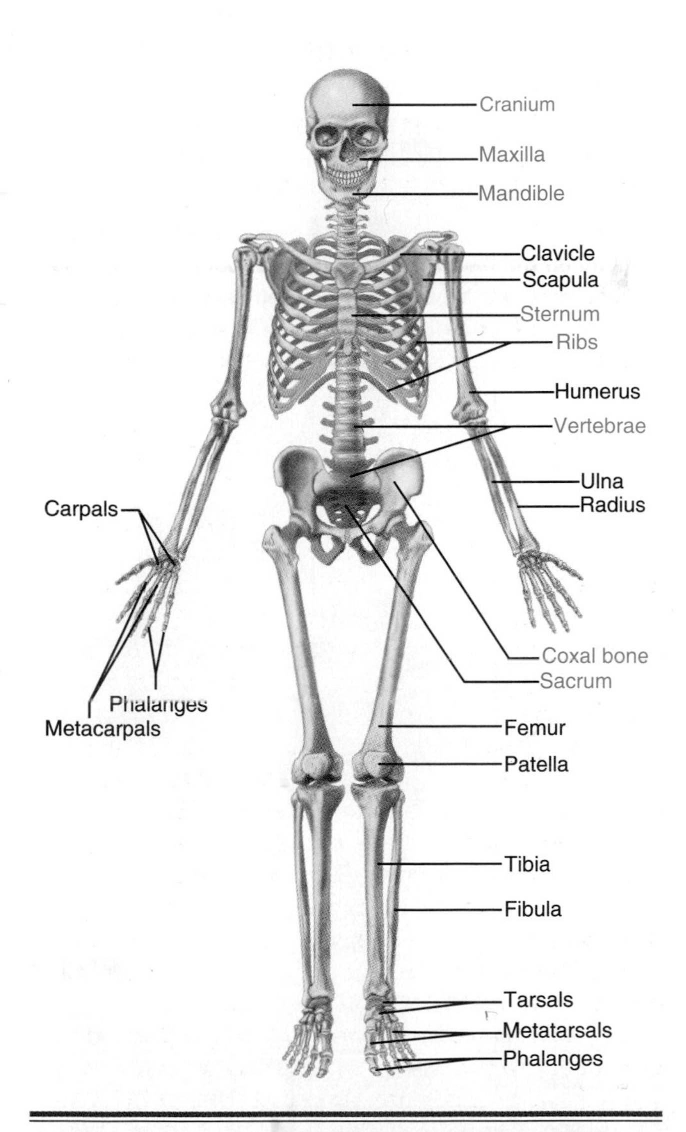

Figure 5.6. The human skeleton. Bones of the axial division are indicated by blue labels and bones of the appendicular skeleton are indicated by black labels.

protected by a bony cage comprised of ribs, the sternum, and vertebrae. Third, because of the way that the bony elements of the skeleton are joined together at joints, the presence of the skeleton permits flexible movement of most parts of the body. This is particularly true of the hands, feet, legs, and arms.

The skeleton is organized into the *axial skeleton* and the *appendicular skeleton.*

Axial skeleton forms the midline of the body

The **axial skeleton** consists of the skull, vertebral column, ribs, and sternum.

The skull: Cranial and facial bones The human **skull** comprises over two dozen bones that protect the brain and form the structure of the face. Figure 5.7 illustrates some of the more important bones of the skull.

The *cranial bones* are flat bones in the skull that enclose and protect the brain. Starting at the front of the skull, the *frontal bone* comprises the forehead and the upper ridges of the eye sockets. At the upper left and right sides of the skull are the two *parietal bones,* and forming the lower left and right sides are the two *temporal bones*. Each temporal bone is pierced by an opening into the ear canal that allows sounds to travel to the eardrum. Between the frontal bone and the temporal bones is the *sphenoid* bone, which forms the back of both eye sockets. The *ethmoid* bone contributes to the eye sockets and also helps support the nose.

Curving underneath to form the back and base of the skull is the *occipital bone*. Near the base of the occipital bone is a large opening called the *foramen magnum* (Latin for "great opening"). This is where the vertebral column connects to the skull and the spinal cord enters the skull to communicate with the brain.

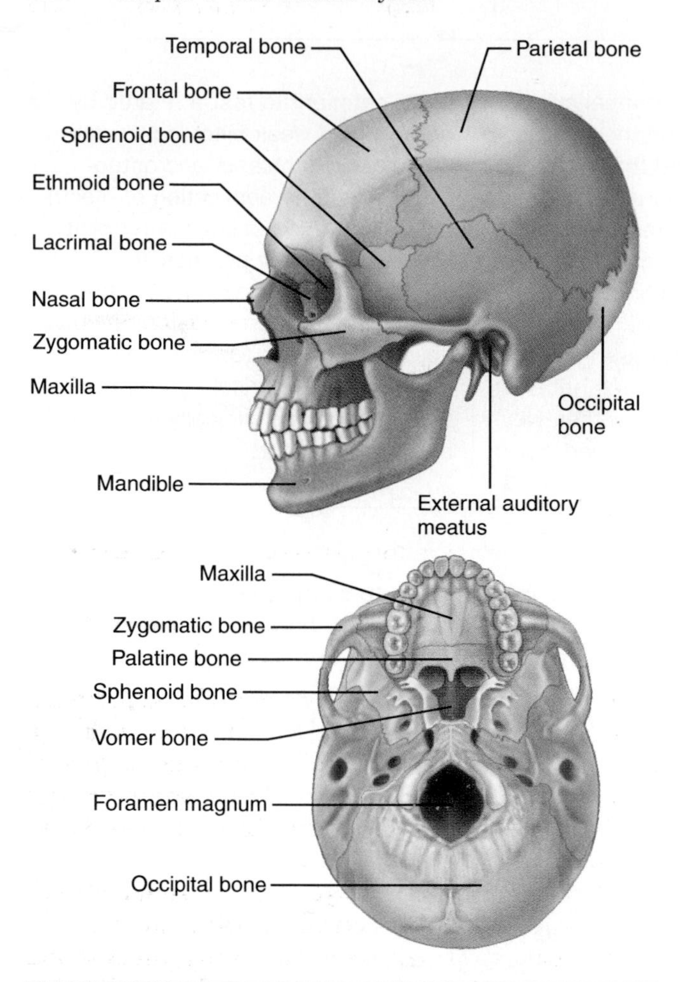

Figure 5.7. The human skull. Except for the mandible, which has a hinged joint with the temporal bone, the bones of the skull are joined tightly together. Their function is protection, not movement.

The *facial bones* comprise the front of the skull. On either side of the nose are the two *maxilla* (maxillary) bones, which form part of the eye sockets and contain the sockets that anchor the upper row of teeth. The hard palate (the "roof" of the mouth) is formed by the maxilla bones and the two *palatine bones*. Behind the palatine bones is the *vomer bone,* which is part of the nasal septum that divides the nose into left and right halves. The two *zygomatic bones* form the cheekbones and the outer portion of the eye sockets. The two small, narrow *nasal bones* underlie only the upper bridge of the nose; the rest of the fleshy protuberance we call our nose is made up of cartilage and other connective tissue. Part of the space formed by the maxillary and nasal bones is the nasal cavity. The small *lacrimal bones,* at the inner eye sockets, are pierced by a tiny opening through which the tear ducts drain tears from the eye sockets into the nasal cavity.

The *mandible,* or lower jaw, contains the sockets that house the lower row of teeth. All the bones of the skull are joined tightly together except for the mandible, which attaches to the temporal bone by a joint that permits a substantial range of motion. This allows us to speak and chew.

Several of the cranial and facial bones contain air spaces called **sinuses,** which make the skull lighter and give the human voice its characteristic tone and resonance. Each sinus is lined with tissue that secretes mucus, a thick, sticky fluid that helps trap foreign particles in incoming air. The sinuses connect to the nasal cavity via small passageways through which the mucus normally drains. However, if you develop a cold or respiratory infection, the tissue lining your sinuses can become inflamed and block these passages. Sinus inflammation is called "sinusitis." If fluid accumulates inside the sinuses, the resulting sensation of pressure may give you a "sinus headache."

The vertebral column: The body's main axis The **vertebral column** (the backbone or spine) is the main axis of the body (Figure 5.8). It supports the head,

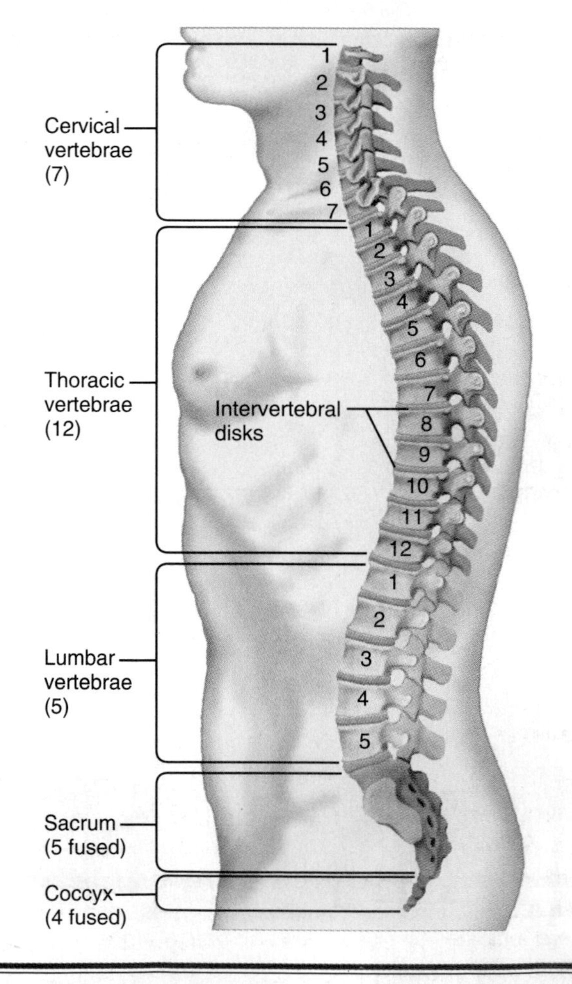

Figure 5.8 The vertebral column. Vertebrae are named and numbered according to their location. The vertebral column is moderately flexible because of the presence of joints and intervertebral disks.

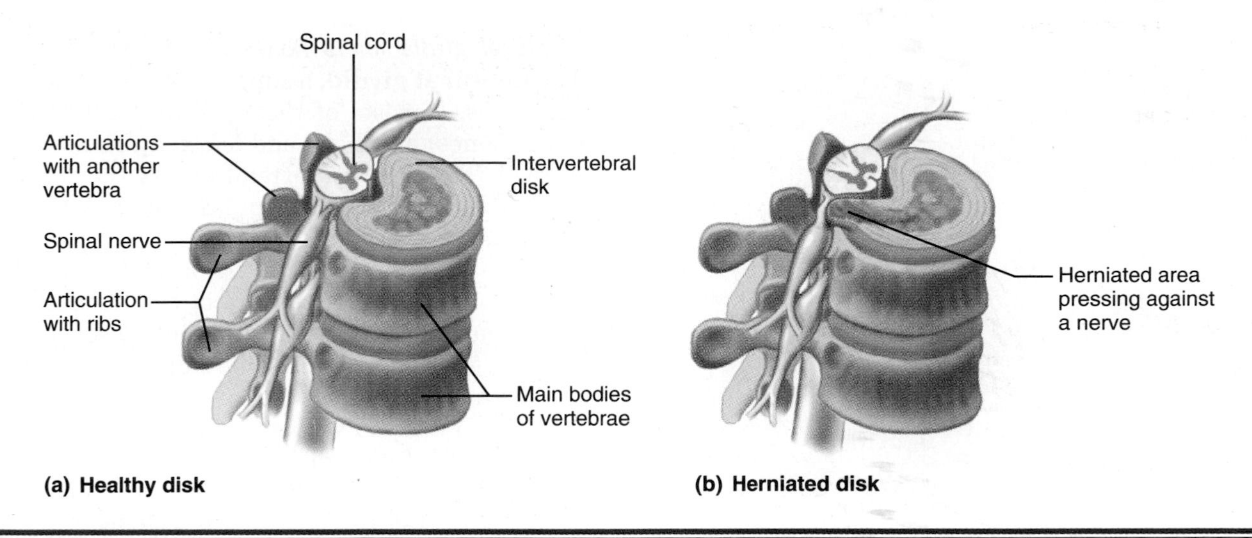

Figure 5.9 Vertebrae. (a) Two healthy vertebrae with their intervertebral disks. **(b)** A herniated disk.

protects the spinal cord, and serves as the site of attachment for the four limbs and various muscles. It consists of a column of 33 irregular bones called *vertebrae* (singular *vertebra*) that extends from the skull to the pelvis. When viewed from the side the vertebral column is somewhat curved, reflecting slight differences in structure and size of vertebrae in the various regions.

We classify the vertebral column into five anatomical regions:

- Cervical (neck)—seven vertebrae
- Thoracic (the chest or thorax)—12 vertebrae
- Lumbar (the lower portion or "small" of the back, which forms the lumbar curve of the spine)—five vertebrae
- Sacral (in the sacrum or upper pelvic region)—In the course of evolution, the five sacral vertebrae have become fused.
- Coccygeal (the coccyx or tailbone)—four fused vertebrae. The coccyx is all that remains of the tails of our ancient ancestors. It is an example of a *vestigial* structure, meaning one that no longer has any function.

A closer look at vertebrae (Figure 5.9) shows how they are stacked on each other and how they are joined. Vertebrae share two points of contact called *articulations* located behind the main body. There are also articulations with the ribs. The spinal cord passes through a hollow cavity between the articulations and the main body. Neighboring vertebrae are separated from each other by a flat, platelike **intervertebral disk** of fibrocartilage. The intervertebral disk serves as a shock absorber that protects the delicate vertebrae from the impact of walking, jumping, and other movements. In conjunction with the joints, the disks also permit a limited degree of movement. This lends the vertebral column greater flexibility, allowing us to bend forward, lean backward, and rotate the upper body.

An especially strong impact or sudden movement can compress an intervertebral disk, forcing the softer center to balloon outward, press against spinal nerves, and cause intense back pain. This condition is referred to as a "herniated" or "slipped disk" (Figure 5.9b), and it occurs most often in the lumbar vertebrae. Occasionally the disk may even rupture, releasing its soft pulpy contents. The pain that accompanies a herniated disk can be alleviated by surgery to remove the damaged disk, relieving the pressure against the nerve. However, surgical correction of a herniated disk reduces spinal flexibility somewhat.

Generally the vertebral column does an effective job of shielding the spinal cord. However, injury to the vertebral column can damage the cord or even sever it, resulting in partial or complete paralysis of the body below that point. Persons with suspected vertebral injuries should not be moved until a physician can assess the situation, because any twisting or bending could cause additional, perhaps permanent damage to the spinal cord. You may have noticed that when athletes are injured on the field, they are instructed to lie absolutely still until a trainer and physician have examined them thoroughly.

The ribs and sternum: Protecting the chest cavity Humans have twelve pairs of **ribs** (Figure 5.10). One end of each rib branches from the thoracic region of the vertebral column. The other ends of the upper seven pairs attach via cartilage to the **sternum,** or

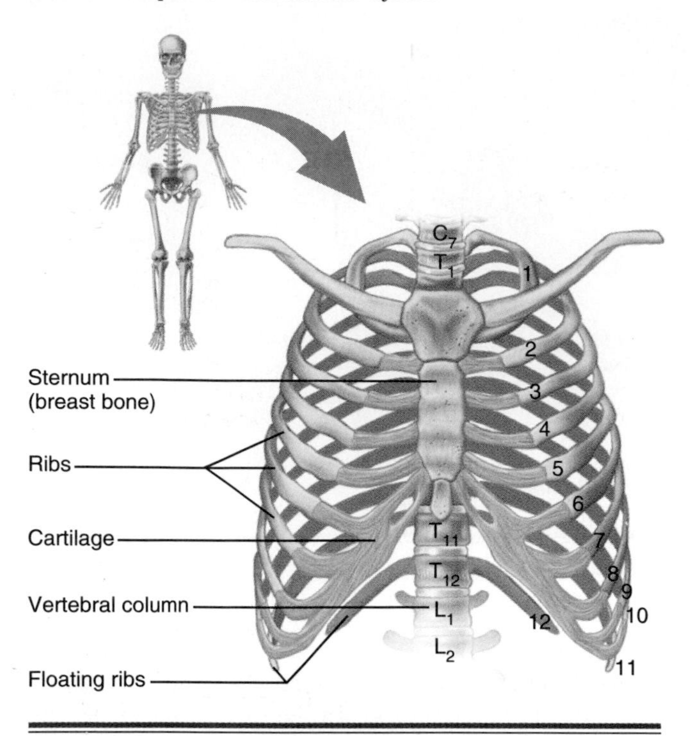

Figure 5.10 Ribs. The twelve pairs of ribs are numbered according to their attachment to the thoracic vertebrae. Only the first seven pairs attach directly to the sternum.

breastbone, a flat blade-shaped bone comprised of three separate bones that fuse during development. Rib pairs 8–10 are joined to the seventh rib by cartilage, and thus attach indirectly to the sternum. The bottom two pairs of ribs are called *floating ribs* because they do not attach to the sternum at all.

The ribs, sternum, and vertebral column form a protective *rib cage* that surrounds and shields the heart, lungs, and other organs of the chest (thoracic) cavity. They also help us breathe, because muscles between the ribs lift them slightly during breathing, expanding the chest cavity and inflating the lungs. The base of the sternum is connected to the diaphragm, a muscle that is important to breathing (see Chapter 10, The Respiratory System).

Recap **The skull protects the brain, the vertebral column protects the spinal cord and supports the appendicular skeleton, and the rib cage protects the organs of the chest cavity.**

Appendicular skeleton: Pectoral girdle, pelvic girdle, and limbs

Those parts of the body that attach to the axial skeleton are called appendages, from the Latin word meaning "to hang upon." The second division of the human skeleton, the **appendicular skeleton,** includes our arms and legs and their attachments to the trunk, the pectoral and pelvic girdles.

Pectoral girdle lends flexibility to the upper limbs

The **pectoral girdle,** a supportive frame for the upper limbs, consists of the right and left **clavicles** (collarbones) and right and left **scapulas** (shoulder blades). The clavicles extend across the top of the chest and attach to the scapulas, triangular bones in the upper back.

The arm and hand consist of 30 different bones (Figure 5.11). The upper end of the **humerus,** the long bone of the upper arm, fits into a socket in the scapula. The other end of the humerus meets with the **ulna** and **radius,** the two bones of the forearm, at the elbow. If you've ever hit your elbow and experienced a painful tingling, you know why this area is nicknamed the "funny bone"; you've just struck the ulnar nerve that travels along the elbow.

The lower ends of the ulna and radius meet the *carpal* bones, a group of eight small bones that makes up the wrist. The five *metacarpal* bones form the palm of the hand, and they join with the 14 *phalanges,* which form the fingers and thumb.

The pectoral girdle and arms are particularly well adapted to permit a wide range of motion. They connect to the rest of the body via muscles and tendons—a relatively loose method of attachment. This structure gives our upper body a degree of dexterity unsurpassed among large animals. We can rotate our upper arms almost 360 degrees—a greater range of movement than with any other joint in the body. The upper arm can rotate in roughly a circle, the arm can bend in one dimension and rotate, and the wrist and fingers can all bend and rotate to varying degrees. We also have an "opposable thumb," meaning we can place it opposite our fingers. The opposable thumb has played an important role in our evolutionary history, as it makes it easier to grasp and manipulate tools and other objects.

We pay a price for this flexibility, because freedom of movement also means relatively little stability. If you fall on your arm, for example, you might dislocate your shoulder joint or crack a clavicle. In fact, the clavicle is one of the most frequently broken bones in the body.

Although our upper limbs are well adapted to a wide range of movements, too much of one kind of motion can be harmful. Repetitive motions—performing the same task over and over—can lead to health problems called overuse or repetitive stress syndromes. Depending on the part of the body that is overused, these injuries can take many forms. A well-known repetitive stress syndrome is *carpal tunnel syndrome,* a condition often due to repetitive typing at a computer keyboard. The carpal bones of the wrist are held together by a sheath of connective tissue. The blood vessels, nerves, and tendons to the hand and fingers pass through the sheath via the

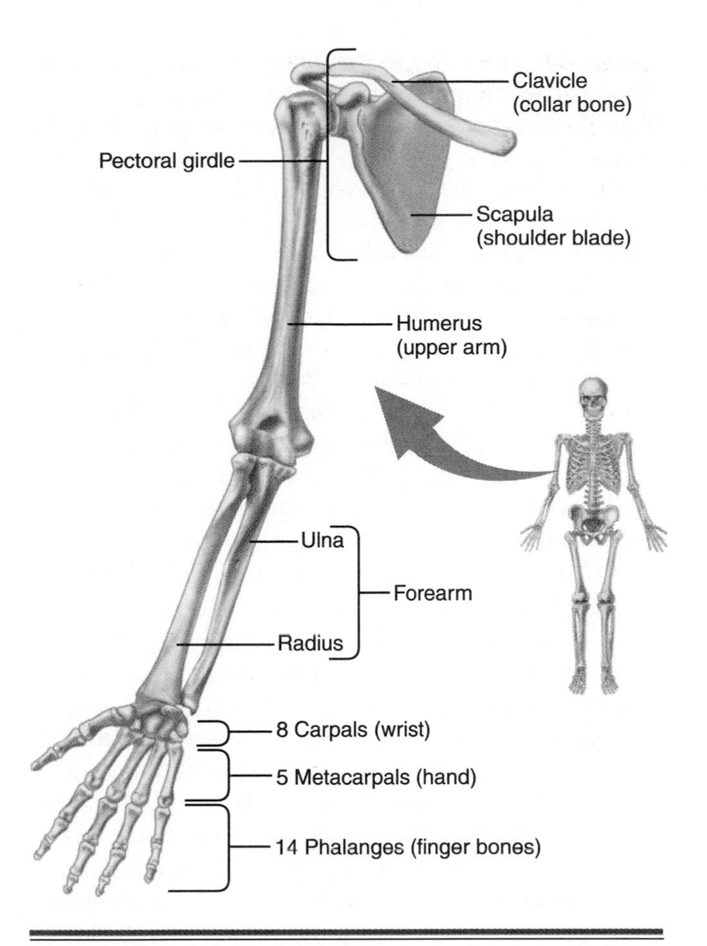

Figure 5.11 Bones of the right side of the pectoral girdle and the right arm and hand.

"carpal tunnel." Overuse of the fingers and hands produces swelling and inflammation of the tendons, which causes them to press against the nerve supplying the hand. The result may be pain, tingling, or numbness in the wrist and hand. Mild episodes of carpal tunnel syndrome respond to rest and pain relievers. Severe cases can be treated with surgery to relieve the pressure.

Pelvic girdle supports the body The **pelvic girdle** consists of the two **coxal bones** and the sacrum and coccyx of the vertebral column (Figure 5.12). The coxal bones attach to the sacral region of the vertebral column in back, then curve forward to meet in front at the *pubic symphysis,* where they are joined by cartilage. You can feel the upper curves of the coxal bones (the iliac region) as your hipbones. Together, these structures form the pelvis.

The primary function of the pelvic girdle is to support the weight of the upper body against the force of gravity. It also protects the organs inside the pelvic cavity and serves as a site of attachment for the legs. The structure of the pelvic girdle reflects a trade-off between dexterity and stability. Partly because the pelvic girdle and lower limbs are larger

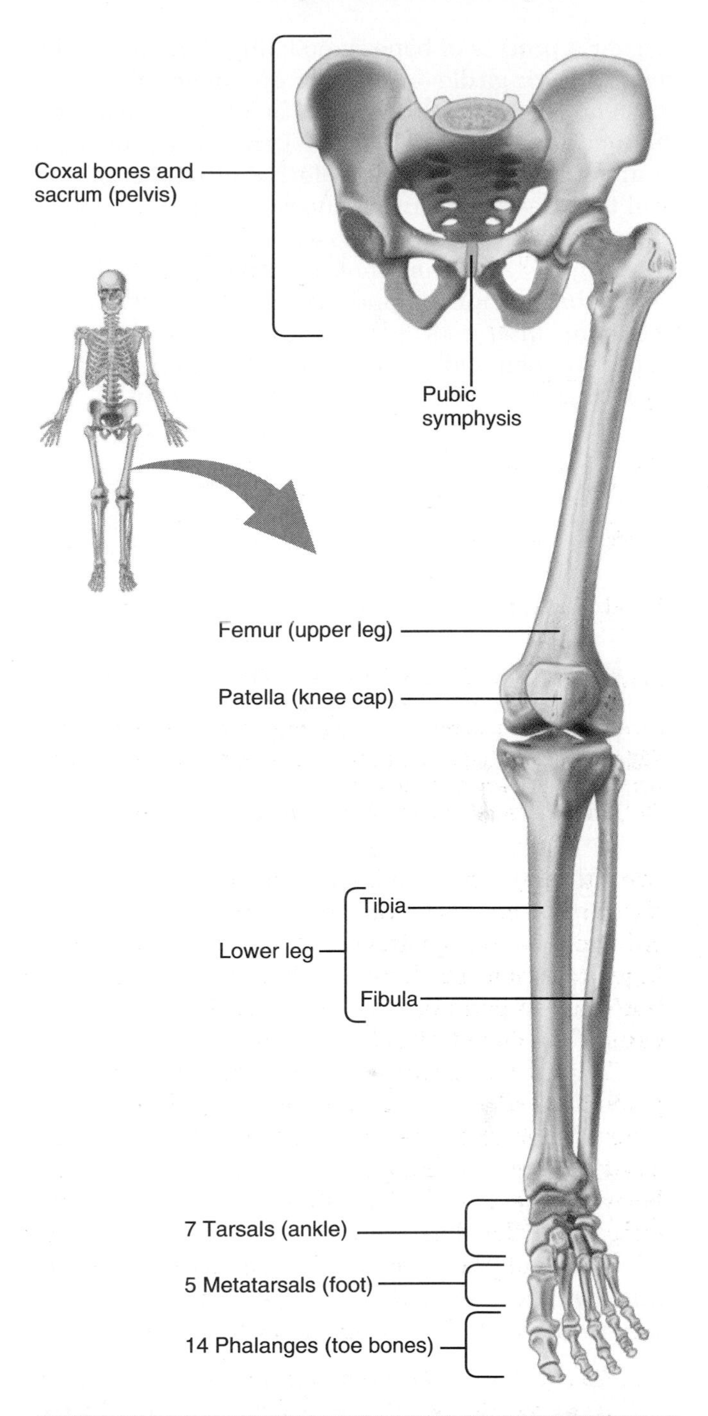

Figure 5.12 Bones of the pelvic girdle and the left leg and foot.

and more firmly connected to the rest of the body than the pectoral girdle and upper limbs, the lower limbs are less dexterous than the upper limbs.

In adult women the pelvic girdle is broader and shallower than it is in men, and the pelvic opening is wider. This allows for safe passage of a baby's head during labor and delivery. These characteristic differences appear during puberty when a woman's body begins to produce sex hormones. The sex hormones trigger a process of bone remodeling that shapes the female pelvic girdle to adapt for pregnancy and birth.

The **femur** (thighbone) is the longest and strongest bone in the body. When you jog or jump your femurs are exposed to forces of impact of several tons per square inch. The rounded upper end of each femur fits securely into a socket in a coxal bone, creating a stable joint that effectively supports the body while permitting movement. The lower end of the femur intersects at the knee joint with the larger of the two bones of the lower leg, the **tibia,** which in turn makes contact with the thinner **fibula.** The *patella,* or kneecap, is a triangle-shaped bone that protects and stabilizes the knee joint.

At the ankle, the tibia and fibula join with the seven *tarsal* bones that make up the ankle and heel. Five long bones, the *metatarsals,* form the foot. The 14 bones of the toes, like those of the fingers, are called *phalanges*.

Recap **The pectoral girdle and upper limbs are capable of a wide range of motions (dexterous movement). The pelvic girdle supports the body's weight and protects the pelvic organs. The lower limbs are stronger but less dexterous than the upper limbs.**

5.5 Joints form connections between bones

We now turn to the structures and tissues that hold the skeleton together while still permitting us to move about freely: joints, ligaments, and tendons. **Joints,** also called articulations, are the points of contact between bones. Ligaments and tendons are connective tissues that stabilize many joints.

Joints vary from immovable to freely movable

Joints vary considerably from basically immovable to freely moveable structures. Types of joints include fibrous, cartilaginous, and synovial joints.

Fibrous joints are immovable. At birth, the flat bones in a baby's skull are separated by relatively large spaces filled with fibrous connective tissue. These "soft spots," called *fontanels,* enable the baby's head to change shape slightly so that it can squeeze safely through the mother's pelvic opening during childbirth. The presence of joints also allows for brain growth and development after birth. During childhood these fibrous joints gradually harden. By the time we reach adulthood, the joints have become thin lines, or sutures, between skull bones. These immovable joints firmly connect the bones that protect and stabilize the skull and brain.

Cartilaginous joints, in which the bones are connected by hyaline cartilage, are slightly movable, allowing for some degree of flexibility. Examples include the cartilaginous joints that connect the vertebrae in the backbone, and those that attach the lower ribs to the sternum.

The most freely movable joints are **synovial joints,** in which the bones are separated by a thin fluid-filled cavity. The two bones of a synovial joint are fastened together and stabilized by ligaments. The interior of the cavity is lined with a *synovial membrane,* which secretes *synovial fluid* to lubricate and cushion the joint. To reduce friction even further, the articulating surfaces of the two bones are covered with a tough but smooth layer of hyaline cartilage. Together, the synovial membrane and the surrounding hyaline cartilage constitute the *joint capsule*.

How do joints resist wear and tear?

Spotlight on Structure & Function

Different types of synovial joints permit different kinds of movements. A *hinge joint,* such as the knee and elbow, gets its name because it allows movement in one plane like the hinges on the door. Figure 5.13 illustrates a human knee joint. The knee joint is strong enough to withstand hundreds of pounds of force, yet it is flexible enough to swing freely in one direction. To reduce friction, there are small disks of cartilage on either side of the knee called *menisci* (singular, meniscus). The knee joint also includes 13 small sacs of fluid, called *bursae* (singular, bursa), for additional cushioning. The entire joint is wrapped in strong ligaments that attach bone to bone and tendons that attach bone to muscle. Note the two *cruciate ligaments* that join the tibia to the femur bone. The anterior cruciate ligament is sometimes injured when the knee is hit with great force from the side.

A second type of synovial joint, a *ball-and-socket joint,* permits an even wider range of movement. Examples include the joint between the femur and the coxal bone, and between the humerus and the pectoral girdle. In both cases, the rounded head of the bone fits into a socket, allowing movement in all planes. ■

Figure 5.14 illustrates the different types of movements made possible by hinge and ball-and-socket joints. Note that you can rotate your arm and your leg because the shoulder and hip are ball-and-socket joints, but you cannot rotate the hinge joint in your knee.

Thanks to its design, a synovial joint can withstand tremendous poundings day after day, year after year without wearing out. But where does it get its strength? For that we turn to ligaments, tendons, and muscles. As we have seen, the bones of a synovial joint are held tightly together by ligaments. They are stabilized even more by **tendons,** another

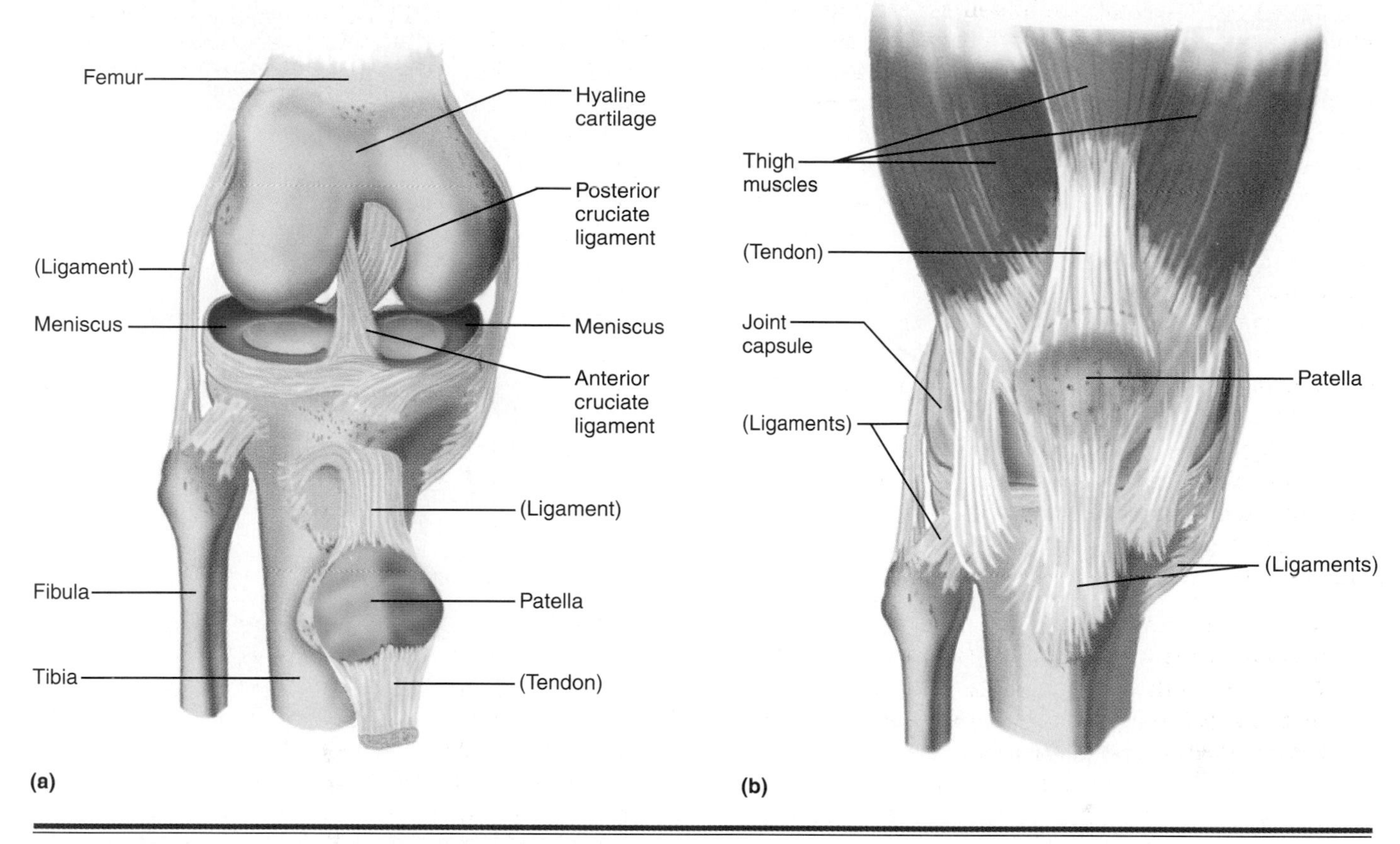

Figure 5.13 The knee joint is a hinged synovial joint. **(a)** A cutaway anterior view of the right knee with muscles, tendons, and the joint capsule removed and the bones pulled slightly apart so that the two menisci are visible. **(b)** A view of the knee with muscles, tendons, and ligaments in their normal position surrounding the intact joint capsule. The combination of ligaments, tendons, and muscles holds the knee tightly together.

type of tough connective tissue, which join the bones to muscles. Both ligaments and tendons contain collagen arranged in parallel fibers, making ligaments and tendons as strong and flexible as a twisted rope of nylon. Muscle contraction increases the tension of the tendons, further tightening and strengthening the joint.

Recap **Joints are the points of contact between bones. Fibrous joints are immovable in adults, cartilaginous joints permit some movement, and synovial joints are highly movable. Synovial joints are held together by ligaments and lubricated by synovial fluid. Their type of structure determines the type of motion they permit.**

5.6 Diseases and disorders of the skeletal system

In this chapter we have already discussed several health conditions related to the skeletal system, including fractures, carpal tunnel syndrome, and osteoporosis. Now we look at several more.

Try It Yourself

Stabilizing a Joint

To appreciate the role ligaments, tendons, and muscles play in stabilizing a joint, try this simple experiment. Sit in a low chair, stretch your leg in front of you with your heel resting on the floor, and relax your muscles. Try to move your kneecap (patella) from side to side gently with your hand. Notice how easily you can shift it out of position. Now, without changing position, tense the muscle of your thigh. Can you still move your patella?

The patella is attached to the tibia by a ligament and to the muscles of the thigh by a tendon. When you contract your thigh muscles the patella remains in position because contraction puts tension on the tendon and the ligament. The increased tension holds the patella firmly in place. If you move your hand to just below the kneecap, you can feel the tightening of the patellar ligament as you alternately contract and relax your thigh muscle.

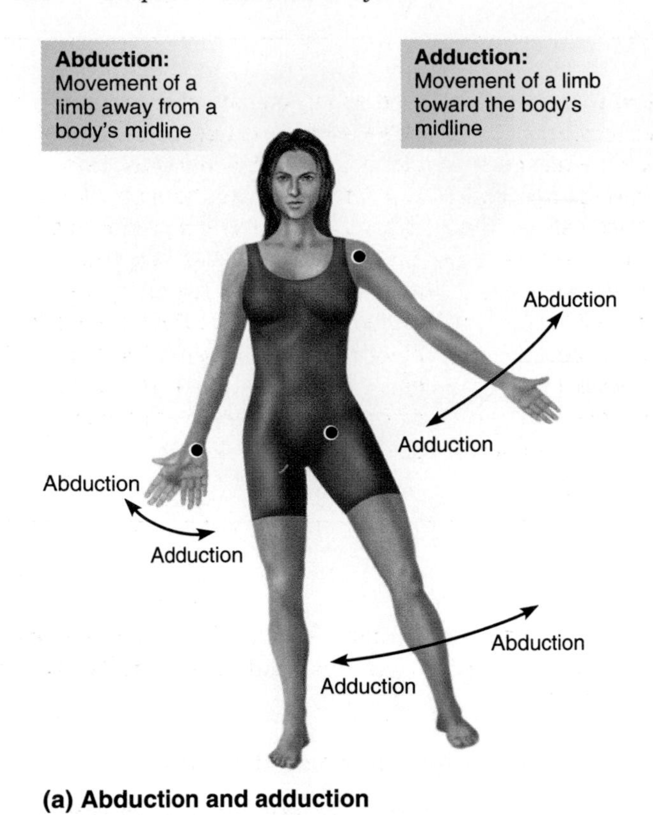

(a) Abduction and adduction

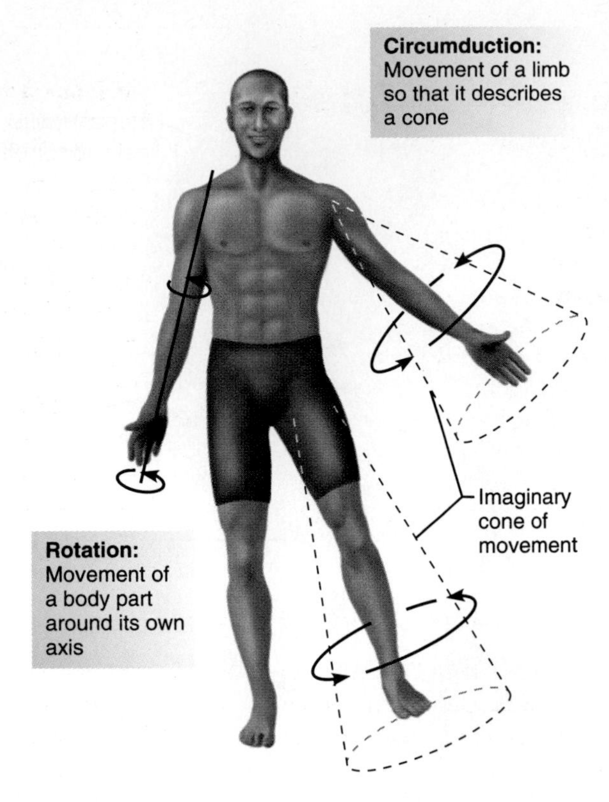

(b) Rotation and circumduction

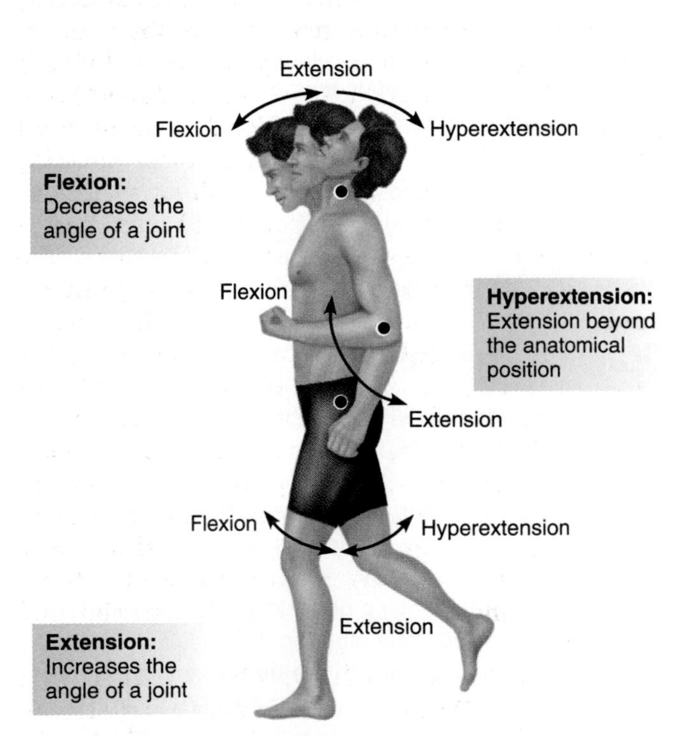

(c) Flexion, extension, and hyperextension

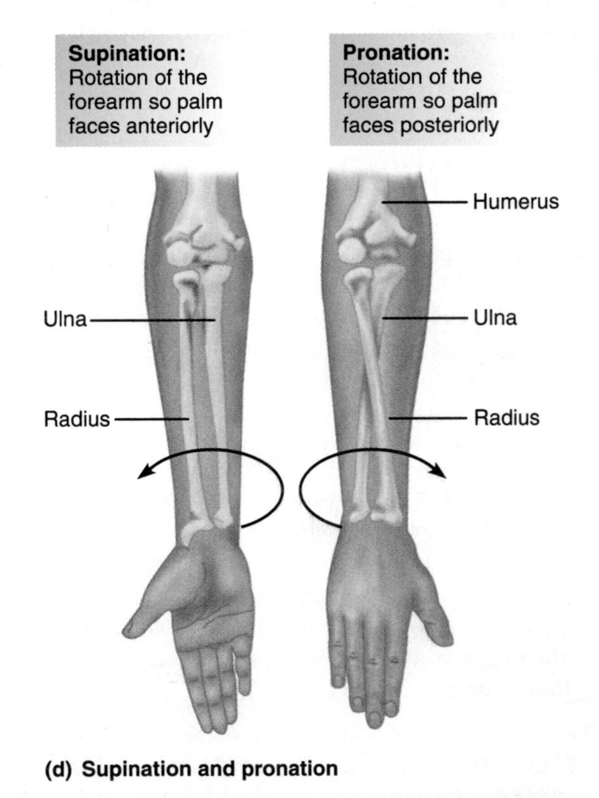

(d) Supination and pronation

Figure 5.14 Types of movements made possible by synovial joints.

Health Watch

Treating a Sprained Ankle

For a severe sprain, many physicians advise the frequent application of cold to the sprained area during the first 24 hours, followed by a switch to heat. Why the switch, and what is the logic behind the timing of cold versus heat?

Very early on, the biggest problem associated with a sprain is damage to small blood vessels and subsequent bleeding into the tissues. Most of the pain associated with a sprain is due to the bleeding and swelling, not the damage to ligaments themselves. The immediate application of cold constricts blood vessels in the area and prevents most of the bleeding. The prescription is generally to cool the sprain for 30 minutes out of every hour or 45 minutes every hour and a half. In other words, keep the sprain cold for about half the time, for as long as you can stand it. The in-between periods assure adequate blood flow for tissue metabolism.

After 24 hours there shouldn't be any more bleeding from small vessels. The damage has been minimized, so now the goal is to speed the healing process. Heat dilates the blood vessels, improves the supply of nutrients to the area, and attracts blood cells that begin the process of tissue repair.

Sprains mean damage to ligaments

A sprain is due to stretched or torn ligaments. Often it is accompanied by internal bleeding with subsequent bruising, swelling, and pain. The most common example is a sprained ankle. Sprains take a long time to heal precisely because the ligaments themselves have very few cells and a poor blood supply. Minor sprains, in which the ligaments are only stretched, usually mend themselves with time. If a large ligament is torn completely, it generally will not heal by itself. Surgery may be required to remove a torn ligament. Sometimes the joint can be stabilized with a piece of tendon or by repositioning other ligaments. Torn ligaments in the knee are particularly problematic because they often leave the knee joint permanently unstable and prone to future injuries.

Bursitis and tendinitis are caused by inflammation

Bursitis and tendinitis refer to inflammation of the bursae or tendons following injury. You will learn more about the inflammatory process in Chapter 9. Causes of bursitis and tendinitis may include tearing injuries to tendons, physical damage caused by blows to the joint, and even some bacterial infections. Like ligaments, tendons and the tissues lining the bursae are not well supplied with blood vessels, so they do not repair quickly. Treatment usually involves applying cold during the first 24 hours followed by heat, resting the injured area, and taking pain-relieving medications. "Tennis elbow" is a painful condition caused by either bursitis or tendinitis. Other common locations for pain include the knees, shoulder, and the Achilles tendon that pulls up the back of the heel.

Arthritis is inflammation of joints

By their nature, joints are exposed to high compressive forces and are prone to excessive wear caused by friction. "Arthritis" is a general term for joint inflammation. The most common type of arthritis is *osteoarthritis,* a degenerative ("wear-and-tear") condition that affects about 20 million Americans, most over age 45. In osteoarthritis the cartilage covering the ends of the bones wears out. With time the bone thickens and may form bony spurs, which further restrict joint movement. The result is increased friction between the bony surfaces, and the joint becomes inflamed and painful. Over-the-counter medications can reduce the inflammation and pain, and surgical joint replacements for severe osteoarthritis are fairly routine today. Injections of hyaluronic acid, a component of hyaline cartilage, can also reduce arthritic knee pain. Many physicians advise people with osteoarthritis to exercise regularly, which helps preserve the joints' healthy range of motion. Several promising new treatments to reduce joint inflammation are still in the experimental stage.

Osteoarthritis should not be confused with *rheumatoid arthritis.* Rheumatoid arthritis also involves joint inflammation, but it is caused by the body's own immune system, which mistakenly attacks the joint tissues. We'll take a closer look at disorders of this type in Chapter 9 (the Immune System).

Chapter Summary

The skeletal system consists of connective tissue

- Connective tissues of the skeletal system are bones, ligaments, and cartilage.
- Bone is a living tissue comprised of cells and extracellular material.
- Ligaments, comprised of dense fibrous connective tissue, attach bones to each other.
- Cartilage forms the intervertebral disks and lines the points of contact between bones.

Bone development begins in the embryo

- By about two months of fetal development, rudimentary models of bones have been formed from cartilage.
- Throughout the rest of fetal development and on into childhood, bone-forming cells called osteoblasts replace the cartilage model with bone.
- Growth in the length of long bones centers around growth plates in each epiphysis.

Mature bone undergoes remodeling and repair

- Bone undergoes replacement throughout life.
- Bones can change shape over time, depending on the forces to which they are exposed.
- The process of bone repair includes: (1) the formation of a hematoma, (2) the formation of a fibrocartilage callus that binds the broken ends together, and (3) the eventual replacement of the callus with new bone.

The skeleton protects, supports, and permits movement

- The axial skeleton is represented by the skull, the vertebral column, the sternum, and the ribs
- In the vertebral column, intervertebral disks of fibrocartilage absorb shock and permit limited movement.
- The appendicular skeleton includes the pectoral girdle, the pelvic girdle, and the upper and lower limbs.

Joints form connections between bones

- Three types of joints connect bones: fibrous, cartilaginous, and synovial.
- Synovial joints are designed for movement without friction. They are lined with a synovial membrane and lubricated by synovial fluid.

Diseases and disorders of the skeletal system

- Sprains are the result of stretched or torn ligaments. Bursitis and tendinitis are caused by injuries to the bursae and tendons.
- Arthritis is a general term for joint inflammation.
- Osteoarthritis is a condition in which the cartilage covering the ends of the bones wears out and joint friction increases.

Terms You Should Know

appendicular skeleton, 116
axial skeleton, 113
bone, 106
cartilage, 108
central (Haversian) canal, 107
chondroblast, 108
clavicle, 116
compact bone, 106
growth plate, 109
intervertebral disk, 115
joint, 118
ligament, 108
osteoblast, 109
osteoclast, 109
osteocyte, 106
osteon, 106
osteoporosis, 110
spongy bone, 106
tendon, 118

Test Yourself

Matching Exercise

_____ 1. Connective tissue structures that attach bone to bone
_____ 2. This bone marrow stores fat
_____ 3. This bone marrow produces blood cells
_____ 4. Young bone-forming cell
_____ 5. Bone-resorbing cell
_____ 6. Cartilage that comprises the vertebral discs
_____ 7. Cartilage that covers the ends of bones in joints
_____ 8. Cartilage that lends structural support to the outer ear
_____ 9. Name for the lower jaw
_____10. Type of joint that is highly moveable

a. osteoclast
b. yellow bone marrow
c. fibrocartilage
d. elastic cartilage
e. ligaments
f. mandible
g. red bone marrow
h. synovial
i. hyaline cartilage
j. osteoblast

Fill in the Blank

11. A long bone has a shaft called the diaphysis and an enlarged knob at each end called an ________________.
12. Mature bone cells (osteocytes) receive nutrients from blood vessels that pass through a ________________ ____________ within a cylindrical structure called an osteon.
13. ________________ *bone* is less dense than *compact bone*.
14. Based on shape, bones are classified as either long, short, ____________, or irregular.
15. During embryonic development, cartilage-forming cells called ________________ create a cartilage model of the future bones.
16. The __________ skeleton forms the midline of the body.
17. The __________ girdle anchors the lower limbs.
18. The __________ girdle supports the upper limbs.
19. A __________ joint such as the knee permits movement in only one plane.
20. ____________ arthritis is caused by the body's own immune system.

True or False?

21. Bones continue to lengthen throughout life.
22. In mature bone, approximately one-third of the matrix is osteoid and two-thirds is crystals of hydroxyapatite.
23. Bone strength can be improved by regular weight-bearing exercise.
24. In humans, all twelve pairs of ribs attach to the sternum, or breastbone.
25. The joints between adjacent vertebrae are cartilaginous joints.

Concept Review

1. List the five functions of bone.
2. Define osteoid.
3. Explain how the two growth plates in a long bone account for the ability of a long bone to lengthen.
4. Explain what might cause a long bone to slowly change shape over many years.
5. Describe how osteocytes embedded in solid bone tissue receive their nourishment and get rid of wastes.
6. Name the three anatomical regions of the vertebral column that are above the *sacral* and *coccygeal* regions.
7. Explain why it is important that someone who may have suffered an injury to the vertebral column not be moved until a medical assessment can be made.
8. Describe the features of synovial joints that reduce friction and prevent the joint from wearing out prematurely.
9. Distinguish between *flexion* and *extension*.
10. Define "arthritis."

Apply What You Know

1. Compare/contrast swimming and running as forms of exercise training in terms of how they might affect muscle mass, bone mass, and the possibility of injuries to joints.
2. The administration of growth hormone is sometimes used clinically to stimulate growth in unusually short children who are growth hormone deficient. However, growth hormone is ineffective in unusually short but otherwise normal adults. What accounts for the difference?
3. Give a possible explanation for why the skin heals more quickly than ligaments following a cut or injury. Can you relate your explanation to other tissues or organs? In other words, what do you think might be factors that affect rates of healing?

Case in Point

Will Onions Prevent Osteoporosis?

A group of Swiss researchers reported recently that onions might be able to reduce the risk of osteoporosis in humans. The researchers removed the ovaries from female rats and then fed them diets supplemented with one gram per day of certain dried vegetables. One gram per day of dried onions for only 10 days produced a significant reduction in the rate of bone resorption, a measure of bone loss. In fact, onion was slightly more effective at reducing bone resorption than the hormone calcitonin, which is often used to prevent osteoporosis in postmenopausal women. The authors suggest that if onions and other vegetables have the same effects in humans as they do in rats, then incorporating certain vegetables into the diet might be an inexpensive way to prevent osteoporosis in humans.

QUESTIONS:

1. If the researchers were interested in osteoporosis in humans, why do you suppose their experiments were conducted on rats?
2. What are some of the differences between rats and humans that you think might be relevant to interpretation of this study?

INVESTIGATE:

Muhlbauer, R.C. and F. Li. Effect of Vegetables on Bone Metabolism. *Nature* 401:343–344, Sept. 23, 1999.

References and Additional Reading

Diamond, J. Building to Code. *Discover* May, 1993, p. 92. (Why are bones and other body parts as strong as they are?)

Shipman, P., A. Walker and D. Bichell. *The Human Skeleton*. Cambridge: Harvard University Press, 1985. 343 pp. (Photographs, fine line drawings, and good written commentary about bones and the human skeleton.)

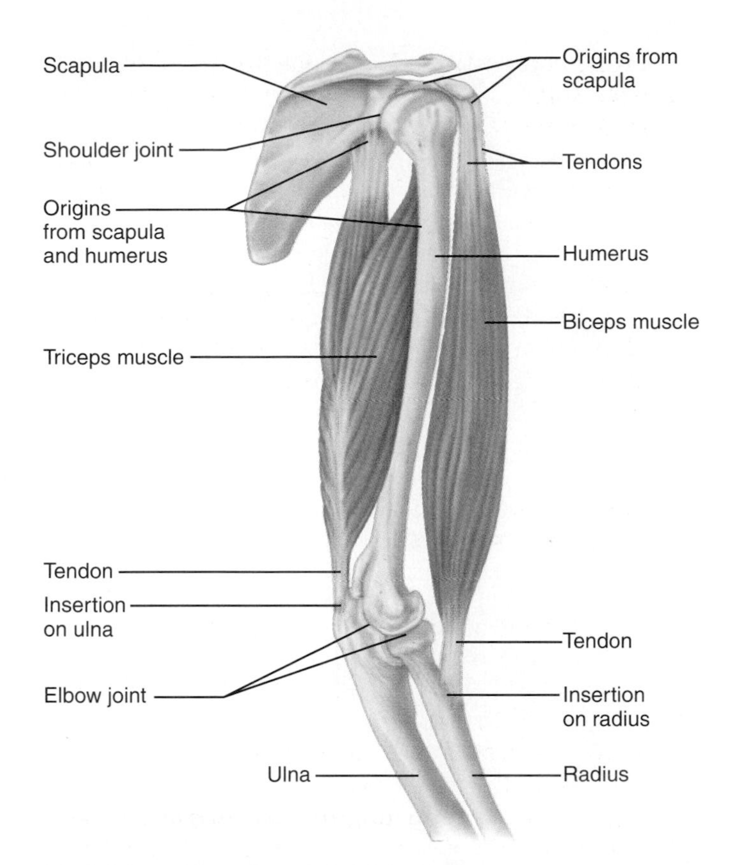

(a) Origin and insertion of skeletal muscle

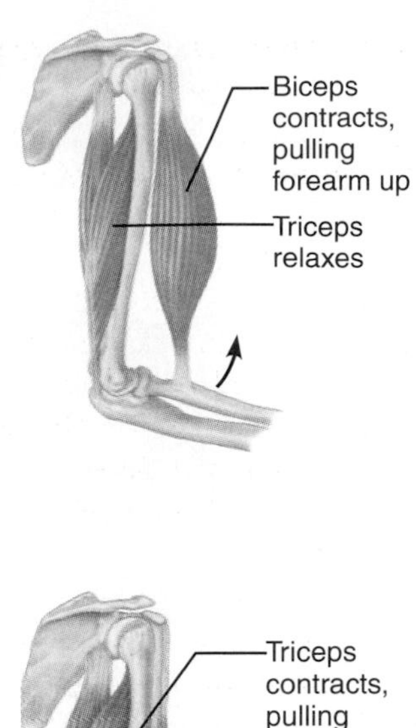

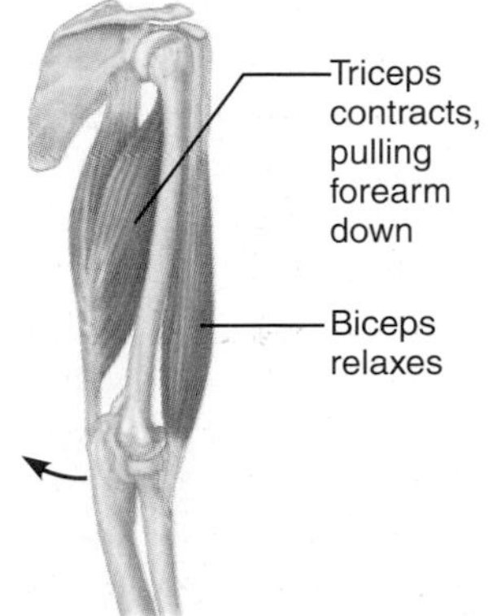

(b) Movement of the forearm

Figure 6.1 Movement of bones. (a) The point of attachment of a muscle to the stationary bone is its origin; the point of attachment to the moveable bone is its insertion. **(b)** Antagonistic muscles produce opposite movements. When the biceps contracts and the triceps relaxes, the forearm bends. The forearm straightens when the biceps relaxes and the triceps contracts.

smile). Muscles join to the skeleton in such a way that each individual muscle produces a very specific movement of one bone relative to another. The skeleton is a complex set of levers that can be pulled in many different directions by contracting or relaxing skeletal muscles. One end of a skeletal muscle, called its **origin,** joins to a bone that remains relatively stationary. The other end of the muscle, called its **insertion,** attaches to another bone across a joint. When the muscle contracts, the insertion is pulled toward the origin. The origin is generally closer to the midline of the body and the insertion is farther away. ■

Figure 6.1b shows how the two antagonistic muscles of the upper arm, the biceps and triceps, oppose each other to bend (flex) and straighten (extend) the forearm. When the triceps muscle relaxes and the biceps contracts, it pulls on the forearm and flexes it. When the biceps relaxes and the triceps contracts, it pulls the forearm down, extending it again.

Hundreds of muscles, each controlled by nerves and acting either individually or in groups, produce all possible human motions. Figure 6.2 summarizes some of the major muscles of the body and their actions.

Recap **When a muscle contracts (shortens), its insertion is pulled toward its origin.**

A muscle is composed of many muscle cells A single *muscle* (sometimes referred to as a "whole muscle") is a group of individual muscle cells, all with the same origin and insertion and all with the same function. In cross section (Figure 6.3), a muscle appears to be arranged in bundles called *fascicles,* each enclosed in a sheath of fibrous connective tissue. Each fascicle contains anywhere from a few dozen to thousands of individual muscle cells, also known as *muscle fibers.* The outer surface of the muscle is covered by several layers of fibrous connective tissue as well. Connective tissue supports the bundles and holds the whole muscle together. At the ends of the muscle all of the connective tissue sheaths come together, forming the tendons that attach the muscle to bone.

Individual muscle cells are tube-shaped, larger, and usually longer than most other human cells. Some muscle cells are only a millimeter in length, whereas others may be as long as 30 centimeters—roughly the length of your thigh muscle. Taking a

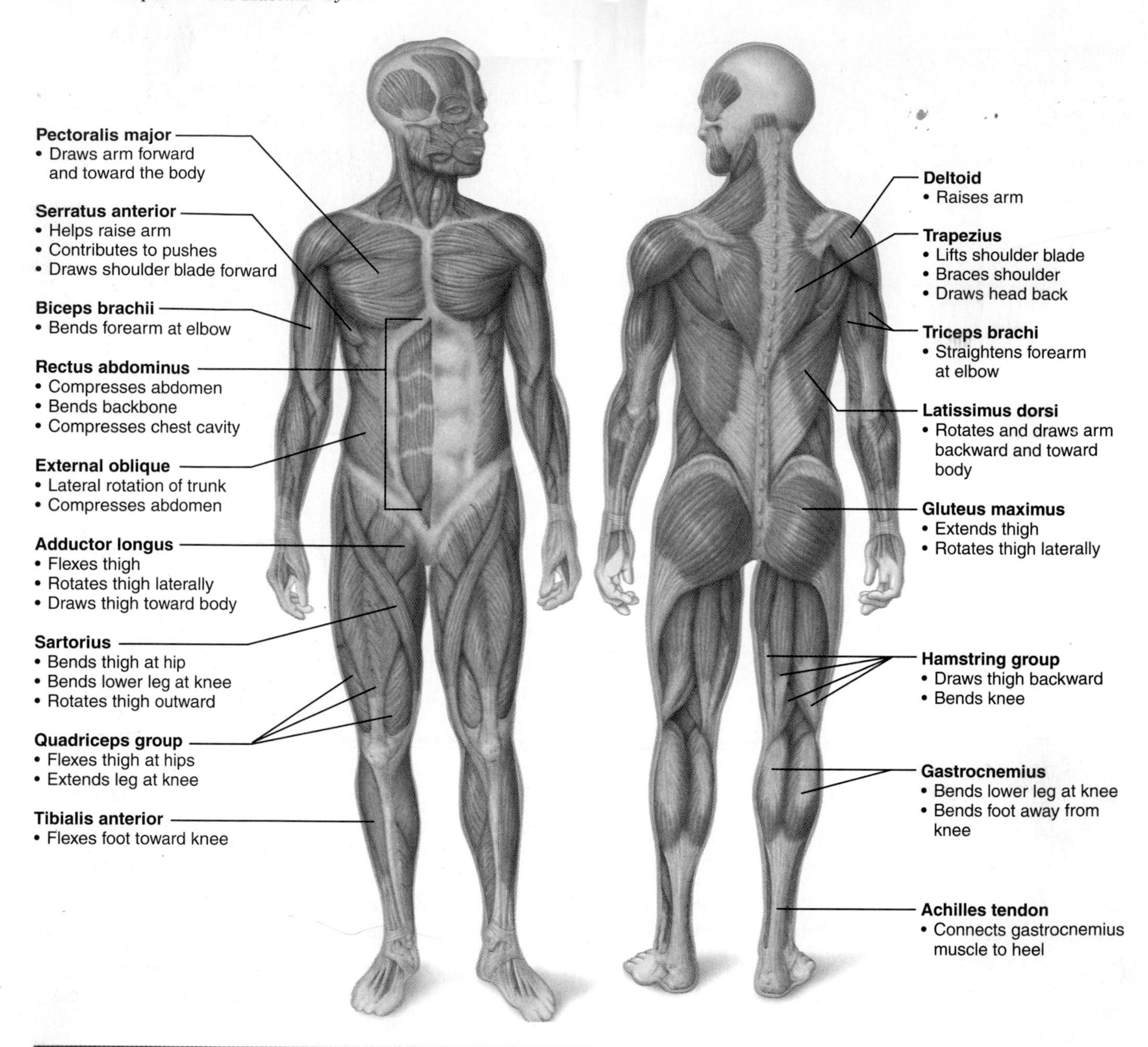

Figure 6.2 Major skeletal muscle groups and their functions.

closer look at a single muscle cell (Figure 6.4), we see that each cell contains more than one nucleus. The nuclei are located just under the cell membrane because nearly the entire interior of the cell is packed with long cylindrical structures arranged in parallel, called **myofibrils.** The myofibrils are packed with contractile proteins called *actin* and *myosin,* discussed below. When myofibrils contract (shorten), the muscle cell also shortens.

The contractile unit is a sarcomere

Spotlight on Structure & Function

Looking still closer at a single myofibril, we see a striated (or banded) appearance that repeats at regular intervals. Various elements of the pattern stand out, but the one that is important for our discussion is a dark line called the *Z-line* (Figure 6.5). A segment of a myofibril from one Z-line to the next is called a **sarcomere.** A single myofibril of one muscle cell in your biceps muscle may contain over 100,000 sarcomeres arranged end to end. The microscopic shortening of these 100,000 sarcomeres all at once is what produces contraction (shortening) of the muscle cell and of the whole muscle. Understanding muscle shortening, then, is simply a matter of understanding how a single sarcomere works.

A sarcomere consists of two kinds of protein filaments. Thick filaments composed of a protein called

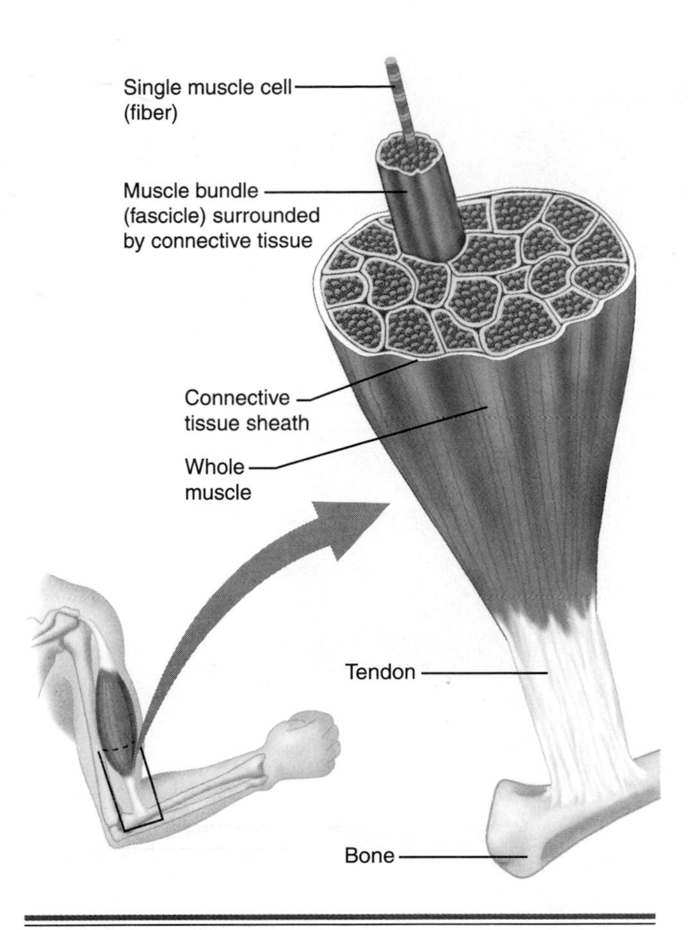

Figure 6.3 Muscle structure. A muscle is arranged in bundles called fascicles, each composed of many muscle cells and each surrounded by a connective tissue sheath. Surrounding the entire muscle is another connective tissue sheath. The connective tissues join to become the tendon, which attaches the muscle to bone.

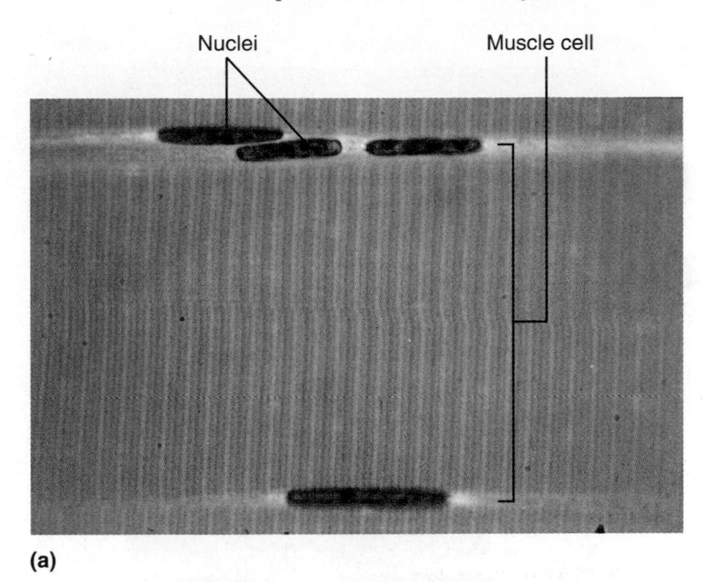

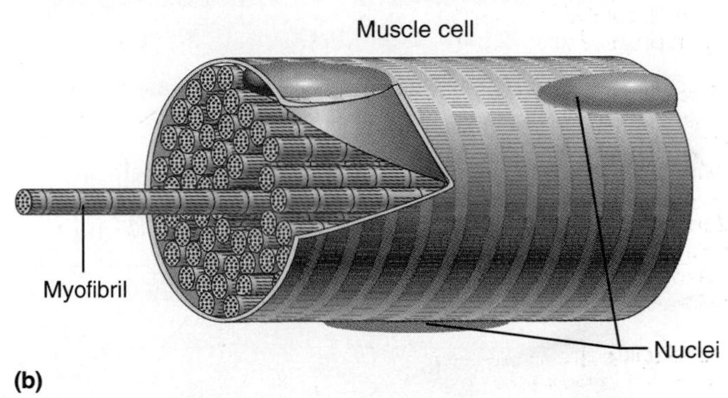

Figure 6.4 Muscle cells. (a) A photograph (×2,000) of portions of several skeletal muscle cells. **(b)** A single muscle cell contains many individual myofibrils and has more than one nucleus.

myosin are interspersed at regular intervals witthin filaments of a different protein called **actin.** Notice that the actin filaments are structurally linked to the Z-line and that myosin filaments are located entirely within sarcomeres, stretching between two different actin filaments. As we will see, muscle contractions depend on the interaction between these actin and myosin filaments. ■

Recap **A muscle is composed of many muscle cells (fibers) arranged in parallel, each containing numerous myofibrils. The contractile unit in a myofibril is called a sarcomere. A sarcomere contains thick filaments of a protein called myosin and thin filaments of a protein called actin.**

6.2 Individual muscle cells contract and relax

During a muscle contraction each sarcomere shortens just a little. Subtle though this action seems, it is also powerful. The contraction of an entire skeletal muscle depends on the simultaneous shortening of the tiny sarcomeres in its cells. There are four keys to understanding what makes a skeletal muscle cell contract and relax:

- A skeletal muscle cell must be activated by a nerve. It will not contract on its own.
- Nerve activation increases the concentration of calcium in the vicinity of the contractile proteins.
- The presence of calcium permits contraction. The absence of calcium prevents contraction.
- When a muscle cell is no longer stimulated by a nerve, contraction ends.

Let's look at each point in more detail.

Nerves activate skeletal muscles

Skeletal muscle cells are stimulated to contract by certain nerve cells called *motor neurons.* The motor neurons secrete a chemical substance called *acetylcholine (ACh).* Acetylcholine is a *neurotransmitter,* a

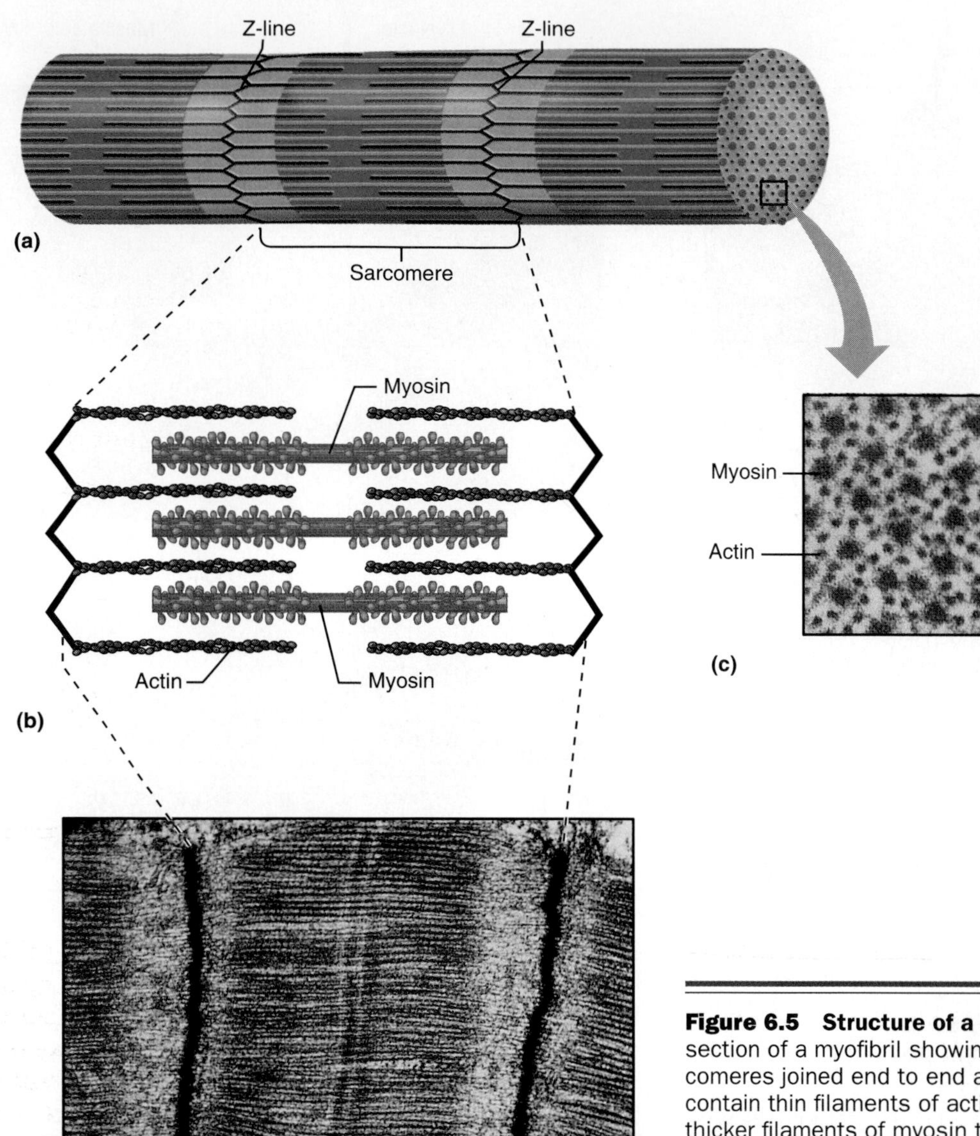

Figure 6.5 Structure of a myofibril. (a) A closer view of section of a myofibril showing that it is composed of sarcomeres joined end to end at the Z-line. **(b)** Sarcomeres contain thin filaments of actin that attach to the Z-lines and thicker filaments of myosin that span the gap between actin molecules. **(c)** An electron micrograph cross section of a sarcomere in a region that contains both actin and myosin. **(d)** An electron micrograph (×38,000) of a longitudinal section of a sarcomere.

chemical released by nerve cells that has either an excitatory or inhibitory effect on another excitable cell (another nerve cell or a muscle cell). In the case of skeletal muscle, acetylcholine excites (activates) the cells.

The junction between a motor neuron and a skeletal muscle cell is called the **neuromuscular junction.** When an electrical impulse traveling in a motor neuron arrives at the neuromuscular junction, acetylcholine is released from the nerve terminal (Figure 6.6). The acetylcholine diffuses across the narrow space between the neuron and the muscle cell and binds to receptor sites on the muscle cell membrane. The binding of acetylcholine to the receptors causes the muscle cell membrane to generate an electrical impulse of its own that travels rapidly along the cell membrane in all directions. In addition, tube-like extensions of the cell membrane called *T tubules* (the "T" stands for transverse) transmit the electrical impulse deep into the interior of the cell. The function of the T tubules is to get the electrical impulse to all parts of the cell as quickly as possible.

***Recap* Motor neurons stimulate skeletal muscles by releasing acetylcholine at the neuromuscular junction.**

Activation releases calcium

As shown in Figure 6.6, T-tubules are in close contact with a series of membrane-bound chambers called the **sarcoplasmic reticulum** (*sarco-* is derived from a Greek word for "flesh" or "muscle"). The sarcoplasmic reticulum is similar to every other

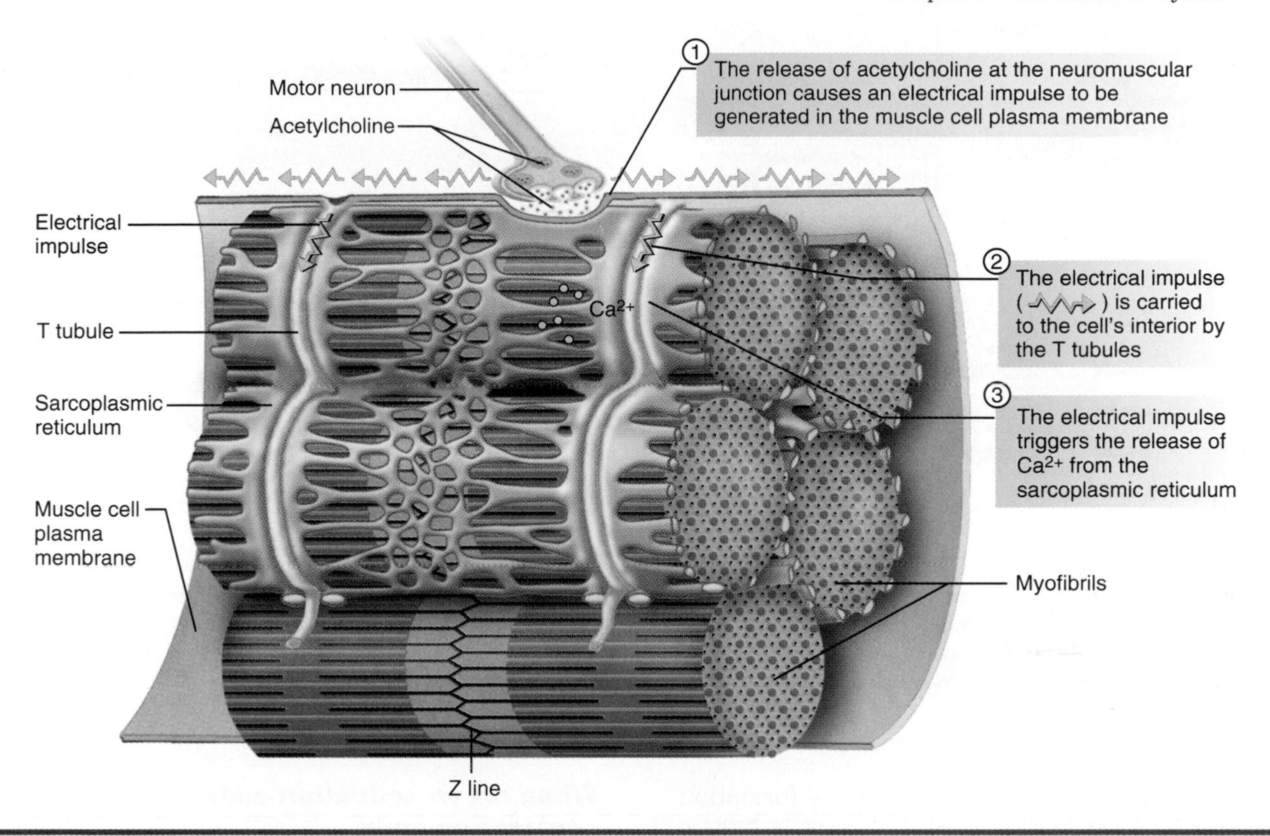

Figure 6.6 How nerve activation leads to calcium release within a muscle cell.

cell's smooth endoplasmic reticulum except that its shape is different, in part because it must fit into the small amount of space in the cell not occupied by myofibrils. The primary function of the sarcoplasmic reticulum is to store ionic calcium (Ca^{2+}).

Inside the muscle cell, an electrical impulse races down the T tubules to the sarcoplasmic reticulum. The arrival of an electrical impulse triggers the release of calcium ions from the sarcoplasmic reticulum. The calcium diffuses into the cell cytoplasm and then comes in contact with the myofibrils, where it sets in motion a chain of events that leads to contraction.

Calcium initiates the sliding filament mechanism

Muscles contract when sarcomeres shorten, and sarcomeres shorten when the thick and thin filaments slide past each other. The mechanism whereby the filaments slide past each other is known as the **sliding filament mechanism** of contraction. Taking a closer look at the arrangement of thick and thin filaments in a single sarcomere (Figure 6.7), we see that every thin filament consists of two strands of actin molecules spiraling around each other. Thick filaments are comprised of many individual molecules of myosin. Myosin molecules are shaped somewhat like a golf club, with a long shaft and a rounded head. Myosin shafts form the main part of the thick filaments. The heads stick out to the side, nearly touching the thin filaments of actin. When a muscle is relaxed, the myosin heads do not quite make contact with the thin filaments, however. Contraction occurs when the myosin heads make contact with the thin filaments, forming a *cross-bridge* between the two filaments. The formation of a cross-bridge causes the head to bend relative to the shaft, pulling the actin molecules toward the center of the sarcomere. The processes of cross-bridge formation and bending (the molecular events of contraction) require energy.

But what initiates the process of contraction? Put another way, what prevents contraction from occurring all the time? The answer is that contraction is inhibited unless calcium is present. An even closer look at a section of myosin and actin (Figure 6.8) shows why. Closely associated with the actin filaments are two other protein molecules called *troponin* and *tropomyosin* that together form the *troponin-tropomyosin protein complex*. In the absence of calcium, the troponin-tropomyosin complex interferes with the myosin binding sites on the actin molecule. Following an electrical impulse, calcium released from the sarcoplasmic reticulum binds to troponin, resulting in a shift in the position of the troponin-tropomyosin protein complex that exposes

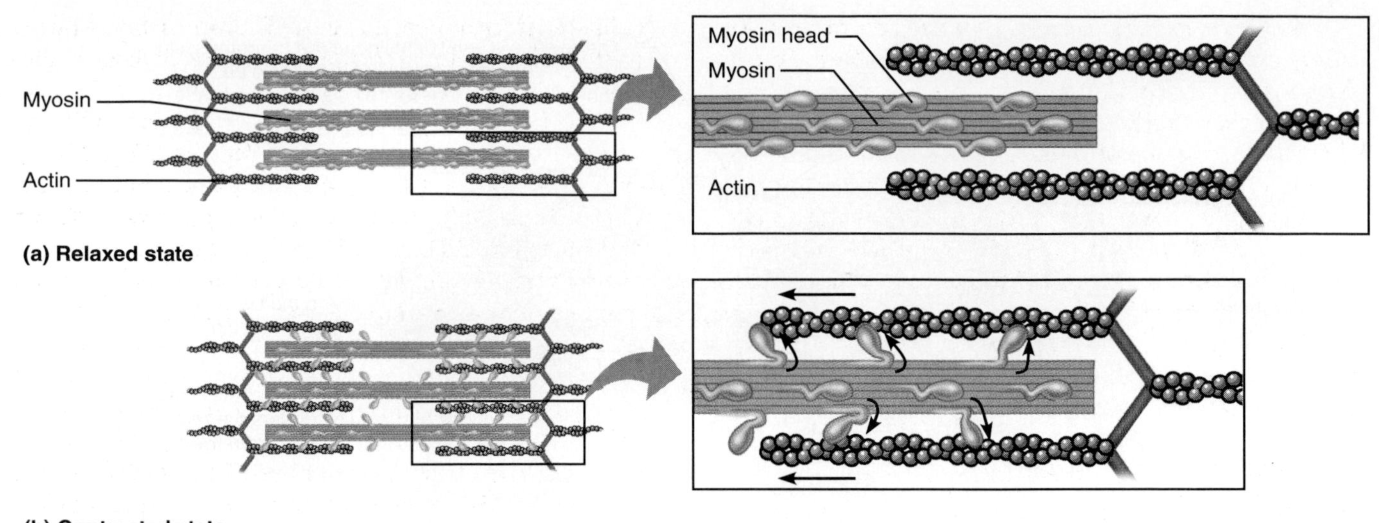

Figure 6.7 Sliding filament mechanism of contraction. (a) In the relaxed state the myosin heads do not make contact with actin. **(b)** During contraction the myosin heads form cross-bridges with actin and bend, pulling the actin filaments toward the center of the sarcomere.

the myosin binding sites and permits the formation of cross-bridges. At this point the myosin heads form cross-bridges with actin, undergo a bending process, and physically pull the actin filaments toward the center of the sarcomere from each end. With thousands of myosin cross-bridges doing this simultaneously, the result is a sliding movement of the thin filaments relative to the thick ones and a shortening of the sarcomere. As hundreds of thousands of sarcomeres shorten, individual muscle cells, and ultimately the whole muscle, shorten as well.

> ***Recap*** **The release of calcium from the sarcoplasmic reticulum triggers contraction. Calcium allows cross-bridges to form between myosin and actin, which leads to contraction by a sliding filament mechanism.**

When nerve activation ends, contraction ends

Relaxation of a muscle cell occurs when nerve activity ends. In the absence of nerve activity, no more calcium is released from the sarcoplasmic reticulum. The calcium released as a result of prior electrical impulses is transported back into the sarcoplasmic reticulum by active transport, which requires energy in the form of ATP. As the calcium concentration in the myofibrils falls, the troponin-tropomyosin complex shifts back into its original position, preventing the binding of the myosin cross-bridges to actin. The sarcomere stretches passively to its original resting length, and the muscle cell relaxes.

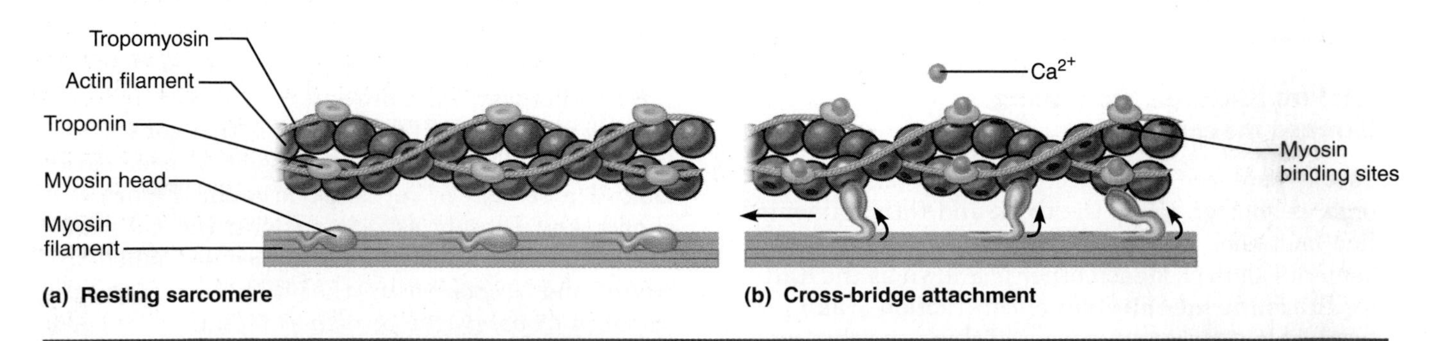

Figure 6.8 Role of calcium in contraction. (a) In the absence of calcium the muscle is relaxed because the myosin heads cannot form cross-bridges with actin. **(b)** When calcium binds to troponin it causes a shift in the troponin-tropomyosin complex, allowing cross-bridges to form.

Any factor that interferes with the process of nerve activation can disrupt muscle function. In the disorder *myasthenia gravis,* the body's immune system attacks and destroys acetylcholine receptors on the cell membrane of muscle cells. Affected muscles respond only weakly to nerve impulses or fail to respond at all. Most commonly impaired are the eye muscles, and many people with myasthenia gravis experience drooping eyelids and double vision. Muscles in the face and neck may also weaken, leading to problems with chewing, swallowing, and talking. Medications that facilitate the transmission of nerve impulses can help people with this condition.

Muscles require energy to contract and to relax

Muscle contraction requires a great deal of energy. Like most cells, muscle cells use ATP as the energy source. In the presence of calcium, myosin acts as an enzyme, splitting ATP into ADP and inorganic phosphate and releasing energy to do work. The energy is used to "energize" the myosin head so that it can form a cross-bridge and undergo bending. Once the bending has occurred another molecule of ATP binds to the myosin, which causes the myosin head to detach from actin. As long as calcium is present, the cycle of ATP breakdown, attachment, bending, and detachment is repeated over and over again in rapid succession. The result is a shortening of the sarcomere.

At the end of the contractile period (when nerve impulses end), energy from the breakdown of ATP is used to transport calcium back into the sarcoplasmic reticulum so that relaxation can occur. However, a second requirement for relaxation is that an intact molecule of ATP must bind to myosin before myosin can finally detach from actin. This last role of ATP is the explanation for *rigor mortis* (Latin, meaning "rigid death"), in which a body becomes stiff between the time period of about four hours and several days after death. Shortly after death calcium begins to leak out of the sarcoplasmic reticulum, causing muscle contraction. The contractions use up the available ATP, but after death the ATP cannot be replenished. In the absence of ATP, the myosin heads cannot detach from actin, and so the muscles remain "locked" in the contracted state. Eventually the stiffness of rigor mortis decreases as the muscle cells themselves degenerate.

What causes rigor mortis?

Muscle cells obtain ATP from several sources Muscle cells store only enough ATP for about 10 seconds worth of maximal activity. Once this is used up the cells must produce more ATP from other energy sources, including creatine phosphate, glycogen, glucose, and fatty acids.

An important pathway for producing ATP involves creatine phosphate (creatine-P), a high-energy molecule with an attached phosphate group. Creatine phosphate can transfer a phosphate group and energy to ADP and therefore create a new ATP quickly. This reaction is reversible: if ATP is not needed to power muscle contractions, the excess ATP can be used to build a fresh supply of creatine phosphate, which is stored until needed. In recent years creatine phosphate loading has become common among body builders and athletes, particularly those who need short-term bursts of power. Creatine phosphate also seems to improve muscle performance in certain neuromuscular diseases, such as ALS (amyotropic lateral sclerosis).

The combination of previously available ATP plus stored creatine phosphate produces only enough energy for up to 30–40 seconds of heavy activity. Beyond that, muscles must rely on stored glycogen, a complex sugar (polysaccharide) composed of many smaller molecules of glucose. For the first three to five minutes of sustained activity, a muscle cell draws on its internal supply of stored glycogen. Glucose molecules are removed from the glycogen and their energy is used to synthesize ATP. Part of the process of the breakdown of glucose can be done without oxygen (called anaerobic metabolism) fairly quickly, but it only yields two ATP molecules per glucose molecule and has the unfortunate side effect of producing lactic acid that can make muscles sore.

The most efficient long-term source of energy is the aerobic metabolism of glucose, fatty acids, and other high-energy molecules such as lactic acid. Aerobic metabolism, as you already know, takes place in mitochondria and requires oxygen. The next time you engage in strenuous exercise, notice that it may take you a few minutes to start breathing heavily. The increase in respiration indicates that aerobic metabolism is now taking place. Until aerobic metabolism kicks in, however, your cells are relying on stored ATP, creatine phosphate, and anaerobic metabolism of glycogen. Weight lifters can rely on stored energy because their muscles perform for relatively short periods. Long-distance runners start out by depending on stored energy, but in less than a minute they are relying almost exclusively on aerobic metabolism. If they could not, they would collapse in exhaustion. Table 6.1 summarizes energy utilization by muscle.

Table 6.1 Energy sources for muscle

Energy Source	Quantity	Time Course of Use	Comments
Stored ATP	Stored only in small quantities	About 10 seconds	ATP is the only direct energy source. It must be replenished by the other energy sources.
Stored creatine phosphate	Three to five times the amount of stored ATP	About 30 seconds	Converted quickly to ATP.
Stored glycogen	Variable. Some muscles store large quantities	Primarily used during heavy exercise within the first 3–5 minutes	ATP yield depends on whether oxygen is available. One glucose molecule (derived from stored glycogen) yields only two ATP in the absence of oxygen, but 36 ATP in the presence of oxygen.
Aerobic metabolism	Not a stored form of energy. Oxygen and nutrients (glucose and fatty acids) are constantly supplied by the blood	Always present. Increases dramatically within several minutes of onset of exercise, when blood flow and respiration increase	High yield. Complete metabolism of one glucose molecule yields 36 ATP molecules.

After you finish exercising, note that you continue to breathe heavily for a period of time. These rapid, deep breaths help reverse your body's **oxygen debt,** incurred because your muscles used more ATP early on than was being provided by aerobic metabolism. The additional ATP was produced by anaerobic metabolism, with the subsequent buildup of lactic acid. After exercise, oxygen is still needed to metabolize the lactic acid by aerobic pathways and to replenish energy stores. The ability of muscle tissue to accumulate an oxygen debt and then repay it later allows muscles to perform at a near-maximal rate even before aerobic metabolism has had a chance to increase. The increase in aerobic metabolism takes several minutes because it requires both an increase in respiration and an increase in blood flow to the tissues.

Muscle **fatigue** is defined as a decline in muscle performance during sustained exercise. The most common cause of fatigue is a lack of sufficient ATP to meet metabolic demands. However, fatigue can also be caused by psychological factors, including discomfort or the boredom of repetitive tasks.

Recap **Energy from the breakdown of ATP is required for contraction and for calcium transport. ATP is produced from metabolism of creatine phosphate and glycogen stores within the muscle, and from glucose and fatty acids obtained from the blood.**

6.3 The activity of muscles can vary

The general functions of muscles are to move body parts or to maintain a certain body position. How well they carry out their functions depends on a number of factors, including whether bones actually move or not, the degree of nerve stimulation, the type of muscle fiber, and the degree to which exercise has improved muscle mass and/or aerobic capacity.

Isotonic versus isometric contractions: Movement versus static position

Most types of exercise include a combination of two different types of muscle contractions, called isotonic and isometric contractions.

Isotonic ("same" + "strength" or "tone") contractions occur whenever a muscle shortens while maintaining a constant force. An example of an isotonic contraction would be when you generate enough muscle force to actually move an object or part of the skeleton. How heavy the object is doesn't matter, as long as parts of the skeleton actually move. It could be just your empty hand, a pencil, a book, or a 100-pound barbell.

In *isometric* ("same" + "length") contractions, force is generated, muscle tension increases, and the muscle may even shorten a little as tendons are stretched slightly, but bones and objects do not move. As a result, isometric contractions do not cause body movement. Examples would be when you tighten your abdominal muscles while sitting still, or when you try to lift a weight too heavy to lift. Isometric contractions help to stabilize the skeleton. In fact you are doing isometric muscle contractions all the time when you are standing up, just to maintain an upright position. If you doubt it, think about how quickly you would fall down if you were to faint and lose control over your skeletal muscles. Isometric contractions are a useful way to strengthen muscles.

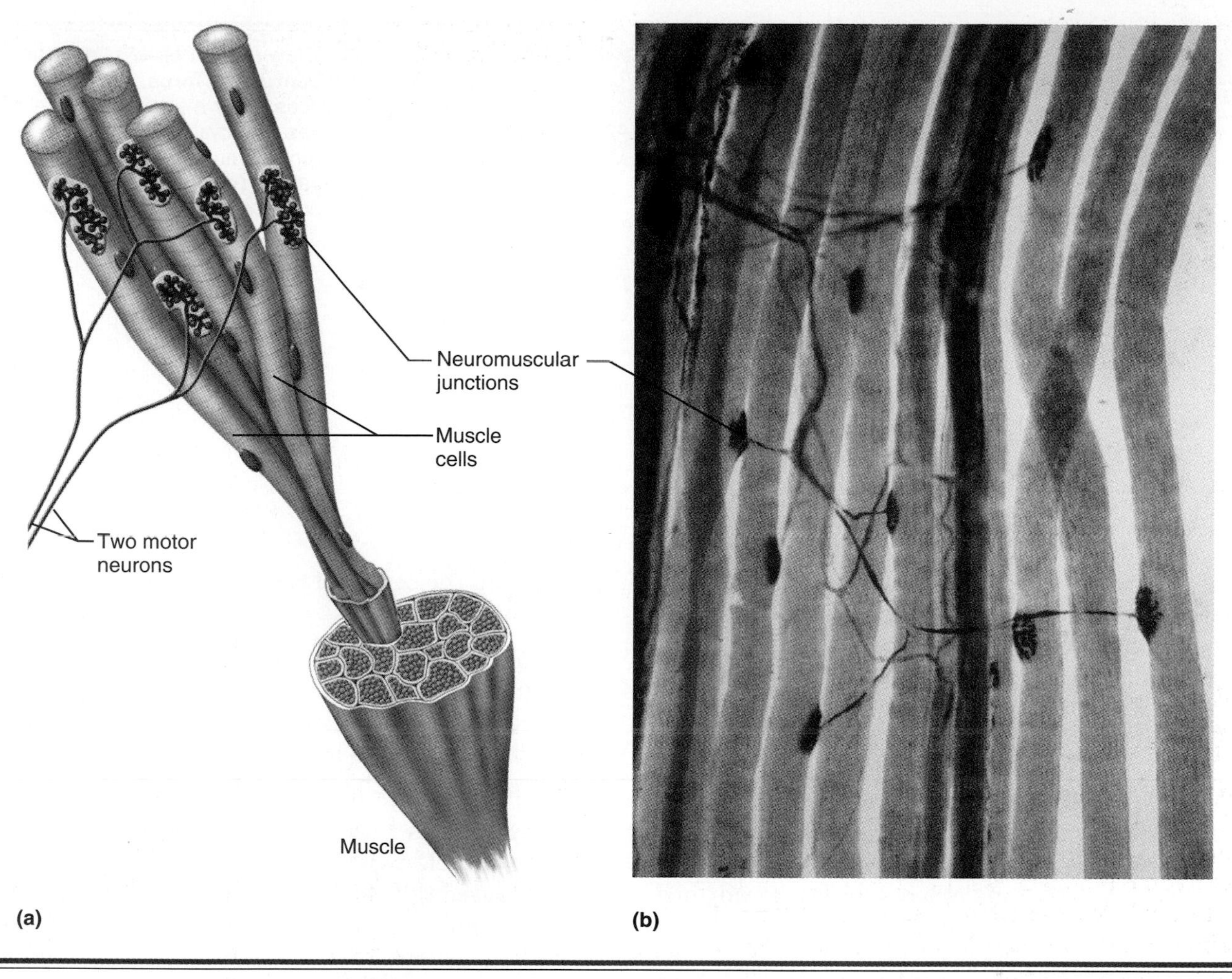

Figure 6.9 Motor units. (a) A motor unit consists of a motor neuron and all of the muscle cells it controls. Any one muscle cell is controlled by only one motor neuron, but a motor neuron controls more than one muscle cell. **(b)** A photograph (×200) of several muscle cells showing branches of the motor neuron and neuromuscular junctions.

The degree of nerve activation influences force

A single muscle may consist of thousands of individual muscle cells. The individual cells in any muscle are organized into groups of cells that all work together. Each group of cells is controlled by a single nerve cell called a *motor neuron* (because it affects movement). The motor neuron and all of the muscle cells it controls are called a **motor unit** (Figure 6.9). A motor unit is the smallest functional unit of muscle contraction, because when the motor neuron is activated all the muscle cells in that motor unit are activated together.

Our strength and ability to move effectively depend on how forcefully our muscles contract. The mechanical force that muscles generate when they contract is called **muscle tension.** How much tension is generated by a muscle depends on three factors:

1. The number of muscle cells in each motor unit (motor unit size)
2. The number of motor units active at any one time
3. The frequency of stimulation of individual motor units

Motor unit size can vary widely from one muscle to the next. The number of muscle cells per motor unit is a trade-off between brute strength and fine control. Larger motor units generate more force but offer less control. In the thigh muscle, where strength is more important than fine control, a single motor unit may consist of as many as a thousand muscle cells. In muscles of the eye, where fine control is essential, a motor unit may consist of only ten muscle cells.

According to the **all-or-none principle,** muscle cells are completely under the control of their motor neuron. Muscle cells never contract on their own.

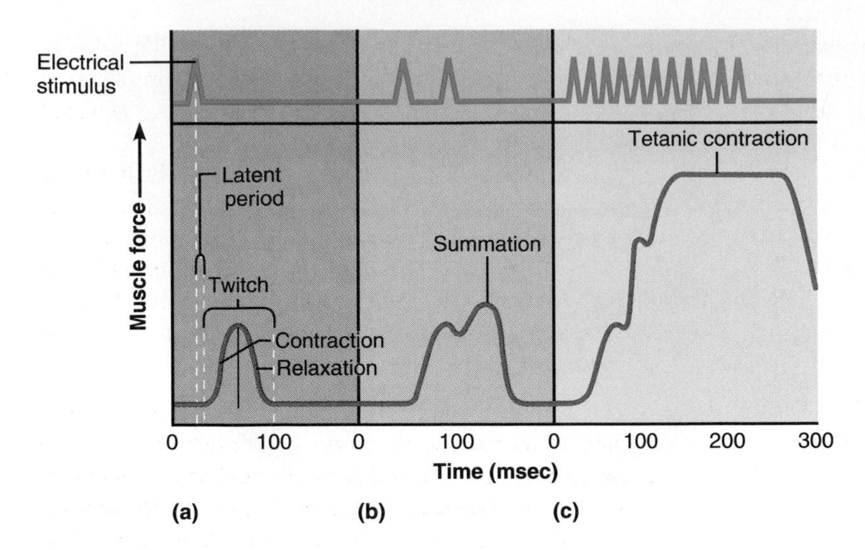

Figure 6.10 How frequency of stimulation affects muscle contractile force. (a) A single stimulus produces (after a short latent period) a contraction/relaxation cycle called a twitch. **(b)** More than one stimulus in a short time may produce summation. **(c)** Many stimuli close together may produce a tetanic contraction (tetanus).

For an individual muscle cell, there is no such thing as a half-hearted contraction, and there is no such thing as disobeying an order. Muscle cells always respond with a complete cycle of contraction and relaxation (called a **twitch**) every time they are stimulated by an electrical impulse from their motor neuron.

Although individual motor units either are contracting or are relaxed, whole muscles generally maintain an intermediate level of force known as *muscle tone*. Muscle tone exists because at any one time, some of the muscle's motor units are contracting while others are relaxed. The second factor that affects overall muscle force, then, is the number of motor units active at any one time. Increasing tone (or force) by activating more motor units is called **recruitment.** The maintenance of muscle tone depends on the nervous system.

The third factor that affects force generation by a muscle is the frequency of stimulation of individual motor units. To understand how frequency of stimulation influences force, we need to take a closer look at what happens when a muscle cell is stimulated by its motor neuron. Although we cannot study the contraction of single muscle cells in the laboratory easily, the study of whole muscles has revealed some important findings regarding the timing of the relationship between a stimulus and a twitch. A laboratory recording of muscle activity, called a *myogram,* reveals that the stimulus-twitch relationship has three stages (Figure 6.10):

- *Latent period* (the time between stimulation and the start of contraction). This is the time it takes for the nerve impulse to travel to the sarcoplasmic reticulum, for calcium to be released, and for the myosin heads to bind to the actin filaments.
- *Contraction* (the time during which the muscle actually shortens). Actin filaments are pulled toward the center of the sarcomere and myofibrils shorten.
- *Relaxation* (muscle returns to its original length). Calcium is transported back into the sarcoplasmic reticulum, the troponin-tropomyosin complex shifts back into its original position, and the sarcomere stretches passively to its original length.

A key point is that the contraction-relaxation cycle of the muscle twitch lasts longer than the stimulus that caused the contraction in the first place. If additional stimuli arrive at the muscle before the muscle has had a chance to transport calcium back into the sarcoplasmic reticulum and relax completely, then the total force produced becomes greater than the force produced by one twitch alone. In effect, the force becomes greater because more calcium is present. Increasing muscle force by increasing the rate of stimulation of motor units is called **summation** (Figure 6.10b).

There is a limit to summation, however. If stimulation becomes so frequent that the muscle cannot relax at all, it will remain in a state of maximum contraction called *tetanus* or a *tetanic contraction.* On a myogram, tetanus appears as a straight horizontal line representing the fusion of the peaks and valleys of individual twitches (Figure 6.10c). A tetanic contraction may lead eventually to muscle fatigue.

Recap **The number of muscle cells in a motor unit varies depending on whether the muscle's function is fine control or brute force. Activation of more motor units and/or increasing the frequency of stimulation of motor units produces greater force.**

Slow-twitch versus fast-twitch fibers: Endurance versus strength

As we have seen, all muscle cells can obtain ATP through both aerobic and anaerobic pathways. Humans have two types of skeletal muscle fibers, called slow-twitch and fast-twitch fibers. The distinction is based on how quickly they can utilize ATP to produce a contraction, and whether they use primarily aerobic or anaerobic metabolic pathways. Most muscles contain a mixture of both slow-twitch and fast-twitch fibers. The ratio of fiber types in any one muscle depends primarily on the function of the muscle.

Spotlight on Structure & Function

Slow-twitch fibers break down ATP slowly and so they contract slowly. They tend to make ATP as they need it by aerobic metabolism. Slow-twitch fibers contain many mitochondria and are well supplied with blood vessels, so they draw more blood and oxygen than fast-twitch fibers. They store very little glycogen because they can obtain glucose and fatty acids quickly from the blood. They store oxygen, however, in a molecule called myoglobin. The ability to maintain a temporary store of oxygen reduces the slow-twitch fiber's need for oxygen from the bloodstream. This is especially important during the early phases of an increase in activity, before blood flow to the muscle has increased. Myoglobin and the presence of numerous blood vessels make slow-twitch fibers reddish in color, so they are sometimes called "red" muscle.

Fast-twitch fibers can contract more quickly than slow-twitch fibers because they break down ATP more quickly. They have fewer mitochondria, fewer blood vessels, and little or no myoglobin compared to slow fibers, so they're called "white" muscle. Fast-twitch fibers store large amounts of glycogen and tend to rely heavily on creatine phosphate and anaerobic metabolism for quick bursts of high energy. Their contractions are rapid and powerful but cannot be sustained for long periods of time. Fast-twitch fibers depend on aerobic mechanisms for any activity that is sustained, but they have the capability of using anaerobic mechanisms for brief periods when bursts of power are needed. During periods of anaerobic activity they tend to accumulate lactic acid, which causes them to become fatigued quickly. ■

Which type of fiber is better? It depends on the activity involved. Because slow-twitch fibers offer more endurance, they are most useful for steady activities such as jogging, swimming, and biking. Slow-twitch fibers are also important for maintaining body posture. Many of the muscles of the leg and back, for example, contain a high percentage of slow-twitch fibers because they must contract for long periods to support us when we stand. Fast-twitch fibers are more often used to power brief, high-intensity activities such as sprinting for short distances, lifting weights, or swinging a tennis racquet. Muscles in our hands, for example, contain a high proportion of fast-twitch fibers, allowing the muscles to contract quickly and strongly when necessary.

The percentage of slow- and fast-twitch fibers varies not only from muscle to muscle but from person to person. The percentages are determined in part by inheritance, and they can influence athletic ability. For example, most world-class marathoners have a higher-than-average percentage of slow-twitch fibers in their legs.

Recap **Most muscles contain a combination of slow-twitch and fast-twitch fibers. Slow-twitch fibers rely on aerobic metabolism and are most useful for endurance. Fast-twitch fibers are capable of short bursts of high-intensity work and are most useful where strength is required.**

Exercise training affects muscle mass, strength, and endurance

Despite the fact that part of your athletic potential might be influenced by inheritance, a consistent, planned program of physical exercise (sometimes called *exercise training*) can improve your strength, endurance, and skill at any athletic endeavor (Figure 6.11). Whether it is primarily strength or endurance that is improved by exercise training depends on the type and intensity of training. The two primary types of exercise training are strength (resistance) training and aerobic (endurance) training.

Strength training involves doing exercises that strengthen specific muscles, usually by providing

(a)

(b)

Figure 6.11 The effects of strength training versus aerobic training. **(a)** A weight lifter. **(b)** A marathon runner.

CurrentIssue

What Are the Consequences of Using Anabolic Steroids?

Seventeen-year-old Aaron is just finishing his junior year of high school and has been thinking about college. He really wants a chance to play football in the PAC 10 league, but he won't be able to afford college unless he gets a scholarship. Aaron is 6′3″ and his league's all-star tight end, but at 170 pounds he's concerned that he may not be big enough or strong enough to make it at the college level. No matter how hard he trains he just can't seem to gain weight. He's heard that anabolic steroids will help him bulk up and is wondering if he should try them. Several of the better players are taking anabolic steroids without telling the coaches and have offered to help him get some if he wants them.

Yesterday Aaron heard a public service announcement warning of dangers of anabolic steroid abuse. But his friends aren't abusing them, are they? They've all gained muscle mass and are among the best athletes on the team. What should he do?

What are Anabolic Steroids?

The term **anabolic steroids** applies to over 100 compounds, all of which are related to the male sex steroid hormone testosterone. Like testosterone, anabolic steroids increase muscle mass (called an *anabolic* effect) and contribute to the development and maintenance of masculine characteristics (called an *androgenic* effect). The proper term for these drugs is "anabolic/androgenic" steroids, but they are generally referred to only as anabolic steroids. People take them for their anabolic effect and for an improvement in athletic performance. Most anabolic steroids have at least some androgenic effects as well.

Anabolic steroids were developed in the 1930s when it was discovered that they would facilitate the growth of skeletal muscle in rats. They have several legitimate medical uses (by prescription only), including the treatment of conditions in which the testes do not produce enough testosterone for normal growth, and the muscle wasting that sometimes accompanies HIV infection or other diseases.

Closely related to the anabolic steroids are certain dietary supplements, including dehydroepiadrosterone (DHEA) and androstenedione (street name Andro). These are legally available over the counter. Baseball slugger Mark McGwire was taking Andro in the year he hit his record number of home runs. Whether these compounds are powerful enough to have much effect on muscle growth is uncertain, but users take them in the belief that they do.

Anabolic Steroid Use May Be Rising

Anabolic steroids are illegal without a prescription. Nevertheless, they are surprisingly easy to get and their use appears to be on the rise. A 1999 survey reported that 2.7% of 10th-graders have taken anabolic steroids at least once, up from 2.0% in 1998. In addition, the percentage of 12th-graders who believe that taking anabolic steroids causes "great risk" to health declined from 68% in 1998 to 62% in 1999. Anabolic steroid use is higher among males than among females, but is growing more rapidly among females. Information on anabolic steroid use by college and professional athletes is unreliable because these drugs are banned by sports federations. Athletes are reluctant to admit to their use and generally don't keep a record of how much and how often they use them. Well-known athletes who have used them at one time or another include

Sprinter Ben Johnson was stripped of his gold medal in the 1988 Olympic Games after testing positive for anabolic steroids.

Arnold Schwarzenegger, sprinter Ben Johnson, wrestler Hulk Hogan, and football player Lyle Alzado (it is not known whether his brain tumor was related to the use of anabolic steroids).

Anabolic steroids are available in either oral or injectable forms or as gels or creams. Typically they are taken in repetitive cycles of 6–12 weeks of steadily increasing doses with periods of non-use in between, a method called "pyramiding." Many users take more than one anabolic steroid at a time, a technique known as "stacking." Because they are not approved and the user typically cannot get good information on what doses to use, the doses used tend to vary widely. The doses used by athletes can be 10 to 100 times higher than the prescription doses used medically. The perceived benefits of pyramiding and stacking have not been proven scientifically.

Do anabolic steroids work? The answer is probably yes, at least for the huge doses being taken by athletes, but there is a wide gap between what is popularly believed and what has been scientifically proved. Only about half of the available scientific studies report a significant improvement in athletic performance. The studies are often hard to compare because of differences in the forms or doses of anabolic steroids used, the intensity and duration of training, and the type of exercise. Importantly, most scientific studies tend to be shorter in duration and use doses of anabolic steroids that are less than a tenth of those used by many athletes.

In sharp contrast to the scientific studies, many athletes and body builders who have used very high doses of anabolic steroids believe strongly that anabolic steroids helped them gain body mass, made them stronger, and improved their ability to train at high intensities. The need for an athletic scholarship, the desire to win or to look better, and perhaps subtle and not-so-subtle pressures from peers, parents, and coaches to perform well, all may contribute to the decision to use anabolic steroids.

Risks Associated with Anabolic Steroids

Authorities are worried that abuse of anabolic steroids may pose a significant health risk. Users, on the other hand, either do not believe that anabolic steroids are harmful or are willing to accept the risks in favor of the perceived benefits. What health risks are known or suspected? What is known about the effects of anabolic steroid use at the dosages typically taken by athletes?

First, anabolic steroids have androgenic side effects in both sexes. Side effects in males include gynecomastia (enlargement of the breasts), shrinkage of the testicles, and male-pattern baldness. Masculinization occurs in females. Breast size and body fat decrease, the voice deepens, and women may lose scalp hair but gain body hair. Acne and oily hair and skin may be seen in both sexes. Some of these changes, including gynecomastia in males and hair loss in both sexes, are irreversible.

Second, anabolic steroid use/abuse is associated with irritability and aggressive behavior. Some steroid users report that they feel more aggressive and commit more aggressive acts while taking anabolic steroids (the so-called "'roid rage"). Not all studies have been able to document these effects.

Third, anabolic steroid use/abuse is associated with an increased risk of heart attacks, strokes, and severe liver diseases, including liver cancer. These reported associations come from case reports rather than carefully controlled scientific studies. Although the number of case reports might suggest that the incidence of these life-threatening complications is low, it is very likely that the effects of steroid use/abuse may be under-recognized or under-reported, particularly since these diseases tend to come later in life.

Fourth, anabolic steroids have been shown to cause premature death in mice. In one study, male mice were exposed for one-fifth of their normal life span to doses of anabolic steroids comparable to low and high doses taken by athletes. Although there were no significant differences in the death rates of the young rats during the period of treatment, over 50% of the rats treated with the high dose of anabolic steroids were dead by 20 months of age, compared with only 12% of the control animals.

Some people argue that although anabolic steroid use *is associated with* an increased risk of heart disease, strokes, and liver problems, it has never been proven scientifically that anabolic steroids actually *cause* these diseases. While this argument may be true, the close association between anabolic steroid use and increased risk of disease is still a legitimate cause for concern because it indicates that there *could* be a cause/effect relationship. A carefully controlled study of the effects of anabolic steroids in humans might help answer the question of risk. Like the mouse study described above, such a study would have a control group (persons given a placebo) and an experimental group (persons given the doses of steroids users/abusers typically take). However, such a study may never be undertaken, because one of the basic tenets of ethical scientific research is that humans should not be used in experiments that may harm or injure the subject.

In the future, new information regarding the risks associated with anabolic steroids is likely to come from either controlled experiments in animals or retrospective population studies of humans who have taken anabolic steroids. Although animal studies can be carried out quickly, it may be decades before retrospective studies clarify the full extent of the risks associated with anabolic steroid use/abuse in humans. A person who chooses to use/abuse anabolic steroids today may be conducting a potentially dangerous experiment with his or her life.

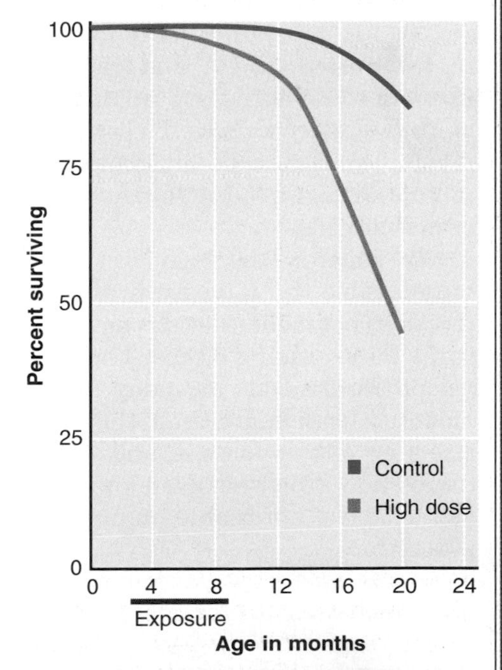

Survival curves for mice treated for 6 months with a placebo (control group), and a high dose of anabolic steroids, comparable to the high doses taken by some athletes.

Health Watch

The Benefits of Regular Exercise

Humans were designed to move, and our bodies thrive on muscular activity. Thus it's not surprising that regular exercise is good for our health. People who stay active experience lower rates of heart disease, lung disease, cancer, arthritis, diabetes, and other health-threatening conditions. The more active we stay throughout our lives, the better off we are.

Exercise Builds Muscle Tissue and Aerobic Capacity

Exercise training cannot change the speed at which a particular muscle cell contracts. Whether a particular fiber is "slow" or "fast" is determined by its internal biochemistry, not by training. Nevertheless, exercise can improve muscle strength because it increases muscle mass, and, as you have learned, the type of exercise can influence which fiber type increases in mass the most. Exercise can also boost endurance by increasing aerobic capacity. As an added benefit, exercise also strengthens bones and increases their density.

Exercise Slows the Effects of Aging

Over our lifetimes we all experience changes in our muscular system. With advancing age the body produces less acetylcholine and the number of motor neurons declines. Decreased nerve stimulation leads to a gradual decrease in size, endurance, and strength of skeletal muscles. Skeletal muscle mass starts to shrink around age 30, so that on average we will have lost about 30% of our muscle mass by age 80. Some muscle cells die; others shrink as the number of myofibrils in them declines.

These changes gradually alter our endurance and strength, not to mention our appearance. They may also be partly to blame for the weight gain that annoys many of us as we get older. Every pound of skeletal muscle metabolizes over 30 calories a day. Less muscle tissue leaves more calories available to be stored as fat tissue.

Regular exercise can slow many of these age-related developments, even if it cannot stop them altogether. It also helps you burn calories more efficiently.

Why do weight lifters "bulk up" but cross-country runners do not?

some type of resistance that makes them work harder. Strength training is generally short, intense exercise such as weight lifting using free weights or weight machines. It builds more myofibrils, particularly in fast-twitch fibers, and causes the fast-twitch fibers to store more glycogen and creatine phosphate as quick energy sources. This increases the size of individual muscle cells and builds muscle mass and muscle strength, but it does not increase the number of muscle cells.

In general, the heavier the weight used, the more visible the increase in muscle size. However, this does not mean that strength training will necessarily build bulging biceps. The extent of muscle development depends on many factors, including the amount of resistance used, the duration of the exercise, how often you do it, and your own genetic predisposition. However, even low to moderate weights can lead to noticeable improvements in muscle strength.

Aerobic training involves activities in which the body increases its oxygen intake to meet the increased demands for oxygen by muscles. Whereas resistance training strengthens muscles, aerobic training builds endurance. With aerobic training the number of blood capillaries supplying muscle increases. In addition, the number of mitochondria in muscle cells and the amount of myoglobin available to store oxygen both increase. The muscle fibers themselves do not increase much in mass, nor do they increase in number. Aerobic exercise also improves the performance of the cardiovascular and respiratory systems. Less intense than strength training but carried out for prolonged periods, aerobic exercises include jogging, walking, biking, and swimming.

The usual guideline for aerobic training is to perform an activity that raises your heart rate to its "target heart rate" for at least 20 minutes, three times a week or more. During aerobic exercise we breathe faster and more deeply to promote adequate flow of oxygen from the lungs into the bloodstream. The heart beats more rapidly and more powerfully to raise blood flow to hard-working muscles. Metabolic activity within muscle cells increases, but the carbon dioxide and other waste products of cell metabolism are quickly removed from the body by the increased blood flow.

It's a good idea to combine any athletic activity with *stretching exercises*. Gentle stretching before exercise increases your heart rate gradually, pumping additional blood to your muscles and preparing you for more strenuous exertion. This lowers your risk of sprains and pulled muscles. After exercising, let your heart rate and breathing return gradually to normal as you walk slowly and do more stretching. Regular stretching improves joint mobility and range of motion. Whenever you stretch, do it gradu-

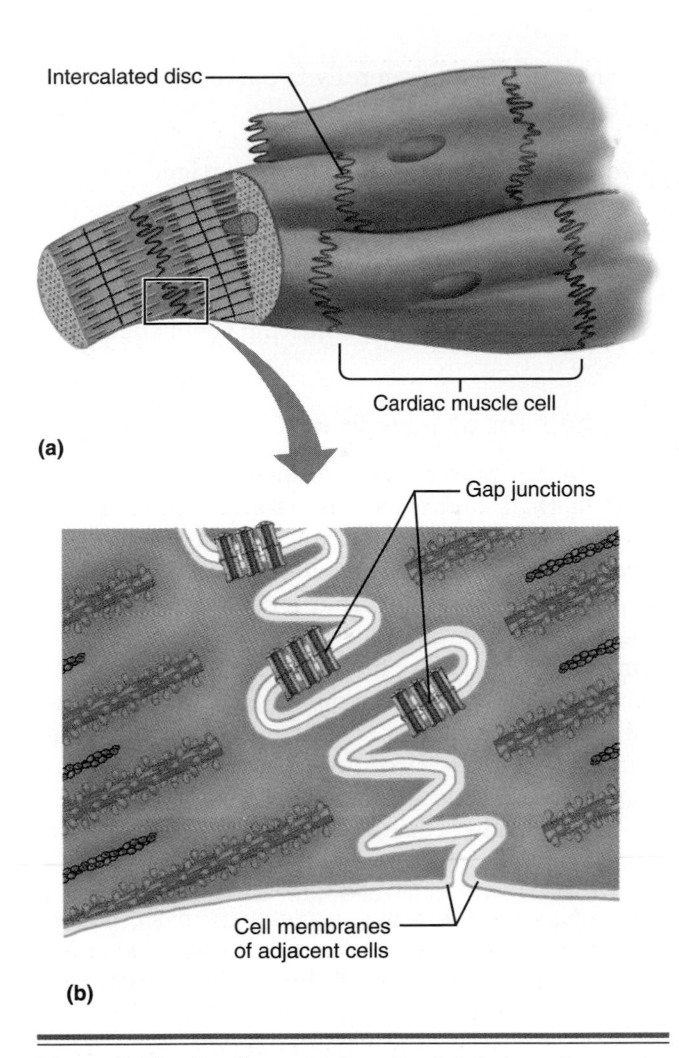

Figure 6.12 Cardiac muscle cells. (a) A view of several adjacent cardiac muscle cells showing their blunt shape and the intercalated discs that join them together. **(b)** A closer view showing that intercalated discs are bridged by gap junctions that permit direct electrical connections between cells.

ally and hold each position for 30 seconds. You should feel a gentle pull in your muscles, but not pain. Try not to bounce, because abrupt stretches could cause your muscles to contract quickly in response, increasing the risk of injury.

Recap **Exercise increases aerobic capacity, muscle mass, and muscle strength, but does not increase the number of muscle cells. The type of exercise training determines whether it is primarily strength or aerobic capacity that is improved.**

6.4 Cardiac and smooth muscles have special features

Most of the overall muscle mass of the body is skeletal muscle. Nevertheless, both cardiac and smooth muscle have unique features that suit them ideally for their roles in the body. All three types of muscle tissue were described briefly in Chapter 4. Here we look at some of the special attributes of cardiac and smooth muscle that set them apart from skeletal muscle.

How cardiac and smooth muscles are activated

Cardiac and smooth muscle are called *involuntary muscle* because we generally do not have voluntary control over them. Both cardiac and smooth muscles can contract entirely on their own, in the absence of stimulation by nerves. Although all cardiac muscle cells are capable of beating spontaneously and establishing their own cycle of contraction/relaxation, those with the fastest rhythm are called *pacemaker cells* because the rest of the cells follow their faster pace. Cardiac muscle cells are joined at their blunt ends by structures called *intercalated discs* (Figure 6.12). The intercalated discs contain gap junctions that permit one cell to electrically stimulate the next one. In effect, the pacemaker cells dictate the rate of contraction of the whole heart because their pace drives the slower cells.

Smooth muscle cells are also joined by gap junctions that permit the cells to activate each other, so that the whole tissue contracts together in a coordinated fashion. In contrast, skeletal muscle cells are only activated by motor neurons. This is why the

Try It Yourself

Calculating Your Target Heart Rate

Your target heart rate for aerobic exercise is considered to be a range that is 65 percent to 85 percent of your maximum attainable heart rate. The general formula for maximum attainable heart rate (without pushing yourself to the maximum just to find out) is 220 minus your age. To calculate your target heart rate, then, subtract your age from 220 and then multiply the result by 0.65 and 0.85 to find the lower and upper limits of your target range. By these calculations, the target range for a 20-year-old is between 130 and 170 beats per minute. When you do aerobic exercise, you should try to keep your heart rate within the range that is right for you. If you're below the minimum you're probably loafing. If you're above it you could be placing undue strain on your heart and muscles.

Notice that the maximum attainable heart rate (and the target heart rate) decreases with age. Consult your doctor about the best degree of exertion for your fitness level, particularly if you're new to exercise.

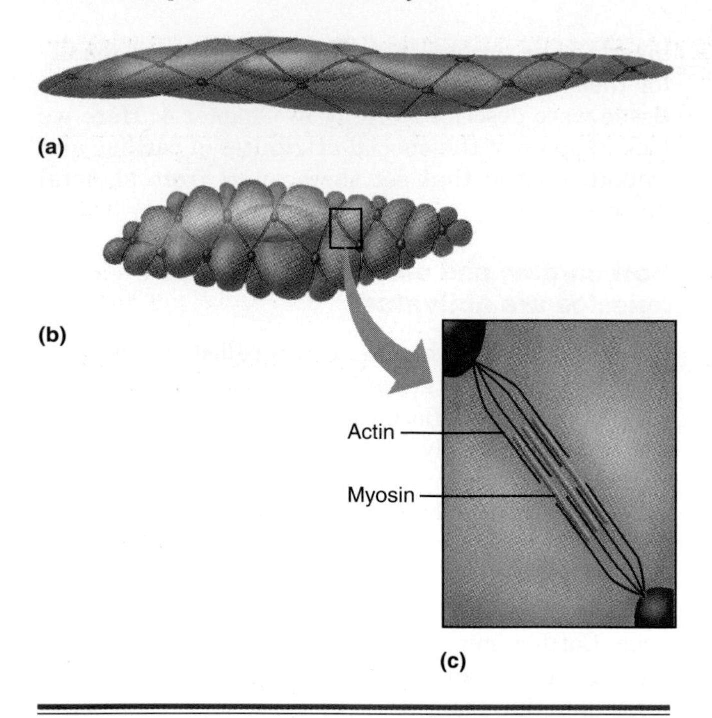

Figure 6.13 Smooth muscle. (a) Relaxed state. **(b)** Contracted state. The crisscross arrangement of bundles of contractile filaments causes the cell to become shorter and fatter during contraction. **(c)** A closer view showing how actin filaments are attached to cell membrane proteins.

skeletal muscles of a person with a severed spinal cord are completely paralyzed below the point of injury.

Even though cardiac and smooth muscle cells are capable of contracting without nerves, they do respond to stimulation by nerves as well. In both types of muscle the nerves belong to the autonomic nervous system (Chapter 11). The effect of nerve stimulation may be either inhibitory or stimulatory. Changes in both inhibitory and stimulatory nerve activity to the heart are responsible for the increase in your heart rate when you exercise, for example. Nerve stimulation can also change the contractile force of smooth muscle.

Recap **Unlike skeletal muscle, both cardiac and smooth muscle can contract in the absence of nerve stimulation. However, both also are capable of responding to nerves.**

Speed and sustainability of contraction

In terms of speed and sustainability of contraction, skeletal muscle is the fastest, cardiac muscle is of moderate speed, and smooth muscle is very slow.

Cardiac muscle cells go through rhythmic cycles of contraction and relaxation. The relaxation periods are necessary periods of rest so that the muscle doesn't fatigue.

Smooth muscle generally is partially contracted all the time. This makes it ideally suited for situations in which contractions need to be sustained. Nevertheless, they almost never fatigue because they contract so slowly that they always use ATP more slowly than they can make it. Smooth muscle is a key player in the homeostatic regulation of blood pressure because it can maintain the diameter of blood vessels indefinitely, adjusting them slightly as necessary.

Arrangement of myosin and actin filaments

Like skeletal muscle, cardiac muscle has a regular array of thick and thin filaments arranged in sarcomeres, so it too is called striated muscle. In contrast, the thick and thin filaments in smooth muscle are arranged in bundles that attach at various angles to the cell membrane. When the thick and thin filaments slide past each other, the points of attachment of the filaments are pulled toward each other and the cell gets shorter and fatter (Figure 6.13). Because it lacks the sarcomere arrangement of filaments, smooth muscle lacks the striated appearance of skeletal and cardiac muscle, which is why it is called "smooth."

Table 6.2 summarizes important features of the three types of muscle.

Recap **Cardiac muscle contracts and relaxes rhythmically, whereas smooth muscle can sustain a contraction indefinitely. Cardiac muscle has thick and thin filaments arranged in sarcomeres, whereas the filaments in smooth muscle are arranged in bundles that crisscross the cell.**

6.5 Diseases and disorders of the muscular system

Throughout this chapter we have discussed a number of musculoskeletal health conditions. We'll finish by looking at several more.

Muscular dystrophy

Serious diseases of muscle are relatively uncommon, but foremost among them is **muscular dystrophy.** The term muscular dystrophy actually applies to several different hereditary diseases of muscle (recall that "dystrophy" means "abnormal growth"). In *Duchenne muscular dystrophy,* a single defective gene results in the lack of a particular muscle cell protein. The normal gene, when present, directs the cell to produce a protein called dystrophin that is part of the muscle cell membrane. The function of dystrophin is to limit the inflow of calcium into muscle

Table 6.2 Characteristics of skeletal, cardiac, and smooth muscle

Characteristics	Skeletal Muscle	Cardiac Muscle	Smooth Muscle
Location	Attached to bones (skeleton)	Found only in the heart	Found in the walls of blood vessels and in the walls of organs of the digestive, respiratory, urinary, and reproductive tracts
Function	Movement of the body. Prevention of movement of the body	Pumping of blood	Control of blood vessel diameter. Movement of contents in hollow organs
Anatomical description	Very large, cylindrical, multinucleated cells arranged in parallel bundles	Short cells with blunt, branched ends. Cells joined to others by intercalated discs and gap junctions	Small, spindle-shaped cells joined to each other by gap junctions
Initiation of contraction	Only by a nerve cell	Spontaneous (pacemaker cells), modifiable by nerves	Some contraction always maintained. Modifiable by nerves
Voluntary?	Yes	No	No
Gap junctions?	No	Yes	Yes
Speed and sustainability of contraction	Fast—50 msec. (0.05 seconds). Not sustainable	Moderate—150 msec. (0.15 sec.). Not sustainable	Slow—1–3 seconds Sustainable indefinitely
Likelihood of fatigue	Varies widely depending on type of skeletal muscle and work load	Low. Relaxation between contractions reduces the likelihood	Generally does not fatigue
Striated?	Yes	Yes	No

cells through calcium "leak" channels. Persons with muscular dystrophy lack dystrophin, and as a result too much calcium leaks into the muscle cell through the leak channels. The high intracellular calcium concentration activates enzymes that damage muscle proteins and ultimately may kill the cell. The result is a loss of muscle fibers and muscle wasting. Eventually much of the muscle mass is replaced with fibrous connective tissue. Many people with muscular dystrophy die before age 30, usually because of failure of the heart muscle or the skeletal muscles responsible for breathing. At the moment there is no cure, though it is an area of intense research interest and progress is being made on several fronts.

Tetanus

A disease called tetanus is caused by a bacterial infection. Generally the infection is acquired by a puncture wound to a muscle. The bacteria produce a toxin that results in overstimulation of the nerves controlling muscle activity, resulting in tetanic contractions. The toxin affects a variety of skeletal muscles, but especially those of the jaws and neck. Jaw muscles may contract so forcefully that the jaw seems locked shut, which is why this disorder is nicknamed "lockjaw." Untreated, tetanus may lead to death due to exhaustion or respiratory failure. The disease is called tetanus because this is the technical term for a maximal (tetanic) muscle contraction. However, most maximal muscle contractions are not the result of disease.

Muscle cramps

What causes muscle cramps?

Muscle cramps occur when a muscle contracts uncontrollably. Generally they occur after heavy exercise, and are accompanied by pain. Cramps may be caused by ATP depletion, dehydration, or an ion imbalance. They may also be caused by a buildup of lactic acid when the exercise is so heavy that the muscle is forced to resort to anaerobic metabolism. Generally they can be soothed by increasing the circulation to the affected muscle through gentle stretching and massage.

Pulled muscles

Pulled muscles, sometimes called torn muscles, result from stretching a muscle too far, causing some of the fibers to tear apart. Internal bleeding, swelling, and pain often accompany a pulled muscle.

Health Watch

Muscle Soreness

Most of us are familiar with the feeling of stiffness and soreness that occurs 1–2 days after participating in an unfamiliar form of exertion. The soreness is actually due to microscopic muscle damage in the form of torn myofibrils throughout the muscle. It is thought that the damage occurs because in the absence of regular use, some sarcomeres become unable to contract as well as they should. They become like old, stiff rubber bands. During exercise these old or damaged sarcomeres become stretched passively to the point that the thick and thin filaments no longer overlap, and they become permanently damaged. The damaged sarcomeres disintegrate over several days and are removed as part of the repair process. As the sarcomeres disintegrate, however, they release chemical substances that cause the feeling of soreness. With time, the damaged sarcomeres are removed completely and new sarcomeres take their place.

Once muscles have become accustomed to a particular exercise, damage (and the accompanying soreness) no longer occurs. To minimize muscle damage and soreness, any new exercise activity should be undertaken in moderation for the first few days.

Fasciitis

Fasciitis involves inflammation of a layer of fascia, usually from straining or tearing the fascia surrounding a muscle. Most often it affects the sole of the foot (plantar fasciitis), where it is a common cause of heel pain. Like tendons and ligaments, fascia mend slowly. Treatment includes resting the area and protecting it from pressure. Injections of corticosteroid drugs can relieve severe pain.

Recap **Muscular dystrophy is an inherited disease in which the absence of a single protein causes an abnormal leak of calcium into muscle cells. Ultimately the leak of calcium damages muscle cell proteins and kills muscle cells.**

Chapter Summary

Muscles produce movement or generate tension

CD Muscular System/Anatomy Review: Skeletal Muscle Tissue

- All muscles produce movement or maintain position by contracting (shortening in length).
- The ways in which skeletal muscles attach to the skeleton determine what particular motion they cause.
- Within a single myofibril in a single muscle cell (fiber), thousands of contractile units called sarcomeres are arranged end to end.

Individual muscle cells contract and relax

CD Muscular System/Contraction of Motor Units

CD Muscular System/Sliding Filament Theory

- Skeletal muscle cells contract only when activated by their motor nerve.
- Motor nerve activation causes calcium to be released from the sarcoplasmic reticulum of the muscle cell.
- In the presence of calcium, the thick (myosin) and thin (actin) filaments slide past each other and the sarcomere shortens.
- ATP supplies the energy for the entire contraction/relaxation process.

The activity of muscles can vary

- An *isotonic contraction* occurs when a muscle shortens while maintaining a constant force; an *isometric contraction* occurs when tension is generated but the bones do not move.
- A motor unit is a single motor neuron and all of the skeletal muscle cells that it controls.
- The force generated by a muscle depends on the number of muscle cells in each motor unit, how many motor units are active at any one moment, and the frequency of stimulation of motor units.
- Muscle strength and endurance are dependent on the ratio of slow-twitch to fast-twitch fibers in the muscle and on the type and amount of exercise training.

Cardiac and smooth muscles have special features

CD Muscular System/Anatomy Review: Skeletal Muscle Tissue

- Cardiac and smooth muscles do not attach to bones.

- Both cardiac and smooth muscles can contract spontaneously, and both can be influenced by nerves of the autonomic nervous system.
- Cardiac muscle contracts rhythmically, with a period of relaxation between each contraction. Smooth muscle can maintain at least some contractile force indefinitely.

Diseases and disorders of the muscular system

- Muscular dystrophy is caused by inheritance of an abnormal gene.
- Pulled (torn) muscles occur when a muscle is stretched too far.

Terms You Should Know

actin, 129	recruitment, 136
all-or-none-principle, 135	sarcomere, 128
fatigue, 134	sarcoplasmic reticulum, 130
motor neuron, 135	sliding filament mechansim, 131
motor unit, 135	summation, 136
myosin, 131	twitch, 136
neurotransmitter, 130	
oxygen debt, 134	

Test Yourself

Matching Exercise

_____ 1. Rapid contraction and relaxation of skeletal muscles to generate heat
_____ 2. Muscle of the upper arm that flexes the forearm
_____ 3. The end of the muscle that attaches to stationary bone
_____ 4. The end of the muscle that attaches to bone that moves
_____ 5. Protein comprising thick filaments in a sarcomere
_____ 6. Protein comprising thin filaments in a sarcomere
_____ 7. A decline in muscle performance during sustained exercise
_____ 8. A type of skeletal muscle fiber capable of explosive bursts of power for short periods of time.
_____ 9. Type of muscle found in the walls of blood vessels
_____10. The mechanical force that muscles generate when they contract

a. insertion
b. actin
c. fast-twitch
d. biceps
e. smooth
f. origin
g. shivering
h. tension
i. fatigue
j. myosin

Fill in the Blank

11. Muscles over which we do not have conscious control are called _____________ muscles.
12. Muscle groups that work together to produce the same movement are called _____________ muscles.
13. A segment of a myofibril from one Z-line to the next is called a _____________.
14. Motor neurons release a chemical neurotransmitter substance called _____________ when a nerve impulse arrives at the neuromuscular junction.
15. When calcium binds to _____________ the troponin-tropomyosin complex shifts position, permitting the myosin heads to form cross-bridges with actin.
16. In an _____________ contraction, bones move relative to each other.
17. Activating more motor units in order to generate more muscle force is called _____________.
18. Cardiac muscle cells with the fastest spontaneous rate of contraction are called _____________ cells.
19. The cells of both cardiac and smooth muscle are joined to each other by _________ junctions that permit the cells to activate each other.
20. Another name for the disease called tetanus is _____________.

True or False?

21. Skeletal muscle cells (fibers) contain numerous cylindrical structures called myofibrils arranged in parallel.
22. *Rigor mortis* occurs after a spinal cord injury.
23. The time between stimulation of a muscle and the start of a contraction is called the latent period.
24. Strength training can markedly increase the number of muscle cells in a muscle.
25. All types of muscles (skeletal, cardiac, and smooth) have a striated appearance.

Concept Review

1. List the three types of muscle tissue.
2. Identify the three primary functions of muscle.
3. Describe the roles of calcium in muscle contraction.
4. Explain what causes *rigor mortis*.
5. Discuss some possible reasons for muscle fatigue.
6. Define summation, and explain why it occurs

when a muscle is stimulated rapidly and repetitively.

7. Explain why a spinal cord injury in the neck completely paralyzes the skeletal muscles of the limbs, whereas the cardiac muscle of the heart still beats rhythmically.
8. Compare/contrast how a constant degree of moderate tension, or tone, is maintained by a skeletal muscle that maintains posture versus a smooth muscle that maintains blood vessel diameter.
9. Define a motor unit, and describe how motor unit size and the number of motor units in a muscle affect muscle strength and fine motor control.
10. List the sources of energy that a muscle cell may use to make more ATP, both from within and from outside the cell.

Apply What You Know

1. Why do you think it is generally accepted medical practice to get bedridden patients up and walking as soon as possible?
2. In what ways would you expect the training regimen for a sprinter to be different from that of a marathon runner, and why?
3. Why do you suppose that heart muscle relies almost exclusively on aerobic metabolism?

Case in Point

Dietary Creatine Supplementation Enhances Athletic Performance

It is becoming increasingly clear that dietary creatine supplementation can increase the performance of certain athletes. In a recent study, researchers at the University of Memphis reported that creatine supplementation improved both weight lifting and sprint performance of 28 weight-matched Division 1A football players. Other studies report that creatine loading enhances performance in sprinters, swimmers, and weight lifters. In contrast, creatine loading has little effect on athletes who engage in submaximal endurance sports. The rationale behind creatine loading is that creatine is converted to creatine phosphate, an energy storage molecule, in muscle. Indeed, studies have shown that creatine loading can increase muscle creatine phosphate content by as much as 30%.

Creatine is an over-the-counter supplement, ranking near vitamins or carbohydrate loading for safety. You get plenty of it naturally in steak. Creatine supplementation has the approval of various sports associations and is acceptable even for Olympic athletes. The side effects of creatine loading appear to be minor, including only a tendency toward fluid retention and a small amount of weight gain.

Because it appears to be both effective and safe, dietary creatine supplementation is becoming common among athletes in certain high-intensity, short-duration sports.

QUESTIONS:

1. From what you know of muscle metabolism, explain the observation that the athletes who benefit the most from creatine loading are those engaged in sports characterized by intermittent, near-maximal strength activities, such as football, sprinting, and weight lifting.
2. Apparently there is an upper limit to the benefits of additional creatine loading. For example, taking five times the recommended dose of creatine does not improve athletic performance any further. Propose a tentative explanation for this observation.

INVESTIGATE:

Haff, G.G., and J.A. Potteiger. Creatine Supplementation for the Strength/Power Athlete. *Strength and Conditioning* 19 (6):72–74, Dec. 1997.

Plisk, S.S. and R.B. Kreider. Creatine Controversy? *Strength and Conditioning Journal* 21(1):14–23, Feb. 1999.

References and Additional Reading

Marieb, Elaine N. *Human Anatomy and Physiology*. 5th ed. San Francisco: Benjamin Cummings, an imprint of Addison Wesley Longman, 2001. (Good overall coverage of human anatomy and physiology.)

McArdle, William D., Frank I. Katch and Victor L. Katch. *Essentials of Exercise Physiology*. Philadelphia: Lea & Febinger, 1994. 563 pp. (In-depth coverage for the student interested in exercise and human performance.)

Yesalis, Charles E. *Anabolic Steroids in Sport and Exercise*. Champaign, IL: Human Kinetics Publishers, 2000. (A readable and well-referenced text on this subject.)

CHAPTER

BLOOD

How are blood cells **produced by your body?**

Why does your blood clot **under certain conditions?**

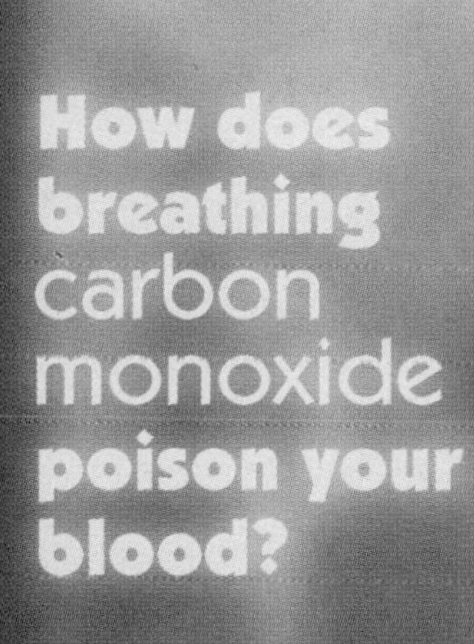

How does breathing carbon monoxide **poison your blood?**

What is your blood type, **and why is it important?**

Why are leukemia patients sometimes given bone marrow transplants?

ALL CELLS IN THE BODY MUST OBTAIN NUTRIENTS AND GET RID of wastes. How do they do that? Diffusion to and from the fluid that surrounds them is only part of the answer. If every cell drew its nutrients from extracellular fluid and there were no way to replenish them, the cells would soon starve. If every cell dumped its waste into the extracellular fluid and those wastes were not promptly removed, the cells would die in a sea of toxic waste. What cells need is a system for keeping the oxygen content, the supply of nutrients, the concentrations of wastes, and the concentrations of every essential molecule and atom within acceptable limits. What they need is a system to maintain homeostasis of the extracellular fluid. In humans, that system is the circulatory system.

The **circulatory system** consists of the heart, the blood vessels, and the blood that circulates through them. As shown in Figure 7.1, the circulatory system plays a central role in supplying all cells with what they need and removing the substances they no longer need. The pathways of the circulatory system ensure that blood flows throughout the entire body, bringing the necessary raw materials to the extracellular fluid surrounding every living cell and removing the waste. It picks up nutrients from the digestive system, exchanges gases with the respiratory system, and carries wastes and excess water and salts to the urinary system for removal from the body. It also carries some metabolic wastes to the liver for removal. Whenever a substance is transported over any distance within the body, the circulatory system is at work. Closely coupled to the circulatory system, indeed often considered part of it, is another system of fluid-filled vessels called the *lymphatic system.* The lymphatic system will be examined in further detail in Chapter 9 in connection with the immune system.

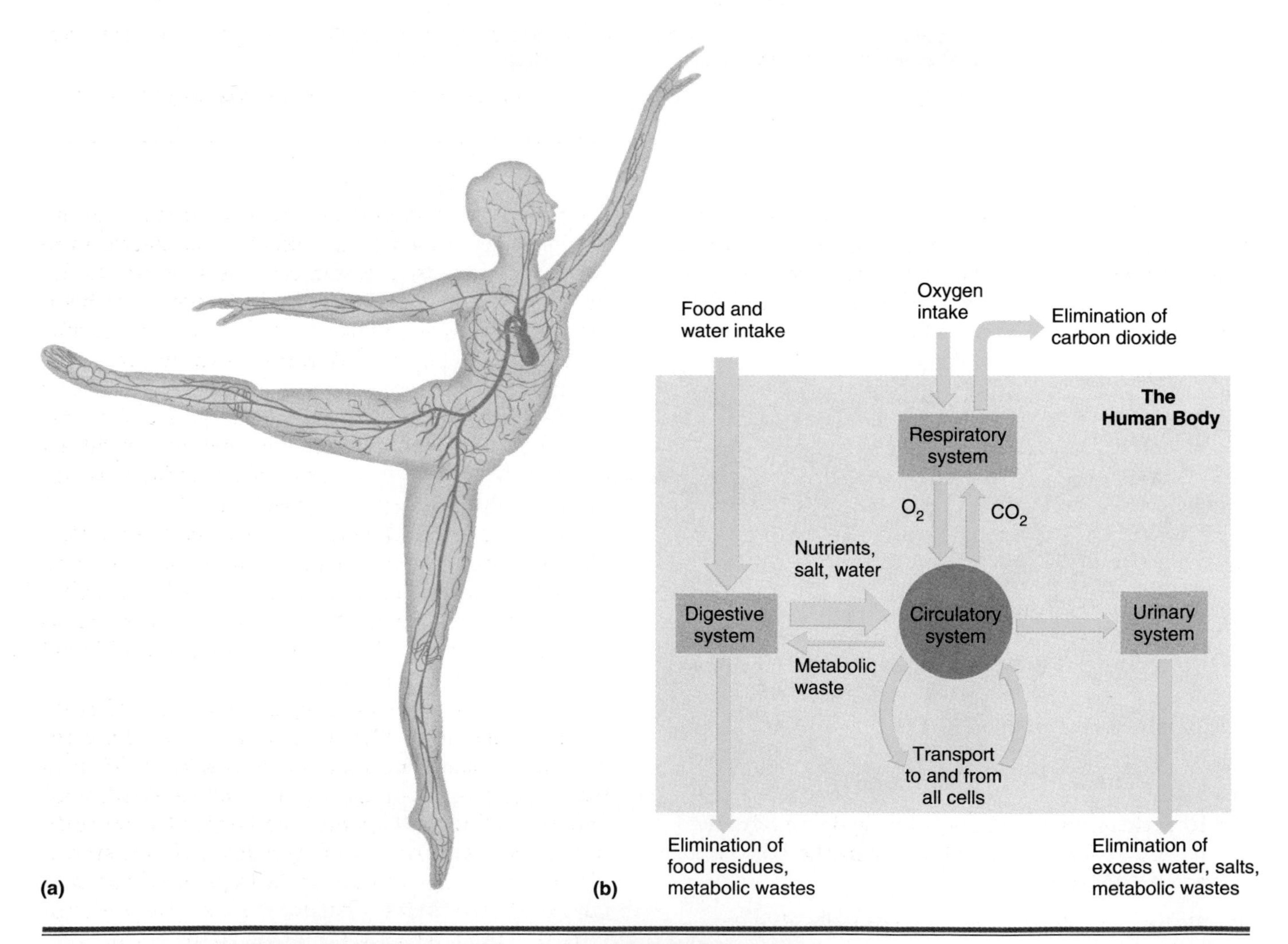

Figure 7.1. The transport role of the circulatory system. (a) An artist's rendition of the cardiovascular system, emphasizing the extensive branching of the system so that every living cell can be served. **(b)** The cardiovascular system serves homeostasis by transporting nutrients and carrying off wastes from all parts of the body.

Table 7.1 Composition of blood

Blood Component	Examples and Functions
Formed Elements (45%)	
Red blood cells	Transport oxygen to body tissues; transport carbon dioxide away from tissues.
White blood cells	Defend body against invading organisms, abnormal cells.
Platelets	Take part in blood clotting as part of the body's defense mechanisms.
Plasma (55%)	
Water	The primary constituent of blood plasma.
Electrolytes (ions)	Sodium, potassium, chloride, bicarbonate, calcium, hydrogen, magnesium, others. Ions contribute to the control of cell function and volume, to the electrical charge across cells, and to the function of excitable cells (nerve and muscle). All ions must be kept at their normal concentrations for homeostasis to occur.
Proteins	Albumins maintain blood volume and transport electrolytes, hormones, and wastes. Globulins serve as antibodies and transport substances. Clotting proteins contribute to blood clotting.
Hormones	Insulin, growth hormone, testosterone, estrogen, others. Hormones are messenger molecules, transporting information and signals throughout the body.
Gases	Oxygen is needed for metabolism; carbon dioxide is a waste product of metabolism. Both are dissolved in plasma as well as carried by RBCs.
Nutrients and wastes	Glucose, urea, heat, many others. Nutrients, raw materials, and wastes (including heat) are transported by blood throughout the body.

We will discuss the heart and blood vessels as well as the transport functions of the lymphatic system in Chapter 8. In this chapter we concentrate on the composition and crucial functions of blood, the fluid that circulates within the heart and blood vessels.

7.1 The components and functions of blood

Blood is a liquid consisting of specialized cells and cell fragments suspended in a watery solution of molecules and ions. Blood carries out three crucial tasks for the body:

- **Transportation:** Blood transports all substances needed anywhere by the body, including oxygen from the lungs, nutrients from the digestive system, and hormones from the endocrine glands. Blood also transports the products of cell metabolism away from body tissues to organs that eliminate them from the body.
- **Regulation:** Blood helps to regulate body temperature, the volume of water in the body, and the pH of body fluids.
- **Defense:** Blood contains specialized defense cells that help protect against infections and illness, and it has the ability to prevent excessive blood loss through the clotting mechanism.

Together, these functions are crucial for maintaining homeostasis. Blood is so effective at performing these functions that, so far, scientists' efforts to develop an artificial blood substitute have not been very successful. If someone needs blood, a transfusion of human blood is often the only solution.

Adult men average 5 to 6 liters of blood (about 1.5 gallons), and adult women average 4 to 5 liters. Differences between men and women reflect differences in body size. In general, blood represents approximately 8% of your body weight.

Blood is a bit thicker and stickier than water. This is because some components of blood are denser (heavier) than water and because blood is roughly five times more *viscous* (thicker—viscosity is a measure of resistance to flow). The old saying that blood is thicker than water is true.

Despite its uniform color, blood carries a rich array of components. Table 7.1 summarizes the components of blood and their functions. They fall into two major categories: the liquid component (plasma) and the cellular component or **formed elements** (*red cells, white cells,* and *platelets*). If you spin a blood sample in a centrifuge (a high-speed rotation device that mimics and magnifies gravitational forces), formed elements will sink to the bottom of a test tube because they are denser than plasma (Figure 7.2). Red blood cells, representing the bulk of the formed elements, settle to the bottom. White blood cells and platelets appear just above red blood cells in a thin, grayish-white layer.

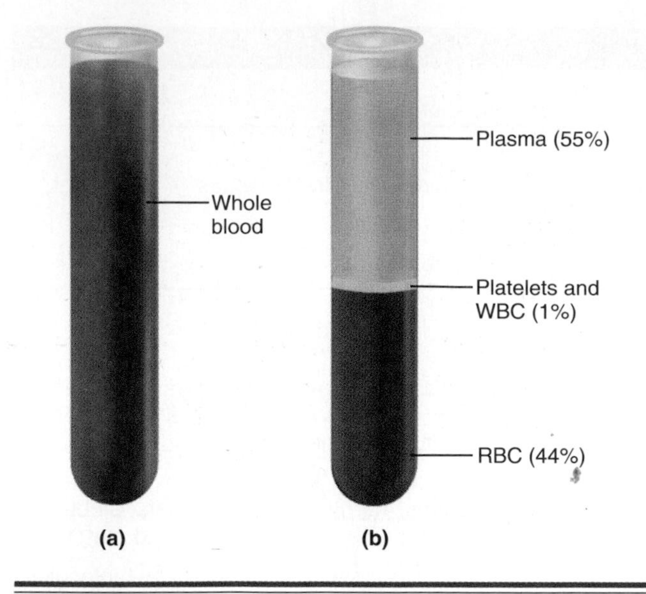

Figure 7.2 **(a)** Whole blood. **(b)** Centrifugation of a blood sample causes the formed elements to settle to the bottom.

Recap **About 8% of our body weight is blood. Blood maintains homeostasis by (1) transporting nutrients and exporting wastes, (2) regulating temperature and water volume, and (3) carrying molecules that fight disease and promote healing.**

Plasma consists of water and dissolved solutes

The top layer of a centrifuged blood sample, representing about 55% of the total volume, consists of a pale yellow liquid called **plasma.** Plasma is the transport medium for blood cells and platelets. About 90% of plasma is water. The rest is dissolved proteins, hormones, more than 100 different small molecules (including amino acids, fats, small carbohydrates, vitamins, and various waste products of metabolism), and ions.

The largest group of solutes in plasma consists of **plasma proteins,** which serve a variety of functions. Important plasma proteins include albumins, globulins, and clotting proteins.

Nearly two-thirds of plasma proteins are **albumins,** which primarily serve to maintain the osmotic relationship between blood and the interstitial fluid and thus ensure proper water balance. Manufactured in the liver, albumins also bind to certain molecules and drugs (such as bilirubin, fatty acids, and penicillin) and assist in their transport in blood.

Globulins (designated alpha, beta, and gamma) are a diverse group of proteins that transport various substances in the blood. Many beta globulins bind to lipid (fat) molecules, such as cholesterol. When a protein attaches to one of these molecules, it creates a complex called a *lipoprotein.* Two medically important lipoproteins are the low-density lipoproteins (LDLs) and high-density lipoproteins (HDLs), and medical exams often include taking a blood sample to measure their relative proportions. The LDLs are sometimes called "bad cholesterol," because high blood levels of these lipoproteins are associated with increased risk of cardiovascular health problems. High blood levels of HDLs often indicate a lower risk of cardiovascular disease. Lipoproteins and the health implications of blood cholesterol levels are discussed in Chapter 8.

Gamma globulins function as part of the body's defense system, helping to protect against infections and illness. We'll take a closer look at them in Chapter 9.

Clotting proteins, a third group of plasma proteins, play an important role in the process of blood clotting. As we see later in this chapter, blood clotting minimizes blood loss and helps maintain homeostasis after injury.

Other plasma components In addition to plasma proteins, plasma transports a variety of other molecules, including ions (also called electrolytes), hormones, nutrients, waste products, and gases. Electrolytes such as sodium and potassium contribute to the control of cell function and cell volume. Hormones, which are chemical "messengers" from the endocrine system, transport information and signals through the body. Nutrients such as carbohydrates, amino acids, vitamins, and other substances are absorbed from the digestive tract or produced by cells' metabolic reactions. Waste products, the byproducts of these reactions, include carbon dioxide, urea, and lactic acid. Gases dissolved in plasma include oxygen, which is necessary for metabolism, and carbon dioxide, a waste product of metabolism.

Recap **About 55% of whole blood consists of plasma. Plasma is mostly water and serves as the medium in which all other blood components are dissolved and transported. Plasma proteins include albumins, globulins, and clotting factors.**

Red blood cells transport oxygen and carbon dioxide

Just under half of the volume of whole blood consists of its formed elements. The most abundant are **red blood cells (RBCs),** also called **erythrocytes** ("red cells" in Greek). Red blood cells function primarily as carriers for oxygen and carbon dioxide. Each cubic millimeter of blood contains approximately 5 million red blood cells. They give blood its color and are the major reason why it is viscous.

Spotlight on Structure & Function

Red blood cells offer a great example of how structure serves function. Red blood cells are small, flattened, doughnut-shaped disks whose centers are thinner than their edges (Figure 7.3). This is an unusual shape among human cells, but it has several advan-

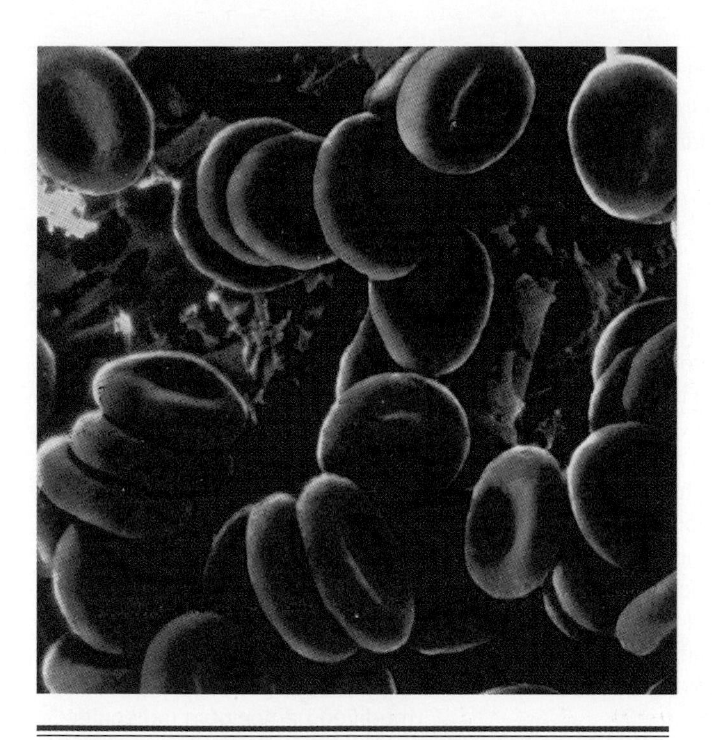

Figure 7.3 Red blood cells. Note that their flattened, biconcave shape gives them a sunken appearance.

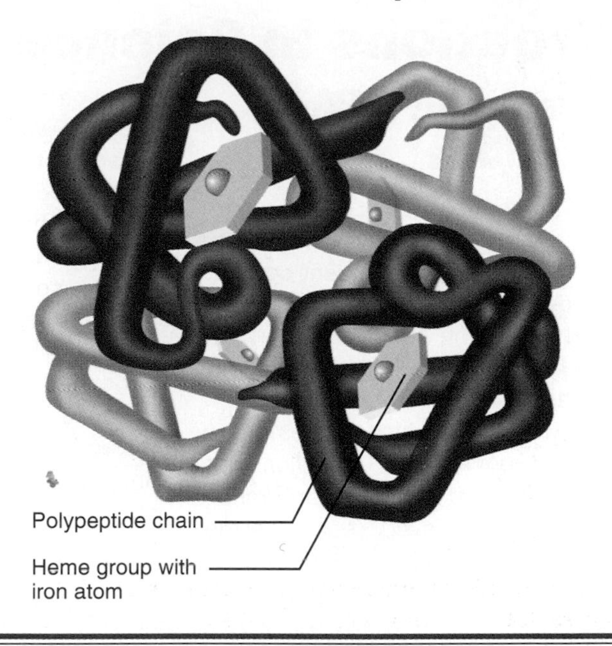

Figure 7.4 A hemoglobin molecule. Hemoglobin consists of four polypeptide chains folded together, each with a heme group containing a single iron atom. There are nearly 300 million of these molecules in every red blood cell.

tages for RBCs. It makes them flexible, so they can bend and flex to squeeze through tiny blood vessels. It also means that no point within an RBC's cytoplasm is ever far from the cell surface, which facilitates the process of gas exchange.

Red blood cells are highly specialized for the transport of oxygen. Mature RBCs have no nucleus and essentially no organelles. They are essentially fluid-filled bags of plasma membrane, crammed with nearly 300 million molecules of an oxygen-binding protein called **hemoglobin.** Hemoglobin consists of four polypeptide chains, each containing a *heme* group (Figure 7.4). At the center of each heme group is an iron atom, which can readily form a bond with an oxygen molecule. In total, a single red blood cell can carry up to 1.2 billion molecules of oxygen. RBCs lack mitochondria and generate ATP by anaerobic pathways, so they don't consume any of the oxygen they carry, they just transport it. ■

Several factors influence the binding of hemoglobin to oxygen. Hemoglobin binds oxygen most efficiently when the concentration of oxygen is relatively high and the pH is fairly neutral. These are precisely the conditions that prevail in the lungs. In the lungs, oxygen diffuses into blood plasma and then into red blood cells, where it attaches readily to the iron atoms in hemoglobin. The binding of O_2 by hemoglobin removes some of the O_2 from the plasma, making room for more O_2 to diffuse from the lungs into the plasma. Hemoglobin with four oxygen molecules attached, called *oxyhemoglobin,* has a characteristic bright red color.

The bond hemoglobin forms with oxygen must be temporary so the oxygen can be released to the cells that need it. In body tissues that are using oxygen in the course of their metabolic activities, the concentration of dissolved oxygen and the pH are both lower. Under these conditions, hemoglobin readily releases oxygen into body tissues, making it available to cells. Increased body heat will also increase the rate at which hemoglobin releases oxygen. Hemoglobin that has given up its oxygen is called *deoxyhemoglobin.* Deoxyhemoglobin is characteristically dark purple, but because venous blood returning from the cells contains a mixture of oxyhemoglobin and deoxyhemoglobin, venous blood generally has a dark red or maroon color that is in between red and purple.

Hemoglobin also transports some carbon dioxide, a waste product of cellular metabolism. In tissues, where carbon dioxide levels are high, about 25% of the CO_2 binds to hemoglobin (at different sites than O_2). In the lungs, CO_2 detaches from hemoglobin and is eliminated through respiration. See Chapter 10 for more about gas transport, including how the rest of the CO_2 is transported.

Recap **The ability of RBCs to transport nearly all of the oxygen and some of the carbon dioxide in blood depends on hemoglobin, the primary protein within red blood cells. Each molecule of hemoglobin contains four atoms of iron that can bind reversibly to oxygen.**

Directions in Science

Developing a Blood Substitute

Blood transfusions first became common in the 1800s. However, transfusion reactions were frequent because nobody knew that there were different blood types. The ABO blood types were discovered in 1901 and the rhesus factor was discovered in 1940, after which transfusions became fairly safe. Currently, over 100 million units of blood or blood products are transfused worldwide every year. Nevertheless, there are causes for concern about the world's blood supply:

- Receiving blood is associated with risk. Although the risk may be small, there is still the very rare chance of contracting a disease or of having a transfusion reaction.
- Blood may not always be available. Blood donations are falling worldwide. Whole blood can only be stored for about a month (although some blood products can be stored longer if frozen). Some places in the world do not have the facilities or the infrastructure to collect blood donations effectively.
- Blood is expensive. According to the Red Cross, it costs over $200 per unit to collect and test blood. That's expensive for a poor person in an underdeveloped nation.

Scientists have long dreamed of a liquid that could substitute for human blood. The ideal blood substitute would be easily and inexpensively produced. It would be safe. It would last for at least a month in the body (time for the body to replace most of the cells that had initially been lost). And it would be able to be stored for a year or more so that it could be shipped and used anywhere in the world.

Actually, the word "substitute" is not entirely accurate because none of the current products even attempt to replace or substitute for the defensive (immune) or clotting functions of blood. Most of the current effort is directed at finding a substitute for the oxygen-carrying capacity of blood, the one function of blood that is of immediate concern following blood loss. Work is proceeding on two very different types of products:

- **Perfluorochemicals.** Perfluorochemicals are a class of inert, nontoxic liquids that are known to be able to dissolve up to 50 times more oxygen than plasma. They are easy to produce in the laboratory and can be sterilized so that there is no chance of infection. They do not mix well with water, however, so it is difficult to get enough of them into solution to do much good. Current efforts are underway to mix them with lipids in order to form microscopic droplets that might be injected, but so far they have not proven to be of significant clinical use.
- **Cross-linked hemoglobin.** Because it is the red blood cell membrane and antibodies in plasma that cause transfusion reactions, the first attempt to duplicate the oxygen-carrying capacity of blood was to try to administer pure hemoglobin that had been extracted from blood. Unfortunately, without the surrounding RBC membrane the natural hemoglobin tends to split in half. The smaller half-molecules are toxic because they can plug up and damage the kidneys. Current efforts are underway to try to cross-link hemoglobin molecules together, forming large *polyhemoglobin* complexes that don't come apart and are too large to cause kidney damage. Because these modified hemoglobin molecules still require hemoglobin derived from RBCs and because human blood is in short supply, some researchers are working with bovine (cattle) blood as their raw material. Other researchers are attempting to genetically engineer bacteria so that they produce hemoglobin molecules that will not come apart.

Although the efforts so far have not yielded a completely effective blood substitute, slow but steady progress is being made. There is a strong possibility that within a decade or two at least some blood substitutes will be available for use in certain circumstances.

Hematocrit reflects oxygen-carrying capacity

The percentage of blood that consists of red blood cells is called the **hematocrit.** Hematocrit is an important medical concept because it is a relative measure of the oxygen-carrying capacity of blood.

Normally, hematocrit levels range from 43% to 49% in men, 37% to 43% in women. An unusual hematocrit is sometimes a reason for concern. A low hematocrit may signal *anemia* or other disorders of inadequate red blood cell production (see Section 7.4: Blood Disorders). A high hematocrit can also be risky, because excessive red blood cells thicken blood and increase the risk of blood clots. In rare cases a high hematocrit could signal *polycythemia,* a disorder of the bone marrow characterized by an unusually large number of red cells. Polycythemia results in increased volume and thickening of the blood, sometimes leading to headaches, blurred vision, and high blood pressure.

Some shifts in hematocrit are normal and temporary. For example, if you visit the mountains on your next vacation and stay for at least several

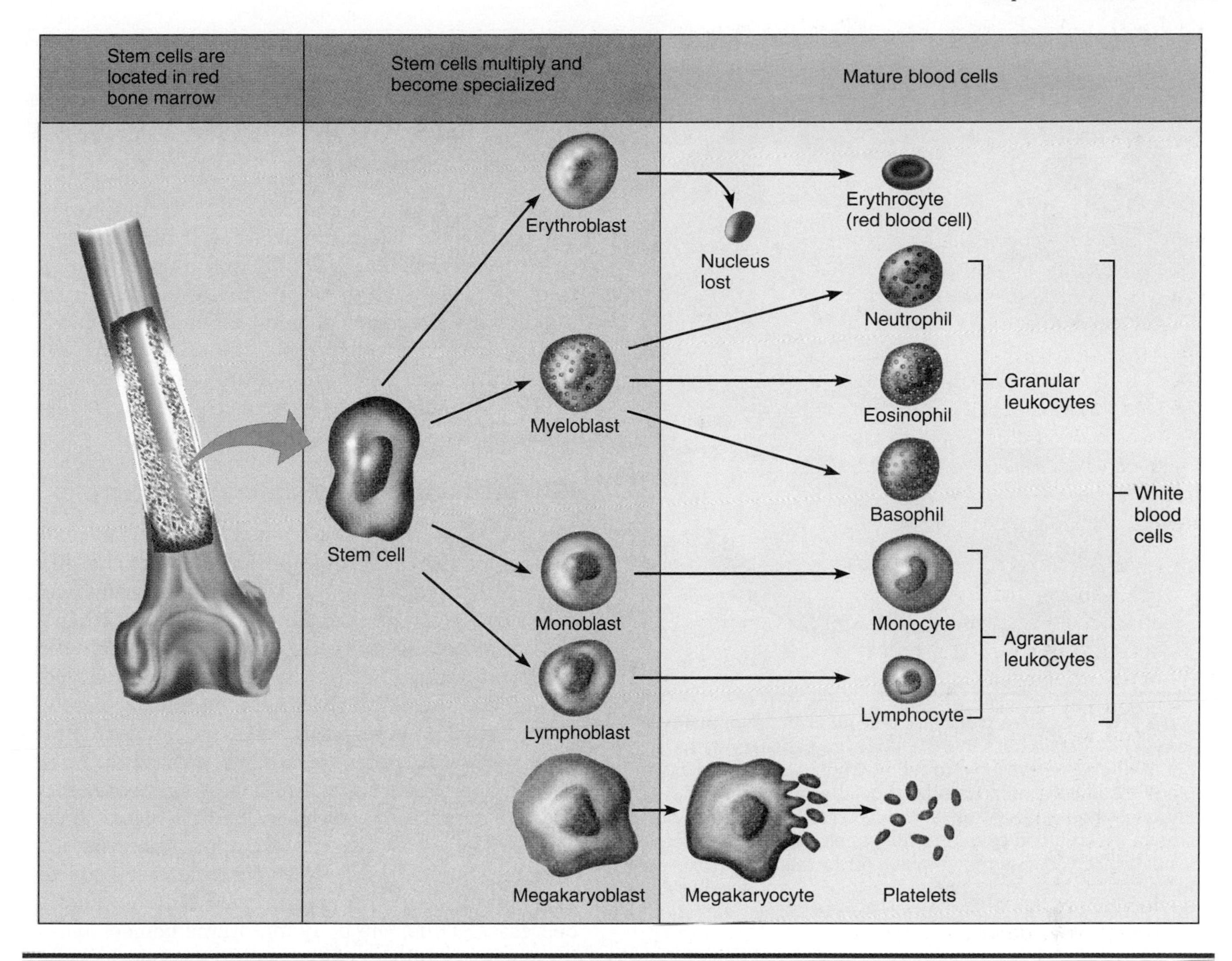

Figure 7.5 The production of blood cells and platelets. Blood cells have short life spans and must be continually replaced. Stem cells in the bone marrow continuously divide and give rise to a variety of types of blood cells.

weeks, your hematocrit will rise to compensate for lower levels of oxygen in the air you breathe. This is part of the normal homeostatic regulation of the oxygen-carrying capacity of the blood. After you return to your usual altitude, your hematocrit will also return to normal.

All blood cells and platelets originate from stem cells

How are **blood cells** produced by your body?

All blood cells and platelets originate from cells in the red bone marrow. These cells, called **stem cells,** divide repeatedly throughout our lives, continually producing immature blood cells. These immature cells develop into platelets and the various types of mature red and white blood cells described in Figure 7.5.

RBCs have a short life span

Some stem cells develop into immature cells called *erythroblasts* ("red" + "immature"). Erythroblasts become filled with hemoglobin and develop into mature RBCs, or erythrocytes, in about a week. As they mature, these cells lose their nucleus and organelles, and so they cannot reproduce. Thus, all new RBCs must originate from dividing stem cells. Because they lack a nucleus and therefore cannot perform many standard cell activities (such as producing new proteins and phospholipids to renew their cell membranes), they tend to become worn out rather quickly. Red blood cells live for only about 120 days, but during that time they make nearly 3,000 round trips a day, ferrying O_2 from the lungs to the tissues and CO_2 from the tissues back to the lungs. Because they live for such a short time, red blood cells must

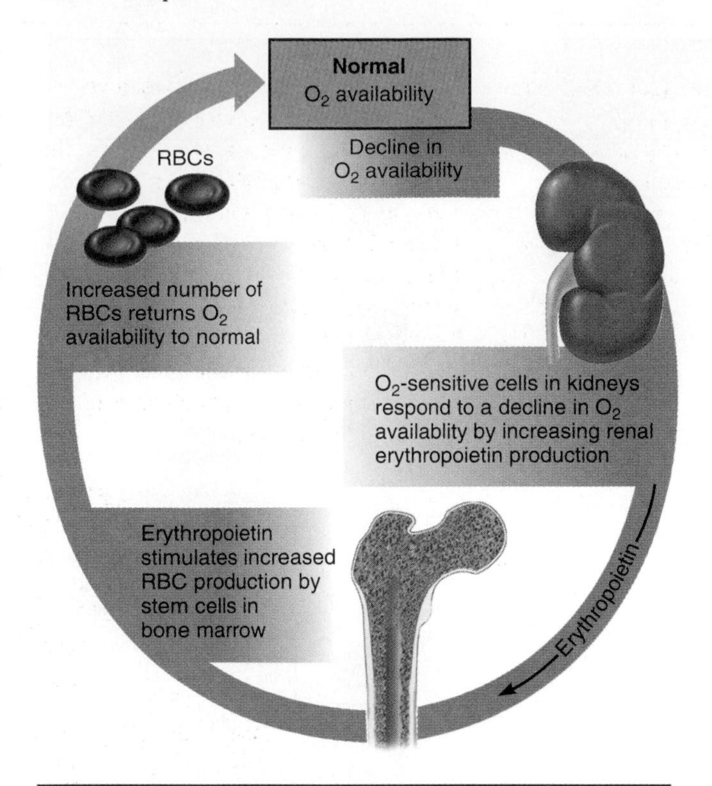

Figure 7.6 Negative feedback control of the availability of oxygen. Certain cells in the kidney are sensitive to the amount of oxygen available to them. When oxygen availability falls these cells produce erythropoietin, a hormone that stimulates the bone marrow to produce red blood cells. The increase in red blood cells returns oxygen availability toward normal, which reduces the stimulus for further erythropoietin secretion. Ultimately, homeostasis of oxygen availability is achieved.

be produced throughout life at the incredible rate of more than two million per second just to keep the hematocrit constant.

Old and damaged RBCs are removed from the circulating blood and destroyed in the liver and spleen. Large white blood cells called **macrophages** destroy damaged RBCs by **phagocytosis,** a process by which extracellular materials are surrounded and transported away. They engulf damaged RBCs and dismantle their component hemoglobin molecules. The body dismantles the four peptide chains into their constituent amino acids and recycles the amino acids into new proteins. The iron atoms of the heme groups are returned to the red bone marrow where they are used again in the production of new hemoglobin for new red blood cells. The heme groups (minus the iron) are converted to a yellowish pigment called *bilirubin* by the liver. If you've ever noticed how a bruise slowly changes color as it heals, from purple to blue to green to yellow, you have observed the chemical breakdown of the heme groups to bilirubin at the site of damage. Under normal circumstances, when hemoglobin is broken down in the liver, bilirubin mixes with bile secreted during digestion and passes into the intestines. This pigment contributes to the characteristic colors of urine and feces.

When the liver fails to secrete bilirubin into the bile properly or when the bile duct from the liver to the intestines is blocked, bilirubin may accumulate in blood plasma. High circulating levels of bilirubin make skin and mucous membranes look yellowish and even turn the whites of the eyes yellow. This condition is called *jaundice* (from *jaune,* French for yellow). Jaundice may also be caused by an increase in the rate of RBC breakdown.

RBC production is regulated by a hormone

Regulation of RBC production is a negative feedback control loop that maintains homeostasis. The number of RBCs is not regulated (there are no cells capable of counting the number of RBCs), only their effect—their ability to transport oxygen. Certain cells in the kidneys monitor the availability of oxygen. Low oxygen availability signals the cells to secrete a hormone called **erythropoietin.** Once released into the bloodstream, erythropoietin is carried to the red bone marrow, where it stimulates the stem cells to produce more red blood cells. When the blood oxygen-carrying capacity rises to an appropriate level as monitored by kidney cells, the cells cut back on their production of erythropoietin, and RBC production decreases. Thus, the body maintains homeostasis of oxygen availability by adjusting the production rate of the RBCs that transport it (Figure 7.6).

Some people with certain kidney diseases do not produce enough erythropoietin to be able to regulate their RBC production properly. Fortunately, erythropoietin is now available commercially and can be administered to stimulate their red cell production. Some athletes have been known to abuse commercial erythropoietin by injecting it to increase their RBC production and thus their blood oxygen-carrying capacity. This practice, called *blood doping,* can have serious consequences. Excess red blood cells make blood more viscous, and the dehydration that follows strenuous exercise can concentrate the blood even more. This increases the risk of blood clots and stroke.

Recap **All blood cells and platelets are derived from stem cells in red bone marrow. RBCs have a short life span and must be continually replaced. The rate of production of RBCs is stimulated by the hormone erythropoietin, which is produced in the kidneys.**

Health Watch

Donating and Receiving Blood

A rapid loss of 30% or more of the blood volume strains the body's ability to maintain blood pressure and tissue oxygenation. When this happens, the patient's very survival may depend on receiving blood donated by other persons. Approximately 4 million people a year receive donated blood as part of their medical care. According to the American Red Cross, 12 million blood donations are given every year. Most persons who donate blood get nothing more (and nothing less) than the satisfaction of knowing they have helped someone in need.

To donate blood you must be 17 years or older and weigh at least 110 pounds. You'll be given a physical examination and asked for your health history, including a confidential questionnaire about your sexual history and recent international travel. This is not done to embarrass you, but to ensure that it is safe for you to give blood and that your blood will be safe for others.

The blood withdrawal procedure itself takes about 10–20 minutes and is relatively painless; a needle will be inserted into an arm vein. All needles used are brand-new and sterile—you cannot catch AIDS or any other blood-borne disease by donating blood. Afterwards you'll be encouraged to drink and eat something and advised to avoid rigorous physical exercise for the rest of the day, but that's about it. This is not the best day to go hang-gliding or mountain climbing, but just about anything less strenuous is okay.

Most donors are only allowed to donate one unit of blood (one pint, or about 10% of your blood volume). This is not enough to affect you adversely. Usually the donated blood volume is replaced within several hours by whatever fluids and foods you drink and eat. The liver replaces the lost plasma proteins within a couple of days, and although full replacement of the RBCs may take a bit longer (perhaps a month or so), a reduction of only 10% of the normal number of RBCs is inconsequential.

What happens to the blood you donate? Sometimes it will be stored as whole blood, but more often it will be separated into three components: packed cells, platelets, and plasma. Each may be given to different recipients, meaning that your single "gift of life" can benefit several people.

Receiving blood is also relatively safe, but it's not as risk-free as donating blood. On rare occasions (and getting rarer every day) recipients of blood contract blood-borne diseases. For some blood-borne diseases, such as HIV, there are screening tests, but new diseases or new strains of diseases can arise for which tests are not yet routinely performed. You may have heard of someone who died as the result of a transfusion, such as the late tennis player Arthur Ashe, who contracted the AIDS virus from a blood transfusion he received during cardiac surgery, before the screening test for HIV had been developed. Today all blood donations undergo extensive testing, including screening for HIV. According to the American Red Cross, the risk of contracting HIV from a blood transfusion is now about one in 676,000. Balance that against the risk of not receiving a blood transfusion when it may save your life, and the risk seems small.

To learn more about donating blood, contact your local chapter of the American Red Cross.

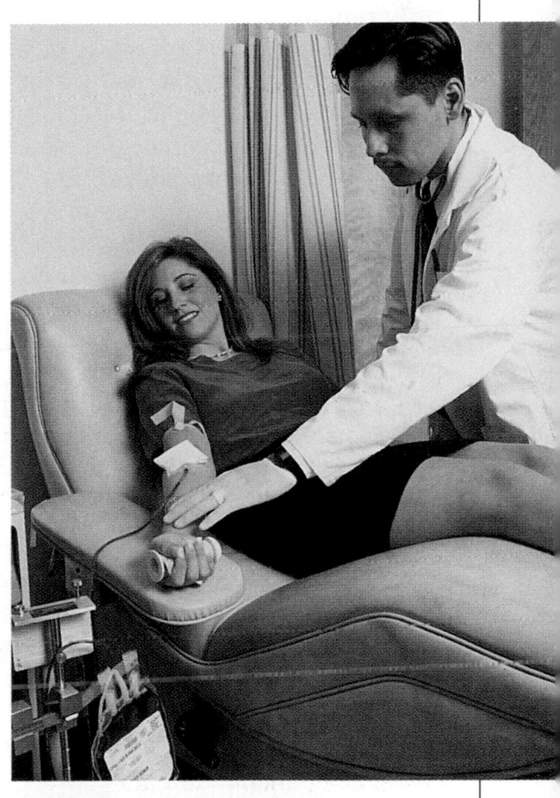

Donating blood. The procedure takes only a short time and is relatively painless.

White blood cells defend the body

Approximately 1% of whole blood consists of **white blood cells (WBCs** or **leukocytes).** Larger than red blood cells, they are also more diverse in structure and function. They have a nucleus but no hemoglobin, and because they are translucent they are difficult to identify under the microscope unless they have been stained. They are also relatively rare—each cubic millimeter of blood contains only about 7,000 of them, and there is only one WBC for every 700 RBCs. White blood cells play a number of crucial roles in defending against disease and injury.

Like red blood cells, white blood cells also arise from stem cells in the red bone marrow. As shown in Figure 7.5, stem cells produce immature blood cells that develop into the various WBCs. There are two major categories of white blood cells: **granular leukocytes** (granulocytes) and **agranular leukocytes** (agranulocytes). Both types contain granules (actually vesicles) in their cytoplasm that are filled with proteins and enzymes to assist their defensive work. Their names come from the fact that these granules are more visible in the granular leukocytes upon staining.

Once produced, most WBCs mature in the red bone marrow. Agranular leukocytes called T lymphocytes are the exception; they travel in the bloodstream to the thymus gland, where they mature. Most WBCs have an even shorter life span than red blood cells. Many granular leukocytes die within a few hours to nine days, probably due to injuries sustained while fighting invading microorganisms. Their agranular cousins usually live longer. Monocytes may survive for several months, lymphocytes for several days to many years. Dead and injured WBCs are continuously removed from the blood by the liver and spleen.

Circulating levels of white blood cells rise whenever the body is threatened by viruses, bacteria, or other challenges to health. Each type of WBC seems able to produce chemicals that circulate in blood back to the bone marrow, where they stimulate the production of leukocyte reinforcements.

Red cells remain in the blood, but some white cells leave the vascular system and circulate in the tissue fluid between cells or in the fluid in the lymphatic system. This is because they are able to change their shape, allowing them to squeeze between the cells that comprise the capillary walls. Because white blood cells are part of the body's defense system, they are discussed in more detail in Chapter 9 (The Immune System). Here we only describe the important characteristics of each type.

Granular leukocytes: Neutrophils, eosinophils, and basophils The granular leukocytes include neutrophils, eosinophils, and basophils. These names are based on their staining properties:

- **Neutrophils,** the most abundant type of granulocyte, comprise about 60% of WBCs. (Their name—which means "neutral-loving"—reflects the fact that their granules do not significantly absorb either a red or blue stain.) The first white blood cells to combat infection, neutrophils surround and engulf foreign cells through phagocytosis. They especially target bacteria and some fungi, and their numbers can rise dramatically during acute bacterial infections such as appendicitis or meningitis.
- **Eosinophils** make up a relatively small percentage (2–4%) of circulating white blood cells. (Their name comes from their tendency to stain readily with an acidic red stain called eosin.) Eosinophils have two important functions. One is to defend the body against large parasites such as worms (hookworms, tapeworms, flukes, and pinworms, among others). These parasites are too big to be surrounded and engulfed through phagocytosis. Instead, clusters of eosinophils surround each parasite and bombard them with digestive enzymes. Eosinophils' second function involves releasing chemicals that moderate the severity of allergic reactions.
- **Basophils,** the rarest white blood cells, account for only 0.5% of leukocytes. (They are named for their tendency to stain readily with basic blue stains.) The granules in basophils' cytoplasm contain histamine, a chemical that initiates the inflammatory response. When body tissues are injured, basophils secrete histamine, causing adjacent blood vessels to release blood plasma into the injured area. The plasma brings in nutrients, various cells, and chemicals to begin the tissue repair process. The swelling, itching, and redness associated with inflammation may not feel pleasant, but they are part of the immune system's defenses against molecules that are perceived as threatening.

Agranular leukocytes: Monocytes and lymphocytes The agranular leukocytes include monocytes and lymphocytes. The largest WBCs, **monocytes,** make up about 5% of circulating white blood cells. They can filter out of the bloodstream and take up residence in body tissues, where they differentiate into the macrophages that engulf invaders and dead cellular debris by phagocytosis. They also stimulate lymphocytes to defend the body. Monocytes seem especially active during chronic infections, such as tuberculosis, and against viruses and certain bacterial parasites.

Lymphocytes total about 30% of circulating white blood cells. They are found in the bloodstream, tonsils, spleen, lymph nodes, and thymus gland. They are classified into two types, *B lymphocytes* and *T lymphocytes* (or B cells and T cells). B lymphocytes give rise to *plasma cells* that produce *antibodies*, specialized proteins that defend against microorganisms and other foreign invaders. T lymphocytes target and destroy specific threats such as bacteria, viruses, and cancer cells. Both play a crucial role in the body's immune system, and we will look at them in more detail in Chapter 9.

Recap **White blood cells (leukocytes) defend the body against disease and injury. The activities of various leukocytes include participating in the response to tissue injury, producing antibodies, engulfing entire foreign cells, and releasing enzymes to attack foreign organisms too large to be engulfed whole.**

Platelets are essential for blood clotting

Less than 1% of whole blood consists of platelets. **Platelets** are actually small cell fragments, not complete cells, and they play an essential role in the process of blood clotting. Platelets are derived from megakaryocytes, which are large cells derived from stem cells in the bone marrow. However, unlike RBCs and WBCs, megakaryocytes never circulate. Platelets are small pieces of megakaryocyte cytoplasm and cell membrane. Megakaryocytes produce about 200 billion platelets a day; a single cubic millimeter of blood contains anywhere from 250,000 to 500,000 of these tiny cell fragments. Platelets last only about five to nine days. Their production is regulated by a recently discovered hormone called *thrombopoietin,* but how thrombopoietin secretion is regulated is still unknown.

If a blood vessel is injured and leaks blood, platelets play an important role in its repair. We examine this process in the next section.

7.2 Hemostasis: Stopping blood loss

One of the most important properties of the circulatory system is the ability to limit blood loss following injury. **Hemostasis,** the natural process of stopping the flow or loss of blood, proceeds in three stages: (1) *vascular spasm*, or intense contraction of blood vessels in the area, (2) formation of a *platelet plug,* and (3) *blood clotting,* also called *coagulation* (Figure 7.7). Once blood loss has stopped, tissue repair can begin.

Vascular spasms constrict blood vessels to reduce blood flow

When a blood vessel is damaged, smooth muscle in its wall undergoes spasms—intense contractions that constrict the vessels. If the vessels are medium-sized to large, the spasms reduce immediate outflow of blood, minimizing the damage in preparation for later steps in hemostasis. If the vessels are small, the spasms press the inner walls together and may even be able to stop the bleeding entirely. Vascular spasms generally last for about half an hour, long enough for the next two stages in hemostasis to occur.

Platelets stick together to seal a ruptured vessel

Normally, platelets circulate freely in blood. However, when the lining of a blood vessel breaks, exposing underlying proteins in the vessel wall, platelets swell, develop spiky extensions, and begin to clump together. They also become sticky and start adhering to the walls of the vessel and to each other. More platelets congregate and undergo these same changes.

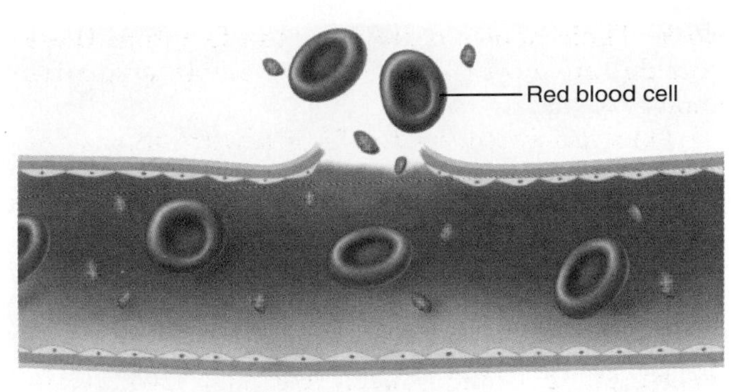

① **Vessel injury.** Damage to a blood vessel exposes the vessel muscle layers and the tissues to blood.

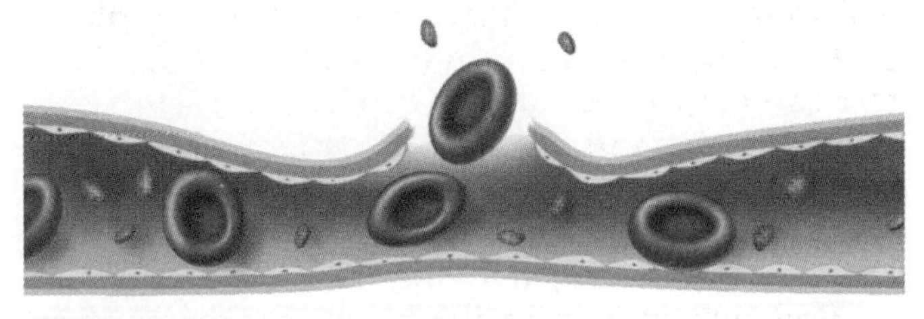

② **Vascular spasm.** The blood vessel contracts, reducing blood flow.

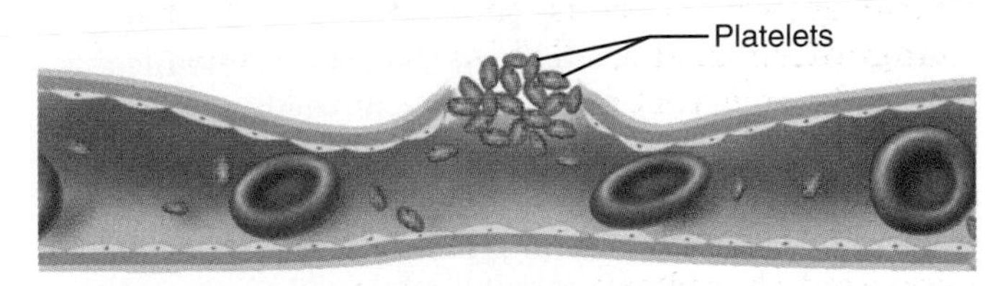

③ **Platelet plug formation.** Platelets adhere to each other and to the damaged vessel.

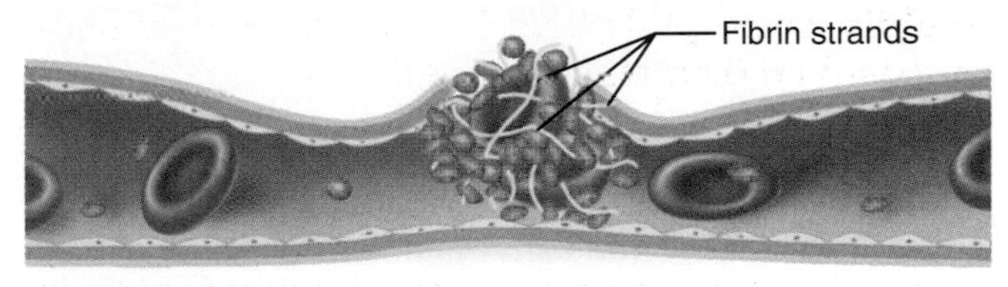

④ **Clot formation.** Soluble fibrinogen forms an insoluble mesh of fibrin, trapping RBCs and platelets.

Figure 7.7 The stages of hemostasis.

The result is a platelet plug that seals the injured area. If the rupture is fairly small, a platelet plug may be able to close it within several seconds. This may be enough to stop the bleeding. If damage is more severe, blood clotting will occur.

A blood clot forms around the platelet plug

The third stage in hemostasis is the formation of a blood clot, during which the blood changes from a

Figure 7.8 Magnified view of a developing clot, showing red blood cells trapped in a network of fibrin fibers.

liquid to a gel. This involves a series of chemical reactions that ultimately produce a meshwork of protein fibers within the blood. At least 12 substances, known as "clotting factors," participate in these reactions. We will focus on three substances involved: prothrombin activator, thrombin, and fibrinogen.

Damage to blood vessels stimulates the vessels and nearby platelets to release a substance called *prothrombin activator.* This activates the conversion of *prothrombin,* a plasma protein, into an enzyme called **thrombin.** This reaction requires the presence of calcium ions (Ca^{2+}). Thrombin in turn facilitates the conversion of a soluble plasma protein, **fibrinogen,** into long insoluble threads of a protein called **fibrin.** The fibrin threads wind around the platelet plug at the wound site, forming an interlocking net of fibers that traps and holds platelets, blood cells, and other molecules against the opening.

Why does your blood clot under certain conditions?

The mass of fibrin, platelets, and trapped red blood cells represents the initial *clot* that reduces the flow of blood at the site of injury (Figure 7.8). This initial fibrin clot can form in less than a minute. Shortly thereafter, platelets in the clot start to contract, tightening the clot and pulling the vessel walls together. Generally the entire process of blood clot formation and tightening takes less than an hour.

If any step in this process is blocked, even a minor cut or bruise can become life threatening. Consider *hemophilia,* an inherited condition caused by a deficiency of one or more clotting factors. People with the most common form of the condition, hemophilia A, lack a specific protein known as clotting factor VIII. When a vessel is breached, blood clots slowly or not at all. Even if the skin is not broken, severe bruising can spread into joints and muscles. Fifty years ago most people with hemophilia did not survive to adulthood, but today many bleeding episodes can be controlled by administering factor VIII. Because clotting factor VIII was initially purified from donor blood, which could not be tested for the HIV virus, in the past some hemophiliacs contracted AIDS. Today all blood is screened for the HIV virus. In addition, the use of donor blood as a source of factor VIII has decreased because genetic engineering techniques have made it possible to produce factor VIII in the laboratory.

Certain medications can also interfere with hemostasis. If you cut yourself after recently taking aspirin, for example, you may notice that you bleed more than usual. This is because aspirin blocks platelet clumping and slows the formation of a platelet plug. If you are planning to have surgery, your doctor will probably advise you to avoid taking aspirin for at least 7–10 days ahead of time.

Recap **Damage to blood vessels causes the vessels to spasm (contract) and the nearby platelets to become sticky and adhere to each other, limiting blood loss. In addition, a series of chemical events takes place that causes the blood in the area to clot, or coagulate (form a gel).**

7.3 Blood types determine blood compatibility

Blood transfusions—the administration of blood directly into the bloodstream of another person—may seem like a miracle of modern medicine, but the concept is not new. For over a century, physicians have tried to counteract severe blood loss by transfusing blood from one living person into another. Sometimes these attempts were successful. More often they were not, resulting in severe illness or even death for the recipient. Why did they save some lives but not others?

Today we know the success or failure of blood transfusions depends largely on blood type. If you ever donate or receive blood, you will undergo testing to determine your blood type. If you receive blood from someone who does not belong to a compatible type, you could suffer a severe reaction.

To understand the concept of blood typing, we must first be familiar with *antigens* and *antibodies.* Our cells have certain surface proteins that our im-

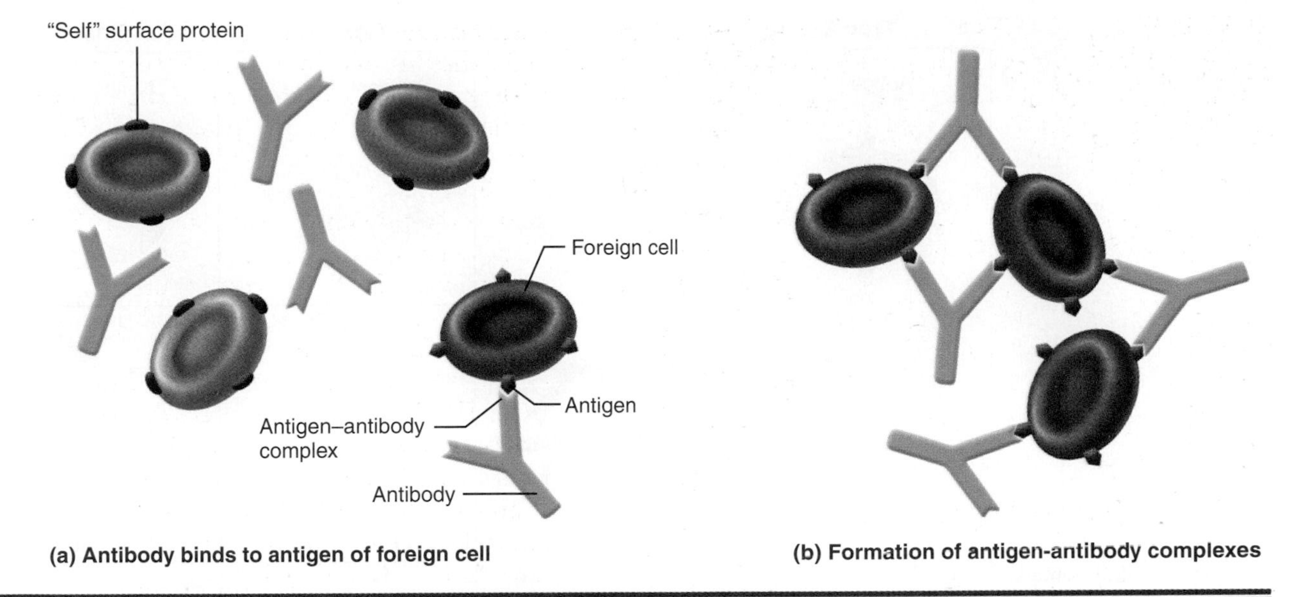

Figure 7.9 How antibodies recognize and inactivate foreign cells. (a) Three "self" cells, circulating antibodies, and one foreign cell (indicated by the darker color). The antibody ignores the "self" surface proteins but binds to the antigen of the foreign cell, forming an antigen-antibody complex. **(b)** The formation of antigen-antibody complexes causes foreign cells to clump together, inactivating them.

mune system can recognize and identify as "self"—in other words, belonging to us. These are like passwords that cause our immune system to ignore our own cells. Foreign cells carry different surface proteins, which the immune system recognizes as "nonself." An **antigen** (*anti* means "against," and the Greek word *gennan* means "to generate") is a nonself cell protein that stimulates the immune system of an organism to defend the organism. As part of this defense, the immune system produces an opposing protein called an **antibody** ("against" plus "body").

Produced by lymphocytes, antibodies belong to the class of plasma proteins called gamma globulins, mentioned earlier. Antibodies mount a counterattack on antigens they recognize as nonself. There are many antibodies, each one specialized to attack one particular antigen. This response has been compared to a lock and key: only a specific antibody key can fit a specific antigen lock. Antibodies float freely in the blood and lymph until they encounter an invader with the matching antigen. They bind to the antigen molecule to form an antigen-antibody complex that marks the foreigner for destruction. The formation of an antigen-antibody complex often causes the foreign cells to clump together (Figure 7.9). Some antibodies also inactivate foreign cells by preventing them from entering human cells.

Antigens and antibodies are discussed in more detail in Chapter 9. For now, let's look at how their interactions relate to blood type and blood transfusions.

Recap **Different blood types are determined by the presence of surface antigens on red blood cells and the presence of antibodies to surface antigens other than their own.**

ABO blood typing is based on A and B antigens

Like other cells, red blood cells have surface proteins that allow the body to identify them as "self." **Blood type** refers to a system for classifying blood according to differences in the surface antigens of red blood cells. The interactions between these antigens, and the development of antibodies against the antigens of foreign red blood cells, underlie the reactions that can occur after blood transfusions.

Red blood cells are classified according to the **ABO system,** in which nearly all individuals belong to one of four types: A, B, AB, or O. Type A blood has A antigens, type B blood has B antigens, type AB blood has *both* A and B antigens, and type O blood has neither (think of the O as a "zero"). In addition, all individuals have circulating antibodies (and the ability to make more antibodies) against any surface antigens different from their own; type A blood has type B antibodies, type B blood has type A antibodies, type O blood has both type A and B

What is your blood type, and why is it important?

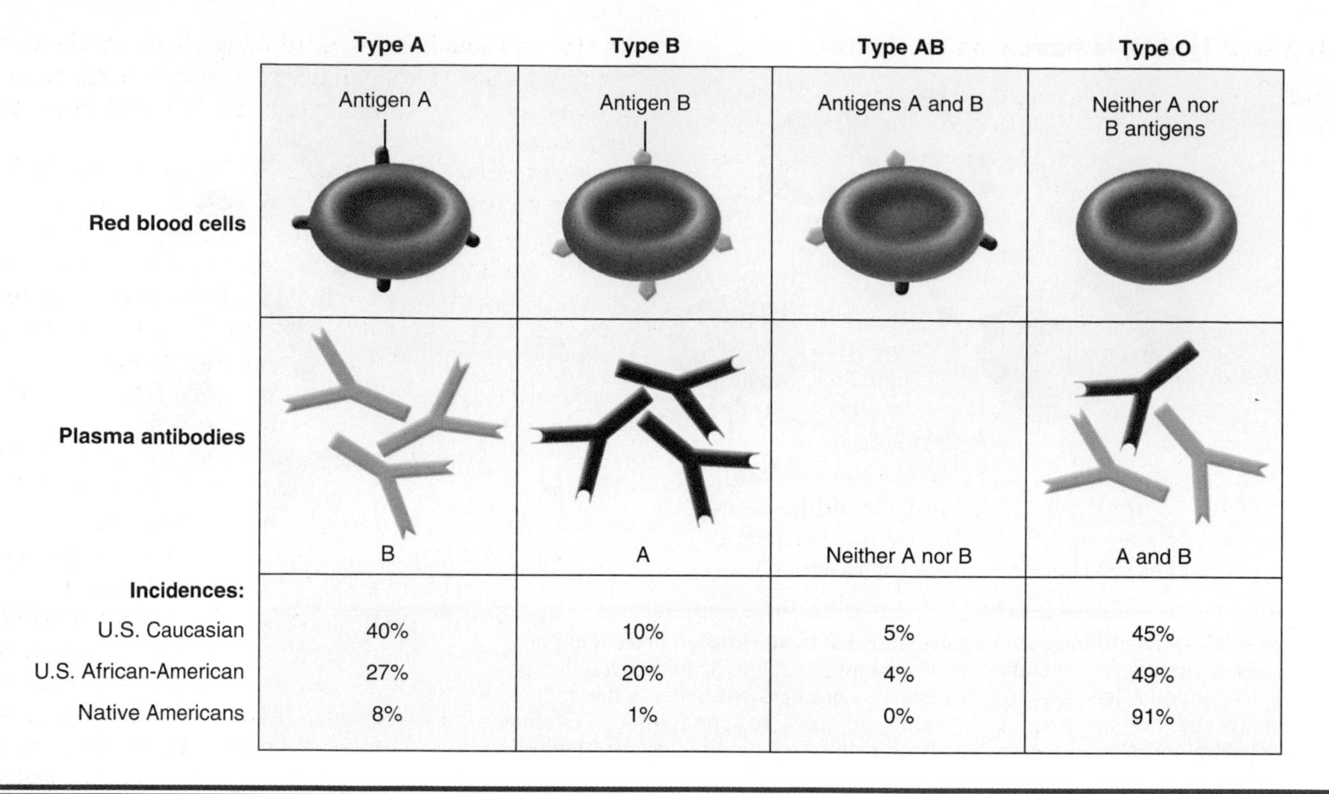

Figure 7.10 Characteristics of the four major blood types of the ABO typing system, showing their RBC surface antigens, antibodies, and relative incidences among various populations.

antibodies, and type AB blood has neither antibody. Figure 7.10 shows these various blood types and also indicates the relative incidences of each type in various populations. The antibodies appear early in life, regardless of whether or not a person has ever received a blood transfusion. These antibodies will attack red blood cells with foreign antigens, damaging them and causing them to *agglutinate*, or clump together (see Figure 7.11). If agglutination is extreme the clumps may block blood vessels, causing organ damage or even death. In addition, hemoglobin released by damaged red blood cells can block the kidneys, leading to kidney failure. Any adverse effect of a blood transfusion is called a *transfusion reaction*.

If you have type A blood, you are restricted to receiving transfusions of either type A or type O blood because neither of them has a foreign (type B) antigen. A transfusion of type B or type AB blood would provoke your antibodies to mount an attack against the B antigen of the donated RBCs, causing them to agglutinate. Similarly, if you're type B, you cannot receive any blood with type A antigens (A or AB). People with type AB blood can generally receive transfusions not only from other AB individuals but from all three of the other blood types as well. People with type AB blood, however, can donate only to other type AB individuals. Type O persons can give blood to anyone, but they can receive blood only from type O. Notice that it is the antibodies of the *recipient* that generally cause the transfusion reaction. Though the donor blood may have antibodies against the recipient's RBCs, transfusion reactions usually don't occur because the volume of blood given is generally small compared to the volume of the recipient's blood.

Recap **Types A, B, AB, and O blood are identified by the presence (or absence) of type A and/or type B surface antigens on their red blood cells.**

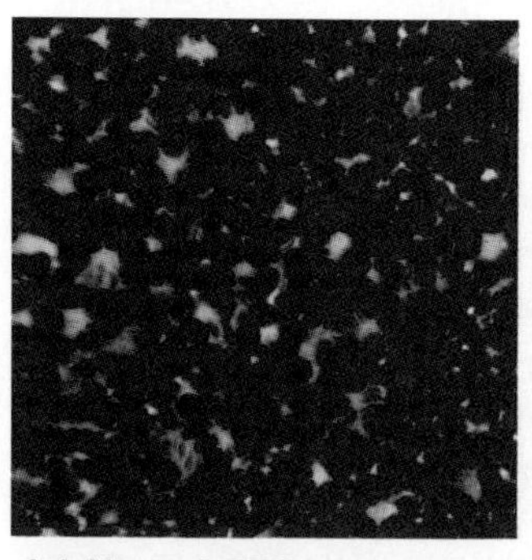

(a) Normal RBCs

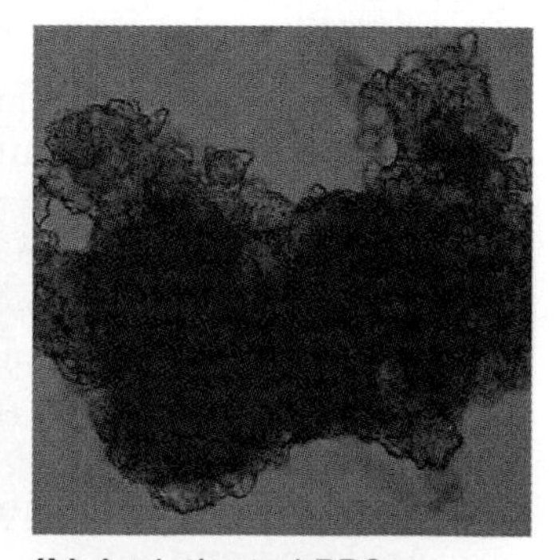

(b) Agglutinated RBCs

Figure 7.11 Agglutination.

Rh blood typing is based on Rh factor

Another red blood cell surface antigen, called **Rh factor** because it was first discovered in Rhesus monkeys, is also important in blood transfusions. Approximately 85% of Americans are *Rh-positive,* meaning they carry the Rh antigen on their red blood cells. Fifteen percent are *Rh-negative* and do not have the antigen.

People who are Rh-positive do not have Rh antibodies. Normally, people who are Rh-negative do not have Rh antibodies either, but they have inherited the ability to produce them over a period of several months after being exposed to the Rh antigen for the first time. If you're Rh-negative and receive a transfusion of Rh-positive blood only once, you'd be okay. Nothing would happen except that your body would produce Rh antibodies in preparation for the next time. But the *second* time you get Rh-positive blood, a severe transfusion reaction could occur because then your antibodies would be ready for it.

In general, blood typing involves determining your ABO type and the presence or absence of the Rh factor. For example, if your blood type is "B-pos" (B+), you are type B and positive for the Rh factor. If you are "O-neg" (O−), you are type O and negative for the Rh factor. AB+ individuals were once called *universal recipients* because they can generally receive blood from any other type. Type O− individuals were called *universal donors* because their blood can usually be donated to any other type. However, because transfusion reactions do occasionally occur unexpectedly, the terms are now considered outdated.

In addition to the A, B, and Rh antigens, scientists have identified about 400 additional blood antigens, most of which are not common. To be safe, however, laboratories generally do a "type and cross-match" on blood to ensure that only donor units with the best match of all possible antigens are administered to every recipient. The type and cross-match test involves mixing small samples of the donor blood and the potential recipient's blood together and examining them for clumping. If agglutination does not occur, the bloods are assumed to be a good match.

Rh factor can affect pregnancy

The Rh factor is a particular concern for Rh-negative women who wish to have children. If an Rh-negative woman becomes pregnant by an Rh-positive man, the child may be Rh-positive. Some of the fetus's Rh-positive blood cells can leak into the mother's blood during late pregnancy, childbirth, or even during a miscarriage or induced abortion, causing the mother to start producing Rh antibodies. The antibodies usually do not affect the first pregnancy because, as we saw earlier, it takes time for the Rh antibodies to be produced.

But difficulties arise because the principal soluble antibodies are small enough to cross the placenta. In general, this is healthy because the mother's antibodies can protect the fetus before it has developed its own antibodies. However, if the Rh-negative woman becomes pregnant with a *second* Rh-positive child, the Rh antibodies she developed after conceiving her first Rh-positive child may enter the fetus's bloodstream, attack the child's red blood cells, and cause the equivalent of a transfusion reaction in the child (Figure 7.12). The result is *hemolytic disease of the newborn* (HDN), a potentially life-threatening disease characterized by reduced numbers of mature red blood cells and elevated blood levels of bilirubin.

To prevent this reaction, an Rh-negative mother will be given an injection of Rh antibodies (RhoGAM) during pregnancy. The injected antibodies will quickly destroy any of the child's Rh-positive red blood cells in the woman's body before her body has a chance to begin producing her own Rh antibodies. The injected antibodies will disappear in a short time.

In addition to its important medical applications, blood typing has many other uses. Because blood types are inherited, anthropologists can track early population migrations by tracing inheritance patterns. Blood typing is also used in criminal investigations to compare the blood of victims and perpetrators, and to eliminate or identify suspects on the basis of matching antigens. DNA tests can be done on blood samples to help determine paternity.

Recap **In addition to ABO type, all persons are classified according to the presence or absence of another red blood cell surface antigen called the Rh factor. Rh factor antibodies can cause a serious immune reaction of the mother to her own fetus under certain circumstances.**

7.4 Blood disorders

Carbon monoxide poisoning: CO competes with oxygen

How does breathing carbon monoxide poison your blood?

Carbon monoxide (CO) is a colorless, odorless, highly poisonous gas produced by burning carbon-containing substances such as natural gas, gasoline, wood, and oil. It has a high affinity for hemoglobin and it competes with O_2 for hemoglobin binding sites. Carbon monoxide binds to hemoglobin almost 200 times more

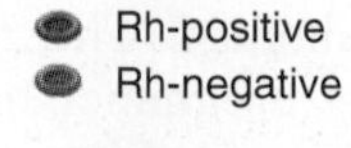

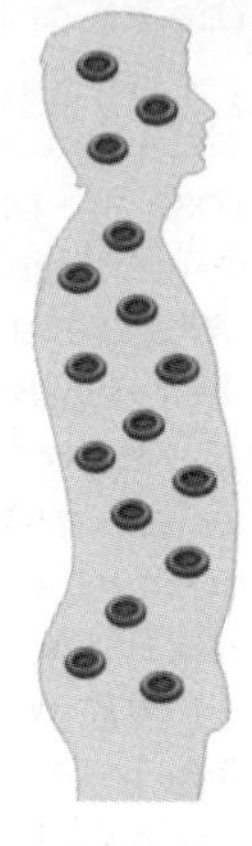

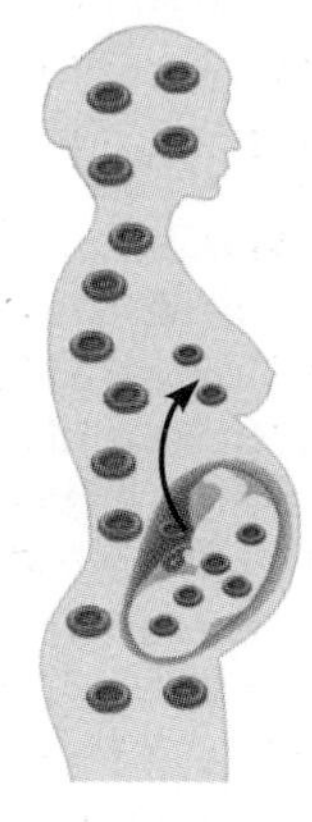

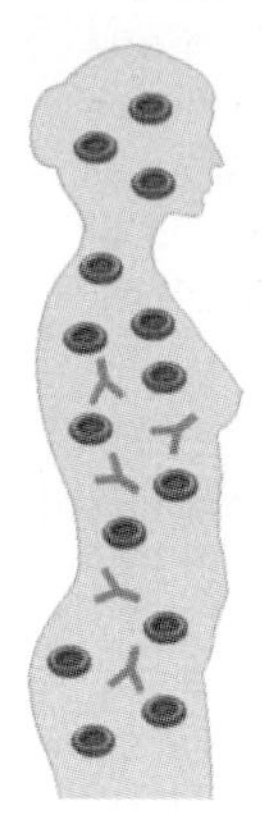

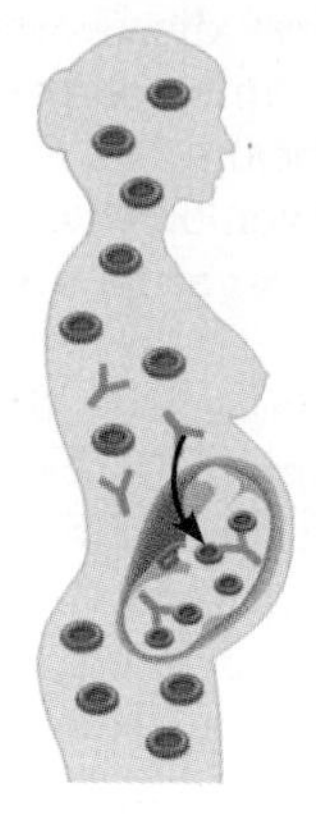

(a) When an Rh-positive man (shown here) fathers a child by an Rh-negative woman (next panel), the fetus may inherit the Rh-positive antigen.

(b) At birth, a small amount of fetal blood enters the mother's circulation.

(c) Over the next several months the woman develops antibodies against the Rh antigen.

(d) When the woman becomes pregnant with her second Rh-positive fetus, her antibodies cross the placenta and attack the fetus's red blood.

Figure 7.12 How Rh-factor incompatibility can affect a fetus.

tightly than oxygen and, once bound, remains attached for many hours. Even trace amounts significantly reduce the oxygen-transporting capacity of blood. High concentrations reduce the oxygen-carrying capacity of hemoglobin so much that body tissues, including those of the brain, literally starve.

Early symptoms of carbon monoxide poisoning include flushed skin (especially of the lips), dizziness, headache, nausea, and feeling faint. Continued exposure leads to unconsciousness, brain damage, and death. It's vital to get away from the gas and into fresh air as soon as possible. People with severe cases should be taken to a hospital, where pressurized oxygen or transfusions of red blood cells can be administered to increase blood oxygen levels.

Common sources of carbon monoxide include industrial pollution, automobile exhaust, furnaces and space heaters, and cigarette smoke. Every year people die from CO poisoning. Fortunately, taking a few precautions can prevent it. For instance, you can install a carbon monoxide detector in your home, and keep your furnace and your car in good repair. Never run a car's engine in an enclosed space. Don't smoke—cigarette smoke contains measurable amounts of carbon monoxide that are delivered directly to your lungs and blood. As much as possible, avoid spending a lot of time in smoke-filled rooms and high-traffic urban areas.

Anemia: Reduction in blood's oxygen-carrying capacity

Anemia is a general term for reduction in the oxygen-carrying capacity of blood. It may be due to below-normal levels of hemoglobin, reduced numbers of red blood cells, or production of abnormal hemoglobin. Some women may become slightly anemic because of heavy menstrual flow. All causes of anemia produce similar symptoms: pale skin, headaches, fatigue, dizziness, and difficult breathing. Major types of anemia include:

- **Iron-deficiency anemia,** in which insufficient iron results in fewer hemoglobin molecules per red blood cell. Iron-deficiency anemia is the most common form of anemia worldwide.
- **Aplastic anemia,** in which the bone marrow does not produce enough stem cells.
- **Hemorrhagic anemia,** caused by extreme blood loss such as with injuries, bleeding ulcers, and certain parasites, especially the parasite that causes malaria.
- **Pernicious anemia,** resulting from a deficiency of vitamin B_{12} absorption by the digestive tract. B_{12} is important for the production of normal red blood cells.

- **Sickle-cell anemia,** an inherited disorder in which the red blood cells assume a sickle shape when the oxygen concentration is low. Their abnormal crescent shape makes it harder for the cells to travel through small blood vessels and encourages their early destruction by the body. Prevalence of sickle-cell anemia is highest in Africans who live near the equator and in African Americans, for reasons that will be discussed in Chapter 19.

Treatment varies according to the specific cause of anemia. Generally, iron-deficiency cases are the easiest to treat.

> ***Recap*** **Carbon monoxide binds with hemoglobin more readily than oxygen, severely reducing the oxygen-carrying capacity of hemoglobin. Other circumstances, such as iron deficiency, can reduce the oxygen-carrying capacity of blood, resulting in various forms of anemia.**

Leukemia: Uncontrolled production of white blood cells

Leukemia refers to any of several types of blood cancers. Their common characteristic is uncontrolled proliferation of abnormal or immature white blood cells in the bone marrow. Overproduction of abnormal WBCs crowds out the production of normal white blood cells, red cells, and platelets. Huge numbers of leukemia cells enter and circulate in the blood, interfering with normal organ function.

There are two major categories of leukemia: acute, which develops rapidly, and chronic. Both are thought to originate in the mutation of a white blood cell (a change in genetic structure) that results in uncontrolled cell division, producing billions of copies of the abnormal cell. Possible causes for the original mutation include viral infection or exposure to radiation or harmful chemicals. Genetic factors may also play a role.

Leukemia can produce a wide range of symptoms. Tissues may bruise easily due to insufficient production of platelets. Anemia can develop if the blood does not contain enough red blood cells. Bones may feel tender because the marrow is packed with immature white blood cells. Some people experience headaches or enlarged lymph nodes.

Treatment can cure leukemia in some cases and prolong lives in others. Treatment generally involves radiation therapy and chemotherapy to destroy the rapidly proliferating cancer cells. This kills the normal stem cells as well, so transplants of bone marrow tissue are required to provide new stem cells. As with blood transfusions, all bone marrow tissue must undergo testing to make sure the donor's antigens are compatible with those of the patient.

Why are leukemia patients sometimes given bone marrow transplants?

Multiple myeloma: Uncontrolled production of plasma cells

Like leukemia, *multiple myeloma* is a type of cancer. In this case it is abnormal plasma cells in the bone marrow that undergo uncontrolled division. Plasma cells are a type of lymphocyte responsible for making a specific antibody. The proliferating plasma cells manufacture too much of an abnormal, frequently incomplete antibody, impairing production of other antibodies and leaving the body vulnerable to infections. Bones become tender as healthy bone marrow is crowded out by malignant plasma cells. Levels of calcium in the blood soar as bone tissue is destroyed. Treatment includes anticancer drugs and radiation therapy.

> ***Recap*** **Leukemia and multiple myeloma are blood cell cancers. They arise as abnormal cells in the bone marrow multiply in huge numbers.**

Mononucleosis: Contagious viral infection of lymphocytes

Mononucleosis is a contagious infection of lymphocytes in blood and lymph tissues caused by the Epstein-Barr virus, a relative of the virus that causes herpes. Most common during adolescence, "mono" is nicknamed the "kissing disease" because it's frequently spread through physical contact.

Symptoms of mononucleosis can mimic those of flu: fever, headache, sore throat, fatigue, swollen tonsils and lymph nodes. A blood test reveals increased numbers of monocytes and lymphocytes. Many of the lymphocytes enlarge and look more like monocytes, which is how the disease got its name. There is no known cure for mononucleosis, but almost all patients recover on their own within four to six weeks. Extra rest and good nutrition helps the body overcome the virus.

Blood poisoning: Infection of blood plasma

Normally blood is well defended by the immune system. Occasionally microorganisms invade the blood, overwhelm these defenses, and multiply rapidly in blood plasma. The organisms may be poisonous themselves, or they may secrete poisonous chemicals as by-products of their metabolism. The result is *blood poisoning,* also called *septicemia* or toxemia.

Blood poisoning may develop from infected wounds (especially deep puncture wounds), severe burns, urinary system infections, or major dental procedures. To help prevent it, wash wounds and

burns thoroughly with soap and water. Consult your doctor immediately if you experience flushed skin, chills and fever, rapid heartbeat, or shallow breathing. Antibiotic drugs are usually very effective against blood poisoning.

Thrombocytopenia: Reduction in platelet number

Thrombocytopenia is a reduction in the number of platelets in the blood. This can happen for a number of reasons, such as viral infection, anemia, leukemia, other blood disorders, exposure to x-rays or radiation, even as a reaction to certain drugs. Sometimes platelet levels decline for no apparent reason, in which case they often rise again after several weeks.

Symptoms include easy bruising or bleeding, nosebleeds, bleeding in the mouth, blood in urine, and heavy menstrual periods. Treatment of the underlying cause generally improves the condition. If it persists, surgical removal of the spleen will often help.

Recap **Mononucleosis and blood poisoning are types of blood infection. Thrombocytopenia, a disease of too few platelets, is characterized by easy bleeding or bruising.**

Chapter Summary

The components and functions of blood

- Blood consists of formed elements and plasma. Blood has transport, regulatory, and protective functions.
- Plasma contains numerous plasma proteins involved in transport, regulation of water balance, and protection. It also contains ions, hormones, nutrients, wastes, and gases.
- Erythrocytes (RBCs) are highly specialized for the transport of oxygen, but they also transport some carbon dioxide.
- Hemoglobin is the primary protein in red blood cells and gives blood its oxygen-carrying capacity.
- The formed elements of blood all originate from stem cells in red bone marrow.
- Leukocytes (WBCs) defend the body against disease and the effects of injury.
- RBCs and WBCs have short life spans and must continually be replaced. RBC production is stimulated when the body detects low oxygen levels in the blood.
- Platelets are cell products that participate in blood hemostasis.

Hemostasis: Stopping blood loss

- Hemostasis is a three-phase process that prevents blood loss through damaged vessels. The phases are (1) vascular spasm, (2) the formation of a platelet plug, and (3) blood clotting.
- During the formation of a blood clot, substances released by damaged blood vessels cause soluble proteins called fibrinogen to become insoluble protein threads called fibrin. The threads form an interlocking mesh of fibers, trapping blood cells and sealing ruptured vessels.

Blood types determine blood compatibility

- Successful transfusion of blood from one person into another depends upon compatibility of their blood types, which is determined by antibodies in plasma and surface antigens on red blood cells.
- Blood types are classified primarily on the basis of the ABO system and the presence or absence of the Rh factor.
- Rh factor in particular can affect certain pregnancies adversely.

Blood disorders

- Carbon monoxide poisoning occurs when carbon monoxide gas binds to hemoglobin, preventing the hemoglobin from binding sufficient oxygen.
- Anemia is a reduction in blood oxygen-carrying capacity for any number of reasons, including insufficient red blood cell production and excessive blood loss.
- Leukemia is a form of cancer characterized by uncontrolled production of abnormal leukocytes (white blood cells).
- Mononucleosis is a contagious viral disease of the lymph.
- Blood poisoning is a general term for infection of blood plasma by various microorganisms.

Terms You Should Know

anemia, 162
blood type, 159
erythrocyte (RBC), 150
erythropoietin, 154
fibrin, 158
hematocrit, 152
hemoglobin, 151
hemostasis, 157
leukocytes (WBC), 155
phagocytosis, 154
plasma, 150
plasma proteins, 150
platelets, 157
Rh factor, 161
stem cell, 153

Test Yourself

Matching Exercise

_____ 1. The practice by some athletes of injecting erythropoietin to increase RBC production
_____ 2. Oxygen-binding protein in RBC
_____ 3. A gas transported by red blood cells
_____ 4. Term for the percentage of the blood that is red blood cells
_____ 5. Process whereby white blood cells surround and engulf other cells
_____ 6. Another name for white blood cells
_____ 7. Hormone that regulates the rate of production of platelets
_____ 8. Another term for blood poisoning
_____ 9. Procedures done before blood is transfused into a person
_____10. Most abundant white blood cells

a. phagocytosis
b. carbon dioxide
c. septicemia
d. hemoglobin
e. leukocytes
f. neutrophils
g. blood doping
h. type and cross-match
i. thrombopoietin
j. hematocrit

Fill in the Blank

11. Granular and agranular leukocytes are distinguished on the basis of their _____________ properties.
12. The soluble protein that precipitates to become fibrin in the presence of thrombin is called _____________.
13. _____________ is an inherited condition caused by a deficiency of one or more clotting factors.
14. Antibodies attach to surface proteins on cells called _____________.
15. _____________ stain best with an acidic red stain.
16. Hemoglobin is broken down by the liver to a yellowish pigment called _____________.
17. An _____________ is an immature red blood cell.
18. All blood cells originate from _________ cells.
19. Contraction of blood vessels in a region of damage is called a vascular _________.
20. Type _____ blood has neither A nor B surface antigens on its RBCs but both A and B antibodies in its plasma.

True or False?

21. Deoxyhemoglobin has a bright red color.
22. Type A blood has type A surface antigens on its RBCs and antibodies against type B surface antigens.
23. Rh-factor is found only in rhesus monkeys.
24. Red blood cells live only about a week.
25. During clotting, blood changes from a liquid to a gel.

Concept Review

1. Describe the functions of blood.
2. Show by a diagram (or explain) how the production of red blood cells is regulated in order to maintain homeostasis of blood oxygen-carrying capacity.
3. How are damaged or dead RBCs removed from the blood, and what happens to the hemoglobin they contained?
4. Describe the difference between the actions of neutrophils and eosinophils.
5. Describe how hemostasis proceeds.
6. List the four ABO blood types. For each one, list its red blood cell surface antigen(s) and plasma antibody(ies).
7. Explain why an Rh-negative woman who has an Rh-positive child will be given an injection of RhoGAM immediately after birth of the child.

Apply What You Know

1. Clarence Smith, age 35, is sent by his physician for a blood test. The lab results indicate his white blood cell count (number of WBC per ml of blood) is 18,000. The typical WBC count for a man his age is 6,000–9,000, meaning Clarence's white blood cell count is considerably higher than normal. What might this mean to the physician?
2. One treatment for certain types of leukemia is to try to kill all of the stem cells in bone marrow through radiation and chemotherapy and then to give a bone marrow transplant from another person (the donor). Can just anyone be the donor? Who is most likely to be a good donor? Explain.
3. Explain how it is possible that on rare occasions even O-neg blood could produce a transfusion reaction in a patient.

Case in Point

Training at High Altitude Enhances Athletic Performance

The University of Texas Southwestern Medical Center studied the performance of 39 elite college male and female runners before and after they lived for 28 days at high altitudes. Researchers found that living at moderately high altitudes (2500 m), combined with lower-altitude training, resulted in significantly greater improvement in oxygen uptake and performance over equivalent sea-level training for most athletes. The athletes who responded best to this "high-low" training model were found to have a significantly larger increase in erythropoietin concentration after 30 hours at altitude compared to the athletes who did not respond to the high-low training method.

QUESTIONS:

1. To what might you attribute the greater oxygen uptake and performance of the athletes who responded best to the high-low training method?
2. Would a measurement of the athletes' hematocrit levels be pertinent to this study? Why or why not?

INVESTIGATE:

Boutellier, U., et al. 1990. Aerobic performance at altitude: Effects of acclimatization and hematocrit. *International Journal of Sports Medicine, 11:* S21–S26.

Chapman, R. F., J. Stray-Gundersen, and B. D. Levine. 1998. Individual variation in response to altitude training. *Journal of Applied Physiology, 85:* 1448–1456.

References and Additional Reading

Becker, R.C. Antiplatelet Therapy. *Scientific American Science & Medicine* 3(4):12, July/Aug. 1996. (An article about how platelet aggregation is controlled.)

Radetsky, P. The Mother of All Blood Cells. *Discover,* March 1995, p. 86. (An article about the apparent isolation of human stem cells and the controversy over who owns the rights to them.)

Thomas, E.D. Hematopoietic Stem Cell Transplantation. *Scientific American Science & Medicine* 2(5):38, Sept./Oct. 1995. (Discusses the use of stem cells for bone marrow transplants.)

CHAPTER 8

HEART AND BLOOD VESSELS

Why do we need a cardiovascular system?

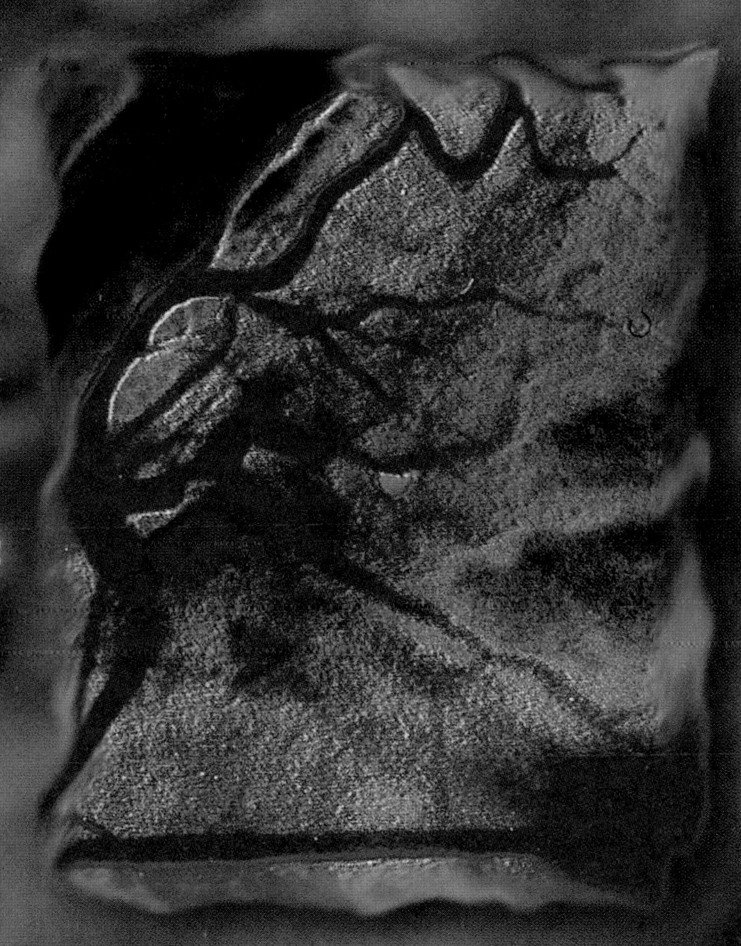

What's so important about maintaining a constant blood pressure?

How does your heart **beat on its own, independent of your** brain?

How can you reduce your risk of cardiovascular disease?

THE HEART IS A PUMP, BUT IT'S A VERY SPECIAL PUMP INDEED. The heart is constructed entirely of living cells and cellular materials, yet it is capable of greater reliability than some of the best pumps ever built by humans. It can easily withstand 80–100 years of continuous service without ever stopping for repairs. The heart is also fully adjustable on demand, over a range of about 5–25 liters of blood per minute.

The heart and blood vessels are known collectively as the **cardiovascular system** ("cardio" comes from the Greek word for "heart," and "vascular" derives from the Latin word for "small vessel"). The heart provides the power to move the blood, and the vascular system represents the network of branching conduit vessels through which the blood flows. The cardiovascular system is essential to life because it supplies every region of the body with just the right amount of blood. It is essential to the maintenance of homeostasis. We consider the blood vessels first.

Why do we need a cardiovascular system?

8.1 Blood vessels transport blood

A branching network of blood vessels transports blood to all parts of the body. The network is so extensive that if the body's blood vessels were laid end-to-end they would stretch 60,000 miles!

We classify the body's blood vessels into three major types: *arteries, capillaries,* and *veins.* Thick-walled arteries transport blood to body tissues under high pressure. Microscopic capillaries participate in exchanging solutes and water with the cells of the body. Thin-walled veins store blood and return it to the heart. Figure 8.1 illustrates the structures of each type of blood vessel, to be described in more detail below.

Arteries transport blood away from the heart

As blood leaves the heart it is pumped into large, muscular, thick-walled **arteries.** Arteries transport blood away from the heart. The larger arteries have a thick layer of muscle because they must be able to withstand the high pressures generated by the heart. Arteries branch again and again, so that the farther blood moves from the heart, the smaller in diameter the arteries become.

Spotlight on Structure & Function

Large and medium-sized arteries are like thick garden hoses, stiff yet somewhat elastic (distensible). Arteries stretch a little in response to high pressure but are still strong enough to withstand high pressures year after year. The ability to stretch under pressure is important because a second function of arteries is to store the blood that is pumped into them with each beat of the heart and then provide it to the capillaries (at high pressure) even between heartbeats. The elastic recoil of arteries is the force that maintains the blood pressure between beats. Think of the arteries as analogous to a city's water system of branching, high-pressure pipes that provide a nearly constant water pressure to every home.

The structure of the walls of large and medium-sized arteries is ideally suited for their functions. The vessel wall is a sandwich of three distinct layers surrounding the *lumen,* or hollow interior of the vessel:

1. The thin inner layer, the *endothelium,* is a layer of flattened, squamous epithelial cells. It is a continuation of the lining of the heart. The flattened cells fit closely together, creating a slick surface that keeps friction to a minimum and promotes smooth blood flow.
2. Just outside the lumen is a layer comprised primarily of smooth muscle with interwoven elastic connective tissue. In most arteries this is the thickest layer of the three. Contraction of the smooth muscle of large and medium-sized arteries stiffens the arteries, but it generally does not constrict them enough to alter blood flow. The elastic tissue makes large and medium-sized arteries slightly distensible so they can stretch passively to accommodate the blood that enters with each heartbeat.
3. The outermost layer of large and medium-sized arteries consists of a tough supportive layer of connective tissue, primarily collagen. This sturdy casing anchors vessels to surrounding tissues and helps protect them from injury. ■

The fact that arteries are constantly under high pressure places them at risk of damage. If the endothelium becomes damaged, blood may seep through the injured area and work its way between the two outer layers, splitting them apart. The result is an *aneurysm*, or ballooning of the artery wall. Some aneurysms cause the smooth muscle and endothelial layers to bulge inward as they develop, narrowing the lumen enough to reduce blood flow to an organ or region of the body. Doctors can sometimes detect such aneurysms with a stethoscope (an instrument for listening to sounds inside the body) because flowing blood produces characteristic sounds as it passes through the narrowed area. Other aneurysms cause the outer connective tissue layer to bulge outward. These are especially dangerous because they may be completely symptomless until they rupture or "blow out," causing massive internal bleeding and often death. If you've ever seen a water line burst, you know how quickly it can be devastating. Some

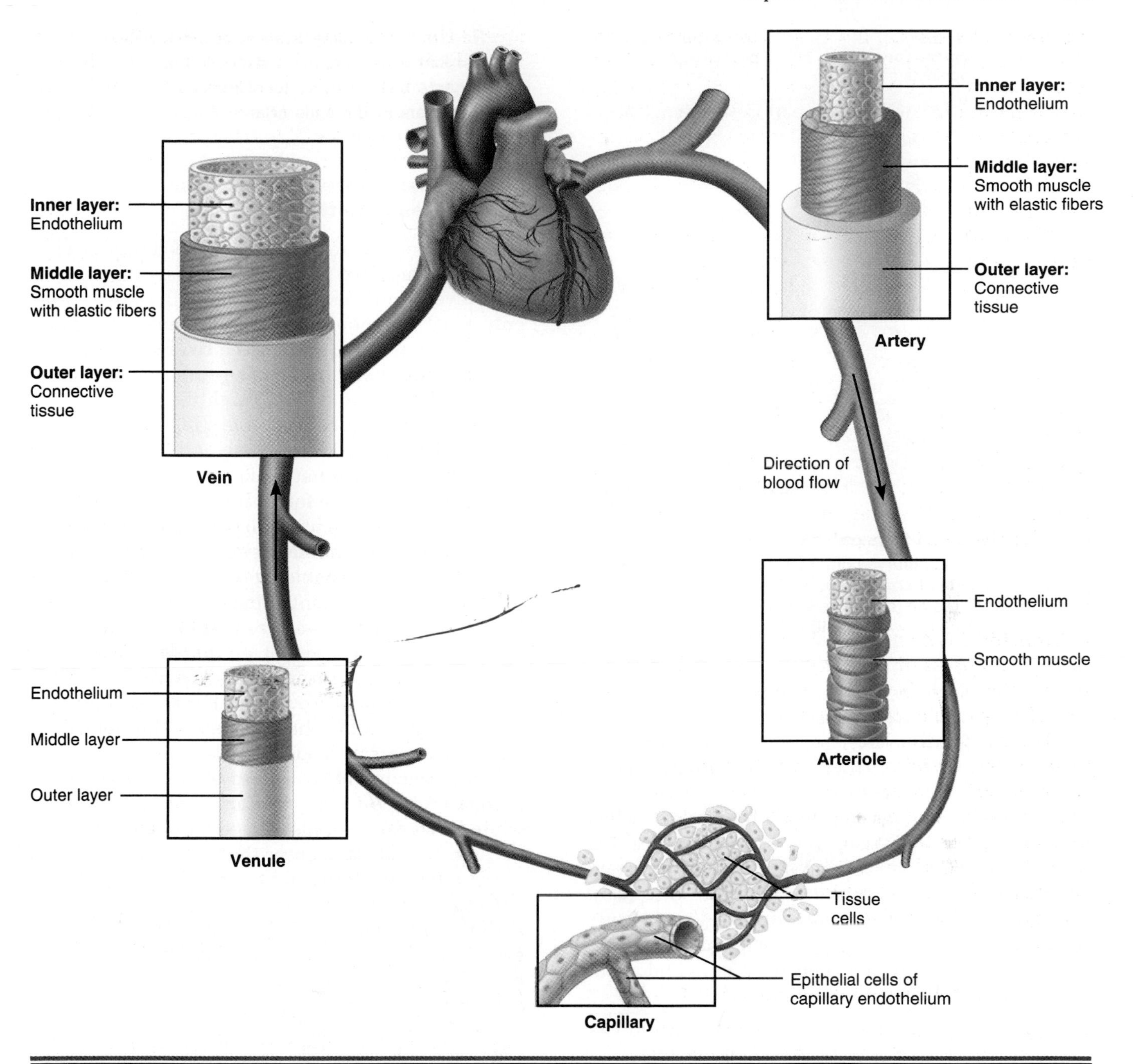

Figure 8.1 The structures of blood vessels in the human body.

aneurysms can be surgically repaired if they are accessible and detected before they rupture.

Arterioles and precapillary sphincters regulate blood flow

Eventually blood reaches the smallest arteries, called **arterioles** (literally, "little arteries"). The largest artery in the body, the *aorta,* is about 2.5 centimeters (roughly one inch) wide. In contrast, arterioles have a diameter of 0.3 millimeters or less, about the width of a piece of thread.

Spotlight on Structure & Function

By the time blood flows through the arterioles, blood pressure has fallen considerably. Consequently, arterioles can be simpler in structure. Generally they lack the outermost layer of connective tissue, and their smooth muscle layer is not as thick. In addition to blood transport and storage, arterioles have a third function not shared by the larger arteries: They help regulate the amount of blood that flows to each capillary. They do this by contracting or relaxing the smooth muscle layer, which has the effect of altering the diameter of the arteriole lumen.

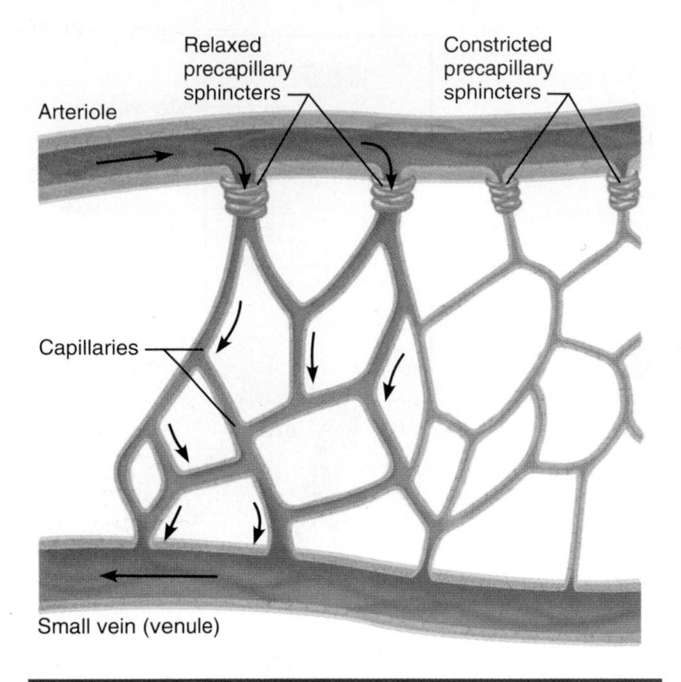

Figure 8.2 Precapillary sphincters control the flow of blood into individual capillaries. In this diagram the two precapillary sphincters on the right are vasoconstricted, reducing flow in that region. Arrows indicate direction of blood flow.

Right where an arteriole joins a capillary is a band of smooth muscle called the **precapillary sphincter** (Figure 8.2). The precapillary sphincters serve as gates that control blood flow into individual capillaries.

Contraction of vascular smooth muscle is called *vasoconstriction*. Vasoconstriction of arterioles and precapillary sphincters reduces their diameters and so reduces blood flow to the capillaries. Conversely, relaxation of vascular smooth muscle is called *vasodilation*. Vasodilation of arterioles and precapillary sphincters increases their diameters and so increases blood flow to the capillaries. ■

A wide variety of external and internal factors can produce vasoconstriction or vasodilation, including nerves, hormones, and conditions in the local environment of the arterioles and precapillary sphincters. If you go outside on a cold day, you may notice your fingers start to look pale. This is because vasoconstriction produced by nerves is narrowing your vessels to reduce heat loss from your body. On the other hand, hot weather will make your skin appear flushed as vasodilation occurs to speed up heat loss and cool you off. Emotions can also have an impact: Vasodilation is partly responsible for the surge in blood flow that causes the penis and clitoris to become erect when we are sexually aroused. Later in this chapter we will talk more about how the cardiovascular system is regulated in order to maintain homeostasis.

Recap **A branching system of thick-walled arteries distributes blood to every tissue in the body. Arterioles regulate blood flow to local regions, and precapillary sphincters at the ends of arterioles regulate flow into individual capillaries.**

Capillaries: Where blood exchanges substances with tissues

Arterioles connect to the smallest blood vessels, **capillaries,** which are thin-walled vessels that branch extensively into body tissues. Capillaries average only about ten-thousandths of a millimeter in diameter—not much wider than the red blood cells that pass through them. In fact, they are so narrow that RBCs often have to pass through them in single file or even bend to squeeze through (Figure 8.3).

Every living cell in the body is near at least one capillary. Extensive networks of capillaries, called *capillary beds,* can be found in all areas of the body, which is why you are likely to bleed no matter where you cut yourself. Capillaries' branching design and thin, porous walls enable blood to exchange oxygen, carbon dioxide, nutrients, and waste products with tissue cells. Capillary walls consist of a single layer of endothelium comprised of epithelial cells. Microscopic pores pierce through this layer, and the cells are separated by narrow slits. These openings are large enough to allow the exchange of fluid and other materials between blood and the interstitial fluid, yet small enough to retain red blood cells and most plasma proteins in the capillary (Figure 8.4). Some white blood cells can also squeeze between the cells in capillary walls and enter the tissue spaces.

In effect, capillaries function as biological strainers that permit selective exchange of substances with the extracellular fluid (the fluid that surrounds every living cell). In fact, capillaries are the *only* blood vessels that can exchange materials with the extracellular fluid.

Figure 8.5 illustrates the general pattern of how water and substances move across a capillary. At the beginning of a capillary, fluid is filtered out of the vessel into the extracellular fluid, accompanied by oxygen, nutrients, and raw materials needed by the cell. The filtered fluid is essentially like plasma except that it contains very little protein because most protein molecules are too large to be filtered. Filtration of fluid is caused by the blood pressure generated by the heart. Waste materials such as carbon dioxide and urea diffuse out of the cells and back into the blood.

Most of the filtered fluid is reabsorbed by diffusion back into the last half of the capillary before it joins a vein. The force for this reabsorption is the presence of protein in the blood but not in the extracellular fluid. In other words, water diffuses from an area

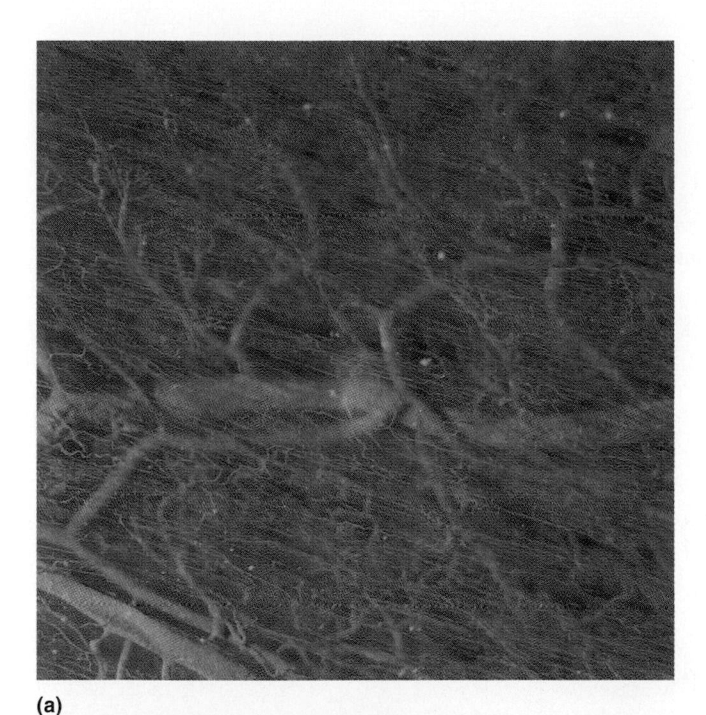

(a)

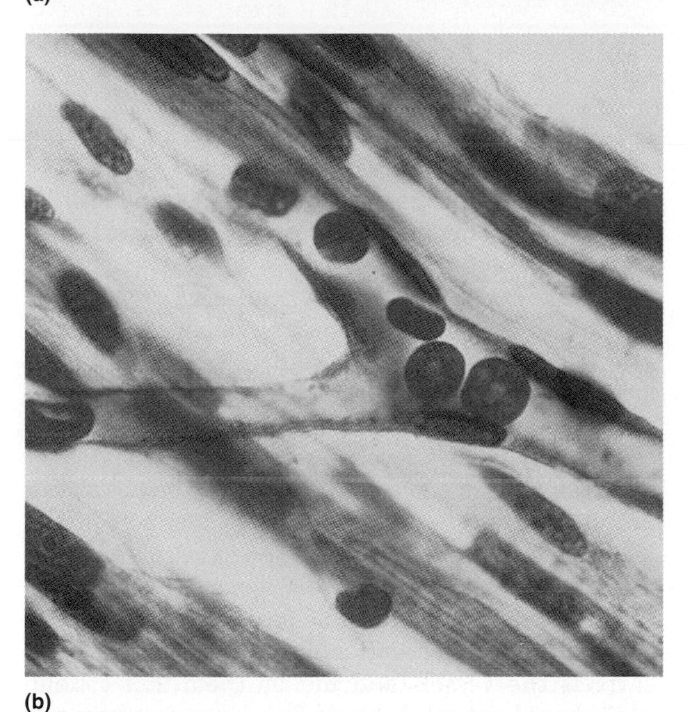

(b)

Figure 8.3 Capillaries. (a) A medium-magnification view showing a rich network of capillaries surrounding and interconnecting small arteries and veins. **(b)** A higher magnification showing a single branching capillary. Notice the red blood cells traveling single-file in the capillary.

of high water concentration (extracellular fluid) to an area of lower water concentration (blood plasma). However, the diffusional reabsorption of water does not quite match the pressure-induced filtration of water, so a small amount of filtered fluid remains in the extracellular space as excess extracellular fluid.

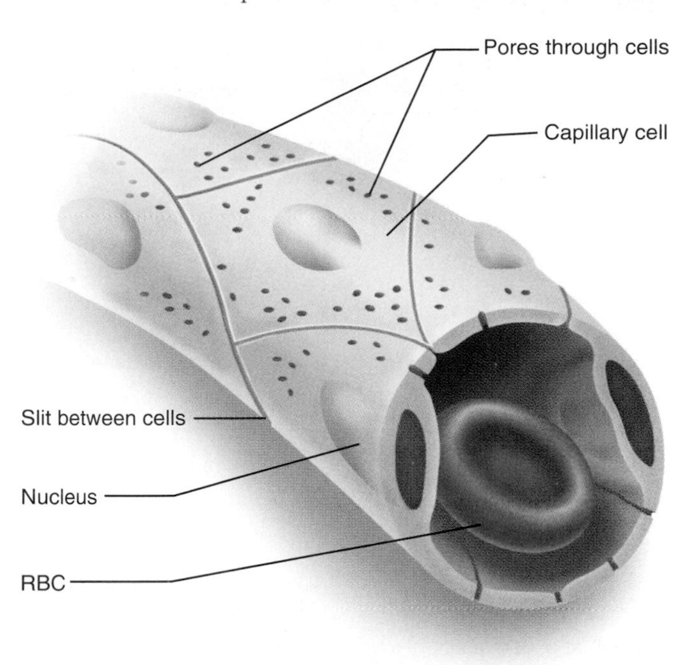

Figure 8.4 The structure of a capillary. A capillary is just a single layer of squamous epithelial cells. Pores through the cells and slits between the cells allow water and small solutes to cross, but not proteins and red blood cells.

Lymphatic system helps maintain blood volume

Although the imbalance between the amount of plasma fluid filtered by the capillaries and the amount reabsorbed is not large, over the course of a day it would amount to about two or three liters. This excess plasma fluid must be returned to the cardiovascular system somehow, or soon all the plasma would end up in the extracellular fluid.

The excess plasma fluid is picked up by a system of blind-ended vessels called *lymphatic capillaries,* which branch throughout our body tissues and are part of the *lymphatic system.* The lymphatic system also picks up a few objects in the extracellular fluid that are too large to diffuse into capillaries. These include lipid droplets absorbed during digestion and invading organisms. Lymphatic capillaries transport the excess extracellular fluid and other objects to larger lymphatic vessels, which eventually return the fluid (called *lymph*) to veins near the heart. Along the way the lymphatic system intercepts the invading microorganisms.

We will say more about the lymphatic system in Chapter 9 when we discuss the immune system. For now, just be aware that the lymphatic system, though technically not part of the cardiovascular system, plays a vital role in maintaining the proper volumes of blood and extracellular fluid.

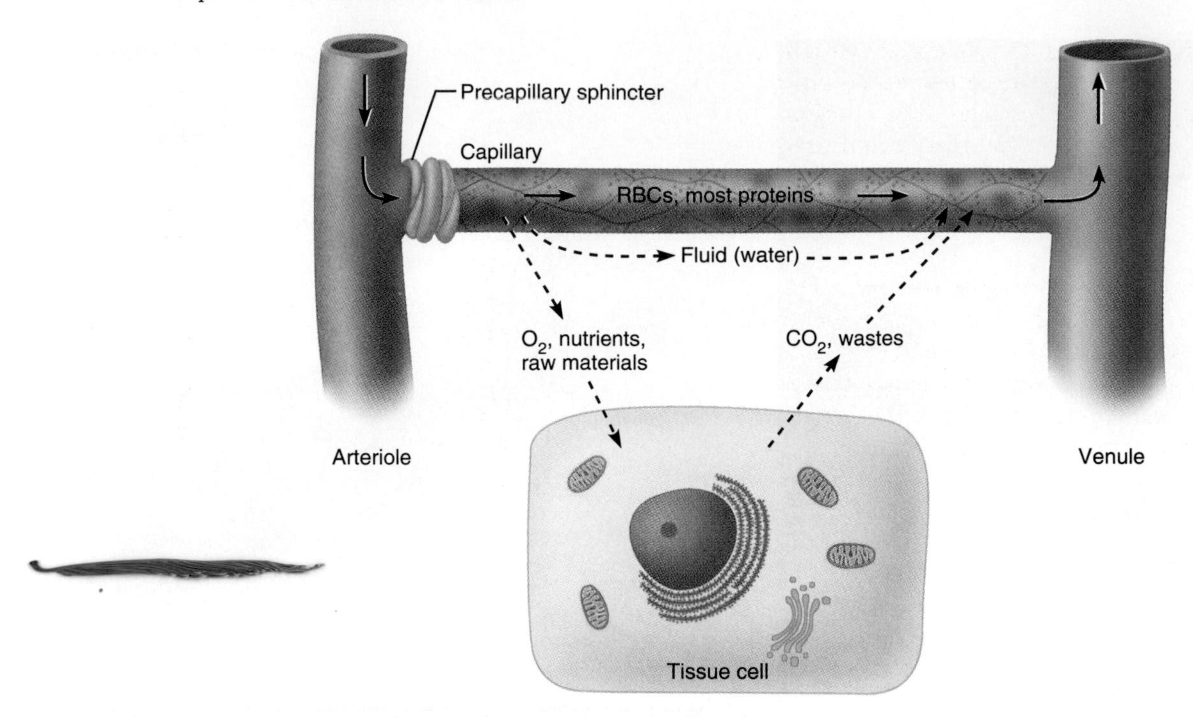

Figure 8.5 The general pattern of movement between capillaries, the extracellular fluid, and cells. For simplicity, only a single tissue cell is shown, but a single capillary may supply many nearby cells.

Recap **Capillaries consisting of just a single layer of cells are the only blood vessels that can exchange materials with the extracellular fluid. The lymphatic system removes excess fluid.**

Veins return blood to the heart

From the capillaries, blood flows back to the heart through *venules* (small veins) and **veins.** Like the walls of arteries, the walls of veins consist of three layers of tissue. However, the outer two layers of the walls of veins are much thinner than those of arteries. Veins also have larger lumens (are larger in diameter) than arteries.

The anatomical differences between arteries and veins reflect their functional differences. As blood moves through the cardiovascular system, the blood pressure becomes lower and lower. The pressure in veins is only a small fraction of the pressure in arteries, so veins do not need nearly as much wall strength (provided by muscle and connective tissue) as arteries. The larger diameter and high distensibility of veins allows them to stretch (like thin balloons) to accommodate large volumes of blood at low pressures.

In addition to their transport function, then, veins serve as a blood volume reservoir for the entire cardiovascular system. Nearly two-thirds of all the blood in your body is in your veins. Thanks to their blood reservoir function, even if you become dehydrated or lose a little blood, your heart will still be able to pump enough blood to keep your blood pressure fairly constant.

The distensibility of veins can lead to problems in returning blood to the heart against the force of gravity. When you stand upright, blood tends to collect in the veins of your legs and feet. People who spend a lot of time on their feet may develop *varicose veins,* permanently swollen veins that look twisted and bumpy from pooled blood. Varicose veins can appear anywhere, but they are most common in the legs and feet. In severe cases the skin surrounding veins becomes dry and hard because the tissues are not receiving enough blood. Often varicose veins can be treated by injecting an irritating solution that shrivels the vessels and makes them less visible. This should not affect blood flow because surrounding undamaged veins will take over and return blood to the heart.

Fortunately, there are three mechanisms that assist the veins in returning blood to the heart: (1) contractions of skeletal muscles, (2) one-way valves inside the veins, and (3) movements associated with breathing. Let's look at each in turn.

Skeletal muscles squeeze veins On their path back to the heart, veins pass between many skeletal muscles. As we move and these muscles contract and relax, they press against veins and collapse them, pushing blood toward the heart. You may have noticed that you tire more easily when you stand still than when you walk around. This is because walking improves the return of blood to your heart and prevents fluid accumulation in your legs. It also increases blood flow and the supply of energy to your leg muscles.

One-way valves permit only one-way blood flow Most veins contain valves consisting of small folds of the inner layer that protrude into the lumen. The structure of these valves allows blood to flow in one direction only: toward the heart. They open to permit blood to move toward the heart and then close to stop it from flowing back. Together, skeletal muscles and valves form what is called the "skeletal muscle pump" (Figure 8.6). Once blood has been pushed toward the heart by skeletal muscles or drained in that direction by gravity, it cannot drain back again because of these one-way valves.

Pressures associated with breathing push blood toward heart The third mechanism that assists blood flow involves pressure changes in the thoracic (chest) and abdominal cavities during breathing. When we inhale, abdominal pressure increases and squeezes abdominal veins. At the same time, pressure within the thoracic cavity decreases, which dilates thoracic veins. The result is to push blood from the abdomen into the chest and toward the heart. This effect is sometimes called the "respiratory pump."

> ***Recap*** **The thin-walled veins return blood to the heart and serve as a volume reservoir for blood. Return of blood to the heart is assisted by: (1) contractions of skeletal muscles, (2) one-way valves inside veins, and (3) pressure changes associated with breathing.**

8.2 The heart pumps blood through the vessels

The **heart** is a muscular, cone-shaped organ slightly larger than your fist, located in the thoracic cavity between the lungs and behind the sternum (breastbone). As we saw in Chapter 6, the heart consists mostly of a special type of muscle called cardiac muscle. Unlike skeletal muscle, cardiac muscle does not connect to bone. Instead, it pumps ceaselessly in a squeezing motion to propel blood through the blood vessels.

Your heart pumps about 75 times every minute—and this does not include the times it speeds up to supply extra blood during exertion or stress. It never rests for more than 2/3 of a second. Over 70 years, this adds up to about 2.8 billion heartbeats, truly an impressive record for any muscle. Under normal circumstances the heart's rate of pumping is controlled

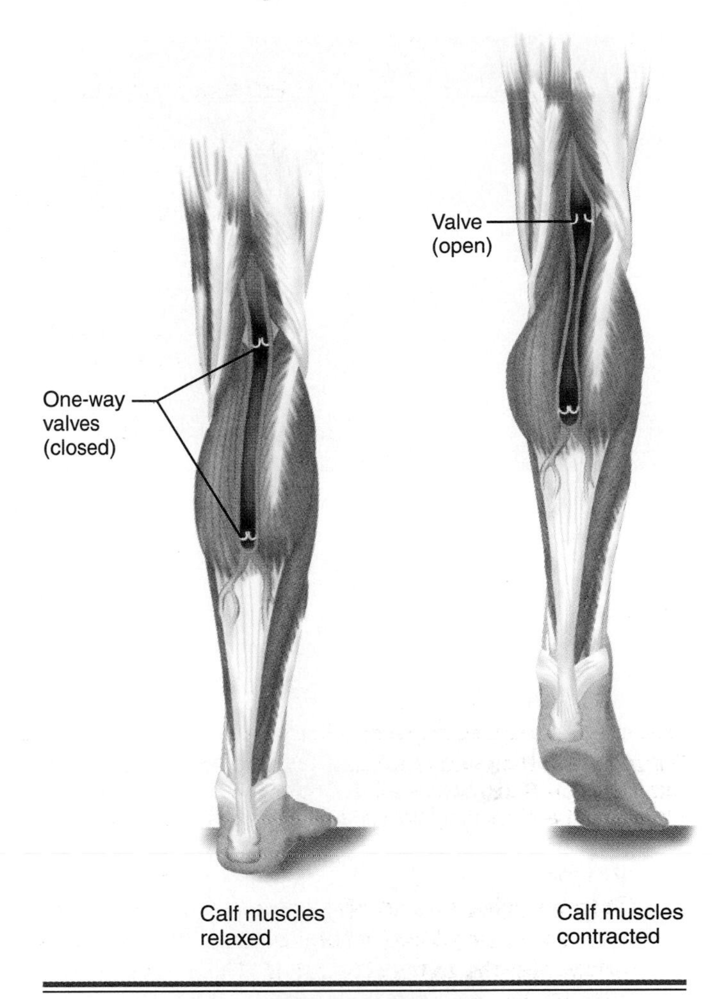

Figure 8.6 The so-called "skeletal muscle pump." With the calf muscle relaxed, blood accumulates in the vein because of the force of gravity. Backflow is prevented by one-way valves. When the calf muscle contracts, skeletal muscles press on the vein, increasing blood pressure in the vein momentarily and forcing blood toward the heart through a one-way valve. Backflow is still prevented by the one-way valve below the muscle.

Try It Yourself

Demonstrating That Veins Store Blood

To demonstrate the capacity of veins to store blood and how the forces of gravity can distend veins, try the following demonstration. (This works best if you have prominent veins on the back of your hand; if you don't, find a friend who does.)

Let your arm hang down by your side, below your waist, for about 30 seconds. Notice that the veins on the back of your hand swell, indicating stored blood. Now raise your hand above your head and hold it there. You can watch the veins almost disappear over the first 10 seconds as the blood in them drains back to the heart.

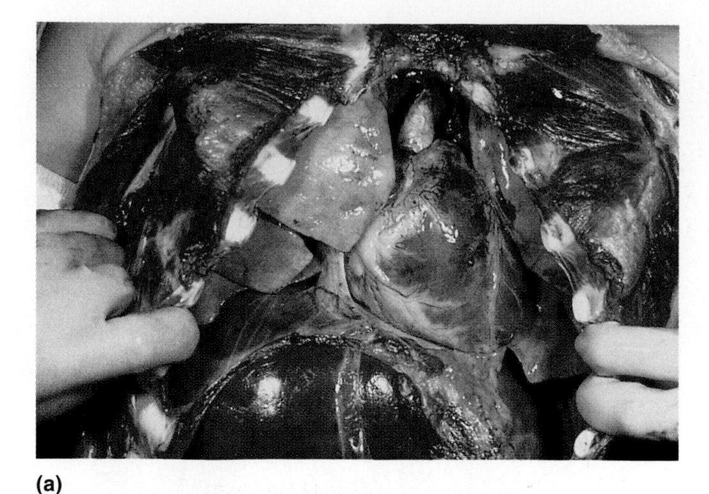

(a)

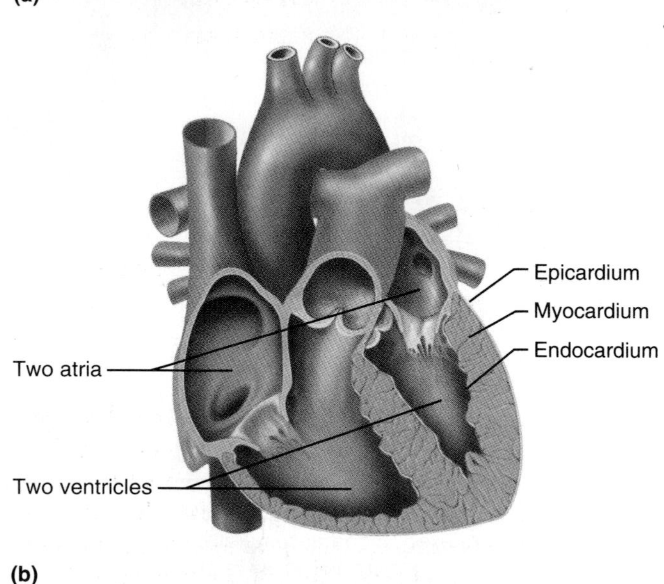

(b)

Figure 8.7 The heart. (a) Photograph of a heart in its natural position in the chest. The sacklike pericardium has been cut away. **(b)** Schematic cross section of a heart showing that it is composed of three layers: a thin outer epicardium, a thick layer of muscle called the myocardium, and an inner lining called the endocardium.

by the brain, but it can also beat on its own without any instructions from the brain at all.

The heart is mostly muscle

Looking at the heart from the outside, we see that it is enclosed in a tough fibrous sac called the *pericardium.* The pericardium protects the heart, anchors it to surrounding structures, and prevents it from overfilling with blood. Separating the pericardium from the heart is a space called the *pericardial cavity.* The pericardial cavity contains a film of lubricating fluid that allows the heart and the pericardium to glide smoothly against each other when the heart contracts.

In cross-section we see that the walls of the heart consist of three layers: the epicardium, myocardium, and endocardium (Figure 8.7). The outermost layer, the **epicardium,** is a thin layer of epithelial and connective tissue. The middle layer is the **myocardium.** This is a thick layer consisting mainly of cardiac muscle that forms the bulk of the heart. The myocardium is the layer that contracts every time the heart beats. Recall from Chapter 6 that the structure of cardiac muscle cells allows electrical signals to flow directly from cell to cell. This means an electrical signal in one cardiac muscle cell can spread to adjacent cells, enabling large numbers of cells to contract as a coordinated unit. Every time the myocardium contracts, it squeezes the chambers inside the heart, pushing blood outward into the arteries. The innermost layer of the heart, the **endocardium,** is a thin endothelial layer resting on a layer of connective tissue. The endocardium is continuous with the endothelium that lines the blood vessels.

Occasionally one of the layers of the heart wall becomes inflamed. These conditions are named according to the location of the problem; "-itis" is a general word-ending that means inflammation. Thus *pericarditis* refers to inflammation of the pericardium, *endocarditis* to an inflamed endocardium, and so on. A variety of factors can lead to inflammation in or around the heart wall, including infections, cancer, injuries, or complications from major surgery. Depending on the underlying cause, many cases respond well to antibiotic and anti-inflammatory drugs.

Recap **The heart wall consists of three layers. The outer epicardium is a thin layer of epithelial and connective tissue; the middle myocardium consists largely of cardiac muscle and is the layer that contracts. The inner endocardium is a layer of endothelium continuous with the endothelium lining blood vessels.**

The heart has four chambers and four valves

Taking a closer look at the details of the structure of the heart, we see that it consists of four separate chambers (Figure 8.8). The two chambers on the top are the **atria** (singular *atrium*), and the two more muscular bottom chambers are the **ventricles.** A muscular partition called the **septum** separates the right and left sides of the heart.

Blood returning to the heart from the body's tissues enters the heart at the *right atrium.* From the right atrium, the blood passes through a valve into the *right ventricle.* The right ventricle is more muscular than the right atrium because it pumps blood at considerable pressure through a second valve and into the artery leading to the lungs.

Blood returning from the lungs to the heart enters the *left atrium* and then passes through a third

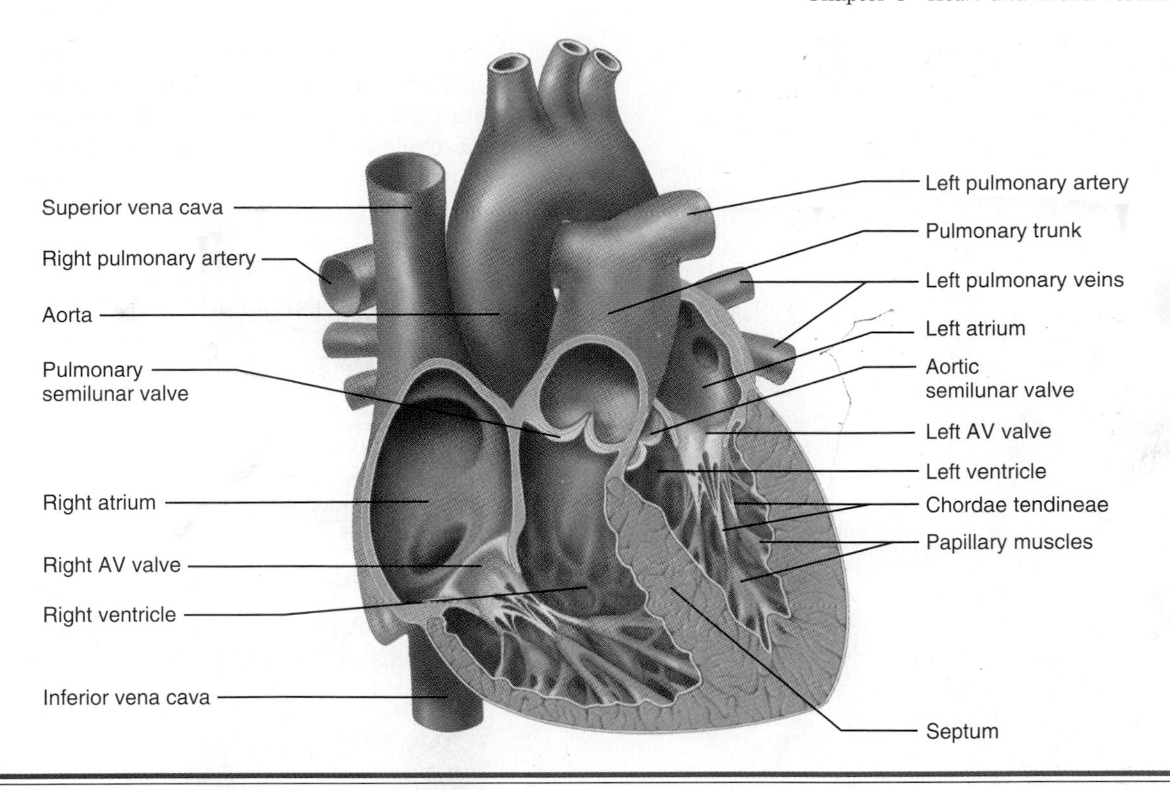

Figure 8.8 A view of the heart showing major blood vessels, chambers, and valves.

valve into the *left ventricle*. The very muscular left ventricle pumps blood through a fourth valve into the body's largest artery, the aorta. From the aorta blood travels through the arteries and arterioles to the systemic capillaries, venules, veins, and then back to the right atrium again.

Spotlight on Structure & Function The left ventricle is the most muscular of the heart's four chambers because it must do more work than any other chamber. The left ventricle must generate pressures higher than aortic blood pressure in order to pump into the aorta (we'll see how high aortic pressure is in a minute). The right ventricle has a thinner wall and does less work because the blood pressure in the arteries leading to the lungs is only about one-sixth that of the aorta.

Four heart valves enforce the heart's one-way flow pattern and prevent blood from flowing backward. The valves open and shut passively in response to changes in the pressure of blood on each side of the valve. The right and left **atrioventricular (AV) valves** located between the atria and their corresponding ventricle prevent blood from flowing back into the atria when the ventricles contract. The AV valves consist of thin connective tissue flaps (cusps) that project into the ventricles. The right AV valve has three flexible flaps, which is why it is called the *tricuspid valve*. The left AV valve has two flaps, so it is referred to as the *bicuspid* or *mitral valve*. These valves are supported by strands of connective tissue called *chordae tendineae* that connect to muscular extensions of the ventricle walls called *papillary muscles*. Together, the chordae tendineae and papillary muscles prevent the valves from everting (opening backward) into the atria when the ventricles contract. ■

Two **semilunar valves** (the pulmonary and the aortic) prevent backflow into the ventricles from the main arteries leaving the heart when the heart relaxes. Each semilunar valve consists of three pocket-like flaps. Their name reflects the half-moon shape of these flaps (*semi* means "one half," *luna* comes from the Latin word for "moon").

Next, let's follow the overall pattern of blood flow through the body.

> ***Recap*** **The heart is comprised primarily of cardiac muscle. It contains four chambers that pump blood to the lungs and the rest of the body simultaneously. Four one-way valves permit blood flow in one direction only.**

Pulmonary circuit provides for gas exchange

Figure 8.9 shows the general structure of the entire cardiovascular system. Note that the heart is pumping blood through the lungs (the **pulmonary circuit**) and through the rest of the body to all the cells

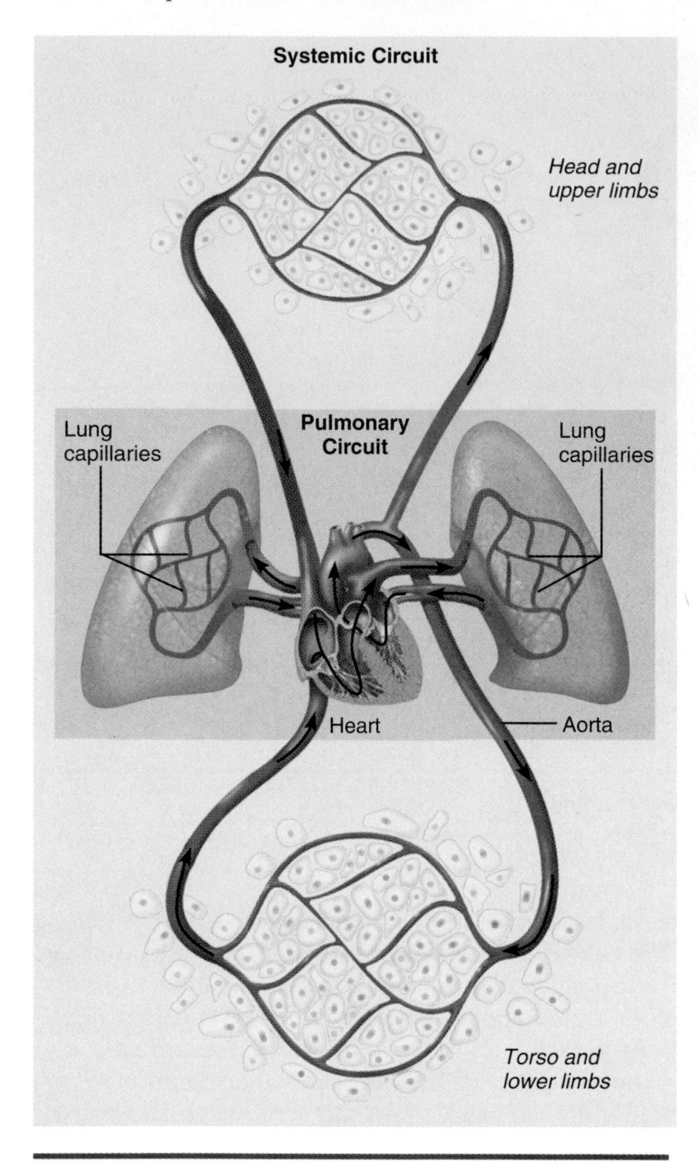

Figure 8.9 A schematic representation of the human cardiovascular system showing the separate pulmonary and systemic circuits. The systemic circuit includes the circulations of the head, torso, limbs, and internal organs, including the heart and some tissues in the lungs.

(the **systemic circuit**) simultaneously. Each circuit has its own set of blood vessels. Let's follow the pulmonary circuit first:

1. When blood returns to the heart from the veins, it enters the right atrium. The blood that returns to the heart is *deoxygenated*—it has given up oxygen to tissue cells and taken up carbon dioxide.
2. From the right atrium, blood passes through the right atrioventricular valve into the right ventricle.
3. The right ventricle pumps blood through the **pulmonary semilunar valve** into the **pulmonary trunk** (the main pulmonary artery) leading to the lungs. The pulmonary trunk divides into the right and left pulmonary arteries, which supply the right and left lungs, respectively.
4. At the pulmonary capillaries, blood gives up carbon dioxide and receives a fresh supply of oxygen from the air we inhale. It is now *oxygenated.*
5. The freshly oxygenated blood flows into the **pulmonary veins** leading back to the heart. It enters the left atrium and flows through the left atrioventricular valve into the left ventricle.

Note that deoxygenated and oxygenated blood never mix. Before it travels from the right side to the left side of the heart, blood must first pass through the lungs via the pulmonary circuit.

Systemic circuit serves the rest of the body

When blood enters the left ventricle, it begins the *systemic circuit,* which takes it to the rest of the body.

1. The left ventricle pumps blood through the **aortic semilunar valve** into the **aorta,** the largest artery.
2. From the aorta, blood travels through the branching arteries and arterioles to the capillaries, where it delivers oxygen and nutrients to all of the body's tissues and organs and removes waste products. Even some tissues of the lungs receive their nutrient blood supply from the systemic circulation.
3. From the capillaries, blood flows to the venules, veins, and then back again to the right atrium.

Figure 8.10 shows some of the major arteries and veins of the human body. Arteries and veins serving the same vascular region often (but not always) have the same name and generally are located very near to each other. For example, a common iliac artery supplies blood to each leg and a common iliac vein returns blood to the heart. On the other hand, carotid arteries supply the head but jugular veins return the blood from the head.

As you might expect from such a hard-working muscle, the heart requires a great deal of oxygen and nutrients to fuel its own operations. Although the heart represents only about 1/200 of your body's weight, it requires roughly 1/20 of your total blood flow at rest. And although the heart is almost continuously filled with blood, the myocardium is too thick to allow adequate diffusion of oxygen and nutrients from the blood passing through. Thus the heart has its own set of blood vessels called the **coronary arteries** that supply the heart muscle (Figure 8.11). The coronary arteries branch from the aorta just above the aortic semilunar valve and encircle the heart's surface (the word coronary comes from

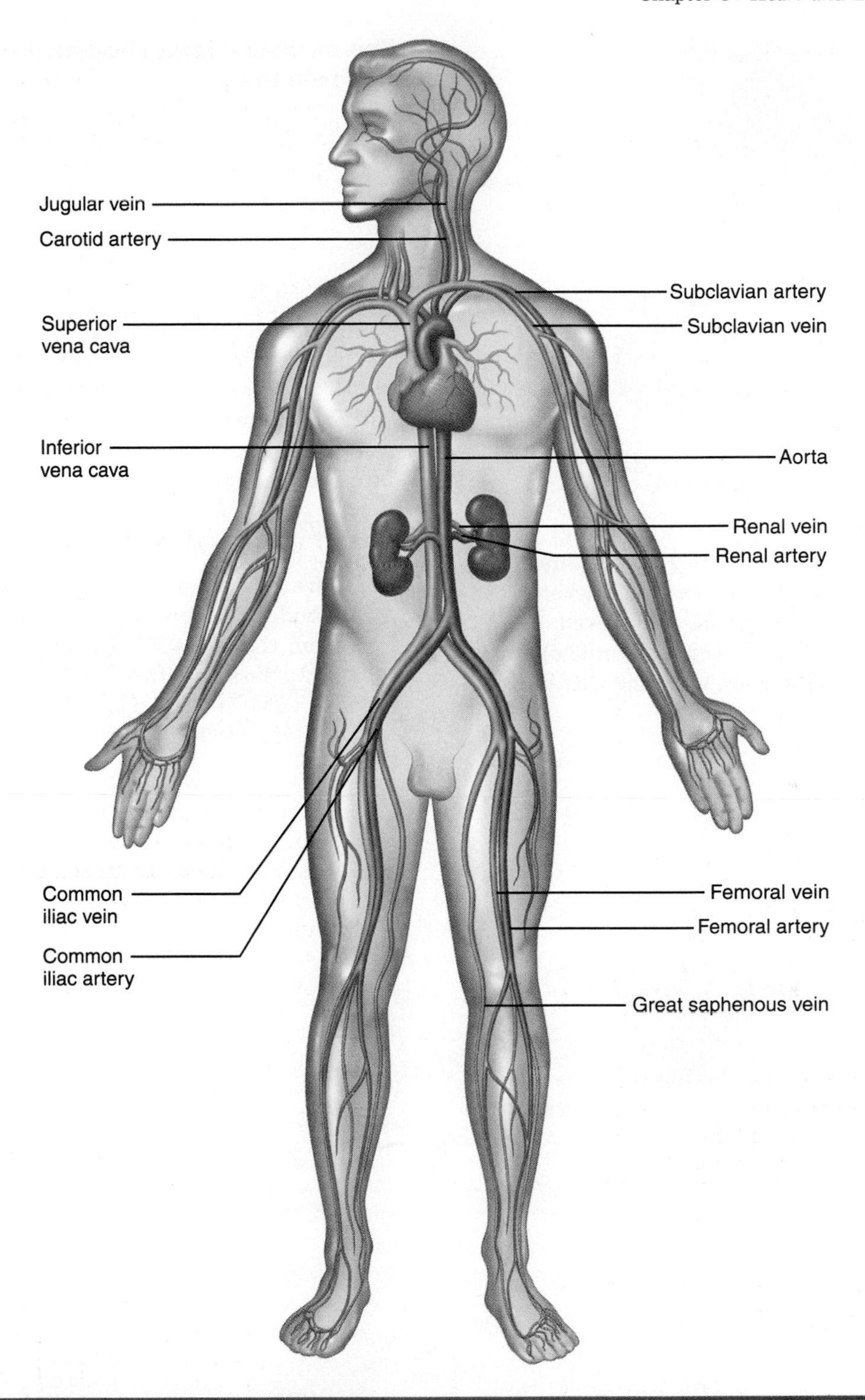

Figure 8.10 Some of the major arteries and veins in the human body. For simplicity, the lungs and most of the internal organs have been omitted.

the Latin *corona,* meaning encircling like a crown). From the surface they send branches inward to supply the myocardium. *Cardiac veins* collect the blood from the capillaries in the heart muscle and channel it back to the right atrium.

The coronary arteries are relatively small in diameter. If they become partially or completely blocked, perhaps due to atherosclerosis, serious health problems can result. Later in this chapter we will look at what happens when circulation to the heart is impaired.

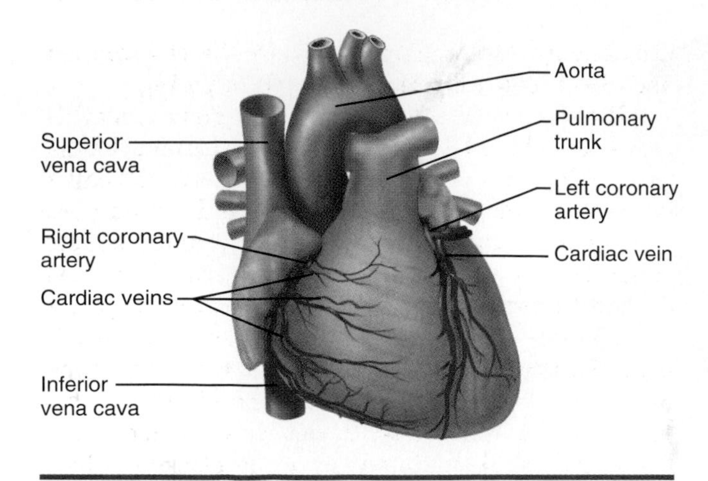

Figure 8.11 Blood vessels of the heart.

Recap **The right side of the heart (right atrium and right ventricle) pumps blood to the lungs, where blood is oxygenated and carbon dioxide is removed. The left side of the heart (left atrium and left ventricle) pumps blood to the rest of the body, supplying nutrients and removing wastes.**

The cardiac cycle: The heart contracts and relaxes

The pumping action of the heart is pulsatile rather than continuous, meaning that it delivers blood in separate and distinct pulses. A complete cardiac cycle involves contraction of the two atria, which force blood into the ventricles, followed by contraction of the two ventricles, which pump blood into the pulmonary artery and the aorta. The term **systole** refers to the period of contraction, and **diastole** refers to the period of relaxation. The entire sequence of contraction and relaxation is called the **cardiac cycle.**

Every cardiac cycle consists of three steps, as shown in Figure 8.12. Starting with the heart in a relaxed state:

1. *Atrial and ventricular diastole:* Both atria and both ventricles are relaxed. At this point pressure inside the heart chambers is lower than blood pressure in the veins. Blood returning from the veins fills the right and left atria and flows passively through the open atrioventricular valves into the two ventricles. Both semilunar valves are closed.

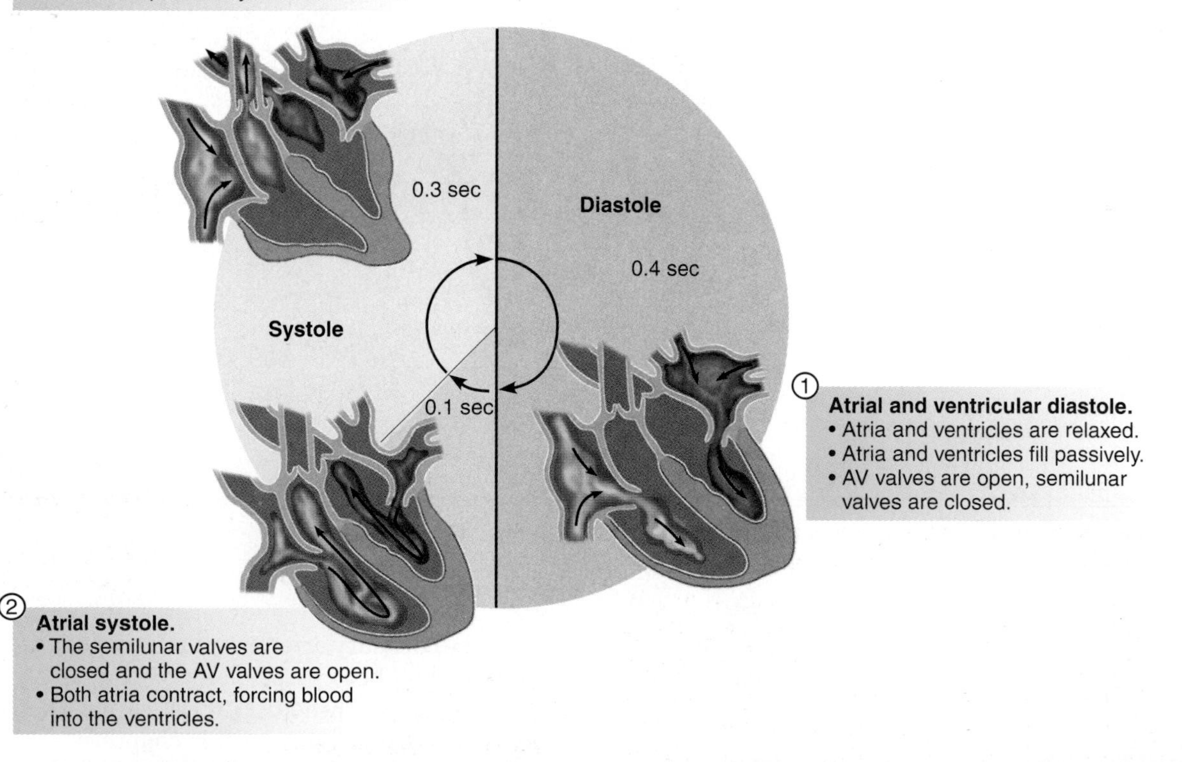

Figure 8.12 The cardiac cycle.

2. *Atrial systole:* Contraction of the heart begins with the atria. During atrial systole both the atria contract, raising blood pressure in the atria and giving the final "kick" that fills the two ventricles to capacity. Atrial systole also momentarily stops further inflow from the veins. Both atrioventricular valves are still open, and both semilunar valves are still closed.
3. *Ventricular systole.* The contraction that began in the atria spreads to the ventricles, and both ventricles contract simultaneously. The rapidly rising ventricular pressure causes the two AV valves to close, preventing blood from flowing backward into the veins. The pressure within the ventricles continues to rise until it is greater than the pressure in the arteries, at which point the pulmonary and aortic semilunar valves open. Blood is forcibly ejected into the pulmonary trunk and aorta.

Atrial and ventricular systole last for only about a third of a second and then the heart relaxes (diastole begins again). As soon as ventricular pressures fall below arterial pressures during early diastole, the pulmonary and aortic semilunar valves close, preventing backflow of arterial blood. Once ventricular pressure falls below blood pressure in the veins, the AV valves open again and another cardiac cycle begins.

As each surge of blood enters the arteries during systole, the artery walls are stretched by the higher pressure to accommodate the extra volume. Then they recoil passively during diastole as blood continues to flow out of them through the capillaries. You can feel this cycle of rapid expansion and recoil in the wall of an artery if it's located close to the skin's surface. This is called a *pulse.* Good places to detect a pulse are the radial artery (inside the wrist, just below the base of the thumb) and the carotid artery (on either side of the neck).

A complete cardiac cycle occurs every 0.8 seconds or so. These cycles repeat, from birth to death, without ever stopping. Atrial systole lasts about 0.1 second, ventricular systole about 0.3 second. During the remaining 0.4 second, the heart relaxes in diastole.

Heart sounds reflect closing heart valves

There is probably no more basic rhythm to which humans respond than the familiar "lub-DUB—lub-DUB" of the heart beating. We probably experience it, at least subconsciously, even before we are born. These *heart sounds* reflect events that occur during the cardiac cycle—specifically the closing of the heart valves. The "lub" signals the closure of the two AV valves during ventricular systole. The slightly louder "DUB" occurs when the aortic and pulmonary semilunar valves close during ventricular diastole. The sounds are actually due to vibrations in the heart chambers and blood vessels caused by the valves closing.

Blood flows silently as long as it flows smoothly. However, if blood encounters an obstruction, the disturbed flow can create unusual heart sounds called *murmurs.* Many murmurs result from incomplete closing of the heart valves due to unusually shaped valve flaps or stiffening of the flap tissue. If a valve does not close completely, some blood will be forced through it during the cardiac cycle, creating a swishing noise that can be detected through a stethoscope. Murmurs are not necessarily a sign of disease, but physicians can diagnose a variety of heart conditions, including leaking or partially blocked valves, from their sound and timing. Even serious murmurs can often be treated with surgery to replace the defective valve with an artificial valve (Figure 8.13).

Recap **Each cardiac cycle is a repetitive sequence of contraction (systole) and relaxation (diastole). The familiar "lub-DUB" of the heartbeat is caused by the closing of heart valves during the cycle.**

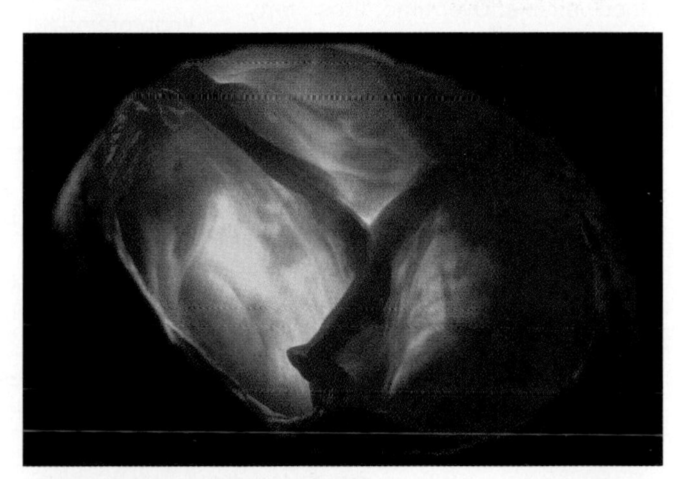

(a) A pulmonary semilunar valve.

(b) An artificial heart valve.

Figure 8.13 Heart valves.

Health Watch

Cholesterol and Atherosclerosis

A major cause of damage to blood vessels is **atherosclerosis,** the buildup of fatty deposits on the inner walls of arteries. In Western nations, atherosclerosis is a key factor in many cases of cardiovascular disease.

How do these deposits form in the first place? There is evidence that cholesterol plays a key role in the development of atherosclerosis. Recall from Chapter 2 that cholesterol is a lipid and an important component of our body's cells. In itself, cholesterol is not bad; in fact, cells require a certain amount of it. However, cholesterol molecules can become risky if they start to adhere to the interior of blood vessels. Over time, these fatty deposits snag other molecules floating by in the blood—calcium, blood cells, and additional cholesterol molecules. These deposits, called *plaques,* continue to enlarge, narrow the lumen, and reduce blood flow.

Atherosclerotic plaques injure the endothelium that lines blood vessels, and calcium deposits stiffen the vessel walls, increasing arterial resistance to blood flow. (This is why atherosclerosis is also called "hardening of the arteries.") Left untreated, atherosclerosis can contribute to heart attacks, strokes, aneurysms, and peripheral vascular disease.

Risk factors for atherosclerosis include factors that contribute to an elevated blood cholesterol (obesity, sedentary lifestyle, and a high-fat diet), smoking, and untreated diabetes and hypertension. Recent evidence indicates that a virus of the herpes type may be involved in some cases. Heredity plays a role, because atherosclerosis often runs in families. Gender is a consideration; before age 45, men have a 10 times greater risk than women. However, the risk rises for women after menopause.

As mentioned above, a major risk factor for atherosclerosis is high levels of cholesterol in the blood. However, cholesterol does not dissolve easily in plasma because it is a lipid and therefore relatively insoluble in water. Most cholesterol molecules in the blood are bound to certain proteins. Together, the protein and cholesterol complex is called a *lipoprotein,* of which there are several types.

From a health standpoint, two important lipoproteins are the low-density and high-density varieties. Low-density lipoprotein complexes (LDLs) are considered "bad" because they carry cholesterol to all cells of the body and make it available to them when they need it. LDL cholesterol may also contribute to plaque formation.

On the other hand, high-density lipoproteins (HDLs) are considered a good sign because they specifically target the cholesterol for removal. HDLs in the blood are picked up by the liver. There the cholesterol is detached from the protein and mixed with bile, which is secreted into the small intestine. Some of the cholesterol in bile is excreted from the body, although some is also reabsorbed to be used again.

Health professionals recommend having your cholesterol measured periodically. Lab results typically indicate LDL, HDL, and total cholesterol levels. These numbers are classified as:

LDL:
- Optimal: Below 130

HDL:
- Optimal: 40 and higher

Total cholesterol:
- Optimal: Below 200
- Borderline: 200–239
- High: 240 and higher

Optimal readings mean less risk of atherosclerosis and related cardiovascular problems.

Over time, all of us develop some degree of atherosclerosis. This is part of the aging process, and it begins in childhood and develops gradually throughout life. However, lifestyle can make a big difference in how rapidly atherosclerosis develops and whether it becomes severe. Maintaining a healthy weight and getting regular exercise will benefit your blood vessels, as will eating a healthy diet and not smoking. There are also medications available to lower cholesterol levels. At the end of this chapter, we look at what you can do to lower your risk of atherosclerosis as well as other cardiovascular conditions.

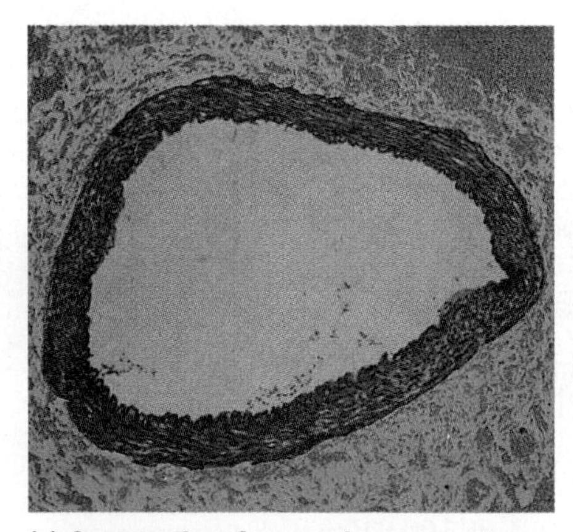

(a) Cross section of a normal coronary artery.

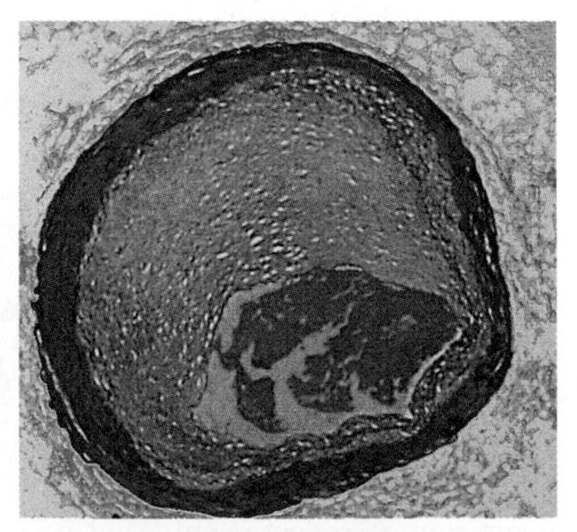

(b) A coronary artery narrowed by atherosclerotic plaque.

Atherosclerosis.

Cardiac conduction system coordinates contraction

How does your heart beat on its own, independent of your brain?

The coordinated sequence of the cardiac cycle is due to the **cardiac conduction system,** a group of specialized cardiac muscle cells that initiate and distribute electrical impulses throughout the heart. These impulses stimulate the heart muscle to contract in an orderly sequence that spreads from atria to ventricles. The cardiac conduction system consists of four structures: sinoatrial node, atrioventricular node, atrioventricular bundle and its two branches, and Purkinje fibers.

The stimulus that starts a heartbeat begins in the **sinoatrial (SA) node,** a small mass of cardiac muscle cells located near the junction of the right atrium and the superior vena cava. The SA node emits an electrical impulse that travels across both atria like ripples on a pond, stimulating waves of contraction (Figure 8.14). (Recall from Chapter 6 that cardiac muscle cells are connected by intercalated discs and gap junctions that let electrical signals flow directly from one cell to the next.) The SA node is properly called the **cardiac pacemaker** because it initiates the heart beat. However, the cardiac pacemaker can be influenced by the brain to speed up or slow down, as we'll see in a minute.

The electrical impulse traveling across the atria eventually reaches another mass of muscle cells called the **atrioventricular (AV) node,** located between the atria and ventricles. The muscle fibers in this area are smaller in diameter, causing a slight delay of approximately 0.1 second, which temporarily slows the rate at which the impulse travels. This delay gives the atria time to contract and empty their blood into the ventricles before the ventricles contract.

From the AV node, the electrical signal sweeps to a group of conducting fibers in the septum between the two ventricles called the **atrioventricular (AV) bundle.** These fibers branch and extend into **Purkinje fibers,** smaller fibers that carry the impulse to all cells in the myocardium of the ventricles. Because the electrical impulse travels down the septum to the lower portion of the ventricles and then spreads rapidly upward through the Purkinje fibers, the lower part of the ventricles contract before the upper part. This lower-to-upper squeezing motion pushes blood into the pulmonary trunk and aorta.

Electrocardiogram records heart's electrical activity

Because the body is largely water and water conducts electrical activity well, we can track the electrical activity of the heart as weak differences in voltage at the surface of the body. An **electrocardiogram (ECG or EKG)** is a record of the electrical impulses in the cardiac conduction system. An ECG involves placing electrodes on the skin at the chest, wrists, and ankles. The electrodes transmit the heart's electrical impulses, which are recorded as a continuous line on a screen or moving graph.

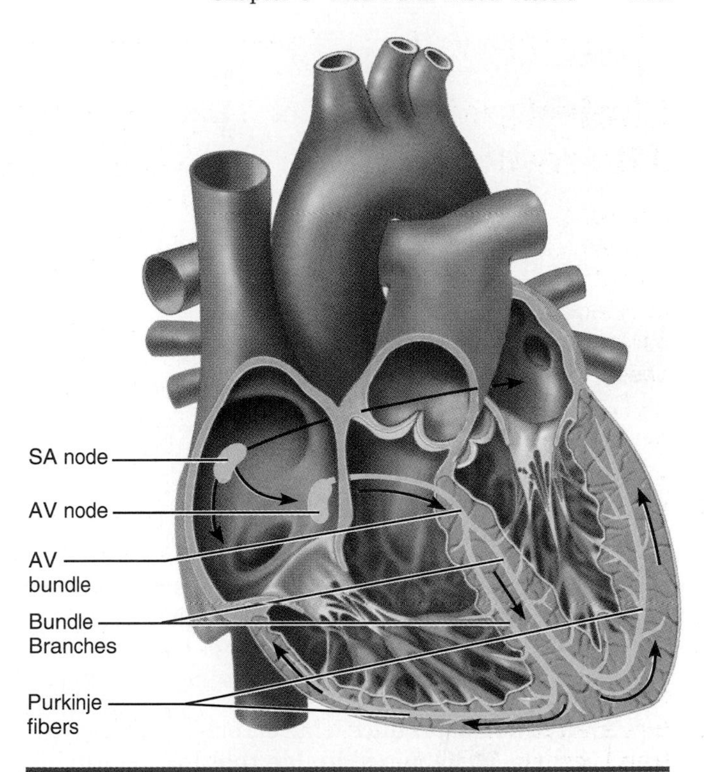

Figure 8.14 The cardiac conduction system. Electrical activity of the heart normally starts at the SA node, spreads across the atria to the AV node, and then progresses down the AV bundle and its branches to the Purkinje fibers.

A healthy heart produces a characteristic pattern of voltage changes. A typical ECG tracks these changes as a series of three formations: P wave, QRS complex, and T wave. First is the small P wave, representing the electrical impulse traveling across the atria. Second is the QRS complex, representing the spread of the electrical impulse down the septum and around the two ventricles in the Purkinje fibers. It occurs just before the ventricles start to contract. Finally, the T wave is the result of the *end* of the electrical activity in the ventricles.

If something goes wrong with the cardiac conduction system or if the heart muscle becomes damaged, abnormal heart electrical impulses and contractions may occur. An abnormality of the rhythm or rate of the heartbeat is called an *arrhythmia.* Arrhythmias take many forms. Occasional skipped heartbeats, for example, are fairly common and usually of no consequence. On the other hand, a type of rapid irregular ventricular contraction known as *ventricular fibrillation* (or "V-fib") is the leading cause of cardiac death in otherwise healthy people.

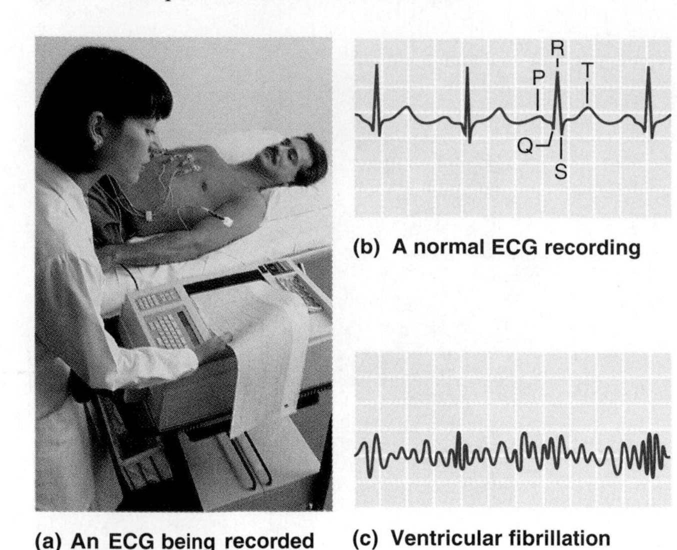

(a) An ECG being recorded (b) A normal ECG recording (c) Ventricular fibrillation

Figure 8.15 **The ECG is a tool for diagnosing heart arrhythmias.**

Arrhythmias produce characteristic ECG tracings, and the ECG is a valuable tool for identifying their cause, type, and location. Figure 8.15 shows a patient undergoing an ECG, a normal ECG recording, and a recording from a patient in ventricular fibrillation.

Ventricular fibrillation is treated by "cardioversion," in which a strong electrical current is applied to the chest to eliminate the abnormal fibrillating pattern and allow the normal sinus rhythm (originated from the SA node) to be restored. Less life-threatening arrhythmias can sometimes be treated with medications. In some cases an artificial pacemaker (a small generating unit that automatically stimulates the heart at set intervals) can be surgically implanted under the chest skin to normalize the heart rate.

Recap **Contraction of the heart is coordinated by modified cardiac muscle cells that initiate and transmit electrical impulses through a specialized conduction system. An electrocardiogram (ECG) is a recording of the heart's electrical activity taken from the surface of the body.**

8.3 Blood exerts pressure against vessel walls

Blood pressure is the force that blood exerts on the wall of a blood vessel as a result of the pumping action of the heart.

Blood pressure is not the same in all blood vessels. Figure 8.16 compares the pressures in the various segments of the vascular system. You can see from the highs and lows shown here that pressure is pulsatile in the arteries; that is, it varies with each beat of the heart. The highest pressure of the cycle, **systolic pressure,** is the pressure reached during ventricular systole when the ventricles contract to eject blood from the heart. The lowest pressure, **diastolic pressure,** occurs during ventricular diastole when the ventricles relax. Arteries store the energy generated by the heart during systole, and during diastole they use that stored energy to supply blood to the tissues.

Maintenance of arterial blood pressure is crucial to drive the flow of blood throughout the body and all the way back to the heart. Recall that fluid always

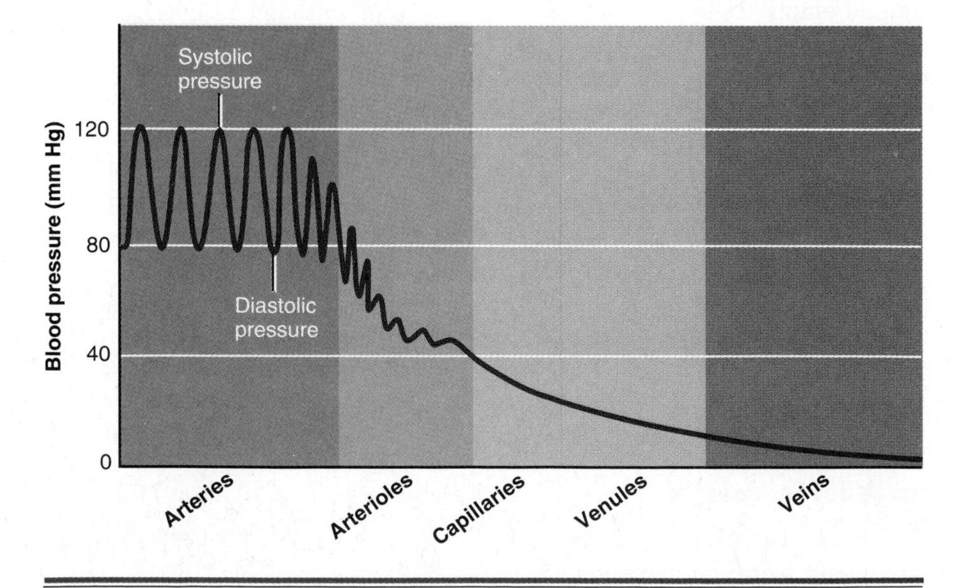

Figure 8.16 **Blood pressure in different segments of the vascular system.**

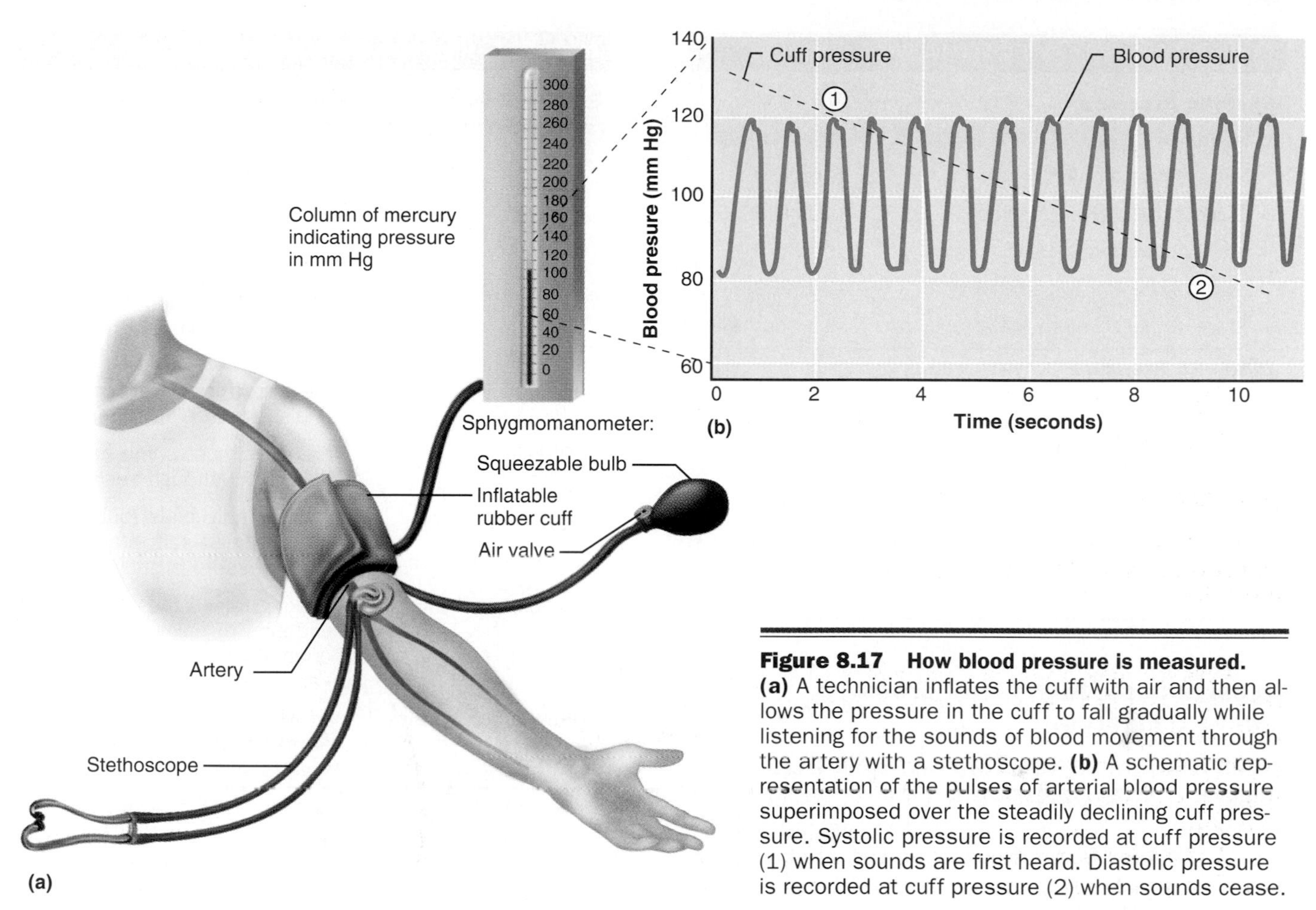

Figure 8.17 How blood pressure is measured. **(a)** A technician inflates the cuff with air and then allows the pressure in the cuff to fall gradually while listening for the sounds of blood movement through the artery with a stethoscope. **(b)** A schematic representation of the pulses of arterial blood pressure superimposed over the steadily declining cuff pressure. Systolic pressure is recorded at cuff pressure (1) when sounds are first heard. Diastolic pressure is recorded at cuff pressure (2) when sounds cease.

flows from a region of high pressure toward a region of lower pressure. By the time it reaches the capillaries, blood flow is steady rather than pulsatile, and pressure continues to fall as blood flows through venules and veins. The differences in the blood pressure of arteries, capillaries, and veins are what keep blood moving through the body.

When health professionals measure your blood pressure, however, they are assessing only the pressure in your main arteries. From a clinical standpoint, blood pressure gives valuable clues about the relative volume of blood in the vessels, the condition or stiffness of the arteries, and the overall efficiency of the cardiovascular system. Trends in blood pressure over time are a useful indicator of cardiovascular changes.

Blood pressure is measured with a *sphygmomanometer* ("sphygmo" comes from the Greek word for "pulse;" a manometer is a device for measuring fluid pressures). An inflatable cuff is placed over the brachial artery in your upper arm and connected to a pressure-measuring device (Figure 8.17). When the cuff is inflated to a pressure above systolic pressure, blood flow through the brachial artery stops because the high cuff pressure collapses the artery. The cuff is then deflated slowly while a health professional listens for the sounds of blood flowing in your artery with a stethoscope. As soon as pressure in the cuff falls to below the peak of systolic pressure, some blood spurts briefly through the artery during the high point of the pressure pulse, making a characteristic light tapping sound that is audible through the stethoscope. The cuff pressure at which this happens is recorded as systolic pressure. As the cuff continues to deflate, eventually blood flow through the artery becomes continuous and the tapping sound ceases. The point where the sound disappears is your diastolic pressure.

From this procedure, you will be given two numbers corresponding to your systolic and diastolic pressures. These represent the high and low points of blood pressure during the cardiac cycle. Blood pressure is recorded as mm Hg (millimeters of mercury, because early equipment used a glass column filled with mercury to measure the pressure). Table 8.1 presents the systolic and diastolic readings health professionals use to classify normal and high blood pressure.

Table 8.1 Systolic and diastolic blood pressures

Systolic Pressure	
Normal	129 and lower
High-Normal (Borderline)	130–139
Stage 1 (Mild Hypertension)	140–159
Stage 2 (Moderate Hypertension)	160–179
Stage 3 (Severe Hypertension)	180–209
Stage 4 (Very Severe Hypertension)	210 and higher
Diastolic Pressure	
Normal	85 and lower
High-Normal (Borderline)	85–89
Stage 1 (Mild Hypertension)	90–99
Stage 2 (Moderate Hypertension)	100–109
Stage 3 (Severe Hypertension)	110–119
Stage 4 (Very Severe Hypertension)	120 and higher

Hypertension: High blood pressure can be dangerous

Blood pressure above normal is called **high blood pressure** or **hypertension** ("hyper" comes from the Greek for "excess"). Hypertension is a significant risk factor for cardiovascular disease because it may damage the heart and blood vessels. The greater the pressure, the greater the strain on the body. Blood vessels react to the pounding by becoming hardened and scarred, which makes them less able to stretch during systole. The changes in blood vessels create greater resistance to blood flow. This stresses the heart because the work it must do is directly proportional to the arterial pressure against which it must pump.

Hypertension is called "the silent killer" because usually it has no symptoms. The American Heart Association estimates that approximately 50 million Americans have hypertension and a third of them don't even realize it. If left untreated, hypertension increases the risk of serious health problems such as heart attack, heart failure, stroke, kidney damage, even damage to the tissues inside the eyes.

What causes hypertension? Many times it happens because blood vessels become narrowed from atherosclerosis. Certain other factors also increase the risk, as summarized in Table 8.2. The only sure way to diagnose it is to have your blood pressure measured.

Blood pressure varies from minute to minute even in healthy individuals. Simply getting up in the morning raises it, as do exercise, emotions, smoking cigarettes, eating, drinking, and many other factors.

Table 8.2 Risk factors for hypertension

Risk Factor	Comments
Heredity	Family history of hypertension raises risk
Age	Blood pressure tends to rise throughout life
Race	African Americans have twice the incidence found in Caucasian Americans and Asian Americans
Sex	Males are more likely than females to develop hypertension
Obesity	The heart must pump harder to push blood through vessels
High salt intake	In some individuals (but not others) a high salt intake raises blood pressure slightly
Smoking	Smoking raises the blood concentration of epinephrine, a hormone that stimulates the heart
Sedentary lifestyle	Not well understood. May be due to higher blood lipids or weight gain
Persistent emotional stress	Emotional stress activates portions of the nervous system that elevate pressure
Diabetes mellitus	Diabetics have a higher incidence of hypertension, for reasons not yet known
Heavy alcohol consumption	Mechanisms unknown
Oral contraceptives and certain medications	Mechanisms vary by medication

Even having your blood pressure measured can make you nervous enough for your blood pressure to rise—a situation that health professionals call "white coat hypertension." This is why physicians generally have you sit quietly while measuring your blood pressure.

If hypertension is suspected, your physician will probably measure your blood pressure on at least three different occasions before making a firm diagnosis. True hypertension is a *sustained* elevation in blood pressure above normal levels—a resting systolic pressure above 140 mm Hg or a resting diastolic pressure above 90 mm Hg. Even blood pressure that is consistently high-normal (borderline) may carry a slightly higher risk of health complications.

Generally, if systolic pressure is high, diastolic pressure will be too. However, sometimes systolic pressure can register at above-normal levels while diastolic pressure remains normal, a condition called *isolated systolic hypertension.* Most common in older adults, it is diagnosed as a systolic pressure of 160 mm Hg or higher with a diastolic reading of less than 90 mm Hg. Like the more common form of hypertension, isolated systolic hypertension is associated with increased health problems.

At the end of this chapter we discuss what you can do to lower your risk of hypertension as well as other cardiovascular conditions. If hypertension does develop, however, there are a number of medical treatments available to lower blood pressure to a healthy level. It is important for people who take antihypertensive drugs to take them consistently because even though hypertension has no symptoms, it still increases the risk of related health problems.

Hypotension: When blood pressure is too low

It is also possible for blood pressure to fall below normal levels, a condition called *hypotension* ("hypo" comes from the Greek for "under"). Generally hypotension is a problem only if blood pressure falls enough to reduce blood flow to the brain, causing dizziness and fainting. Drops in blood pressure can follow abrupt changes in position, such as standing up suddenly. Other causes include severe burns or injuries involving heavy blood loss.

Recap **Blood pressure is the force that blood exerts on the wall of a blood vessel. It is measured as two numbers corresponding to systolic and diastolic pressures. Hypertension (high blood pressure) is a serious risk factor for cardiovascular disease and other health problems.**

8.4 How the cardiovascular system is regulated

The overall function of the cardiovascular system is to provide every cell with precisely the right blood flow to meet its needs at all times. This might seem like a complicated process because different types of tissues have different needs and their requirements may change at different times. In principle, however, the system is rather simple.

What's so important about maintaining a constant blood pressure?

Consider for a minute how your community provides water to every house according to the needs of the occupants. Basically, municipal water systems provide constant water *pressure* to every house. With steady water pressure available, each household simply adjusts the *flow* of water into the house by turning faucets on and off.

The human cardiovascular system is based on similar principles. Consider the following key points:

- Homeostatic regulation of the cardiovascular system centers on *maintaining a constant arterial blood pressure.*
- A constant arterial blood pressure is *achieved* by regulating the heart rate and force of contraction (adjusting flow *into* the arteries) and by regulating the diameters of all the body's arterioles as a whole (adjusting overall flow *out* of the arteries).
- With arterial blood pressure held relatively constant, local blood flows are adjusted to meet local requirements.

As we will see, this overall regulation is achieved with nerves, hormones, and local factors coupled to metabolism.

Baroreceptors maintain arterial blood pressure

There is probably no more important regulated variable in the entire body than the regulation of blood pressure. Without a relatively constant blood pressure as the driving force for supplying blood to the capillaries, homeostasis simply would not be possible. Of course blood pressure does vary, but always within a range consistent with being able to provide blood to all cells.

If we are to be able to regulate blood pressure, the body must have some way of measuring it. In fact, several of the large arteries, including the aorta and two **carotid arteries,** have certain regions called **baroreceptors** (Gr. "baro" denotes pressure). The baroreceptors regulate arterial blood pressure in the following manner:

1. When blood pressure rises, arterial blood vessels are stretched passively.
2. Stretch of baroreceptors in the carotid arteries and aorta causes them to send signals via nerves to an area of the brain called the "cardiovascular center."
3. The cardiovascular center responds by sending signals via nerves to the heart and blood vessels.
4. The effect on the heart is to lower heart rate and the force of contraction. This reduces **cardiac output,** the amount of blood that the heart pumps into the aorta each minute.
5. The effect on arterioles is vasodilation, an increase in arteriole diameter. Vasodilation increases blood flow through all tissues.

6. The net effect of both reducing cardiac output and increasing flow through the tissues is to return arterial pressure to normal.

Exactly the opposite sequence of events occurs when arterial pressure falls *below* normal. When pressure falls and the arteries stretch less than normal, the baroreceptors send fewer nerve signals to the brain. The brain correctly interprets this as a fall in pressure and sends nerve signals that increase cardiac output and constrict arterioles, raising arterial blood pressure again.

All day long—every time you stand up, sit down, get excited, run for the bus—your blood pressure fluctuates up or down. But it is quickly brought back within normal range by a negative feedback loop initiated by baroreceptors.

> ***Recap*** **Homeostatic regulation of the cardiovascular system centers on maintaining a relatively constant arterial blood pressure, making blood available to all cells as needed. Arterial pressure is sensed by baroreceptors located in the carotid arteries and aorta.**

Nerves and a hormone adjust cardiac output

As mentioned before, cardiac output is the amount of blood pumped into the aorta in one minute. We calculate cardiac output by multiplying *heart rate* (number of heartbeats per minute) by *stroke volume* (volume of blood pumped out with each heartbeat). For a healthy adult at rest, the heart rate averages about 75 beats per minute and the stroke volume averages about 70 milliliters per beat. This yields a cardiac output of 5250 milliliters per minute: 5.25 liters or approximately 5 quarts.

Our bodies can adjust cardiac output over a fairly wide range according to the need to regulate arterial blood pressure, as described earlier. Exercise and strong emotions can raise heart rate (and cardiac output) as well.

Regulation of cardiac output is centered in an area of the brain called the *medulla oblongata,* which is where the cardiovascular center is located. The medulla oblongata receives nerve signals from the baroreceptors and also from areas of the brain involved in emotions and then sends nerve signals to the heart in two sets of nerves. *Sympathetic nerves* stimulate the heart and cause it to beat faster, and *parasympathetic nerves* inhibit the heart and cause it to beat more slowly.

The hormone *epinephrine* (also called adrenaline) stimulates the heart as well. Epinephrine is secreted by the adrenal glands in times of stress or fear.

The sympathetic and parasympathetic nerves also help to maintain blood pressure by regulating the overall diameter of the body's arterioles. In general, the sympathetic nerves constrict blood vessels and raise blood pressure, and the parasympathetic nerves dilate blood vessels and lower blood pressure.

Local requirements dictate local blood flows

With arterial blood pressure held relatively constant, the flow of blood through each precapillary sphincter (and hence each capillary) can be adjusted according to need, just like turning a faucet on or off. How is the flow adjusted? When a particular area is metabolically active, such as when a muscle is contracting, it consumes more oxygen and nutrients. Increased metabolism also raises the production of carbon dioxide and other waste products. One or more of these changes associated with increased metabolism cause precapillary sphincters to vasodilate, increasing flow. Scientists do not yet know the precise mechanisms of this vasodilation or the identity of all the chemical substances that influence blood flow, but we do know that it occurs. Figure 8.18 shows how an increase in cellular metabolism might increase the local blood flow to that region.

Look at the overall process and you can see how together these control mechanisms deliver blood efficiently to all tissues. First, nerves and hormones adjust cardiac output and the rate of blood flow through the vascular system as a whole, in an effort to maintain relatively constant arterial blood pressure. Second, any cell or part of the body that is not receiving enough blood can override the system locally and get exactly what it needs—no more, no less.

Some tissues and organs need a more consistent blood supply than others. That too is taken into account by our control system. Consider what would happen if you lost half of your blood volume due to injury, producing a precipitous fall in blood pressure. The negative feedback control of arterial blood pressure would stimulate your heart and constrict your arterioles, reducing flow to most organs in an effort to raise blood pressure. However, organs whose survival depends critically on a constant blood supply (your brain and heart, for example) can override the generalized vasoconstriction with their local control mechanisms. Organs whose metabolic activities are not tied to their immediate survival (such as your kidneys and digestive tract) just remain vasoconstricted for awhile. In effect, the limited available blood supply is shunted to essential organs.

You may wonder why, if these homeostatic controls are working properly, people develop hypertension in the first place. Every time blood pressure went up a little, wouldn't feedback mechanisms bring it back to normal? This is the great mystery of

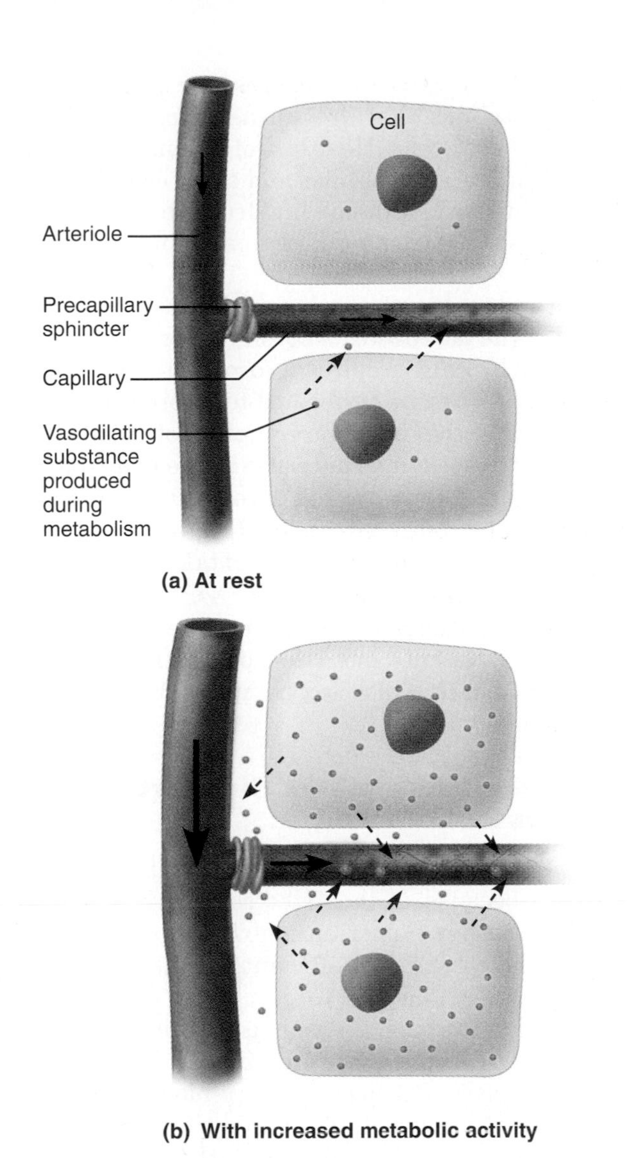

Figure 8.18 Theoretical representation of how the increased production of a vasodilating substance during metabolism could regulate local flow. (a) At rest, very little of the vasodilating substance would be produced and flow would be minimal. **(b)** With increased metabolic activity, the presence of more of the substance in the extracellular space near an arteriole and precapillary sphincter would cause the vessels to vasodilate, increasing flow and washing out the substance.

hypertension. Although scientists don't fully understand it yet, for some reason the body seems to adjust to high pressure once it has been sustained for a long time. Apparently the feedback system slowly resets to recognize the higher pressure as normal. This may be because hypertension tends to develop slowly over decades, so there is never a single defining moment when pressure registers as too high.

Recap **Two opposing sets of nerves (sympathetic and parasympathetic) and a hormone (epinephrine) adjust cardiac output and arteriole diameters to maintain arterial blood pressure. Local factors regulate blood flow into individual capillaries by altering the diameters of precapillary sphincters.**

8.5 Cardiovascular disorders: A major health Issue

Disorders affecting the cardiovascular system are a major health issue in Western countries. Cardiovascular disorders cause more than 900,000 deaths per year in North America alone. They are the number one killer in the U.S., far ahead of the number two killer, cancer (approximately 500,000 deaths per year).

We have already considered two important cardiovascular disorders, atherosclerosis and hypertension. In the rest of this chapter we look at several other disorders, including angina pectoris, heart attack, congestive heart failure, embolism, and stroke. Finally, we examine what you can do to reduce your own risk of cardiovascular health problems.

Angina pectoris: Chest pain warns of impaired blood flow

As a hard-working muscle, the heart requires a constant source of blood. Normally the coronary arteries and their branches provide all the blood the heart needs, even during sustained exercise. However, if these arteries become narrowed (usually due to atherosclerosis), blood flow to the heart may not be sufficient for the heart's demands. This may lead to *angina pectoris* (sometimes just called angina), a sensation of pain and tightness in the chest. Often angina is accompanied by shortness of breath and a sensation of choking or suffocating (*angina* comes from the Latin word for "strangling;" *pectoris* refers to the chest). Many angina episodes are triggered by physical exertion, emotional stress, cold weather, or eating heavy meals because the heart requires more blood and oxygen at these times.

Angina is uncomfortable but usually temporary. Stopping to rest and taking several deep breaths can often relieve the discomfort. However, angina should never be ignored because it is a sign of insufficient circulation to the heart. *Angiography* is a procedure that enables blood vessels to be visualized after being filled with a contrast medium (a substance that is opaque to x-rays). Angiography allows health professionals to take x-ray pictures of blood vessels (called *angiograms*) and assess their condition (Figure 8.19).

Certain medications can increase blood flow to the heart muscle. Another treatment is *balloon angioplasty*, which involves threading a slender flexible

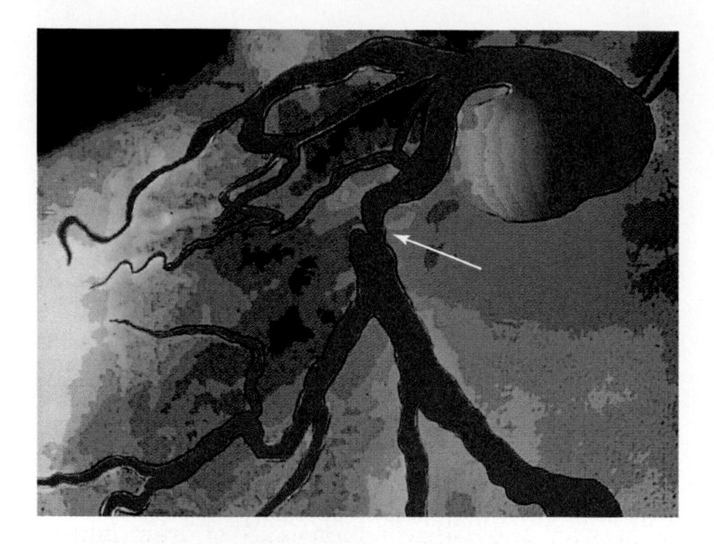

Figure 8.19 A coronary angiogram. A contrast medium (similar in function to a dye) is injected into the coronary arteries, causing the coronary arteries to become visible on an x-ray photograph. Note the area of narrowing of the coronary artery indicated by the arrow.

tube with a small balloon attached into the blocked artery. When the balloon reaches the narrowest point of the vessel it is inflated briefly so that it presses against the fatty plaques that cause the vessel lumen to be narrowed, flattening them and widening the lumen. Balloon angioplasty has a high success rate, although in some cases the vessel narrows again over time, requiring repeat treatments.

Heart attack: Permanent damage to heart tissue

If blood flow to the heart is impaired long enough, the result is a **heart attack**—sudden death of an area of heart tissue due to oxygen starvation. (The clinical term is *myocardial infarction;* infarction refers to tissue death from inadequate blood supply.) Many people who suffer a heart attack have a previous history of angina pectoris. Symptoms of heart attack include feelings of intense chest pain, nausea, and the sensation of a heavy object on the chest, making it hard to breathe. Pain may radiate to the left arm, jaw, back, or upper abdomen.

By definition, a heart attack means that part of the heart has been permanently injured. Because the body cannot replace cardiac muscle cells, any damage to them can impair the heart's ability to function. Most heart attack fatalities occur because of a serious heart arrhythmia called *ventricular fibrillation.*

Prompt treatment is crucial for recovery from a heart attack. If a heart attack is even suspected, the person should be rushed immediately to a hospital. The diagnosis of a heart attack is generally made on the basis of the ECG and the presence in the blood of certain enzymes that are released from dead and damaged heart cells. Health professionals can often control cardiac arrhythmias and other complications and administer clot-dissolving drugs to unblock vessels. The sooner treatment begins the more successful it is likely to be.

Later (or even before a heart attack has occurred, if the narrowing of coronary vessels has been diagnosed by angiography), a *coronary artery bypass graft (CABG)* can be performed to improve coronary blood flow. In this procedure, a piece of blood vessel is removed from somewhere else in the body (often a leg vein is used) and grafted onto the blocked artery to bypass the damaged region (Figure 8.20). Over time the grafted vein thickens and takes on the characteristics of an artery.

Thanks to these and other treatments, the heart attack survival rate has risen dramatically. Eighty percent of heart attack survivors are back at work within three months.

Congestive heart failure: The heart becomes less efficient

Generally our bodies maintain adequate arterial pressure because of the tight control mechanisms described earlier. Sometimes, however, the heart becomes weaker and less efficient at pumping blood. When the heart begins to pump less blood, blood backs up in the veins (the venous system) *behind* the failing heart. As a result, pressure in the veins and capillaries rises. Because capillary blood pressure is now higher than usual, more fluid than usual filters out of the capillaries and into the extracellular space. When that happens a person is said to have **congestive heart failure.** ("Congestive" refers to the buildup of extracellular fluid).

People with congestive heart failure get out of breath when walking or climbing even a short flight of stairs. They may even have trouble breathing while lying down because the horizontal position results in even higher venous pressure and fluid accumulation in the lungs. Other symptoms include swollen ankles and legs, swollen neck veins, and weight gain from the extra fluid.

There are several reasons why a heart might begin to fail. Aging is one factor, but certainly not the only one because older people do not necessarily develop heart failure. Other possible causes include past heart attacks, leaking heart valves, heart valves that fail to open normally, uncontrolled hypertension, or serious arrhythmias. Lung conditions such as emphysema can raise blood pressure in the lungs and strain the heart.

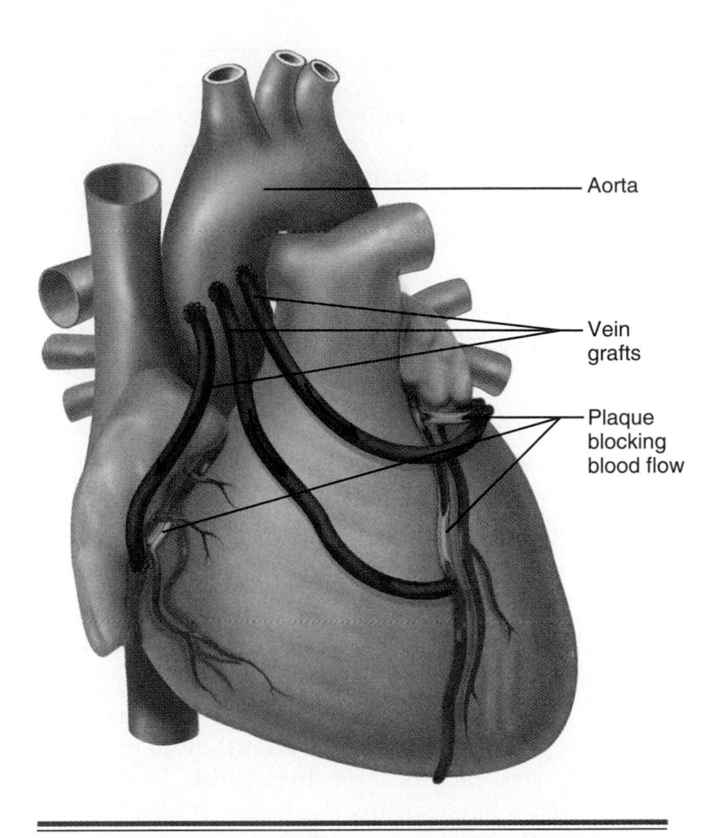

Figure 8.20 Coronary artery bypass grafts. This is an example of a "triple CABG," in which three vein grafts have been placed to bypass three areas of atherosclerotic plaque. Additional blood flow to the area beyond the restricted artery is indicated by arrows.

Treatments for congestive heart failure focus on improving cardiac performance and efficiency while preventing the accumulation of extracellular fluid. Regular mild exercise promotes more efficient blood flow, and frequent resting with the feet elevated helps fluid drain from leg veins. Physicians may prescribe diuretics (drugs that help the body get rid of excess fluid), vasodilating drugs to expand blood vessels, or medications to help the heart muscle beat more forcefully.

Embolism: Blockage of a blood vessel

Embolism refers to the sudden blocking of a blood vessel by material floating in the bloodstream. Most often the obstacle (an *embolus*) is a blood clot that has broken away from a larger clot elsewhere in the body (often a vein) and lodged in an artery at a point where arterial vessels branch and get smaller in diameter. Other possible emboli include cholesterol deposits, tissue fragments, cancer cells, clumps of bacteria, or bubbles of air.

Embolism conditions are named according to the area of the body affected. A *pulmonary embolism* blocks an artery supplying blood to the lungs, causing sudden chest pain and shortness of breath. A *cerebral embolism* impairs circulation to the brain, possibly causing a stroke. These conditions are emergencies that require immediate treatment. An *embolectomy* (surgery to remove the blockage) may be possible. Clot-dissolving drugs may also help.

Stroke: Damage to blood vessels in the brain

The brain requires a steady blood supply—about 15% of the heart's output at rest—to function normally. Any impairment of blood flow to the brain rapidly damages brain cells. A **stroke** (*cerebrovascular accident*) represents damage to part of the brain caused by an interruption to its blood supply. In effect, it is the brain equivalent of a heart attack. Strokes are the most common cause of brain injury and a leading cause of death in Western nations.

Many strokes are caused by cerebral embolism —a blood clot that lodges and blocks blood flow through a vessel serving the brain. Some clots form in brain arteries that are already narrowed by atherosclerosis. A stroke can also result from rupture of a vessel in the brain. Probably the greatest risk factor for stroke is untreated hypertension.

Symptoms of strokes appear suddenly and vary according to the area of the brain affected. They may include weakness or paralysis on one side of the body, fainting, inability to speak or slurred speech, difficulty in understanding speech, impaired vision, nausea, or a sudden loss of coordination.

Immediate medical care is crucial. If the stroke resulted from an embolism, clot-dissolving drugs or an embolectomy can save a life. Eliminating the clot as quickly as possible (within minutes or hours) can limit the area of damage and reduce the severity of the permanent injury. If a ruptured artery is responsible, health providers may be able to surgically drain the excess blood.

Some people recover well from a stroke or suffer only minor effects, but others do not recover completely despite intensive physical therapy over many months. The reason for the poor recovery rate is that the body does not grow new nerve cells to replace damaged ones. However, rehabilitation with skilled health professionals can help many stroke patients improve their ability to function. Recovery involves retraining nerve pathways that already exist so they can take over the functions of damaged nerve cells.

Recap **Cardiovascular disorders are the number one killer in the U.S. Most disorders are caused either by conditions that result in failure of the heart as a pump or by conditions in which damage to blood vessels restricts flow or ruptures vessels.**

Health Watch

Smoking and Cardiovascular Disease

We think of smoking as a leading cause of lung disease, and indeed it is. You might be surprised to learn, however, that smokers also have higher rates of heart attack and stroke. For example, smokers have more than twice the risk of suffering a heart attack than nonsmokers, and if they do have a heart attack they are almost four times as likely to die from it. The American Heart Association calls smoking the leading preventable cause of premature death in this country today.

Why the association between smoking and cardiovascular disease? One reason is that smokers tend to have higher levels of LDL cholesterol, the type of cholesterol that leads to atherosclerosis. Smokers also are more likely to develop hypertension, possibly because tobacco smoke stimulates the release of epinephrine, a hormone that increases the speed and force of the heartbeat. Tobacco smoke contains carbon monoxide, which, as we saw in Chapter 7, binds to hemoglobin and reduces the amount of oxygen that reaches body tissues.

Smoking may also make other risk factors even riskier. For example, some studies indicate that female smokers who take oral contraceptives may have an increased incidence of cardiovascular disease. The American Heart Association recommends that women who take the Pill should not smoke. Consult your physician for the latest information.

8.6 Reducing your risk of cardiovascular disease

How can you reduce your risk of cardiovascular disease?

Cardiovascular disorders are among the most preventable of chronic health conditions. Although there are some factors beyond your control—such as sex, race, age, and genetic inheritance—your lifestyle choices can also affect your risk. Things you can do to reduce your risk include:

- Don't smoke, or if you do, quit. Some researchers think secondhand smoke poses a risk as well.
- Watch your cholesterol levels. Cardiovascular risk rises with the blood cholesterol level. There is also evidence that high cholesterol increases risk even more when it is combined with other factors such as hypertension and tobacco smoke.
- Keep moving (Figure 8.21). Regular, moderate exercise lowers the risk of cardiovascular disease. This is not surprising because the heart is, after all, a muscle. Physical activity tends to lower blood pressure and cholesterol and makes it easier to maintain a healthy body weight. Always consult your physician before starting an exercise program.
- If your blood pressure is on the high side, seek treatment. As discussed earlier, untreated hypertension damages blood vessels and increases the workload on the heart.

In addition to these major risks, at least three other factors are associated with cardiovascular disease, although the precise link has not yet been determined. This is why doctors recommend the following:

- Maintain a healthy weight. It's not clear how obesity contributes to cardiovascular problems, but overweight people have a higher rate of heart disease and stroke even if they do not have other risk factors. One hypothesis is that increased weight strains the heart. It also has adverse effects on other risk factors such as blood cholesterol and hypertension.
- Keep diabetes under control. Diabetes mellitus is a disorder of blood sugar levels. Untreated diabetes damages blood vessels, but effective treatments reduce cardiovascular damage significantly. (For more on diabetes, see Chapter 13.)

Figure 8.21 Moderate, regular exercise improves cardiovascular performance and lowers the risk of cardiovascular disease.

Directions in Science

What's the Best Way to Exercise Your Heart?

The heart is a muscle, and it makes sense that it could benefit from regular exercise like any other muscle. But how much exercise, and what kind of exercise, is the right way to keep your heart in shape?

Until the mid-1990s the prescription called for fairly vigorous exercise that would raise heart rate substantially for at least 20 to 30 minutes at least three times a week. The theory was that the exercise needed to increase heart rate significantly or the heart wasn't really being pushed. Just performing daily activities wasn't considered sufficient—you needed to take a serious jog, go swimming or biking, or take an aerobics class. This kind of exercise takes considerable effort, not only to do the activity but to arrange the time and place for it. Many of us don't get that kind of exercise on a regular basis, and we feel guilty about it.

Then, in 1993, a panel of experts convened by the Centers for Disease Control and Prevention (CDC) and the American College of Sports Medicine met to consider these questions. The panel concluded that exercise did not need to be strenuous—even moderate activities such as housework, walking, or gardening could provide benefits that are comparable to more strenuous, sustained exercise. The panel also reported that these activities could be extended over the course of a day rather than concentrated into one intense session. The media greeted this report very positively, perhaps because most of us already do the kinds of activities proposed by the panel. Many of us breathed a collective sigh of relief. We could continue with our usual lifestyle, secure in the knowledge that we were being heart-healthy.

More recently, however, the debate has re-resurfaced. Several researchers now argue that the data are not as clear-cut as we were led to believe. They claim the benefits of moderate exercise are being overemphasized to make it more acceptable to the public, and the available data do not convincingly support the notion of breaking up exercise into small chunks of time throughout the day.

The issue is not trivial, for the wrong recommendation could lead to a false sense of security in the public. It could affect the course of cardiovascular disease in millions of people. Indeed, some scientists estimate that inactivity is a contributing factor in a third of all deaths from heart disease.

Different studies, different conclusions. Why? One reason is that it's virtually impossible to tell whether a heart is "athletic" or not. To determine how much exercise is appropriate for the average person in a large population, scientists need to rely on statistics—the relationship, for example, between the amount of exercise that different groups of people get versus the number of people in each group who suffer a heart attack. Heart attacks are sufficiently rare, however, that it is hard to include enough people in each exercise group to produce reliable numbers. Consequently, scientists sometimes choose to rely on indirect measures of cardiovascular fitness, such as blood cholesterol level or body fat percentage, because everyone has a blood cholesterol level and some fat tissue. This approach assumes, of course, that blood cholesterol levels or amounts of fat correlate directly and accurately with cardiovascular fitness. They may not in all cases.

Finally, there's the problem of determining how much exercise a person actually gets. Information about the amount and severity of exercise is usually obtained through questionnaires or interviews. If you were to ask a hundred people if they exercise "strenuously," do you think they would all interpret the word strenuously the same? Self-reports are notoriously unreliable because they are inconsistent between individuals, making the data highly inaccurate.

It may well be that more intense exercise does yield greater cardiovascular benefits, but the question is still unresolved. We can expect scientists to continue debating these issues and refining exercise recommendations over the next decade or so. As always when you read about scientific studies, be prepared to evaluate new claims. Examine the data they put forward, and ask yourself if the data were collected properly and whether they support the conclusions. Look for underlying assumptions that might affect the study's outcome, and assess counterarguments by other scientists to see if they have a valid point. Be willing to change your mind on the basis of well-supported information.

In the meantime, continue to exercise. It's hard to imagine that moderate exercise in any form could do anything but help your heart.

- Avoid chronic stress. Again the mechanism is unclear, but there is an association between a person's perceived stress and behavior patterns and the development of cardiovascular disease. Stress may also affect other risk factors, for example, how much a smoker smokes or whether a person starts smoking.

Recap **You can reduce your risk of developing cardiovascular disease by not smoking, exercising regularly, watching your weight and cholesterol, and avoiding prolonged stress. If you have diabetes and/or hypertension, try to keep these conditions under control.**

Chapter Summary

Blood vessels transport blood

CD Cardiovascular System/Anatomy Review: Blood Vessel Structure and Function

- The primary function of blood vessels is to bring blood into close proximity of all living cells.
- Thick-walled arteries transport blood to the capillaries at high pressure.
- Small arterioles and precapillary sphincters regulate the flow of blood into each capillary.
- Thin-walled capillaries are the only vessels that exchange fluids and solutes with the extracellular fluid.
- Distensible venules and veins store blood at low pressure and return it to the heart.

The heart pumps blood through the vessels

CD Cardiovascular System/Anatomy Review: The Heart

CD Cardiovascular System/Cardiac Cycle

CD Cardiovascular System/Cardiac Output

- The heart is composed primarily of cardiac muscle. Structurally, it consists of four separate chambers and four one-way valves. Its primary function is to pump blood.
- The heart pumps blood simultaneously through two separate circuits: the pulmonary circuit, where blood picks up oxygen and gets rid of carbon dioxide, and the systemic circuit, which supplies the rest of the body's cells.
- The heart contracts and relaxes rhythmically. Contraction is called systole and relaxation is called diastole.
- The coordinated contraction of the heart is produced by a system of specialized cardiac muscle cells that initiate and distribute electrical impulses throughout the heart muscle.
- An electrocardiogram, or ECG, records electrical activity of the heart from the surface of the body. An ECG can be used to diagnose certain cardiac arrhythmias and disorders.

Blood exerts pressure against vessel walls

- The heart generates a blood pressure, and the arteries store the blood under pressure during diastole.
- Systolic and diastolic arterial blood pressures can be measured with a sphygmomanometer.
- High blood pressure, called hypertension, is a major risk factor for cardiovascular disease.

How the cardiovascular system is regulated

CD Cardio Output

- The most important controlled variable in the cardiovascular system is arterial blood pressure.
- Arterial blood pressure is monitored by stretch receptors located in certain large arteries.
- Cardiac output and the diameters of the arterioles are regulated (controlled) in order to keep arterial blood pressure relatively constant.
- With pressure held constant, local blood flows can be adjusted according to the metabolic needs of the tissues and cells in that area of the body.

Cardiovascular disorders: A major health issue

- The heart muscle is always working. Impairment of blood flow to the heart can lead to a sense of pain and tightness in the chest (angina pectoris) and/or permanent damage to heart tissue (myocardial infarction, or heart attack).
- Slowly developing, chronic failure of the heart as a pump can lead to excessive extracellular fluid, a condition known as congestive heart failure.
- An embolism is the sudden blockage of a blood vessel by any object.
- Strokes, also called cerebrovascular accidents, can be caused by either embolisms or rupture of blood vessels. The result is damage to a part of the brain when its blood supply is interrupted.

Reducing your risk of cardiovascular disease

- Your chances of developing a cardiovascular disorder depend on certain risk factors. Some risk factors you cannot change, whereas others depend on the choices you make in life.

Terms You Should Know

artery, 168
atrium, 174
AV valves, 175
capillary, 170
diastole, 178
myocardium, 174
precapillary sphincter, 170
pulmonary circuit, 175
semilunar valves, 175
sinoatrial (SA) node, 181
systemic circuit, 176
systole, 178
vein, 172
ventricle, 174

Test Yourself

Matching Exercise

_____ 1. Blood vessels that store most of the blood volume in the body

_____ 2. Blood vessels with a thick muscle layer able to withstand high pressures

_____ 3. Blood vessels that exchange materials with the extracellular fluid

_____ 4. Blood vessels whose diameters can be adjusted to regulate blood flow

_____ 5. The innermost layer of tissue of the heart; it lines the heart chambers

_____ 6. Inflammation of the sac surrounding the heart

_____ 7. Period of relaxation of all four chambers of the heart

_____ 8. Highest pressure of the arterial pressure cycle

_____ 9. Stretch receptors that monitor arterial pressure

_____10. A heart attack

a. arterioles
b. endocardium
c. arteries
d. pericarditis
e. systolic pressure
f. baroreceptors
g. veins
h. myocardial infarction
i. diastole
j. capillaries

Label the following components of the heart (11–20):

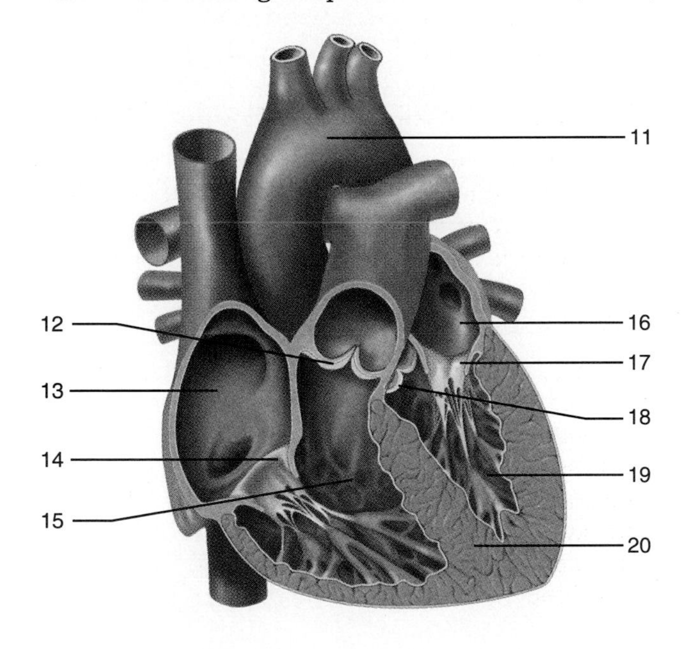

Fill in the Blank

21. The largest artery in the body is called the __________.
22. A blood clot that breaks off from a vein in the leg and lodges in a pulmonary artery is called a __________ __________.
23. A __________ is a device used for measuring blood pressure.
24. The amount of blood that the heart pumps every minute is called the __________ __________.
25. Damage to or blockage of a blood vessel in the brain may lead to a __________.

Concept Review

1. Why is it so important that the heart always get a consistent and adequate blood supply?
2. Describe how arterial blood pressure is measured by the body.
3. Describe the function of the aortic semilunar valve.
4. Compare the functions of the pulmonary circuit and the systemic circuit.
5. Describe the cardiac cycle of relaxation and contraction, and explain what causes each of the two heart sounds during the cycle.
6. Name the site in the heart that normally initiates the electrical activity that leads to a contraction.
7. Of what functional value is the distensibility of veins? In other words, why not have them just as thick and stiff as arteries?
8. Name the one hormone that has a stimulatory effect on the heart (increases heart rate).

Apply What You Know

1. Regarding the cardiac cycle shown in Figure 8.12, explain why no blood is ejected into the aorta during the very first part of ventricular systole.
2. During exercise, arterial blood pressure changes very little. However, cardiac output may double and blood flow to exercising muscle may go up 10-fold, while at the same time the blood flow to kidneys may decline by nearly 50%. Explain possible mechanisms that might account for these very different changes.
3. When a coronary artery bypass graft (CABG) is performed, the vessels used for the bypass grafts are usually veins taken from the patient's legs. Over time the grafted veins take on many of the characteristics of arteries; that is, they become thicker and stiffer. What might this suggest about the possible cause(s) of the structural differences between arteries and veins? Hypothesize what might happen if you took a section of artery and implanted it into the venous circulation.

Case in Point

Heart Attacks: Worse for Women?

Younger women are usually considered unlikely prospects for serious heart attacks. Recent studies cast doubt on this assumption.

Men's heart attacks are frequently caused by an artery that has become completely blocked. Heart attacks in men are often accompanied by the classic signs of crushing chest pain that radiates to the left shoulder or jaw, often accompanied by nausea and sweating. They also usually produce characteristic ECG patterns and angiograms. In contrast, women's heart attacks often result from less significant artery blockage, according to a recent study by researchers at St. Luke's-Roosevelt Hospital in New York. The symptoms of a heart attack in women are often as subtle as shortness of breath and overwhelming fatigue. Perhaps for that reason, women take longer to call an ambulance during a heart attack.

Despite their seemingly less dramatic symptoms, female heart attack patients are more apt to suffer complications such as coronary-artery spasms. In addition, women under age 50 who suffer a heart attack are more likely to die from it than men their age—6.1% compared to 2.9%. In the 50–54 age group, 7.4% of female heart attack patients die afterward, compared to 4.1% of men.

QUESTIONS:

1. Why do you think female heart attack patients have a higher mortality rate than male patients?
2. Among younger heart attack victims, women are more likely than the men to have other preexisting health conditions (such as congestive heart failure and diabetes). How do you think this might contribute to the different outcomes between women and men?

INVESTIGATE:

Ron Winslow, "Heart Disease Affects Women, Men Differently." *Wall Street Journal,* July 22, 1999, p. B2.

Robert Bazell, "Heart Attacks More Deadly for Women." MSNBC News, http://www.msnbc.com/news/292252.asp

References and Additional Reading

Centers for Disease Control and Prevention and American College of Sports Medicine, in cooperation with the President's Council on Physical Fitness and Sports. *Summary Statement: Workshop on Physical Activity and Public Health.* Indianapolis, IN: July 29, 1993. (A panel of experts' report, 1993, regarding exercise and cardiovascular disease.)

Cowley, Geoffrey. The Heart Attackers. *Newsweek,* Aug. 11, 1997, pp. 54–60. (A popular press update of recent findings concerning the causes of heart attacks.)

Dickman, Steven. Mysteries of the Heart. *Discover,* July 1997, pp. 117–119. (A short article discussing the possibility that a virus contributes to coronary artery disease.)

Williams, Paul T. Relationship of Distance Run per Week to Coronary Heart Disease Risk Factors in 8283 Male Runners: The National Runners' Health Study. *Archives of Internal Medicine* 157:191–198, 1997. (A research publication with an alternative view on exercise and cardiovascular disease, written for professionals in the field.)

CHAPTER 9

THE IMMUNE SYSTEM AND MECHANISMS OF DEFENSE

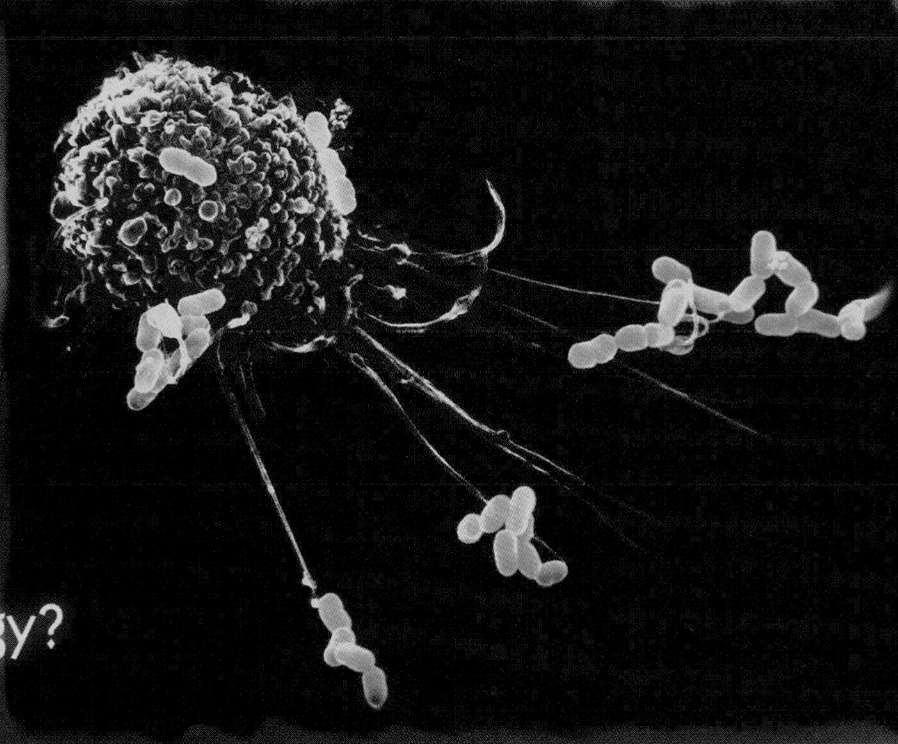

When is a fever beneficial?

What causes an allergy?

How do immunizations prevent diseases?

How does an HIV infection lead to AIDS?

THE WORLD IS FULL OF LIVING ORGANISMS TOO SMALL TO BE seen with the naked eye. Called *microorganisms* or sometimes just *microbes,* they're found on doorknobs, the money we handle, and our clothes. They're in the food we eat and the air we breathe. Most are harmless—indeed, some are highly beneficial. Only a relatively small percentage cause disease, but they account for many important human disorders and contribute to human suffering worldwide.

These challenges to health come from outside the body, but others come from within. Mutations may disrupt the process of normal cell division in our bodies. Abnormal cells can develop due to inherited factors, environmental pollutants, and many other reasons. Although we still have much to learn about what causes cells to become abnormal, we do know that our bodies dispatch many of them before they can develop into cancer.

So why aren't we constantly ill? For this we can thank our bodies' defense mechanisms, which detect and ward off many of these threats before they have a chance to do us harm. They include:

1. *Physical and chemical barriers* that prevent many harmful substances from entering the body in the first place (such as skin and stomach acid).
2. *Nonspecific mechanisms* that help the body respond to all kinds of tissue damage (such as phagocytic white blood cells).
3. *Specific defense mechanisms* that recognize and kill specific microorganisms and abnormal cells (such as specific immune responses).

All three mechanisms work cooperatively and simultaneously to protect us. They involve a wide variety of cells, proteins, chemicals, and organs. The lymphatic system plays a crucial role, and in this chapter we will look at its tissues and functions. These mechanisms also involve the **immune system,** a complex group of cells, proteins, and structures of the lymphatic and circulatory system.

What are they defending us against? We start by looking at microorganisms and their impact on human health.

9.1 Pathogens cause disease

Many microorganisms are beneficial. Some microorganisms break down raw sewage and cause the decomposition of dead animals and plants, thereby playing an essential role in the recycling of energy and raw materials. Other bacteria living inside our bodies help us by controlling population levels of more harmful organisms. A few bacteria even produce vitamins in our bodies. Microorganisms also assist us in producing a variety of commercial products, including antibiotic drugs, hormones, vaccines, and foods ranging from sauerkraut to soy sauce. Life as we know it would not be possible without these little organisms.

However, some microorganisms cause cell death and disease. These microbes, called **pathogens** (Gr. *pathos* disease + *gennan* to produce), invade or poison living cells. Some pathogens release enzymes or toxins that damage cells, causing them to rupture and release raw materials the invaders can use. Others enter cells and consume the raw materials that they find there, draining the cell of what it needs for itself.

It is impossible to put into a single category the pathogens that live within or on our bodies. They include bacteria, viruses, fungi, and a few protozoa. Larger parasites, including various worms, can also be pathogens, although this is relatively rare in developed countries. Because bacteria and viruses are by far the most numerous and problematic pathogens in Western nations, we'll focus on them as we describe the ways we protect and defend ourselves.

Bacteria: Prokaryotic pathogens

As you learned in Chapter 3, all organisms except bacteria are called *eukaryotes.* The defining feature of the cells of eukaryotes (called eukaryotic cells) is that they have a membrane-bound nucleus that surrounds and encloses the cell's DNA. They also have well-organized internal structures called organelles (the mitochondria, golgi apparatus, endoplasmic reticulum, and various vesicles) that are also enclosed in thin membranes.

Bacteria (singular *bacterium*), on the other hand, are all single-celled organisms called *prokaryotes* (prokaryotic cells). Bacteria lack a nucleus and membrane-bound organelles. Their DNA is a single closed loop within the cytoplasm that is attached to the plasma membrane. Their ribosomes are smaller than ours, and they float freely in the cytoplasm. The outer surface of bacteria is covered by a rigid cell wall that surrounds the plasma membrane and gives bacteria their distinctive shapes, which include spheres, rods, and spirals (Figure 9.1).

Judging by their variety and numbers, bacteria are among the most successful organisms on earth. Although they are smaller than the typical cell of a eukaryote, their small size is actually an advantage. Like all living organisms, bacteria need energy and raw materials to maintain life and to grow and divide. Their small size means that they have a high surface-to-volume ratio, a decided advantage when it comes to getting raw materials and getting rid of wastes by diffusion.

Like our own cells, bacteria use ATP as a direct energy source and amino acids for making proteins. They store energy as carbohydrates and fats. Where do they obtain those raw materials? Anywhere they

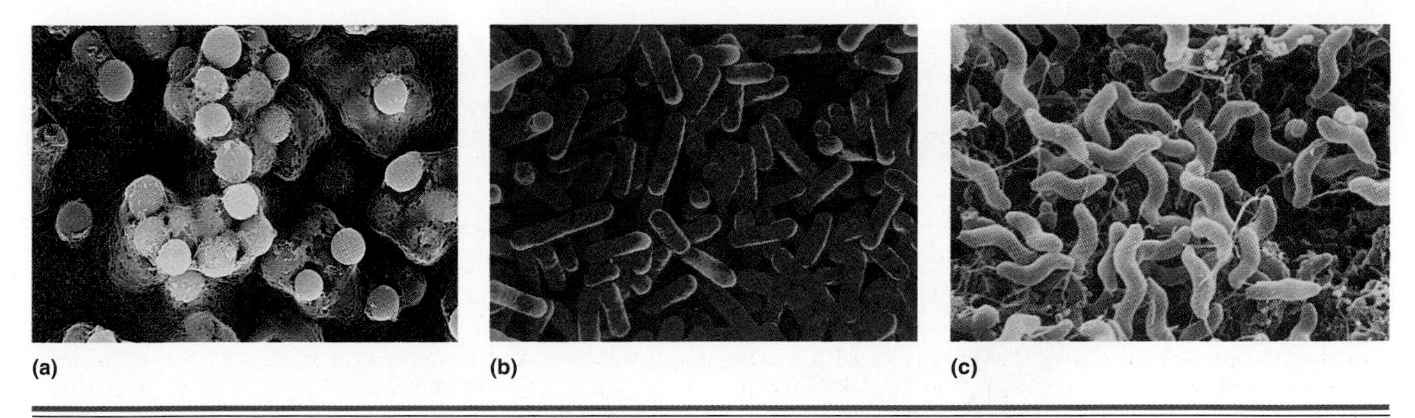

Figure 9.1 Electron micrographs of the three common shapes of bacteria. (a) SEM (×2,000) of *Streptococcus,* a spherical bacterium that causes sore throats. **(b)** SEM (×2,000) of *Legionella pneumophila,* the rod-shaped bacterium responsible for Legionnaire's disease. **(c)** SEM (×12,000) of *Campylobacter jejuni,* a spiral-shaped bacterium that causes food poisoning.

can. Some bacteria draw minerals and nutrients from the earth, atmosphere, and dead cells, but others turn to other living organisms as their primary source. It makes perfect sense, for living cells represent a concentrated supply of the raw materials and energy needed for life's processes.

Pathogenic bacteria are responsible for pneumonia, tonsillitis, tuberculosis, botulism, toxic shock syndrome, syphilis, Lyme disease, and many other diseases. Although we concentrate on pathogenic organisms in this chapter, keep in mind that most bacteria are harmless, and many are even beneficial.

> ***Recap*** **Bacteria are single-celled organisms. Like all cells, they draw their energy and raw materials from their environment. Pathogenic bacteria get the materials they need from living cells, damaging or killing the cells in the process.**

Viruses: Smallest infectious agents

Viruses are the smallest known infectious agents. They may be only one-hundredth the size of a bacterium and one-thousandth the size of a typical human cell (Figure 9.2). Structurally a virus is very simple, consisting solely of a piece of genetic material (either a double strand of DNA or a single strand of RNA) surrounded by a protein coat. They have no organelles, they don't grow, and they can't reproduce on their own.

Biologists are divided on whether to call viruses "living" or not. Viruses seem to perform no observable activity associated with life when they are not in contact with another living cell. However, once they make contact with a living cell they take it over and use the cell's organelles to make more viruses.

Viruses have several ways of gaining entry into living cells. Most viruses that infect human cells are taken into the cell by endocytosis and released into the cytoplasm. The virus then makes its way to the cell nucleus, where it releases its genetic material for incorporation into the cell's genetic material. The presence of the viral genetic material causes the cell to begin producing thousands of copies of the virus instead of carrying out its own metabolic activities. Sometimes the newly formed viruses are released by a type of budding from the cell membrane while the cell is still alive. In other cases the cell becomes so packed with viruses that it dies and bursts, releasing a huge number of viruses all at once.

Diseases caused by viruses range from serious—AIDS, hepatitis, encephalitis, rabies—to annoying—colds, warts, chicken pox. Viral infections can be minor for some people but serious for others. An otherwise healthy person may be ill for a few days with a viral infection, whereas someone who is very young, very old, or in poor health may die.

> ***Recap*** **A virus consists of a single strand of DNA or RNA surrounded by protein. Viruses use their DNA or RNA to force a living cell to make more copies of the virus.**

Transmissibility, mode of transmission, and virulence determine health risk

Some pathogens are clearly more risky to human health than others. Factors that determine the danger of a particular pathogen include transmissibility (how easily it is passed from one person to another), mode of transmission (how it is transmitted), and virulence (how damaging the disease is that it causes).

For instance, the viruses that cause the common cold are easily transmitted from hands to mucus membranes, as well as in the fluid particles present in a sneeze. However, these viruses tend not to be very virulent. The HIV virus that causes AIDS, on the other hand, is only moderately transmissible. Its mode of transmission is limited to exchange of body fluids (blood, semen, breast milk, or vaginal secretions).

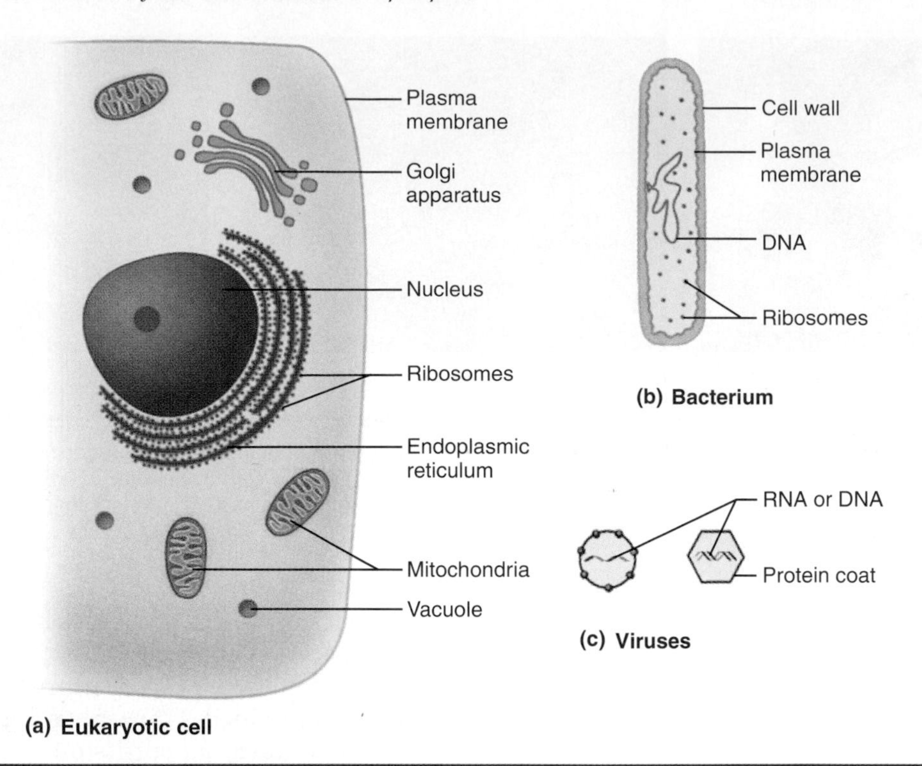

Figure 9.2 Size and structural differences between a eukaryotic cell, a bacterium (prokaryotic cell), and viruses. (a) Eukaryotic cells have a membrane-bound nucleus and well-defined membrane-bound organelles. **(b)** Bacteria have a single strand of DNA and free-floating small ribosomes within the cytoplasm. Their plasma membrane is surrounded by a rigid cell wall. **(c)** Viruses consist of a protein coat surrounding either RNA or DNA.

However, the HIV virus is tremendously virulent, and this is what makes it so dangerous.

Imagine what would happen if a disease as transmissible as the flu and as virulent as AIDS were to arise in the human population. In fact, there have been such diseases, and they have caused deadly epidemics. Between 1348 and 1350, bubonic plague, a bacterial infection, killed an estimated 25–40% of the European population. A 1918 outbreak of influenza killed more than 20 million people worldwide.

Pathogens continue to challenge human defenses. A prime example is the Ebola virus that arose in Africa in 1976 and is still present in the world today. It is one of the most virulent pathogens known, killing more than 80% of an exposed population in less than two weeks.

9.2 The lymphatic system defends the body

In the last chapter we mentioned that the **lymphatic system** is closely associated with the cardiovascular system. The lymphatic system performs three important functions:

- It helps maintain the volume of blood in the cardiovascular system.
- It transports fats and fat-soluble vitamins absorbed from the digestive system.
- It defends the body against infection and injury.

In the previous chapter we mentioned briefly that the lymphatic system helps to maintain blood volume and interstitial fluid volume by returning excess fluid that has been filtered out of the capillaries back to the cardiovascular system. We will discuss its role in transporting fats and vitamins when we describe the digestive system. In this chapter we turn to the third function: the role of the lymphatic system in protecting us from disease and injury.

Lymphatic vessels transport lymph

Figure 9.3 shows the structure of the lymphatic system. The lymphatic system begins as a network of small, blind-ended *lymphatic capillaries* in the vicinity of the cells and blood capillaries.

Spotlight on Structure & Function

Lymph capillaries have wide spaces between overlapping cells. Their structure allows them to take up substances

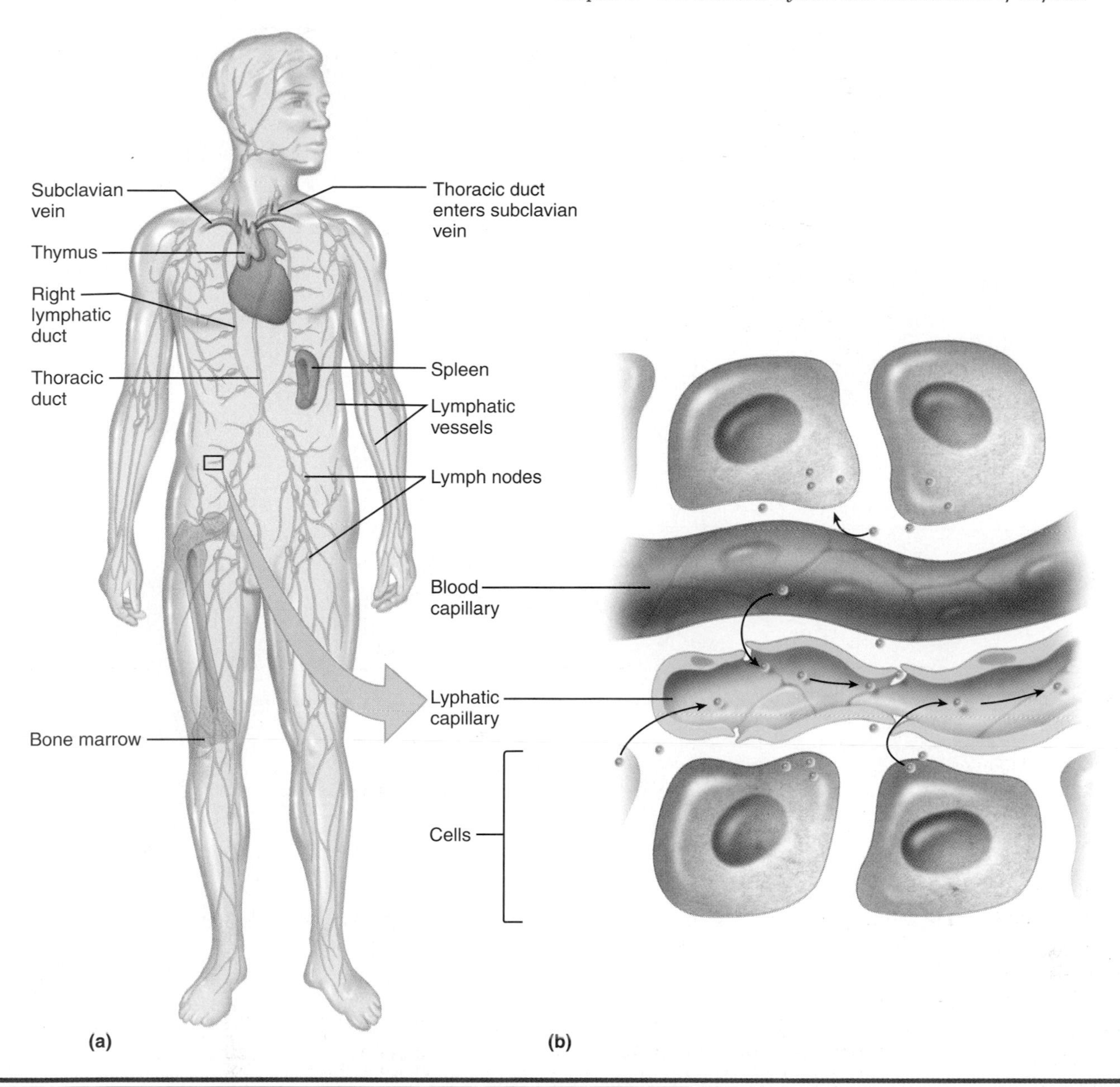

Figure 9.3 The lymphatic system. (a) The lymphatic system consists of lymphatic vessels, lymph nodes, the spleen, the thymus, and the tonsils and adenoids (tonsils and adenoids not shown in this figure). **(b)** Lymphatic vessels begin as blind-ended capillaries throughout the body.

(including bacteria) that are too large to reenter a blood capillary.

The fluid in the lymphatic capillaries is *lymph,* a milky body fluid that contains white blood cells, proteins, fats, and the occasional bacterium. Lymphatic capillaries merge to form the *lymphatic vessels.* Located at intervals along the lymphatic vessels are small organs called **lymph nodes,** described below. Like veins, lymphatic vessels have walls consisting of three thin layers, and they contain one-way valves to prevent back-flow of lymph. Also like veins, skeletal muscle contractions and pressure changes in the chest during respiration help keep the lymph flowing. The lymphatic vessels merge to form larger and larger vessels, eventually creating two major lymphatic ducts: the *right lymphatic duct* and the *thoracic duct.* The two lymph ducts join the subclavian veins near the shoulders, thereby returning the lymph to the cardiovascular system. ■

In addition to the lymphatic vessels, important structures of the lymphatic system include lymph nodes, spleen, thymus gland, and tonsils and adenoids (Figure 9.4).

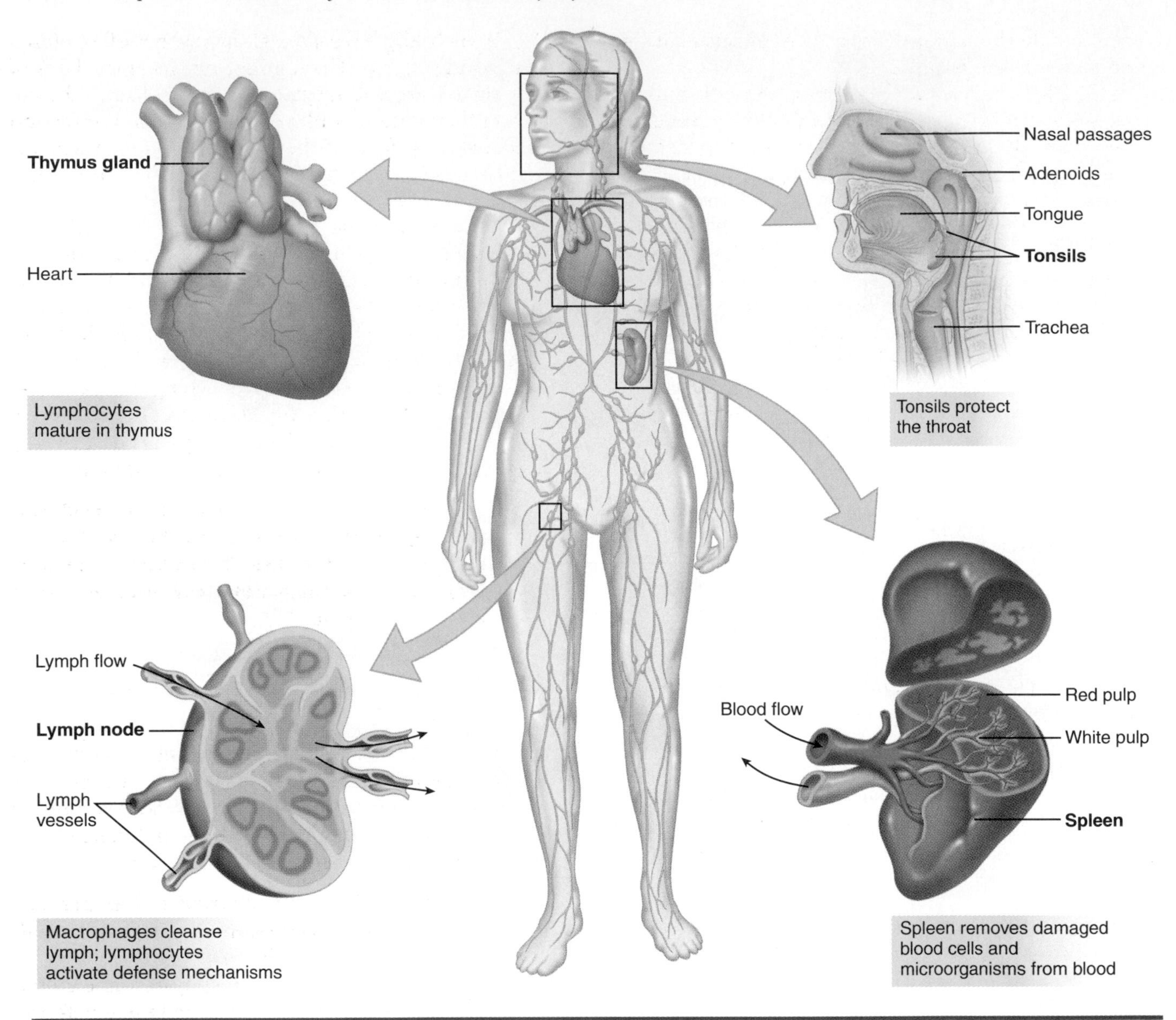

Figure 9.4 Structures and functions of the various components of the lymphatic system.

Lymph nodes cleanse the lymph

Lymph nodes remove microorganisms, cellular debris, and abnormal cells from the lymph before returning it to the cardiovascular system. There are hundreds of lymph nodes, clustered in the areas of the digestive tract, neck, armpits, and groin. They vary in diameter from about 1 millimeter to 2.5 centimeters. Each node is enclosed in a dense capsule of connective tissue pierced by lymphatic vessels. Inside each node are connective tissue and two types of white blood cells known as macrophages and lymphocytes.

The lymphatic vessels carry lymph into and out of each node. Valves within these vessels ensure that lymph flows only in one direction. As the fluid flows through a node, the macrophages destroy foreign cells by phagocytosis, and the lymphocytes activate other defense mechanisms. The cleansed lymph fluid flows out of the node and continues on its path to the veins.

Recap **The lymphatic system helps protect us from disease and injuries. As lymph flows through the lymph nodes, macrophages and lymphocytes within the nodes identify microorganisms and remove them.**

The spleen cleanses blood

The largest lymphatic organ, the **spleen,** is a soft, fist-sized mass located in the upper-left abdominal cavity. The spleen is covered with a dense capsule of connective tissue interspersed with smooth muscle cells. Inside the organ are two types of tissue, called red pulp and white pulp.

The spleen has two main functions: It controls the quality of circulating red blood cells by removing the old and damaged ones, and it helps fight infection. The *red pulp* contains macrophages that scavenge and break down microorganisms as well as old and damaged red blood cells and platelets. The cleansed blood is then stored in the red pulp. The body can call on this reserve for extra blood in case of blood loss, a fall in blood pressure, or whenever it needs extra oxygen-carrying capacity. In fact, splenic contraction during vigorous exercise is one of the causes of the sharp pain in your left side known as a "stitch." The *white pulp* contains primarily lymphocytes. Notice that the main distinction between spleen and lymph nodes is *which* fluid they cleanse—the spleen cleanses the blood, and the lymph nodes cleanse lymph. Together, they keep the circulating body fluids relatively free of damaged cells and microorganisms.

A number of diseases, such as tuberculosis and leukemia, cause the spleen to enlarge. The swollen spleen can sometimes be felt as a lump in the upper-left abdomen. A strong blow to the abdomen can rupture the organ, causing severe internal bleeding. In this case, a *splenectomy*—surgical removal of the spleen—may be necessary to forestall a fatal hemorrhage. We can live without a spleen because its functions are shared by the lymph glands, liver, and red bone marrow. However, people who have had a splenectomy are often a little more vulnerable to infections.

Thymus gland hormones cause T lymphocytes to mature

The **thymus gland** is located in the lower neck, behind the sternum and just above the heart. Encased in connective tissue, the gland contains lymphocytes and epithelial cells. The thymus gland secretes two hormones, thymosin and thymopoietin, that cause certain lymphocytes called *T lymphocytes* (T cells) to mature and take an active role in specific defenses.

The size and activity level of the thymus gland varies with age. It is largest and most active during childhood. During adolescence it stops growing and then slowly starts to shrink. By that time our defense mechanisms are typically well established. In old age the thymus gland may disappear entirely, to be replaced by fibrous and fatty tissue.

Tonsils protect the throat

The *tonsils* are masses of lymphatic tissue near the entrance to the throat. Lymphocytes in the tonsils gather and filter out many of the microorganisms that enter the throat in food or air.

We actually have several tonsils, not all of which are readily visible. The familiar tonsils at the back of the throat are the largest and most often infected. When they become inflamed, the resulting infection is called *tonsillitis*. If the infection becomes serious, the tissues can be surgically removed in a procedure known as a *tonsillectomy*.

Lymphatic tissue called the *adenoids* (or *pharyngeal tonsils*) lie at the back of the nasal passages. The adenoids tend to enlarge during early childhood, but in most people they start to shrink after age 5 and usually disappear by puberty. In some cases they continue to enlarge and obstruct airflow from nose to throat. This can cause mouth breathing, a nasal voice, and snoring. Like other tonsils, if the adenoids grow large enough to cause problems, they can be surgically removed. This procedure is called *adenoidectomy*.

Recap **The spleen removes damaged red blood cells and foreign cells from blood. The thymus gland secretes hormones that help T lymphocytes mature. Cells in the tonsils gather and remove microorganisms that enter the throat.**

9.3 The body has three lines of defense

The lymphatic system works with other body systems to protect us against pathogens and cellular changes. As mentioned earlier, our bodies employ three different mechanisms, or lines of defense, each with an increasing level of sophistication:

1. *Physical and chemical barriers* that prevent many pathogens from entering the body at all are the first line of defense.
2. Pathogens that get past the physical and chemical barriers encounter a second line of defense,

Try It Yourself

Detecting Swollen Lymph Nodes

The next time you get a bad cold or the flu, you may notice that your lymph nodes become swollen and tender (try feeling the ones in your neck and under your lower jaw). This is a healthy response to the pathogen, because it indicates that you have an infection and your lymphatic system is taking the appropriate action. The entry of the pathogen into your lymphatic system has triggered a massive explosion in the lymphocyte population, and your lymph nodes are swollen with dividing lymphocytes, clumps of virus particles, and pathogens under attack by lymphocytes. When the infection subsides, your lymph nodes should return to their normal size.

nonspecific defense mechanisms that help the body respond to all kinds of tissue damage. These mechanisms include a series of actions that clear injured areas of debris and damaged cells. At the same time, the body's defense cells seek out and kill the invading pathogens.
3. Finally, *specific defense mechanisms* enable the body to recognize, respond to, and kill specific viruses, bacteria, and foreign tissues. This third line of defense is sophisticated weaponry indeed. Specific defense mechanisms are the basis of immunity from future disease.

Let's look at each type of protection in more detail.

9.4 Physical and chemical barriers: Our first line of defense

For all organisms there are certain conditions that are ideal for growth and reproduction, others that are unfavorable, and still others that are fatal. Our bodies are equipped with physical and chemical barriers that create an inhospitable environment for many invading microorganisms. Physical barriers can render the invaders helpless, sweep them off body surfaces, or wash them harmlessly away. Chemical defenses engage in a type of chemical warfare by producing substances that slow the growth of many pathogens or kill them outright. Table 9.1 summarizes our physical and chemical barriers.

Skin: An effective deterrent

Spotlight on Structure & Function

The most important barrier against entry of any organism into our bodies is our skin. Skin has three key attributes that make it such an effective barrier: its structure, the fact that it is being constantly replaced and/or repaired, and its acidic pH.

As we saw in Chapter 4, the outermost layers of the skin's epidermis consist of dead, dried-out epithelial cells. These cells contain a fibrous protein called keratin, which is also a primary component of fingernails and hair. Once the cells have died and the water has evaporated, the keratin forms a dry, tough, somewhat elastic barrier to the entry of microorganisms. A second key feature of skin is that it is constantly being replaced throughout life. Dead cells shed from the surface are replaced by new cells arising from the base of the epidermis. Even if pathogens manage to collect on the skin surface, this constant shedding effectively removes many of them. Finally, healthy skin has a pH of about 5 to 6. This relatively low (acidic) pH makes skin a hostile environment for many microorganisms.

For further proof of skin's effectiveness as a barrier to infection, look at what happens when you get a tiny cut or scratch in your skin. If the damage reaches the moist layers of living cells underneath the skin, you may see signs of infection in the area within a few days. One of the most critical problems in treating patients with extensive burns is the infections that often result from the loss of the barrier function of skin. ■

Other physical and chemical barriers

Most successful pathogens enter the body at places where we do not have skin. They enter through the mucous membranes that line the digestive, urinary, respiratory, and reproductive tracts, where they can take advantage of moist surfaces in direct contact with living cells. A few organisms can even enter through the moist tissue covering the eyes.

Areas covered by mucous membranes are often well defended in other ways, however. Impediments to the entry of pathogens include tears, saliva, ear-

Table 9.1 The first line of defense: Physical and chemical barriers

Barrier	Functions
Skin	Forms durable barrier; constant cell shedding removes pathogens; acidic pH repels microorganisms.
Tears	Lubricate and wash eyes. Contain lysozyme.
Saliva	Lubricates mouth and rinses particles into digestive tract. Contains lysozyme.
Earwax	Traps particles and microorganisms.
Digestive acids	Hydrochloric acid kills most pathogens that enter stomach.
Mucus	Traps pathogens so they can be swept away by cilia and then removed by swallowing or sneezing.
Vomiting	Removes toxic and infective agents from the stomach.
Urination	Slightly acidic; flushes out urinary tract.
Defecation	Removes pathogens from digestive tract.
Resident bacteria	Control harmful organisms by competing against them; may produce substances that alter body's pH.

wax, digestive and vaginal acids, and mucus; the ability to remove microorganisms by vomiting, urination, and defecation; and even competition created by nonpathogenic bacteria that normally live in (and on) the body.

Tears, saliva, and earwax Although we may not think of tears as a defense mechanism, they perform a valuable service by lubricating the eyes and washing away particles. Tears and saliva both contain **lysozyme,** an enzyme that kills many bacteria. In addition, saliva lubricates the delicate tissues inside the mouth so they do not dry out and crack. It also rinses food and microorganisms safely from the mouth into the stomach. Earwax traps small particles and microorganisms.

Mucus Mucus is a thick, gel-like material secreted by cells at various surfaces of the body, including the lining of the digestive tract and the branching airways of the respiratory system. Microorganisms that come into contact with the sticky mucus become mired and cannot gain access to the cells beneath. In addition, the cells of the airways have tiny hairlike projections, called cilia, that beat constantly in a wavelike motion to sweep mucus upward into the throat. There we get rid of the mucus by coughing or swallowing it. Sometimes we remove mucus and microorganisms by sneezing, which is also one of the primary ways we pass microorganisms to other people (Figure 9.5).

Digestive and vaginal acids Swallowed microorganisms encounter hydrochloric acid, a digestive acid secreted by the stomach that is strong enough to kill nearly all pathogens that enter the digestive tract. A few microbes can survive it; one strain of bacteria, *Helicobacter pylori*, has actually evolved to thrive in the highly acidic environment of the stomach. *Helicobacter pylori* is now known to contribute to many cases of stomach ulcers (see Chapter 14). Bacteria in the vaginas of adult women encounter secretions that are fairly acidic (low pH).

Vomiting, urination, and defecation Vomiting, urination, and defecation wash microorganisms away. Vomiting, though perhaps unpleasant, is certainly an effective way of ridding the body of toxic or infected stomach contents. Generally speaking, urine contains very few bacteria because the flushing action of urination normally prevents microorganisms from reaching the bladder. If microorganisms do get that far, the flushing action of urination tends to keep their populations low.

Because urine is usually slightly acidic, it is a hostile environment for most microorganisms. However, it can become even more acidic depending on diet. Some physicians advise patients with bladder or urethra infections to drink large amounts of cranberry juice. The tart juice makes urine more acidic, inhibiting microorganism growth and reproduction, and the additional fluid increases urine production and flushes the microorganisms away.

Figure 9.5 Sneezing removes mucus and microorganisms from the body. Sneezing also promotes the transmission of certain respiratory infections from person to person.

The movement of feces and the act of defecation also help remove pathogenic microorganisms from the digestive tract. When we become ill, the muscles in the intestinal wall may start to contract more vigorously, and the intestine may secrete additional fluid into the feces. The result is *diarrhea*—increased fluidity, frequency, or volume of bowel movements. Unpleasant though diarrhea may be, mild cases serve a useful function by speeding the process of defecation and the rate at which pathogens are expelled.

Resident bacteria Certain strains of bacteria normally live in the mucous membranes lining the vagina and the gastrointestinal tract. They help control population levels of more harmful organisms by competing successfully against them for food. Their presence may also make the body less vulnerable to pathogens. For example, *Lactobacillus* bacteria in the vagina produce a substance that lowers vaginal pH to levels that many fungi and bacteria cannot tolerate.

Recap **Various physical and chemical barriers create an inhospitable environment for pathogenic microorganisms. Skin is a dry outer barrier. Tears, saliva, earwax, and mucus trap microorganisms or wash them away. Stomach acid kills swallowed microorganisms; urination, defecation, and vomiting forcibly expel them; and resident harmless bacteria compete with pathogens for food.**

9.5 Nonspecific defenses: The second line of defense

If pathogens manage to breach our physical and chemical barriers and start to kill or damage cells, we have a problem of a different sort. Now the body must actively seek out the pathogens and get rid of them. It must also clean up the injured area and repair the damage.

Our second line of defense includes a varied group of defense mechanisms. We refer to them as *nonspecific* because they do not target specific pathogens. Instead, they occur in response to all types of health challenges without discriminating between them. Table 9.2 summarizes nonspecific defenses, which include phagocytes, natural killer cells, the inflammatory response, the complement system, interferons, and fever.

Phagocytes engulf foreign cells

As noted in Chapter 7, *phagocytes* are white blood cells that destroy foreign cells through the process of **phagocytosis.** As illustrated in Figure 9.6:

1. A phagocyte surrounds and engulfs a foreign cell or cell fragment.
2. The phagocyte contains the invader within a cytoplasmic vesicle.
3. The vesicle containing the bacterium fuses with lysosomes.
4. Powerful digestive enzymes dissolve the foreign cell's membranes.
5. The phagocyte discards the wastes and debris that remain from the foreign cell.

Recall from Chapter 7 that some white blood cells can filter through the walls of blood vessels into tissue spaces, attracted by substances released by injured cells at the site of the infection. Other phagocytes remain in connective tissues of the lymph nodes, spleen, liver, lungs, and brain.

Neutrophils are the first white blood cells to respond to infection. They digest and destroy bacteria and some fungi in the blood and tissue fluids. Other white blood cells known as monocytes leave the vascular system, enter the tissue fluids, and develop into **macrophages** (from the Greek for "large eater") that can engulf and digest large numbers of foreign cells, especially viruses and bacterial parasites. Macrophages serve a cleanup function by scavenging old blood cells, dead tissue fragments, and cellular debris. They also release chemicals that stimulate the production of more white blood cells. Technically, macrophages are no longer white blood cells because they are no longer in blood.

When invaders are too big to be engulfed and digested by phagocytosis, other white blood cells called **eosinophils** take action. Eosinophils cluster around large parasites such as flukes and pinworms and bombard them with digestive enzymes. Eosinophils also engulf and digest certain foreign proteins.

When the body is actively fighting an infection, the mortality rate of white blood cells rises dramatically. Tissue fluid, dead phagocytes and microorganisms, and cellular debris accumulate at the infection site, producing a characteristic discharge called *pus*. If pus becomes trapped and cannot drain, the body may wall it off with connective tissue. The result is an *abscess*. Common places for abscesses to form include the breast (mastitis), the gums (dental abscesses), and more rarely in the liver or brain. Many abscesses subside after being drained, while others may require antibiotic drugs or surgical removal.

Recap **Nonspecific defense mechanisms involve a general attack against all foreign and damaged cells. Neutrophils and macrophages engulf and digest bacteria and damaged cells. Eosinophils bombard larger organisms (too large to be engulfed) with digestive enzymes.**

Table 9.2 The second line of defense: Nonspecific defenses

Defense	Functions
Phagocytes	Neutrophils and macrophages engulf and digest foreign cells. Eosinophils bombard large parasites with digestive enzymes and phagocytize foreign proteins.
Natural killer cells	Release chemicals that disintegrate cell membranes of tumor cells and virus-infected cells.
Inflammatory response	Four components: redness, warmth, swelling, and pain. Attracts phagocytes and promotes tissue healing.
Complement system	A group of proteins that assists other defense mechanisms. Enhances inflammation and phagocytosis, kills pathogens.
Interferons	Stimulate the production of proteins that interfere with viral reproduction.
Fever	Modest fever makes internal environment less hospitable to pathogens, fosters ability to fight infcctions.

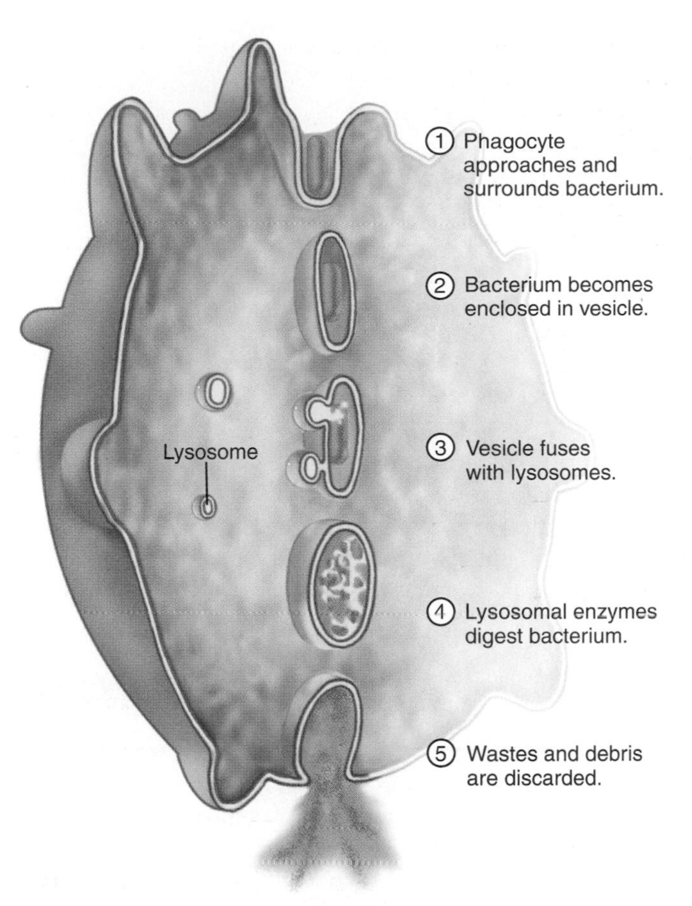

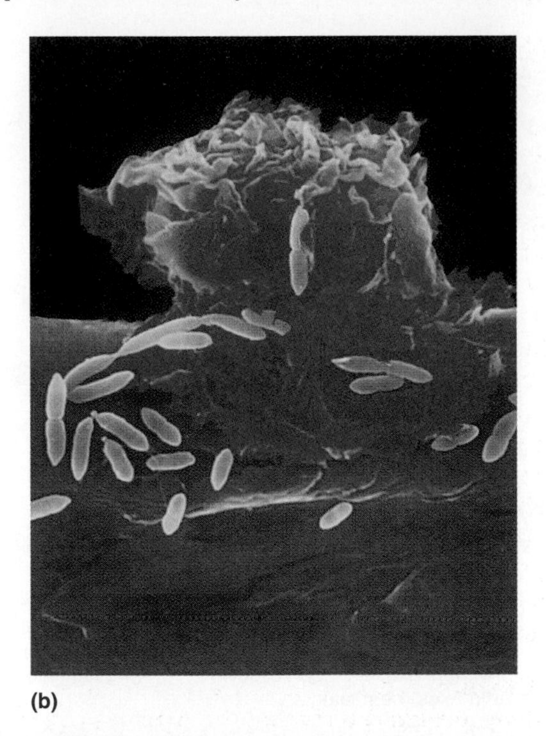

Figure 9.6 Phagocytosis. (a) Steps in the process. **(b)** An electron micrograph of a phagocytic cell engulfing a bacterium.

The inflammatory response: Redness, warmth, swelling, and pain

Any type of tissue injury—whether infection, burns, irritating chemicals, or physical trauma—triggers a series of related events collectively called the **inflammatory response** or *inflammation.* Inflammation has four outward signs: redness, warmth, swelling, and pain. While these may not sound like positive developments, the events that cause these signs prevent the damage from spreading, dispose of cellular debris and pathogens, and set the stage for tissue repair mechanisms.

The inflammatory response starts whenever tissues are injured (Figure 9.7). The release of chemicals from damaged cells is the signal that sounds the alarm for the process to begin. These chemicals stimulate **mast cells,** which are connective tissue cells specialized to release **histamine.** Histamine promotes vasodilation of neighboring small blood vessels. White blood cells called **basophils** also secrete histamine.

Recall that most white blood cells are too large to cross capillary walls. As histamine dilates blood vessels, however, the endothelial cells in vessel walls pull slightly apart and the vessels become more permeable. This allows additional phagocytes to squeeze through capillary walls into the interstitial fluid. There they attack foreign organisms and damaged cells. After destroying pathogens, some phagocytes travel to the lymphatic system, where their presence activates lymphocytes to initiate specific defense mechanisms as discussed later.

Vasodilation brings more blood into the injured area, making it red and warm. The rising temperature increases phagocyte activity. The increased leakiness of capillary walls allows more fluid to seep into tissue spaces, causing swelling. The extra fluid dilutes pathogens and toxins and brings in clotting proteins that form a fibrin mesh to wall off the damaged area from healthy tissue. As a bonus, the fluid carries in extra oxygen and nutrients to promote tissue healing and carries away dead cells, microorganisms, and other debris from the area.

Swollen tissues press against nearby nerve endings. This swelling, plus the sensitizing effects of inflammatory chemicals, creates the sensation of pain that accompanies inflammation. However, even pain can be positive. The discomfort hinders active movement and forces the injured person to rest, facilitating the healing process.

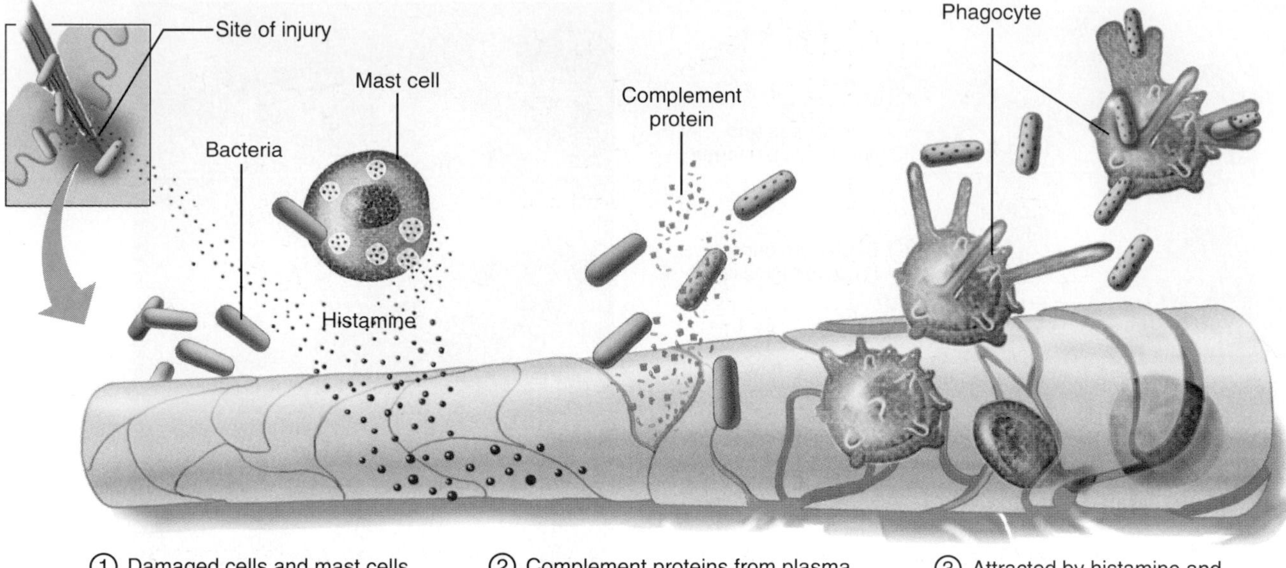

Figure 9.7 The inflammatory response.

Natural killer cells target tumors and virus-infected cells

Natural killer (NK) cells are a group of white blood cells (lymphocytes) that destroy tumor cells and cells infected by viruses. They apparently recognize the cells by certain changes that take place in their plasma membranes. The name "natural killer" reflects the fact that NK cells are nonspecific killers, unlike other killer cells discussed later in this chapter that target only specific enemies.

NK cells are not phagocytes. Instead, they release chemicals that break down their targets' cell membranes. Soon after an NK attack, the cell's membrane develops holes and its nucleus rapidly disintegrates. NK cells also secrete substances that enhance the inflammatory response.

Complement system assists other defense mechanisms

The **complement system,** or *complement,* comprises a group of at least 20 plasma proteins that circulate in the blood and complement, or assist, other defense mechanisms. Normally these proteins circulate in an inactive state. When activated by the presence of an infection, however, they become a potent defense force. Once one protein is activated it activates another, leading to a cascade of reactions. Each protein in the complement system can activate many others, creating a powerful "domino effect."

Some activated complement proteins join to form large protein complexes that create holes in bacterial cell walls. Fluids and salts leak in through these holes, until eventually the bacterium swells and bursts (Figure 9.8). Other activated complement proteins bind to bacterial cell membranes, marking them for destruction by phagocytes. Still others stimulate mast cells to release histamine or serve as chemical attractants to draw additional phagocytes to the scene of infection.

Interferons interfere with viral reproduction

One of the most interesting defense mechanisms is an early warning system between virus-infected and still-healthy cells. As mentioned earlier, viruses cannot reproduce on their own. Instead, they invade body cells and use the cells' machinery to make more viruses.

When a cell becomes infected by viruses, however, it secretes a group of proteins called **interferons.** Interferons diffuse to nearby healthy cells, bind to their cell membranes, and stimulate the healthy cells to produce proteins that interfere with the synthesis of viral proteins, making it harder for the viruses to infect the protected cells.

Interferons are now being produced in pharmaceutical laboratories. At least one interferon protein *(alpha interferon)* has shown promise against certain viral diseases, including a form of leukemia, genital warts, and hepatitis B.

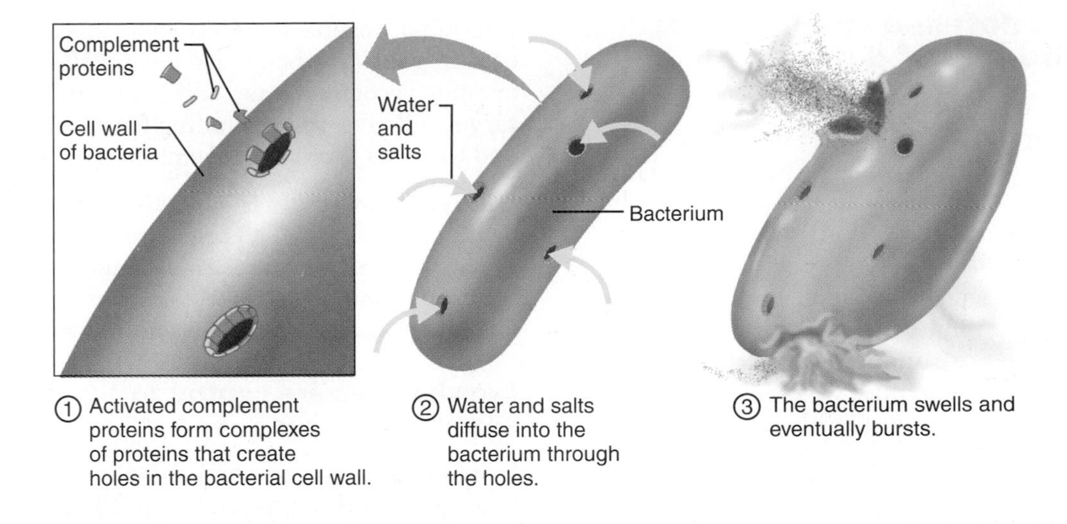

① Activated complement proteins form complexes of proteins that create holes in the bacterial cell wall.

② Water and salts diffuse into the bacterium through the holes.

③ The bacterium swells and eventually bursts.

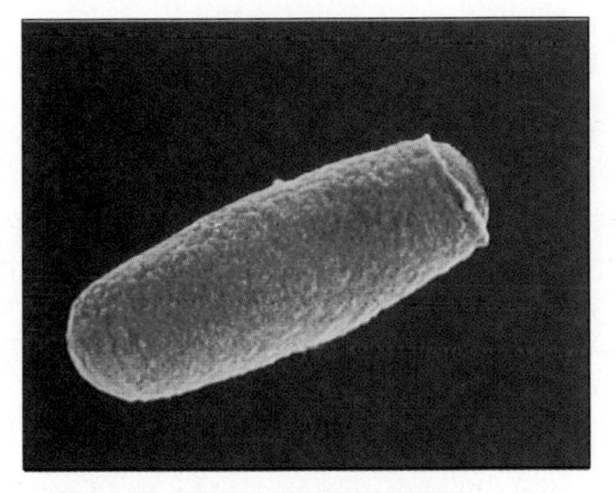

Photomicrograph of an intact bacterium

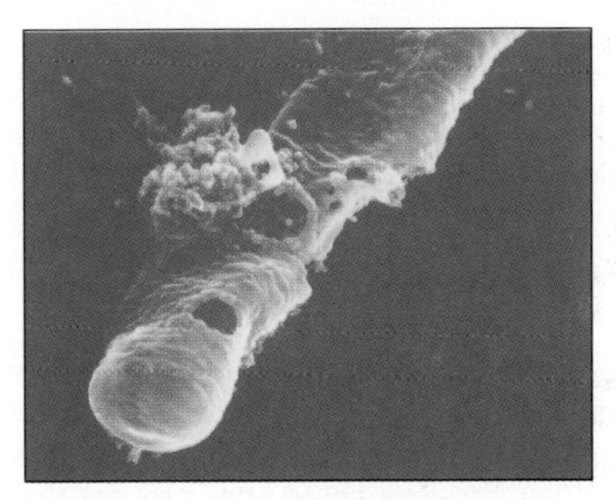

A bacterium after lysis by activated complement proteins

Figure 9.8 How activated complement proteins kill bacteria.

Fever raises body temperature

When is a fever beneficial?

A final weapon in our second line of defense is **fever,** an abnormally high body temperature. Typically, your body's "thermostat" is set to approximately 98.6°F (37°C), with a normal range of about 97 to 99°F (36 to 37.2°C). When macrophages detect and attack bacteria, viruses, or other foreign substances, they release certain chemicals into the bloodstream. These chemicals, collectively called *pyrogens,* cause the brain to reset your thermostat to a higher temperature.

There is a tendency to treat all fevers as if they were a problem. But a modest fever may be beneficial because it makes our internal environment less hospitable to pathogens and fosters the body's ability to fight infections. Fever increases the metabolic rate of body cells, speeding up both defense mechanisms and tissue repair processes. When the infection is gone, the process reverses. Macrophages stop releasing pyrogens, the thermostat setting returns to normal, and your fever "breaks."

On the other hand, high fevers can be dangerous. Remember from Chapter 2 that the chemical bonds that give a protein its shape are relatively weak. Consequently, the shape (and function) of some proteins can be affected by high temperatures. It's a good idea to monitor the course of any fever, particularly in children and older adults. Health professionals recommend seeking medical advice for any fever that lasts longer than two days or rises above 100°F.

Recap **The inflammatory response attracts phagocytes and promotes tissue healing. Natural killer cells release chemicals that disintegrate cell membranes of tumor cells and virus-infected cells. The complement system assists other defense mechanisms, interferons interfere with viral reproduction, and a modest fever enhances our ability to fight infections.**

9.6 Specific defense mechanisms: The third line of defense

Even if foreign cells manage to bypass physical and chemical barriers and overcome nonspecific defenses, they must still cope with the body's third line of defense, the most sophisticated weapon of all. The *immune system* comprises cells, proteins, and the lymphatic system, all working together to detect and kill particular pathogens and abnormal body cells. The activities of the immune system are collectively called the **immune response.** Because the immune system targets specific enemies, we refer to these operations as *specific defense mechanisms.*

The immune response has three important characteristics:

- It recognizes and targets specific pathogens or foreign substances.
- It has a "memory," the capability to store information from past exposures so it can respond more quickly to later invasions of the same pathogen.
- It protects the entire body; the resulting immunity is not limited to the site of infection.

The key to specific defenses is the body's ability to distinguish between its own cells and those of foreign invaders. Among its own cells, it must also be able to distinguish between those that are healthy, those that are abnormal (such as cancer cells), and those that are dead or dying.

The immune system targets antigens

An **antigen** is any substance that mobilizes the immune system and provokes an immune response. Most antigens are large molecules not normally found in the body, such as foreign proteins and large polysaccharides.

Whole cells have surface proteins that can act as antigens under the right circumstances. Your cells have a unique set of proteins on their surfaces that your immune system uses as a way of recognizing that the cells belong to you. These *self markers* are known as **major histocompatibility complex (MHC) proteins.** Your MHC proteins are unique to you by virtue of your unique set of genes. Normally they signal your immune system to bypass your own cells. They are a sort of password, the equivalent of a cellular fingerprint. Your immune system "reads" the password and leaves your cells alone.

However, the same MHC proteins that define your cells as belonging to you would be read as *nonself markers* in another person. In other words, your MHC proteins represent antigens that would provoke an immune response in another person. The cells of pathogens and even abnormal and cancerous cells in your own body also have MHC proteins that are not recognized as "self." The immune system targets all antigens, including foreign MHC proteins and the cells that carry them, for destruction.

Recap **An antigen is any substance that provokes an immune response. All cells of an organism have markers called MHC proteins that identify the cells as "self." Nonself MHC proteins are antigens, and as such they activate the immune system.**

Lymphocytes are central to specific defenses

Lymphocytes play crucial roles in our specific defense mechanisms. As we saw in Chapter 7, lymphocytes are white blood cells originating from stem cells in bone marrow. They are fairly small white blood cells with a single nucleus that fills nearly the entire cell. They total about 30% of circulating WBCs. Lymphocytes are found in the bloodstream, tonsils, spleen, lymph nodes, and thymus gland.

There are two types: B lymphocytes and T lymphocytes, also called **B cells** and **T cells.** (Their names are based on where in the body they mature: B cells in bone marrow, T cells in the thymus gland.) Although both types of lymphocytes can recognize and target antigen-bearing cells, they go about this task in different ways.

B cells are responsible for **antibody-mediated immunity.** B cells produce **antibodies**—proteins that bind with and neutralize specific antigens. They release antibodies into the lymph, bloodstream, and tissue fluid, where they circulate throughout the body. Antibody-mediated immunity works best against viruses, bacteria, and foreign molecules that are soluble in blood and lymph.

T cells are responsible for **cell-mediated immunity,** which depends on the actions of several types of T cells. Unlike B cells, T cells do not produce antibodies. Instead, some T cells directly attack foreign cells that carry antigens. Other T cells release proteins that help coordinate other aspects of the immune response, including the actions of T cells, B cells, and macrophages. Cell-mediated immunity protects us against parasites, bacteria, viruses, fungi, cancerous cells, and cells perceived as foreign (including, unfortunately, transplanted tissue—see Section 9.9 on "Tissue Rejection"). T cells can identify and kill infected human cells even before the cells have a chance to release new bacteria or viruses into the blood.

B cells: Antibody-mediated immunity

In adults, B cells mature in the bone marrow. As they mature, they develop unique surface receptors

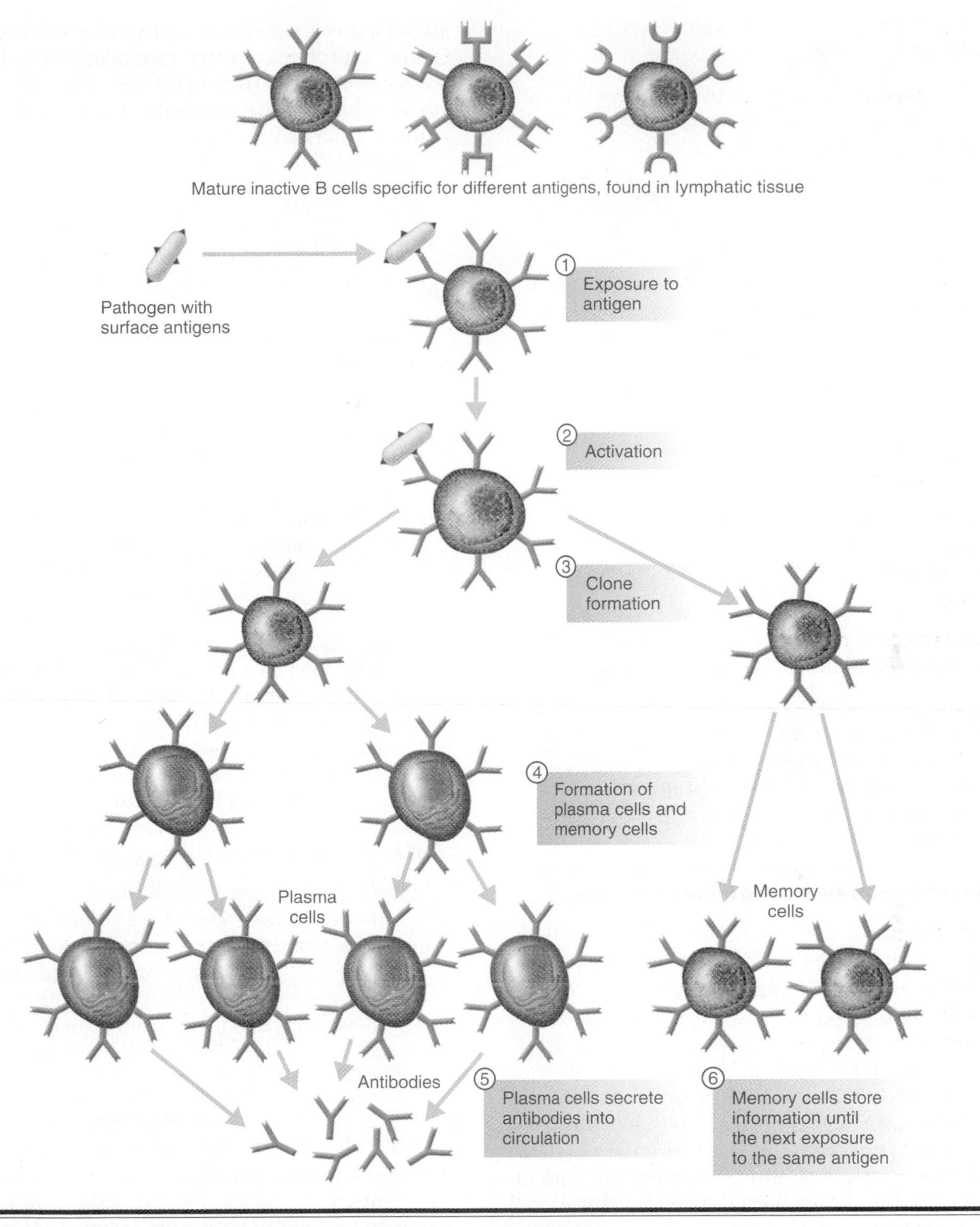

Figure 9.9 The production of antibodies by B cells. ① The surface antigen of a pathogen binds to the matching receptor on a mature, inactive B cell in lymphatic tissue. ② The B cell becomes activated and grows in size. ③ The enlarged cell begins to divide rapidly, forming a clone. ④ Some of the clone cells become plasma cells; others become memory cells. ⑤ Plasma cells secrete antibodies into the lymph fluid. ⑥ Memory cells lie in wait for the next exposure to the antigen.

(with the same structure as an antibody) that allow them to recognize specific antigens. Then they travel in the bloodstream to the lymph nodes, spleen, and tonsils, where they remain inactive until they encounter a foreign cell with that particular antigen. When a B cell with just the right surface receptor encounters the appropriate antigen, its surface receptors bind to the antigen (Figure 9.9). This activates

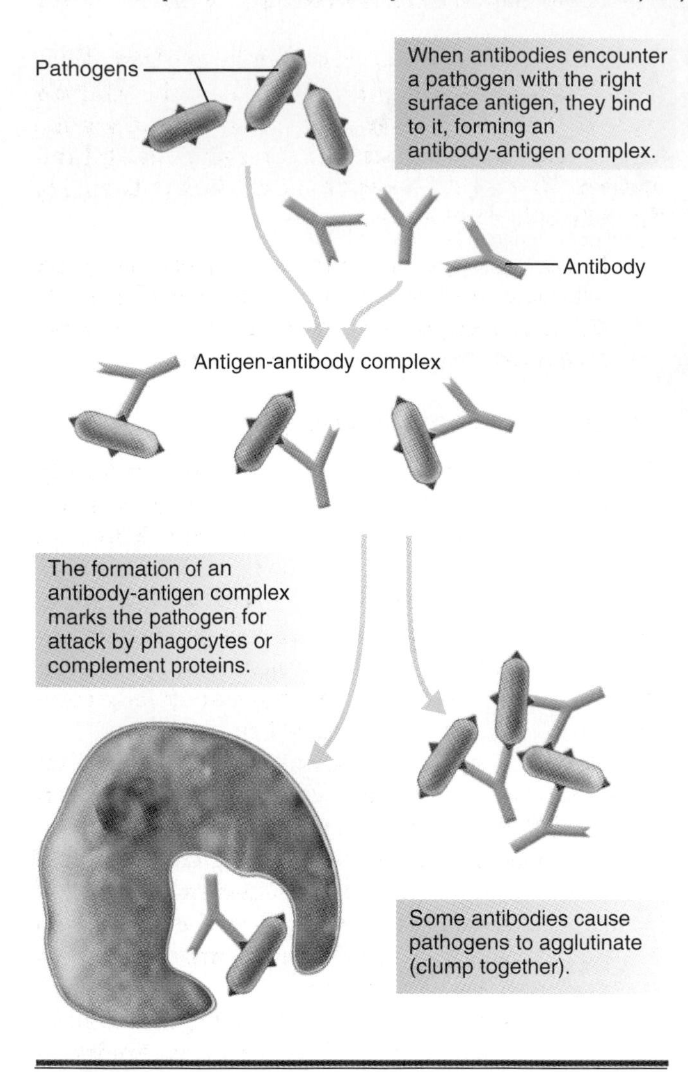

Figure 9.10 How antibodies inactivate pathogens.

the B cell to grow and then multiply rapidly, producing more B cells exactly like the original and bearing the same surface receptors. The resulting group of identical cells, all descended from the same cell, is called a *clone.*

Although the B cells themselves tend to remain in the lymphatic system, most of the cells of the clone are called **plasma cells** because they begin to secrete their antibodies into the lymph fluid and ultimately into the blood plasma. A typical plasma cell can make antibody molecules at a staggering rate—about 2,000 molecules per second. A plasma cell maintains this frantic pace for a few days and then dies, but its antibodies continue to circulate in blood and lymph.

When the antibodies encounter matching antigens, they bind to them and create an **antigen-antibody complex.** Antibodies specialize in recognizing certain proteins; thus one particular antibody can bind to one particular antigen (Figure 9.10). This binding marks the antigen (and the foreign cell that carries them) for destruction by activating complement proteins or attracting phagocytes. Some antibodies also inactivate pathogens by causing their cells to agglutinate, preventing them from entering human cells and causing disease.

Some of the clone cells become **memory cells,** long-lived cells that remain inactive until that same antigen reappears in the body at some future date. The memory cells store information about the pathogen so that if there is a second exposure, the immune response will be even faster than the first time. Upon exposure, the memory cells quickly become plasma cells and start to secrete antibodies. These memory cells are the basis for long-term immunity.

> ***Recap*** **Lymphocytes called B cells are responsible for producing antibodies against specific antigens. When activated by first exposure to an antigen, B cells quickly produce a clone of identical antibody-producing B cells called plasma cells. They also produce a few long-lived memory cells that remain inactive until the next exposure to the same antigen.**

The five classes of antibodies

Antibodies belong to the class of blood plasma proteins called gamma globulins. Because they play such a crucial role in immunity, they are often called **immunoglobulins (Ig).** There are five classes of immunoglobulins, each designated by a different letter: IgG, IgM, IgA, IgD, and IgE. Each type has a different size, location in the body, and function:

- **IgG** (75% of immunoglobulins): This is the most common class. Found in blood, lymph, intestines, and tissue fluid, IgG are long-lived antibodies that activate the complement system and neutralize many toxins. They are the only antibodies that cross the placenta during pregnancy and pass on the mother's acquired immunities to the fetus.
- **IgM** (5–10%): These antibodies are the first to be released during immune responses. Found in blood and lymph, they activate the complement system and cause foreign cells to agglutinate. ABO blood cell antibodies belong to this class.
- **IgA** (15%): These antibodies enter areas of the body covered by mucous membranes, such as the digestive, reproductive, and respiratory tracts. There they neutralize infectious pathogens. They are also present in the mother's milk and are transmitted to the infant during breast feeding.
- **IgD** (less than 1%): IgD antibodies are in blood, lymph, and B cells. Their function is not clear, but they may play a role in activating B cells.

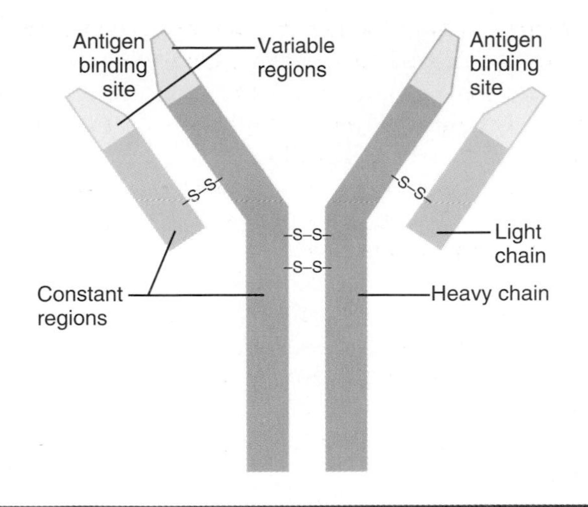

Figure 9.11 Structure of an antibody (or B cell surface receptor). Light and heavy chains are linked by disulfide (S-S) bonds. Only the lighter-colored ends of each chain vary. These variations are what determine the specificity of each antibody for only one antigen.

- **IgE** (approximately 0.1%): The rarest immunoglobulins, IgE antibodies are in B cells, mast cells, and basophils. They activate the inflammatory response by triggering the release of histamine. They are also the trouble-makers behind allergic responses (covered in section 9.10).

Antibodies' structure enables them to bind to specific antigens

An antigen provides all the information the immune system needs to know about a foreign substance. Essentially, antigens are the locks on an enemy's doors. The immune system can identify the lock, produce the antibody key, and then send in immune system cells to open the door and neutralize the invader. How does this happen?

Spotlight on Structure & Function Regardless of their class, antibodies share the same basic structure, and this structure enables them to neutralize nonself antigens. As an example, let's look at the structure of IgG, the most common immunoglobulin. As shown in Figure 9.11, each antibody (or surface receptor, if it is attached to a B cell) consists of four polypeptide chains arranged in a "Y" shape. Two of these chains are large, so they are called *heavy chains*. The other two are shorter and are called *light chains*. All four chains are linked by covalent bonds between sulfur atoms.

Each of the four chains has a *constant region* at one end, where the sequence of amino acids is fixed, and a smaller *variable region* at the other end, where the sequence varies. The constant regions are similar for all antibodies in one class, although they differ from those of other classes. The tips of the variable regions form antigen-binding sites with a unique shape that fits only one specific antigen. The IgG immunoglobulin has two such binding sites, but some of the other classes have more. ■

Recap **There are five classes of antibodies (also called immunoglobulins), each with a characteristic size, shape, and function. Within each class, a particular antibody fits only one specific antigen.**

T cells: Cell-mediated immunity

The T cells responsible for cell-mediated immunity develop from stem cells in bone marrow but mature in the thymus gland. T cells directly attack foreign cells that carry antigens, or release substances that enhance the immune response. The basic difference between B and T cells is that B cells produce circulating antibodies against an antigen, whereas T cells recognize the presence of an antigen-antibody complex, attach to it, and kill the cell carrying it.

Like B cells, T cells have receptors. However, these receptors do not allow them to recognize antigens (and the pathogens to which they are attached) directly. Instead, T cells react to fragments of antigens. Thus it is necessary for a certain amount of antigen processing to occur before T cells can do their job. The antigen must be presented to them in a form that they can recognize.

Certain cells—including macrophages and activated B cells—fulfill this role by acting as **antigen-presenting cells (APCs)** that engulf foreign particles, partially digest them, and display fragments of the antigens on their surfaces (Figure 9.12). After an APC engulfs a pathogen, it partially degrades the antigen inside a vesicle. The vesicle containing antigen fragments joins with another vesicle containing MHC molecules, the molecules that become the cell's self markers. The MHC binds to the antigen pieces and moves to the cell surface, to be displayed as an antigen-MHC complex. Essentially, the cell "presents" a sample of the antigen for T cells to recognize, along with its own cell-surface self marker.

In the thymus gland, under the influence of the hormone thymosin, immature lymphocytes develop into mature but inactive T cells. T cells carry specialized surface receptors that enable them to identify and bond to the antigen fragments displayed by APCs. During maturation they also develop one of two sets of surface proteins, CD4 or CD8. These proteins determine which type of T cell they will become.

Types of T cells include helper, cytotoxic, memory, and suppressor T cells. Figure 9.13 summarizes the development and activation of all four types of T cells.

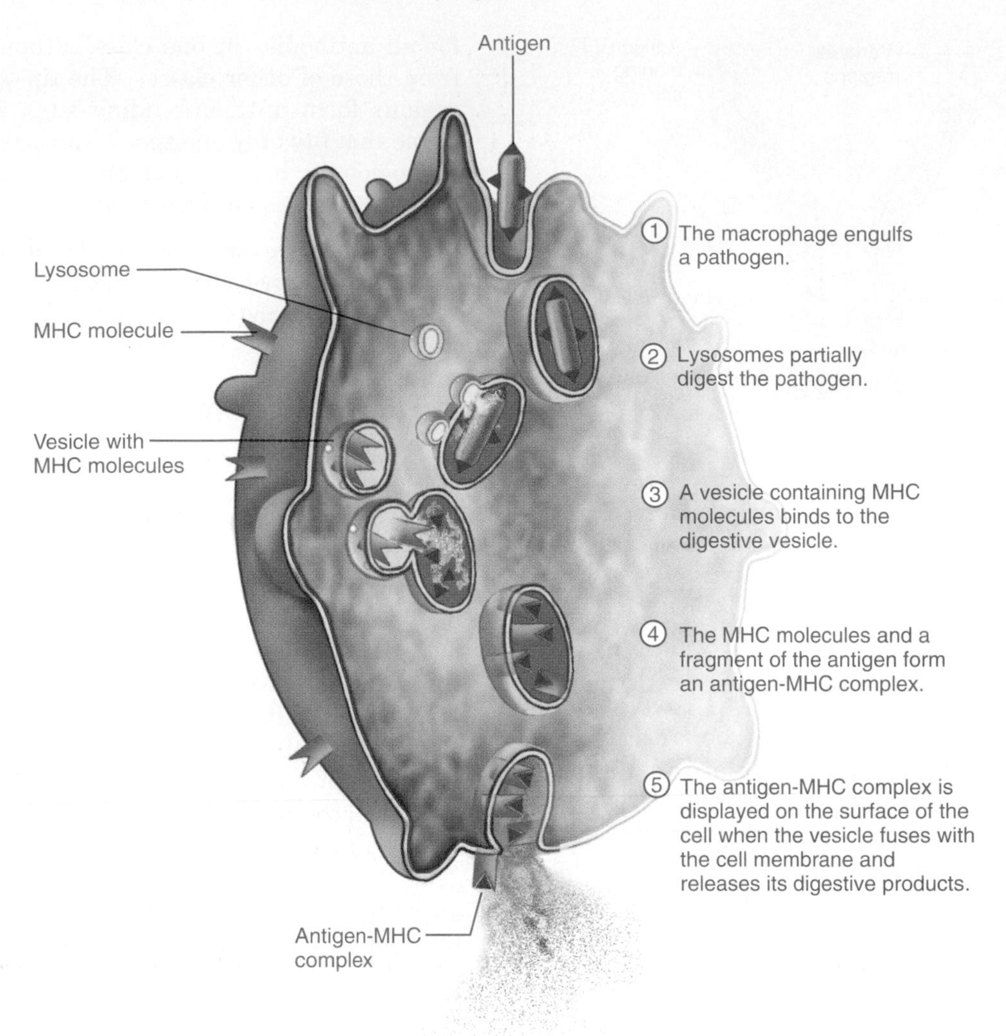

Figure 9.12 How a macrophage acts as an antigen-presenting cell (APC).

Helper T cells stimulate other immune cells When a T cell with CD4 receptors encounters an APC displaying an antigen-MHC complex, it differentiates into a **helper T cell.** The new helper T cell undergoes mitosis (see Chapter 17), quickly producing a clone of identical T cells. Because all the cells in the clone carry the same receptors, they all recognize the same antigen.

Many cells in the clone begin secreting a class of signaling molecules called *cytokines.* Among them are *lymphokines,* proteins that enhance inflammation and the immune response by stimulating the actions of T cells and macrophages, and *interleukins,* substances that promote development of other immune cells. Lymphokines released by helper T cells stimulate a host of defense mechanisms: phagocytes, natural killer cells, and T cells with CD8 receptors. They attract other types of white blood cells to the area, enhancing nonspecific defenses. They even activate B cells, creating an important link between antibody-mediated and cell-mediated immunity.

The role of helper T cells is crucial. Indeed, without them the entire immune response would be severely impaired or nonexistent because they direct or enhance the activities of so many other cells in the immune system. The reason AIDS is so devastating is that the virus that causes AIDS destroys helper T cells and weakens the body's ability to mount a cell-mediated immune response. (See Section 9.11, "Immune Deficiency—The Special Case of AIDS").

Cytotoxic T cells kill abnormal and foreign cells When a mature T cell with CD8 receptors meets an APC that displays an antigen-MHC complex, it is activated to produce a clone of **cytotoxic T cells** (also called *killer T cells* or T8s). These are the only T cells that directly attack and destroy other cells.

Once activated, cytotoxic T cells roam through the body. They circulate through blood, lymph, and lymphatic tissues in search of cells that display the antigens they recognize. They migrate to a tumor or site of infection, where they release chemicals that

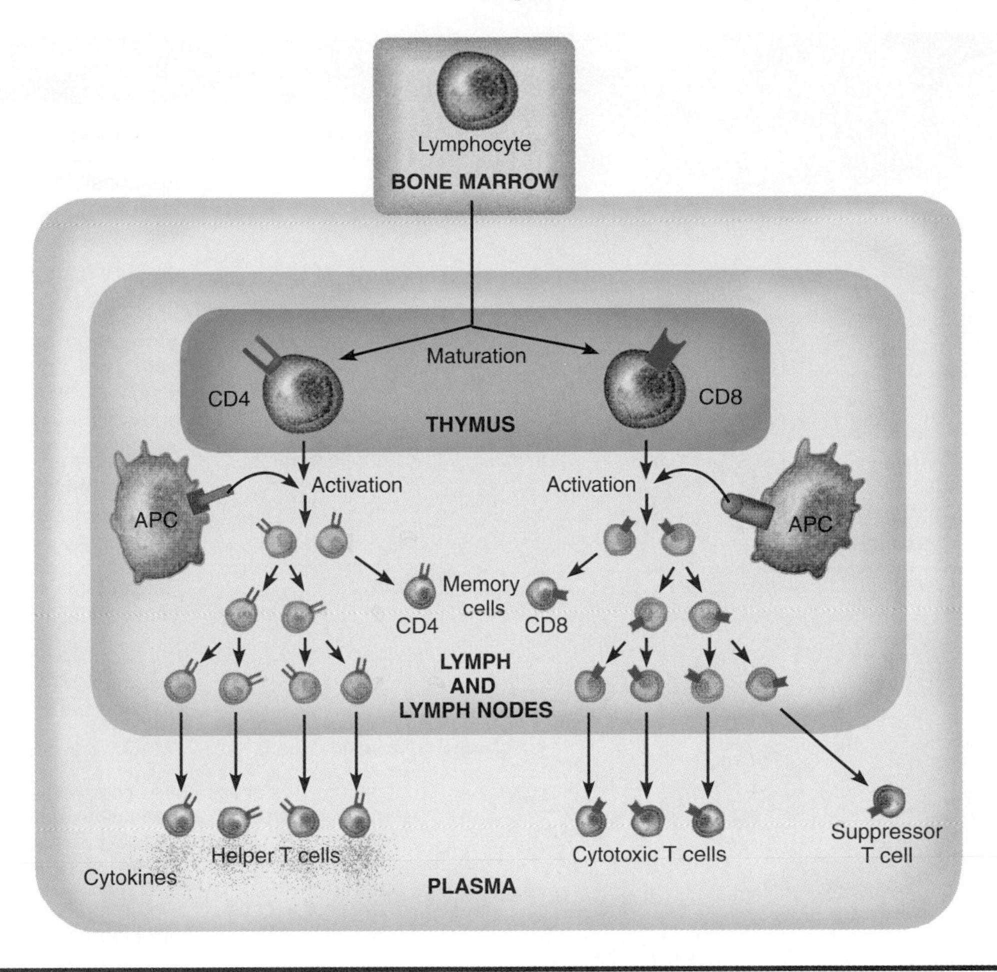

Figure 9.13 The formation and activation of T cells. Activation begins when mature but inactive T cells come in contact with an antigen-presenting cell displaying the appropriate antigen fragment.

are toxic to abnormal cells (*cytotoxic* literally means "cell-poisoning"). Figure 9.14 illustrates cytotoxic T cells in action. When a cytoxic T cell locates and binds to a target cell, a protein called *perforin* is released from secretory vesicles in the cytotoxic cell and inserted into the plasma membrane of the target cell. The cytotoxic T cell then detaches from the target cell and goes off in search of other prey. The individual molecules of perforin assemble themselves into large pores, or holes, in the target cell membrane. Water and salts enter through the holes, and eventually the target cell swells and bursts.

Other lymphokines secreted by cytotoxic T cells include lymphotoxin, gamma interferon, and tumor necrosis factor. *Lymphotoxin* destroys foreign cells by fragmenting their DNA. *Gamma interferon* stimulates macrophages to become more active. *Tumor necrosis factor* exerts anticancer effects that are not fully understood.

Several promising medical treatments involve harnessing the defensive capabilities of lymphokines. Genetically engineered gamma interferon is used to treat the chronic viral disorder hepatitis C. In other experiments, researchers removed cytotoxic T cells from cancer patients' tumors and altered the T cells to boost their production of tumor necrosis factor. The patients then received doses of the enhanced cells to stimulate their bodies' natural defenses. Another type of interferon has been moderately successful for treating multiple sclerosis.

Memory T cells reactivate during later exposures Some helper T cells and cytotoxic T cells become long-lived **memory cells** that retain receptors for the antigen that originally stimulated their production. If that antigen is presented to them again, the memory cells are reactivated. Some form new helper T cells that multiply quickly to marshal an immune response. Others form a new army of cytotoxic T cells to attack and destroy. Like memory B cells, memory T cells are an important factor that distinguishes specific defenses from nonspecific defense mechanisms.

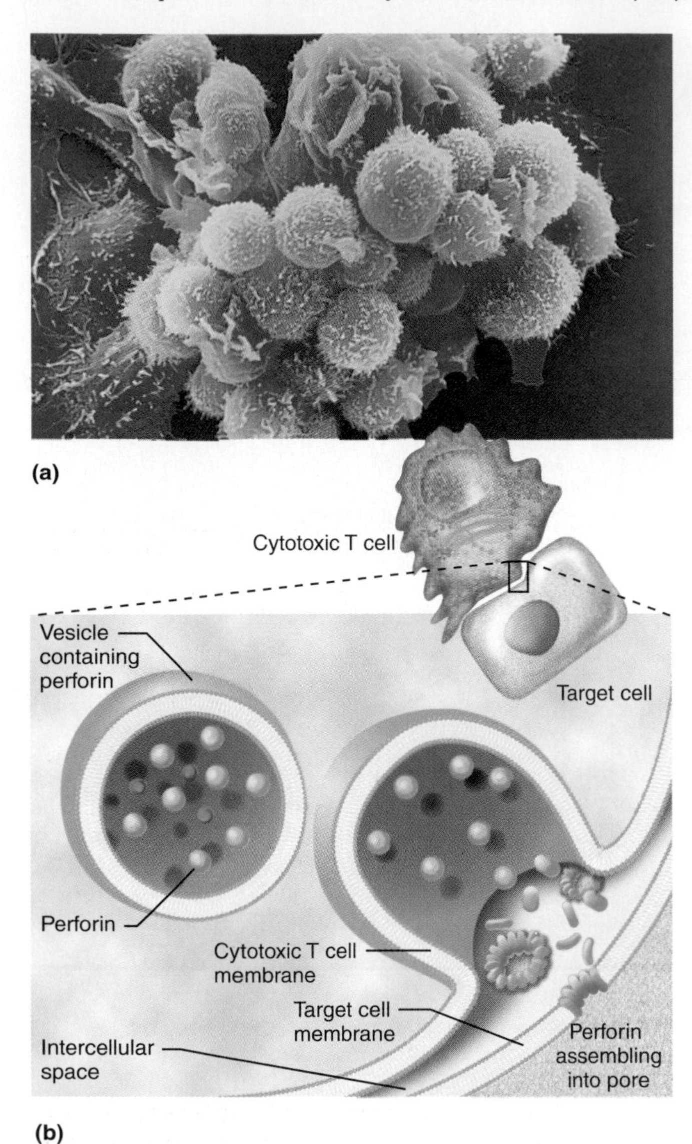

Figure 9.14 Cell-mediated immunity in action. (a) Cytotoxic T cells attaching to a target cell. **(b)** How cytotoxic T cells kill target cells. When the T cell attaches to the target, vesicles within the T cell fuse with the T cell membrane and release perforin molecules. The perforin molecules create holes in the target cell membrane, leading to destruction of the cell. The action of perforin is like one of the actions of complement proteins that circulate in blood (review Figure 9.8). By direct contact, however, the T cell can deliver a high concentration of perforins directly to its target.

Suppressor T cells may suppress immune response Some studies indicate there is another type of T cell, a **suppressor T cell,** that is thought to develop from cytotoxic T cells and that suppresses the immune response after an antigen has been successfully neutralized. Its role is to prevent immune system activity from going out of control and wreaking havoc on healthy cells. It may also help control autoimmune reactions, in which the body targets its own tissues.

Table 9.3 Cells and proteins involved in specific defenses

Cell/Protein	Functions
B cells	Responsible for antibody-mediated immunity (AMI), or humoral immunity.
Plasma cells	Produce and secrete specific antibodies.
Memory B cells	Store information about pathogen. Upon subsequent exposure, become plasma cells and secrete antibodies.
Immunoglobulins	Five classes of antibodies. Have antigen-binding sites with a unique shape that fits one specific antigen.
T cells	Responsible for cell-mediated immunity (CMI).
Helper T cells	Enhance specific and nonspecific defenses by stimulating other immune cells.
Cytotoxic T cells	Attack and destroy abnormal cells.
Memory T cells	Reactivate upon later exposure to the same antigen to form helper and cytotoxic T cells.
Suppressor T cells	May suppress immune response after an antigen has been destroyed.

The concept of suppressor T cells remains controversial. In practice they are difficult to study and little is known about them. Some researchers believe that lymphokines released by T cells are actually responsible for suppressing the immune response when it is no longer needed.

Table 9.3 summarizes the various cells and proteins involved in specific defense mechanisms.

Recap **Lymphocytes that mature in the thymus gland are called T cells. Helper T cells enhance specific and nonspecific defenses by stimulating other immune cells. Cytotoxic T cells attack abnormal and foreign cells. Memory T cells store information until the next exposure to the same antigen, and suppressor T cells may suppress the immune response after an antigen has been neutralized.**

9.7 Immune memory creates immunity

When you are first exposed to an antigen, your immune system protects you with the wealth of defense mechanisms described so far. Your first exposure to this particular antigen generates a **primary immune response.** As we have seen, this involves: recognition of the antigen, and production and proliferation of B and T cells.

Typically the primary immune response has a lag time of three to six days after the antigen first appears. During this period B cells specific to that antigen multiply and develop into plasma cells. Antibody concentrations rise, typically reaching their peak about 10–12 days after first exposure. Then they start to level off (Figure 9.15).

However, as you have learned, B and T cells create a population of memory cells. The presence of these memory cells is the basis for **immunity** from disease. (The Latin word *immunis* means "safe" or "free from.") Subsequent exposure to the pathogen elicits a **secondary immune response** that is faster, longer lasting, and more effective than the first. Within hours after second exposure to an antigen, memory cells bind to the pathogen. New armies of T and plasma cells form, and within a few days antibody concentrations rise rapidly to much higher levels than in the primary response. Notice that antibody levels remain much higher in the body after second exposure.

Memory cells are long-lived, and many retain their ability to generate a secondary immune response over a lifetime. The secondary immune response can be so effective that you're not even aware you've been exposed to the pathogen a second time. At worst, you may experience only a fleeting sensation of feeling unwell. Other memory cells, such as the ones for the bacterial infection that causes tetanus, need to be reactivated every ten years or so.

Given this immunity, though, why is it possible to get a cold or the flu over and over, sometimes several times a year? One reason is that there are more than 100 different viruses that can cause colds and flu. Even if your latest respiratory ailment feels like the previous one, it may actually be due to an entirely different pathogen. Furthermore, the viruses that cause colds and flu evolve so rapidly that they are essentially different each year. Their antigens change enough that each one requires a different antibody, and each exposure triggers a primary response. Rapid evolution is their survival mechanism. Our survival mechanism is a good immune system.

Recap **First exposure to a specific antigen generates a primary immune response. Subsequent exposure to the same antigen elicits a secondary immune response that is faster, longer lasting, and more effective.**

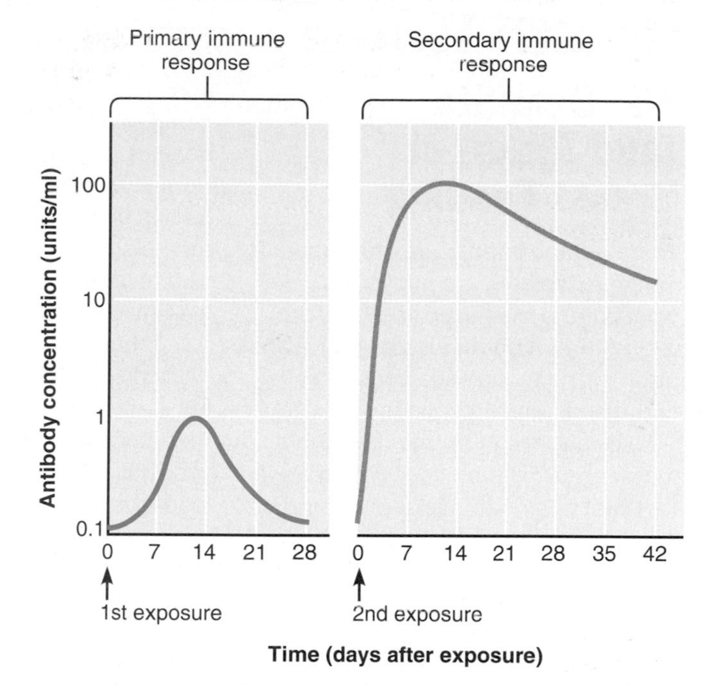

Figure 9.15 The basis of immunity. The antibody response to a first exposure to antigen (the primary immune response) declines in about a month. A second exposure to the same antigen results in a response that is more rapid in onset, larger, and longer-lasting. On this graph the abscissa scale is broken because it may be months or even years before second exposure. Note that the concentration of antibody during a secondary response may be 100 times higher than the primary response.

9.8 Medical assistance in the war against pathogens

Our natural defenses against pathogens are remarkable. Nevertheless, we humans have taken matters into our own hands by developing the science of medicine. We have been able to conceive of and produce other sophisticated weaponry to help us combat pathogens. Important milestones in human health include the development of vaccines, passive immunization, monoclonal antibodies, and antibiotics.

Active immunization: The most effective weapon against viruses

How do immunizations prevent diseases?

It is said that an ounce of prevention is worth a pound of cure, and this is certainly true when dealing with viruses. The best weapon against viruses is to give the body a dose of the viral antigen in advance so the immune system will mount a primary immune response against it. Then, if exposure to that virus does occur, the body is

Health Watch

The Benefits (and Risks) of Breast-Feeding

When human infants are born, their immune systems are not yet fully developed and their intestines are not yet mature. As a result, newborns are often prone to diarrhea caused by recurring bacterial infections.

Breast milk from the newborn's mother may be beneficial. When an infant first begins to suckle, the mother secretes a sort of "premilk" called colostrum. Colostrum is rich in antibodies and immune cells, including phagocytes that participate in nonspecific defense mechanisms and lymphocytes of the specific immune system. Together these provide some immunity for the infant until its immune system can develop its own antibodies (about a month after birth). Human milk continues to provide antibodies and immune cells throughout the nursing period, though the concentration of these substances in milk is less than it is in colostrum.

The natural role of human milk in delivering antibodies to the infant has led some immunologists to suggest that mothers should be immunized by vaccination against certain childhood diseases. The mother would then produce specific antibodies for the infant against those diseases. Others have suggested putting antibodies directly into infant formulas or cow's milk for babies who are not breast-fed.

There are other benefits to human milk as well. All of us harbor beneficial bacteria in our digestive systems, and human milk encourages the growth of these good bacteria. Human milk is also more easily digested by human infants than cow's milk.

There is also a slight risk associated with human milk that might not be associated with cow's milk. "Street drugs," prescription medicines, and chemicals in the mother's diet can, under certain circumstances, end up in human milk. Examples include heroin, cocaine, alcohol, caffeine, and Tagamet® (an anti-ulcer medication). HIV (the virus that causes AIDS) can also be transmitted in human milk. With a thorough physical exam and medical history, however, the risks associated with breast-feeding can be kept to a minimum.

already primed with the appropriate antibodies and memory cells. The immune system can react swiftly with a secondary response, effectively shielding you from the discomfort and danger of disease symptoms.

The process of activating the body's immune system in advance is called **active immunization.** This involves administering an antigen-containing preparation called a **vaccine.** Some vaccines are produced from dead or weakened pathogens or their antigens. An example is the oral polio vaccine (the Sabin vaccine), made from weakened poliovirus. Other vaccines are made from organisms that have been genetically altered to produce a particular antigen.

Of course, vaccines have limitations. First, there are issues of safety, time, and expense. Weakened but not dead pathogens generally make better vaccines because they elicit a greater immune response. However, a vaccine that contains weakened pathogens has a slight potential to cause disease itself. This has happened, although rarely, with the polio vaccine. It takes a great deal of time, money, and research to verify the safety and effectiveness of a vaccine. Second, a vaccine only confers immunity against one pathogen, so a different vaccine is needed for every virus. This is why doctors may recommend getting a flu vaccine each time a new flu strain appears (nearly every year). Third, vaccines are not particularly effective after a pathogen has struck; that is, they do not cure an already existing disease.

Nonetheless, vaccines are an effective supplement to our natural defense mechanisms. Active immunization generally produces long-lived immunity that can protect us for many years. The widespread practice of vaccination has eliminated smallpox, once a major scourge of human beings, and has greatly reduced many other childhood diseases such as polio, measles, and whooping cough.

In the U.S., immunization of adults has lagged behind that of children. It's estimated that more than 50,000 Americans die each year from infections that could have been prevented with timely vaccines, including pneumonia, hepatitis, and influenza.

Passive immunization can help against existing infections

There is a second weapon in the war against pathogens. A procedure called **passive immunization** involves administering prepared antibodies (gamma globulin serum) for a specific pathogen. The antibodies are obtained from a human or animal donor who already has immunity against that illness.

Passive immunization has the advantage of being somewhat effective against an existing infection. It can be administered to prevent illness in someone who has been unexpectedly exposed to a pathogen, and it confers at least some short-term immunity. However, protection is not as long-lasting as active immunization because the antibodies disappear

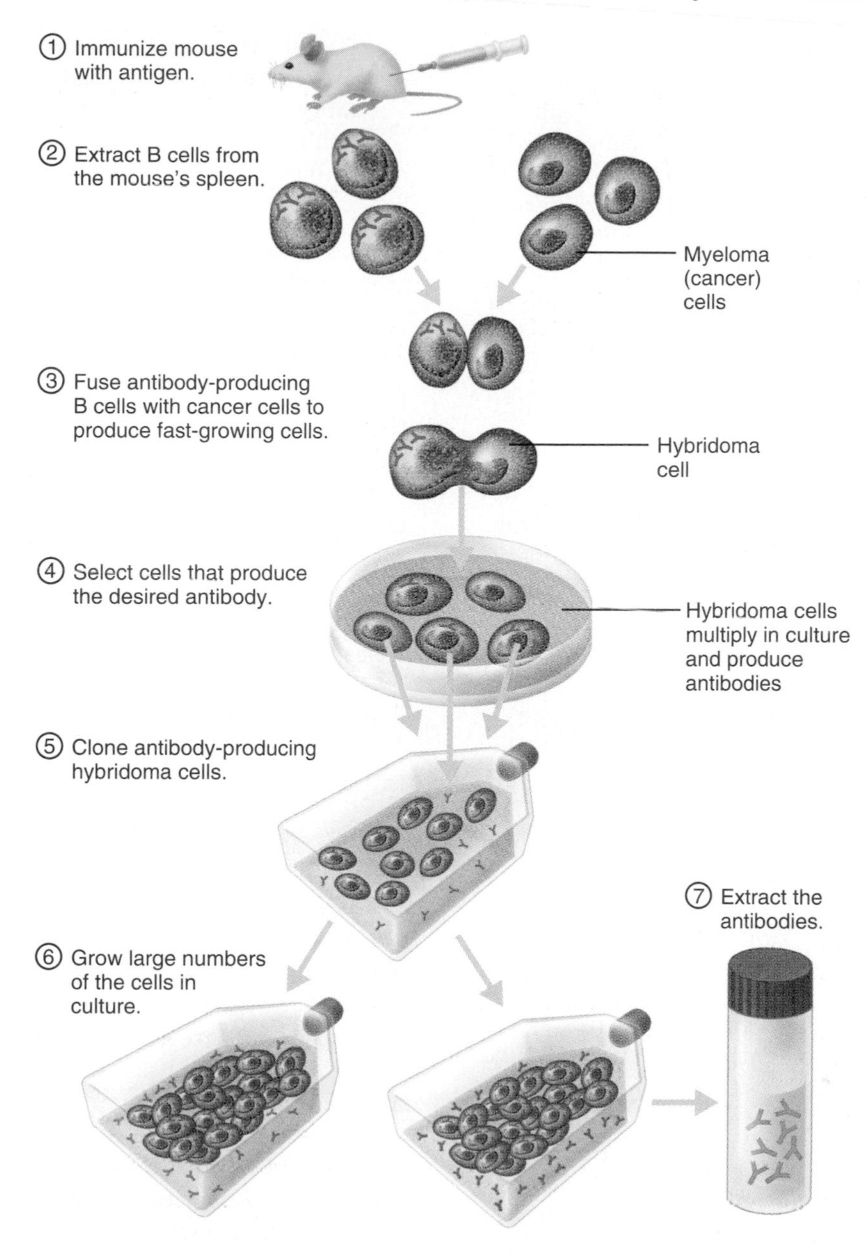

Figure 9.16 How monoclonal antibodies are produced.

from the circulation with time. The person's own B cells are not producing these antibodies, so memory cells for this pathogen do not develop and long-term immunity does not occur.

Passive immunization has been used effectively against certain common viral infections, including those that cause hepatitis B and measles, bacterial infections such as tetanus, and Rh incompatibility. Passive immunization of the fetus and newborn also occurs naturally across the placenta and via breast-feeding.

Recap **Active immunization with a vaccine (a preparation of weakened or dead pathogens) produces a primary immune response and readies the immune system for a secondary immune response. Administration of prepared antibodies (passive immunization) can be effective against existing infections but does not confer long-term immunity.**

Monoclonal antibodies: Laboratory-created for commercial use

In addition to providing passive immunity, antibody preparations are proving useful in research, testing, and cancer treatments. **Monoclonal antibodies** are antibodies produced in the laboratory from cloned descendants of a single hybrid B cell. As such, monoclonal antibodies are relatively pure preparations of antibodies specific for a single antigen.

Figure 9.16 summarizes a technique for preparing monoclonal antibodies. Typically, B cells are removed from a mouse's spleen after the mouse has been immunized with a specific antigen to stimulate B cell production. The B cells are fused with myeloma (cancer) cells to create new *hybridoma* (hybrid cancer) cells that have desirable traits of both parent cells: they each produce a specific antibody, and they proliferate with cancer-like rapidity. As these hybridoma cells grow in culture, those that produce the desired antibody are separated out and cloned, producing millions of copies. The antibodies they produce are harvested and processed to create preparations of pure monoclonal antibodies. (The term *monoclonal* means these antibody molecules derive from a group of cells cloned from a single cell.)

Monoclonal antibodies have a number of commercial applications, including home pregnancy tests, screening for prostate cancer, and diagnostic testing for hepatitis, influenza, and HIV/AIDS. Monoclonal antibody tests tend to be more specific and more rapid than conventional diagnostic tests. Researchers are experimenting with bonding monoclonal antibodies to anticancer drugs, to deliver the drugs directly to cancerous cells and spare healthy tissue.

Antibiotics combat bacteria

Literally, "antibiotic" means "against life." **Antibiotics** are drugs that kill bacteria or inhibit their growth. The first antibiotics were derived from extracts of molds and fungi, but today most antibiotics are synthesized by pharmaceutical companies.

There are hundreds of antibiotics in use today, and they work in dozens of ways. In general, they take advantage of the differences between bacteria and human cells, including: (1) bacteria have a thick cell wall (our cells do not have a thick cell wall), (2) bacterial DNA is not safely enclosed in a nucleus (ours is), (3) bacterial ribosomes are smaller than ours, and (4) their rate of protein synthesis is very rapid as they grow and divide. For example, penicillin blocks the synthesis of bacterial cell walls, and streptomycin inhibits bacterial protein synthesis by altering the shape of bacterial ribosomes.

Some antibiotics combat only certain types of bacteria. Others, called broad-spectrum antibiotics, are effective against several groups of bacteria. By definition, however, antibiotics are ineffective against viruses. Recall that viruses do not reproduce on their own; they only replicate once they are inside living cells. Using antibiotics to fight viral infections such as colds or the flu is wasted effort and contributes to the growing health problem of bacterial resistance to antibiotics.

Recap **Monoclonal antibodies are laboratory-produced antibody preparations that are specific for a single antigen. Most antibiotics kill bacteria by interfering with bacterial protein synthesis or bacterial cell wall synthesis.**

9.9 Tissue rejection: A medical challenge

Because the goal of the immune system is to protect the body from invasion by nonself cells, it is not surprising that it attacks foreign tissues, even those that are human, with vigor. This phenomenon is called *tissue rejection*. In Chapter 7 we saw that a type of tissue rejection called a transfusion reaction can be fatal. Red blood cells carry a dozen or so self markers, but most body cells have far more. Therefore, it is even more difficult to match tissues between donor and recipient than it is to match blood types.

Surgical techniques for performing many organ transplants are really not that difficult. Historically, the major stumbling block to widespread transplantation of most organs has been the effectiveness of the recipient's immune response. In the normal immune response, cytotoxic T cells will swiftly attack and destroy the foreign cells.

Before a transplant is even attempted, then, the donor's and recipient's ABO and other blood group antigens must first be determined. Next, donor and recipient tissues are tested to compare MHC antigens, because cytotoxic T cells target foreign MHC proteins. The closer the relationship between donor and recipient the better, because their MHC antigens are likely to be similar. Although successful transplants can be done between unrelated people, at least a 75% match between tissues is essential. After surgery the patient must take *immunosuppressive drugs* that block the immune response, such as corticosteroid drugs to suppress inflammation or cytotoxic medications that kill rapidly dividing cells (to block activated lymphocytes).

Immunosuppressive therapy can dramatically prolong the lives of transplant patients, but it brings other risks. An impaired immune system cannot protect the body as effectively against pathogens and abnormal cells, so these patients are vulnerable to infections and certain cancers. The key to a successful transplant is to suppress the immune system enough to prevent rejection while preserving as much immune function as possible. Antibiotic drugs can help control infections as they arise.

In recent years, three factors have made organ transplants more successful than ever before: (1) im-

Health Watch

Bacterial Resistance to Antibiotics

When antibiotics first became available in the 1940s, they were hailed as a breakthrough in medicine. Indeed, many people today owe their lives to these drugs.

But there is a downside: indiscriminate use of antibiotics encourages the proliferation of antibiotic-resistant bacteria. A few of these are dangerous pathogens. As of now, certain strains of *Staphylococcus aureus,* a bacterium that causes blood poisoning and wound infections, are resistant to all but one antibiotic. That antibiotic too is beginning to fail. If we are not careful, we may soon face a deadly pathogen with no defenses other than our natural mechanisms.

Most bacteria are not harmful to humans. Some are quite useful. Because the bacteria in (and on) your body compete for resources just like all other organisms occupying the same ecological space, good bacteria can limit the growth of harmful bacteria. But consider what happens when you take an antibiotic to kill a particular pathogen: you destroy the helpful bacteria as well as the pathogens. If the drug kills most of the bacteria but not every single one, those that survive are the ones most resistant to the antibiotic's effects. With the total population of bacteria down, these antibiotic-resistant survivors face reduced competition. They thrive and multiply. Any time we take an antibiotic, then, the percentage of resistant bacteria in the remaining population increases and thus the antibiotic will be less effective the next time around.

We didn't realize this 50 years ago. Because antibiotics were so magically effective, there was (and still is) a societal tendency to use them indiscriminately. In the U.S. alone we manufacture over 50 million pounds of antibiotics annually, about half of which end up in livestock feed or sprayed on fruit trees. In terms of medical use, researchers at the Centers for Disease Control and Prevention estimate that fully one-third of all antibiotic prescriptions given to non-hospitalized patients in the U.S. are not needed. Many patients demand them, believing they can do no harm, and physicians may prescribe them even though they suspect the disease is viral in origin and will not respond to antibiotic treatment.

Another problem involves failure to complete the full treatment. Many antibiotics are prescribed as a 7- to 10-day course of treatment in order to kill all the bacteria. Many people feel better within three days when most (but by no means all) of the bacteria have been killed, so they stop taking the drug and save the leftover pills for a future time when they feel sick again. They have inadvertently selected for survival the more antibiotic-resistant bacteria!

What can be done to preserve the effectiveness of antibiotics? Here are some tips:

- Don't ask for antibiotics for a cold, flu, or other viral illness. Take them only when they are needed and when a physician prescribes them.
- Always complete the full course of treatment as prescribed—even if you believe that you are back to normal.
- Use antibiotic ointments on cuts and abrasions, but consider reducing your use of antibacterial hand creams, soaps, and laundry detergents.
- Encourage farmers and orchard owners to reduce their use of antibiotics in cattle feed and on fruit trees, bearing in mind that these groups have legitimate reasons of their own for antibiotic use.
- Support research efforts to find new classes of antibiotics so older ones can be used more sparingly.

The development of resistance to antibiotics is a worldwide issue that affects us all. Used properly and judiciously, however, antibiotics will remain in our antibacterial arsenal for a long time to come.

provements in immunosuppressive drugs, (2) better techniques for cross-matching (or "typing") tissue, and (3) national sharing of information and donor organs via organ-bank systems. The latter allow patients to receive the best matches possible regardless of where they live. These improvements have made organ transplants a viable option for many people. Still in the development phase are attempts to "mask" foreign antigens so that they do not provoke an immune response.

Recap **The major obstacle to organ transplants is the recipient's immune response, as cytotoxic T cells usually attack all foreign cells.**

9.10 Inappropriate immune system activity causes problems

Conditions characterized by inappropriate immune system activity include allergies and autoimmune disorders (such as lupus erythematosus and rheumatoid arthritis).

Allergies: A hypersensitive immune system

Many of us—perhaps 10% of North Americans—suffer from allergies. Some are relatively minor; others are quite severe. Examples include hay fever, poison ivy, and severe reactions to specific foods or drugs. Some allergic reactions can even send you to the hospital.

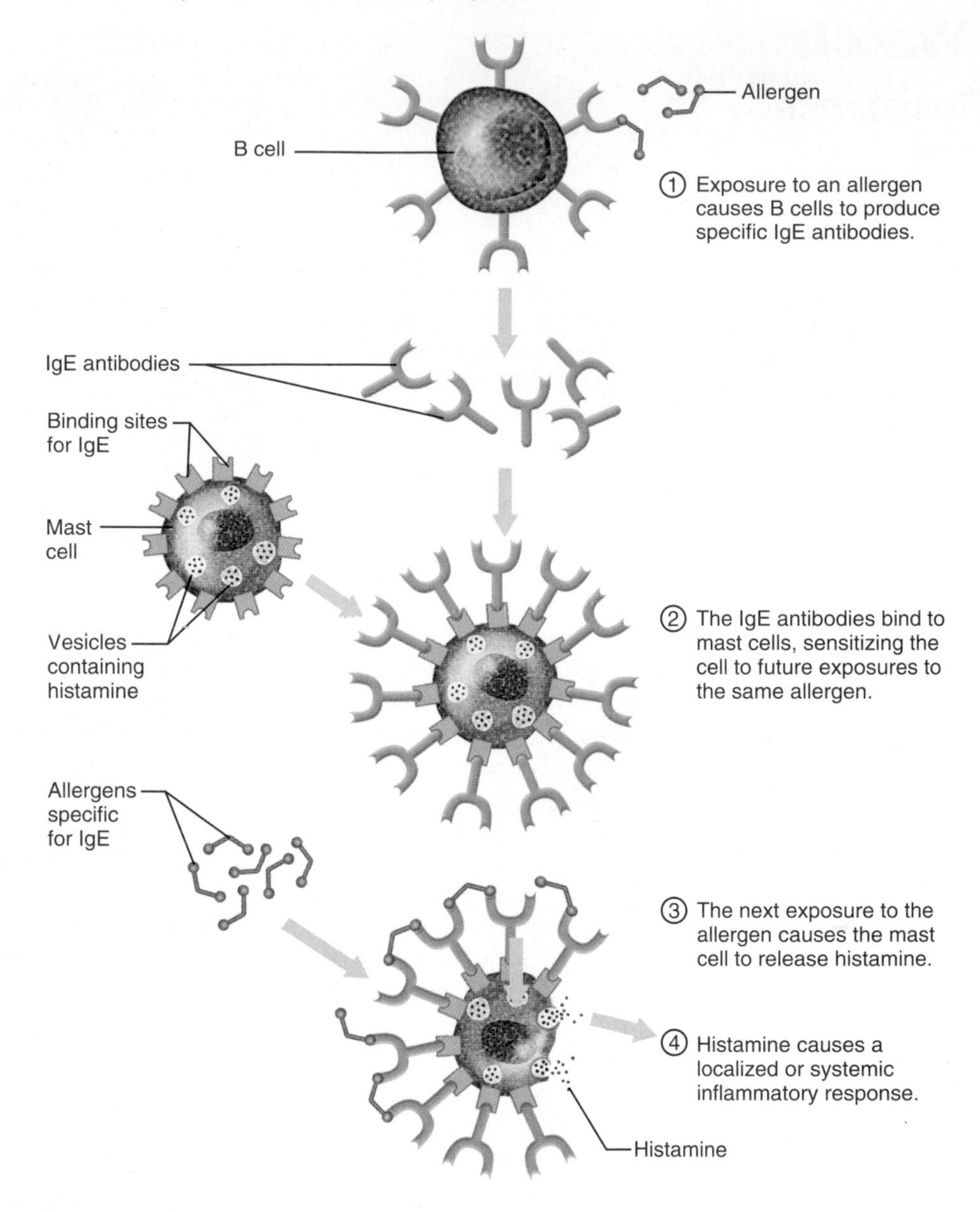

Figure 9.17 How an allergic reaction develops.

An **allergy** is an inappropriate response of the immune system to an **allergen** (any substance that causes an allergic reaction). The key is "inappropriate": the allergen is not a dangerous pathogen, but the body reacts as if it is.

What causes an allergy?

Recall that there are five classes of immunoglobulins. As shown in Figure 9.17, the culprits involved in allergic reactions are those in the IgE group. At some point, exposure to an allergen triggers a primary immune response, causing B cells to produce specific IgE antibodies. Some of these IgE antibodies bind to cell surface receptors of mast cells and basophils in tissues.

When the same allergen enters the body a second time, it binds to the IgE antibodies, causing the mast cells to release histamine. The result is an allergic reaction, a typical inflammatory response that includes warmth, redness, swelling, and pain in the area of contact with the allergen. Histamine also increases secretion of mucus in the region. Every time it is exposed to this allergen the body reacts as if there is an injury or infection, even though there isn't.

Some allergens, such as poison ivy, affect only the areas exposed. At least 85% of the U.S. population is sensitive to poison ivy. The allergen is a chemical in the plant's sap. Other allergens, including food allergens and bee sting venom, are absorbed or injected into the bloodstream. These substances are carried quickly to mast cells throughout the body, including in the respiratory, digestive, and circulatory systems. Such allergens often elicit a *systemic response,* meaning they affect several organ systems. Systemic responses include constriction of smooth muscle in the lungs and digestive system and dilation of blood vessels.

Symptoms of a severe systemic allergic reaction can include difficulty breathing (caused by constricted airways), severe stomach cramps (muscle contractions), swelling throughout the body (increased capillary permeability), and circulatory collapse with a life-threatening fall in blood pressure (dilated arterioles). This is known as *anaphylactic shock.* Anyone who appears to be suffering from anaphylactic shock should be rushed to a hospital because the reaction can be fatal. Because the symptoms often begin suddenly, doctors advise people with a history of strong allergic reactions to carry an emergency kit with them at all times. The kit contains a self-injecting hypodermic syringe of epinephrine, a hormone that dilates the airway and constricts peripheral blood vessels, preventing shock.

Most allergies, however, are more annoying than dangerous. Antihistamines—drugs that block the effects of histamine—are often effective treatments for mild to moderate reactions. Allergy shots can help by causing the body to produce large numbers of IgG antibodies, which combine with the allergen and block its attachment to IgE.

Recap **An allergy is an inappropriate inflammatory response. Allergens that produce a systemic response can rapidly affect the ability to breathe and maintain blood pressure.**

Autoimmune disorders: Defective recognition of "self"

On rare occasions, the immune system's remarkable ability to distinguish self from nonself fails. When that happens, the immune system may produce antibodies and cytotoxic T cells that target its own cells. Conditions in which this happens are called **autoimmune disorders.**

Approximately 5% of adults in North America, two-thirds of them women, have some type of autoimmune disorder. We don't yet know all the details of how these diseases arise. In some cases, certain antigens simply are never exposed to the immune system as it undergoes fetal development. These antigens were never programmed into the system as self, so when tissue damage exposes them, the mature immune system responds as if they are foreign. In other cases, antibodies produced against a foreign antigen may cross-react with the person's own tissues.

At the moment there are no cures for autoimmune disorders. Treatments include therapies that depress the body's defense mechanisms and relieve the symptoms. These conditions include a wide range of diseases, including multiple sclerosis, a progressive disorder of the central nervous system (see Chapter 11) and juvenile (Type I) diabetes mellitus, which targets cells in the pancreas (see Chapter 13). Below we look at two more conditions: lupus erythematosus and rheumatoid arthritis.

Lupus erythematosus: Inflamed connective tissue *Lupus erythematosus* (or *lupus*) is an autoimmune disorder in which the body attacks its own connective tissue. One type of lupus, discoid lupus erythematosus, targets exposed areas of the skin. A more serious form, systemic lupus erythematosus, can affect many other body systems, including the kidneys and joints.

Lupus often starts as a red skin rash on the face or head. Other symptoms include fever, fatigue, joint pain, and weight loss. Spreading inflammation can lead to osteoarthritis (see Chapter 5), pericarditis (see Chapter 8), or pleurisy (inflammation of the lining of the lungs).

Lupus affects nine times as many women as men. Typically it occurs during childbearing age and is more common in certain racial groups such as African Americans, West Indians, and Chinese. Medications can reduce the inflammation and improve the symptoms.

Rheumatoid arthritis: Inflamed synovial membranes Rheumatoid arthritis is a type of arthritis involving inflammation of the *synovial* membrane that lines certain joints (see Chapter 5). Fingers, wrists, toes, or other joints become painful and stiff. Without treatment, synovial fluid accumulates and joints swell. Neutrophils, lymphocytes, and other inflammatory cells migrate into the joint. Eventually the synovial membrane thickens, scar tissue forms, and the bones of the joint may fuse (Figure 9.18).

Pain-relieving medications can help many people with rheumatoid arthritis, as can regular mild exercise and physical therapy to improve the range of motion. Powerful drugs that neutralize chemicals in the inflammatory response can prevent joints from becoming deformed. Surgery to replace damaged joints with artificial joints can restore the ability to move and prevent painful disabilities.

Recap **An autoimmune disorder occurs when the immune system fails to distinguish self from nonself cells and begins to attack the body's own cells. Examples of autoimmune disorders are lupus erythematosus (inflammation of connective tissue) and rheumatoid arthritis (inflammation of synovial membranes).**

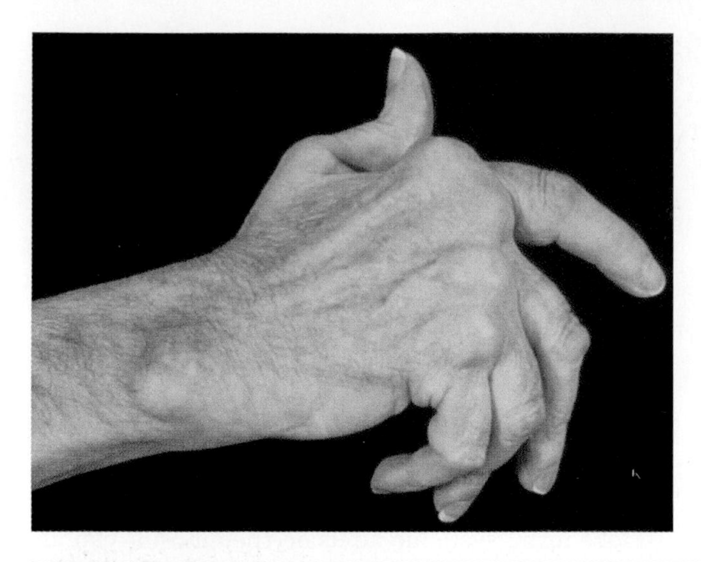

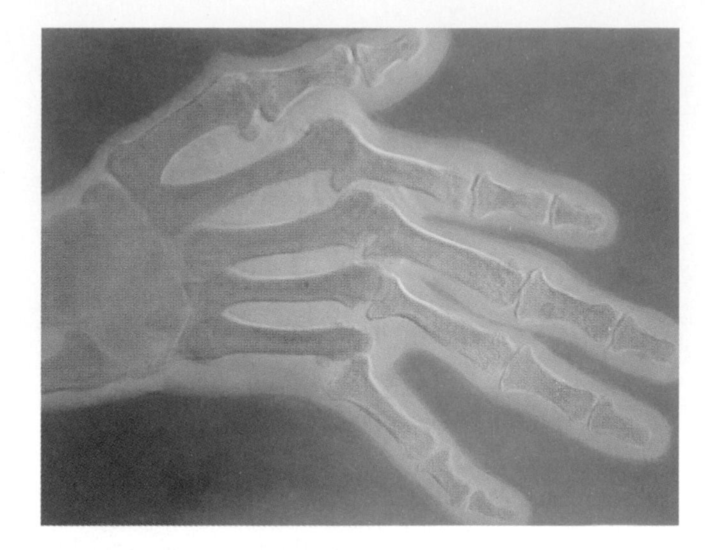

Figure 9.18 Rheumatoid arthritis.

9.11 Immune deficiency: The special case of AIDS

Immune deficiency is a general term for an immune system that is not functioning properly. One immune deficiency disease is *severe combined immunodeficiency disease (SCID)*. For the rare person who inherits SCID, even a minor infection can become life-threatening. People with SCID have too few functional lymphocytes to defend the body against infections.

By far the most common and best-known severe immune deficiency condition is **AIDS (acquired immune deficiency syndrome).** A *syndrome* is a medical term for a group of symptoms that occur together, and *acquired* means that one catches it—in this case by becoming infected with the virus called **HIV (human immunodeficiency virus).**

HIV targets helper T cells of the immune system

Figure 9.19 shows the structure of HIV. The virus consists of nothing more than single-stranded RNA and enzymes, wrapped in two protein coats and a phospholipid membrane with protein spikes. There is no nucleus and no organelles. Like other viruses, HIV infects by entering a cell and using the cell's machinery to reproduce. HIV targets helper T cells, gaining entry by attaching to CD4 receptors.

HIV belongs to a particular class of viruses called *retroviruses* that have a unique way of replicating (Figure 9.20). They enter the helper T cell after attaching to its CD4 receptor. Retroviruses use their RNA as a template (a pattern). Using their own enzymes, they force the host cell to make a complementary single strand of DNA, and from it a second strand of DNA complementary to the first. The new double-stranded DNA fragment is then inserted into the cell's DNA. The cell, not recognizing the DNA as foreign, uses it to produce more viral RNA and proteins, generating thousands of new viruses at a time. The sheer magnitude of the viral replication so saps the T cell's energy that eventually it dies. The viral copies move on to infect other helper T cells.

HIV is transmitted in body fluids

HIV is a fragile virus that cannot survive dry conditions for even a short period of time. This means that it cannot be transmitted through the air, by casual contact, or by doorknobs or toilet seats. HIV can be transmitted in at least four body fluids—blood, semen, breast milk, and vaginal secretions. HIV is not known to be transmitted by contact with urine,

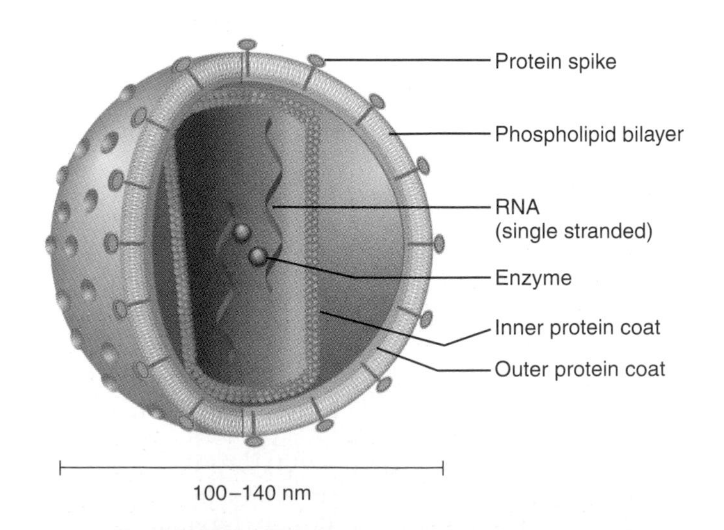

Figure 9.19 The structure of HIV.

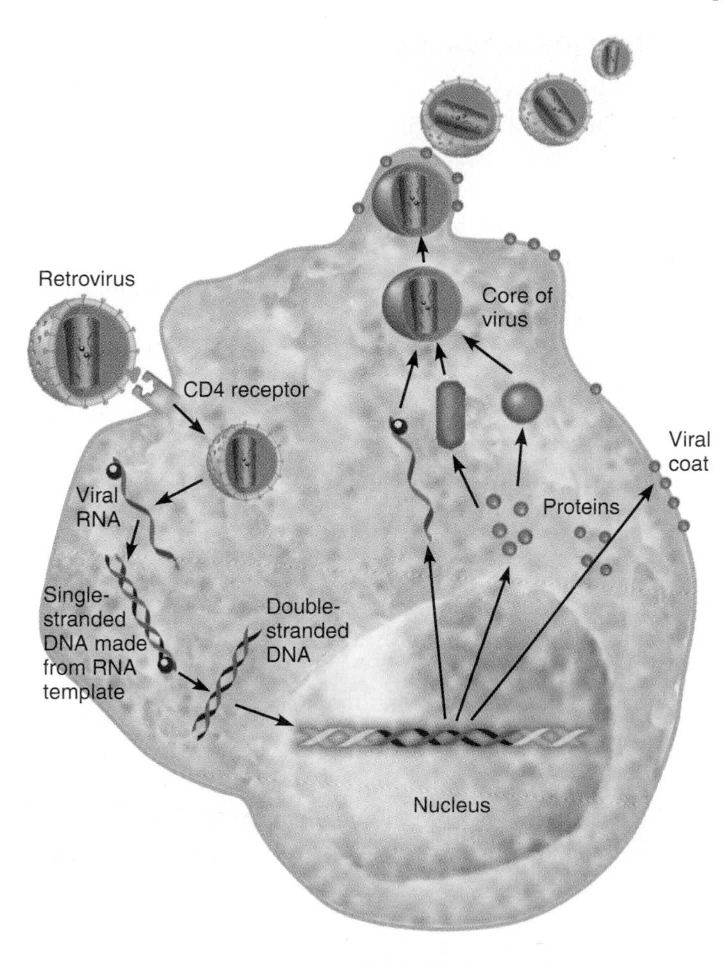

Figure 9.20 How HIV replicates. After binding to a helper T cell's CD4 receptor, the virus enters the cell and uses its own enzymes to force the cell to make single- and then double-stranded DNA. The DNA is inserted into the cell's own DNA within the nucleus, where it directs the cell to make more protein coat and viral RNA and protein core. The coats and cores join, exiting the cell as new viruses.

feces, saliva, perspiration, tears, or nasal secretions unless there is blood in these fluids.

The most common ways that HIV is transmitted is via either sexual contact or contaminated hypodermic needles. In addition, infected mothers can pass it to their children during pregnancy, labor and delivery, or when breast-feeding. It has been estimated, for example, that a healthy newborn who is breast-fed by an infected mother has about a 20% chance of contracting HIV from its mother's milk. Up until the mid-1980s a significant number of cases of HIV infection were acquired from blood transfusions. However, the risk of acquiring HIV from a blood transfusion has been greatly reduced since 1985 when HIV testing of blood became routine.

Because transmission of HIV generally requires direct contact of bodily fluids, HIV is considered relatively hard to transmit from person to person (at least when compared to such common diseases as the flu). However, unlike the flu, HIV is extremely virulent if left untreated. Over 90% of all U.S. citizens diagnosed with AIDS by 1988 are now dead.

AIDS develops slowly

How does an HIV infection lead to AIDS?

The symptoms of HIV infection and the subsequent development of full-blown AIDS follow a slow time course that can be divided into three phases (Figure 9.21).

Phase I Phase I lasts anywhere from a few weeks to a few years after initial exposure to the virus. During this time there may be typical flulike symptoms of swollen lymph nodes, chills and fever, fatigue, and body aches. The immune system's T cell population may decline briefly then rebound as the body begins to produce more cells and antibodies against the virus. The presence of antibodies against the virus is the basis for a diagnosis of HIV infection, and a person having these antibodies is said to be "HIV-positive." However, he or she will not have the disease syndrome called AIDS yet.

Unless they suspect they have been exposed to HIV, most people would not associate Phase I symptoms with HIV infection and so would not think to be tested. The antibodies do not destroy the virus entirely because many of the virus particles remain inside cells, where antibodies and immune cells cannot reach them.

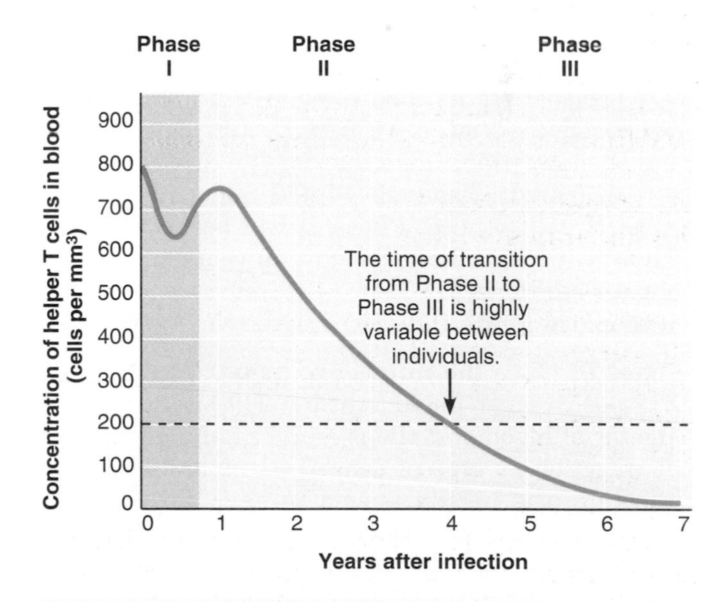

Figure 9.21 Time course of the progression toward AIDS after HIV infection. Phase I is characterized by flulike symptoms followed by apparent recovery. During **Phase II** there is a slow decline in the number of T cells. **Phase III** (AIDS) is characterized by a T cell count of less than 200/mm^3, opportunistic infections, and certain cancers.

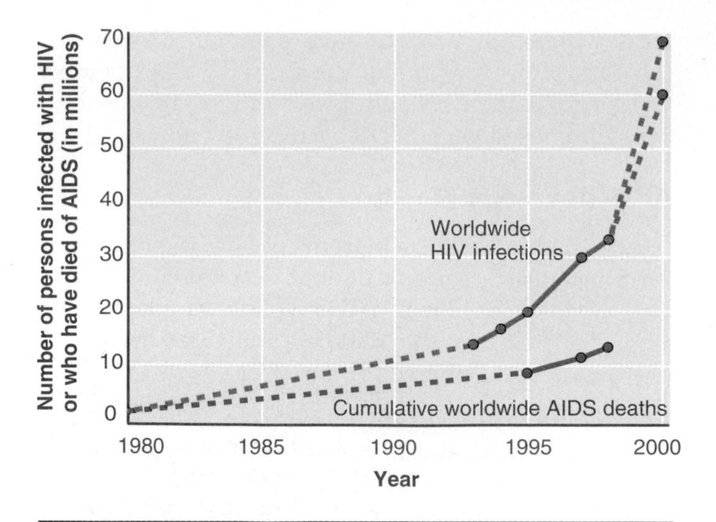

Figure 9.22 Estimated cumulative worldwide HIV infections and cumulative worldwide deaths from AIDS. Dashed lines represent data that were unavailable or are projections into the future.

Phase II In phase II the virus begins to do its damage, wiping out more and more of the helper T cells. The loss of T cells makes the person more vulnerable to *opportunistic infections*, meaning infections that take advantage of the weakened immune system to establish themselves in the body. During phase II a person *may* have persistent or recurrent flulike symptoms, but on the other hand they may have no symptoms at all if they do not have an opportunistic infection. If they have not been tested for HIV antibodies they may not even know they are infected.

Two-thirds to three-quarters of all people who test positive for antibodies to HIV do not exhibit symptoms associated with AIDS. During Phases I and II, many people pass the virus on to others without realizing it. Those they infect may transmit the virus to still others. Phase II can occur within 6 months or may take as long as 10 years or more to progress to Phase III. Left untreated, only about 5% of people in Phase II do not progress to Phase III.

Phase III Once the number of helper T cells of an HIV-positive person falls below 200 per cubic millimeter of blood *and* the person has an opportunistic infection or type of cancer associated with HIV infection, the person is said to have AIDS. Infections and cancers associated with AIDS include pneumonia, meningitis, tuberculosis, encephalitis, Kaposi's sarcoma, and non-Hodgkin's lymphoma, among others.

Notice that AIDS may not appear until years after initial HIV infection. Untreated AIDS is nearly always fatal.

The AIDS epidemic: A global health issue

AIDS was first described in 1981 when the Centers for Disease Control in Atlanta began to notice a disturbing similarity between cases involving a strange collection of symptoms (this illustrates the advantage of having a central clearinghouse for information). It is now believed that HIV first infected humans in the 1960s in Africa after "jumping species" from other primates to humans.

The worldwide damage done by HIV so far is truly astonishing. By the end of 1992 there were already over 14 million HIV-infected people worldwide (although accurate numbers are hard to come by). According to the World Health Organization (WHO), more than 33 million people worldwide are living with HIV infection or AIDS. Fully 65% of them reside in sub-Saharan Africa, where the disease is thought to have originated. The areas of the world where the AIDS epidemic is increasing most rapidly include sub-Saharan Africa, South and Southeast Asia, and Eastern Europe. Current estimates are that 60–70 million persons will be infected with HIV by the year 2000 (Figure 9.22).

The statistics for AIDS in the U.S. are easier to come by and more accurate because all 50 states require diagnosed cases of AIDS to be reported. However, only slightly more than half the states require that infections of HIV be reported, so any numbers for HIV infection are an estimate at best. Figure 9.23 shows the newly diagnosed cases of AIDS each year and deaths per year since 1981.

In the U.S., at least, it appears that the number of new cases and deaths from AIDS have stabilized. Part of the decline in death rates is due to new AIDS treatment drugs that suppress the HIV infection and keep infected people alive much longer. Whether this will just delay the deaths is still uncertain. However, what is happening in the U.S. is not necessarily good news for the rest of the world. The treatment of AIDS is very expensive (over ten thousand dollars per year per person), leaving little hope that those currently infected with HIV in Africa or Asia will ever live to see effective treatment.

Although the overall death rate from AIDS may be declining slightly in this country, the number of diagnosed cases among teenagers and young adults continues to rise. AIDS is now a major killer of Americans ages 25 to 44.

Risky behaviors increase your chances of getting AIDS

Because your behavior will ultimately determine your risk for AIDS, it's worth knowing what consti-

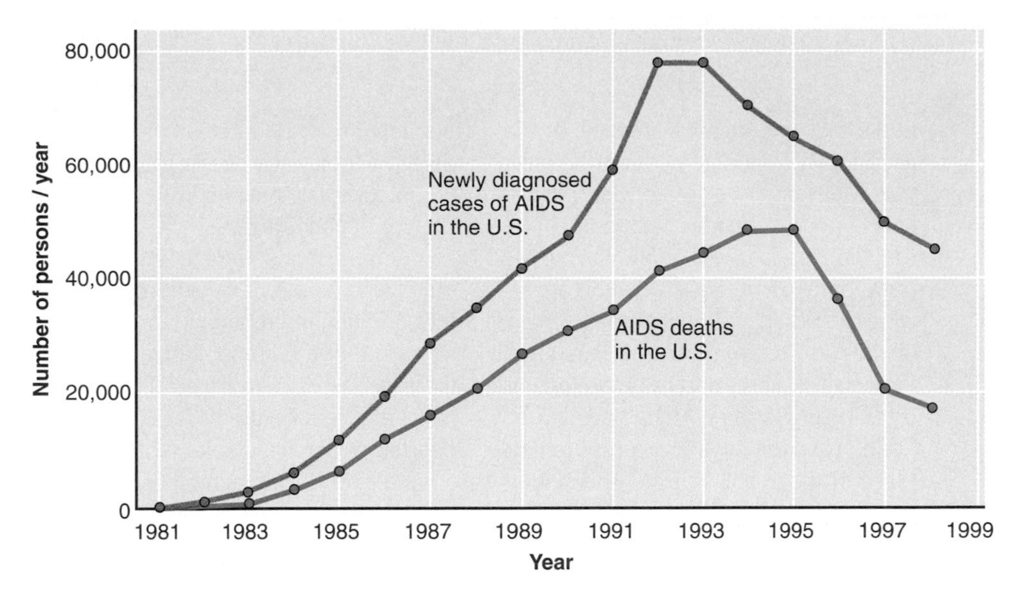

Figure 9.23 The number of newly diagnosed cases of AIDS and the number of deaths from AIDS in the United States for each year since the disease was first reported.

SOURCE: Centers for Disease Control and Prevention, HIV/AIDS Surveillance Report, 1999: vol. 11(1), pp 32, 36; 1997: vol. 9(2), p 19.

tutes "risky behavior" in terms of HIV infection. Table 9.4 shows how HIV is contracted in the U.S. among young people in the 20–24 age group.

By far the most risky behaviors are sharing needles during intravenous drug use, or (if you are male) engaging in sex with other males. Some men mistakenly believe that they cannot get AIDS if their only risky behavior is heterosexual penile-vaginal intercourse. They are wrong. The fact that a low percentage of men contract HIV by heterosexual contact may be due to much higher infection rates in men by drug use or homosexual activity, rather than to a low risk of infection by heterosexual contact per se. The risk of infection by heterosexual contact, especially among women, is on the rise. Note the particularly high risk for women who have sex with HIV-infected men.

Sex can be safer

Given the risks of contracting HIV/AIDS, it would be wise to consider making sex as safe as possible. Below are several *safer sex guidelines* and some of the evidence for them. Note that current suggestions for improving the safety of sex are no guarantee, which is why they are called recommendations for "safer" rather than "safe" sex.

- **Abstain from sex.** Eliminating sexual contact eliminates risk, and some people choose abstinence for ethical as well as practical reasons. However, abstinence may not be considered desirable or practical. In that case, your safest course is to follow as many of the following guidelines as possible.

Table 9.4 AIDS cases in 20–24 year-olds in 1997, by sex and exposure category

Male exposure category:	Percent of total
Men who have sex with other men	55
Injecting drug use	11
Men who have sex with other men and injecting drug use	4
Hemophilia/coagulation disorder	2
Heterosexual contact	7
Receipt of blood transfusions, blood component, or tissue	0
Risk not specified	21
Male total:	**100**
Female exposure category:	
Injecting drug use	1
Hemophilia/coagulation disorder	0
Heterosexual contact:	
Sex with injecting drug user	11
Sex with bisexual male	3
Sex with HIV-infected person, risk not specified	37
Receipt of blood transfusion, blood components, or tissue	1
Risk not reported or identified	34
Female total:	**100**

CurrentIssue

AIDS: An African Crisis, A World Problem

An entire continent is slowly dying. Africa, the birthplace of the human race, is on the verge of collapse. How bad is it, did it happen, how will it affect us, and what (if anything) are we willing to do about it?

In Africa, AIDS is Out of Control

More than 65 percent of the world's HIV-infected people live in sub-Saharan Africa. Researchers estimate that 30 million people in southern Africa are already infected with HIV, with 10 more becoming infected every minute. Life expectancy in the area is projected to decline to just 45 years of age by 2010.

Patterns of AIDS infection and transmission in Africa differ from those of industrialized countries. In the United States, more men than women are infected with HIV, which is often transmitted via homosexual sex. In Africa, an estimated 55 percent of HIV-infected persons are women who have contracted the virus heterosexually. Studies in several African nations have found that females aged 15–19 are five to six times more likely to be infected than males their age. One report ascribes this disparity to older HIV-infected men who coerce or pay impoverished girls for sex in the mistaken belief that sex with a virgin will cure AIDS. Of course, these girls may transmit the virus to their children.

The problem of AIDS in Africa is made worse by a host of other problems. Many sub-Saharan African nations are plagued by severe political, economic, and social instability. Of the area's 42 nations, almost a third are at war. Many economies are weak, sanitation is poor, and even rudimentary medical services are often lacking. Malnutrition and outright starvation are widespread. In this environment, the social fabric of some societies is breaking down. Approximately 11 million children in sub-Saharan Africa have already been orphaned by AIDS, war, or famine. Many of these children are abandoned and impoverished, fending for themselves without adult supervision.

Best for Us, Or Best for Humanity?

What should we do? Do we come first, or do we (as individuals or as a society) have a responsibility to all of humanity? We need to examine where we stand on these questions before we determine the possible actions we might take.

If we come first, then we can probably continue to reduce the risk of contracting AIDS in this country through concerted public education efforts, drug research, and expensive drug treatments. If the goal were only to reduce AIDS in this country, then money spent in this country would probably be as effective as money spent elsewhere. But we will probably not be able to eliminate AIDS in this country because of the prevalence of world-wide travel. Eliminating or severely containing any disease in the future, including AIDS, will require a concerted worldwide effort.

Even if we wished to remain self-absorbed, however, could we ignore the growing AIDS problem in Africa and elsewhere? The answer is almost certainly no. Eventually, an out-of-control worldwide AIDS epidemic would affect our security and our economy, as desperate people take desperate measures. Like it or not, we live in a global village. Recognizing this, the U.S. sometimes chooses to get involved in other regions out of self-interest, if not for humanitarian reasons.

What of the humanitarian issue? Is there an issue of equity or fairness in who gets treatments for AIDS and who does not? Can we just let large populations die, or do we as members of the human species have a responsibility toward other less fortunate human populations? In this regard, the AIDS crisis is just one specific example of a disparity in health and/or economic well-being between populations.

What Can Be Done?

If we do choose to take action, any proposed solution to the growing AIDS crisis in Africa must start with the recognition that the problems in Africa are different than the problems in North America. Meaningful solutions will require concerted efforts on several fronts and at all levels, from personal actions we might take as individuals, to international efforts by consortiums of nations. The possibilities might include (but are not limited to):

- Improving the delivery of health care in Africa
- Providing accurate information on HIV transmission, including helping people understand how their sexual practices may contribute to their risk of infection and to the spread of the disease
- Continuing to seek AIDS treatments and HIV infection preventative measures that are inexpensive enough to be used effectively in poor countries

- Taking political or social action to enable, encourage, or force drug companies to provide AIDS treatment drugs more cheaply
- Providing economic assistance and encouraging political stability in the region

Most of these general approaches are not likely to be easy or cheap. For example, improving the delivery of health care in sub-Saharan Africa may be difficult in rural areas without roads and bridges and in nations with an inadequate number of trained medical personnel. Regarding AIDS specifically, there is the additional problem of even identifying and tracking those with the disease. As discussed in the chapter, people in North America who are suspected of having AIDS generally are tested for HIV, and health professionals maintain careful records documenting their symptoms and treatments. In contrast, in sub-Saharan Africa there is no requirement that suspected AIDS patients be tested for HIV, and health professionals often are unable to monitor AIDS cases effectively or even to pay for the tests required to diagnose AIDS. Other health problems, such as malnutrition, tuberculosis, and malaria, are rampant in the region, and some of their symptoms—weight loss, fever, fatigue—resemble those of AIDS.

What about educational efforts? The kinds of educational efforts used in this country may not work in countries in which the media (radio, newspapers, and television) are not as well developed as they are here. Furthermore, social norms are different and misinformation and denial are widespread. Many researchers advocate developing strategies to change men's sexual behavior. Others propose the creation of broad-based programs of sex education for both boys and girls. Providing male and female condoms at low cost and encouraging their use might also help.

Finally, there is the big hurdle of the cost of AIDS treatment drugs. Recent advances in AIDS treatment depend on new drugs that support the immune system and control HIV infection. These medications can cost $1,000 per month—more than many Africans earn in a year. Pharmaceutical managers point out that a company spends an average of $500 million to develop and test just one new drug. High prices for new AIDS drugs are a direct consequence of the enormous research and development costs that are required by U.S. law to ensure the drugs' safety and effectiveness. In other words, because they have been proven so safe and so effective, they are also out of reach of all but the citizens of the richest nations. At some point we may wish to ask whether this is fair and equitable to all of the world's people. Would we be willing to accept lower development costs (and therefore less assurance of safety and effectiveness in this country) in order to make drugs more widely available to those in need around the world? Should we set a different standard for drugs destined for export than we accept here?

In addition to the costs of research and development, the high cost of drugs is also a consequence of U.S. patent law, which seeks to prevent companies in other countries from copying drugs developed in the U.S and selling them overseas at lower prices. Firms in India, Thailand, and Brazil can manufacture and sell AIDS medicines at one-tenth of the U.S. price, but they do not have to recover development costs. In addition to selling the low-cost drugs abroad, they smuggle them into North America. Pharmaceutical firms protest that, by cutting into profits, the knockoffs endanger future research and the development of even better treatments. They have struck back by taking legal action against distributors and purchasers of illegal AIDS medications. South Africa attempted to manufacture less-expensive copies of U.S. medicines for its own use, but backed down after the federal government threatened to retaliate with trade sanctions. Do you think we should enforce U.S. patent law in foreign countries? Are we willing to alter our patent laws for humanitarian goals?

The problem of AIDS is likely to remain with us for some time, unless a very cheap and effective treatment or preventative measure is discovered soon. We can begin to deal with it now or we can deal with it later.

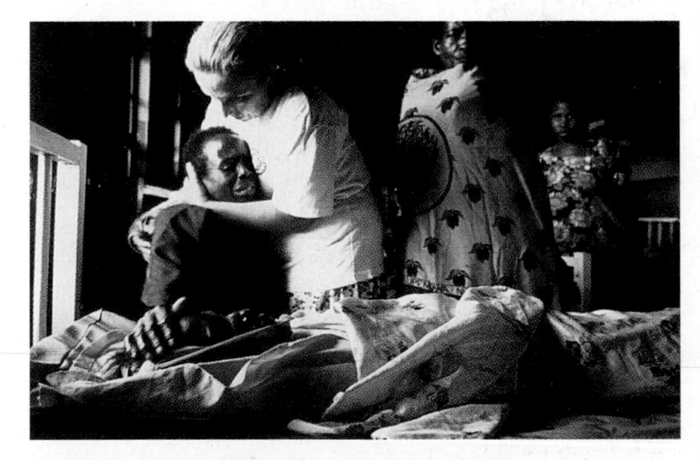

Facilities for treating AIDS in Africa are often inadequate.

- **Reduce the number of sexual partners.** Evidence shows that this is particularly effective among homosexual men. The evidence is not yet conclusive regarding the relationship between HIV infection and number of partners in heterosexual transmission, perhaps because the median number of partners is generally less. Nevertheless, it makes sense that fewer partners translates to less risk.

- **Choose a sexual partner with low-risk behavior.** This is especially important. A partner who has a history of injecting drugs or a man who has sex with other men is a high-risk partner. One study reported that choosing a partner who is not in any high-risk group lowers the risk of AIDS almost 5,000-fold.

- **Avoid certain high-risk sexual practices.** Any sexual behavior that increases the risk of direct blood contact, such as anal-genital sex, should be considered risky. Oral-genital sex is less risky but not risk-free.

- **Use latex or polyurethane condoms or other barriers.** In laboratory tests, latex or polyurethane condoms prevent the passage of HIV. How effective they are in actual use is not certain, but as they are about 90% effective as a birth control method, a reasonable hypothesis (as yet untested) is that they reduce the risk of HIV by about the same amount too. Condoms are effective only if used consistently and correctly. Natural skin condoms are not as safe as latex condoms. The use of dental dams is recommended for oral-vaginal sex.

- **Use nonoxynol 9, a spermicidal agent.** In laboratory tests, nonoxynol 9 has been shown to inhibit the growth of HIV. Be aware, however, that the evidence is conflicting regarding whether nonoxynol 9 offers any protection during sexual contact.

- **Get tested (and have your partner tested).** However, you should know that the HIV tests currently available are designed to detect the human *antibodies* to HIV, not the virus itself. It can take up to six months after exposure to HIV for enough antibodies to be produced for the antibody test to become positive. For this reason, retesting at least six months after the last possible exposure is a good idea. Evidence shows that close to 90% of HIV-infected persons will show a positive antibody test for HIV within six months. A negative test at six months reduces risk substantially, but it does not provide complete assurance. A few individuals may not test positive for three or more years.

New treatments offer hope

As yet there is no cure for AIDS, but researchers are investigating more than a hundred drugs to treat the condition. New antiviral drugs show promise in combating HIV. Some (AZT, ddI, ddC, d4T, and 3TC) inhibit the enzymes the virus needs in order to replicate. Others, such as ritonavir and saquinavir, are protease inhibitors (protease is an enzyme required to assemble viral proteins).

In addition to treating full-blown AIDS, doctors are starting to treat people who are HIV-infected even though they do not yet show the symptoms of AIDS. This approach may prevent the widespread destruction of the immune system that precedes the onset of symptoms. Many health professionals advise people to be tested for HIV so asymptomatic cases can be detected and treated early.

Some researchers believe that safe and effective vaccines offer the only real hope of conquering HIV. However, the production of vaccines is complicated by the fact that HIV mutates rather quickly and there are already several strains of the virus, each of which would need a separate vaccine. In addition, HIV is so dangerous that it could be risky to produce vaccines from whole but weakened viruses. If we had to choose, would we prefer a safe vaccine that doesn't protect against HIV very well, or an effective vaccine that might rarely *cause* HIV infection in the vaccinated person?

Recap **AIDS (Acquired Immune Deficiency Syndrome) is a devastating disorder of the immune system caused by a virus (HIV) that attacks helper T cells. HIV is transmitted in body fluids, typically through sexual contact, blood transfusions, contaminated needles, or breast-feeding.**

Chapter Summary

Pathogens cause disease

- Pathogens include bacteria, viruses, fungi, protozoa, and worms.
- Bacteria have very little internal structure and are covered by a rigid outer cell wall.
- Viruses cannot reproduce on their own. Viral reproduction requires a living host cell.
- The danger from a particular pathogen depends on *how* it is transmitted, how *easily* it is transmitted, and how *damaging* the disease is that it causes.

The lymphatic system defends the body

- The lymphatic system consists of vessels, lymph nodes, the spleen, the thymus gland, and the tonsils.
- The lymphatic system helps protect us against disease.
- Phagocytic cells in the spleen, lymph nodes, and tonsils engulf and kill microorganisms.
- The thymus gland secretes hormones that help T lymphocytes mature.

The body has three lines of defense

- The three lines of defense against pathogens and cellular damage include: (1) physical and chemical barriers, (2) nonspecific defense mechanisms, and (3) specific defense mechanisms.

Physical and chemical barriers: The first line of defense

- Skin is an effective barrier to the entry of microorganisms.
- Vomiting, defecation, and urination physically remove microorganisms after entry.
- Tears, saliva, mucus, and earwax trap organisms and/or wash them away.
- Digestive acid in the stomach kills many microorganisms.

Nonspecific defenses: The second line of defense

- *Phagocytes* surround and engulf microorganisms and damaged cells.
- *Natural killer cells* kill their targets by releasing damaging chemicals.
- Circulating proteins of the *complement system* either kill microorganisms directly or mark them for destruction.
- *Interferons* are proteins that interfere with viral reproduction.
- *Fever* raises body temperature, creating a hostile environment for some microorganisms.

Specific defense mechanisms: The third line of defense

- Cells of the immune system are able to distinguish foreign or damaged cells from our own healthy cells.
- All cells have cell-surface markers called *MHC proteins* that identify the cells as "self."
- An *antigen* is a substance that stimulates the immune system and provokes an immune response.
- B cells produce *antibodies* against foreign antigens.
- T cells of several types release chemicals that enhance the immune response and kill foreign cells directly.

Immune memory creates immunity

- Information about an antigen is stored in memory cells after first exposure.
- The second exposure to the antigen produces a much greater immune response than the first.
- The rapidity of the second response is the basis of immunity from disease.

Medical assistance in the war against pathogens

- Vaccines immunize the body in advance against a particular disease.
- Injected antibodies provide temporary immunity and are of some benefit against an existing infection.
- Monoclonal antibodies are used primarily in medical tests.
- Antibiotics are effective against bacteria, but not against viruses.

Tissue rejection: A medical challenge

- The phenomenon of tissue rejection is a normal consequence of the body's ability to recognize self from nonself.
- Immunosuppressive drugs, the ability to test for various antigens, and organ donor matching programs have increased the success rate of organ transplants in humans.

Inappropriate immune system activity causes problems

- Allergies occur when the immune system responds excessively to foreign particles that are not otherwise harmful.

- Autoimmune disorders develop when a person's immune system attacks the person's own cells as if they were foreign.

Immune deficiency: The special case of AIDS

- AIDS is caused by a virus (HIV).
- The disease (AIDS) can take years to develop after initial HIV infection, but it is nearly always fatal.
- Worldwide, the number of cases of HIV infection and of AIDS are still rising rapidly.
- The chances of contracting AIDS can be reduced (but never eliminated completely) by practicing "safer" sex.

Terms You Should Know

AIDS (acquired immune deficiency syndrome), 222
antibiotics, 218
antibody, 208
antibody-mediated immunity, 208
antigen, 208
bacteria, 196
B cell, 208
cell-mediated immunity, 208
HIV (human immunodeficiency virus), 222
inflammatory response, 205
monoclonal antibodies, 217
pathogen, 196
primary immune response, 215
secondary immune response, 215
T cell, 208
vaccine, 216
virus, 197

Test Yourself

Matching Exercise

_____ 1. A type of cell with no nucleus and very little internal structure
_____ 2. An enzyme in saliva that kills some bacteria
_____ 3. A phagocytic white blood cell best described as a "large eater"
_____ 4. B cells that produce antibodies for secretion into the lymph and plasma
_____ 5. A "self" marker group of proteins found on the surface of every cell
_____ 6. Memory cells are responsible for this phenomenon
_____ 7. A type of cell that secretes histamine
_____ 8. The word for the type of proteins to which antibodies belong
_____ 9. Organ in which damaged red blood cells are removed from blood
_____10. A particular virus that specifically targets helper T cells

a. mast cell
b. spleen
c. major histocompatibility complex (MHC)
d. immunoglobulins
e. prokaryote
f. plasma cells
g. immunity
h. macrophage
i. HIV
j. lysozyme

Fill in the Blank

11. A disease-causing microorganism is called a ______________.
12. Virus-infected cells secrete a group of proteins called ______________ that improve the ability of nearby cells to resist the viral infection.
13. Cells that engulf and digest foreign particles and then display fragments of the foreign antigen on their surfaces to alert T cells are called ______________.
14. Lymphocytes called T cells mature in the ______________.
15. Drugs used to kill bacteria are called ______________.
16. Masses of lymphatic tissue near the entrance to the throat and at the back of the nasal passages are called ______________ and ______________.
17. A substance that causes an allergic reaction is called an ______________.
18. An antigen stimulates B cells to produce a specific ______________.
19. Defective recognition of "self" by the immune system results in an ______________ disorder.
20. Antibodies prepared from a clone of cells derived from a single B cell are called ______________ antibodies.

True or False?

21. A mild fever may be beneficial in fighting an infection.
22. Rheumatoid arthritis is an autoimmune disorder.
23. Passive immunization protects against future infections longer than active immunization.
24. AIDS generally develops in the first week after HIV infection.
25. Left untreated, AIDS is almost always fatal.

Concept Review

1. List the three basic structural forms of bacteria.
2. Describe what is unusual about viruses that cause some people to consider them not "living."
3. Name two areas of the body where acidity discourages bacterial infections.
4. Describe the three functions of the lymphatic system.
5. List the four components of an inflammatory response.
6. Describe in general terms the distinction between *nonspecific* and *specific* defense mechanisms.
7. Show by a drawing how antibodies might cause agglutination of a group of bacteria that carry the complementary antigen.
8. Describe how cells that belong to a particular individual are identified so that the individual's immune system doesn't attack them.
9. Explain how cytotoxic T cells kill target cells.
10. Describe how a vaccine produces immunity from a specific disease.

Apply What You Know

1. Why do you think that, in an otherwise healthy person, acute illnesses due to respiratory infections (colds and the flu) are so much more common than skin infections?
2. When you get a minor infection in a small cut in the skin, it sometimes speeds the healing process to soak it in hot water. How might the heat help?
3. Explain why antibiotics don't work against viruses.
4. Although it is possible to create vaccines against bacteria, antibiotics are more commonly used. What might be the potential advantages and/or disadvantages of a vaccine versus an antibiotic against bacteria?

Case in Point

Interspecies Transplants: From Pigs to People?

In this chapter, we see that organ transplants can save lives. Unfortunately, there are many more people who need transplants than there are available human organs. Thousands of patients die every year while waiting in vain for a donor. Researchers are developing ways to transplant living tissue from pigs and other animals. Interspecies transplants can buy valuable time for desperately ill patients until human tissue becomes available.

Transplantation of pig tissue, while promising, also creates certain problems. The inevitable tissue rejection can often be overcome with immunosuppressive drugs and genetic modification of the pig tissue to incorporate human genes. Another issue involves potential infection by microorganisms in pig tissue. One microbe, porcine endogenous retrovirus (PERV), is embedded in pigs' DNA but apparently kept under control by the animals' immune defenses. Some researchers worry that transplanted PERV could spread to human cells and start dangerous new epidemics. A recent study of 160 patients who had received pig tissue grafts several years before found them free of PERV infection. Although this is encouraging, researchers caution that PERV is only one of many animal viruses.

QUESTIONS:

1. Do you feel the benefits of pig tissue transplants outweigh the risks? Explain your answer.
2. Some animal rights activists oppose interspecies transplants as unethical because the pigs' tissue must be altered with human genes. What is your opinion?

INVESTIGATE:

"Xenotransplantation: No PERVersion." *The Economist,* August 21, 1999, p. 71.

Stephen Moore, "Novartis and CDC Support Case for Pig-Tissue Grafts in Humans." *Wall Street Journal,* August 20, 1999, p. B10.

References and Additional Reading

Garrett, Laurie. *The Coming Plague: Newly Emerging Diseases in a World Out of Balance.* New York: Penguin Books, 1995. 750 pp. (A compelling look at the potential for worldwide epidemics of communicable diseases.)

Levy, Stuart B. The Challenge of Antibiotic Resistance. *Scientific American,* March 1998. (A good popular scientific press review of this subject.)

Preston, Richard. *The Hot Zone.* New York: Anchor Books, 1994. 411 pp. (A gripping true account of an outbreak of Ebola, one of the most dangerous viruses in the world.)

Nevid, Jeffrey S. *Choices: Sex in the Age of STDs.* Boston: Allyn and Bacon, 1998. 180 pp. (A practical guide to making responsible sexual decisions.)

Newman, Jack. How Breast Milk Protects Newborns. *Scientific American,* December 1995. (Cells and antibodies in breast milk protect newborns until their immune system develops.)

Shilts, Randy. *And the Band Played On: Politics, People, and the AIDS Epidemic.* New York: St. Martin's Press, 1987. 630 pp. (The story of the very earliest years of the AIDS epidemic in America.)

CHAPTER 10

THE RESPIRATORY SYSTEM: EXCHANGE OF GASES

Why do we die within minutes without oxygen?

How does air move **into and out of our lungs?**

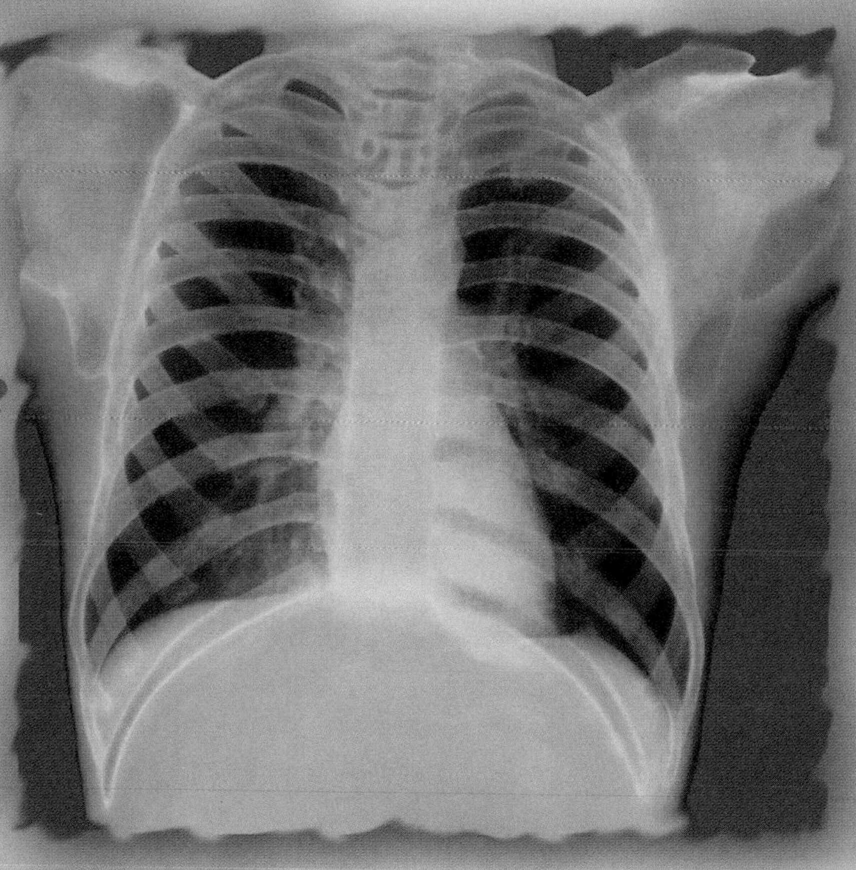

How is our breathing **regulated?**

Why are red blood cells essential to respiration?

Why is the respiratory system so prone to infections?

YOUR DINNER COMPANION IS TURNING BLUE. ONE MINUTE AGO he was laughing and drinking and spearing his steak with enthusiasm, and now he can't breathe or talk, and is frantically pulling at his collar.

Choking is an emergency, but one that is seldom handled by emergency medical personnel. The victim's fate is almost always decided before professional help can arrive. Unless someone in the immediate vicinity can quickly intervene to relieve the choking, there's a good chance the victim won't survive.

A person can live for days without nutrients or water but will die within minutes if denied oxygen. Of all the exchanges of materials between an animal and its environment, the exchange of respiratory gases is almost always the most urgent.

What is the physical basis of this urgency? All cells use energy and certain raw materials in order to survive, grow, and reproduce. To do this, cells rely on aerobic metabolism, meaning that they use up oxygen and create carbon dioxide as a waste product. As discussed in Chapter 3, oxygen is the final electron acceptor in cellular respiration, essential for the production of ATP for most cells. Without oxygen, essential cellular processes cannot continue. Removal of carbon dioxide from the tissues is a pressing concern as well, as its accumulation in the body has many harmful effects.

Why do we die within minutes without oxygen?

The primary function of the **respiratory system** is to exchange these gases (oxygen and carbon dioxide) with the air. We extract oxygen from the air we breathe, and when we exhale we get rid of carbon dioxide, the waste product of our metabolism. The oxygen in the air we breathe comes from plants. Plants absorb carbon dioxide through small pores in their leaves and use it in their own energy-producing process called *photosynthesis*. In the process, they produce oxygen for their (and coincidentally our) use. (We'll further consider this and other examples of the delicate balance of life on Earth in Chapter 22, on ecosystems.)

10.1 Respiration takes place throughout the body

The term **respiration** encompasses four processes:

- *Breathing* (also called **ventilation**): the movement of air into and out of the lungs
- *External respiration:* the exchange of gases between inhaled air and blood
- *Internal respiration:* the exchange of gases between the blood and tissue fluids
- *Cellular respiration:* the process of using oxygen to produce ATP within cells. Cellular respiration generates carbon dioxide as a waste product.

Breathing is facilitated by the respiratory system and its associated bones, muscles, and nerves. External respiration takes place within the lungs, and internal respiration and cellular respiration take place in the tissues throughout the body.

The respiratory system in humans and most animals has another function in addition to gas exchange, and that is the production of sound (vocalization). The production of sound is an important mechanism used by infants to signal their adult caregivers. Sound production has, of course, also proved valuable in information exchange and cultural development, and thus contributes to the long-term survival of our species.

10.2 The respiratory system consists of upper and lower respiratory tracts

The respiratory system (shown in Figure 10.1) consists of (1) a system of passageways for getting air to and from the lungs and (2) the lungs themselves, where gas exchange actually occurs. Also important to respiration are the bones, muscles, and components of the nervous system that cause air to move into and out of the lungs.

For the sake of convenience, the respiratory system can be divided into the upper and lower respiratory tracts. The *upper respiratory tract* comprises the nose (including the nasal cavity) and pharynx—structures above the "Adam's apple" in your neck. The *lower respiratory tract* starts with the larynx and includes the trachea, the two bronchi that branch from the trachea, and the lungs themselves.

The upper respiratory tract filters, warms, and humidifies air

When you inhale, air enters through your nose or your mouth (Figure 10.2). Your *nose* is to be appreciated, as it does more than serve as a passageway for respiration. The nose also:

- Contains receptors for the sense of smell
- Filters inhaled air and screens out some foreign particles
- Moistens and warms incoming air
- Provides a resonating chamber that helps give your voice its characteristic tone

The visible portion of the nose is known as the **external nose.** The internal portion of the nose is

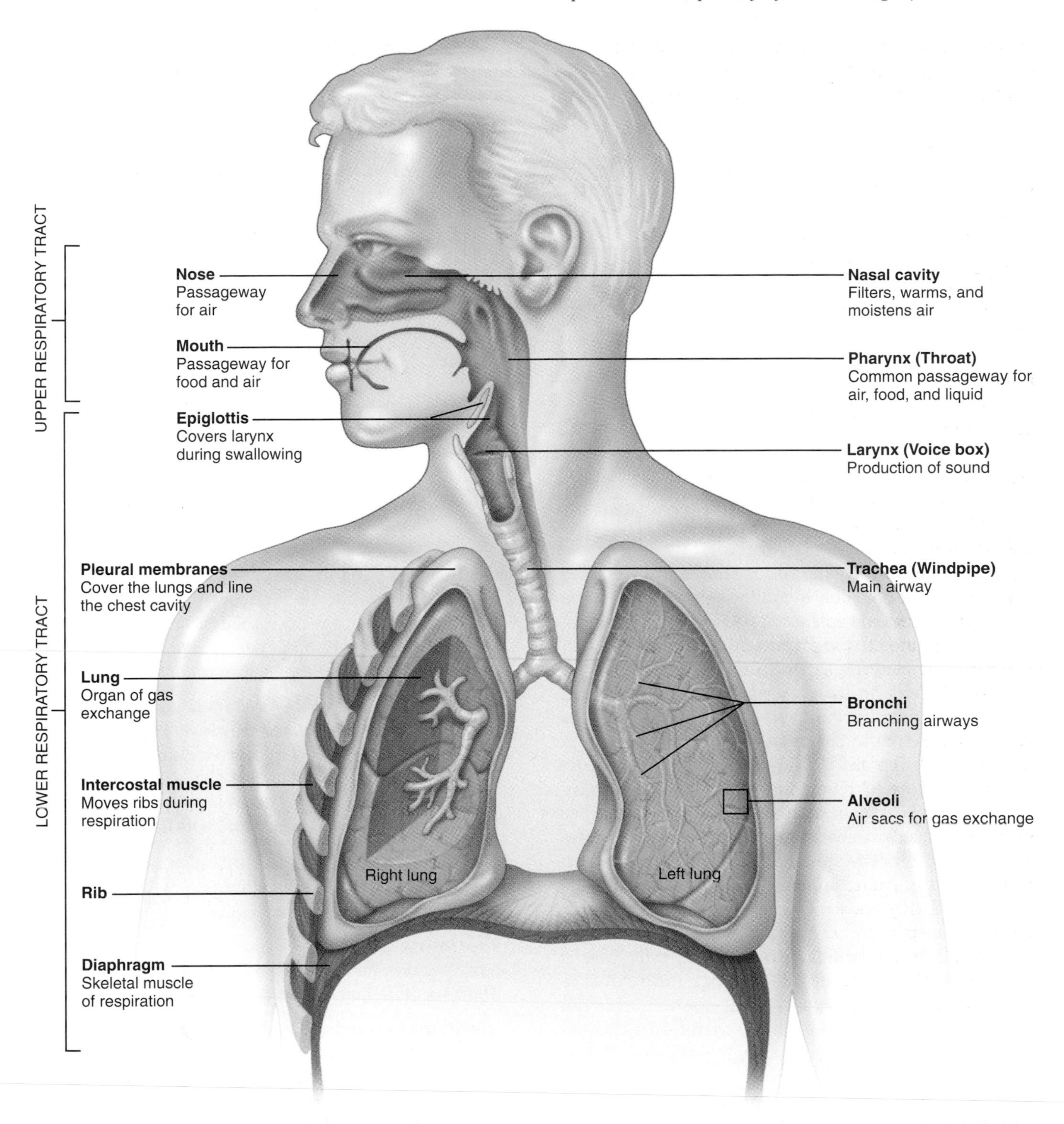

Figure 10.1 The human respiratory system. The functions of each component are listed along with the anatomical structures.

called the **nasal cavity.** The external nose consists of cartilage in the front and two nasal bones behind the cartilage. The nose varies in size and shape from person to person, primarily as a result of individual differences in the cartilage tissue.

The external nose and nasal cavity are divided into two chambers by the nasal septum. Air enters through the nostrils, the two openings at the base of the external nose, where it is partially filtered by nose hairs, then flows into the nasal cavity. This cavity is

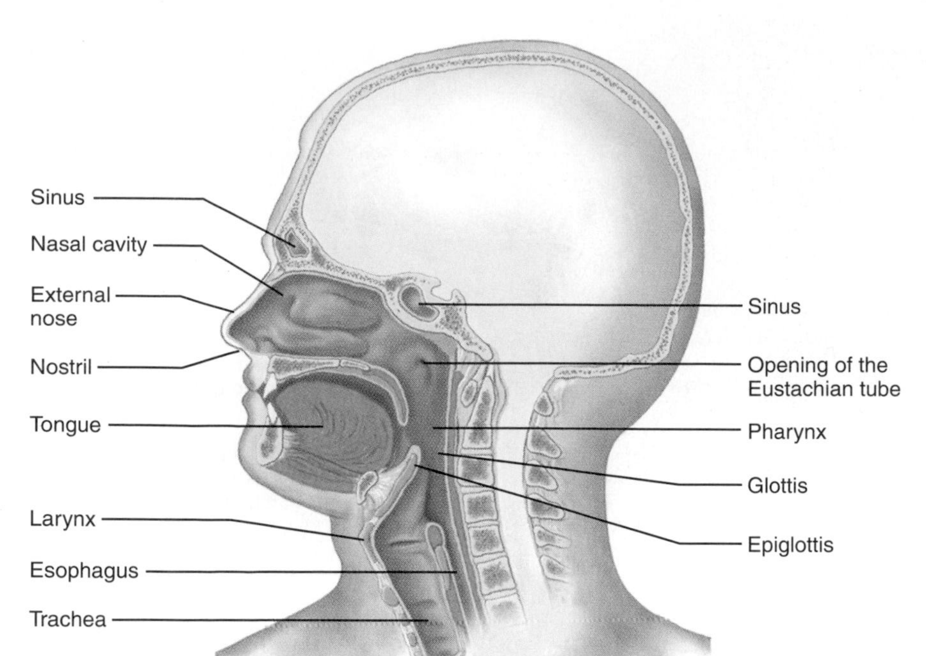

Figure 10.2 Components of the upper respiratory tract as seen in a lateral view of the head and neck.

lined with moist epithelial tissue that is well supplied with blood vessels. The blood vessels help to warm incoming air and the epithelial tissue secretes mucus, which humidifies the air. The epithelium is also covered with tiny hairlike projections called cilia.

The mucus in the nasal cavity traps dust, pathogens, and other particles in the air before they get any farther into the respiratory tract. The cilia beat in a coordinated motion, creating a gentle current that moves the particle-loaded mucus toward the back of the nasal cavity and pharynx. There we cough it out, or swallow it to be digested by powerful digestive acids in the stomach. Ordinarily we are unaware of our nasal cilia as they carry on this important task. However, exposure to cold temperatures can slow down their activity, allowing mucus to pool in the nasal cavity and drip from the nostrils. This is why your nose "runs" in cold weather.

Recall from Chapter 5 that air spaces inside the skull called *sinuses* are also lined with tissue that secretes mucus and helps trap foreign particles. The sinuses drain into the nasal cavity via small passageways. Two tear ducts, carrying fluid away from the eyes, drain into the nasal cavity as well. This is why excess production of tears, perhaps due to strong emotions or irritating particles in your eyes, also makes your nose "runny."

Incoming air next enters the **pharynx** (throat), which connects the mouth and nasal cavity to the larynx (voice box). The upper pharynx extends from the nasal cavity to the roof of the mouth. Into it open the two *Eustachian tubes* that drain the middle ear cavities and equalize air pressure between the middle ear and outside air. The lower pharynx is a common passageway for both food and air. Food passes through on its way to the esophagus, and air flows through to the lower respiratory tract.

> ***Recap*** **The respiratory system is specialized for the exchange of oxygen and carbon dioxide with the air. The upper respiratory tract filters, warms, and humidifies the air.**

The lower respiratory tract exchanges gases

The lower respiratory tract includes the larynx, the trachea, the bronchi, and the lungs with their bronchioles and alveoli (Figure 10.3).

The larynx produces sound The **larynx,** or voice box, extends for about five cm (2 inches) below the pharynx. The larynx serves to:

- Maintain an open airway
- Route food and air into the appropriate channels
- Assist in the production of sound

Spotlight on Structure & Function

The larynx contains two important structures: the epiglottis and the vocal cords (Figure 10.4). The **epiglottis** is a flexible flap of cartilage located at the opening to the larynx. When air is flowing into the larynx, the epiglottis remains open. But when we swallow food or liquids, the epiglottis tips to block the opening temporarily. This "switching mechanism" routes

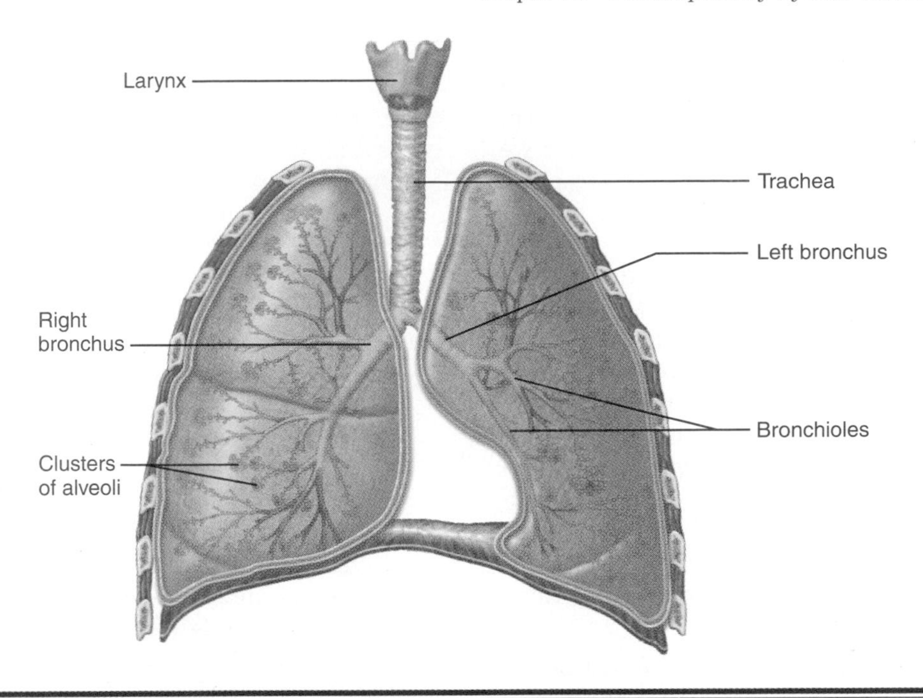

Figure 10.3 Components of the lower respiratory tract. Individual alveoli are too small to be seen clearly in this figure.

food and beverages into the esophagus and digestive system, rather than into the trachea. This is why it is impossible to talk while you are swallowing.

The **vocal cords** consist of two folds of connective tissue that extend across the airway. They surround the opening to the airway, called the **glottis.** The vocal cords are supported by ligaments and enclosed within a cartilaginous structure nicknamed the "Adam's apple."

We produce most sounds by vibration of the vocal cords, although we can also make a few sounds by moving our tongue and teeth. The tone of the sounds

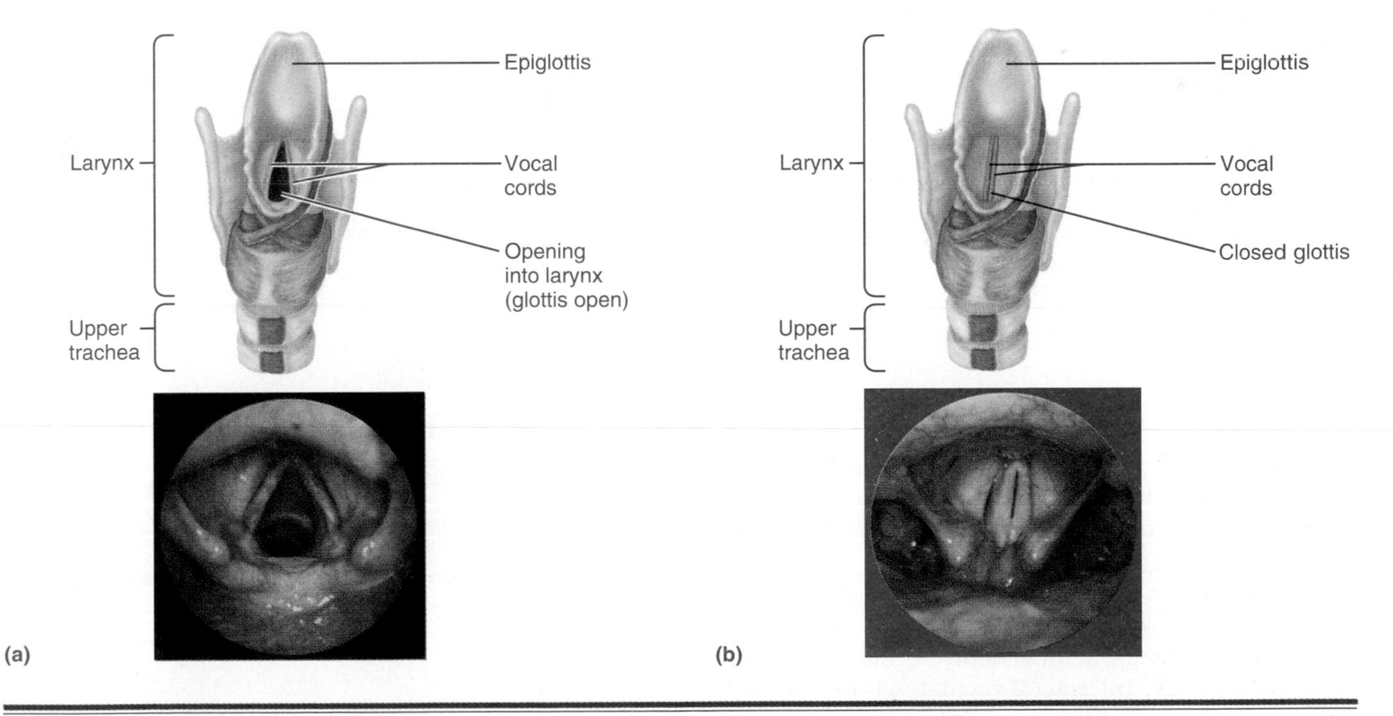

Figure 10.4 Structures associated with the production of sound. (a) Position of the vocal cords during quiet breathing. **(b)** Position of the vocal cords during sound production.

produced by the vocal cords depends on how tightly the vocal cords are stretched, which is controlled by skeletal muscle. When we are not talking, the vocal cords are relaxed and open (Figure 10.4a). When we start to talk they stretch tightly across the tracheal opening, and the flow of air past them causes them to vibrate (Figure 10.4b).

Like any string instrument, cords that are relatively short yield higher-pitched tones than longer cords do. Also, the tighter the vocal cords stretch, the higher the tone they produce. Men tend to have deeper voices than women and a more prominent Adam's apple, due to testosterone that causes the larynx to enlarge at puberty. We can exert some control over the volume and pitch of the voice by adjusting the tension on our vocal cords. (You may have noticed that when you're nervous, your voice becomes higher pitched.) The resulting vibrations travel through the air as sound waves. ■

Most of us can be recognized by the distinctive quality of our voices. In addition to the shape and size of vocal cords, individual differences in voices are determined by many components of the respiratory tract and mouth, including the pharynx, nose and nasal cavity, tongue, and teeth. Muscles in the pharynx, tongue, soft palate, and lips cooperate to create recognizable sounds. The pharynx, nose, and nasal sinuses serve as resonating chambers to amplify and enhance vocal tone.

Speech has played an important role in our evolutionary history and the development of human culture. Although sound production does not play a role in the homeostasis of respiratory gases, animals have evolved to take advantage of the energy available in the moving air to accomplish this important function.

Recap **The larynx maintains an open airway, routes food and air into the appropriate channels, and produces sound. Sound is produced by vibration of the vocal cords as air passes through the glottis.**

The trachea transports air As air continues down the respiratory tract, it passes to the **trachea,** the "windpipe" that extends from the larynx to the left and right bronchi. The trachea consists of a series of C-shaped, incomplete rings of cartilage held together by connective tissue and muscle. As shown in Figure 10.5, each cartilage ring extends only three-quarters of the circumference of the trachea. The rings of car-

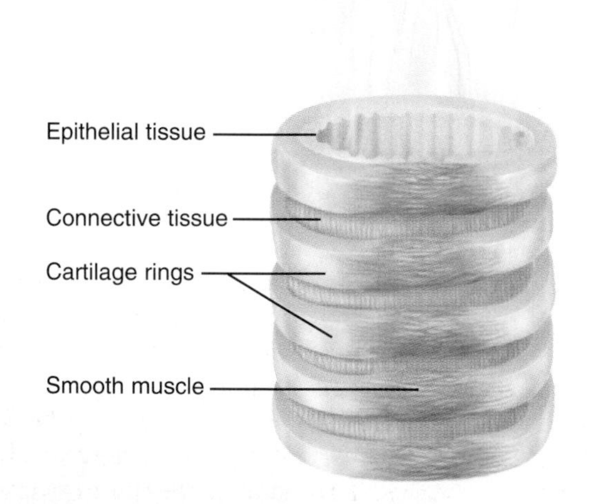

(a) Trachea during normal breathing

(b) Trachea during a cough

Figure 10.5 The trachea. The trachea consists of smooth muscle and layers of epithelial and connective tissue held open by tough, flexible C-shaped bands of cartilage. **(a)** Relaxed state. The maximum diameter facilitates air movement in and out. **(b)** During the cough reflex, the smooth muscle contracts briefly, reducing the diameter of the trachea. Combined with contraction of the abdominal muscles, this increases the velocity of air movement, forcibly expelling irritants or mucus in the trachea.

Health Watch

Saving Lives with the Heimlich Maneuver

Every year thousands of people choke to death when a piece of food lodges in the larynx or trachea. Many of these deaths could be prevented by the Heimlich maneuver.

The Heimlich maneuver is a method for dislodging objects that completely block the airway. It is an emergency procedure for saving a life. Do not try it on someone who is having difficulty breathing but is still getting some air in and out—you could shift the obstacle to a position where it does block the airway completely. People whose airways are completely blocked will appear agitated. They may signal that they are choking by grasping their throat or gesturing with their hands, but they will not be able to speak.

Without startling the person, approach from behind. Make a fist of one hand with your thumb inward, place it below the rib cage but above the navel, and grasp your fist with the other hand. Then pull sharply upward and inward.

The Heimlich maneuver should be done firmly and abruptly to force the obstruction up the trachea. It's best to learn the technique from a qualified instructor, because you need to exert enough force to push the diaphragm and lungs upward, creating air pressure in the lungs to expel the object. A little light pressure won't help. Furthermore, the person may panic and struggle against you. However, if done correctly, the Heimlich maneuver can save a life.

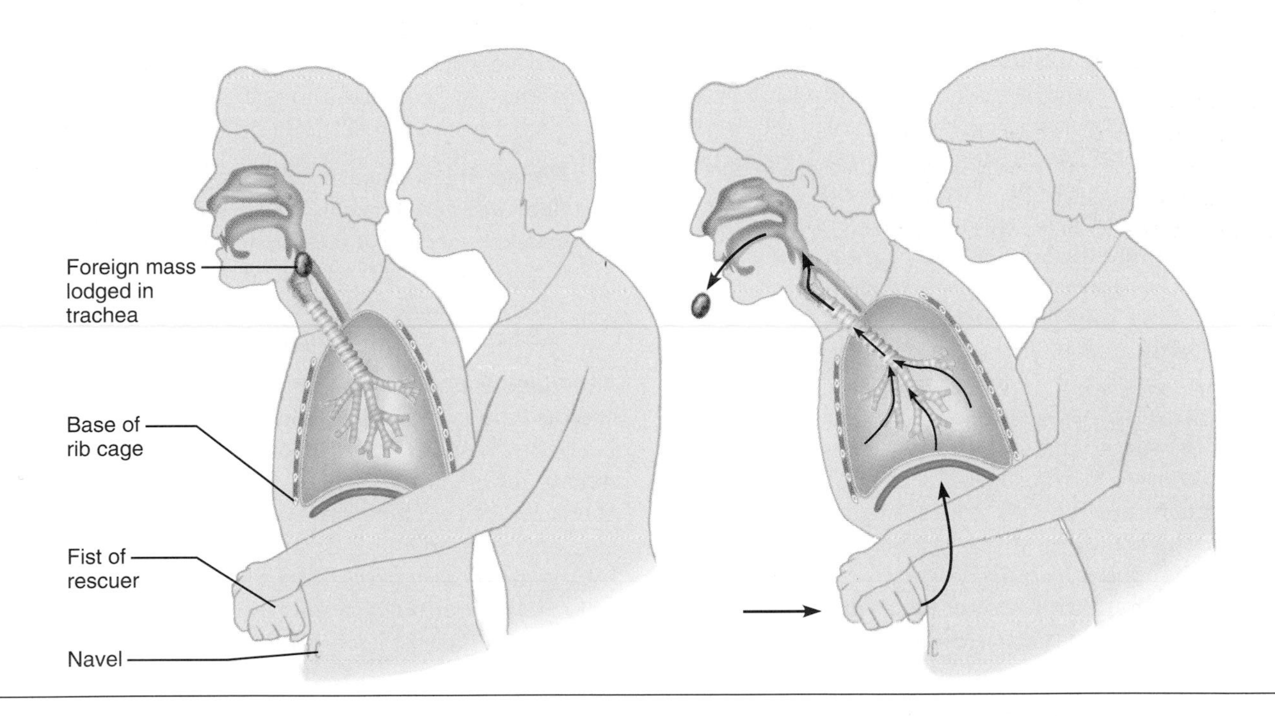

tilage keep the trachea open at all times, but because they are not complete circles they permit the trachea to change diameter slightly when we cough or breathe heavily.

Like the nasal cavity, the trachea is lined with cilia-covered epithelial tissue that secretes mucus. The mucus traps foreign particles and the cilia move them upward, away from the lungs.

If a foreign object lodges in the trachea, respiration is interrupted and choking occurs. If the airway is completely blocked, death can occur within minutes. Choking often happens when a person carries on an animated conversation while eating. Beyond good manners, the risk of choking provides a good reason not to eat and talk at the same time.

Choking typically stimulates receptors in the throat that trigger the *cough reflex.* This is a sudden expulsion of air from the lungs in an attempt to dislodge foreign material (Figure 10.5b). If the object blocks the airway completely before the person has finished inhaling, there may not be much air in the lungs. This will make the obstacle more difficult to remove. If the object blocks airflow only partially, it may be possible to dislodge it by inhaling slowly, then coughing.

Bronchi branch into the lungs The trachea branches into two airways called the right and left **bronchi** (singular *bronchus*) as it enters the lung cavity (refer to Figure 10.3). Like the branches of a tree, the two bronchi divide into a network of smaller and smaller

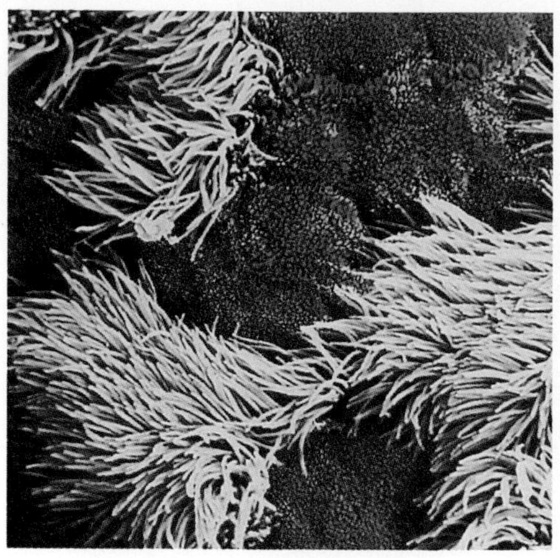

(a)

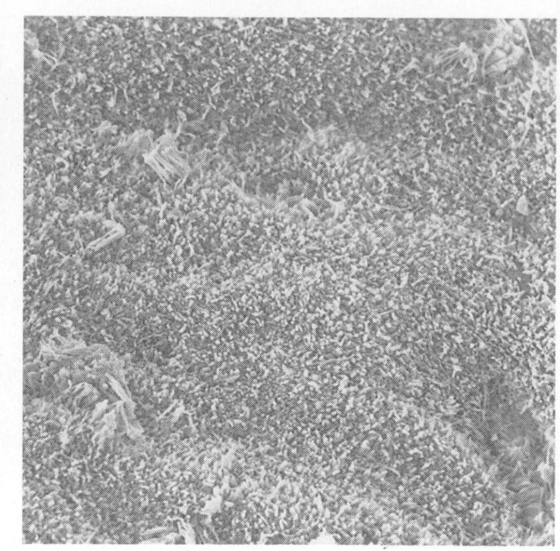

(b)

Figure 10.6 Effects of smoking on the cilia of the airways. (a) The inner surface of an airway of a normal person, showing normal ciliated epithelial cells with their clumps of hairlike cilia. The cells without cilia are the mucus-secreting cells. **(b)** The inner surface of an airway of a smoker, showing destruction of the cilia.

bronchi. The walls of bronchi contain fibrous connective tissue and smooth muscle reinforced with cartilage tissue. As the branching airways get smaller and smaller the amount of cartilage declines. By definition, the smaller airways that lack cartilage are called **bronchioles.** The smallest bronchioles are 1 mm or less in diameter and consist primarily of a thin layer of smooth muscle surrounded by a small amount of elastic connective tissue.

The bronchi and bronchioles have several other functions in addition to air transport. They also clean the air, warm it to body temperature, and saturate it with water vapor before it reaches the delicate gas-exchange surfaces of the lungs. The air is warmed and humidified by contact with the moist surfaces of the cells lining the bronchi and bronchioles. With the exception of the very smallest bronchioles, the bronchi and bronchioles are lined with ciliated epithelial cells and occasional mucus-secreting cells. These cells function to trap particles and then sweep them upward toward the pharynx so that they can be swallowed.

Tobacco smoke contains chemicals and particles that irritate the respiratory tract. Mucus production increases in response, but the smoke impairs the activity of the cilia. With continued smoking, the cilia become destroyed, and mucus and debris from the smoke accumulate in the airway. "Smoker's cough" refers to the violent coughing necessary to dislodge the mucus from the airway. Mucus pooling leads to frequent infections because pathogens and irritants remain in the respiratory tract. It also increases the risk of bronchitis, emphysema, and lung cancer (Figure 10.6) (see "Disorders of the Respiratory System" later in this chapter).

From the nose and mouth to the tiniest bronchioles in the lungs, none of the airways we have described so far participate in gas exchange. Essentially, they are all tubes for getting air to the lungs where gas exchange actually occurs.

Recap **The trachea, or "windpipe," branches into the right and left bronchi, which further subdivide into smaller bronchi and then bronchioles. Like other areas of the respiratory tract, the bronchi and bronchioles filter, warm, and humidify the incoming air before it reaches the gas exchange structures in lungs.**

The lungs are organs of gas exchange The **lungs** are organs consisting of supportive tissue enclosing the bronchi, bronchioles, blood vessels, and the areas where gas exchange occurs. They occupy most of the thoracic cavity. There are two lungs, one on the right side and one on the left, separated from each other by the heart (Figure 10.7). The shape of the lungs follows the contours of the rib cage and the thoracic cavity. The base of each lung is broad and shaped to fit against the convex surface of the diaphragm.

Each lung is enclosed in two layers of thin epithelial membranes called the **pleural membranes.** One of these layers represents the outer lung surface and the other lines the thoracic cavity. The pleural membranes are separated by a small space, called the **pleural cavity,** that contains a very small amount of watery fluid. The fluid reduces friction between the pleural membranes as the lungs and chest wall move during breathing. Inflammation of the pleural membranes, a condition called pleurisy, can result in a reduced secretion of pleural fluid, an increase in friction, and pain during breathing. Pleurisy can be a symptom of pneumonia (see "Disorders of the Respiratory System").

Each lung consists of several lobes, three in the right lung and two in the left. Each lobe contains a branching tree of bronchioles and blood vessels. The

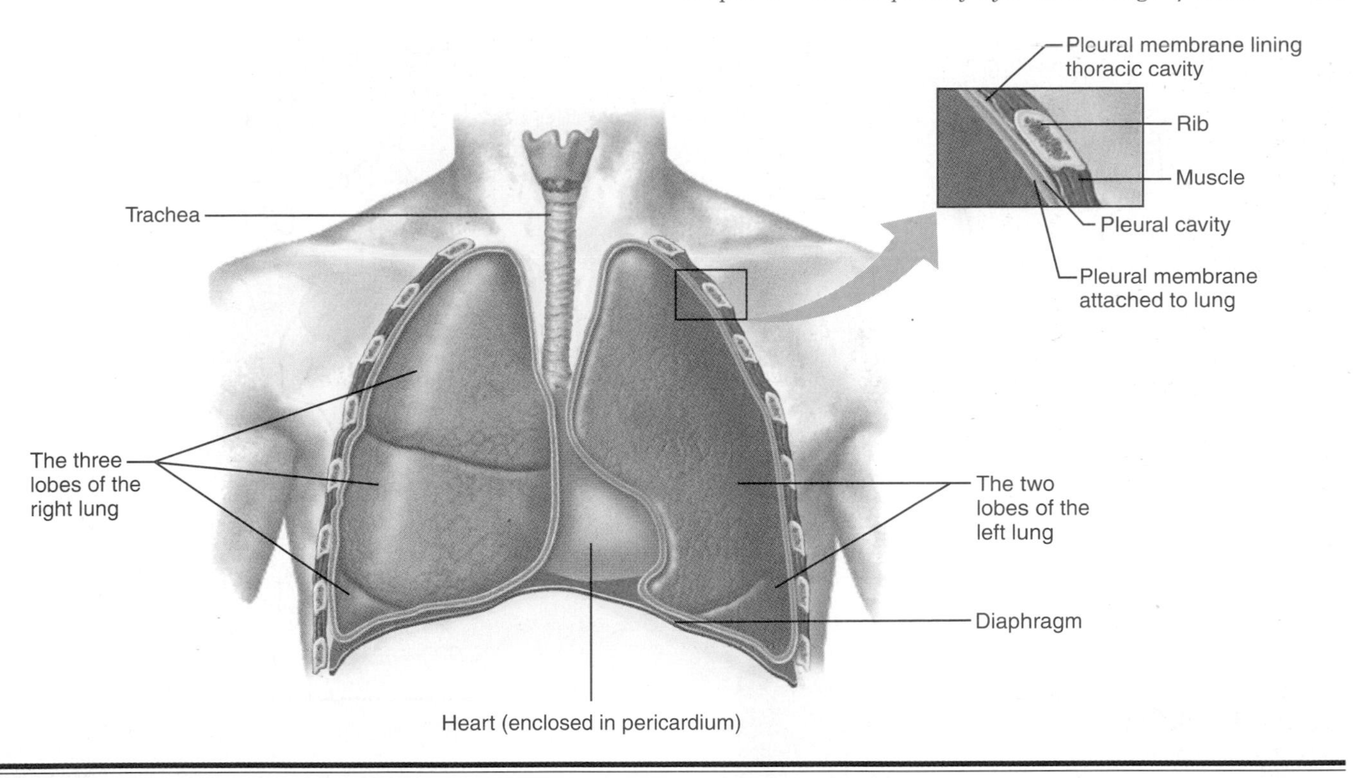

Figure 10.7 The lungs, the pleural membranes, and the pleural cavity. The pleural cavity has been expanded so that it can be seen in the drawing. In reality, it is no more than a very thin, watery space that reduces friction between the two pleural membranes.

lobes can function fairly independently of each other, so it is possible to surgically remove a lobe or two without totally eliminating lung function.

Gas exchange occurs in alveoli If you could touch a living lung, you would find that it is very soft and frothy. In fact, most of it is air. The lungs are basically a system of branching airways that end in 300 million tiny air-filled sacs called **alveoli** (singular *alveolus*). It is here that gas exchange takes place (Figure 10.8). Alveoli are arranged in clusters at the end of every terminal bronchiole, like grapes clustered on a stem. A single alveolus is a thin bubble of living squamous epithelial cells only one cell thick. Their combined surface area is nearly 800 square feet, approximately 40 times the area of our skin. This tremendous surface area and the thinness of the squamous type of epithelium facilitates gas exchange with nearby capillaries.

Within each alveolus, certain epithelial cells secrete a lipoprotein called *surfactant* that coats the interior of the alveoli and reduces surface tension. Surface tension is due to the attraction of water molecules toward each other (review Chapter 2). Without surfactant, the force of surface tension could collapse the alveoli. This can occur in infants who are born prematurely, because the surfactant-secreting cells in their lungs are underdeveloped. Called *infant respiratory distress syndrome,* the condition is treated with surfactant replacement therapy.

Pulmonary capillaries bring blood and air into close contact

Spotlight on Structure & Function Recall from Chapter 8 that the right ventricle of the heart pumps deoxygenated blood into the pulmonary artery. The pulmonary artery divides into smaller and smaller arteries and arterioles, eventually terminating in a capillary bed called the *pulmonary capillaries.* In the pulmonary capillaries, blood comes into very close contact with the air in the alveoli. Only two living cells (the squamous epithelial cell of the alveolus and the cell of the capillary wall) separate blood from air at this point. ■

Note that blood always travels through the pulmonary capillaries, where gas exchange takes place, every time it passes through the heart. A series of venules and veins collects the oxygenated blood from the pulmonary capillaries and returns the blood to the left side of the heart. From there it can be transported to all parts of the body.

Recap **The lungs are organs containing a branching system of bronchi and bronchioles, blood vessels, and 300 million alveoli. Gas exchange occurs between the alveoli and pulmonary capillaries.**

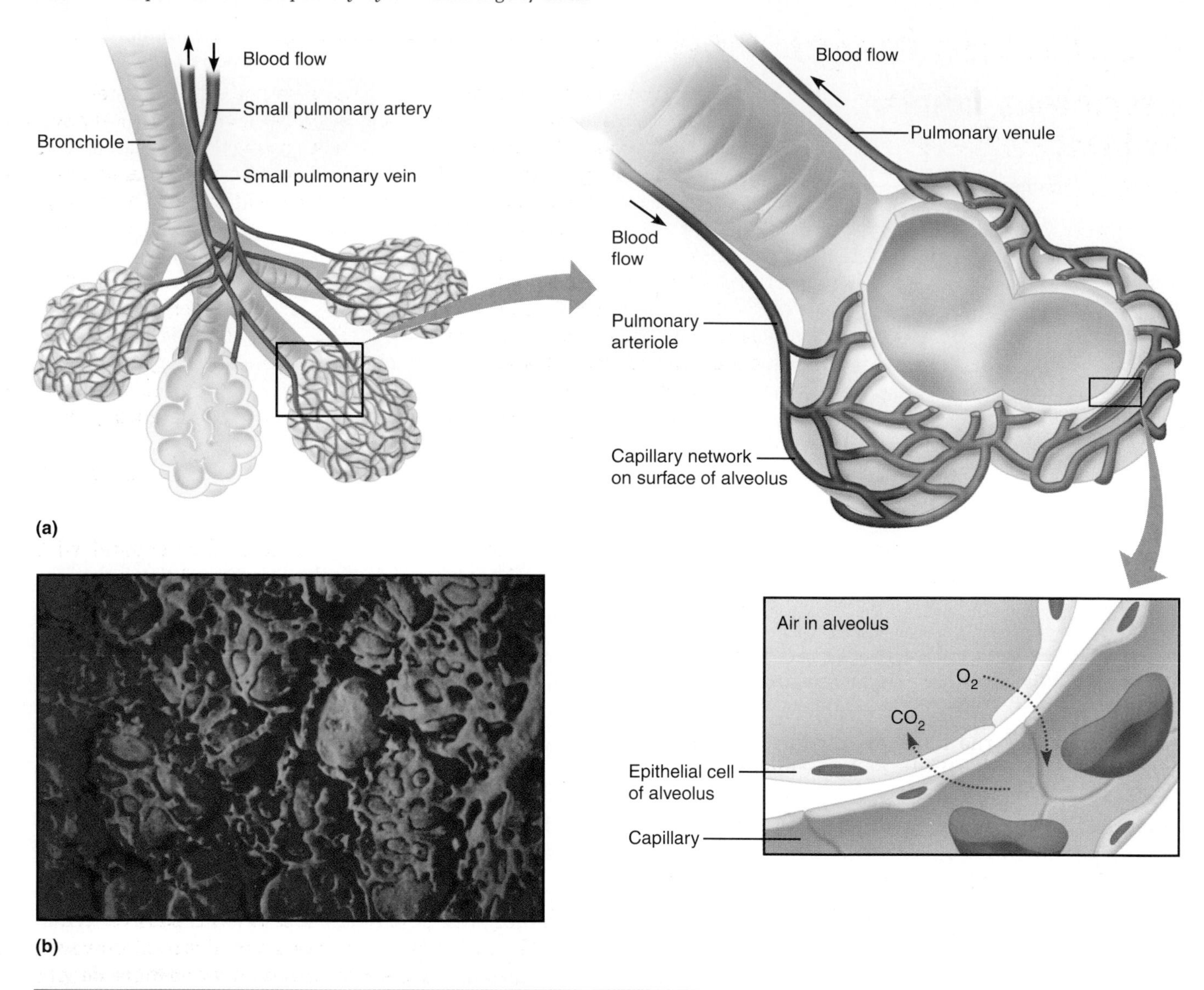

Figure 10.8 Gas exchange between the blood and alveoli. (a) Bronchioles end in clusters of alveoli, each surrounded by capillaries. CO_2 and O_2 are exchanged across the capillary and alveolar walls by diffusion. **(b)** Photo of the surface of alveoli covered with capillaries.

10.3 The process of breathing involves a pressure gradient

Breathing involves getting air into and out of the lungs in a cyclic manner, and that requires muscular effort. However, the lungs themselves don't have any skeletal muscle tissue. The lungs expand passively because the surrounding bones and muscles expand the size of the chest cavity.

The bones and muscles of respiration include the ribs, the **intercostal muscles** between the ribs, and the main muscle of respiration called the **diaphragm,** a broad sheet of muscle that separates the thoracic cavity from the abdominal cavity (review Figure 10.1). The intercostal muscles and the diaphragm are skeletal muscles.

Inspiration brings in air, expiration expels it

To understand why air moves into and out of the lungs in a cyclic manner, we need to understand some general principles of gas pressure and of how gases move:

- Gas pressure is caused by colliding molecules of gas.
- When the volume of a closed space *increases,* the molecules of gas in that space are farther away

Directions in Science

Respiratory Inhalers as Drug Delivery Systems

One challenge that drug companies face is getting their products into patients. Drugs containing peptides and proteins are especially challenging because they generally are broken down by digestive acids in the stomach before they can be absorbed. Injections are another alternative, but they are inconvenient, painful, and expensive.

In recent years many companies have been investigating the possibility of using the lungs as a drug delivery system. One advantage of the lungs is that they have a large surface area and a thin barrier to diffusion. Furthermore, they don't contain all those protein-digesting enzymes found in the stomach. Asthmatics have been using inhalers for years to administer drugs that open the bronchi and bronchioles. Could the respiratory tract be used to deliver medications aimed at other body systems as well?

Currently pharmaceutical researchers are developing delivery systems in which medications are first dissolved in liquid, then inhaled as a fine mist of liquid droplets. An allergy medication that is a steroid can now be administered this way. Other researchers are experimenting with the inhalation of dry powders. One of the first major drugs to be delivered by one of these methods may be insulin, because the potential market is large (over 1 million diabetics require daily injections of insulin).

from each other, and the pressure inside the space *decreases*. Conversely, when the volume in a closed space *decreases*, the gas pressure *increases*.

- Gases flow from areas of *higher* pressure to areas of *lower* pressure.

As we have seen, the lungs are air-filled structures consisting almost entirely of bronchioles, alveoli, and blood vessels. Lacking skeletal muscle, they cannot expand (increase in volume) or contract (decrease in volume) on their own. The lungs expand and contract only because they are compliant (stretchable) and because they are surrounded by the pleural cavity, which is airtight and sealed. If the volume of the pleural cavity expands, the lungs will expand with it.

Inspiration (inhalation) pulls air into the respiratory system as lung volume expands, and *expiration* (exhalation) pushes air out as lung volume declines again. Let's look at a cycle of inspiration and expiration, starting from the relaxed state at the end of a previous expiration (Figure 10.9):

1. *Relaxed state.* At rest, both the diaphragm and the intercostal muscles are relaxed. The relaxed diaphragm appears dome-shaped.
2. *Inspiration.* As inspiration begins, the diaphragm contracts, flattening it and pulling its center downward. At the same time, the intercostal muscles contract, pulling the ribs upward and outward. These two actions of skeletal muscle increase the volume of the pleural cavity and lower the pressure within the pleural space. Because the lungs are elastic and the pressure around them has just fallen relative to the atmosphere, they expand with the pleural cavity. Expansion of the lungs reduces air pressure within the lungs relative to the atmosphere, allowing air to rush in.
3. *Expiration.* Eventually the muscle contractions end. As the muscles relax the diaphragm returns to its domed shape, the ribs move downward and inward, and the pleural cavity becomes smaller. The rest of the process reverses as well. The lungs become smaller, so pressure within the lungs rises relative to the atmosphere and air flows out.

How does air move into and out of our lungs?

During quiet breathing, inspiration is active (requiring muscular effort) and expiration is passive. When we are under physical or emotional stress, however, we need to breathe more frequently and more deeply. At this point both inspiration and expiration may become active. We can take bigger breaths because additional rib cage muscles raise the rib cage higher. As we exhale deeply, abdominal muscles contract and push the diaphragm even higher into the thoracic cavity, and the inner intercostal muscles contract to pull the rib cage downward. These events combine to increase the speed and force of respiration.

We also exhale forcibly when we sneeze or cough. During sneezing and coughing the abdominal muscles contract suddenly, raising abdominal pressure. The rapid increase in abdominal pressure pushes the relaxed diaphragm upward against the pleural cavity, forcing air out of the lungs.

Recap **Inspiration is an active process (requiring energy) that occurs when the diaphragm and intercostal muscles contract. Normally expiration is passive, but it can become active when we forcibly exhale, cough, or sneeze.**

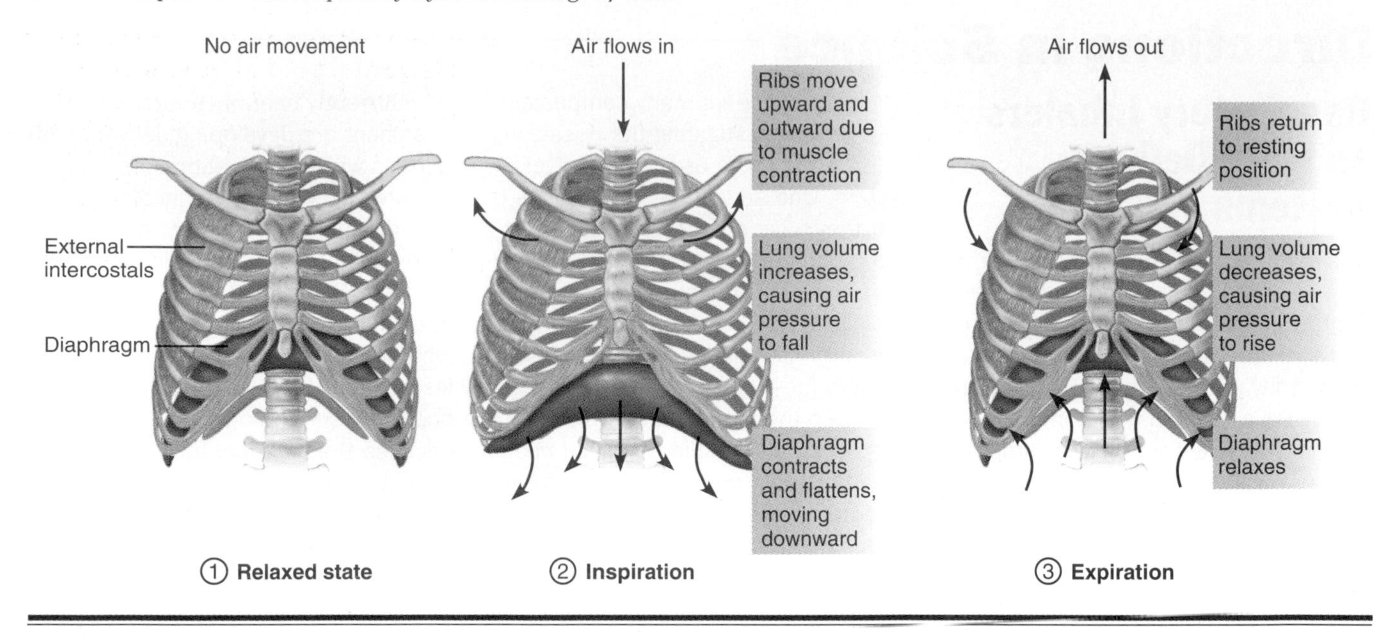

Figure 10.9 The respiratory cycle. During inspiration, the ribs move upward and outward, enlarging the thoracic cavity. During expiration, the muscles relax and the ribs return to their relaxed position.

Lung volumes and vital capacity measure lung function

At rest, you take about 12 breaths every minute. Each breath represents a **tidal volume** of air of approximately 500 milliliters, or about a pint. On average, only about 350 ml of each breath actually reach the alveoli and become involved in gas exchange. The other 150 ml remain in the airways, and because it does not participate in gas exchange, this air is referred to as *dead space volume.*

The maximal volume that you can exhale after a maximal inhalation is called your **vital capacity.** Your vital capacity is about 4,800 ml, almost ten times your normal tidal volume at rest. The amount of additional air that can be inhaled beyond the tidal volume (about 3,100 ml) is called the *inspiratory reserve volume,* and the amount of the air that we can forcibly exhale beyond the tidal volume (about 1,200 ml) is the *expiratory reserve volume.* No matter how forcefully you exhale, some air always remains in your lungs. This is your *residual volume,* approximately 1,200 ml.

Measurements of lung capacity are performed with a device called a *spirometer* (Figure 10.10). The measurements are made by having a person breathe normally into the device (to measure tidal volume) and then take a maximum breath and exhale it forcibly and as completely as possible (to measure vital capacity). Lung volumes and rates of change of volume are useful in diagnosing various lung diseases. For example, emphysema is a condition in which a loss of elasticity of the smaller airways causes them to collapse during expiration, impairing the ability to exhale naturally. A spirometer recording of someone with emphysema might show a prolonged period of expiration after a maximum inspiration because of resistance to air outflow.

Recap **Although we normally take breaths of about 500 ml (tidal volume), our vital capacity is about 4,800 ml. Some air, called the residual volume, remains in the lung even at the end of expiration.**

10.4 Gas exchange and transport occur passively

So far, we have focused on the first process in respiration: breathing. Once air enters the alveoli, gas exchange and transport occur. In this section we review some basic principles governing the diffusion of gases to set the stage for our discussion of external and internal respiration (the second and third processes of respiration). We also describe how the gases are transported by blood. The fourth process of respiration, cellular respiration, is the use of oxygen by cells in the production of energy. Cellular respiration was described in Chapter 3.

Gases diffuse according to their partial pressures

The Earth is surrounded by an atmosphere of gases. Like liquids, gases have mass and are attracted to the earth by gravity. Though it doesn't really feel like we are pressed down by a heavy weight of gases,

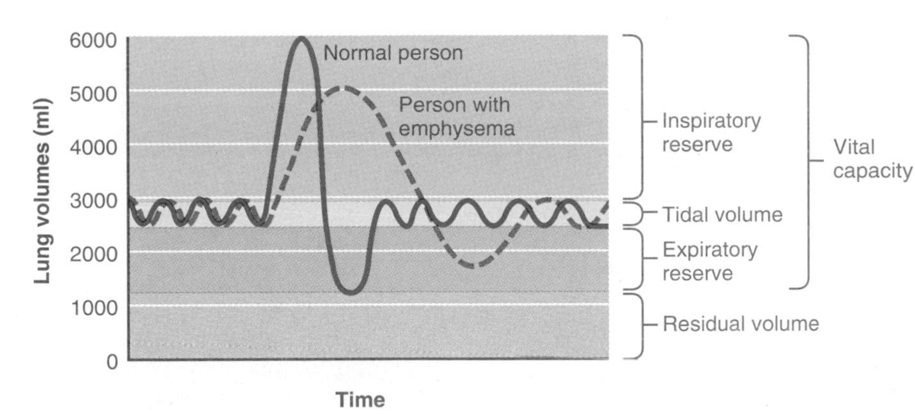

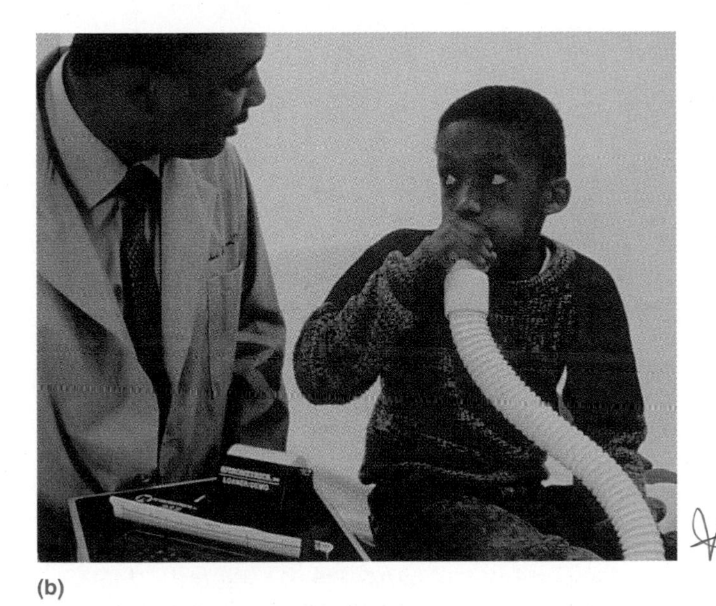

Figure 10.10 Measurement of lung capacity. (a) A recording of lung capacity. After several normal breaths, the person inhales and then exhales maximally. The volumes indicated by the solid line are for a normal person. The dashed line is typical of a patient with emphysema. **(b)** A patient having his lung capacity determined with a spirometer.

in fact the atmosphere (air) exerts a total atmospheric pressure at sea level of about 760 mm Hg (millimeters of mercury). A normal atmospheric pressure of 760 mm Hg means that the pressure of the atmosphere will cause a column of mercury in a vacuum to rise 760 mm, or about 2 feet. The pressure seems like zero to us because the pressure inside our lungs is the same as atmospheric pressure, at least when we are resting between breaths.

The primary gases of the earth's atmosphere are nitrogen (78%) and oxygen (21%), with a trace amount of carbon dioxide (about 0.04%) and less than one percent of all other gases combined. In a mixture of gases, each gas exerts a **partial pressure** that is proportional to its percent of the total gas composition. Partial pressure is represented by "P" and, like atmospheric pressure, it is measured in mm Hg. The pressure of the atmosphere is thus the sum of the partial pressures of each of the gases found in the atmosphere at sea level. Because we know the percentages of each gas in our atmosphere, we can find the partial pressure of each. For example, the partial pressure of oxygen (P_{O_2}) in air is about 160 mm Hg (760 mm Hg × 0.21).

Because partial pressures in a mixture of gases are directly proportional to concentrations, a gas will always diffuse down its *partial pressure gradient,* from a region of higher partial pressure to a region of lower partial pressure. As we shall see, the exchanges of O_2 and CO_2, both between the alveoli and the blood and between the blood and the tissues, are purely passive. No consumption of ATP is involved; changes in partial pressures are entirely responsible for the exchange and transport of these gases.

External respiration: The exchange of gases between air and blood

A comparison of the partial pressures of O_2 and CO_2 in inhaled air, in the alveoli, and in the blood in the lungs illustrates how external respiration takes place (Figure 10.11). As stated above, the partial pressure of the oxygen (P_{O_2}) in the air we breathe is about 160 mm Hg. As there is very little CO_2 in the air we breathe, the partial pressure of inspired CO_2 is negligible. The partial pressures of alveolar air are not, however, the same as those of inspired air. This is because only about 1/8 of the air is actually exchanged with each breath, so most of the air in the lungs is actually "old" air that has already undergone some gas exchange. Consequently, the partial pressures of O_2 and CO_2 in the alveoli average about 104 and 40 mm Hg, respectively.

When venous (deoxygenated) blood with a P_{O_2} of only 40 mm Hg and a P_{CO_2} of 46 mm Hg arrives at the pulmonary capillaries from the pulmonary arteries, O_2 diffuses from the alveoli into the capillaries and CO_2 diffuses in the opposite direction. As a result, the P_{O_2} of oxygenated (arterial) blood leaving the lungs rises to 100 mm Hg and the P_{CO_2} falls to 40 mm Hg. The oxygenated blood is carried in the pulmonary veins to the heart and then throughout the body in the arterial blood vessels. The CO_2 that

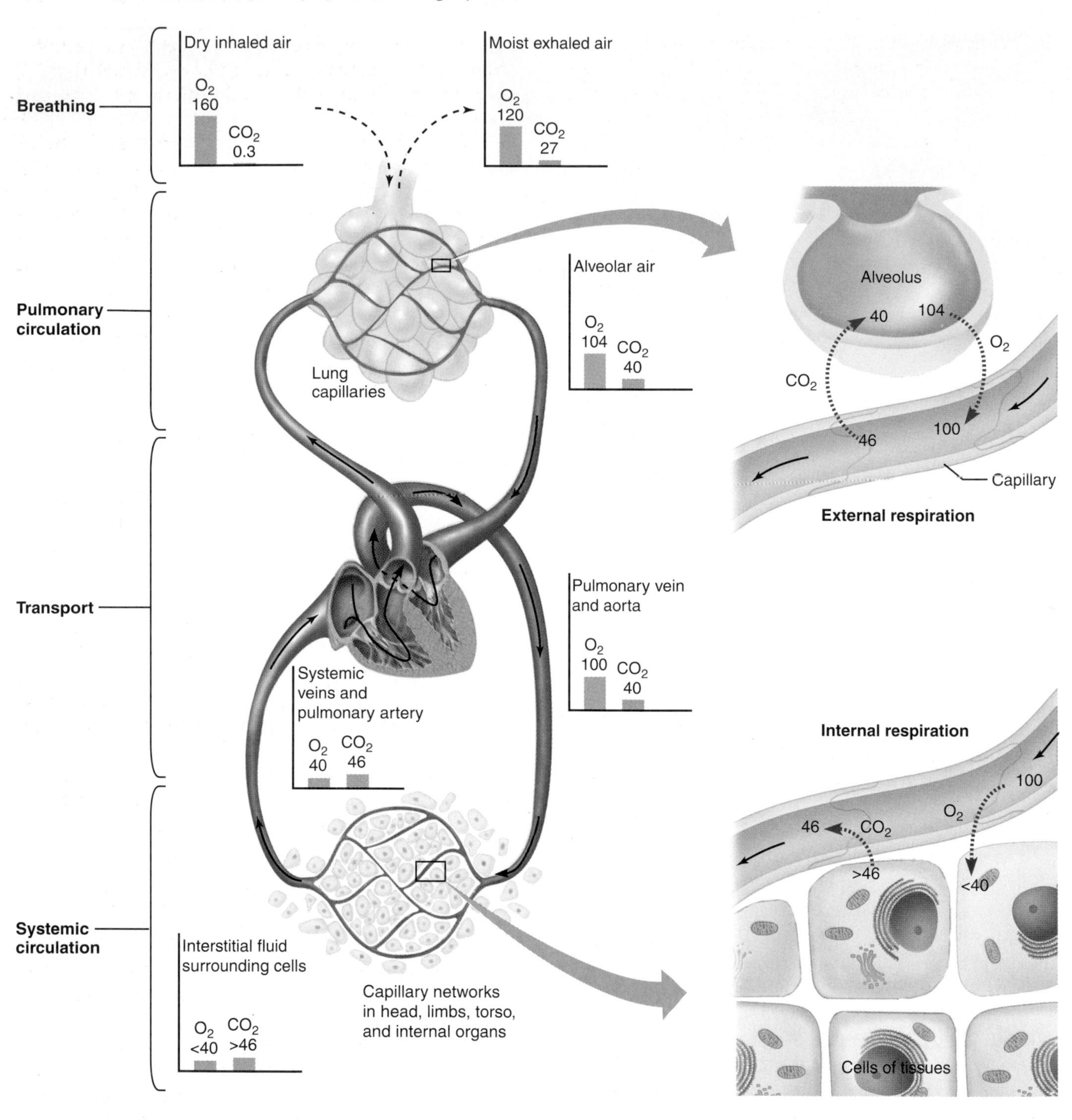

Figure 10.11 Partial Pressures. All partial pressures are expressed in units of mm Hg. Differences in partial pressures account for the diffusion of O_2 and CO_2 between the lungs and blood, and between blood and the body's tissues.

diffuses into the alveoli is exhaled, along with some water vapor.

Notice that the definition of venous blood is that it is deoxygenated, not that it happens to be in a vein. The pulmonary arteries transport venous blood to the lungs, and the pulmonary veins transport arterial blood to the heart.

Internal respiration: The exchange of gases with tissue fluids

The body's cells get their supply of O_2 for cellular respiration from the interstitial fluid that surrounds them. Because the cells are constantly drawing oxygen from the interstitial fluid (again, by diffusion), the interstitial fluid P_{O_2} is usually quite a bit lower than that of arterial blood (less than 40 mm Hg). As blood enters the capillaries, then, O_2 diffuses from the capillaries into the interstitial fluid, replenishing the O_2 that has been used by the cells. CO_2 diffuses in the opposite direction: from the cell into the interstitial fluid and then into the capillary blood.

Both external and internal respiration occur entirely by diffusion. The partial pressure gradients that permit diffusion are maintained by breathing, blood transport, and cellular respiration. The net effect of all of these processes is that homeostasis of the concentrations of O_2 and CO_2 in the vicinity of the cells is generally well maintained.

Recap **In a mixture of gases such as air, each gas exerts a partial pressure. Gases diffuse according to differences in their partial pressures. Diffusion accounts for both external and internal respiration.**

Hemoglobin transports most oxygen molecules

Our discussion of external and internal respiration has ignored one important aspect of the overall subject of gas exchange, and that is how the gases are transported between the lungs and tissues in the blood. Recalling from Chapter 7 that an important function of blood is to transport oxygen from the lungs to body tissues, we next examine oxygen transport.

Oxygen is transported in blood in two ways: either it is bound to hemoglobin (Hb) in red blood cells, or it is dissolved in blood plasma (Figure 10.12). The presence of hemoglobin is absolutely essential for the adequate transport of O_2 because O_2 is not very soluble in water. Only about 2 percent of all O_2 is dissolved in the watery component of blood known as blood plasma. Most of it—98 percent—is taken out of the watery component by virtue of its binding to hemoglobin molecules. Without hemoglobin, the tissues would not be able to receive enough oxygen to sustain life.

Why are red blood cells essential to respiration?

As we saw in Chapter 7, *hemoglobin* is a large protein molecule consisting of four polypeptide chains, each of which is associated with an iron-containing heme group that can bind oxygen. Because there are four heme groups, each hemoglobin molecule can bind four oxygen molecules at a time, forming **oxyhemoglobin** (HbO_2). We can represent this reaction as:

$$\underset{\text{hemoglobin}}{Hb} + \underset{\text{oxygen}}{O_2} \rightarrow \underset{\text{oxyhemoglobin}}{HbO_2}$$

This reaction is reversible and highly dependent on the partial pressures of O_2 in plasma. When the P_{O_2} rises (in the lungs), oxygen attaches to hemoglobin and is transported in arterial blood. When the P_{O_2} falls (at the tissues), oxygen detaches from hemoglobin.

Several other factors affect O_2 attachment to hemoglobin as well. Hemoglobin binds O_2 most efficiently in conditions of fairly neutral pH and relatively cool temperatures—similar to conditions existing in the lungs. Body regions having warmer temperatures and lowered pH—such as in body tissues—reduce hemoglobin's affinity for binding O_2. Consequently, O_2 and hemoglobin tend to combine in the lungs, facilitating the transport of oxygen to the tissues, and to detach in body tissues, making O_2 available to cells.

Most CO_2 is transported in plasma as bicarbonate

Cellular metabolism in body tissues continuously produces carbon dioxide as a waste product. As we learned in Chapter 7, one function of blood is to transport CO_2 away from tissues and back to the lungs, where it can be removed from the body.

Because the partial pressure of CO_2 is higher in the tissues than it is in blood, CO_2 readily diffuses from tissues into the bloodstream. Once in the blood, CO_2 is transported in three ways: dissolved in blood plasma, bound to hemoglobin, or in the form of bicarbonate (see Figure 10.12).

Only about 10 percent of the CO_2 remains dissolved in blood plasma. Another 20 percent binds with hemoglobin to form **carbaminohemoglobin** ($HbCO_2$). This reaction is represented as:

$$\underset{\text{hemoglobin}}{Hb} + \underset{\text{carbon dioxide}}{CO_2} \rightarrow \underset{\text{carbaminohemoglobin}}{HbCO_2}$$

Hemoglobin can transport O_2 and CO_2 molecules simultaneously because the two gases attach to different sites on the hemoglobin molecule. O_2 combines with heme, CO_2 with globin.

About 70% of all of the CO_2 produced by the tissues is converted to bicarbonate prior to transport. To produce bicarbonate, CO_2 combines with water (H_2O) to become carbonic acid (H_2CO_3). This first reaction is catalyzed by an enzyme called *carbonic anhydrase*. The carbonic acid immediately breaks apart into bicarbonate and hydrogen ions:

$$\underset{\text{carbon dioxide}}{CO_2} + \underset{\text{water}}{H_2O} \rightarrow \underset{\text{carbonic acid}}{H_2CO_3} \rightarrow \underset{\text{bicarbonate}}{HCO_3^-} + \underset{\text{hydrogen}}{H^+}$$

The formation of bicarbonate from CO_2 occurs primarily inside the red blood cells because this is where the carbonic anhydrase enzyme is located. However, most of the bicarbonate quickly diffuses

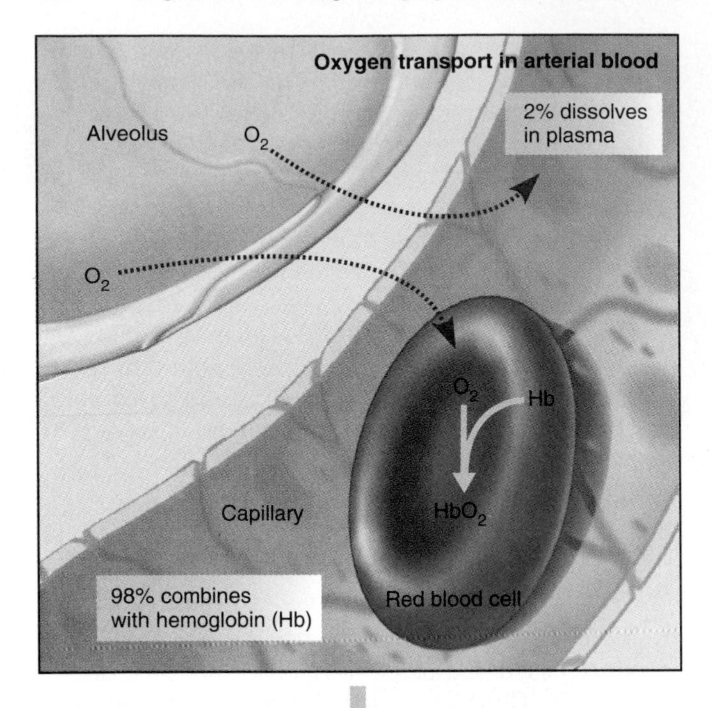

Blood is transported from lungs to tissue

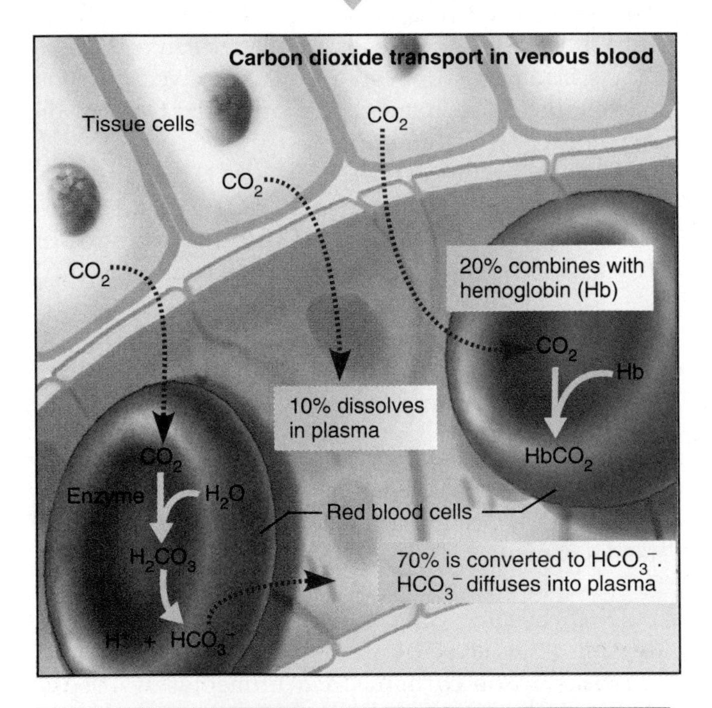

Figure 10.12 How oxygen and carbon dioxide are transported in blood. Approximately 98% of all the O_2 transported to the tissues by arterial blood is bound to hemoglobin within red blood cells. Most of the CO_2 transported to the lungs by venous blood is converted to HCO_3^- within red blood cells and then transported in plasma as dissolved HCO_3^-. The rest is either dissolved in plasma as CO_2 or transported within the red blood cell bound to hemoglobin.

out of red blood cells and is transported back to the lungs dissolved in plasma.

Some of the hydrogen ions formed during the formation of bicarbonate stay inside the red blood cells and bind to hemoglobin. Their attachment to hemoglobin weakens the attachment between hemoglobin and oxygen molecules and causes hemoglobin to release more O_2. This is the explanation for the effect of pH on oxygen binding described above. The overall effect is that the presence of CO_2 (an indication that cellular metabolism has taken place) actually enhances the delivery of O_2 to the very sites where it is most likely to be needed.

At the lungs, dissolved CO_2 diffuses out of the blood and into the alveolar air. The loss of CO_2 from the blood plasma causes the P_{CO_2} to fall, which in turn causes the chemical reaction that formed bicarbonate in the first place to reverse:

$$\underset{\text{bicarbonate}}{HCO_3^-} + \underset{\text{hydrogen}}{H^+} \rightarrow \underset{\text{carbonic acid}}{H_2CO_3} \rightarrow \underset{\text{carbon dioxide}}{CO_2} + \underset{\text{water}}{H_2O}$$

Thus, as CO_2 is removed by breathing, the bicarbonate and hydrogen ions formed in the peripheral tissues to transport CO_2 are removed as well.

Recap **Nearly all of the oxygen in blood is bound to hemoglobin in red blood cells. Most carbon dioxide produced in the tissues is transported in blood plasma as bicarbonate.**

10. 5 The nervous system regulates breathing

As mentioned earlier, breathing depends on the contractions of skeletal muscles. Recall from Chapter 5 that skeletal muscles are activated only by motor neurons. It follows, then, that respiration is controlled by the nervous system. The nervous system regulates the rate and depth of breathing in order to maintain homeostasis of the concentrations of certain key variables, most notably the concentrations of CO_2, H^+, and O_2. In addition, we can exert a certain amount of conscious control over breathing if we wish.

A respiratory center establishes rhythm of breathing

The basic cyclic pattern of inspiration and expiration and the rate at which we breathe are established in an area near the base of the brain called the *medulla oblongata* (Figure 10.13). Within this area, called the **respiratory center,** groups of nerve cells automatically generate a cyclic pattern of electrical impulses

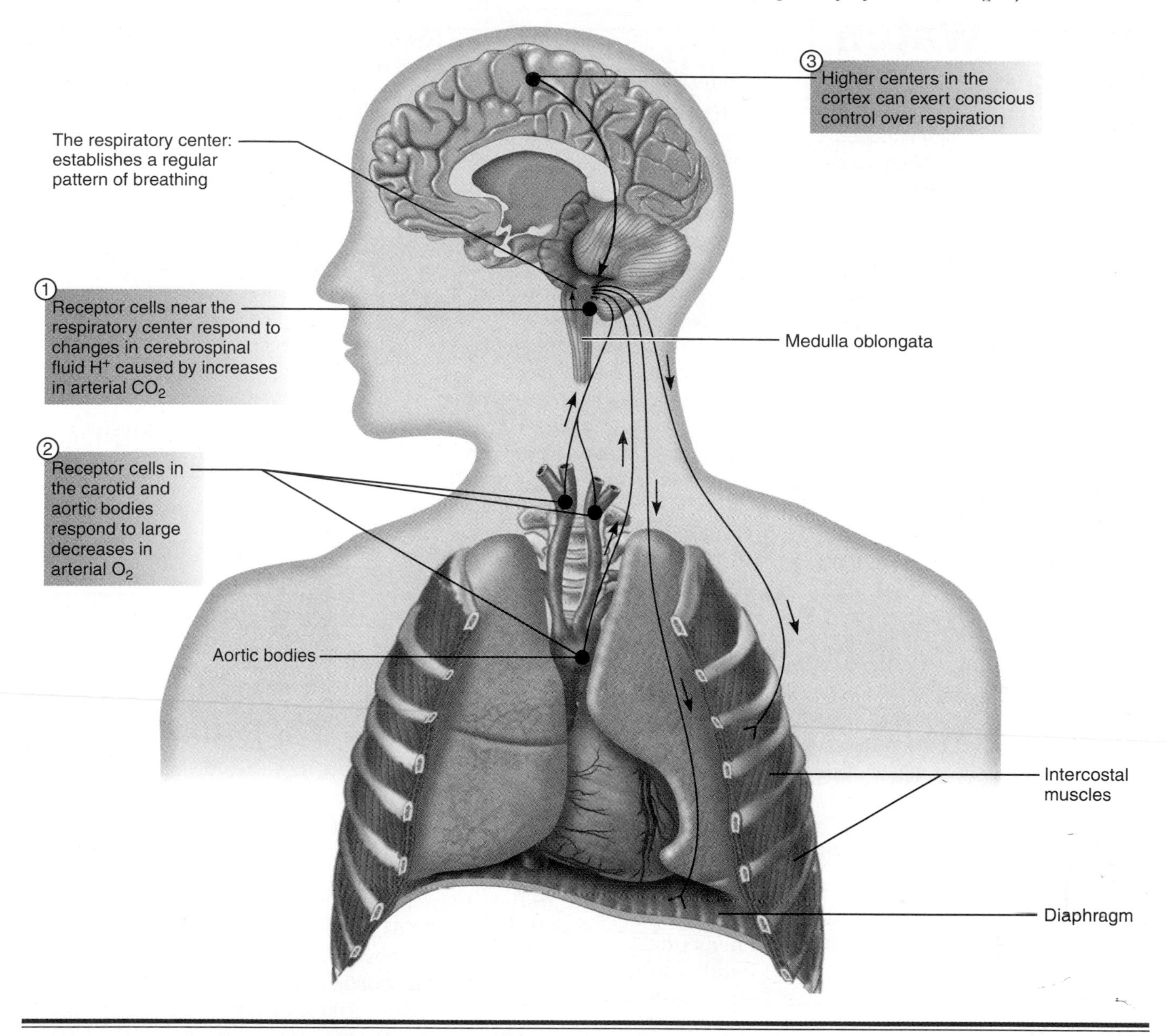

Figure 10.13 Regulation of breathing. The basic pattern of breathing is established by a respiratory center within the medulla oblongata. Normally, the rate and depth of respiration are set primarily to regulate blood CO_2 levels, but respiration can also be stimulated by a substantial drop in blood O_2 levels. In addition, respiration can be controlled (at least for short periods of time) by conscious control.

every 4–5 seconds. The impulses travel along nerves to the diaphragm and the intercostal muscles and stimulate those muscles to contract. As these respiratory muscles contract the rib cage expands, the diaphragm is pulled downward, and we inhale. As inhalation proceeds, the respiratory center receives sensory input from stretch receptors in the lungs. These receptors monitor the degree of inflation of the lungs and serve to limit inhalation and initiate exhalation. When nerve impulses from the respiratory center to the muscles end, the respiratory muscles relax, the rib cage returns to its original size, the diaphragm moves upward again, and we exhale.

Any disorder that interferes with the transmission of these nerve impulses can affect breathing. Consider amyotrophic lateral sclerosis (ALS; also known as Lou Gehrig's disease for the famous baseball player who succumbed to it). In ALS the nerves to skeletal muscle become damaged and no longer conduct impulses properly. Over time the skeletal

Health Watch

Altitude Sickness

Air gets "thinner" at high altitudes, meaning that the atmospheric pressure is lower. Atmospheric pressure at sea level is about 760 mm Hg; at the top of Pikes Peak (elevation 14,110 ft.) it falls to 450 mm Hg. Most people who go from sea level to 14,000 feet without taking several days to get used to the change will experience a reaction called altitude sickness. Symptoms include headache, fatigue, dizziness, heart palpitations, loss of appetite, and nausea. Some people begin to experience mild symptoms at around 6,000 feet.

What causes altitude sickness? As the P_{O_2} in inspired air falls, there is a corresponding decline in the P_{O_2} in blood. This is one of the few situations in which the arterial P_{O_2} falls enough to stimulate the O_2 receptor cells in the carotid and aortic bodies. Activation of these cells stimulates breathing and the person *hyperventilates* (breathes faster and more deeply) in an attempt to get P_{O_2} back toward normal. However, hyperventilation causes the person to blow off more CO_2 than usual, and the arterial blood P_{CO_2} falls to below its normal value of 40 mm Hg. The low P_{CO_2} causes a decline in the H^+ concentration as well. The combination of the low P_{O_2}, low P_{CO_2}, and the low H^+ concentration (a condition called *alkalosis*) causes the symptoms of altitude sickness.

Altitude sickness can usually be prevented by ascending gradually—going up 2,000 to 3,000 feet, then stopping for a day or two before ascending further. If symptoms do develop, they generally disappear in 2–4 days as the body adjusts to higher elevations. First, the kidneys excrete less H^+ than usual in the urine, which gradually returns the H^+ concentration to normal. Second, the bone marrow produces more red blood cells and many tissues grow more capillaries and mitochondria. This increases aerobic efficiency to compensate for a scarcity of O_2.

muscles, including the diaphragm and intercostal muscles, weaken and waste away from lack of use. Most ALS patients die within 5 years of initial diagnosis. Although ALS is not a respiratory condition *per se*, the immediate cause of death is usually respiratory failure.

Chemical receptors monitor CO_2, H^+, and O_2 levels

The rate and depth of the breathing pattern is modified, or regulated, in order to maintain homeostasis. Under normal circumstances the regulation of breathing centers on maintaining homeostasis of CO_2, H^+, and O_2, with the main emphasis on CO_2.

How is our breathing regulated?

The sensory mechanisms for detecting changes in CO_2 levels are actually indirect. Rather than detecting changes in CO_2 levels, certain cells in the medulla oblongata can detect changes in the H^+ concentration of the cerebrospinal fluid (the extracellular fluid around the cells in the brain). This is pertinent to control of CO_2 because, as you may recall, a rise in CO_2 concentration is accompanied by a rise in the hydrogen ion concentration according to the following reaction:

$$CO_2 + H_2O \rightarrow H_2CO_3 \rightarrow HCO_3^- + H^+$$

Thus, when the P_{CO_2} of arterial blood rises, the concentration of H^+ in cerebrospinal fluid also rises. Receptor cells detecting an elevated H^+ concentration transmit signals to the respiratory center, causing it to increase the rate of the cyclic pattern of impulses and the number of impulses per cycle. As a result we breathe more frequently and more deeply, exhaling more CO_2 and lowering blood levels of the gas back to normal. Our normal arterial P_{CO_2} is maintained at about 40 mm Hg because the normal regulation of respiration keeps it there. Any rise above 40 mm Hg will stimulate breathing and any fall below 40 mm Hg will inhibit it.

Certain other receptor cells respond to blood P_{O_2} rather than P_{CO_2}. These receptors are located in small structures associated with the carotid arteries (the large arteries to the head) and the aorta, called the *carotid* and *aortic bodies*. Normally, P_{O_2} in arterial blood is about 100 mm Hg. If P_{O_2} drops below about 80 mm Hg, these receptors signal the respiratory center to increase the rate and depth of respiration in order to raise blood O_2 levels back toward normal. The carotid and aortic bodies also can be stimulated by an increase in the blood concentrations of H^+ and CO_2 if they are great enough.

It is important to note that the receptors for O_2 in the carotid and aortic bodies only become activated when arterial P_{O_2} falls by at least 20 percent. In contrast, even a 2–3 percent change in P_{CO_2} will stimulate respiration. Under normal circumstances, then, respiration is controlled entirely by the receptors that regulate CO_2 by responding to cerebrospinal fluid H^+. It just so happens that the regulation of CO_2 also keeps O_2 within the normal range. In other words, the rate and depth of normal breathing is set by the need to get rid of CO_2, not to obtain O_2.

There are circumstances, however, when the O_2 receptors do come into play. These include certain disease states, drug overdoses, and breathing at high altitudes where the partial pressure of O_2 is much lower than normal.

We can exert some conscious control

Finally, there is another way to regulate breathing, and that is by conscious control. Conscious control resides in higher brain centers, most notably the cortex. The ability to modify our breathing allows us to speak and sing. You can choose to hold your breath for a minute or even to hyperventilate (breathe rapidly) for a short while.

But if you don't think those automatic regulatory mechanisms are powerful, just try to hold your breath indefinitely. You will find your conscious control overpowered by the automatic regulatory mechanisms described above.

Recap **The respiratory center in the brain establishes a regular pattern of cyclic breathing. The rate and depth of breathing are then adjusted by regulatory mechanisms that monitor arterial concentrations of CO_2, H^+, and O_2. Conscious control can modify regulatory control but cannot override it completely.**

10.6 Disorders of the respiratory system

There are numerous causes of disorders of respiration. Among them are conditions that reduce air flow or gas exchange, infections by microorganisms, cancer, diseases of other organs such as heart failure, and even genetic diseases.

Reduced air flow or gas exchange impedes respiratory function

Respiration depends on the flow of air between the atmosphere and the alveoli and upon the diffusional exchange of gases across the alveolar and capillary walls. Any factor that impairs these activities will impede respiratory function.

Asthma: Spasmodic contraction of bronchi **Asthma** is characterized by spasmodic contraction of the bronchi. The spasms partially close the bronchi, making breathing difficult. It is a recurrent, chronic lung disorder that affects over 14 million people in North America. Its incidence is on the rise both in the U.S. and around the world (see Table 10.1).

Symptoms of an asthma "attack" include coughing while exercising, shortness of breath, wheezing, and a sense of tightness in the chest. People with asthma often wheeze when they exhale. The symptoms can be triggered by any number of causes, including viruses, air particles, allergies (such as allergies to pollen, house dust, and animal fur), exercise (especially in cold temperatures), tobacco smoke, and air pollution. The symptoms may come and go.

Drugs are available to dilate the bronchi (bronchodilators) and restore normal breathing. Treatments also focus on preventing attacks by isolating the cause and avoiding it when possible.

Emphysema: Alveoli become permanently impaired **Emphysema** is a chronic disorder involving damage to the alveoli. It begins with destruction of connective tissue in the smaller airways. As a result the airways are not as elastic, do not stay open properly, and have a tendency to collapse during expiration. The high pressures in the lungs caused by the inability to exhale naturally through the collapsed airways eventually damage the fragile alveoli. The result is a permanent reduction in the surface area available for diffusion, and eventually breathlessness and reduced capacity to exchange gases across the lung.

At least one form of emphysema is inherited, but most cases are associated with smoking or long-term exposure to air pollutants. The difference between asthma and emphysema is that asthma is an episodic, recurrent condition of increased airway resistance that largely goes away between episodes, whereas emphysema involves permanent damage to airways that eventually destroys alveoli.

Bronchitis: Inflammation of the bronchi **Bronchitis** refers to inflammation of the bronchi, resulting in a persistent cough that produces large quantities of

Table 10.1 Asthma Statistics

Americans with asthma:	14.6 million[1]
Children with asthma:	4.8 million children under age 18[1]
Rise in prevalence of asthma:	75% increase in reported cases since 1980[2]
Asthma deaths:	More than 5,000 each year[3]
Asthma-related hospitalizations:	466,000 in 1994[2]
Emergency room visits for asthma:	1.9 million in 1995[2]
Health care costs for asthma care:	Estimated at more than $6 billion a year[4]
Missed school days:	Estimated at more than 10 million a year[5]

SOURCES: 1. *Vital and Health Statistics,* December 1995; 10(193): Table 62. 2. "Surveillance for asthma—United States 1960–1995," *Morbidity and Mortality Weekly Report,* April 24, 1998; 47(SS-1). 3. *Monthly Vital Statistics Report,* August 14, 1997; 46(1): Table 6. 4. "HHS Targets Efforts on Asthma," Department of Health and Human Services, May 21, 1998. 5. "Asthma: A Concern for Minority Populations," National Institute of Allergy and Infectious Diseases, January, 1997.

phlegm. Bronchitis may be acute (comes on suddenly and clears up within a couple of weeks) or chronic (persists over a long period and recurs over several years). Both forms are more common in smokers and in people who live in highly polluted areas.

Symptoms include wheezing, breathlessness, and a persistent cough that yields yellowish or greenish phlegm. Sometimes there is fever and a feeling of discomfort behind the sternum. Acute bronchitis can be treated by humidifying the lungs (using a home humidifier or inhaling steam), drinking plenty of fluids, and taking antibiotics if the infection is bacterial in origin. People with chronic bronchitis may need further testing to rule out other health conditions. Bronchodilators can widen the bronchi; oxygen may be prescribed to raise blood oxygen levels.

Microorganisms can cause lung disease

The lungs are particularly prone to infections. One of the reasons is that the lungs are moist, warm, and covered in a thin layer of fluid, exactly the conditions favored by microorganisms. Besides the common cold and the flu (both caused by viruses), more serious diseases caused by microorganisms include pneumonia, tuberculosis, and botulism.

Why is the respiratory system so prone to infections?

Pneumonia: Infection inflames the lungs In **pneumonia** the lungs become inflamed from an infection, usually caused by viruses or bacteria. The alveoli secrete excess fluid, impairing the exchange of oxygen and carbon dioxide. Symptoms typically include fever, chills, shortness of breath, and a cough that produces yellowish-green phlegm and sometimes blood. Some people experience chest pain when breathing due to inflammation of the membranes that line the chest cavity and cover the lungs.

In North America, pneumonia ranks among the top 10 causes of death, primarily because it is a frequent complication of many serious illnesses. Treatment depends on the microorganism involved. In severe cases, oxygen therapy and artificial ventilation may be necessary. However, most people who develop pneumonia recover completely within a few weeks.

Tuberculosis: Bacterial infection scars the lungs **Tuberculosis (TB)** is an infectious disease caused by the bacterium *Mycobacterium tuberculosis*. People pass the infection in airborne droplets by coughing or sneezing. The bacteria enter lungs and multiply to form an infected "focus." In most cases, the immune system fights off the infection, although it may leave a scar on the lungs. In perhaps 5% of cases, however, the infection spreads via lymphatic vessels to the lymph nodes and may enter the bloodstream. Sometimes the bacteria become dormant for many years, then reactivate later to cause more lung damage.

Major symptoms include coughing (sometimes bringing up blood), chest pain, shortness of breath, fever, night sweats, and appetite and weight loss. A chest x-ray usually reveals lung damage, perhaps cavities in the lungs or old infections that have healed. A skin test called the tuberculin test can indicate whether someone has been exposed to the infection.

A century ago, tuberculosis was a major cause of death worldwide. With the development of antibiotics the incidence of tuberculosis declined precipitously, and most patients in industrialized countries now recover fully. The disease remains a major health problem in undeveloped nations, however, and recently the incidence of tuberculosis has increased in industrialized countries as well. Many authorities attribute the increased prevalence of tuberculosis in developed nations to immigration of people from underdeveloped nations. A compounding factor is that some strains of tuberculosis are becoming resistant to the antibiotics.

Botulism: Poisoning by bacterial toxin *Botulism* is a form of poisoning caused by a bacterium, *clostridium botulinum*, occasionally found in inadequately cooked or preserved foods. The bacterium produces a powerful toxin that blocks the transmission of nerve signals to skeletal muscles, including the diaphragm and intercostal muscles.

Symptoms of botulism poisoning usually appear 8 to 36 hours after eating the contaminated food. They can include difficulty swallowing and speaking, double vision, nausea, and vomiting. If not treated, botulism can be fatal due to paralysis of the respiratory muscles.

Lung cancer is caused by proliferation of abnormal cells

Cancer is the uncontrolled growth of abnormal cells. In the lung, cancerous cells crowd out normal cells and eventually impede normal function. Lung cancer can impair not only the movement of air in the airways, but also the exchange of gases in the alveoli and the flow of blood in pulmonary blood vessels (Figure 10.14). It accounts for one third of all cancer deaths in the United States.

Lung cancer takes years to develop and is strongly associated with smoking. More than 90 percent of lung cancer patients are current or former smokers, and some of those who don't smoke were

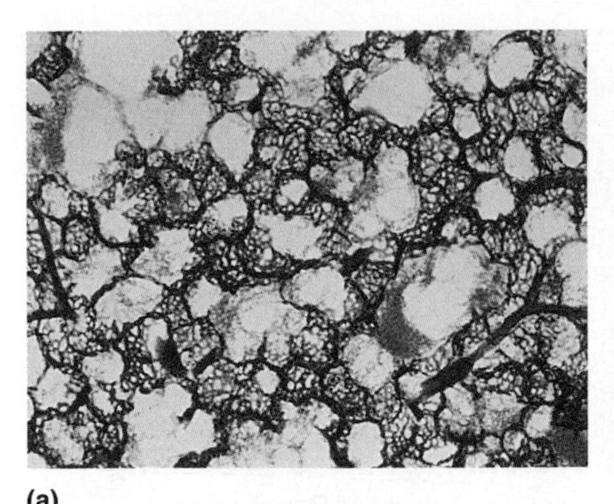

(a)

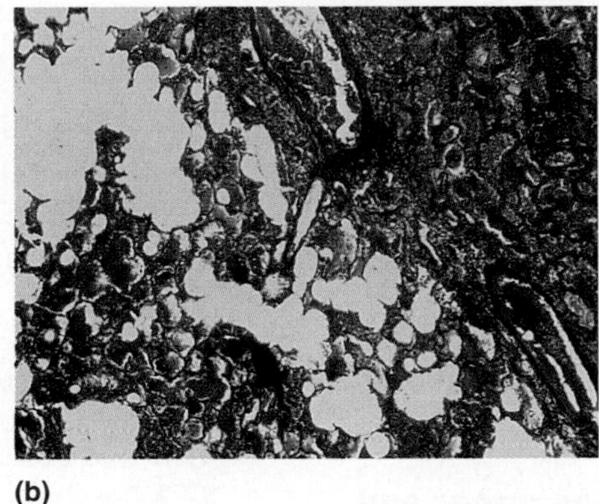

(b)

Figure 10.14 Lung cancer. **(a)** Normal lung alveoli and surrounding capillaries. **(b)** A section through a cancerous lung showing alveolar destruction and capillary bleeding (dark red color).

exposed to secondhand smoke from other people. Two other important causes are radon gas (formed in the earth's rock as a breakdown product of uranium) and chemicals in the workplace, the best known of which is asbestos.

Symptoms of lung cancer include chronic cough, wheezing, chest pain, and coughing up blood. Most people don't go to the doctor for every trivial cough, but any cough that is persistent or accompanied by chest pain—especially coughs that bring up blood—should be checked out immediately. If cancer is caught early it may be possible to remove the cancerous lobe or region of the lung, or to eradicate the tumor with radiation therapy and chemotherapy.

Lung cancer is highly preventable. It is a good idea to not smoke, know the conditions of your work environment, and, depending on where you live, have your home inspected for radon gas. For more information on all types of cancer, see Chapter 18.

Congestive heart failure impairs lung function

In Chapter 7 we discussed *congestive heart failure* as a cardiovascular condition in which the heart gradually becomes less efficient. Even though it starts as a heart disorder, it can end by impairing lung function as well.

Recall that in congestive heart failure the heart begins to fail as a pump. When the left side of the heart fails, blood backs up in the pulmonary blood vessels behind the left side of the heart. The result is a rise in blood pressure in the pulmonary vessels. When pulmonary capillary pressure increases, the balance of physical pressure and osmotic forces across the capillary wall favors fluid loss from the capillary. As a result, fluid builds up in the interstitial spaces between capillaries and alveoli and sometimes within alveoli themselves. This reduces diffusion of gases because the diffusional distance is increased. Treatments focus on reducing this fluid buildup by helping the body get rid of fluid and improving the heart's pumping action.

Cystic fibrosis is an inherited disease

Cystic fibrosis is an inherited condition in which a single defective gene causes the mucus-producing cells in the lungs to produce a thick, sticky mucus. The disease affects other organ systems as well. For more on this subject, see the Cystic Fibrosis Health Watch in Chapter 19. In the lungs, the abnormally thick mucus impedes air flow and also provides a site for the growth of bacteria. People with cystic fibrosis tend to get frequent infections of the airways. Treatment of the disease includes consistent physical therapy to try to dislodge the mucus and keep the airways open. Several promising new drugs are now on the market for this disease.

Recap **The lungs are prone to damage by environmental pollutants, tobacco smoke, and infections by microorganisms. Cases of both asthma and tuberculosis are on the rise.**

Chapter Summary

Respiration takes place throughout the body

- Respiration encompasses four processes: breathing, external respiration, internal respiration, and cellular respiration.
- External respiration occurs in the lungs; internal respiration and cellular respiration take place in the tissues.

The respiratory system includes upper and lower respiratory tracts

CD Respiratory System/Anatomy Review: Respiratory Structures

- The respiratory system includes the air passageways to the lungs and the lungs themselves.
- Bones and skeletal muscles support the respiratory system and participate in breathing.
- The upper respiratory tract consists of the nose and pharynx. The upper respiratory tract filters, warms, and humidifies the air we breathe.
- The lower respiratory tract consists of the larynx, the trachea, the main bronchi, and the lungs.
- Within the lungs, the bronchi branch many times, becoming smaller airways called bronchioles that end in air-filled sacs called alveoli.
- The tremendous surface area of the alveoli, coupled with the thinness of the respiratory membrane, facilitates gas exchange with the pulmonary capillaries.

The process of breathing involves a pressure gradient

CD Respiratory System/Pulmonary Ventilation

- Inspiration occurs as the lungs expand due to the action of the diaphram and the intercostal muscles; expiration occurs when these muscles relax.
- When the lungs expand, the pressure within them falls relative to atmospheric pressure and air rushes in; during expiration the lungs become smaller and increasing pressure within them forces air out.
- During normal breathing, inspiration is active (requiring energy) and expiration is passive.
- Normally we breathe at about 12 breaths per minute with a tidal volume of 500 ml per breath.
- Vital capacity is the maximum amount of air a person can exhale after a maximal inhalation.

Gas exchange and transport occurs passively

CD Respiratory System/Gas Exchange

- The diffusion of a gas is dependent on a partial pressure gradient, which is equivalent to a concentration gradient.
- External and internal respiration are both processes that occur entirely by diffusion.
- Nearly all (98%) of the oxygen transported by blood is bound to hemoglobin in red blood cells.
- Although some carbon dioxide is transported as dissolved CO_2 or is bound to hemoglobin, most CO_2 (70%) is converted to bicarbonate and then transported in plasma.

The nervous system regulates breathing

CD Respiratory System/Control of Respiration

- A respiratory center in the medulla oblongata of the brain establishes a regular cycle of inhalation and exhalation.
- Under normal conditions, the rate and depth of breathing is adjusted primarily to maintain homeostasis of arterial blood P_{CO_2}.
- Regulation of respiration by O_2 only comes into play when the P_{O_2} concentration falls by more than 20%, such as in disease states or at high altitude.
- We can exert some conscious control over breathing.

Disorders of the respiratory system

CD Respiratory System/Resistance and Lung Compliance

- Asthma is episodic, spasmodic contractions of the bronchi that impede air flow.
- Emphysema is a chronic disorder characterized by high resistance to air flow and destruction of alveoli.
- The lungs are prone to infections because their surface is kept moist and warm in order to facilitate gas exchange. TB is an infectious disease caused by a bacterium. In pneumonia, infected alveoli secrete excess fluid, impairing gas exchange.
- Lung disease can be a secondary condition resulting from impairment of another organ, as in congestive heart failure.

Terms You Should Know

alveoli, 241
asthma, 251
bronchi, 239
bronchioles, 240
bronchitis, 251
carbaminohemoglobin, 247
diaphragm, 242
emphysema, 251
epiglottis, 236
glottis, 237
intercostal muscles, 242
larynx, 236
lungs, 240
oxyhemoglobin, 247
partial pressure, 245
pharynx, 236
pleural cavity, 240
respiratory center, 248
tidal volume, 244
trachea, 238
tuberculosis, 252
ventilation, 234
vital capacity, 244
vocal cords, 237

Test Yourself

Matching Exercise

_____ 1. The components of the respiratory system located above the "Adam's apple."
_____ 2. Movement of air into and out of the lungs.
_____ 3. Passageway for air between the middle ear and the upper part of the pharynx.
_____ 4. The opening between the vocal cords.
_____ 5. The air-filled sacs that are the sites of gas exchange between air and blood.
_____ 6. Name for the volume of a normal breath at rest.
_____ 7. Maximum amount of air we can inhale and then exhale in a single breath.
_____ 8. Respiratory disorder characterized by spasmodic contraction of bronchi.
_____ 9. Inherited disorder characterized by the buildup of mucus in the airways.
_____10. Inflammation of the lungs caused by an infection.

a. alveoli
b. glottis
c. vital capacity
d. breathing
e. tidal volume
f. asthma
g. upper respiratory tract
h. pneumonia
i. eustachian tube
j. cystic fibrosis

Fill in the Blank

11. The flexible flap of cartilage that closes off the airway during swallowing is called the _____________.
12. The single airway between the larynx and the two main bronchi is called the _____________.
13. Layers of epithelial tissue that cover the outer surface of the lung and line the chest cavity are called _____________ membranes.
14. Approximately _____% of the air we breathe is oxygen.
15. Hemoglobin bound to oxygen is called _____________.
16. Hemoglobin bound to carbon dioxide is called _____________.
17. The _____________ _____________ establishes a regular, rhythmic cycle of breathing.
18. During inspiration, the dome-shaped _____________ contracts and flattens.
19. The _____________ and _____________ bodies contain sensory receptors that monitor blood P_{O_2}.
20. The lower portion of the _____________ is a common passageway for food and air that leads to the larynx (passageway for air only) and the esophagus (passageway for food only).

True or False?

21. Under normal conditions of quiet breathing at rest, both inspiration and expiration are active processes, requiring contraction of opposing skeletal muscle groups.
22. At rest we normally breathe at a rate of about 12 breaths per minute.
23. About 98% of all oxygen carried in arterial blood is bound to hemoglobin.
24. Under normal conditions, the rate and depth of breathing are adjusted by a homeostatic control mechanism for CO_2, not O_2.
25. The inner surfaces of the airways (trachea, bronchi, and bronchioles) are lined with smooth muscle.

Concept Review

1. List at least three functions of your nose.
2. Explain why men's voices are lower than women's.
3. Explain why smokers sometimes have a chronic cough.
4. Distinguish between *bronchi* and *bronchioles*.
5. Describe what happens to gas pressure in a closed container when the size of the container is enlarged.
6. Explain what causes the lungs to inflate, in light of the fact that the lungs themselves do not have any skeletal muscle.
7. Define *partial pressure* and explain its importance to gas diffusion.
8. Explain why the partial pressure of oxygen in the alveoli is always lower than the partial pressure of oxygen in the air we breathe.
9. Explain (or use a diagram to show) how bicarbonate is formed from dissolved CO_2 in the region of metabolizing tissues.
10. Describe how heart failure can lead to a decrease in the ability of O_2 and CO_2 to diffuse between the blood and the alveoli.

Apply What You Know

1. Why would a person administering a breathalyzer test for alcohol ask the person being tested to give one deep breath rather than many shallow ones?

2. You are having dinner with a friend when she becomes agitated and says very clearly and distinctly, "I think I'm choking. Please do something!" Should you attempt the Heimlich maneuver on her? Why or why not?
3. Would you guess that there is nitrogen gas in the blood? If you were to assume that nitrogen gas is completely inert and not used by the body, what do you think its partial pressure would be in blood?
4. How do you suppose a pathologist could tell whether a baby had been born stillborn or whether it died after birth, after it had begun breathing?

Case in Point

Why is the Incidence of Asthma Rising?

In a 1999 government publication, the CDC (Centers for Disease Control and Prevention) reported that more than 5% of the U.S. population has asthma, and that the numbers are growing. The American Lung Association has reported that asthma prevalence has increased from 34 cases/1000 population in 1982 to 56 cases/1000 in 1994. What might be causing this rapid increase in the incidence of asthma?

Several recent studies may provide a clue. A 1999 study in Aberdeen, Scotland, examined 133 adults for signs of respiratory illness and asthma, all of whom had participated in a similar study as children 30 years earlier. A higher incidence of asthma was noted in the second study. The investigators found a significant association between asthma occurrence in these adults and family histories of asthma/allergies, indicating that there is a genetic component to the disease. However, they also found a high prevalence of IgE antibody responsiveness and wheezing in the children of those subjects who had no previous family history of allergies. Although increased IgE responsiveness and wheezing by themselves are not proof of asthma, the investigators suggest that changes in the environment may be contributing to the observed incidence of asthma.

In another 1999 study, researchers from several countries in Europe studied questionnaires about respiratory symptoms provided by over 15,000 people in 12 industrialized countries. They found that farmers, painters, plastic workers, and cleaners were at particularly high risk for asthma.

QUESTIONS:

1. What questions would you raise before accepting the conclusions of the Aberdeen study?

2. How could occupation be a factor in asthma incidence? How would you set up a study to find out?

INVESTIGATE:

Christie, G.L., P.J. Helms, D.J. Gooden, S.J. Ross, J.A.R. Friend, J.S. Legge, N.E. Haites, and J.G. Douglas. Asthma, Wheezy Bronchitis, and Atopy across Two Generations. *Am. J. Resp. Crit. Care Med.* 159:125–129, 1999.

Kogevinas, M., J.M. Anto, J. Sunyer, A. Tobias, H. Krumhout, P. Burney, and the European Community Respiratory Health Survey Study Group. Occupational asthma in Europe and other industrialised areas: a population-based study. *Lancet* 353:1750–1754, 1999.

References and Additional Reading

Buckley, Christopher. *Thank You for Smoking*. New York: HarperPerennial, 1994. 272 pp. (Fiction. A humorous satirical view of smoking from the perspective of a fictional tobacco lobbyist.)

Gibbs, T.W. Breath of Fresh Liquid. *Scientific American* 280(2):40–41, 1999. (Brief research news article about breathing liquids instead of air.)

Levitzky, Michael G. *Pulmonary Physiology*. 5th ed. New York: McGraw-Hill, 1999. (A good all-around reference resource for human respiratory system function.)

Service, Robert. Drug Delivery Takes a Deep Breath. *Science* 277:1199–1200, Aug. 29, 1997. (Brief research news article about delivery of drugs via the respiratory system.)

CHAPTER 11

THE NERVOUS SYSTEM—INTEGRATION AND CONTROL

What triggers a nerve cell to send an impulse?

How fast can a nerve impulse travel?

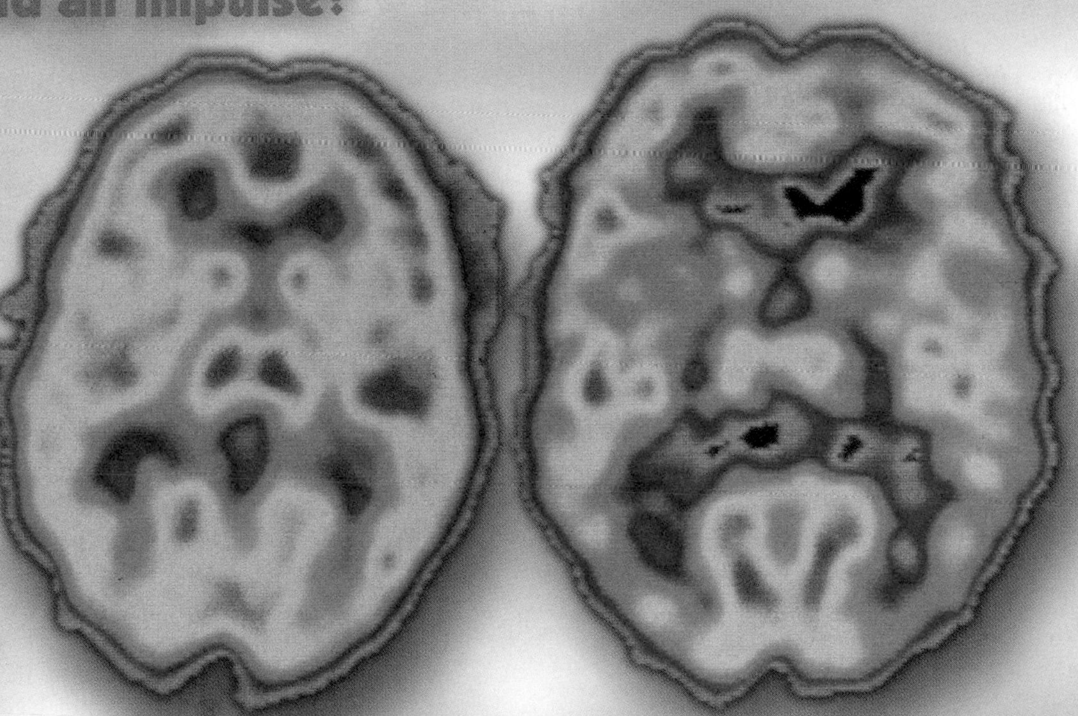

What part of the brain controls the body's automatic functions?

How do psychoactive drugs work?

How are memories stored in the brain?

CONSIDER THE FOLLOWING EXAMPLE: YOU ARE CROSSING THE street. It is cold but you are not wearing your coat, and you are hungry. You see a donut shop, you smell donuts in the air, and you hear the sound of your shoes on the pavement. Suddenly you hear other sounds—the screech of a skidding truck behind you and the blaring of the truck's horn. Quickly you whirl around, glance at the approaching truck, and leap to safety on the sidewalk.

Your nervous system is constantly receiving input from all kinds of sources. There's a huge amount of information to be sorted through, including data in your large memory banks about the probable meaning of all these sounds and other stimuli. The nervous system integrates this seemingly unrelated information quickly and enables you to react in time. How is this done?

Three characteristics of the nervous system are:

1. The nervous system requires information, which it receives from our senses.
2. The nervous system integrates information. *Integration* is the process of taking different pieces of information from many different sources and having it make sense as a whole. It involves sifting through mountains of data and coming up with a plan or course of action.
3. The nervous system is very fast. It can receive information, integrate it, and produce a response within tenths of a second. In this chapter we learn about the special properties of the nervous system that give it such speed.

Most of the functions of the nervous system are completely automatic and can be carried on simultaneously without requiring our attention or conscious decision making. However, the nervous system can also bring selected information to the level of conscious awareness. Working with the endocrine system (Chapter 13), the nervous system maintains homeostasis. It also allows us to feel emotions, to be aware of ourselves, and to exert conscious control over the incredible diversity of our physical movements.

11.1 The nervous system has two principal parts

The nervous system comprises the **central nervous system (CNS)** and the **peripheral nervous system (PNS).** The CNS consists of the brain and the spinal cord. It receives, processes, stores, and transfers information. The PNS represents the components of the nervous system that lie outside the CNS. The PNS has two functional subdivisions: The **sensory division** of the PNS carries information *to* the brain and spinal cord, and the **motor division** carries information *from* the CNS (Figure 11.1).

The motor division of the peripheral nervous system is further subdivided along functional lines. The *somatic division* of the PNS controls skeletal muscles, and the *autonomic division* of the PNS controls smooth muscles, cardiac muscles, and glands. In turn, the autonomic division has two subdivisions called the *sympathetic* and *parasympathetic* divisions. In general, the actions of the sympathetic and parasympathetic divisions oppose each other. They work antagonistically to accomplish the automatic, subconscious maintenance of homeostasis.

We examine each of these components in more detail later in this chapter. But first, we discuss the structure and function of *neurons* and of the *neuroglial cells* that accompany and support them.

Recap **The nervous system has two major subdivisions: the central nervous system (CNS), consisting of the brain and spinal cord, and the peripheral nervous system (PNS), which includes all parts of the nervous system that lie outside the CNS. The PNS has a somatic division, which controls skeletal muscles, and an autonomic division, which controls smooth muscles, cardiac muscles, and glands. The autonomic division has both sympathetic and parasympathetic divisions.**

11.2 Neurons are the communication cells of the nervous system

Neurons are cells specialized for communication. They generate and conduct electrical impulses, also called *action potentials*, from one part of the body to another. The longest neurons extend all the way from your toes to your spinal cord. These are single cells, remember.

There are three types of neurons in the nervous system, as shown in Figure 11.2:

1. **Sensory neurons** of the PNS are specialized to respond to a certain type of stimulus, such as pressure or light. They transmit information about this stimulus to the CNS in the form of electrical impulses. In other words, sensory neurons provide input to the CNS.
2. **Interneurons** within the CNS transmit impulses between components of the CNS. Interneurons receive input from sensory neurons, integrate this information, and influence the functioning of other neurons.
3. **Motor neurons** of the PNS transmit impulses away from the CNS. They carry the nervous system's output, still in the form of electrical impulses, to all of the tissues and organs of the body.

All neurons consist of a cell body, one or more dendrites, and an axon. As in all cells, the main body of

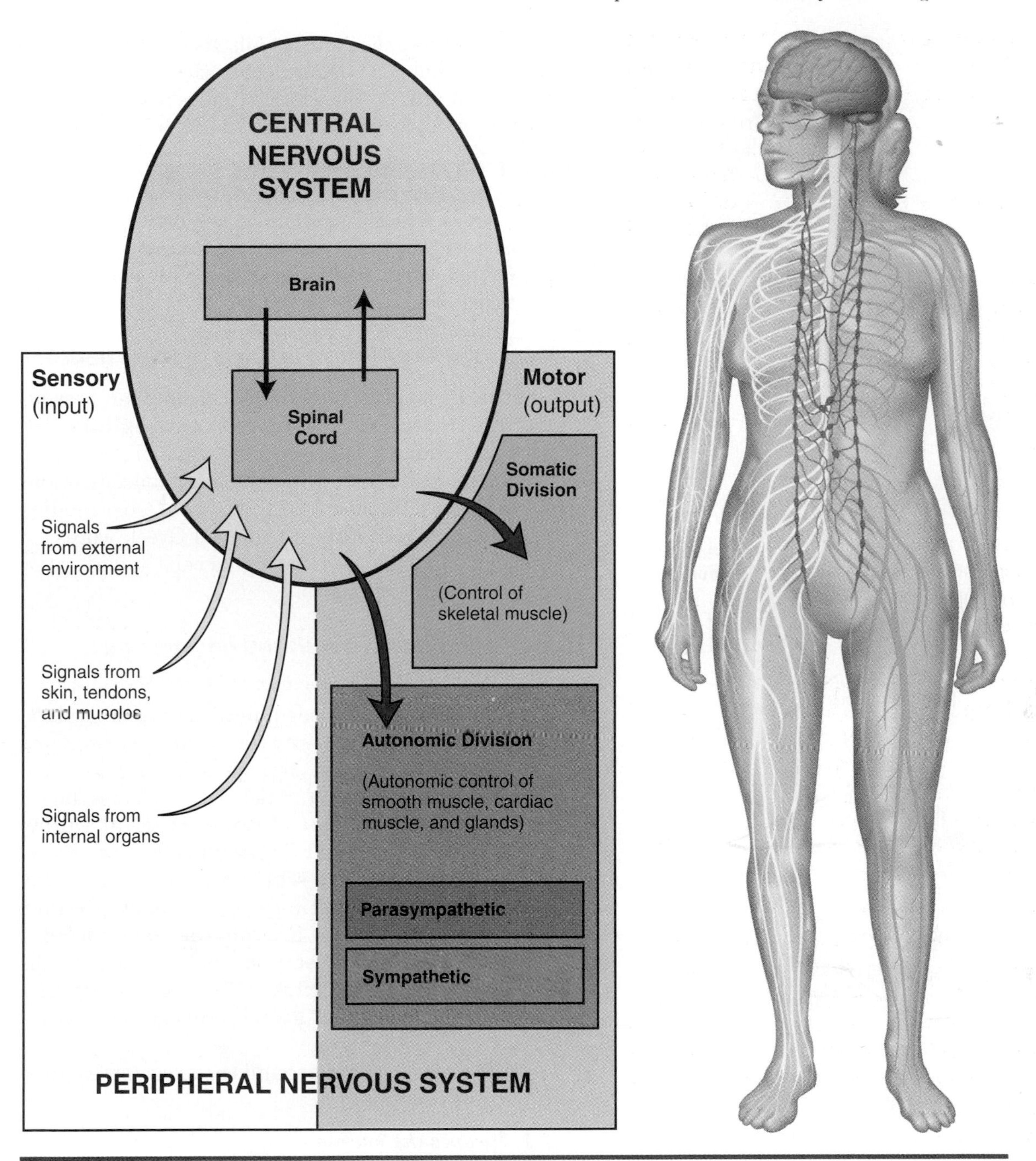

Figure 11.1 Components of the nervous system. The CNS receives input from the sensory component of the PNS, integrates and organizes the information, and then sends output to the periphery via the motor components of the PNS.

a neuron is called the **cell body.** The nucleus with its content of DNA is located in the cell body, as are the mitochondria and other cell organelles.

Slender extensions of the cell body, called **dendrites,** receive information from receptors or incoming impulses from other neurons. Interneurons and motor neurons have numerous dendrites that are fairly short and extend in many directions from the cell body. Sensory neurons are an exception, for their dendrites connect directly to an axon.

An **axon** is a long, slender tube of cell membrane containing a small amount of cytoplasm. Axons are specialized to conduct electrical impulses. Axons of sensory neurons originate from a dendrite, whereas the axons of interneurons and motor neurons originate from a cone-shaped area of the cell body called the *axon hillock.* At its other end, the axon branches into slender extensions called *axon terminals.* Each axon terminal ends in a small rounded tip called an *axon bulb.*

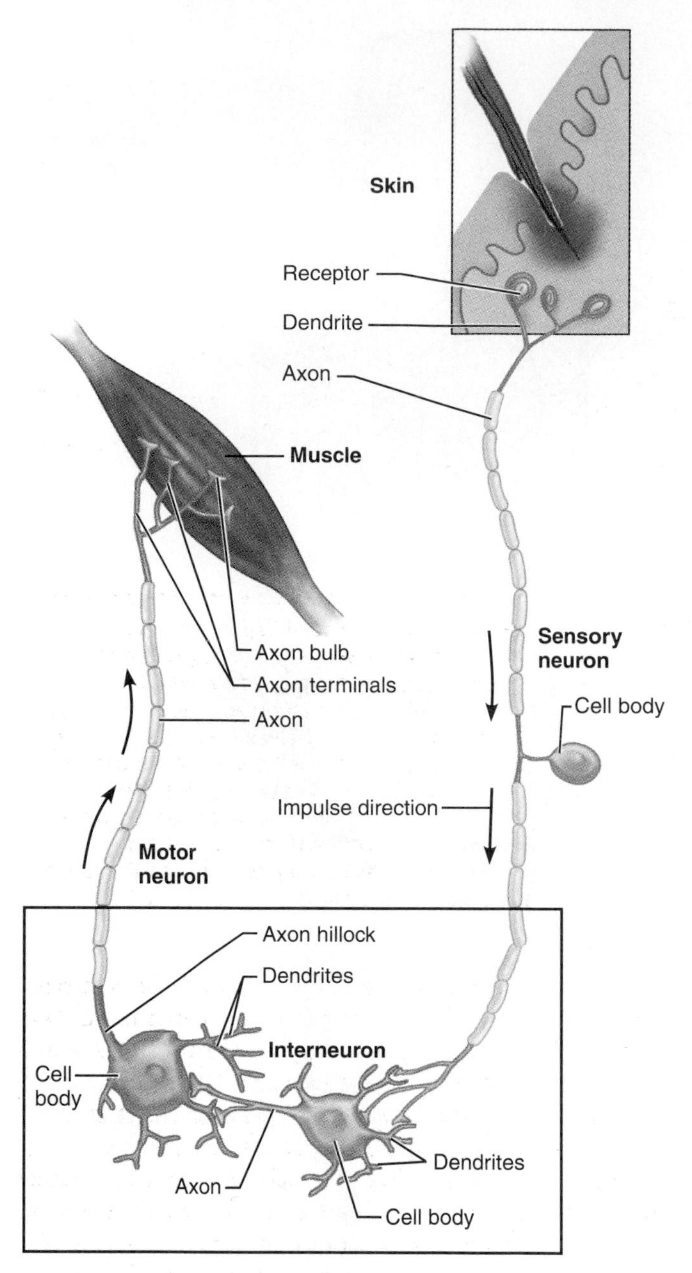

Figure 11.2 Types of neurons in the nervous system. The flow of information begins at a receptor near a dendritic ending of a sensory neuron and ends at the axon bulbs of a motor neuron. In this example the receptors are in skin. Interneurons are located entirely within the central nervous system and typically have short axons.

Typically, an interneuron or motor neuron receives incoming information from other neurons at its dendrites or cell body. If the incoming information is of the right kind and is strong enough, the neuron responds by generating an electrical impulse of its own at its axon hillock. In contrast, in a sensory neuron the impulse is initiated where the dendrite joins the axon. The impulse is then transmitted from one end of the axon to the other, bypassing the cell body entirely. We will talk more about different types of receptors in Chapter 12 (Sensory Systems).

Recap **Neurons generate and transmit electrical impulses from one part of the body to another. Sensory neurons transmit impulses to the CNS. Interneurons transmit impulses between components of the CNS. Motor neurons transmit impulses away from the CNS to muscles and glands.**

Neuroglial cells support and protect neurons

Only about 20 percent of cells in the human nervous system are neurons. The rest are **neuroglial cells,** which provide physical support and protection to neurons and help maintain healthy concentrations of important chemicals in the fluid surrounding them. (*Neuroglia* derives from the Greek words for "neuron" and "glue.") Neuroglial cells do not generate or transmit impulses.

A myelin sheath insulates neurons and speeds impulses

Spotlight on Structure & Function

In the peripheral nervous system, many neuron axons are enclosed and protected by specialized neuroglial cells called *Schwann cells*. Schwann cells produce a fatlike insulating material called **myelin.** During development, individual Schwann cells wrap themselves around a short segment of an axon many times as a sort of insulating blanket, creating a shiny white protective layer around the axon called a **myelin sheath.** Between adjacent Schwann cells are short uninsulated gaps called *nodes of Ranvier* where the surface of the axon is still exposed (Figure 11.3).

The myelin sheath around the axon serves three important functions:

1. *It saves the neuron energy.* The insulating layer of myelin prevents some of the slow inward leak of sodium and outward leak of potassium that would otherwise occur. These leaks normally have to be replaced by active transport processes requiring energy (see Section 11.3).
2. *It speeds up the transmission of impulses.* Because myelin insulates a segment of the axon electrically in the same way that rubber insulates an electrical wire, electrical impulses traveling down the axon are forced to jump rapidly from one uninsulated node of Ranvier to the next. This "leaping" pattern of conduction along myelinated neurons is called *saltatory conduction* (from *saltare*, Latin for "dance"). Impulses travel at a speed of only about 5 mph (2.3 meters per second) along unmyelinated

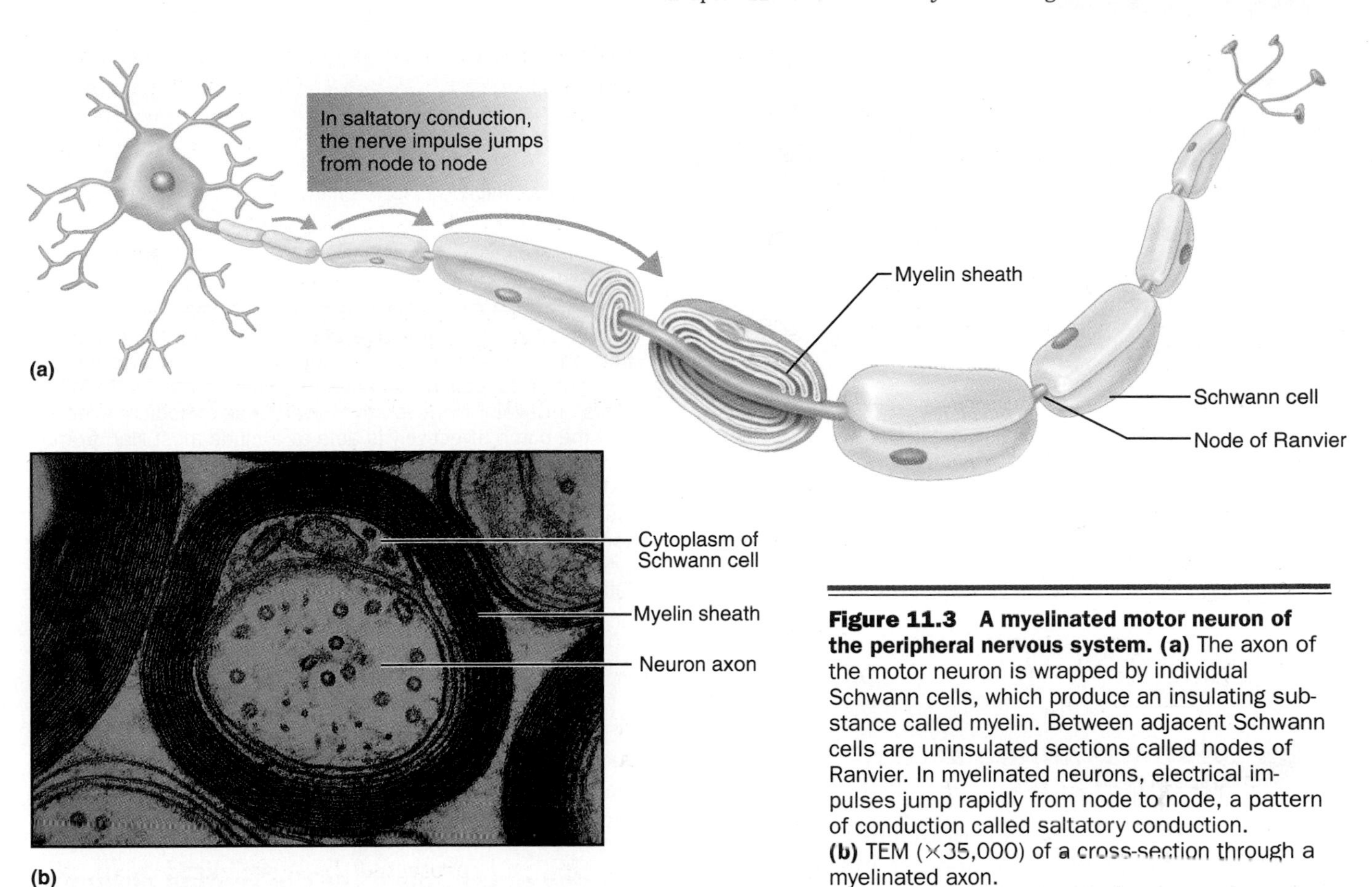

Figure 11.3 A myelinated motor neuron of the peripheral nervous system. (a) The axon of the motor neuron is wrapped by individual Schwann cells, which produce an insulating substance called myelin. Between adjacent Schwann cells are uninsulated sections called nodes of Ranvier. In myelinated neurons, electrical impulses jump rapidly from node to node, a pattern of conduction called saltatory conduction. **(b)** TEM (×35,000) of a cross-section through a myelinated axon.

neurons. In contrast, they travel at approximately 250 mph (approximately 110 meters per second) in myelinated neurons.

3. *It helps damaged or severed axons of the peripheral nervous system regenerate.* If a neuron axon is severed, the portion of the axon distal to the cell body may degenerate. However, the axon can regrow through the channel formed by the sheath, eventually reconnecting with the cell that it serves. Depending on the length of the axon, the regeneration process can take anywhere from a few weeks to more than a year. On average, an axon in the peripheral nervous system grows by nearly 1/16 inch (approximately 1.5 millimeters) per day. ■

In the central nervous system, protective sheaths of myelin are produced by another type of neuroglial cell called an *oligodendrocyte.* Unlike the sheath formed by Schwann cells, however, the sheath formed by oligodendrocytes degenerates once the axon it protects is destroyed. Consequently, neurons of the central nervous system do not regenerate after injury. This is why spinal cord injuries and CNS disorders such as multiple sclerosis result in a permanent change or loss of function (see "Disorders of the Nervous System" at the end of this chapter).

In the autoimmune disorder *multiple sclerosis (MS),* the sheaths of myelinated neurons in the brain and spinal cord become progressively damaged until they form hardened (sclerotic) scar tissue. These sclerotic areas can no longer insulate the neurons effectively, which disrupts and slows the transmission of impulses. People with MS can experience a variety of symptoms depending on which areas of the CNS are damaged. Common symptoms include muscle weakness, visual impairment, and urinary incontinence. The disease usually appears between 20 and 40 years of age, and although it is generally progressive, it may go through periods of remission.

A similar but rarer disease is amyotrophic lateral sclerosis (ALS). In ALS the sclerotic areas typically begin in regions of the spinal cord involved in the motor control of skeletal muscle. The primary symptom of ALS is progressive weakening and wasting of skeletal muscle tissue, most notably the diaphragm and intercostal muscles. As mentioned in Chapter 10 (The Respiratory System), people with ALS usually die from respiratory failure.

Recap **Neuroglial cells support and protect neurons. Neuroglial cells called Schwann cells (in the PNS) and oligodendrocytes (in the CNS) form myelin sheaths that protect axons and speed transmission of impulses.**

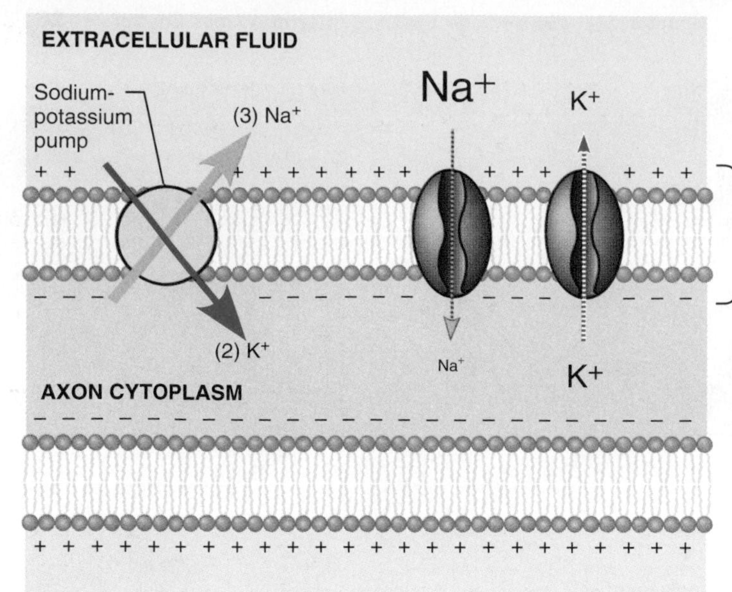

Figure 11.4 Maintenance of the resting membrane potential. The sodium-potassium pump actively transports three Na^+ out of the cell in exchange for two K^+. Most of the K^+ leaks out again. However, the inward leak of sodium is so slow that the pump effectively is able to exclude most Na^+ from the cell. The exclusion of positive ions (Na^+) from the cell creates a voltage difference (an electrical potential) across the cell membrane of about −70 mV (inside negative). The relative concentrations of Na^+ and K^+ in the extracellular fluid and the cytoplasm are indicated by the sizes of the letters associated with the diffusion channels.

11.3 Neurons initiate action potentials

The function of a neuron is to transmit information from one part of the body to another in the form of electrical impulses. Next, we look at how a neuron initiates and conducts an impulse.

Sodium-potassium pump maintains resting potential

Recall from Chapter 3 that the plasma membrane of all cells in the body (including neurons) is selectively permeable. In fact it is considerably more permeable to potassium than to sodium. Plasma membranes also have an active transport protein called the *sodium-potassium pump* that transports three sodium ions (Na^+) out of the cell in exchange for two potassium ions (K^+). The primary function of the sodium-potassium pump is to maintain cell volume by maintaining unequal distributions of sodium and potassium across the cell membrane. But because the sodium-potassium pump transports three Na^+ out for every two K^+ in, the net effect is removal of positive charges from the cytoplasm. In addition, because K^+ then leaks out more easily than Na^+ leaks in, even more positive charges leave the cytoplasm. The end result is that the interior of the cell is electrically negative compared to the exterior.

The slight difference in voltage (electrical potential) between the inside and outside of a cell is called the **resting potential.** When a neuron is not transmitting an impulse its resting potential is about −70 millivolts (mV). Figure 11.4 illustrates the role of the sodium-potassium pump and the processes of diffusion of Na^+ and K^+ (through channels) in maintaining the resting potential.

Graded potentials alter the resting potential

The resting potential of a neuron undergoes a change when an impulse arrives from another neuron. Every incoming impulse produces, in a series of steps to be discussed in a moment, a small transient change in the resting potential in a local area of the membrane. Depending on the type of signal and its strength, the change might be to *depolarize* the membrane (move the voltage closer to zero), or *hyperpolarize* it (make it even more negative). These transient local changes in resting potential are called **graded potentials** because they can vary in size. They occur only at a single region on the membrane, fading away at increasing distances from that region like the ripples made by a raindrop on water. A neuron may receive hundreds of incoming signals from other neurons, so hundreds of these graded potentials may be occurring at the same time.

A key feature of graded potentials is that they can sum in space and time, meaning that many incoming signals from other neurons produce a bigger change in membrane potential than just one impulse alone. This is known as **summation.**

Recap **A neuron's resting potential of about −70 millivolts is maintained by the constant action of the sodium-potassium pump. Impulses arriving from other neurons can cause small, local changes in the neuron's membrane potential called graded potentials.**

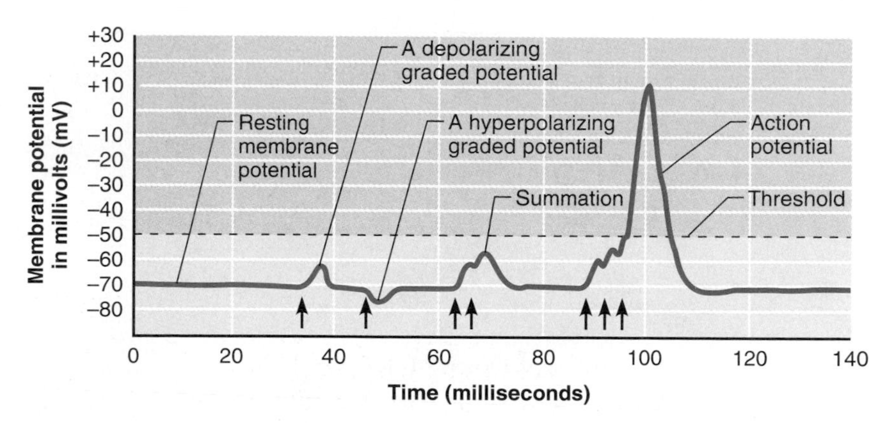

Figure 11.5 The resting membrane potential, graded potentials, and an action potential in a neuron. Stimulation by other neurons is indicated by the arrows. Note that graded potentials can either depolarize or hyperpolarize the membrane and that they can summate. If the sum of all graded potentials is sufficient to depolarize the membrane to threshold, an action potential occurs.

Action potential is a sudden reversal of membrane voltage

If the sum of all graded potentials is sufficiently strong to reach a certain triggering membrane voltage called the **threshold,** an *action potential* results. An **action potential** is a sudden, temporary reversal of the voltage difference across the cell membrane. Once an action potential is initiated, it sweeps rapidly down the axon at a constant amplitude and rate of speed until it reaches the axon terminals.

What triggers a nerve cell to send an impulse?

Figure 11.5 depicts the resting membrane potential, graded potentials and how they summate, and an action potential.

Functionally, an action potential (impulse) is the only form in which information is transmitted long distance by the nervous system. An action potential is like a single monotonal "beep" in a touch-tone phone or a single on-off pulse of current. Its meaning is provided only by where it came from and where it goes—that is, by how the nervous system is wired.

Action potentials occur because the axon membrane contains *voltage-sensitive ion channels* that open and close sequentially once threshold has been reached. An action potential occurs as a sequence of three events: (1) depolarization, (2) repolarization, and (3) reestablishment of the resting potential (Figure 11.6).

1. *Depolarization: sodium moves into the axon.* When threshold is exceeded, Na^+ channels in the axon's membrane open briefly and Na^+ ions diffuse rapidly into the cytoplasm of the axon. This influx of positive ions causes *depolarization,* meaning that the membrane potential shifts from negative (−70 mV) to positive (about +30 mV).
2. *Repolarization: potassium moves out of the axon.* After a short delay the Na^+ channels close automatically. But the reversal of the membrane polarity triggers the opening of potassium channels. This allows more K^+ ions than usual to diffuse rapidly out of the cell. The loss of positive ions from the cell leads to *repolarization,* meaning that the interior of the axon becomes negative again.
3. *Reestablishment of the resting potential.* Because the potassium channels are slow to close, there is a brief overshoot of membrane voltage during which the interior of the axon is slightly hyperpolarized. Shortly after the potassium channels close, the resting potential is reestablished by the normal activity of the sodium-potassium pumps, and the axon is prepared to receive another action potential. The entire sequence of three steps takes about 2 milliseconds.

While an action potential is underway, an axon is physically incapable of generating another action potential under any circumstances. This period is called the *absolute refractory period.* The absolute refractory period is followed by a brief *relative refractory period* during which it is harder than usual to generate the next action potential. The presence of these refractory periods ensures that action potentials always travel in one direction only, because an area of the axon that has just experienced an action potential cannot produce another one right away.

Whether or not a neuron is generating action potentials, the sodium-potassium pumps continue to operate to maintain the normal concentrations of Na^+ and K^+ and the normal resting potential. The actual number of Na^+ and K^+ ions that diffuse across the membrane during a single action potential is so small compared to the activity of the sodium-potassium pumps that an axon can transmit many action potentials in a short period of time without adversely affecting the subsequent resting membrane potential.

Action potentials are all-or-none and self-propagating

As we have seen, an action potential does not occur unless a certain threshold membrane voltage is reached. Once it is triggered, however, it always looks

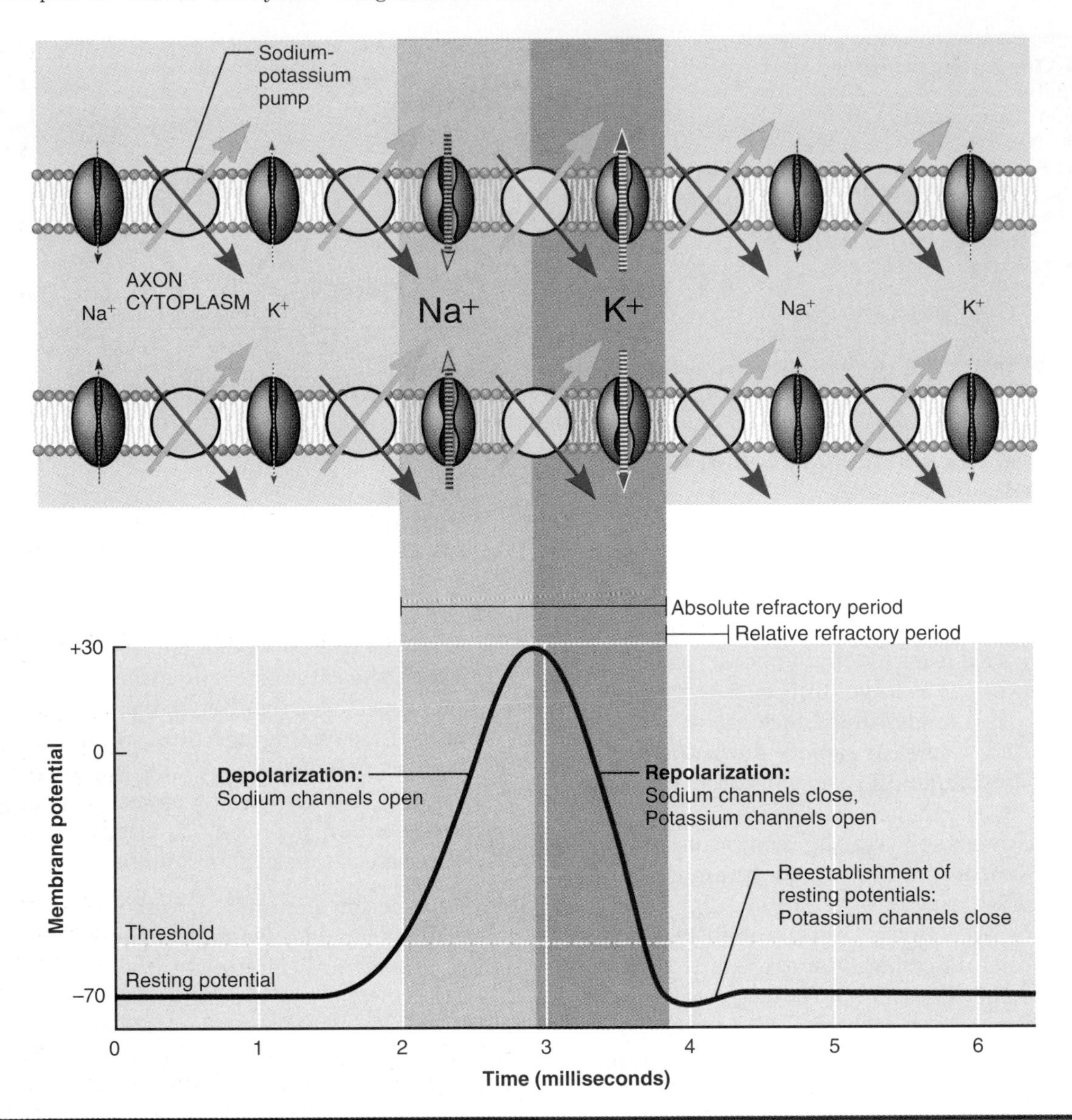

Figure 11.6 Propagation of an action potential. This is a graph of changes in potential over time recorded from a single point on the axon. During the depolarization phase, voltage-sensitive sodium channels open briefly and sodium diffuses rapidly into the axon, depolarizing the membrane. After a brief period, the sodium channels close, potassium channels open, potassium diffuses outward, and the membrane repolarizes. Throughout all phases of the action potential, the sodium-potassium pump continues to transport sodium and potassium so that the proper ion concentrations are either maintained or reestablished.

exactly the same in form and voltage. Even if the graded stimulus that causes it is significantly higher than the threshold level, the form and amplitude of the action potential is always the same. Consequently an action potential is an "all-or-none" phenomenon: either it occurs, or it does not occur. It's like firing a gun. A certain amount of pressure—the threshold level—is required to make the gun fire. Pressing the trigger too lightly will not make it fire, and pressing it harder than necessary will not cause the bullet to leave the gun any faster.

An action potential is also a self-propagating event, meaning that it continues to create itself in the next region of the axon. This is because the local spread of electrical current in the area that is currently undergoing an action potential causes threshold to be reached in the area next to it, opening the voltage-sensitive sodium channels there as well. The action potential sweeps down the axon like a wave, at a constant rate of speed and amplitude.

How fast can a nerve impulse travel?

Although the speed of travel of an action potential is always constant for a given neuron, the speed in different neurons can vary from about 5 mph up to 250 mph. Action potential speed is increased either by increasing the diameter of the axon or by the presence of a myelin sheath. As we saw earlier, action potentials travel faster along myelinated neurons because they jump from one node of Ranvier to the next.

Besides the generation of a single action potential, the only additional information that can be transmitted by a neuron is how frequently action potentials are generated per unit of time. Within a single neuron, action potential frequencies generally range from near zero to perhaps several hundred per second.

Recap **The sum of all graded potentials may cause the initiation a self-propagating, all-or-none action potential in a neuron. An action potential involves three events: depolarization, repolarization, and reestablishment of the resting potential.**

11.4 Information is transferred from a neuron to its target

Once an action potential reaches the axon terminals of a neuron, the information inherent in it must be converted to another form for transmittal to its target (muscle cell, gland cell, or another neuron). In essence, the action potential causes the release of a chemical that crosses a specialized junction between the two cells called a **synapse.** This chemical substance is called a **neurotransmitter** because it transmits a signal from a neuron to its target. The entire process of transmission from a neuron to its target is called *synaptic transmission*. You have already seen how synaptic transmission occurs between a neuron and skeletal muscle (Chapter 6, The Muscular System). Now we'll see how it occurs between neurons.

Neurotransmitter is released

Figure 11.7 illustrates the structure of a typical synapse and the events that occur during synaptic transmission. At a synapse, the *presynaptic membrane* is the cell membrane of the neuron that is sending the information. The *postsynaptic membrane* refers to the membrane of the cell that is about to receive the information. The small fluid-filled gap that separates the presynaptic and postsynaptic membranes is the *synaptic cleft*.

The axon terminal of the presynaptic neuron ends in an axon bulb that contains neurotransmitter stored in membrane-bound vesicles. The events that occur during synaptic transmission follow a particular pattern:

1. An action potential arrives at the axon bulb, causing calcium (Ca^{2+}) channels in the presynaptic membrane to open. Ca^{2+} diffuses into the axon bulb.
2. The presence of Ca^{2+} causes the vesicles to fuse with the presynaptic membrane and release their contents of neurotransmitter directly into the synaptic cleft. Because the synaptic cleft is so narrow (about 0.02 micrometer, or one-millionth of an inch), the neurotransmitter molecules reach the postsynaptic membrane by diffusion.
3. Molecules of neurotransmitter bind to receptors on the postsynaptic membrane. This causes certain chemically gated channels, such as those for sodium, to open.
4. Sodium ions diffuse inward, producing a graded depolarization of the postsynaptic membrane in the area of the synapse.

Note that graded potentials are caused by the opening of *chemically* sensitive ion channels (the chemical being a neurotransmitter), rather than *voltage*-sensitive ones.

Neurotransmitters exert excitatory or inhibitory effects

The response of the postsynaptic cell to neurotransmitter depends on several factors, including the type of neurotransmitter, how concentrated it is in the synaptic cleft, and the types of receptors and chemically sensitive ion channels in the postsynaptic membrane.

To date, scientists have identified more than 50 chemicals that can function as neurotransmitters. All are stored in vesicles within the axon bulb and released into the synaptic cleft in response to an action potential. All produce a graded potential in the postsynaptic cell, though the effects may be quite different because they can affect different ion channels.

Neurotransmitters are classified as excitatory, inhibitory, or both. *Excitatory* neurotransmitters depolarize the postsynaptic cell, causing threshold to be approached or exceeded. Therefore, excitatory neurotransmitters encourage the generation of new impulses in the postsynaptic neuron. *Inhibitory* neurotransmitters cause the postsynaptic cell to hyperpolarize, meaning that the cell's interior becomes even more negative than before. Hyperpolarization makes it harder for threshold to be reached, so inhibitory neurotransmitters tend to prevent the generation of action potentials in the postsynaptic neuron. Some neurotransmitters can be either excitatory or inhibitory, depending on the type of receptor with which they bind on the postsynaptic membrane.

The effects of neurotransmitters are relatively short-lived because the neurotransmitter remains in the synaptic cleft for only a short time. Prompt removal of neurotransmitter is important because it allows neural signals to be terminated almost as

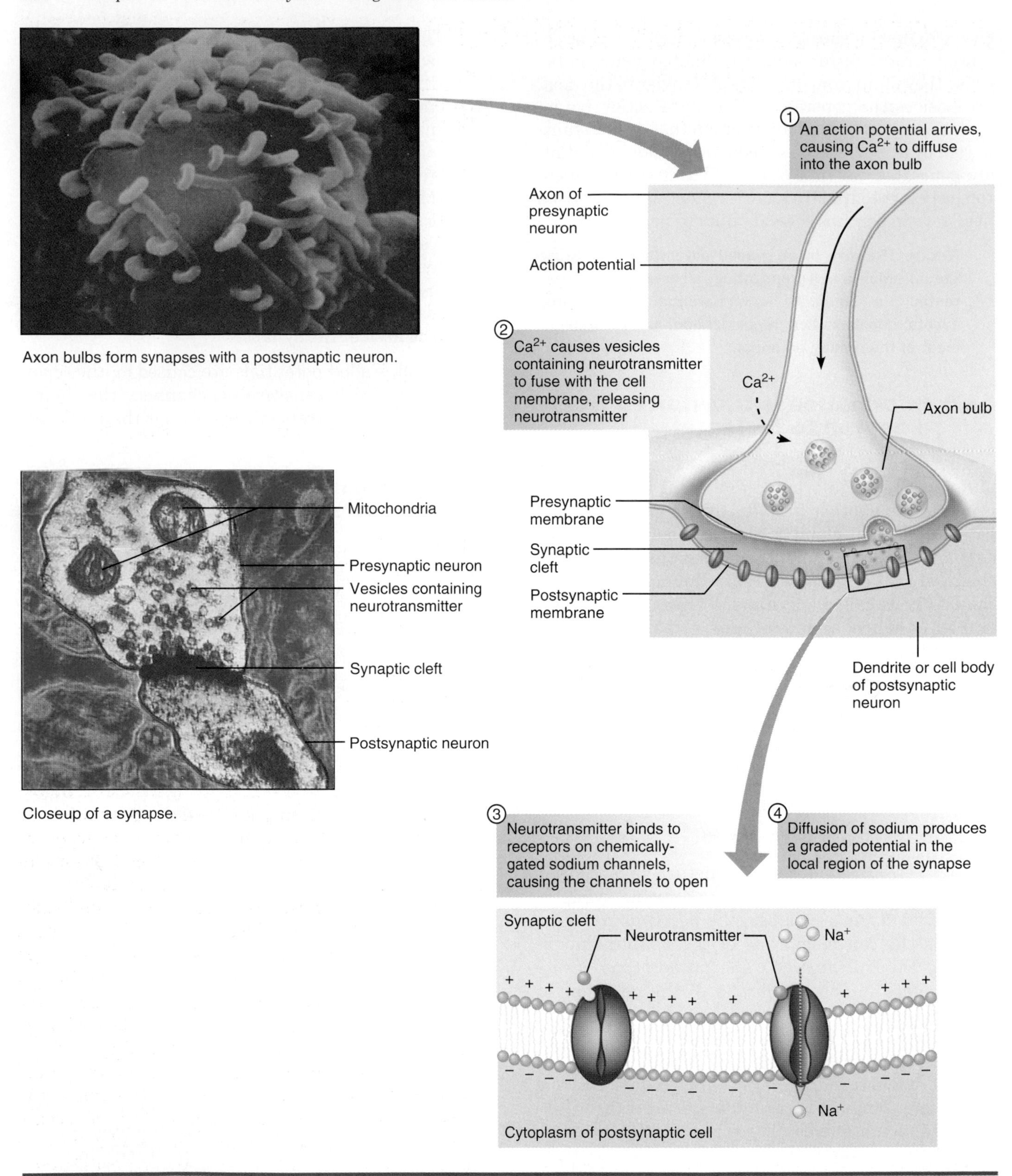

Figure 11.7 Summary of synaptic transmission.

rapidly as they are initiated. Only then can the next message be received and recognized. The neurotransmitter may be removed from the synaptic cleft in any of three ways: (1) it may be taken back up again by the presynaptic neuron and repackaged into membrane-bound vesicles, to be used again; (2) it may be destroyed by enzymes in the synaptic cleft; or (3) it may diffuse away from the synaptic cleft into

Table 11.1 Actions of Common Neurotransmitters

Neurotransmitter	Sites Where Released	Principal Actions
Acetylcholine	Brain Neuromuscular junctions Autonomic nervous system	Excitatory on skeletal muscles Excitatory or inhibitory on internal organs
Norepinephrine	Areas of brain and spinal cord Autonomic nervous system	Excitatory or inhibitory, depending on receptors Plays a role in emotions
Serotonin	Areas of brain Spinal cord	Usually inhibitory Involved in moods, sleep cycle, appetite
Dopamine	Areas of brain Parts of peripheral nervous system	Excitatory or inhibitory, depending on receptors Plays a role in emotions
Glutamate	Areas of brain Spinal cord	Usually excitatory Major excitatory neurotransmitter in brain
Endorphins	Many areas in brain Spinal cord	Usually inhibitory Natural opiates that inhibit pain
Gamma-aminobutyric acid	Areas of brain Spinal cord	Usually inhibitory Principal inhibitory neurotransmitter in brain
Somatostatin	Areas of brain Pancreas	Usually inhibitory Inhibits release of growth hormone

the general circulation, where it will ultimately be destroyed.

Table 11.1 summarizes the actions of these and several other common neurotransmitters.

Recap **A neuron transmits information to another cell across a synapse. At a synapse, neurotransmitter released from a presynaptic neuron diffuses across the synaptic cleft and binds to receptors on the postsynaptic membrane, opening ion channels and causing a graded potential in the postsynaptic cell.**

Postsynaptic neurons integrate and process information

The conversion of the signal from electrical (an action potential) to chemical (neurotransmitter) allows the postsynaptic cell to do a lot of integration and information processing. This is because a single all-or-none action potential in the presynaptic neuron does *not* always cause a corresponding all-or-none action potential in the postsynaptic neuron, as it does if the target is a muscle cell. A single action potential produces only a small graded potential in the postsynaptic neuron, and a lot of these graded potentials generally are required for threshold to be reached.

One way for threshold to be reached in a postsynaptic neuron is for the presynaptic neuron (or a few neurons) to increase their frequency of stimulation, sending lots of action potentials in a short time (hundreds per minute). In addition to this effect of stimulus intensity, whether threshold is reached in the postsynaptic neuron depends on: (1) how many other neurons synapse with it, and (2) whether those synaptic connections are stimulatory or inhibitory. Typically, one neuron receives input from many neurons, a phenomenon known as *convergence*. In turn, its action potential will go to many other neurons, a phenomenon known as *divergence* (Figure 11.8).

Though it would be incorrect to say that neurons "interpret" information—that would imply that a neuron has a brain—the effect of convergence is that neurons may integrate and process millions of simultaneous incoming stimulatory and inhibitory signals before generating and transmitting their own action potentials. They also reroute the information to many different destinations. Individual neurons do not see, smell, or hear, yet their combined actions allow us to experience those complex sensations.

Consider again the example given at the beginning of this chapter, in which you were able to ignore certain incoming signals—the smell of donuts, the sound of your shoes on pavement, the sensation of being cold—and focus on others, such as the noises made by the truck and its horn. You tapped your memory banks for the probable meaning of these sounds (danger approaching!), and then coordinated your muscles to turn around, look at the truck, and jump out of the way—all in a fraction of a second. This is a complex set of events, and it is all done with neurons, action potentials, and neurotransmitters.

Muscle cells are also target cells of presynaptic neurons. As you learned in Chapter 6, however, skeletal muscle cells do not process and integrate information. There are two reasons for this. First, a muscle cell receives input from only one neuron. There are no converging neural inputs to process.

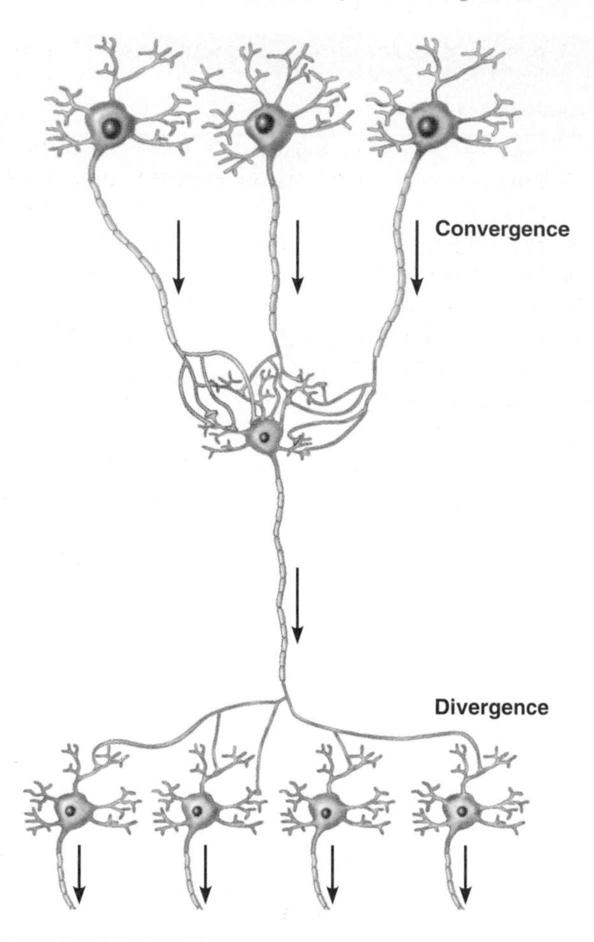

Figure 11.8 Neural information processing by convergence and divergence. Convergence occurs when several presynaptic cells form synapses with a single postsynaptic cell. Divergence occurs when one nerve cell provides information to a variety of postsynaptic cells.

Second, the neuromuscular junction is quite large, with lots of points of contact between neuron and muscle cell. Consequently, threshold is always reached in a skeletal muscle cell every time the motor neuron sends even a single action potential. Functionally, what this means is that the nervous system has absolute and final control over the activity of skeletal muscle cells.

> ***Recap*** **Postsynaptic neurons integrate and process incoming signals. Information processing occurs because of the complex patterns of divergence and convergence between neurons, and because many graded potentials must sum in order to produce an action potential.**

11.5 The PNS relays information between tissues and the CNS

Now that we have seen how neurons generate and transmit action potentials, we will look at how this information travels throughout the nervous system. As stated earlier, the nervous system has two major subdivisions: the central nervous system (CNS), which includes the spinal cord and brain, and the peripheral nervous system (PNS), which consists of nerves that transmit information between body tissues and the CNS.

Nerves carry signals to and from CNS

Action potentials are carried to and from the central nervous system along information pathways in the peripheral nervous system called nerves. **Nerves** are cablelike bundles of many neuron axons all wrapped together in a protective connective tissue sheath like a large communications cable. The peripheral nervous system consists of cranial nerves and spinal nerves. The function of each nerve depends on where it originates and the organs or tissues to which it travels.

There are 12 pairs of **cranial nerves,** which connect directly with the brain. Cranial nerves carry action potentials between the brain and the muscles, glands, and receptors of the head, neck, and thoracic and abdominal cavities.

There are 31 pairs of **spinal nerves,** which connect with the spinal cord. Each spinal nerve attaches to the spinal cord via two short branches of the spinal cord called the dorsal root and the ventral root. The *dorsal root* contains the sensory neurons that transmit incoming action potentials from body tissues to the spinal cord. The *ventral root* contains motor neurons that carry action potentials away from the spinal cord to the rest of the body. Thus each spinal nerve conveys both sensory and motor information.

> ***Recap*** **In the peripheral nervous system, 12 pairs of cranial nerves transmit action potentials between the brain and the head, neck, and thoracic and abdominal cavities. The 31 pairs of spinal nerves transmit action potentials to and from the spinal cord to all regions of the body.**

Sensory neurons provide information to the CNS

The nervous system must have good information if it is to take appropriate action. That information arrives at the CNS as action potentials traveling in sensory neurons located throughout the body. Because information provided by sensory neurons to the CNS may affect the motor output of both the somatic and autonomic divisions of the PNS, the sensory (input) component of the PNS cannot be classified as belonging specifically to either the somatic or autonomic motor (output) divisions.

Somatic division controls skeletal muscles

Recall that the motor (or output) neurons of the peripheral nervous system are organized into the somatic division and the autonomic division. The **somatic division** controls voluntary and involuntary skeletal muscle movement. Somatic motor neurons transmit information from the spinal cord to skeletal muscles.

As we have seen, a single somatic motor neuron controls several skeletal muscle cells (together they form a *motor unit*), and every time the neuron sends an action potential down its axon, all the muscle cells of that motor unit experience an action potential and contract. Motor neurons are activated either by conscious control from the brain or by an involuntary response called a *reflex*.

Spinal reflexes are involuntary responses that are mediated primarily by the spinal cord and spinal nerves, with little or no involvement of the brain. Figure 11.9 illustrates two such spinal reflexes. Stepping on a sharp object, for example, activates pain receptors, producing action potentials in sensory neurons traveling to your spinal cord. The sensory neurons stimulate interneurons within the spinal cord, which in turn transmit signals to motor neurons in your leg, causing you to flex the appropriate muscles to lift your foot. This is called the *flexor (withdrawal) reflex*. Meanwhile, other interneurons that cross over to the opposite side of the spinal cord are also stimulated by the sensory neuron. These interneurons stimulate motor neurons to extend, rather than flex, certain skeletal muscles in your other leg. After all, it wouldn't do to flex both legs and fall on the sharp object! This second reflex is called a *crossed extensor reflex*.

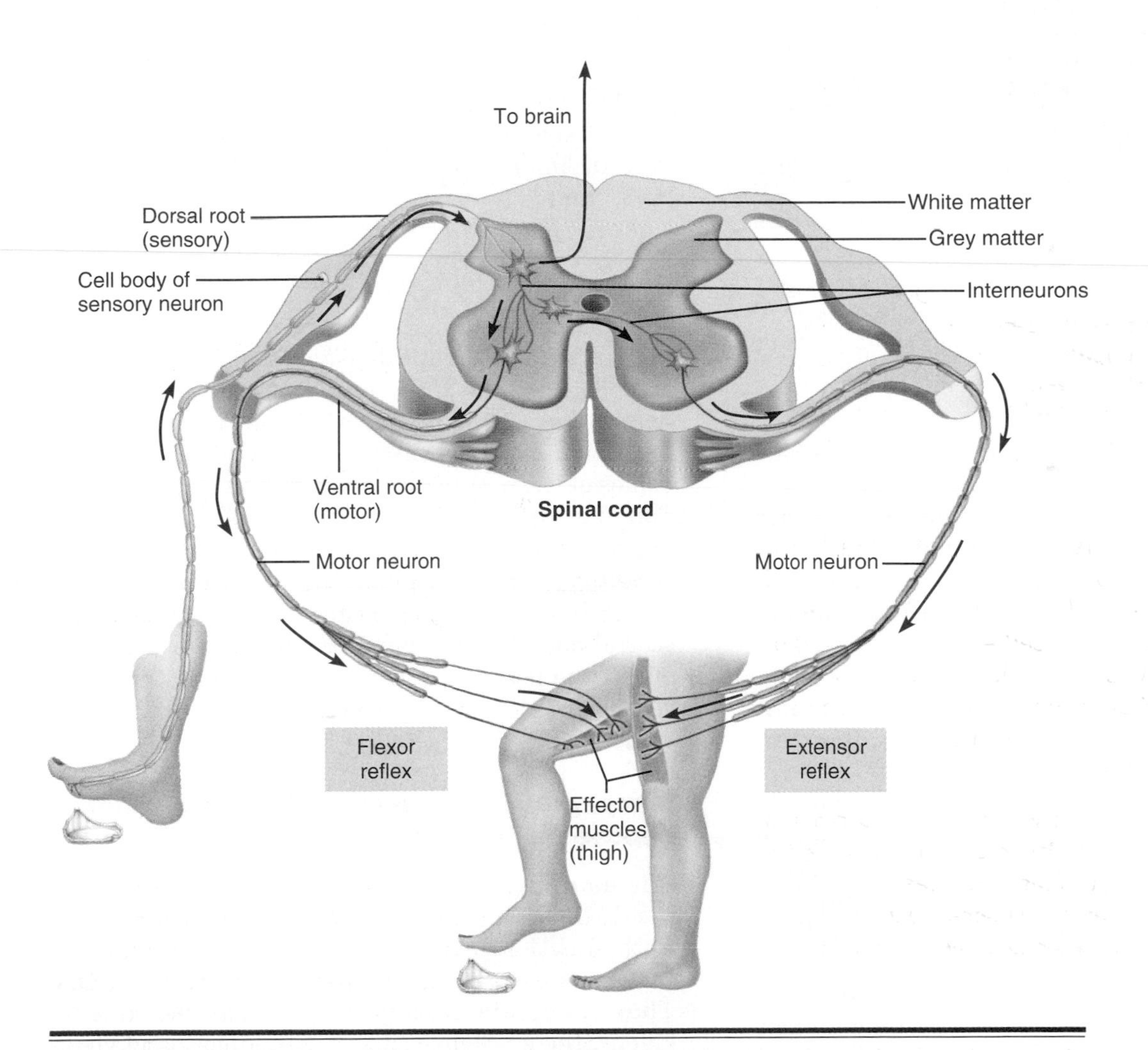

Figure 11.9 Spinal reflexes. In this example of withdrawal and crossed extension reflexes, a painful stimulus activates sensory nerves in a foot. Sensory nerves stimulate interneurons within the spinal cord that activate motor neurons to muscles on both sides of the body. Note that the motor effects are different on the two sides of the body: withdrawal (flexion) on the stimulus side and extension on the opposite side. Other interneurons relay the sensory information on to the brain.

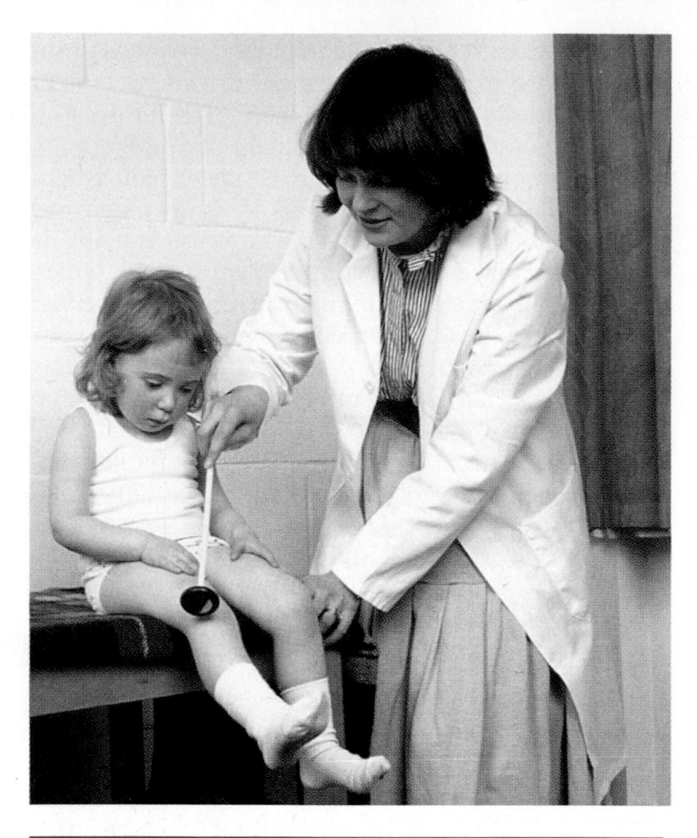

Figure 11.10 The patellar (knee-jerk) reflex. The physician tests the integrity of the patellar reflex to verify that certain sensory and motor nerves, the spinal cord, areas of the brain, and skeletal muscle are all functioning normally.

Note that these events did not require conscious thought. A simple spinal reflex is processed primarily at the level of the spinal cord, allowing a rather complex series of actions to occur quickly. Some sensory information does ascend to the brain as well, allowing the brain to integrate the sensory information from your foot with all the other information it is receiving and determine your next course of action (such as sitting down and examining your foot). But most of the spinal reflex itself occurs without any involvement of the brain.

Another type of spinal reflex is a *stretch reflex.* In a stretch reflex, stretch receptors in a skeletal muscle stimulate sensory nerves, which carry the signals to the spinal cord. In the spinal cord, motor neurons to the stretched muscle are activated. At the same time, motor neurons to any antagonistic muscles are inhibited so that contraction will be effective. Stretch reflexes play an important role in maintaining upright posture and movement, as they allow us to stand and move without having to concentrate on our actions.

Stretch reflexes associated with posture and movement are often modified slightly by signals from higher brain centers. Physicians can test such reflexes to determine whether the spinal cord and areas of the brain involved in stretch reflexes are intact and functioning properly. An example is the well-known "knee jerk" reflex. When the patellar tendon is tapped lightly below the kneecap, the resulting slight stretch of the thigh muscles initiates a spinal reflex that ultimately results in contraction of the thigh muscles and an upward movement of the foot and lower leg (Figure 11.10). If the neural pathways are not intact or if the brain is not modifying the reflex properly, the reflex may be either less vigorous or more vigorous than normal.

Recap **Sensory neurons provide information (input) to the CNS. The somatic output division of the PNS controls voluntary (conscious) and involuntary (reflex) skeletal muscle movement.**

Autonomic division controls automatic body functions

Motor neurons carrying information to smooth or cardiac muscle and to all other tissues and organs except skeletal muscle belong to the **autonomic division.** *Autonomic* derives from the Greek words for "self" and "law." The autonomic division of the peripheral nervous system carries signals from the CNS to the periphery that control "automatic" functions of the body's internal organs. Many of these activities occur without our ever being aware of them.

Unlike the somatic division in which one neuron extends from the CNS to the target cell, the autonomic division requires two neurons to transmit information from the CNS to the target cell. The cell bodies of the first neurons (called *preganglionic neurons*) lie within the central nervous system. The axons of many of these neurons go to structures called *ganglia* (plural of ganglion) that lie outside the CNS. A ganglion is a cluster of neuron cell bodies of the second neurons (called *postganglionic neurons*). Axons of the postganglionic neurons extend to the far reaches of the body to communicate with internal glands and organs.

Sympathetic and parasympathetic divisions oppose each other

There are two types of motor neurons in the autonomic division: sympathetic and parasympathetic. Both function automatically, usually without our conscious awareness, to regulate body functions. They innervate smooth muscle, cardiac muscle, glands, the epithelial cells of the kidneys, and virtually every internal organ except skeletal muscle, bone, and the nervous system itself.

In most organs, the actions of the sympathetic and parasympathetic divisions have very opposite effects, as shown in Figure 11.11. Which division is predominating at any one time will determine what happens in your body. It is as if you were driving a car with one foot on the gas pedal and one on the brake—you could speed up by hitting the gas pedal, releasing the brake, or both. The fact that there are two opposing types of motor neurons probably indicates the importance of tight regulatory control to the body.

Sympathetic division arouses body for "fight or flight" Preganglionic motor neurons of the **sympathetic division** originate in the thoracic and lumbar regions

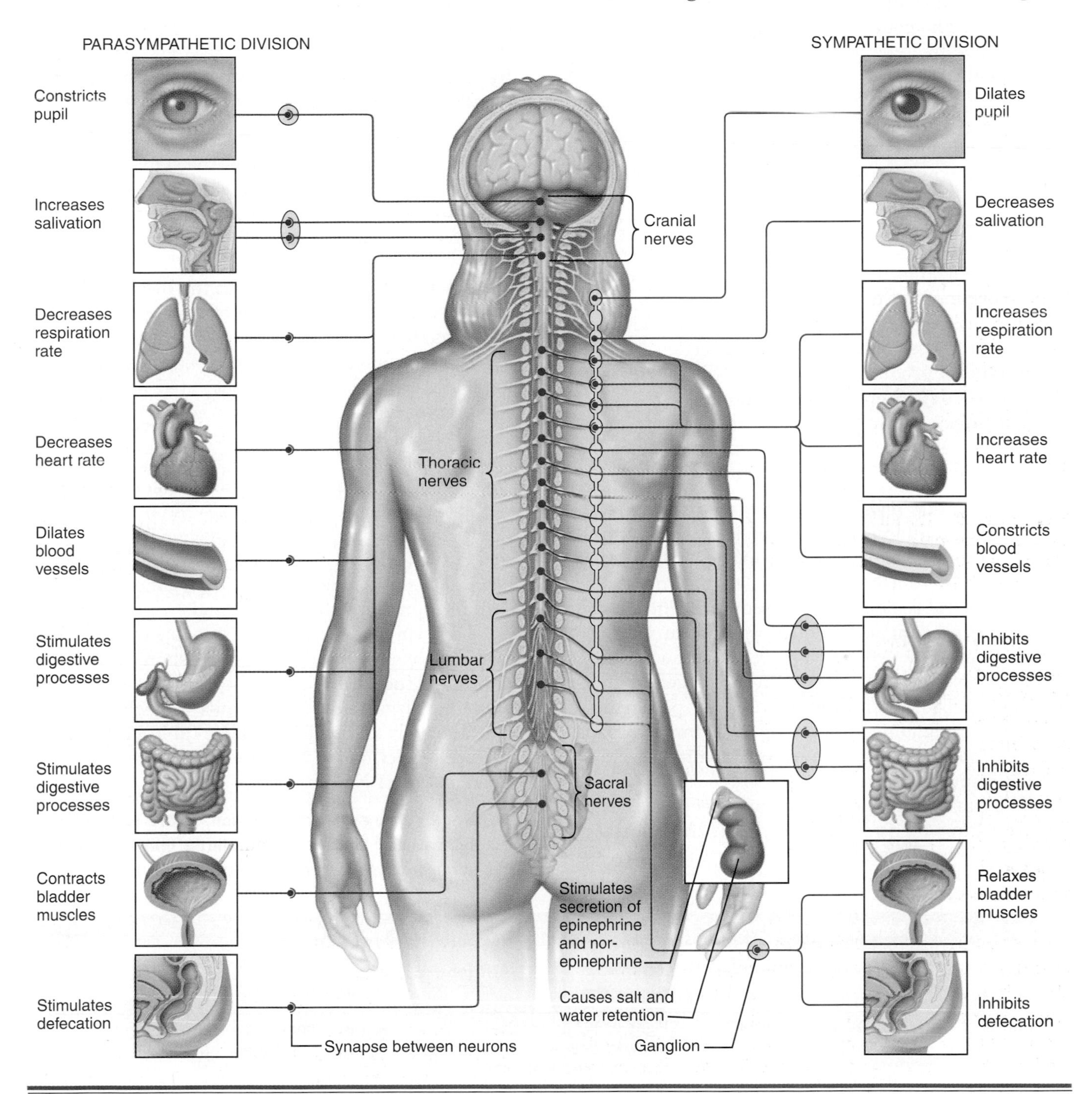

Figure 11.11 The sympathetic and parasympathetic divisions of the autonomic nervous system. The sympathetic diversion is associated with "fight or flight" responses. The parasympathetic division predominates in the body's relaxed state. The nerves of the sympathetic division arise in the thoracic and lumbar segments of the spinal cord, while the nerves of the parasympathetic division arise from cranial and sacral segments. Ganglia (groups of nerve cell bodies) are shown in yellow.

of the spinal cord. Many of their axons extend to a chain of sympathetic ganglia that lie alongside the spinal cord, where they synapse with the postganglionic cells. Because these ganglia are connected to each other, the sympathetic division tends to produce unified response in all organs at once. Axons of the postganglionic neurons branch out to innervate all of the organs. The neurotransmitter released by sympathetic postganglionic neurons in the target organs is norepinephrine.

The sympathetic division transmits signals that prepare the body for emergencies and situations requiring high levels of mental alertness or physical activity, such as play and sexual activity. It increases heart rate and respiration, raises blood pressure, and dilates the pupils—effects that help you detect and respond quickly to changes in your environment. The sympathetic division also reduces blood flow to organs that do not help you cope with an immediate crisis, such as the intestines and kidneys, and inhibits less important body functions such as digestion and production of saliva (this is why your mouth feels dry when you are anxious).

The effects of the sympathetic division are sometimes called the "fight-or-flight response," reflecting its association with emergency situations.

Parasympathetic division predominates during relaxation Preganglionic neurons of the **parasympathetic division** originate either in the brain, becoming part of the outflow of certain cranial nerves, or from the sacral region of the spinal cord. The ganglia where the preganglionic neurons synapse with the postganglionic neurons is generally some distance from the CNS and may even be in the target organ itself. The neurotransmitter released by parasympathetic postganglionic neurons at their target organs is acetylcholine, the same neurotransmitter used at the neuromuscular junction.

The parasympathetic division predominates in situations in which you are relaxed. It transmits signals that lower heart rate and respiration, increase digestion, and permit defecation and urination, among other actions. The parasympathetic division exerts calming, restorative effects that counteract the fight-or-flight stimulation of the sympathetic division. Parasympathetic nerves are also responsible for the vasodilation that causes erection in males and swelling of the labia and erection of the clitoris in females.

Sympathetic and parasympathetic divisions work antagonistically to maintain homeostasis

As we have seen, the actions of the sympathetic and parasympathetic divisions oppose each other. They work antagonistically to accomplish the automatic, subconscious maintenance of homeostasis. Both systems are intimately involved in feedback loops that control homeostasis throughout the body.

An example is the control of blood pressure. Recall (Chapter 8) that certain sensory nerves are activated by baroreceptors. When blood pressure is too high, the baroreceptors are activated and the sensory nerves send more frequent action potentials to the brain. The brain integrates this information with the other information it is receiving. Then it decreases the number of action potentials traveling in sympathetic nerves and increases the number of action potentials traveling in parasympathetic nerves serving the heart and blood vessels. As a result, heart rate slows, blood vessels dilate, and blood pressure falls until it is back to normal.

The sympathetic and parasympathetic divisions can also cause a widespread response, such as arousal of the entire body. They can even carry out many localized homeostatic control mechanisms simultaneously and independently. Table 11.2 summarizes the functions of the somatic and autonomic divisions of the PNS.

Recap **The autonomic division of the PNS includes sympathetic and parasympathetic components that serve internal organs and regulate automatic body**

Table 11.2 The Somatic and Autonomic Divisions of the PNS

Comparison	Somatic Division	Autonomic Division	
		Sympathetic	Parasympathetic
Functions	Serves skeletal muscles	Fight or flight; arouses body to deal with situations involving physical activity, mental alertness	Relaxes body; promotes digestion and other basic functions
Neurotransmitter	Acetylcholine	Norepinephrine	Acetylcholine
Number of neurons required to reach the target cell	One	Two (preganglionic and postganglionic)	Two (preganglionic and postganglionic)

functions in order to maintain homeostasis. The sympathetic division predominates in times of high activity or stress, and the parasympathetic division predominates during rest.

11.6 The brain and spinal cord comprise the CNS

A neuron does not think or love, yet humans do. Scientists are still trying to unravel the biology of emotion and thought, but we do know that they originate in the central nervous system, specifically in the brain.

The central nervous system, consisting of brain and spinal cord, is "central" both in location and action. It is where integration, or processing, of information occurs. It is where information is received and where complex outputs are initiated. It's where billions of action potentials traveling in millions of neurons all come together as a conscious thought. That is not a very satisfactory explanation for "thinking," but it may be the best we can do.

Bone, meninges, and the blood-brain barrier protect the CNS

The central nervous system is vital to proper functioning of the entire body, so it is no surprise that it is well protected against physical injury and disease.

First, the CNS is protected by bone. The brain is encased in the skull, and the spinal cord is enclosed in a hollow channel within the vertebrae. This shell of bone around the central nervous system helps shield it from physical trauma.

Second, the CNS is enclosed by three membranes of connective tissue, called **meninges,** named (from outermost to innermost layers) the *dura mater,* the *arachnoid,* and the *pia mater.* These three meninges protect the neurons of the CNS and the blood vessels that service them.

Third, the CNS is bathed in its own special liquid, called **cerebrospinal fluid,** which fills the space between the arachnoid and the pia mater. In addition to serving as a liquid shock absorber around the brain and spinal cord, cerebrospinal fluid tends to isolate the central nervous system from infections. Inflammation of the meninges due to a bacterial or viral infection is called *meningitis.* Meningitis can be serious because it may spread to the nervous tissue of the CNS.

Cerebrospinal fluid is somewhat similar to the extracellular fluid that bathes all cells, but it does not exchange substances as freely with blood. It is secreted from specialized capillaries into four hollow cavities (called *ventricles*) in the brain. From the ventricles, it flows through a system of fluid-filled cavities and canals, bathing and surrounding the neurons in the brain and spinal cord (Figure 11.12).

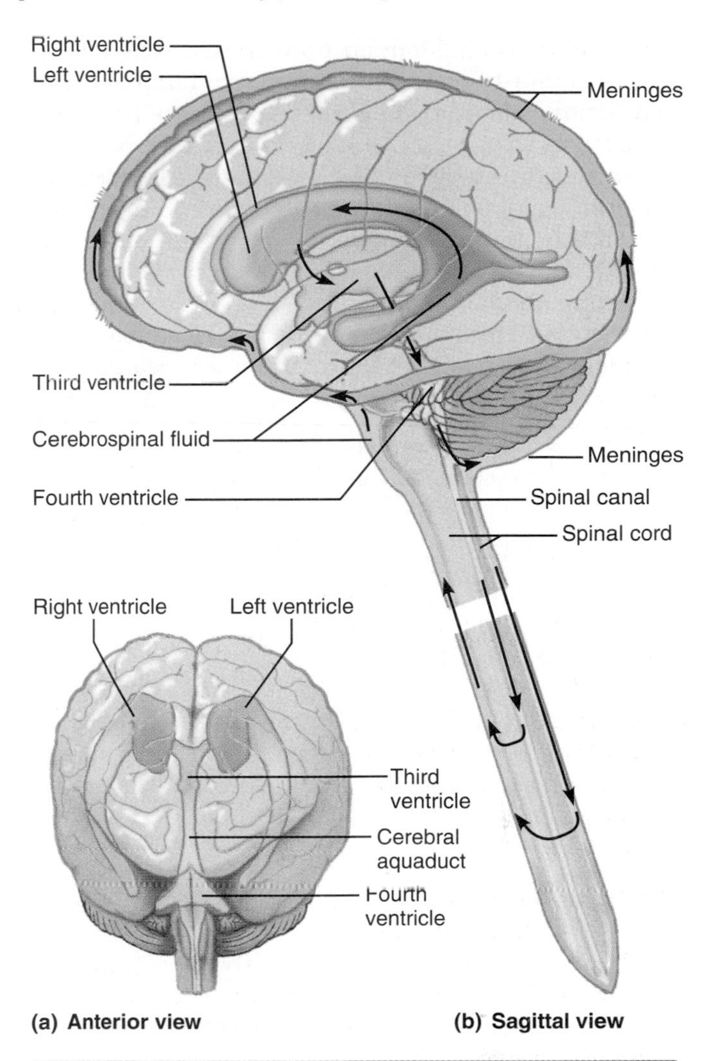

Figure 11.12 The ventricles of the brain and circulation of cerebrospinal fluid. Cerebrospinal fluid is secreted into the four brain ventricles (left, right, third, and fourth). From there it flows through a series of cavities, bathing the brain and spinal cord.

Spotlight on Structure & Function

Most capillaries in the body are fairly leaky because there are narrow slits between adjacent capillary cells. However, the cells of the capillaries that produce cerebrospinal fluid are fused tightly together. Consequently, substances must pass through the capillary cells rather than between them in order to get from blood to cerebrospinal fluid. Lipid-soluble substances such as oxygen and carbon dioxide can diffuse easily through the lipid bilayer of the cell membrane. Certain important molecules such as glucose are actively transported. Most charged molecules and large objects, however, such as proteins, viruses, and bacteria, generally are prevented from entering the cerebrospinal fluid.

This functional barrier between blood and brain is called the **blood-brain barrier.** Thanks to its protection, bacterial and viral infections of the brain and spinal cord are rare. Isolation of the CNS from infections is important because these illnesses are generally serious and difficult to treat. Many of our best antibiotics are not lipid-soluble, which means they cannot cross the blood-brain barrier and so are ineffective against CNS infections. Lipid soluble substances that cross the blood-brain barrier rather easily include alcohol, caffeine, nicotine, cocaine, and general anesthetics. ■

Recap **The CNS consists of the brain and spinal cord. It is well protected by bone, the meninges, and cerebrospinal fluid. The capillaries that produce cerebrospinal fluid have special properties that constitute a functional blood-brain barrier.**

Spinal cord relays information

The **spinal cord** serves as the superhighway for action potentials traveling between the brain and the rest of the body. As we have seen, the spinal cord can also process and act on certain information on its own via spinal reflexes, without necessarily involving the brain. However, its power to act independently is limited, and such actions never reach the level of conscious awareness.

Approximately the diameter of your thumb, the spinal cord extends from the base of the skull to the area of about the second lumbar vertebra, or about 17 inches (Figure 11.13). It is protected by the vertebral column. The outer portions of the spinal cord consist primarily of bundles of axons, which in the CNS are called *nerve tracts* rather than nerves. Because these axons are generally myelinated, giving them a whitish appearance, the areas of the cord occupied by these ascending (sensory) and descending (motor) nerve tracts are called *white matter.* Neurons of the peripheral nervous system enter and leave the spinal cord at regular intervals via the dorsal (sensory) and ventral (motor) horns that fuse to form spinal nerves.

Near the center of the spinal cord is a region occupied primarily by the cell bodies, dendrites, and axons of neurons of the CNS, and also neuroglial cells. These structures are not myelinated, so the area they occupy is referred to as *gray matter.* Within

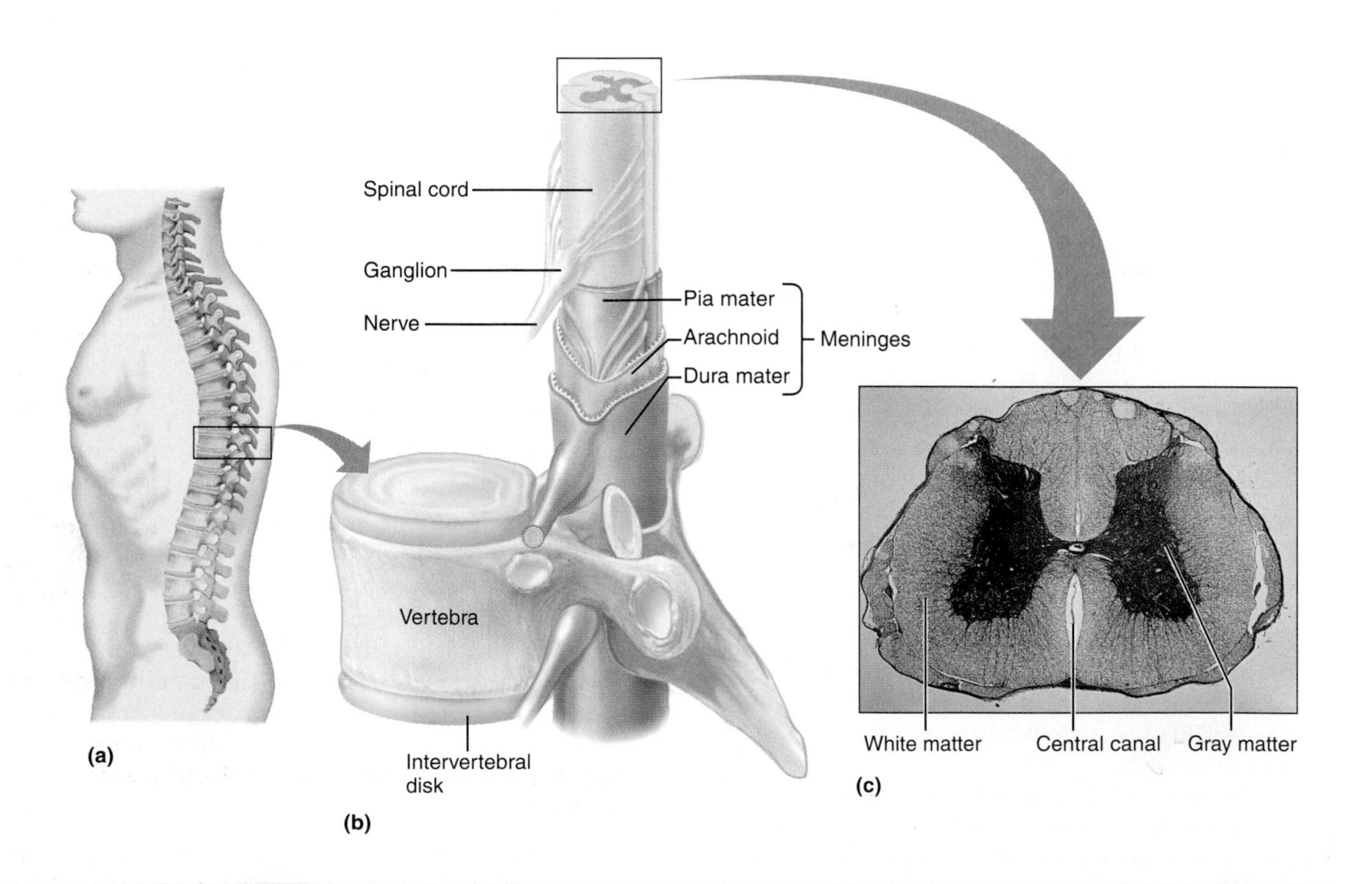

Figure 11.13 The human spinal cord. (a) The spinal cord lies within the vertebral column. **(b)** A closer look at the spinal cord and its relationship with a vertebra. **(c)** A transverse slice of the cord, showing white and grey matter.

The limbic system is involved in emotions and memory The *limbic system* refers to a group of neuronal pathways that connect parts of the thalamus, hypothalamus, and inner portions of the cerebrum. The limbic system is associated with emotions and memory, and we will look at it in more detail later in this chapter when we discuss memory and behavior.

The cerebrum deals with higher functions In humans, the large **cerebrum** is the most highly developed of all brain regions. It deals with many of the functions that we identify with being human: language, decision making, and conscious thought.

The cerebrum (Figure 11.15) consists of left and right *cerebral hemispheres* joined in the middle by a network of nerve tracts called the *corpus callosum.* The corpus callosum enables the two hemispheres to share sensory and motor information. The outer layer of the cerebrum is called the **cerebral cortex.** The cerebral cortex is primarily gray matter, consisting of CNS neurons with unmyelinated axons and their associated neuroglial cells. Numerous folds in its surface give it an appearance similar to a walnut. The inner portion of the cerebrum is mostly white matter containing myelinated nerve axons that connect lower brain areas to the cerebral cortex.

The structure of the cerebrum—an inner section of ascending and descending axons and an outer layer of cells—makes it ideally suited to direct incoming information to the proper region of the brain, integrate it and process it at the highest level, and then route outgoing motor activity to the appropriate areas of the body.

> ***Recap*** **The forebrain contains the hypothalamus, thalamus, limbic system, and cerebrum. Structures in the forebrain contribute to the maintenance of homeostasis: They receive, process and transfer information; control emotions and memory; and are responsible for language, decision making, and conscious thought.**

A closer look at the cerebral cortex Although the cerebral cortex is only about 6 millimeters thick (about 1/4 inch), it contains roughly 50 to 80 billion neuron cell bodies. It is the center for integrating and interpreting sensory information, but it is also much more than that. The cerebral cortex is responsible for memory storage, abstract thought, conscious awareness, and conscious control of skeletal muscle.

All of these functions are location-specific. Functionally, the cortex is divided into four lobes (Figure 11.15b). At the back of the brain, the *occipital lobe* processes visual information. The *temporal lobe* interprets auditory information, comprehends spoken and written language, and is responsible for perceptual judgement. The *parietal lobe* receives and interprets sensory information from the skin, such as temperature, touch, pain, and pressure. Finally, toward the front of the cortex is the *frontal lobe,* perhaps the most highly developed of all brain regions. The frontal lobe initiates motor activity and is also responsible for speech and conscious thought. All four lobes are involved in memory storage.

Notice that the region for speech (frontal lobe) is anatomically separate from the region for under-

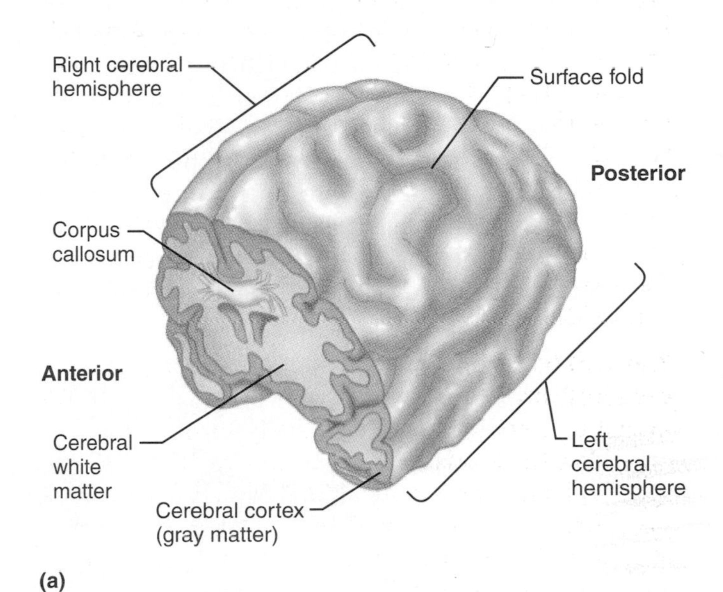

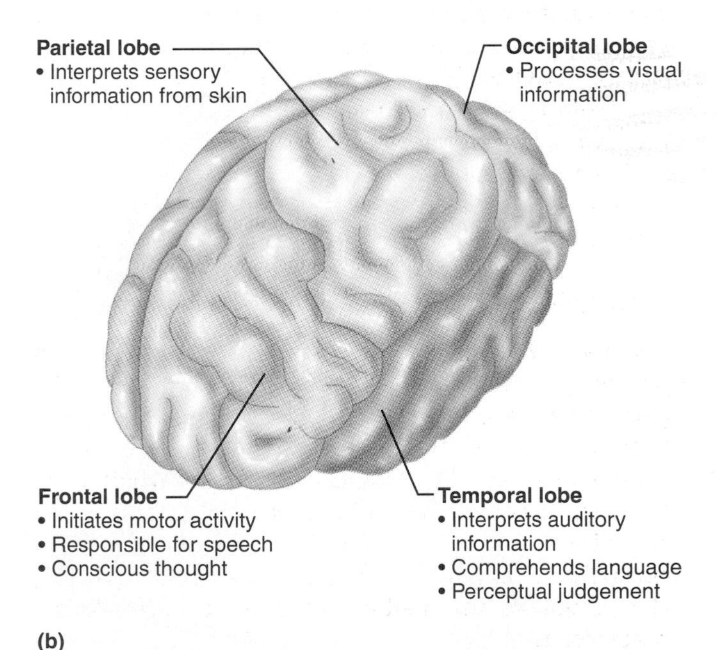

Figure 11.15 The cerebrum. (a) Gray matter (the cerebral cortex) consists of interneurons that integrate and process information. White matter consists of ascending and descending nerve tracts. The two separate hemispheres are joined by the corpus callosum. **(b)** The functions of the four lobes of the cerebral cortex are location specific.

standing spoken language (temporal lobe). The reasons for this separation are not known.

Spotlight on Structure & Function Two areas of the brain that have been studied extensively are the region in the parietal lobe that receives sensory input from the skin (called the *primary somatosensory area*) and the corresponding region of the frontal lobe that initiates motor activity (called the *primary motor area*). Researchers have found very specific regions within each of these areas corresponding to specific parts of the body (Figure 11.16). Body parts that are extremely sensitive, such as the lips and genitals, involve more neurons and consequently larger portions of the primary somatosensory area, whereas less sensitive body parts involve fewer neurons and smaller portions. Similarly, body parts that can perform intricate and precise movements, such as the fingers, involve more neurons in the primary motor area than less dexterous regions.

The location-specific nature of information processing explains why damage to different brain regions can cause such varied, seemingly contradictory effects. This is often apparent in stroke victims. One stroke patient might be unable to speak yet still be fully capable of understanding speech. Another could have trouble controlling basic skeletal muscle movements while abstract thought processes remain unimpaired. Physicians use their understanding of these known location-specific functions to diagnose which areas of the brain have been affected. ■

Recap **The cerebral cortex is divided into four lobes called the occipital, temporal, parietal, and frontal lobes, each with location-specific functions. For example, the occipital lobe processes visual information whereas the temporal lobe processes auditory information.**

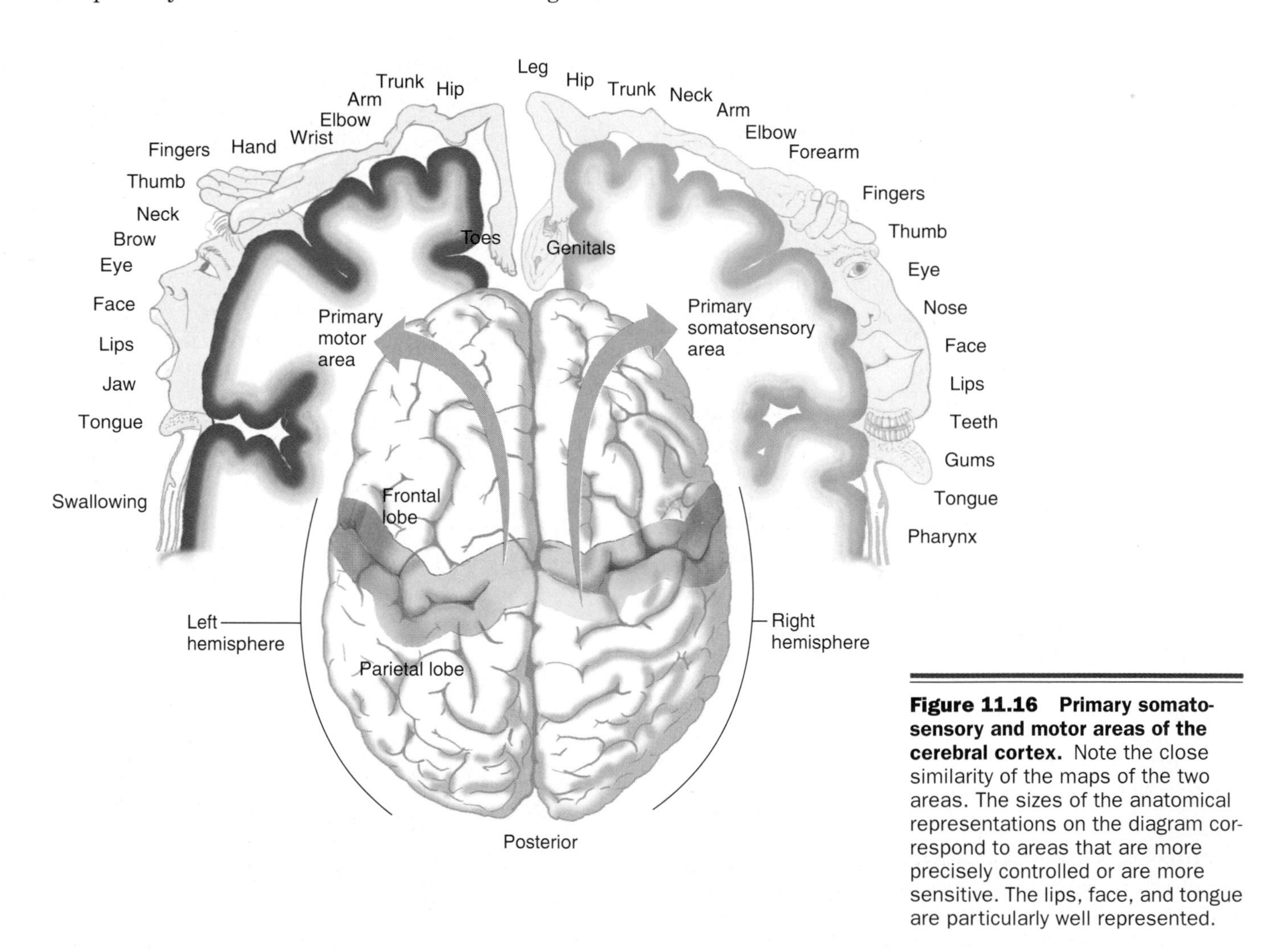

Figure 11.16 Primary somatosensory and motor areas of the cerebral cortex. Note the close similarity of the maps of the two areas. The sizes of the anatomical representations on the diagram correspond to areas that are more precisely controlled or are more sensitive. The lips, face, and tongue are particularly well represented.

11.8 Brain activity continues during sleep

Why do we sleep? This may seem like a simple question, but in fact scientists are still looking for the answer. The obvious hypothesis is we do it to rest. However, brain activity continues even during deep sleep.

Levels of sleep and wakefulness are controlled by the *reticular activating system (RAS),* a group of neurons in the reticular formation. Some neurons of the RAS transmit a steady stream of action potentials to the cerebrum, keeping us awake and alert. A "sleep center" in the RAS releases the neurotransmitter serotonin, which induces sleep by inhibiting the neurons that arouse the brain. Norepinephrine, secreted from another area in the brain, inhibits the effects of serotonin and makes us feel more alert.

We can study what happens to the brain during sleep with *electroencephalograms (EEGs),* recordings of the brain's electrical activity as measured at the body's surface. As Figure 11.17 shows, EEG readings for different states of sleep show characteristic patterns:

- Stage 1: Stage 1 is a transitional phase between wakefulness and sleep. The pupils gradually constrict, breathing slows, and heart rate slows. On an EEG, Stage 1 appears as random small waves, only slightly different from wakefulness.
- Stage 2: During this phase, skeletal muscles relax, and eye and body movements usually cease. On an EEG, Stage 2 produces sharp waves known as "sleep spindles."
- Stage 3: Sleep deepens as heart rate and respiration slow even more. On an EEG, this stage and Stage 4 produce large, slow oscillations, which is why they are sometimes called "slow wave sleep."
- Stage 4: Sleepers are difficult to awaken; heart rate and respiration are at their slowest and body temperature falls.
- REM (rapid eye movement) sleep: This is when we dream. Heart rate, respiration, and blood flow to the brain increase. Most of the body's muscles go limp, preventing us from responding

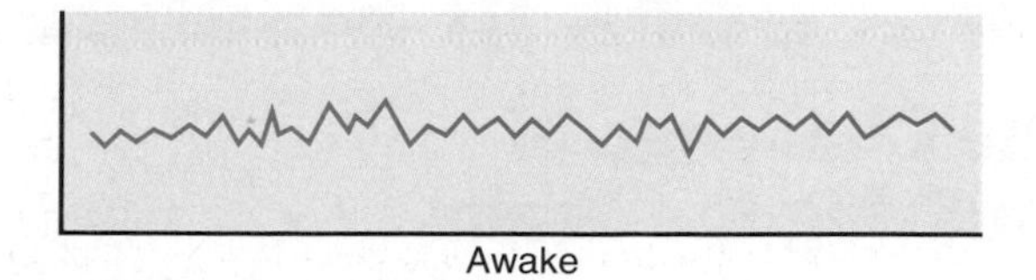

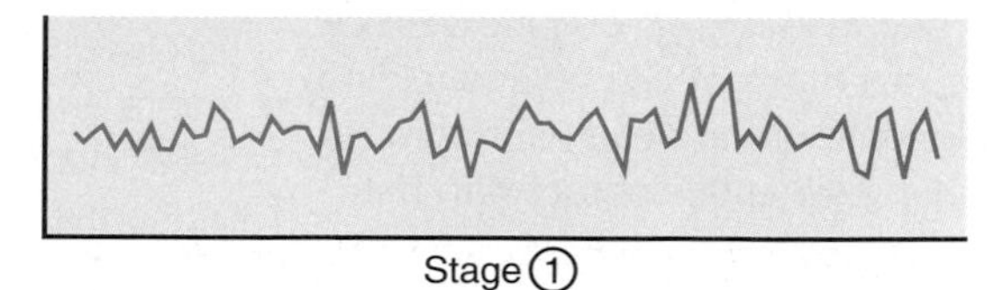

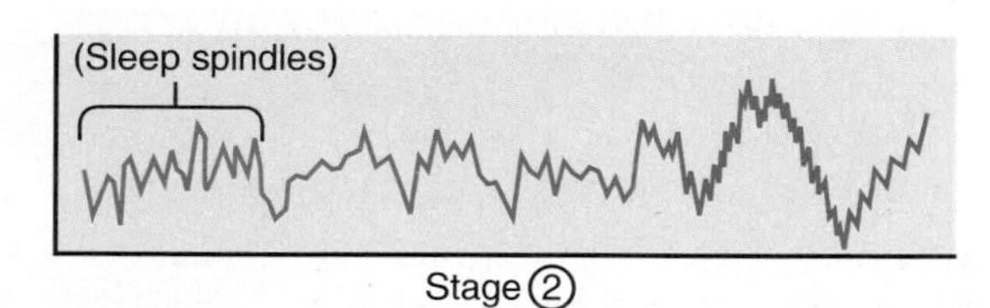

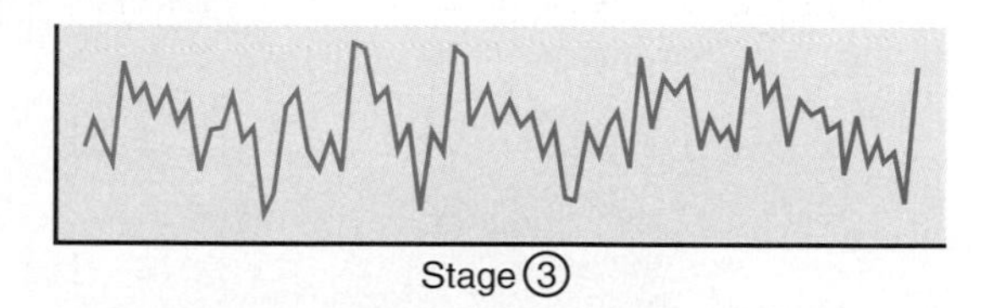

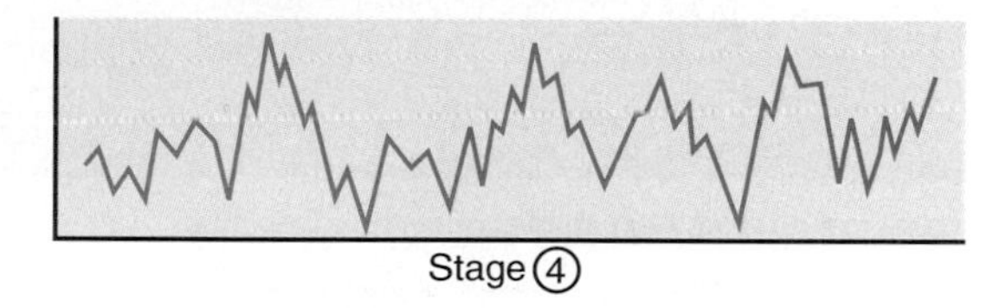

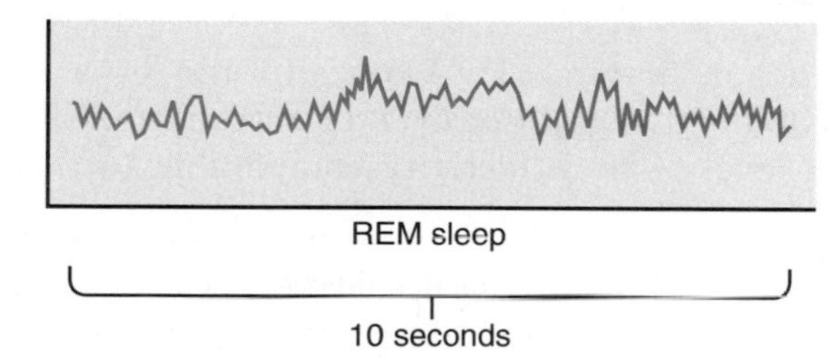

(a)

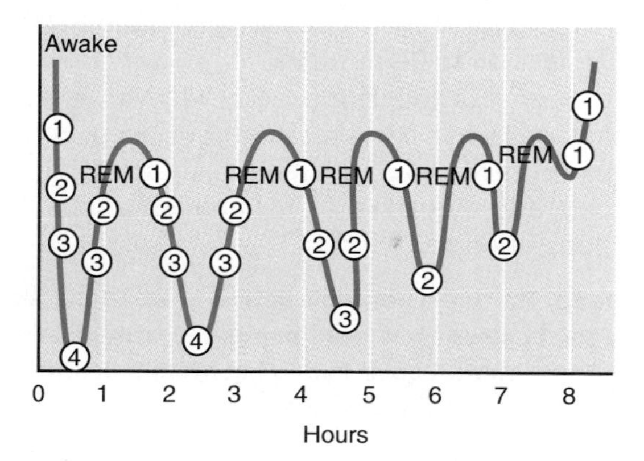

(b)

Figure 11.17 Brain activity patterns during wakefulness and sleep. (a) EEG recordings in an awake person and during stages 1–4 and REM sleep. Note the progressively larger waves during stages 2–4 and the similarity between REM sleep and wakefulness. **(b)** Patterns of sleep.

Directions in Science

The Challenge of Studying the Brain

The human brain is a difficult organ to study because it is so complex. Every region is a mass of cell bodies, axons, and dendrites, and each neuron may be making synaptic contact with millions of other neurons located anywhere in the body.

How did we learn what we do know about the human brain? Basic information about neurons and action potentials was derived from experiments on laboratory animals. Squid were chosen for the experiments because their axons are 10 times larger in diameter than ours, making them easier to study. In addition, the experiments could be done at room temperature in seawater. Thanks to squid, we now know that action potentials are nearly a universal form of information code in the animal kingdom.

Squid axons are often used in studies of nerve conduction.

Our knowledge of location-specific brain functions also owes a great deal to laboratory animals. Researchers can cut through or remove a particular brain region, and then observe effects on the animals' behavior. From rats, we have learned that breathing and blood pressure are controlled by respiratory and cardiovascular centers located in the medulla oblongata. Researchers have identified visual and auditory processing centers by training rats and pigeons to respond to stimuli, either to obtain food or avoid punishment. In other experiments, areas of anesthetized animals' skin were stimulated while recording from neurons in the spinal cord and brain, in order to "map" where the action potentials traveled. In fact, we know far more about a rat's brain than we do about our own!

Complex human emotions and behaviors have proven much harder to study. We certainly can't ask a rat whether it feels love or rage. Furthermore, the cerebral cortex is better developed in humans, so knowledge gained from laboratory animals might not be applicable in any case. Histori-

to our dreams. Paradoxically, the eyes move rapidly under closed eyelids. EEG patterns show a level of brain activity comparable to that of wakefulness.

Although a typical night's sleep involves cycling through these phases, the length of time spent in each stage can vary. REM sleep returns about every 90 minutes throughout the night and lasts a little longer each time. Toward morning a single REM period may last up to 20 minutes.

Although it is not clear exactly why we sleep, it is clear that we need a certain amount of sleep. Persons deprived of REM sleep become moody, irritable, and depressed, and may even suffer hallucinations if deprived long enough.

Recap **The brain remains active even during sleep. Stages of sleep have been correlated with brain wave patterns. The reticular activating system controls levels of sleep and wakefulness.**

11.9 The limbic system is the site of emotions and basic behaviors

The control of emotions and basic patterns of behavior resides in the **limbic system** (Figure 11.18). *Limbic* means "border." The name was originally used to describe those structures that bordered the basal regions of the cerebrum. As we learned about the function of these structures, however, the name came to describe all of the neuronal structures that, together, control emotional behavior and motivational drives. When different areas of the limbic system are stimulated, we experience strong emotions such as fear, anger, sorrow, or love. These reactions, both pleasant and unpleasant, influence our behavior.

You may have noticed that odors have the power to evoke emotions and vivid recollections. This is because pathways for the sense of smell also pass through the limbic system, creating a strong association between olfactory information and memory.

cally, researchers have had to rely on observation of the rare human patient with a particular area of brain damage.

Consider the classic case of Phineas Gage. On September 13, 1848, the 25-year-old Gage, a railroad construction foreman, was tamping explosives into a hole. Suddenly the explosives detonated, driving the tamping rod—42 inches long and 1 inch thick—into his left cheek and out the top of his skull. The rod landed some distance away.

Gage was able to walk away from the accident with help from his crew, and, though temporarily stunned, he remained fully conscious and able to talk. After the injury healed, he even returned to his railroad job. However, it soon became clear that his personality had changed dramatically. Once affable, easygoing, and polite, he became irritable, bad-tempered, and profane. The efficient, reliable foreman was transformed into an undependable nuisance who could no longer hold a steady job. In short, he seemed to have lost all social inhibitions, and his friends said he was "no longer Gage." Eventually, his employer had to let him go, and he was reduced to earning his living as a circus freak.

Although tragic for him, Gage's accident provided some crucial evidence for medical scientists. For the first time, scientists recognized that entire patterns of behavior might be controlled by specific areas of the brain. We now know that one function of the frontal lobe is to modulate our basic animal behaviors so our actions are rational and socially acceptable.

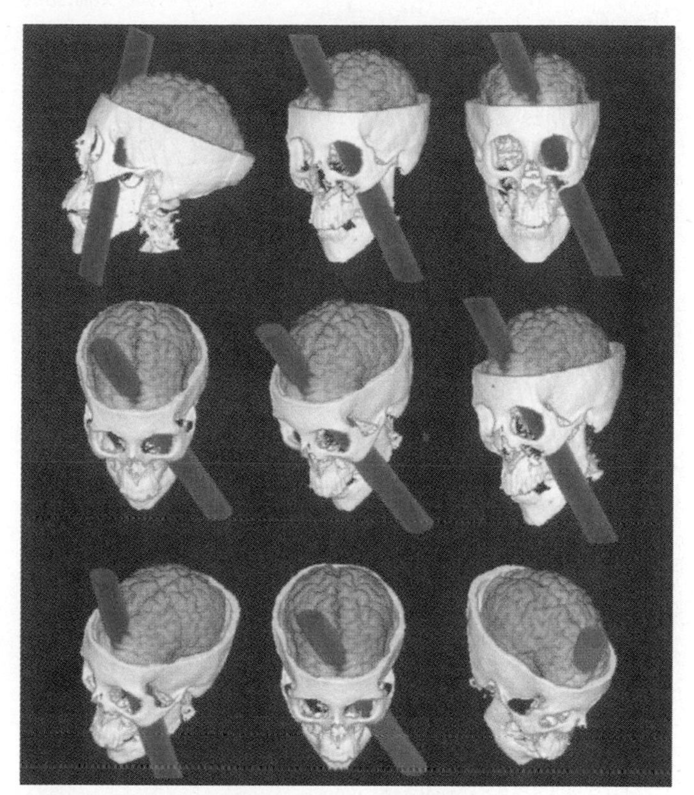

A three-dimensional digital image of a reconstruction of the skull of Mr. Phineas Gage.

Recall that the hypothalamus monitors and regulates homeostasis of many basic bodily functions. The hypothalamus also serves as a gateway to and from the limbic system and so it exerts some control over basic self-gratifying behaviors, such as satisfying our hunger, thirst, and sexual desire. The thalamus receives and organizes information regarding these behaviors and transmits it to the cerebrum. The cerebrum integrates this information with the other information it is receiving, and then adjusts behavior accordingly.

Thus the intensity with which we feel emotions and the extent to which we act on them depends on modification by the cerebrum. To appreciate the importance of cerebral modification, just imagine what would happen if our most basic urges were no longer influenced by social rules or codes of conduct!

Recap **A collection of structures bordering the cerebrum called the limbic system controls our emotions and behavior center. The activities of the limbic system are monitored by the hypothalamus and modified by the cerebrum.**

11.10 Memory involves storing and retrieving information

How are **memories** stored in the brain?

Memory involves storing information and retrieving it later as needed. Memory has two stages: *short-term* ("working") *memory* involves retrieval of information that was stored within the past few hours, and *long-term memory* is the ability to retrieve information days or years later.

There is a clear difference between how the brain manages the two stages of memory. Short-term memory occurs in the limbic system. When you receive new sensory information (such as reading an unfamiliar telephone number), the stimulus triggers a quick burst of action potentials in the limbic system. You can remember the number for a few seconds or minutes, because neurons that have just fired are easier to fire again in the short term. The pathway becomes entrained, if only for a brief time. If the number is not important, you quickly forget it

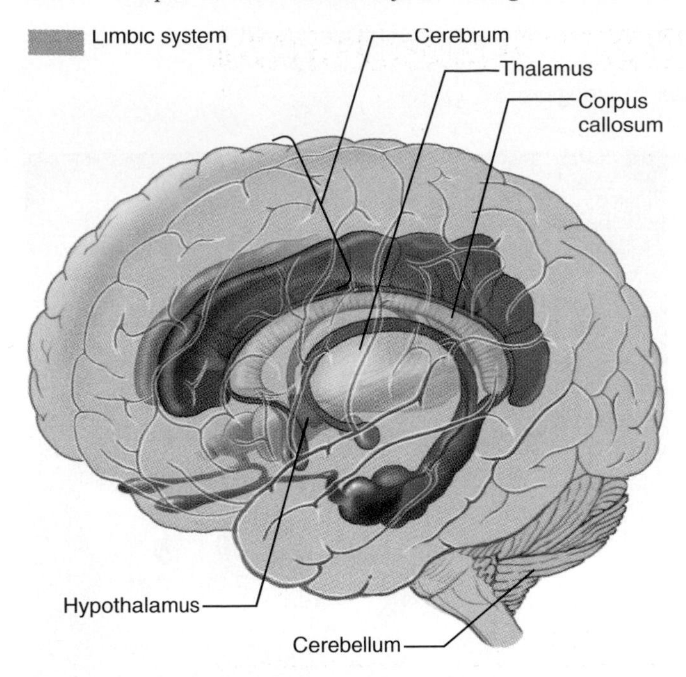

Figure 11.18 The limbic system. The limbic system is a collection of neuronal areas near the base of the cerebrum that are involved in emotions, basic behaviors, and short-term memory.

because the enhancement of the limbic pathway is not long lasting. That particular piece of information does not move into long-term memory.

However, if the number is important to you, as evidenced by the fact that you say it out loud several times, write it down and then read it to yourself, or use it repeatedly over several weeks, the information may be transmitted to your cerebral cortex for storage in long-term memory centers. During this process, neurons undergo a permanent chemical or physical change. Evidence from experimental animals suggests that long-term memory storage creates additional synapses between connecting neurons, enabling those circuits to be activated more easily in the future.

Recap **Short-term memory involves quick bursts of action potentials in the limbic system. Long-term memory resides in the cerebral cortex and involves permanent changes in neurons and synapses.**

11.11 Psychoactive drugs affect higher brain functions

A drug is any substance introduced into the body for the purpose of causing a physiologic change. **Psychoactive drugs** are those that affect states of consciousness, emotions, or behavior—in other words, the higher brain functions. Some psychoactive drugs are legal, such as alcohol and nicotine. Others are not legal but have found their way into the very fabric of our society.

Most drugs, including psychoactive substances, were originally discovered or designed to treat disease. The nonmedical use of drugs is euphemistically called "recreational" use. Table 11.3 lists statistics for recreational drug use in the 18–25 age group.

How do psychoactive drugs work?

All psychoactive drugs are able to cross the blood-brain barrier (not all drugs are able to do so). Their common mode of action is that they influence the concentrations or actions of brain neurotransmitters. Recall that normally the effects of neurotransmitters are short-lived because the neurotransmitter remains in the synaptic cleft for only a brief time. Psychoactive drugs either bind to neurotransmitters or affect their release, action, or re-uptake (reabsorption) in some way. This changes normal patterns of electrical activity in the brain.

Try It Yourself

Improve Your Memory

Can you improve your ability to learn and remember new information? Yes! Memory specialists suggest the following techniques to take advantage of the brain's storage and retrieval mechanisms:

- Concentrate on what you are studying. This may seem obvious, but paying attention increases brain activity and boosts epinephrine levels, promoting consolidation of information into long-term memory.
- Minimize interference. Studying in a noisy, crowded environment will distract you and impair your ability to concentrate.
- Allow sufficient time to study. Leisurely study extending over several days or weeks is more effective than frantic "cramming" the night before a test.
- Break down large amounts of information into smaller topics. Give yourself time to review each topic, and take a break in between.
- Rephrase material in your own words. Don't repeat what you are told or read exactly; restate the information in a way that makes sense to you personally.
- Test yourself. Create outlines or diagrams. Define key terms before looking up their definitions. Use practice questions when they are available.

Note that many of these techniques are based on repeating and actively thinking about what you study. As we have seen, short-term memory involves quick bursts of action potentials. Every time you read, think about, or test yourself on a concept, more neurons fire. By studying new material actively and repeatedly, you trigger more action potentials and improve long-term retention.

Table 11.3 Where Do You Stand Among Your Peers?*

Drug	Lifetime	Past Year	Past Month
Marijuana and hashish	44.0	23.8	13.2
Cocaine	10.2	4.7	2.0
Crack	3.0	1.3	0.6
Inhalants	10.8	3.0	1.0
Hallucinogens	16.3	6.9	2.3
PCP	2.3	0.5	0.1
LSD	13.9	4.6	0.9
Heroin	1.3	0.9	0.4
Nonmedical use of any prescription-type stimulant, sedative, tranquilizer, or analgesic	12.7	6.7	2.9
Alcohol	83.8	75.3	60.0
Cigarettes	68.5	44.7	38.3
Smokeless tobacco	23.4	9.7	6.1

*This table lists the percentage of 18–25-year-olds who reported having used these psychoactive drugs in their lifetime, the past year, or the past month.

SOURCE: Estimates are preliminary data from the National Household Survey on Drug Abuse, 1996, of the Substance Abuse and Mental Health Services Administration, Department of Health and Human Services.

Depending on which neurotransmitter they influence and where they act, these drugs can exert a variety of effects. Many impair muscle coordination and artificially alter mood.

Psychoactive drugs can lead to *psychological dependence,* meaning that users begin to crave the feelings associated with a drug and alter their behavior in order to obtain it. Eventually *tolerance* develops—they need to use more of it, or use it more frequently, to achieve the same effect. Tolerance occurs because the liver produces enzymes that detoxify many drugs, and repeated intake causes circulating levels of these detoxifying enzymes to rise. Soon the original dose is no longer enough to create the desired effect, or sometimes any effect at all.

Tolerance leads to **addiction,** the need to continue obtaining and using a substance despite one's better judgment and good intentions. Essentially, choice is lost. Addiction has serious negative consequences for both emotional and physical health. Eventually, most addicts find it impossible to concentrate and hold a steady job. At the same time, they must spend large amounts of money to support their habit. Personal relationships suffer, and many addicts lose their families, their employment, and their physical health. In addition, they face the probability of **withdrawal,** the symptoms that occur if the drug is removed suddenly.

Cocaine and nicotine are both highly addictive. On the other hand, alcohol is addictive in some people but not in others, for reasons we do not yet fully understand.

Recap **The common mode of action of psychoactive drugs is that they alter either the concentration or action of brain neurotransmitters. Some lead to psychological dependence, tolerance, and addiction.**

11.12 Disorders of the nervous system

Because the nervous system is so complex and controls so many bodily functions, disorders of the nervous system can be particularly debilitating. We have already discussed stroke in the context of cardiovascular disease. Here we discuss several classes of disorders that affect the nervous system, including trauma, infections, disorders of neural or synaptic transmission, and brain tumors.

Trauma

Trauma (physical injury) to either the brain or spinal cord can be particularly dangerous because the brain controls so many bodily functions and the spinal cord is the information pathway to virtually all the organs.

Concussion disrupts electrical activity A violent blow to the head or neck can cause *concussion*, a brief period of unconsciousness in which electrical activity of brain neurons is briefly disrupted. After

Health Watch

Cocaine

The basic mechanism of action of cocaine is that it blocks the re-uptake of the neurotransmitters dopamine and norepinephrine back into the presynaptic bulb. As a result, more of these two neurotransmitters remain in the synaptic cleft for longer periods of time. The ultimate effect is excessive and continued stimulation of postsynaptic neurons that are stimulated by these neurotransmitters.

Dopamine is one of the most important neurotransmitters in areas of the brain associated with pleasure. Cocaine stimulates those regions, initially allowing the user to experience pleasurable effects for a longer-than-normal period of time.

However, the body is very adaptive, even to the point of adjusting production of a neurotransmitter if there seems to be too much of it. Feedback mechanisms that regulate dopamine production adapt to its oversupply by cutting back. This reaction may begin after only one use. Over time, as levels of dopamine decline, more cocaine is required to maintain even a normal amount of neurotransmitter in the synaptic cleft. As dopamine supplies continue to fall, cocaine addicts become incapable of experiencing pleasure without the drug. Eventually they cannot even maintain a normal emotional state, and severe anxiety and depression set in. They lose weight, have trouble sleeping, and suffer frequent infections. Babies born to women addicted to crack, a ready-to-smoke form of cocaine, are often brain damaged and dangerously thin.

Cocaine also blocks the re-uptake of norepinephrine, the main neurotransmitter of the sympathetic division. Cocaine overstimulates the heart, raises blood pressure, and can lead to irregular heartbeats, seizures, strokes, and heart attacks. Even healthy young adults have died suddenly from cocaine-induced cardiac arrest.

Although scientists understand the basic mechanism of cocaine's action at the level of the synapse, it is still unclear why cocaine and especially crack cocaine are so addictive. The subject of psychoactive drug addiction continues to be the focus of considerable research.

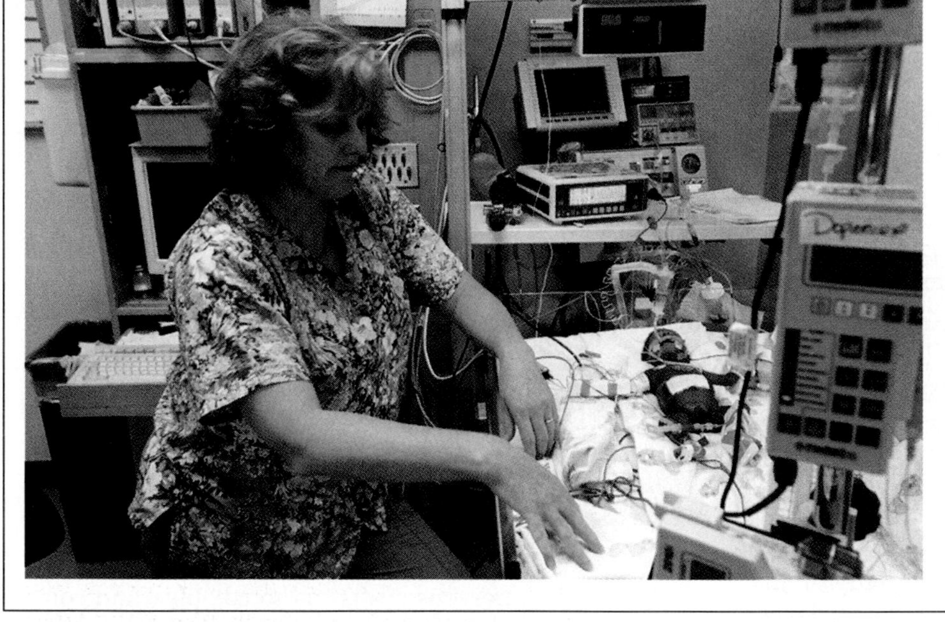

This baby was born to a woman who smoked crack throughout her pregnancy.

the person "comes to," he or she may experience blurred vision, confusion, and nausea and vomiting. Although most concussions are not associated with permanent injury, one of the biggest dangers is a *subdural hematoma*, or hemorrhage (bleeding) into the space between the meninges.

Because the brain is encased in a rigid skull, any bleeding increases pressure within the skull. Eventually the trapped blood presses against healthy brain tissue and disrupts function. Symptoms can include drowsiness, headache, and growing weakness of one side of the body. Depending on how rapidly bleeding occurs, symptoms may take weeks or even months to appear. Sometimes surgeons drill through the skull to drain the hematoma, relieve the pressure, and repair damaged blood vessels.

Spinal cord injuries impair sensation and function Violent trauma to vertebrae may press against the spinal cord, tear it, or even sever it completely. Damage to the spinal cord usually leads to impaired sensation and body function below the injured area, so the extent of injury depends on the site of damage. Injuries below the neck can cause paraplegia (weakness or paralysis of the legs, and sometimes of part of the trunk). Injuries to the neck may lead to quadriplegia (weakness or paralysis of the legs, arms, and trunk). The higher the spinal cord injury, the more likely it is to be fatal. Spinal cord injuries may also cause problems with bladder or bowel control because the autonomic division no longer functions properly.

Researchers are investigating many exciting treatments for spinal cord injuries. Experiments that are helping injured lab animals to move again include transplanting neurons to grow over the injured spinal cord, and developing antibodies to fight inhibitors that currently stop damaged neurons from regenerating. Some of these treatments may become available for human testing in a few years.

Infections

The brain and spinal cord are spared most of the infections that affect the rest of the body because of the blood-brain barrier. Nevertheless, certain viruses and bacteria do occasionally infect the brain or spinal cord.

Encephalitis: Inflammation of the brain Encephalitis refers to inflammation of the brain, often caused by a viral infection. Frequent sources of encephalitis include the herpes simplex virus (which causes cold sores) and a virus transmitted through mosquito bites. HIV (human immunodeficiency virus) infection is responsible for a growing number of cases.

Symptoms of encephalitis can vary, but often include headache, fever, and fatigue, progressing to hallucinations, confusion, and disturbances in speech, memory, and behavior. Some patients also experience epileptic seizures. Treatment varies depending on the cause, but antiviral drugs may help.

Meningitis: Inflammation of the meninges Meningitis involves inflammation of the meninges, usually due to viral or bacterial infection. Symptoms include headache, fever, nausea and vomiting, sensitivity to light, and a stiff neck. Mild cases may resemble the flu.

Viral meningitis tends to be relatively mild and usually improves within a few weeks. Bacterial meningitis can be life threatening and requires antibiotics.

Rabies: Infectious viral disease Rabies is an infectious viral brain disease of mammals, especially dogs, wolves, raccoons, bats, skunks, foxes, and cats. It is transmitted to humans by direct contact, usually through a bite or a lick over broken skin. The virus makes its way through the sensory neurons in the region of the bite to the spinal cord and then to the brain, where it multiplies and kills cells.

Signs of rabies include swollen lymph glands, painful swallowing, vomiting, choking, spasms of throat and chest muscles, fever, and mental derangement. The symptoms generally appear one to six months after initial infection, followed by coma and death within two to twenty days.

If an animal has bitten you, wash the wound thoroughly and consult a physician. If possible, have the animal tested for rabies. If there is any possibility it could be rabid, your doctor will advise you to be immunized, because untreated rabies may be fatal. The rabies vaccine is most effective if administered within a few days after the bite, but it may still work when given weeks or months later.

Disorders of neural and synaptic transmission

The nervous system depends on action potentials traveling in neurons and the chemical transfer of signals from one neuron to another across a synapse. It should come as no surprise, then, that problems involving neural or synaptic transmission can cause disease. The symptoms of these disorders depend on which nerves and/or neurotransmitters are affected.

Epilepsy: Recurring episodes of abnormal electrical activity *Epilepsy* is characterized by recurring bouts of abnormal brain electrical activity that disrupt normal functioning. Fatigue, stress, or patterns of flashing lights may trigger attacks. Generally the person experiences a *seizure,* which involves jerky body movements and loss of consciousness. The patient recovers fully after a short period of unconsciousness. Some cases of epilepsy result from infections (such as meningitis or encephalitis), head injuries, brain tumors, or stroke, but often the cause is not known. Epilepsy usually appears during childhood or adolescence.

Parkinson's disease: Loss of dopamine-releasing neurons *Parkinson's disease* is a progressive, degenerative disorder that strikes nearly 50,000 people a year in North America, most over age 55. Symptoms include stiff joints and muscle tremors in the hands and feet. Eventually people with Parkinson's lose mobility; they may also become mentally impaired and depressed.

We now know that Parkinson's disease is caused by degeneration of dopamine-releasing neurons in the area of the midbrain that coordinates muscle movement. The shortage of dopamine impairs the ability to perform smooth, coordinated motions. The symptoms can be improved by taking L-dopa, a drug that the body converts into dopamine, but this does not slow the loss of neurons. As the disease progresses, even L-dopa becomes ineffective.

However, there is an experimental and highly controversial treatment for patients with Parkinson's disease that might replace lost cells: receiving transplants of fetal neurons directly into the brain. Experiments with animals have already shown that transplanted fetal neurons will survive and grow in the recipient animal. In humans, the fetal cells would undoubtedly come from aborted fetuses, which is why the technique generates such controversy.

Health Watch

Cell Phones and Brain Tumors: Is There a Connection?

In 1993, as cell phones were coming into widespread use, a lawsuit was filed by a Florida man who claimed that his wife died from a brain tumor caused by her cell phone. His allegation had an immediate, but temporary, negative effect on cell phone purchase and use. The original lawsuit and several others that followed failed to prove any connection between cell phones and cancer, and the furor died out.

Then, in 1997, a study published in a scientific journal again suggested a possible link between cell phones and cancer. The study was done in a strain of mice highly susceptible to tumors who were exposed to high radiofrequency fields in the laboratory.

To date, most of the evidence suggests there is no danger from cell phone use. However, now that there are so many cell phone users worldwide, some health officials think the issue should be studied carefully and on a large scale. Accordingly, in 1998 the World Health Organization initiated a study in eight countries to determine whether there is a link between cell phones and brain tumors. It will be several years at least before the results are in.

In the meantime, be aware of the results of another study: People who use cell phones while driving have roughly the same risk of an accident as if they were legally drunk.

Alzheimer's disease: Shortage of acetylcholine In contrast to Parkinson's disease, which affects primarily physical activity, *Alzheimer's disease* is a disorder of mental impairment, especially impairment of memory. One in every ten people over age 65 has the disorder.

Alzheimer's disease begins with memory lapses and progresses to severe memory loss, particularly for recent events; long-term memory is affected more slowly. Patients can also become disoriented, demented, and develop personality changes. Eventually they can no longer function independently.

The disease primarily affects neurons in the limbic system and frontal lobe, which use acetylcholine as their neurotransmitter. As the disease progresses the supply of acetylcholine becomes insufficient for normal neurotransmission. The study of Alzheimer's-affected brains has revealed definite anatomic changes, including abnormal, tangled neurons and unusually large deposits of an abnormal protein called *beta amyloid*. Along with these anatomic changes, metabolic activity in the frontal and temporal lobes declines, as do levels of the enzyme that produces acetylcholine (Figure 11.19). Drugs that enhance the brain's production of acetylcholine can slow the progression of the disease, but as yet there is no cure.

The cause of Alzheimer's is still unknown, but possibilities include infections and impaired circulation to the brain. There appears to be a genetic component, as a family history of the disease increases the risk. Recent studies suggest that an abnormal gene on chromosome 21 for the beta amyloid precursor protein may be involved.

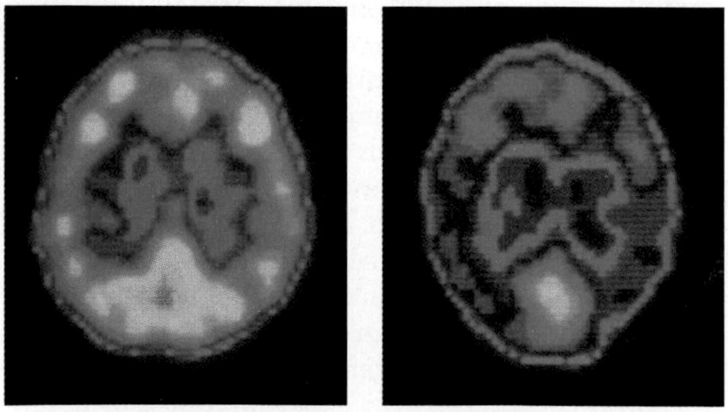

(a) (b)

Figure 11.19 Changes in brain metabolic activity in Alzheimer's disease. A positron emission tomography (PET) scan reveals differences in metabolic activity of tissue. Reds and yellows represent high metabolic activity, blues and greens represent lower activity. **(a)** A PET scan of a normal person. **(b)** A PET scan of a patient with Alzheimer's disease.

Brain tumors: Abnormal growths

A *brain tumor* is an abnormal growth in or on the brain. Brain tumors are not necessarily cancerous, but even noncancerous growths can cause problems if they press against neighboring tissue and increase pressure within the skull. Over time, the rising pressure disrupts normal brain function. Depending on the location and size of the tumor, symptoms include headache, vomiting, visual impairment, and confusion. Some people also experience muscle weakness, difficulty speaking, or epileptic seizures.

Some brain tumors originate from the brain itself, but most are seeded by cancer that spreads from other parts of the body. For more information on cancer, see Chapter 18.

Some brain tumors can be surgically removed. If tumors are inaccessible or too large to be removed, radiation and chemotherapy treatments may shrink them or limit their growth.

Recap **Disorders of the nervous system are often serious because of the central role of the nervous system in regulating other body systems. Common disorders include physical trauma and infections, as well as acquired or genetic diseases that disrupt neural transmission in some way, such as epilepsy, Parkinson's disease, and Alzheimer's disease.**

Chapter Summary

The nervous system has two principal parts

- The central nervous system receives, processes, and stores information, and the peripheral nervous system provides sensory input and motor output.
- The PNS has somatic and autonomic divisions.
- The autonomic division of the PNS has sympathetic and parasympathetic divisions.

Neurons are the communication cells of the nervous system

CD Nervous System/Anatomy of a Neuron

- Neurons generally consist of a cell body, several short extensions called dendrites, and a long thin extension called an axon.
- Neurons are supported and protected by nonneural cells called neuroglial cells.

Neurons initiate action potentials

CD Nervous System/The Membrane Potential
CD Nervous System/Ion Channels
CD Nervous System/Action Potential

- Incoming signals from other neurons produce short-lived graded potentials in a neuron that alter its resting potential slightly.
- If the sum of all graded potentials exceeds threshold, voltage-sensitive sodium channels open briefly and an action potential is initiated.
- The three components of an action potential are: (1) depolarization, in which sodium channels open briefly and then close, (2) repolarization, in which potassium channels open briefly and then close, and (3) reestablishment of the resting potential by the continuous action of the sodium-potassium pump.

Information is transferred from a neuron to its target

CD Nervous System/Anatomy Review: Synapses
CD Nervous System/Synaptic Transmission

- An action potential in a presynaptic neuron causes ion channels for Ca^{2+} to open in the membrane of the axon bulb.
- Ca^{2+} diffuses into the axon bulb and causes a chemical neurotransmitter to be released into the synaptic cleft.
- The neurotransmitter binds to receptors on the postsynaptic cell membrane, causing a graded potential in the postsynaptic cell.
- Information processing depends on differences in stimulus intensity and on converging and diverging neural connections.

The PNS relays information between tissues and the CNS

- The peripheral nervous system consists of 12 pairs of cranial nerves and 31 pairs of spinal nerves. Each nerve may contain a mixture of sensory, somatic motor, and autonomic motor nerve axons.
- The somatic motor division of the peripheral nervous system controls skeletal muscle.
- The autonomic motor division of the peripheral nervous system controls automatic functions and helps to maintain homeostasis.

The brain and spinal cord comprise the CNS

- The brain and spinal cord are protected by bone, by three layers of connective tissue called meninges, and by cerebrospinal fluid.
- The spinal cord consists of ascending (going to the brain) and descending (coming from the brain) nerve tracts, plus neurons and neuroglial cells of the CNS.

The brain processes and acts on information

- The brain receives sensory information, integrates and stores it, and generates the appropriate response.
- The medulla oblongata contains areas that control blood pressure and respiration.
- The cerebellum coordinates basic and certain skilled body movements.
- The hypothalamus coordinates certain automatic functions and participates in the maintenance of homeostasis.
- The cerebrum is involved in language, decision making, and conscious thought. It is also the site of long-term memory storage.

Brain activity continues during sleep

- The brain exhibits characteristic cyclic patterns of activity during sleep.
- The functional importance of sleep is unknown.

The limbic system is the site of emotions and basic behaviors

- The limbic system is the center of emotions such as fear, anger, sorrow, and love and basic behaviors such as seeking food or sexual gratification.

Memory involves storing and retrieving information

- Short-term memory resides in the limbic system. Long-term memory involves permanent changes to neurons in the cerebral cortex.

Psychoactive drugs affect higher brain functions

- Psychoactive drugs act either by altering the concentrations of brain neurotransmitters or by altering the responses of postsynaptic neurons to them.
- Addiction to psychoactive drugs is characterized by the compulsion to use the drug, often at increasing doses, despite good intentions and despite negative consequences.

Disorders of the nervous system

- Head trauma can be dangerous because any bruising or bleeding within the confined space of the skull presses against the brain.
- Spinal cord injury can interrupt neural signals traveling up and down the spinal cord, resulting in a loss of sensation and paralysis of muscle.
- Rabies is an infectious viral disease that is transmitted by the bite of an infected animal. The virus travels to spinal cord and brain via the axon of a sensory nerve.
- Alzheimer's disease is characterized by a decline in the amount of the neurotransmitter acetylcholine in the limbic system and frontal lobe. Although the cause of Alzheimer's is unknown, the disease is associated with deposits of an abnormal protein in the brain.

Terms You Should Know

action potential, 263
autonomic division, 270
axon, 259
blood-brain, barrier, 274
central nervous system (CNS), 258
cerebellum, 276
cerebral cortex, 275
cerebrospinal fluid, 273
cerebrum, 277
convergence, 267
dendrite, 259
depolarization, 263
divergence, 267
forebrain, 276
graded pontential, 262
hindbrain, 275
interneuron, 258
midbrain, 276
motor neuron, 258
myelin sheath, 260
nerve, 268
neuroglial cell, 260
neuron, 258
neurotransmitter, 265
peripheral nervous system (PNS), 258
pons, 276
repolarization, 263
resting potential, 262
sensory neuron, 258
somatic division, 269
spinal reflex, 269
synapse, 265

Test Yourself

Matching Exercise

_____ 1. Type of neuron providing information to the central nervous system
_____ 2. Type of neuron carrying the output of the central nervous system to the tissues and organs of the body
_____ 3. A neuron located entirely within the central nervous system
_____ 4. A type of neuroglial cell that wraps itself around the axon of a neuron of the peripheral nervous system as a sort of insulating blanket
_____ 5. A small change in membrane potential caused by the opening of chemically gated ion channels
_____ 6. The difference in voltage across the cell membrane of a resting cell
_____ 7. An all-or-none, self-propagating electrical impulse
_____ 8. A cablelike bundle of neuron axons in the peripheral nervous system
_____ 9. A network of nerve tracts joining the right and left cerebral hemispheres
_____10. A recording of brain electrical activity

a. interneuron
b. action potential
c. motor neuron
d. nerve
e. corpus callosum
f. sensory neuron
g. electroencephalogram (EEG)
h. graded potential
i. Schwann cell
j. resting potential

Fill in the Blank

11. Interneurons and motor neurons receive information from other neurons either at their ____________ or at their cell bodies.

12. Axons branch into slender extensions called ______ ______________, each of which ends with an axon bulb.
13. Approximately 80% of the cells of the nervous system are ______________ cells, which provide support and protection to neurons.
14. There is a brief ______________ ______________ following an action potential during which a neuron cannot initiate another action potential.
15. The junction between two neurons is called a ______________.
16. The ______________ nervous system consists of the brain and spinal cord.
17. Activation of the ______________ division of the autonomic nervous system raises heart rate; activation of the ______________ division lowers heart rate.
18. The ______________ lobe of the cerebral cortex initiates motor activity and is responsible for speech and conscious thought.
19. The symptoms that occur when a psychoactive drug is no longer available to an addict are known as ______________.
20. ______________ is an infectious viral brain disease that is transmitted by the bite of an infected animal.

True or False?

21. Myelinated axons conduct action potentials more slowly than unmyelinated axons.
22. The repolarization phase of an action potential is due entirely to the sodium-potassium pump.
23. Sensory nerves provide input to the central nervous system only in the form of action potentials.
24. Short-term memory resides in the limbic system.
25. The muscle tremors of Parkinson's disease are caused by a shortage of the neurotransmitter dopamine in the area of the brain that coordinates muscle movement.

Concept Review

1. Distinguish between oligodendrocytes and Schwann cells in terms of where they are located and how they affect nerve regeneration.
2. Explain what is meant by summation of graded potentials and explain how summation may lead to an action potential.
3. Describe in terms of the opening and closing of ion channels and the subsequent diffusion of ions how the depolarization and repolarization phases of an action potential take place.
4. Describe in general terms the primary functions of the somatic and autonomic motor divisions of the peripheral nervous system.
5. Distinguish between gray matter and white matter in the spinal cord.
6. List the three major divisions of the brain and identify their general functional roles.
7. Explain what is meant by the statement that the functions of the cerebral cortex are "location specific."
8. Define *REM sleep* and indicate the physiological changes that might be observed during REM sleep.
9. Describe why a subdural hematoma is so much more dangerous than a hematoma in a muscle (known commonly as a bruise).
10. Explain why an injury to the spinal cord may impair sensations and motor movement below, but not above, the point of injury.

Apply What You Know

1. Why do you suppose the treatment for rabies must be begun even before symptoms have appeared in order for treatment to be successful?
2. Explain in terms of brain anatomy and function how it would be possible for a person to remember whole events in the distant past yet not be able to recall what they had for breakfast.
3. What do you suppose would happen to a person's behavior and emotional expression if the neural connections between the limbic system and the cerebral cortex were severed?

Case in Point

Evidence That the Human Brain May Regenerate

Over the years one of the most persistent hypotheses in neurobiology has been that the adult human brain does not produce new nerve cells. That hypothesis has now been challenged. In a study reported in the November 1998 issue of *Nature Medicine*, U.S. and Swedish researchers have found that certain cells in a small region of the limbic system called the *hippocampus* continue to divide into adulthood. The researchers were able to obtain the brains of five deceased cancer patients who had received a drug used to identify dividing cells. The drug clearly identified certain cells in the hippocampus as having been formed recently by cell division.

The two hippocampi (one in each brain hemisphere) contribute to the limbic system's role in learning and memory storage. The findings raise the intriguing possibility of future treatments for learning deficits and memory loss if researchers are able to find a way to stimulate the growth of neurons in intact humans.

QUESTIONS:

1. How do these findings affect the hypothesis that the adult human brain does not produce new nerve cells?
2. Which do you think would be most affected by disease of the hippocampi: short-term memory or long-term memory? Explain.

INVESTIGATE:

Eriksson, P.S., E. Perfilieva, T. Bjork-Eriksson, A.-M. Alborn, C. Nordborg, A. Peterson, and F.H. Gage. Neurogenesis in the adult human hippocampus. *Nature Medicine* 4:(11), 1998.

References and Additional Reading

Cowley, Geoffrey, and Anne Underwood. Our Latest Health Obsession. *Newsweek* June 15, 1998, pp. 49–54. (An article in the popular press about how memory works.)

Damasio, H., T. Grabowski, R. Frank, A.M. Galaburda, and A.R. Damasio. The Return of Phineas Gage: Clues About the Brain from the Skull of a Famous Patient. *Science* 264:1102–1105, 1994. (A reanalysis of Mr. Phineas Gage's 150-year-old skull using modern techniques.)

Pinker, Steven. *How the Mind Works*. New York: W.W. Norton & Company, Inc., 1997. 660 pp. (A modern examination of the human mind by a leading cognitive scientist. Difficult but informative reading.)

CHAPTER 12

SENSORY MECHANISMS

Why do we see in color?

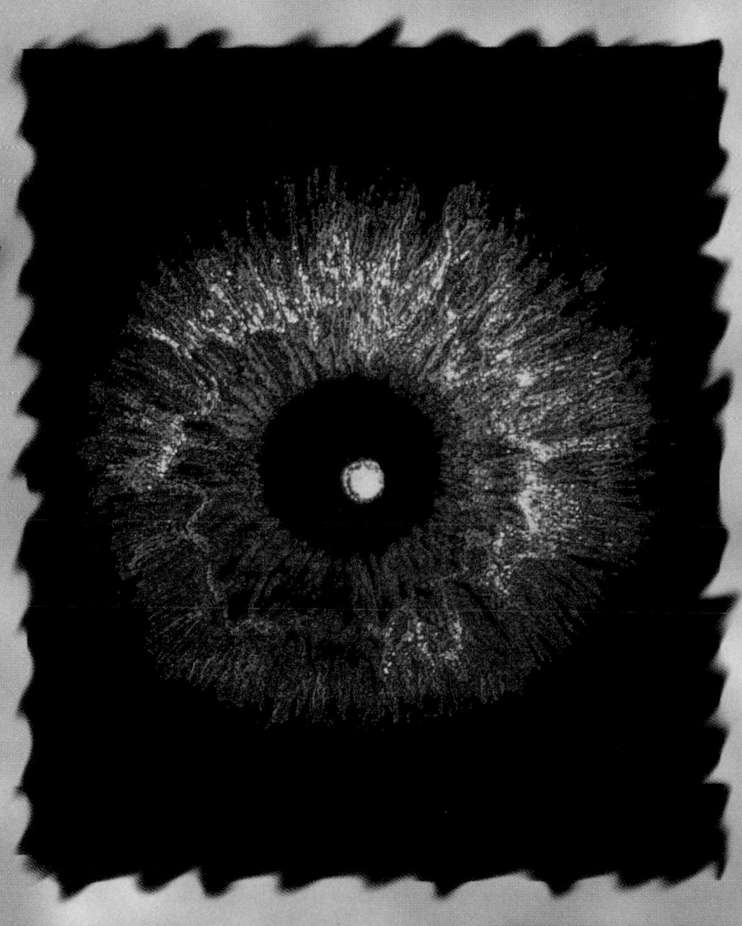

Why do your ears "pop" **when you change altitude quickly?**

How do your eyes adjust **to different** light intensities?

What is the value or purpose of pain?

How can you know body position **with your eyes closed?**

LOOK AROUND YOU AT THE COMPLEX AND COLORFUL WORLD you inhabit. Then after reading this paragraph, close your eyes and concentrate for a moment on what you can hear. Identify the sources of any sounds you hear and try to judge their direction and distance from you. Describe any odors that you smell, and make a guess at the temperature in the room. Now concentrate on your own body. Move your arms and legs about, and then identify what type of clothing you are wearing. Finally, visualize in your mind the precise position of every part of your body, right down to your fingers and toes.

This exercise demonstrates that the body's sensory mechanisms are constantly providing the brain with detailed information about the world around you, and even about the body itself. In this chapter we'll learn how different kinds of sensory information are received by the body, how they are converted to nerve impulses, and how the nerve impulses are transmitted to the brain in such a way that the brain can make sense of it all.

12.1 Receptors receive and convert stimuli

Sensory input that causes some change within or outside the body is called a **stimulus** (plural *stimuli*). The stimulus is a form of energy such as heat, physical pressure, or sound waves. A **receptor**—a structure specialized to receive certain stimuli—accepts the stimulus and converts its energy into another form.

Some receptors evolved from dendritic structures that are part of a sensory neuron. They convert the stimulus into a graded potential that, if it is powerful enough, initiates an impulse within the sensory neuron. Other receptors are part of cells that produce graded potentials and release a neurotransmitter, stimulating a nearby sensory neuron. In the end, the effect is the same—generation of an impulse in a sensory neuron.

When the central nervous system receives these impulses, we often experience a **sensation,** meaning we become consciously aware of the stimulus. A sensation is different from **perception,** which means understanding what the sensation means. As an example, hearing the sound of thunder is a sensation; the belief that a storm is approaching is a perception.

Recap **Receptors receive stimulus energy and convert it into action potentials. A sensation is the conscious awareness of a stimulus.**

Receptors are classified according to stimulus

Receptors are classified according to the type of stimulus energy they convert:

- **Mechanoreceptors** respond to forms of mechanical energy, such as waves of sound, changes in fluid pressure, physical touch or pressure, stretching, or forces generated by gravity and acceleration.
- **Thermoreceptors** respond to heat or cold.
- **Pain receptors** respond to tissue damage or excessive pressure or temperature.
- **Chemoreceptors** respond to the presence of chemicals in the nearby area.
- **Photoreceptors** respond to light.

Many of these receptors contribute to sensations, although some (such as joint receptors) contribute no more than a general sense of where our limbs are located. Table 12.1 lists examples of each type of receptor and the sensations they give us.

A few receptors are "silent" in the sense that we are not consciously aware of their actions. These receptors function in negative feedback loops that maintain homeostasis inside the body. They include stretch receptors for monitoring and regulating blood pressure and fluid volumes, and chemoreceptors for regulating the chemical composition of our internal environment. They will not be discussed in detail here, as we discuss most of them in the context of the homeostatic control of various organ systems.

Recap **Receptors are classified according to the type of stimulus they convert. The five classes of receptors are mechanoreceptors, thermoreceptors, pain receptors, chemoreceptors, and photoreceptors.**

The CNS interprets nerve impulses based on origin and frequency

How can cells in the central nervous system interpret some incoming impulses as images and others as sounds? How do they distinguish a loud sound from a soft one and bright lights from dim?

As we saw in the last chapter, nerve impulses are transmitted from receptors to specific brain areas. For example, impulses generated by visual stimuli travel in sensory neurons whose axons go directly to brain regions associated with vision. All incoming impulses traveling in these neurons are interpreted as light, regardless of how they were initiated. This is why, when you are hit in the eye, you may "see stars." The blow to your eye triggers impulses in visual sensory neurons. When these signals arrive in the visual area of the brain, they are interpreted as light.

Stronger stimuli activate more receptors and trigger a greater frequency of impulses in sensory neurons. In effect, the central nervous system gets all the information it needs by monitoring where impulses originate and how frequently they arrive.

Table 12.1 Types of receptors

Type of Receptor	Examples	Sensation
Mechanoreceptors		
Touch, pressure	Unencapsulated dendritic endings, Meissner corpuscles, Pacinian corpuscles, Ruffini endings, Merkel disks	Touch and pressure on skin or body hair
Vibration	Pacinian corpuscles	Vibration on skin
Stretch	Muscle spindles	Muscle length
	Tendon receptors	Tendon tension
	Joint receptors	Joint position
Hearing	Hair cells in cochlea of the inner ear	Vibration of fluid originating from sound waves
Balance	Hair cells in semicircular canals of the inner ear	Rotational movement of the head
	Hair cells in the utricle and saccule of the inner ear	Static position and linear acceleration and deceleration of the head
Thermoreceptors		
Temperature	Receptors in unencapsulated dendritic endings in skin	Heat or lack of heat (cold)
Pain receptors		
Pain	Receptors in unencapsulated dendritic endings in skin and internal organs	Pain caused by excessive pressure, heat, light, or chemical injury
Chemoreceptors		
Taste	Taste receptors in taste buds	Salty, sweet, sour, and bitter tastes
Smell	Olfactory receptors of nasal passages	Over a thousand different chemical smells
Photoreceptors		
Vision	Rods of the retina	Light (highly sensitive black and white vision)
	Cones of the retina	Light (color vision)

Some receptors adapt to continuing stimuli

The central nervous system can ignore one sensation to concentrate on others. For example, you ignore the feel of your clothing when you are more interested in other things. Some sensory inputs are ignored after awhile due to **receptor adaptation,** in which the sensory neuron stops sending impulses even though the original stimulus is still present. To demonstrate, try touching a few hairs on your forearm. Notice that the initial feeling of light pressure goes away within a second or so, even though you maintain the same amount of pressure.

Receptors in the skin for light touch and pressure adapt rather quickly. This confers a survival advantage, because they can keep the central nervous system informed of *changes* in these stimuli, without constantly bombarding it with relatively unimportant stimuli. Olfactory (smell) receptors also adapt rapidly. Olfactory adaptation can be hazardous to people if they are continually exposed to low levels of hazardous chemicals that they can no longer smell.

Other receptors—pain receptors, joint and muscle receptors that monitor the position of our limbs, and essentially all of the silent receptors involved in homeostatic feedback control loops—adapt slowly or not at all. Lack of adaptation in these receptors is also important to survival. Persistent sensations such as pain alert us to possible tissue damage from illness or injury and prompt us to take appropriate action. Persistent activity of silent receptors is essential to the ability to maintain homeostasis.

Recap **Sensations are transmitted to specific brain areas. Stronger stimuli activate more receptors and trigger a greater frequency of action potentials in sensory neurons. Some receptors adapt rapidly; others adapt slowly or not at all.**

Somatic sensations and special senses provide sensory information

The sensations (or senses) provided by receptors are categorized as either *somatic* or *special.* The **somatic sensations** originate from receptors present at more than one location in the body (*soma* is the Greek word for "body"). The somatic sensations include temperature, touch, vibration, pressure, pain, and awareness of body movements and position. The five **special senses** (taste, smell, hearing, balance, and vision) originate from receptors that are restricted to special areas of the body, such as the ears and eyes. The special senses deliver highly specialized information about the external world.

We'll look at the somatic sensations first and then discuss the five special senses in detail.

12.2 Somatic sensations arise from receptors throughout the body

The somatic sensations of temperature, touch, pressure, vibration, pain, and awareness of body movements and position are essential to help us coordinate muscle movements, avoid danger, and maintain body temperature. Some somatic sensations contribute to pleasurable feelings as well, such as our response to the gentle touch of a loved one.

Receptors that detect the somatic sensations are located throughout the body in skin, joints, skeletal muscles, tendons, and internal organs. Sensory neurons linked to these receptors send their impulses to the *primary somatosensory area* of the parietal lobe of the cerebral cortex (review Figure 11.16). As you already know, body parts with the greatest sensory sensitivity, such as the mouth and fingers, involve more neurons in the somatosensory area, whereas less-sensitive body parts involve fewer neurons.

The somatosensory area processes the information, integrates it, and sends it to the nearby *primary motor area* in the frontal lobe. Impulses are then generated in motor (output) neurons of the peripheral nervous system.

Mechanoreceptors detect touch, pressure, and vibration

Receptors for sensing touch, pressure, and vibration are called mechanoreceptors. Though they may take several different forms, all mechanoreceptors are the modified dendritic endings of sensory neurons. Any force that deforms the plasma membrane of the dendritic ending produces a typical graded potential. If the graded potential (or the sum of multiple graded potentials) is large enough to exceed threshold, the sensory neuron will initiate an impulse.

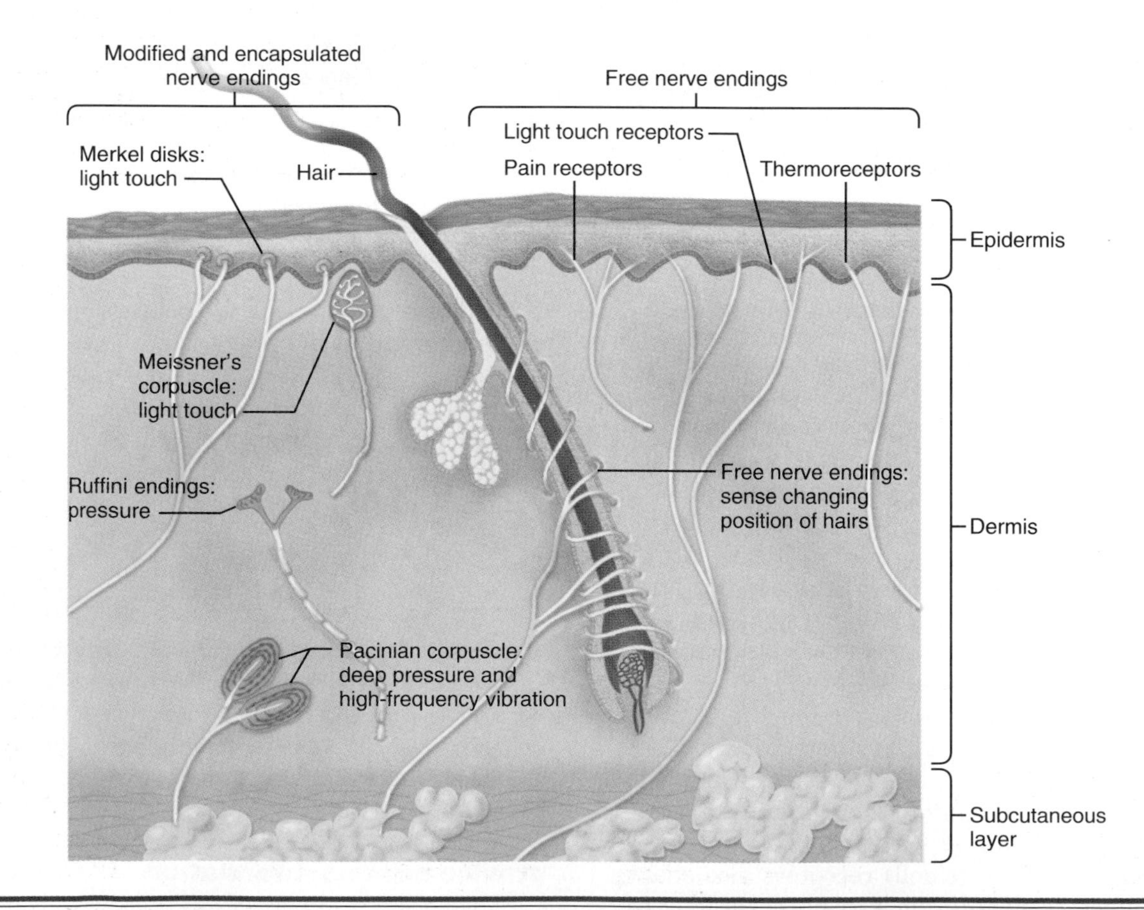

Figure 12.1 Sensory receptors in skin. Thermoreceptors, some of the receptors for light touch, and the receptors for pain are the free endings of sensory neurons. Other receptors for touch, pressure, and vibration are modified or encapsulated endings of sensory neurons.

The various receptors for touch, pressure and vibration differ in their locations, the intensity of the stimulus that must be applied in order to generate an impulse (pressure is a stronger mechanical force than light touch), and the degree to which they adapt. Vibration-sensitive receptors adapt so quickly that it takes a rapidly changing physical deformation (many on-and-off cycles per second) to keep them stimulated. Vibration-sensitive receptors are particularly useful for providing information about potentially harmful flying insects because they respond to the kinds of vibrations generated by insects' wings.

Figure 12.1 depicts the skin with several types of receptors for detecting somatic sensations. We list them here:

- *Unencapsulated dendritic endings:* Naked dendritic endings of sensory neurons around hairs and near the skin surface that signal pain, light pressure, and changes in temperature.
- *Merkel disks:* Detect light touch and pressure. Merkel disks are modified unencapsulated dendritic endings.
- *Pacinian corpuscles:* There are several kinds of dendritic endings that are encapsulated (enclosed) in epithelial or connective tissue. The best known, called Pacinian corpuscles, are encapsulated receptors located in the dermis that respond to either deep pressure or high-frequency vibration. Figure 12.2 shows how encapsulation enables Pacinian corpuscles to adapt to a continuously applied stimulus.
- *Meissner's corpuscles:* Encapsulated touch receptors located closer to the skin surface than Pacinian corpuscles. They detect the beginning and the end of light pressure and touch.
- *Ruffini endings:* Encapsulated receptors that respond continually to ongoing pressure.

Recap **Somatic sensations include temperature, touch, pressure, vibration, pain, and awareness of body movements and position. Mechanoreceptors that detect touch, pressure, and vibration include a variety of unencapsulated and encapsulated dendritic endings.**

Mechanoreceptors indicate limb position, muscle length, and tension

How can you know body position with your eyes closed?

With your eyes closed, try to describe the positions of your arms and legs. You can probably describe their position rather well. You may even be able to tell which muscle groups are contracting. You do this with a variety of mechanoreceptors in joints (for joint position), in skeletal muscles (for muscle length), and in tendons (for tension).

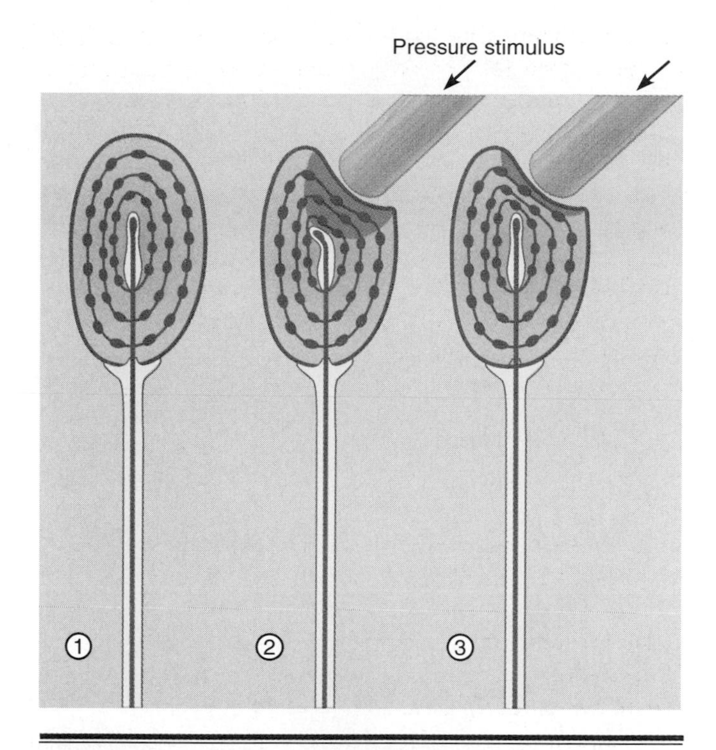

Figure 12.2 The Pacinian corpuscle. The encapsulation structure of the Pacinian corpuscle permits the mechanoreceptor to adapt rapidly to a continuously applied stimulus. (1) The Pacinian corpuscle in the unstimulated state. (2) The initial application of a stimulus causes mechanical bending of the receptor ending of the nerve, leading to excitation of the receptor. (3) With continuous application of the stimulus, the layers of encapsulation slip past each other, dissipating the energy generated by the stimulus and permitting the receptor ending to return to its initial shape.

Perhaps the best known of these mechanoreceptors are the specialized structures for monitoring muscle length, called **muscle spindles.** For most joints, muscle length determines joint position because of the way the muscle is attached to the bones.

A muscle spindle is a small bundle of modified skeletal muscle cells located within a skeletal muscle (Figure 12.3). Muscle spindles are innervated by sensory nerves whose dendritic endings are mechanoreceptors that respond to stretch (lengthening). One type of dendritic ending wraps around the middle of a muscle spindle cell. Another has several branches that attach nearer the end of the muscle spindle cell. When the whole muscle is stretched, the receptors attached to the muscle spindle cells are stretched as well. Mechanical distortion of the mechanoreceptors produces local graded potentials in the dendritic endings and (if threshold is passed) an action potential.

This is exactly what happens in the classic patellar stretch reflex, described in the last chapter. When a physician taps your patellar tendon (at the front of your knee) with a hammer, the patellar tendon and quadriceps (thigh) muscle are stretched, stretching the muscle spindle as well. Stretch of the muscle

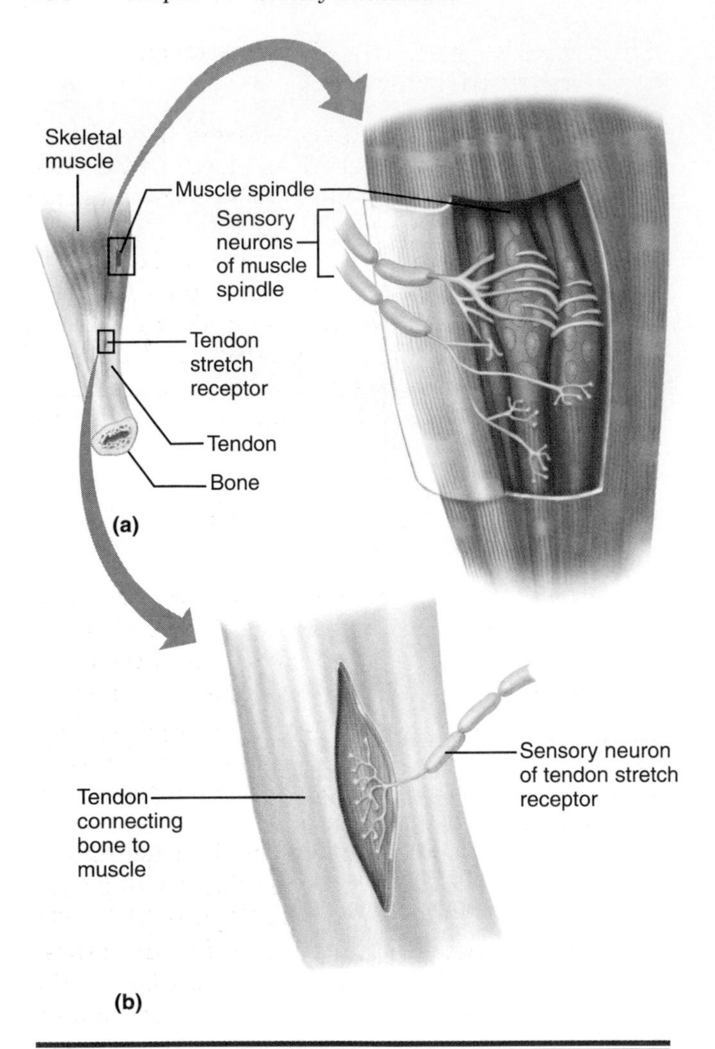

Figure 12.3 Muscle spindles and tendon receptors. **(a)** Muscle spindles are a group of modified muscle cells that are innervated by sensory neurons. Passive stretch of a muscle stretches the muscle spindle, stimulating mechanoreceptors in the nerve endings of the sensory neurons. Conversely, muscle contraction shortens the muscle spindle and reduces muscle spindle mechanoreceptor stimulation. **(b)** Tendon stretch receptors respond to tension in tendons. When a muscle contracts and also when it is stretched passively, tension on the tendon increases, causing activation of the tendon stretch receptors.

causes sensory neurons innervating the muscle spindle to transmit action potentials to the spinal cord. Within the cord, motor nerves to the quadriceps are activated, causing the quadriceps to contract. All this occurs in a fraction of a second.

Tendons, which connect muscle to bone, also have mechanoreceptors. Tendon receptors respond to tension, but they cannot distinguish between tension produced by passive stretch and tension produced by active muscle contraction. Tendon receptors may play a protective role in preventing injury by extremely high tensions. For example, if you were to fall out of a tree but land on your feet, the impact when you hit the ground would produce high levels of tension, activating your tendon stretch receptors. The sensory information they provide inhibits muscle contraction, so you would crumple to the ground rather than stay upright. This reflex-induced collapse protects your muscles and bones from tearing and breaking, respectively.

Recap **Mechanoreceptors in skeletal muscle, tendons, and joints respond to changes in muscle length, tendon tension, and joint position. They provide information about the position of our limbs and the amount of tension on muscles.**

Thermoreceptors detect temperature

Thermoreceptors for heat and cold near the skin's surface provide useful information about the external environment. These receptors adapt quickly, allowing us to monitor changes in temperature accurately and yet adjust sensory input so it becomes more bearable. For example, when you step into a hot shower the water may seem uncomfortably warm at first, but you soon become used to it.

Other thermoreceptors located in abdominal and thoracic organs throughout the body monitor *core* (internal) *temperature.* In keeping with their role in maintaining homeostasis of core temperature, these thermoreceptors do not adapt rapidly.

Recap **Thermoreceptors near the skin surface monitor the temperature of the external environment. Thermoreceptors in internal organs maintain homeostasis of body temperature.**

Pain receptors signal discomfort

What is the value or purpose of pain?

As much as we dislike feeling pain, the ability to perceive pain is essential for survival. Pain warns us to avoid certain stimuli and informs us of injuries. Pain receptors are unencapsulated endings that respond to injury from excessive physical pressure, heat, light, or chemicals. Pain is described as either fast or slow, depending on its characteristics.

Fast pain, also called sharp or acute pain, occurs as soon as a tenth of a second after the stimulus. Receptors for fast pain generally respond to physical pressure or heat and usually are located near the surface of the body. They inform us of stimuli to be avoided—for example, stepping on a nail or touching a hot burner on the stove. The reflex withdrawal response to fast pain is rapid and strong.

In contrast, *slow pain* generally arises from muscles or internal organs. Slow pain, which may not appear until seconds or even minutes after injury, is due to activation of chemically sensitive pain receptors by chemicals released from damaged tissue. Slow pain from internal organs is often perceived as originating from an area of the body completely removed from

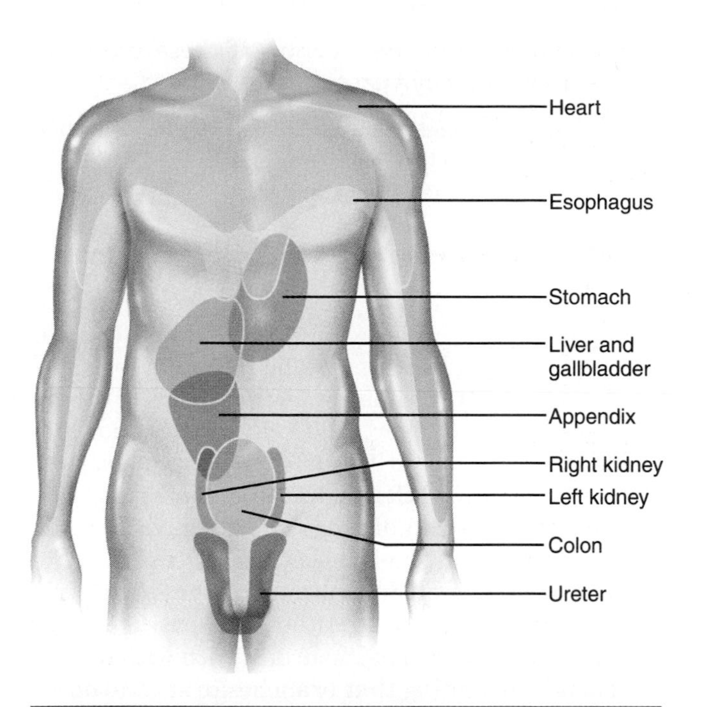

Figure 12.4 Locations of referred pain from various internal organs. Note that the areas of overlap for esophageal and heart pain are so extensive that it is often difficult to tell them apart by location.

the actual source. This phenomenon is called *referred pain,* and it happens because action potentials from internal pain receptors are transmitted to the brain by the same spinal neurons that transmit action potentials from pain receptors in the skin and skeletal muscles. The result is that the brain has no way of knowing the exact source of the pain, so it assigns the pain to another location.

Referred pain is so common that physicians use it to diagnose certain disorders of internal organs. For example, a heart attack usually manifests itself as pain in the left shoulder and down the left arm. Figure 12.4 shows where pain from internal organs may be felt. The close association between areas of referred pain from the heart and esophagus is why it can be difficult, based on sensations alone, to distinguish a heart attack from the less serious problem of esophageal spasm.

Pain receptors generally do not adapt. From a survival standpoint this is beneficial, because the continued presence of pain reminds us to avoid doing whatever causes it. However, it also means that people with chronic diseases or disabilities can end up experiencing constant pain.

Recap **Fast pain arises from receptors near the body surface, while slow pain originates in muscles or internal organs. Pain is important to survival because it warns us of injury.**

12.3 Taste and smell depend on chemoreceptors

We turn now to the five special senses: taste, smell, hearing, balance, and vision. As mentioned earlier, these senses originate from receptors located in special areas of the body.

The first senses we consider, taste and smell, are both due to the action of chemoreceptors. The rich fragrance of a rose, the less pleasing odor of a skunk, the tart taste of an orange—all depend on our ability to convert a chemical stimulus into the only language the central nervous system can understand: action potentials.

Taste: Chemoreceptors bind with dissolved substances

Spotlight on Structure & Function

The function of our **taste buds** is to convert the presence of a chemical signal into action potentials. A taste bud is a cluster of about 25 *taste cells* and 25 *supporting cells* that support and insulate the taste cells from each other. We have approximately 10,000 taste buds, mostly on the tongue, although a few are scattered throughout the mouth.

The tongue has a rough texture because its surface is composed of numerous *papillae* (small projections) surrounded by deep folds (Figure 12.5). The taste buds are located at the surface of many of these folds. The exposed tip of a taste cell has *taste hairs* that project into the mouth. The taste hairs contain chemoreceptors, which are cell membrane proteins specific for a certain chemical (called the "tastant").

When we eat or drink, tastants dissolve in our saliva and bind to the chemoreceptor. The binding of tastant to chemoreceptor causes the taste cell to release a neurotransmitter, which stimulates a nearby sensory neuron to produce action potentials. Note that the taste cell itself is not a sensory neuron. It takes two cells (the taste cell and a sensory neuron) to convert the chemical stimulus into a nerve impulse, but the effect is ultimately the same as if it were one cell. ■

The four major categories of taste are sweet, sour, salty, and bitter. The receptors for the four different taste qualities tend to be concentrated in different areas of the tongue. Receptors for sweet are located on the tip of the tongue, sour on the sides, salty on the tip and sides, and bitter at the back. Very few receptors are located on the center of the tongue.

To gain the maximum taste sensation, we move a substance around in our mouths before swallowing. On the other hand, if we don't want to taste something, we swallow quickly and pass it over the center of the tongue. The combined sensitivities of our taste

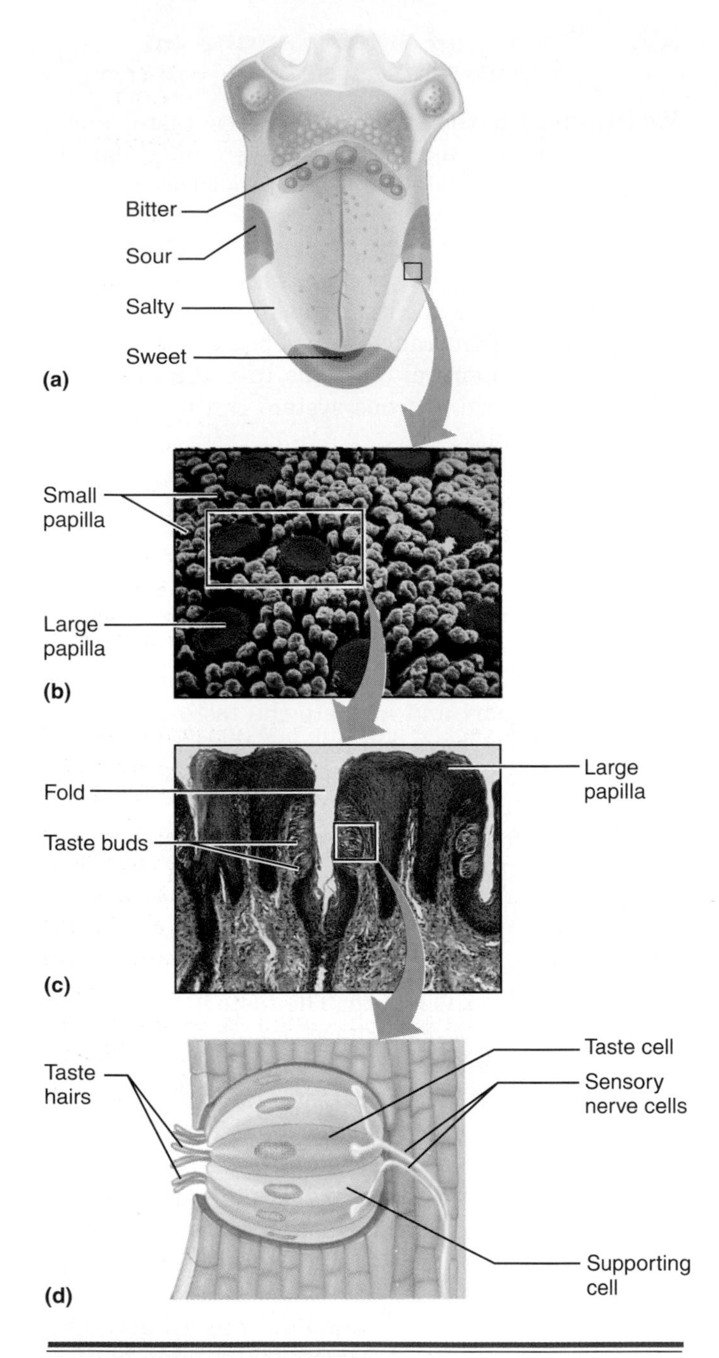

Figure 12.5 Locations and structure of the receptors for taste. (a) Locations of the four primary taste categories on the tongue. **(b)** A view (approximately ×75) of the surface of the tongue showing that it is covered with large and small papillae. **(c)** A sagittal section (approximately ×200) through several large papillae showing the location of the taste buds along their sides, just beneath the tongue's surface. **(d)** A taste bud is composed of a group of taste cells and supporting cells. Each taste cell connects to a sensory neuron.

receptors allow us to distinguish hundreds of different flavors, far beyond the basic four categories.

It is probably no accident that bitter taste receptors are located at the very back of the tongue. Bitterness is often a sign of an inedible or even poisonous substance, and the location of these receptors makes it difficult to bypass them.

> ***Recap*** **Taste buds contain chemoreceptor cells that bind dissolved chemicals. The four basic taste qualities are sweet, salty, sour, and bitter.**

Smell: Chemoreceptors bind with odorants

The sense of smell also relies on a chemoreceptor mechanism. There are two fundamental differences between taste and smell. (1) The receptors for smell are located on true sensory neurons, and (2) there are receptors for over 1,000 different "odorant" chemicals, as opposed to just four tastant classes. With these 1,000 different receptor types and our sense of taste we can distinguish as many as 10,000 different taste and smell sensations.

Odors are detected by **olfactory receptor cells** located in the upper part of the nasal passages. The receptor cells are sensory neurons, each with a modified dendritic ending that branches to several *olfactory hairs* (Figure 12.6). The olfactory hairs extend into a layer of mucus covering the surface of the nasal passages. The mucus, produced by nearby olfactory glands, keeps these hairs from drying out.

Gaseous and airborne odorants enter the nasal passages, become dissolved in the mucus, bind to chemoreceptors located on the olfactory hairs, and cause the olfactory receptor cell to generate impulses. Olfactory receptor cells synapse with olfactory neurons in a nearby area of the brain called the *olfactory bulb,* where information is partially integrated and then passed to higher brain centers.

Our sense of smell complements our sense of taste because when food is chewed it releases chemicals that come into contact with olfactory receptors. The combined sensations create an effect that is interpreted at higher brain levels. Children learn this early by experience—they hold their noses when asked to eat foods they hate (Figure 12.7). When we have a cold, most foods taste less appetizing. In part this is because increased production of mucus prevents odorant molecules from reaching olfactory receptors, impairing our sense of smell.

> ***Recap*** **Our sense of smell augments and complements our sense of taste, allowing us to experience over 10,000 different combinations of taste and smell.**

12.4 Hearing: mechanoreceptors detect sound waves

Sounds are waves of compressed air. The *loudness* of a sound is determined by the amplitude of the waves and is measured in units called decibels (db). The threshold of human hearing (barely audible) is set at

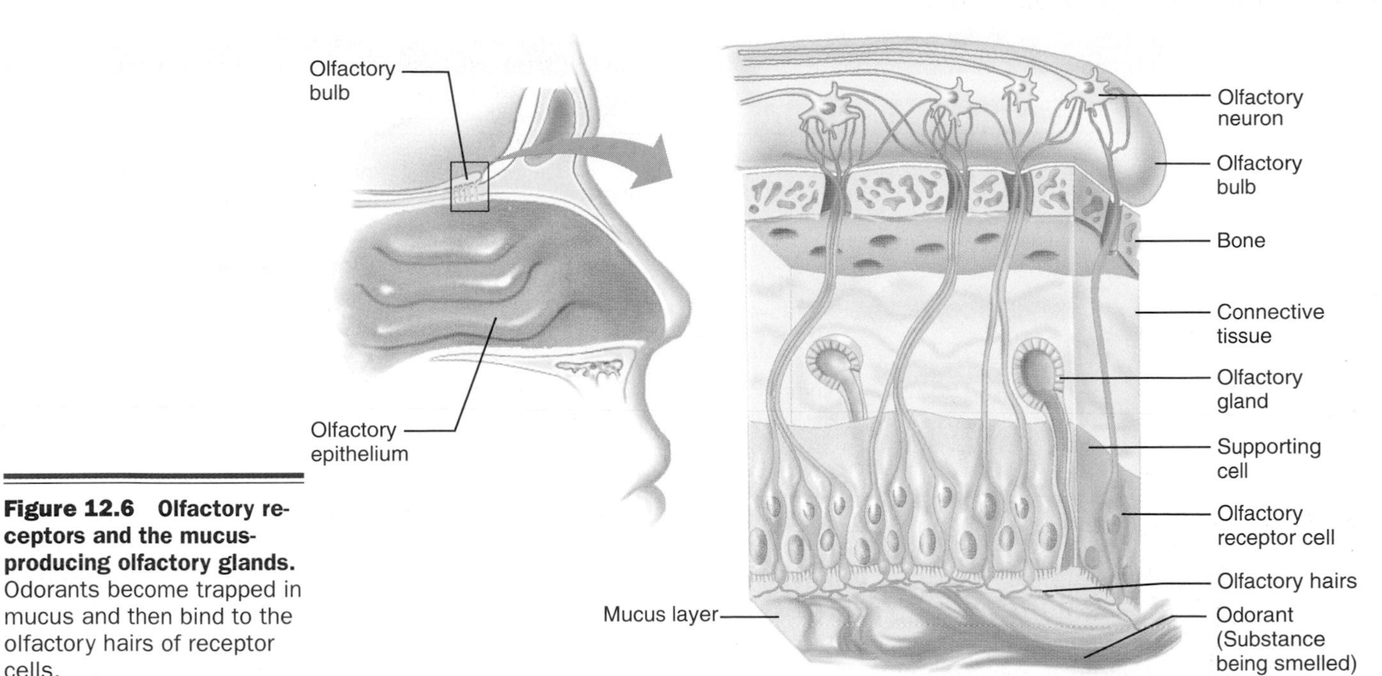

Figure 12.6 Olfactory receptors and the mucus-producing olfactory glands. Odorants become trapped in mucus and then bind to the olfactory hairs of receptor cells.

0 db, and each 10db increase represents a perceived doubling in loudness. Severe hearing loss can result from exposure to sounds louder than 90db. Table 12.2 lists the decibels associated with some common sounds.

The *tone* of a sound is determined by its frequency, the number of wave cycles that pass a given point per second (cycles/sec). Frequency is expressed in *hertz* (Hz). Higher tones are those of a higher frequency (Figure 12.8). The frequency range for normal human hearing is from 20–20,000 Hz.

When waves of sound strike the ear they cause vibration-sensitive mechanoreceptors deep within the ear to vibrate. The vibration causes the production of impulses that travel to the brain for interpretation.

The outer ear channels sound waves

The human ear consists of three functionally different regions: the outer ear, the middle ear, and the inner ear (Figure 12.9).

The **outer ear** consists of the *pinna,* or visible portion of the ear, and the *auditory canal.* Sound waves arrive at the pinna and are directed into the auditory canal, which channels them to the **eardrum (tympanic membrane)**. The eardrum serves as a partition between the outer and middle ears.

The middle ear amplifies sound

Spotlight on Structure & Function

The **middle ear** consists of an air-filled chamber within the temporal bone of the skull, bridged by three small bones called the *malleus* ("hammer"), *incus* ("anvil"), and *stapes* ("stirrup"). When sound waves strike the eardrum (called the tympanic membrane), it moves back and forth slightly (vibrates). Vibration of the eardrum makes the malleus, incus, and stapes vibrate. The stapes touches a smaller membrane, called the *oval window,* causing it to vibrate too.

All the vibrating force of the much larger eardrum is now concentrated on the smaller oval window. In effect, the middle ear amplifies the sound waves several-fold. Amplification is essential because the oval window must vibrate with sufficient force to produce pressure waves in the watery fluid of the inner ear. ■

Figure 12.7 Smell and taste. Olfactory receptors contribute to our sense of taste. This child would prefer not to taste at all.

Table 12.2 Common noises and your hearing

Sound	Decibels	How It Affects Your Hearing
Mechanoreceptors		
Rustling leaves	20	Does not cause hearing loss
Normal conversation	50–60	Does not cause hearing loss
Vacuum cleaner	70	Does not cause hearing loss
Power lawn mower	85–100	Cumulative exposure causes hearing loss; newer mowers may be quieter
Live rock concert	110–130	Definite risk of permanent hearing loss
Jackhammer	120–150	Definite risk of permanent hearing loss
Shotgun	140	One exposure likely to cause permanent hearing loss
Jet engine	150	One exposure likely to cause permanent hearing loss

The air-filled middle ear is kept at atmospheric pressure by the **eustachian (auditory) tube,** a narrow tube that runs from the middle ear chamber to the throat. Although the tube is normally held closed by the muscles along its sides, it opens briefly when you swallow or yawn to equalize air pressure in the middle ear with that of the surrounding atmosphere. When you change altitude quickly, your ears "pop" as the eustachian tube opens briefly and air pressures are suddenly equalized. When you have a cold or allergy, inflammation and swelling may prevent the eustachian tube from equalizing pressures normally, which is why airplane trips can be painful when you have a cold or allergies. Children tend to get a lot of "earaches" (middle ear infections) via the eustachian tube until the skull elongates, changing the angle of the eustachian tube to the ear.

Why do your ears "pop" when you change altitude quickly?

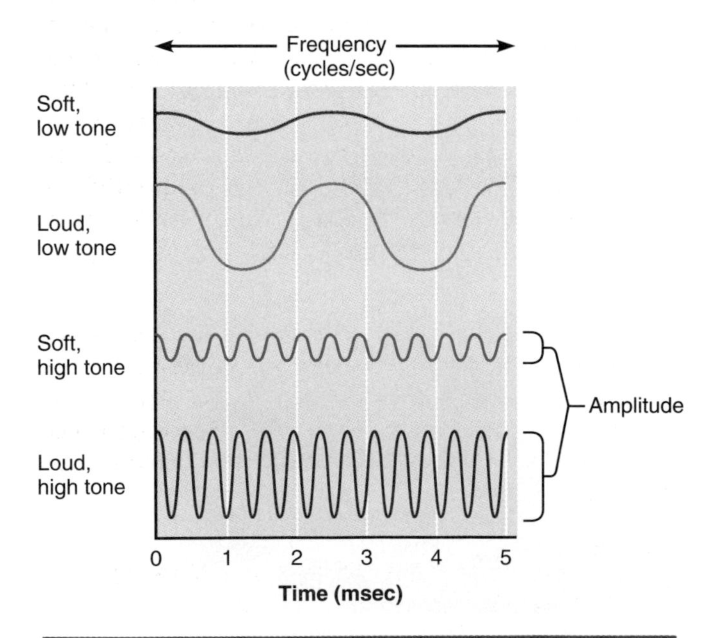

Figure 12.8 Properties of sound waves. Intensity (loudness) is determined by sound wave amplitude. Tone is determined by sound wave frequency.

The inner ear sorts and converts sounds

The **inner ear** sorts sounds by tone and converts them into impulses. It consists of the **cochlea,** where sound is converted, and the *vestibular apparatus* consisting of the vestibule and the three semicircular canals, which does not contribute to hearing at all. We will discuss the vestibular apparatus later in connection with the sense of balance.

The cochlea is a coiled structure shaped a bit like a snail. Figure 12.10a shows what the cochlea would look like if it were uncoiled. It is a tapered tube containing two interconnected outer canals called the *vestibular canal* and the *tympanic canal* surrounding a third closed fluid-filled space called the *cochlear duct*. The base of the cochlear duct is formed by the **basilar membrane.** The basilar membrane supports a population of about 15,000 **hair cells,** the mechanoreceptor cells of the ear. Hair cells have hairlike projections that are embedded in an overhanging structure called the **tectorial membrane** (the Latin word for "roof" is *tectum*), which is composed of a firm, gelatinous, noncellular material. Together, the hair cells and the tectorial membrane are called the *organ of Corti,* representing the organ that converts pressure waves to impulses.

When sound waves strike the oval window, the oval window generates pressure waves in the watery fluid in the vestibular canal. These waves travel around the cochlea to the tympanic canal. Eventually the waves strike another membrane called the *round window,* which bulges inward and outward in synchrony with the oval window, dissipating some of the pressure.

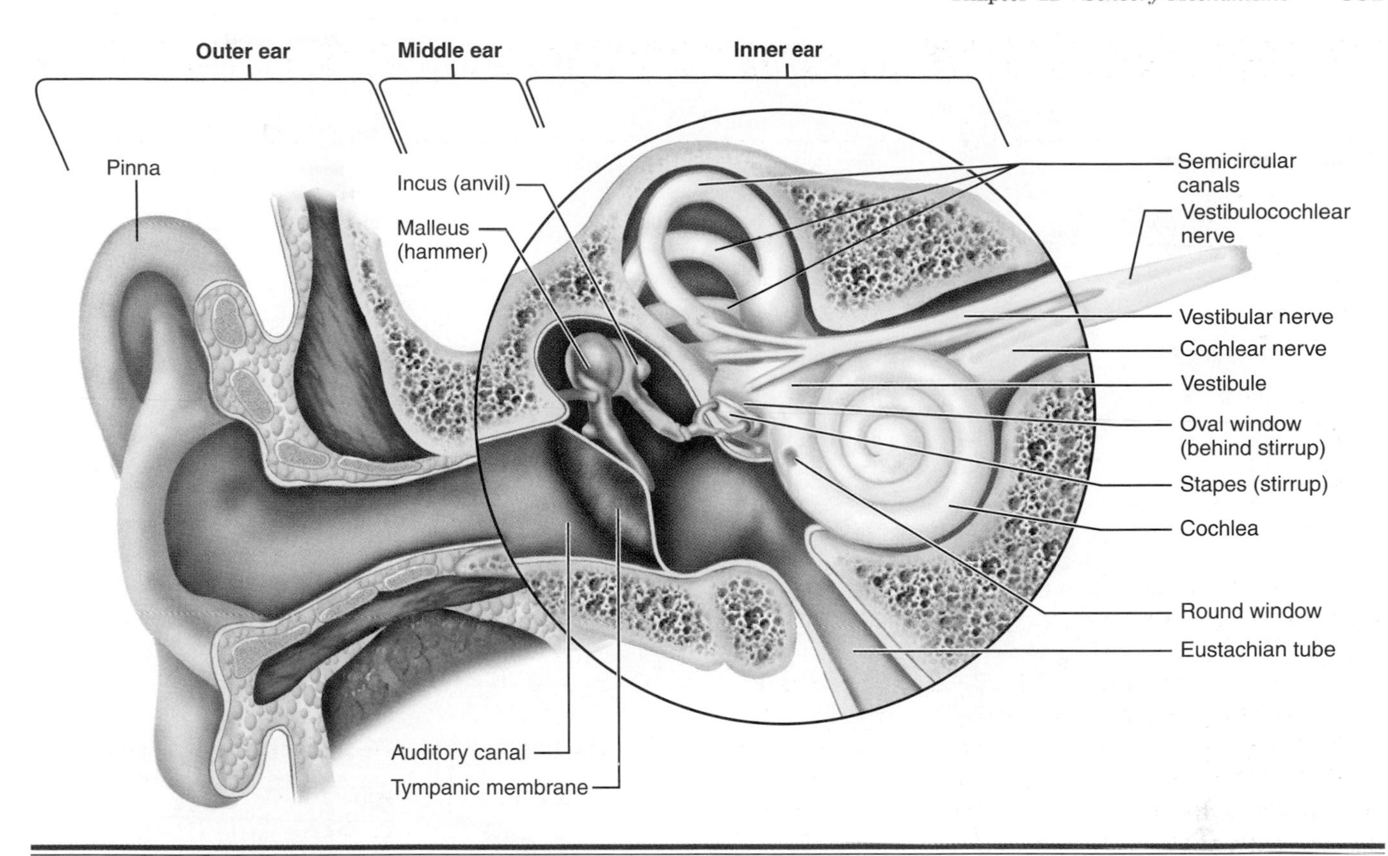

Figure 12.9 Structure of the human ear. The outer ear receives and channels sound, the middle ear amplifies sound, and the cochlea of the inner ear converts the sound to action potentials. The vestibule and the three semicircular canals are responsible for balance.

Some of the pressure waves take a shortcut from the vestibular canal to the tympanic canal via the cochlear duct. These pressure waves cause the basilar membrane to vibrate like the strings of a musical instrument. The slight movement of the basilar membrane causes the hair cells above the membrane to vibrate as well. Because the hairs of the hair cells are embedded in the less moveable tectorial membrane, vibration of the basilar membrane causes the hairs to bend. This physical movement of the hairs causes the hair cells to release a neurotransmitter, which in turn generates impulses in nearby sensory neurons. The impulses travel in neurons of the auditory nerve, which joins the vestibular nerve to become the *vestibulocochlear nerve* leading to the brain.

Vibrations of fluid in the vestibular and tympanic canals are of the same frequency as the original sound waves, which may be of many different tones at once. But because the fibers of the basilar membrane are different lengths at different regions of the cochlea, sound waves are converted to vibrations of the basilar membrane in a location-specific manner according to tone. High-pitched tones vibrate the basilar membrane closer to the base of the cochlea (near the oval and round windows), whereas low-pitched tones vibrate the basilar membrane closer to the cochlear tip. The identical impulses generated in sensory neurons at these different locations are interpreted by the brain as sounds of specific tones on the basis of where they originated on the basilar membrane. This is yet another example of the location-specific nature of information processing by the brain.

Recap **The outer ear receives and directs sound to the middle ear. The three bones of the middle ear amplify the sound and pass it to the cochlea. Vibrations of fluid in the cochlea cause the hairs of receptor hair cells to bend, generating action potentials in nearby neurons.**

12.5 The inner ear plays an essential role in balance

Maintaining your balance against gravity requires integration of multiple sensory inputs. Those sensory

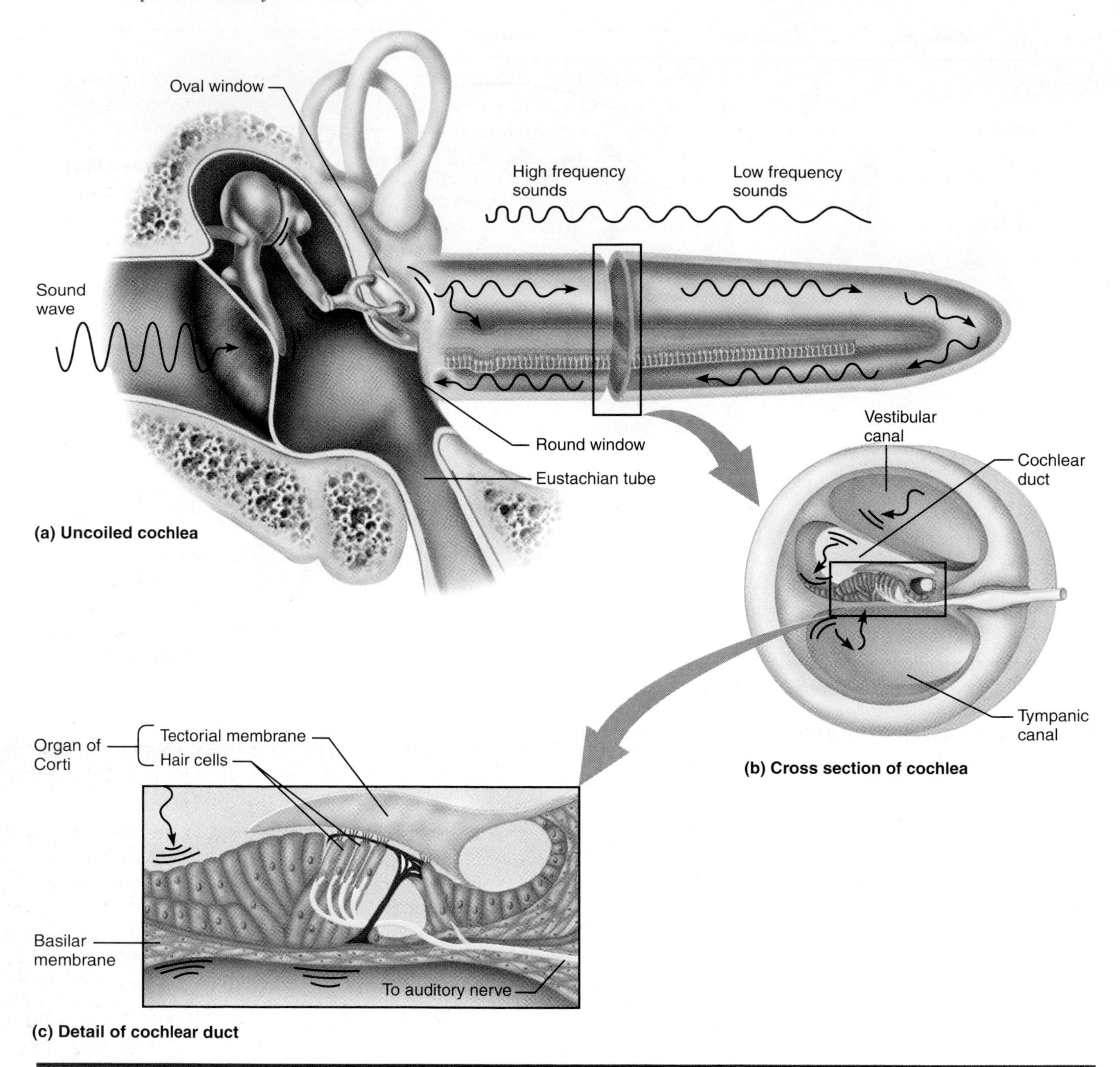

Figure 12.10 Structure and function of the cochlea. (a) The cochlea as it might appear if it were uncoiled. Higher frequency sounds are converted to impulses nearer to the oval window, whereas lower frequency sounds are converted near the tapered tip. **(b)** A cross section through the cochlea, showing how pressure waves passing from vestibular canal to tympanic canal through the cochlear duct cause the basilar membrane to vibrate. **(c)** A section through part of the central structure shows the hair cells with their hairs embedded in the tectorial membrane. Vibration of the basilar membrane causes hairs to bend, ultimately generating impulses in sensory neurons.

inputs include signals from the joint receptors, muscle spindles, and tendon receptors previously described as well as special structures associated with the inner ear. Even visual input is involved. The inner ear structures, described next, provide information about rotational movement, position, and linear acceleration of the head.

Looking again at the inner ear (Figure 12.11a), we see that next to the cochlea is the **vestibular apparatus,** a system of fluid-filled canals and cham-

Health Watch

Hearing Loss

Impaired hearing is not the same as having cancer or AIDS. Nevertheless, a loss of hearing can certainly affect your quality of life.

More and more people, especially musicians and others who are exposed to loud sounds on the job or in their hobbies, suffer from at least partial hearing loss. Over one-third of adults in North America will experience significant declines in hearing capacity by retirement age. We tend to think of being "hard of hearing" as part of the aging process precisely because it is so common, but it does not have to be that way. Evidence suggests that most hearing loss results from cumulative exposure to loud sounds over our lifetime. Often the decline in hearing progresses so slowly that we hardly notice it until the damage has already occurred.

Sound loudness is measured in decibels. The rustling of leaves might be 20 decibels, a conversation about 50 decibels, a motorcycle 85–100 decibels, and a loud rock concert nearly 130 decibels. Any noise over 80 decibels is loud enough to produce at least some hearing loss (see Table 12.2).

Because most of us don't think in decibels, here's an easy rule of thumb. If you must raise your voice to be heard by someone nearby, the sound around you is probably over 80 decibels. If you have to shout or scream or can't be heard at all, it's too loud.

Loud sounds impair our hearing because the excessive amplitude of the sound waves vibrates the hair cells too strongly and breaks the delicate hairs. Your body may be able to repair minor damage, but repeated severe damage can kill the entire hair cell. Hair cells are not replaced, so when you lose them they are gone forever.

Deafness caused by damage to hair cells is called *nerve deafness* because sounds cannot be converted into impulses in sensory nerves. Another type of hearing loss, *conduction deafness*, results from damage to the eardrum or the bones of the middle ear. This impedes the transfer of sound waves to the inner ear so that sounds never reach the hair cells. Conduction deafness is frequently due to arthritis of the middle ear bones.

Hearing loss may be partially corrected with a hearing aid, which amplifies incoming sounds, but even amplification won't be very effective if hair cells are missing. Some people benefit from a cochlear implant, which consists of a tiny microprocessor that converts sound waves into electrical signals.

What can you do to protect your hearing? Health professionals suggest these guidelines:

- Wear earplugs when using noisy equipment. This includes lawn mowers, chain saws, rug cleaning machines, and any other loud indoor or outdoor devices.
- Earplugs are also a good idea around recreational vehicles such as snowmobiles and motorcycles.
- Turn down the volume on headphones, boom boxes, and stereo speakers.
- Be cautious about attending rock concerts and other noisy events. Repeated exposure to loud music will definitely damage your hearing.

As a simple rule, if you notice that you suffer a temporary change in hearing after certain activities (temporary hearing loss, ringing in your ears, or a feeling of "fullness" in your ears), don't put yourself repeatedly in that situation.

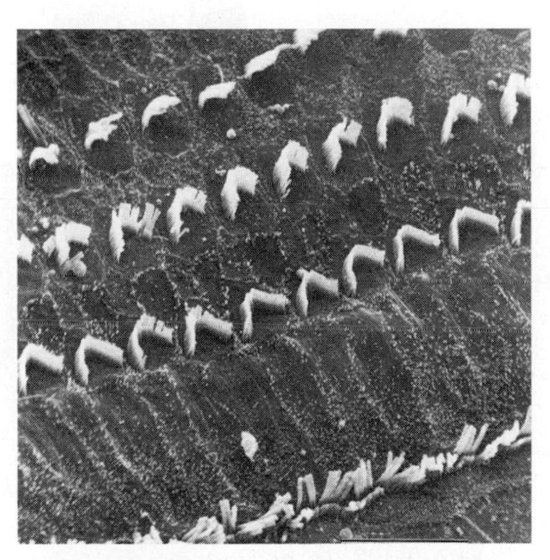

(a) Normal healthy hair cells of the cochlea of a guinea pig.

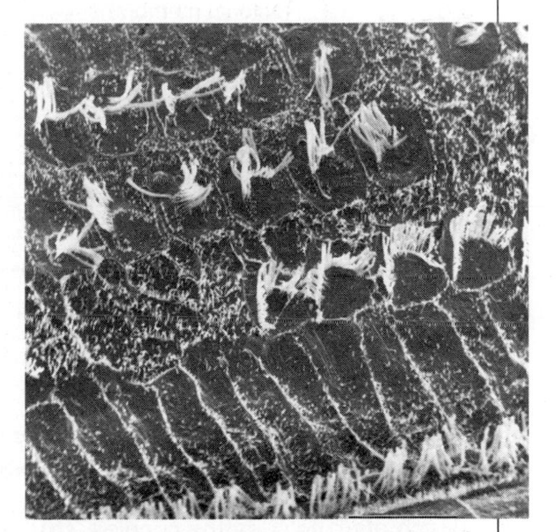

(b) Hair cells after 24 hours of exposure to loud sounds.

bers. The vestibular apparatus consists of three *semicircular canals* for sensing rotational movement of the head and body, and an area called the *vestibule,* which senses static (nonmoving) position and linear acceleration and deceleration.

The semicircular canals and the vestibule have mechanoreceptor-type hair cells embedded in a gel-like material. When the head moves or changes position, the gel also moves, although a bit more slowly because of inertia. The movement of the gel bends the hairs, which causes the hair cells to release a neurotransmitter that ultimately generates impulses in sensory neurons of the vestibular nerve.

Sensing rotational movement

The three semicircular canals are fluid-filled tubes of bone. Near the end of each canal is a bulging region

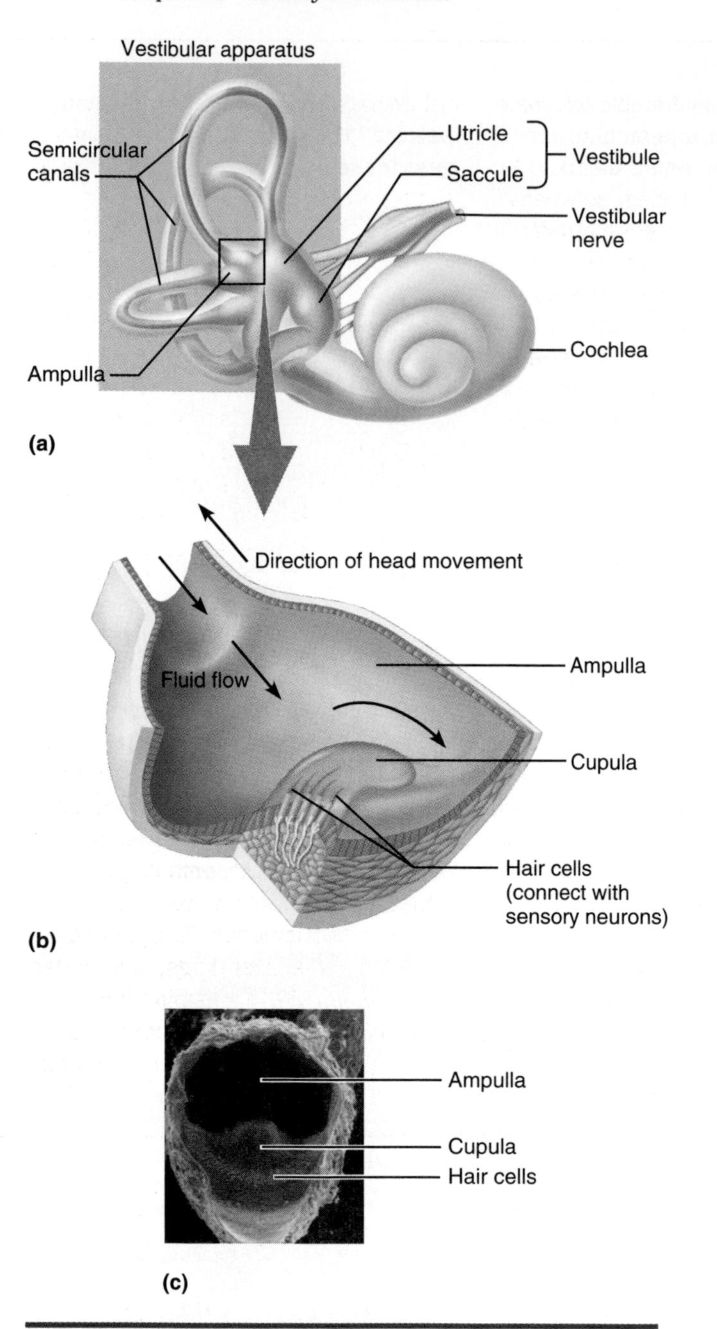

Figure 12.11 The vestibular apparatus. (a) The vestibular apparatus (the organ of balance) is part of the inner ear, which also includes the cochlea. **(b)** Cutaway view of the ampulla of a semicircular canal, showing the hair cells embedded in the base of the gel-like cupula. Rotational movement of the head causes fluid movement in the opposite direction in the semicircular canal. The fluid pushes against the cupula, bending it and causing activation of the mechanoreceptor hairs of the hair cells. The hair cells release a neurotransmitter that stimulates sensory neurons of the vestibular nerve. **(c)** Scanning EM of a cupula.

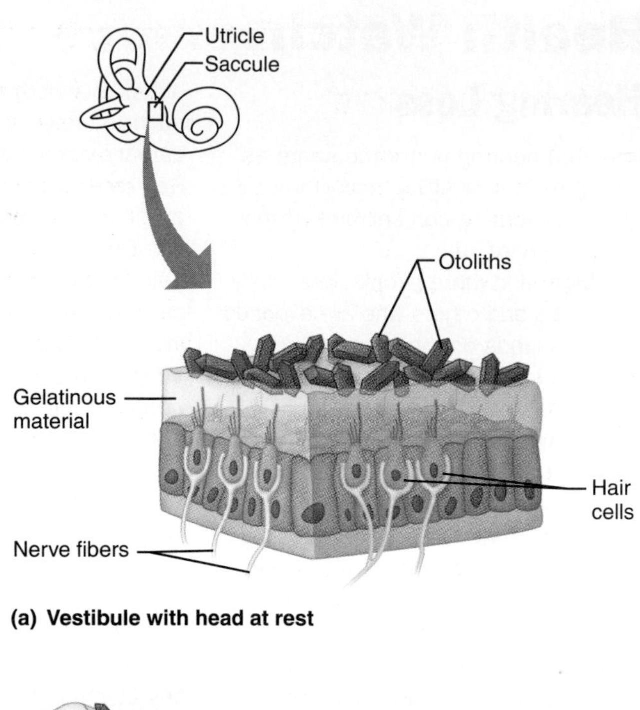

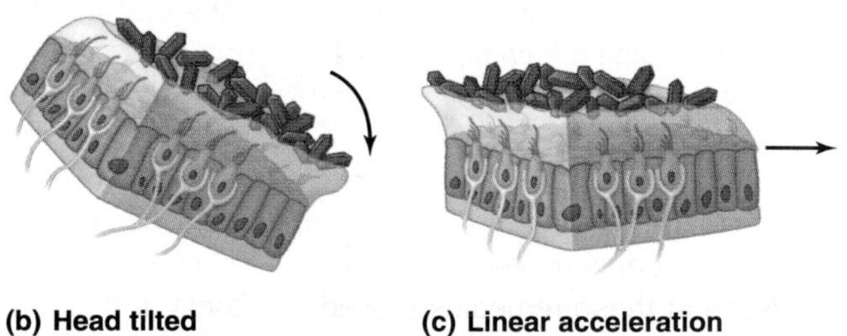

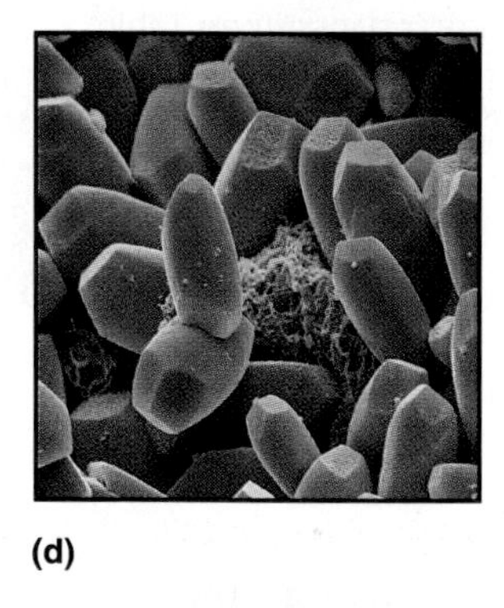

Figure 12.12 How the utricle and saccule sense static head position and linear acceleration. (a) A cutaway view of the vestibule with the head at rest. **(b)** A change of head position causes a shift of position of crystals of bonelike material called otoliths, which ultimately stimulates hair cells. **(c)** Linear acceleration also produces a shift of position of the otoliths. **(d)** Scanning EM of otoliths.

called the *ampulla* containing a soft, gel-like dome called the *cupula* (Figure 12.11b). Hairs of mechanoreceptor sensory neurons are embedded in the cupula.

When you turn your head, the fluid in your semicircular canal moves more slowly than your head because of inertia. The shifting position of the fluid bends the soft cupula and the hairs within it, producing impulses. The structure of the hairs enables them to detect the direction of the movement, in addition to movement in general. Because there are

Directions in Science

Motion Sickness

One of the more curious phenomena related to sensory mechanisms is the unpleasant sensation known as "motion sickness." Motion sickness is not a disease but a physiological response to abnormal body motion and visual experience. Nearly everyone is susceptible to some degree. Symptoms of motion sickness include nausea and vomiting, cold sweats, pale skin, hyperventilation, and headache.

The history of our theories about motion sickness shows how scientific beliefs change as we gain new knowledge and discard old hypotheses. Until the late 19th century, scientists believed motion sickness was caused by excessive motion of the stomach and its contents (a plausible enough hypothesis, given that nausea and vomiting are prominent symptoms). Treatments ranged from wearing a specially designed motion sickness belt to a variety of medicines and diets. None of these remedies proved effective.

A breakthrough occurred in 1881, when scientists discovered the function of the vestibular apparatus. Subsequent research led to a new hypothesis: that motion sickness was due to overstimulation of the vestibular apparatus.

A hypothesis must be discarded or at least modified if it can be contradicted by convincing evidence. In this case, in the past 50 years we have learned that motion sickness is not just due to overstimulation of the vestibular apparatus. In fact, it also depends on sensory inputs from muscle, joint, and visual receptors. These findings have broadened the earlier vestibular apparatus explanation into what is now called the "sensory conflict" hypothesis.

According to the sensory conflict hypothesis, motion sickness occurs when the central nervous system receives conflicting sensory inputs from the vestibular apparatus, the eyes, and various receptors in muscles, tendons, and joints throughout the body. The symptoms result from the inability of the CNS to integrate conflicting sensory signals into an understandable whole.

The adaptive value of these symptoms, if there is one, is not clear. One untested hypothesis is that feeling ill immobilizes us, thus protecting us from the risk of falling or being injured.

The sensory conflict hypothesis is strengthened by the fact that we do not get motion sickness from voluntary motions such as running. Voluntary motions are initiated by the brain and therefore presumably are understood by the brain. The theory also seems to fit with what we know about adaptation. The symptoms of motion sickness fade over time even though the motions that cause it continue, perhaps because the brain comes to accept the sensory inputs as normal.

Scientific discovery relies on curiosity as well as careful scientific method. Breakthroughs are sometimes achieved by following through on chance observations that might just as easily be ignored. Such is the case of the first effective treatment for motion sickness. In 1949 a woman who was taking the drug dimenhydrinate for a skin condition told her physician that she seemed to be immune to motion sickness. Intrigued, he reported the observation to others.

Subsequent tests with large groups of individuals proved that it was a general phenomenon. Today, dimenhydrinate and similar drugs, sold under brand names like Dramamine and Scopolamine, are still the most effective remedies against motion sickness.

three semicircular canals and each is oriented on a different plane, together they can detect any and every possible rotational movement of the head.

Sensing head position and acceleration

To sense the position of the head relative to the earth, we need a mechanism that is sensitive to gravity. The two fluid-filled chambers of the vestibule, called the *utricle* and the *saccule,* each contain mechanoreceptor hair cells, gel, and hard crystals of bone-like material called **otoliths** (literally, "ear stones") (Figure 12.12).

The otoliths are embedded in the gel near the gel's surface, unattached to the bone wall of the vestibule. The otoliths are indeed like little rocks, and they are considerably heavier than the gel itself. When the head tips forward, for instance, the otoliths slide toward the center of the gravitational

Try It Yourself

A Model of Fluid Movement in the Semicircular Canals

You can easily demonstrate the principle of fluid movement in the semicircular canals relative to movement of the head. Fill a bucket half-full of water and place a floating object in it. Now quickly rotate the bucket a half-turn or so. Notice that the object (and the water) tend to move more slowly than the bucket, especially if the movement of the bucket is rapid. The water and the object appear to be flowing in the opposite direction from the bucket, but in fact they are just slow to get started. When you stop the bucket, the opposite happens. The water and the object are now in motion and so they continue to spin for a short time.

Table 12.3 Structures and functions of the ear

Region of Ear	Functions	Structures	
Outer ear	Receives and channels sound	Pinna	
		Auditory canal	
Middle ear	Amplifies sound	Eardrum	Stapes
		Malleus	Oval window
		Incus	Eustachian tube
Inner ear	Sorts sounds by tone and converts them into impulses	Cochlea	Hair cells
		Vestibular canal	Tectorial membrane
		Tympanic canal	Organ of Corti
		Cochlear duct	Round window
		Basilar membrane	
	Senses rotational movements and converts them into impulses	Semicircular canals	Cupula
		Ampulla	Hair cells
	Senses static (nonmoving) position and linear acceleration and deceleration; converts them into impulses	Vestibule	Otoliths
		Utricle	Hair cells
		Saccule	

pull, causing the gel to slip as well and bending the hairs of hair cells (Figure 12.12a, b). Again, this produces impulses in nearby sensory neurons that are transmitted to the brain for interpretation.

The utricle and saccule respond to linear acceleration and deceleration as well as static position. When you are in a car that is accelerating rapidly, you feel like you are being pushed backward in your seat. This is the force of acceleration acting against your own inertia. Likewise, your otoliths are being pushed backward, causing the hair cells to bend (Figure 12.12c).

The otoliths respond only to acceleration and deceleration (a change in linear velocity), and not to a constant rate of speed in one direction.

Table 12.3 summarizes the structures and functions of the ear.

Recap **Rotational movement of the head is sensed as the result of movement of fluid in the three semicircular canals of the inner ear. Head position and linear acceleration are sensed as the result of movement of otoliths (ear stones) in the utricle and saccule of the inner ear.**

12.6 Vision: Detecting and interpreting visual stimuli

Light is a form of electromagnetic radiation that travels in waves at a speed of 186,000 miles per second. Our sense of vision enables us to detect light from objects both nearby and distant, from the table lamp next to you to a star light-years away.

In order to interpret visual stimuli, we must first collect and focus light onto specialized cells called *photoreceptors.* Photoreceptors are located in our eyes, the organs that allow us to receive and process visual information.

Structure of the eye

The structure and functional components of the eye are illustrated in Figure 12.13 and summarized in Table 12.4. A tough outer coat known as the *sclera,* or "white of the eye," covers the outer surface except in the very front, where it is continuous with the clear **cornea.**

Light passes through the cornea and a fluid-filled space called the *aqueous humor* that nourishes and cushions the cornea and lens. Light then either strikes the **iris,** a colored, disk-shaped muscle that determines how much light enters the eye, or passes through the **pupil,** the adjustable opening in the center of the iris. Light that passes through the pupil strikes the **lens,** a transparent, flexible structure attached to a ring of circularly arranged smooth muscle, called the *ciliary muscle,* by connective tissue fibers.

After passing through the *vitreous humor,* the fluid-filled main chamber of the eye, light encounters the layers at the back and sides of the eye. This is the **retina,** which comprises primarily photoreceptor cells, neurons, and a few blood vessels. Between the retina and the sclera lies the *choroid,* consisting of pigmented cells and blood vessels. The pigmented

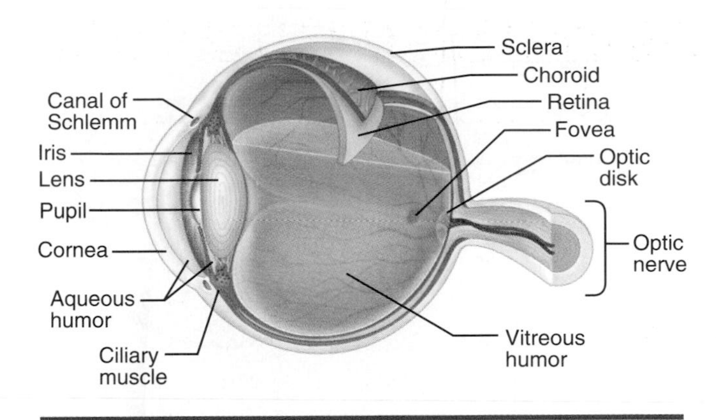

Figure 12.13 Structure of the eye. Light enters through the transparent cornea and passes through aqueous humor, the pupil, the lens, and vitreous humor before striking the retina at the back of the eye. The fovea is the area of the retina with the highest visual acuity. The nerves and blood vessels exit the eye through the optic disk.

cells absorb light not sensed by photoreceptors so the image does not become distorted by reflected light, and the blood vessels nourish the retina. At the back of the eyeball is the *optic nerve,* which carries information to the thalamus, to be forwarded to the visual cortex for interpretation. Finally, skeletal muscles surround the eye and control its movements, so we can choose exactly where to look.

Two particular sites on the retina deserve special mention. The *fovea centralis* (or fovea) is a small, funnel-shaped pit where the concentration of photoreceptors is higher than anywhere else in the eye. When we look directly at an object, we are focusing the image directly on the fovea. The *optic disk* is the area where the axons of the optic nerve exit the eye, so there are no photoreceptors there at all. The optic disk leaves us with a "blind spot" in each eye.

Recap **Light passes through the clear cornea, the adjustable opening in the iris called the pupil, the lens, and the aqueous fluid of the eye before striking the photoreceptors in the retina.**

Regulating the amount of light and focusing the image

Spotlight on Structure & Function

The iris adjusts the amount of light entering the eye with two sets of smooth muscle. When bright light strikes the eye, contraction of muscle arranged circularly around the pupil causes the pupil to contract. Otherwise, the intensity of daylight would overwhelm our photoreceptors and temporarily blind us. In dim light, contraction of smooth muscle arranged radially around the pupil causes the pupil to dilate.

Nerves control each set of muscles. When examining an unconscious or injured person, a physician may shine a light in each eye. The pupils should contract in response to light and dilate when the light is removed. Pupils that are "fixed" (unresponsive to light) are a bad sign, possibly indicating more widespread failure of the nervous system.

Light entering the eye is focused by the cornea and the lens. The cornea, which is curved, is actually responsible for bending most of the incoming light. However, the curvature of the cornea is not adjustable. Our ability to regulate the degree to which incoming light is bent, and therefore our ability to change focus between near and far objects, is accomplished solely by adjusting the curvature of the lens. This is done by the ciliary muscle.

When the ciliary muscle contracts, the inner radius of the muscle shrinks, reducing the tension on the fibers attached to the lens. This allows the lens to bulge, and we focus on a near object. When the ciliary muscle relaxes, the ring of muscle increases the tension on the lens, stretching and flattening it and bringing more distant objects into focus (Figure 12.14). **Accommodation** refers to the adjustment of lens curvature so we can focus on either near or far objects. ■

The light rays from each point of an object are bent and focused so that the image created on the retina is inverted (upside down). However, the brain interprets the image as right side up.

Table 12.4 Parts of the eye and their functions

Eye structure	Functions
Sclera	Covers and protects eyeball
Cornea	Bends incoming light (focuses light)
Aqueous humor	Nourishes and cushions cornea and lens
Iris	Adjusts amount of incoming light
Lens	Regulates focus
Ciliary muscle	Adjusts curvature of the lens
Vitreous humor	Transmits light to retina
Retina	Absorbs light and converts it into impulses
Rods	Photoreceptors responsible for black-and-white vision in dim light
Cones	Photoreceptors responsible for color vision and high visual acuity
Fovea	Contains greatest concentration of photoreceptors
Optic disk	"Blind spot" where optic nerve exits eye
Optic nerve	Transmits impulses to the brain
Choroid	Nourishes retina and absorbs light not absorbed by retinal photoreceptors

Try It Yourself

Find Your Blind Spot

You can locate the blind spot in each eye with the following demonstration. Cover your left eye and hold the image four inches in front of your face. Focus on the square with your right eye, then move the book slowly away until the circle disappears. The circle disappears because at this distance from your eye, its image is striking the blind spot of your right eye.

Try the same demonstration with your left eye by focusing on the circle with your right eye closed. In this case the square will disappear when the image strikes the blind spot of your left eye.

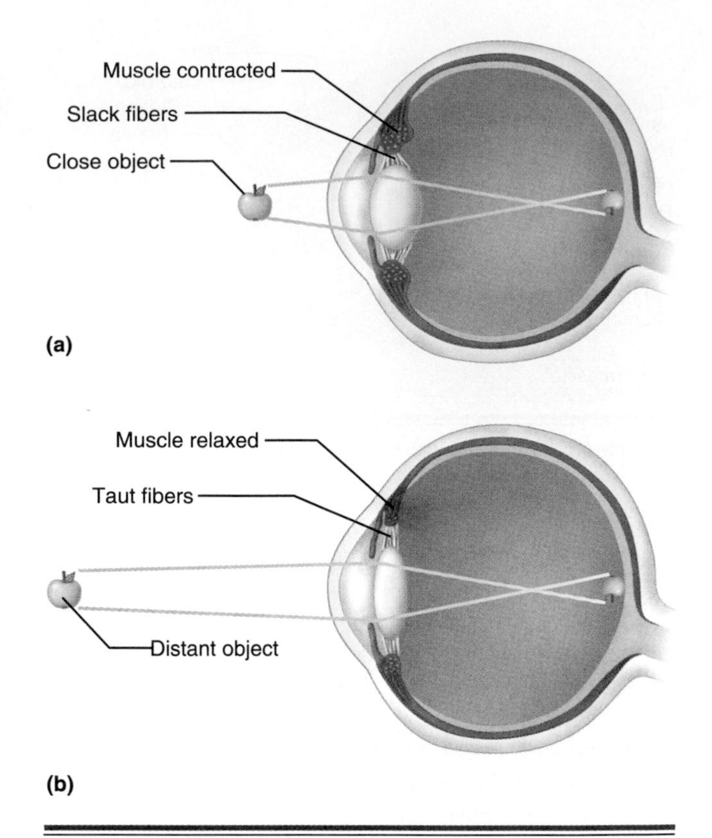

Figure 12.14 Accommodation. **(a)** For near objects, contraction of ciliary muscles reduces the tension on the fibers and allows the lens to bulge. **(b)** For focus on distant objects, relaxation of ciliary muscles increases tension on the fibers, which pull on the lens and cause it to flatten. Notice that the image created on the retina is upside down.

As we get older, the lens tends to stiffen and cannot resume a bulging shape even when the ciliary muscles are maximally contracted. The result is *presbyopia,* the inability to focus on nearby objects, which generally appears after age 40. People with presbyopia must hold reading material farther away to see the print clearly. Presbyopia can be easily corrected with reading glasses.

Eyeball shape affects focus

Differences in the shape of the eyeball can affect the ability to focus properly (Figure 12.15).

Myopia is a common inherited condition in which the eyeball is slightly longer than normal (more rarely, myopia occurs when the ciliary muscle contracts too strongly). Even when the lens is flattened maximally, distant objects focus in front of the retina. People with myopia can see nearby objects, but distant objects are out of focus, which is why it is called *nearsightedness.* Concave lenses, which bend incoming light so it focuses on the retina, can correct nearsightedness.

Less common is the opposite condition, *hyperopia (farsightedness),* which occurs when the eyeball is too short and nearby objects focus behind the retina. Farsighted people can see distant objects clearly, but nearby objects are out of focus. Convex lenses correct farsightedness.

Astigmatism refers to blurred vision caused by irregularities of the shape of the cornea or lens. The result is that light is scattered and may not focus evenly on the retina. Astigmatism can be corrected with specially ground lenses that exactly compensate for the irregularities of the cornea and lens.

Recap **The iris regulates the amount of light entering the eye, and the cornea and the lens adjust the focus. Problems that affect the ability to focus include presbyopia, nearsightedness, farsightedness, and astigmatism.**

Light is converted into action potentials

Like other sensory organs, the eyes convert a stimulus (light) into impulses. In the eye this process occurs in the retina. The unique structure of the retina allows us to see in color, adapt to varying light intensities, and perceive images of the world around us.

Figure 12.16 shows a closeup of the retina, which consists of four layers:

1. The outermost layer consists of pigmented cells that, along with the choroid, absorb light not captured by the photoreceptor cells.
2. Next is a layer of photoreceptor cells called *rods* and *cones* because of their shapes.
3. The rods and cones synapse with the third layer, a layer of neurons called *bipolar cells.*

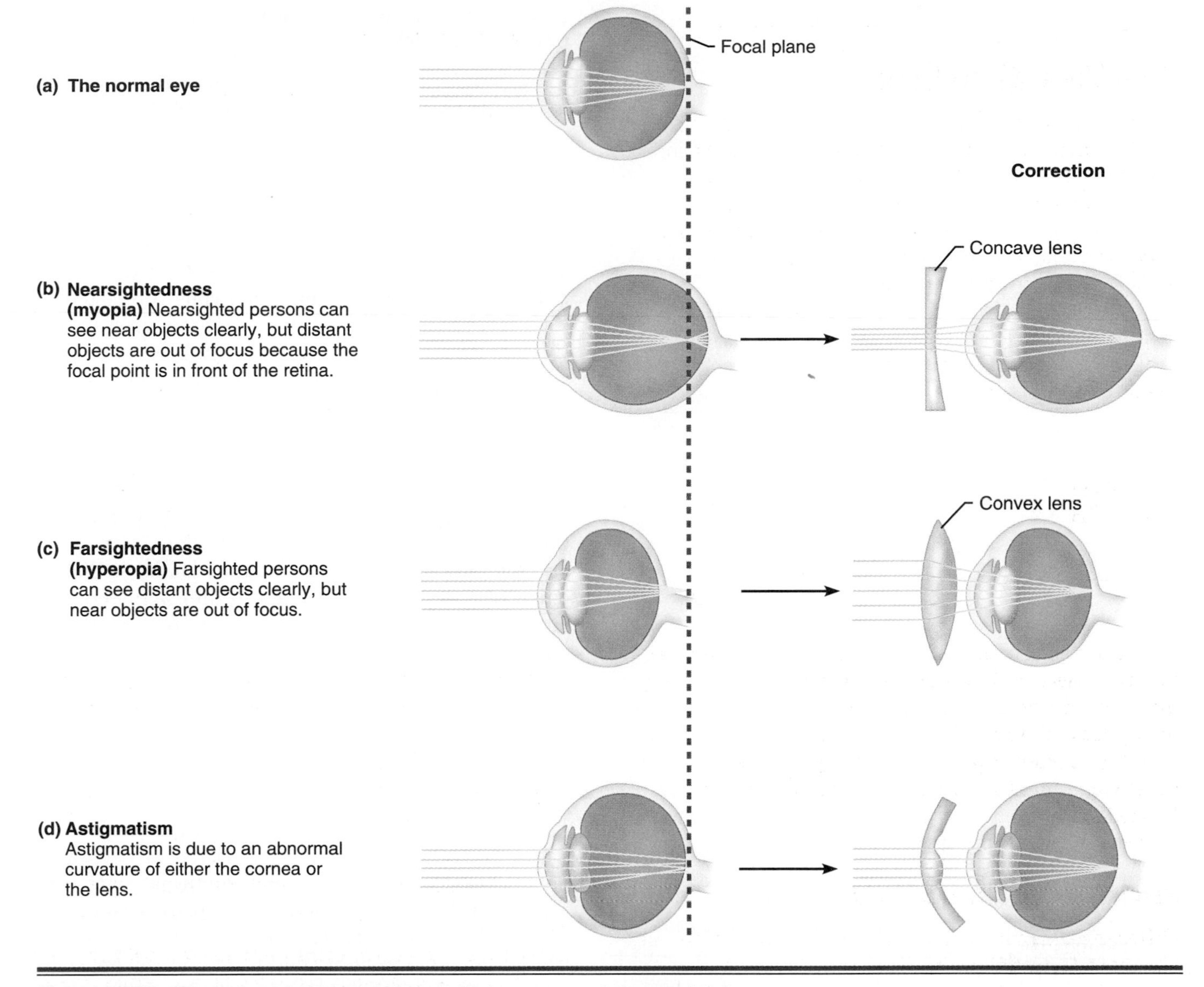

Figure 12.15 Examples of abnormal vision. All three types of abnormal vision are correctable.

Bipolar cells partially process and integrate information and then pass it on to the fourth layer.

4. The innermost layer consists of ganglion cells. Ganglion cells are also neurons. Their long axons become the optic nerve going to the brain.

Note that light must pass through the third and fourth layers to reach the photoreceptor cells, but this does not significantly hinder the ability of the photoreceptor cells to receive and convert light.

Rods and cones respond to light

Figure 12.17 takes a closer look at the photoreceptor cells, **rods** and **cones.** One end of the cell consists of a series of flattened disks arranged to form either a rod- or cone-shaped structure, as the names imply. The flattened disks contain numerous molecules of a particular protein called a *photopigment.*

A photopigment protein has the unique ability to change its shape when it is exposed to energy in the form of light. The change in photopigment molecule shape causes the photoreceptor (rod or cone) to close some of its sodium channels and reduce the amount of neurotransmitter it normally releases. Because the neurotransmitter released by rods or cones normally inhibits the bipolar cells, light ultimately increases the activity of the bipolar cells, which in turn activate the ganglion cells.

There are approximately 120 million rods and 6 million cones, but only 1 million ganglion cells with axons going to the brain. Clearly, a great deal of convergence and summation of signals occurs at each level of neuron transfer.

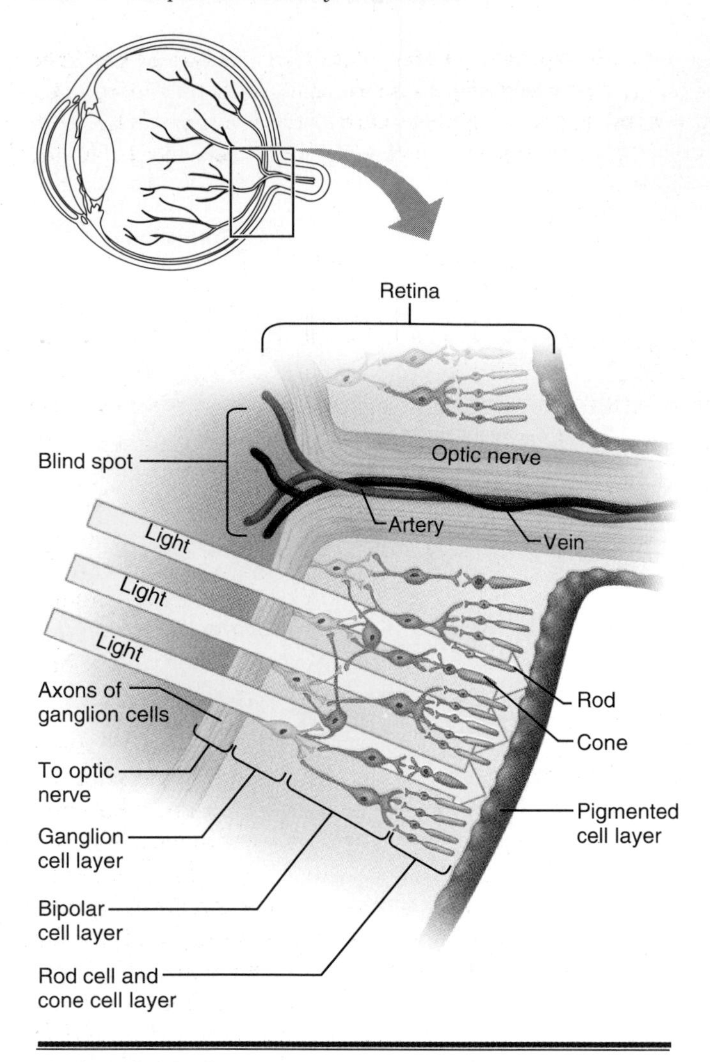

Figure 12.16 Structure of the retina. Light must pass between the axons of the ganglion cells and through the ganglion and bipolar cell layers to reach the photoreceptor cells, the rods and cones.

Rods provide vision in dim light

Rods all have the same photopigment, called **rhodopsin.** As we have seen, there are about 20 times more rods than cones. If we imagine that all 120 million rods converge on only half of the ganglion cells (one half million), there would be 240 rods converging on a single ganglion cell. This means that the ganglion cells supplied by rods are highly sensitive to dim light.

Rods function in dim light, but they do not detect color or detailed images. They give us only black-and-white vision, which is why objects appear gray and a bit blurred at dusk or dawn.

Rods and cones are not distributed evenly on the retina. Regions of the retina farthest away from the fovea have the highest ratio of rods to cones. If you want to see a dim star at night, don't look directly at it—look off to the side just a little.

Cones provide color vision and accurate images

Why do we see in color?

Not all animals perceive color, but humans do. We are able to see colors because we have three different kinds of cones: red, green, and blue. Each contains a photopigment that absorbs the energy of red, green, or blue light particularly well.

Our ability to distinguish a variety of colors is due to the way the brain interprets the ratios of impulses coming from the ganglion cells connected to the three types of cones. When all three types are activated by all different wavelengths, we perceive white light. The perception of black is no light at all.

Cones also are responsible for visual acuity (the most accurate and detailed vision). A typical ganglion cell may receive input from only a few dozen cones at most. In the fovea, where all of the photoreceptor cells are cones and visual acuity is highest, the connections may even be one-to-one.

You may have noticed that your ability to distinguish between colors declines in dim light. For example, at dusk it becomes difficult to tell whether a dark-colored car is green or red. The photopigments in cones are about 300 times less sensitive to light than the rhodopsin in rods, so cone vision is the first to fail as the light fades.

Visual receptors adapt

How do your eyes adjust to different light intensities?

From your own experience you know that your vision will adapt to changing light conditions over several seconds or minutes. Adaptation seems to take longer when going from bright to dim light than it does in the opposite direction. Adaptation depends on rapid adjustment of the pupil by the iris and on adaptation by the rods.

When rhodopsin absorbs light, this "uses up" the photopigment temporarily. Light energy actually breaks the rhodopsin into two molecules. These molecules can be resynthesized into rhodopsin again, but only over a period of minutes.

When you have been out in bright light, most of the rhodopsin in the light-sensitive rods has been broken down. Then, when you enter a dim room and the cones no longer are functioning, there is not enough intact rhodopsin for good vision in dim light. With time (5–15 minutes) the rhodopsin is resynthesized and available again.

Conversely, when you go out into sunlight from a dimly lit room, the light seems very bright because you have the maximal amount of photopigment available in both rods and cones. Most of the rhodopsin in the rods is quickly used up, however, and you revert to the use of primarily cones in bright light.

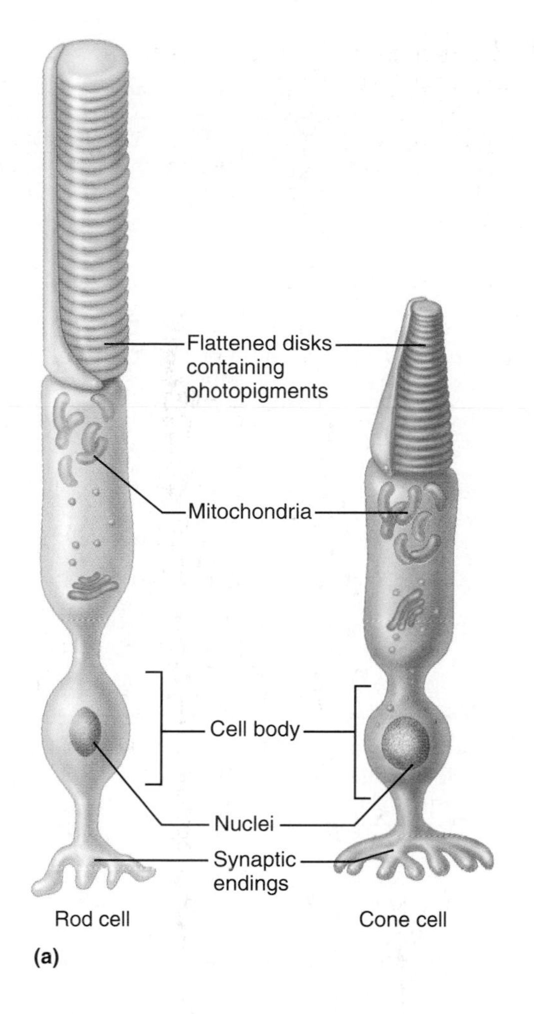

Figure 12.17 The photoreceptor cells, rods and cones. **(a)** The structure of rod and cone cells. The photopigments in the flattened disks are sensitive to light. **(b)** An SEM of rods and cones.

Recap **Light energy alters the structure of photopigments—rhodopsin in rods, and red, blue, or green photopigments in cones. Rods provide vision in dim light. Cones are responsible for color vision and detailed images.**

12.7 Disorders of sensory mechanisms

Disorders of the eyes

Retinal detachment: Retina separates from choroid If the thin retina is torn, the vitreous humor may leak through the tear and lift the retina from the choroid. The most common cause of *retinal detachment* is a blow to the head. The detached region of the retina loses most of its blood supply and its ability to focus an image properly. Symptoms include flashes of light, blurred vision, or loss of peripheral vision. Retinal detachment is an emergency, but prompt surgery can usually repair the damage.

Glaucoma: Pressure inside the eye rises As mentioned previously, a clear fluid called aqueous humor fills the chamber between the cornea and the lens. Aqueous humor is produced by capillaries located near the ciliary muscles of the eye. Its function is to bring nutrients and oxygen to the lens and cornea and to carry away wastes. Normally, aqueous humor is drained from the eye and returned to the venous blood by a vessel called the canal of Schlemm (review Figure 12.13). *Glaucoma* is a condition in which this drainage vessel becomes blocked. The excess fluid increases pressure inside the eye and compresses blood vessels supplying the retina. Cells of the retina or optic nerve may die, gradually impairing vision and ultimately, if not treated, leading to blindness. Glaucoma usually has no symptoms, and most patients do not realize they have it until they have already lost some vision. However, it can be detected with a simple test, and if caught early can be controlled with drugs or surgery.

Cataracts: The lens becomes opaque *Cataracts* are a decrease in the normal transparency of a lens. This condition normally develops late in life. The treatment, which is usually successful, involves removing the lens surgically and replacing it with an artificial lens.

Color blindness: Inability to distinguish the full range of colors *Color blindness* is a general term for the inability to distinguish the full range of colors. Most forms of color blindness are caused by deficient numbers of particular types of cones. Less common are conditions in which one of the three cones is completely missing.

People with red-green color blindness, the most common form, are deficient in either red cones or green cones. They have trouble distinguishing between red and green, or they perceive them as the same color. The color they see is red or green, depending on which type of cone is missing.

The inability to perceive any color at all is quite rare in humans because it only occurs when two of the three cones are completely missing.

Color blindness is often inherited. Red-green color blindness is an X-linked recessive trait that is more common in men than in women (see Chapter 19, Genetics and Inheritance). It affects approximately 8 percent of Caucasian males but only 1 percent of Caucasian females, and is less common in other races. Traffic lights always have the red ("stop") light at the top and the green ("go") light at the bottom so that colorblind persons can tell them apart by position. Color blindness is easily tested with a series of colored test plates (Figure 12.18).

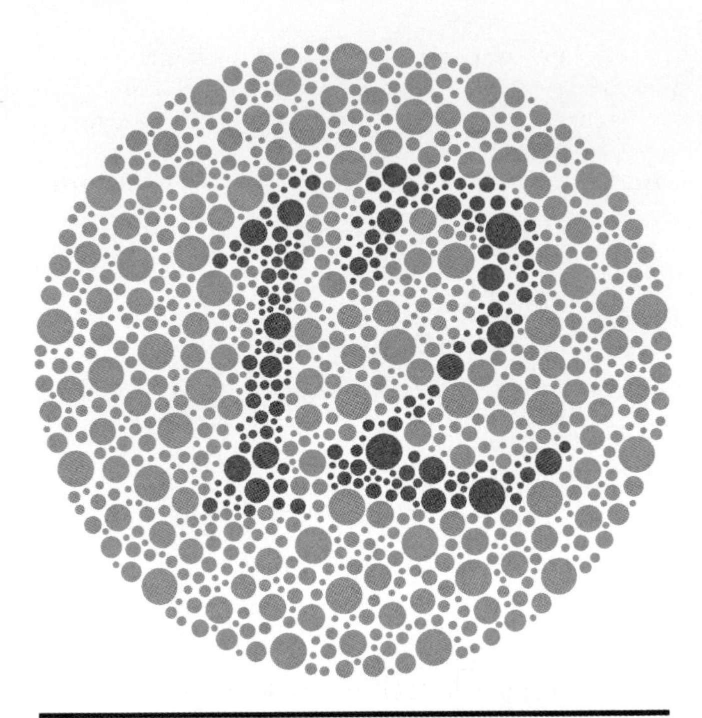

Figure 12.18 Test plates for diagnosing color blindness. A person with red-green color blindness would not be able to see the number 12 in this test plate. The original plates should be used for accurate testing of color blindness.

Disorders of the ears

In addition to hearing loss discussed in the Health Watch, disorders of the ears include otitis media and Meniere's syndrome.

Otitis media: Inflammation of the middle ear A common cause of earache is *otitis media,* in which the middle ear becomes inflamed. Usually this results from an upper respiratory tract infection that extends up the eustachian tube. The tube may become blocked and trap fluid in the middle ear. Otitis media usually responds to antibiotics.

Meniere's syndrome: Inner ear condition impairs hearing and balance Meniere's syndrome is a chronic condition of the inner ear. The cause is unknown, but it may be due to excess fluid in the cochlea and semicircular canals. Symptoms include repeated episodes of dizziness and nausea accompanied by progressive hearing loss. Balance is usually affected to the point that patients find it difficult to stay upright. Mild cases can be treated with motion sickness medications. Severe cases may benefit from surgery to drain excess fluid from the inner ear.

Recap **A detached retina is a retina that has separated from the underlying choroid. Glaucoma is a condition in which fluid pressure inside the eye rises, constricting the blood vessels that normally supply the living cells. Earaches are most often caused by infections of the middle ear.**

Chapter Summary

Receptors receive and convert stimuli

- Receptors convert the energy of a stimulus into impulses in sensory neurons.
- The brain interprets incoming impulses on the basis of where the sensory neurons originate.
- Some types of receptors adapt to continuous stimulation, whereas others (most notably pain receptors) do not.
- The somatic senses (temperature, touch, vibration, pressure, pain, and awareness of body position and movements) arise from receptors throughout the body.
- The five special senses (taste, smell, hearing, balance, and vision) arise from only certain specialized areas of the body.

Somatic sensations arise from receptors located throughout the body

- The skin contains receptors for heat, cold, vibration, pressure, light touch, and pain. Receptors for temperature and for pain are also located in some internal organs.

- Receptors in joints sense joint position.
- Muscle spindles sense muscle length, and tendon receptors sense tendon (and muscle) tension.

Taste and smell depend on chemoreceptors

- Chemoreceptors for taste are located in structures called taste buds, most of which are on the tongue.
- The receptors for our sense of smell (called olfactory receptors) are located in our nasal passages. We have over 1,000 different olfactory receptor types.

Hearing: Mechanoreceptors detect sound waves

- The outer ear channels sound waves to the eardrum, the middle ear amplifies the sound waves, and the cochlea of the inner ear converts sound waves to impulses.
- Sound waves of different tones are converted to impulses at different locations in the cochlea.

The inner ear plays an essential role in balance

- The vestibular apparatus of the inner ear consists of three semicircular canals and the vestibule.
- The semicircular canals are structures that sense rotational movement of the head on three planes.
- Structures of the vestibule sense static (nonmoving) head position and linear acceleration.

Vision: Detecting and interpreting visual stimuli

- The eye is the organ of vision, with structures for moving the eye (muscles), regulating the amount of light entering the eye (iris), focusing the visual image (lens), and converting light to impulses (retina).
- The photoreceptor cells in the retina are called rods and cones.
- Rods are 20 times more numerous than cones. They give us vision in dim light, but not in color and not with the highest detail.
- The three types of cones have different photopigments, so each type responds best to a different wavelength (color) of light. Cones are most effective in bright light, and they give us color vision and the sharpest image.
- Photoreceptors adapt to conditions of varying light intensities (brightness). It takes about 15 minutes to fully adapt to very dim light.

Disorders of sensory mechanisms

- Common eye disorders include retinal detachment, glaucoma, cataracts, and color blindness.
- Inflammation of the middle ear is a common cause of earache.

Terms You Should Know

accommodation, 307
basilar membrane, 300
cochlea, 300
cone, 309
cornea, 306
eardrum (tympanic membrane), 299
eustachian tube, 300
hair cell, 300
iris, 306
lens, 306
muscle spindle, 295
olfactory receptor cell, 298
otolith, 305
photoreceptor, 306
receptor adaptation, 293
retina, 306
rhodopsin, 310
rods, 309
somatic sensations, 293
special senses, 294
taste buds, 297
vestibular apparatus, 302

Test Yourself

Matching Exercise

____ 1. Sensory input that causes a change inside or outside the body
____ 2. A structure specialized to receive a certain stimulus
____ 3. Specialized structure for monitoring muscle length
____ 4. A cluster of about 50 receptor cells for taste
____ 5. A chemical that produces a particular odor or smell
____ 6. A unit of sound loudness
____ 7. Gelatinous structure of the cochlea in which the hairs of hair cells are embedded.
____ 8. Hard crystals of bonelike material that are important to our ability to sense position of the head
____ 9. Another word for nearsightedness
____10. A protein in the photoreceptor cells of the eye that changes shape when exposed to light.

a. taste bud
b. decibel
c. receptor
d. odorant
e. tectorial membrane
f. myopia
g. stimulus
h. photopigment
i. otoliths
j. muscle spindle

Fill in the Blank

11. Some sensory inputs are ignored after a while because of receptor ________.
12. Thermoreceptors near the skin's surface monitor temperature in the external environment; thermoreceptors deep within the body monitor ______ temperature.
13. Both taste and smell depend on the class of receptors called ________.
14. The four taste categories are ______, ______, ______, and ______.
15. Odors are detected by ________ receptors located in the nasal passages.
16. The eardrum is also called the ________ ________.
17. The photoreceptor cells of the eye are called ______ and ______.
18. Light is bent and focused on the retina by the cornea and the lens, but only the ______ has an adjustable curvature.
19. Blurred vision caused by irregularities of shape of either the cornea or the lens is called ________.
20. An eye condition called ________ is due to a decrease in the normal transparency of the lens.

True or False?

21. Visual and auditory sensory input are both received by the brain in exactly the same form—as action potentials (impulses).
22. Visual and auditory sensory input travel to the brain in the same nerve.
23. Rotational movement of the head is sensed by the three semicircular canals of the inner ear.
24. Rods are responsible for color vision.
25. Color blindness is more common in males than in females.

Concept Review

1. List the five classifications of receptors in terms of types of stimulus they convert.
2. Describe what is meant by a "silent" receptor.
3. Explain why you may "see stars" when you are hit in the eye.
4. Name the five "special senses."
5. Compare/contrast fast pain, slow pain, and referred pain.
6. Describe the overall function of the three bones (the malleus, the incus, and the stapes) of the middle ear.
7. The eye contains both skeletal muscle and smooth muscle. Describe where each is located and indicate their functions.
8. Compare/contrast rods and cones.
9. Explain why it may take up to 15 minutes to become fully adapted to very dim light conditions.
10. Explain what causes glaucoma.

Apply What You Know

1. What might be the benefit of slowly adapting or nonadapting taste receptors for bitterness?
2. With his/her eyes closed, would an astronaut in outer space be able to detect lateral movement of the head (would the semicircular canals be functioning normally)? Would he/she be able to detect static head position (would the utricle and saccule be functioning normally)? Explain.
3. Why do you suppose that you are not normally aware of the blind spot in each eye?
4. What would be a possible explanation for why dogs are completely colorblind (they see only in black and white)?

Case in Point

Humans Respond to Silent Olfactory Sensory Signals

Most studies on how women and men respond to each other are focused on behavioral or visual cues. Yet some cues may involve olfactory signals, called pheromones, of which we are not even consciously aware. Prime candidates for pheromones are *androstenol* and its breakdown product called *androstenone*, both found in male sweat, and a female vaginal secretion called *copuline*. Researchers reported in 1996 that when 290 young women were asked to smell samples of androstenone, most reacted negatively unless they were ovulating, at which time they found the smell neutral. Similarly, when males were asked to rate women's pictures for beauty, the ratings went up when they were also exposed to air samples containing copulines. Studies such as these suggest that smell could play a key part in sexual attractiveness and receptivity.

QUESTIONS:

1. What might be the ultimate purpose, or evolutionary advantage, of olfactory sexual cues?

2. If you could do research in this area, what would you like to know? What specific question would you choose to address?

INVESTIGATE:

Holden, Constance, editor. Sex and Olfaction. *Science* 273 (July 19, 1996), p. 315.

References and Additional Reading

Karb, Claudia. Our Embattled Ears. *Newsweek* Aug. 25, 1997, pp. 75–76. (A popular press article on hearing loss.)

CHAPTER 13

THE ENDOCRINE SYSTEM

Why do only certain cells **respond to hormones?**

Can hormone therapy prevent dwarfism?

How does the endocrine system respond to fear **or** danger?

AT ABOUT AGES 11–13, SOMETHING STRANGE AND WONDERFUL happens to the human body and mind. The body grows rapidly and secondary sex characteristics develop. Many boys get interested in girls, and many girls get interested in boys with an intensity that sometimes seems to border on obsession, when only months before they couldn't care less. At that age, neither sex can articulate why it happens. It just does.

Welcome to the *endocrine system,* an internal system of powerful chemicals that controls a host of bodily functions. In addition to sexual maturation and sexual desire, the endocrine system regulates salt and water balance, blood pressure, stress responses, digestion, cellular metabolism, the production of red blood cells, and overall organ growth and development.

13.1 The endocrine system regulates body functions with hormones

The **endocrine system** is a communication system that uses chemical messengers rather than nerve impulses. The chemical messengers of the endocrine system are called **hormones.** The defining features of hormones are that they (1) come from *endocrine glands,* (2) circulate in the bloodstream, and (3) act only upon specific cells in the body. **Endocrine glands** are groups of cells specialized to secrete hormones into the extracellular space (*endocrine* means "secreted internally"). From the extracellular space, the hormones make their way into the blood or lymphatic fluid. There are approximately 50 different hormones circulating in the human bloodstream, and new ones are still being discovered.

Hormones are blood-borne units of information, just as nerve impulses are units of information carried in nerves. Some hormones participate in feedback control loops regulating various bodily functions. We need these hormones to maintain homeostasis. Other hormones produce specific effects, such as contraction of the uterus at birth, growth during childhood, and the development of sexual characteristics at puberty. We need specific hormones to be able to carry out each of these very specific functions.

The various tissues and organs that make up the endocrine system and the most important hormones they secrete are shown in Figure 13.1. The many functions of these hormones are discussed in future sections of this chapter or in other chapters where appropriate.

Why do only certain cells respond to hormones?

The endocrine system has certain characteristics that set it apart from the nervous system as a communications system:

1. *The endocrine system has universal access to nearly every living cell.* This attribute gives the endocrine system a distinct advantage over the nervous system. Access is virtually guaranteed because almost every cell in the body is near a blood vessel. With the endocrine system there is no need for the direct connections (the hard wiring, so to speak) required by the nervous system.
2. *Each hormone acts only on certain cells.* As all hormones circulate together in the same well-mixed blood, how can one hormone specifically regulate red blood cell production while another regulates blood calcium concentration? The answer is that each hormone acts only on a certain group of cells, called its **target cells.** All other cells are unresponsive to the hormone.
3. *Only certain cells respond because only those cells have the right receptors.* Only the hormone's target cells have the appropriate receptor to fit a particular hormone. When a hormone binds to its receptor, a change occurs within the target cell. The cell may grow, divide, or change its metabolism in some way. All other cells of the body fail to respond to the hormone because they lack the appropriate receptor. As an analogy, consider that your car is a cell, the car's ignition switch is the receptor, and the car key is the hormone. Your car key fits only your car (the "target" car) and no other. Fitting your key into the ignition and turning it causes the car to start. Gas is consumed, engine parts rotate, and heat is generated.
4. *Endocrine control tends to be slower than nervous system control.* There is a price for using the cardiovascular system as the message delivery system, and that is a reduction in speed—a lengthening of the response time. Nerve impulses can travel from head to toe in a fraction of a second. It takes longer to secrete a hormone into the bloodstream, have the hormone circulate to its target, and exert its effect. At the very fastest, this takes 20 seconds or so.

Given this difference in speed, it is not surprising that reflexes that prompt us to avoid a hot flame are controlled by the nervous system and not the endocrine system. On the other hand, endocrine communication is highly effective for longer-term controls, such as regulating

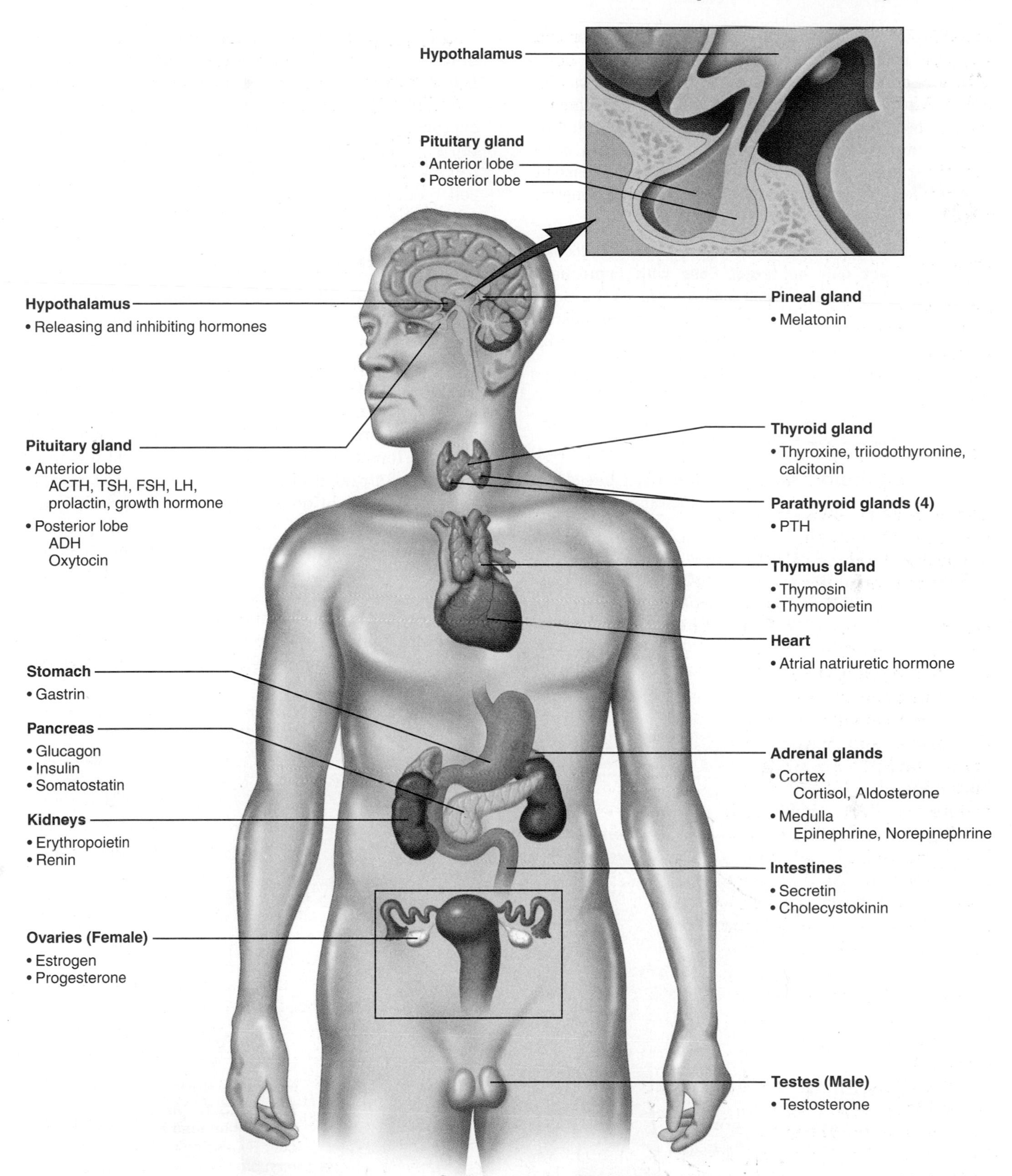

Figure 13.1 Components of the human endocrine system. Note that some organs with non-endocrine primary functions (the heart, stomach, intestines, and kidneys) produce and secrete hormones as secondary functions.

blood pressure, producing red blood cells, and controlling the onset of puberty.

5. *The endocrine and nervous systems can (and often do) interact with each other.* The timing of growth and sexual maturation, for example, involves a complex sequence of changes in both neural and endocrine signals, and the release of some hormones depends on input from sensory neurons.

Recap **Hormones secreted by glands of the endocrine system act only on target cells with appropriate receptors. Hormones reach their targets via the circulatory system, making endocrine system control slower than nervous system control. The two systems frequently interact.**

13.2 Hormones are classified as steroid or nonsteroid

Hormones fall into two basic categories based on their structure, which determines their mechanism of action.

Spotlight on Structure & Function

Steroid hormones are structurally related to cholesterol; in fact all of them are synthesized from cholesterol and all are lipid-soluble. **Nonsteroid hormones** consist of, or at least are partly derived from, the amino acid building blocks of proteins. In general, they are lipid-insoluble. The differences in lipid solubility explain most of the important differences in how the two categories of hormones work. Steroid hormones enter the cell, bind to an intracellular receptor, and activate genes that produce new proteins. Nonsteroid hormones bind to receptors on the cell's surface. Their binding either opens or closes cell membrane ion channels or activates enzymes within the cell. ■

Steroid hormones enter target cells

Recall that the cell membrane is primarily composed of a bilayer of phospholipids. Because they are lipid-soluble, steroid hormones cross the target cell membrane and even the nuclear membrane rather easily by diffusion. Once inside the cell, they bind to specific receptors floating within the nucleus or cytoplasm, forming a hormone-receptor complex.

The formation of a hormone-receptor complex starts a chain of events within the cell that begins with either activation or inhibition of a specific gene on the DNA and ends with a change in the production of a specific protein. The protein then carries out the cellular response to the hormone, whatever it might be. Figure 13.2 illustrates this process for a receptor in a target cell nucleus.

As we will see in Chapter 17, activation or inhibition of specific genes regulates essentially every-

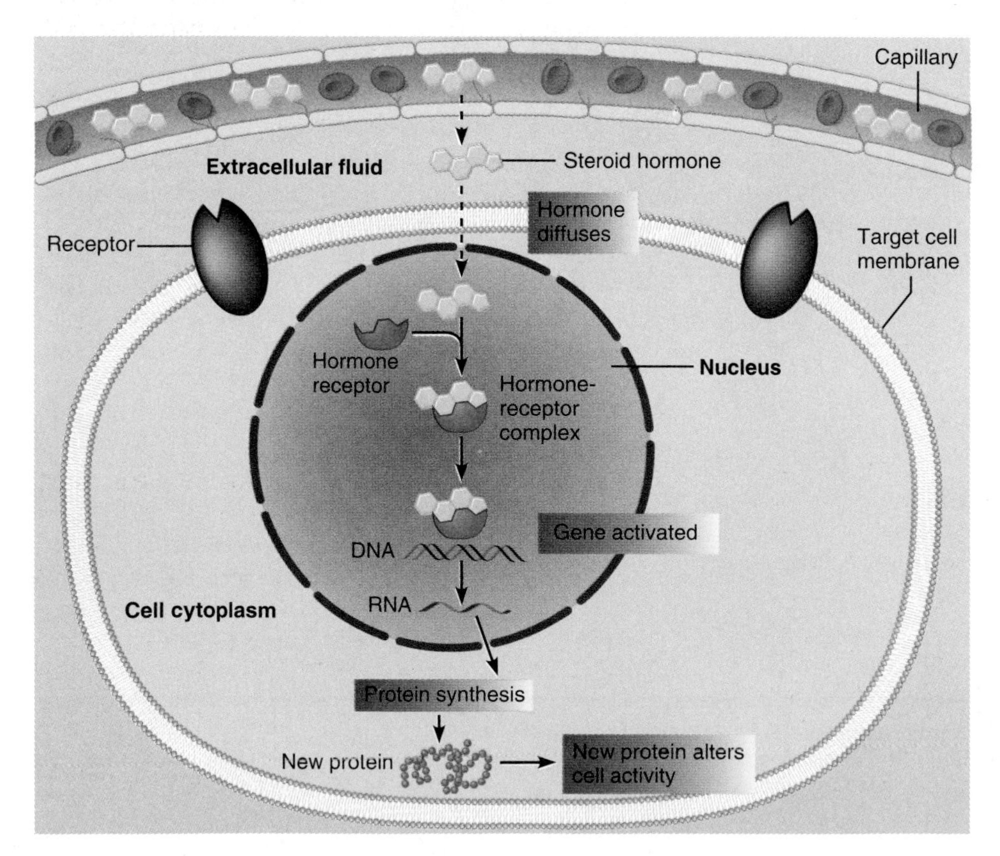

Figure 13.2 Mechanism of steroid hormone action on a target cell. Bypassing receptors in the cell membrane, a lipid-soluble steroid hormone diffuses across the cell and nuclear membranes into the nucleus, where it binds to a hormone receptor that activates a gene. Gene activation results in the production of a specific protein.

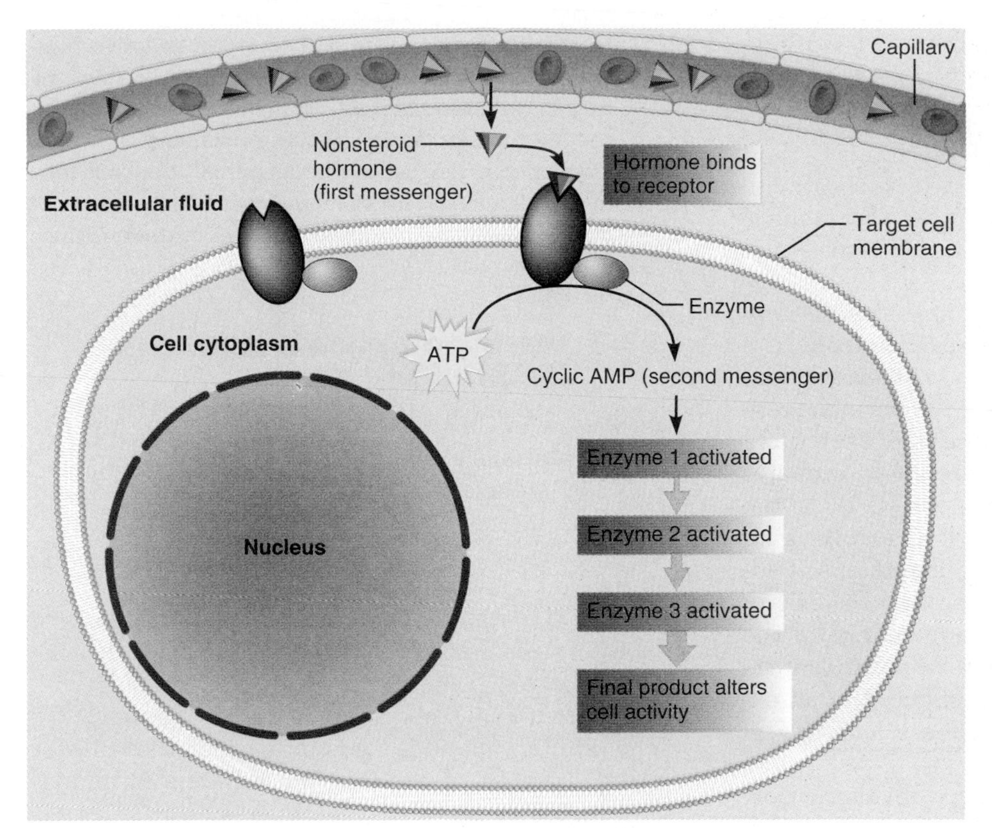

Figure 13.3 Mechanism of nonsteroid hormone action on a target cell. The nonsteroid hormone binds to a receptor in the cell membrane, leading to the conversion of ATP to cyclic AMP (the second messenger) within the cell. Cyclic AMP initiates a cascade of enzyme activations, amplifying the original hormonal signal and generating a cellular response.

thing a cell does, because most genes are the precise recipes for a single protein. The cell's proteins, in turn, comprise its enzymes and many of its structural units as well. A few genes regulate yet other genes, which in turn produce proteins.

Steroid hormones tend to be slower acting than nonsteroid hormones because it can take minutes or even hours to produce a complete protein starting from the entry of the hormone into the cell.

Nonsteroid hormones bind to receptors on target cell membranes

Because they are not lipid-soluble, nonsteroid hormones cannot enter the target cell directly. Instead they bind to a receptor located on the outer surface of the cell membrane. The receptors are part of the integral protein molecules in the cell membrane (Chapter 3). The binding of the hormone to the receptor causes some change in the protein or an associated protein, which in turn produces a change within the cell. The effect of a nonsteroid hormone is analogous to turning on the lights in a room by flipping a switch located in the hall outside.

The actual effect of the binding of the hormone to the receptor may be to trigger the opening or closing of a membrane ion channel, similar to the action of a neurotransmitter. This changes the membrane potential or alters ionic concentrations within the cell.

More commonly, the binding of hormone to receptor converts an inactive molecule within the cell into an active molecule. The activated molecule is called a **second messenger,** because it carries the message provided by the hormone (the "first messenger") without the hormone ever entering the cell. A common second messenger is cyclic AMP, produced within the cell from ATP when the hormone binds to the receptor. The cyclic AMP second messenger activates an enzyme, starting a cascade in which the first enzyme activates another, which in turn activates another, and so on (Figure 13.3). At each stage the activation is amplified, so that even a small amount of hormone can have a major effect on the eventual cellular change that it produces.

Amplification of the original signal is a common feature of both steroid and nonsteroid hormone actions. If the hormone were a lighted match, the effect of the hormone would be the equivalent of a bonfire.

Most nonsteroid hormones tend to be faster acting than steroid hormones because they activate proteins (enzymes) that already exist within the cell but in inactive form. Nonsteroid hormones can activate an entire cascade of existing enzymes within seconds or minutes.

Hormones participate in negative feedback loops

As messenger molecules, some hormones participate in internal homeostatic control mechanisms and control vital physiological processes. It is essential to regulate carefully the rate at which each hormone is secreted, so that its concentration in blood is just right to carry out its function.

As you learned in Chapter 4, homeostasis is generally maintained by a *negative feedback loop*. In a negative feedback loop involving a hormone the endocrine gland is the control center, the hormone represents the pathway between the control center and the effectors, and the effectors are the hormone's target cells, tissues, or organs (Figure 13.4). As the figure shows, a negative feedback loop involving an endocrine gland and a hormone is a stable, self-adjusting mechanism for maintaining homeostasis of the controlled variable, because any change in the controlled variable sets in motion a response that reverses that change. (You were introduced to an example of a hormonal negative feedback loop in Figure 7.6.)

Not all negative feedback loops involving hormones are quite as simple as the one depicted in Figure 13.4. In some cases the real control center is the brain, which activates an endocrine gland via nerves. Nevertheless the effect is the same—reversal of the initial change in the controlled variable.

A few hormones are secreted in response to specific environmental cues or for specific purposes, such as to avoid danger or initiate puberty. This type of secretion is not part of a negative feedback loop. We turn now to specific endocrine glands and their hormones.

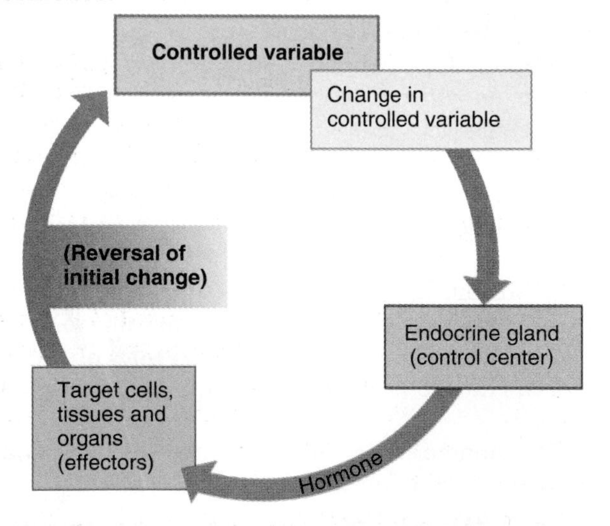

Figure 13.4 A negative feedback loop involving a hormone. In response to a change in the controlled variable, the endocrine gland releases a hormone that acts on target cells to return the controlled variable to its normal state.

Recap **Many hormones participate in negative feedback loops that maintain homeostasis of certain controlled variables. Steroid hormones enter the target cell, activate or inhibit specific genes, and cause the production of new proteins. Nonsteroid hormones bind to a cell membrane receptor that either opens or closes membrane ion channels or activates enzymes within the cell.**

13.3 The hypothalamus and the pituitary gland

As we saw in Chapter 11 (Nervous System), the **hypothalamus** is a small region in the forebrain that plays an important role in homeostatic regulation. It monitors internal environmental conditions such as water and solute balance, temperature, and carbohydrate metabolism.

The hypothalamus also produces hormones and monitors the **pituitary gland,** a small endocrine gland located beneath the hypothalamus and connected to it by a stalk of tissue (review Figure 13.1). The pituitary gland is sometimes called the "master gland" because it secretes eight different hormones and regulates many of the other endocrine glands (Table 13.1).

The pituitary gland consists of two lobes: a *posterior lobe* that appears to be a bulging extension of the connecting stalk and a larger, more distinct *anterior lobe.*

The posterior pituitary stores ADH and oxytocin

Spotlight on Structure & Function The interaction between the hypothalamus and posterior lobe of the pituitary gland is an example of the close working relationship between the nervous and endocrine systems. Cells in the hypothalamus called **neuroendocrine cells** essentially function as both nerve cells and endocrine cells, because they generate nerve impulses and release hormones into blood vessels.

Figure 13.5 illustrates the posterior pituitary lobe and its connections to the hypothalamus. Neuroendocrine cells with cell bodies in the hypothalamus have axons that extend down the stalk connecting to the posterior pituitary. The neuron cell bodies make either antidiuretic hormone (ADH) or oxytocin—both nonsteroid hormones—in the hypothalamus and then send the hormones down the axon for storage in axon endings in the posterior pituitary.

When the hypothalamus is stimulated to release the hormones, the neuroendocrine cells send impulses down the axon, causing packets of hormone to be released into nearby capillaries. The hormones circulate to their target cells. ■

Table 13.1 Hormones of the Pituitary Gland

Pituitary Lobe	Hormone	Abbreviation	Main Targets	Primary Actions
Posterior				
	Antidiuretic hormone	ADH	Kidneys	Reduces amount of water lost in urine
	Oxytocin	—	Uterus	Induces uterine contractions
			Mammary glands	Induces ejection of milk from mammary glands
Anterior				
	Adrenocorticotropic hormone	ACTH	Adrenal cortex	Stimulates adrenal cortex to release glucocorticoids
	Thyroid stimulating hormone	TSH	Thyroid gland	Stimulates synthesis and secretion of thyroid hormones
	Follicle stimulating hormone	FSH	Ovaries	In females, stimulates egg maturation and the secretion of estrogen
			Testes	In males, stimulates the formation of sperm
	Luteinizing hormone	LH	Ovaries	In females, stimulates ovulation (egg release) and the secretion of progesterone
			Testes	In males, stimulates testerone secretion
	Prolactin	PRL	Mammary glands	Stimulates the development of mammary gland cells and production of milk
	Growth hormone	GH	Most cells	Stimulates growth in young individuals; plays multiple roles in cell division, protein synthesis, and metabolism in adults

ADH regulates water balance Antidiuretic hormone reduces the amount of water lost in urine, helping to regulate the body's overall water balance. The primary target cells for ADH are in the kidneys, where it causes changes in cell permeability to water as part of the mechanism for conserving water (Chapter 15).

When the hypothalamus detects low concentrations of water (equivalent to a high concentration of solute particles) in the blood, neuroendocrine cells release ADH from the posterior pituitary. ADH circulates in blood to the kidneys, where it stimulates the reabsorption of water. When water concentrations rise to normal levels, the hypothalamus is no longer stimulated to release ADH, and ADH secretion declines.

Oxytocin causes uterine contractions and milk ejection The primary target cells for oxytocin are in the uterus and the mammary glands (breasts) of pregnant and lactating females, respectively. Oxytocin is secreted at high concentrations during childbirth and also when an infant nurses at its mother's breast (Figure 13.6). Its actions are stimulation of contractions of the uterus during childbirth, and ejection of milk into the ducts of the mammary glands during suckling.

The stimulation of oxytocin secretion by nursing is an example of a *neuroendocrine reflex* in which a nervous system stimulus (activation of sensory mechanoreceptors in the mother's breast) is responsible for secretion of a hormone. When a baby suckles, sensory receptors in the nipple are stimulated. Impulses are transmitted to neuroendocrine cells in the hypothalamus, which release oxytocin from the posterior pituitary. The hormone circulates to breast tissue, binds with receptors, and causes contractions of smooth muscle that eject milk.

Oxytocin is also present in males. In both sexes it is thought to play a role in sexual arousal and the feeling of sexual satisfaction after orgasm.

Recap **ADH acts on the kidneys to regulate water balance. Oxytocin stimulates uterine contractions and milk ejection in pregnant and lactating females, respectively. In both sexes, oxytocin contributes to feelings of sexual satisfaction.**

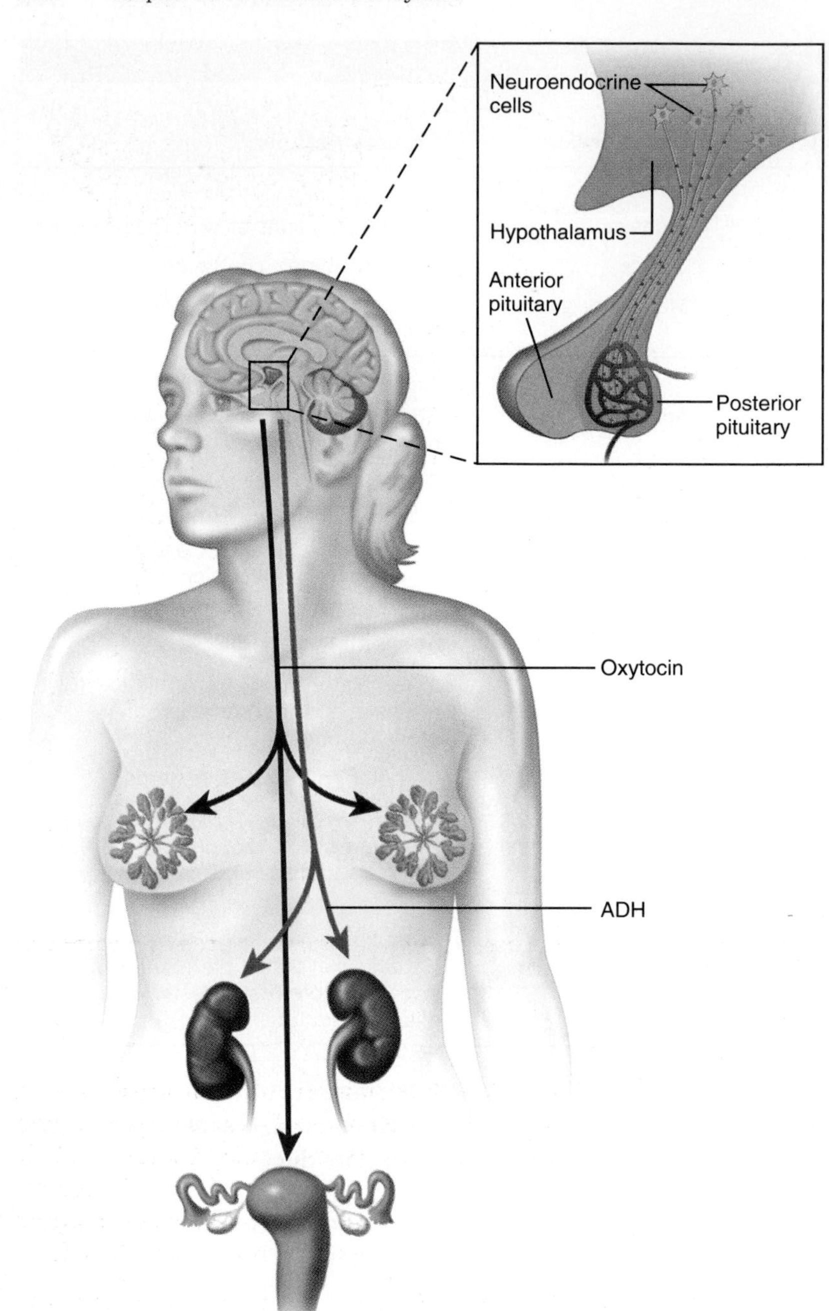

Figure 13.5 Posterior pituitary lobe and hypothalamus. Individual neuroendocrine cells of the hypothalamus and the posterior pituitary secrete either ADH or oxytocin. The hormones enter the blood supply of the posterior pituitary and make their way to the general circulation. ADH acts on the kidneys to regulate water balance in the body. Oxytocin primarily acts on the mammary glands and uterus muscles of lactating and pregnant women, respectively.

The anterior pituitary produces six key hormones

The anterior lobe of the pituitary gland produces several hormones. Six key anterior pituitary hormones include:

- Adrenocorticotropic hormone (ACTH)
- Thyroid stimulating hormone (TSH)
- Follicle stimulating hormone (FSH)
- Luteinizing hormone (LH)
- Prolactin (PRL)
- Growth hormone (GH)

Each anterior pituitary hormone is produced and secreted by a separate cell type, and each hormone is regulated by separate mechanisms. The first four hormones (ACTH, TSH, FSH, and LH) act by stimulating the release of yet other hormones from other endocrine glands.

The secretion of anterior pituitary hormones depends partly on the hypothalamus, but the mechanism differs from that of the posterior pituitary. The connection between the hypothalamus and anterior pituitary is endocrine, not neural, and relies on releasing and inhibiting hormones.

Releasing hormones stimulate the release of anterior pituitary hormones, and *inhibiting hormones*

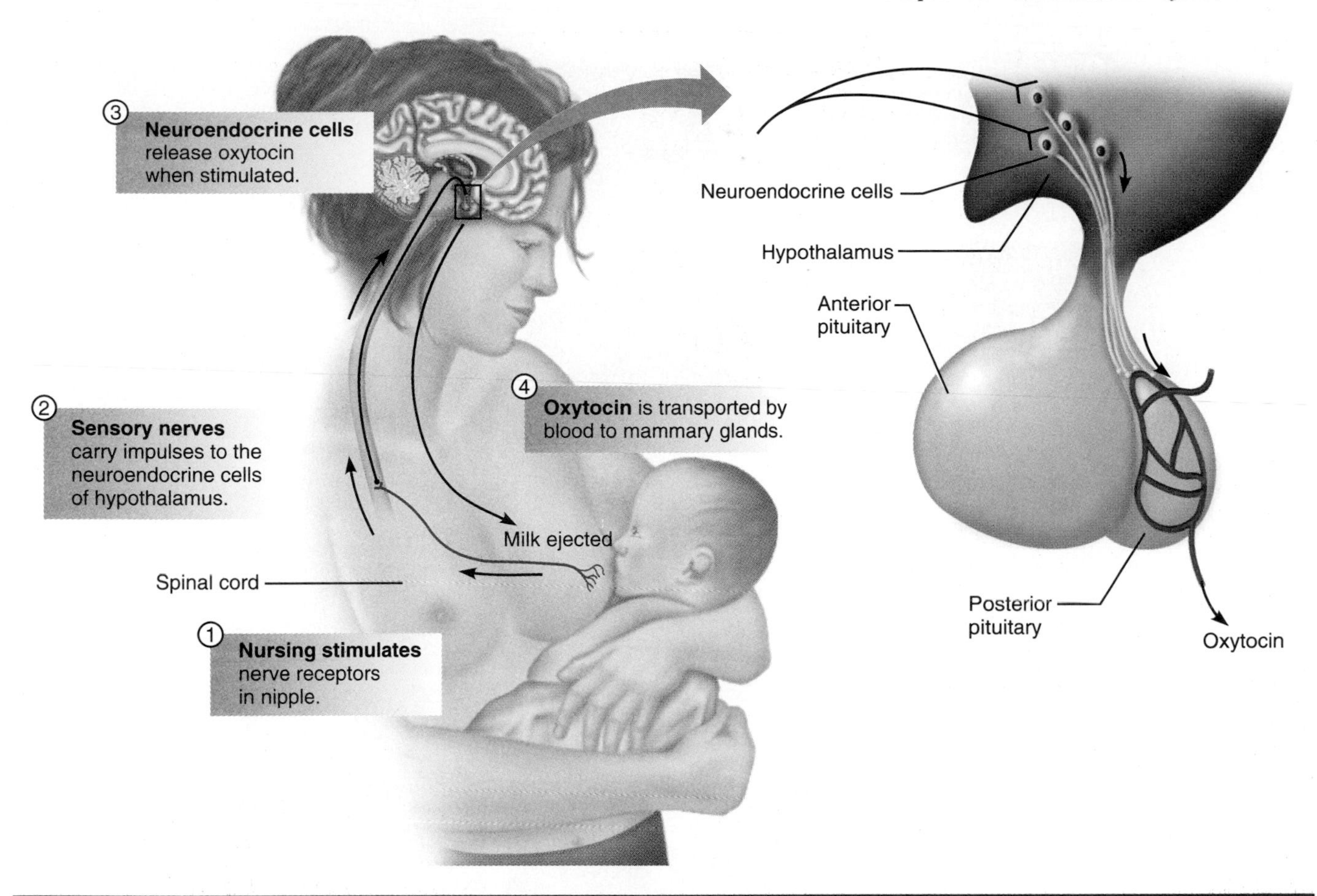

Figure 13.6 The control of oxytocin secretion by nursing. The act of suckling triggers a series of events that ultimately eject milk from the nipple.

suppress the release of some anterior pituitary hormones. Both releasing and inhibiting hormones are secreted by neuroendocrine cells in the hypothalamus (Figure 13.7). Both are produced in miniscule quantities. But because they are secreted into the *pituitary portal system,* a special blood supply that runs directly between the hypothalamus and anterior pituitary, their concentrations are sufficient to stimulate the anterior pituitary gland cells.

For example, the release of TSH is stimulated by another hormone called thyroid stimulating hormone-releasing hormone (TSH-RH). Gonadotropin-releasing hormone (GnRH) stimulates the release of FSH and LH, which are called "gonadotropins" because they stimulate the gonads (reproductive organs).

ACTH stimulates the adrenal cortex The names of hormones generally offer a clue to their function or site of action. The target cells for ACTH are located in the outer layer, or cortex, of the adrenal gland (*adrenocortico-*). The suffix *-tropic* means "acting upon."

ACTH stimulates the adrenal cortex to release another group of hormones called glucocorticoids, which are steroid hormones involved in stress-related conditions and the control of glucose metabolism (see "Adrenal Cortex" later in this chapter). ACTH secretion is regulated by a negative feedback loop involving a releasing hormone from the hypothalamus and the concentration of glucocorticoids in blood.

TSH acts on thyroid gland TSH stimulates the thyroid gland to synthesize and release thyroid hormones. There is a TSH releasing hormone from the hypothalamus, mentioned earlier, called TSH-RH.

FSH and LH stimulate the reproductive organs FSH and LH stimulate the growth, development, and function of the reproductive organs in both males and females. In females, FSH induces the development of eggs and also stimulates the secretion of the ovarian hormone estrogen. LH promotes ovulation (egg release) and the secretion of the ovarian hormone progesterone after ovulation. In males, FSH induces the development of sperm and LH stimulates the production of the hormone testosterone by the testes.

FSH and LH are not produced until about age 10–13. The first production of these two hormones is

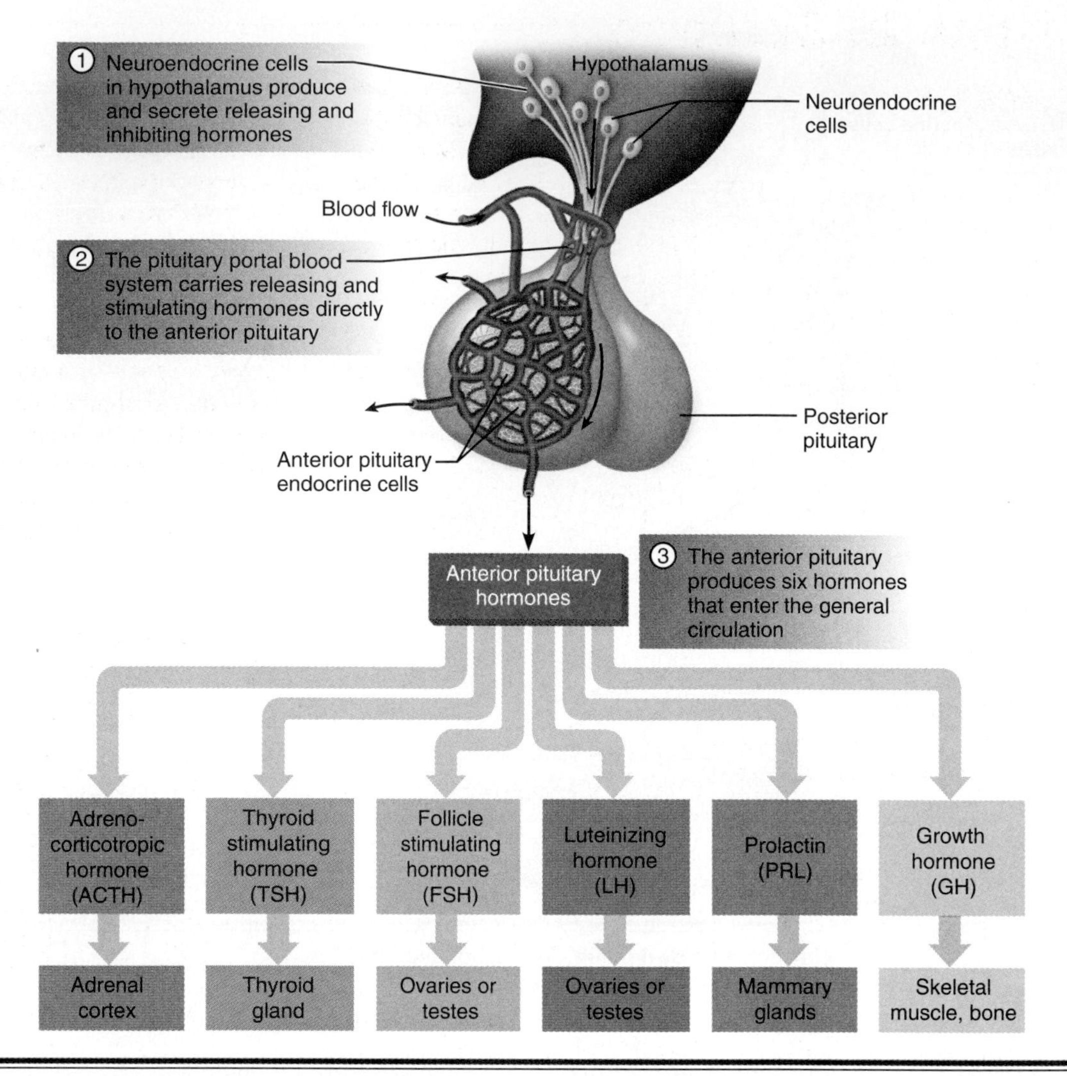

Figure 13.7 The relationship between the hypothalamus and the anterior pituitary gland. Neuroendocrine cells in the hypothalamus secrete releasing hormones and inhibiting hormones into a system of capillaries that goes directly to the anterior pituitary. Specific hypothalamic hormones inhibit or stimulate each anterior pituitary hormone.

what initiates sexual maturation and the development of secondary sex characteristics (puberty).

Prolactin (PRL): Mammary glands and milk production Prolactin's primary function is to stimulate the development of mammary gland cells and the production of milk. Prolactin concentrations in blood rise late in pregnancy in preparation for the arrival of the infant. The increase occurs because two other hormones, estrogen and progesterone, cause the hypothalamus to step up production of *prolactin-releasing hormone*. When estrogen and progesterone levels decline after birth, prolactin-releasing hormone and prolactin levels are sustained by a sensory reflex associated with nursing.

When a woman is not in late pregnancy or breast-feeding, prolactin secretion is suppressed by *prolactin-inhibiting hormone,* also from the hypothalamus. Prolactin is present in males but its function is unknown.

Growth hormone (GH): Widespread effects on growth The effects of growth hormone are so widespread that it is difficult to define a specific target cell or action. The effects of GH are most obvious in bone and muscle. In general, GH influences cells in ways that promote growth. Specific actions include stimulating protein synthesis and cell division, and using body fat as a source of energy.

GH is present throughout life, but most of its growth-promoting effects occur during childhood and adolescence. At these times it causes a dramatic increase in the mass and length of muscles and bones. Abnormalities of GH concentration during development can lead to permanent changes in the final height of an adult. In adults it continues to contribute to the regulation of metabolism by promoting the utilization of fats and amino acids as cellular fuel and by helping to maintain blood glucose levels.

Directions in Science

Making Human Hormones

Before the 20th century, hormones simply were unavailable and people who suffered from endocrine disorders often died young. The first hormone to become available in amounts sufficient for use in humans was insulin. The discovery and purification of insulin in the 1920s was a major breakthrough, for in those days Type I diabetics usually died within a year of diagnosis.

However, there were problems with early preparations of insulin. It required many pounds of pancreatic beef tissue to obtain enough insulin for even one person. In addition, although the preparations were relatively pure, enough contaminants still remained to cause adverse reactions in some patients.

Since that time, several scientific advances have resulted in the production of many hormones in essentially pure form. Insulin, erythropoietin, growth hormone, melatonin, epinephrine, and norepinephrine, for example, are now all readily available. These advances include:

- Biochemical techniques. We knew about the existence of hormones and their functions before we understood their structure. Only after crude preparations of hormone had been created in nearly pure form did it become possible to analyze their structure. Even then the procedures for determining structure were laborious and slow. Today, automated amino acid sequencing machines can determine the structure of a peptide in a matter of hours.
- Molecular biology of the cell. Advances in molecular biology enable us to understand the structure of genes and how the cell uses them to produce specific proteins. In recent years, scientists have learned how genes are regulated—that is, what turns them on and off so they make the right proteins at the right time.
- Recombinant DNA technology. Scientists have learned how to insert small segments of DNA from one species of animal or plant into the DNA of a different species. That is, they can get DNA to recombine across species. The technique, called *recombinant DNA technology,* has been used to insert segments of human DNA into the DNA of other organisms, such as the bacterium *E. coli.* If the bacteria can be made to activate the gene, they will produce the hormone. And because bacteria can be grown easily in large vats, this results in a nearly limitless supply of the hormone. This technique works best with smaller hormones because the genes that encode for them are smaller and thus easier to insert into a bacterium.

One of the first human hormones to be produced by bacteria by recombinant DNA techniques was insulin, followed closely by human growth hormone and erythropoietin. The use of recombinant DNA technology to solve problems in medicine and agriculture is called *genetic engineering.* We talk more about genetic engineering in Chapter 19.

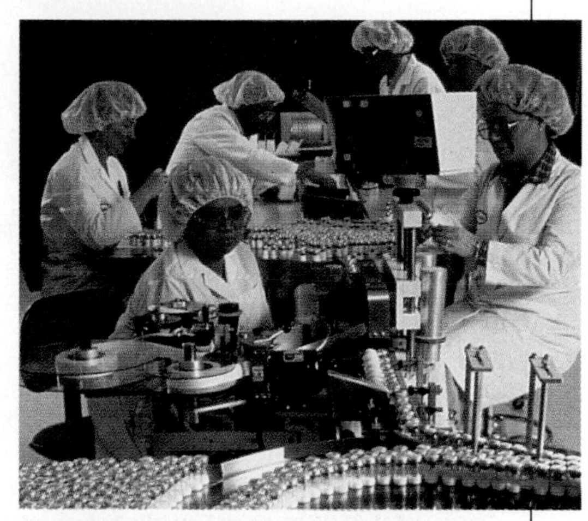

Production of human growth hormone. The workers in this photograph are completing the process of manufacturing human growth hormone by recombinant techniques.

Recap **The six hormones produced by the anterior pituitary are ACTH, TSH, FSH, LH, prolactin (PRL), and GH. The release of each hormone is controlled in part by a releasing or inhibiting hormone from the hypothalamus.**

Pituitary disorders: Hypersecretion or hyposecretion

Many pituitary disorders (and many endocrine disorders in general) are chronic conditions of too much hormone *(hypersecretion)* or too little hormone *(hyposecretion).* Some of them are inherited conditions. To illustrate, we will focus on hypersecretion and hyposecretion of a posterior pituitary hormone (ADH) and an anterior pituitary hormone (GH).

Hypersecretion of ADH is called *syndrome of inappropriate ADH secretion* (SIADH). Recall that ADH causes the kidneys to retain water, and so people with SIADH retain too much water for their state of water balance. The symptoms are often fairly general (such as headache and vomiting). Many underlying disorders can cause SIADH, ranging from HIV infection and brain injuries to pneumonia or chronic lung conditions. Some drugs may also increase ADH production and cause the condition.

On the other side of the coin, a disorder called *diabetes insipidus* is characterized by the inability to save water properly. Diabetes insipidus can be caused either by lack of ADH or by lack of the receptors for ADH in kidney cells. Some cases of diabetes insipidus are caused by head trauma or brain surgery that disrupts the normal production of ADH. Symptoms of diabetes insipidus include frequent urination and a

Figure 13.8 Effect of growth hormone on body growth. The tall person exhibits the traits of gigantism caused by excessive secretion of growth hormone during childhood. The short person has pituitary dwarfism, the result of too little growth hormone during childhood.

Figure 13.9 Acromegaly, the result of excessive growth hormone during adulthood. If excessive amounts of growth hormone are secreted after the epiphyseal plates have closed, acromegaly results. In this disorder the bones and muscles of the face, hands, and feet gradually enlarge, progressively altering the person's appearance.

high fluid intake. People with diabetes insipidus must drink enormous amounts of water just to keep up with the amount of water lost by their kidneys.

Note that an endocrine control loop can be disrupted not only by the lack of a hormone but also by the lack of target cell receptors for that hormone. Either way, the system does not function effectively. Endocrine disorders due to defects in target cell receptors are quite common and are often inherited.

Recall that growth hormone is intimately involved in the growth and development of children. For that reason, abnormalities of growth hormone secretion are most damaging when they occur before puberty. Hyposecretion of growth hormone during childhood can result in *pituitary dwarfism*. Conversely, hypersecretion of growth hormone during childhood and adolescence produces *gigantism* (Figure 13.8).

Dwarfism can be prevented if children are diagnosed early and treated with growth hormone throughout childhood. However, once adulthood is reached, dwarfism cannot be overcome by administering growth hormone. The reason is that during puberty the sex steroid hormones cause the cartilaginous growth plates at the ends of long bones to be replaced by bone. After that, growth in bone length is no longer possible.

Can hormone therapy prevent dwarfism?

In adults, an excess of growth hormone over long periods of time produces a condition called *acromegaly* (Figure 13.9). In acromegaly the bones and muscles of the face, hands, and feet (bones not affected by sex steroids) enlarge, changing patients' facial features but not necessarily their height. Excessive production of growth hormone in adults usually results from a tumor of the pituitary or hypothalamus. Unfortunately, such growth hormone-secreting tumors often go undiagnosed prior to the development of acromegaly because the symptoms may mimic those of type II diabetes. The definitive diagnosis of excessive growth hormone secretion is made by measurement of growth hormone levels.

13.4 The pancreas secretes glucagon, insulin, and somatostatin

The **pancreas** is both an endocrine gland (secreting hormones into the blood) and exocrine gland (secreting enzymes, fluids, and ions into the digestive tract to aid in digestion). In this chapter we deal with its endocrine function.

The endocrine cells of the pancreas are located in small clusters of cells scattered throughout the pancreas called the *islets of Langerhans* (an "islet" is a "little island"), or just the pancreatic islets. The pancreatic islets contain three different types of cells and produce three different hormones, all of them having to do with the regulation of blood sugar (glucose). One hormone raises blood sugar, the second lowers it, and the third inhibits the secretion of the other two hormones. Specifically:

1. Alpha cells secrete **glucagon,** which raises blood sugar. As blood glucose levels decline between meals, glucagon is secreted into the bloodstream. In the liver, glucagon causes the breakdown of glycogen (a storage form of sugar) to glucose and also promotes the production of glucose from the metabolism of amino acids. The net result is that glucagon raises blood glucose levels. Glucagon is particularly important between meals for increasing the supply of glucose to cells that need it.
2. Beta cells secrete **insulin,** which lowers blood sugar. After a meal, blood glucose levels rise as sugars are absorbed from the digestive tract. The high glucose concentration stimulates the beta cells to secrete insulin into the blood, where it does the opposite of glucagon. Insulin promotes the entry of glucose into cells of the liver, muscle, and fat tissue. It also promotes the storage of sugar by stimulating the synthesis of proteins, fats, and glycogen and by inhibiting the breakdown of protein. In effect, insulin lowers blood glucose levels. It is particularly important for finding a place to store all the glucose that you take in with a meal. Figure 13.10 illustrates the interactions between insulin and glucagon in the homeostatic control of blood glucose concentrations.
3. Delta cells secrete *somatostatin*. In the pancreas, somatostatin inhibits the secretion of glucagon and insulin. Somatostatin is secreted from several different locations in the body, and its effects vary. For example, somatostatin secreted by the hypothalamus serves as the inhibitory hormone for growth hormone secretion. In the digestive tract, somatostatin helps regulate several digestive secretions. The example of somatostatin reinforces the principle that the function of a hormone in any particular tissue or organ depends on the types of target cells present in those locations.

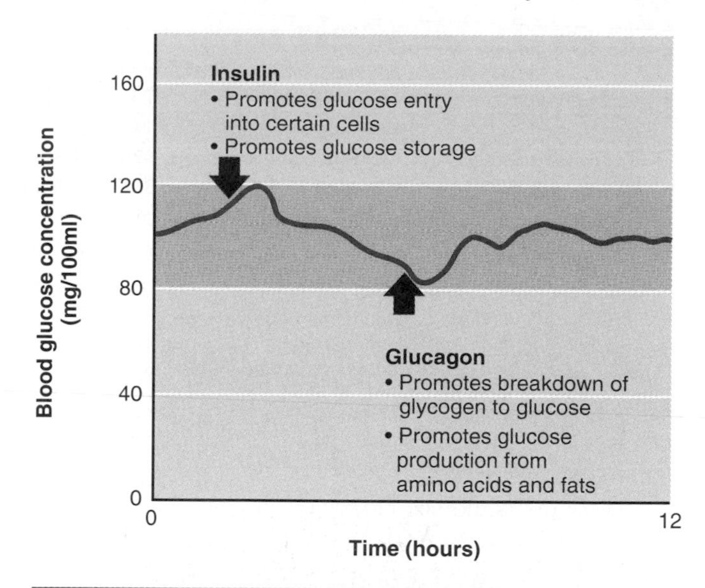

Figure 13.10 Roles of insulin and glucagon in regulating blood glucose. The normal range for blood glucose concentration is between 80 and 120 mg/100 mL. As blood glucose concentrations approach the outer limits of the normal range, the secretion of either insulin or glucagon (arrows) increases to keep blood glucose within the normal range.

Recap **Hormones secreted by the pancreas include glucagon, insulin, and somatostatin. Glucagon raises blood glucose levels by increasing production of glucose from glycogen and the metabolism of amino acids. Insulin lowers glucose levels by facilitating glucose uptake and storage. Somatostatin inhibits the secretion of both glucagon and insulin.**

13.5 The adrenal glands comprise the cortex and medulla

The **adrenal glands** are two small endocrine organs located just above the kidneys. Each gland has an outer layer, the *adrenal cortex,* and an inner core, the *adrenal medulla*.

The adrenal cortex: Glucocorticoids and mineralocorticoids

The **adrenal cortex** produces small amounts of the sex hormones estrogen and testosterone (most of the sex hormones are produced by the ovaries and testes, to be discussed later) and two classes of steroid hormones called *glucocorticoids* and *mineralocorticoids*. Their names give a clue to their functions: glucocorticoids help regulate blood glucose levels, and mineralocorticoids help regulate the minerals sodium and potassium.

Health Watch

Diabetes

Diabetes mellitus, or **diabetes,** as it is more commonly known, is a disorder of blood sugar regulation involving the hormone insulin. Diabetes mellitus should not be confused with *diabetes insipidus,* an entirely different disorder caused by a deficiency of antidiuretic hormone. Diabetes insipidus is generally referred to by its full name. When people say "diabetes" they mean diabetes mellitus.

Diabetes is the sixth leading cause of death in North America, affecting over 15 million people. Approximately a third of the people who have it are not even aware of it.

Two Types of Diabetes

There are two distinct forms of diabetes, known as Type I and Type II. Their common feature is an inability to get glucose into cells so it may be used as a source of cellular energy. As a result the cells literally starve—even though blood glucose levels may be above normal.

- **Type I diabetes** is caused by the failure to produce enough insulin. Type I diabetes is also called *insulin-dependent diabetes* because the person depends on daily doses of insulin for survival, or *juvenile-onset diabetes* because it generally appears during childhood or adolescence. Because insulin is the only hormone that facilitates the entry of glucose into cells, lack of insulin leads to very high blood glucose levels, especially after meals.

  Type I diabetes represents about 5–10% of all cases, and it is the more dangerous of the two types. Most cases of Type I diabetes appear to have an autoimmune cause: the person's own immune system attacks and destroys the beta cells of the pancreas.
- **Type II diabetes** (also called *non-insulin-dependent* or *maturity-onset* diabetes), represents 90–95% of all cases. It is slower in onset (generally first appearing after age 45) and usually less severe. Insulin levels may be low, normal, or even higher than normal. The hallmark of Type II diabetes is *insulin resistance,* meaning that the cells fail to respond adequately to insulin even when it is present at high levels.

  Although we do not know fully what causes Type II diabetes, risk factors include obesity and lack of regular exercise. One popular theory is that our modern high-sugar, high-calorie diets contribute to its development by signaling the pancreas to secrete more insulin to maintain blood glucose homeostasis. Over time, the high insulin levels cause the cells to become less sensitive to insulin. As a result, blood glucose levels continue to rise, stimulating the pancreas even more, until the pancreas just "wears out."

Both types of diabetes can lead to other health problems if blood sugar levels are not controlled properly. In particular, chronic untreated diabetes may damage blood vessels and nerves.

Vascular damage is most prevalent in the retina of the eyes and in the kidneys. Indeed, untreated diabetes is the leading cause of new cases of blindness in most age groups; this condition is called *diabetic retinopathy.* Uncontrolled diabetes is also the leading cause of kidney failure, accounting for nearly 40% of all cases, and increases the risk of heart disease and stroke approximately two- to four-fold.

Complications of diabetes are directly related to elevation of blood sugar levels over many years. This is why it is so important for diabetics—both Type I and Type II—to keep their blood sugar within normal limits.

Symptoms of Diabetes

Warning signs of Type I diabetes include frequent urination and unusual thirst, extreme hunger and unusual

Glucocorticoids: Cortisol The adrenal cortex produces a group of glucocorticoids with nearly identical structures. **Cortisol** accounts for approximately 95% of these glucocorticoids. Cortisol assists glucagon in maintaining blood glucose levels during prolonged fasting by promoting the utilization of fats and by increasing the breakdown of protein to amino acids in muscle. The liver can then metabolize the free amino acids to yield glucose. In addition to its role in the control of blood glucose, cortisol is also important in suppressing inflammatory responses that occur following infection or injury.

Cortisol secretion is stimulated by ACTH from the anterior pituitary gland, which in turn is stimulated by ACTH-releasing hormone (ACTH-RH) from the hypothalamus. Normally the blood levels of cortisol are kept fairly constant by a negative feedback loop in which a decline in the blood concentration of cortisol below a certain point stimulates the secretion of both ACTH-RH and ACTH (Figure 13.11). The rise in ACTH increases the adrenal secretion of cortisol, returning cortisol to normal. Conversely, a rise in plasma cortisol inhibits both ACTH-RH and ACTH secretion, and the subsequent decline in ACTH ultimately limits the rise in cortisol.

Decreases in blood glucose concentration as well as injury and emotional stress will raise plasma cortisol concentrations significantly. All of these stimuli cause the hypothalamus to release more ACTH-RH, which raises the ACTH level and ultimately the level of cortisol. The elevated cortisol concentration helps weight loss, fatigue, and irritability. The disease is named, in fact, for its urinary symptoms. *Diabetes mellitus* literally means "high urine flow rate with a sweet taste." The blood level of glucose rises to the point that the kidneys cannot reabsorb all the glucose that is filtered. When that happens, glucose appears in the urine and draws water with it. Loss of water in urine leads to dehydration and extreme thirst.

Type II diabetics may have any of these symptoms. Many also experience frequent infections, blurred vision, bruises or cuts that are slow to heal, and tingling or numbness in the feet or hands.

Treatments

The treatment of diabetes depends on whether it is Type I or Type II.

Type I diabetics need daily injections of insulin to stay alive. However, even daily injections cannot keep blood glucose levels within the normal range all the time, because a single injection does not mimic the cyclic periods of insulin secretion that follow each meal. As a result, Type I diabetics may still develop complications over the long term. New therapies are aimed at providing true feedback control of insulin availability within the body (see below).

Type II diabetes can often be managed just by careful control of caloric intake and a regular program of exercise. Cutting caloric intake reduces the need for insulin, and the lower the insulin levels are the more likely it is that cellular sensitivity to insulin will return toward normal.

Exercise lowers blood glucose in both Type I and Type II diabetics because exercising muscles can take up glucose even without the help of insulin. Some Type II diabetics take *oral hypoglycemic drugs,* pills that lower blood sugar by stimulating the beta cells to secrete more insulin than they would on their own. However, the use of oral hypoglycemic drugs is controversial because they may further weaken the beta cells. Exercise and weight control are considered better alternatives.

Science Is Making Progress

The treatment of Type I diabetes is an area where scientific research may be about to pay off. Promising techniques include:

- Transplanting a pancreas from a human cadaver. Although technically feasible, the main problems are cost and availability. The cost can run to $100,000 and there is only one donor available for every 10–15 diabetics who might benefit.
- Injecting purified human pancreatic islets. Researchers are working on techniques for isolating human pancreatic islets and injecting them into the circulation of the liver. A few patients have benefited from this technique, but so far results have been short-lived (three years or less).
- Implanting a semipermeable device containing pancreatic islets. A hot field of research is the development of devices that enclose pancreatic islets within a semipermeable membrane to allow free exchange of nutrients, insulin, and glucose without contact with immune cells. These membranes could house pancreatic cells from another species (pigs have been proposed) without fear of immune system rejection. The device would offer true feedback control of insulin secretion.
- Developing a true artificial pancreas. Some researchers envision a nonliving implantable mechanical device. So far, the stumbling block has been the development of a sensor for continuously monitoring blood glucose that is sufficiently small, reliable, and accurate.

Other companies are developing devices for inhaling insulin. Watch for these and other advances to replace insulin injections in the future.

to restore the blood glucose concentration to normal and makes more amino acids available to repair damaged tissues. It also reduces inflammation by decreasing fluid loss from capillaries.

Cortisol and cortisol-like drugs can be administered topically to treat inflammation or injury. When used for prolonged periods, however, these drugs suppress the immune system.

Mineralocorticoids: Aldosterone The other hormones produced by the adrenal cortex are the mineralocorticoids, the most abundant of which is **aldosterone.** Aldosterone is the hormone primarily responsible for regulating the amounts of sodium and potassium in the body. With ADH, it helps maintain body water balance.

Adrenal secretion of aldosterone increases whenever the total amount of sodium and water in the body is too low or when there is too much potassium present (see Chapter 15 for the details). In turn, aldosterone acts on cells in the kidneys to promote sodium reabsorption and potassium excretion, precisely what you would expect it to do if it were part of a homeostatic control loop for these ions.

The adrenal medulla: Epinephrine and norepinephrine

The **adrenal medulla** produces the nonsteroid hormones **epinephrine** (adrenaline) and **norepinephrine** (noradrenaline). These hormones play roles in metabolism and controlling blood pressure and heart activity.

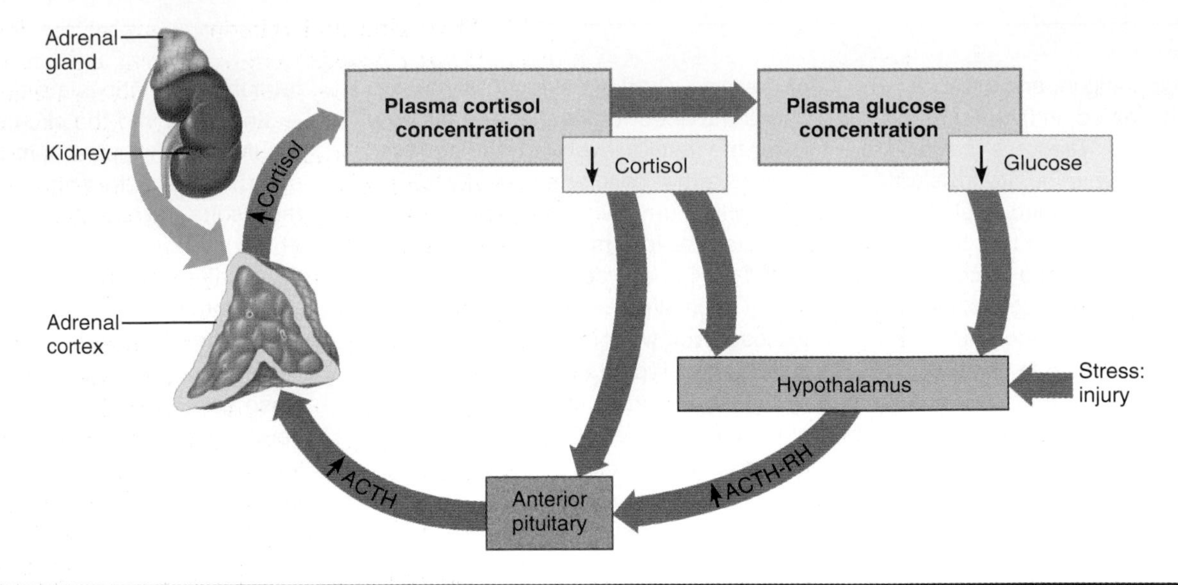

Figure 13.11 Negative feedback loop regulating the secretion of cortisol. A decrease in plasma cortisol concentration acts on the hypothalamus and directly on the anterior pituitary to increase the secretion of ACTH-RH and ACTH, respectively. In addition, a decrease in plasma glucose concentration, stress, or injury all can raise cortisol levels by increasing the hypothalamic secretion of ACTH-RH.

We mentioned epinephrine and norepinephrine in Chapter 11 (on the nervous system) because they are also neurotransmitters. Recall that the distinction between a neurotransmitter and a hormone is in how it functions, not its structure. When norepinephrine and epinephrine are released into the blood and act on target cells, they are functioning as hormones.

Like the hypothalamus and the posterior pituitary, the adrenal medulla is really a neuroendocrine organ. Epinephrine and norepinephrine are synthesized and stored in sympathetic neurons that end in the adrenal medulla. When the sympathetic division of the autonomic nervous system is activated, it stimulates the release of these hormones (Figure 13.12).

Once released, the two hormones participate in a wide range of actions that are essentially like the "fight or flight" responses of the sympathetic division. They raise blood glucose levels, increase heart rate and force of contraction, increase respiration rate, and constrict or dilate blood vessels in many organs, all actions that prepare the body for emergency activity.

You may have noticed that approximately 30 seconds after you become afraid or are aware of a dangerous situation, such as avoiding an auto accident, your heart starts to pound, you breathe faster, and you feel "pumped up," alert, and a little shaky. These changes may occur even though the danger has already passed. You are experiencing a massive adrenal release of norepinephrine and epinephrine from your adrenal glands in response to stimulation of your sympathetic nervous system. The adrenal response occurs about half a minute after the event, because it takes that long for the two hormones to be released, circulate to all tissues, bind to receptors on their target cells, and initiate the appropriate intracellular cascade of events. The response is also short-lived, lasting only a couple of minutes at most. Norepinephrine and epinephrine from the adrenal glands enhance the action of the sympathetic nervous system because the hormones reach more cells and they are released in large amounts.

How does the the endocrine system respond to fear or danger?

Recap **The main secretory products of the adrenal cortex are cortisol and aldosterone. Cortisol raises blood glucose levels and suppresses inflammatory responses, and aldosterone regulates water balance by promoting sodium reabsorption. The hormones secreted by the adrenal medulla (epinephrine and norepinephrine) contribute to the "fight or flight" response initiated by the sympathetic nervous system.**

13.6 Thyroid and parathyroid glands

The thyroid and parathyroid glands are anatomically linked. The **thyroid gland** is situated just below the larynx at the front of the trachea, and the two lobes of the thyroid gland wrap part of the way around the trachea (Figure 13.13). The four small **parathyroid glands** are embedded in the back of the thyroid.

The thyroid and parathyroid glands are also functionally linked; both help to regulate calcium balance. In addition, the thyroid has a separate and important role in controlling metabolism.

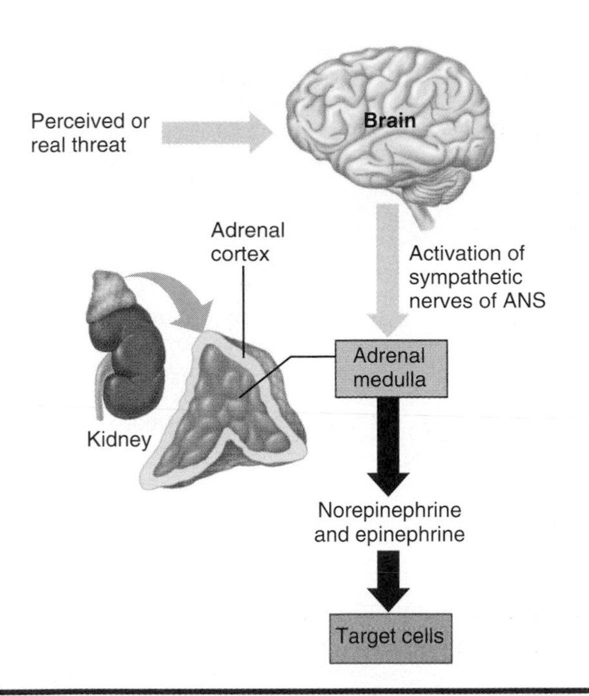

Figure 13.12 Secretion of norepinephrine and epinephrine by the adrenal medulla. The secretion of these two hormones is closely linked with the activity of the sympathetic division of the autonomic nervous system (ANS). Neural pathways are noted in yellow, hormonal pathways in blue.

The two main hormones produced by the thyroid gland are thyroxine and calcitonin. The parathyroid glands produce parathyroid hormone.

The thyroid gland: Thyroxine speeds cellular metabolism

The thyroid gland produces two very similar hormones called *thyroxine* (T_4) and *triiodothyronine* (T_3). Thyroxine and triiodothyronine are identical except that thyroxine contains four molecules of iodine (hence the abbreviation T_4) whereas triiodothyronine contains only three. The thyroid gland secretes mainly thyroxine, but eventually most of it is converted to the more active T_3 form of the hormone in the blood. Because they are so similar we'll consider them together as thyroxine.

Structurally, thyroxine is not a steroid hormone. Nevertheless, it acts like a steroid because it is lipid-soluble. It crosses the cell and nuclear membranes, binds to nuclear receptors, and activates genes as its mode of action. Those genes carry the code for the various enzymes that regulate the rate of cellular metabolism.

Thyroxine regulates the production of the high-energy molecule ATP from glucose in nearly all body cells. When thyroxine increases, basal metabolic rate (BMR) also increases. More energy is used, more work is done, and more heat is produced. Decreasing the blood concentration of thyroxine reduces all of these activities.

Secretion of thyroxine is stimulated by TSH from the anterior pituitary, which in turn is stimulated by TSH-RH from the hypothalamus. Conditions that raise the body's energy requirements, such as exposure to cold temperatures, cause an increase in TSH-RH, TSH, and ultimately in thyroxine. As the concentration of thyroxine rises, however, it exerts a negative feedback effect on the hypothalamus and the anterior pituitary to reduce the secretion of TSH-RH and TSH (Figure 13.14).

Iodine deficiency can cause goiter Iodine is required to produce active thyroid hormones. In fact, the main reason we need iodine in our diet is to ensure

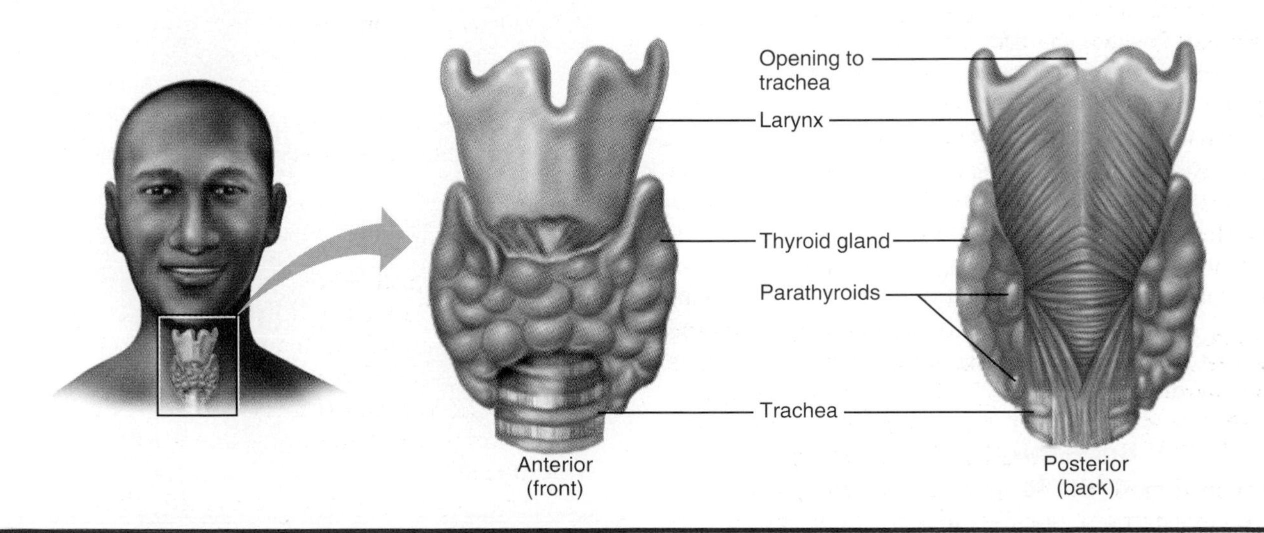

Figure 13.13 The thyroid and parathyroid glands.

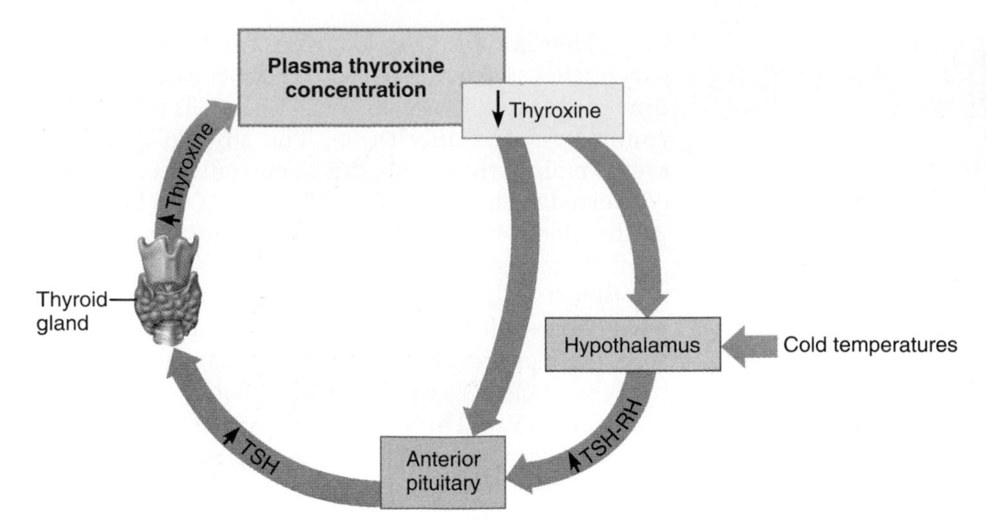

Figure 13.14 Negative feedback control of thyroxine secretion. A fall in plasma thyroxine concentration causes the hypothalamus and anterior pituitary to increase their secretion of TSH-RH and TSH, respectively. The increase in TSH stimulates thyroid gland production and secretion of thyroxine, returning the plasma concentration to normal. Conversely, a rise in thyroxine causes a fall in TSH-RH and TSH. Cold temperatures alter the feedback loop (raise plasma thyroxine concentration) by increasing hypothalamic secretion of TSH-RH.

adequate production of thyroxine. Iodine deficiency can result in a thyroid-deficiency disease. When thyroxine is absent or abnormally low, the normal feedback inhibitory controls on the hypothalamus and pituitary are missing, so the hypothalamus and pituitary continue to secrete large quantities of TSH-RH and TSH unchecked.

The high TSH levels stimulate the thyroid gland to grow to enormous size in an effort to get the thyroid to make more hormone, which it cannot do because it lacks iodine. An enlarged thyroid, as a result of iodine deficiency or other factors, is called a *goiter* (Figure 13.15). Goiter is less common today in industrialized nations because iodine is added to table salt. However, it is still a public health problem in certain areas of the world, most notably in Africa, China, and mountainous regions where the rain has leached iodine from the soil.

Calcitonin lowers blood calcium levels **Calcitonin,** the other main hormone of the thyroid gland, is produced by a separate group of thyroid cells. Calcitonin lowers the blood calcium concentration because it increases the deposition of calcium in bone. Calcitonin is part of a negative feedback control mechanism for calcium in which high blood calcium levels stimulate calcitonin release, promoting the deposition of calcium in bone, and low blood calcium levels inhibit calcitonin release, promoting removal of calcium from bone. Calcitonin is particularly important for the growth and development of bones in children. Once adulthood is reached, the responsiveness to calcitonin declines.

Parathyroid hormone (PTH) raises blood calcium levels

The parathyroid glands produce only one hormone, **parathyroid hormone (PTH),** which:

1. Removes calcium and phosphate from bone
2. Increases absorption of calcium by the digestive tract
3. Causes the kidneys to retain calcium and excrete phosphate

All of these actions tend to raise the blood calcium concentration (and lower blood phosphate concentration), opposing the action of calcitonin. The secretion of PTH is stimulated by low blood calcium

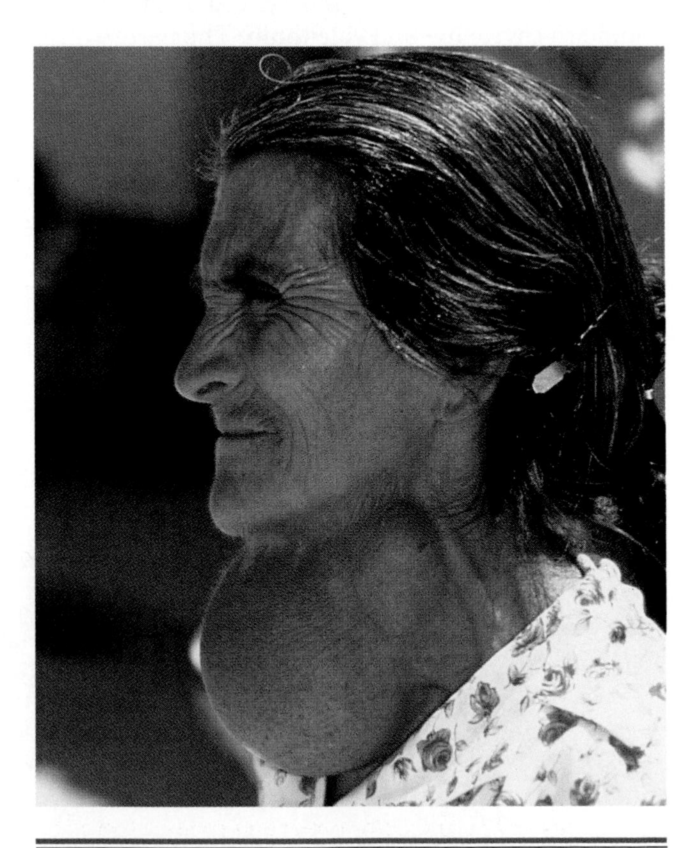

Figure 13.15 A goiter caused by dietary iodine deficiency.

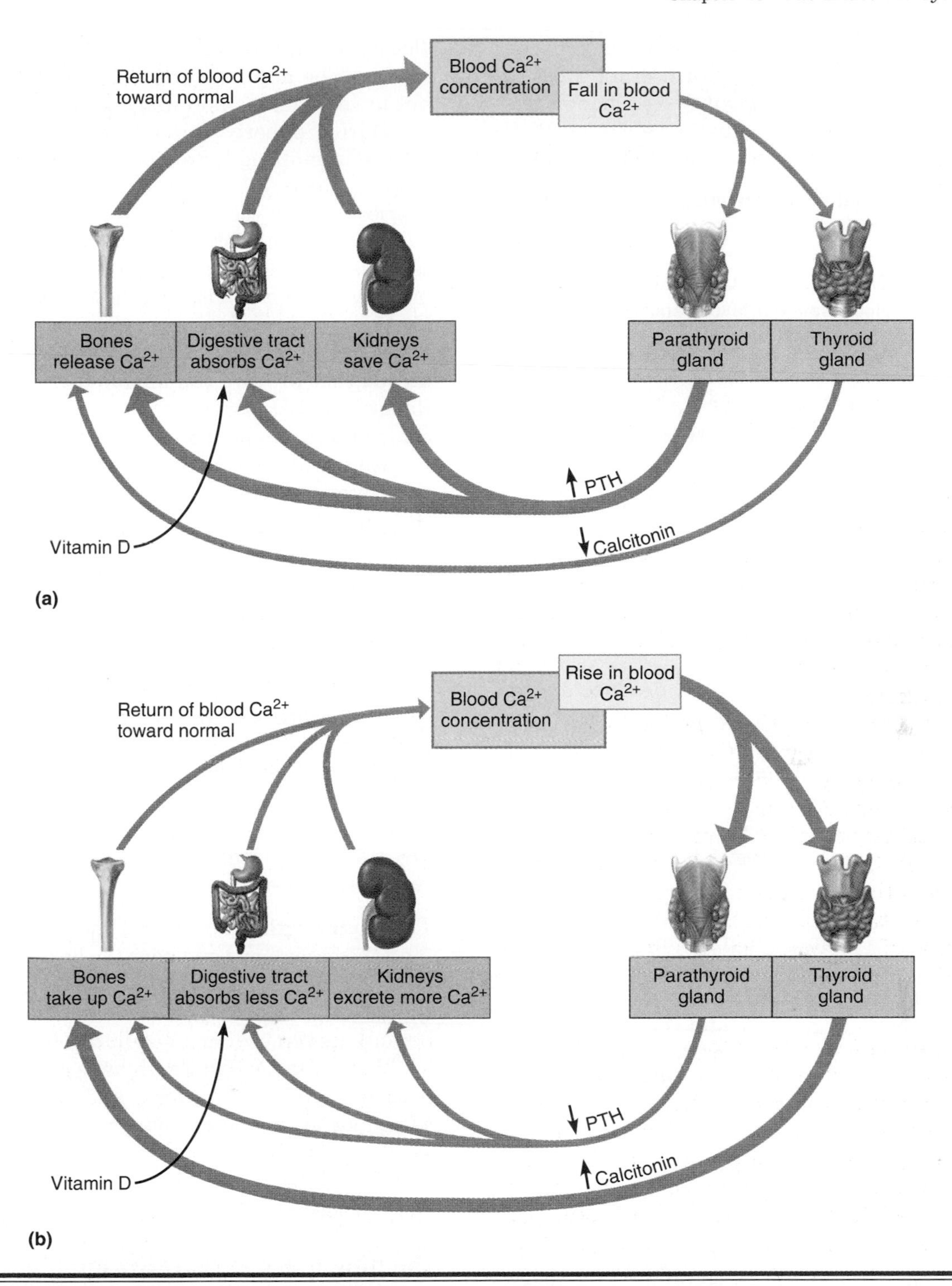

Figure 13.16 The homeostatic regulation of blood calcium concentration. (a) Responses to a fall in blood calcium. **(b)** Responses to a rise in blood calcium. The importance of calcitonin is greatest in children, declining once adulthood is reached.

levels and inhibited by high blood calcium levels. As Figure 13.16 illustrates, the negative feedback control of blood calcium concentration depends on two hormones (PTH and calcitonin) and three effector organs (bone, the digestive tract, and the kidneys), all acting together. The control of blood calcium concentration also depends on activated Vitamin D, which enhances calcium absorption from the digestive tract.

Because PTH plays the dominant role in controlling calcium balance once adulthood is reached, disorders of PTH secretion can significantly affect calcium homeostasis in adults. Over long periods of time (years), excessive PTH can remove enough calcium from bones to make them brittle and weak, a condition known as *osteodystrophy* (see Chapter 5—The skeletal system). One cause of elevated PTH concentrations and osteodystrophy is kidney failure.

Recap **Production of the thyroid gland hormones (T_4 and T_3) requires iodine. T_3 and T_4 speed cellular metabolism and have widespread effects on growth in children and BMR in adults. Calcitonin (from the thyroid) lowers blood calcium levels and PTH (from the parathyroids) raises blood calcium.**

13.7 Testes and ovaries produce sex hormones

The human *gonads* are the testes of males and the ovaries of females. These organs are responsible for the production of sperm and eggs, respectively. Both organs are also endocrine glands as well, in that they produce the steroid sex hormones.

The testes produce testosterone

The testes, located in the scrotum, produce *androgens,* the male sex hormones. The primary androgen produced by males is *testosterone*.

Between birth and puberty, the testes produce little if any testosterone. But during puberty, the anterior pituitary gland releases luteinizing hormone (LH), which stimulates the testes to begin producing male sex hormones.

Testosterone has many functions in males. It regulates the development and normal functioning of sperm, the male reproductive organs, and male sex drive. It is also responsible for the spurt of bone and muscle growth at puberty and the development of male secondary sex characteristics such as body hair and a deepening voice. LH, FSH, and testosterone continue to maintain the male reproductive system after puberty.

The adrenal glands in both sexes produce a small amount of a testosterone-like androgen called *dihydroepiandrosterone* (DHEA). DHEA has no demonstrable effect in males because they have an abundance of the more powerful testosterone. In females, DHEA from the adrenal glands is responsible for many of the same actions as testosterone in males, including enhancement of female pubertal growth, the development of axillary (armpit) and pubic hair, and the development and maintenance of the female sex drive.

The ovaries produce estrogen and progesterone

The ovaries, located in the abdomen, produce the female sex hormones known collectively as *estrogens*. The two main estrogens are *estrogen* and *progesterone*. (The adrenal cortex also produces estrogen, but in very small amounts.)

During puberty, the anterior pituitary starts to release luteinizing hormone (LH) and follicle stimulating hormone (FSH). These hormones stimulate the ovaries to begin secreting estrogen and progesterone.

Estrogen initiates the development of female secondary sex characteristics such as breast development, widening pelvis, and distribution of body fat. Both estrogen and progesterone regulate the *menstrual cycle,* the monthly changes in the tissue of the uterus that prepare the female body for pregnancy. Release of LH, FSH, estrogen, and progesterone continues throughout a woman's reproductive years.

We will discuss male and female hormones and reproductive systems in greater detail in Chapter 16 (Reproductive Systems).

13.8 Other glands and organs also secrete hormones

In addition to the glands described so far, there are several other lesser endocrine glands, and also organs with glands within them whose functions have been (or will be) described in other chapters. Among these are the thymus and pineal glands, the heart, the digestive system, and the kidneys.

Thymus gland hormones aid the immune system

The **thymus gland** is located between the lungs, behind the breastbone and near the heart. Sometimes it is considered part of the lymphatic system. For years the function of the thymus gland went unrecognized because its removal in an adult has little effect.

We now know that the thymus gland secretes two peptide hormones, *thymosin* and *thymopoietin,* that are important in helping lymphocytes of the immune system to mature into T cells. The thymus is relatively important in children, but it shrinks and reduces its secretion in adults, especially after age 30.

The pineal gland secretes melatonin

The **pineal gland** is a pea-sized gland located deep within the brain, in the roof of the third ventricle. Its name derives from the fact that it is shaped like a small pine cone (*pineas* in Latin). More than 200 million years ago our pineal gland was a photosensitive area, or "third eye," located near the skin's surface. Although it is now shielded from light by the skull, it still retains its photosensitivity because it receives input indirectly from the eyes via the optic nerve and nerve pathways in the brain.

The pineal secretes the hormone **melatonin** (not to be confused with the skin pigment, melanin) in a cyclic manner, coupled to the daily cycle of light and dark. Melatonin is sometimes called the "hormone of darkness" because its rate of secretion rises nearly 10-fold at night and then falls again during daylight. Its secretion appears to be regulated by the absence or presence of visual cues. During the day, nerve impulses from the retina inhibit its release.

We are just beginning to understand what melatonin does. Researchers believe that it may be important in synchronizing the body's rhythms to the daily light/dark cycle (the *circadian cycle*). Some scientists suggest that declining sensitivity to melatonin may contribute to the onset of puberty. Evidence in favor of this hypothesis is that destruction of the pineal gland by disease is associated with early puberty. Other proposed roles include inducing sleep, slowing the aging process, and enhancing immunity.

Melatonin is readily available over the counter as a pill. Some people take it as a natural sleep aid, although its effectiveness is controversial.

Table 13.2 lists all of the primary endocrine organs other than the hypothalamus and the pituitary, and summarizes their targets and hormonal actions.

Endocrine functions of the heart, the digestive system, and the kidneys

The heart, the digestive system, and the kidneys all have functions unrelated to the endocrine system. Nevertheless, all three of these organs secrete at least one hormone. The hormones they secrete and the functions of the hormones are listed below and summarized in Table 13.3.

- *Atrial natriuretic hormone (ANH)* is a peptide (nonsteroid) hormone secreted by the atria of the heart that helps regulate blood pressure. When blood pressure rises, ANH increases the rate at which sodium and water are excreted in urine. This decreases blood volume and lowers blood pressure. We discuss how ANH acts on the kidneys in Chapter 15.
- *Gastrin, secretin,* and *cholecystokinin* are all hormones secreted by the digestive system. They have effects on the stomach, pancreas, and gall bladder (see Chapter 14, The Digestive System and Nutrition).
- *Erythropoietin* and *renin* are secreted by the kidneys. Erythropoietin stimulates the production of red blood cells in bone marrow (see Chapter 7, The Circulatory System: Blood). Renin ultimately stimulates aldosterone secretion and constricts blood vessels (Chapter 15, The Urinary System).

13.9 Other chemical messengers

Some chemical messengers can function in ways similar to hormones, but they are not considered hormones because they are not secreted into the blood. Their actions are primarily local (near their site of release). You already know about neurotransmitters, one group of chemical messengers that act on nearby cells.

Such nonhormonal chemical messengers include histamine, prostaglandins, nitric oxide, and various growth factors and growth-limiting factors, plus dozens of other molecules about which we still know very little (Table 13.4). Most of these substances are short-lived because they are either quickly destroyed or taken back up by the cells that produced them.

Histamine is important in inflammation

We discussed **histamine** in Chapter 9 (Immune System). Mast cells release histamine into the local interstitial fluid in response to tissue injury or the presence of an allergen. In turn, the histamine increases local secretion of mucus by mucus-secreting cells, dilates blood vessels, and increases the leakiness of capillaries, all part of a constellation of symptoms called inflammation.

Histamine is largely responsible for the runny nose of a cold and the nasal congestion of allergies. This is why over-the-counter drugs for colds and allergies generally contain an antihistamine compound.

Prostaglandins: Local control of blood flow

Prostaglandins are a group of chemicals derived from a fatty acid precursor. Their name comes from the fact that they were first discovered in semen and were thought to originate from the prostate gland, but they have since been found in nearly every tissue in the body.

Prostaglandins have a variety of functions, including local control of blood flow. Some prostaglandins constrict blood vessels; others dilate blood vessels. They also contribute to the inflammatory response and are involved in blood clotting at the site of an injury.

Nitric oxide has multiple functions

Nitric oxide (NO) is a gas that has only recently received attention as a chemical messenger. NO was originally called "endothelial-derived relaxing factor" because an extract of the endothelial cells lining blood vessels caused blood vessels to relax, or dilate. We now know that NO is produced in many other tissues as well.

Table 13.2 Hormones of Endocrine Glands Other Than the Hypothalamus and Pituitary

Gland	Hormone(s)	Main Targets	Primary Actions
Pancreas			
	Glucagon	Liver	Raises blood sugar levels
	Insulin	Liver, muscle, adipose tissue	Lowers blood sugar levels
	Somatostatin	Pancreas, hypothalamus, digestive tract	Inhibits release of glucagon, insulin, growth hormone, and digestive secretions
Adrenal cortex			
	Glucocorticoids (including cortisol)	Most cells	Raise blood glucose levels and promote breakdown of fats and proteins; suppress inflammation
	Mineralocorticoids (including aldosterone)	Kidneys	Regulate body water balance by promoting sodium reabsorption, potassium excretion
Adrenal medulla			
	Epinephrine and norepinephrine	Liver, muscle, adipose tissue	Raise blood sugar; increase heart rate and force of contraction; dilate or constrict blood vessels, increase respiration
Thyroid			
	Thyroxine (T_4) and triiodothyronine (T_3)	Most cells	Regulate the rate of cellular metabolism; affect growth and development
	Calcitonin	Bone	Lowers blood calcium levels
Parathyroid glands			
	Parathyroid hormone (PTH)	Bone, digestive tract, kidneys	Raises blood calcium levels
Testes			
	Testosterone	Most cells	Development of sperm, male secondary sex characteristics, and reproductive structures
Ovaries			
	Estrogen, progesterone	Most cells	Development of female secondary sex characteristics and reproductive structure; regulate menstrual cycle
Thymus			
	Thymosin, thymopoietin	Lymphocytes	Help lymphocytes mature, especially in children
Pineal			
	Melatonin	Many cells	Synchronizes body rhythms; may be involved in onset of puberty

Nitric oxide functions include regulating local blood flow in many tissues of the body, controlling penile erection, and regulating smooth muscle contraction in the digestive tract. It is also used as a "chemical warfare" agent against bacteria by macrophages, and it interferes with clotting mechanisms. It even functions as a neurotransmitter in the brain.

Growth factors regulate tissue growth

Chemical messenger molecules called **growth factors** modulate development of specific tissues at the local level. The list includes nerve growth factor, epidermal growth factor, fibroblast growth factor, platelet-derived growth factor, tumor angiogenesis growth factors, insulin-like growth factor, and many

Table 13.3 Hormones of the Digestive Tract, Kidneys, and Heart

System or Organ	Hormone(s)	Main Targets	Primary Actions
Digestive tract			
	Gastrin, secretin, cholecystokinin	Stomach, pancreas, gall bladder	Stimulate activities of the stomach, pancreas, and gall bladder
Kidneys			
	Erythropoietin	Bone marrow	Stimulates red blood cell production
	Renin	Adrenal cortex, blood vessels	Stimulates aldosterone secretion, constricts blood vessels
Heart			
	Atrial natriuretic hormone (ANH)	Kidney, blood vessels	Increases sodium excretion, dilates blood vessels

more. Growth factors influence when a cell will divide and even in what direction it will grow. For example, nerve growth factors help to determine the direction of neuron axon growth such that nerves develop properly.

Recap **Various chemical messengers (histamine, prostaglandins, nitric oxide, and growth factors) act primarily locally. Among their actions are regulation of local blood flows, participation in local defense responses, and regulation of growth of specific types of tissue.**

13.10 Disorders of the endocrine system

Because the endocrine system is one of the two primary systems for controlling bodily functions, any disruption to it can have dramatic and widespread effects on the body. We have already discussed several endocrine disorders earlier in this chapter. Here we look at hypothyroidism, hyperthyroidism, Addison's disease, and Cushing's syndrome.

Hypothyroidism: Underactive thyroid gland

Hypothyroidism refers to underactivity of the thyroid gland (hyposecretion of thyroid hormones). Mild cases of hypothyroidism may not have any signs, but more severe deficiencies can cause a variety of symptoms.

In children, insufficient thyroxine production for any reason can slow body growth, alter brain development, and delay the onset of puberty. Left untreated, this can result in *cretinism,* a condition of mental retardation and stunted growth.

In adults, insufficient thyroxine can lead to *myxedema,* a condition characterized by edema (swelling) under the skin, lethargy, weight gain, low BMR, and low body temperature. Hyposecretion of the hormone can be treated with thyroxine pills.

Hyperthyroidism: Overactive thyroid gland

Hyperthyroidism involves an overactive thyroid gland and hypersecretion of thyroid hormones. Too much thyroxine increases BMR and causes hyperactivity, nervousness, agitation, and weight loss.

The most common form of hyperthyroidism is *Graves' disease,* an autoimmune disorder in which a person's own antibodies stimulate the thyroid to produce too much thyroxine. Graves' disease is often accompanied by protruding eyes caused by fluid accumulation behind the eyes (Figure 13.17).

Table 13.4 Other Chemical Messengers

Messenger(s)	Primary Actions
Histamine	Important in inflammatory response; increases secretion of mucus, dilates blood vessels, increases leakiness of capillaries
Prostaglandins	Control local blood flow; constrict blood vessels, dilate blood vessels, contribute to inflammatory response and blood clotting
Nitric oxide (NO)	Regulates local blood flow, controls penile erection, regulates smooth muscle contraction in digestive tract, acts as a "chemical warfare" agent against bacteria, interferes with clotting mechanisms, acts as a neurotransmitter in brain
Growth factors	Modulate growth of tissues at local level

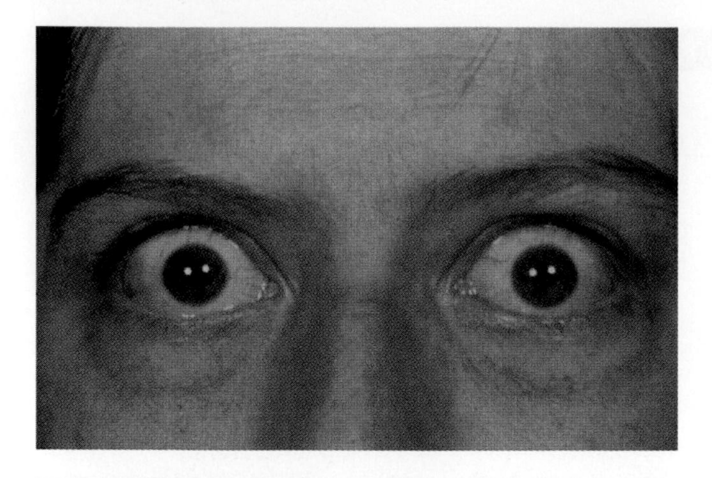

Figure 13.17 Graves' disease. Note the protruding eyes.

Addison's disease: Too little cortisol and aldosterone

Addison's disease is a condition caused a failure of the adrenal cortex to secrete sufficient cortisol and aldosterone. Lack of cortisol lowers blood glucose levels, and lack of aldosterone lowers blood sodium. Addison's disease tends to develop slowly, with chronic symptoms of fatigue, weakness, abdominal pain, weight loss, and characteristic darkening of the skin in pressure areas such as the palms. It can be successfully treated with medications that replace the missing hormones.

Cushing's syndrome: Too much cortisol

The symptoms of Cushing's syndrome are due to the exaggerated effects of too much cortisol, the main effects of which are: (1) excessive production of glucose from glycogen and protein, and (2) retention of too much salt and water. Blood glucose concentration rises and there is a decrease in muscle mass because of the utilization of protein to make glucose. Some of the extra glucose is converted to body fat, but only in certain areas of the body, including the face, abdomen, and the back of the neck. Symptoms of Cushing's syndrome include muscle weakness and fatigue, edema (swelling due to too much fluid), and high blood pressure. The disease can be caused by tumors of the adrenal gland or the ACTH-secreting cells of the pituitary. It can also be caused by the excessive use of cortisol and cortisol-like drugs (cortisone, prednisone, dexamethasone, and others) to control chronic inflammatory conditions such as allergies and arthritis.

Chapter Summary

The endocrine system regulates body functions with hormones

- The endocrine system is a collection of glands that secrete hormones into the blood.
- Hormones are chemical messengers secreted by the endocrine system that circulate in the blood and act on specific target cells.

Hormones are classified as steroid or nonsteroid

- Steroid hormones are lipid-soluble, hence they enter the target cell in order to function.
- Nonsteroid hormones generally are not lipid-soluble. They bind to receptors on the outer surface of the cell membrane of the target cell.

The hypothalamus and the pituitary gland

- The hypothalamus of the brain and the pituitary gland of the endocrine system link the nervous and endocrine systems.
- The anterior lobe of the pituitary produces and secretes six hormones (ACTH, TSH, FSH, LH, PRL, and GH). The first four listed stimulate the release of other hormones from other endocrine glands.
- The posterior lobe secretes two hormones (ADH and oxytocin) that are produced by the hypothalamus.
- Abnormalities of ADH secretion lead to disorders of water balance.
- Hypersecretion of growth hormone in a child leads to gigantism; hyposecretion causes dwarfism. In an adult, hypersecretion of growth hormone causes acromegaly.

The pancreas secretes glucagon, insulin, and somatostatin

- Within the pancreas, clusters of cells called the pancreatic islets produce and secrete three hormones (glucagon, insulin, and somatostatin).
- Glucagon raises blood glucose levels, insulin lowers blood glucose levels, and somatostatin inhibits the secretion of glucagon and insulin.

The adrenal glands comprise the cortex and medulla

- The outer zone of the adrenal gland, called the cortex, secretes steroid hormones that affect glucose metabolism (glucocorticoids) and salt balance (mineralocorticoids).
- The inner zone of the adrenal gland, called the medulla, is actually a neuroendocrine organ. The adrenal medulla secretes epinephrine and norepinephrine into the blood.

Thyroid and parathyroid glands

- The thyroid gland produces thyroxine (T_4), triiodothyronine (T_3), and calcitonin.
- The four small parathyroid glands produce PTH, which, along with calcitonin, is responsible for the regulation of blood calcium concentration.

Testes and ovaries produce sex hormones

- The primary male sex hormone is testosterone, produced by the testes.
- The ovaries produce the two female sex hormones, estrogen and progesterone.

Other glands and organs also secrete hormones

- The thymus gland produces two hormones (thymosin and thymopoietin) that help lymphocytes mature into T cells.
- Melatonin from the pineal gland synchronizes the body's daily rhythms to the daily light/dark cycle.
- The heart secretes atrial natriuretic hormone, the kidneys secrete renin and erythropoietin, and the digestive tract secretes gastrin, secretin, and cholecystokinin.

Other chemical messengers

- Some substances not defined as hormones (because they are not specifically secreted into the bloodstream) nevertheless serve as chemical messengers.
- Most of these chemical messengers tend to act locally, near their site of secretion. They include histamine, prostaglandins, nitric oxide, and growth factors.

Disorders of the endocrine system

- Because the endocrine system controls so many body functions, disruptions can have widespread effects.
- Hypothyroidism can lead to cretinism (stunted growth and mental retardation) in children and myxedema in adults. Graves' disease is the most common form of hyperthyroidism.
- Insufficient cortisol and aldosterone can cause Addison's disease, and Cushing's syndrome results from hypersecretion of cortisol.

Terms You Should Know

adrenal glands, 327
aldosterone, 329
calcitonin, 332
cortisol, 328
endocrine gland, 316
epinephrine, 329
glucagon, 327
hormone, 316
hypothalamus, 320
insulin, 327
neuroendocrine cells, 320
nonsteroid hormone, 318
norepinephrine, 329
parathyroid gland, 330
parathyroid hormone, 332
pineal gland, 334
pituitary gland, 320
second messenger, 319
steroid hormone, 318
target cell, 316
thymus gland, 334
thyroid gland, 330

Test Yourself

Matching Exercise

____ 1. A type of hormone that initiates a cascade of enzymes within the cell
____ 2. A type of hormone that activates or inhibits a gene
____ 3. Sometimes functions as a second messenger within a cell
____ 4. Region of the forebrain that produces the releasing and inhibiting factors associated with the anterior pituitary hormones
____ 5. Pituitary hormone associated with feelings of sexual satisfaction
____ 6. Anterior pituitary hormone that stimulates the development of mammary gland cells and the production of milk late in pregnancy
____ 7. A glucocorticoid produced by the adrenal cortex
____ 8. Hormone from the pineal gland, available over the counter as a sleep aid
____ 9. A chemical messenger once called "endothelial-derived relaxing factor" that dilates blood vessels and participates in penile erection
____10. One of two hormones from the kidneys

a. cyclic AMP
b. hypothalamus
c. prolactin
d. cortisol
e. nonsteroid

f. melatonin
g. erythropoietin
h. steroid
i. nitric oxide
j. oxytocin

Fill in the Blank

11. A (steroid/nonsteroid) hormone binds to a receptor within the target cell.
12. A (steroid/nonsteroid) hormone binds to a receptor on the target cell's cell membrane.
13. The two hormones secreted from the posterior lobe of the pituitary gland are ________ and ________.
14. A ________ cell functions as both a neuron and an endocrine cell.
15. The two anterior pituitary hormones that stimulate the reproductive organs in both men and women are ______ and ______.
16. Four ions of ________ are required for the synthesis of each thyroxine molecule.
17. The primary mineralocorticoid hormone produced by the adrenal cortex is ________.
18. The hormones PTH and calcitonin both participate in the homeostatic regulation of the blood concentration of ________.
19. Cushing's syndrome and Addison's disease are both disorders of the cortex of the ________ glands.
20. The hormone properly called ________ is also commonly known as adrenaline.

True or False?

21. Testosterone is produced and secreted by the anterior pituitary gland.
22. Insulin and glucagon are produced and secreted by different types of cells in the pancreas.
23. The heart secretes a hormone called atrial natriuretic factor.
24. A disease called goiter can be caused by insufficient iodine in the diet.
25. A disorder called Cushing's syndrome can be produced in otherwise healthy people by the overuse of cortisol or cortisol-like drugs.

Concept Review

1. List the six hormones secreted by the anterior lobe of the pituitary gland.
2. Explain why it is important that withdrawal of your hand from a hot flame depends on the nervous system and not the endocrine system.
3. Name one of the two places in the endocrine system that one might find neuroendocrine cells.
4. Define when you would call norepinephrine a neurotransmitter and when you would call it a hormone.
5. Explain why a 30-year-old short person cannot be made taller by the administration of growth hormone.
6. Describe the function of the two hormones from the thymus gland. Why is this function likely to be more important in children than in adults?
7. Explain how the pineal gland can be sensitive to light, even though it is located deep within the brain.
8. Describe why substances such as histamine and nitric oxide, although they are chemical messengers, are not considered true hormones.
9. Name the hormone that is released from the posterior pituitary of lactating females when the infant nurses.
10. What does the suffix *-tropic* mean?

Apply What You Know

1. The statement was made early in the chapter that "the endocrine system has universal access to nearly every living cell." The word "nearly" implies that there must be an exception. Which cells, tissues, or organs would you guess are most likely *not* exposed to most hormones? Explain.
2. If the biological effectiveness of a hormone were greatly reduced because its target cells lacked the appropriate receptors, what would you expect the hormone concentration to be: high, normal, or low? Explain. (HINT: it may help to draw the components of a negative feedback loop.)
3. Explain why an injection of epinephrine to combat an acute immune reaction (such as an allergic response to a bee sting) has an almost immediate effect, whereas an injection of a steroid hormone can take hours to have an effect.

Case in Point

Hormone Treatments for the Healthy?

In this chapter we saw that treating children with growth hormone can prevent the endocrine disorder called pituitary dwarfism. But what about children who are simply shorter than average—should they be treated with growth hormone too?

An increasing number of parents say yes, and they are asking their doctors to prescribe growth hormone for children who are shorter than their peers. These parents believe that people who are shorter than average often encounter discrimination, and they view the treatments as enhancing their children's quality of life. Some health professionals agree that lack of height can be a social drawback and psychological barrier. It's estimated that at least 10 percent of growth hormone prescriptions in the U.S. are written to treat short stature rather than a clinical disorder. Parents in other countries, such as Canada and Thailand, are also requesting this treatment for their children.

Although popular, this use of growth hormone remains controversial. Some physicians worry that its long-term effects are unknown. Others question the wisdom of using medications to alter a condition that isn't really a health problem. There may be a link between growth hormone treatments and leukemia, but cases are rare and the evidence is inconclusive.

A recent study at Stanford University monitored 80 youngsters who started taking the hormone between the ages of 7 and 10, and found that 61 percent of the girls and 51 percent of the boys exceeded their expected adult height by more than two inches. However, the treatments had no effect at all on 20 percent of the children.

QUESTIONS:

1. Do you think this is a valid use for growth hormone? Explain your answer.
2. In rare cases, children taking growth hormone may experience side effects such as joint pain, a diabetes-like rise in blood glucose, and gigantism. Explain why these side effects might occur.

INVESTIGATE:

White, G. B. Human Growth Hormone: The Dilemma of Expanded Use in Children. *Kennedy Institute of Ethics Journal* 3: 401–409, 1993.

References and Additional Reading

Atkinson, Mark A. and Noel K. Maclaren. What Causes Diabetes? *Scientific American* July 1990, pp. 62–71. (A review of the autoimmune basis of type I diabetes.)

Lacy, Paul E. Treating Diabetes with Transplanted Cells. *Scientific American* July 1995, pp. 50–58. (A discussion of recent advances and future directions in the treatment of type I diabetes.)

Hadley, M.E. *Endocrinology*. 5th ed. Prentice Hall, 1999. (A standard textbook of endocrinology.)

CHAPTER 14

THE DIGESTIVE SYSTEM AND NUTRITION

What are the components of saliva?

What should a healthy diet **contain?**

Why doesn't the stomach digest itself?

Why is it so difficult to lose weight **and** keep **it off?**

STOP FOR A MOMENT AND MAKE A LIST OF WHAT YOU ATE FOR lunch—a hamburger, fries, and a soft drink? Pizza and bottled water? A salad, an apple, and milk?

All cells need **nutrients**—substances in food that are required for growth, reproduction, and the maintenance of health. Quite simply, the cells of your body must live on whatever you choose to eat. Right now the nutrients in your last meal (as well as nutrients stored in your body from previous meals) are being used to fuel cellular activities, build cell components, and serve other vital functions. Your cells draw these nutrients from your blood, and your blood obtains them from your digestive system or from nutrient storage pools.

In many ways the digestive system is a highly efficient disassembly line. It takes in food and processes it, breaking it into small pieces and digesting the fragments with enzymes and strong chemicals. Then it passes the mixture down the line, absorbing its nutrients along the way, and stores the leftover waste until it can be eliminated from the body.

Does what you eat matter? Yes! Poor nutrition is associated with diseases ranging from cancer to cavities. Good nutrition, on the other hand, improves

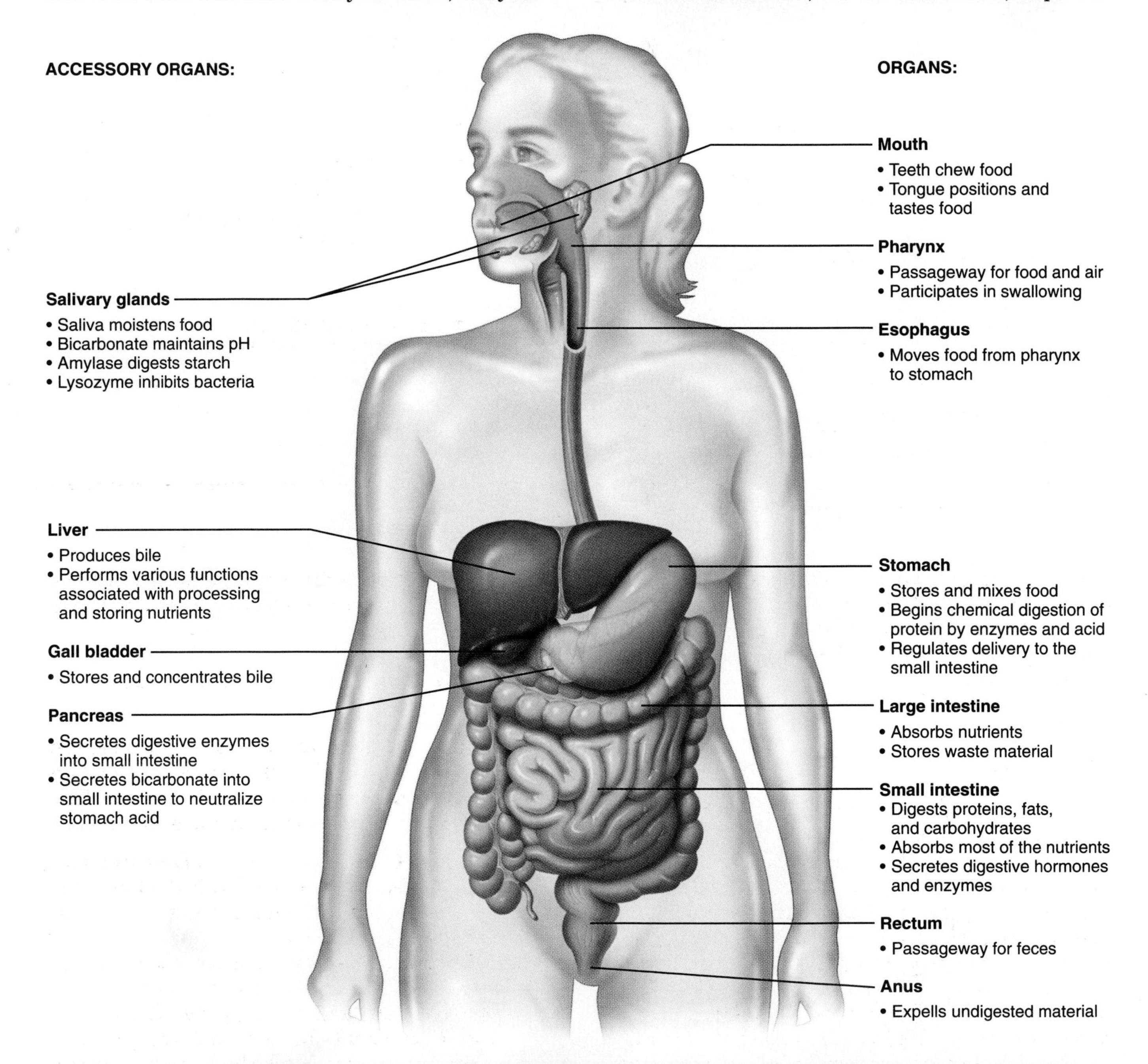

Figure 14.1 Organs and accessory organs of the digestive system and their functions.

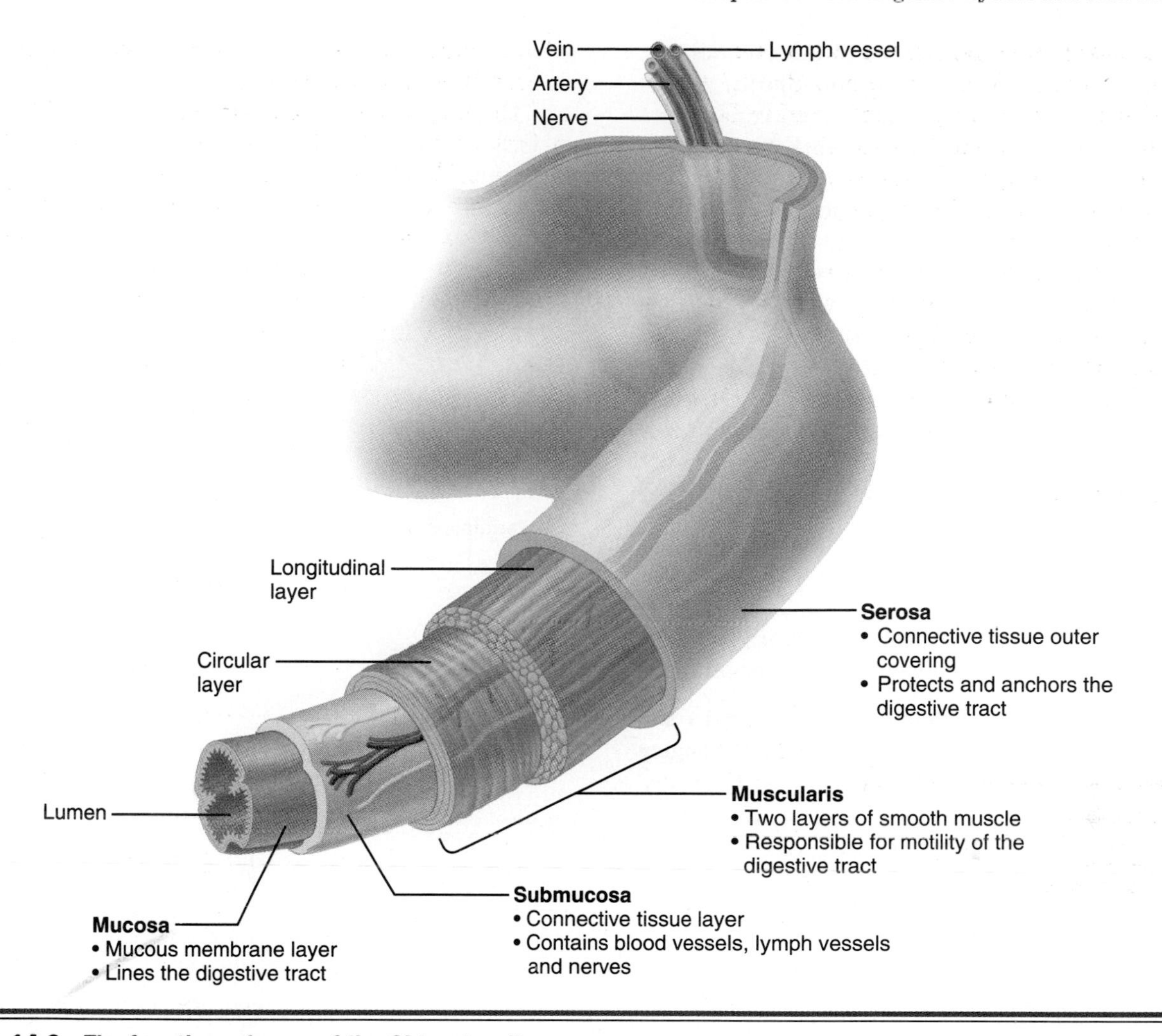

Figure 14.2 The four tissue layers of the GI tract wall.

overall health and lowers the risk of health problems. In this chapter we follow the digestive process and the fate of your food. Then we discuss how you can apply this information to promote a healthy diet and body weight.

14.1 The digestive system brings nutrients into the body

The **digestive system** consists of all the organs that share the common function of getting nutrients into the body. It includes a series of hollow organs extending from the mouth to the anus: the mouth, pharynx, esophagus, stomach, small intestine, large intestine, rectum, and anus. These organs comprise a hollow tube called the **gastrointestinal (GI) tract** ("gastric" is the adjective for stomach). The space within this hollow tube—the area through which food and liquids travel—is called the *lumen.*

The digestive system also includes four **accessory organs**—the salivary glands, liver, gallbladder, and pancreas. Figure 14.1 summarizes the functions of the organs and accessory organs of the digestive system.

The walls of the GI tract are composed of four layers

Spotlight on Structure & Function

From the esophagus to the anus, the walls of the GI tract share common structural features (Figure 14.2). The walls of the GI tract consist of four layers of tissue:

1. **Mucosa:** The innermost tissue layer (the mucous membrane in contact with the lumen) is the *mucosa.* All nutrients must cross the mucosa to enter the blood.
2. **Submucosa:** Next to the mucosa is a layer of connective tissue containing blood vessels, lymph vessels, and nerves, called the *submucosa.* Components of food that are absorbed across the mucosa enter the blood and lymph vessels of the submucosa.

3. **Muscularis:** The third layer of GI tract tissue, called the *muscularis,* is responsible for *motility,* or movement. The muscularis consists of two or three sublayers of smooth muscle. In general, the fibers of the inner sublayer are oriented in a circular fashion around the lumen, whereas those in the outer sublayer are arranged lengthwise, parallel to the long axis of the digestive tube. The exception is the stomach, which has a diagonal (oblique) sublayer of muscle inside the other two.
4. **Serosa:** The outermost layer of the GI tract wall is the *serosa.* The serosa is a thin connective tissue sheath that surrounds and protects the other three layers and attaches the digestive system to the walls of the body cavities. ■

Some of the organs of the GI tract are separated from each other by thick rings of circular smooth muscle called *sphincters.* When these sphincters contract they can close off the passageway between organs.

Recap **The digestive system consists of organs and accessory organs that share the function of bringing nutrients into the body. The wall of the GI tract consists of four tissue layers: the mucosa, the submucosa, the muscularis, and the serosa.**

Five basic processes accomplish digestive system function

In practical terms, the digestive system is a huge disassembly line that starts with huge chunks of raw material (we call it food) and takes them apart so that the nutrients in food can be absorbed into the body. The digestive system accomplishes this task with five basic processes:

1. *Mechanical processing and movement.* Chewing breaks food into smaller pieces, and two types of movement (motility) mix the contents of the lumen and propel it forward.
2. *Secretion.* Fluid, digestive enzymes, acid, alkali, bile, and mucus are all secreted into the GI tract at various places. In addition, several hormones that regulate digestion are secreted into the bloodstream.
3. *Digestion.* The contents of the lumen are broken down both mechanically and chemically into smaller and smaller particles, culminating in nutrient molecules.
4. *Absorption.* Nutrient molecules pass across the mucosal layer of the GI tract and into the blood or lymph.
5. *Elimination.* Undigested material is eliminated from the body via the anus.

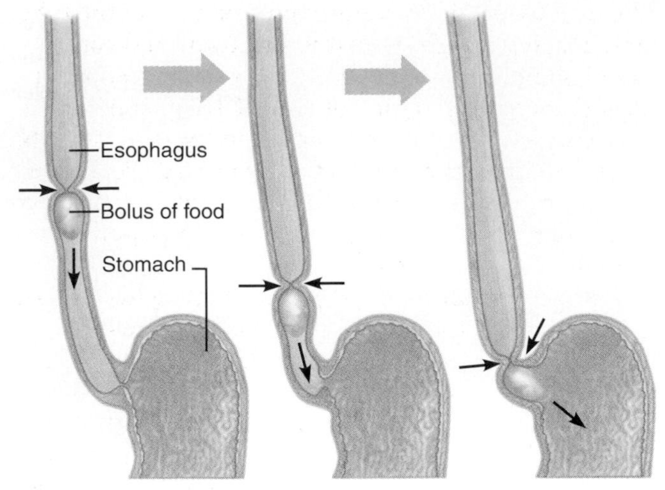

(a) Peristalsis

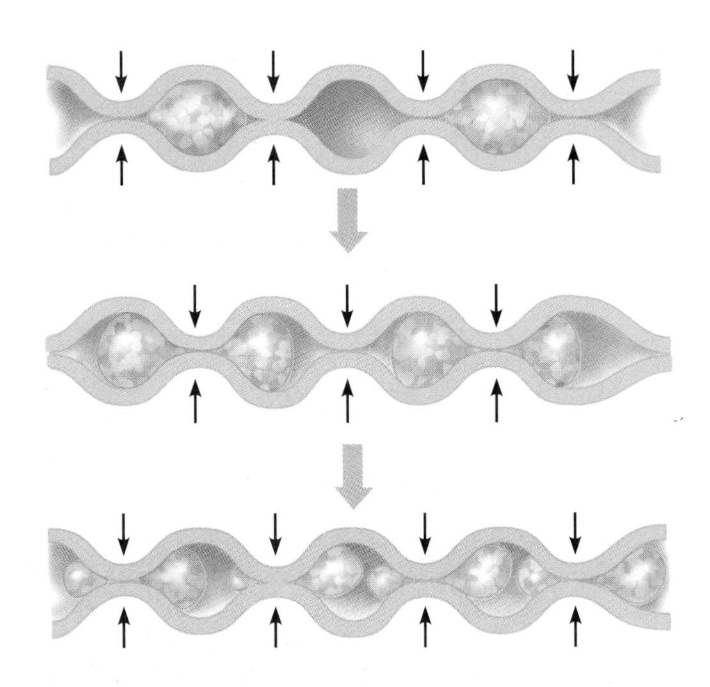

(b) Segmentation

Figure 14.3 Motility of the gastrointestinal tract. **(a)** Peristalsis in the esophagus. The three sequential diagrams show how peristalsis moves the contents of the lumen onward. **(b)** Segmentation in the small intestine. Segmentation mixes the contents of the lumen.

Two types of motility aid digestive processes

The smooth muscle of the GI tract produces two kinds of motility, called peristalsis and segmentation. The functions of peristalsis and segmentation are quite different.

Peristalsis propels food forward (Figure 14.3a). Peristalsis begins when a lump of food (called a bolus) stretches a portion of the GI tract, causing the smooth muscle in front of the bolus to relax and the muscle behind it to contract. The contractions push the food forward, stretching the next part of the tube and causing muscle relaxation in front and contraction behind. The peristaltic wave of contraction ripples through the organs of the GI tract, mixing the contents of the stomach and pushing the contents of the esophagus and intestines forward. Peristalsis occurs in all parts of the GI tract, but is especially prevalent in the esophagus where it transports food rapidly to the stomach.

Segmentation mixes food (Figure 14.3b). In segmentation, short sections of smooth muscle contract and relax in seemingly random fashion. The result is a back-and-forth mixing of the contents of the lumen. Food particles are pressed against the mucosa, enabling the body to absorb their nutrients. Segmentation occurs primarily in the small intestine as food is digested and absorbed.

14.2 The mouth processes food for swallowing

The **mouth,** or oral cavity, is the entrance to the GI tract. Digestion begins in the mouth with the process of chewing, which breaks food into smaller and smaller chunks. Essentially, the mouth functions as an effective food processor.

Teeth bite and chew food

The *teeth* chew food into pieces small enough to swallow. There are four types of teeth, each specialized for a different purpose (Figure 14.4a). The sharp-edged incisors cut food and the pointed canines tear it. The flat surfaces of the premolars and molars are well adapted to grinding and crushing food.

Children have only 20 teeth. They develop by about age 2 and are gradually replaced by permanent teeth. Permanent teeth usually develop by late adolescence, except for the wisdom teeth (third molars), which generally appear by age 25. Most adults have 32 permanent teeth.

Each tooth (Figure 14.4b) consists of a visible region called the *crown* and a region below the gum line called the *root.* The crown is covered by a layer of enamel, an extremely hard nonliving compound of calcium and phosphate. Beneath the enamel is a bonelike living layer called *dentin.* The soft innermost *pulp cavity* contains the blood vessels that supply the dentin, as well as the nerves that cause so much pain when a tooth is infected or injured. The entire tooth sits in a socket in the jaw bone that is lined with periodontal membrane.

Our mouths contain large numbers of bacteria that flourish on the food that remains between the teeth (Figure 14.4c). During their metabolism these bacteria release acids that can dissolve enamel, creating cavities or *dental caries.* If not treated, cavities deepen, eroding the dentin and pulp cavity and causing a toothache. Tooth decay may inflame the soft gum tissue (*gingiva*) around the tooth, causing *gingivitis.* Decay that inflames the periodontal membrane leads to *periodontitis.* However, good dental hygiene—including regular exams and teeth cleaning—can prevent most dental problems from becoming serious.

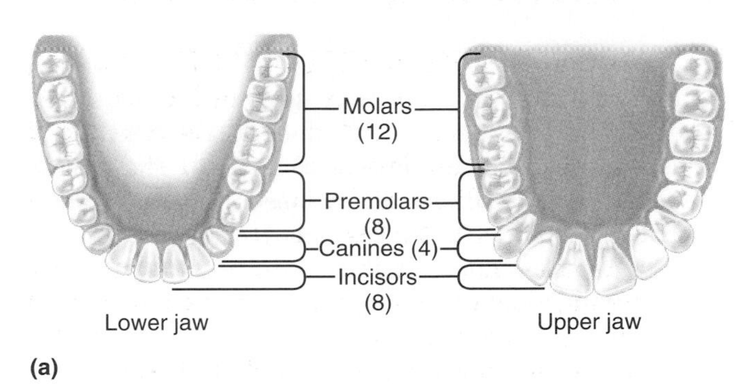

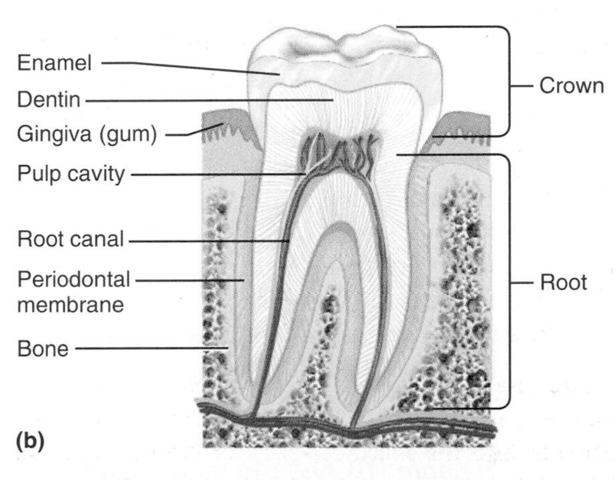

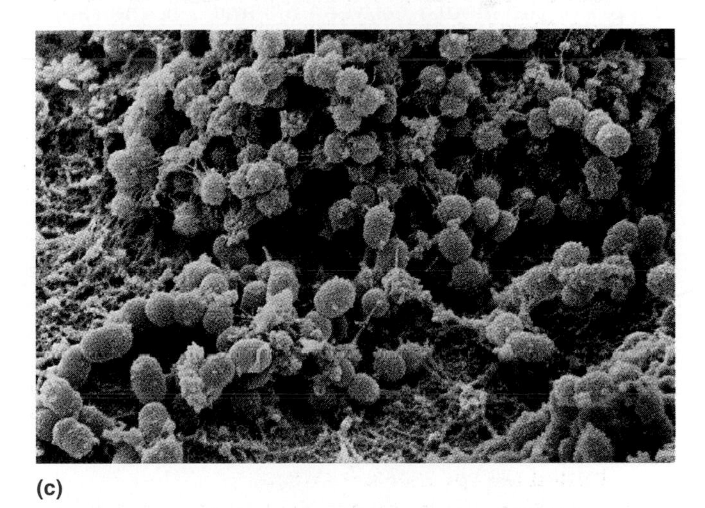

Figure 14.4 Teeth. (a) Locations and types of adult human teeth. **(b)** Anatomy of a tooth. **(c)** Bacteria on a tooth's surface (approx. ×8,000).

The tongue positions and tastes food

Chewing would be inefficient without the muscular tongue, which positions food over the teeth and mashes it against the roof of the mouth. The tongue consists of skeletal muscle enclosed in mucous membrane, so we have voluntary control over its movements. The tongue contributes to the sense of taste (Chapter 12) and also is important for speech.

Saliva begins the process of digestion

Three pairs of **salivary glands** produce a watery fluid called *saliva.* The *parotid gland* lies near the back of the jaw, and the smaller *submandibular* and *sublingual glands* are located just below the lower jaw and below the tongue, respectively (Figure 14.5). All three glands connect to the oral cavity via ducts.

Saliva moistens food, making it easier to chew and swallow. Saliva contains four ingredients, each with an important function. One is *mucin,* a mucus-like protein that holds food particles together so they can be swallowed more easily. An enzyme called *salivary amylase* begins the process of digesting carbohydrates. *Bicarbonate* (HCO_3^-) in saliva maintains the pH of the mouth between 6.5 and 7.5, the range over which salivary amylase is most effective. Salivary bicarbonate may also help protect your teeth against those acid-producing bacteria. Saliva also contains small amounts of an enzyme called *lysozyme* that inhibits bacterial growth.

What are the components of saliva?

Recap **The four kinds of teeth (molars, premolars, canines, and incisors) mechanically digest chunks of food. Salivary glands secrete saliva, which moistens food, begins the chemical digestion of carbohydrates, maintains the pH of the mouth, and protects the teeth against bacteria.**

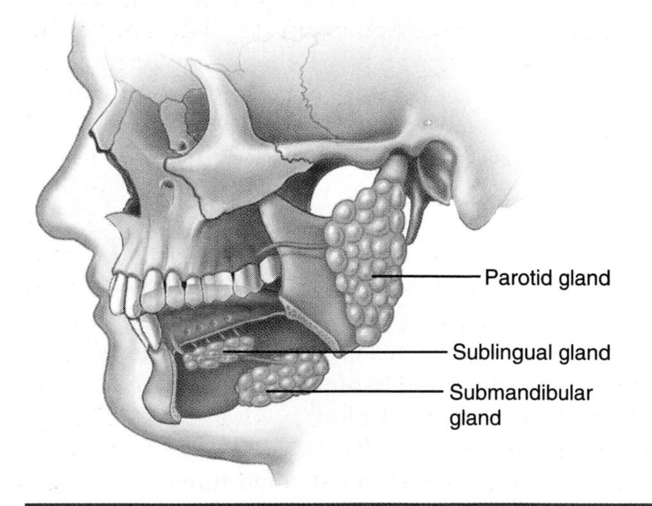

Figure 14.5 The salivary glands. The three pairs of glands deliver saliva to the mouth via salivary ducts.

14.3 The pharynx and esophagus deliver food to the stomach

After we have chewed our food and mixed it with saliva, the tongue pushes it into the **pharynx,** or throat, for swallowing. The act of swallowing involves a sequence of events that is coordinated with a temporary halt in breathing.

First, voluntary movements of the tongue and jaws push a bolus of food into the pharynx. As we saw in Chapter 10, the pharynx is a common passageway for food and air (Figure 14.6a). The presence of the food stimulates receptors in the pharynx that initiate the second, involuntary phase, the "swallowing reflex." The soft palate rises to close off the passageway into the nasal cavities and the larynx rises slightly. The epiglottis bends to close the airway to the trachea temporarily, so we do not inhale the food. Meanwhile, the tongue pushes the food back even farther, sliding it past the epiglottis and into the esophagus (Figure 14.6b). Once started, the swallowing reflex is involuntary and cannot be stopped. The initiation of swallowing is our last voluntary act until defecation.

Just beyond the pharynx is the **esophagus,** a muscular tube consisting of both skeletal and smooth muscle that connects the pharynx to the stomach. The lining of the esophagus produces lubricating mucus that helps food slide easily. The presence of a bolus of food in the esophagus initiates peristaltic contractions of smooth muscle that push the bolus to the stomach. Occasionally thick foods like mashed potatoes or peanut butter stick briefly to the mucosa of the esophagus, causing a painfully strong reflex contraction behind the food. If the contraction does not dislodge the food, drinking liquid may help.

Gravity assists peristalsis in propelling food. However, peristaltic contractions enable the esophagus to transport food even against gravity, such as when we are lying down.

The *lower esophageal sphincter,* located at the base of the esophagus, opens briefly as food arrives and closes after it passes into the stomach. The sphincter prevents reflux of the stomach's contents back into the esophagus. Occasionally this sphincter malfunctions, resulting in the backflow of acidic stomach fluid into the esophagus. This condition, known as *acid reflux,* is responsible for the burning sensation known as "heartburn." Acid reflux becomes more common with weight gain, pregnancy, and age. Occasionally it indicates a *hiatal hernia,* a

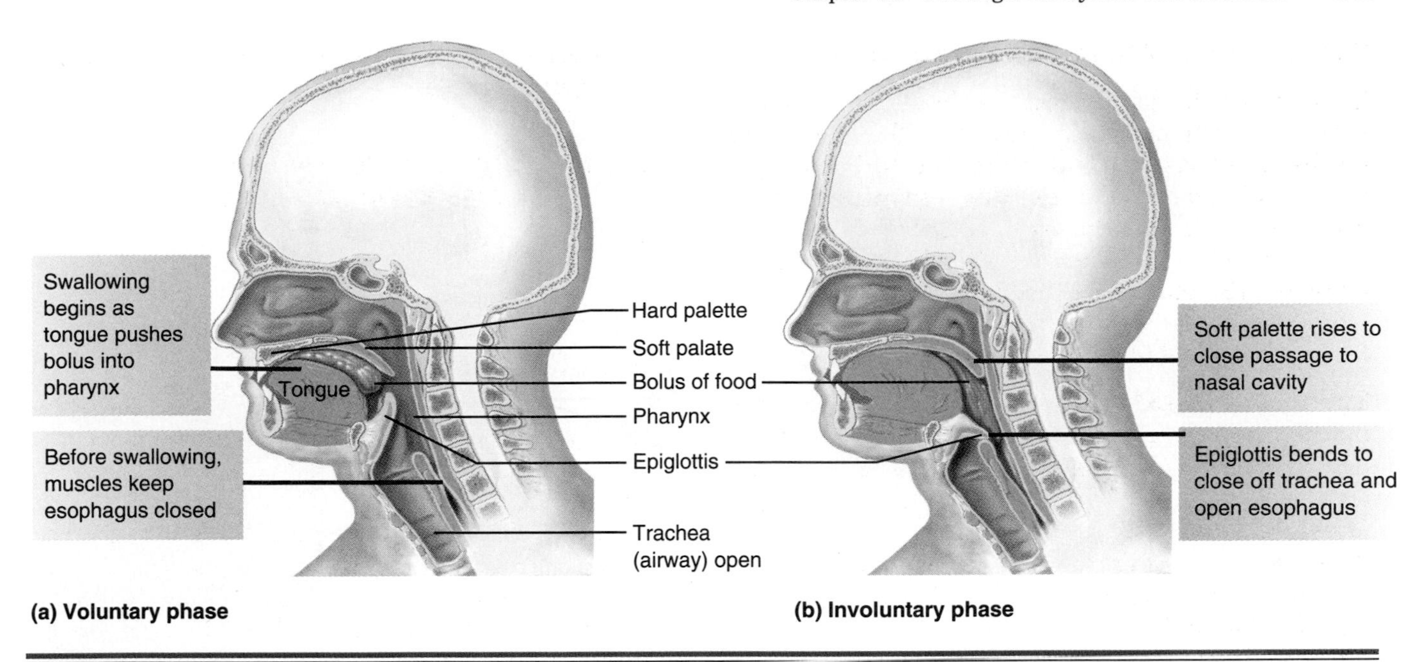

Figure 14.6 Swallowing. (a) During the voluntary phase the tongue pushes a bolus of food back against the soft palate. **(b)** During the involuntary (reflex) phase the force of the tongue causes the soft palate to rise, closing off the nasal passage. The epiglottis bends to close off the airway and the bolus slides past the epiglottis and into the esophagus.

condition in which part of the stomach protrudes upward into the chest through an opening (hiatus) in the diaphragm muscle. Mild or temporary acid reflux is usually not serious and often improves with weight loss. However, prolonged acid reflux may cause esophageal ulcers because stomach acid can erode the mucosa of the esophagus.

> ***Recap*** **Swallowing begins with voluntary movements of the tongue; the presence of food initiates an involuntary swallowing reflex. Peristalsis and gravity transfer food through the esophagus to the stomach.**

14.4 The stomach stores, digests, and regulates food delivery

The **stomach** is a muscular, expandable sac that performs three important functions:

1. **Food storage.** Humans tend to eat meals several times a day. The stomach stores the food until it can be digested and absorbed. The stomach shrinks when empty, then expands to 1–3 liters capacity when we eat.
2. **Digestion.** The stomach is more than a storage bag, however. Chemical digestion occurs because exocrine glands in the stomach add strong acid and protein-degrading enzymes to the stomach contents. Muscle contractions mix food with these secretions and push it into the small intestine.
3. **Regulation of delivery.** The stomach regulates the rate at which food is delivered to the small intestine.

Gastric juice breaks down proteins

The walls of the stomach consist of the usual four layers: mucosa, submucosa, muscularis, and serosa (Figure 14.7a). A closer look at the mucosal layer reveals millions of small openings called *gastric pits* that lead to **gastric glands** below the surface (Figure 14.7b, c). Some of the cells lining the glands secrete either hydrochloric acid (HCl) or mucus, but most secrete *pepsinogen,* a large precursor molecule that becomes a protein-digesting enzyme called **pepsin** once it is exposed to acid in the stomach. Collectively, the HCl, pepsinogen, and fluid that are secreted into the glands are known as **gastric juice.**

Typically the stomach produces one to two liters of gastric juice per day, most of it immediately after meals. The pepsin and acid in gastric juice dissolve the connective tissue in food and digest proteins and peptides into amino acids so they can be absorbed in the small intestine. The acid in gastric juice gives the stomach an acidic pH of approximately 2. The watery mixture of partially digested food and gastric juice that is delivered to the small intestine is called **chyme.** Between the stomach and the small intestine is the *pyloric sphincter,* which regulates the rate of transport of chyme into the small intestine.

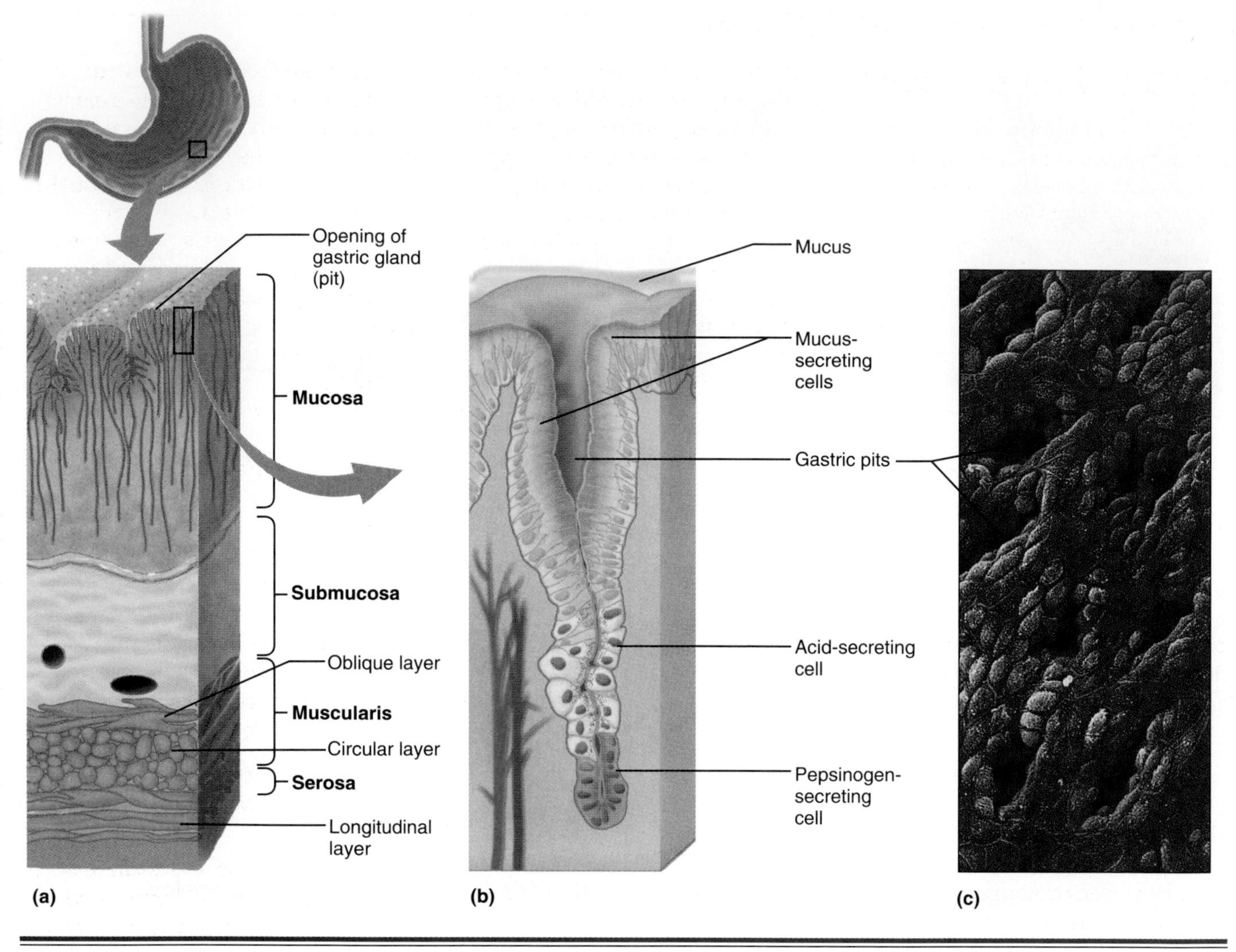

Figure 14.7 The stomach. **(a)** The structure of the stomach wall. **(b)** An enlargement of the mucosa showing the gastric glands lined with epithelial cells. **(c)** Gastric pits as seen from the lumen of the stomach (approximately ×3,700).

If gastric juice is powerful enough to digest proteins, why doesn't it digest the stomach too? The reason is that some of the cells lining the stomach and the gastric glands continuously produce a protective barrier of mucus. Normally the stomach contents are in contact with mucus, not living cells. If the mucus layer becomes damaged, however, this leaves the underlying tissue vulnerable. An open (sometimes bleeding) sore called a **peptic ulcer** (from the Greek word *peptein,* "digest") may form. Peptic ulcers occasionally occur in the esophagus and upper part of the small intestine as well.

Why doesn't the stomach digest itself?

In addition to pepsinogen, HCl, and mucus, some mucosal cells secrete *intrinsic factor,* a protein that binds to vitamin B_{12} so it can be absorbed in the small intestine. Finally, certain cells in the gastric glands secrete a hormone called *gastrin* into the bloodstream. We'll talk about gastrin when we discuss the regulation of the digestive process.

Recap **The stomach stores food, digests it, and regulates its delivery to the small intestine. Gastric juice dissolves connective tissue, large proteins, and peptides in food.**

Stomach contractions mix food and push it forward

While your stomach is empty, tonic muscle contractions keep it small. When you eat a meal, tonic contractions cease and the stomach relaxes and stretches to accommodate the food. Stretching signals peristalsis to increase.

Each wave of peristalsis starts at the lower esophageal sphincter and moves toward the pyloric sphincter, becoming stronger as it proceeds. The peristaltic wave pushes the chyme forward and then,

Directions in Science

Treating Peptic Ulcers

Peptic ulcers afflict four to five million Americans each year. Until the early 1980s doctors believed that peptic ulcers were caused by excess production of stomach acid, triggered by stress, diet, alcohol, and smoking. Many sufferers had to undergo surgery, and research focused on developing drugs that block gastric acid production. By the mid-1990s two drugs, Tagamet® and Zantac®, were prescribed for nearly 90% of all peptic ulcer patients, generating huge profits for the pharmaceutical companies that held their patents.

By the time the two drugs had been in use for 15 years, ulcer surgeries had declined by nearly 60%—an astounding success by any measure. There was just one problem: neither Tagamet® nor Zantac® prevented the recurrence of ulcers. Ulcer relapse rates were as high as 95% over two years, meaning that patients had to treat new ulcers over and over again.

Meanwhile, in 1982 a group of Australian medical researchers discovered that the mucosal lining of ulcer patients' stomachs was infected with a previously unknown bacterium, which they named *Helicobacter pylori*. They proposed that *H. pylori* actually causes peptic ulcers and that antibiotics would cure them.

So why did it take until 1995 for an antibiotic treatment for *H. pylori* to be approved? Some researchers point to the pharmaceutical industry, which at first showed little interest in the bacterium. This may be because the drugs required to treat it, a combination of bismuth and two antibiotics, were already available and relatively inexpensive (less than $200 per person for a complete 1-week course of treatment). Another reason for the delay was healthy scientific skepticism. The *H. pylori* theory represented a major departure from the status quo, and the scientific community expects any new discovery to be repeated and confirmed by other researchers.

Recent research does confirm the *H. pylori* hypothesis. One study reports that peptic ulcers recur in only 8% of ulcer patients treated with antibiotics that eradicate *H. pylori*, compared to 86% in patients not receiving the antibiotics. Indeed, the antibiotic regimen could well cure ulcers *and* save $600 to $800 million per year in total health-care costs!

And what about the role of stomach acid, so long believed to be the culprit? It turns out that acidity is a necessary condition for ulcers to form but does not cause them. Stomach acidity is normal, in fact, in most ulcer patients. Tagamet® and Zantac® help because they reduce stomach acid and soothe existing sores, but continuing *H. pylori* infection allows ulcers to redevelop.

when there is nowhere else for it to go, backward again in a squeezing, mixing motion (Figure 14.8). Each contraction propels about a tablespoon of chyme into the small intestine before the pyloric sphincter closes momentarily. A peristaltic contraction occurs every 15–25 seconds. Have you ever noticed your stomach gurgling at this rate after a meal? You're hearing peristalsis in action.

It takes two to six hours for the stomach to empty completely after a meal. Most forceful when the stomach is full, peristalsis declines as the stomach empties. Chyme with a high acid or fat content stimulates the release of hormones that slow stomach peristalsis, giving the small intestine more time to absorb the nutrients.

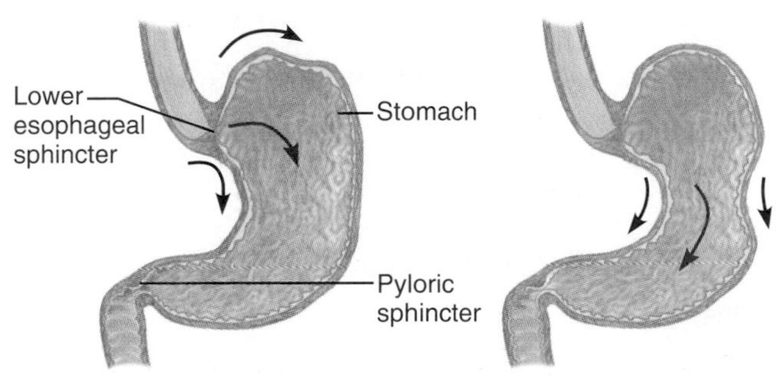

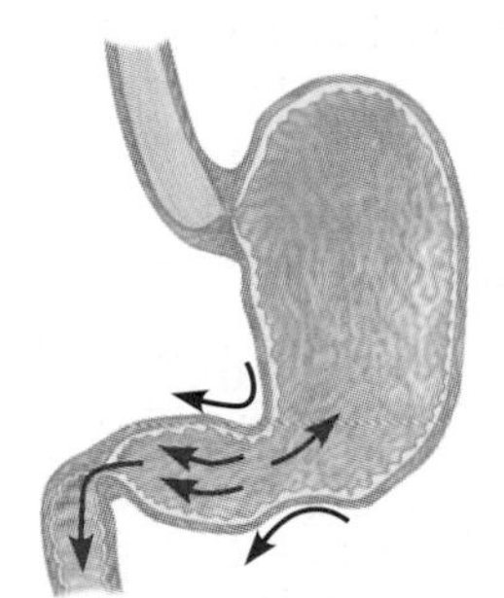

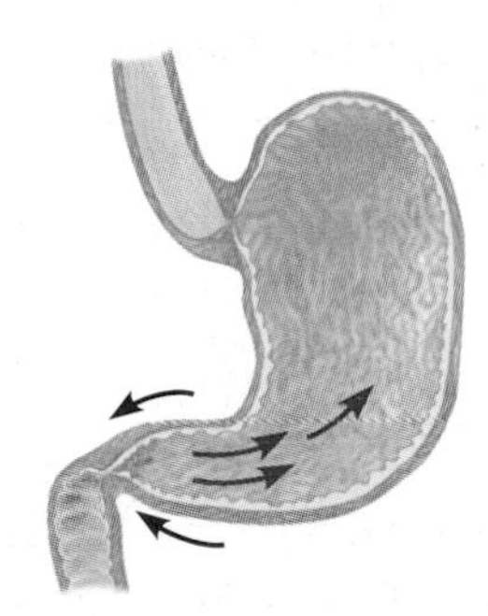

Key: ⟶ = Location and direction of peristalsis
⟶ = Movement of stomach contents

Figure 14.8 Peristalsis of the stomach. A peristaltic wave of contraction occurs in the stomach approximately every 15–25 seconds. Peristalsis mixes the contents of the stomach and forces a small amount of chyme into the small intestine with contraction.

The stomach essentially does not absorb nutrients because it lacks the required cellular transporting mechanisms and its inner lining is coated with mucus. Exceptions to this rule are alcohol and aspirin, both of which are small lipid-soluble substances that can cross the mucus barrier and be absorbed into the bloodstream directly from the stomach. When the stomach contains food, alcohol is absorbed more slowly. Drinking "on an empty stomach" hastens absorption and its intoxicating effects.

> ***Recap*** **The presence of food stretches the stomach and increases peristalsis. Peristaltic contractions mix the chyme and push it gradually into the small intestine.**

14.5 The small intestine digests food and absorbs nutrients

The process of digestion continues in the **small intestine,** so named because it is smaller in diameter (about 2.5 centimeters or 1 inch) than the final segment of the digestive tract, the large intestine. The small intestine has two major functions:

1. **Digestion:** The stomach partially digests proteins to smaller peptides, under the influence of strong acids and pepsin. Protein digestion continues in the small intestine, but here we also digest carbohydrates and lipids. Digestion of proteins, carbohydrates, and lipids in the small intestine involves neutralizing the highly acidic gastric juice and adding additional digestive enzymes from the intestine and pancreas.
2. **Absorption:** Eventually the proteins, carbohydrates, and lipids in food are broken down to single amino acids, monosaccharides, fatty acids, and glycerol. These are small enough to be transported across mucosal cells into the blood. Approximately 95% of the food you eat is absorbed in the small intestine.

The small intestine has a large surface area for absorption

The small intestine consists of three different regions. The first region, the **duodenum,** is only about 10 inches long, but it is here that most of the digestion takes place. The products of digestion are absorbed primarily in the other two segments, the *jejunum* and the *ileum,* which together are about 10 feet long.

Spotlight on Structure & Function The structure of the small intestine wall makes it well suited for absorption (Figure 14.9). The mucosa contains large folds covered with microscopic cytoplasmic projections called **villi** (singular *villus*). Each epithelial cell of the villi has dozens of even smaller projections called *microvilli.* Combined, the folds, villi, and microvilli enlarge the surface area of the small intestine by more than 500 times, increasing its ability to absorb nutrients. At the center of each

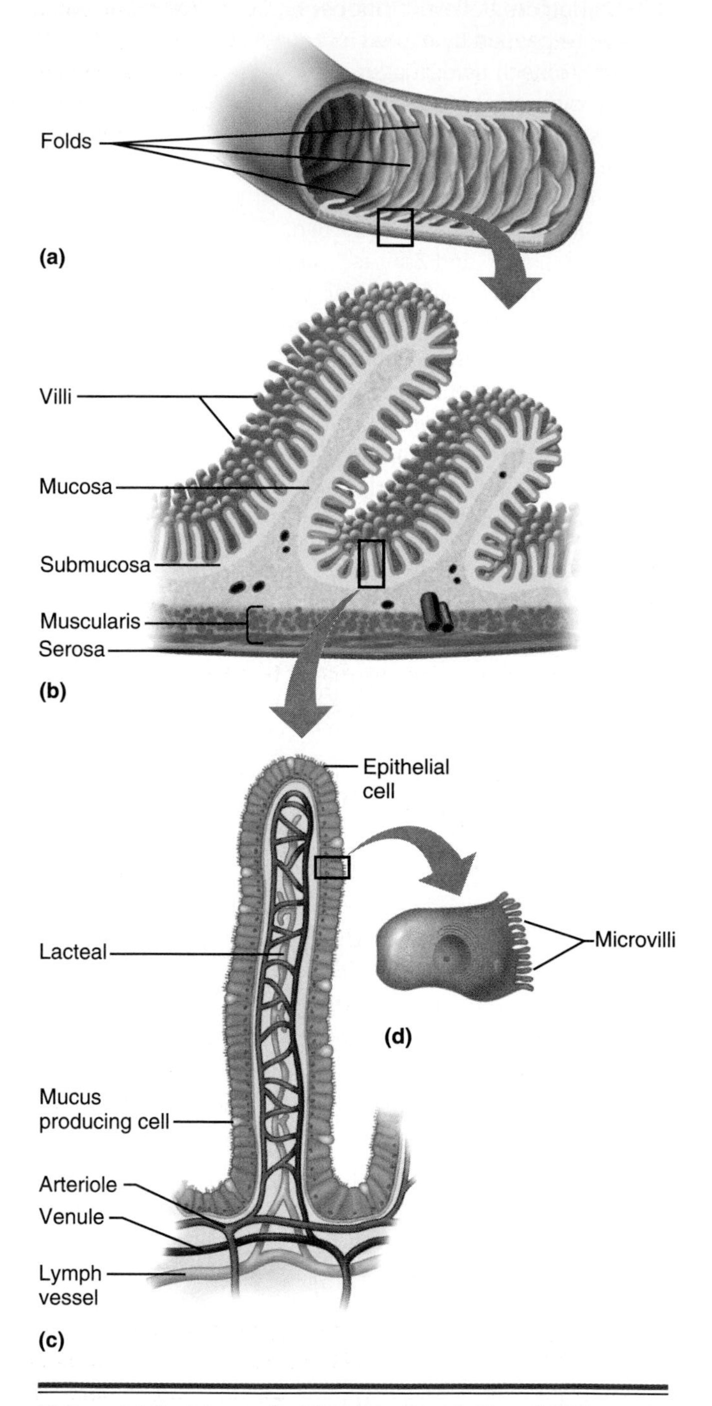

Figure 14.9 The wall of the small intestine. (a) The wall of the small intestine contains numerous folds that increase its surface area. **(b)** Each fold is covered with smaller folds called villi. **(c)** A close-up view of a villus showing the single layer of mucosal cells covering the surface and the centrally located lymph vessel (lacteal) and blood vessels. **(d)** A single cell with its outer membrane of microvilli.

villus are capillaries and a small lymph vessel called a **lacteal,** which transport nutrients to larger blood vessels and lymph vessels. The villi and microvilli give the mucosal surface a velvety appearance, which is why it is sometimes called the "brush border." ■

> ***Recap*** **The small intestine has two major functions: (1) digesting proteins, carbohydrates, and lipids; (2) absorbing approximately 95% of the nutrients we consume. Projections called villi in the mucosa increase the small intestine's surface area for absorption.**

14.6 Accessory organs aid digestion and absorption

As mentioned earlier, the digestive system has four accessory organs: salivary glands, pancreas, gallbladder, and liver. We have already discussed the salivary glands. Now, before discussing absorption in more detail, we need to look at the roles of the other three accessory organs.

Pancreas secretes enzymes and $NaHCO_3$

The pancreas, an elongated organ that lies just behind the stomach, has both endocrine and exocrine functions (Figure 14.10). As we saw in the last chapter, as an endocrine gland it secretes hormones that regulate blood glucose levels. In this chapter we focus on its exocrine role. The pancreas produces and secretes:

1. **Digestive enzymes:** Pancreatic enzymes include *proteases* (enzymes that digest proteins), such as trypsin, chymotrypsin and carboxypeptidase; *pancreatic amylase,* which continues the digestion of carbohydrates only partially accomplished by salivary amylase; and *lipase,* a lipid-digesting enzyme.
2. **Sodium bicarbonate ($NaHCO_3$):** functions to neutralize stomach acid. This is important because unlike pepsin, which is effective in acid conditions, pancreatic enzymes work best at a more neutral pH.

Two *pancreatic ducts* deliver these secretions to the duodenum, where they facilitate the process of digestion. Table 14.1 lists major digestive enzymes in the gastrointestinal tract.

> ***Recap*** **The pancreas assists digestion by producing (1) digestive enzymes and (2) sodium bicarbonate to neutralize stomach acid and make enzymes more effective.**

Liver produces bile and performs many other functions

The **liver** is a large organ located in the upper right abdominal cavity. The liver performs many significant functions, some of which are associated with digestion.

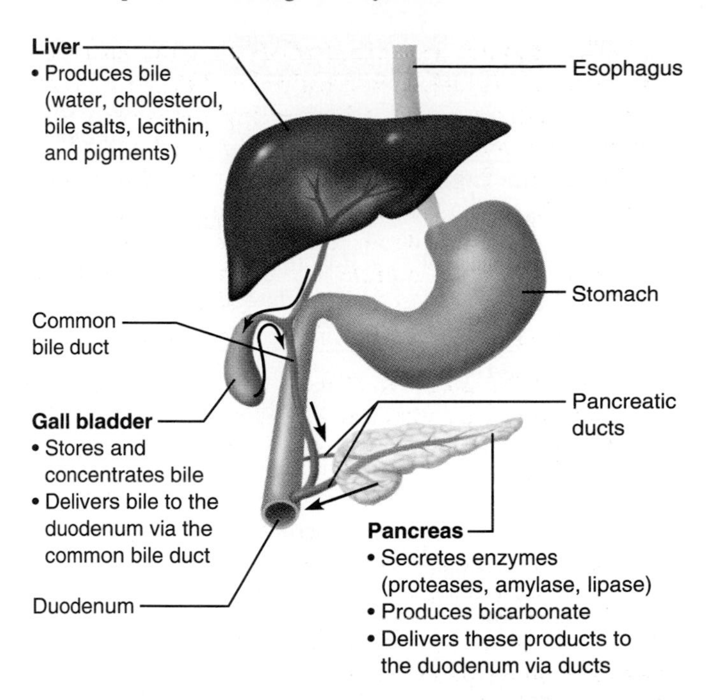

Figure 14.10 Locations and digestive functions of the liver, gallbladder, and pancreas. These three accessory organs are connected to the small intestine via ducts. The common bile duct joins one of the two pancreatic ducts just prior to where the latter enters the duodenum. Arrows indicate direction of movement of secretory products.

In terms of digestion, the liver's primary function is to facilitate the digestion of lipids via the production of **bile.** Bile is a watery mixture of bicarbonate, cholesterol, *bile salts* derived from cholesterol, and pigments derived from the breakdown of hemoglobin. The bile salts and lecithin in bile *emulsify* lipids, that is, they break them into smaller and smaller droplets. Eventually the droplets are small enough to be digested by lipases (lipid-digesting enzymes) from the pancreas and intestine.

An important feature of the vascular anatomy of the GI tract is that all of the nutrient-rich venous blood leaving the GI tract goes directly to the liver via the **hepatic portal system** before entering the general circulation (Figure 14.11). A *portal system* carries blood from one capillary bed to another: in this case it carries blood from the digestive organs to the liver (*hepatos* is the Greek word for "liver"). The smaller veins of the hepatic portal system join together to become the *hepatic portal vein,* which enters the liver. Therefore the liver is ideally located to begin processing and storing nutrients for the body just as soon as digestion and absorption have occurred. After passing through the liver, blood from the GI tract enters the general circulation via two *hepatic veins.*

Table 14.1 Major enzymes of digestion

Enzyme	Source	Where Active	Substance Digested	Breakdown Products
Carbohydrate Digestion				
Salivary amylase	Salivary glands	Mouth	Polysaccharides	Disaccharides
Pancreatic amylase	Pancreas	Small intestine	Polysaccharides	Disaccharides
Intestinal enzymes	Small intestine	Small intestine	Disaccharides	Monosaccharides (e.g., glucose)
Protein Digestion				
Pepsin	Stomach	Stomach	Proteins	Peptides
Trypsin	Pancreas	Small intestine	Proteins	Peptides
Chymotrypsin	Pancreas	Small intestine	Proteins	Peptides
Carboxypeptidase	Pancreas	Small intestine	Peptides	Amino acids
Intestinal enzymes	Small intestine	Small intestine	Peptides	Amino acids
Lipid Digestion				
Lipase	Pancreas	Small intestine	Tryglycerides	Free fatty acids, glycerol

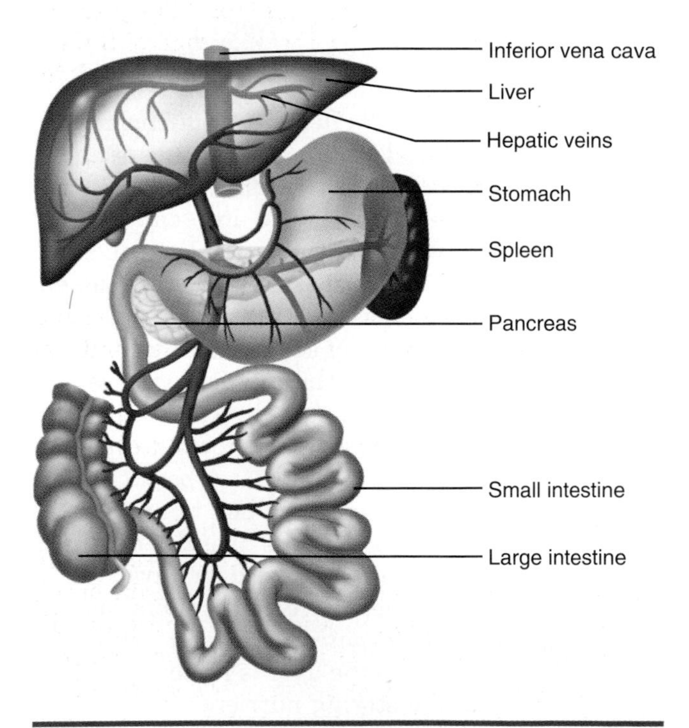

Figure 14.11 The hepatic portal system. The hepatic portal system carries nutrient-rich venous blood from the stomach, intestines, pancreas, and spleen directly to the liver before it is returned to the general circulation. Only the first part of the large intestine is shown.

The liver serves a number of other functions that maintain homeostasis. The liver:

- Stores fat-soluble vitamins (A, D, E, and K) and iron
- Stores glucose as glycogen after a meal, and converts glycogen to glucose between meals
- Manufactures plasma proteins, such as albumin and fibrinogen, from amino acids
- Synthesizes and stores some lipids
- Inactivates many chemicals, including alcohol, hormones, drugs, and poisons
- Converts ammonia, a toxic waste product of metabolism, into less toxic urea
- Destroys worn-out red blood cells

Because of its central role in so many different functions, liver injury can be particularly dangerous. Overexposure to toxic chemicals, medications, or alcohol can damage the liver because it takes up these substances to "detoxify" them, killing some liver cells in the process. Long-term exposure, such as prolonged alcohol abuse, can destroy enough cells to permanently impair liver function, a condition known as *cirrhosis*.

The gallbladder stores bile until needed

The bile produced by the liver flows through ducts to the **gallbladder.** The gallbladder concentrates bile by removing most of the water and stores it until after a meal, when it is secreted into the small intestine via the bile duct, which joins the pancreatic duct.

Recap **Bile is produced by the liver and stored in the gallbladder until after a meal. The liver also produces plasma proteins; inactivates toxic chemicals; destroys old red blood cells; stores vitamins, iron, and certain products of metabolism; and performs other functions important for homeostasis.**

14.7 The large intestine absorbs nutrients and eliminates wastes

By the time the contents of the digestive tract reach the large intestine, most nutrients have been absorbed. The **large intestine** absorbs the last of the remaining water, ions, and nutrients and stores the now nearly solid waste material until it can be eliminated.

The large intestine is larger in diameter than the small intestine (about 6.5 centimeters or 2.5 inches) but shorter (approximately 1.5 meters [5 feet] compared to 3 meters). It begins at a pouch called the *cecum,* which receives the chyme from the small intestine (Figure 14.12). A small fingerlike pouch, the *appendix,* extends from the cecum. The appendix has no known digestive function, but we become acutely aware of its presence if it becomes inflamed or infected, a condition called *appendicitis.*

From the cecum the large intestine continues as the **colon.** The colon is subdivided into sections called the *ascending colon,* the *transverse colon,* and the *descending colon,* the names reflecting their positions in the abdominal cavity. The short *sigmoid colon* joins the **rectum,** the passageway through which the waste materials, now called *feces,* pass as they are eliminated.

In addition to indigestible material, the feces contain about 30% bacteria by weight. Many strains of bacteria thrive on the leftover material in the colon that we cannot digest. Some of these bacteria release byproducts that are useful to us, such as vitamin K, which is important for blood clotting. They also produce less helpful substances such as intestinal gas, a by-product of their metabolism as they break down food.

During *defecation* feces pass from the rectum, through the anal canal, and out of the body through the **anus,** the terminal opening of the digestive tract. The neural *defecation reflex* controls this process. Normally the anus is kept closed by tonic contraction of the *internal anal sphincter,* a ring of smooth muscle. The presence of feces stretches the rectum, stimulating sensory receptors that send impulses to the spinal cord. Nerves signal the internal anal sphincter to relax and the rectum to contract, expelling the feces.

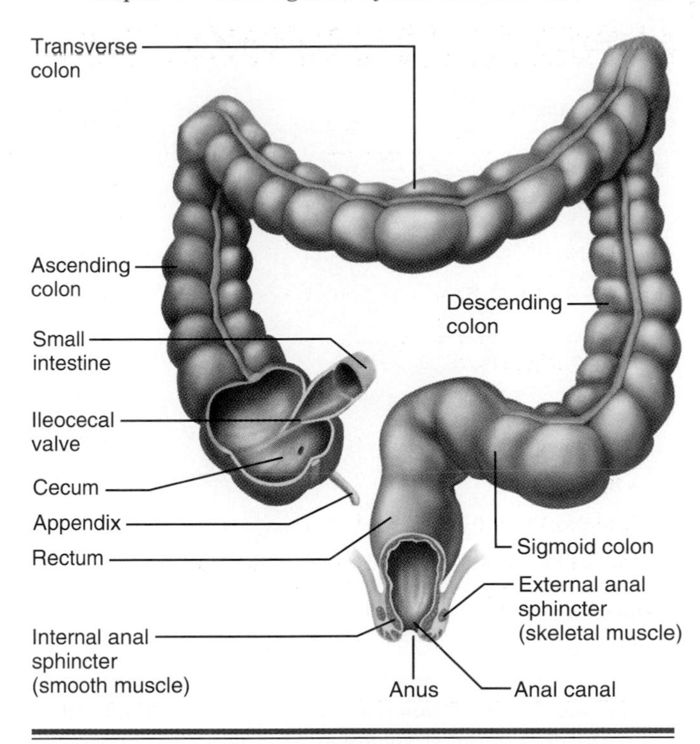

Figure 14.12 The large intestine. The large intestine begins at the cecum and ends at the anus.

We can prevent defecation by voluntarily contracting the *external anal sphincter,* a ring of skeletal muscle under our conscious control. "Potty training" of children involves learning to control the external anal sphincter. Until that time children defecate whenever the defecation reflex occurs.

Recap **The large intestine absorbs water, ions, and some nutrients, and stores feces until defecation occurs.**

14.8 How nutrients are absorbed

Once food has been digested, how are its nutrients absorbed into the body? The mechanism depends on the type of nutrient.

Proteins and carbohydrates are absorbed by active transport

In the small intestine, enzymes from the pancreas and enzymes secreted by the mucosal layer of the stomach and from the small intestine itself break down proteins into amino acids, completing the digestion of proteins that was begun in the stomach. The amino acids are then actively transported into the mucosal cells. Eventually they move by facilitated diffusion out of the mucosal cells and make their way into the capillaries (Figure 14.13).

Carbohydrate digestion begins in the mouth, where salivary amylase breaks down polysaccharides into disaccharides (review Table 14.1). It is completed in the small intestine with the addition of pancreatic

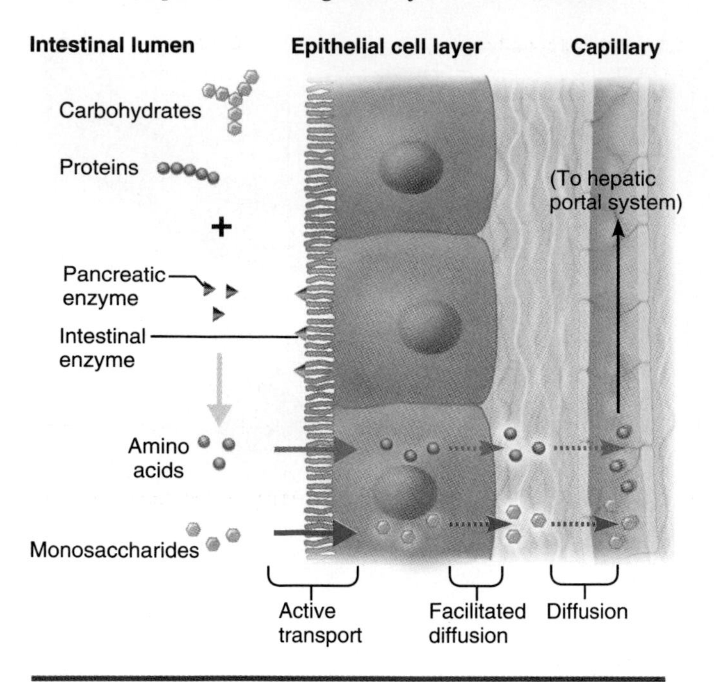

Figure 14.13 Digestion and absorption of proteins and carbohydrates in the small intestine. Digestion depends on enzymes from the pancreas and enzymes attached to the surface of the epithelial cells of the intestine. The amino acid and carbohydrate products of digestion cross the epithelial cell layer of the mucosa and enter a capillary for transport to the liver.

amylase and enzymes from the small intestine. Together these enzymes break the remaining carbohydrates into monosaccharides (simple sugars such as glucose). Monosaccharides follow transport pathways that are similar to those for amino acids. However, they utilize different active transport proteins for the active transport step.

Recap **Proteins and carbohydrates are absorbed from the lumen of the small intestine by active transport processes, then move by facilitated diffusion into capillaries.**

Lipids are broken down, then reassembled

Recall that bile salts emulsify lipids into small droplets, which are then digested by pancreatic and intestinal lipases. The products of lipid digestion are fatty acids and monoglycerides. Because these substances are not water soluble, if left alone they would float on top of the watery chyme and remain unavailable to cells. Instead, the fatty acids and monoglycerides cling to bile salts, cholesterol, and other substances, forming small droplets called *micelles*. Micelles are small enough to come into close contact with mucosal cells, allowing the fatty acids and monoglycerides to diffuse out of the micelles and into the cells (Figure 14.14).

Once inside the cells, the fatty acids and monoglycerides recombine into triglycerides. Clusters of triglycerides are then coated with proteins to form water-soluble droplets called *chylomicrons*. Chylomicrons are released from the cell by exocytosis. However they are too large to enter the capillaries directly. Instead, they enter the more permeable lacteals and travel in the lymph vessels until the lymph is returned to the venous blood vessels near the heart.

Water is absorbed by osmosis

Most of the water we ingest is absorbed in the small intestine by osmosis, right along with all of the nutrients being absorbed. The ability of the small intestine to absorb almost limitless quantities of water explains why you do not experience diarrhea every time you drink a lot of fluid.

Water reabsorption continues in the large intestine, but here the capacity for water reabsorption is more limited. Conditions that cause the small intestine to deliver too much chyme to the large intestine or that speed up the rate of movement through the large intestine can lead to *diarrhea* (watery feces). A common cause of diarrhea is a bacterial infection in the small intestine.

The opposite problem is *constipation,* in which chyme remains in the colon so long that too much water is reabsorbed. The feces become dry and hard, making defecation difficult. Constipation can result from stress, lack of exercise, or insufficient fiber (indigestible material) in the diet.

Vitamins and minerals follow a variety of paths

How vitamins are absorbed depends on whether they are fat soluble or water soluble. Fat-soluble vitamins dissolve in the micelles and are absorbed by diffusion across the lipid membrane of the mucosal cell layer, just like the components of lipids. Water-soluble vitamins are absorbed either by active transport or diffusion through channels or pores. Minerals (ions) such as sodium, potassium, calcium, phosphate, sulfate, and magnesium are electrically charged and hence not lipid soluble. These are either actively transported or are absorbed by diffusion via specific transport proteins, pores, or channels.

In addition to the nutrients we ingest in our food, the body digests and reabsorbs the components of the digestive secretions themselves. Nearly 9 liters of gastric juice, pancreatic juice, digestive enzymes, and bile are produced every day. The water and minerals in the digestive secretions are reabsorbed by the normal mechanisms for these nutrients. Enzymes are digested to their component amino acids and the amino acids are then reabsorbed. Bile salts are reabsorbed, returned to the liver, and used again.

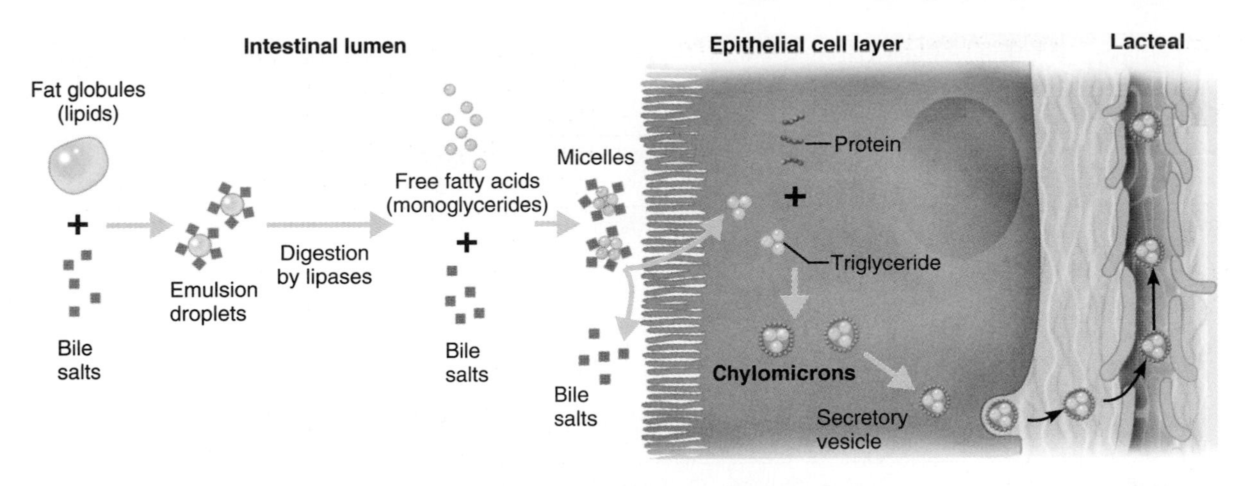

Figure 14.14 Digestion and absorption of fats in the small intestine. Bile salts emulsify large fat droplets into smaller droplets so that lipases can digest the fats into free fatty acids and monoglycerides. The free fatty acids and monoglycerides join with bile salts to become micelles. Micelles near the cell's surface release their free fatty acids and monoglycerides, which then diffuse across the cell membrane. In the cell the free fatty acids and monoglycerides are resynthesized to triglycerides and coated with protein. The protein-coated triglycerides are called chylomicrons. Chylomicrons exit the cell by exocytosis and enter a lacteal for transport to the circulation.

Recap **The components of lipid digestion are transported to the mucosa in micelles, diffuse into the cell, and recombine into lipids within the cell. Then they are coated with protein to become chylomicrons that enter the lymph. The digestive system also absorbs water, vitamins, minerals, and digestive secretions.**

14.9 Endocrine and nervous systems regulate digestion

The digestive system is most active when food or chyme is present and fairly inactive when it is not. Regulation involves altering the motility and secretions of various organs so that each operates as efficiently as possible.

Most regulatory mechanisms operate to maintain a constant internal environment. Regulation of the digestive system, on the other hand, promotes rapid, efficient digestion and absorption of whatever is delivered to the system, regardless of homeostasis. The digestive process actually alters the internal environment temporarily because all those absorbed nutrients enter the blood in a relatively short time—a few hours.

Regulation depends on volume and content of food

The endocrine system and nervous system regulate digestion according to both the volume and content of food (Table 14.2). Because most digestion and absorption occur in the stomach and small intestine, most regulatory processes involve those organs.

When the stomach stretches to accommodate food, neural reflexes increase stomach peristalsis and secretion of gastric juice. Stretching and the presence of protein stimulate release of the hormone **gastrin,** which stimulates the release of more gastric juice.

When chyme arrives at the small intestine, stretching of the duodenum increases segmentation to mix the chyme. The duodenum also secretes two hormones into the bloodstream: secretin and cholecystokinin. Acid in chyme triggers release of **secretin,** which stimulates the pancreas to secrete water and bicarbonate to neutralize acid. Fat and protein stimulate release of **cholecystokinin (CCK),** which signals the pancreas to secrete more digestive enzymes. CCK and stretching of the duodenum also stimulate the gallbladder to contract and release bile.

Secretin, CCK, and stretching of the small intestine inhibit stomach motility and stomach secretions. In other words, if chyme flows too quickly from the stomach, the small intestine will slow stomach activity accordingly.

Gastrin and a neural reflex involving stretching of the stomach increase motility of the large intestine after eating. This is why people often feel the urge to defecate after their first meal of the day.

Table 14.2 Nervous and endocrine systems regulation of the digestive processes

Organ	Nervous System	Endocrine System
Stomach	Stretching increases peristalsis secretion of gastric juice	Stretching and protein trigger release of gastrin, which stimulates gastric glands
Small intestine	Stretching increases segmentation Stretching inhibits stomach motility and stomach secretions	Acid stimulates release of secretin, which stimulates pancreas to secrete bicarbonate, neutralizing acid Fat and protein stimulate release of CCK, which stimulates pancreas to secrete enzymes and the gallbladder to release bile CCK and secretin inhibit stomach motility and stomach secretions
Large intestine	Stomach stretching increases motility	Gastrin increases motility

Recap **The endocrine and nervous systems regulate digestion based on content and volume of food. Regulatory mechanisms include neural reflexes involving organ stretching, and release of the hormones gastrin, secretin, and cholecystokinin.**

Nutrients are utilized or stored until needed

Once all those nutrients have been absorbed, the body must utilize them. Will they be consumed for immediate energy, stored until later, or combined with more nutrients to create other molecules? Regulation of organic metabolism involves interactions between virtually all organs in the body. Key organs are the pancreas with its two main endocrine secretions (insulin and glucagon) and the liver with its multiple roles in overall metabolism.

Figure 14.15 summarizes how the body utilizes organic molecules. Three concepts are evident from this figure:

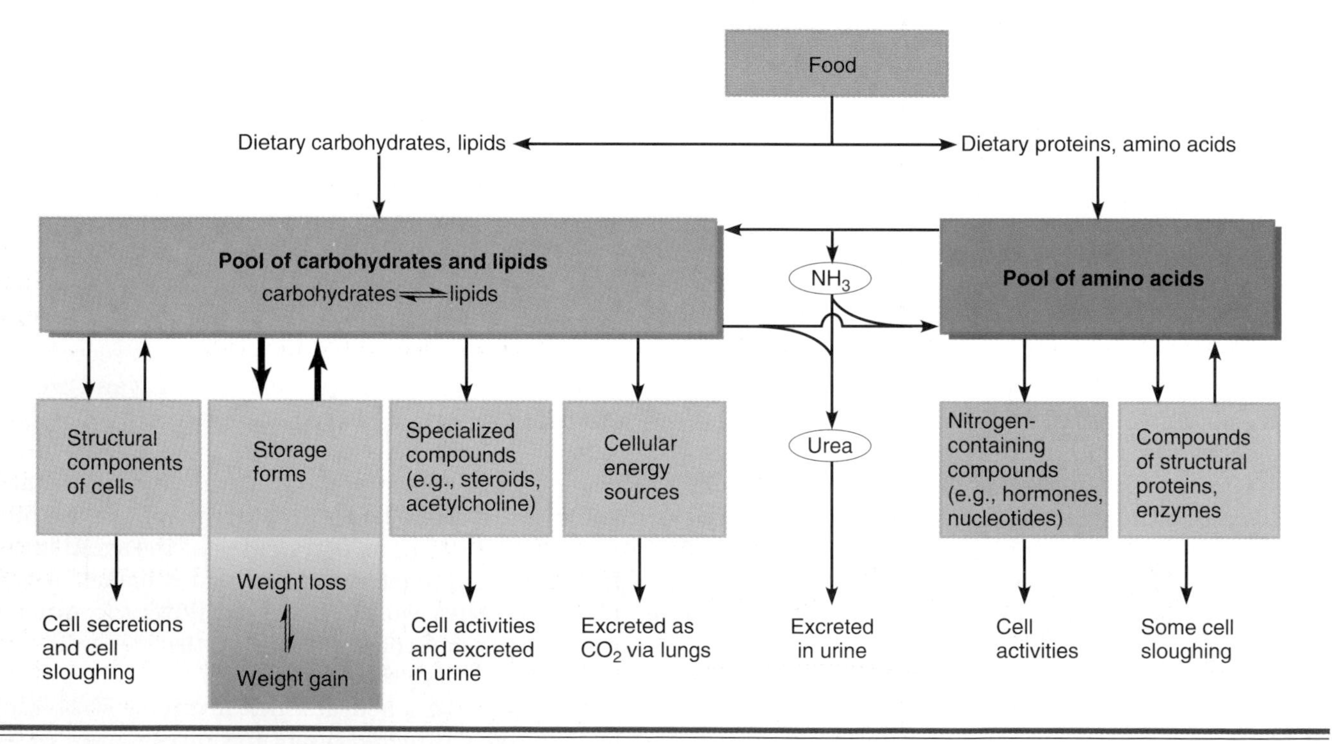

Figure 14.15 The pathways of organic metabolism. When energy intake is greater than utilization the excess is stored, often as fat. When energy intake is less than utilization the storage forms are used to produce additional energy.

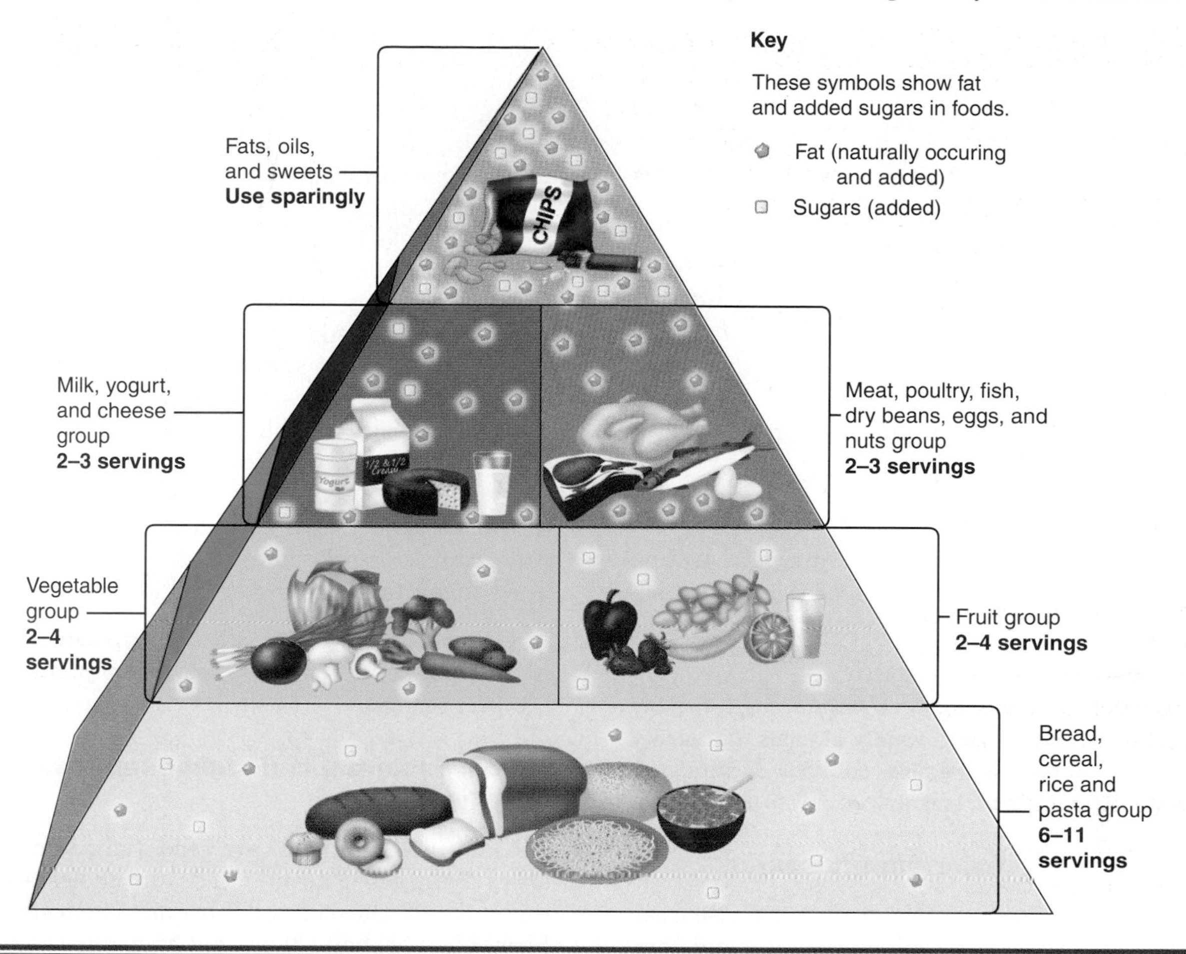

Figure 14.16 The Food Guide Pyramid. The Food Guide Pyramid is a visual representation of the recommended dietary guidelines of the United States Department of Agriculture (USDA) in conjunction with the U.S. Department of Health and Human Services (USDHHS). The pyramid shows the recommended weighting of six different food groups, taking into account both nutritional needs and concerns about the effects of diet on nutrition-related disease.

1. Depending on which molecules are in short supply and which are in excess at the moment, there can be a great deal of interconversion of one to another. This concept was first introduced in connection with cellular metabolism in Chapter 3. Lipids, carbohydrates, and proteins can all be converted to storage forms of lipid or carbohydrate, then recycled according to the body's needs.
2. When we consume more energy-containing nutrients than we use, our bodies store the excess for future use. Over time, these mounting reserves can increase body weight.
3. When we consume fewer energy-containing nutrients than we use, the body draws on these storage forms of energy to make up the difference. If we do this on a regular basis, we lose weight.

In the rest of this chapter we discuss nutrition, energy balance, and how to maintain a healthy body weight.

Recap **Lipids, carbohydrates, and amino acids can all be converted to storage forms of lipid or carbohydrate. They can be re-created from stored forms according to the body's needs at that moment.**

14.10 Nutrition: You are what you eat

Nutrition is the interaction between an organism and its food. Because nearly all nutrients enter the body through the digestive system, it is fair to say that "you are what you eat." What nutrients do we actually need, and why?

Figure 14.16 presents one answer as provided by the U.S. Department of Agriculture (USDA) and the Department of Health and Human Services. This is the *Food Guide Pyramid,* which divides foods into six different groups and shows the recommended weighting of each.

What should a healthy diet contain?

Other recommendations for a healthy diet include:

- Eat a variety of foods.
- Maintain a healthy weight.
- Eat plenty of fruits, vegetables, and grain products.
- Choose a diet low in cholesterol and saturated fat.
- Use sugar in moderation.
- Consume salt and sodium in moderation.
- If you drink alcoholic beverages, do so in moderation.

We consider the three basic nutrient groups first—carbohydrates, lipids, and proteins—before discussing vitamins, minerals, and the importance of fiber.

Recap **Good nutrition includes maintaining a healthy weight while consuming a variety of foods. Eat plenty of fruit, vegetables, and grains; cut back on saturated fat; and use sugar, salt, and alcohol only in moderation.**

Carbohydrates: A major energy source

Carbohydrates are one of the body's main sources of energy, and many nutritionists recommend that they represent approximately 60 percent of the Calories we consume. (A Calorie is a measure of energy, discussed in more detail in Section 14.11.)

Carbohydrates may be either simple or complex (see Table 14.3). *Simple carbohydrates* (sugars) are found in many natural foods such as fruit and honey. Refined sugars, such as granulated sugar and corn syrup, have had most other nutrients removed, so they are less nutritious than their natural counterparts. *Complex carbohydrates,* such as starch or glycogen, consist of many sugar units linked together. Whole foods containing complex carbohydrates are better for us than refined sugars because they release sugars more slowly and also contribute fiber, vitamins, and minerals. In the body, stored starch and glycogen are broken down to glucose, one of the premier sources of energy.

It is difficult to tell exactly how much refined sugar we eat because so much is disguised on food labels as "corn sweeteners," "dextrose," or "fructose" (levulose). The USDA estimates the average North American consumes more than 120 pounds of refined sugar annually—over 2 pounds a week.

Table 14.3 Good sources of natural carbohydrates

Carbohydrates	
Simple	**Complex**
Fruit	Starchy vegetables
Citrus	Potatoes
Berries	Corn
Melons	Peas
Unsweetened fruit juice	Beans
Honey	Grains
	Whole-grain breads and cereals
	Whole-grain pasta
	Rice
	Oatmeal
	Wheat bran

Recap **Carbohydrates are a good source of energy. Natural sugars and complex carbohydrates are the best sources.**

Lipids: Essential cell components and energy sources

Lipids (including fats) are essential components of every living cell. Phospholipids and cholesterol make up most of the cell membrane. Cholesterol also forms the backbone of steroid hormones and is used to synthesize bile. Fat stores energy, cushions organs, insulates the body under the skin, and stores several vitamins.

Most of the fats in food are triglycerides, which consist of three fatty acids attached to a glycerol molecule. As you learned in Chapter 2, fats are classified according to the ratio of hydrogen to carbon atoms in their fatty acids:

1. *Saturated fats:* There are two hydrogen atoms for every carbon atom in the fatty acid tails of saturated fats. They tend to be solids at room temperature. They are found in meat and dairy products and in a few plant sources such as coconut and palm kernel oil.
2. *Unsaturated fats:* There are fewer than two hydrogen atoms for every carbon atom in the fatty acid tails of unsaturated fats. Unsaturated fats are likely to be liquids (oils) at room temperature. Examples include plant oils—such as olive, safflower, canola, and corn—and fish oil.

Although lipids are essential to human health, most of us consume far more than we need. Nutritionists recommend limiting lipids to 10–30% of Calories consumed per day, but the average North American diet derives more than 35 percent of its Calories from fat. In addition to watching the amount of fat we eat, it's important to pay attention to the type. Diets high in saturated fat and cholesterol are a risk factor for cardiovascular disease and certain cancers. On the other hand, some unsaturated fats appear to decrease the risk of heart disease, perhaps by lowering blood cholesterol levels.

The liver can synthesize many lipids, including cholesterol. The body cannot, however, produce all of the fats it requires, and so these *essential fatty acids* must be consumed as food. An example is *linoleic acid,* present in corn and olives. Here again we do not need much: a teaspoon of corn or olive oil, or its equivalent in food, satisfies our daily requirement.

Recap **Some essential fatty acids must be consumed to meet the body's nutritional requirements. Saturated fats are found mostly in meat and dairy products; unsaturated fats include plant and fish oils. Unsaturated fats are a healthier choice.**

Complete proteins contain every amino acid

Like lipids, **proteins** are vital components of every cell. They constitute the enzymes that direct metabolism, they serve as receptor and transport molecules, and they make up our muscle fibers. A few are hormones.

Despite their variety, all proteins are composed of just 20 different *amino acids.* The body is capable of making 12 of these amino acids. The eight that the body cannot produce (isoleucine, leucine, lysine, methionine, phenylalanine, threonine, tryptophan, and valine) are called *essential amino acids* because they must be ingested in food. Two more that the body can make (histidine and arginine) are sometimes considered essential in children because their rapidly growing bodies cannot synthesize them fast enough.

A *complete protein* contains all 20 of the amino acids in proportions that meet our nutritional needs. Most animal proteins are complete, but nearly all plant proteins lack one or more of the essential amino acids. Vegetarians must be careful to choose the right combinations of plant-based foods to obtain all amino acids. As Figure 14.17 shows, only a few plant foods offer complete proteins. A combination of foods from the three right-most columns of Figure 14.17 can prevent protein deficiency.

Approximately 15% of our Calories should come from protein. In many parts of the world protein deficiency is a serious health problem. Every enzyme has a unique amino acid sequence, so even if only one amino acid is missing in the diet, the body may be unable to produce crucial enzymes at the proper time of development. Protein deficiencies during pregnancy and childhood can retard growth and alter physical and mental performance. Unfortunately, foods with complete proteins such as meat and milk are often expensive in poor societies, and many people remain uninformed about ways to achieve a balanced diet of plant proteins.

Recap **The body needs 20 amino acids but cannot synthesize eight of them, called essential amino acids. Complete proteins contain all 20 amino acids. Most animal proteins are complete but most plant proteins are incomplete.**

Vitamins are essential for normal function

In addition to carbohydrates, lipids, and proteins, humans need **vitamins,** a group of at least 13 other chemicals that are essential for normal functioning. The body can produce only a few vitamins; our skin synthesizes vitamin D when exposed to sunlight, and bacteria living in the colon manufacture vitamins K, B_6, and Biotin. We must obtain all others from our food and absorb them in the digestive tract. Table 14.4 lists the 13 vitamins along with their sources, functions, and signs of deficiency or excess.

Vitamins fall into two groups: fat soluble and water soluble. The distinction affects how a particular vitamin is absorbed and stored and how steady a supply of it you need. Fat-soluble vitamins are absorbed more readily if there is fat in the diet. They tend to be absorbed along with the components of fat and they are stored in fat tissue and released as needed. Body stores of some fat-soluble vitamins may last for years, so it is rarely necessary to take daily supplements.

Water-soluble vitamins are absorbed more readily than fat-soluble vitamins, but they are stored only briefly and rapidly excreted in urine. Thus we need to consume foods containing water-soluble vitamins on a regular basis.

Minerals: Elements essential for body processes

Minerals are the atoms of certain chemical elements that are also essential for body processes. They are the ions in blood plasma and cell cytoplasm (sodium, potassium, chloride, and many others). They represent most of the chemical structure of bone (calcium, phosphorus, and oxygen). They also contribute to the activity of nerves and muscles (sodium, potassium, and calcium), among many other functions. Twenty-one minerals are considered essential for animals. Nine of the 21 (arsenic, chromium, cobalt, manganese, molybdenum, nickel, selenium, silicon, and vanadium) are called *trace minerals* because they represent less than 0.01% of

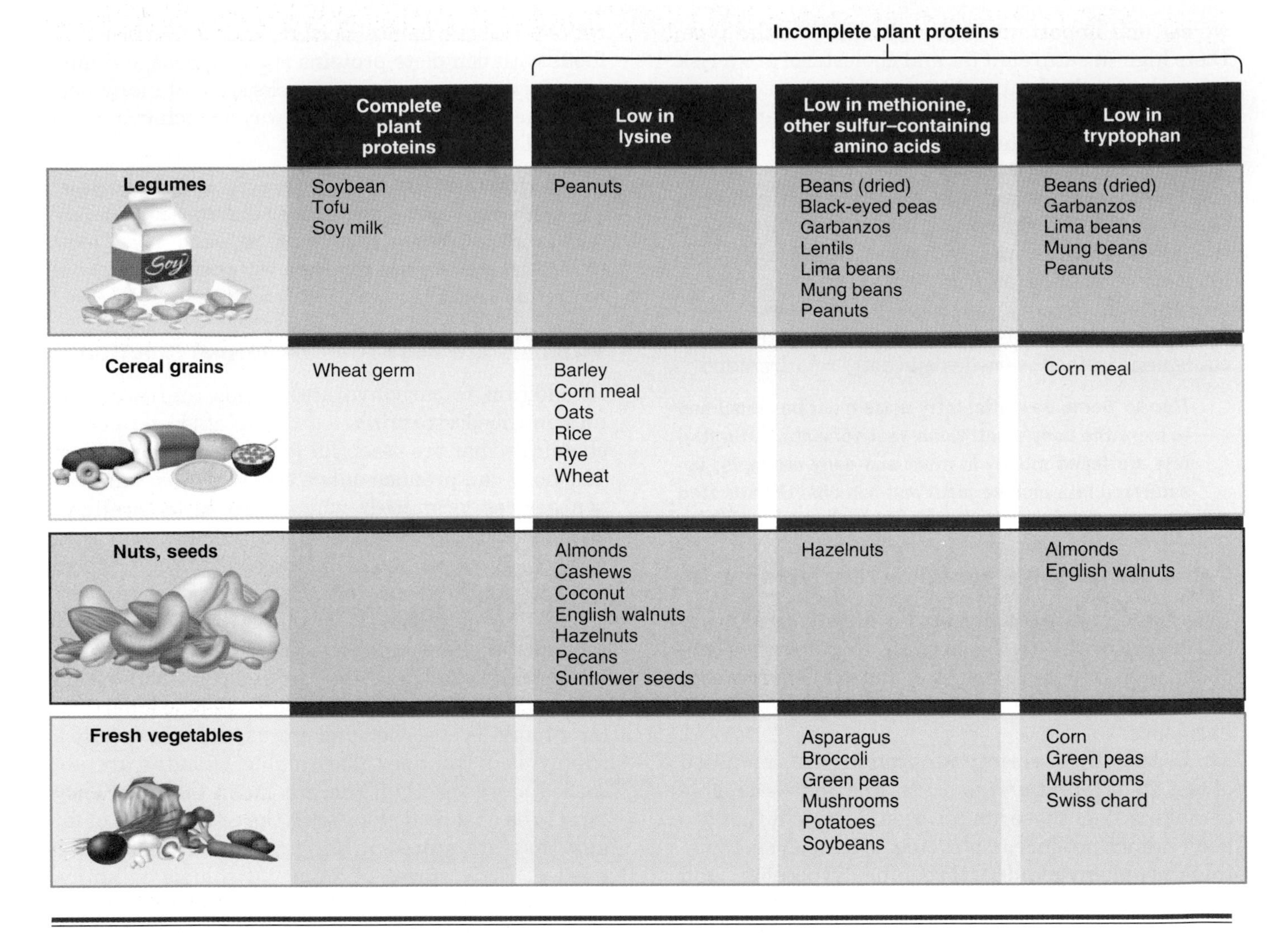

		Incomplete plant proteins		
	Complete plant proteins	Low in lysine	Low in methionine, other sulfur–containing amino acids	Low in tryptophan
Legumes	Soybean Tofu Soy milk	Peanuts	Beans (dried) Black-eyed peas Garbanzos Lentils Lima beans Mung beans Peanuts	Beans (dried) Garbanzos Lima beans Mung beans Peanuts
Cereal grains	Wheat germ	Barley Corn meal Oats Rice Rye Wheat		Corn meal
Nuts, seeds		Almonds Cashews Coconut English walnuts Hazelnuts Pecans Sunflower seeds	Hazelnuts	Almonds English walnuts
Fresh vegetables			Asparagus Broccoli Green peas Mushrooms Potatoes Soybeans	Corn Green peas Mushrooms Swiss chard

Figure 14.17 Getting the essential amino acids from plant foods. Most plants are low in one or more of the essential amino acids. A vegetarian should take care to learn which foods can be eaten in combination to ensure an adequate intake of all essential amino acids.

your body weight. The 12 most important minerals are listed in Table 14.5.

How much do we need of each vitamin and mineral? The National Research Council publishes the current best estimate, the **Recommended Daily Allowance (RDA).** Most healthy people can achieve the RDA without taking supplements—if they eat a balanced diet of whole foods. That's a big "if" for some of us, considering our diets. Scientific studies suggest that supplements may also benefit certain groups, such as newborns, the elderly, or people taking medications that interfere with nutrient absorption. Calcium is sometimes advised for women to prevent the postmenopausal bone loss of osteoporosis.

Never take massive doses of any vitamin or mineral unless prescribed by your doctor. As the "excess" columns show, more is not always better.

Fiber benefits the colon

Fiber—found in many vegetables, fruit, and grains—is indigestible material. Even though our bodies cannot digest it, we need a certain amount of it in our diets. Fiber is beneficial because it makes feces bulky and helps them pass more efficiently through the colon.

A low-fiber diet can lead to chronic constipation, *hemorrhoids* (swollen veins in the lining of the anus, often caused by straining during defecation), and a disorder called *diverticulosis* (see Section 14.12 on "Disorders"). A low fiber diet has also been associated with a higher risk of colon cancer, perhaps because cancer-causing substances remain in the colon longer (see Chapter 18 on Cancer). Doctors recommend eating 20–35 grams of fiber every day—considerably more than the average 12–17 grams that most North Americans consume.

Table 14.4 Vitamins

Vitamin	Common Sources	Primary Functions	Signs of Severe, Prolonged Deficiency	Signs of Extreme Excess
Fat-Soluble Vitamins				
A	Fortified dairy products, egg yolk, liver, yellow fruits, yellow and dark green leafy vegetables	Synthesis of visual pigments and the development of bones and teeth	Night blindness, dry scaly skin, increased susceptibility to infections	Damage to bone and liver, blurred vision
D	Fish oil, fortified milk, egg yolk; also produced by skin	Promotes absorption of calcium, required for healthy bones and teeth	Bone deformities in children, bone weakening in adults	Calcium deposits in tissues, leading to cardiovascular and kidney damage
E	Vegetable oils, nuts, whole grains	Thought to be an antioxidant that prevents cell membrane damage	Damage to red blood cells and nerves	Generally nontoxic
K	Green vegetables, also produced by bacteria in colon	Formation of certain proteins involved in blood clotting	Defective blood clotting, abnormal bleeding	Liver damage and anemia
Water-Soluble Vitamins				
B_1 (thiamine)	Whole grains, legumes, eggs, lean meats	Coenzyme in carbohydrate metabolism	Damage to heart and nerves, pain	None reported
B_2 (riboflavin)	Dairy products, eggs, meat, whole grain, green leafy vegetables	Coenzyme in carbohydrate metabolism	Skin lesions	Generally nontoxic
Niacin	Nuts, meats, grain, green leafy vegetables	Coenzyme in carbohydrate metabolism	Causes a disease called pellagra (damage to skin, digestive tract, and nervous system)	Skin flushing, potential liver damage
B_6	High protein foods in general; also made by colonic bacteria	Coenzyme in amino acid metabolism	Muscle, skin, and nerve damage, anemia	Poor coordination, numbness in feet
Folic acid	Green vegetables, nuts, grains, legumes, orange juice	Coenzyme in nucleic acid and amino acid metabolism	Anemia, digestive disturbances	Masks vitamin B_{12} deficiency
B_{12}	Animal products	Coenzyme in nucleic acid metabolism	Anemia, nervous system damage	Thought to be nontoxic
Pantothenic acid	Widely distributed in foods, especially animal products	Coenzyme in glucose metabolism, fatty acid and steroid synthesis	Fatigue, tingling in hands and feet, intestinal disturbances	Generally nontoxic except occasional diarrhea
Biotin	Widely distributed in foods; also made by colonic bacteria	Coenzyme in amino acid metabolism, fat and glycogen formation	Scaly skin (dermatitis)	Thought to be nontoxic
C (absorbic acid)	Fruits and vegetables, especially citrus fruits, cantaloupe, broccoli, and cabbage	Antioxidant; needed for collagen formation; important for bone and teeth formation; improves iron absorption	A disease called scurvy, poor wound healing, impaired immune responses	Digestive upsets

Table 14.5 Major minerals in the human body

Mineral	Common Sources	Primary Functions	Signs of Severe, Prolonged Deficiency	Signs of Extreme Excess
Calcium (Ca)	Dairy products, dark green vegetables, legumes	Bone and teeth formation, nerve and muscle action, blood clotting	Decreased bone mass, stunted growth	Kidney damage, impaired absorption of other minerals
Chloride (Cl)	Table salt	Role in acid-base balance, acid formation in stomach, body water balance	Poor appetite and growth, muscle cramps	Contributes to high blood pressure in susceptible people
Copper (Cu)	Seafood, nuts, legumes	Synthesis of hemoglobin, melanin	Anemia, bone and blood vessel changes	Nausea, liver damage
Fluorine (F)	Fluoridated water, tea, seafood	Maintenance of teeth and perhaps bone	Tooth decay	Digestive upset, mottling of teeth
Iodine (I)	Iodized salt, marine fish and shellfish	Required for thyroid hormone	Enlarged thyroid (goiter), metabolic disorders	Goiter
Iron (Fe)	Green leafy vegetables, whole grains, legumes, meats, eggs	Required for hemoglobin, myoglobin, and certain enzymes	Iron deficiency anemia, impaired immune function	Acute: shock and death. Chronic: liver and heart failure
Magnesium (Mg)	Whole grains, green leafy vegetables	Coenzyme in several enzymes	Nerve and muscle disturbances	Impaired nerve function
Phosphorus (P)	Meat, poultry, whole grains	Component of bones, teeth, phospholipids, ATP, nucleic acids	Loss of minerals from bone, muscle weakness	Depressed absorption of some minerals, abnormal bone deposition
Potassium (K)	Widespread in the diet, especially meats, grains	Muscle and nerve function; a major component of intracellular fluid	Muscle weakness	Acute: cardiac arrhythmias and death, Chronic: muscle weakness
Sodium (Na)	Table salt; widespread in the diet	Muscle and nerve function; major component of body water	Muscle cramps	High blood pressure in susceptible people
Sulfur (S)	Meat, dairy products	Component of many proteins	None reported	None reported
Zinc (Zn)	Whole grains, meats, seafood	Component of many enzymes	Impaired growth, reproductive failure, impaired immune function	Acute: nausea, vomiting, diarrhea. Chronic: anemia and impaired immune function

Recap **All minerals and nearly all vitamins must be obtained from food. For most healthy people a balanced diet, including adequate fiber, is the best way to achieve the RDA.**

14.11 Weight control: Energy consumed versus energy spent

The body requires energy to fuel metabolic processes and other activities. When we digest nutrients, the energy they contain can be put to use.

Energy is measured in units called calories. A *calorie* is the amount of energy needed to raise the temperature of 1 gram of water by 1° Celsius. Because this is not much energy in biological terms, scientists use a larger unit to measure the nutrient content of food and the energy used to perform biological activities. This is a **kilocalorie,** commonly referred to as a capital "C" **Calorie,** which is 1,000 calories. When people "count calories" they really mean the big Calories, and that is the unit we use in this chapter.

BMR: Determining how many Calories we need

To maintain a stable body weight, the number of Calories we consume must equal the number we use. So how do we determine the number of Calories we utilize?

You can estimate your daily energy needs by determining your **basal metabolic rate (BMR),** the energy your body needs to perform essential activities such as breathing and maintaining organ function. Factors that influence the BMR include:

- **Gender and body composition:** BMR is higher in males. Muscle tissue consumes more energy than fat tissue; because men generally have more muscle than women, they have a higher BMR. Also, highly muscular men and women have a faster BMR than their less-muscular counterparts.
- **Age:** BMR declines over time.
- **Health:** Some health conditions (such as fever, infections, and hyperthyroidism) increase BMR. Other conditions (such as hypothyroidism) lower it.
- **Stress:** Norepinephrine and epinephrine raise BMR.
- **Food intake:** Eating increases the metabolic rate, whereas fasting and extreme dieting decrease it. This is why crash diets often fail to lower body weight permanently. Although weight may fall temporarily, a slower BMR makes it difficult to keep the pounds from returning.
- **Genetics:** Rates of metabolism vary between individuals, independent of the factors listed above. Genetics plays a strong role in determining your BMR, although precisely how has not been fully determined.

Of course, your body performs many activities in addition to basic metabolic functions. However, the BMR accounts for a surprisingly large proportion of Calories used, as much as two-thirds of daily utilization. The rest go to fuel physical activity.

Energy balance and body weight

Healthy weight control involves balancing energy intake against energy expenditure. Note that the three classes of nutrients have quite different Caloric contents. Fat contains about nine Calories per gram, but carbohydrates and proteins contain only four. Thus, although fat is a good energy source, a high-fat diet can easily contain more Calories than we need. This is one reason why controlling fat intake is so important.

When we eat more Calories than we use, the excess energy is stored in specialized cells as fat. The number of fat cells a person has is largely determined by the time he or she is an adult. This may be why some people find losing weight so difficult and others do not. Research suggests that overweight people seem to have more fat cells than normal-weight individuals, so when they diet and shrink their fat reserves in each cell, their bodies respond as if they are starving. Dieting is difficult for chronically overweight persons because they are constantly fighting the body's own weight-control system, which responds as if their excess weight were normal.

Why is it so difficult to lose weight and keep it off?

Recap **Weight control involves balancing energy consumed in food against energy spent. Calculating your BMR helps you estimate how many Calories you need each day.**

Physical activity: An efficient way to use calories

Although the BMR stays fairly constant, the amount of exercise we get can dramatically change the total number of Calories we burn each day. Table 14.6 compares the number of calories spent per hour by a 100-, 150-, or 200-pound person performing various types of exercise. Note that heavier people do more work per hour for the same level of activity. To lose one pound of fat, we must spend about 3,500 Calories.

Try It Yourself

Calculating Your BMR

1. Calculate your weight in kilograms: Divide the number of pounds by 2.2 = ________
2. If you are male: Multiply your weight × 1.0 = ________
3. If you are female: Multiply your weight × 0.9 = ________
4. This number approximates the number of Calories you consume per hour. Multiply this number by 24 to estimate how many Calories you need each day to support basic metabolic functions = ________

For example:

- A 200-pound man (91 kg) has a BMR of 91 Calories/hour. His metabolic activities consume approximately 2,184 Calories per day.
- A 130-pound woman (59 kg) has a BMR of 53 Calories/hour. Her metabolic activities consume approximately 1,272 Calories per day.

Table 14.6 Approximate number of calories burned per hour by various activities

Activity	100 Lb person	150 Lb person	200 Lb person
Bicycling, 6 mph	160	240	312
Bicycling, 12 mph	270	410	534
Jogging, 5.5 mph	440	660	962
Jogging, 10 mph	850	1,280	1,664
Jumping rope	500	750	1,000
Swimming, 25 yds/min	185	275	358
Swimming, 50 yds/min	325	500	650
Walking, 2 mph	160	240	312
Walking, 4.5 mph	295	440	572
Tennis (singles)	265	400	535

SOURCE: American Heart Association

The best approach to weight loss is a gradual one. Nutritionists recommend reducing Caloric intake by a small amount each day while gradually increasing physical activity.

Of course, exercise has many benefits in addition to weight loss. It improves cardiovascular health, strengthens bones, tones muscles, and promotes a general sense of well-being. Your best strategy is a healthful diet combined with moderate exercise.

Healthy weight improves overall health

Why do we worry about Caloric intake and body weight, anyway? A primary reason is that body weight is linked to health status. The connection is strong enough that government groups and insurance companies regularly publish charts, like Table 14.7, that define a healthy weight range for various heights. Because they are based on population statistics, tables like this are not intended to provide definitive information about weight status for any one individual. Notice the chart does not consider gender or body frame, and it classifies age into only two groups.

Overweight is generally defined as 15–20% above ideal body weight, and *obese* is defined as more than 20% above it. People who weigh 10% less than ideal are considered *underweight;* this condition is less common in developed countries.

Recap **Increasing physical activity is an efficient way to increase Calorie expenditure. The best strategy for losing weight combines a healthful diet with moderate regular exercise.**

Table 14.7 Height and weight table

	Weight (lb)			
Height	19–34 years		Over 35 years	
	Midpoint	Range	Midpoint	Range
5'0"	112	97–128	123	108–138
5'1"	116	101–132	127	111–143
5'2"	120	104–137	131	115–148
5'3"	124	107–141	135	119–152
5'4"	128	111–146	140	122–157
5'5"	132	114–150	144	126–162
5'6"	136	118–155	148	130–167
5'7"	140	121–160	153	134–172
5'8"	144	125–164	158	138–178
5'9"	149	129–169	162	142–183
5'10"	153	132–174	167	146–188
5'11"	157	136–179	172	151–194
6'0"	162	140–184	177	155–199
6'1"	166	144–189	182	159–205
6'2"	171	148–195	187	164–210
6'3"	176	152–200	192	168–216
6'4"	180	156–205	197	173–222
6'5"	185	160–211	202	177–228
6'6"	190	164–216	208	182–234

NOTE: Within each range, the higher weights generally apply to men and the lower weights to women. Higher weights for people over age 35 reflect research that suggests people can carry more weight as they age without additional risk to their health.

SOURCE: U.S. Department of Agriculture and the U.S. Department of Health and Human Services, Home and Garden Bulletin No. 232, *Nutrition and Your Health: Dietary Guidelines for Americans,* 4th ed. Washington, D.C.: U.S. Government Printing Office, 1995.

14.12 Disorders of the digestive system

Although common, many digestive problems are not necessarily life threatening. One of the most common conditions worldwide is *food poisoning,* caused by food and beverages contaminated with bacteria or their toxic products. Diarrhea and vomiting often accompany food poisoning. *Food allergies* can also cause diarrhea, vomiting, and generalized allergic responses throughout the body. Common food allergens include shellfish, wheat, peanuts, and eggs.

Disorders of the GI tract

Lactose intolerance: Difficulty digesting milk A common disorder of digestion and absorption is *lac-*

tose intolerance. Human infants are born with the enzyme lactase in their small intestines so they can digest lactose, the primary sugar in milk. However, many adults gradually lose the enzyme, and with it their ability to digest the lactose in milk. The symptoms of lactose intolerance include diarrhea, gas, bloating, and abdominal cramps. Diarrhea occurs because the undigested lactose causes fluid to be retained in the digestive tract. The gas, bloating, and abdominal cramps are due to bacterial fermentation of the readily available lactose, which produces gases.

Lactose-intolerant people can eat cheese or yogurt, as the lactose in these milk products has been already digested. Lactose-free milk is also available.

Diverticulosis: Weakness in the intestinal wall *Diverticulosis* gets its name from its characteristic diverticula (small sacs produced when the mucosal lining of the large intestine protrudes through the other layers of the intestinal wall). Most people with diverticulosis do not experience discomfort, but sometimes the diverticula become infected or inflamed. Antibiotics usually resolve the problem.

Diverticulosis is relatively common in Western countries, perhaps due to inadequate dietary fiber. A low-fiber diet produces smaller feces, narrowing the colon and making its contractions more powerful. This increases the pressure on colon walls, forcing weak areas outward and forming diverticula.

Colon polyps: Noncancerous growths A polyp is a noncancerous growth that projects from a mucous membrane. Polyps can develop in many areas of the body, including the colon. The majority of colon polyps do not become cancerous, but because most colon cancers start as polyps, doctors recommend removing them. Polyps can be detected and removed in an outpatient procedure called a *colonoscopy,* in which a flexible fiberoptic scope is inserted into the colon through the anus.

Disorders of the accessory organs

Hepatitis: Inflammation of the liver *Hepatitis* refers to inflammation of the liver, generally caused by viruses or toxic substances. Researchers have identified at least five viruses that cause hepatitis. The most common are hepatitis A, B, and C.

Hepatitis A is transmitted by contaminated food or water and causes a brief illness from which most people recover completely. A vaccine is now available for hepatitis A. Although the vaccine is not considered necessary for most residents of industrialized nations, it may be a good idea for travelers to underdeveloped countries.

Hepatitis B travels in blood or body fluids, so it is usually passed via contaminated needles, blood transfusions, or sexual contact with infected individuals. Approximately 300,000 new cases are diagnosed each year. If not treated, hepatitis B can lead to liver failure. Symptoms include jaundice (the skin takes on a yellowish color due to accumulated bile pigments in the blood), nausea, fatigue, abdominal pain, and arthritis. There is a vaccine for hepatitis B. Federal law requires all health care workers to be vaccinated.

Hepatitis C is also transmitted in infected blood, usually through contaminated needles or blood transfusions. By 1992 U.S. blood banks began routine testing for hepatitis C, so the risk of contracting it from transfusions has fallen. However, researchers estimate that as many as four million Americans may already be infected, and many of them have no symptoms. Hepatitis C may remain dormant for years but still damage the liver. Severe cases can lead to chronic hepatitis, cirrhosis, or liver cancer. Doctors recommend testing for anyone who has had kidney dialysis (see Chapter 15, Urinary System) or an organ transplant before 1992. Also at risk are those who have injected drugs or had sexual contact with a hepatitis C carrier.

Gallstones can obstruct bile flow The gallbladder normally concentrates bile about 10-fold by removing about 90% of the water. Excessive cholesterol in the bile may precipitate out of solution with calcium and bile salts, forming hard crystals called *gallstones.* Only about 20% of gallstones ever cause problems, but if the crystals grow large enough they can obstruct bile flow and cause intense pain, especially after a meal. Treatments include drugs to dissolve the crystals, ultrasound vibrations or laser treatments to break the stones apart, or surgery to remove the gallbladder.

Malnutrition: Too many or too few nutrients

Malnutrition refers to nutritional conditions in which human development and function are compromised by an unbalanced or insufficient diet. Malnutrition can be caused either by *overnutrition* or *undernutrition* (Figure 14.18). In developed countries, overnutrition has led to an increase in obesity. Worldwide, however, undernourishment is far more common. The United Nations estimates that 800 million people, or about 13% of the world's population, are undernourished.

Deficiencies of one or more nutrients can produce specific effects. For example, years of vitamin A deficiency leads to eye damage and night blindness.

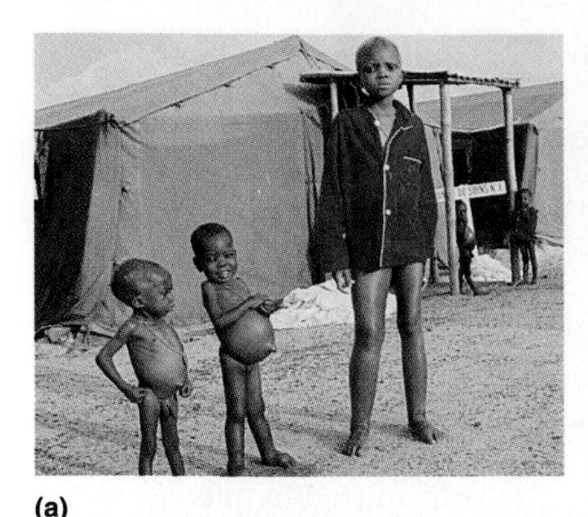

(a)

(b)

Figure 14.18 Malnutrition. Malnutrition can take many forms. **(a)** Starvation. The swollen bellies are typical of a marked deficiency of protein in the diet. **(b) Obesity.** Obesity is defined as a body weight more than 20% above ideal weight

In children, undernutrition stunts growth and increases susceptibility to infection. Severe undernutrition, or *starvation,* is still the leading cause of malnutrition worldwide. Nearly 20 million people, most of them children, die every year of starvation or related diseases.

Recap **Disorders of the GI tract and accessory organs include lactose intolerance, diverticulosis, colon polyps, gallstones, and hepatitis. Malnutrition can be caused by over- or undernutrition. Whereas starvation is the leading cause of malnutrition in underdeveloped countries, obesity is increasing in industrialized nations.**

Eating disorders: Anorexia nervosa and bulimia

Several conditions involving digestion and nutrition are really not digestive disorders at all but eating disorders involving the nervous system. Although eating disorders occur around the world, they are most common in women living in industrialized Western countries. Two examples are anorexia nervosa and bulimia.

Anorexia nervosa is a condition in which a person diets excessively or stops eating altogether, even to the point of starvation and death. Symptoms include:

- Refusal to maintain healthy body weight; people with anorexia often weigh less than 85% of their ideal weight.
- Intense fear of gaining weight, even though underweight.
- Distorted perception or preoccupation with body weight or shape.
- In premenopausal women, the absence of at least three consecutive menstrual cycles. Severe undernutrition interferes with the hormonal cycles of menstruation.

Male anorexics also experience hormonal abnormalities.

Bulimia is a binge-and-purge condition in which someone eats and deliberately vomits or takes other steps to minimize the Calories ingested. Symptoms of bulimia are:

- Recurrent episodes of binge eating. An episode of binge eating involves both (1) eating large amounts of food and (2) feeling a lack of control over eating.
- Taking recurrent inappropriate steps to prevent weight gain, such as self-induced vomiting; misusing laxatives, diuretics, enemas, or other medications; fasting; or exercising excessively.
- Binge eating and compensatory behaviors that occur, on average, at least twice a week for three months.
- Preoccupation with body shape and weight. However, unlike anorexics, some bulimics maintain a normal weight.

Both anorexia and bulimia play havoc with the body and mind. Anorexics become malnourished and suffer insomnia, hair loss, fatigue, and moodiness. Over time, they lose bone mass and develop osteoporosis. Repeated trauma to a bulimic's digestive system leads to ulcers, chronic heartburn, and rectal bleeding. Recurrent vomiting damages gums, erodes tooth enamel, and makes salivary glands swell, giving a chipmunk-like appearance.

Doctors are not sure what causes eating disorders, but there appear to be deep-rooted psychological and cultural factors. Many people with eating disorders also suffer from depression and anxiety. Because these conditions are complex, multidisciplinary treatment seems most effective, with a team of professionals who can address the patient's medical, psychiatric, dental, psychological, and nutritional needs.

Health Watch

Obesity: A Worldwide Epidemic?

In a world where nearly 15% of the population is undernourished, it seems ironic that obesity is on the rise. Indeed, the World Health Organization calls obesity a worldwide epidemic.

It's hard to get good statistics for obesity because there is no consistency regarding how, when, or whether the numbers are collected. The few reliable trend analyses available are startling. In the U.S., obesity increased from 14.5% of the population in 1980 to 22.5% by the late 1990s. In Canada between 1978 and 1982, obesity prevalence rose from 6.8% to 12.0% in men and from 9.6% to 14.0% in women. Obesity has also increased substantially in Australia and East Germany but apparently only slightly in China and Japan. For most of the world, accurate numbers are simply not available.

Why are health officials concerned? Study after study reveals a direct correlation between body weight and the incidence of certain diseases, including heart disease, diabetes, and cancer. The trend is so strong that even populations of normal-weight individuals show a higher risk compared to groups of people who weigh slightly below average. In other words, any increase in weight above fairly skinny is accompanied by an increased risk of disease for the population as a whole.

Note that we are talking about group effects based on statistics. The numbers do not take into account individual variations in body frame or amount of exercise. Indeed, some health officials worry that because obese people tend to be more sedentary, the real culprit may be physical inactivity rather than body weight per se. There is some evidence for this. In one study, lean but physically unfit men showed twice the death rate of men who were overweight but physically fit.

Why has obesity increased over the past 20 years? Our body weight is determined by both *internal* factors (such as genetic makeup), and *external* factors in the environment (such as activity level and availability of food). There's a lot of interest in genetic factors these days, especially so-called "fat genes" that might make it harder to maintain a healthy weight.

But one point is clear: genetics cannot account for an entire population's weight gain because genes don't evolve that fast. It took us millions of years to become what we are, and our collective gene pool can't be changed in just 20 years. A population-wide increase in obesity, then, can be blamed squarely on the environment.

So what's different about our environment? Around 1980, modernization produced an environment in many industrialized countries that favors an imbalance between Caloric intake and Caloric expenditure. Computers, remote controls, cars for nearly everyone, the ability to watch sports on television rather than participate in them—all combine to produce a more sedentary lifestyle. At the same time, food became relatively cheap, available, and energy-dense.

Throughout our millions of years of evolution, survival depended on the ability to defend against famine and weight loss, not against excessive food and overeating. We're not good at voluntarily regulating food intake, and indeed our modern environment seems to actively discourage it. For example, as a child were you ever told to "clean your plate"? Does your fast-food restaurant routinely ask whether you want to "super-size" your meal?

Another factor: energy density of food. Foods that are high in fat are energy-dense, because fat packs over twice the Calories of protein or carbohydrates for the same weight. Experiments show that subjects tend to eat a constant weight of food, regardless of whether or not the food is high in fat.

If we want to reduce population-wide obesity, we must change our environment. Studying children to see how eating patterns develop and how they could be modified, educating consumers to limit portion sizes at home and in restaurants, making the environment more conducive to exercise—all would help. Are health officials and the general public prepared to take these steps to reduce future health costs? That is another question.

Chapter Summary

The digestive system brings nutrients into the body

- The digestive system consists of the gastrointestinal (GI) tract and four accessory organs: the salivary glands, liver, gall bladder, and pancreas.
- The five basic processes of the digestive system are motility, secretion, digestion, absorption, and excretion.
- Two types of motility in the GI tract are peristalsis, which propels food forward, and segmentation, which mixes the contents.

The mouth processes food for swallowing

- The 32 teeth of adults include incisors, canines, premolars, and molars. Teeth cut, tear, grind, and crush food into smaller pieces.
- Saliva moistens food, begins the digestion of starch, and helps to protect against bacteria.

The pharynx and esophagus deliver food to the stomach

- Swallowing is a reflex that is initiated by voluntary movements of the tongue. Once started, swallowing is involuntary.
- The sole function of the esophagus is to get food from the mouth to the stomach.
- The lower esophageal sphincter prevents reflux of stomach contents.

The stomach stores, digests, and regulates food delivery

- The stomach stores ingested food until it can be delivered to the small intestine.
- Glands in the mucosa of the stomach secrete gastric juice into the lumen, beginning the process of protein digestion.
- Peristalsis of the stomach mixes the food and pushes it toward the small intestine.

The small intestine digests food and absorbs nutrients

- Digestion occurs primarily in the first part of the small intestine, called the duodenum.
- The jejunum and ileum of the small intestine absorb most of the products of digestion.
- The small intestine has a very large surface area as the result of folds, villi, and microvilli.

Accessory organs aid digestion and absorption

- The pancreas secretes fluid containing bicarbonate and digestive enzymes into the small intestine.
- The liver produces bile and participates in homeostasis in a variety of ways.
- All of the venous blood from the GI tract is routed directly to the liver.
- The gallbladder stores bile from the liver and concentrates it by removal of most of the water.

The large intestine absorbs nutrients and eliminates wastes

- The large intestine absorbs nearly all the remaining nutrients and water, as well as vitamin K produced by bacteria. It also stores wastes.
- Defecation is generally controlled by a neural reflex, but it can be overridden by conscious control.

How nutrients are absorbed

CD **Fluid Balance/Introduction to Body Fluids**

- Amino acids and simple sugars are actively transported into the mucosal cells that line the small and large intestines.
- The products of fat digestion enter the mucosal cells by diffusion and then reform into triglycerides. The triglycerides are coated with protein and then make their way into the lymph vessels for transport to the blood.
- Water is reabsorbed by osmosis.
- Vitamins and minerals follow a variety of specific pathways.

Endocrine and nervous systems regulate digestion

- The volume and content of food play a large part in regulating digestive processes.
- Stretching of the stomach causes stomach peristalsis to increase and increases the secretion of gastric juice.
- Stretching of the small intestine inhibits gastric motility, increases intestinal segmentation, and causes the secretion of two digestive enzymes, secretin and cholecystokinin.
- Acid in the small intestine triggers the secretion of pancreatic juice containing bicarbonate.

Nutrition: You are what you eat

- Good nutrition requires a variety of foods weighted toward grain products, fruits, and vegetables.
- The human body needs certain nutritional components that it cannot make, including a few

fatty acids, eight amino acids, 13 vitamins, and all essential minerals.

Weight control: Energy consumed versus energy spent

- The basal metabolic rate (BMR) represents the daily energy needs of the body for all essential activities except physical activity.
- Fat contains over twice as many Calories per gram as carbohydrate or protein.
- A person must use 3,500 Calories more than he or she ingests to lose a pound of body fat.

Disorders of the digestive system

- Lactose intolerance is caused by the lack of the enzyme (lactase) that normally digests lactose.
- Hepatitis, or inflammation of the liver, can be caused by several different viruses.
- Starvation is the most common form of malnutrition in the world.

Terms You Should Know

accessory organs, 345	large intestine, 355
anorexia nervosa, 368	mucosa, 345
basal metabolic rate, 365	muscularis, 346
bile, 353	nutrient, 344
bulimia, 368	pancreas, 353
Calorie, 364	pepsin, 349
cholecystokinin, 357	peptic ulcer, 350
chyme, 349	peristalsis, 346
colon, 355	Recommended Daily Allowance, 362
complete protein, 361	rectum, 355
esophagus, 348	salivary glands, 348
essential amino acids, 361	secretin, 357
gallbladder, 354	segmentation, 347
gastrin, 357	serosa, 346
gastrointestinal tract, 345	small intestine, 352
hepatic portal system, 353	stomach, 349
hepatitis, 367	submucosa, 345
	villi, 352
	vitamins, 361

Test Yourself

Matching Exercise

_____ 1. Ionic component of saliva, pancreatic juice, and bile
_____ 2. Organ of the digestive system that produces intrinsic factor
_____ 3. Structure that closes off the tracheal airway when we swallow
_____ 4. Vitamin K can be synthesized in this organ
_____ 5. The only digestive organ that is prone to the development of stones
_____ 6. The first and shortest region of the small intestine
_____ 7. Mineral element needed for the production of hemoglobin
_____ 8. One of two hormones secreted by the small intestine
_____ 9. Hormone secreted by the stomach
_____10. A fat-digesting enzyme

a. epiglottis
b. gall bladder
c. duodenum
d. secretin
e. bicarbonate
f. iron
g. stomach
h. lipase
i. gastrin
j. large intestine

Fill in the Blank

11. The four accessory organs of the digestive system are the salivary glands, the pancreas, the ________, and the gallbladder.
12. All nutrients must cross the innermost layer of the GI tract, called the ________, in order to enter the vascular or lymphatic systems.
13. The four types of adult teeth are incisors, ________, premolars, and molars.
14. Reflux of stomach acid into the esophagus can lead to a painful condition known as ________.
15. The lining of the stomach is protected from the potentially damaging effects of stomach acid by a layer of ________.
16. The removal of too much water from the feces results in difficulty defecating, a condition known as ________.
17. Glycogen and starch belong to the nutrient group known as ________.
18. Calcium, sodium, and potassium belong to the nutrient group known as ________.
19. In order to maintain a constant body weight, our energy ________ must exactly balance our energy expenditure.
20. ________ is a binge-and-purge eating disorder.

True or False?

21. Saliva contains fat-digesting enzymes.
22. The primary type of motility of the esophagus is peristalsis.
23. The pancreas secretes digestive enzymes and bicarbonate into the small intestine.

24. Only the large intestine absorbs ingested water.
25. The gallbladder concentrates bile by removing water from it.

Concept Review

1. Describe the one general function of the digestive system.
2. Name the five basic processes associated with carrying out that function.
3. Describe the location and function of the pyloric sphincter.
4. Compare/contrast *peristalsis* and *segmentation*.
5. Indicate how many teeth adult humans have.
6. List the functions of the stomach.
7. Indicate where in the GI tract most of the absorption of nutrients occurs.
8. Name the major nutrient group (carbohydrates, lipids, or proteins) that contains the most Calories per gram.
9. Indicate what fraction of our daily Caloric expenditure is accounted for by our basal metabolic rate (BMR).
10. Explain why it is harder to get an adequate supply of all amino acids from a vegetarian diet than from meat.

Apply What You Know

1. A meal containing a lot of fat takes longer to be emptied out of the stomach than a low-fat meal. Why?
2. Why do you suppose that heartburn is more likely to occur in pregnant women during the third trimester than in nonpregnant women?
3. In 1996 Procter & Gamble began to market a Calorie-free fat substitute called Olean® (olestra) that could be used to deep-fry foods such as potato chips. Olestra is a fat-soluble molecule that is neither digested nor absorbed. At the time there was concern that one side effect of eating too much Olean® might be vitamin deficiencies. What types of vitamins do you think might be affected by the use of Olean®, and how might they be affected?

Case in Point

Do Weight Loss Dietary Supplement Products Work?

Herbal dietary supplements are hot products these days, and none are hotter than those that profess to help people lose weight. According to a *Newsweek* magazine article, sales of one such product, called Metabolife, approached 1 billion dollars in 1999. Dietary supplements designed for weight loss generally contain a natural stimulant called ephedra (closely related to the drug ephedrine) plus over 15 other natural ingredients, including caffeine. Ephedra is thought to suppress appetite, and both ephedra and caffeine speed up the nervous system and increase the basal metabolic rate (BMR).

But do the weight loss products really work, and perhaps more importantly, how safe are they? The short answer is that in most cases we just don't know. Potential buyers should be aware that these products have not been approved by the Food and Drug Administration (FDA) and may not have been tested extensively. According to *Newsweek*, "Under a federal law passed in 1994, naturally occurring substances can be sold freely as dietary supplements unless they're proved dangerous." In contrast, drugs created by the pharmaceutical industry must undergo extensive testing for both safety and effectiveness before they're FDA-approved. In other words, the burden of proof that is applied to drugs does not apply to herbal dietary supplements. At least at the moment, in the case of dietary supplements it's "buyer beware."

QUESTIONS:

1. Explain how herbal dietary supplements such as Metabolife are thought to assist with weight loss.
2. With a classmate, list some pros and cons of regulating the herbal supplements industry.

INVESTIGATE:

Cowley, G. and J. Reno. Mad About Metabolife. *Newsweek* Oct. 4, 1999, 52–53.

References and Additional Reading

Alper, Joseph. Ulcers as an Infectious Disease. *Science* 260: 159–160, 1993. (A short research news update on the subject.)

Christian, Janet L. and Janet L. Greger. *Nutrition for Living*. Redwood City, California: Benjamin/Cummings, 1994. 612 pp. (A textbook of nutrition for the interested student.)

Kiberstis, Paula A. and Jean Marx, editors. Regulation of Body Weight. *Science* 280:1363–1390, 1998. (A compilation of four news articles and five research articles representing some of the latest information on this subject.)

U.S. Department of Agriculture and the U.S. Department of Health and Human Services, Home and Garden Bulletin No. 232, *Nutrition and Your Health: Dietary Guidelines for Americans,* 4th ed. Washington, D.C.: U.S. Government Printing Office, 1995. (The U.S. government dietary guide, published every 5 years.)

CHAPTER 15

THE URINARY SYSTEM

How do the kidneys get rid of toxic substances?

How is urine formed **from blood fluid?**

How do coffee **or** alcohol **affect your body's water balance?**

How is urination controlled?

DROP BY DROP—SLOWLY BUT STEADILY, THE KIDNEYS GO about the business of making urine. Minutes or even hours pass as the urinary bladder slowly fills. Finally, although you hadn't been thinking about it before, the thought comes to your mind that you need to find a bathroom, and soon. Before long you can think of little else.

We are aware of our heartbeat, consciously develop muscle tone, and are often preoccupied by our stomach. But do we ever give our kidneys a second thought? They are forgotten organs, tucked away at the back of the abdominal cavity. With urine as their primary product, the kidneys may not seem to serve a very glamorous function, but in fact they play a vital role in maintaining the constancy of the internal environment. Urine itself serves no purpose; it is simply the end product of the regulation of the internal environment, the waste that is discarded.

As a recurrent theme in physiology, homeostasis has been discussed in terms of the regulation of body temperature, blood pressure, blood gases, and many other controlled variables in connection with the functions of other organ systems. This chapter describes the central role of the urinary system in maintaining the constancy of the composition and volume of the body fluids.

15.1 The urinary system contributes to homeostasis

Excretion refers to processes that remove wastes and excess materials from the body. Let's briefly review the excretory organs and systems involved in managing metabolic wastes and maintaining homeostasis of water and solutes (Figure 15.1). The digestive system provides the body with nutrients and water and excretes food residues. The lungs take in oxygen and excrete carbon dioxide gas. The skin gets rid of heat and is also a site of water and salt loss, especially when getting rid of heat quickly. The liver destroys or inactivates numerous substances but has a limited capacity to actually excrete them; the only available mechanism is to secrete them in bile, some of which is excreted with feces.

Because the excretory capacity of the other organs is limited, the urinary system has primary responsibility for homeostasis of water and most of the solutes in blood and other body fluids. The **urinary system** consists of the organs (kidneys, ureters, bladder, and urethra) that produce, transport, store, and excrete urine.

Urine is essentially water and solutes. Among the solutes excreted in urine are excess elements and ions, drugs, vitamins, toxic chemicals, and waste

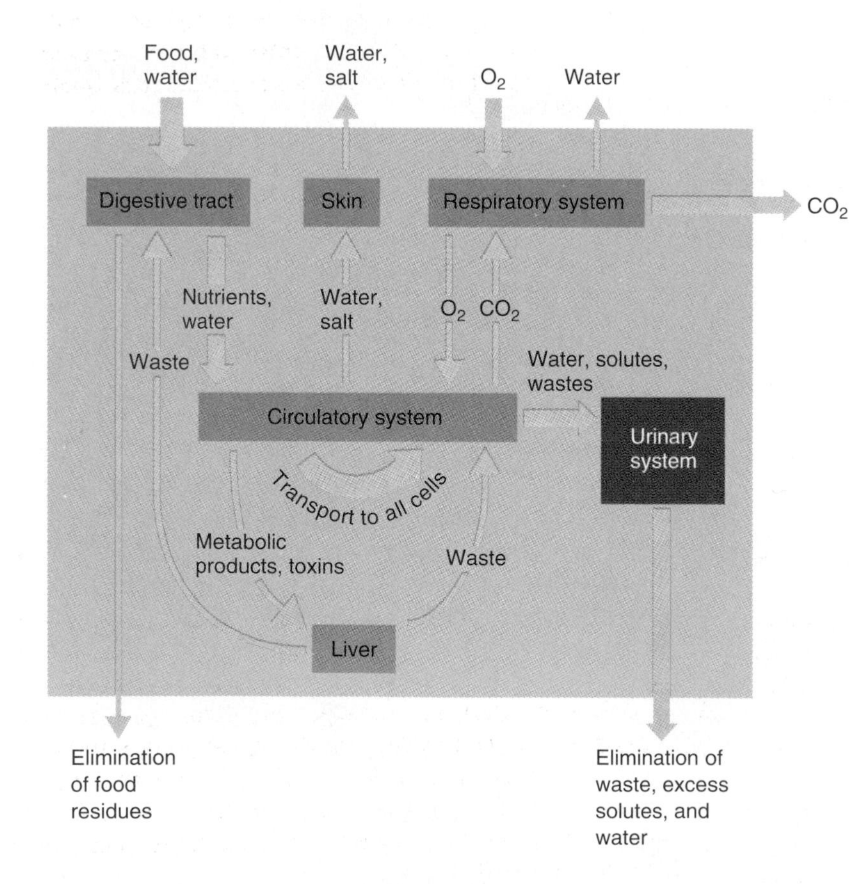

Figure 15.1 Organ systems involved in the removal of wastes and the maintenance of homeostasis of water and solutes. With the large brown box representing the body, this diagram maps the inflow and outflow of key compounds we consume. The kidneys of the urinary system are the organs primarily responsible for the maintenance of homeostasis of water and solutes and for the excretion of most waste products.

products produced by the liver or by cellular metabolism. Some substances, such as water and sodium chloride (salt), are excreted to regulate body fluid balance and salt levels. About the only major solutes *not* excreted by the kidneys under normal circumstances are the three classes of nutrients (carbohydrates, lipids, and proteins). The kidneys keep these nutrients in the body for other organs to regulate.

The urinary system regulates water levels

Water is the most abundant molecule in your body, representing about 60% of body weight. Normally you exchange about 2½ liters of water per day with the external environment (Table 15.1). We consume most of our water in food and beverages, but we also produce about 300 milliliters as part of cellular metabolism. Meanwhile, we lose water through evaporation from lungs and skin, and through defecation.

Water intake can vary tremendously, whereas water loss from evaporation stays relatively constant. It's up to the urinary system to excrete the excess. To ensure healthy water balance in the body, the urinary system can adjust water output over a wide range, from about one-half liter per day to nearly one liter *per hour.*

The urinary system regulates nitrogenous wastes and other solutes

Even though many solutes in the body are essential for life, we continually acquire more of them than we can use. The primary solutes excreted by the urinary system are nitrogenous wastes, excess ions, and trace amounts of other substances, as described below.

Nitrogenous wastes are formed during the metabolism of proteins. The major nitrogenous waste product in urine is **urea.** The metabolism of protein initially liberates ammonia (NH_3). Ammonia is quite toxic to cells; however it is quickly detoxified by the liver. In the liver, two ammonia molecules are combined with a molecule of carbon dioxide to produce a molecule of urea ($H_2N\text{-}CO\text{-}NH_2$) plus a molecule of water. Although far less toxic than ammonia, urea is also dangerous in high concentrations. A small amount of urea appears in sweat, but most of it is excreted by the urinary system.

Dozens of different ions are ingested with food or liberated from nutrients during metabolism. The most abundant ions in the body are sodium (Na^+) and chloride (Cl^-), which are important in determining the volume of the extracellular fluid, including blood. The volume of blood, in turn, affects blood pressure. Other important ions include potassium (K^+), which maintains electrical charges across membranes; calcium (Ca^{2+}), important in nerve and muscle activity; and hydrogen (H^+), which maintains acid-base balance. The rate of urinary excretion of each of these ions is regulated by the kidneys in order to maintain homeostasis.

Trace amounts of many other substances are excreted in proportion to their daily rate of gain by the body. Among them are *creatinine,* a waste product that is produced during the metabolism of creatine phosphate in muscle, and various waste products that give the urine its characteristic yellow color.

Table 15.1 Sources of water gain and loss per day

Water Gain (ml/day)		Water Loss (ml/day)	
Drinking fluids	1,000	Urine	1,500
Water in food	1,200	Evaporative loss (lung)	500
Metabolic production	300	Evaporative loss (skin)	350
		Feces	150
TOTAL	2,500	TOTAL	2,500

Recap **The urinary system maintains a constant internal environment by regulating water balance and body levels of nitrogenous wastes, ions, and other substances. It filters metabolic wastes from the blood and excretes them in urine. The major nitrogenous waste product is urea.**

15.2 Organs of the urinary system

The organs of the urinary system include the kidneys, ureters, urinary bladder, and urethra (Table 15.2).

Kidneys: The principal urinary organs

The main organs of the urinary system are the two **kidneys.** The kidneys are located on either side of the vertebral column, near the posterior body wall (Figure 15.2a). Each kidney is a dark reddish-brown organ about the size of your fist and shaped like a kidney bean. A *renal artery* and a *renal vein* connect each kidney to the aorta and inferior vena cava, respectively (*renal* comes from the Latin word *ren,* kidney).

Seen in longitudinal and cross section (Figure 15.2b), each kidney consists of inner pyramid-shaped zones of dense tissue (called renal pyramids) that constitute the **medulla,** and an outer zone called the **cortex.** At the center of the kidney is a hollow space, the *renal pelvis,* where urine collects after it is formed.

Table 15.2 Functions of organs in urinary system

Organ	Urinary functions	Other homeostatic functions
Kidneys	Regulate body water volume Regulate body concentrations of inorganic ions Remove metabolic wastes (primarily urea) from blood	Help regulate blood volume and blood pressure Assist in controlling body salt balance Help regulate acid-base balance and blood pH Control production of red blood cells Activate vitamin D
Ureters	Transport urine to bladder	
Urinary bladder	Stores urine until excretion	
Urethra	Transports urine to body surface	

A closer look at a section of the renal cortex and medulla reveals that it contains long, thin, tubular structures called *nephrons* (Figure 15.2c). Nephrons share a common final section called the *collecting duct,* through which urine produced by the nephrons is delivered to the renal pelvis.

Besides being the primary organs of the urinary system, the kidneys have many other functions related to maintaining homeostasis, which are discussed later in this chapter.

Ureters transport urine to the bladder

The renal pelvis of each kidney is continuous with a **ureter,** a muscular tube that transports urine to the bladder. Peristaltic waves of smooth muscle contraction, occurring every 10–15 seconds, move urine along the 10-inch length of the ureters to the bladder.

Urinary bladder stores urine

The **urinary bladder** stores urine. The bladder consists of three layers of smooth muscle lined on the inside by epithelial cells. Typically the bladder can hold about 600–1,000 ml of urine, though volumes that large may feel uncomfortable. Women generally have a smaller bladder capacity than men because their bladders are slightly compressed by the nearby uterus (Figure 15.3).

Urethra carries urine from the body

During urination urine passes through the **urethra,** a single muscular tube that extends from the bladder to the body's external opening. Until then the bladder is prevented from emptying by the *internal urethral sphincter,* where the bladder joins the urethra, and the *external urethral sphincter* farther

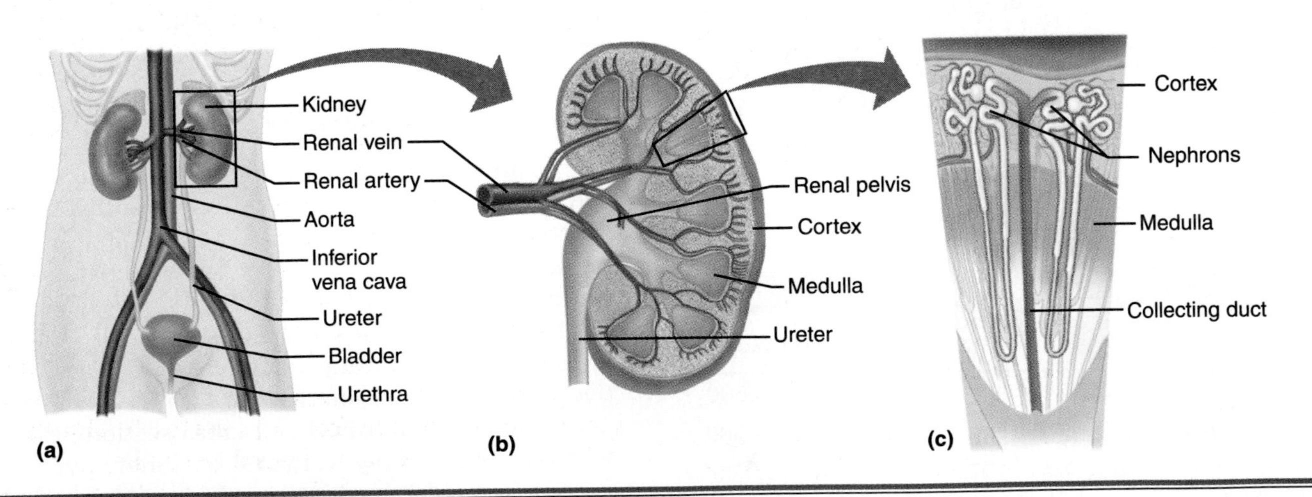

Figure 15.2 The human urinary system. (a) Locations of the components of the urinary system within the body. **(b)** Internal structure of a kidney. **(c)** The cortex and medulla of the kidney are composed of numerous nephrons.

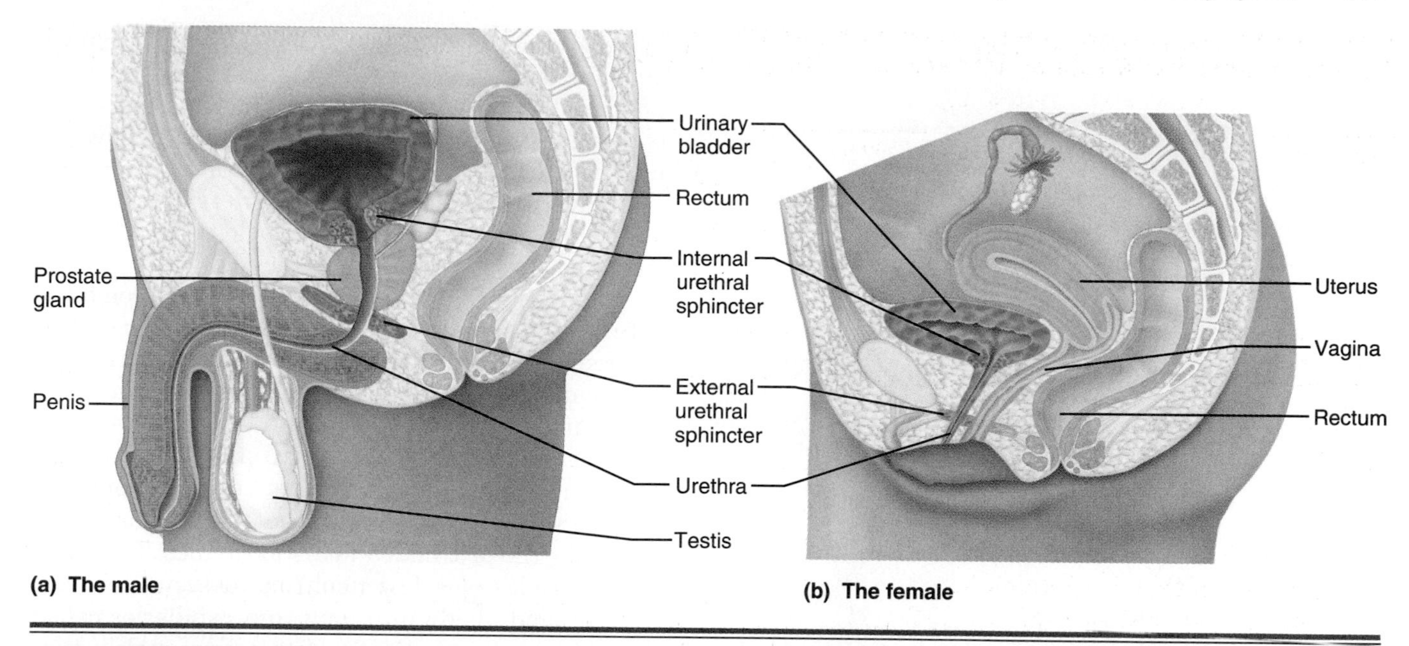

Figure 15.3 The positions of the bladder, the urethra, and associated organs in males and in females. (a) The male. **(b)** The female.

down the urethra. The urethra is about 8 inches long in men and about 1.5 inches long in women.

> ***Recap*** **Organs of the urinary system include the kidneys, ureters, bladder, and urethra. The kidneys are the principal urinary organs, although they have several homeostatic functions as well. The ureters transport urine to the bladder, where it is stored until carried by the urethra to the body's external opening.**

15.3 Nephrons produce urine

Each kidney contains approximately a million small functional units called **nephrons** (from the Greek word for kidney, *nephros*). An individual nephron consists of a thin hollow tube of epithelial cells, called a *tubule,* plus the blood vessels that supply the tubule. The function of the nephron is to produce urine. However, nephrons don't just pick the waste molecules out of blood and excrete them. Instead, they remove about 180 liters of fluid from the blood every day (about 2½ times your body weight) and then return almost all of it to the blood, leaving just a small amount of fluid behind in the tubule to be excreted as urine. The formation of urine would be similar to cleaning your room by taking everything out of the room and then putting it all back except for the dust and waste paper you want discarded.

The tubule filters fluid and reabsorbs substances

Spotlight on Structure & Function

Figure 15.4 depicts the tubular portion of a nephron. The nephron begins with a cup of tissue that looks like a deflated ball with one side pushed in, called the **glomerular capsule** (sometimes called **Bowman's capsule**). The glomerular capsule surrounds and encloses a network of capillaries called the **glomerulus,** which is part of the blood supply of the nephron.

The process of urine formation begins when plasma fluid is filtered out of the capillaries of the glomerulus and into the space between the two layers of the glomerular capsule. From the glomerular capsule the tubule continues as a long, thin tube with four distinct regions: proximal tubule, loop of Henle, distal tubule, and collecting duct. The **proximal tubule** (*proximal* meaning it is the part nearest the glomerular capsule) starts at the glomerular capsule and ends at the renal medulla. The hairpin-shaped **loop of Henle** (loop of the nephron) extends deep into the medulla as the *descending limb,* and then loops back up to the vicinity of the glomerular capsule as the *ascending limb.* After it passes the glomerular capsule, the tubule is called the **distal tubule** (*distal* means it is more distant from the glomerular capsule than the proximal tubule). Finally, the distal tubules of up to a thousand nephrons join together to become a **collecting duct.** The collecting duct extends from the cortex to the renal pelvis, where the final urine is deposited.

As we will see, the tubule and collecting duct are not just passive conduits. As urine flows through them, their cells modify it substantially by reabsorbing water and ions. Reabsorption prevents the loss of valuable substances from the body and adjusts the composition of blood and body fluids. ■

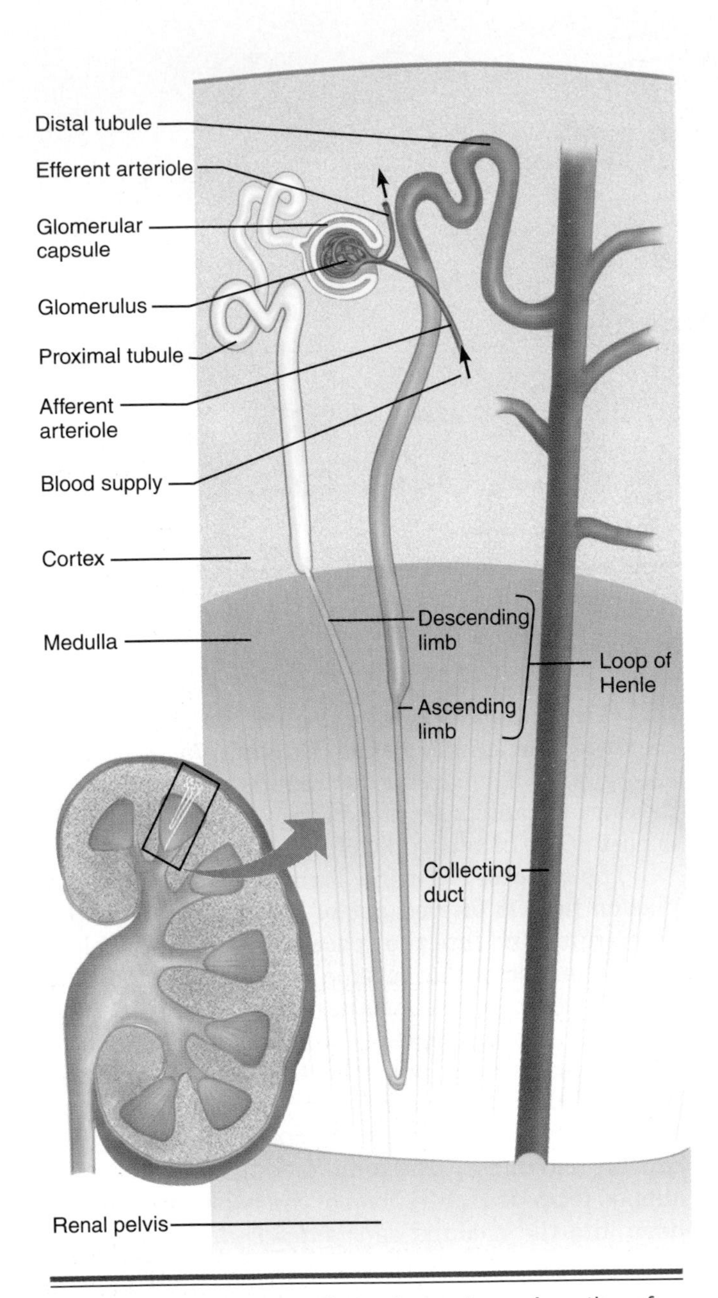

Figure 15.4 Tubular regions of a nephron. A portion of the glomerular capsule has been cut away to expose the blood vessels (the glomerulus) enclosed by the glomerular capsule. The proximal tubule begins at the glomerular capsule and ends where the tubule narrows and enters the medulla. The loop of Henle consists of descending and ascending limbs; it ends where the tubule passes the glomerular capsule. The distal tubule connects to a collecting duct, which is shared by many nephrons.

Special blood vessels supply the tubule

Figure 15.5 provides a closer look at the blood vessels serving the nephrons. The renal artery supplying a kidney branches many times to serve those million nephrons. Ultimately, every nephron is supplied by a single arteriole, called the **afferent arteriole** (*afferent* means "directed toward"). The afferent arteriole enters a glomerular capsule and then divides many times to become the network of capillaries that constitutes the glomerulus. Here plasma fluid and solutes are filtered from the blood into the capsular space.

The glomerular capillaries rejoin to become the **efferent** ("directed away") **arteriole,** which carries filtered blood from the glomerulus. The efferent arteriole divides again into another capillary network that surrounds the proximal and distal tubules in the cortex, called the **peritubular capillaries** (*peri-* means "around"). The peritubular capillaries remove water, ions, and nutrients, which are reabsorbed by the proximal and distal tubules. The efferent arterioles of a few nephrons descend into the medulla and divide into long thin capillaries called the **vasa recta** (Latin for "straight vessels") that supply the loop of Henle and collecting duct. Eventually the filtered blood flows into progressively larger veins that become the single renal vein leading to the inferior vena cava.

> ***Recap*** **A nephron is the functional unit of a kidney. A nephron tubule consists of the glomerular capsule, where fluid is filtered, and four regions in which the filtrate is modified before it becomes urine: proximal tubule, loop of Henle, distal tubule, and collecting duct. Blood flows to the glomerular capsule via the renal artery and afferent arterioles. Peritubular capillaries carry the blood to the proximal and distal tubules, and vasa recta supply the loops of Henle and collecting ducts.**

15.4 Formation of urine—filtration, reabsorption, and secretion

As we have noted, the urinary system regulates the excretion of water and ions to achieve homeostasis of fluid volume and composition. It excretes certain wastes while retaining precious nutrients. How do the kidneys select what to retain and what (and how much) to excrete in urine?

The formation of urine involves three processes:

1. *Glomerular filtration:* the movement of a protein-free solution of fluid and solutes from the glomerulus into the glomerular capsule
2. *Tubular reabsorption:* the return of most of the fluid and solutes back into the peritubular capillaries or vasa recta
3. *Tubular secretion:* the addition of certain solutes from the peritubular capillaries or vasa recta into the tubule

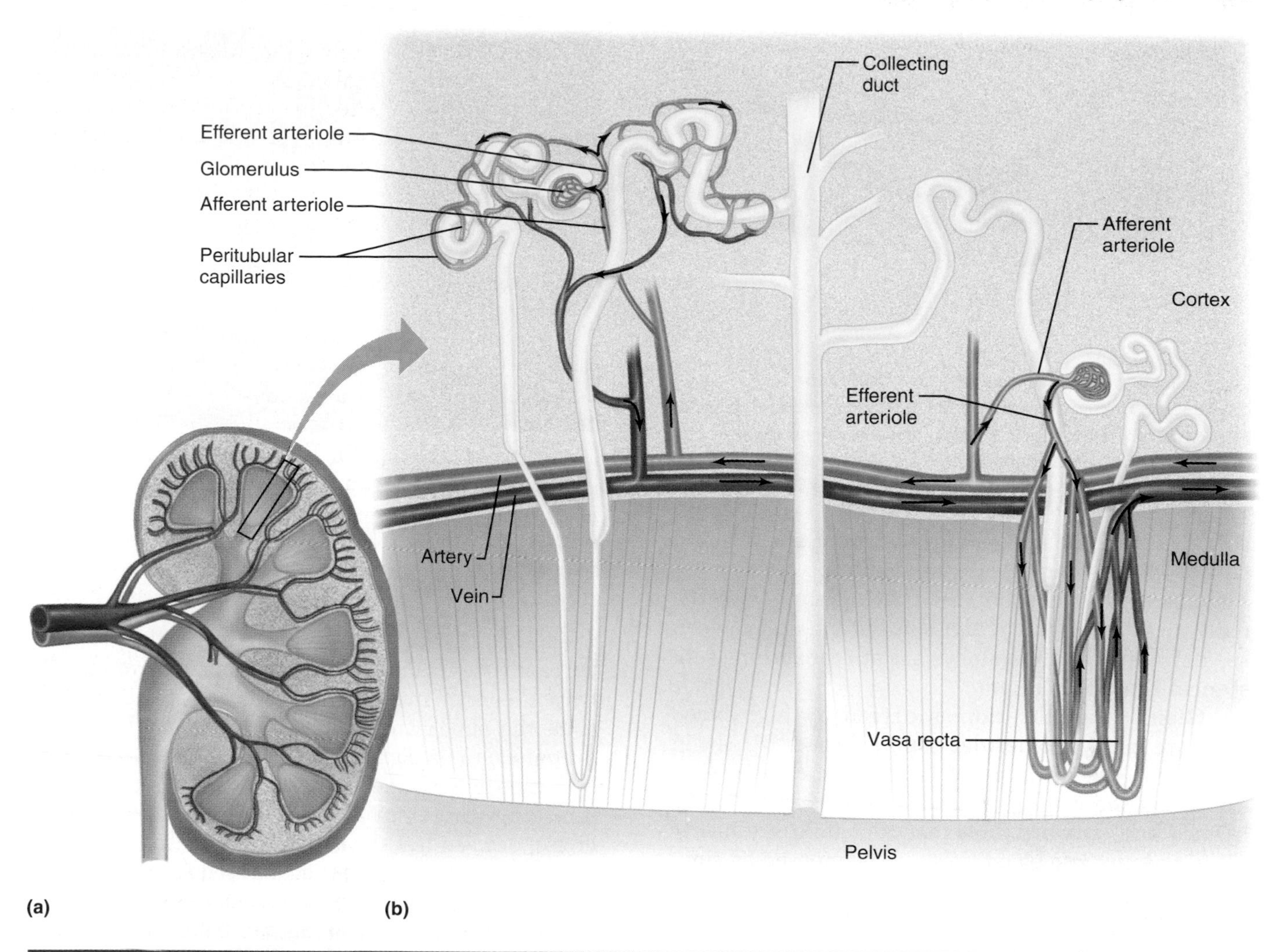

Figure 15.5 Relationships between the tubular and vascular components of nephrons. (a) The renal artery and renal vein branch many times to deliver blood to and return blood from the lobes of the kidney. **(b)** The efferent arteriole of most nephrons, such as the one shown here on the left, divide to become the peritubular capillaries, which supply proximal and distal tubules in the cortex. In some nephrons, such as the one on the right, the efferent arteriole descends into the medulla to become the vasa recta, which supply loops of Henle. Nephrons are so closely packed together that these two nephron tubules would actually share peritubular capillaries and vasa recta.

The fluid and solutes that remain in the tubule constitute the urine, which is eventually excreted. Figure 15.6 summarizes the relationship between these three processes. As the figure indicates, the amount of any substance excreted in urine is equal to the amount filtered, minus the amount reabsorbed back into the blood, plus the amount secreted into the tubule.

How is urine formed from blood fluid?

Glomerular filtration filters fluid from capillaries

Urine formation starts with **glomerular filtration,** the process of filtering a large quantity of protein-free plasma fluid from the glomerular capillaries into the glomerular capsule. The capillary walls of the glomerulus are highly specialized for filtration—so porous that they are nearly 100 times more permeable to water and small solutes than are most other capillaries (Figure 15.7). However, they are also highly selective, in that they are *less* permeable than other capillaries to large proteins and whole cells. As a result, the fluid that filters into the glomerular capsule, called the *glomerular filtrate,* contains water and all of the small solutes (ions, nutrients, small proteins, and urea) in the same concentrations as in blood plasma, but it contains no large proteins, red blood cells, or white blood cells.

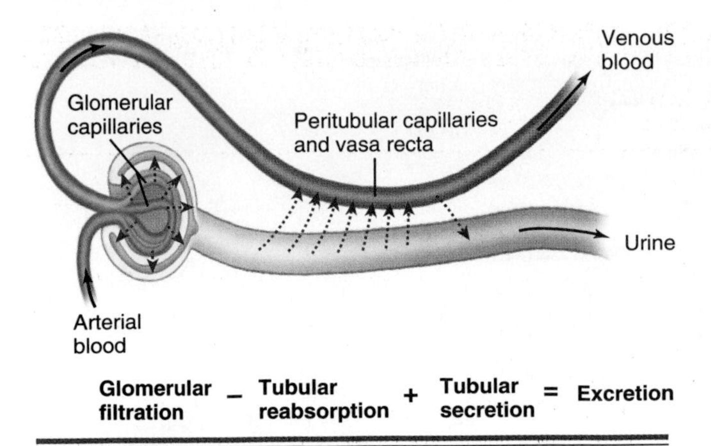

Figure 15.6 Schematic representation of the three processes that contribute to the formation of urine.

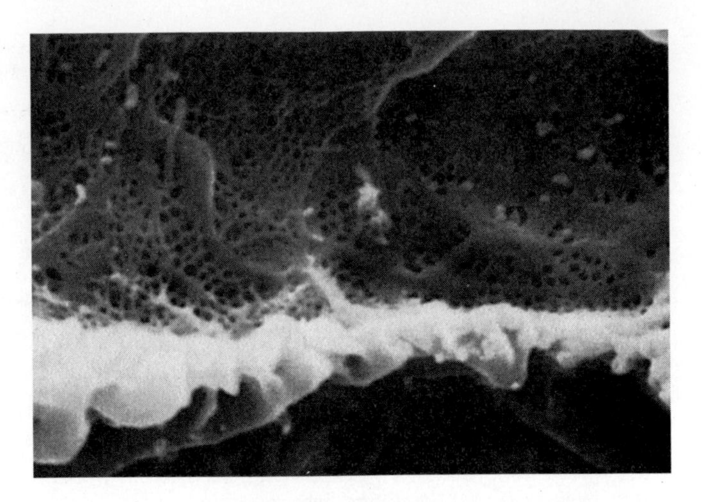

Figure 15.7 The porous structure of glomerular capillaries. A highly magnified view of the inner surface (blood side) of a single glomerular capillary reveals its porous, sievelike structure.

Together, the two kidneys produce about half a cup of glomerular filtrate per minute, a whopping 180 liters per day. Contrast that amount with the daily urine excretion of about 1.5 liters, and you can see how efficient the kidneys are at reabsorbing substances that are filtered.

Glomerular filtration is driven by high blood pressure in the glomerular capillaries. As we saw in Chapter 8, the filtration of fluid out of capillaries is driven by the capillary blood pressure. Usually the inward and outward movements of fluid are nearly equal because the blood pressure is opposed by similar osmotic pressure generated by the proteins in blood. The difference is that in the kidneys, the blood pressure in the glomerular capillaries is about twice that of any other capillary. This high pressure forces large amounts of glomerular filtrate into the glomerular capsule. The kidneys themselves do not have to expend any energy to produce the filtrate.

The rate of filtration is regulated in two different ways:

1. Under resting conditions, pressure-sensitive cells in the arterioles and flow-sensitive cells in the tubule walls can release chemicals to adjust the diameter of the afferent arterioles. These feedback mechanisms maintain a relatively constant rate of glomerular filtration, allowing the kidneys to carry out their regulatory functions.
2. During times of stress (such as after an injury or while running a marathon), blood flow to the kidneys falls substantially as blood is redistributed to more critical organs. The sympathetic division of the autonomic nervous system constricts afferent and efferent arterioles, reducing blood flow and the rates of glomerular filtration and urine formation. The kidneys are unharmed because they do not need a high blood flow to survive, and the body can cut back on urine production temporarily without ill effects.

If the delicate sievelike structure of the glomerular capillaries is disrupted, proteins may cross the filtration barrier into the tubule fluid. This is *proteinuria,* the appearance of protein in urine. Persistent proteinuria is a sign of glomerular damage, perhaps by toxins or sustained high blood pressure. However, temporary proteinuria can occur after strenuous exercise, even in healthy people. The reason for this is unknown, but it may be that reduced renal blood flow during exercise allows blood to remain in contact with the glomerular barrier longer, letting more plasma proteins leak through. Exercise-induced proteinuria is not dangerous and goes away within a day or so.

Recap **Glomerular filtration separates plasma fluid and small solutes from larger proteins and blood cells. High blood pressure in the glomerular capillaries drives this process.**

Tubular reabsorption returns filtered water and solutes to blood

Tubular reabsorption, the second step in urine formation, returns filtered water and solutes from the tubule into the blood of the peritubular capillaries or vasa recta. As filtrate flows through the tubule, major nutrients are almost completely reabsorbed—all of the filtered glucose, amino acids, and bicarbonate, and more than 99% of the water and sodium. About 50% of the urea is also reabsorbed. The final urine contains just enough water and sodium to balance the daily net gain from all other sources, the equivalent of all the urea produced in one day's metabolism, and trace amounts of other ions and wastes. Some waste products, such as creatinine, are not reabsorbed at all once they are filtered (Table 15.3).

Table 15.3 Amounts of various substances filtered and excreted or reabsorbed

Substance	Amount Filtered per Day	Amounted Excreted in Urine per Day	Proportion Reabsorbed
Water	180 L	1–2 L	99%
Sodium (Na^+)	620 g	4 g	99%
Chloride (Cl^-)	720 g	6 g	99%
Potassium (K^+)	30 g	2 g	93%
Bicarbonate (HCO_3^-)	275 g	0	100%
Glucose	180 g	0	100%
Urea*	52 g*	26 g	50%
Creatinine*	1.6 g*	1.6 g	0

*The plasma concentrations of these substances and the amounts filtered per day will vary with diet, age, and level of physical activity.

Most tubular reabsorption occurs in the proximal tubule, leaving the fine-tuning and regulation of reabsorption to the more distal regions. A good example is water. The proximal tubule always reabsorbs about 65–70% of the water and the loop of Henle reabsorbs another 25%, both completely unregulated. The distal tubule and collecting duct together reabsorb less than 10%, yet it is here that water excretion is effectively regulated.

Spotlight on Structure & Function

The process of reabsorption requires that substances cross the layer of epithelial cells of the tubules to reach blood capillaries. Like the epithelial cells of the digestive tract, the cell membrane on the lumen side of the tubule has a "brush border" of microvilli to increase the surface area for reabsorption (Figure 15.8). The process of reabsorption begins in the proximal tubule with the active transport of sodium from inside the cell toward the peritubular capillaries that surround the tubule. This is an active process that requires energy in the form of ATP. Other steps in the reabsorptive process, diagrammed in Figure 15.9, follow from this one event:

1. The active transport of sodium out of the tubular cell decreases the intracellular concentration of sodium, setting up a driving force for sodium to diffuse into the cell by facilitated transport.
2. Because sodium is a positively charged ion, its movement drives the diffusion of chloride across the tubular cell (negative charges must follow positive ones to maintain electrical neutrality).
3. The transport of solutes (Na^+ and Cl^-) across the tubular cell creates a concentration gradient for the diffusion of water, so water is also reabsorbed.
4. The inward carrier-mediated facilitated transport of sodium across the luminal cell membrane provides the energy for the secondary active transport of glucose and amino acids into the cell. Once inside the cell, the glucose and amino acids diffuse into the interstitial fluid and eventually into the peritubular capillaries.

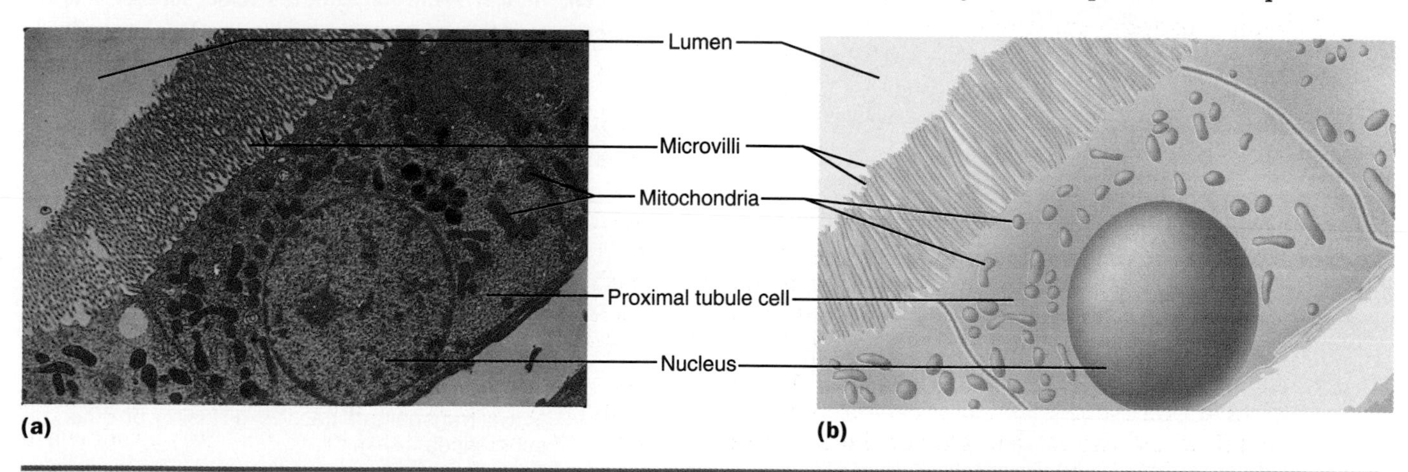

Figure 15.8 The proximal tubule. (a) TEM (×19,000) of a section of a proximal tubule, showing the microvilli that form the brush border of the inner (luminal) surface. **(b)** A diagram showing the major features of the photograph.

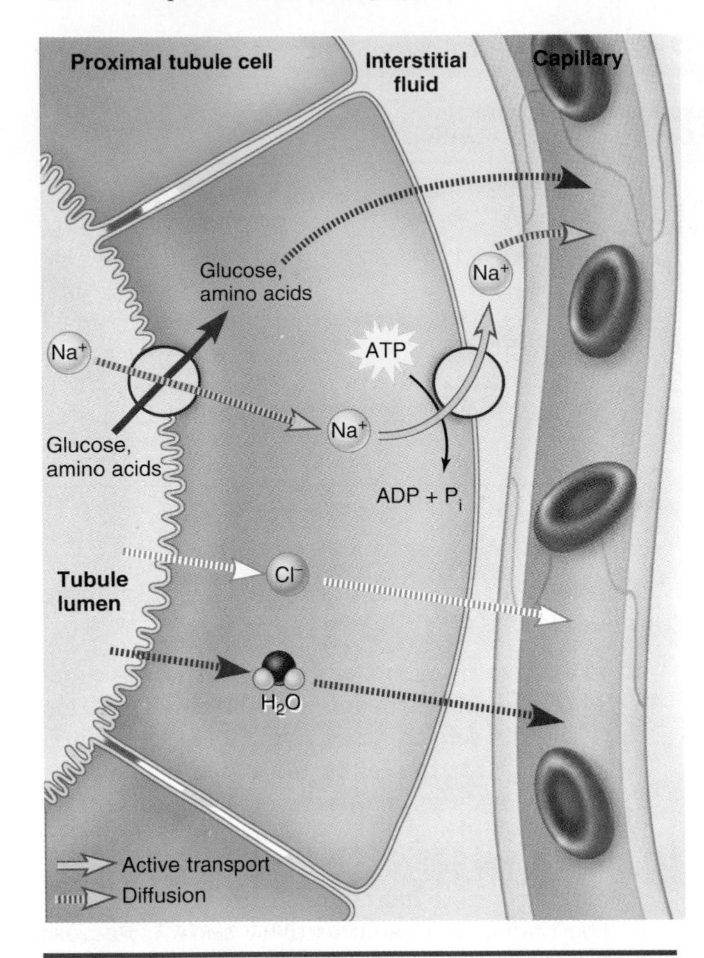

Figure 15.9 Basic mechanisms of reabsorption in the proximal tubule. The key to the entire process is the active transport of sodium (Na^+) across the tubular cell membrane on the capillary side of the cell. This step requires energy in the form of ATP. The active transport of sodium keeps the intracellular sodium concentration low, which permits sodium to diffuse into the cell from the luminal side. Negatively charged chloride ions follow in order to maintain electrical neutrality, and water follows because the reabsorption of sodium and chloride produces an osmotic driving force for the diffusion of water. Finally, the facilitated diffusion of sodium across the luminal membrane is used as an energy source for the reabsorption of glucose and amino acids.

Note that the reabsorption of water, salt, glucose, and amino acids depends on just one metabolic energy-using process: the primary active transport of sodium out of the tubular epithelial cell. Although the amounts of the various substances reabsorbed are different, reabsorption in the more distal regions of the tubule still relies on the active transport of sodium as the energy-utilizing step. ■

Tubular secretion removes other substances from blood

How do the kidneys get rid of toxic substances?

A few substances undergo **tubular secretion**—that is, they move from the capillaries (either peritubular capillaries or vasa recta) into the tubule to be excreted. Tubular secretion may occur by either active transport or passive diffusion, depending on the specific substance being secreted.

Tubular secretion is critical for removing or regulating levels of certain chemicals, including toxic and foreign substances and drugs. The list of substances secreted into the tubule includes penicillin, cocaine, marijuana, many food preservatives, and some pesticides.

Some substances that occur naturally in the body are secreted just to maintain normal homeostatic levels. The proximal tubule secretes hydrogen (H^+) and ammonium (NH_4^+) to regulate the body's acid-base balance, and the distal tubule secretes potassium (K^+) to maintain healthy levels of that mineral. Table 15.4 summarizes all three processes in urine formation.

Recap **During tubular reabsorption, nearly all the filtered water and sodium and all the major nutrients are reabsorbed from the nephron tubule. The process begins with the active transport of sodium across the cell membrane located on the capillary side of the tubular cell. Tubular secretion removes toxic, foreign, and excess substances from the capillaries. It is essential to**

Table 15.4 Steps in the formation of urine

Step	Function	Process
Glomerular filtration	Separates plasma fluid and small solutes from larger proteins and blood cells	High blood pressure in glomerular capillaries forces fluid through capillary walls into the glomerular capsule
Tubular reabsorption	Returns filtered water and nearly all major nutrients to blood	Primary active transport of Na^+ across cell membrane drives diffusion of Cl^- and water; provides energy for reabsorption by secondary active transport or passive diffusion of other substances
Tubular secretion	Removes harmful or excess substances from blood	Substances move by active transport or passive diffusion from capillaries into tubule

the regulation of acid-base balance, potassium balance, and the excretion of certain wastes.

15.5 The kidneys can produce dilute or concentrated urine

Your kidneys are capable of producing urine that is either more dilute or more concentrated than plasma. In other words, the kidneys can conserve water when it is in short supply and get rid of it when there is too much. The ability to do this depends on a high concentration of solutes in the renal medulla, coupled with the ability to alter the permeability of the collecting duct to water.

Producing dilute urine: Excreting excess water

If you drink a large glass of water quickly, the water will be absorbed by your digestive system and enter your blood. This will increase your blood volume, and decrease the concentration of ions in blood and body fluids. Most of the water enters your cells. To prevent osmotic swelling and damage to cells, the kidneys adjust the process of urine formation in such a way that you reabsorb less water and produce a dilute urine. Figure 15.10 illustrates the process by which this happens.

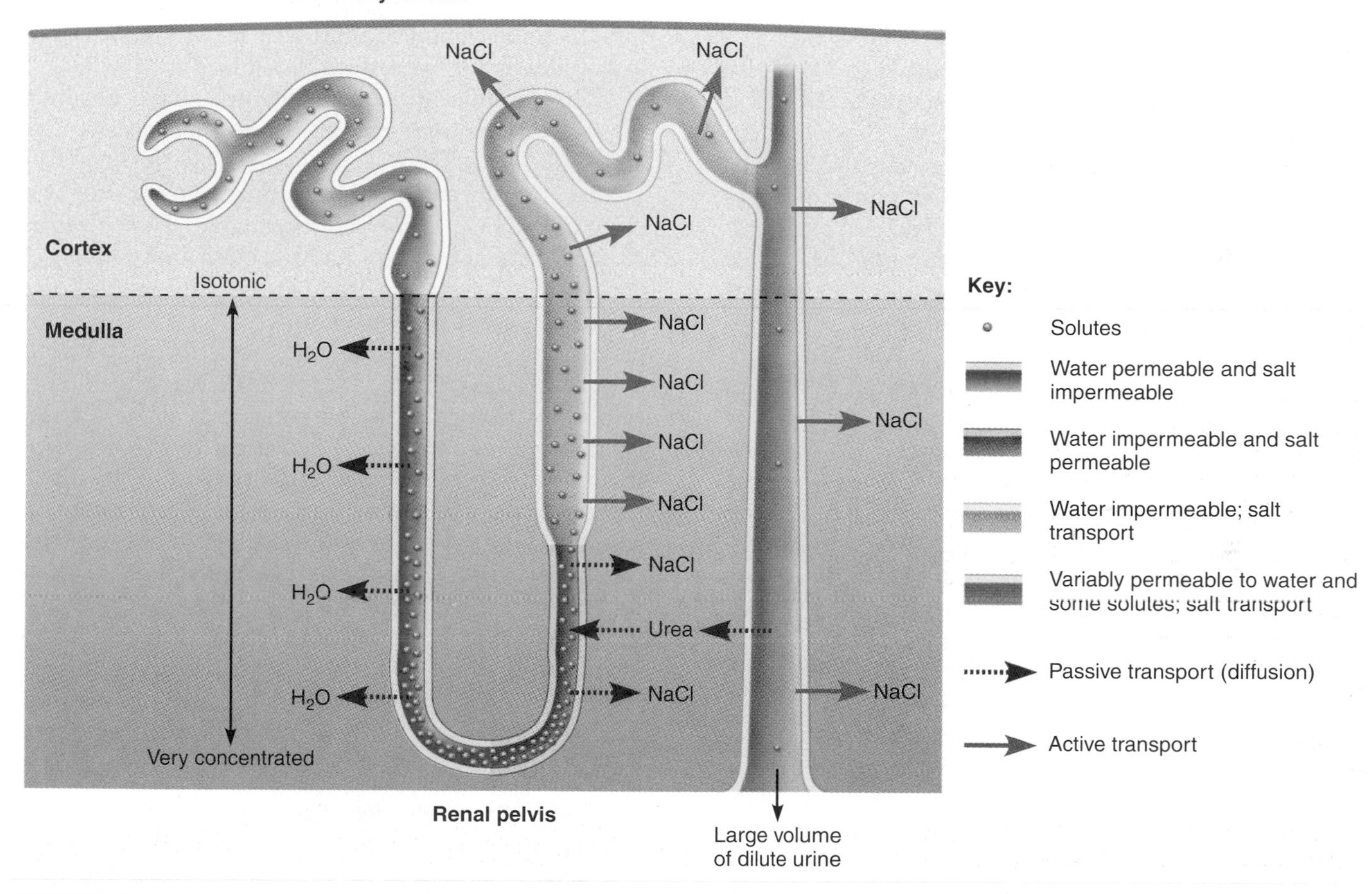

Figure 15.10 The formation of dilute urine. Urine volume and concentration are regulated by controlling the water permeability of the collecting duct as it passes through the medulla. The medulla has an increasing total solute concentration gradient from top to bottom. The concentration gradient is produced by passive and active transport of salt from the ascending limb of the loop of Henle and by passive transport (diffusion) of urea from the collecting duct. Starting at the loop of Henle, the descending limb of the loop of Henle is permeable to water, so water is reabsorbed as the fluid descends. In the thin part of the ascending limb, the tubule becomes impermeable to water but permeable to salt, so salt is reabsorbed passively but not water. In the thick part of the loop of Henle and also in the distal tubule, salt is actively transported, but since water cannot follow (these regions remain impermeable to water) the tubular fluid becomes more dilute. Water permeability of the collecting duct varies; when it is impermeable, as shown here, water is not reabsorbed. The water is excreted and the urine becomes even more dilute as salt reabsorption continues.

Spotlight on Structure & Function

The process of forming a dilute urine begins at the descending limb of the loop of Henle. The descending limb is highly permeable to water. Because the solute concentration is higher in the interstitial fluid than in the lumen, water diffuses out of the loop, into the interstitial fluid, and eventually into the blood. As a result, the remaining fluid becomes more concentrated.

When fluid turns the hairpin corner of the loop, however, the permeability characteristics of the tubule change. The first part of the ascending limb is impermeable to water but permeable to NaCl and urea. Because so much water was removed in the descending limb, sodium concentration in the ascending limb is now higher than in the interstitial fluid of the medulla. With the change in permeability, NaCl can now diffuse out of the ascending limb and into the interstitial fluid of the medulla.

At the same time that salt is diffusing out of the ascending limb of the loop of Henle, urea is diffusing inward from the interstitial fluid. As Figure 15.10 shows, the concentration of urea in the deepest (innermost) portion of the medulla is actually quite high because urea is being reabsorbed from the collecting duct. In other words, urea recycles—from collecting duct, to ascending limb of the loop of Henle, and back to the collecting duct again. This recycling pattern allows the medullary interstitial concentration of urea to become quite high compared to blood plasma.

The final portion of the ascending limb of the loop of Henle becomes impermeable to salt and urea as well as water, so no passive diffusion occurs. However, this region of the ascending limb actively transports sodium and chloride into interstitial fluid by a process that requires energy in the form of ATP. Consequently the interstitial fluid gains solutes while the fluid remaining in the tubule becomes more dilute. The process of active salt reabsorption without the reabsorption of water continues in the distal tubule and collecting duct. By the time the urine reaches the end of the collecting duct it is a very dilute solution of a lot of water, a little urea, and very little salt. You may need to urinate frequently when you are forming a very dilute urine because your kidneys are producing over a liter of urine per hour. ■

Recap **Production of dilute urine depends on the permeability of the loop of Henle. The interstitial fluid gains solutes while water remains in the loop, forming highly diluted urine.**

Producing concentrated urine: Conserving water

Sometimes our problem is too little water, perhaps due to perspiration or living in an arid climate. Less water in the blood means lower blood volume, declining blood pressure, and a risk of dehydration for body cells. Again, your kidneys compensate by reabsorbing more water. The result is a more concentrated urine.

We urinate less when we are dehydrated because the kidneys are reabsorbing more of the glomerular filtrate water than usual. The process of reabsorbing water is regulated by **antidiuretic hormone (ADH)** from the posterior pituitary gland (see Chapter 12, The Endocrine System). ADH increases the permeability of the collecting duct to water. The formation of concentrated urine follows a process similar to that just described for the formation of a dilute urine until the urine reaches the collecting duct. In the presence of ADH, more water is reabsorbed because it diffuses out of the collecting duct toward the high solute concentration in the medulla. Consequently only a small volume of urine will be excreted, and it will be highly concentrated (Figure 15.11).

The production of concentrated urine depends on a **countercurrent exchange** mechanism. Because of the hairpin arrangement of the loops of Henle, fluid flows in opposite directions in the two sides of the loop (*countercurrent* flow). The countercurrent direction of flow and the close anatomical association between the two sides of the loop allows events happening on one side of the loop to influence conditions on the other, and even for water and solutes to be *exchanged* between the two sides if permitted by permeability characteristics of the tubule. Countercurrent exchange is critical to the development of the high concentration gradient of the inner medulla because it permits a small difference in composition between the tubular fluid and the medullary interstitial fluid to be multiplied severalfold along the length of the loop (from the top of the medulla to the bottom). Countercurrent flow and exchange in the vasa recta is also important, because it removes solutes and water from the medulla without dissipating the medullary solute concentration from top to bottom (Figure 15.12).

Recap **The production of concentrated urine is determined by plasma concentrations of ADH and countercurrent mechanisms in the loops of Henle and vasa recta. In the presence of ADH, most of the water is reabsorbed from the collecting duct, leaving a small volume of concentrated urine.**

15.6 Urination depends on a reflex

Urination depends on a neural reflex called the *micturition reflex*. This involves the two urethral sphincters and the bladder. The internal urethral sphincter, which consists of smooth muscle, is an extension of the smooth muscle of the bladder. It remains con-

How is urination controlled?

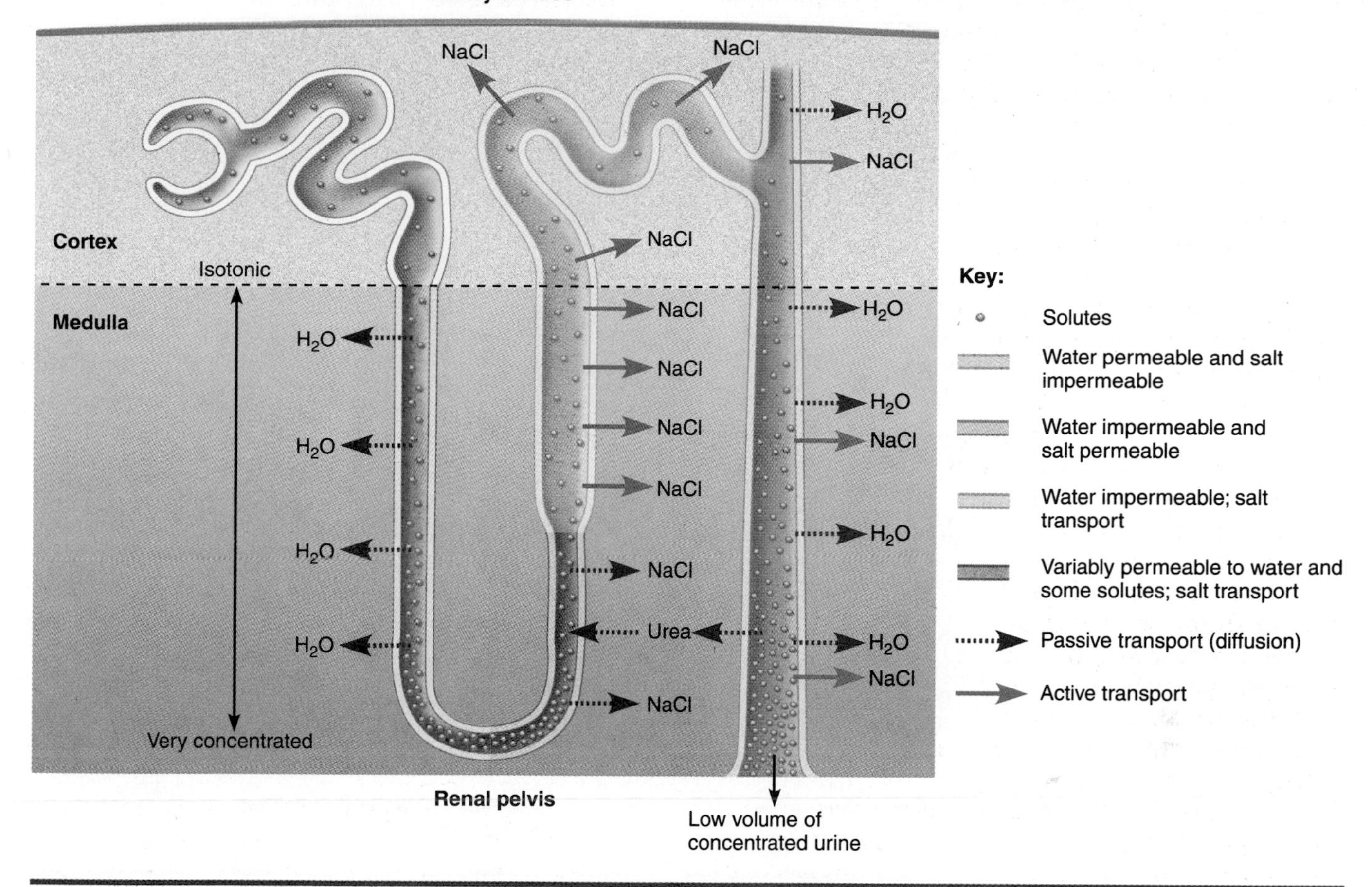

Figure 15.11 The formation of concentrated urine. In the presence of the hormone ADH the collecting duct becomes permeable to water. Consequently most of the water in the collecting duct is reabsorbed by passive diffusion as the fluid in the collecting duct passes through the medulla with its high solute concentration.

tracted unless the bladder is emptying. The external urethral sphincter is skeletal muscle, and therefore it is under voluntary control.

Normally the external urethral sphincter is kept closed by tonic activity of somatic motor neurons controlled by the brain. As the bladder fills with a cup (roughly 250 ml) or more of urine, it starts to stretch. Stretching stimulates sensory nerves that send signals to the spinal cord. Nerves in the spinal cord initiate an involuntary (parasympathetic autonomic) reflex that contracts the smooth muscle of the bladder and relaxes the internal urethral sphincter. Stretch receptor input also goes to the brain, which decreases the activity of somatic motor neurons to the external urethral sphincter, allowing it to relax so urine can flow through.

Humans don't usually have the opportunity to urinate whenever the urge strikes. Fortunately the brain can voluntarily override the micturition reflex by increasing the activity of the somatic nerves that control the external urethral sphincter. This delays urination, for extended periods if necessary. When we finally decide to urinate, the somatic nerve activity is shut off, the external urethral sphincter relaxes, and urination occurs. Voluntary prevention of urination becomes increasingly difficult as the bladder approaches maximum capacity.

Recap **Urination depends on the neural micturition reflex; bladder stretching initiates involuntary relaxation of the internal urethral sphincter. The brain can override the reflex by voluntary contraction of the external urethral sphincter.**

15.7 The kidneys maintain homeostasis in many ways

The kidneys are much more than simply the organs that make urine. As mentioned earlier, they play other important roles in the body as well. The kidneys help to:

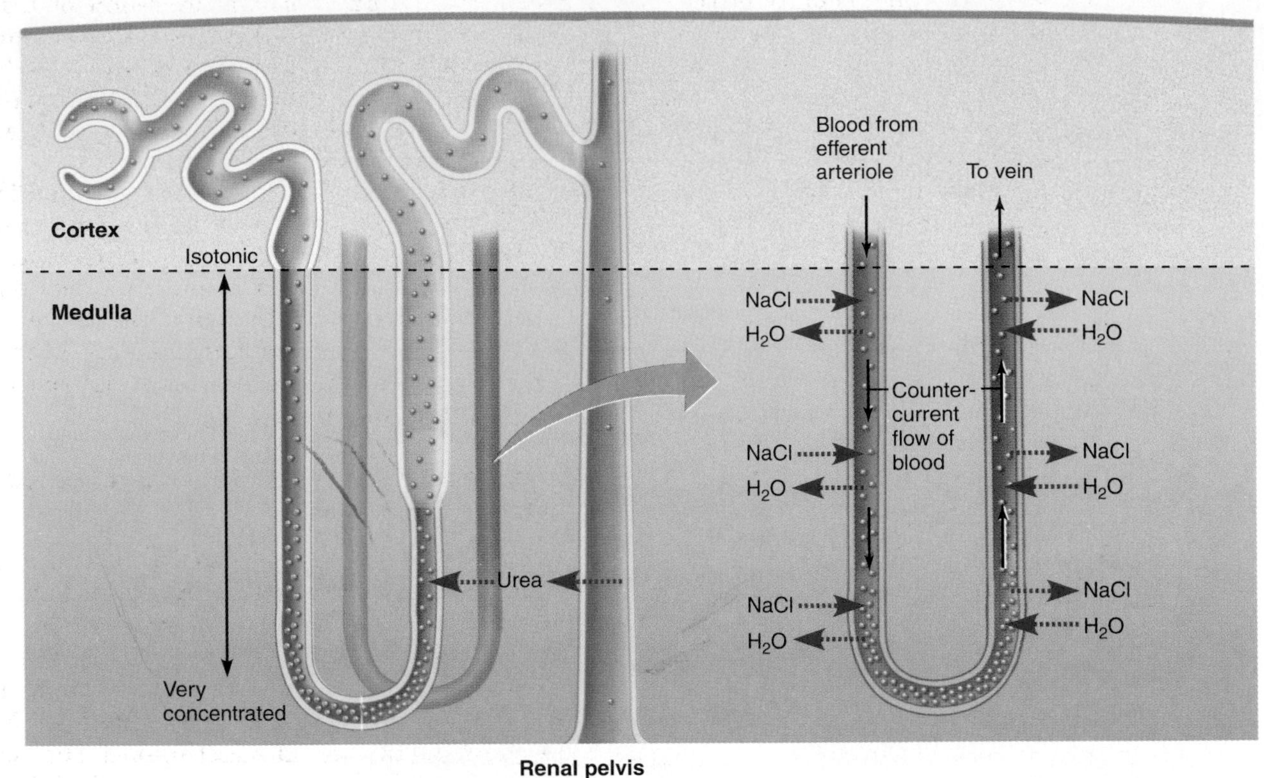

Figure 15.12 Countercurrent exchange in the vasa recta. The vasa recta are closely associated with the tubular segments in the medulla. The vasa recta remove the water and solutes that accumulate in the medulla as a result of reabsorption from the loop of Henle and the collecting duct. Like most capillaries, the vasa recta are highly permeable to water and solutes and rapidly equilibrate with their surrounding fluid. The countercurrent flow of blood in the two parallel vessels of the vasa recta permits the rapid exchange of water and solutes by diffusion, at the same time preserving the medullary concentration gradient from top to bottom.

- Maintain water balance, which adjusts blood volume and blood pressure
- Regulate salt balance, which also controls blood volume
- Maintain acid-base balance and blood pH
- Control production of red blood cells
- Activate an inactive form of vitamin D

The kidneys are responsible for maintaining homeostasis (by excretion) of virtually every type of ion in the body, considering that ions cannot be created or destroyed by metabolic processes. Because the kidneys secrete hormones, they are also endocrine organs. And by chemically altering the vitamin D molecule so it can function as a vitamin, they play a role in metabolism as well. The common theme in these activities is the maintenance of homeostasis.

Water balance determines blood volume and blood pressure

As mentioned earlier, blood volume is determined in part by how we retain or excrete water. Increases in blood volume raise blood pressure, and decreases in blood volume lower it, so it is critical that the blood volume (and water volume) be maintained.

The maintenance of water balance is a function shared by the kidneys, the hypothalamus of the brain, and the posterior pituitary gland of the endocrine system. Water balance is achieved by a negative feedback loop that regulates the solute concentration of the blood (Figure 15.13).

When the concentration of blood solutes rises (too little water compared to the amount of solutes in the blood), the ADH-producing neurons in the hypothalamus cause ADH to be secreted from the posterior pituitary into the blood. ADH circulates to the kidney, where it increases the permeability of the collecting duct to water. This allows more water to be reabsorbed, reducing the amount of water excreted in the urine.

In addition to its effect on ADH, a rise in the blood solute concentration triggers a sensation of thirst, encouraging us to drink more water. The increased water intake combined with the reduced loss of water in the urine lowers the blood solute concentration toward normal again.

Normally there is always some ADH present in the blood. This allows the control mechanism to respond to excess water, too. When there is too much water and blood solute levels fall, the hypothalamus signals the posterior pituitary to reduce its secretion of ADH. The fall in ADH concentration makes the renal collecting duct less permeable to water, and so the excess water will be excreted. We also feel less thirsty when we already have more than enough water.

The term *diuresis* refers to a high urine flow rate. A *diuretic* is any substance that increases the formation and excretion of urine. Diuretic drugs, such as furosemide (Lasix®), are prescribed to reduce blood volume and blood pressure in certain patients, such as people with congestive heart failure or hypertension.

How do coffee or alcohol affect your body's water balance?

Caffeine is a mild diuretic because it inhibits sodium reabsorption, and as sodium is excreted it takes water with it. Alcohol is also a diuretic, although by a different mechanism: it inhibits ADH release. The result is that the renal tubules become less permeable to water and more water is excreted. If you drink excessive amounts of alcohol you may feel dehydrated and thirsty the next day.

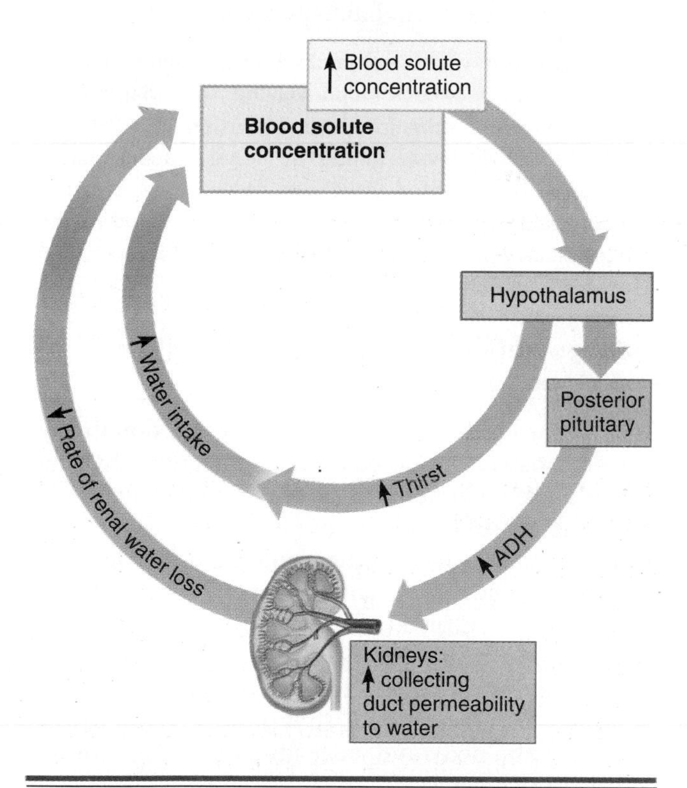

Figure 15.13 Negative feedback loop for the control of blood solute concentration. A rise in blood solute concentration causes an increase in thirst and an increase in ADH secretion. ADH causes the kidneys to form a more concentrated urine and decrease their excretion of water. An increase in water intake coupled with a reduction in water loss from the kidneys dilutes the blood and returns blood solute concentration toward normal. The control loop responds equally well to a fall in blood solute concentration (not shown).

Aldosterone, renin, and ANH control blood volume

The control of blood volume also depends on maintaining the body's salt balance, which in turn depends on three hormones: aldosterone, renin, and atrial natriuretic hormone (ANH).

Aldosterone, a steroid hormone from the adrenal gland, increases the reabsorption of Na^+ across the distal tubule and collecting duct. High concentrations of aldosterone cause nearly all of the filtered sodium to be reabsorbed, so that less than 50 mg (0.05 gram) per day appears in the urine. Low levels of the hormone allow as much as 20–25 grams of sodium to be excreted each day. When we consider that the average North American consumes about 10 grams of sodium daily, we see that aldosterone provides more than enough control over sodium excretion.

Given what aldosterone does, we might think its concentration is controlled by blood sodium levels. This is only partly true, for sodium concentration is only a weak stimulus. The strongest stimulus for aldosterone secretion is a decrease in blood volume (or blood pressure), and in particular how a change in blood volume affects another hormone, *renin.*

Renin is technically an enzyme that functions like a hormone. It is secreted by a group of afferent arteriole cells that are part of a structure located very near the glomerulus called the **juxtaglomerular apparatus** (Figure 15.14). The juxtaglomerular apparatus is composed of the renin-secreting afferent arteriole cells, the cells of the distal tubule that are in close contact with them, and some other renal cells located between them.

Figure 15.15 illustrates the negative feedback mechanism for the control of blood volume. The steps in this mechanism are as follows:

- Decreases in blood volume or pressure stimulate the juxtaglomerular apparatus within the kidneys to secrete renin.

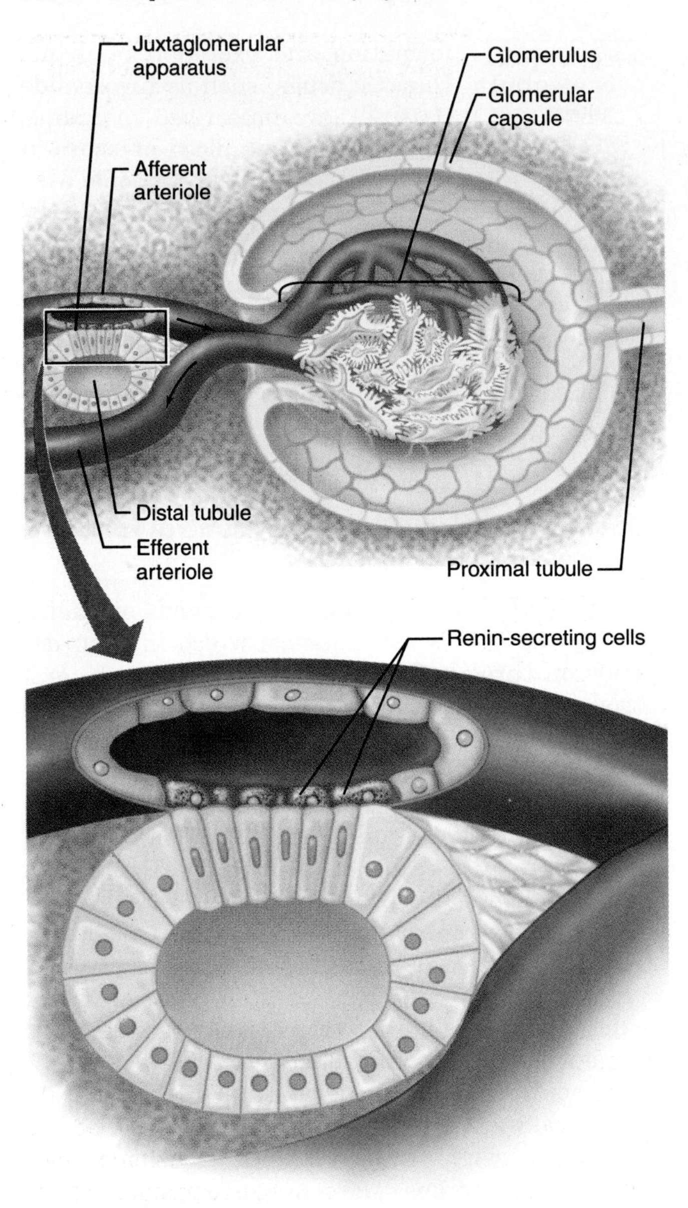

Figure 15.14 The juxtaglomerular apparatus. The juxtaglomerular apparatus is a region of contact between the afferent arteriole and the distal tubule. The afferent arteriolar cells of the juxtaglomerular apparatus synthesize and secrete renin.

- In the blood, renin cleaves a small peptide of 10 amino acids called *angiotensin I* from a large inactive protein molecule, *angiotensinogen,* produced by the liver. Angiotensin I loses two more amino acids to become *angiotensin II,* a highly active and important regulatory hormone.
- Angiotensin II stimulates the adrenal cortex to secrete aldosterone.
- Aldosterone increases the rate of sodium reabsorption by the distal tubules and collecting ducts.
- As blood sodium levels rise, water is retained and blood volume rises. Water is retained along with sodium because of ADH and its control over solute concentrations, as discussed previously. Thus renin and aldosterone (and ADH) all interact to control blood volume and blood pressure.

Renin regulates blood pressure in another way as well: angiotensin II is a potent vasoconstrictor that narrows the diameter of blood vessels. An increase in renin secretion can raise arterial blood pressure within minutes, long before it has influenced aldosterone secretion and blood volume.

There is yet another controller of sodium excretion, a peptide hormone called *atrial natriuretic hormone (ANH).* (A "natriuretic" is a substance that increases the excretion of Na^+.) The atria of the heart secrete ANH in response to stretch (which indicates increased blood volume). After circulating to the kidneys, ANH inhibits Na^+ reabsorption in the distal tubule and collecting duct, and so causes Na^+ excretion to increase. In effect, ANH has an effect opposite to that of renin and aldosterone.

Recap **The water concentration of blood is maintained by a feedback loop involving ADH. Rising blood solute levels stimulate the secretion of ADH, and falling solute levels inhibit it. Renal reabsorption of sodium is regulated largely by aldosterone, which is controlled by the secretion of renin when blood volume (or blood pressure) declines. The hormone ANH stimulates sodium excretion.**

Kidneys help maintain acid-base balance and blood pH

Recall that *acids* are molecules that can donate hydrogen ions (H^+), and *bases* are molecules that can accept a hydrogen ion. Many of the body's metabolic reactions generate H^+. If not eliminated, these acids would accumulate—a dangerous situation, because blood pH must remain in a relatively narrow range of 7.35 to 7.45. If blood pH strays by more than a few tenths of a unit beyond this range, it could be fatal.

The kidneys contribute to the maintenance of acid-base balance, a role they share with various buffers in the body and with the lungs. The kidneys do this in two ways:

- The kidney tubules secrete H^+ into the tubular lumen. Most of the secreted H^+ is used in the process of reabsorbing (conserving) all of the bicarbonate (HCO_3^-) that was filtered by the glomerulus. Once the reabsorption of HCO_3^- is complete, the remaining H^+ is buffered by negatively charged ions in the urine (phosphate and sulfate) and eventually excreted.

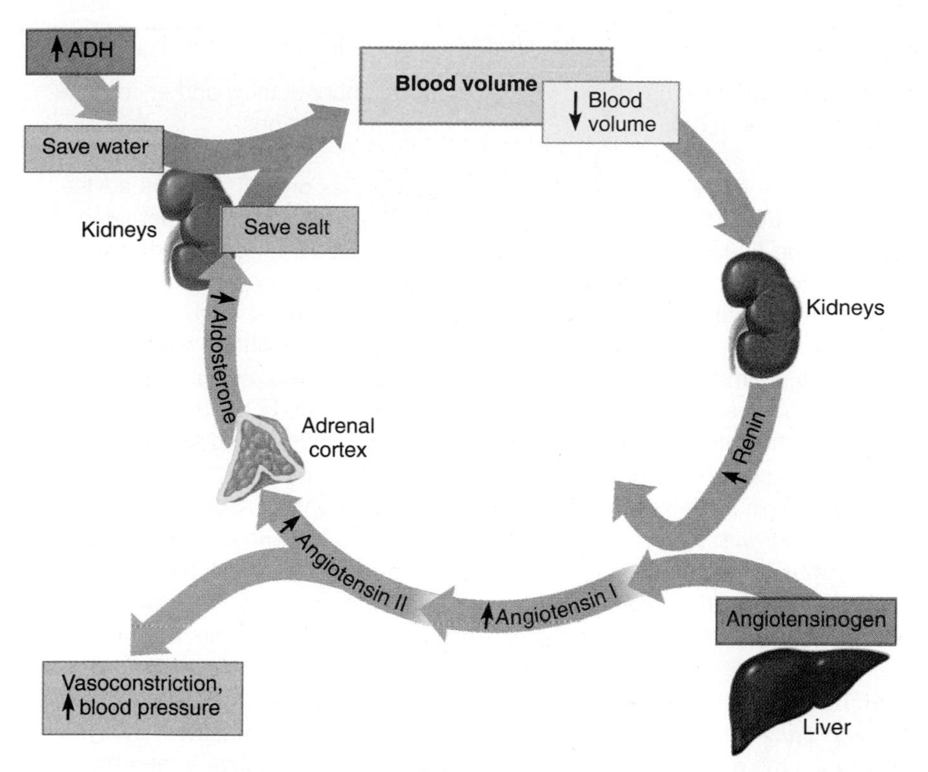

Figure 15.15 Regulation of blood volume by renin, aldosterone, and ADH. A decline in blood volume causes the kidneys to secrete renin. Renin activates a cascade of events that ultimately causes the kidneys to save salt and water, returning blood volume to normal.

- Kidney tubule cells produce ammonia (NH_3) as part of their normal metabolic activity. The NH_3 combines with H^+ within the cell to become ammonium (NH_4^+), an acid. The NH_4^+ is then secreted into the renal tubule and excreted.

Ordinarily our diets provide us with about 50–60 millimoles/day of excess H^+ that must be excreted if we are to stay in balance. The rest of our daily production of H^+ results from the creation of CO_2 during metabolism, which is handled by the lungs. The renal excretion of H^+ with sulfate, phosphate, and ammonia exactly offsets the excess acid intake per day.

Erythropoietin stimulates production of red blood cells

Recall from Chapter 7 (Blood) that the kidneys regulate the production of red blood cells in the bone marrow. They do this via secretion of a hormone, **erythropoietin.**

When certain cells scattered throughout the kidneys sense a fall in the amount of oxygen available to them, they release erythropoietin, which then stimulates the bone marrow to produce more red blood cells. The presence of more red blood cells increases the oxygen-carrying capacity of the blood. Once oxygen is again being delivered normally, the cells stop activating erythropoietin and RBC production slows. The activation of erythropoietin by certain kidney cells constitutes a negative feedback loop for the control of oxygen availability to all cells, because it keeps the oxygen-carrying capacity of the blood within normal limits.

Try It Yourself

Demonstrating the Water-Retaining Power of Salt

The amount of water your body retains depends more on the amount of salt in your diet than on how much water you drink. To demonstrate this, try the following experiment.

On a normal day between meals, empty your bladder. Wait half an hour, then urinate again. This time, collect and measure the volume of urine you produced. This is your baseline rate of urine production. Then drink a quart of water quickly and measure your urine output for four more consecutive half-hour periods. Subtract the baseline amount of urine from each time period to find out how much of the water you excreted in two hours.

On a separate day repeat the experiment, but this time dissolve about 5 grams (approximately 3/4 of a measuring teaspoon) of salt in the water first. It may help to dissolve the salt in most of the quart of water, drink it, and then drink the remaining water to remove the very salty taste from your mouth. You should find that you excrete a lot less of the extra water in the next two hours than you did the first time.

Certain people—for example those with high blood pressure or congestive heart failure—may need to retain less water for health reasons. Because of the water-retaining power of salt, such people generally are advised to restrict their salt intake, not their water intake.

Health Watch

Urinary Tract Infections

The phrase **urinary tract infection (UTI)** refers to the presence of microbes in the urine or an infection in any part of the urinary system. Symptoms include swelling and redness around the urethral opening, a burning sensation or pain while urinating, difficulty urinating, bed-wetting, even low back pain. Severe infections can produce visible blood and pus in urine. Depending on where they are located, UTIs include *urethritis* (inflammation of the ureter), *cystitis* (inflammation of the bladder), *pyelitis* (inflammation of the renal pelvis), and *pyelonephritis* (inflammation of the kidney tissues). Of these, pyelonephritis is potentially the most dangerous—if left untreated it may permanently damage the kidneys and impair renal function.

Most UTIs are caused by bacteria that make their way up the urethra from the body surface. Frequent culprits include *E. coli*, which normally inhabits the colon, and sexually transmitted microbes such as *chlamydia* and the organism that causes gonorrhea. From the urethra and bladder, microbes can travel up the ureters to the kidneys. This is why it is important to treat all urinary tract infections immediately. Most cases can be cured with antibiotics.

Cystitis is more common in women because their urethras are much shorter, allowing organisms to reach the bladder more easily. However, men get UTIs too. In males, a swollen prostate gland (an organ that surrounds the male urethra) may restrict urinary outflow and encourage bacterial infections.

Personal habits can significantly cut your risk of UTIs. Doctors advise urinating frequently; don't let urine stay in your bladder for extended periods. Urinating immediately after sex helps flush out microbes. And, although it may seem self-evident, good personal hygiene can also reduce the risk.

A common home remedy that is based on sound principles is to drink lots of fluids, especially those with a high acid content such as cranberry juice. This has two benefits: first, you urinate more. Second, because the kidneys excrete the extra acid into the urine, the urine pH falls from its normal 6.5 to near 5.5. Most microbes do not thrive under such acid conditions. However, no home remedy is a substitute for medical treatment.

Kidneys activate vitamin D

Vitamin D is important for absorbing calcium and phosphate from the digestive tract and developing healthy bones and teeth. We absorb some vitamin D from food, but the body also manufactures vitamin D in a three-step process involving the skin, liver, and kidneys.

Vitamin D synthesis begins when ultraviolet rays in sunlight strike a steroid molecule in skin that is similar to cholesterol, producing an inactive form of vitamin D. The inactive vitamin D is transported to the liver where it is chemically altered, then carried to the kidneys where it is converted to a usable form. This is why people with long-standing kidney failure often suffer from vitamin D deficiency.

Recap **The kidneys help maintain the body's acid-base balance and blood pH by reabsorbing all filtered bicarbonate and by excreting H^+. Decreased oxygen delivery to the kidneys triggers the release of erythropoietin, which stimulates the production of red blood cells by bone marrow. Synthesis of vitamin D involves the skin, liver, and kidneys.**

15.8 Disorders of the urinary system

The good news about urinary system disorders is that the kidneys have an enormous reserve of function. You can survive quite nicely on one kidney, and even on one kidney that is only functioning half as well as it should. Nevertheless, sometimes urinary system problems do occur. They include kidney stones, acute and chronic renal failure, and urinary tract infections.

Kidney stones can block urine flow

Sometimes minerals in urine crystallize in the renal pelvis and form solid masses called *kidney stones*. Most stones are less than 5 mm in diameter and are excreted in urine with no trouble. Others may grow larger and obstruct a ureter, blocking urine flow and causing great pain. Kidney stones can be removed surgically or crushed with ultrasonic shock waves, after which the pulverized fragments can be excreted with less pain.

Acute and chronic renal failure impair kidney function

Kidney impairments that are short-term and possibly correctable are called *acute renal failure*. Conditions that might impair kidney function temporarily include sustained decreases in blood pressure to below the pressure required for filtration, large kidney stones in the renal pelvis, infections, transfusion reactions, burns, severe injuries, and toxic drugs or chemicals. A surprising number of common medications are toxic to the kidneys if taken too long or at

Directions in Science

Is It Safe to Donate a Kidney?

The 35,000 people on waiting lists for kidney transplants almost certainly have parents, siblings, and/or children. Relatives make the best donors because they are genetically closer than unrelated individuals. One of the most difficult decisions those relatives will ever have to make is whether to donate one of their kidneys if it turns out to be a good match.

The burning question that any potential donor would ask, of course, concerns safety. The surgery itself is very safe, but what about the long-term consequences—is it possible to live a long and healthy life with just one kidney? The answer has changed several times since live kidney donations became possible, an example of how our knowledge has progressed over time.

During the 1960s, when renal transplants were first performed, donating a kidney seemed perfectly safe. After all, the reasoning went, we can function on as little as 25% of normal renal capacity. Furthermore, when one kidney is removed the remaining organ grows substantially, and within a year renal function achieves 60%–75% of the function of two kidneys. For several years, nearly all transplanted kidneys came from living, related donors.

In the early 1980s the story changed dramatically. A prominent group of researchers from Harvard University questioned the long-term safety of donating a kidney. Based on findings in rats, they proposed that reducing the kidney mass might, under poor dietary conditions (such as a very high-protein diet), cause renal damage and a slow decrease of renal function over time. As the debate raged in academic circles, kidney transplants from living donors came to a virtual halt.

To resolve the controversy, scientists needed reliable long-term data on people with only one kidney. Researchers began to comb the country for past donors, but finding and convincing them to be examined wasn't the only problem. The other was that transplants had only been performed for about 20 years, a relatively short time. To address the time issue, some researchers studied people who had had a kidney removed in childhood.

By the 1990s the data began to come in. Study after study revealed that the subjects retained good renal function after losing a kidney. In addition, they did not show increased rates of high blood pressure or other kidney-related diseases. Today we again believe (based on more data than we had originally) that kidney donation is safe. Consequently, the number of living kidney donations is again on the rise.

Were the Harvard researchers "wrong"? Absolutely not! They conducted valuable experiments in rats that caused them to formulate an important hypothesis about the possible risks of kidney donation in humans. Thanks to their efforts and follow-up from other researchers, today's kidney donors can enjoy greater confidence than ever.

high doses. The kidneys are particularly vulnerable because of their reabsorptive efficiency, which tends to concentrate harmful substances in the urine. The toxicity problem may be prevented by drinking lots of water and excreting a more dilute urine.

Chronic renal failure, also known as "end-stage renal disease" (ESRD), is defined as long-term irreversible damage leading to at least a 60% reduction in functioning nephrons and failure of the kidneys to function properly. People with chronic renal failure may have less than 10% of the normal filtering capacity of the kidneys, and some essentially have no renal function at all.

Any of the factors that cause acute renal failure can, if not corrected, progress to chronic failure. In addition, approximately 40% of people with type I (insulin-dependent) diabetes eventually develop renal failure, especially if the diabetes has not been well controlled. Approximately 13 million Americans have some form of chronic renal failure.

Recap **The kidneys have a huge reserve of function. However, acute or chronic renal failure can result from prolonged changes in blood pressure, disease, large kidney stones, transfusion reactions, burns, injuries, toxic substances, and other conditions such as diabetes.**

Dialysis cleanses the blood artificially

The body cannot replace irreparably damaged nephrons. This leaves two options for people with severe renal failure: dialysis or a kidney transplant.

Dialysis attempts to duplicate the functions of healthy kidneys. A dialysis technique that can be done at home without a kidney machine is called *continuous ambulatory peritoneal dialysis (CAPD).* In this procedure, dialysis fluid (a fluid similar to extracellular fluid) is placed directly into the peritoneal cavity through an access port permanently implanted in the abdominal wall. The fluid is left in the peritoneal cavity for several hours, during which time it exchanges wastes and ions with the capillaries. Then the fluid is drained and discarded.

If done on a regular basis, CAPD can be modestly effective. It is convenient and allows freedom of movement. However, there is a risk of infection because of the access port through the abdominal wall.

CurrentIssue

How Should Scarce Organs Be Allocated?

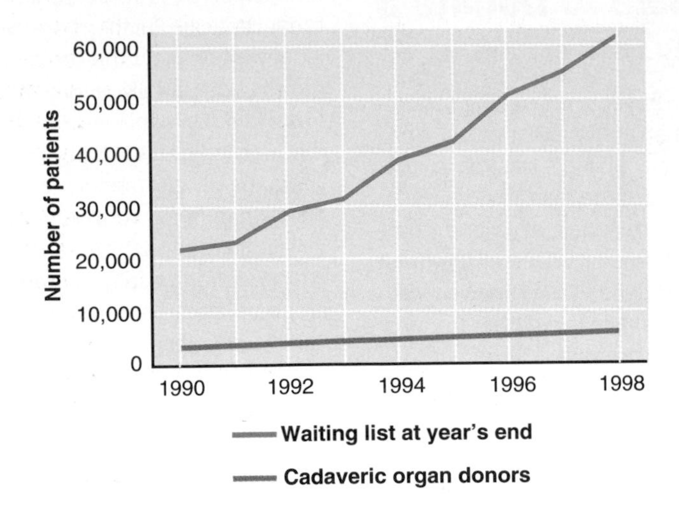

Every 14 minutes another person is added to the growing list of over 70,000 patients who are waiting for an organ transplant. Within the next year, 6,000 of the people on the list will die. Only 6,000 cadaver (deceased person) organ donors are available each year, not nearly enough to meet the ever-increasing demand. Over 47,000 people are waiting for a kidney; another 15,000 are waiting for a liver, 4,000 for a heart, 4,000 for a lung, and about 1,000 for a pancreas.

Given this huge difference between organ supply and need, how is it decided who gets the organs that do become available? Will you be able to get an organ if you ever need it? For those on the list right now, the answers to these questions are a matter of life for one person, death for another. Three key facts determine organ availability to for particular patient:

1. No matter how rich you are, in the U.S.A. you can't buy an organ. It's against the law.
2. A living donor may donate certain organs to a specific patient. A live donor may donate a kidney (because he/she has two) or part of a lung or liver. Patient and donor must have the same blood type and are tested for six key tissue antigens that determine the closeness of the immunological "match." The more antigens that match, the less likely it is that the patient will reject the "foreign" organ. Organs from living donors usually come from a close relative because they are more likely to be a good match.
3. If you cannot find your own living donor, or if you need a heart or a pancreas, the only source is an organ from a cadaver. Organs may be harvested from a cadaver only when it is the express wish of the deceased or (in the absence of knowledge of the deceased's wishes) permission is granted by relatives. Persons who wish to donate their organs after their death can make their wishes known by signing and carrying a uniform donor card issued by most Departments of Motor Vehicles or available on the Web. Although national statistics aren't available, it is estimated that fewer than a third of all persons over 18 have signed such a card.

How Are Scarce Cadaver Organs Currently Allocated?

With all those people on waiting lists, what is the process for distributing the organs of a cadaveric donor? Is the current allocation system equally fair to all patients?

All cadaveric organs are procured and allocated to patients by 65 independent regional *organ procurement organizations* (OPOs), each linked to a geographical region or group of transplant centers. A private non-profit corporation called the United Network for Organ Sharing (UNOS) links the OPOs together and maintains the national lists of potential organ transplant recipients. UNOS and the various OPOs operate according to federal rules established by the Department of Health and Human Services (DHHS). In essence, the system is operated by the private sector, but is watched over by the government.

According to the current DHHS rules, a patient who needs an organ places him/herself on the transplant list of one of the regional OPOs, generally the one that covers the region in which he/she lives. When a cadaveric organ becomes available in a particular OPO it is tested for blood type and (for some organs) the six antigens. Perfect matches—the same blood type and the same six antigens—are so rare that a patient anywhere in the country has first priority for a perfectly matched organ, provided there is time to transport either the organ or the patient.

For all partial matches and for organs that do not need to be matched except for blood type, such as livers, the organ is first made available to all patients on the list in the OPO that procured the organ. Only when none of the patients on the waiting list within the regional OPO is considered a suitable candidate for the organ will the organ be made available to patients of other OPOs. The amount of money a patient has is not supposed to be a consideration, but patients who are willing to pay for the transplant themselves can put themselves on the transplant lists of several different OPOs at once (insurance generally only pays for a transplant in the patient's primary OPO).

In the past few years a heated debate has developed over the fairness of the current system. Opponents of the current system argue that:

- The current system results in disparities in waiting times between the various OPOs. For example, the average waiting time for a liver ranges from four months to over 2½ years, depending on the OPO.
- Organs do not always go to the patients who need them the most (the sickest first) because they can be offered first to *all* possible patients in the OPO that procured the organ.
- The current system may discriminate against minorities and the poor who need organs, because they live in regions that tend to have lower rates of organ donation.

In the face of these criticisms, in 1998 the U.S. Department of Health and Human Services (DHHS) attempted to change the rules governing the operation of UNOS. The new rule would have required that organs be allocated according to common criteria of medical need, not geographical location. The proposed rule, however, itself created a firestorm of controversy, particularly from smaller OPOs with higher rates of organ donation. Critics of the proposed rule change argued that:

- With geography no longer a factor, the country's largest hospitals and transplant centers would sign up the most patients on their waiting lists. With the most patients as potential matches, they would then get most of the donated organs (and make the most money). Smaller transplant centers might close. In the long run competition might be reduced and this would not be in the best interest of patients.
- Organ donations might decline when people realized that their organs would not be used for patients in their region.
- Patient survival rates would decline if the sicker patients were always given higher priority for organs.

In response, in late 1998 Congress suspended implementation of the new DHHS rule pending further study by the National Academy of Sciences. Their report, completed in 1999, found that:

- At least for some organs, under the current regional allocation system less severely ill patients were more likely to receive transplants in some OPOs, as critics of the current regional allocation system claimed.
- Evidence from the few instances of voluntary organ sharing by several OPOs did not indicate that widescale sharing would result in either the closing of smaller transplant centers or cause organ donation rates to decline.
- There was no evidence that minorities and poor patients did (or would) suffer adversely under either allocation system.

As a result of their study, the commission recommended that the new rules be implemented as soon as possible. Nevertheless the controversy continues, and as of June 2000 Congress still had not enacted the proposed new rule into law.

Developments to Watch

- It's illegal to sell an organ in this country. Organs reportedly have been legally sold, however, in India, the Philippines, and China. Critics charge that the sale of organs exploits desperately poor people. Nevertheless, a bill was introduced in the House of Representatives in 2000 that would give a $10,000 tax credit to a family who donated a person's organs at the time of his/her death.
- A switch to "presumed consent" laws could dramatically increase organ donations. The current U.S. policy, called "family consent," requires clear evidence of donor consent, or family consent after death if the potential donor had not made his/her wishes known. Singapore, Austria, and Belgium, and Spain now presume that organs are available unless the donor (or the family) explicitly states otherwise. In Belgium the number of kidney donations doubled in the first three years after the law's passage.
- Animal organs may soon be used for human transplantation. An organ transplant from another species is called a xenograft. Researchers are experimenting with pigs that have been genetically altered with the goal of producing organs that cause a less vigorous immune response from a human recipient.
- Some researchers hypothesize that undifferentiated human stem cells could be coaxed into growing and differentiating into entire organs suitable for transplantation. Such a possibility is still a long way off.

In the meantime several states have taken matters into their own hands. By late 1998 at least five states (Kentucky, Louisiana, Oklahoma, South Carolina, and Wisconsin) had passed laws mandating that organs donated in their states must be offered first to in-state patients. Whether these laws will stand up under challenge is an open question. The scramble for scarce organs continues.

In another type of dialysis, called **hemodialysis,** the patient's blood is circulated through an artificial kidney machine consisting of semipermeable membranes in contact with a large volume of clean fluid (Figure 15.16). Metabolic wastes and excess ions diffuse from the blood into the dialysis fluid, which is discarded. The procedure is costly (over $50,000 per year) and time-consuming. Typically the patient must go to a medical site called a dialysis unit for treatment several times each week, with each treatment taking a whole morning or afternoon.

Kidney transplants are a permanent solution to renal failure

Most people currently on renal dialysis will remain on dialysis for the rest of their lives. Although dialysis is a life-saving technique, it is not the perfect solution. It is difficult, if not impossible, to achieve complete homeostasis of all ions and wastes by artificial means. Furthermore, dialysis does not replace the renal hormones. The best hope for many chronic renal failure patients is to receive a living kidney from another person, but there are 47,000 people on waiting lists to get one.

When kidney transplants began in the 1960s, the biggest challenge was to find a good immunological match so the recipient's body would not reject the foreign kidney. Often close relatives were asked to consider donating a kidney. With the advent of better drugs to suppress the immune system, improved tissue matching techniques, national data banks of patients awaiting kidneys, and rapid jet transport, it is now feasible for the donor kidney to come from a cadaver. The main bottleneck now is that not enough people have offered to donate their organs after death. Thousands of people currently on waiting lists will die before they can receive a kidney.

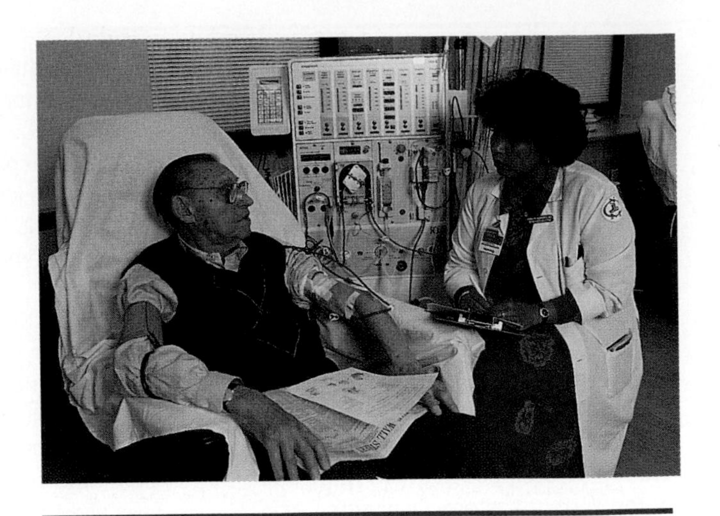

Figure 15.16 **A patient undergoing hemodialysis.**

What can you do to help? Specify your willingness to make an anatomical gift by completing an organ donor card. Many states' drivers licenses have a donor card on the back.

> ***Recap*** **CAPD is an at-home technique in which fluid is placed in the peritoneal cavity and replaced at regular intervals. Hemodialysis cleanses the blood by means of an artificial kidney machine consisting of a semipermeable membrane and clean fluid. Kidney transplants are the best hope for people in renal failure. Currently there is a severe shortage of kidneys available for transplant.**

Chapter Summary

The urinary system contributes to homeostasis

CD Fluid Balance/Water Homeostasis

- The kidneys, the lungs, the liver, and the skin all participate in the maintenance of homeostasis of the internal environment.
- The kidneys are the primary regulators of water balance and most excess solutes, including especially inorganic ions and urea.

Organs of the urinary system

CD Urinary System/Anatomy Review

- The urinary system consists of those organs that produce, transport, store, and excrete urine. The urinary system includes the kidneys, the ureters, the bladder, and the urethra.
- Functions of the kidneys include regulation of the volume and composition of body fluids, excretion of wastes, regulation of blood pressure, regulation of the production of red blood cells, and the activation of vitamin D.

Nephrons produce urine

CD Urinary System/Glomular Filtration

- The functional unit of the kidneys is the nephron. Each nephron consists of a tubular component and the blood vessels that supply it.
- The tubular components of a nephron are the glomerular capsule, proximal tubule, loop of Henle, distal tubule, and collecting duct. The collecting duct is shared by many nephrons.
- A tuft of capillaries called the glomerulus is enclosed within each glomerular capsule. Peritubu-

lar capillaries supply proximal and distal tubules, and vasa recta supply loops of Henle and collecting ducts.

Formation of urine—filtration, reabsorption, and secretion

CD Fluid Balance/Introduction to Body Fluids

CD Fluid Balance/Water Homeostasis

- The formation of urine involves three processes: glomerular filtration, tubular reabsorption, and tubular secretion.
- Approximately 180 liters per day of protein-free plasma fluid is filtered across the glomerulus and into the tubular space within the glomerular capsule. Filtration is driven by high blood pressure in the glomerular capillaries.
- Ninety-nine percent of all filtered water and salt and all of the filtered bicarbonate, glucose, and amino acids are reabsorbed in tubular reabsorption. The active transport of sodium provides the driving force for the reabsorption of nearly all other substances.
- Tubular secretion is a minor process relative to reabsorption but is critical for the regulation of acid-base balance and for the removal of certain toxic wastes.

The kidneys can produce dilute or concentrated urine

CD Fluid Balance/Water Homeostasis

- The ability of the kidneys to form either a dilute or a concentrated urine depends on the high solute concentration in the renal medulla and the ability to alter the permeability of the collecting duct to water.
- Dilute urine is formed in the absence of the hormone ADH. In the absence of ADH, reabsorption of salt without reabsorption of water continues in the collecting duct.
- Concentrated urine is formed when ADH increases the collecting duct's permeability to water, allowing water to diffuse toward the high solute concentration in the medulla.
- The ability of the kidneys to produce a concentrated urine is dependent upon a countercurrent mechanism that exists in the hairpin arrangements of the loops of Henle and the vasa recta.

Urination depends on a reflex

- Urination is caused by the micturition reflex, a neural reflex initiated by stretch of the bladder.
- Urination can be prevented by higher (voluntary) neural signals from the brain.

The kidneys maintain homeostasis in many ways

CD Fluid Balance/Water Balance

- Water balance is maintained by a negative feedback loop involving ADH. The main stimulus for the secretion of ADH is an increase in the solute concentration of the blood.
- Blood volume is regulated by maintenance of the body's salt balance, which is controlled primarily by a negative feedback loop influenced by two hormones: renin from the kidney and aldosterone from the adrenal cortex.
- The kidneys secrete (and excrete) H^+ and NH_4^+ in amounts equal to the net gain of acid per day (other than as CO_2). They also secrete H^+ as part of the mechanism for the reabsorption of all filtered HCO_3^-.
- The kidneys synthesize and secrete erythropoietin, the hormone responsible for the regulation of red blood cell production.
- The kidneys are required for the activation of vitamin D.

Diseases of the Urinary System

- The kidneys are vulnerable to damage by toxic substances, infections, sustained decreases in blood pressure, or blockage of urinary outflow.
- Nephrons that are damaged beyond repair are not replaced.
- Hemodialysis is an artificial procedure for cleansing the blood of wastes and excess solutes.
- Renal transplantation is technically easy and usually quite successful, but there is a shortage of donor kidneys.

Terms You Should Know

ADH, 384
collecting duct, 377
cortex, 375
countercurrent mechanism, 384
distal tubule, 377
glomerular capsule, 377
glomerular filtration, 379
glomerulus, 377
hemodialysis, 394
kidneys, 375
loop of Henle, 377
medulla, 375
nephron, 377
peritubular capillaries, 378
proximal tubule, 377
renal failure, 390
renin, 387
tubular reabsorption, 380
tubular secretion, 382
urea, 375
ureter, 376
urethra, 376
vasa recta, 378

Test Yourself

Matching Exercise

_____ 1. Structure that surrounds and encloses the glomerulus
_____ 2. All the glomerular capsules and all proximal and distal tubules are located in this region of the kidney
_____ 3. Blood vessel that leads directly to either peritubular capillaries or vasa recta
_____ 4. Long, thin blood vessels in the renal medulla
_____ 5. Capillaries that receive most of the fluid and water reabsorbed from the renal tubules
_____ 6. Tuft of capillaries where filtration occurs
_____ 7. Structure that transports urine from a kidney to the bladder
_____ 8. Structure that transports urine from the bladder to the exterior of the body
_____ 9. The last region of the renal tubule, shared by many nephrons
_____10. The unit of function in a kidney

a. collecting duct
b. ureter
c. vasa recta
d. urethra
e. renal cortex
f. glomerulus
g. glomerular capsule
h. peritubular capillaries
i. nephron
j. efferent arteriole

Fill in the Blank

11. The secretion of the hormone __________ is stimulated by a drop in oxygen delivery to the kidney.
12. It is believed that the cells responsive to changes in solute concentration that alter the secretion of ADH are located in the ________ of the brain.
13. ________ is a hormonelike enzyme secreted by afferent arteriole cells of the juxtaglomerular apparatus.
14. The primary nitrogenous waste excreted in the urine is ________.
15. The hollow space in the interior of the kidney where urine collects after it is formed is called the renal ________.
16. The first region of the renal tubule, called the __________ tubule, receives the filtrate from the glomerular capsule.
17. [Filtration] minus [reabsorption] plus [secretion] equals [__________].
18. A short-term and possibly reversible impairment of renal function is called __________ __________ __________.
19. The active transport of __________ out of the renal tubular cell is ultimately responsible for the tubular reabsorption of sodium, chloride, water, and all amino acids and monosaccharides.
20. The hormone ADH increases the permeability of the collecting duct to ________.

True or False?

21. The interstitial fluid in the renal medulla has the same solute concentration as interstitial fluid in the renal cortex.
22. The kidneys play a role in activating the vitamin D that is made within the body.
23. The kidneys cannot contribute to acid-base balance because they do not excrete CO_2.
24. Diabetes can lead to chronic renal failure.
25. Currently there is a shortage of kidneys available for transplant.

Concept Review

1. List the sites of water gain and loss from the body.
2. Name the primary nitrogenous waste excreted by the kidneys.
3. Name the two urethral sphincters controlling micturition (urination), and indicate which one is under voluntary control.
4. Indicate where vasa recta are located, and list the two tubular regions of the nephron they serve.
5. Describe the driving force responsible for filtration of fluid from the glomerular capillaries into the tubular space within the glomerular capsule.
6. Name the one region of the tubule where water permeability is *regulated*.
7. Indicate the primary stimulus for the secretion of the hormone ADH.
8. Describe the function of the hormone aldosterone.
9. Describe how chronic renal failure differs from acute renal failure.
10. Discuss the treatment options available to someone in chronic renal failure (end-stage renal disease.)

Apply What You Know

1. What do you suppose might happen to hematocrit (the fraction of the blood that is red blood cells) if you were to live at altitudes of greater than 10,000 feet for three months? Explain.

2. Are you working your kidneys harder than normal when you eat a high-salt diet or drink lots of fluid? Explain your reasoning.
3. Explain the mechanism for why one feels thirsty after heavy exercise accompanied by sweating.

Case in Point

Renal Failure Patients are at Risk for Certain Cancers

Over the past quarter-century the number of Americans with chronic renal failure, also called end-stage renal disease (ESRD), has increased dramatically, from approximately 10,000 people in 1973 to about 284,000 today. During this period the National Institute of Diabetes and Digestive and Kidney Diseases (NIDDKD) has collected and analyzed data on the risk factors, incidence, and outcomes of the condition.

Among the NIDDKD findings: there is a strong correlation between diabetes mellitus, hypertension, and kidney problems. In fact, uncontrolled diabetes is the leading cause of renal failure, and uncontrolled hypertension runs a close second. Together these two health conditions account for over 60 percent of all new cases of renal failure every year. The third most important cause of renal failure is glomerulonephritis, or inflammation of the glomerulus. Frequent urinary tract infections also increase the risk of kidney trouble. The NIDDKD recommends that all people with these risk factors should have their urine tested for the presence of protein.

As we learned in the chapter, one treatment for chronic renal failure patients is long-term dialysis. However, dialysis is not a perfect solution. A recent study funded by the NIDDKD and the Italian Association for Cancer Research found that chronic renal failure patients on dialysis experience higher-than-average rates of kidney cancer—2,053 diagnosed cases compared to 570 expected cases—and bladder cancer—1,646 diagnosed cases versus 1,100 expected. The increased risk occurs in both men and women, and appears to be consistent for patients in the United States, Europe, Australia, and New Zealand.

QUESTIONS:

1. Based on what you learned in this and other chapters, explain why uncontrolled hypertension might damage the kidneys.
2. Explain why clinicians test for the presence of protein in urine—what would protein indicate about kidney function? Would you expect urine protein levels to be higher or lower than normal?
3. Speculate upon why renal failure patients undergoing dialysis might have an increased risk of kidney and bladder cancer.

INVESTIGATE:

Allen Nissenson and Richard Rettig. Medicare's End-Stage Renal Disease Program: Current Status and Future Prospects. *Health Affairs* (January 1, 1999) p. 161.

More than One in Four Adults at Risk for Kidney Failure. *Tufts University Health & Nutrition Letter* (June 1, 1999), p. 8.

References and Additional Reading

Caplan, Arthur L. and David L. Coelho. *The Ethics of Organ Transplants: The Current Debate.* New York: Prometheus Books, 1998. (An up-to-date discussion of this subject.)

O'Brien, Mary Elizabeth. *The Courage to Survive: The Life Career of the Chronic Dialysis Patient.* New York: Grune & Stratton, 1983. (A look at the physical, psychological, and social aspects of long-term hemodialysis.)

Smith, Homer W. *From Fish to Philosopher.* Boston: Little, Brown and Company, 1953. 264 pp. (A classic book in renal physiology that describes the evolution of the vertebrate kidney.)

Valtin, Heinz and James Schafer. *Renal Function, 3rd ed.* Boston: Little, Brown and Company, 1995. 314 pp. (A comprehensive reference text of renal function and the maintenance of body fluids.)

CHAPTER 16

REPRODUCTIVE SYSTEMS

How many sperm does a single ejaculation **contain?**

How long can sperm live **inside a woman's body?**

Which methods of birth control **are most effective?**

How do male and female sexual responses **compare?**

How can you protect yourself against sexually transmitted diseases?

MOST OF THE ORGAN SYSTEMS WE HAVE CONSIDERED SO FAR in this book are not exactly hot topics in popular media. How many songs have you heard praising the urinary system or the digestive process? Certainly some of the subjects in this chapter, such as differences in male and female anatomy and how males and females are aroused sexually, get more than their share of attention—in songs, books, movies, even Web sites. The beauty and mystery of sex are a source of endless fascination to us.

On a practical level, sexual attractiveness and sexual arousal are normal events in the human reproductive process—a process that requires the **reproductive systems.** The male and female reproductive systems consist of the tissues and organs that participate in creating a new human being. At first glance, it may seem that the male and female reproductive systems are completely dissimilar. After all, the external and internal structures look different from each other. Even secondary sexual characteristics, such as body form, muscle mass, hair patterns, and distribution of body fat, don't look the same. From an evolutionary standpoint, there must have been a tremendous advantage to having the male and female reproductive systems evolve in this manner. We'll touch on the advantages of sexual reproduction in Chapter 21 when we consider evolution.

However, there are also similarities between male and female reproductive systems. Both systems consist of primary reproductive organs, accessory glands, and ducts. Both men and women experience similar phases of sexual desire. Each system has highly specific functions that allow it to complement the other perfectly. And ultimately, both systems share the same goal: to join a sperm and an egg in an environment (the woman's body) that enables the fertilized egg to become an individual.

16.1 Male reproductive system delivers sperm

The male reproductive system evolved to deliver the male reproductive cells, called **sperm,** to the female. As we will see, the chance of any one sperm fertilizing a female reproductive cell (an **egg**) is astronomically low. The reproductive strategy of the male, then, is to deliver huge numbers of sperm at a time to improve the odds that one will reach the egg.

Testes produce sperm

Figure 16.1 illustrates the organs of the male reproductive system. Although we are accustomed to thinking of men as having external *genitalia* (reproductive organs), only the penis and scrotum are visible.

The sites of sperm production in the male are the paired **testes** (singular, *testis*). The testes differentiate from the tissues that eventually become the abdominal wall early in embryonic development. Shortly before birth they descend into the **scrotum,** an outpouching of skin and smooth muscle. The scrotum regulates the temperature of the developing sperm within the testes. Sperm develop best at tem-

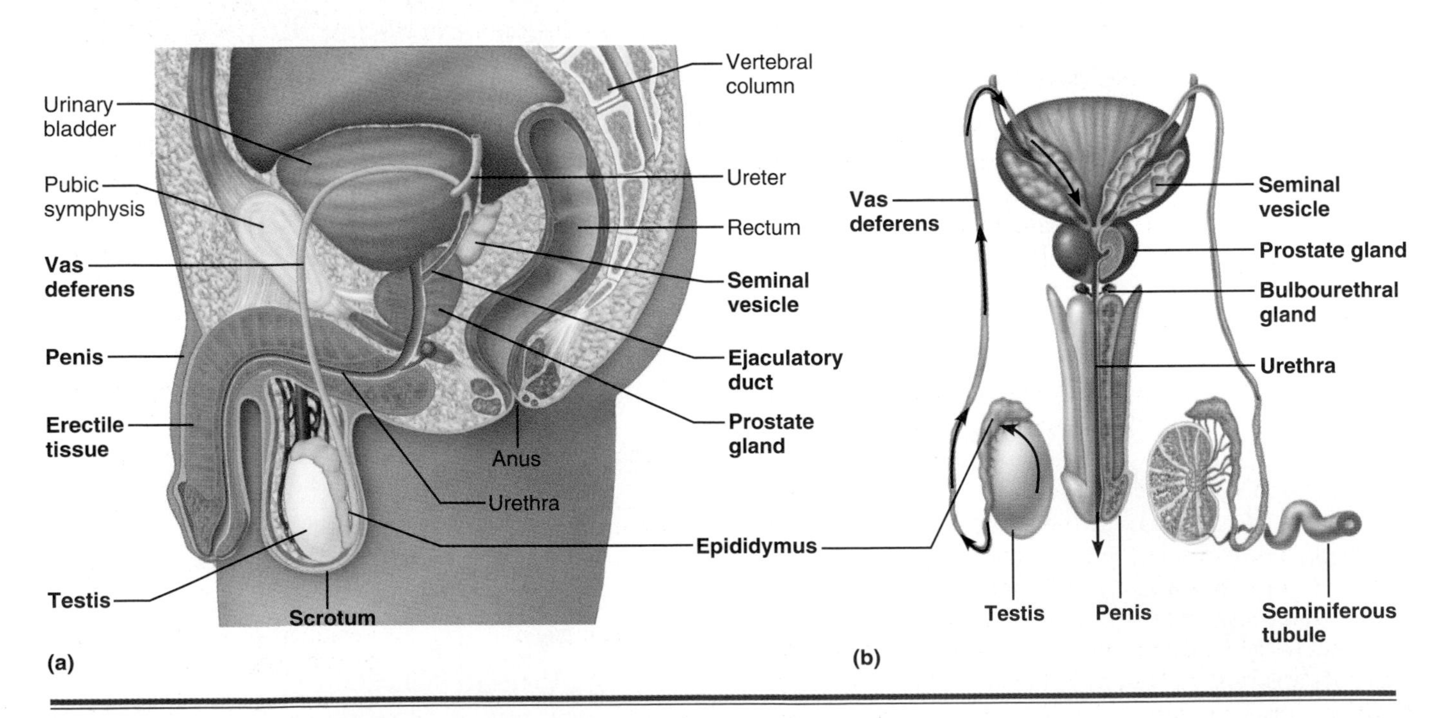

Figure 16.1 The male reproductive system. (a) A sagittal cutaway view. **(b)** A posterior view of the reproductive organs and the bladder.

peratures a few degrees cooler than body temperature. When the temperature within the scrotum falls below about 95°F the smooth muscle of the scrotum contracts, bringing the testes closer to the warmth of the body. When the temperature warms to above 95°F the smooth muscle relaxes and the scrotum and testes hang away from the body.

Each testis is only about 2 inches long, but it contains over 100 yards of tightly packed **seminiferous tubules** where the sperm form. The several hundred seminiferous tubules join to become the **epididymis,** a single coiled duct just outside the testis. The epididymis joins the long **ductus (vas) deferens,** which eventually connects to the **ejaculatory duct.** The newly formed sperm are not fully mature (they cannot yet swim) when they leave the seminiferous tubules. Their ability to swim develops in both the epididymis and the ductus deferens. The epididymis and the ductus deferens also store the sperm until ejaculation.

When the male reaches sexual climax and ejaculates, rhythmic contractions of smooth muscle propel the sperm through the short **ejaculatory duct** and finally through the urethra, which passes through the penis. (As we saw in Chapter 15, the urethra is a pathway for urine as well.)

The **penis** is the male organ of sexual intercourse. Its function is to deliver sperm internally to the female, far away from the harsh external environment. The penis contains erectile tissues that permit **erection**—an increase in length, diameter, and stiffness of the penis that facilitates its entry into the female vagina.

Accessory glands help sperm survive

Sperm must endure the rapid journey from epididymis to vagina and then continue to live within the woman for many hours. To improve their chances of survival, the male delivers sperm in a thick, whitish mixture of fluids called **semen.** In addition to sperm, semen contains secretions from the seminal vesicles, the prostate gland, and the bulbourethral gland.

Spotlight on Structure & Function

The **seminal vesicles** produce *seminal fluid,* a watery mixture containing fructose and prostaglandins that represents about 60% of the volume of semen. Fructose, a carbohydrate, provides the sperm with a source of energy. The prostaglandins are thought to induce muscle contractions in the female reproductive system that help sperm travel more effectively. The **prostate gland** contributes an alkaline fluid. The female's vagina is generally fairly acid (pH 3.5–4.0), which helps prevent infections, but is too acid for sperm. The prostate secretions temporarily raise vaginal pH to about 6, optimal for sperm. Finally, the **bulbourethral glands** secrete mucus into the urethra during sexual arousal. The mucus provides lubrication for sexual intercourse and also washes away traces of acidic urine in the urethra before the sperm arrive. To enhance the possibility of sperm survival, the sperm are not mixed with seminal and prostate fluids until the moment of ejaculation. ■

Table 16.1 summarizes the organs and glands of the male reproductive system.

Recap **The male reproductive system consists of the testes where sperm are produced, a series of ducts and accessory glands, and the penis. Semen consists of sperm and three glandular secretions that provide energy and the proper pH environment for the sperm and also lubrication for sexual intercourse.**

Sperm production requires several cell divisions

Let's take a closer look at the process of sperm production. As mentioned, it takes place in the seminiferous tubules within each testis (Figure 16.2a). Toward the outer surface of each tubule are undifferentiated cells (cells that have not yet become any kind of specialized cell) called *spermatogonia* (singular, *spermatogonium*). Spermatogonia are *diploid* cells, meaning they have 46 chromosomes, the normal human number.

Table 16.1 Summary of the male reproductive organs and glands

Organ	Function
Testis (2)	Produces sperm, testosterone, and inhibin
Scrotum	Keeps the testes at the proper temperature
Epididymis (2)	Site of sperm maturation and storage
Ductus deferens (2)	Duct for sperm maturation, storage, and transport
Ejaculatory duct (2)	Duct for transporting sperm and glandular secretions
Penis	Erectile organ of sexual intercourse
Accessory Glands	
Seminal vesicle (2)	Secretes fructose and most of the seminal fluid
Prostate gland	Secretes watery alkaline fluid to raise vaginal pH
Bulbourethral gland (2)	Secretes lubricating mucus

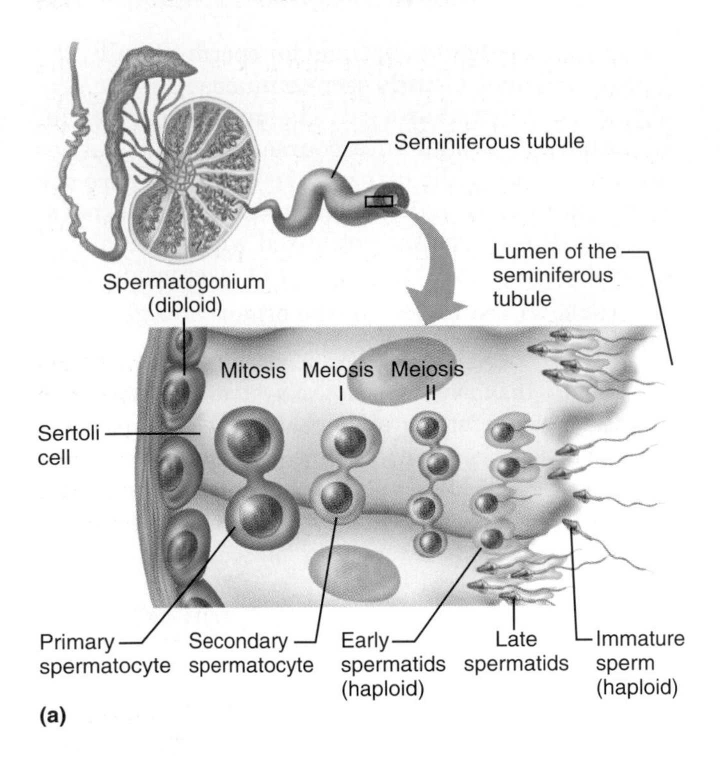

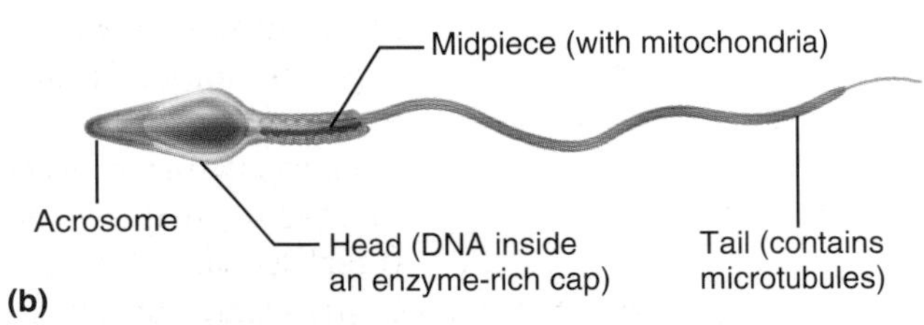

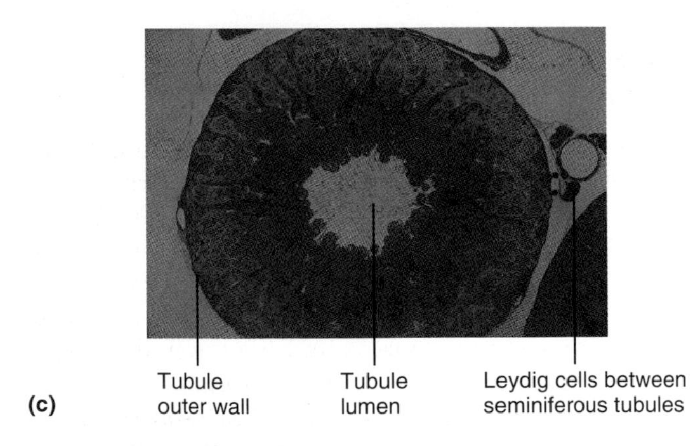

Figure 16.2 Sperm formation in a seminiferous tubule. **(a)** Sperm originate from undifferentiated spermatogonia located near the outer surface of the tubule. The production of sperm is supported by Sertoli cells. **(b)** A mature sperm. **(c)** A cross section through a seminiferous tubule showing the location of the testosterone-producing Leydig cells.

The formation of both male sperm and female eggs involves a series of cell divisions called *mitosis* and *meiosis*. In the next chapter, we discuss meiosis and mitosis in more detail. The important point to remember here is that in order to produce the **gametes** (sperm and eggs), the diploid number of chromosomes must be halved. Gametes are called *haploid* cells (from the Greek *haploos,* "single") because they contain only 23 chromosomes. This ensures that when a sperm and egg unite, the embryo will have the proper (diploid) number of chromosomes again.

Spermatogonia divide several times during the course of sperm development. The first division, by mitosis, ensures a constant supply of *primary spermatocytes,* each with the diploid number of chromosomes. Primary spermatocytes then undergo a series of two cell divisions during meiosis to become *secondary spermatocytes* and finally *spermatids.* Spermatids, which are haploid cells, mature slowly to become the male gametes, or sperm. The entire process of sperm formation and maturation takes about 9–10 weeks. During this time the cells are nourished by the large **Sertoli cells** that comprise most of the bulk of the seminiferous tubules.

A mature sperm has a head, midpiece, and tail (Figure 16.2b). The head contains the DNA, organized into chromosomes. The head is covered by an *acrosome,* a cap containing enzymes that help the sperm penetrate the egg. The sperm's whiplike tail enables it to move, and the mitochondria in the midpiece produce energy to power the tail.

Every day, from puberty to old age, tens of millions of sperm form in the testes, although the number tapers off somewhat in later years. A typical ejaculation may contain 100–300 million sperm, approximately 10% of which are already nonviable for a variety of reasons. If an egg is fertilized, it will be by only one sperm.

How many sperm does a single ejaculation contain?

Recap **The haploid sperm form continuously in the seminiferous tubules from undifferentiated diploid cells called spermatogonia. Millions of sperm form every day throughout a man's life, and a typical ejaculation contains up to 300 million.**

Testosterone, LH, and FSH regulate male reproductive activity

Male reproductive capacity is regulated by three hormones: *testosterone, luteinizing hormone (LH),* and *follicle stimulating hormone (FSH).* Historically, the names for LH and FSH came from their functions in the female, but they are active hormones in the male as well.

Testosterone is a steroid hormone produced in the testes and secreted into the blood by **Leydig cells** located between the seminiferous tubules (Figure 16.2c). Testosterone controls the growth and function of the male reproductive tissues, stimulates

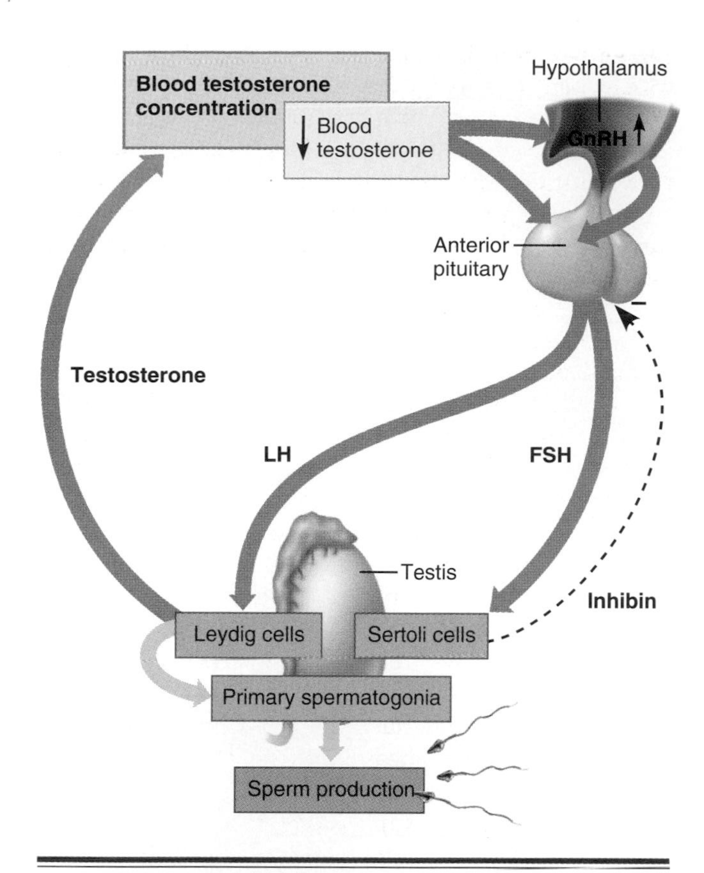

Figure 16.3 Negative feedback control of blood testosterone concentration and sperm production. Sperm production is kept fairly constant by a feedback mechanism that controls blood testosterone concentration. When testosterone concentration falls below normal the hypothalamus secretes more GnRH. The anterior pituitary is stimulated both by GnRH and by the decline in testosterone to produce LH and FSH. LH stimulates testosterone production (returning the blood levels to normal), and testosterone stimulates sperm production by the testes. FSH enhances further sperm production via stimulation of the Sertoli cells, which can inhibit FSH secretion via the production of another hormone called inhibin (dashed line and negative sign indicates inhibition).

aggressive and sexual behavior, and causes the development of secondary sexual characteristics at puberty, such as facial hair and deepening voice. Within the testes, testosterone stimulates the undifferentiated spermatogonia to begin dividing and the Sertoli cells to support the developing sperm, and thus it determines the rate at which sperm form.

A negative feedback loop involving the hypothalamus and the pituitary maintains a fairly constant blood concentration of testosterone, and thus a consistent rate of sperm production (Figure 16.3). When blood concentrations of testosterone fall, the hypothalamus secretes gonadotropin-releasing hormone (GnRH), which stimulates the anterior pituitary gland to release LH and FSH. (Recall from Chapter 12 that LH and FSH are called *gonadotropins* because they stimulate the reproductive organs—the *gonads.*) LH stimulates the secretion of testosterone. The role of FSH is less clear, but it seems to enhance sperm formation by stimulating the Sertoli cells.

If the concentration of testosterone becomes too high, it inhibits the secretion of GnRH and also LH. In addition, if too many sperm are produced, the Sertoli cells secrete a hormone called *inhibin* that directly inhibits the secretion of FSH.

Recap **Testosterone stimulates the growth and function of the male reproductive system, aggressive and sexual behavior, and development of secondary sexual characteristics. Blood levels of testosterone are regulated by a negative feedback loop involving GnRH from the hypothalamus and LH from the anterior pituitary.**

16.2 Female reproductive system produces eggs and supports pregnancy

The female reproductive system evolved to perform more functions than that of the male. The female system hoards the precious eggs, typically releasing only one or perhaps two in any given month. In addition, because it can never be known for certain if an egg will be fertilized, the female reproductive system must go through cyclic changes to prepare for the possibility of fertilization. If fertilization does occur, the female system adapts to pregnancy and supports the developing fetus.

Ovaries release oocytes and secrete hormones

Figure 16.4a illustrates the organs of the female reproductive system. The primary female reproductive organs are the two **ovaries,** which release the female gametes, immature eggs called **oocytes,** at regular intervals during the reproductive years. The ovaries also secrete the female sex steroid hormones **estrogen** and **progesterone.** Like the male testes, the ovaries differentiate from the tissues destined to become the abdominal wall early in fetal development, but unlike the male they are eventually housed inside the abdominal cavity.

Once released, the oocyte is swept into the open end of an **oviduct** (also called a *Fallopian tube* or *uterine tube*), which leads from the ovary to the uterus. The open end of the oviduct has fingerlike projections called *fimbriae* that help move the oocyte into the oviduct. If fertilization by a sperm occurs, it usually takes place in the upper third of the oviduct. The unfertilized oocyte or fertilized egg moves down the oviduct over the course of about 3–4 days to the uterus.

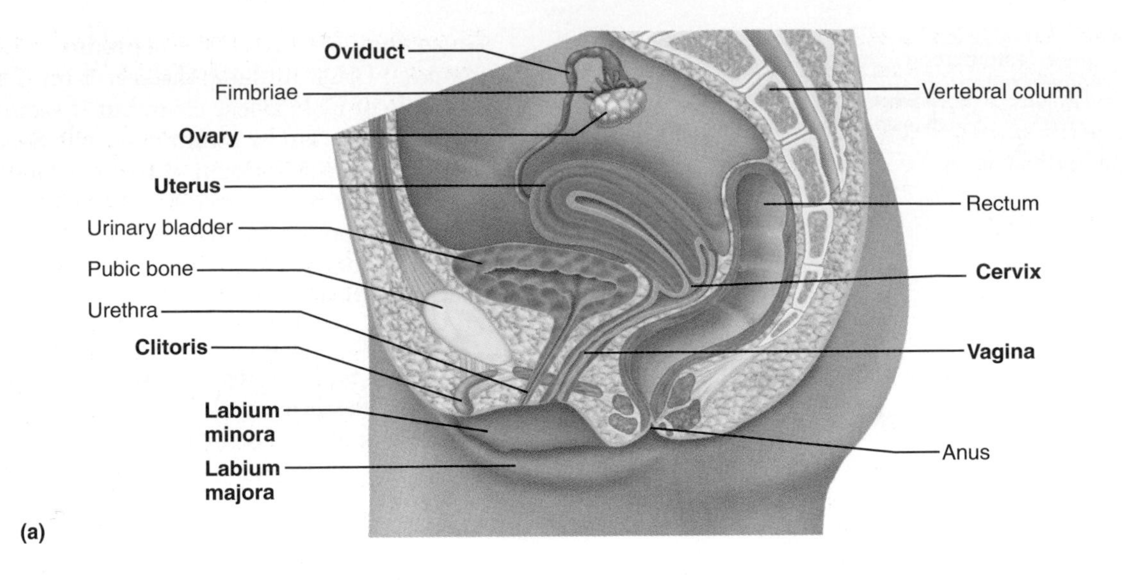

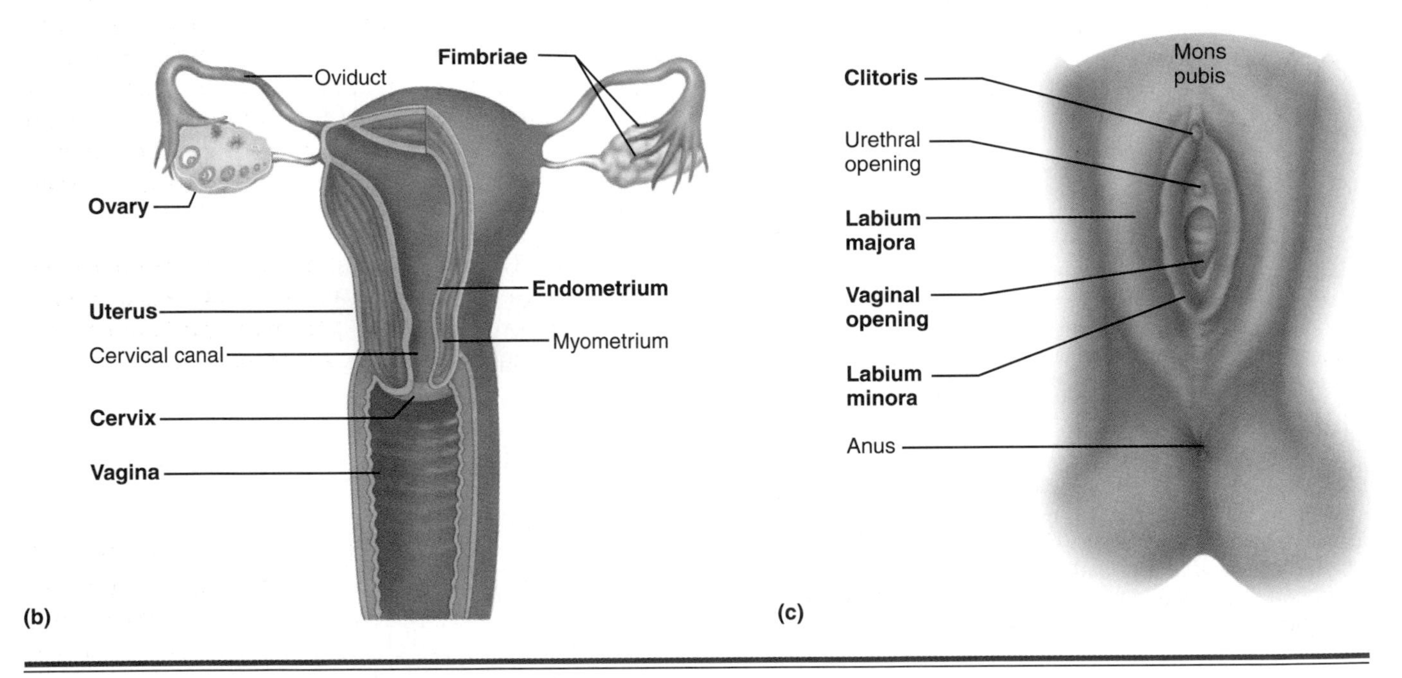

Figure 16.4 The female reproductive system. (a) Sagittal section showing the components of the system in relation to other structures. **(b)** Anterior view of the internal organs. **(c)** The female external genitalia. The labia have been pulled to the sides to expose the vaginal and urethral openings.

The uterus nurtures the fertilized egg

Spotlight on Structure & Function

The **uterus** is a hollow pear-shaped organ where the fertilized egg grows and develops. The walls of the uterus comprise two layers of tissue (Figure 16.4b).

The inner layer of the uterus, called the **endometrium,** is a lining of epithelial tissue, glands, connective tissue, and blood vessels. If an egg has been fertilized it will attach to the endometrium in a process called **implantation.** After implantation, the endometrium helps to form the *placenta,* a common tissue between the uterus and the fetus. The placenta provides nourishment to the developing fetus and also provides for waste removal and gas exchange.

The outer layer of the uterus, or *myometrium,* consists of thick layers of smooth muscle. The myometrium stretches during pregnancy to accommodate the growing fetus. It also provides the muscular force to expel the mature fetus during labor.

The narrow lower part of the uterus is the **cervix.** An opening in the cervix permits sperm to enter the uterus and allows the fetus to exit during birth. ■

Recap **The ovaries secrete estrogen and progesterone, store immature oocytes, and release generally one oocyte at a time at intervals of about 28 days. The oocyte travels through the oviduct to the uterus, where implantation occurs if the egg has been fertilized.**

The vagina: Organ of sexual intercourse and birth canal

The cervix joins the **vagina,** a hollow muscular organ of sexual intercourse and also the birth canal. Glands within the vagina produce lubricating secretions during sexual arousal. A thin ring of tissue called the *hymen* may partially cover the opening to the vagina. The hymen usually is ruptured by the first sexual intercourse, by inserting tampons to absorb menstrual flow, or by vigorous physical activity.

The vagina is continuous at the body surface with the female external genitalia, collectively called the **vulva** (Figure 16.4c). An outer, larger pair of fat-padded skin folds called the **labia majora** (in Latin, "major lips") surround and enclose the **labia minora,** a highly vascular but smaller pair of folds. The **clitoris** is a small organ partly enclosed by the labia minora that is an important contributor to the female sexual response. The clitoris and the penis originate from the same tissue during early development, and like the penis, the clitoris contains erectile tissue that is highly sensitive to stimulation. Between the clitoris and the vaginal opening lies the urethral opening. Table 16.2 summarizes the main components of the female reproductive system.

Mammary glands nourish the infant

Human breasts contain **mammary glands,** modified sweat glands that technically are part of the skin, or integumentary system. Though breasts are not required to produce a viable fetus, they undergo developmental changes during puberty in women and they support the infant's development, so we consider them here.

A *nipple* at the center of each breast contains smooth muscles that can contract, causing the nipple to become erect in response to sexual stimulation, cold temperatures, or a nursing infant. Surrounding the nipple is the pigmented *areola.* Internally, the breasts contain the mammary glands that are specialized to produce milk (Figure 16.5). The mammary glands consist of hundreds of small milk-producing lobules. Contractile cells around each lobule allow the milk to be released, and ducts deliver it to the nipple. Most of the breast consists of adipose tissue, so breast size does not indicate the potential for milk production. A network of fibrous connective tissue supports the glands and adipose tissue.

At puberty, female mammary glands enlarge under the control of the hormone estrogen. Both estrogen and progesterone prepare the glands for **lactation,** or the production of milk, late in pregnancy. At birth, prolactin stimulates lactation and oxytocin stimulates the contractions that release milk during nursing. In males the mammary glands are vestigial, meaning they have no function.

Table 16.2 Summary of the components of the female reproductive system

Component	Function
Ovary (2)	Site of storage and development of oocytes
Oviduct (2)	Duct for transporting oocyte from ovary to uterus; also site of fertilization if it occurs
Uterus	Hollow chamber in which embryo develops
Cervix	Lower part of the uterus that opens into the vagina
Vagina	Organ of sexual intercourse; produces lubricating fluids; also the birth canal
Clitoris	Organ of sexual arousal

Recap **The vagina is the female organ of sexual intercourse and the birth canal; around its opening are the structures of the vulva. Mammary glands are accessory organs that produce and store milk.**

16.3 Menstrual cycle consists of ovarian and uterine cycles

Every month, the ovaries and uterus go through a pattern of changes called the **menstrual cycle.** The menstrual cycle lasts about 28 days and is controlled by hormones. Menstrual cycles begin at puberty and generally continue throughout the reproductive years, except during pregnancy.

A complete menstrual cycle consists of two linked cycles, the *ovarian cycle* and the *uterine cycle.* Together, they periodically release oocytes and prepare the uterus to receive a fertilized egg.

The ovarian cycle: Oocytes mature and are released

The **ovarian cycle** is a regular pattern of growth, maturation, and release of oocytes from the ovary. At birth, a female has approximately one million *primary oocytes* already formed and stored in each ovary, and no more are ever produced. Each primary oocyte has already developed partway through meiosis at birth, but at birth all further development halts until after puberty. By puberty 85% of them have been resorbed, leaving only about 300,000 in both ovaries.

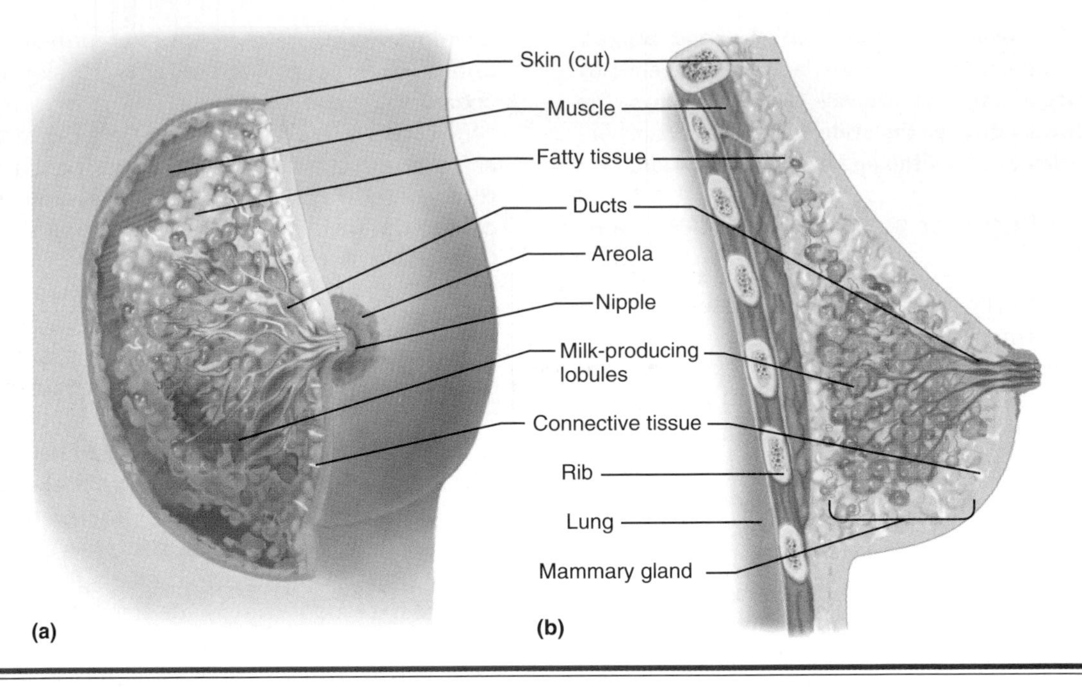

Figure 16.5 The female breast. (a) Front view and **(b)** side view, showing the anatomical relationships between the milk-producing mammary glands, adipose tissue, and muscle.

Each month perhaps a dozen primary oocytes start the development process, but typically only one will complete it. Only about 400 to 500 oocytes will ever be released during a woman's lifetime, a number that contrasts sharply with the 100–300 million sperm released in a single ejaculation.

Figure 16.6 illustrates the events of the ovarian cycle:

1. A primary oocyte is surrounded by a layer of *granulosa cells* that nourish it. Together, a primary oocyte and its granulosa cells constitute an immature **follicle.** At the beginning of the ovarian cycle, GnRH from the hypothalamus raises concentrations of the anterior pituitary hormones FSH and LH. FSH stimulates the follicle to grow and enlarge. As it develops the follicle begins to change form.
2. The granulosa cells begin to divide and to secrete a sugary material (glycoproteins) that becomes the *zona pellucida,* a noncellular coating around the oocyte.
3. A fluid-filled space called the *antrum* develops within the follicle, and some of the granulosa cells begin to secrete estrogen and progesterone.
4. The primary oocyte completes stage I of meiosis to yield a *secondary oocyte* and a nonfunctional *polar body*. The mature follicle, sometimes called a "Graafian follicle," consists of a secondary oocyte, a polar body, and numerous granulosa cells, all surrounded by ovarian connective tissue cells.
5. Midway through the ovarian cycle the rising levels of estrogen produced by the follicle stimulate the pituitary to release a surge of LH. The LH surge triggers **ovulation**—the follicle wall balloons outward and ruptures, releasing the secondary oocyte with its polar body, zona pellucida, and surrounding granulosa cells into the extracellular fluid.

   At this point the secondary oocyte enters the oviduct and is swept toward the uterus (and any sperm that may be swimming toward it) by peristaltic contractions of the oviduct. If fertilization occurs, the secondary oocyte quickly goes through the second stage of meiosis to produce a mature egg, or *ovum,* which unites with the sperm. Chapter 20 discusses the developmental events that follow the union of sperm and egg.
6. But the ovarian cycle is not over yet. A **corpus luteum** forms from what is left of the ruptured follicle (the name "luteinizing hormone" comes from the fact that LH causes the corpus luteum to form). The corpus luteum has an important function: it secretes large amounts of progesterone and estrogen. These hormones prepare the endometrium of the uterus for the possible arrival of a fertilized egg that would need the proper environment for further development. The high levels of progesterone and estrogen also feed back to the hypothalamus and inhibit FSH secretion, temporarily preventing other follicles from developing.

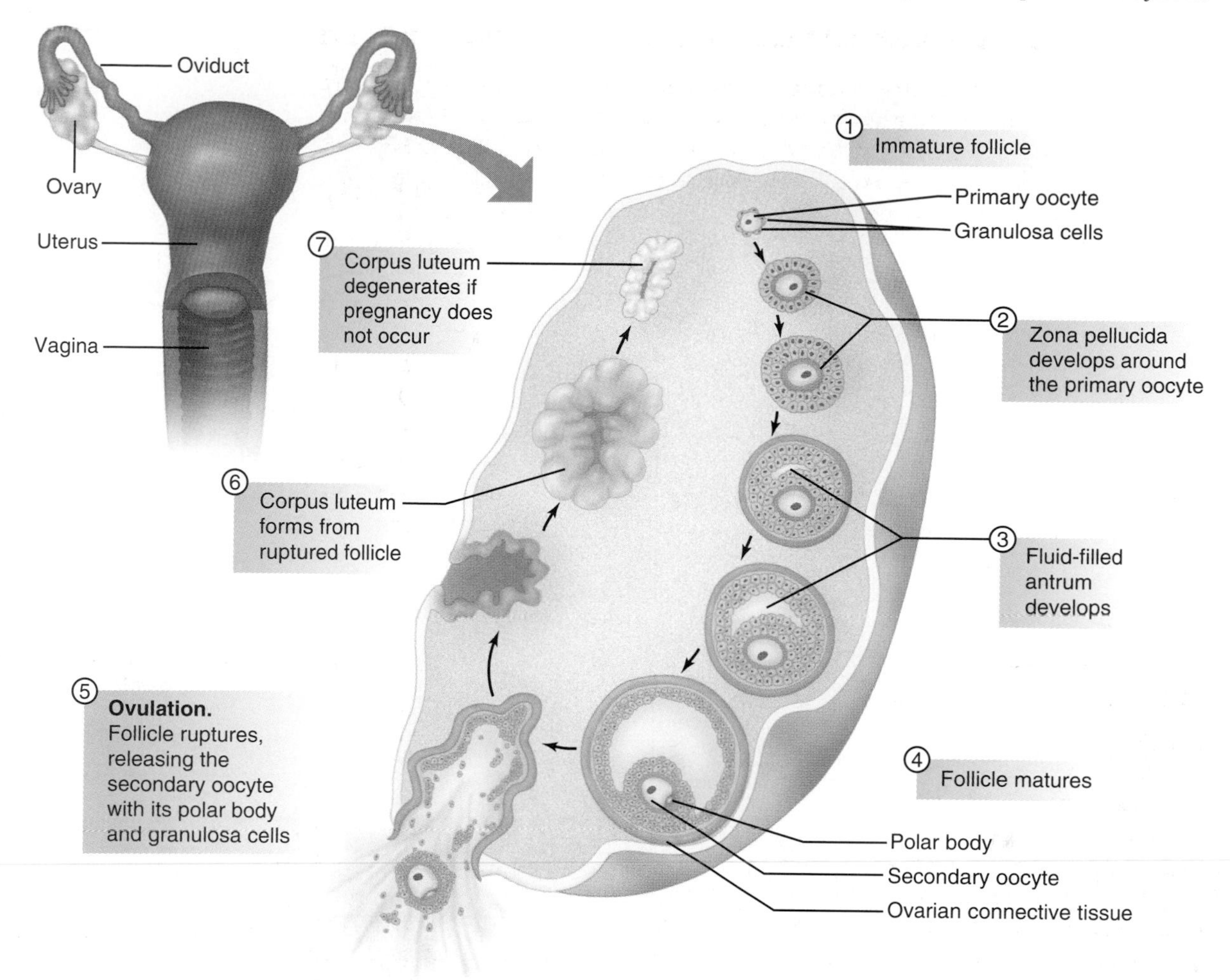

Figure 16.6 The ovarian cycle. Approximately a dozen primary follicles start this process each month, but generally only one completes it. For any particular primary follicle the events take place in one location, but for clarity the events are shown as if they migrate around the ovary in a clockwise fashion.

7. If fertilization does not occur, the corpus luteum degenerates in about 12 days. At this point estrogen and progesterone levels fall rapidly and the ovarian cycle starts again.

If pregnancy does occur, a layer of tissue derived from the embryo called the *chorion* begins to secrete a hormone called *human chorionic gonadotropin* (hCG). The detection of hCG in urine is the basis of home-use pregnancy test kits. hCG maintains the ability of the corpus luteum to produce progesterone and estrogen for another 9–10 weeks. By 10 weeks the developing placenta secretes enough progesterone and estrogen to support the pregnancy, and the corpus luteum degenerates. The high levels of estrogen and progesterone that are maintained throughout pregnancy ensure that ovulation does not occur during pregnancy. You will learn more about pregnancy and fetal development in Chapter 20.

Recap **During the ovarian cycle, a primary oocyte within a developing follicle divides once to form a secondary oocyte. The follicle ruptures, releases the oocyte, and forms the corpus luteum that secretes progesterone and estrogen.**

The uterine cycle prepares the uterus for pregnancy

Spotlight on Structure & Function

The **uterine cycle** is a series of structural and functional changes that occur in the endometrium of the uterus as it prepares each month for the possibility that a fertilized egg may arrive. The uterine cycle is linked to the ovarian cycle. Because the beginning of the cycle is defined as the first day of menstruation, we begin there. Note that although a complete cycle generally lasts about 28 days, many women have cycles that are shorter or longer. Figure 16.7 summarizes

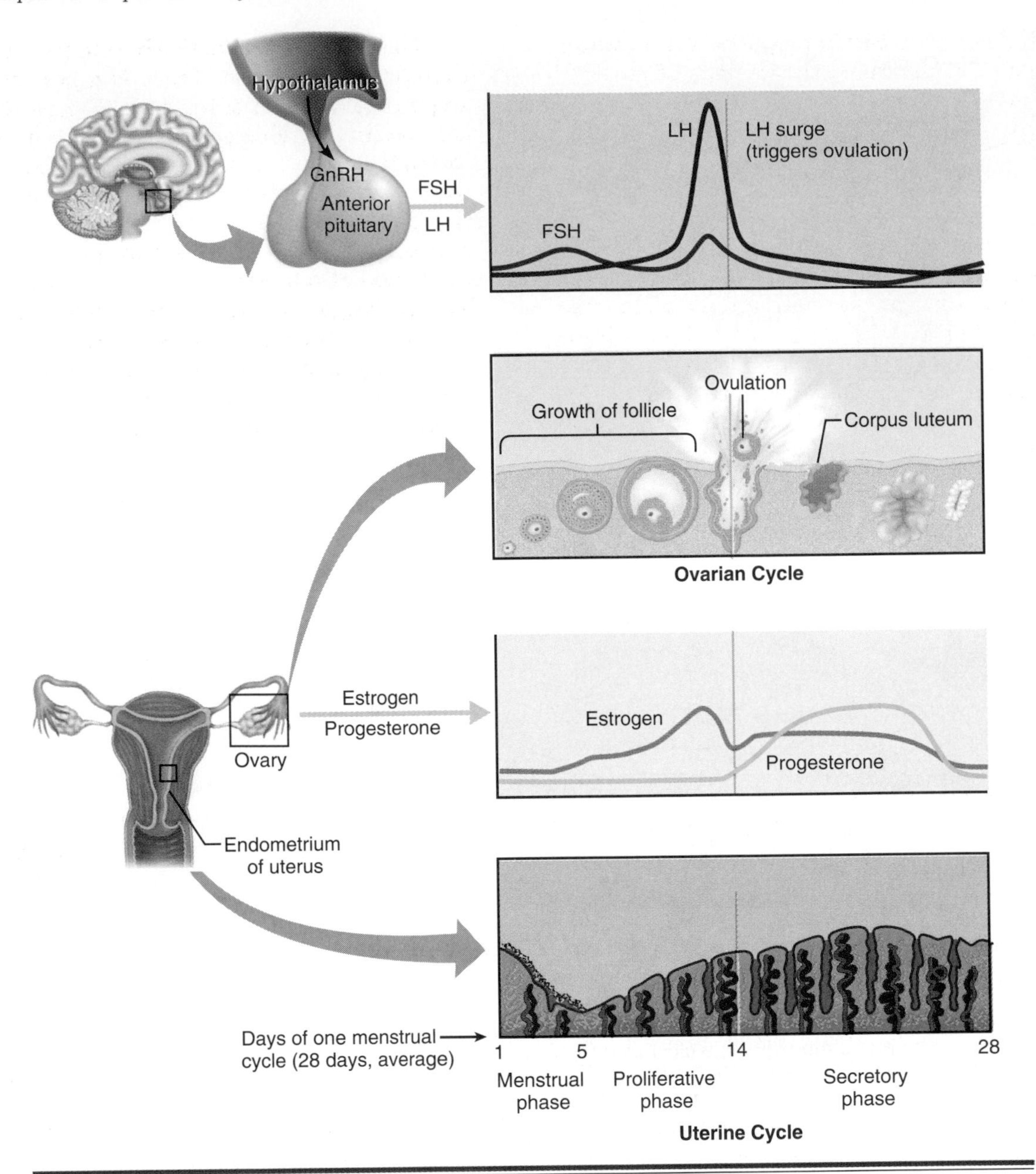

Figure 16.7 The menstrual cycle. The menstrual cycle consists of linked ovarian and uterine cycles. The ovarian cycle is a series of changes associated with oocyte maturation, which depends on the blood levels of FSH and LH. The uterine cycle involves changes in the endometrial lining of the uterus, which depend on the blood levels of estrogen and progesterone.

the ovarian, uterine, and hormonal events of a single menstrual cycle.

- *Days 1–5: Menstrual phase.* In the absence of the arrival of a fertilized egg from the previous cycle, body levels of estrogen and progesterone decline precipitously. Without these hormones the uterus cannot maintain its endometrial lining, which had thickened to prepare for a potential pregnancy. **Menstruation** is the process by which the endometrial lining disintegrates and its small blood vessels rupture. The shed tissue and blood pass through the vagina and are excreted. The period of visible menstrual flow, which lasts about 3–5 days, is called the *menstrual period*.

- *Days 6–14: Proliferative phase.* Triggered by the rising levels of estrogen and progesterone from the next developing follicle, the endometrial lining begins to proliferate. It becomes thicker, more vascular, and more glandular. In addition, the lining of the cervix produces a thin, watery mucus that facilitates the passage of sperm, if present, into the uterus from the vagina.

- *Ovulation.* Ovulation occurs midway through the cycle, around the fourteenth day.
- *Days 15–28: Secretory phase.* Production of estrogen and especially progesterone by the corpus luteum causes the endometrium to continue proliferating, until it becomes almost three times thicker than it was after menstruation. Uterine glands mature and begin producing glycogen as a potential energy source for the embryo. The cervical mucus becomes thick and viscous, forming a "cervical plug" that prevents sperm and bacteria from entering the uterus. These structural changes in the endometrium prepare the uterus for a developing embryo.

If fertilization does not occur, degeneration of the corpus luteum at about 12 days after ovulation results in declining estrogen and progesterone levels, and another menstrual cycle begins. ■

The menstrual cycle is not a pleasant experience for some women, to put it mildly. It involves rapid hormonal and physical changes that can sometimes cause pain or mood changes.

Premenstrual syndrome (PMS) is a group of symptoms often associated with the premenstrual period, including food cravings, mood swings, anxiety, back and joint pain, water retention, and headaches. It generally develops in the second half of the cycle, after ovulation, and lasts until menstruation begins. Exercise and medications that reduce pain and water retention can help.

Painful menstruation is known as *dysmenorrhea.* The endometrium of the uterus produces prostaglandins, which can trigger contractions and cramping of the uterine smooth muscle. Aspirin and ibuprofen may offer some relief from dysmenorrhea because these drugs inhibit prostaglandin formation. Although ovulation itself is usually not uncomfortable, some women report they experience a brief sharp pain when the follicle ruptures.

***Recap* Rising levels of progesterone and estrogen cause the endometrium to proliferate. If pregnancy does not occur, hormone levels fall and the endometrial layer disintegrates and is shed, a process known as menstruation.**

Cycles in hormone levels maintain menstrual cycle

The menstrual cycle is almost unique among regulatory events, precisely because it *cycles* rather than maintaining hormones at a constant level.

The female reproductive hormones cycle through these changes because there is a *positive feedback* step in the first half of the cycle (Figure 16.8). Recall that positive feedback accentuates an initial change instead of opposing it. The positive feedback step in the menstrual cycle is that rising levels of estrogen in the first half of the cycle (the proliferative phase) trigger a surge in LH, both by stimulating LH directly and also by stimulating the release of GnRH. The LH surge goes on to cause ovulation.

Because the rising levels of estrogen come from a maturing follicle, this makes sense; ovulation is triggered when a follicle approaches maturity, not before. The positive feedback ensures that ovulation will

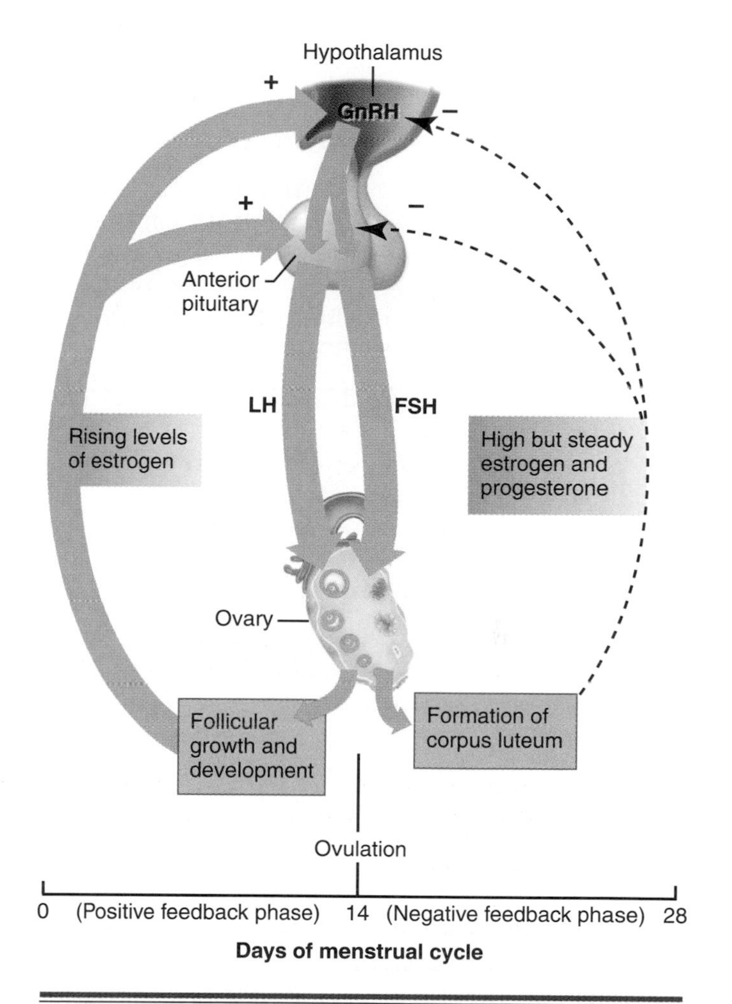

Figure 16.8 Regulation of the menstrual cycle. The cyclic nature of the menstrual cycle is due to a positive feedback loop in the first half of the cycle, in which LH and FSH from the anterior pituitary gland stimulate estrogen production by the developing follicle. The estrogen further stimulates LH secretion. Eventually, the rise in estrogen concentration triggers the sharp surge in LH concentration that causes ovulation. A negative feedback loop prevails in the second half of the cycle because the high but fairly steady levels of estrogen and progesterone produced by the corpus luteum inhibit the secretion of LH and FSH. Once the corpus luteum regresses, the cycle repeats.

occur at fairly regular intervals, as the other events in the cycle take a certain amount of time. The LH surge is short-lived because, in the second half of the cycle (the secretory phase), high but steady levels of estrogen combined with progesterone inhibit LH and FSH in the typical negative feedback fashion. This prevents a second ovulation until progesterone and estrogen decline. Then estrogen begins to rise rapidly again during the proliferative phase of the next cycle.

Recap **Ovulation is triggered by a surge of LH, which in turn is caused by the positive feedback effect of rising levels of estrogen from the maturing follicle. During the second half of the menstrual cycle, sustained high levels of estrogen and progesterone inhibit further ovulation.**

16.4 Human sexual response, intercourse, and fertilization

How do male and female sexual responses compare?

The human sexual response is a series of events that coordinate sexual function to accomplish intercourse and fertilization. The response can have up to four phases in both men and women:

1. *Excitement:* Increased sexual awareness and arousal
2. *Plateau:* Intense and continuing arousal
3. *Orgasm:* The peak of sexual sensations
4. *Resolution:* Arousal subsides

The male sexual response

In men, excitement is accompanied by pleasurable sensations from the penis and by an erection. Erection occurs when neural activity relaxes (dilates) arterioles leading into vascular compartments within the penis. This causes the compartments to fill with blood, and as they fill they press on the veins draining the penis and reduce the rate of venous outflow. As more blood enters the penis and less blood leaves it, the penis swells, lengthens, and stiffens. Heart rate and breathing increase, as does the tone of skeletal muscle. Sights, sounds, memories, and other psychological stimuli, kissing or other physical contact, and especially physical contact with the penis all contribute to excitement. Arousal continues during the plateau phase, which can last for a few seconds or many minutes.

At some point sexual arousal becomes so great that **orgasm** occurs. An orgasm is a brief, intensely pleasurable reflex event that accomplishes **ejaculation,** or the expulsion of semen. During ejaculation, sympathetic nerve activity contracts smooth muscle in the seminal vesicles, epididymis, bulbourethral gland, and prostate to move glandular secretions into the urethra. The internal urethral sphincter closes tightly to prevent urine from leaking into the urethra, and the skeletal muscles at the base of the penis contract rhythmically, forcing semen out of the urethra in several spurts. A typical ejaculation ejects about 3–4 cc (a teaspoon or more) of semen.

During the resolution phase, the erection subsides and breathing and heart rate return to normal. Generally a refractory period follows, during which another erection and orgasm cannot be achieved. The refractory period may last for several minutes up to several hours.

The female sexual response

Women experience the same four phases of the sexual response. As in men, the excitement phase is triggered by sights, sounds, psychological stimuli, and physical stimulation, especially of the clitoris and breasts. Stimulation initiates neural reflexes that dilate blood vessels in the labia, clitoris, and nipples. During arousal the vagina and the area around the labia secrete lubricating fluids. Without these lubricating fluids intercourse can be difficult and sometimes painful for both partners.

The female orgasm, like that of the male, consists of rhythmic and involuntary muscular contractions. Contractions of the uterus and vagina are accompanied by intense feelings of pleasure and sensations of warmth and relaxation. The rate of orgasm is more variable in females than in males. Some women do not experience orgasms frequently, whereas others experience multiple orgasms within one period of sexual contact. Female orgasm is not required for fertilization or pregnancy, nor is arousal for that matter.

Recap **Women and men experience the same four phases of sexual responsiveness. Sexual arousal in the male results in penile erection that leads to orgasm and ejaculation. Females experience sexual arousal and pleasurable orgasms marked by rhythmic muscular contractions.**

Fertilization: One sperm penetrates the egg

The term for sexual intercourse is **coitus** (*coitio* means "coming together"). During coitus, pelvic thrusts stimulate the penis, leading ultimately to male orgasm and ejaculation. The semen is deposited inside the vagina, near the cervix.

The several hundred million sperm in semen still have a long way to go, however, before one of them encounters an egg. Random swimming takes some sperm though the cervical opening, through the uterus, and up the two oviducts. The journey can take hours or even a day or more. Fertiliza-

How long can sperm live inside a woman's body?

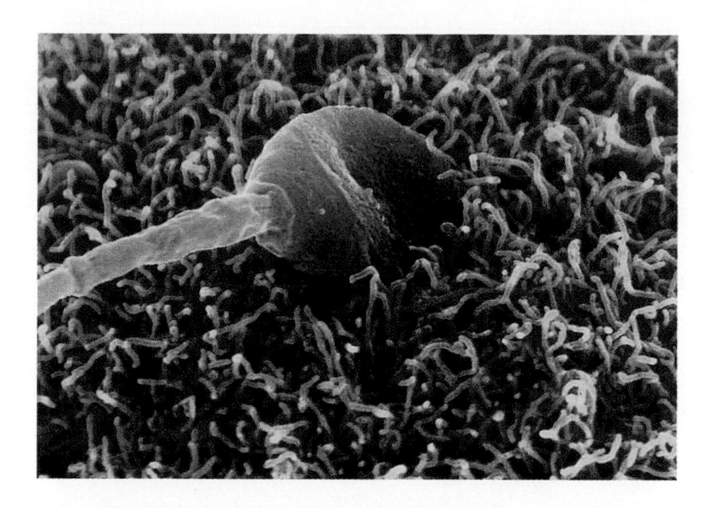

Figure 16.9 SEM (×1,650) of a sperm in contact with an egg. Only one sperm will fertilize the egg.

tion occurs when and if a sperm encounters an oocyte in an oviduct and penetrates the zona pellucida (Figure 16.9). Some sperm may be viable for up to five days in the female reproductive tract, although most do not last more than two days.

Recap **Ejaculation deposits several hundred million sperm in the vagina. Fertilization of the egg by a single sperm occurs within five days, if it occurs at all.**

16.5 Birth control methods: Controlling fertility

Which methods of birth control are most effective?

There are many reasons why couples or individuals might choose to exercise control over fertility, and many ways to influence reproductive outcome. Birth control methods include abstinence, surgical sterilization, hormonal methods, intrauterine devices, diaphragms and cervical caps, chemical spermicides, condoms, morning-after pills, and natural techniques such as the rhythm method and withdrawal.

As shown in Table 16.3, birth control methods vary widely in effectiveness. Furthermore, some help protect against sexually transmitted diseases and others do not.

Abstinence: Not having intercourse

The only completely effective method of birth control is to not have sexual intercourse at all. Abstinence, as it is called, works for many people for short periods of time, and throughout life for others. Abstinence is natural in that no intervention or artificial method need be employed. When abstinence is not the option of choice, there are still many ways to reduce the probability of pregnancy.

Surgical sterilization: Vasectomy and tubal ligation

Both men and women may opt for surgical sterilization (Figure 16.10). Male surgical sterilization is called *vasectomy.* It can generally be performed in a doctor's office under local anesthesia. The doctor makes small incisions in the scrotum, locates each ductus deferens, and ties them in two places. The segments between the two ties are then removed, preventing sperm from reaching the urethra.

After a vasectomy the testes continue to produce testosterone and sperm, and sexual interest and all secondary sexual characteristics remain unchanged. With no access to the urethra, however, the sperm eventually are resorbed, just as they are if ejaculation never takes place. Generally it is impossible to detect a change in semen volume per ejaculation. If done properly, surgical male sterilization is close to foolproof. However, to be on the safe side, the surgeon will probably recommend doing a sperm count several months after the operation.

Female sterilization is accomplished by a *tubal ligation.* The doctor makes a small incision in the woman's abdominal wall, locates each oviduct, and ties (ligates) each at two sites. The sections between the ties are cut away, leaving the oocyte no way to reach the uterus or sperm to reach the oocyte. In a newer version of a tubal ligation, called a *hysteroscopy,*

Table 16.3 Approximate failure rates of various contraceptive methods*

Contraceptive method	Average failure rate (annual pregnancies per 100 women)
None attempted	85–90
Douche	Greater than 30
Withdrawal	20–30
Rhythm method	20–30
Chemical contraceptives (spermicides)	About 20
Diaphragm or cervical cap with spermicide	10–15
Male or female condoms	About 10
Intrauterine device	Less than 5
Oral contraceptives	Approximately 2
Injected or implanted contraceptives	Less than 1
Vasectomy or tubal ligation	Nearly 0
Abstinence	0

*The numbers should not be taken as absolutes since effectiveness will vary with the degree of proper use. The list should be viewed as an approximate rank order in terms of failure rate.

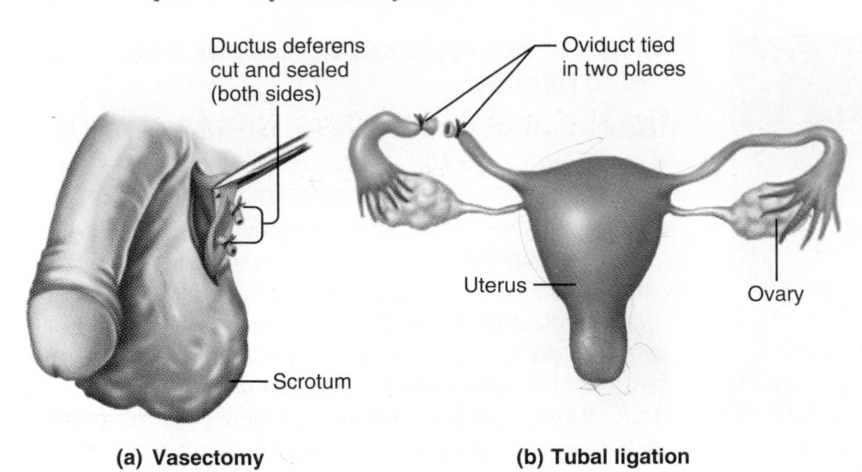

Figure 16.10 Surgical sterilization. (a) In a vasectomy, small incisions are made in the scrotum, and each ductus deferens is tied in two places. **(b)** In a tubal ligation, an incision is made in a woman's abdominal wall, the two oviducts are located, and each is tied in two places.

a flexible tube is inserted through the vagina and uterus and into the oviduct. The tube emits electrical current that seals shut the oviduct. Both tubal ligation and hysteroscopy are highly effective but require surgery that is slightly more difficult than in the male. In addition, tubal ligation may leave two small but possibly visible scars.

A drawback to surgical sterilization in both men and women is that although it *may* be possible to reverse, sterilization generally is permanent. Sterilization should be carefully considered before it is chosen as a method of birth control.

Hormonal methods: Pills, injections, and implants

Manipulation of hormone levels provides reasonably effective and safe methods of birth control. The most common is oral contraceptives, or *birth control pills*. Although the exact mix of hormones in the pills varies, they all administer synthetic progesterone and estrogen in amounts that inhibit the release of FSH and LH. That way, follicles do not mature and ovulation does not take place. It is important to take a pill at the same time every day; if not taken regularly, ovulation (and pregnancy) may occur.

Roughly a third of all women in the U.S. use birth control pills for at least some period of their lives. Oral contraceptives have several side effects, some beneficial and others harmful. Some women report that oral contraceptives reduce cramps and menstrual flow, and they may offer some protection against cancers of the ovaries and uterus. On the negative side, oral contraceptives can cause acne, headaches, fluid retention, high blood pressure, or blood clots. Women who smoke have a greater risk of blood clots and vascular problems from birth control pills.

To avoid the drawbacks of taking—or forgetting to take—a daily pill, some women opt for hormone injections or implants. *Depo-Provera* is a progesterone injection that lasts three months, and *Norplant* is a slow-release form of progesterone that is implanted under the skin of the upper arm. The *Norplant* implant lasts for five years, but can be removed sooner if desired. Injections and implants can also have side effects, including weight gain, headaches, and irregular menstrual periods.

Hormonal methods of contraception have the advantage of not interfering with sexual activity or foreplay in any way. However, none of them protect against sexually transmitted diseases.

Recap **Abstinence and male or female surgical sterilization are highly effective methods of birth control. Hormonal methods—pills, implants, and injections—are also relatively effective but can have side effects.**

IUDs are inserted into the uterus

Intrauterine devices, or IUDs, are small pieces of plastic or metal that are inserted into the uterus by a health care provider. They should only be removed by a health care provider.

IUDs create a mild chronic inflammation that either prevents fertilization or prevents the fertilized egg from implanting in the uterine wall. On the positive side, they are relatively effective and, like hormonal methods, do not interfere with sexual activity. However, risks include uterine cramping and bleeding, infection, and possible damage to the uterus. IUDs offer no protection from sexually transmitted diseases.

Diaphragms and cervical caps block the cervix

Diaphragms and *cervical caps* are devices made of latex that a woman inserts into her vagina. They cover the cervical opening and prevent sperm from entering the uterus.

Both types of devices must be fitted initially to the user's cervix. A disadvantage is that they must be inserted shortly before intercourse and then removed sometime later, possibly interfering with sexual activity. However, they are fairly effective, especially if inserted correctly and used with chemical spermicides. Diaphragms and cervical caps do not protect against sexually transmitted disease.

Chemical spermicides kill sperm

Chemical *spermicides,* which destroy sperm, are available in a wide variety of forms—foams, creams, jellies, and douches. Their effectiveness varies; all reduce the risk of pregnancy, but none are 100% effective. Most are inserted into the vagina shortly before intercourse, except douches, which are designed to be used after intercourse (and are less effective). Some kill organisms that cause sexually transmitted diseases; others do not.

Condoms trap ejaculated sperm

Despite their decidedly low-tech approach (a condom is just a balloon, really), *condoms* remain one of the most popular forms of birth control. They are fairly effective if used properly. Combining them with a chemical spermicide makes them even more effective. They also offer some protection against sexually transmitted diseases.

The male condom is a soft sheath, made of latex or animal membrane, that encloses the penis and traps ejaculated sperm. There is also a female condom that is essentially a polyurethane liner that fits inside the vagina. Condoms made of latex (but not animal membrane) have the tremendous advantage of offering some protection against disease, including AIDS.

The female condom along with several other contraceptive devices may be seen in Figure 16.11.

Recap **Physical barriers (diaphragms, cervical caps, and condoms) and chemical spermicides are moderately effective, and a few afford some protection against disease. IUDs, although more effective against pregnancy, are also more risky.**

Natural alternatives: The rhythm method and withdrawal

Some people prefer to rely on the only two natural methods besides abstinence, the rhythm method and withdrawal. The *rhythm method* takes advantage of the fact that fertilization is possible for only a limited time in each menstrual cycle, defined by the life span of sperm and the length of time an oocyte can be fertilized after ovulation.

The rhythm method requires that a woman avoid intercourse for about an 8-day span every month in the middle of her cycle, from about five days before ovulation to three days after ovulation. It is possible to determine the day of ovulation by carefully monitoring body temperature, because the temperature rises slightly—several tenths of a degree—at ovulation and stays elevated for about three days. However, it is difficult to know which are the five days *preceding* ovulation, because obviously it has not yet occurred. The rhythm method is less effective than many other birth control techniques.

Also less effective is the *withdrawal* method. In principle, the man withdraws his penis from the vagina at just the right moment before ejaculation. In practice it takes skill and commitment on the part of the male and a lot of trust on the part of the female. In addition, sperm can be flushed into the ejaculatory duct even before ejaculation, so there is a possibility of fertilization even if the man does not ejaculate.

Recap **The rhythm method and withdrawal are among the least effective forms of birth control, but if performed correctly they at least reduce the risk of pregnancy.**

Try It Yourself

Do Natural Membrane Condoms Leak More Than Latex Condoms?

If used correctly, latex condoms can help protect against disease. Is the same true of natural membrane condoms? You can test both types of condoms for leakiness.

Dissolve ordinary clothing dye in a quart of warm water. Fill latex and natural membrane condoms with equal volumes of the dye solution, tie them tightly, and rinse them thoroughly. Place each condom in a separate glass of water and observe for several hours.

Within an hour, some dye will leak through the natural membrane condoms, but not through the latex condoms. If you leave the condoms in the water overnight you will also see that the natural membrane condoms swell, indicating that water is moving into them by osmosis. You can conclude that natural condoms are indeed more leaky than latex condoms, at least to water and dye.

But what of the hypothesis that natural membrane condoms are less protective against disease? Your findings do not prove the disease hypothesis because in your experiment you only observed the movement of water and dye, neither of which cause disease. However, your findings certainly would *support* the hypothesis. In fact, research indicates that some viruses *can* pass through natural membrane condoms.

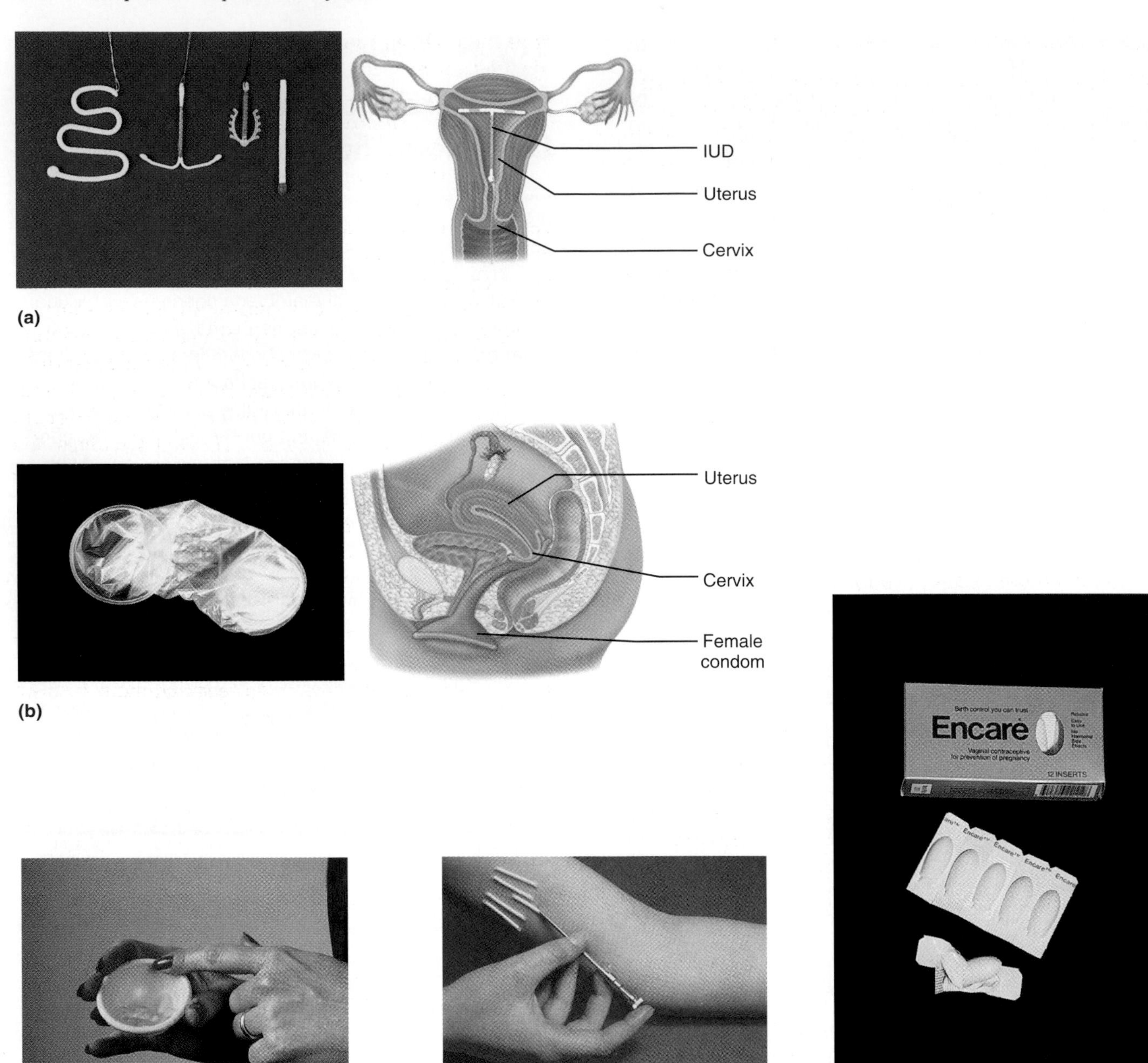

Figure 16.11 Some birth control methods. (a) An intrauterine device and its position in the uterus. **(b)** A female condom and how it is placed in the vagina. **(c)** A diaphragm and spermicidal jelly. **(d)** Contraceptive implants. **(e)** A spermicide.

Morning-after pills prevent pregnancy from continuing

When control of fertility fails, other methods can prevent pregnancy from continuing. One pill, Preven, can be taken up to 72 hours *after* intercourse. Available only by prescription, Preven is a high progesterone and estrogen pill that is thought to work by preventing ovulation if it has not already occurred. It is commonly called an "emergency contraception" pill to distinguish it from another morning-after drug, RU-486, which blocks the action of progesterone and makes the endometrium unprepared to support a developing embryo. RU-486 quickly became known as the "abortion pill," which may be one reason why it is controversial in the U.S. Despite the controversy, RU-486 was finally approved for sale in this country in September of 2000.

Abortion: Terminating pregnancy

Abortion refers to elective termination of a pregnancy. The procedure can be performed in several different ways, including vacuum suction of the uterus, surgical scraping of the uterine lining, or infusion of a strong saline solution that causes the fetus to be rejected. Sometimes abortions are performed in cases of incest or rape, or when continuing the pregnancy would endanger the mother's health. Some women choose to have an abortion if medical tests reveal a nonviable fetus.

Abortions are highly controversial. Some people support a woman's right to choose her own reproductive destiny, whereas others support an embryo's right to life. The beliefs on both sides are often based on strongly held moral and religious convictions.

Recap **Morning-after pills can be taken after intercourse to forestall ovulation or prevent a pregnancy from continuing. Abortion is a controversial procedure that terminates a pregnancy electively.**

The future in birth control

Several new forms of birth control are still in the research and development stage. Possibilities include a male birth control pill that reduces sperm production and several vaccines for women. For instance, a vaccine is being developed to immunize women against human chorionic gonadotropin, the hormone that maintains the corpus luteum so implantation can occur. Also under development is a vaccine against sperm.

16.6 Infertility: Inability to conceive

If both partners are healthy, pregnancy is usually not difficult to achieve. As Table 16.3 shows, the annual pregnancy rate without any birth control averages 85–90 percent. Timing can improve the odds. If a couple wants to increase their chance of having a child, for example, they might choose to have intercourse near the middle of the menstrual cycle, around the time of ovulation.

But some couples find it difficult to become pregnant. Next we describe problems that can impair fertility and solutions that are available.

Infertility can have many causes

A couple is considered **infertile** if they fail to achieve pregnancy after a year of trying. By this definition, about 15% of all couples in the U.S. are infertile. Infertility is about equally attributable to men and women.

Male infertility results from insufficient numbers of normal, healthy sperm. The chance of any one sperm achieving reproductive success is so low that men whose sperm count is below 60 million per ejaculation are, for all practical purposes, infertile. Reasons for a low sperm count include too little testosterone, immune disorders that attack the sperm, radiation, certain drugs such as anabolic steroids, and diseases such as mumps and gonorrhea.

Sperm function best at temperatures slightly cooler than body temperature. One simple measure that may raise the viable sperm count is to switch from tight-fitting underwear, which hold the testicles close to the body, to looser-fitting boxer shorts.

Causes of female infertility are more variable. One of the most common causes of female infertility is *pelvic inflammatory disease* (PID), a general term for any extensive bacterial infection of the internal female reproductive organs. PID that involves the oviducts can result in scar tissue that seals the oviducts shut, preventing passage of either eggs or sperm (Figure 16.12). Abnormal production of LH or FSH may limit the development of oocytes into follicles. Irregular menstrual cycles can make it difficult to time the period of ovulation. Some women produce strongly acid vaginal secretions or thick cervical mucous that damage sperm or make it difficult for them to move toward the egg. Uterine tumors may interfere with implantation.

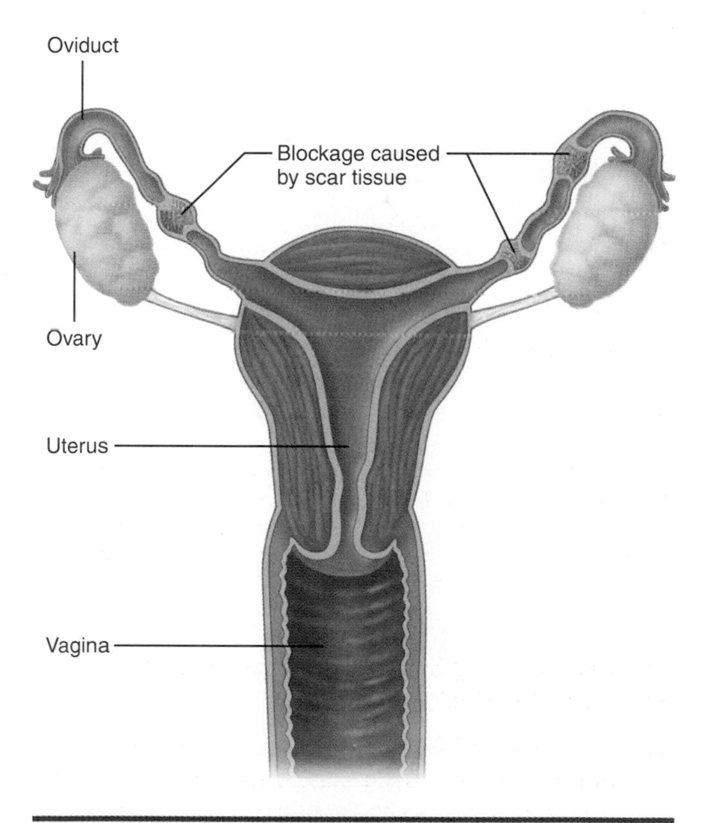

Figure 16.12 Pelvic inflammatory disease can lead to sterility in females. Bacteria that migrate from the vagina into the oviducts can cause inflammation of the oviducts, leading to scarring and either partial or complete closure of the oviduct.

One to three percent of women develop *endometriosis,* a condition in which endometrial tissue migrates up the oviduct during a menstrual period and implants on other organs such as the ovaries, the walls of the bladder, or the colon. There it grows, stimulated each month by the normal hormonal cycle. Endometriosis can cause pain and infertility, but can sometimes be corrected by surgery, drugs, or hormone therapy.

Female reproductive capacity is more affected by age than male reproductive capacity. By the mid 40s, women begin to run out of oocytes. In addition, the ovaries become less responsive to LH and FSH, leading eventually to menopause and the end of menstrual cycles. Oocytes ovulated later in life are also more likely to have been damaged by years of exposure to radiation, chemicals, and disease.

A common cause of a failure to achieve reproductive success even when the couple is fertile is spontaneous abortion, or *miscarriage,* defined as the loss of a fetus before it can survive outside the uterus. Perhaps as many as one third of all pregnancies end in miscarriages, some of them so quickly that a woman has no idea she ever conceived.

Recap **Male infertility is caused by a low sperm count or no sperm. Causes of female infertility are more variable, including failure to ovulate, damage to oviducts, pelvic inflammatory disease, secretions that impair sperm function, uterine tumors, endometriosis, age-related changes, and miscarriages.**

Enhancing fertility

Several options are available for infertile couples who wish to have a child. None guarantee success and most are expensive. It is not unusual for couples to spend tens of thousands of dollars, even $200,000 or more, in the process of becoming biological parents. Techniques include artificial insemination, *in vitro* fertilization, gamete intrafallopian transfer, and fertility-enhancing drugs. Some couples choose to hire a surrogate mother.

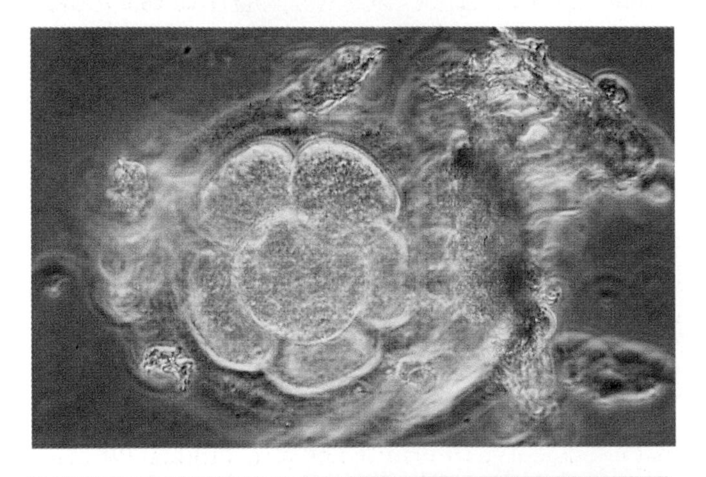

Figure 16.13 A human embryo at the eight-cell stage, approximately 2½ days after *in vitro* fertilization.

Artificial insemination By far the easiest, most common, and least expensive technique is *artificial insemination:* Sperm are placed with a syringe into the vagina or uterus, as close to ovulation as possible. Artificial insemination is the method of choice when the man's sperm count is low, because his sperm can be collected over a period of time and all delivered at once. It is also a good option for men who produce no sperm, in which case the woman may receive sperm from an anonymous donor who has sold his semen to a "sperm bank," and for single women who make the personal choice to become a parent. Over 20,000 babies are born in the U.S. each year as a result of artificial insemination.

In vitro *fertilization* *In vitro fertilization* (IVF) essentially means fertilization in a test tube, outside the body (*vitreus* is Latin for "glass"). A needle, guided by ultrasound, is used to harvest immature eggs from the woman. The eggs mature in laboratory glassware under sterile conditions and then sperm are added. Two to four days later, when it is clear that fertilization has occurred because cells are dividing (Figure 16.13), the embryo is inserted into the woman's uterus through her cervix. If the embryo implants in the endometrium, pregnancy and fetal development are generally normal.

IVF is effective when infertility results from damaged oviducts or female secretions that prevent contact between sperm and egg. If the woman cannot produce viable eggs due to an ovarian problem, IVF can be done with an egg donated by another woman. It is even possible to freeze embryos so they can be used later, without having to collect new eggs and start over again.

IVF results in a live birth only about 20% of the time, so some couples try it several times before achieving pregnancy. To improve the chances of pregnancy, it is not uncommon to transfer up to three embryos at once. Consequently, multiple births are common among women undergoing IVF.

GIFT and ZIFT *Gamete intrafallopian transfer (GIFT)* attempts to simplify the IVF procedure and improve the odds of success. In GIFT, unfertilized eggs and sperm are placed directly into an oviduct (Fallopian tube) through a small incision in the woman's abdomen. In a variation of this method, *zygote intrafallopian transfer (ZIFT),* the egg is fertilized outside the woman's body and then inserted in the same way (a zygote is a fertilized egg).

Directions in Science

Assisted Reproductive Technologies

July 25, 1978 is the birthday of Louise Joy Brown, the first human being conceived *in vitro,* or outside the human body. When two British physicians announced the birth of the first "test-tube baby," it changed reproductive choices forever and created a new industry known as "assisted reproductive technologies" (ART).

The first *in vitro* fertilization procedure in the U.S. took place in 1982 in Norfolk, Virginia. By 1995, just 17 years after Louise's birth, there were 281 ART clinics (or fertility clinics, as they are more commonly known) in America. These clinics are where couples go for *in vitro* fertilization, gamete intrafallopian transfer (GIFT) and zygote intrafallopian transfer (ZIFT).

The discovery that a human being could be conceived outside the human body startled the public at first, and the media paid it a great deal of attention. There were cries of "playing God," and the subject was debated by ethicists, physicians, and religious leaders. The outcry from strong lobbyist groups made the federal government decidedly uneasy about taking sides. As a result the government banned the use of federal research dollars for ART research. The ban still stands today. However, the government stopped short of declaring *in vitro* fertilization illegal because there seemed no good reason to do so.

Inevitably, these decisions created the ART industry. Public outcry soon died down, and infertile couples began clamoring for the technology. The ART clinics arose in response. However, none of the ART clinics in the U.S. receive federal support for research or infertility treatments, which is why infertile couples may spend thousands of dollars in the process of having a child. The user pays, not the taxpayer.

Some scientists argue that the government's hands-off approach has caused the U.S. to lag behind other countries in reproductive technologies, because the fledgling ART clinics have had to develop and advance the technologies themselves. This is debatable.

In the end the government did get involved. With hundreds of private clinics performing ART, legislators recognized a need to regulate the industry. The U.S. Congress passed the Fertility Clinic Success Rate and Certification Act, which requires that the Centers for Disease Control and Prevention (the CDC in Atlanta) publish pregnancy success rates for all ART clinics in the U.S. The CDC lists the names and locations of every clinic and publishes their individual success rates, broken down by type of procedure.

According to the CDC, in one year the 281 ART clinics make more than 59,000 attempts to produce a pregnancy. About 70% are *in vitro* fertilizations with subsequent transfer to the uterus through the cervix. The overall success rate in producing a live birth is 19.6% for all techniques and clinics combined, and therefore there are nearly 12,000 live births as a result of assisted reproductive technologies in one year alone. Expect the success rate to get even better in the future.

Fertility-enhancing drugs Medications are available to boost the production of developing eggs. Sometimes these drugs are tried before IVF, or given to women who are preparing for IVF so several eggs may be harvested at once. Occasionally multiple pregnancies can occur as a result of the medications themselves. When you hear news reports of multiple births involving four or more babies, it is almost certain the mother was taking fertility-enhancing drugs.

Surrogate mothers Some couples choose to pay another woman, called a surrogate mother, to become pregnant and bear a baby for them. Depending on the arrangement, the prospective parents may contribute sperm, eggs, or both.

Surrogate motherhood raises a number of difficult legal issues that should be explored before entering into an agreement. The surrogate must confirm legally that she will give up the baby for adoption to the parents. However, there have not yet been enough court cases to define legal precedents should conflicts arise between surrogate mother and prospective parents.

Recap **A variety of methods are available to improve fertility; the choice depends on what is causing the infertility. Options include artificial insemination, *in vitro* fertilization, gamete intrafallopian transfer, zygote intrafallopian transfer, fertility-enhancing drugs, and hiring a surrogate mother.**

16.7 Sexually transmitted diseases: A worldwide concern

Disorders of the reproductive systems fall into four classes: (1) infertility, (2) complications of pregnancy, (3) cancers and tumors of the reproductive organs, and (4) sexually transmitted diseases. We have already considered infertility and its solutions. We discuss potential complications of pregnancy in Chapter 20, and we cover cancer in detail in Chapter 18. In the rest of this chapter, we focus on sexually transmitted diseases.

Sexually transmitted diseases (STDs) are transmitted by sexual contact, whether genital, oral-genital, or anal-genital. STDs are on the rise worldwide, apparently because the age of sexual contact is declining and birth control enables people to be sexually active without fear of pregnancy. In the U.S. alone, over 12 million people contract an STD each year. A quarter of them are teenagers.

Not only are STDs embarrassing socially, but some cause serious organ damage and a few are deadly. Many don't actually affect the reproductive organs, even though they are transmitted sexually. Some STDs are difficult if not impossible to cure once you have them, so an ounce of prevention is definitely worth a pound of (attempted) cure.

Now that we have studied the human reproductive systems, you can see why STDs are so common. Most disease organisms cannot live long outside a warm, moist environment. That's why their most common sites of entry are those that are warm and moist, such as the digestive tract, respiratory tract, and, of course, the reproductive organs during sexual contact. Some organisms, such as the virus that causes AIDS, travel in body fluids. Others, like the pubic lice known as "crabs," simply take the opportunity afforded by intimate contact to hop from one human to another.

STD-causing organisms include viruses, bacteria, fungi, protozoa, and even arthropods. Table 16.4 summarizes common STDs, including causes, symptoms, potential complications, and treatments.

Bacterial STDs: Gonorrhea, syphilis, and chlamydia

Gonorrhea Caused by the bacterium *Neisseria gonorrhoeae,* gonorrhea is easily transmitted—roughly 50% of women and 20% of men contract it after a single exposure. It can be transmitted to the mouth and throat by oral/genital contact, or to the mouth or eyes through hand contact. An infected mother can pass gonorrhea to her child during birth, causing a potentially blinding eye infection.

In men, early symptoms include pus from the penis and painful urination, but the symptoms may disappear even though the disease remains. Women may experience a burning sensation during urination and vaginal discharge that may not appear particularly abnormal. However, nearly 20% of infected men and up to 80% of infected women show no symptoms, at least for periods of time, and so do not seek medical advice. Left untreated, gonorrhea can lead to inflammation, scarring, and sterility.

Although gonorrhea has become less common, it is still with us: nearly 326,000 new cases were diagnosed in 1996 (Figure 16.14). Although gonorrhea generally responds to antibiotics, in recent years antibiotic-resistant strains have emerged.

Syphilis The bacterium *Treponema pallidum* causes syphilis, one of the most dangerous STDs if left untreated. Syphilis develops in three phases, often separated by periods in which the disease seems to disappear. One to eight weeks after infection, the disease enters its *primary phase*. A hard, dry sore filled with bacteria, called a chancre, appears at the infection site. The chancre is not painful and disappears spontaneously after several weeks; infected women may not even notice it.

During the *secondary phase,* which can last several years, the bacteria invade lymph, blood vessels, mucous membranes, bones, and the nervous system. Symptoms include a widespread but non-itchy rash, hair loss, and gray patches representing areas of infection on mucous membranes. Again the symptoms may disappear, and sometimes the symptoms end there. However, some people progress to the *tertiary phase* characterized by widespread damage to the cardiovascular and nervous systems, blindness, skin ulcers, and eventual death. A child of a syphilis-infected woman can be born blind, malformed, or dead because the bacteria cross the placenta into the developing fetus. Syphilis can be treated with penicillin.

Chlamydia An infection by *Chlamydia trachomatis,* chlamydia is now the most prevalent of all STDs, with nearly 500,000 cases reported in 1996. Many of these are college students. Women are diagnosed nearly five times more often than men, leading public health officials to suspect that most men go undiagnosed and the actual number of cases could be even ten times higher than reported.

Symptoms are mild and often go unnoticed. Men may experience a discharge from the penis and a burning sensation during urination, and women

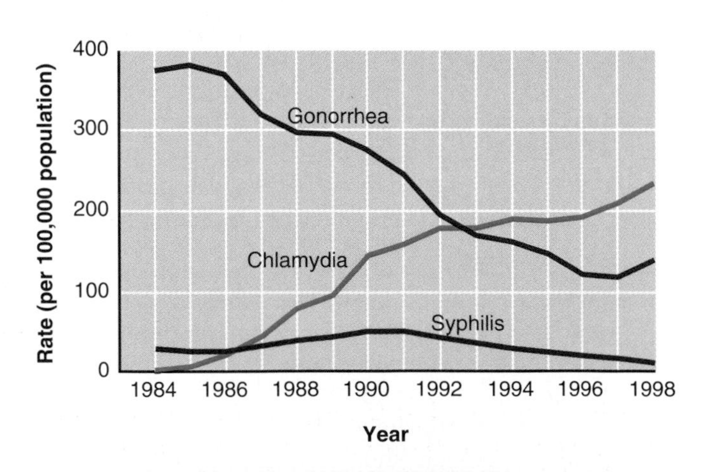

Figure 16.14 Relative rates of infection for the three most common bacterial STDs. Data are expressed in terms of numbers of infection per 100,000 population per year, since 1984. Source: Sexually Transmitted Disease Surveillance 1998. Atlanta: Centers for Disease Control and Prevention, 1999.

Table 16.4 Sexually transmitted diseases

Disease	Organism	Symptoms	Complications	Comments
Gonorrhea	*Neisseria gonorrhoeae* bacterium	Men: Pus from penis, painful urination Women: Painful urination, vaginal discharge	Men: Inflamed testicles and sterility Women: Pelvic inflammatory disease, scarring, sterility	May be asymptomatic, or symptoms may disappear even though disease remains; can be treated with antibiotics
Syphilis	*Treponema pallidum* bacterium	Primary: Chancre at infection site	Secondary: Non-itchy rash, hair loss, gray areas of infection on mucous membranes Tertiary: Widespread damage to cardiovascular and nervous systems, blindness, skin ulcers	Symptoms may disappear spontaneously between phases; can be treated with antibiotics
Chlamydia	*Chlamydia trachomatis* bacterium	Men: Discharge from penis, burning during urination Women: Vaginal discharge, burning, and itching	Pelvic inflammatory disease, urethral infections, sterility, complications of pregnancy	Can be treated with antibiotics
AIDS (also see Chapter 9)	Human immunodeficiency virus (HIV)	Fatigue, fever, and chills, night sweats, swollen lymph nodes or spleen, diarrhea, loss of appetite, weight loss	Infections, pneumonia, meningitis, tuberculosis, encephalitis, cancer	Asymptomatic phase can last for years, even though disease is still contagious; no cure
Hepatitis B (also see Chapter 14)	Hepatitis B virus	Nausea, fatigue, jaundice, abdominal pain, arthritis	Cirrhosis, liver failure	Only STD for which there is a vaccine
Genital herpes	Herpes simplex type 2 virus	Genital blisters, painful urination, fever, swollen lymph nodes in groin	Symptoms disappear during remission, then reappear; contagious during active phases	Drugs available to control active phase
Genital warts	Human papillomavirus (HPV)	Warts on penis, labia, anus, or in vagina or cervix	Risk factor for cancers of cervix or penis	Can be removed by laser surgery, freezing, or drugs; tend to recur
Yeast infections	*Candida albicans*	Men: Discharge from penis, painful urination Women: Vaginal pain, inflammation, cheesy discharge	Can develop in women without sexual contact, but is transmitted sexually to men	Can be treated with antifungal drugs
Trichomoniasis	*Trichomonas vaginalis* protozoan	Men: Discharge from inflammed penis Women: Frothy, foul-smelling vaginal discharge	Recurrent urethral infections	Can be treated with drugs
Pubic lice	*Phthiris pubis*	Intense itching, skin irritation in pubic area; lice are small but visible	Can also be transmitted through infected sheets, clothing	Can be treated with drugs

may have a vaginal discharge and a burning and itching sensation. Both sexes may develop swollen lymph nodes. If not treated, chlamydia can cause pelvic inflammatory disease, complications of pregnancy, and sterility. It can infect a newborn's eyes and lungs at birth, leading to pneumonia.

Recap **Major bacterial STDs include gonorrhea, syphilis, and chlamydia. If not treated, gonorrhea and chlamydia can cause inflammation and sterility; syphilis can cause widespread damage to body systems and eventual death.**

Viral STDs: AIDS, hepatitis B, genital herpes, and genital warts

AIDS *Acquired immune deficiency syndrome (AIDS)* is incurable and one of the most dangerous STDs of our time. It results from infection by the human immunodeficiency virus (HIV), which slowly destroys the immune system.

Early symptoms include fatigue, fever and chills, night sweats, swollen lymph nodes or spleen, diarrhea, loss of appetite, and weight loss. As the immune system grows weaker, patients experience numerous infections and complications. One of the most dangerous features of AIDS is its lengthy asymptomatic stage, which can last for years even though the disease can still be transmitted during this time. AIDS and its global impact were discussed in more detail in Chapter 9 (Immune System).

Hepatitis B Several viruses can cause *hepatitis* (inflammation of the liver). The hepatitis B virus is transmitted in blood or body fluids, and appears to be even more contagious than HIV. For more on hepatitis, see Chapter 14 (Digestive System).

Genital herpes Genital herpes are caused by the *herpes simplex type 2* virus. Over 15% of the population has genital herpes, of which primary symptoms are genital blisters, painful urination, swollen lymph nodes in the groin, and fever. Rupture of the blisters spreads the virus. The symptoms come and go, but once contracted the disease never leaves; it just goes into remission.

As yet there is no cure for genital herpes, but drug treatments are available that can suppress the active phase during which it is contagious. Like so many STDs, genital herpes can infect infants' eyes during birth and cause blindness if not treated. Herpes treatment is an active area of research, so look for a vaccine soon.

Genital warts Genital warts are due to the *human papillomavirus* (HPV). Usually they appear as warts on the penis or labia, in the vagina or cervix, or around the anus. They can be removed by several different methods including laser surgery, freezing with liquid nitrogen, or certain drugs, but they have a tendency to reappear. Research indicates they are a risk factor for cancers of the cervix or penis.

Recap **Two particularly dangerous viral STDs are AIDS and hepatitis B; neither is curable but a vaccine can prevent hepatitis B. Genital herpes may go into remission and genital warts can be removed, but both tend to recur.**

Other STDs: Yeast infections, trichomoniasis, and pubic lice

Yeast infections *Candida albicans* is a yeast (a type of fungus) that is always present in the mouth, colon, and vagina. Normally it is kept in check by competition with other organisms. Sometimes in the vagina, however, the *Candida* population may grow out of control, leading to pain, inflammation, and a thick, cheesy vaginal discharge. A woman with an active yeast infection can pass it to her partner. Infected men sometimes develop a discharge from the penis accompanied by painful urination. It is easily treated by antifungal drugs.

Trichomoniasis Another condition that can cause *vaginitis* (inflammation of the vagina) is *trichomoniasis,* caused by the protozoan *Trichomonas vaginalis.* In women it causes a frothy, foul-smelling vaginal discharge; in men it inflames the penis and causes a discharge. Trichomoniasis can be transmitted even when there are no symptoms, and reinfection is common if partners are not treated simultaneously. It is treatable with drugs.

Pubic lice Pubic lice are tiny arthropods related to spiders and crabs (Figure 16.15). Pubic lice are commonly called "crabs." They live in hair but especially prefer pubic hair, jumping from one host to the next during sexual contact. They can also be transmitted by infected bed-sheets or clothes. The adult lice lay eggs near the base of the hair, and the eggs hatch in a few days. Pubic lice nourish themselves by sucking blood from their host, causing intense itching and skin irritation; the adults are very small but visible with the naked eye. They can be killed with antilice treatments, but all clothes and bedding should be thoroughly washed in hot water to prevent reinfestation.

Recap **Yeast, normally present in the vagina, can multiply and cause a yeast infection; trichomoniasis results from infection by the protozoan Trichomonas vaginalis. Pubic lice are tiny arthropods that are transmitted during intimate contact or by contact with clothes or bedding.**

Figure 16.15 The pubic louse, *Phthirus pubis.* (Approximately ×50)

Protecting yourself against STDs

We've just reviewed a litany of diseases with symptoms ranging from painful urination to brain damage and outcomes ranging from full recovery to death. The good news is that you can probably avoid contracting an STD, but you need to know what to do and, more importantly, be willing to do it.

How can you protect yourself against sexually transmitted diseases?

In Chapter 9 (Immune System) we presented a number of recommendations that can lower your risk of STDs. Many of them are common sense precautions:

- Select your partner wisely. A monogamous relationship with someone you know and trust is your best defense.
- Communicate with your partner. If you have questions or concerns, share them, and be willing to do the same yourself.
- Use suitable barriers, depending on the sexual activity in which you engage. Be aware, though, that barrier methods of birth control are not a cure-all for risky behavior. For example, condoms are only marginally effective for preventing the transfer of HPV and genital herpes, and they will not protect at all against pubic lice.

Finally, know your own health. If you or your partner think you are at risk for or may have a certain disease, get tested. It's confidential, it's not difficult, and it can save you a lot of embarrassment. It may even save your life.

Recap **You can reduce your risk of contracting an STD with a little effort. Choose your partner wisely, use a barrier method of birth control, and, if you suspect you have a disease, get tested promptly.**

Chapter Summary

Male reproductive system delivers sperm

- Sperm are produced in the male reproductive organs, called the testes, and stored in the epididymis and ductus deferens.
- Semen contains sperm and the secretions of three glands: the seminal vesicles, the prostate gland, and the bulbourethral glands.
- Tens of millions of sperm are formed every day throughout the male's adult life. The production of sperm is under the control of three hormones; testosterone, LH, and FSH.

Female reproductive system produces eggs and supports pregnancy

- The female reproductive organs (the ovaries), produce mature oocytes and release them one at a time on a cyclic basis.
- Fertilization of the oocyte (if it is to occur) takes place in the oviduct.
- The fertilized egg makes its way to the uterus, where it implants and begins to develop into a fetus.
- The vagina contains glands that produce lubricating fluid during sexual arousal.
- The hormone estrogen causes the mammary glands to enlarge at puberty.

Menstrual cycle consists of ovarian and uterine cycles

- The cyclic changes in the female reproductive system are called the menstrual cycle. The menstrual cycle consists of an ovarian cycle that produces mature oocytes, and a uterine cycle in which the uterus prepares for pregnancy.
- The menstrual cycle is controlled by the hormones estrogen, progesterone, LH, and FSH. A surge in LH secretion triggers ovulation.

Human sexual responses, intercourse, and fertilization

- Both males and females are aroused by certain stimuli and respond in ways that facilitate intercourse and ejaculation. In males, the penis swells and hardens. In females, glandular secretions provide lubrication.
- Both males and females experience a pleasurable reflex event called an orgasm. Orgasm accomplishes ejaculation, or the expulsion of semen, in the male.
- Sperm deposited in the vagina during intercourse make their way through the cervix and uterus and migrate up the oviducts to the egg. Only one will fertilize the egg.

Birth control methods: Controlling fertility

- The most effective methods of preventing pregnancy are abstinence and female or male surgical sterilization. Sterilization should be considered permanent, though it may be reversible.
- Other effective methods of birth control include manipulation of hormone levels with pills, implants, or injections, and intrauterine devices (IUDs).
- Moderately effective methods of birth control include condoms, cervical caps and diaphragms, and various spermicides. The "natural" methods (the rhythm method and withdrawal) are the least effective.
- A morning-after pill is now available that can be used as an emergency contraceptive up to 72 hours after intercourse.

Infertility: Inability to conceive

- Infertility is defined as the failure to achieve a pregnancy after a year of trying. Nearly 15% of all couples are infertile.
- For male infertility, the primary option is artificial insemination. When the male has a low sperm count but still some viable sperm, the sperm can be concentrated before insemination. In many cases, donor sperm are used.
- Female infertility may be overcome by several methods. In *in vitro* fertilization, egg and sperm are combined under laboratory conditions. The fertilized egg is allowed to develop for several days and then inserted into the uterus. In GIFT, the egg and sperm are inserted directly into an oviduct immediately after collection. In ZIFT, the egg and sperm are combined first, and then the fertilized egg is inserted into the oviduct.

Sexually transmitted diseases: A worldwide concern

- The common feature of sexually transmitted diseases (STDs) is that they are transmitted during sexual contact. Their disease effects are not necessarily on the reproductive system.
- Bacterial STDs include gonorrhea, syphilis, and chlamydia. Syphilis is the most dangerous, chlamydia the most common. All are treatable with antibiotics.
- Viral STDs include HIV (the virus that causes AIDS), hepatitis B virus, and genital herpes and warts. HIV virus is particularly deadly, and there is as yet no cure. There is now a vaccine for hepatitis B virus.
- Avoiding the diseases caused by STDs is a matter of reducing your risk of exposure and paying attention to (and taking responsibility for) your own health.

Terms You Should Know

cervix, 404
clitoris, 405
corpus luteum, 406
diploid, 401
ductus (vas) deferens, 401
egg, 400
endometrium, 404
epididymis, 401
erection, 401
estrogen, 403
follicle, 406
gametes, 402
haploid, 402
implantation, 404
infertile, 415
menstrual cycle, 405
menstruation, 408
oocytes, 403
orgasm, 410
ovarian cycle, 405
ovaries, 403
oviduct, 403
ovulation, 406
penis, 401
progesterone, 403
semen, 401
sperm, 400
testes, 400
testosterone, 402
uterine cycle, 407
uterus, 404
vagina, 405

Test Yourself

Matching Exercise

_____ 1. The primary reproductive organs in males
_____ 2. The primary reproductive organs in females
_____ 3. Rupture of the follicle and release of the oocyte
_____ 4. A sexually transmitted disease caused by *human papillomavirus (HPV)*
_____ 5. Stimulates the secretion of testosterone in males
_____ 6. Stimulates the division of undifferentiated spermatogonia (production of sperm) in males
_____ 7. Causes the LH surge at the midpoint of the menstrual cycle
_____ 8. The other hormone secreted by the corpus luteum, besides estrogen
_____ 9. The production of milk
_____10. Method of male surgical sterilization

a. testosterone
b. ovaries
c. LH
d. lactation
e. estrogen
f. testes
g. genital warts
h. progesterone
i. vasectomy
j. ovulation

Fill in the Blank

11. Newly produced sperm are stored in the ________ and the ductus deferens until ejaculation. They also undergo a maturation process here.
12. The ________ gland in the male produces a watery alkaline fluid that neutralizes the acidic pH of the vagina.
13. The seminal gland produces the carbohydrate ________ as a source of energy for sperm.
14. The inner lining of the uterus, called the ________, undergoes a monthly cycle of growth and then disintegration and shedding each month.
15. After ovulation, the remainder of the follicle becomes the ________, which continues to produce estrogen and progesterone during the second half of the ovarian cycle.
16. After ovulation, the secondary oocyte enters the ________ and begins the journey to the uterus.
17. The ________ is a small organ that contributes to the female sexual response. It originates from the same tissue as the penis early in fetal development.
18. In females, the hormone ________ causes a primary follicle to begin to grow and mature.
19. A common cause of infertility in women is spontaneous abortion, also called a ________.
20. A natural birth control method that requires abstinence for about 8 days a month is called the ________ method.

True or False?

21. Sperm develop best at temperatures several degrees below body temperature.
22. Both male and female orgasms are required for fertilization to take place.
23. Chlamydia, gonorrhea, and syphilis are all bacterial STDs.
24. Pelvic inflammatory disease can lead to female infertility.
25. Sperm and the eggs are diploid cells, like all other body cells.

Concept Review

1. Name the three accessory glands of the male reproductive system.
2. Does a male still produce sperm after a vasectomy? Explain.
3. Explain where fertilization takes place in the female reproductive system.
4. What happens to the remains of the follicle once ovulation has occurred?
5. Describe what is happening in the uterus that leads to menstruation, and explain why menstruation does not occur during pregnancy.
6. How long after intercourse can fertilization occur?
7. Describe how *in vitro* fertilization is done.
8. Name three viral STDs.
9. Describe birth control options, and list those that afford at least some protection against STDs.
10. What factors might go into one's choice of birth control method? (List as many as you can.)

Apply What You Know

1. Male sexual responsiveness and secondary sex characteristics remain unchanged after a vasectomy. Explain why this is so.
2. What sorts of circumstances might cause a woman to choose tubal ligation over birth control pills as a method of birth control? What might cause her to choose birth control pills over tubal ligation?
3. What keeps the ovarian and uterine cycles always exactly in phase with each other, even though the length of the menstrual cycle can vary slightly?
4. If you could ask a potential sexual partner three questions and know that you would get a completely honest answer, what would you ask?

Case in Point

A Birth Control Pill for Men?

It all began with one couple who consulted Dr. Susan Benoff, a fertility researcher at North Shore University Hospital in Manhasset, New York, about their persistent trouble with conceiving a child. Although the man's sperm and the woman's eggs tested normal, repeated attempts at artificial insemination and *in vitro* fertilization failed. Benoff studied their situation and eliminated a number of potential causes. Finally, she took a closer look at nifedipine, the prescription drug the man was taking to treat high blood pressure.

Nifedipine, a calcium-channel blocker, blocks the passage of calcium ions through cell membranes. It and other calcium-channel blockers are a widely prescribed treatment for hypertension and heart disease, taken by millions of people worldwide. In the U.S. alone, they constitute a $4 billion-a-year market. So how could they affect fertility?

When a sperm locates the egg it burrows through the surrounding layer of granulosa cells, binds to the zona pellucida, and releases enzymes that dissolve the zona pellucida. Finally, the sperm cell membrane fuses with the cell membrane of the egg, so that the fertilized egg is surrounded by just one cell membrane. Researchers believe that by blocking the entry of calcium into sperm, drugs such as nifedipine cause the sperm to produce more cholesterol. The extra cholesterol apparently causes changes in the sperm cell membrane that prevent it from binding to the egg.

Benoff estimates that calcium-channel blockers could impair fertility in 95% of male patients. The infertility is reversible; when Benoff's patient switched to another hypertension drug his wife soon became pregnant. Benoff suggests that nifedipine's male infertility side effect, while undesirable for some men, could be an asset for others: the drug could have a promising future as a male birth control pill.

QUESTIONS:

1. Drug companies were not immediately receptive to the idea that calcium channel blockers such as nifedipine might be useful as a male birth control pill. Why do you suppose that is?
2. Do you feel there is a market for a male birth control pill? Why or why not?

INVESTIGATE:

Paul Raeburn. A Promising Pill for Male Birth Control. *Business Week* (February 7, 2000), pp. 95–100.

Maryann Napoli. Calcium Channel Blockers: Causes Male Infertility. *Health Facts* (November 1, 1999).

References and Additional Reading

1995 Assisted Reproductive Technology Success Rates: National Summary and Fertility Clinic Reports. A publication developed and produced by the National Center for Chronic Disease Prevention and Health Promotion of the Centers for Disease Control and Prevention in collaboration with the American Society for Reproductive Medicine; the Society for Assisted Reproductive Technology; and RESOLVE, a national consumer group, December, 1997. (The official report on the U.S. fertility clinics.)

Crooks, R., and K. Baur. *Our Sexuality*. 6th ed. Redwood City, CA: Benjamin/Cummings Publishing, 1996. (An introduction to human sexuality.)

Kaplan, Lawrence J. and Rosemarie Tong. *Controlling Our Reproductive Destiny: A Technological and Philosophical Perspective*. Cambridge, MA.: The MIT Press, 1994. 418 pp. (An interdisciplinary look at birth control and fertility enhancement, including scientific, ethical, legal, and social implications.)

Steptoe, P.C., and R.G. Edwards. Birth after the reimplantation of a human embryo. *Lancet* 2:366, 1978. (Brief announcement of the birth of the first test-tube baby.)

CHAPTER 17

CELL REPRODUCTION AND DIFFERENTIATION

Why is a fetus **more vulnerable to** genetic damage **than an adult?**

How do new cells end up with the same DNA **as other cells?**

How can your body repair mutations **in your DNA?**

How often **do different types of cells divide?**

CELLS REPRODUCE BY DIVIDING IN TWO. INDEED, CELL DIVISION (cell reproduction) is one of the defining features of life itself. For single-celled organisms, cell division is the equivalent of organism reproduction. For multicellular organisms, cell division and growth allow the organism to grow in size.

You started life as just a single cell created by the union of your father's sperm and your mother's egg. By the time you were born you were already composed of over 10 trillion cells, each created by cell division. Cell division continues throughout childhood and adolescence as your body grows and matures. Even after adulthood, some cells continue to divide in order to replace cells that die or become damaged. For instance, your 25–30 trillion red blood cells are completely replaced every 120 days, which works out to an astonishing 175 million cell divisions every minute.

The mechanisms of cell division are the same in all eukaryotes (organisms with cell nuclei) regardless of organism size. In this chapter we see how the cell's genetic material is copied to prepare for cell division, how the cell divides in two, and how cell growth and division are regulated. Finally, we discuss how cells differentiate to take on special forms and functions.

17.1 The cell cycle creates new cells

The creation of new cells from existing cells involves a repetitive sequence of events known as the **cell cycle** (Figure 17.1). The cell cycle consists of two phases, called the *interphase* and the *mitotic phase:*

- **Interphase** is a long growth phase during which the cell grows and DNA is duplicated in preparation for the next division. Most of the growth of the cell occurs during the first part of interphase, called G1 (the G stands for growth). During the S phase (the S stands for synthesis of DNA), the cell's DNA is duplicated and growth continues, although at a slower pace. The cell continues to grow and prepares for cell division during the final growth phase, called G2.

- The **mitotic phase** is a much shorter period during which first the nucleus and then the cytoplasm divide. The mitotic phase consists of (1) *mitosis,* during which the duplicated DNA is divided into two sets and the nucleus divides, and (2) *cytokinesis,* when the cytoplasm divides and two new cells, called "daughter" cells, are formed. The two new cells are called daughter cells regardless of the person's sex.

In mammals, a complete cell cycle takes about 18–24 hours if the cells are actively dividing. DNA replication during the S phase takes about 7–8 hours. Mitosis and cytokinesis generally take only about 30–45 minutes. Cells that are not dividing stop growing during G1 and are said to be in an *arrested G1 phase.*

Recap **Cell reproduction is required for growth and to replace cells throughout life. The cell cycle consists of a long growth phase (interphase), during which the cell's DNA is replicated, and a shorter phase (mitotic phase), during which first the nucleus and then the cell cytoplasm divide.**

17.2 Replication, transcription, and translation—An overview

Recall from Chapter 2 that **DNA (deoxyribonucleic acid)** is a double-stranded string of nucleotides intertwined into a helical shape. Three billion base pairs of human DNA are packed onto 46 separate molecules called **chromosomes** that organize and arrange the DNA within the nucleus. Chromosomes also contain proteins called *histones* that confer a certain structure to the chromosome molecule. The number of chromosomes varies between species of organisms, but it is always the same for all

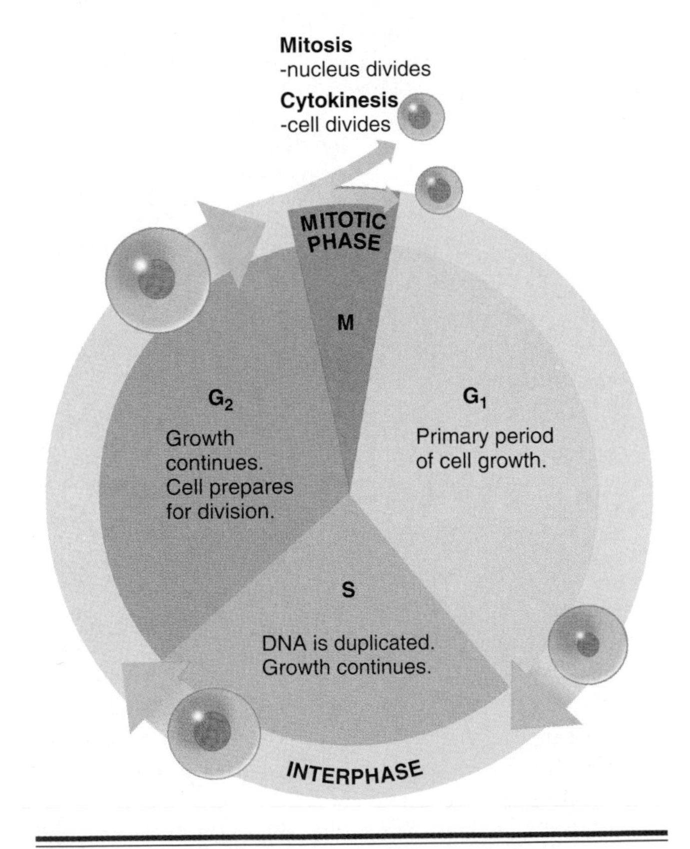

Figure 17.1 The cell cycle. The cell grows throughout all three phases of interphase, though most of the growth is in G1. During the short mitotic phase, first the nucleus divides (mitosis) and then the cell divides (cytokinesis).

of the individuals of a species. Humans have 46 chromosomes in their cell nuclei.

Throughout most of the cell cycle the chromosomes are not visible because they are so long and thin. During interphase they appear only as a diffuse, grainy material known collectively as *chromatin* material. However, just before the nucleus divides the duplicated chromosomes condense briefly into a shorter, compact shape that can be seen under the microscope (Figure 17.2). Chromosomes are moved around a lot during nuclear division, and the compactness of chromosomes at this stage helps prevent breakage of the delicate chromatin strands. Each visible duplicated chromosome consists of two identical *sister chromatids* held together by a *centromere*.

Mitochondria also contain a very small amount of DNA. The single human mitochondrial DNA molecule is not organized into a chromosome and contains only 37 genes, only about 5% of the genes needed for mitochondrial function. Scientists believe that mitochondrial DNA is a holdover from a time in evolutionary history when mitochondria were independent organisms. At some point, mitochondria were incorporated into eukaryotic cells and became dependent on them.

Because DNA represents essentially all instructions for life, every cell in the organism must contain exactly the same set of DNA. It follows that the DNA must be duplicated every time a cell divides. DNA **replication** is the process of copying the DNA prior to cell division.

A **gene** is a short segment of DNA that contains the code, or recipe, for one or more proteins. A single gene is the smallest functional unit of DNA. There are approximately 100,000 genes on the 46 chromosomes, each one located at a precise position on a particular chromosome. The chromosomes are too large to pass through the nuclear pores, so they remain within the nucleus. For a gene to be useful to the cell, its code must first be converted to a form that can leave the nucleus and enter the cytoplasm. This is the function of **ribonucleic acid (RNA).**

Transcription is the process by which the DNA code of a single gene is converted into a complementary single strand of RNA. RNA is small enough to pass through the nuclear pores and enter the cytoplasm. In the cytoplasm, RNA is used as a template to create a precise sequence of amino acids that comprise one particular protein. The process of converting the RNA template code into a particular protein is called **translation.**

Figure 17.3 summarizes replication, transcription, and translation. We will now look more closely at these three processes.

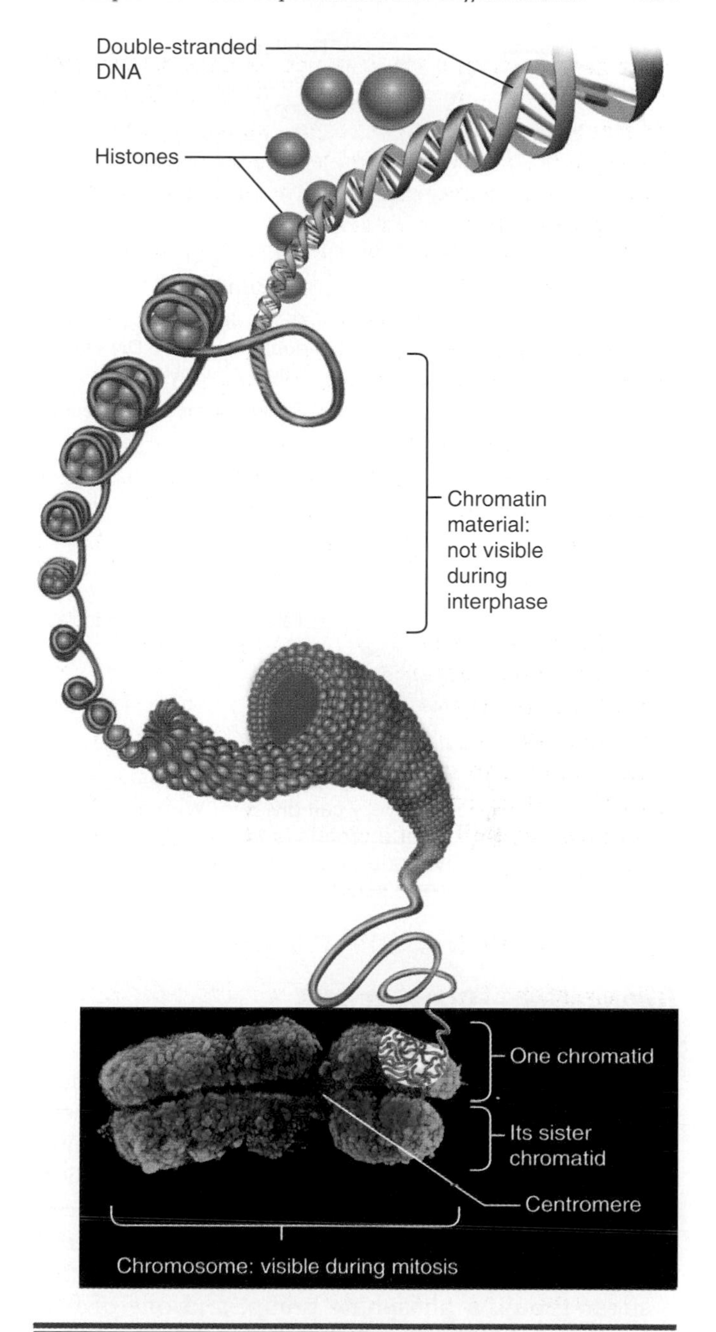

Figure 17.2 The structure of a chromosome. DNA combines with histone proteins to become long thin strands of chromatin material. During most of the cell cycle the chromatin material of a chromosome is not visible within the cell. During mitosis the chromatin material condenses and the duplicated chromosome becomes visible as two identical sister chromatids held together by a centromere. The two sister chromatids separate during nuclear division.

Recap **Human DNA is organized and arranged on 46 chromosomes within the cell nucleus. A gene is the smallest functional unit of DNA. Before a cell divides, its DNA is duplicated during the process of replication.**

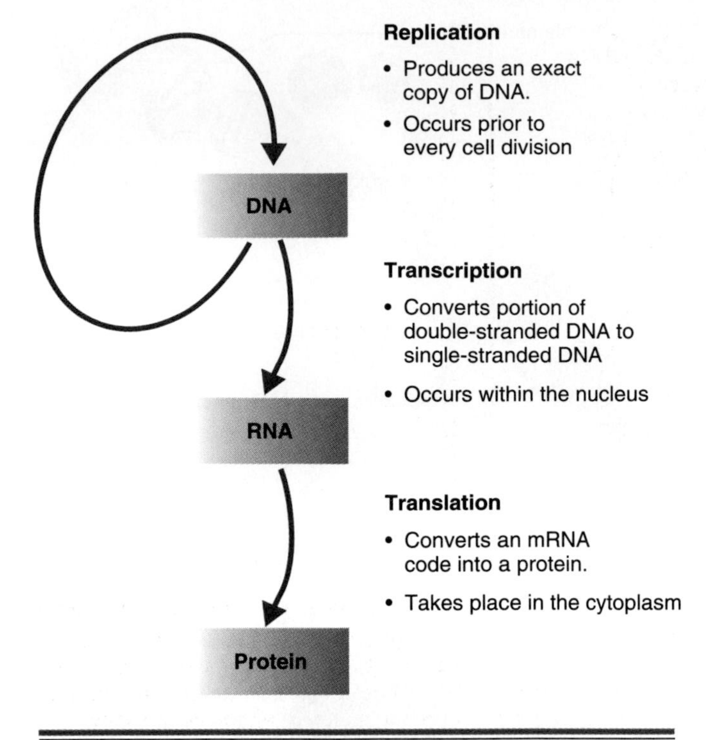

Figure 17.3 Replication, transcription, and translation. DNA is *replicated* before every cell division. Within the nucleus the genetic code for a protein is first *transcribed* into RNA. RNA transfers the code to the cytoplasm, where it is *translated* into the correct sequence of amino acids for the protein.

Replication—Copying DNA before cell division

Spotlight on Structure & Function

Before a cell can divide, its DNA must be duplicated so that the two new cells resulting from the division each have an identical set of genes. This process, called replication, begins with the uncoiling of the DNA helices. Recall from Chapter 2 that the nucleotides in each strand of the double-helical DNA molecule consist of a sugar group, a phosphate group, and one of four bases (adenine, thymine, guanine, or cytosine). The double helix is formed as the bases pair up; thymine with adenine and guanine with cytosine. This precise pairing facilitates the replication of DNA, because upon uncoiling and "unzipping" the two strands, each single strand can serve as a template for the creation of a new complementary strand. The precise nature of the base pairing of the nucleotides (T–A and G–C) ensures that the new complementary strand will be exactly like the original complementary strand (Figure 17.4a). The new nucleotides are positioned and linked together by enzymes called *DNA polymerases*.

The "unzipping" of the DNA strands does not proceed linearly from one end of the DNA molecule to the other. Instead, the DNA strands become detached from one another at multiple locations along the length of the molecule. Were this not so, replication would take too long, considering the huge number of nucleotides involved. Because there are three billion nucleotide pairs in the human set of DNA, six billion nucleotides have to be added to the unzipped single strands in order to replicate the DNA entirely. Even if the cell replicated all 46 DNA molecules simultaneously, it would have to add an average of 5,000 nucleotides per second to each developing new strand of DNA to get the job done during the 7–8 hours of the S-phase. That is an incredibly fast rate of unzipping and complementary base pairing—too fast when accuracy is important!

What actually happens during replication is much more efficient. Certain enzymes bind at various points along the DNA molecule and gently unwind and "unzip" a portion of the DNA, creating a "replication bubble" in the two strands. New complementary strands begin to form at these bubbles. Replication proceeds outward in opposite directions in each strand until the expanding replication bubbles join (Figure 17.4b). Because replication occurs at thousands of sites along the molecule at once, it can conclude much more quickly.

Once replication is complete the two identical sister chromatids remain attached at a single point called the *centromere* (refer back to Figure 17.2). The centromere holds the sister chromatids together until they are physically pulled apart during mitosis. Figure 17.5 illustrates how the structure of a chromosome changes during the various stages of the cell cycle. ■

Recap **During replication, the two strands of DNA unwind and separate from each other. Each strand serves as a template to form a new complementary strand. Enzymes called DNA polymerases add new nucleotides to each of the old strands in order to build two complete molecules of DNA from one.**

Mutations can be helpful or harmful Considering that DNA spells out the instructions for building all the proteins of a complex organism, you might think that it is essential that every copy of it be perfect. It turns out, though, that changes in DNA are necessary in order for evolution to occur. Changes in the DNA of the cells that produce sperm and eggs may be passed to future generations.

Alterations in DNA, called **mutations,** create the differences between individuals and species. Generally speaking, heritable mutations occur very slowly. The DNA of humans and chimpanzees, for example, differ only by about 1% even though our evolutionary paths diverged over 5 million years ago. Two humans differ by perhaps one base pair in a thousand, but that is still three million base pair dif-

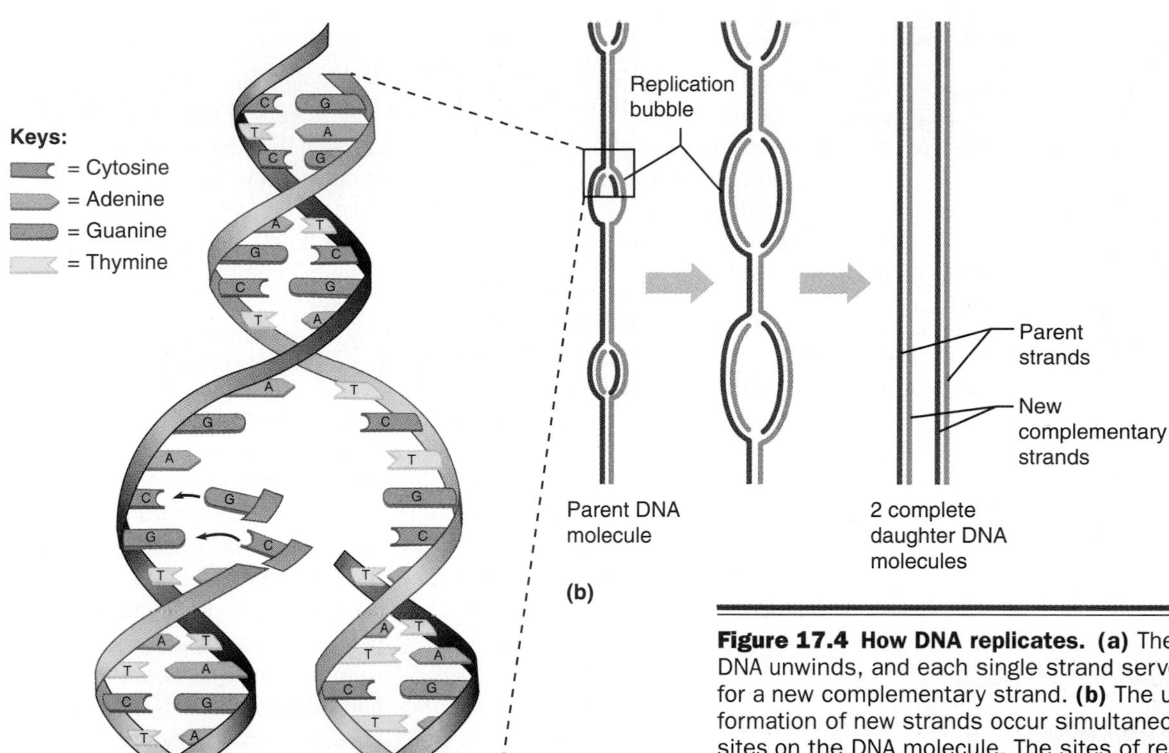

Figure 17.4 How DNA replicates. (a) The double-stranded DNA unwinds, and each single strand serves as a template for a new complementary strand. **(b)** The unwinding and the formation of new strands occur simultaneously at many sites on the DNA molecule. The sites of replication expand outward until they join. For simplicity the two strands are shown as parallel in **(b)**, but in actuality they form a helical shape as shown in **(a)**.

ferences, more than enough to account for all our individual variations.

Mutations may result from mistakes in DNA replication. In addition, chemicals or physical forces can damage a segment of DNA before it is replicated. Unless these errors are detected and corrected before the next replication, they may be passed on to future cells or could even prevent the DNA from being copied at all. In other words, the DNA needs to be in as good a shape as possible before it is replicated.

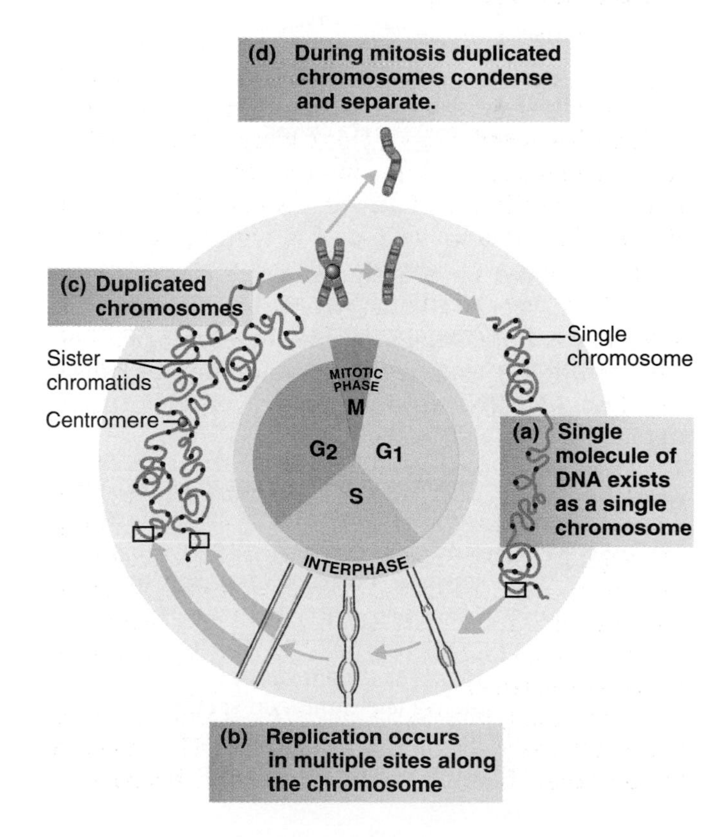

Figure 17.5 The structure of DNA throughout the cell cycle. (a) During G1 the DNA molecule exists as a long, thin chromosome that cannot be seen distinctly under the microscope. **(b)** During the S phase the two strands of the DNA molecule unwind and replicate. **(c)** After the DNA has been replicated, the duplicated chromosome consists of sister chromatids held together by a centromere. **(d)** At the beginning of the mitotic phase, the duplicated chromosome condenses and becomes visible. During mitosis the duplicates are pulled apart, with one going to each daughter cell. When mitosis is complete, the single chromosome again takes on the more dispersed form and is no longer visible.

Directions in Science

DNA Fingerprinting

An exciting new development in science is the ability to identify human beings, whether living or dead, from only a few strands of their DNA. "DNA fingerprinting," as it is called, is really a sophisticated technique for analyzing very small samples of DNA. DNA fingerprinting is the modern sequel to dermal fingerprinting.

Fingerprints were first used as identification during the 1870s in a district of India. In 1880 it was proposed that they could be used to identify criminals, but it was not until the pioneering work of Francis Galton in the 1890s that the technique gained wide acceptance. Galton stressed that it is not the general pattern of a fingertip that is important, but the precise details of its forking and branching. He established two important precepts of fingerprinting:

- A person's fingerprints are stable over time. Galton showed that although fingerprints grow with the individual, they remain unchanged throughout life and even after death.
- Fingerprints are unique to only one person, or at least sufficiently different that they can be reliable evidence.

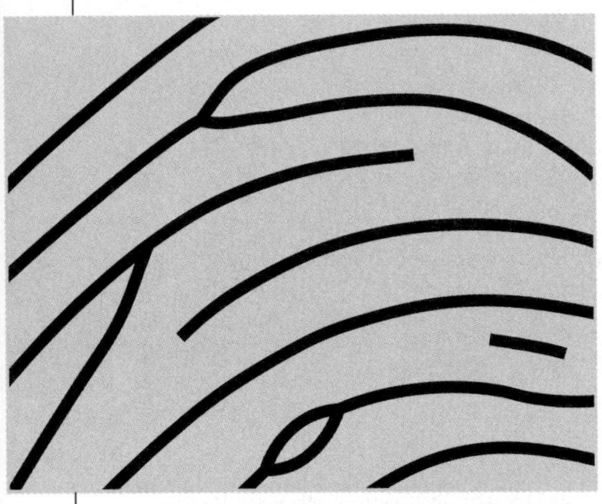

The characteristics of human dermal fingerprints. The pattern of forks and branches is unique to any one individual. This is a re-rendering of a drawing from a work by Sir Francis Galton in 1892.

Proving the second statement was essential to the acceptance of fingerprints as legal evidence. Galton devised and tested an ingenious statistical argument, which led him to claim that the chance of any random fingerprint matching any other fingerprint was one in 64 billion. (At the time the entire world population was only 1.6 billion.) Later he showed that the gross patterns of fingerprints are inherited, but not the minute patterns of forks and branches. Even siblings' fingerprints are unique. Galton's arguments and methods were so thorough that they have withstood the tests of time and courtroom scrutiny.

DNA fingerprinting is a newer and more controversial technique. Court cases using the technique often get bogged down in weeks or months of scientific testimony about the method itself. The potential payoff is that DNA fingerprinting can identify a person when fingerprints are not available, using only a few cells or a strand of DNA. It is based on the following facts:

- The human set of DNA consists of the same genes located at the same places on the same chromosomes. In other words, the genes and their locations are not unique between individuals.
- The nucleic acid sequences of the genes vary slightly between individuals. These slight variations result from millions of years of accumulated mutations.
- Certain enzymes can be used to slice DNA at specific locations.

Specific enzymes are added to a sample of a person's DNA to cut the DNA into small pieces. The sizes and nucleic acid sequences of these pieces are unique for each person. Although scientists don't know how to put the pieces back together, they do know how to line them up by size (using a technique called gel electrophoresis), label them, and photograph the labels. The resulting

When DNA is severely damaged, it may not be possible to repair all the errors. In this case, substantial mutations and even life-threatening cancers may result. For example, skin cancer can be caused by damage to DNA from excessive exposure to sunlight.

How can your body repair mutations in your DNA?

Mechanisms of repair DNA repair mechanisms play a crucial role in the survival of an organism and its species. These mechanisms are quite efficient unless they are overwhelmed by massive damage.

DNA repair involves recognizing the errors, cutting out and replacing the damaged section or incorrect nucleotide base, and reconnecting the DNA backbone. The process utilizes numerous *DNA repair enzymes*. Typically one enzyme might recognize certain types of damage and cut out the impaired nucleotide, another might insert the correct nucleotide, and yet another might reconnect the adjoining ends of the DNA. The repair process is most active in the hour between DNA replication and the beginning of mitosis. Repair prior to cell division ensures that the best possible copy of DNA is passed on to each daughter cell.

DNA repair mechanisms are directed by certain genes that code for the repair enzymes. Thus the genetic code plays a vital role in repairing itself. If the damaged genes are the ones that control the repair process, DNA errors may accumulate more quickly than normal. Researchers now know that mutations of certain genes that direct DNA repair are associ-

"DNA fingerprint" is, at least in theory, unique for every person except identical twins.

But there's a catch. The complete set of human genes is large (3 billion base pairs), and many of the genes are still unknown. Furthermore, DNA is difficult to purify. We may only get a short segment of DNA, which is like having a partial thumbprint. So even when two samples match, defense attorneys still argue (sometimes successfully) that they *could* have come from different individuals, even though the statistical chance of it being true is usually less than one in millions.

On the other side of the coin, a failure of two samples to match is generally considered proof positive that they did not come from the same individual. For that reason the technique can be used effectively to establish nonpaternity, an important issue in child support, custody, and inheritance cases. It has also been used to overturn convictions based on other evidence before DNA fingerprinting became available.

The first conviction in the U.S. based on DNA evidence occurred in a rape trial in Florida in 1987. A year later, the so-called "Forest Hills Rapist" was convicted in a highly publicized trial in New York. In that case a light-skinned Hispanic man stood accused of sexual assault, but the three victims had indicated to police that the rapist was black. DNA fingerprinting played a central role in his conviction because the DNA pattern of sperm found at the crime scenes matched that of his blood. DNA analysis of blood samples also played a prominent role in the criminal and civil trials of O. J. Simpson for the murder of his ex-wife and her friend in 1994.

In legal cases, pretrial hearings about DNA evidence can take weeks and cost taxpayers millions of dollars without permanently settling the issues of reliability and uniqueness. In the O. J. Simpson case, for example, the pretrial defense document challenging the DNA data was over 100 pages long. Mr. Simpson was acquitted in the criminal trial but convicted in the civil trial.

Why is DNA fingerprinting challenged so consistently in the courts when dermal fingerprinting is so well accepted? One reason is that the technique is new. More importantly, it is difficult for a jury to understand. There's something elegant in the visual simplicity of a fingerprint, but for the average juror there's no simple elegance in DNA analysis.

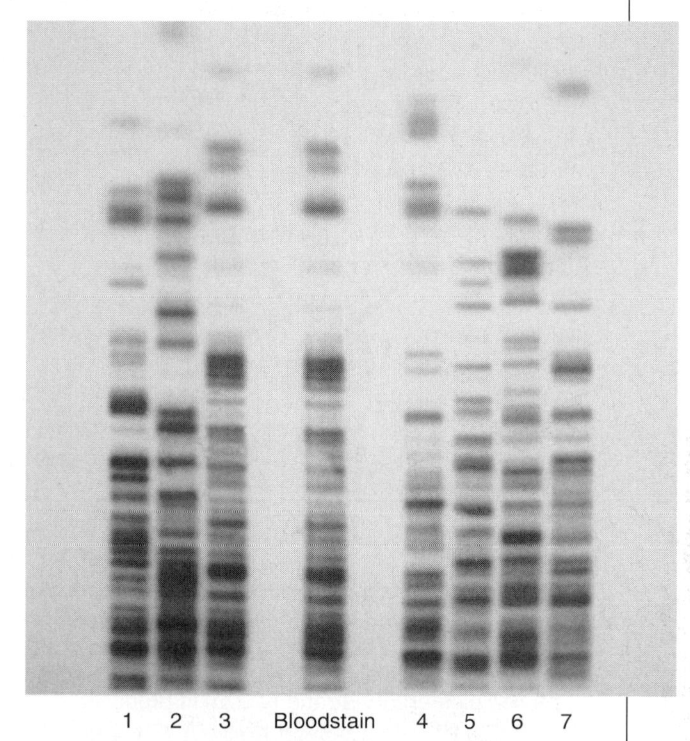

DNA fingerprints. This group of DNA fingerprints includes one from a blood sample found at a crime scene and seven from different persons, including the suspect. Can you identify the match? Photo courtesy of Cellmark Diagnostics, Inc., Germantown, Maryland.

ated with an increased risk of several cancers, including forms of colon and breast cancer.

Recap **Mutations may result from mistakes in DNA replication or from physical or chemical damage. Repair mechanisms remove and replace damaged DNA, if possible, before replication.**

Transcription—Converting a gene's code into RNA

A gene is "expressed" when it is called into action and the protein for which it codes is produced. An essential step in this process is to make a complementary single strand of RNA. Transcription involves converting the DNA code of a single gene into a complementary single strand of RNA.

Transcription is similar to the process of DNA replication, with these exceptions:

1. Only the segment of DNA representing a single gene unwinds, as opposed to the entire molecule.
2. RNA is single-stranded, so that only one of the two strands of DNA actually carries the genetic code specifying the synthesis of RNA.
3. One of the four complementary base pairs of RNA is different from those of DNA; uracil replaces thymine.
4. The sugar group of RNA is ribose rather than deoxyribose.

Once it is fully transcribed the RNA molecule is released from the DNA strand and the two DNA strands entwine around each other again. Figure 17.6a illustrates the process of transcription.

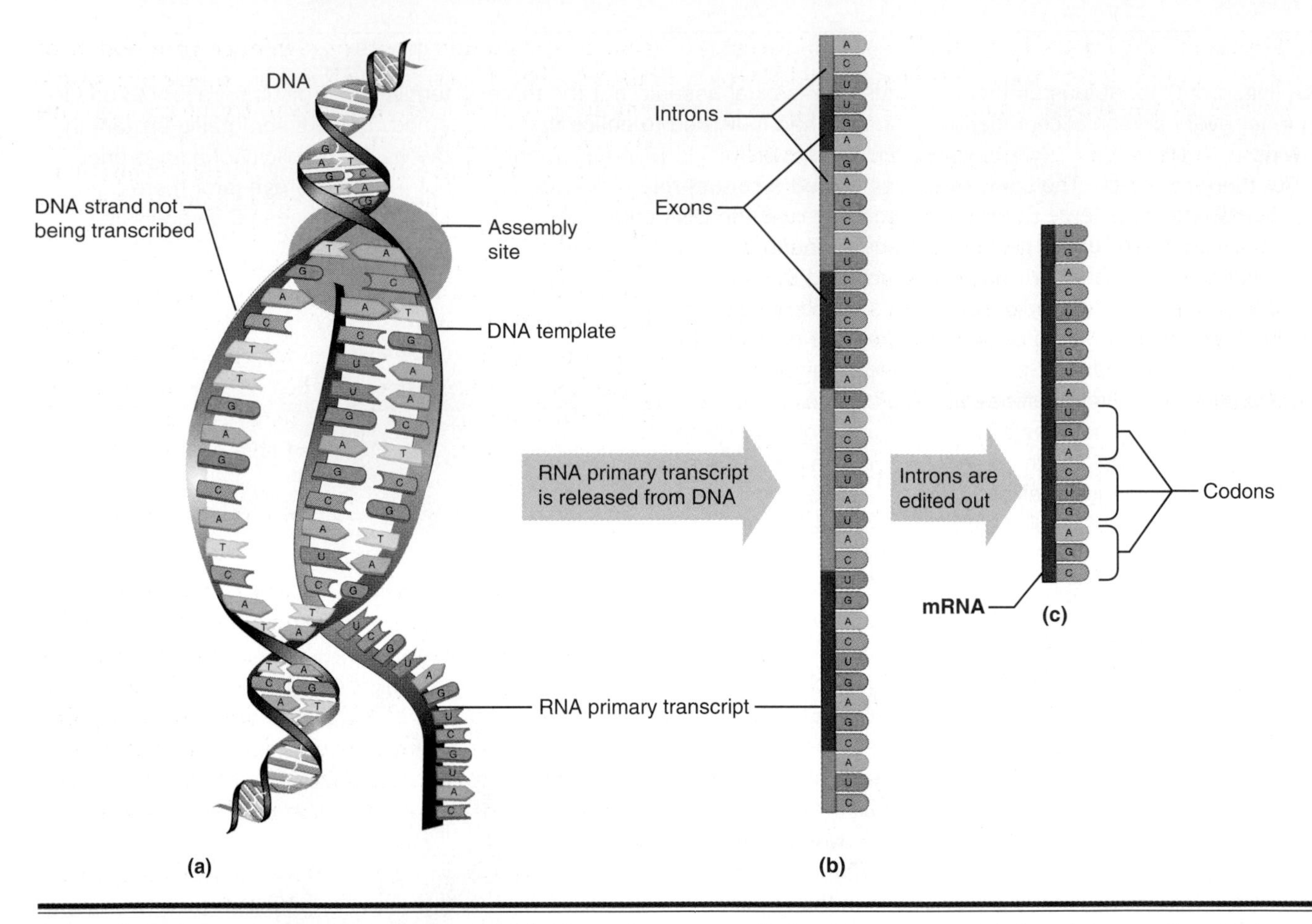

Figure 17.6 Transcription of a gene into mRNA. Transcription takes place in the nucleus. **(a)** The portion of the DNA molecule corresponding to the gene unwinds temporarily, and a complementary strand of RNA is produced from one DNA strand. **(b)** The strand of RNA released from DNA is called a primary transcript. It contains sections that carry genetic information, called exons, and sections that are apparently useless, called introns. **(c)** The introns are edited out by enzymes to produce the final mRNA.

How is the specific gene to be transcribed identified on the chromosome? A unique base sequence called the *promoter* marks the beginning of every gene. An enzyme called **RNA polymerase** attaches to the promoter, starts the DNA unwinding process, and assists in attaching the appropriate RNA nucleotides to the growing chain. The process ends at another base sequence that identifies the end of the gene.

The RNA molecule first transcribed from DNA, called the *primary transcript,* is not yet functional, because most of the DNA base sequences of a gene do not code for anything at all. The meaningless fragments of DNA are called **introns,** and the sequences that carry genetic information are called **exons** (Figure 17.6b). The primary transcript will be "edited" by still other enzymes, which snip out the sections corresponding to the introns. The importance of the introns is unclear, though there is now some evidence that they may allow the exon fragments to be joined in different ways so that a single gene could code for different RNA molecules in different cells. The RNA molecule produced from the primary transcript is called **messenger RNA (mRNA).** It is called "messenger" because it contains a message, in the form of a template, that can be "translated." The message spells out a specific sequence of amino acids that constitute a protein.

The message is encrypted as a *triplet code,* so called because three successive bases of mRNA, called a **codon,** each code for one of the 20 amino acids. Because there are four possible nucleotide bases and the code is a triplet code, there are precisely 64 different possible codons ($4 \times 4 \times 4$). We know exactly which amino acids are specified by the 64 possible codons of mRNA (Figure 17.7). Note two important points:

1. Several different codons can specify the same amino acid because there are more possible codons than there are amino acids. For example, both the UUU and UUC codons code for the same amino acid, phenylalanine (Phe).

2. The codon AUG specifies a "start" code and three others specify "stop" codes. These are needed to specify where to begin and end the protein.

Recap **Transcription converts short segments of DNA representing single genes into a readable and transportable RNA code. Only a portion of the DNA molecule unwinds, and only one strand is used to make single-stranded RNA; the result is an mRNA molecule in which three successive bases, called a codon, represent the code for a particular amino acid.**

Translation—Making a protein from RNA

Translation is the process of using mRNA as a template to convert, or "translate," the mRNA code into a precise sequence of amino acids that compose a specific protein.

The genes of DNA actually produce three different kinds of RNA, each with a specific role in the production of proteins. Only mRNA carries the recipe for a new protein. The other two kinds of RNA are called *transfer RNA* (tRNA) and *ribosomal RNA* (rRNA). Each has an important role in protein synthesis.

Transfer RNA (tRNA) molecules are small molecules that carry the code for just one amino acid. They also carry an **anticodon,** or base triplet that is the complementary sequence to a codon of mRNA. The function of tRNA is to capture single amino acids and then bring them to the appropriate spot on the mRNA chain.

First position	Second position: U	C	A	G	Third position
U	UUU Phe	UCU Ser	UAU Tyr	UGU Cys	U
	UUC Phe	UCC Ser	UAC Tyr	UGC Cys	C
	UUA Leu	UCA Ser	UAA Stop	UGA Stop	A
	UUG Leu	UCG Ser	UAG Stop	UGG Trp	G
C	CUU Leu	CCU Pro	CAU His	CGU Arg	U
	CUC Leu	CCC Pro	CAC His	CGC Arg	C
	CUA Leu	CCA Pro	CAA Gln	CGA Arg	A
	CUG Leu	CCG Pro	CAG Gln	CGG Arg	G
A	AUU Ile	ACU Thr	AAU Asn	AGU Ser	U
	AUC Ile	ACC Thr	AAC Asn	AGC Ser	C
	AUA Ile	ACA Thr	AAA Lys	AGA Arg	A
	AUG Met/start	ACG Thr	AAG Lys	AGG Arg	G
G	GUU Val	GCU Ala	GAU Asp	GGU Gly	U
	GUC Val	GCC Ala	GAC Asp	GGC Gly	C
	GUA Val	GCA Ala	GAA Glu	GGA Gly	A
	GUG Val	GCG Ala	GAG Glu	GGG Gly	G

Figure 17.7 The genetic code of mRNA. Three successive bases represent a codon, and all but three codons encode for an amino acid. The letters correspond to the four RNA bases: uracil (U), cytosine (C), adenine (A), and guanine (G). To read the chart start with the letter in the left-hand column, then move across to one of the letters in the four center columns, then down the final column to the appropriate letter. UAG is the "start" codon and the only codon for the amino acid methionine; there are three "stop" codons.

Once they have been captured and brought to the mRNA molecule, the amino acids must be connected together into one long chain in order to become a protein. This is the function of a ribosome. A **ribosome** consists of two subunits comprised of **ribosomal RNA** (rRNA) and proteins. A ribosome has binding sites for both mRNA and tRNA. It also contains the enzymes capable of connecting the amino acids together. The function of a ribosome, then, is to hold the mRNA and tRNA in place while joining the amino acids.

Translation occurs in three steps, as illustrated in Figure 17.8:

1. *Initiation.* A particular initiator tRNA binds to the smaller of two ribosome subunits and to the mRNA molecule. The tRNA and ribosome subunit move along the mRNA until they encounter a "start" (AUG) codon. At this point they are joined by a larger ribosomal subunit to form the intact ribosome, which holds the mRNA in place while the tRNAs bring amino acids to it.
2. *Elongation.* The chain of amino acids lengthens one amino acid at a time. tRNA molecules carrying the appropriate amino acids bind to the ribosome and to mRNA. The ribosome moves along the mRNA molecule, catalyzing the binding of the amino acids to each other and breaking the bonds that hold the tRNA to the amino acid and mRNA. The tRNA molecule is then released to find another amino acid.
3. *Termination.* There is no tRNA anticodon corresponding to a "stop" codon on mRNA. When a "stop" mRNA codon is encountered, the ribosomal subunits and the newly formed peptide chain detach from the mRNA.

In cooking terms the mRNA brings the recipe, the tRNAs find and deliver the ingredients, and the ribosome is the cook who creates the final product. What happens next to the newly formed peptide chain depends on its function. Some enter the endoplasmic reticulum for further processing, packaging, and shipping, whereas others remain in the cytoplasm. Many become parts of new structural units, organelles, cell membrane components, and enzymes that the cell will need as it grows and then divides into two new cells.

Recap **Translation is a three-step process (initiation, elongation, and termination) by which proteins are assembled according to an mRNA code. The process utilizes tRNA molecules to capture the amino acids in the**

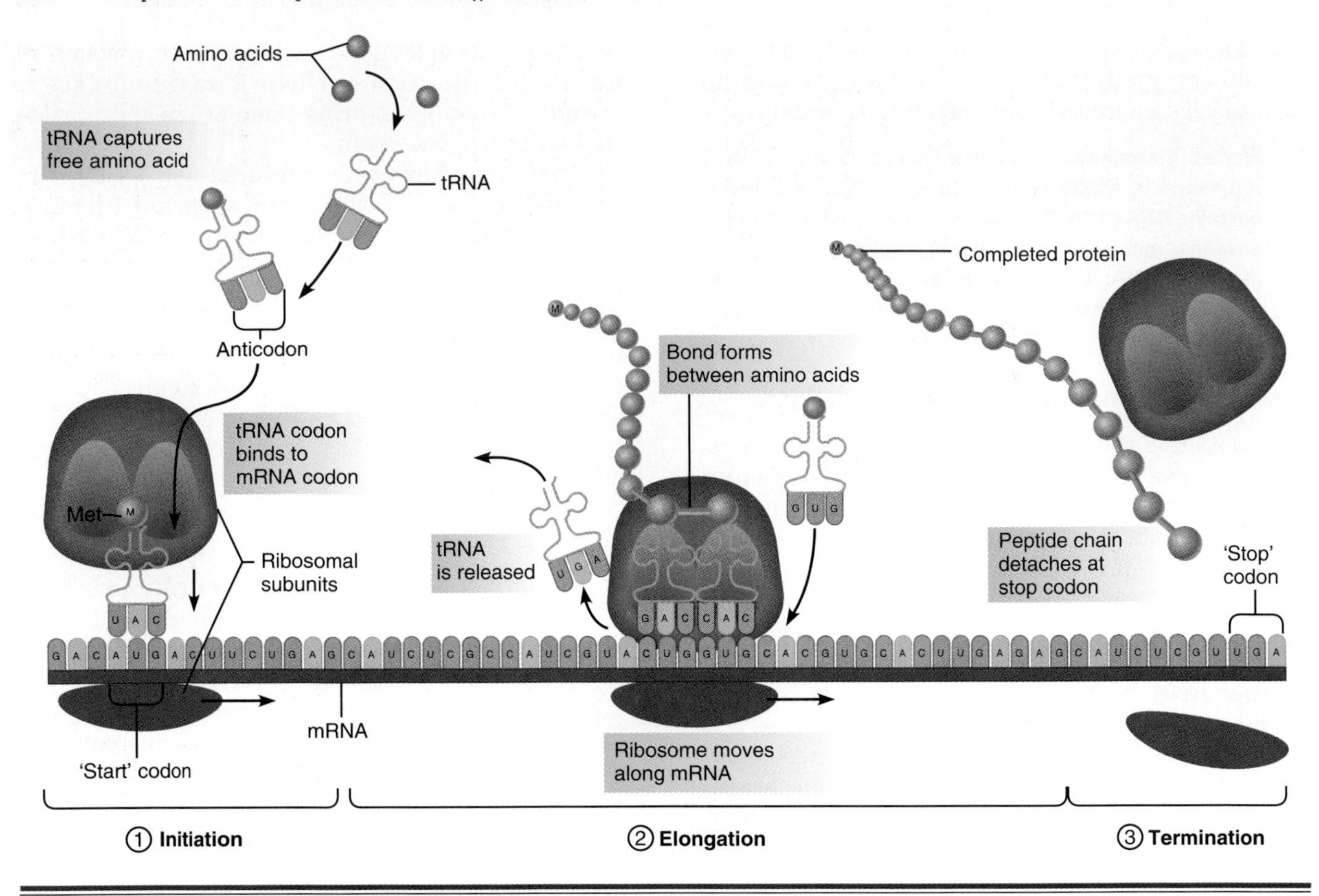

Figure 17.8 The three steps of translation. ① Initiation. A tRNA carrying the "start" anticodon binds to the smaller ribosomal subunit and to the "start" codon of mRNA. The larger ribososmal subunit joins them. ② Elongation. The ribosome moves along the mRNA. tRNA molecules capture free amino acids and bring them to the appropriate codon on the mRNA. The ribosome binds to the tRNA and catalyzes the formation of the bond between successive amino acids. The tRNA is then released to find another amino acid. ③ Termination. When a stop codon is reached, the ribosomal subunits and the newly formed protein detach from the mRNA.

cytoplasm specified by each mRNA codon. The amino acids are connected together on a ribosome, which binds both mRNA and tRNA.

17.3 Cell reproduction—One cell becomes two

Let's review what we have learned so far about the preparations for cell division. DNA is replicated during the S phase of the cell cycle so there are two complete copies. The genes of DNA are used to produce specific proteins. These proteins in turn direct the cell's activities and form many of its structural units as the cell grows.

We turn now to the mitotic phase of the cell cycle, when first the nucleus and then the cell cytoplasm divide. The division of the nucleus is called **mitosis.** It is followed by **cytokinesis,** the division of the cytoplasm.

Mitosis: Daughter cells are identical to the parent cell

How do new cells end up with the same DNA as other cells?

Mitosis is the process of nuclear division in which the sister chromatids of each duplicated chromosome are separated from each other. Following mitosis, each daughter cell has a complete set of DNA and is identical to the parent cell. Mitosis ensures that all cells of a complex organism are endowed with the same set of DNA.

Mitosis (Figure 17.9) includes a sequence of phases that compose most of the mitotic phase of the cell cycle. The phases are defined by the structural changes taking place in the cell as the DNA divides and two new nuclei form. Although each stage has a name and certain defining characteristics, mitosis is really a seamless series of events.

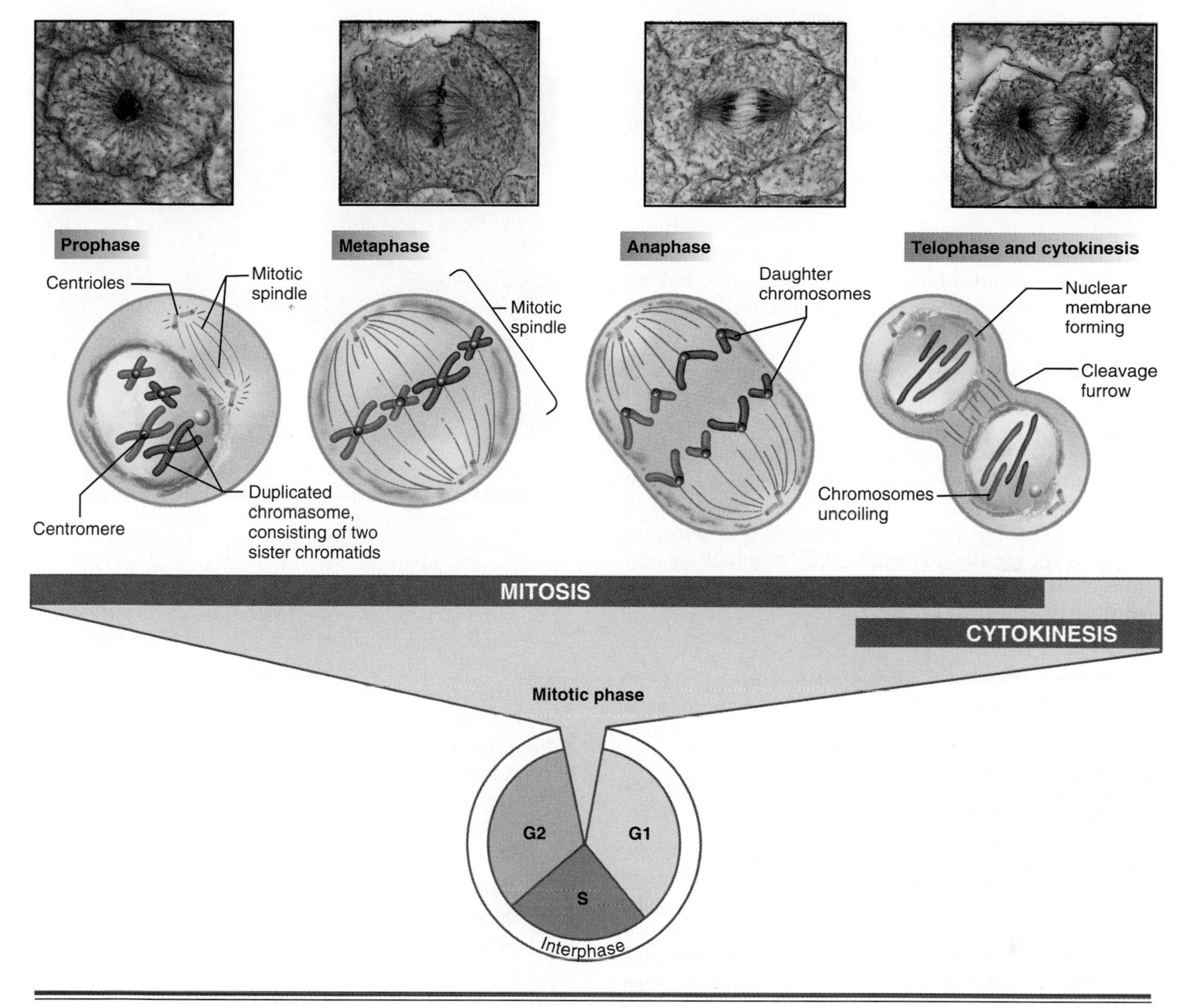

Figure 17.9 Mitosis. Mitosis comprises most of the mitotic phase of the cell cycle, overlapping slightly with cytokinesis, the division of the cell into two cells. The duplicated chromosomes become visible in prophase, line up in metaphase, are pulled apart in anaphase, and uncoil and become surrounded by nuclear membranes in telophase.

Prophase In late G2 of the interphase, the strands of replicated DNA and associated proteins begin to condense and coil tightly. Eventually they become visible as a duplicated chromosome consisting of sister chromatids held together by the centromere. **Prophase** is defined as the point at which the duplicated chromosomes first become visible.

Recall that the cell has a cytoskeleton that helps hold it together. During prophase the tubular elements of the cytoskeleton come apart and are reassembled between the pairs of centrioles. This newly formed parallel arrangement of microtubules, called the **mitotic spindle,** will become the structure that causes the two identical sets of DNA to divide. The pairs of centrioles drift apart and the mitotic spindle lengthens between them. The nuclear membrane dissolves and the centrioles move to opposite sides of the cell. This allows the mitotic spindle to cross through the middle of the cell. The centromere in each duplicated chromosome develops into two separate structures (one for each of the duplicate DNA molecules) that attach to microtubules originating from opposite sides of the cell. Forces within the mitotic spindle draw the duplicated chromosomes toward the center of the cell.

Metaphase During **metaphase,** the duplicated chromosomes align on one plane at the center of the cell. Cells often seem to rest at this stage, and metaphase can occupy as much as 20 minutes of the entire one-hour process of mitosis. In fact, the duplicate DNA molecules are being pulled in opposite directions by equal forces (like a tug-of-war), but no separation occurs because they are still held together by the centromeres.

Anaphase During **anaphase,** the duplicate DNA molecules separate and move toward opposite sides of the cell. It is the shortest phase of mitosis, lasting only a minute or so. The centromere abruptly comes apart, and the now-separated DNA molecules are pulled in opposite directions by the microtubules to which they are attached. The process of separation requires energy in the form of ATP and utilizes certain proteins that act as "motors" to tow the microtubules in opposite directions. Cytokinesis usually begins at this time.

Telophase **Telophase** begins when the two sets of chromosomes arrive at opposite poles of the cell. At this point the mitotic spindle comes apart and new nuclear membranes form around the chromosomes. The chromosomes uncoil and revert to the extended form in which they are no longer under a microscope.

> ***Recap*** **Mitosis is the process by which the cell nucleus containing duplicated chromosomes divides in two. During mitosis, the sister chromatids of a duplicated chromosome are separated, resulting in two daughter cells with complete sets of DNA. The duplicated chromosomes become visible in prophase, align along the center of the cell in metaphase, are pulled apart in anaphase, and move to opposite sides of the cell and become surrounded by new nuclear membranes in telophase.**

Cytokinesis divides one cell into two identical cells

Spotlight on Structure & Function

Cytokinesis is the process by which a cell divides to produce two daughter cells. Certain physical properties of the cell facilitate this process. A contractile ring of protein filaments of the same type found in muscle forms just inside the cell membrane. The contractile ring tightens, forming a cleavage furrow and then pinching the cell in two (Figure 17.10). The ring is perpendicular to the long axis of the mitotic spindle, which ensures that one nucleus will be enclosed in each of the daughter cells.

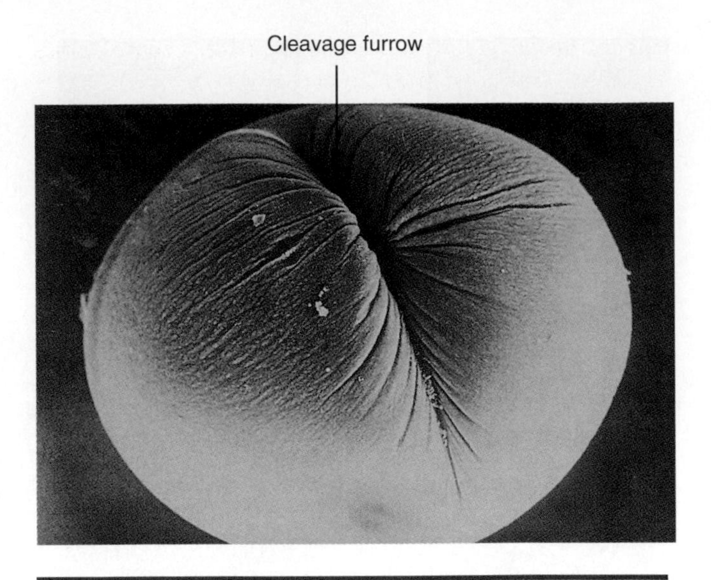

Figure 17.10 Cytokinesis. A cleavage furrow forms as the ring of contractile tissue in the cell tightens.

Cytokinesis is an example of the resourcefulness and efficiency of life's processes. The contractile ring is assembled just before it is needed from the remnants of the cytoskeleton that dissolved during mitosis. After the cell divides the ring itself is disassembled, presumably to form new cytoskeletons for the daughter cells. In effect, cells use a "just-in-time" materials delivery system coupled to an efficient recycling system. Living cells have been doing it long before the business community invented a name for it. ■

> ***Recap*** **During cytokinesis, a contractile ring of microfilaments pinches the cell into two cells. Each daughter cell contains a nucleus identical to that of the parent.**

Mitosis produces diploid cells and meiosis produces haploid cells

In humans, all cells in the body except those developing into sperm or egg cells have 46 chromosomes. Human cells with 46 chromosomes are called **diploid** cells (in Greek, *diploos* means "twofold") because the 46 chromosomes actually represent 23 *pairs* of chromosomes. Diploid cells reproduce by first replicating the 46 chromosomes and then undergoing mitosis (one nuclear division), so that the two daughter cells end up with 46 chromosomes that are identical to that of the parent cell.

In diploid cells, one of each pair of chromosomes came from each parent. There are 22 pairs of **homologous chromosomes** plus the sex chromosomes X and Y. Homologous chromosomes look identical under the microscope, and they have copies of the same genes in the same location (*homologous* means "same form and function").

Sperm and eggs are **haploid** cells, meaning that they have only one set of 23 chromosomes. Haploid sperm and eggs are created by **meiosis,** a sequence of two successive nuclear divisions in which the human genes are mixed, reshuffled, and reduced by half. Once fertilization occurs the fertilized egg and all subsequent cells have the diploid number of chromosomes again.

Meiosis—Preparing for sexual reproduction

The two successive nuclear and cell divisions of meiosis are called *meiosis I* and *meiosis II.* Both meiosis I and II can be subdivided into four stages: prophase, metaphase, anaphase, and telophase. Figure 17.11 illustrates the events that occur in each stage.

Meiosis I Before meiosis begins, the precursor cell undergoes a typical S phase of interphase in which the DNA is replicated and the 46 chromosomes are duplicated. But then something very different happens, compared to mitosis:

1. During prophase of meiosis I the duplicated homologous chromosomes pair up and swap sections of DNA (genes) by a process called **crossing-over.** Consequently, the homologous chromosomes now contain a recombination of both of the person's parent's genes.
2. During the rest of meiosis (metaphase through telophase), the homologous pairs of chromosomes are separated from each other, rather than the duplicates of each pair.

Meiosis II Meiosis II proceeds like mitosis except that the chromosomes are not duplicated again. During meiosis II, the 23 duplicated chromosomes line up and the duplicates are separated from each other. However, because of crossing-over during meiosis I, none of the four haploid daughter cells are exactly alike.

Sex differences in meiosis: Four sperm versus one egg

Spotlight on Structure & Function

There is a fundamental difference in how meiosis proceeds in males and females that has to do with the different functions of sperm and eggs. In males there is only a slim chance that any one sperm will ever reach an egg. It is important, therefore, that lots of sperm be produced. Meiosis in males produces four equal-sized but genetically different sperm from every precursor cell (Figure 17.12).

In females each egg is precious. If fertilized, an egg will need a lot of energy to grow and develop. A large egg with plenty of cytoplasm and lots of organelles has a better chance of surviving the early stages of development. In females, therefore, as much of the cytoplasm as possible is reserved for only one daughter cell at each cell division. The smaller cell produced at each division is called a *polar body.* The polar body produced during meiosis I may or may not divide again during meiosis II, but in any event the two (or three) polar bodies eventually degenerate.

In females, meiosis II is not completed until fertilization. The secondary oocyte released from the ovary stops development at metaphase of meiosis II. Once fertilization occurs the secondary oocyte quickly completes meiosis II and the sperm and egg nuclei then unite. ■

Recap **Meiosis is a two-step process in which the nucleus and cell divide twice, producing sperm or eggs with the haploid number of chromosomes. Crossing over ensures that sperm (and eggs) are genetically different from each other because they contain a mixture of genes from each parent.**

17.4 How cell reproduction is regulated

Although division is the norm for most cell types, not all cells in the body divide at the same rate. Some groups of cells divide rapidly throughout life, whereas others stop after adolescence. Still others reproduce at highly variable speeds, increasing or decreasing their rates of cell division in response to regulatory signals.

Table 17.1 lists examples of variations in the rates of cell division. Although it was once thought that nerve cells do not divide after adulthood, we now know that at least certain nerve cells in the hippocampus do divide on occasion. With regard to muscle, it is interesting to note that most world-class swimmers and gymnasts are younger than world-class runners. One theory is that muscle cells lost through disuse are not replaced. Because we all walk, we may retain our leg muscles throughout life better than our upper body muscles, which do not get as consistent a pattern of exercise. Young swimmers may retain upper body muscle cells longer if they have trained continuously since an early age. Whether muscle cells are ever replaced in adulthood seems to be an open question.

How often do different types of cells divide?

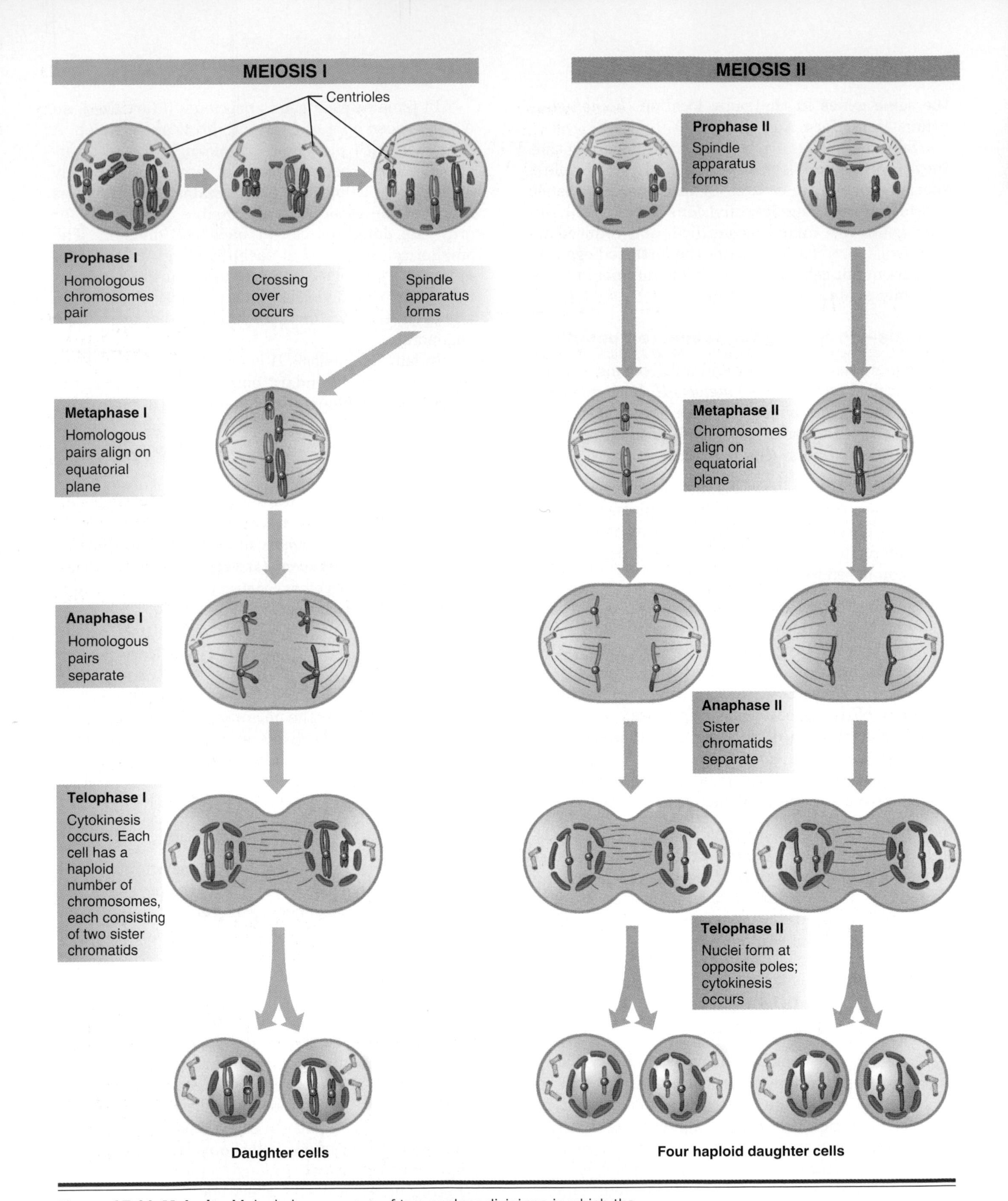

Figure 17.11 Meiosis. Meiosis is a process of two nuclear divisions in which the chromosomal material is reshuffled and reduced by half during the formation of sperm and eggs. The red and blue colors indicate homologous chromosomes that were inherited from each of the person's parents. During prophase of meiosis I, pairs of duplicated homologous chromosomes pair up and exchange sections of DNA by a process known as crossing-over. During the rest of meiosis I, the homologous pairs are separated from each other. In meiosis II the duplicates of each chromosome are separated from each other, just as they are in mitosis. The result is four daughter cells, none of which are alike.

Control of the cell cycle is a hot research topic these days because it may have a lot to do with the development of cancer. Cancer cells divide repeatedly, ultimately destroying the organism in which they live. Although the exact molecular mechanisms regulating cell growth and division are not yet completely known, we are beginning to understand some of the details.

Selective gene expression regulates the cell cycle

The key to understanding both the regulation of the cell cycle and cell differentiation, discussed below, is the concept of selective gene expression. We say a gene is *expressed* when the protein for which it codes is being produced. By *selective gene expression* we mean that in any given cell at any one time, only a few of the 100,000 genes are actually being expressed. Even though every cell has exactly the same set of genes, only a white blood cell expresses the genes that produce antibodies, and only at certain times. Only a muscle cell expresses the gene that codes for myoglobin, the oxygen storage molecule in muscle. Only a cell that is going to divide will express the genes that lead to the formation of a second pair of centrioles, and so on.

Selective gene expression is regulated by certain proteins, called regulatory proteins, whose sole function is to turn certain genes on and off. Some proteins, called *repressors,* prevent gene expression, so a gene can become active only when these repressor proteins are absent. Other proteins are *activators;* their presence is required to activate certain genes. Because repressors and activators are proteins, they are the products of activation of other genes that code for them.

Regulatory genes code for repressor or activator regulatory proteins. **Structural genes** code for an enzyme or structural protein. Figure 17.13 illustrates one way that regulatory genes could control the activity of structural genes.

Regulatory proteins can be deactivated or activated by molecules in their environment, such as concentrations of calcium, hydrogen, or hormones. The influence of environmental factors on gene expression becomes particularly important when we discuss how cells differentiate; that is, how cells take on different forms and functions.

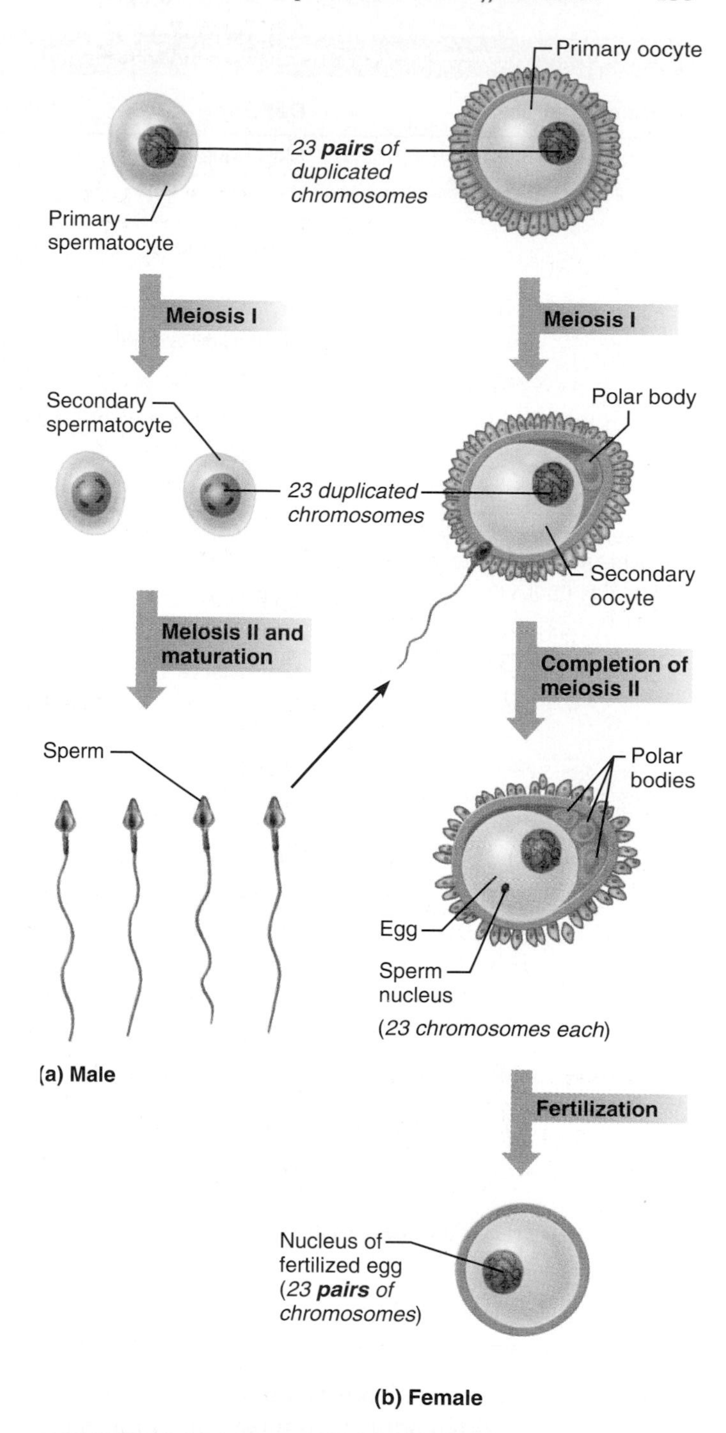

Figure 17.12 Sex differences in meiosis. (a) In males, meiosis produces four different small sperm. **(b)** In females, nearly all of the cytoplasm is reserved for the future egg at each stage of meiosis. Meiosis I in females thus results in a secondary oocyte and a polar body. Meiosis II is not completed in females until fertilization. The polar bodies eventually disintegrate.

Table 17.1 Variation in rates of cell division by human cells

Rate of Division	Cell Type	Comments
Divide constantly and rapidly throughout life	Skin cells	Skin is continually being formed from deep layers. The outermost layer of skin is composed of dead cells; they are constantly being sloughed off.
	Most epithelial cells	Epithelial cells line the inner surfaces of body organs such as the digestive tract and the lungs. These cells are exposed to frequent damage and must be replaced.
	Bone marrow cells	Precursor cells in the bone marrow produce red blood cells (RBCs) and white blood cells (WBCs) throughout life. RBCs live only about 120 days. The rate of WBC production can be increased as part of the immune response.
	Spermatogonia (after puberty)	Spermatogonia divide to produce sperm throughout life in the adult male, though the rate may decline some in old age.
Do not normally divide but can be stimulated to divide under the right circumstances	Liver cells	Liver cells don't normally divide in adulthood but will do so if part of the liver is removed.
	Epithelial cells surrounding the egg	Called granulosa cells, they are quiescent for most of adult life. They begin to divide as the follicle matures prior to ovulation.
Do not divide in adulthood	Most nerve cells	Most human nerve cells apparently do not divide in adulthood, although recently it has been shown that neurons in the hippocampus do.
	Osteocytes	Mature bone cells called osteocytes become trapped within the hard crystalline matrix of bone, with no room for growth or division. Only immature bone cells (osteoblasts) divide.
	Most muscle cells	The commonly held view is that muscle cells do not divide in adulthood. There is some disagreement about this.

Recap **Selective expression of structural genes regulates the cycle of cell growth and reproduction. Regulatory genes code for proteins that either repress or promote structural gene expression.**

Environmental factors can modify the cell cycle

We now know there are at least three times during the cell cycle where regulatory control is exerted:

1. In G1, just before DNA is duplicated. Cells that do not divide in adulthood most commonly stop at this stage
2. Just before mitosis
3. Less often, mitosis may be arrested during metaphase

These are natural stopping points, because to proceed forward will consume energy and result in a substantial structural change to the cell.

Cell cycle stopping points are controlled by regulatory proteins. These proteins, in turn, can be modified by environmental influences such as growth factors, hormones, availability of nutrients, or even the presence of other cells. When cells come into contact with each other, for instance, they release a substance that inhibits cell division. This inhibition is an effective negative feedback mechanism to regulate tissue growth and organ size. As an example, normally adult liver cells are in an arrested G1 state. If, however, part of the liver is removed so that cells no longer touch, the remaining cells begin dividing again.

Recap **There are at least three points in the cell cycle at which cell growth and division are regulated. Regulatory proteins and environmental influences can modify the cell cycle.**

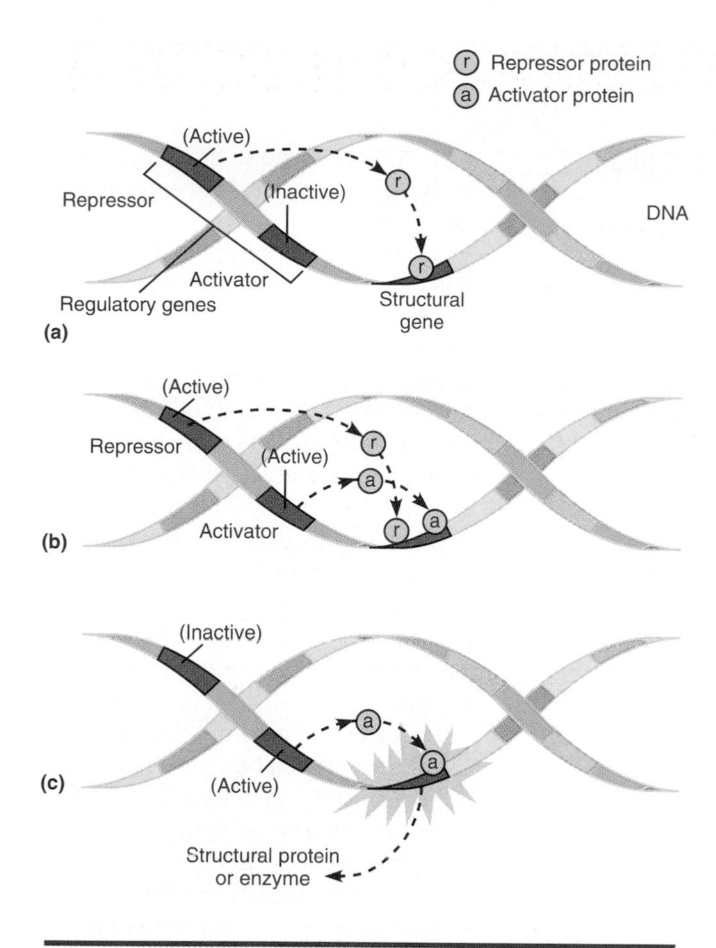

Figure 17.13 Repressor and activator genes control structural genes. (a) When a repressor gene is active, a repressor protein R is produced that inhibits expression of the structural gene. **(b)** The production of an activator protein A by an activator gene cannot activate the structural gene as long as the repressor protein is still present. **(c)** Only in the presence of the activator protein and the simultaneous absence of the repressor protein will the structural gene become active.

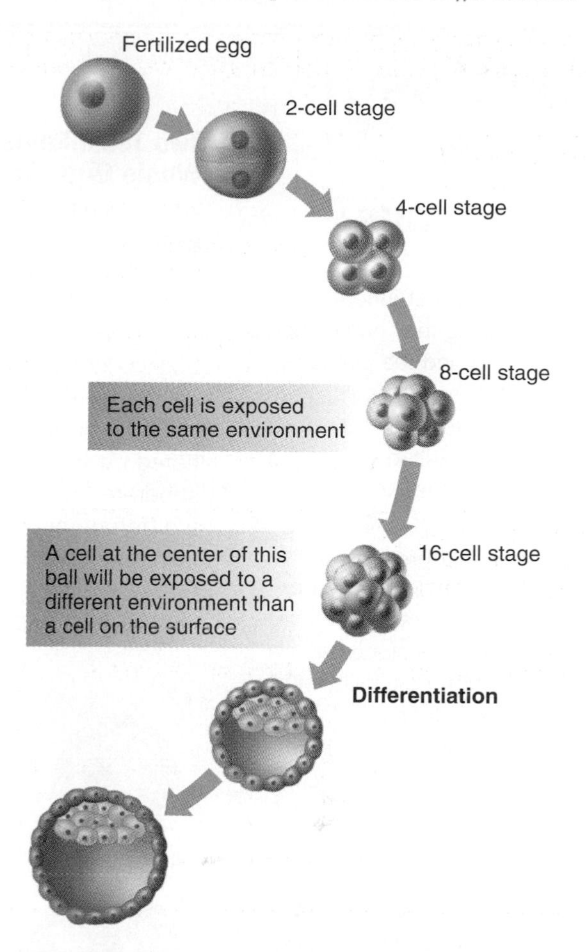

Figure 17.14 How early differentiation might occur. Beyond about the eight-cell stage, the cells begin to be exposed to different environments. Differences in environments may trigger different gene expressions, leading to differentiation.

17.5 Environmental factors influence cell differentiation

All of the cells in your adult body originated from a single cell, the fertilized egg, and all of those cells (except those destined to become your sperm or eggs) have exactly the same set of DNA. Yet your body is composed of lots of different kinds of specialized cells. If all your cells have the same exact set of genes, why don't all your cells stay identical, like identical twins? What causes the cells to change form and function?

Differentiation is the process whereby a cell becomes different from its parent or sister cell. At various developmental stages of life, cells differentiate because they begin to express different genes. With the examples below we'll see that differentiation is probably due to environmental influences.

Differentiation during early development

Let's examine the earliest stage of human development, starting from a fertilized egg. The fertilized egg is a relatively large cell. Soon after fertilization it begins to divide, becoming two cells, then four, then eight. During these early divisions, up to about 16 or 32 cells, the cells do not grow. Consequently each cell gets progressively smaller as a ball of identical cells develops (Figure 17.14).

Spotlight on Structure & Function

Up to about eight cells, each cell is exposed to the same environment—the surrounding fluid plus contact with cells just like themselves. Once they form a tight ball of many cells, however, the situation changes, for now only the cells at the outer surface of the ball will be exposed to the surrounding fluid. The rest are exposed only to the extracellular fluid between the cells. At this point the environment that surrounds a cell at the center of the ball becomes different from

CurrentIssue

Should We Clone Humans?

In 1997, researchers stunned the world by announcing they had cloned a sheep. The media and the public reacted immediately to the news. There were fanciful suggestions that sports teams might clone skilled athletes, parents would clone babies to provide donor organs for a sick child, and grief-stricken widows might seek to clone their spouses from a cell of the deceased. A few years later other scientists revealed they had cloned a monkey, bringing the possibility of human cloning even closer.

Could we actually clone a human being? Should we allow human cloning even if we could? If you and your spouse were infertile, would you want to clone yourself or your spouse? The answer to the first question is that with enough effort, probably yes. The answer to the second question is a societal decision, up to all of us. The answer to the third question would be a personal choice, but whether you will have that choice depends on how society answers the second question.

In the broadest sense, a clone is defined as a copy. The popular media generally thinks of a clone as a copy of an adult organism, but scientists use the term cloning to describe the copying of cells, genes, and molecules as well. Using genetic engineering techniques, scientists now routinely insert human genes into bacteria, yeast, and even large animals in order to produce cloned human hormones in large quantities. Human erythropoietin (to treat anemia), insulin (for diabetes), and growth hormone (for endocrine disorders) are now produced this way. Human cells can also be cloned in order to produce specialized tissues. Scientists who grow layers of skin cells to treat badly burned patients are practicing a form of cloning.

Two Techniques for Cloning Whole Organisms

When it comes to cloning entire organisms, one of two techniques is generally employed: *embryo splitting* and *somatic cell nuclear transfer.*

How does cloning by embryo splitting work? Recall that until the eight-cell stage of development, all of the cells in a fertilized egg are undifferentiated. At this stage it is possible to separate the cells, implant each into a different mother, and create up to eight identical clones of the organism that would have resulted from the original fertilized egg (Figure 1). It is fairly commonplace to produce cloned offspring of valuable farm animals in this manner.

There are, however, limitations to this technology. In embryo splitting the eight clones are made *after* the union of sperm and egg. Clones produced by embryo splitting, while identical to each other, are not exact copies of either parent. In other words, a prized adult animal cannot be copied exactly by embryo splitting, and humans produced in this manner would be no more than identical siblings (up to octuplets).

What about cloning an adult by the technique of somatic cell nuclear transfer? A *somatic cell* is any cell in the body except a germ (sex) cell. Every somatic cell has the same entire set of DNA, so in theory any one body cell could be used to produce a new adult identical to the original. But until the birth of Dolly the sheep it was considered impossible to make a somatic cell essentially "start over" and differentiate as if it were a fertilized egg. The birth of Dolly proved that wrong. Dolly was created when a somatic cell (from the udder) of an adult sheep was fused with another sheep's egg after the egg's nucleus had been removed. (Figure 2). The egg, now containing the DNA of the udder cell, developed into Dolly. In essence, Dolly is a twin of the sheep providing the udder cell.

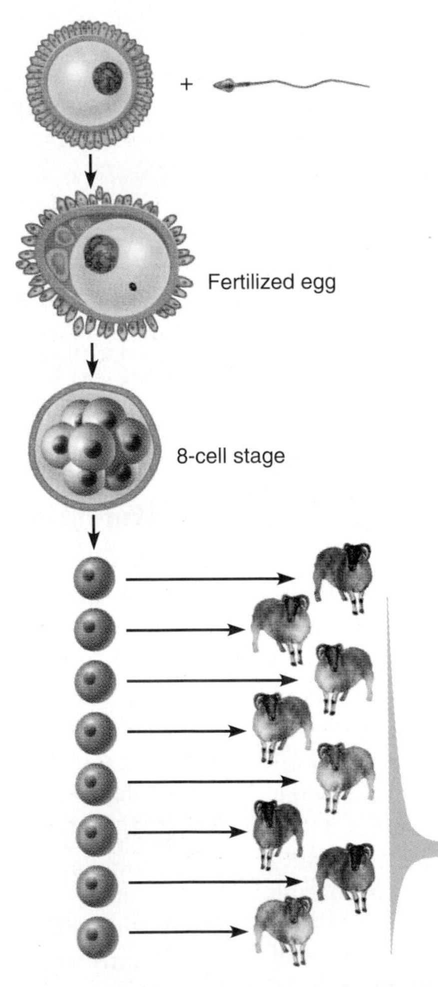

Figure 1. Cloning by embryo splitting. A fertilized egg is formed by the union of a sperm and egg. The egg is allowed to develop to about the eight-cell stage. The cells are separated from each other and implanted into different females. Ultimately they develop into identical individuals.

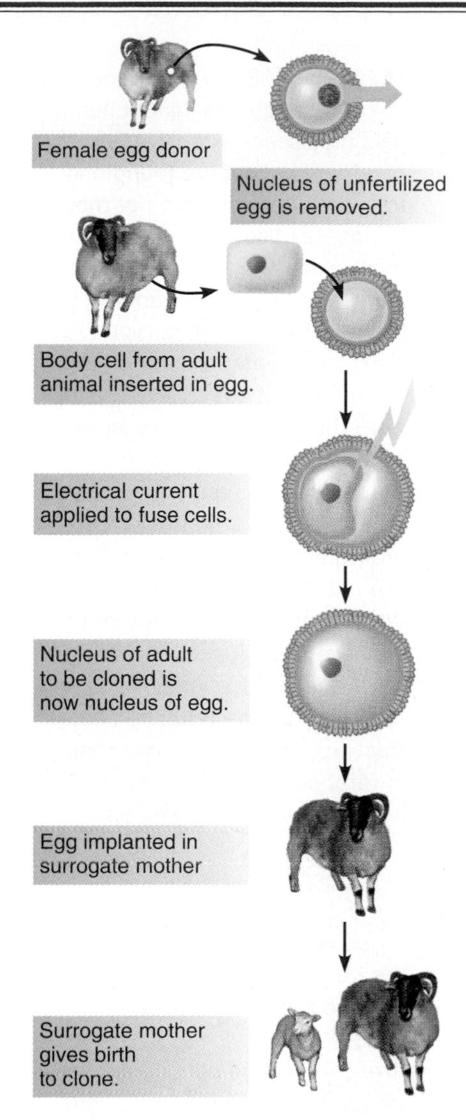

Figure 2. Cloning by somatic cell nuclear transfer.

Biological, Ethical, and Political Issues

While it might be entirely possible right now to clone a human by somatic cell nuclear transfer, most biologists doubt that cloning could ever replace sexual reproduction in the long run. With sexual reproduction, sperm or eggs with damaged chromosomes generally don't survive. Thus, potentially harmful mutations are eliminated from the human gene pool before fertilization takes place. With cloning by somatic cell nuclear transfer, however, any defects in the cell used for cloning would be transferred to the clone. Thus, genetic defects would be passed on indefinitely, until descendants become too damaged to sustain life. On the positive side, from what we know of the rate of evolutionary change, the rate at which genetic defects would accumulate is likely to be slow. Better information about the potential for harmful genes to be passed on during cloning will probably come from further experiments with animals over the next several decades.

The cloning of humans raises fundamental ethical questions that deserve serious debate. Would human cloning lessen the value of a human life, and could a clone be harmed by a loss of individual identity?

Regarding the value of human life, cloning by embryo splitting offers the possibility of making up to eight copies of a human embryo. Would it be ethical to choose only one embryo for implantation and discard the rest? Before you answer, perhaps it would be instructive to find out what happens to extra *non*-identical embryos created by *in vitro* fertilization at assisted reproductive technology clinics (fertility clinics) these days. Cloning by somatic cell nuclear transfer could potentially produce a limitless number of embryos, but the technique is more difficult. The Scottish researchers who produced Dolly implanted 29 early-stage embryos before their single success with Dolly; the other 28 embryos died. Would we be willing to accept such a low success rate with implanted human embryos? And if we *could* produce human clones reliably, should we allow human embryos to be produced solely as tissue or organ donors for people who become seriously ill?

The second ethical issue—a possible loss of human identity—stems from a commonly held belief that our sense of identity comes from our genetic uniqueness. In fact, we are unique primarily because each of us is a separate organism. Consider identical twins with the same set of genes. Though they may look alike, each twin is shaped by his/her own individual experiences, and each is accorded full rights under the law. No one would dare suggest that the second-born twin could be forced to be an organ donor for the first. A more difficult potential problem would be created by cloning by somatic cell nuclear transfer. In this case the cloned "twin" could be of an entirely different age. Would the clone be treated by the legal system as the son or daughter of the somatic cell donor?

Some opponents of cloning argue that clones could be harmed psychologically because they had no choice in being created as clones. Is this a valid concern, considering that no child *ever* has the choice of whether to be born? The counter-argument is that the clones' psychological health would more likely depend on how society handles their uniqueness.

The birth of Dolly caught politicians, the public, and even most of the scientific community by surprise. In the U.S., the initial political reaction was largely negative. The president moved quickly to ban the use of federal funds for cloning experiments, and a few states passed their own laws to ban human cloning entirely. Some professional societies also came out against human cloning, at least until its consequences could be better understood. The Federation of American Societies for Experimental Biology, for example, endorsed a voluntary moratorium on the cloning of humans.

Political actions and voluntary moratoriums will slow the development of cloning techniques because they make it more difficult to obtain funding for research. Cloning could easily be banned permanently in this country by federal law if we so desire. It is unlikely, however, that political actions taken by the U.S. alone will affect human cloning very much in the long run. If human cloning were banned in the U.S., what's to stop scientists from conducting cloning research and/or establishing human cloning clinics in other countries? The issue of cloning will probably remain with us as a biological, ethical, and political dilemma for some time to come.

Health Watch

Reducing the Risk of Birth Defects

Birth defects, the leading cause of infant mortality, account for nearly 20% of all infant deaths today. They also contribute significantly to the one-third of all pregnancies that end in spontaneous abortion. Some birth defects are inherited traits. Others are the direct result of mutations acquired during early development from environmental influences.

"Birth defects" is a general term encompassing a wide range of newborn health problems. Agents that cause general problems with fetal health are *fetotoxic,* meaning that they injure the fetus in some way. The damage may be no more than low birth weight, correctable with time. Other substances are *teratogenic* (meaning literally "to create a monster"). Listed below are agents that are either teratogenic or more toxic to a fetus than they are to an adult.

A young boy born to a mother who took thalidomide early in pregnancy. Thalidomide was taken off the market as soon as its teratogenic effects became known.

- *Chemicals in air, water, and soil.* Lead and mercury are fetotoxic and teratogenic at concentrations below those that affect adults. Both chemicals appear to damage DNA, and prenatal exposure can lead to fetal malformations, mental retardation, and nervous system damage. Many pesticides and a number of industrial solvents can be teratogenic, especially at high concentrations. Two solvents in particular, polychlorinated biphenyls (PCBs) and toluene, are risky. Although this list of chemicals may seem extensive and frightening, most are found only in trace amounts and are not thought to pose a major health hazard in most places.
- *Radiation.* Radiation can injure the developing fetal nervous system. The greatest radiation threat is from radon gas that seeps up from the ground in some areas. Another source is pelvic irradiation as part of cancer treatment. Video display terminals and natural radiation from cosmic rays are not thought to pose any risk. We do not yet know whether microwave radiation is harmful.
- *Intrauterine infections.* Certain infections that the fetus can contract from the mother can result in fetal death, birth defects, or illnesses that persist well past birth. Fetal infections are particularly dangerous during the first trimester. The most damaging are those caused by the HIV virus, the bacterium that causes syphilis, and the rubella (German measles) virus.
- *Prescription and nonprescription drugs.* Some medications, both

the environment of a cell at the surface. Cells at the center may be exposed to a lower O_2 content, high levels of CO_2, a different pH, or may be influenced by products secreted by the other cells. Any one of these different environmental factors could activate different genes, leading to different developmental pathways for the cells at the center versus the cells at the surface.

Notice that no "thought" or feedback control mechanisms are needed to account for early differentiation; it can be accounted for solely by differences in the physical and chemical environment of individual cells. Cell differentiation begins long before homeostatic feedback mechanisms ever develop in the organism. ■

Scientists have known for some time that it is possible to clone (make copies of) an organism by "embryo-splitting"; that is, by splitting the eight-cell ball that develops from a fertilized egg into eight separate cells. At that stage, all eight cells are equally capable of becoming complete organisms, presumably because they have not yet differentiated. After about 16 cells, however, cloning by embryo-splitting no longer works.

Recap **Differentiation is the process whereby cells become different from each other. During early development, environmental influences trigger differentiation.**

Differentiation later in development

Differentiation at later stages is similar in principle to that of earlier stages. Each cell is shaped by two factors: the developmental history of the cells that came before it and its local environment. Genes are turned on at certain stages of development and then turned off, some never to be used again.

prescription and nonprescription, can be hazardous to the fetus. Some drugs have been shown to be teratogenic in animals at high doses, but relatively few are proven human teratogens. One is thalidomide, a mild sedative used briefly in Europe in the 1960s before its dramatic teratogenic effects became apparent. Nearly 10,000 children were affected before thalidomide was banned. The danger of such unforeseen consequences is why prescription drugs are now extensively tested in animals and then tested again in limited trials in patients under strict government supervision before being approved. Only after years of use do they become available on a nonprescription basis. Anticonvulsants, used to treat epilepsy, slightly increase the risk of birth defects. Here is a case where a small risk to the fetus may have to be balanced against the medical needs of the mother.

- *Illegal drugs.* Cocaine and heroin use can lead to fetal death, fetal abnormalities, or low birth weight. Even babies who appear normal may suffer through withdrawal symptoms at birth. The effects of marijuana are difficult to judge because it is seldom used independently of other drugs, alcohol, or cigarettes, but evidence associates it with poor fetal weight gain and behavioral abnormalities in the newborn.
- *Alcohol.* Although legal and common, alcohol is also commonly abused. Babies born to mothers who abuse alcohol may show a characteristic pattern of abnormalities called fetal alcohol syndrome. The syndrome includes a group of minor facial characteristics, heart defects, limb abnormalities, and delays in motor and language development. It can be difficult to diagnose fetal alcohol syndrome because typically the infant does not have all the symptoms. The condition is most likely when the mother abuses alcohol heavily before and during pregnancy, but even smaller amounts of alcohol may harm the fetus.
- *Cigarette smoking.* The best-documented effect of cigarette smoking is fetal growth retardation. Smoking may also contribute to spontaneous abortions. Whether cigarette smoking is teratogenic remains controversial. Cigarette packs are required to carry warning labels describing the health risks in both the U.S. and Canada. Nevertheless, most women who smoked before pregnancy continue to smoke during pregnancy.

The fetus is very vulnerable during the earliest stages of development, but with a little care the most likely outcome of pregnancy can still be a healthy baby. If you become pregnant, doctors advise you to:

- Stop smoking
- Refrain from drinking alcohol
- Review with your physician all drugs you take
- Avoid having X-rays or irradiation treatments if possible. Diagnostic X-rays are now performed with fairly low levels of radiation, but it is still a good idea to keep them to a minimum. If you must have them done, inform health care workers that you are pregnant.

You may also wish to measure radon gas levels in your home. The question of chemicals in our air, water, and soil is a complex issue that we all need to address in our local communities and the larger world in which we live.

For example, early in development some cells begin to express the genes that cause them to become epithelial cells. Others become connective tissue, nerves, or muscle cells. Those that are destined to become muscle tissue differentiate even further, becoming either skeletal, smooth, or cardiac muscle.

In the end, even cells that are closely related in function and structure can differ in hundreds of ways. Each difference in gene expression adds up to produce a very specialized cell. For example, your smooth and skeletal muscle cells differ in their metabolic pathways (how they use energy), speed and strength of contraction, shape and size, and internal arrangement of muscle fibers. They got that way by an ever-increasing cascade of selective gene expressions.

The mechanisms of differentiation explain why the developing fetus is more vulnerable to genetic damage than an adult. Alteration of a gene that plays a critical role in early development can be so disruptive to embryonic differentiation that the embryo may not survive. Later in development the embryo may be able to survive the damage but be left with marked physical deformities. In an adult, alteration of a single gene may be similarly dangerous, or it could have little or no visible effect. It all depends on what the protein encoded by the gene does.

Why is a fetus more vulnerable to genetic damage than an adult?

Recap **Some genes are expressed only at certain stages of development; for this reason, genetic mutations during early development can be particularly damaging. Later in development hundreds of different genes may be expressed by various types of cells.**

Chapter Summary

The cell cycle creates new cells

- Cells reproduce by a repetitive cycle called the cell cycle in which one cell grows and then divides in two.
- The cell cycle has two primary phases: interphase and the mitotic phase. During interphase the cell grows and the DNA is replicated. During the mitotic phase, first the nucleus and then the cell divide in two.

Replication, transcription, and translation—An overview

- Human DNA is packed into 46 separate molecules called chromosomes.
- A gene is the smallest functional unit of a chromosome. A gene contains the code for making a specific protein.
- DNA replication is a process in which the two strands of DNA separate and a new complementary copy is made of each strand.
- DNA is repaired when it is damaged and is checked for errors after it is replicated.
- For a gene to be expressed, the strand of DNA with that gene must be transcribed to create a complementary strand of mRNA. The mRNA leaves the nucleus, attaches to a ribosome, and serves as the template for protein synthesis.
- Translation is the process of making a polypeptide chain (protein) using the mRNA as a template. Three successive bases of mRNA, called a codon, code for a particular amino acid.
- The amino acid building blocks for the polypeptide chain are captured in the cytoplasm by tRNA, brought to the mRNA, and attached to each other by ribosomal enzymes.

Cell reproduction—One cell becomes two

- Mitosis is a sequence of events in which the replicated chromosomes are separated to form two new genetically identical nuclei.
- Cytokinesis is the process whereby the cell divides into two new cells, each with one of the new nuclei produced by mitosis and roughly half of the cell's organelles and mass.
- Meiosis is a sequence of two cell divisions that produces haploid cells. Meiosis only occurs in cells destined to become sperm or egg. Crossing-over during meiosis mixes the genes of homologous chromosomes, and subsequent cell divisions reduce the number of chromosomes by half.

How cell reproduction is regulated

- Cell reproduction is regulated in part by selective gene expression. Selective gene expression is controlled by regulatory genes.
- The cell cycle may be influenced by the physical and chemical environments both inside and outside the cell.

Environmental factors influence cell differentiation

- Differentiation is the process whereby cells become different from each other, acquiring specialized forms and functions.
- Because all cells have the same set of genes, differentiation in the early embryo must be triggered by environmental influences.
- Cell differentiation later in development can be influenced by environmental cues, but it also depends on the developmental history of the cells that preceded it.

Terms You Should Know

anaphase, 436
cell cycle, 426
chromosome, 426
codon, 432
cytokinesis, 434
differentiation, 441
diploid, 436
gene, 427
haploid, 437
homologous chromosomes, 436
interphase, 426
meiosis, 437
messenger RNA (mRNA), 432
metaphase, 436
mitosis, 434
mitotic phase, 426
mutation, 428
prophase, 435
regulatory genes, 439
replication, 427
ribosomal RNA (rRNA), 433
ribosome, 433
structural genes, 439
telophase, 436
transcription, 427
transfer RNA (tRNA), 433
translation, 427

Test Yourself

Matching Exercise

_____ 1. The process of making a protein using the mRNA code as a template
_____ 2. The process of making a single strand of RNA from a short segment of DNA
_____ 3. The process of copying all of the cell's DNA prior to cell division
_____ 4. The process whereby duplicated homologous chromosomes exchange genes during meiosis
_____ 5. Meaningless fragment of DNA in a gene

_____ 6. Fragment of DNA in a gene that carries important genetic information

_____ 7. mRNA base triplet representing the code for a particular amino acid

_____ 8. tRNA base triplet that binds to the base triplet of mRNA

a. exon
b. translation
c. intron
d. codon
e. replication
f. anticodon
g. transcription
h. crossing-over

Fill in the Blank

9. The long phase of the cell cycle during which the cell grows and the DNA is replicated is called the ________________.
10. A short segment of DNA representing the code for a particular protein is called a ____________.
11. An alteration to a cell's DNA is called a ________________.
12. Small molecules of RNA called ______________ bind to specific amino acids and bring them to the ribosomes, where protein synthesis takes place.
13. The process of nuclear division (in all cells except those developing into sperm and eggs) is called ____________.
14. The process of cell division is called ________________.
15. A pair of chromosomes that look alike and have the same genes at the same locations are called ________________ chromosomes.

True or False?

16. Daughter cells are always female cells.
17. Human cells (except those developing into sperm or eggs) have 23 pairs of chromosomes.
18. Sperm and eggs are haploid cells.
19. Cells that are not dividing usually stop growing in G2 of the cell cycle, just before the mitotic phase.
20. The function of a centromere is to keep the sister chromatids of a duplicated chromosome together until they are pulled apart during mitosis.

Concept Review

1. Describe how DNA is replicated before cell division.
2. Explain why it is necessary to transcribe a gene into a complementary strand of RNA before a protein can be made.
3. List the three steps of translation.
4. Name the four phases of mitosis and describe briefly what is happening to the chromosomes during each phase.
5. Explain why only one large egg is formed during meiosis in the female, whereas four sperm are formed during meiosis in the male.
6. Describe what is meant by *selective gene expression* and why it is important to how a cell functions.
7. Explain how factors present in the environment can influence cell differentiation.
8. How do ribosomes contribute to the formation of a protein?
9. Compare/contrast the functions of *regulatory* genes and *structural* genes.
10. How is the primary transcript of mRNA that is released from the DNA molecule different from the final mRNA molecule that is used to make a protein?

Apply What You Know

1. Recall that the base pairing in DNA is always A–T and C–G. Suppose that during a DNA replication a C is accidentally paired with a T. Explain why it is important that the error be detected and corrected before the cell divides again, by illustrating what would happen at the next cell division if the error were not corrected.
2. In reviewing Figure 17.6 you will notice that only one of the two strands of DNA carries the genetic code for making RNA. What do you think is the function of the other strand?
3. The RNA code is a "triplet code," meaning that three bases code for a single amino acid. Would a "doublet code" be sufficient? Explain.

Case in Point

Fragments of DNA May Save Your Skin

Think back to a sunny day when you spent a lot of time outdoors. The exposed areas of your skin might have turned pinkish at first, then become darker red. Your skin probably felt sensitive, and after several days it flaked or peeled. If you were more careful in how much time you spent outdoors each day over a prolonged period, you may have developed a suntan that partially protected you from these effects. What was happening to your skin at the cellular level?

When skin cells are exposed to ultraviolet radiation (whether from natural sunlight or from artificial sunlamps), their DNA may be damaged. Within two to three days the injured cells undergo a characteristic series of events designed to help the cells recover from the damage and prevent future damage. First, the production of DNA repair enzymes increases. These enzymes increase the transcription of more than 20 genes that produce proteins that enhance skin cells' survival. Second, the pigmented cells in skin called melanocytes increase their production of melanin, the skin pigment responsible for a suntan. Melanin partially shields the cell from the damaging effects of UV radiation. Third, cells that are damaged beyond repair are tagged by a protein called Fas ligand. Researchers speculate that Fas ligand marks cells that may contain dangerously mutated DNA, in effect spotlighting them for destruction by the immune system. When your skin peels after a sunburn, your body is shedding potentially harmful cells. Repeated exposure to ultraviolet radiation or even intermittent exposure that causes bad sunburns may overwhelm the repair process and allow DNA mutations to develop into skin cancer.

Researchers at Boston University are working to develop a topical cream containing short fragments of DNA similar to those caused by exposure to ultraviolet radiation. They hypothesize that this preparation could stimulate the melanocytes to produce more melanin (and give us a suntan) without causing the damage to the person's own DNA that typically accompanies exposure to UV light.

QUESTIONS:

1. Although excessive sun exposure at any age can cause skin cancer, frequent exposure and serious sunburns before age 15 seem to be especially risky. Explain why this might be so.
2. Some tanning salons advertise that their sunlamps do not injure skin the way natural sunlight does. Is this claim scientifically valid? Why or why not?

INVESTIGATE:

Gilchrest, B.A., M.S. Eller, A.G. Geller, and M. Yaar. Mechanisms of Disease: The Pathogenesis of Melanoma Induced by Ultraviolet Radiation. *New Engl. J. Med.* 340:1341–1348, 1999.

Hill, L.L., A. Ouhtit, S.M. Loughlin, M.L. Kripke, H.N. Ananthaswamy, and L.B. Owen-Schaub. Fas Ligand: A Sensor for DNA Damage Critical to Skin Cancer Etiology. *Science* 285:898–900, 1999.

Key, S.W., and M. Marble. Skin Cancer Connected to Intermittent Sun Exposures and Sunburns. *Cancer Weekly Plus* May 10, 1999. p. 15.

References and Additional Reading

Becker, Wayne M., Jane B. Reece, and Martin F. Poenie. *The World of the Cell*. Menlo Park, CA.: Benjamin Cummings, 1996. 897 pp. (A comprehensive textbook of cell and molecular biology, for advanced reference only.)

Kolata, Gina. *Clone: The Road to Dolly, and the Path Ahead*. New York: Wm. Morrow & Co., 1997. (An account of how cloning became possible and what cloning may mean for our future.)

Persaud, T.V.N. *Environmental Causes of Human Birth Defects*. Springfield, IL: Charles C. Thomas, Publisher, 1990. 182 pp. (A good summary, with references to the literature for specific substances that cause birth defects.)

Silver, Lee M. *Remaking Eden: Cloning and Beyond in a Brave New World*. New York: Avon Books, 1997. (A provocative account suggesting that cloning will be driven by our desire for new reproductive technologies.)

Watson, James D. *The Double Helix: A Personal Account of the Discovery of the Structure of DNA*. New York: Atheneum, 1968. 226 pp. (A readable and entertaining account of the discovery of the structure of DNA by the Nobel laureate.)

CHAPTER 18

CANCER

Can cancer be cured?

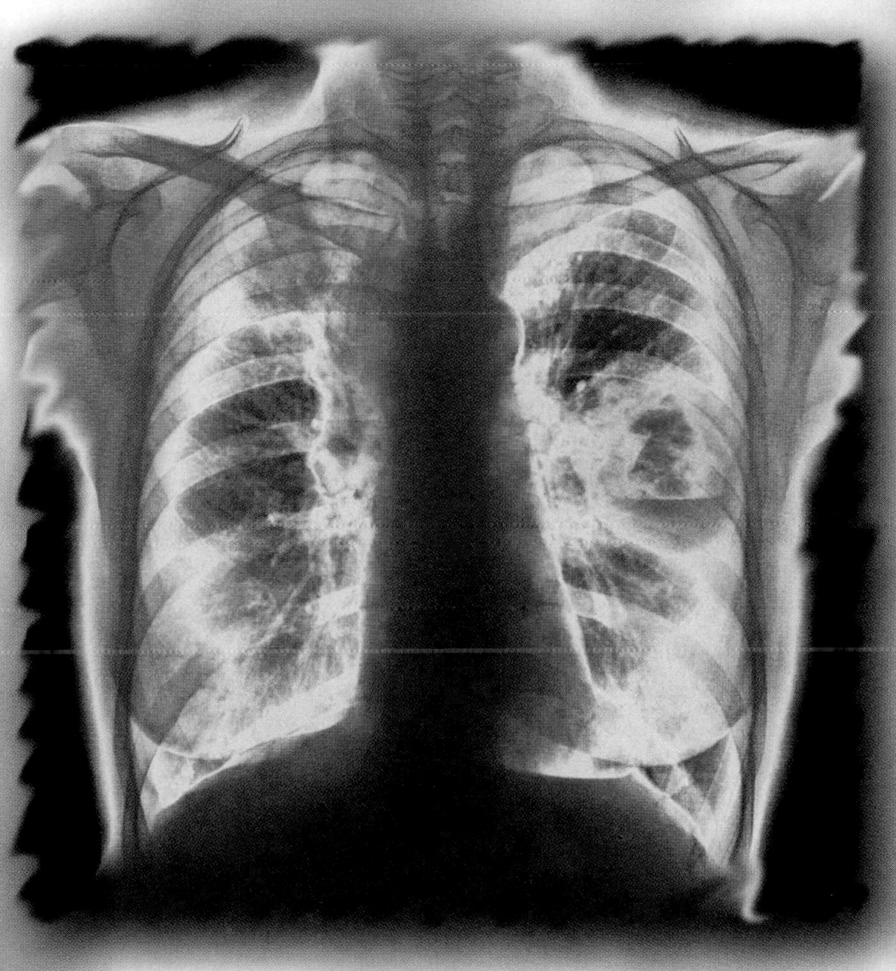

What are the most common cancers?

What can you do to lower **your** cancer risk?

What single substance causes **more than 30 percent of all** cancer deaths?

THE WORD CANCER MAY ITSELF CONJURE UP IMAGES OF PAIN, suffering, and lingering death. We may associate the word with visits to the hospital with our parents when we were children; we may recall years of long-term care of a grandparent at home; we may remember a friend or loved one no longer with us. A diagnosis of cancer sounds like a death sentence.

And yet, people do recover from cancer. There are known risk factors for most types of cancers and things we can do to reduce our risk. There are treatments for many cancers, some of which actually work rather well. The death rate from some cancers is even declining as we learn more about the disease.

There are many different types of cancers, but all are the same in one respect—they all are diseases of cell division and differentiation. Once that is understood, cancer makes some sense. In this chapter we examine the biological basis for how cancer develops and spreads. Then we discuss how cancer is diagnosed and look at new advances in cancer treatment and prevention.

To understand how cancer develops we must first understand two key characteristics of normal cells:

- Normal cells have regulatory mechanisms that keep their rates of cell division appropriately in check. They may divide more frequently at some times than at others, but their rates of cell division are carefully controlled by regulatory mechanisms that may involve an internal clock and various hormones. In addition, their rates of division are generally inhibited by signals from nearby cells.
- Normal cells generally remain in one location throughout their lifetime because they adhere to their nearby neighbors. (Blood cells are an exception, because they must travel throughout the body to carry out their normal function.)

18.1 Tumors can be benign or cancerous

If a cell loses the ability to control its rate of cell division it may begin to divide more frequently than normal. The result is a condition known as **hyperplasia** ("increased" plus Gr. *plasis,* "formation"). Eventually the cells will develop into a discrete mass called a **tumor,** or *neoplasm* (meaning literally "new growth").

Not all tumors are cancers. Tumors that remain in one place as a single well-defined mass of cells are *benign tumors* (from the Latin *benignus,* "well disposed"; Figure 18.1). The cells in a benign tumor still have most of the structural features of the cells from which they originated, and in particular they stay together and may be surrounded by a layer of connective tissue. Benign tumors tend to threaten health only if they become so large that they crowd normal cells, which can happen sometimes in the brain. In most cases surgery can easily remove benign tumors because they are so well contained. Many moles on the skin are a form of benign tumor so inconsequential that usually we don't bother to remove them. Occasionally, a previously benign tumor may change into something more damaging. For example, certain moles have the potential to develop into melanoma, a form of skin cancer.

18.2 Cancerous cells lose control over cell functions

If and when cells progress toward cancer, in addition to dividing more frequently than normal they also undergo a series of structural changes. The nucleus often becomes larger, there is less cytoplasm, and the cells lose their specialized functions and structures. A change for the worse in cell structure is called **dysplasia,** meaning "bad form" (Figure 18.2).

Doctors have learned to recognize characteristic patterns of changes that cells typically go through as they get more and more out of control. Dysplasia is often a sign that tumor cells are "precancerous," meaning they are altering in ways that may herald the possibility of cancer. The tumor itself becomes more and more disorganized as cells pile on top of each other in seemingly random fashion.

A tumor is defined as a **cancer** when at least some of its cells completely lose all semblance of organization, structure, and regulatory control. As

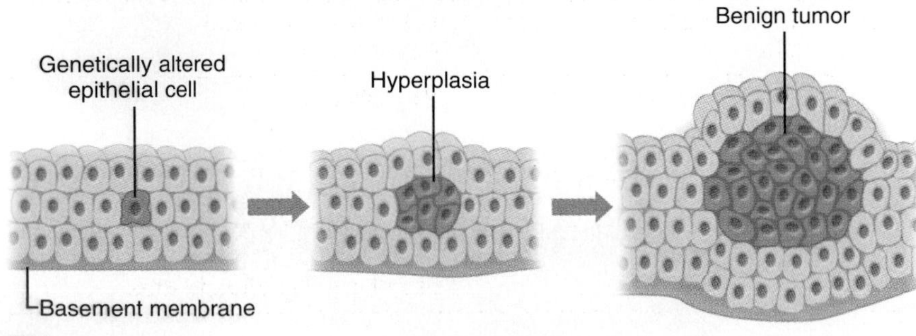

Figure 18.1 Development of a benign tumor. A benign tumor begins when a cell (in this case an epithelial cell) becomes genetically altered and begins to divide more frequently than normal. The mass enlarges but stays well contained. Benign tumors generally are well contained by a capsule of normal cells.

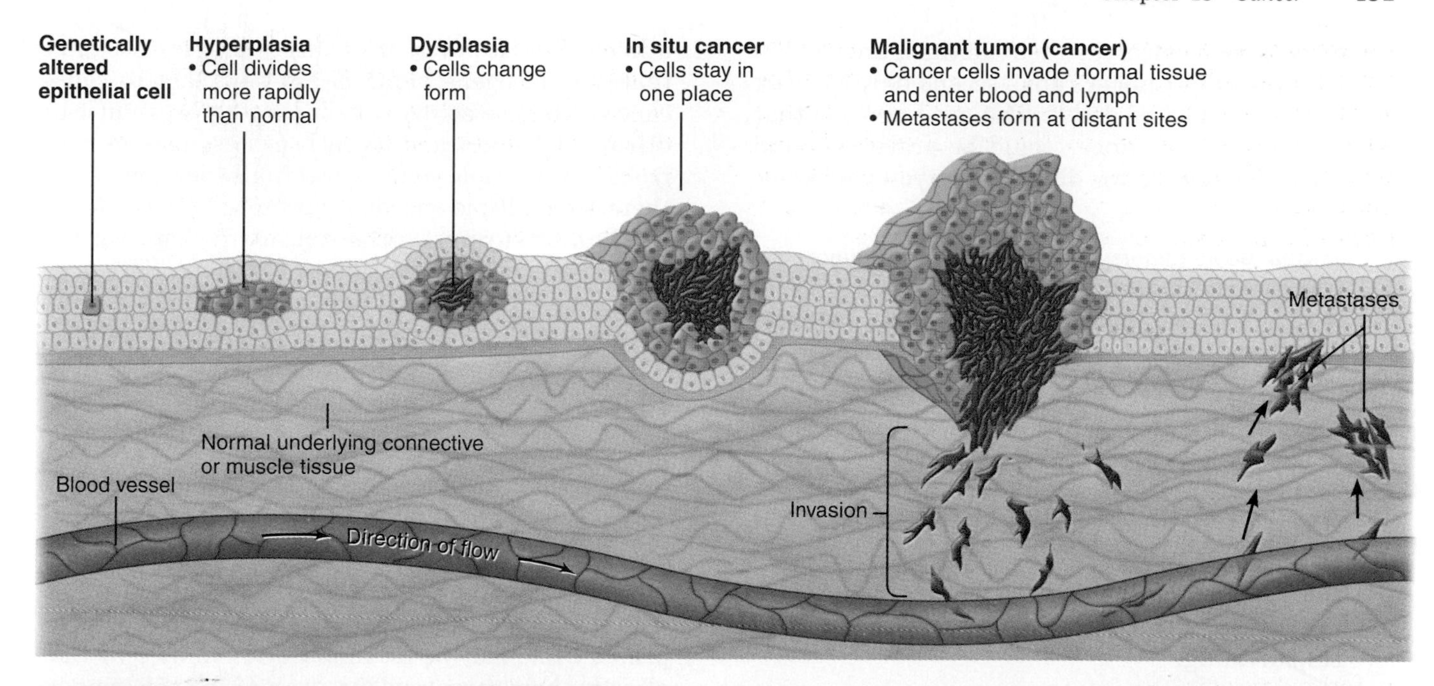

Figure 18.2 The development of a malignant tumor (cancer). A genetically altered cell begins to divide more frequently than normal, resulting in an increased number of cells (hyperplasia). Additional genetic alterations cause some cells to change form (dysplasia) and to lose all semblance of organization or control over cell division and function. As long as the mass stays in one place it is called an *in situ* cancer. However, additional changes in the cells may cause them to break away from the tumor and invade other normal tissues or enter the blood or lymph. These invading cells may set up new colonies of cancer (called metastases) at distant sites. A cancer that invades and metastasizes is called a malignant tumor.

long as the entire tumor remains in one place it is known as an ***in situ* cancer** (tumor). *In situ* cancers often can be successfully removed surgically if they are detected early enough.

If cells in the *in situ* cancer go through additional changes that cause them not to adhere to each other in the normal fashion, they may break away from the main tumor and invade neighboring tissues. Some of these invading cancer cells may even enter the blood or lymph vessels, travel to distant sites in the body, and set up new colonies of cancer, called **metastases.** Cancers that *invade* normal tissues and *metastasize* to distant sites are called **malignant tumors** (from the Latin *malignus,* "ill-disposed"). Table 18.1 compares benign, *in situ,* and malignant tumors.

A malignant tumor is like a forest fire that is out of control and throwing off sparks. The sparks start new fires in distant locations, which also spread uncontrollably. Invasive, metastasizing cancer continues to spread until it completely overruns the functions of tissues, organs, and organ systems. Ultimately it may kill the entire organism, just as a forest fire may not die out until it has consumed the whole forest.

Cancer is extremely difficult to treat, for obvious reasons. One in three people in the United States

Table 18.1 A summary of the characteristics of benign tumors and cancers

	Benign Tumor	*In Situ* Tumor (cancer)	Malignant Tumor (cancer)
Frequency of cell division	More rapid than normal	Rapid and completely out of control	Rapid and completely out of control
Cell structure and function	Slightly abnormal	Increasingly abnormal	Very abnormal
Tumor organization	A single mass generally surrounded by a capsule	Still a single mass, but increasingly disorganized and not always surrounded by a capsule	Some cells break away to invade underlying normal tissues or to enter the blood or lymph. New tumor colonies (metastases) become established at distant sites

will experience cancer in their lifetime, and one in four will die of it. Cancer ranks second only to heart disease as a cause of death. The good news is that most incidences of cancer could be prevented, and later in this chapter we discuss how you can lower your risk.

Recap **Some tumors are benign, but when tumor cells change form dramatically and divide uncontrollably, the tumor is called cancer. Cancer becomes malignant when cells invade and metastasize, starting new tumors at distant sites.**

18.3 How cancer develops

In order for cancer to develop, at least two things must happen simultaneously:

- The cell must grow and divide uncontrollably, ignoring inhibitory signals from nearby cells to stop dividing.
- The cell must undergo physical changes that allow it to break away from surrounding cells.

These events begin when some of the cell's genes become abnormal and no longer function properly.

Mutated or damaged genes contribute to cancer

In the last chapter we learned that *regulatory genes* code for proteins that either activate or repress the expression of *structural genes.* Structural genes, in turn, code for the proteins that control cell growth, division, differentiation, and even cell adhesion (the adherence of cells in the same tissue to each other).

Some regulatory genes code for proteins that cells secrete in order to influence nearby cells. These proteins are called *growth factors* or *growth inhibitors,* depending on which it is that they do. Growth inhibitors are especially important in regulating cell division because they influence cells in a particular tissue to stop dividing when there are enough cells present.

Spotlight on Structure & Function

Under normal circumstances, two types of regulatory genes control the cell's activities: proto-oncogenes and tumor suppressor genes. **Proto-oncogenes** are normal regulatory genes that *promote* cell growth, differentiation, division, or adhesion (*proto,* "first"; *onco-* means "mass, bulk", i.e., a tumor). Mutated or damaged proto-oncogenes that contribute to cancer are called **oncogenes.** Some oncogenes drive the internal rate of cell growth and division faster than normal. Others produce damaged protein receptors that fail to heed inhibitory growth signals from other cells. However, it is important to recognize that one oncogene alone is not sufficient to cause cancer. Because so many cellular processes must be disrupted at once and as each process may be controlled by multiple genes, cancer only develops when there are multiple oncogenes present.

Tumor suppressor genes are regulatory genes that *repress* cell growth, division, differentiation, or adhesion. Under normal circumstances they play an important role in inhibiting cell activities when cells are stressed or when conditions are not right for division. Tumor suppressor genes contribute to cancer when they become damaged, because then the cell functions they normally suppress can occur unchecked.

One tumor suppressor gene, called p53, has received a lot of attention lately. The p53 gene codes for a protein found in most tissues that prevents damaged or stressed cells from dividing. In fact its primary role may be to inhibit division by cells that already show cancerous features. If the p53 gene itself becomes damaged, a variety of cancers may develop more easily. To date, mutated p53 genes have been identified in tumors of the cervix, colon, lung, skin, bladder, and breast. ■

Recap **Regulatory genes called proto-oncogenes and tumor suppressor genes normally control the rate of cell division. When these regulatory genes mutate or become damaged in ways that contribute to cancer, they become oncogenes.**

A variety of carcinogens can cause cancer

Carcinogenesis is the process of transforming a normal cell into a cancerous one. A **carcinogen** is any substance or physical factor that causes cancer. The word carcinogen comes from the recognition by early physicians that certain skin cancers appeared crablike (Greek *karkinos,* "crab"; *genein,* "to produce"). Carcinogenesis involves a number of accumulated changes to a cell's DNA that add up to a cell out of control.

A car analogy will demonstrate how a cell progresses toward cancer. At first your new car runs really well. As time goes on a few parts wear out, but you manage to keep it in good repair with regular service checkups. But as the car gets older, the need for repairs gets more frequent, and you aren't always able to keep up with them. Now the suspension needs replacing, the tires are worn out, the brakes are thin, and the steering wobbles at freeway speeds. No single defect is enough to cause an accident, but together they spell trouble.

The same sort of progression occurs within cells. Through millions of years a number of mutations have accumulated in the set of human genes, some of which you inherited from your parents. Then, over your lifetime, your genes may suffer additional damage from a variety of carcinogens, such as:

- Viruses and bacteria
- Chemicals in the environment
- Tobacco
- Radiation
- Dietary factors and alcohol consumption

Finally, your DNA may be harmed by internal factors, including unrepaired mistakes made during DNA replication and internally produced chemical carcinogens. Let's look at these carcinogens in more detail.

Inherited susceptibility Your parents carry a few of the mutations that have accumulated over millions of years in human genes. These rare heritable mutations may not be enough to guarantee cancer, but they do increase the risk. If six mutations are required to produce cancer, for example, you could inherit five mutations and still be cancer free. This is why, even in families with a history of cancer, some siblings develop the disease and others do not.

The multigene basis of cancer also explains why it is possible to develop cancer even if there is no family history. Remember that we all inherit a slightly different combination of our parents' genes, and if you happen to get mutated genes from both parents, you could be at greater risk than either of them. (We discuss genetic inheritance in more detail in the next chapter.)

Recap **Cancer generally develops only after at least four to six different regulatory genes are transformed into oncogenes. Some oncogenes are inherited.**

Viruses and bacteria Viruses and bacteria contribute to some cancers. Recall from Chapter 9 that viruses reproduce by inserting their DNA into the host's DNA. If the viral sequence of DNA impairs the function of a normal proto-oncogene, perhaps by preventing a regulatory gene from being turned off after it is turned on, this will increase the cancer risk.

Viruses thought to contribute to cancer include the human papillomavirus (cancers of the cervix and penis), hepatitis B virus (liver cancer), HIV (Kaposi's sarcoma), and the Epstein-Barr virus that causes mononucleosis (Hodgkin's disease and non-Hodgkin's lymphoma). The *Helicobacter pylori* bacterium, which causes ulcers, may contribute to stomach cancer as well. Although this list may sound frightening, viruses and bacteria probably account for fewer than 15% of all cancers.

Chemicals in the environment Chemical carcinogens that directly damage DNA are called *initiators*. Other chemicals, called *promoters,* do not damage DNA themselves but increase the potency of other carcinogens. The list of chemical carcinogens in the environment is quite long and includes many chemicals used in (or produced by) industrial processes, such as coal tar, soot, asbestos, benzene, vinyl chloride, and some pesticides and dyes. Table 18.2 lists examples of viral, chemical, and radiation carcinogens, ranked by likelihood of exposure.

Tobacco Perhaps 2% of all cancers are thought to be caused by industrial pollutants. In contrast, tobacco is by far the single most lethal carcinogen in this country, accounting for over 30% of all cancer deaths. Smoking dramatically increases your risk of cancer of the lungs, mouth, pharynx, pancreas, and bladder. Smokers are not the only ones at risk; passive smoking (inhaling secondhand smoke) can cause lung cancer too.

What single substance causes more than 30 percent of all cancer deaths?

Smokeless tobacco (chewing tobacco or snuff) is also a carcinogen. It is a major risk factor for cancers of the mouth, pharynx, and esophagus.

Radiation Sources of radiation include the sun and radioactive radon gas, which is released by rocks, soil, and groundwater at intensities that vary in different geographic areas. Cellular telephones, household appliances, and electric power lines also emit trace amounts of radiation.

From time to time, the carcinogenic potential of radiation receives a great deal of media attention. In truth, all sources of radiation combined cause only about 2% of all cancer deaths, and most of these are due to skin cancer caused by the sun's ultraviolet rays.

The high-frequency ultraviolet B rays in sunlight cause over 80% of all skin cancers, including a particularly dangerous type of cancer of the melanin-producing cells of the skin called **melanoma.** Frequent sunburns, especially during childhood, increase the risk of melanoma. Fair-complexioned people who sunburn quickly are more vulnerable than those who tan easily (Figure 18.3). However, cumulative exposure to ultraviolet radiation, even without burning, raises the risk of skin cancer. Ultraviolet radiation from sunlamps and tanning booths is just as dangerous as radiation from natural sunlight.

Diet How much does diet contribute to cancer? Although difficult to define precisely, its role appears substantial. Dietary factors may be involved in up to 30% of cancers, rivaling smoking. In particular, red meat and saturated animal fat raise the risk of cancers

Table 18.2 Examples of carcinogens

Carcinogen	Source of Exposure or Types of Persons Exposed	Type of Cancer	Exposure of General Population
Tobacco (smoking and smokeless)	Smokers, chewers, people exposed to secondhand smoke	Lung, bladder, mouth, pharynx, pancreas	Common
Diesel exhaust	Buses and trucks, miners, railroad yard workers	Lung	Common
Benzene	Paints, dyes, furniture finishes	Bone marrow (leukemias)	Common
Pesticides	Agricultural workers	Lung	Common
Ultraviolet light	Sunlight	Skin cancers	Common
Ionizing radiation	Radioactive materials, medical and dental procedures	Bone marrow (leukemias)	Common
Human papilloma viruses	Sexually transmitted	Cervix, penis	Common
Hepatitis B virus	Sexually transmitted	Liver	Less common
HIV	Sexually transmitted	Kaposi's sarcoma	Less common
Hydrocarbons in soot, tar smoke	Firefighters, chimney cleaners	Skin, lung	Uncommon
Asbestos	Shipyard, demolition, and insulation workers Brake linings	Lung, epithelial linings of body cavities	Uncommon
Radon	Mine workers Basements of houses	Lung	Uncommon

of the colon, rectum, and prostate gland. Obesity in general is a risk factor for several types of cancer, and a high salt intake may contribute to stomach cancer. Excessive consumption of alcohol is associated with breast, rectal, colon, and liver cancer.

Certain fungi and plants produce carcinogens. One such carcinogen is *aflatoxin,* a byproduct of a fungus that infects ground nuts. Aflatoxin may be present in raw peanut butter. Commercial peanut butters are processed to eliminate the fungus.

Figure 18.3 Risky behavior. Persons with light complexions who were often sunburned as children have an increased risk for skin cancer later in life.

In recent years researchers have become interested in what might be *missing* from the diet that could contribute to cancer. Recent evidence suggests that fruits and vegetables in particular may be mildly protective. A compound in broccoli sprouts has received a lot of attention lately as a cancer risk-reducing agent. Dietary analysis is complicated because there are literally thousands of compounds in the foods we eat.

Internal factors It is difficult to estimate how significant internal factors might be to the development of cancer. In theory, errors introduced during DNA replication and damage caused by certain chemicals within the body could contribute to the disease.

In particular, scientists are studying the effects of *free radicals,* highly reactive fragments of molecules that are produced by the body's biochemical processes. Free radicals are a normal by-product of cellular metabolism, and ordinarily they are detoxified by peroxisomes. However, if detoxification processes become less efficient for any reason, free radicals may accumulate in the body and damage other molecules, including DNA. Some researchers believe that free radicals play a role in the aging process as well.

Certain vitamins known as *antioxidants* appear to neutralize free radicals, although the mechanism by which they do this is still controversial. In particular, vitamins A, C, and E seem to exert antioxidant benefits. Whether antioxidants can have an impact on the development of cancer or the aging process, however, is still an open question.

Recap **Some oncogenes result from damage done to DNA by various carcinogens, including viruses and bacteria, tobacco, radiation, environmental chemicals, dietary factors, and alcohol. Free radicals created during cellular metabolism may also damage DNA.**

Immune system plays an important role in cancer prevention

Spotlight on Structure & Function Under normal circumstances, the immune system plays a critical role in defending against damaged or defective cells, including cancer cells. Recall that all cells display proteins that identify them as "self" so the immune system leaves them alone. As cells change in ways that make them cancerous they may stop displaying "self" proteins, or they may display completely different proteins that mark them as abnormal. This makes them subject to attack and destruction by the immune system, at least as long as it is functioning normally. ■

Suppression of the immune system by drugs, viruses such as HIV, or even mental states of anxiety, stress, or depression may allow certain cancers to develop more easily. Some cancers apparently can suppress the immune system, and others may have mechanisms for disguising themselves from attack by the immune system. Figure 18.4 presents an overview of how cancer develops.

Recap **The immune system normally protects us from cancer cells by killing them before they spread. Immune system suppression allows cancers to develop more easily.**

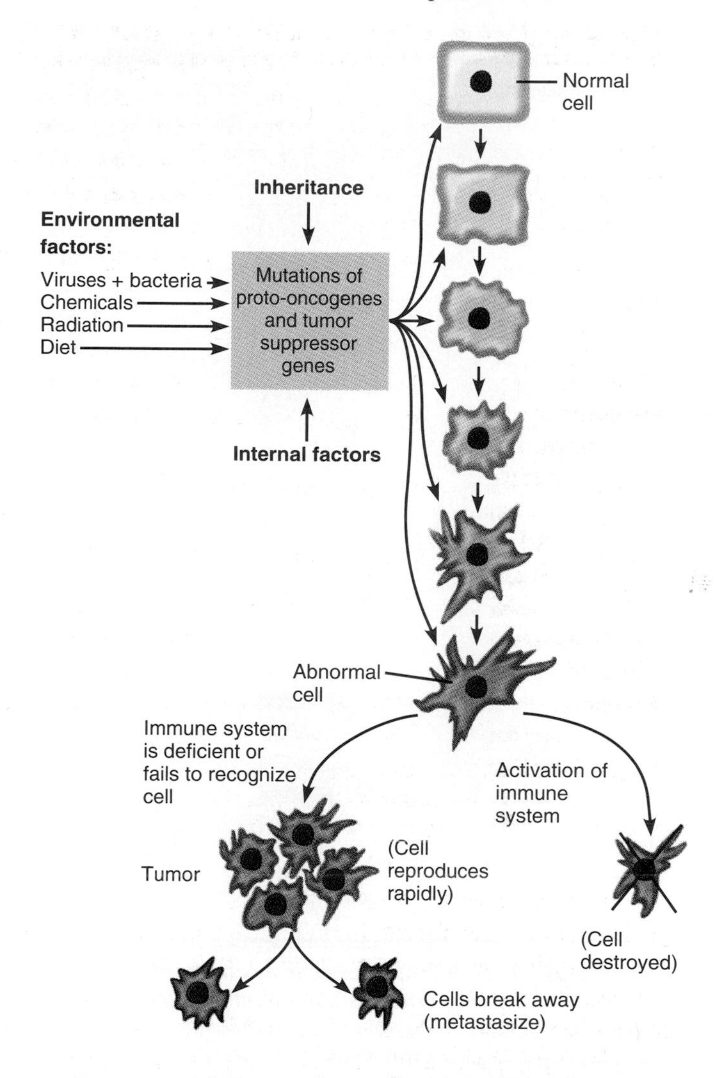

Figure 18.4 Overview of the development of cancer. The progression to cancer requires multiple genetic changes in regulatory genes, some of which may be inherited and some of which may be caused by environmental or internal factors. A cancer cell reproduces rapidly (develops into a cancer) and may invade and colonize distant tissues (metastasize). Suppression of the immune system appears to be involved in many cancers.

18.4 Advances in diagnosis enable early detection

In 1971 President Nixon signed into law the National Cancer Act, declaring "war" on cancer. At that time many researchers hoped we could overcome cancer with early diagnosis and better treatments.

Today, although we have made progress, much more remains to be done. The sobering fact is that after 30 years of concerted effort, the overall death rate from cancer still has not declined. In fact, from 1973 to 1996 it increased by 3.3% (Figure 18.5). The increase was due primarily to a sharp rise in lung cancer among women, probably because more women smoked.

Scientists have learned a tremendous amount about the basic mechanisms of cancer and how to diagnose and treat it effectively. Some researchers believe we are on the verge of seeing a true change in the number of deaths due to cancer.

Early detection (diagnosis) of cancer is an important key to success. Early detection means that therapy can begin sooner, hopefully before the disease has had a chance to metastasize. Prompt treatment can even cure some cancers.

Tumor imaging: X-rays, PET, and MRI

The classic approach to diagnosing a tumor is to view its image, either on an x-ray (Figure 18.6) or with a newer method. Computer-enhanced imaging

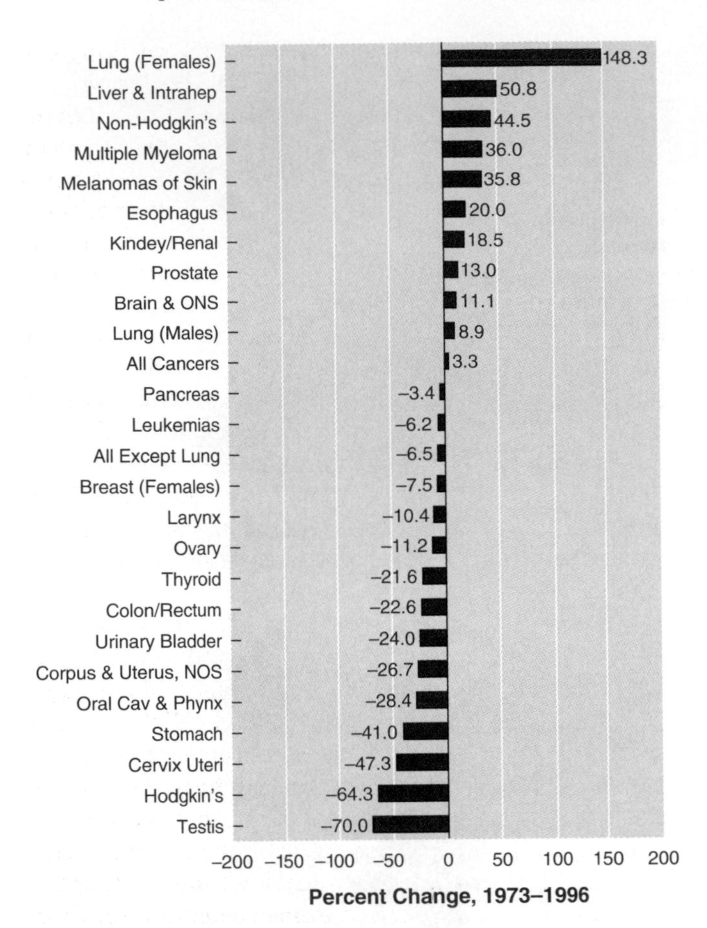

Figure 18.5 Trends in U.S. cancer mortality rates, 1973 to 1996. The figure shows the percent change in the incidence of the listed cancers in 1996 compared to 1973. For example, for every 100 women who developed lung cancer in 1973 in a given population size, 248 women developed lung cancer in 1996 within the same population size. For 100 men who developed testicular cancer in 1973, only 30 men developed testicular cancer in 1996. Abbreviations: ONS (other nervous system); NOS (location not specified).

SOURCE: Ries, L.A.G., C.L. Kosary, B.F. Hankey, B.A. Miller, L. Clegg, and B.K. Edwards (eds). SEER Cancer Statistics Review, 1973–1996, National Cancer Institute, Bethesda, MD, 1999.

is now able to detect changes that visual imaging may miss. Advanced imaging techniques such as *positron-emission tomography* (PET) and *magnetic resonance imaging* (MRI) show physiological differences between tissues in places that cannot be diagnosed effectively with x-rays.

PET scans employ radioactive substances to create three-dimensional images showing the metabolic activity of body structures. In addition to diagnosing cancer, PET scans are used to study blood flow, heart function, and brain activity.

MRI scans use short bursts of a powerful magnetic field to produce cross-sectional images of body structures. For example, MRI detects differences in water and chemical composition between tissues on the basis of changes in magnetic fields. MRI can detect tumors because tumors often have a slightly different consistency and composition than normal tissue. It can detect tumors even when they are hidden behind bone, such as tumors of the brain (Figure 18.7). Unlike x-rays, MRI is not itself an added risk factor for cancer.

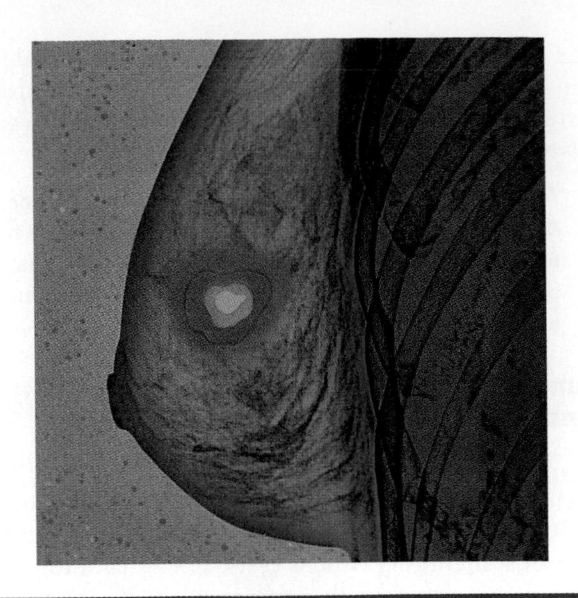

Figure 18.6 An x-ray of the breast, called a mammogram. A tumor is clearly visible in this x-ray.

Genetic testing can identify mutated genes

As we learn more about which human characteristics are transmitted by specific genes, genetic testing for diseases, including cancers, becomes a real possibility. Already hundreds of new genes and their mutated counterparts have been identified, and tests are being devised to detect them.

Such tests are highly controversial, however, and society will have to choose how to use them. One of the concerns about genetic testing is that it may be possible to test for mutated genes that have no cure. Some people believe that such knowledge may not be in the patient's best interest. In addition, it is difficult to convince people that in many cases a mutated gene is nothing more than a risk factor. It does not guarantee development of a disease.

Enzyme tests may detect cancer markers

An enzyme called *telomerase* is rarely found in normal cells but is present in nearly all cancer cells. Some researchers propose that a sensitive test for telomerase in body fluids could detect early stages of cancer. An enzyme test could also be used to screen large populations or monitor the progression of the disease in known cancer patients. Researchers are exploring other potential cancer markers as well.

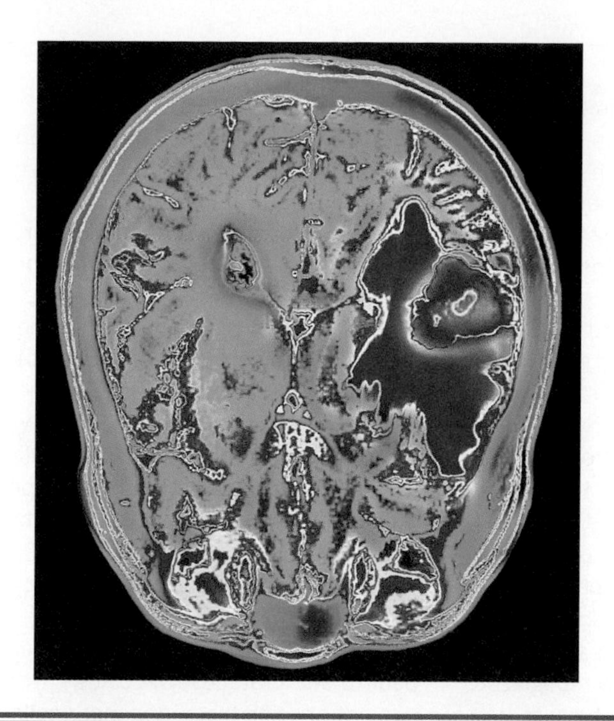

Figure 18.7 An MRI of the brain. The magenta and red area to the right is a brain tumor.

Recap **Early diagnosis and prompt treatment of cancer are vital. Diagnostic techniques include x-rays, positron-emission tomography, magnetic resonance imaging, genetic testing, and enzyme markers.**

18.5 Cancer treatments

Can cancer be cured?

Many cancers are treatable, especially if discovered early. Although some forms of the disease have higher survival rates than others, current treatments cure approximately 50 percent of all cancer cases.

Conventional cancer treatments: Surgery, radiation, and chemotherapy

Classic treatments for cancer include surgery, radiation, chemotherapy, or a combination of two or more. Better imaging techniques have benefited the surgical treatment of cancer, and techniques for focusing external sources of radiation directly onto the cancer have improved as well. These focused sources of intense radiation are designed to damage and kill cancer cells in a specific location. It is sometimes possible to implant radioactive materials in the region of a tumor.

Healthy cells recover from radiation more readily than cancer cells, but radiation therapy generally does injure or kill some normal tissue too. A drawback of both surgery and radiation is that they may miss small groups of metastasized cells, allowing the cancer to reappear later.

Chemotherapy is the administration of cytotoxic (cell-damaging) chemicals to destroy cancer cells. Chemotherapy addresses some of the limitations of surgery and radiation. Some chemotherapeutic drugs act everywhere in the body, and for that reason they must inflict only acceptable amounts of damage to normal cells. Some chemotherapy chemicals stop cells from dividing by interfering with their ability to replicate DNA. Chemicals that inhibit cell division act primarily on cancer cells because cancer cells divide very rapidly. But such chemicals also damage or kill many normal cells that divide rapidly, such as bone marrow cells and cells lining the digestive tract. Damage to cells in the digestive tract is the reason for the nausea experienced by many patients undergoing chemotherapy.

Common side effects of chemotherapy are nausea, vomiting, hair loss, anemia (a reduction in the number of red blood cells), and an inability to fight infections (due to fewer white blood cells). A major problem is that many tumors become resistant to chemotherapeutic drugs, just as bacteria develop resistance to antibiotics. Treatment then becomes a battle of trying to destroy abnormal cells with stronger and stronger chemicals without killing the patient in the process. Sometimes combination therapy, in which surgery is followed by radiation and combinations of chemotherapeutic drugs, works best.

Immunotherapy promotes immune response

Immunotherapy attempts to boost the general responsiveness of the immune system so it can fight cancer more effectively. Recent efforts have focused on finding specific antigen molecules that are present on cancer cells, but not on normal cells, and using them to produce antibodies that target cancer cells for destruction by the immune system. This is an area that is showing particular promise. Taking it one step farther, researchers are working to attach radioactive molecules or chemotherapeutic drugs to the antibodies so treatments could be delivered only to cancer cells and spare normal cells. Someday it may be possible to create vaccines that prevent specific cancers.

Researchers are also testing a skin cancer treatment vaccine created from a patient's own cancer cells. When injected into the patient, the modified cancer cells appear to stimulate the person's immune defenses to recognize and combat the abnormal cells.

Directions in Science

New Weapons in the War on Cancer

When the war on cancer began in 1971, scientists hoped that with enough effort and financial support we could conquer cancer. Has the war been a success? This remains an open question.

Critics would argue that, despite all that spending, deaths from cancer have hardly budged. Supporters counter that certain cancers are much less common than they were 30 years ago and that the knowledge gained about the mechanisms of cancer will pay off shortly.

Both views are probably correct. When the war on cancer began it was understood that cancer is a disease of uncontrolled growth and metastasis. The feeling was that any cancer could be successfully treated (given the limited arsenal available at the time) if detected at an early stage. During those first years, then, researchers focused on early detection and treatment. As a result, two things happened: (1) more cancers are now detected at earlier stages due to increased public awareness, improved screening methods, and better diagnostic tests, and (2) refinements in radiation and chemotherapy have led to better treatment once cancer is detected.

However, it turns out that these improvements only affect some cancers. Death rates from testicular cancer, cervical cancer, and Hodgkin's disease have all plummeted, for example. At the same time, however, deaths from other cancers—notably lung cancer and melanoma—have risen dramatically. Many physicians attribute this increase to behavioral trends such as increased smoking and sunbathing. It is now recognized that other strategies will have to be adopted for those cancers not amenable to early detection or effective treatment.

One of the greatest advances, though it doesn't yet show up in the death statistics, is our improved understanding of how cancer develops and progresses. We now know that cancers generally result from an accumulation of multiple defects, and that by the time the tiniest tumor is detectable it may already be out of control. As a result, in addition to early detection and treatment, we are adding two new weapons to the war on cancer: *prevention* and *protection.*

Prevention aims to reduce exposure to damaging carcinogens in the first place, so cancer never develops. Expect to hear a lot more about risk factors for specific cancers and how you can reduce your risk. Indeed, ultimately we may save more lives by persuading people not to smoke than by any amount of early detection.

Protection is aimed at boosting the body's own defenses once carcinogenic damage has occurred. Expect to see protective vaccines against specific types of cancer cells in the future. Another new area of cancer research seeks to determine the protective effects of certain chemicals and dietary components. Indeed, there is some indication that antioxidant vitamins can reduce the development of some cancers, perhaps by raising cellular concentrations of natural enzymes that protect against free radicals.

"Starving" cancer by inhibiting angiogenesis

Tumors grow and divide rapidly, so they require a great deal of energy. We now know that enlarging tumors promote the growth of new blood vessels (called *angiogenesis*) to serve their energy needs. Without angiogenesis, the tumor would reach about the size of a pea and stop growing. At least a dozen proteins stimulate angiogenesis, some of which are released by tumor cells, and several other proteins are known to inhibit it. Interest is very high in antiangiogenic drugs, which literally starve tumors by limiting their blood supply. Several drugs are already in clinical trials (the early stages of testing in humans).

Molecular treatments target oncogenes

Sophisticated molecular strategies target the oncogenes that lead to cancer. Some researchers are attempting to inactivate specific oncogenes, or the proteins they encode, to slow the runaway cell division. Future techniques may include *gene therapy,* in which defective genes are repaired or replaced with their normal counterparts. A key target for gene therapy is the p53 tumor suppressor gene that, when defective, contributes to many cancers.

Recap **The mainstays of cancer treatment are surgery, radiation, and chemotherapy. Recent advances include immunotherapy, drugs to inhibit angiogenesis, and molecular treatments that focus on specific oncogenes.**

18.6 The ten most common cancers

What are the most common cancers?

Table 18.3 ranks the ten most common cancers by incidence. Each has its own risk factors, warning signs, methods of detection, treatments, and (most important of all) survival rate.

In both men and women, the most frequent cancers are those of the skin, lung, and colon and rectum. In addition, prostate cancer is common in men and breast cancer is common in women.

Table 18.3 The top ten cancers ranked by incidence (numbers of new cases per year)

Cancer	New Cases per Year	Deaths per Year
Skin (melanoma, basal, and squamous cell)	1,300,000	9,600
Breast	182,800 (women) 1,400 (men)	40,800 (women) 400 (men)
Prostate	180,400 (men)	31,900 (men)
Lung	164,100	156,900
Colon and rectum	130,200	47,700
Lymphoma	62,300	27,500
Urinary bladder	53,200	12,200
Uterus	36,100 (women)	6,500 (women)
Leukemia	30,800	21,700
Oral and pharynx	30,200	7,800

SOURCE: American Cancer Society. Cancer Facts & Figures 2000. American Cancer Society, Inc.; 2000.

The most deadly cancers—in terms of the percentage of people diagnosed who will die—are lung and pancreatic cancer. Lung cancer kills more people per year than any other form of the disease.

Which cancers are the most deadly?

Skin cancer: Look for changes in your skin

There are three major types of skin cancer: basal cell, squamous cell, and melanoma.

- Basal cell cancer, the most common, occurs when the basal cells in skin divide abnormally. Basal cell cancer usually appears as a pink or flesh-colored bump with a smooth texture (Figure 18.8a). Sometimes the bump bleeds or crusts. Basal cell cancer rarely metastasizes but should still be removed. If not treated, it can eventually spread to underlying tissues.
- Squamous cell cancer arises from the epithelial cells produced by the basal cells. It consists of pink scaly patches or nodules that may ulcerate and crust (Figure 18.8b). Squamous cell cancer can metastasize, although slowly.
- Melanoma, the least common skin cancer, is also the deadliest. It arises from abnormal melanocytes. Look for dark spots or patches on your skin and evaluate them according to the "ABCD" rule: (1) *Asymmetry*—the two halves of the spot or patch don't match; (2) the *Border* is irregular in shape; (3) the *Color* varies or is intensely black; and (4) the *Diameter* is greater than 6 millimeters, or about the size of a pea (Figure 18.8c). Itchiness, scaliness, oozing, or bleeding may also be significant signs.

Early detection is critical; both basal and squamous cell cancers have a better than 95 percent cure rate if removed promptly. For melanoma, the five-year survival rate (percentage of persons diagnosed who are still alive after five years) is 95% if the melanoma is detected while still localized, but only 16% if it has spread to distant sites.

Breast cancer: Early detection pays off

Although it can occur in men, breast cancer is almost exclusively a cancer of women. It is usually first diagnosed by a mammogram (an x-ray of breast tissue) or when the woman or health care provider detects an abnormal lump during a physical breast exam.

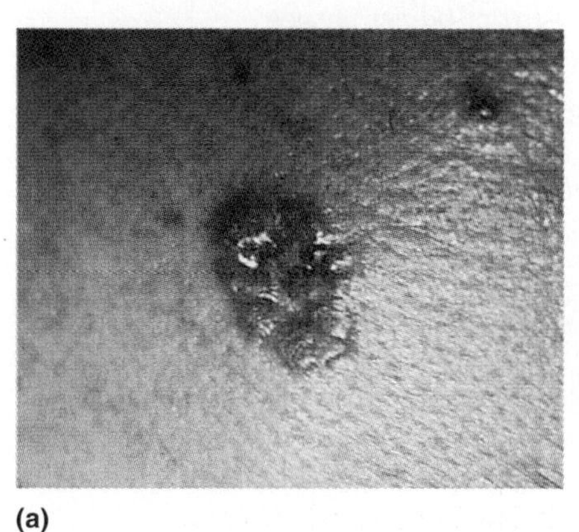

(a)

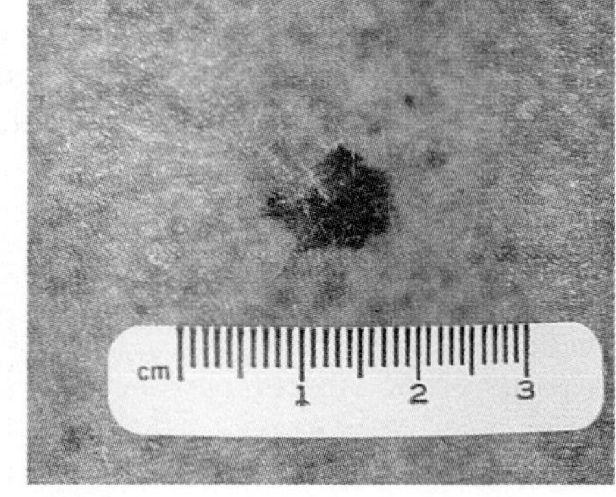

(b)

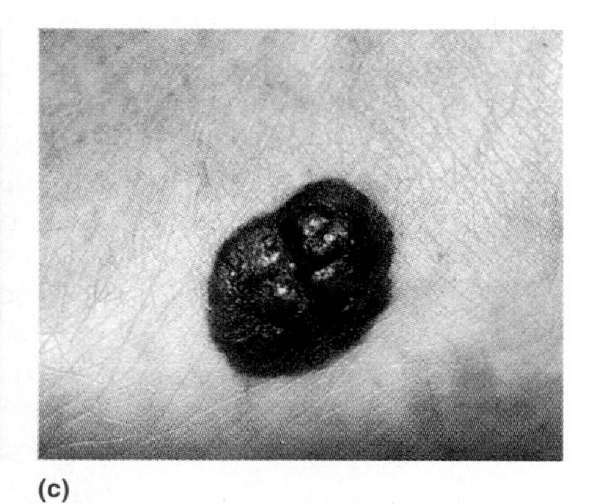

(c)

Figure 18.8 Skin cancers. (a) Basal cell carcinoma arises from basal cells that divide abnormally. **(b)** Squamous cell carcinoma occurs in the squamous epithelial cells derived from basal cells. **(c)** Melanoma, the most malignant skin cancer, arises from melanocytes.

Recent research suggests that there are at least two inherited susceptibility genes for breast cancer. Other risk factors include:

- Age (more common after age 40)
- Early onset of menstruation or late onset of menopause
- Using oral contraceptives
- Taking estrogen supplements after menopause

Environmental and dietary factors may also play a role. Early detection dramatically increases the chance of survival: the five-year survival rate for breast tumors that have not yet metastasized is 97%, but it falls to 21% for women who have distant metastases at diagnosis. Overall, the 15-year survival rate for all diagnoses is now 56%.

Prostate cancer: Most common after age 50

Symptoms of prostate cancer include:

- Difficulty or inability to urinate
- Blood in the urine
- Pain, usually in pelvic area

The biggest risk for prostate cancer is advancing age. Physicians recommend yearly testing for men aged 50 and older (Table 18.4). The cancer can be diagnosed by digital rectal exam or by the PSA (prostate-specific antigen) test, which detects elevated blood levels of a protein produced by the prostate gland.

Treatment options include surgery, radiation therapy, and hormones. Treatment in older men is controversial because prostate cancer generally grows so slowly that it is unlikely to be the cause of death. The five-year survival rate is now 89% and the 15-year survival rate is 50%.

Table 18.4 Recommendations for cancer screening

Cancer	Gender	Test	Age and Frequency	Description
All	Men and women	Cancer-related checkup	20–40: Every 3 years 40+: Every year	General exam by health professional that includes health counseling and screening as appropriate for malignant and nonmalignant diseases
Colon and Rectum	Men and women	Fecal occult blood test	50+: Every 5 years	Checks for hidden blood in feces
		Digital rectal exam	50+: Every 5 years	Physical exam by health professional
		Colonoscopy	50+: Every 5–10 years	Examining interior of colon with flexible fiber-optic scope
Breast	Women	Self-examination	20+: Once per month	See Health Watch Box
		Clinical examination	20–40: Every 3 years 40+: Every year	Physical exam by health professional
		Mammogram	40+: Every year	X-ray of breast tissues
Cervix and Endometrium	Women	Pelvic exam	18+ and all sexually active women: Every year	Physical exam by health professional
		Pap test	18+ and all sexually active women: Every year	Smear of cells taken from cervix by health professional
Endometrium	Women	Endometrial tissue sample	At menopause	Sample of uterine tissue taken by health professional
Prostate	Men	Digital rectal exam	50+: Every year	Physical exam by health professional
		PSA (prostate-specific antigen)	50+: Every year	Blood test that detects elevated levels of PSA
Testicles	Men	Self-examination	18+: Once per month	See Health Watch Box

NOTE: People who have one or more risk factors for any cancers should begin screening at earlier ages and have tests done more often. Consult your physician for advice. Data based on American Cancer Society Prevention and Detection Guidelines.

Lung cancer: Smoking is leading risk factor

The fact is that most cases of lung cancer are preventable, because by far the greatest risk factor is cigarette smoking. All other risk factors, which include exposure to smoke, radiation, and industrial chemicals such as asbestos, are relatively minor by comparison.

Treatment may include surgery, radiation, or chemotherapy. Early detection improves the odds, but the overall five-year survival rate for all stages of diagnosis is still only 14%.

Cancers of colon and rectum: Tests can detect them early

An obvious worrisome sign is blood in the stool or rectal bleeding, though bleeding can also result from noncancerous causes. Risk factors include a family history of colorectal cancer and a low-fiber, high-fat diet with inadequate fruits and vegetables.

Precancerous growths known as *polyps* can be removed surgically. The five-year survival rate for all cases of colorectal cancer has improved to 62% largely due to early detection and treatment. For people with distant metastases at the time of diagnosis, the five-year survival rate drops to just 7%. Here is another case where early detection is extremely beneficial.

Fortunately, several tests are available to diagnose colorectal cancer in early stages. For everyone over age 50, the American Cancer Society recommends regular rectal exams, a yearly test to detect fecal occult (hidden) blood, and a colonoscopy (examining the interior of the colon with a flexible fiberoptic scope) every ten years. People who have any risk factors should begin screening earlier and undergo testing more often.

Lymphoma: Cancers of lymphoid tissues

Lymphoma is the general term for cancers of the lymphoid tissues, the most familiar of which is called Hodgkin's disease. Symptoms include:

- Enlarged lymph nodes
- Fever
- Anemia
- Night sweats

Risk factors are not entirely clear but seem to relate to altered immune function. People at risk include organ transplant recipients, those with HIV infection, and perhaps workers with high occupational exposure to herbicides. Treatments generally are radiation or chemotherapy. Treatments involving specific antibodies to lymphoma cells can be effective. Bone marrow transplant may be necessary after high-dose therapy.

Urinary bladder cancer: Surgery often successful if done early

Blood in the urine is an important sign, though there may be other reasons for it. Smoking doubles the risk of bladder cancer. Living in an urban area or occupational exposure to rubber, leather, or dye industries are also risks. Surgery is almost always successful if the cancer is detected early (94% five-year survival rate), but the survival rate declines dramatically if the cancer has already metastasized to distant sites (7% five-year survival rate).

Cancer of the uterus: Unusual uterine bleeding is major symptom

Cancers of the uterus include cervical cancer and cancer of the endometrium of the main body. The primary symptoms are abnormal bleeding or spotting. Pain is a late manifestation of the disease. Estrogen replacement therapy after menopause is a major risk factor, but progesterone added to estrogen is thought to offset the risk related to estrogen alone. Pregnancy and the use of oral contraceptives seem to offer some protection from endometrial cancer.

Cervical cancer is usually detected by a Pap test (examination of a smear of cervical cells; the name comes from Papanicolaou, the doctor who devised the test). The Pap test is not very effective at detecting cancer of the endometrium, which is generally detected by a pelvic exam. Treatments include surgery, radiation, hormones, or chemotherapy. The five-year survival rate is 96% if the cancer is detected early but declines if the disease has already spread.

Leukemia: Chemotherapy is often effective

Leukemias are cancers of the blood-forming organs characterized by a marked increase in the number of white blood cells (leukocytes) and their precursors in the blood. The causes are largely unknown, but may be linked to Down's syndrome, excessive exposure to ionizing radiation, and benzene. Certain leukemias are caused by a retrovirus, human T-cell leukemia/lymphoma virus-I (HTLV-I).

Although leukemia is commonly thought of as a childhood disease, more adults than children develop leukemias. The symptoms of leukemia are fairly general (fatigue, weight loss, increased incidence of infections), and as a result leukemias are not always diagnosed early. A firm diagnosis is made with a blood test and a bone biopsy.

Health Watch

Examining Yourself for Breast or Testicular Cancer

Breast and testicular cancers can sometimes be detected at an early stage through self-examination techniques that check for unusual lumps. It's a good idea to examine yourself on a regular basis because early detection and treatment greatly enhance the survival rate of both these cancers. Besides, self-examination is quick, easy, private, and free.

Learning a self-examination technique now and practicing it regularly means that you will know how your tissues normally feel. This increases your odds of detecting a suspicious lump if one should develop.

Examining Your Breasts

Examine your breasts at the same time each month. If you are still menstruating, do it after your period when your breasts are least tender. The idea is to know your breasts both visually and internally through physical contact.

A breast self-examination involves pressing on each breast following a regular and complete pattern. Lie down with a pillow or folded towel under your right shoulder (see illustration). Using the pads of the first three fingers of your left hand, check the right breast in every possible lo-

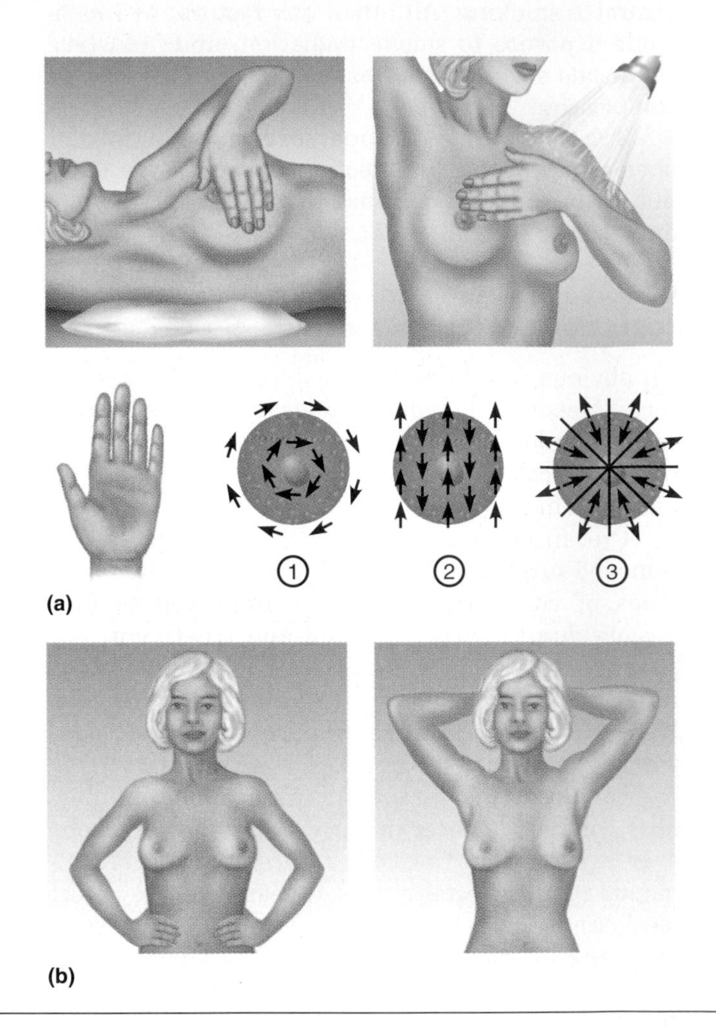

Breast self-examination. (a) Use the pads of the first three fingers of the hand opposite to each breast. Establish a regular and complete pattern of pressure (1, 2, or 3) that works for you. Do the exam while lying down and again in the shower, when your hands are soapy and glide easily over the skin. (b) It is a good idea to examine your breasts visually while standing in front of a mirror.

SOURCE for **(a):** American Cancer Society. *How to Do Breast Self-Examination.* American Cancer Society, Inc., 1997.

The standard treatment for leukemias is chemotherapy. Sometimes chemotherapy is followed by a bone marrow transplant to replace the normal blood-forming stem cells eliminated by chemotherapy. The survival rates for certain types of leukemias have improved over the years, especially among children.

Cancers of the mouth and pharynx: Tobacco is major risk

Symptoms include:

- Sores that bleed and do not heal
- Red or white patches in the mouth
- Lumps or areas of thickened tissue

The major risk factors are the use of tobacco products (smokeless tobacco as well as cigarettes, cigars, or pipes) and excessive consumption of alcohol. Early detection is important and greatly improves survival. Radiation and surgery are the usual treatments.

Recap **Cancers vary widely in terms of incidence, ease of detection and treatment, and death rate. For any cancer the probability of survival is much higher if the disease is detected early, before it metastasizes.**

18.7 Most cancers can be prevented

At present, some cancers are not preventable. This is particularly true of cancers that are linked to the oncogenes we inherit from our parents. For these cases especially, early detection is not only vital but increasingly likely as we learn more about which cancers relate to which oncogenes. True prevention of heritable cancers will have to wait for advances in gene therapy and vaccines.

cation by pressing firmly. Press hard enough that you can feel the underlying tissue.

Establish a regular pattern of pressure (you may wish to use pattern a, b, or c depending on which feels right to you) and use it consistently so you become familiar with how every area of your breast feels. Then examine the left breast by placing the pillow under your left shoulder and using your right hand.

As an added precaution, repeat the process while standing up in the shower. Your soapy, wet hands will glide more easily over your skin, making it easier to tell how your breasts feel.

Finally, be sure to do a mirror examination. Stand in front of a mirror, first with your arms on your hips and then behind your head. Look for any changes in how your breasts look, such as swelling, dimpling of the skin, redness, or changes in the nipples.

Examining Your Testicles

Because testicular cancer accounts for only 1% of all cancers in men, most people don't give it much thought. Nevertheless, testicular cancer is *the* most common cancer among men aged 20–35. In addition, it is one of the most curable cancers if caught early.

It is a good idea to get in the habit of examining your testicles every month. Do the exam after a warm shower, when the skin of the scrotum is fully relaxed. Place your thumbs and the index and middle fingers of both hands gently underneath one testicle with the thumbs on top (see figure). Roll the testicle gently between your thumbs and fingers, feeling for unusual pea-sized lumps on the front or sides. Then examine the other testicle.

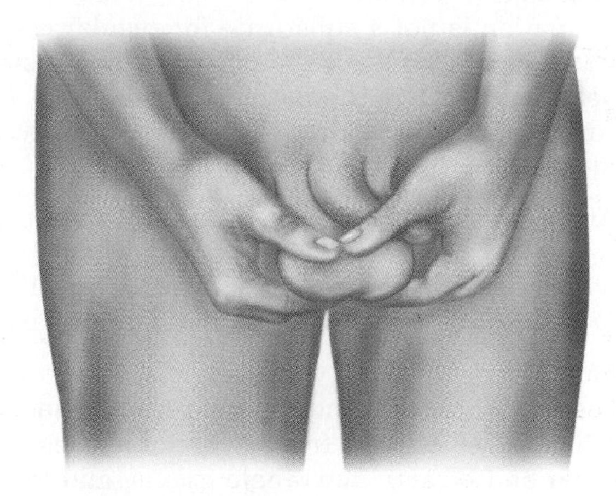

Testicular self-examination. Roll each testicle gently between your thumb and fingers, feeling for unusual lumps.

SOURCE: National Cancer Institute, National Institutes of Health Publication No. 93-2636–*Testicular Self-Examination,* reprinted November 1992.

Be aware that it is not unusual for one testicle to be slightly larger than the other, and do not confuse the epididymus with a lump. The epididymus is the cordlike structure on the top and the back of the testicle that stores and transports sperm.

Follow Up on Suspicious Lumps

If you find anything suspicious, consult your physician immediately. The good news is that many breast and testicular lumps turn out to be noncancerous when further tests are done.

Of course, self-examination should not take the place of regular exams by a health professional. However, when done correctly it can save your life.

However, most incidences of cancer are preventable. Indeed, fully 60% of all cancer cases are thought to be caused by just two factors—smoking and diet. Only a small percentage result from radiation and environmental pollutants.

What can you do to lower your cancer risk?

If we really want to make a dent in the cancer death rate, we need to work on societal issues: encouraging people to exercise, control their weight, eat a healthy diet, limit alcohol, and, above all, not smoke. Researchers are studying what constitutes a healthy diet and why certain dietary components are harmful or beneficial.

What can you do to reduce your own risk of cancer?

- Know your family history. Because cancer generally involves multiple genetic defects, you may have inherited one or more risk factors. Of course, this does not guarantee you will develop cancer. However, until we know of and can test for all such defects you can at least assess your potential by examining your family's history of cancer and discussing it with your physician.
- Get regular medical screenings for cancer. As we've seen, early detection is best, and some slow-growing cancers can be cured if caught early. Your physician may suggest other tests based on your family history, gender, and age, in addition to those described previously in Table 18.4.
- Learn self-examination techniques. Women can examine their breasts, and men their testicles, to check for lumps. All of us should be on the lookout for changes in the skin that might represent

skin cancer (refer to Figure 18.8). While self-examination is not a substitute for regular checkups by a health professional, many cancers are detected early by the patients themselves. The American Cancer Society provides information on self-examination techniques.

- Avoid direct sunlight between 10 a.m. and 4 p.m., wear a broad-brimmed hat, and apply sunscreen with an SPF (sun protection factor) of 15 or greater. Also avoid sunlamps and tanning salons.
- Watch your diet and your weight. There are no absolutes here, but the general recommendation is for a diet high in fruits, vegetables, legumes (peas and beans), and whole grains, and low in saturated fats, red meat, and salt. Unsaturated fats, such as fish and olive oil, appear to be beneficial.
- Don't smoke. Here is a personal choice you can make that will dramatically lower your risk. Smoking is considered *the* single largest preventable risk factor for cancer. It accounts for fully 85% of all lung cancers and also contributes to cancers of the bladder, mouth, and pharynx.
- If you consume alcohol, drink in moderation. One to two alcoholic beverages per day *may* reduce the chance of cardiovascular disease, but the tradeoff is that alcohol consumption appears to raise the risk of certain cancers. It is particularly risky when combined with smoking.
- Seek more information. There's a wealth of information on cancer in general, on specific types of cancer, and on the effects of nutrition and diet. Good resources include your physician, the American Cancer Society, the National Cancer Institute, or any good library.

Recap **Most cancers could be prevented; smoking is by far the leading preventable risk factor. Other prevention strategies include knowing your family history, getting regular medical screenings, learning self-examination techniques, avoiding sunlight and sunlamps, watching your diet and weight, drinking alcohol in moderation if at all, and staying informed.**

Chapter Summary

Tumors can be benign or cancerous

- Normal cells have two key characteristics: (1) their rates of division are kept under control, and (2) most remain in one location in the body.
- A mass of cells that is dividing more rapidly than normal is called a tumor. Some tumors are benign.

Cancerous cells lose control over cell functions

- Cancer develops when cells divide uncontrollably, undergo physical changes, and no longer adhere to each other.
- Eventually some cancer cells invade nearby tissues. Others metastasize, colonizing distant sites.
- Cancer is the second leading cause of death in the U.S.

How cancer develops

- Inheritance of one or more mutated regulatory genes from your parents may increase your risk of developing certain cancers.
- If proto-oncogenes and tumor suppressor genes mutate or become damaged by carcinogens, they can become oncogenes.
- Known carcinogens include some viruses and bacteria, environmental chemicals, tobacco, radiation, and dietary factors including alcohol consumption. Internal factors, such as faulty DNA replication and internally produced chemicals, can also contribute to cancer.
- Cancer is a multifactorial disease. Perhaps six or more oncogenes may need to be present before a cell becomes cancerous.

Advances in diagnosis enable early detection

- Most cancers are diagnosed when a tumor is detected by imaging techniques such as x-rays, PET, or MRI.
- Enzyme tests may also identify cancer.
- Genetic testing can determine the presence of a specific known oncogene, but the presence of the oncogene does not necessarily mean that cancer will develop.

Cancer treatments

- Many cancers are treatable; early diagnosis is important.
- Conventional treatments for cancer include surgical removal of the tumor, radiation, and chemotherapy.

- Newer treatments include immunotherapy to activate the patient's immune system, drugs to inhibit angiogenesis and cut off the tumor's blood supply, and molecular treatments that target specific genes.

The ten most common cancers

- In both sexes, the most frequently occurring cancers are cancers of the skin, lung, and colon and rectum.
- In addition, prostate cancer is common in men; breast cancer is common in women.
- Lung cancer causes more deaths than any other type of cancer.

Most cancers can be prevented

- Some cancers will occur despite our best efforts because we cannot control inherited risk factors.
- However, most cancers could be prevented. The single most effective way to reduce cancer deaths is to reduce the rate of smoking.
- Another important step to reduce your cancer risk is to eat a healthy diet. Eat lots of fruit, vegetables, legumes, and grains; reduce your intake of saturated fat, red meat, and salt.
- Other preventive strategies include knowing your family's health history; examining yourself for cancer and getting regular medical screenings; maintaining a healthy weight; avoiding direct sunlight and sunlamps; drinking alcohol in moderation if at all; and staying informed.

Terms You Should Know

cancer, 450	malignant tumor, 451
carcinogen, 452	melanoma, 453
carcinogenesis, 452	metastasis, 451
chemotherapy, 457	oncogene, 452
dysplasia, 450	proto-oncogene, 452
hyperplasia, 450	tumor, 450
immunotherapy, 457	tumor suppressor gene, 452
leukemia, 461	
lymphoma, 461	

Test Yourself

Matching Exercise

_____ 1. A change for the worse in cell structure
_____ 2. This organism is associated with stomach cancer
_____ 3. An increase in cell number as a result of increased cell division
_____ 4. Regulatory genes that repress cell growth and division
_____ 5. A tumor imaging technique
_____ 6. A carcinogen in some paints and dyes
_____ 7. A tumor that invades normal tissues and spreads to distant organs
_____ 8. A sexually transmitted carcinogen
_____ 9. Mutated or damaged genes that contribute to cancer
_____10. The administration of cytotoxic chemicals to kill cancer cells

a. malignant
b. dysplasia
c. oncogenes
d. hyperplasia
e. x-ray
f. chemotherapy
g. benzene
h. hepatitis B virus
i. *helicobacter pylori*
j. tumor suppressor genes

Fill in the Blank

11. _____________ is the least common form of skin cancer but also the most likely to be fatal.
12. The three standard methods of treating cancer are _____________, _____________, and _____________.
13. The leading cause of cancer deaths is cancer of the _____________.
14. A cancer-producing substance is called a _____________.
15. In males, cancer of the _____________ is very common but grows so slowly that it may not even need to be treated in older patients.
16. Cancer of the bone-forming cells is called _____________.
17. A discrete mass of cells is called a(n) _____________.
18. _____________ are regulatory genes that promote cell growth, differentiation, division, or adhesion.
19. _____________ refers to treatments that improve the general responsiveness of the immune system so it can fight cancer more effectively.
20. The biggest risk factor for lung cancer is _____________.

True or False?

21. Smoking is a risk factor for cancer of the urinary bladder.
22. Radon gas in the environment is the primary cause of skin cancer.
23. Over 20% of all cancers are caused by industrial environmental pollutants.

24. The *Helicobacter pylori* bacterium may contribute to cancer of the stomach.
25. Chemotherapeutic drugs generally kill some normal cells.

Concept Review

1. What are the critical differences between a benign tumor and a malignant tumor?
2. Aside from uncontrolled cell division, what other key change must occur in cancer cells in order for them to metastasize?
3. What does a carcinogen do to a cell?
4. Which causes more cancers—tobacco smoke or environmental pollutants?
5. What causes most skin cancers?
6. Which two dietary factors are associated with an increased risk of cancers of the colon and rectum?
7. What role does the immune system play in permitting or preventing cancers?
8. What is the primary way that a tumor is diagnosed?
9. How could antibodies to cancer cells be used to improve the delivery of chemotherapeutic drugs?
10. Why might antiangiogenic drugs be effective against cancer?

Apply What You Know

1. Why do you suppose that the rate of lung cancer increased much more in women from 1973 to 1996 than in men? What information would you like to have that might support or disprove your hypothesis?
2. Because diesel fuel is a known carcinogen, why do you suppose we still allow its use? If there were a referendum on the ballot in your state to ban the use of diesel fuel entirely in your state, what information would you like to have before deciding how to vote?
3. Suppose that a man who uses smokeless tobacco develops cancer of the larynx. Will the fact that he used smokeless tobacco increase his children's inherited risk of cancer of the larynx? In other words, can environmental causes of cancer be inherited? Explain your reasoning.
4. What do you think might be the possible positive or negative consequences of self-examination for skin cancer?
5. If you can't do anything about the genes you inherit, what good does it do to know your family history for particular cancers?

Case in Point

Saving Lives with DNA Testing

No one doubts the Pap test can be a life saver. Since the 1960s, increasing use of the technique has led to a fivefold drop in cancers of the cervix. However, the test isn't perfect. It can be uncomfortable and it must be performed by a health professional, making it difficult for women who lack access to health care. Furthermore, the Pap test averages a worryingly low sensitivity of 67 percent, meaning it misses approximately 33 percent of cervical cancers.

Is there an alternative? Evidence points to the value of a DNA test that screens for infection by the human papillomavirus (HPV). A 1995 study estimated that HPV was present in 93 percent of cervical cancer cases worldwide. New research using a more precise screening technique has raised that estimate to 99.7 percent.

The DNA test involves swabbing cell samples from the vagina that are then examined in a laboratory for genetic material from any of 13 forms of HPV. More than 70 forms of the virus have been identified, but so far only these 13 have been implicated in cancer.

In a recent study, 1,365 South African women collected their own vaginal cell samples, which were sent to the U.S. for both DNA and Pap test analysis. The DNA test detected 66 percent of the women's cervical cancers, almost equivalent to the Pap test's 68 percent detection rate. When physicians collected the cell samples, DNA analysis was actually more accurate than the Pap test. Another study conducted at clinics in Philadelphia found equivalent detection rates in DNA samples collected by doctors and by patients. The findings of these studies suggest that a simple and inexpensive at-home vaginal cell collection system may have potential as a cervical cancer screening tool.

QUESTIONS:

1. Explain how HPV infection might lead to cancer of the cervix.
2. Approximately $6 billion a year is spent in the United States on Pap testing. Discuss the financial and public health implications of an at-home test for HPV.
3. Most women with HPV infections do not ultimately develop cervical cancer. Based on what you read in this and other chapters, describe factors that could increase or decrease an HPV-positive woman's risk of developing cancer.

INVESTIGATE:

Check, E. An At-Home Lifesaver? *Newsweek* (Jan. 17, 2000), p. 62.

Editorial: Cervical Screening: Making It Better. *The Practitioner* (June 1, 1999), p. 441.

Key, S.W., and M. Marble. Study Shows Human Papillomavirus in 99.7% of All Cervical Cancers. *Cancer Weekly Plus* (August 30, 1999).

References and Additional Reading

Boon, T. Teaching the Immune System to Fight Cancer. *Scientific American* 268(3): 782–89, 1993. (Discusses efforts to improve the immune system response to cancer.)

Cancer Facts and Figures—1998. (A yearly fact booklet published by the American Cancer Society.)

Cavanee, W.K., and R.L. White. The Genetic Basis of Cancer. *Scientific American* 272(3): 72–79, 1995. (Discusses the role of multiple oncogenes in causing cancer.)

Radetsky, P. Magic Missiles. *Discover* March 1993. (Describes the use of antibodies to fight cancer.)

What You Need to Know About Cancer. *Scientific American Special Issue* 275(3): September, 1996. (An entire issue devoted to the subject of cancer, serving as an excellent review.)

CHAPTER 19

GENETICS AND INHERITANCE

What determined whether you are male or female?

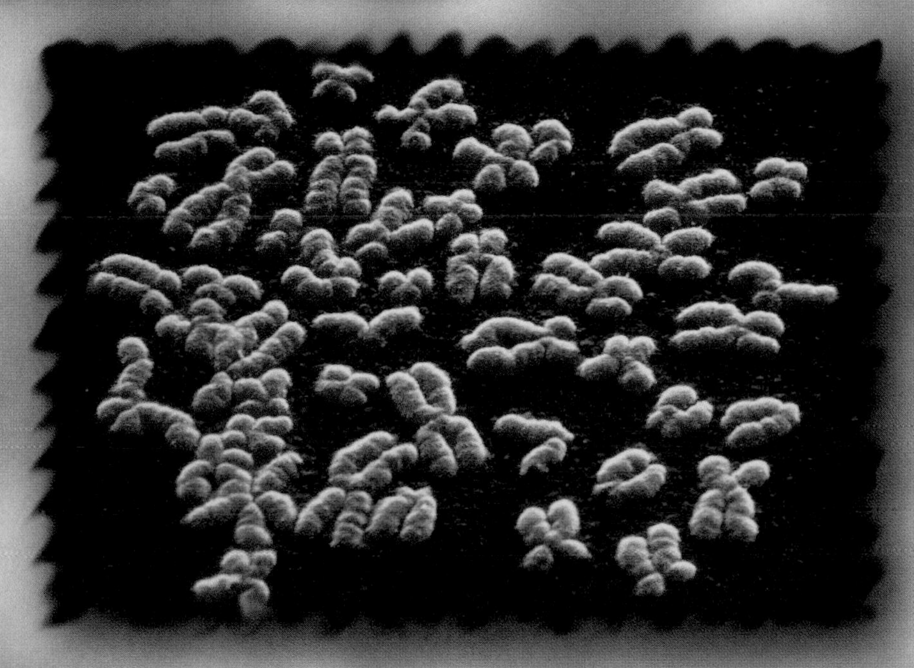

How do genes **affect** behavior?

How can you defend yourself **against** heritable diseases?

Why are men **more likely than women to be** color blind?

AS WE HAVE SEEN, THE INSTRUCTIONS FOR CREATING A NEW human being reside in the DNA within the nucleus of the fertilized egg. Those instructions take the form of 100,000 or so genes that provide the recipes for specific proteins. **Genetics** is the study of genes and their transmission from one generation to the next.

The word **inheritance** often conjures up notions of money or possessions. The broader definition of inheritance, however, is something received from an ancestor or another person. In terms of human biology, everyone gets the same inheritance—one complete set of DNA. So why does one human end up with straight blond hair and blue eyes and another with curly black hair and dark brown eyes?

Although each of us inherits a complete set of DNA, the sets vary ever so slightly. Those variations are enough to account for the differences between us. In this chapter, we explore the patterns of genetics and inheritance that help to make us who we are—influencing our appearance, health, possibly even thoughts and actions. At the end of the chapter, we look at exciting new developments in genetic engineering and their potential to transform our lives.

19.1 Alleles are different forms of homologous genes

Recall from Chapter 17 that DNA is organized into genes, most of which code for proteins, and that the genes are located on 23 pairs of chromosomes (actually 22 pairs of chromosomes called **autosomes** that we all have regardless of gender plus the two nonidentical **sex chromosomes**). We inherit one of each pair of chromosomes from each parent. The members of each autosome pair look alike and they have homologous genes at the same locus, or location. In effect, we get two copies of each gene.

Sometimes homologous genes are not exactly identical, though they are usually close. **Alleles** are different structural and functional variations of homologous genes. Alleles may have one or more bases that are different from each other, and as a result they produce slightly different proteins. In turn, the alteration in protein structure can affect how the protein functions. For example, one allele of a gene for hair pattern produces what is known as a widow's peak whereas the other allele produces a straight hairline.

If both alleles at a particular gene are identical the person is said to be **homozygous** for that gene (*homo-* means "same," and *zygous* means "yoked," or joined together). If the alleles are different the person is **heterozygous.** In some cases there are more than two alleles for a particular gene, although a given individual can only inherit two of them. An example of more than two possible alleles are the alleles for blood type, A,B, and O.

So how did we get different alleles? Alleles probably resulted from millions of years of **mutations,** or changes to the DNA base pair sequence, of cells destined to become sperm or egg. If those mutations were not corrected before cell division and meiosis, and if they did not cause the sperm or egg to be nonviable, they would be passed on as slight variations of the original gene. Because we have no way of knowing which of a pair of alleles was the original form, they are both called alleles of the same gene locus.

A *genome* refers to the complete set of genes in the chromosomes of a particular organism. All the various human genes and their alleles are known collectively as the *human gene pool.* There are enough different alleles among our 100,000 genes to account for the uniqueness of individuals.

The complete set of genes and alleles that you inherited is called your **genotype.** On a practical level it is still not possible to determine your exact genotype because it would require a complete analysis of the structure of your DNA. To analyze the genetic basis of inheritance we must rely on the study of **phenotype,** the observed physical and functional traits that characterize us (Figure 19.1). Phenotype includes such traits as hair and eye color, height and body shape, skin color, whether you can curl your tongue or not, and whether your hair is straight or curly. Your phenotype is determined in part by the genes you inherit from your parents (your genotype), but environmental factors such as how much

Figure 19.1 Human phenotypes. Our phenotype is our unique observable physical and functional traits.

you are exposed to the sun, whether or not you exercise, and how much and what you eat can also influence it.

> ***Recap*** **We inherit either two identical genes or two slightly different alleles of every gene. The sum of our genes is our genotype and the physical and functional expression of those genes is our phenotype.**

19.2 Genetic inheritance follows certain patterns

Punnett square analysis predicts patterns of inheritance

The shorthand way to designate two alleles of the same gene is to assign them upper case (such as A) and lower case (a) letters. Recall that people with the same two alleles of a possible pair (either aa or AA) are homozygous for that allele and people with different alleles (aA or Aa) are heterozygous. A **Punnett square** provides a simple way to view patterns of inheritance for a single pair of alleles and the probability that a particular genotype will be inherited (Figure 19.2).

To create a Punnett square, place the possible alleles of the male gametes (produced during meiosis, Chapter 17) on one axis and the possible alleles of the female gametes on the other. By combining the letters for the sperm and egg in the appropriate squares (the possible combinations that might occur during fertilization), we can see the genotypes of the offspring and the ratios of offspring of each possible type.

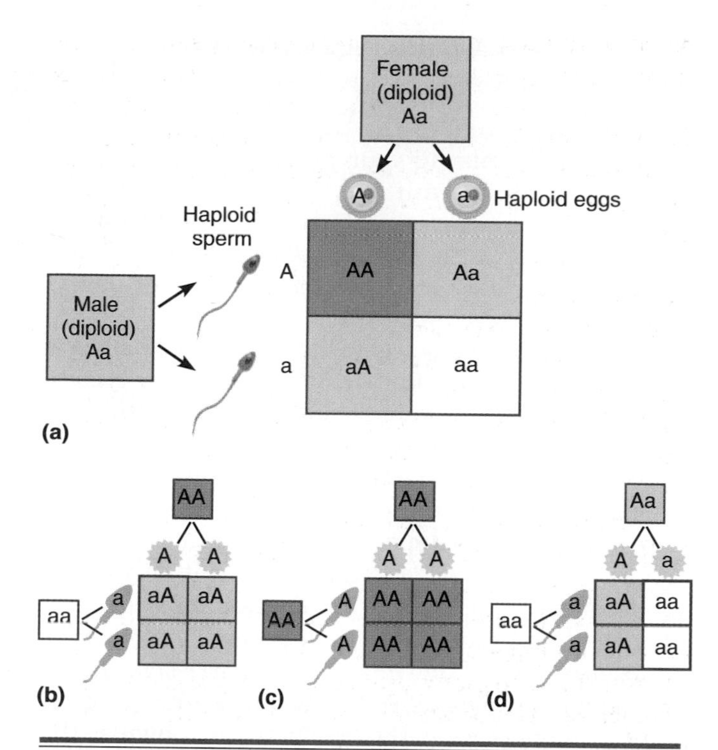

Figure 19.2 Punnett squares. Punnett squares are used to predict the outcomes of a particular combination of parental alleles. **(a)** To construct a Punnett square, the possible combinations of male and female gametes are placed on the two axes, and then the possible combinations of the offspring are plotted in the enclosed squares. This plot shows that in a cross between two heterozygotes only half the offspring will be heterozygotes. **(b, c)** A cross between two homozygotes produces offspring that are all the same genotype as each other, but not necessarily the same genotype as their parents. **(d)** A cross between a homozygote and a heterozygote produces an equal number of offspring of each parent's genotype.

Mendel developed the basic rules of inheritance

The basic rules of inheritance were first worked out in the 1850s by Gregor Mendel, a university-educated Austrian monk who specialized in natural history. Based on his experiments with the inheritance patterns of garden peas, Mendel proposed that there were discrete "factors" of heredity that united during fertilization and then separated again in the formation of sperm and eggs.

It is remarkable how right Mendel was, considering that he knew nothing about DNA, chromosomes, or meiosis. Specifically, he was convinced that organisms inherit two units of each factor, one unit from each parent. Today we know that the factors are genes and that most complex organisms are indeed diploid, that is, they can have two different alleles or two copies of a particular gene.

Based on his experiments, Mendel proposed the *law of segregation:* when the gametes are formed in the parents, these factors (genes) separate from each other so that each sperm or egg gets only one unit. He was right. Today we understand that the sperm and egg are haploid, with only half the number of chromosomes and genes of the parent (review the process of meiosis, Chapter 17).

He also proposed what came to be known as Mendel's *law of independent assortment:* genes for different traits are separated from each other independently during the formation of sperm and egg. On this second point he was only partly right. We now know that independent assortment of genes only fully applies if the genes are located on different chromosomes, as we'll see when we discuss linkage groups.

> ***Recap*** **A Punnett square is a useful device for predicting the ratios of possible offspring genotypes of a particular combination of alleles. Mendel's first rule of inheritance, the law of segregation, still holds. His second rule, the law of independent assortment, holds only for genes located on different chromosomes.**

Dominant alleles are expressed over recessive alleles

In most cases, one of the two alleles dominates in terms of determining your phenotype. In this case one allele is **dominant** and the other is **recessive.** This is the inheritance pattern that Mendel studied with his pea plants. It is customary to denote the dominant allele by an upper case letter and the recessive allele by a lower case letter.

When the presence of just one dominant allele is all that is necessary for the dominant phenotype to be expressed, the allele is said to exert **complete dominance.** The dominant allele codes for a functional protein. If even one dominant allele is present, enough of the protein can be produced to produce the observed phenotype. The recessive allele is usually the mutated form, and as a result it either fails to produce the protein or produces a modified protein that is nonfunctional. The recessive phenotype is observed only if one inherits both recessive alleles.

A good example of complete dominant versus recessive phenotypes is the hair pattern known as the widow's peak (Figure 19.3). The widow's peak results from a dominant allele (W). Anyone who is WW or Ww will have a widow's peak, whereas persons who are homozygous for the recessive allele (ww) will have a straight hairline. Note that only half of the offspring of two heterozygotes, on average, will have the same *genotype* as the parents (Ww), but three quarters of them will have the parental *phenotype.*

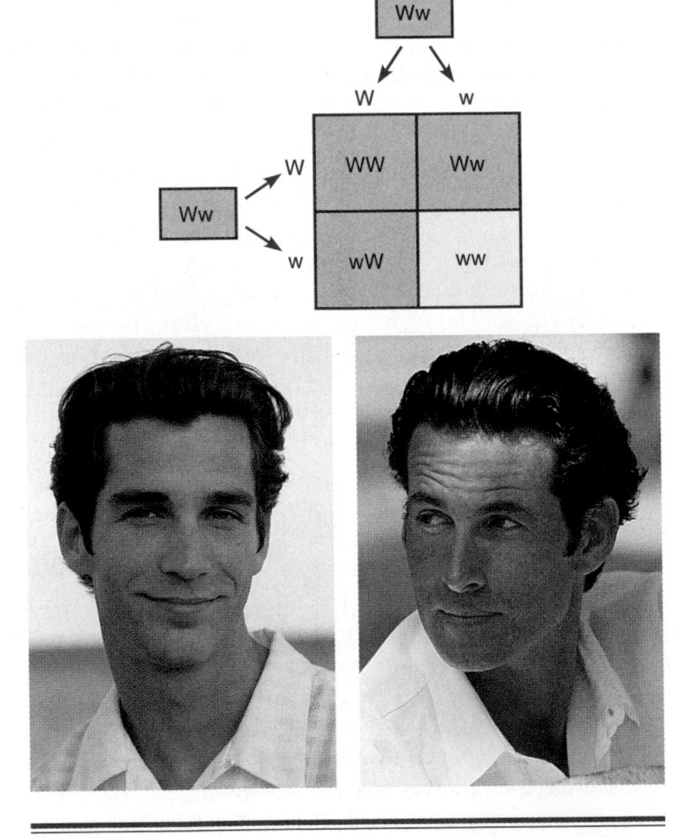

Figure 19.3 Inheritance of dominant/recessive alleles. The allele for the widow's peak (W) is dominant. Only the homozygous recessive genotype (ww) will have a continuous hairline.

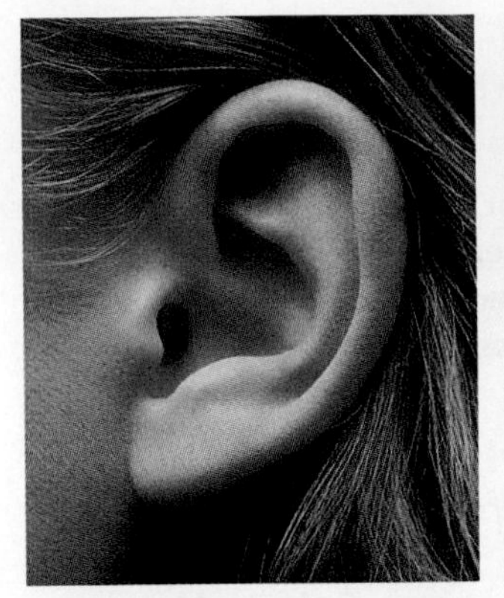
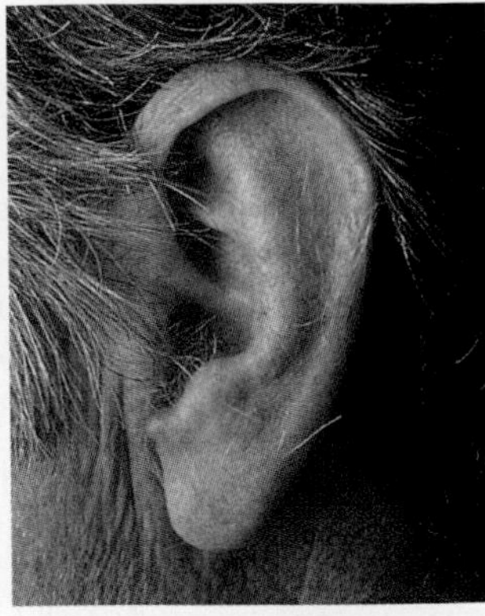

(a) Attached versus unattached earlobes.

(b) Tongue-curling ability.

Figure 19.4 Harmless dominant/recessive traits and alleles. (a) The allele for unattached earlobes is dominant over that for attached earlobes. **(b)** The daughter in this family photo inherited the dominant allele for tongue-curling from her father. Her mother has two of the recessive alleles, and so cannot curl her tongue.

Most recessive alleles, such as this one, are fairly inconsequential, that is, they do not confer a distinct disadvantage to the homozygous recessive person. Many recessive alleles—such as the one that results in a straight hairline, another that causes the earlobes to be attached (Figure 19.4), or another that prevents you from being able to curl your tongue—just happened at some point in our evolutionary history. They remain in the human gene pool because they do no harm.

At the other end of the spectrum, some recessive alleles result in the absence of a functionally important protein. An example is cystic fibrosis, caused by inheriting two recessive alleles. These alleles no

Try It Yourself

Calculating Probability with a Punnett Square

Can you curl your tongue into a U-shape? Can your parents do it? The ability to curl (or roll) the tongue is a dominant allele, so we can calculate the probability that a child will have this phenotype if we know the parents' genotype. Draw a Punnett square to illustrate the combinations of genes that would result if both parents are heterozygous (Tt) for this trait:

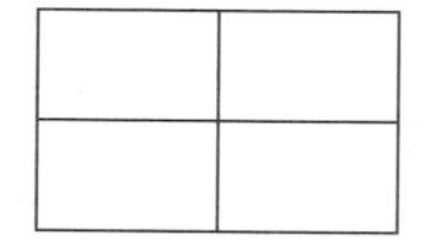

What percentage of these parents' offspring would be able to curl their tongues, and what percentage would be unable to? Able: ____ Unable: ____.

In another family, the father is heterozygous for this trait but the mother cannot curl her tongue. Draw a Punnett square to show the probable genotypes of their children:

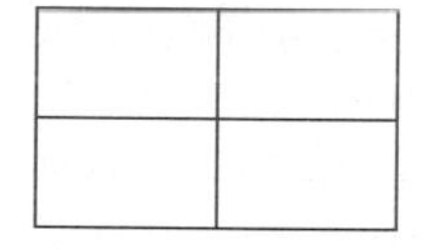

Do you know your parents' tongue-curling ability? If so, draw Punnett squares to illustrate the possibilities for their children's (you and your siblings') genotypes:

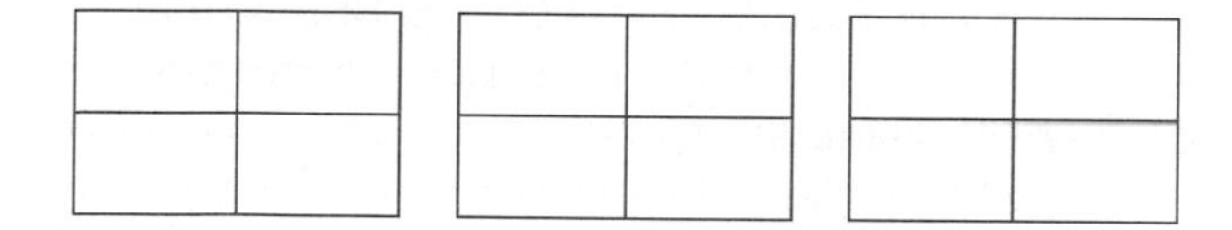

As far as we know, tongue-curling ability confers no evolutionary advantage at all. Its recessive allele remains in the human gene pool because it does not affect our chances of survival.

---

longer code for a specific essential protein because their nucleotide sequence has been altered slightly from the dominant form. In principle, the frequency of truly harmful recessive alleles in a population is kept in check by the occasional premature death of homozygous recessives, but the allele still can persist in the population in heterozygotes.

Recap **Where two alleles of a particular gene exist in the human gene pool, one is usually dominant over the other in terms of phenotypic expression in heterozygotes. Harmless alleles are more likely to remain in the population than harmful ones.**

Incomplete dominance: The heterozygote is an intermediate phenotype

Some alleles do not follow the complete dominance pattern described above. In **incomplete dominance,** the heterozygous genotype results in a phenotype that is intermediate between the two homozygous conditions. The color of the palomino horse, in which neither of the alleles (one for nearly white color and one for chestnut) is dominant, is an example of incomplete dominance (Figure 19.5). An example of incomplete dominance in Caucasian humans is the trait of having straight, wavy, or curly hair.

Recap **In incomplete dominance, the phenotype of the heterozygous genotype is intermediate between the phenotype of either homozygous genotype.**

Codominance: Both phenotypes are equally expressed

In **codominance** the heterozygote exhibits the phenotype of both alleles equally, rather than the intermediate phenotype of incomplete dominance. A good example of codominance is the relationship between the A and B alleles for blood type. Recall that there are three alleles for blood type (A, B, and O) but that a given individual can only inherit two of them. As a result there are six possible genotypes (Table 19.1). However, because alleles A and B are codominant and allele O is recessive to both, there are four possible phenotypes (in this case blood types): types A, B, AB, and O. Type AB represents the codominant heterozygous condition. Blood phenotype has important implications for which types of blood can be given to which patients, as we saw in Chapter 7.

Spotlight on Structure & Function

Sickle-cell anemia is a disease caused by one of two codominant alleles which are involved in the production of hemoglobin for red blood cells. One of the two alleles codes for an altered hemoglobin molecule. Sickle-cell anemia results from the homozygous sickle-cell genotype, designated Hb^SHb^S. Unaffected individuals are designated Hb^AHb^A and the heterozygous condition, called *sickle-cell trait,* is designated Hb^AHb^S. Because the two alleles are codominant, half of the hemoglobin in persons with sickle-cell trait is normal and half is altered.

Recall from Chapter 7 that hemoglobin transports oxygen in the blood. The problem for people with sickle-cell anemia and to a lesser extent those with sickle-cell trait is that exposure of their affected hemoglobin to low oxygen concentration causes the red blood cells to take on a crescent, or sickle, shape

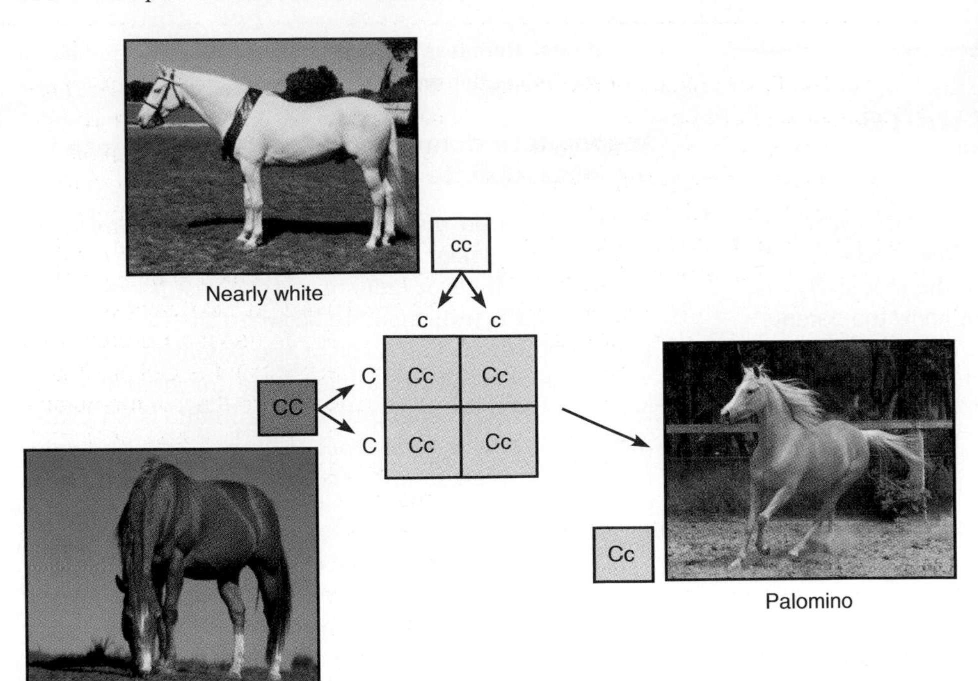

Figure 19.5 Incomplete dominance. The palomino horse represents an intermediate color between the light chestnut and nearly white phenotypes because neither allele is dominant.

(Figure 19.6). As a result the red cells clump together, clogging small blood vessels and blocking blood flow. Eventually lack of blood flow damages tissues and organs.

Many people with sickle-cell anemia do not live into their thirties. People with sickle-cell trait can generally lead normal lives because their condition is intermediate in terms of severity, but they too have occasional problems. Sickle-cell anemia affects primarily Africans (and African-Americans) and Caucasians of Mediterranean descent. In the United States, approximately 1 of every 500 African-Americans is born with the disease.

Why would a deadly condition such as sickle-cell anemia and the less deadly sickle-cell trait be retained in the human population? It turns out that the sickle-cell allele protects against malaria, a deadly parasitic disease common in tropical climates. Malaria kills a million people a year worldwide. Somehow the presence of affected hemoglobin prevents the malaria parasite from entering red blood cells. This means that in malaria-infested areas of the world, heterozygous individuals with the sickle-cell trait (Hb^AHb^S) are actually more likely to survive than are normal individuals. Were it not for malaria, we could surmise that sickle-cell disease and sickle-cell trait would become rare. In portions of East Africa where malaria is still common, nearly 40% of the black population are heterozygous carriers of sickle-cell trait. In the United States, malaria is rare and only about 8% (1 in 12) of African-Americans are carriers. ■

Table 19.1 Blood genotypes and blood types (phenotypes) in humans

Blood Genotype	Blood Type (Phenotype)
AA	A
AO	A
BB	B
BO	B
AB	AB
OO	O

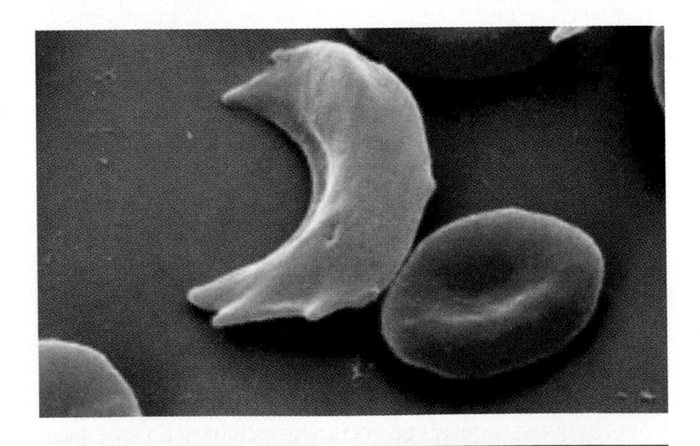

Figure 19.6 Codominance. These red blood cells are from a person with sickle-cell trait. Half of this person's red blood cells are normal, but the other half have a tendency to take on a crescent, sickle shape when exposed to a low oxygen concentration.

Health Watch

Cystic Fibrosis

Cystic fibrosis, the most fatal genetic disease in North America today, affects approximately 30,000 people in the U.S. and another 3,000 in Canada. Approximately one in 29 persons carries an altered gene and about one in 3,300 live births has the disease. The mean age of death for people with cystic fibrosis is currently 31 years, but fortunately that number is rising as we understand the disease better.

A breakthrough came in 1989 when Canadian scientists discovered the cystic fibrosis gene. We now know the disease is caused by an altered allele of a single gene located on chromosome 7. It is a recessive allele, meaning that only people who are homozygous recessive (that is, who inherit altered alleles from both parents) will have the disease.

The allele produces a defective version of a protein that normally aids the transport of chloride across the cell membrane of epithelial cells. Impaired chloride transport affects the secretion of salt and water across the cells that line the ducts in several organs. As a result, organs that normally produce a thin, watery mucus secrete a thick, sticky mucus instead.

The organs most affected are the lungs and the pancreas. In the lungs, the thick mucus clogs the bronchioles and leads to infections that can be fatal. In the pancreas, the mucus blocks ducts and prevents pancreatic enzymes from reaching the intestines. The sweat glands are also affected, but ordinarily this does not cause major physical problems.

Symptoms of cystic fibrosis include persistent coughing, wheezing or pneumonia, excessive appetite but poor weight gain, and foul-smelling, bulky feces. The disease is usually diagnosed by age three. The standard diagnostic test measures the amount of salt in perspiration. A high salt level indicates cystic fibrosis.

Management of cystic fibrosis is aimed at reducing the damage caused by the symptoms, and often it becomes a constant battle throughout life. Vigorous chest physical therapy is usually required on a regular basis to dislodge mucus in the lungs and antibiotics may be needed to treat lung infections. When cystic fibrosis affects digestive processes patients often do not absorb enough nutrients. Careful dietary management coupled with vitamins and enzymes can help.

Several drugs can benefit the condition. In 1993 the Food and Drug Administration (FDA) approved the first new cystic fibrosis drug in 30 years. Called Pulmozyme, the drug thins pulmonary mucus and reduces the number of respiratory infections. In 1997 the FDA approved TOBI (tobramycin), an inhalant antibiotic that delivers a concentrated dose directly to the lungs.

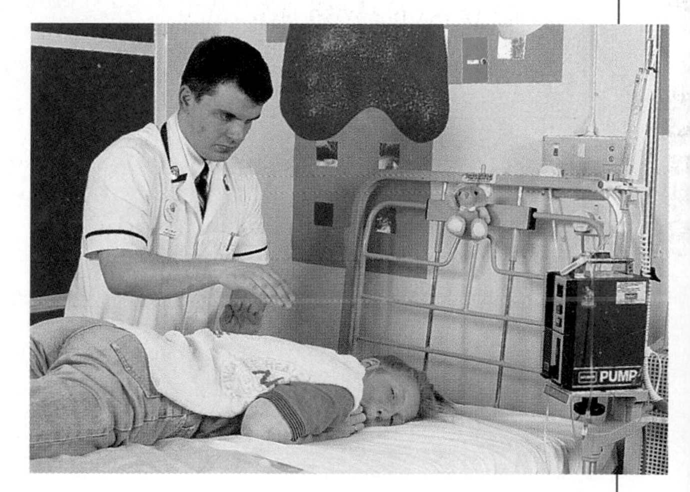

Cystic fibrosis. This teen with cystic fibrosis is receiving vigorous percussion therapy ("clap therapy") to loosen the mucus in the lungs.

With these two advances, cystic fibrosis patients are living longer and spending less time in the hospital.

On the horizon are attempts to treat the causes of cystic fibrosis rather than just the symptoms, and even to cure the disease altogether. We now know that the defective sodium chloride transport protein gets trapped deep within the cell, so it is never inserted into the cell membrane properly. In the early stages of clinical trials are drugs that help the transport protein move to the cell membrane. Other researchers are working on gene therapy techniques that insert copies of the normal gene into viruses, such as those that cause colds, and then deliver the modified viruses into the airway. This is a field that is still in its infancy. However, do not be surprised if cystic fibrosis is cured in our lifetime, or at least treated so effectively that cystic fibrosis patients can live long and healthy lives.

Recap **A-B blood type is an example of codominance. There are three alleles (A, B, and O) for blood type; A and B are codominant and O is recessive. The alleles for sickle cell trait and for normal hemoglobin are another example of codominance. In individuals with just one sickle-cell allele (Hb^AHb^S), half of their hemoglobin is normal and half is abnormal.**

Some phenotypes are influenced by many genes

Many traits result from not just one pair of homologous genes but from many genes acting simultaneously. Inheritance of phenotypic traits that depend on many genes is called **polygenic inheritance.** Eye color, for example, is controlled by three or more genes, yielding a range of different eye color phenotypes from nearly black to light blue (Figure 19.7). Both skin and eye color are determined by the amount

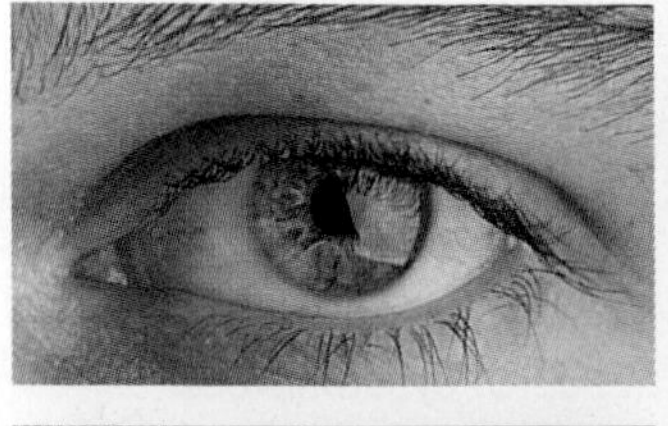
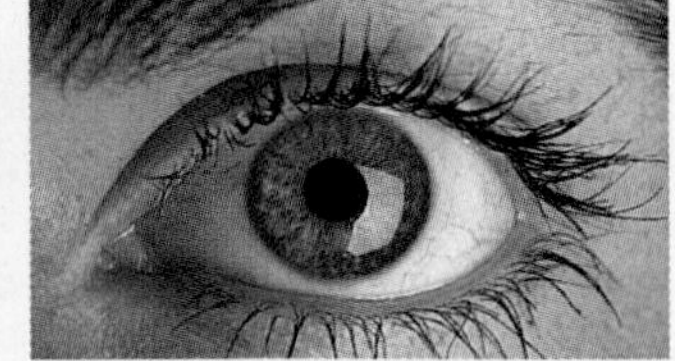
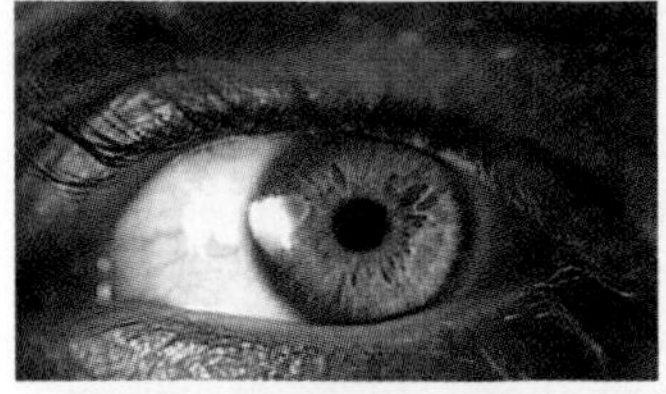
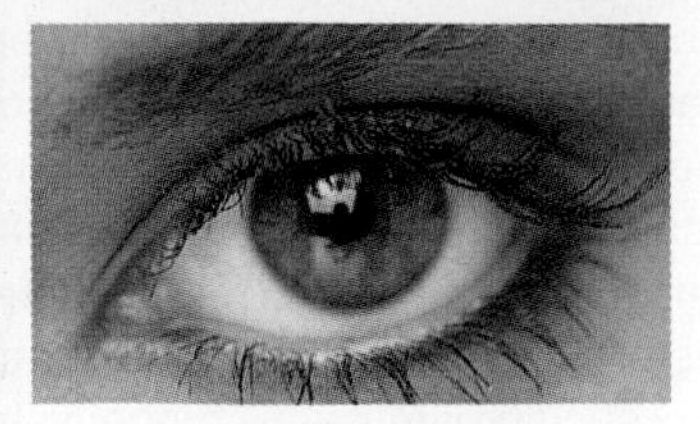
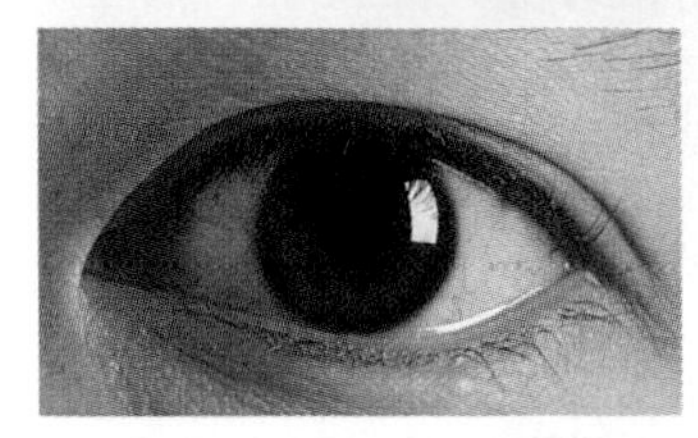

Figure 19.7 Polygenic inheritance. Eye color is determined by three or more genes, not just by one. Different combinations of the alleles of the genes account for a range of eye colors.

of a light-absorbing pigment called melanin in the skin and iris. The striking differences in eye color occur because at the lower range of amounts of melanin we see different wavelengths of light (green, gray, or blue) being reflected from the iris.

Physical characteristics such as height, body size, and shape are also examples of polygenic inheritance. Traits governed by polygenic inheritance generally are distributed in the population as a continuous range of values, with more people at the middle (or median) and fewer people at the extremes. A plot of the numbers of individuals at each value for the trait typically forms a normal (bell-shaped) curve (Figure 19.8).

We are just beginning to understand that many health conditions may be caused or at least influenced by polygenic inheritance. Examples include cancer, high blood pressure, heart disease, and stroke.

Both genotype and the environment affect phenotype

Phenotype is determined in part by our genotype and in part by environmental influences (Figure 19.9). A prime example of an environmental influence is the effect of diet on height and body size. Analysis of human populations reveals a clear trend toward increased height and weight in certain human populations, particularly in developed countries. The changes in height and weight have sometimes occurred within one generation, much too short a time to be due to gene mutations. The trend is primarily due to improvements in diet and nutrition, especially early in childhood.

Except for a few rare genetic diseases such as Huntington disease, our genotype is not the sole determinant of whether or not we will actually develop a heritable disease (that is, have the disease phenotype). Usually we inherit only a slightly increased *risk* of developing the disease, with those risks modifiable by environmental factors and our own actions. If you know that certain heritable conditions or diseases run in your family, it may be beneficial to work toward reducing any environmental risk factors that may be associated with that particular disease. For example, if your family has a history of skin cancer it might be prudent to limit your exposure to the sun; if your family has a history of heart disease due to high cholesterol, have your cholesterol measured on a regular basis and change your diet or begin other corrective measures if your cholesterol is above the normal range.

How can you defend yourself against heritable diseases?

This is really the old nature-versus-nurture question: Are we the way we are because of our genes or because of our environment? The answer is both. Our genes carry the instructions for all our proteins, but the environment can influence how genes are expressed and how those expressions interact. There is no one answer to the nature-versus-nurture dichotomy. But as you cannot change your genes, your best defense against any possible future disease is to take proper care of yourself.

Recap **In polygenic inheritance, many different genes contribute to phenotype. In addition, some phenotypes can be influenced by environmental factors such as diet.**

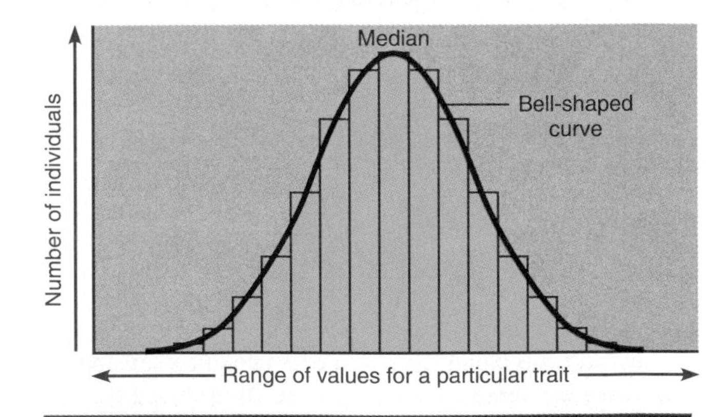

Figure 19.8 Continuous variation in phenotypic traits. Many phenotypic traits appear in the population as a continuous range of values, with more people in the middle than at either extreme. This graph would be typical of a polygenic trait such as body height or the distribution of blood pressures. Measured variables are usually reported as falling within discrete ranges (heights in 2-inch groups, for example), but the actual population is characterized by the continuous bell-shaped curve.

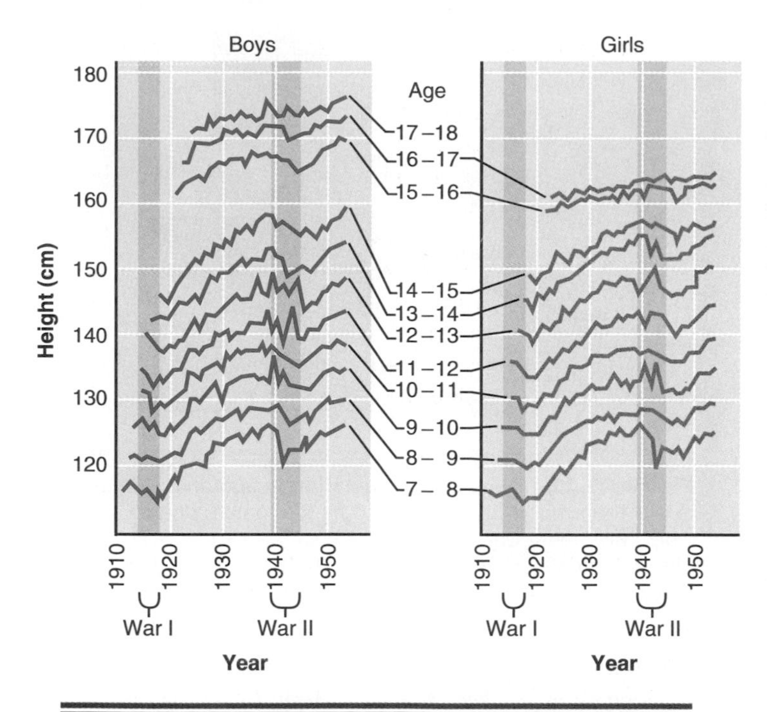

Figure 19.9 Environmental factors may influence phenotype. Sudden changes in environmental factors can cause changes in phenotype over a very short period of time. The average height of schoolchildren in Stuttgart, Germany, aged 7–18 increased slowly between World Wars I and II. During the second world war, however, nutrition was often inadequate and average height decreased significantly.

Linked genes may be inherited together

Recall that we said Mendel's law of independent assortment was not entirely correct. The reason is that the 100,000 genes in the human genome do not float independently in the cell's nucleus. Every homologous pair of genes is part of a homologous pair of chromosomes, and there are only 23 pairs of chromosomes. Independent assortment holds only for genes that are on different chromosomes.

Genes on the same chromosome are called **linked genes** because they are physically linked together. The closer they are located to each other on the chromosome the more likely it is that they will be inherited as a *linkage group*. However, genes on a chromosome are not always inherited together because of crossing-over during meiosis, which reshuffles the genes across homologous chromosomes (review meiosis in Chapter 17). How often two genes on the same chromosome are inherited as a linkage group is directly proportional to how close together they are to each other (the closer they are, the more likely the genes will be linked) and inversely proportional to how often crossing-over occurs (the more that crossing-over occurs, the less likely the genes will be linked).

To get a feel for how often genes on a chromosome might be inherited together, co that the average pair of homologous chromoso may have over 4,000 genes and that crossing-over occurs approximately 30 times per pair of chromosomes. Genes located close to each other, then, may be inherited together (linked) quite often. Scientists have learned that they can map the positions of genes by studying how often particular genes are inherited together, a technique called *linkage mapping*.

Recap **When genes are located close together on the same chromosome, they may be inherited together as a linkage group.**

19.3 Sex-linked inheritance: X and Y chromosomes carry different genes

Chromosomes are only identifiable in cells just prior to cell division. At this time each chromosome can be identified by its characteristic size, centromere location, and distinct banding pattern. A composite display of all of the chromosomes of an organism is called a **karyotype.**

The karyotype shown in Figure 19.10 shows the 23 pairs of human chromosomes. Of these, 22 are matched pairs (autosomes). The last pair of chromosomes, however, do not look alike and do not match up. These are the sex chromosomes, X and Y. The sex chromosomes are not a homologous pair. They look different from each other, and they function differently because they carry different genes. They represent a pair only in that you get one from each parent as a consequence of meiosis. Next we discuss how the differences in sex chromosomes may have originated, and then look at how these differences affect inheritance.

The sex chromosomes may have originated by chance mutations

According to one theory, about 250 million years ago—before mammals branched off from the reptiles—our ancestors had only homologous pairs of chromosomes. One of these pairs, according to the theory, carried a pair of homologous genes that determined sex. The sex genes, like so many other genes that cause cells and tissues to differentiate, may have been switched on and off by environmental cues such as temperature or humidity. In fact, this type of gender determination still occurs in some reptiles today. In contrast, a human embryo's gender is determined solely by which sex chromosome (X or Y) is carried by the male's sperm (Figure 19.11).

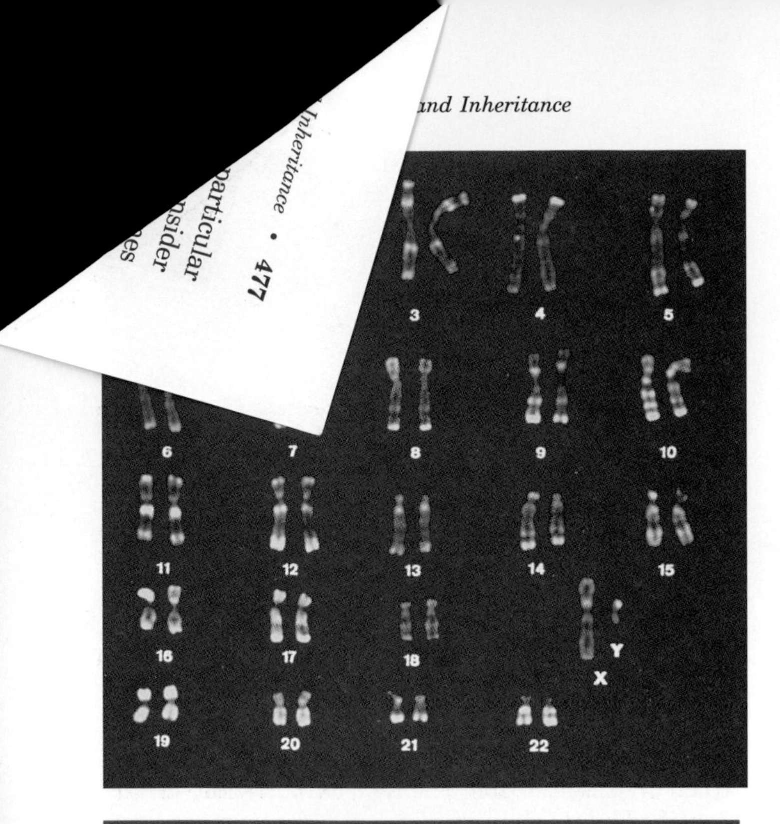

Figure 19.10 The human karyotype. Chromosomes are identified and paired according to their size, centromere location, and characteristic banding patterns. A numbered arrangement of all 23 pairs of human chromosomes is called the human karyotype. Note that the sex chromosomes (X and Y) are quite different from each other. Females are XX and males are XY.

So how did the two sex chromosomes become different? Current thinking is that chance mutations caused the two homologous sex genes to become stuck, one in the "on" position and the other in the "off" position. From this point on, gender would be determined by which of the two chromosomes one inherited and not by environmental cues. Presumably the chromosome with the "on" gene became the Y chromosome (for determining male gender), whereas the chromosome with the "off" gene became the X chromosome.

We now know *how* the sex chromosomes determine gender. Apparently a single gene located on the Y chromosome is activated for only a few days at about six or seven weeks of embryonic development. Its activation leads to the development of the testes and the production of the male hormone testosterone. This is a good example of how a single gene can play a critical role in differentiation, in this case the differentiation of the embryo into a male. In the absence of this male-producing gene on the Y chromosome, other genes on both the X chromosome and the autosomes cause the embryo to develop into a female.

What determined whether you are male or female?

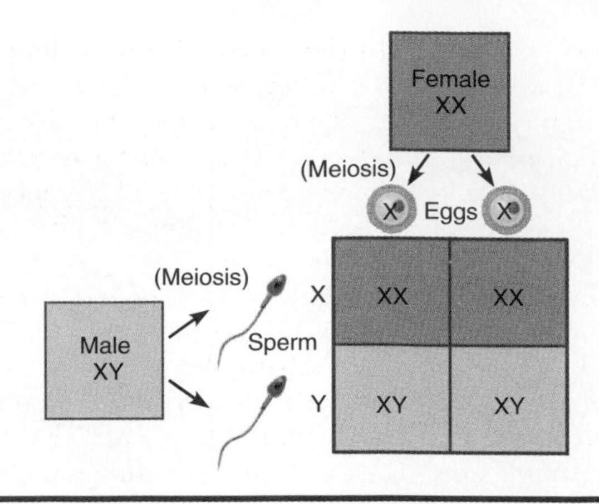

Figure 19.11 Gender determination in humans. The gender of the embryo is determined by the male's sperm, which are about equally distributed with an X or a Y chromosome. Maleness is determined by the inheritance of a Y chromosome.

***Recap* The sex chromosomes evolved during the course of our evolutionary history. Today they are not homologous in structure or function.**

Sex-linked inheritance depends on genes located on sex chromosomes

The hard-wiring of gender determination into a specific chromosome was not without cost, particularly to the male. The advantage of having two homologous chromosomes is that each has a second copy—a backup file, if you will. If something goes wrong with one copy, the other can still carry out its function.

Spotlight on Structure & Function

In sexually reproducing animals in which the Y chromosome determines maleness, females have a homologous pair of sex chromosomes but males do not. This means that males have a greater susceptibility for diseases associated with mutations of the sex chromosomes.

Sex-linked inheritance refers to inheritance patterns that depend on genes located on the sex chromosomes. Sex-linked inheritance is **X-linked** if the gene is located only on the X chromosome and *Y-linked* if the gene is located only on the Y chromosome. There are plenty of examples of X-linked inheritance because many of the genes on the X chromosome are not related to gender at all. However there are essentially no well-documented examples of Y-linked inheritance. The reason is that the Y chromosome is relatively small and most of its genes relate to "maleness." Genes on the Y chromosome apparently influence differentiation of the male sex organs, production of sperm, and the development of secondary sex characteristics, but not much else.

X-linked inheritance In the female, X-linked inheritance behaves just like inheritance of homologous chromosomes with paired homologous genes. Not so in the male, because he inherits only one X chromosome. In the male, X-linked genotype and phenotype are both determined solely by the single X chromosome he inherited from his mother.

The best known example of an X-linked disease is hemophilia, also known as bleeder's disease. Hemophiliacs lack a blood-clotting factor that is controlled by an X-linked gene with two alleles. The recessive allele (X^h) is a mutant that cannot produce the clotting factor, whereas the dominant allele (X^H) produces the clotting factor. Individuals have the disease only if they do *not* have at least one normal dominant allele (that is, if they are an X^hX^h female or an X^hY male).

The inheritance of hemophilia and other X-linked recessive diseases follows this pattern:

- Many more males have the disease than females. Females can be protected by inheriting at least one normal allele (X^H) out of two, whereas males inherit only one allele because they inherit only one X chromosome.
- The disease is passed to sons solely through their mothers even though the mothers may only be *carriers.* Carriers are heterozygous (X^HX^h); therefore, they can pass on the recessive hemophilia allele even though they do not suffer from the disease themselves. Statistically, half of the sons of carrier mothers will have the disease and half of the daughters will be carriers.
- A father cannot pass the disease to a son, but his daughters will all be carriers (unless the mother is also a carrier or has the disease, in which case the daughters may also have the disease).

Why are men more likely than women to be color-blind?

Figure 19.12 is a pedigree chart that illustrates a typical inheritance pattern for hemophilia. Red-green color blindness and muscular dystrophy are also X-linked recessive conditions that follow the inheritance pattern described for hemophilia. Nearly all X-linked diseases are caused by a recessive allele rather than a dominant one. ■

Recap **Sex-linked inheritance refers to patterns of inheritance that depend on genes located on the sex chromosomes.**

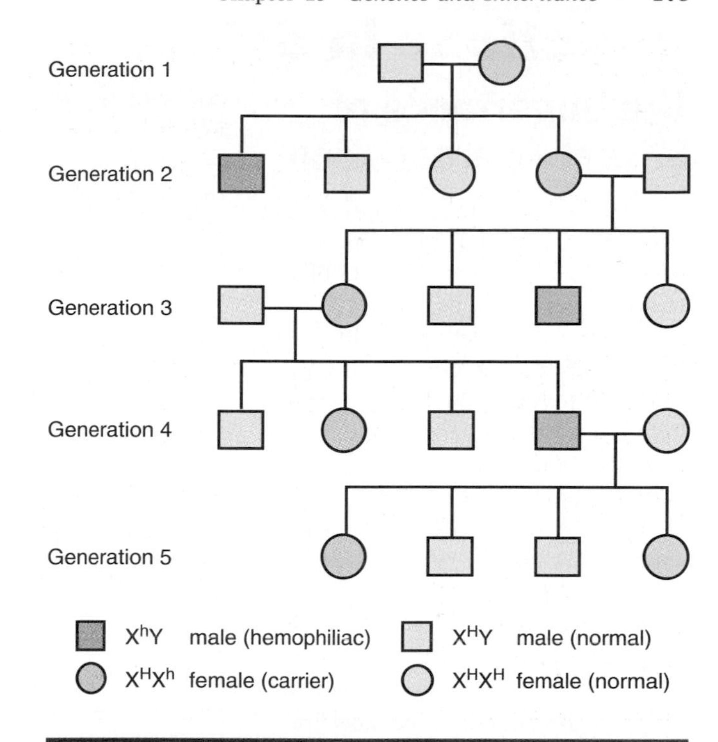

Figure 19.12 A pedigree chart showing a typical inheritance pattern for an X-linked recessive disease. The example depicted here is hemophilia. The normal (X^H) and hemophilia (X^h) alleles are located on the X chromosome. Note that female carriers (genotype X^HX^h) pass the hemophilia allele to half their daughters and the disease to half their sons. Males with the disease pass the hemophilia allele to all of their daughters (if they survive long enough to have children).

Sex-influenced traits are affected by actions of sex genes

A few genes that you might suspect are located on the sex chromosomes are actually located on one of the 22 autosomal pairs of chromosomes. A good example is a gene that influences baldness. The normal allele (b^+) and the allele for baldness (b) can be present in both men and women. However, the baldness allele is recessive in women. Women must be (bb) and probably have other baldness genes or factors as well before their hair loss is significant. Men become very bald if they are (bb) and develop significant pattern baldness even if they are heterozygous (bb^+).

Why the difference between men and women? Testosterone strongly stimulates the expression of the baldness allele, in effect converting it from a recessive to a dominant allele in males. This is an example of a *sex-influenced* phenotypic trait, one not inherited with the sex gene but influenced by the actions of the sex genes. In fact, a number of genes influence the presence and pattern of hair on the head in addition to this particular baldness gene.

Directions in Science

The Importance of Sexual Reproduction

Currently a debate is underway among scientists concerning the adaptive value of sexual reproduction. Until a few years ago many believed that the reshuffling of genes that results from sexual reproduction permits an organism to better adapt to its environment. The purpose of sexual reproduction, they held, was to ensure the long-term evolution of the species in the face of changing environmental conditions. However, we now know that environmental change is a slow process that occurs over thousands and even millions of years, far too slow to require gene reshuffling every generation. In fact, evolution needs to be slow if a species is to remain stable over long periods of time.

In recent years two new theories have emerged: that sex evolved to help complex organisms combat their parasites, and that sex is important for the repair and rejuvenation of DNA. The keeping-up-with-the-parasites hypothesis is called the "Red Queen" theory (after a character in Lewis Carroll's *Through the Looking Glass* who had to run constantly just to stay in the same place). The Red Queen theory holds that long-lived organisms engage in a constant struggle for survival against internal parasites that can evolve much more quickly than they can. Parasitic microorganisms such as bacteria and viruses have generation times (the period of time between parents and offspring) on the order of days or even hours, so their opportunities for recombining the genes in their gene pool to gain an advantage over us are high indeed. Because we recombine genes only about once every 20 years or so, we need to do a good job of reshuffling them maximally when we do.

Evidence to support the Red Queen theory comes from analyzing the frequency of chromosomal crossing-over among different species. Organisms with long generation times typically have more crossing-over than organisms with short generation times. For example, scientists calculate that crossing-over occurs approximately 30 times in each chromosome pair in humans, 10 times in rabbits, and only 3 times in mice. In other words, crossing-over seems to have been encouraged by evolution when reproduction is less frequent. From this, scientists hypothesize that genetic mixing has a short-term benefit. If this is true then sex is very important indeed, for every sexual reproduction cycle immediately doubles the gene pool from which offspring originate. The hypothesized short-term benefit is to combat internal parasites.

There is probably some truth to this, for it's not slow climatic change over millions of years that kills most of us. Before modern medicine, it was syphilis, bubonic plague, polio, meningitis, tuberculosis, even influenza and the common cold. The battle between microorganisms and humans is constant and never won or lost completely by either side. We shuffle our gene pool as quickly as possible just to keep up.

The second theory, also supported by considerable evidence, is that sex ensures the continued health of the organism's DNA. According to this hypothesis sexual reproduction serves to remove mutated DNA from the gene pool. How does this work? Recall that all of the cells of your body (except sperm and eggs), are diploid, meaning that they have two sets of homologous chromosomes. During meiosis the chromosomes separate so that each sperm or egg cell gets only one copy.

Imagine now that a particular gene on one chromosome of any body cell undergoes a mutation. The adult organism can live with this minor mutation because the other allele on the

19.4 Chromosomes may be altered in number or structure

During mitosis of diploid cells and meiosis of the precursor cells of sperm and eggs, duplicated chromosomes must separate appropriately into two different cells. Just as duplication of DNA is not always perfect, neither is separation of the chromosomes.

Failure of homologous chromosomes or sister chromatids to separate properly is called **nondisjunction.** Sometimes both sister chromatids go to one of the two cells formed during mitosis. Occasionally homologous chromosomes or sister chromatids fail to separate during meiosis (Figure 19.13), resulting in an alteration in chromosome number of sperm or egg. Very rarely, a piece of a chromosome may break off and be lost, or it may reattach to another chromosome where it doesn't belong.

For the most part we never see these chromosomal alterations. If mitosis goes awry, the two daughter cells probably just die and their place is taken by other dividing cells. Alterations that arise during meiosis are more serious because they have at least the potential to alter the development of an entire organism. However, it appears that we never see most of these either because so many of the genes on the chromosomes are essential for embryonic development. Any sperm, egg, or developing embryo with errors as great as an extra or missing chromosome is unlikely to survive. Most die early in embryonic development before we are even aware of their presence.

other homologous chromosome can still make the required protein. But during meiosis the two homologous chromosomes end up in different sperm cells (or egg cells). The sperm (or eggs) with just one mutated allele may be unable to function properly and may die before fertilization. Thus sexual reproduction favors sperm and eggs with good, clean, single copies of genes. The mutated genes are continually cast aside (see figure).

Sexual reproduction also ensures that mutated genes are expressed as infrequently as possible. Everyone carries some mutations, but they are so rare that unrelated people generally have mutations on different genes. Generally, even one good copy of a gene pair is sufficient for normal function, but two bad copies spell trouble. The physical changes that become apparent after too much inbreeding show what happens when both alleles of a pair are damaged. Similarly, a very real danger associated with asexual reproduction and cloning is that mutations may accumulate in the DNA.

In all likelihood both theories have merit. In the short term, sexual reproduction helps us keep up in the battle against rapidly evolving parasites. In the long term, it ensures the continued good health of DNA.

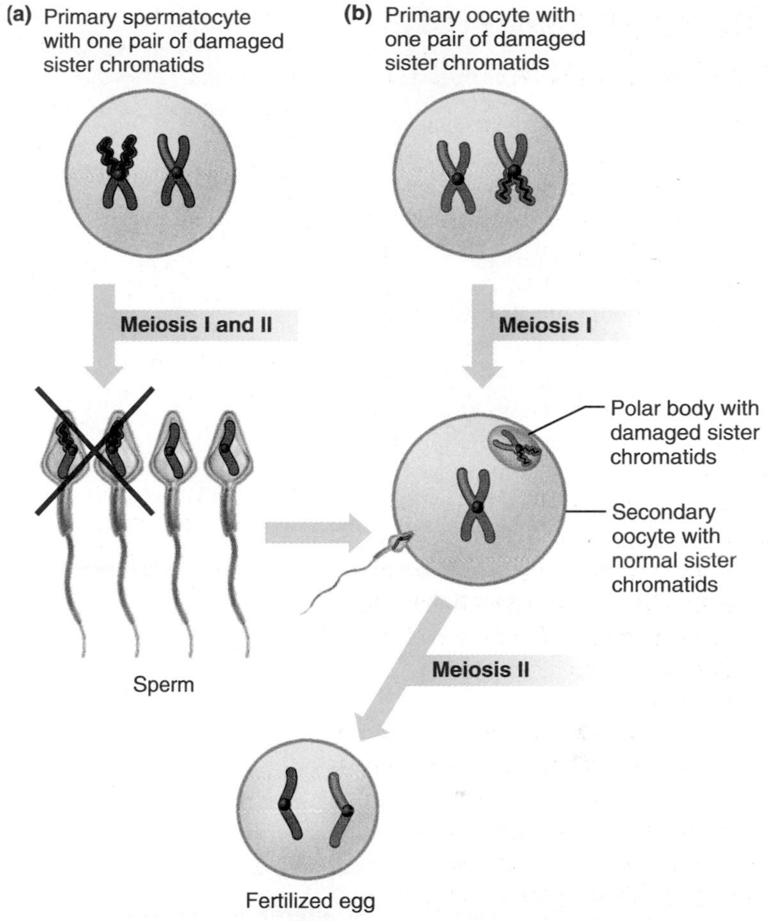

Effect of sexual reproduction on DNA. Only one pair of homologous chromosomes is depicted. Because homologous chromosomes are separated into different sperm (or eggs) and because sperm and eggs with damaged DNA are less likely to survive, sexual reproduction has the effect of winnowing out damaged DNA. In this example the damaged DNA in the female passed into a polar body. An egg with damaged DNA might not have survived.

Down syndrome: Three copies of chromosome 21

A few alterations of autosomal chromosome number result in live births. The most common example is *Down syndrome* (also called trisomy 21), caused by having three copies of chromosome 21 (Figure 19.14). Less common are Edwards syndrome (trisomy 18) and Patau syndrome (trisomy 13).

Down syndrome affects one in every 800 live births in the United States. People with Down syndrome are typically short with round faces and other distinctive physical traits; most are friendly, cheerful, and affectionate persons. They generally are slow to develop mentally and are prone to respiratory complications or heart defects that can lead to death at a young age. However, with improved care many now live well into their adult years.

The risk of a woman giving birth to a Down syndrome baby rises markedly with increasing age. The chance is one in 3,000 for mothers under 30 and one in 17 for mothers over 45. Recent evidence suggests that the frequency of Down syndrome embryos is the same in all age groups of women, but younger women are more likely to naturally abort the embryo.

Some pregnant women elect to undergo fetal testing to detect Down syndrome and other disorders. In amniocentesis, a sample of amniotic fluid containing fetal cells is withdrawn and tested. Another option is chorionic villi sampling, in which a small sample of fetal tissue is suctioned from the placenta for examination. Fetal cells are then examined for the

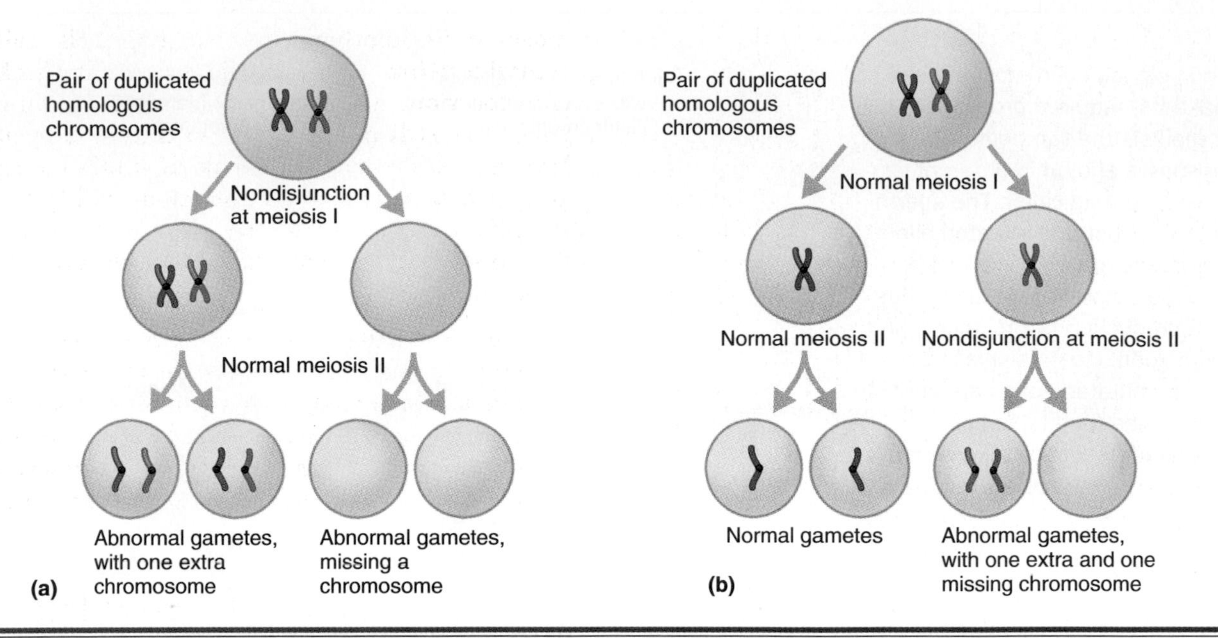

Figure 19.13 Nondisjunction during meiosis. (a) When nondisjunction occurs during meiosis I, homologous chromosomes fail to separate from each other. **(b)** When nondisjunction occurs during meiosis II, sister chromatids fail to separate from each other. In either case, some gametes (sperm or egg) may have an altered number of chromosomes.

presence of third copies of chromosome 21 (Down syndrome), third copies of chromosomes 18 and 13, and for alterations of sex chromosome number. Chromosomal damage can also sometimes be detected.

Recap **Nondisjunction leads to altered numbers of chromosomes in daughter cells. Down syndrome is caused by inheriting three copies of chromosome 21.**

Figure 19.14 An individual with Down syndrome. Down syndrome is caused by inheritance of three copies of chromosome 21.

Alterations of the number of sex chromosomes

Nondisjunction of the sex chromosomes can produce a variety of combinations of sex chromosome number, several of which are fairly common in the human population. In general, an individual with at least one Y chromosome will be essentially a male phenotype. An individual lacking a Y chromosome will be a female phenotype, regardless of the number of X chromosomes. The four most common alterations of sex chromosome number are the following:

XYY—Double-Y syndrome XYY individuals are males who tend to be tall but otherwise fairly normal, except that some show impaired mental function. At one time it was thought that XYY males had a tendency toward violent criminal behavior, but careful study has shown this is not true. Whether their rate of minor, nonviolent criminality tends to be higher than normal is still a disputed issue.

XXY—Klinefelter syndrome People with Klinefelter syndrome are also tall male phenotype. They are sterile, may show mild mental impairment, and may develop enlarged breasts because of the extra X chromosome (Figure 19.15a).

XXX—Trisomy-X syndrome Trisomy-X syndrome individuals are female phenotype. Typically they are nearly normal except for a tendency toward mild mental retardation.

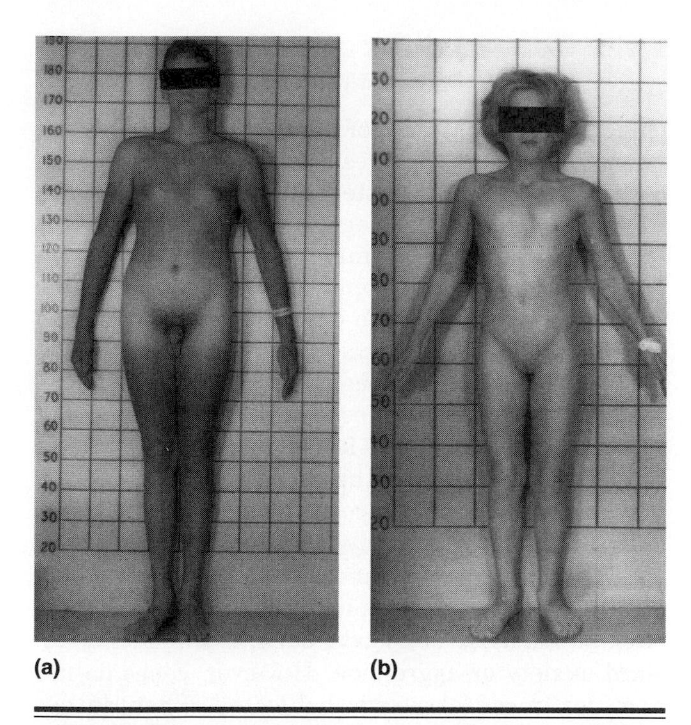

Figure 19.15 Alterations of sex chromosome number. **(a)** An individual with Klinefelter syndrome (XXY). **(b)** An individual with Turner syndrome (XO).

XO—Turner syndrome Individuals with only one X chromosome (XO) are phenotypically female. However, they tend to be short with slightly altered body form and small breasts (Figure 19.15b). Most are not mentally retarded and can lead normal lives, though they are sterile and may have shortened life expectancies. Turner syndrome is relatively rare compared to the other three because the XO embryo is more likely to be spontaneously aborted.

Table 19.2 summarizes the most common alterations of chromosomal number and gives their frequencies of occurrence.

Deletions and translocations alter chromosome structure

A **deletion** occurs when a piece of chromosome breaks off and is lost. In general, we cannot afford to lose genes, and therefore, nearly all deletions are lethal to the sperm, egg, or embryo. There are a few rare conditions in which a chromosome deletion results in a live birth. An example is *cri-du-chat* (French, meaning "cat-cry") syndrome, caused by a deletion from chromosome 5. Babies born with this syndrome are often mentally and physically retarded, with a characteristic kitten-like cry due to a small larynx.

When a piece of a chromosome breaks off but reattaches at another site, either on the same chromosome or another chromosome, it is called a **translocation.** Conceptually you might think that translocations are not a problem because all of the genes are still present, just in different locations. However, translocations result in subtle changes in gene expression and therefore their ability to function. For instance, translocation appears to increase the risk of certain cancers, including one form of leukemia.

Recap **Alterations of sex chromosome number occur when the sex chromosomes fail to separate during meiosis. In general, people with at least one Y chromosome are male phenotype, and people without a Y chromosome are female phenotype. A deletion occurs when part of a chromosome is lost; it rarely results in a live birth.**

19.5 Inherited genetic disorders usually involve recessive alleles

In principle, genetic defects of the cells that become sperm or eggs are transmitted to the next generation. Because there are two copies of each gene in every individual, diseases caused by recessive alleles are often passed on because the recessive allele is masked by the dominant allele in heterozygous individuals. It is also easy to pass on genetic alterations in which the gene is only one of multiple genes controlling a particular function.

Although most mutations result in a recessive allele that is unable to carry out its normal function,

Table 19.2 Some common alterations of chromosome number

Chromosome Genotype	Sexual Phenotype	Common Name	Approximate Frequency Among Live Births
Trisomy 21	Male or female	Down syndrome	1/800
Trisomy 18	Male or female	Edwards syndrome	1/6,000
Trisomy 13	Male or female	Patau syndrome	1/5,000
XYY	Male	Double-Y syndrome	1/2,000
XXY	Male	Klinefelter syndrome	1/2,000
XXX	Female	Trisomy-X syndrome	1/2,000
XO	Female	Turner syndrome	1/10,000

occasionally a mutation leads to an allele that is dominant rather than recessive. Dominant alleles which cause disease are dangerous because even heterozygous individuals will become ill. An example is Huntington disease.

Huntington disease is caused by a dominant-lethal allele

Huntington disease is marked by progressive nerve degeneration leading to physical and mental disability and death. Afflicted individuals begin to show symptoms only in their thirties, and most die in their forties or fifties. Huntington disease follows the simple rules of inheritance of a pair of dominant (HD) and recessive (hd) alleles. It is the *dominant* allele (HD) that causes Huntington disease. Anyone with even one HD allele will ultimately develop the disorder, and all will die of it unless they die of something else first. For this reason HD is called a *dominant-lethal* allele.

Dominant-lethal alleles are by their nature uncommon because they tend to eliminate themselves from the population. Huntington disease is unusual in this regard. The disease remains in the human population because until recently an individual who carried the HD allele had no way of knowing it until after having children. As a result, people whose genome determined that they would get the disease passed the dominant allele on to roughly half their children. Here is an example where genetic testing and counseling could have a dramatic effect on the prevalence of a disease.

Genetic testing for Huntington disease and other inherited disorders is one of the great promises of the *Human Genome Project.* Involving scientists from around the world, this massive endeavor involves locating each of the approximately 100,000 genes on the human genome and identifying their functions. Researchers have already successfully identified the genes responsible for several disorders, including forms of breast cancer, skin cancer, and osteoporosis. This information holds promise for developing ways to prevent new cases of genetic diseases and treating existing cases with new technologies. One possibility might be to correct genetic defects by techniques collectively called *gene therapy.* However, effective gene therapy may be years or decades away.

Recap **Dominant disease-causing alleles are unusual because they tend to be eliminated from the population. Huntington disease is a rare example of a dominant-lethal disease caused by inheriting just one dominant allele.**

19.6 Genes code for proteins, not for specific behaviors

Genes represent the set of instructions for a human being. The actions of genes can be very specific, such as determining hair patterns or earlobe shape. In terms of behavior, the logical conclusion might be that specific genes determine specific behaviors. Indeed, we read in the popular press of a heritable "schizophrenia gene." We hear of studies that show certain groups of criminals may share a common genetic defect; that some people may have a gene that predisposes them to depression; that there may even be a "happiness gene." Is any of this true?

The bottom line is that genes code for specific proteins. Specific genes code for proteins that give you curly hair or freckles. Other genes (and their protein products) may strongly *influence* whole patterns of behavior or moods, such as a tendency toward anxiety or aggression. However, genes do not necessarily cause *specific* behaviors. They do not lead to specific thoughts, and there is no one gene that causes you to turn to the left and another one that causes you to turn to the right. Nor are there any specific genes that cause people to commit specific crimes, as far as we know.

How do genes affect behavior?

The key to understanding how genes influence behaviors lies in understanding the many roles of proteins. Proteins may act as hormones, neurotransmitters, enzymes, or intracellular messengers. They comprise many of the structural components of cells and they control cell functions. Some proteins influence a broad range of other genes and proteins; others have very specific effects within only certain target cells. Together, groups of genes and their protein products apparently do influence broad patterns of behavior, such as feeding, mating, or learning. They may also influence moods. Nevertheless, there is no evidence that specific genes or their protein products cause depression, lead to complete happiness, or produce specific physical behaviors.

Recap **Genes influence patterns of behavior. They do not cause specific behaviors.**

19.7 DNA can be modified in the laboratory

For billions of years nature has been recombining DNA every time an organism reproduces. For thousands of years humans have been domesticating and selectively breeding plants and animals to induce nature to recombine DNA in ways useful to us. In just the last few decades we have learned enough about

DNA to be able to cut, splice, and create huge quantities of it at will, spawning a new field of applied science known as **recombinant DNA technology.**

Recombinant DNA technology is an essential step in **cloning** (making multiple identical copies of) genes and molecules. Another application of recombinant DNA technology involves **genetic engineering,** the planned alteration of the genetic makeup of an organism by modifying, inserting, or deleting genes or groups of genes. In effect, we now have the capacity to develop organisms never before created in nature and perhaps even the ability to modify or fix defective human genes.

The availability of such powerful techniques raises questions about how these new technologies might best be used. These are important decisions that we as a society must make.

Recombinant DNA technology: Cutting, splicing, and copying DNA

Cutting, splicing, and copying (cloning) DNA and the genes it contains requires specialized tools and components, including restriction enzymes, plasmids, DNA ligases, and bacteria. The technique for producing recombinant DNA is shown schematically in Figure 19.16:

1. *Isolate the DNA.* The human DNA of interest and a bacterial *plasmid* (a small circular piece of DNA found in some bacteria) are isolated.
2. *Cut both DNAs with restriction enzymes.* Certain enzymes found in bacteria can be used as "DNA scissors" to cut DNA into fragments. These enzymes are called *restriction enzymes* because they only cut DNA at specific nucleotide sequences called restriction sites.
3. *Mix the human DNA fragments with the cut plasmids.* The human DNA fragments begin to join together by complementary base pairing at the ends of the fragments, which are generally single-stranded.
4. *Add an enzyme to complete the connections.* An enzyme called *DNA ligase* is added. DNA ligase joins the plasmid and human DNA strands together. The circular bacterial plasmid now contains human DNA.
5. *Introduce the new plasmid into bacteria.* Commonly, *E. coli* bacteria are used for this purpose.
6. *Select the bacteria containing the human DNA of interest and allow them to reproduce.* The bacteria will copy any plasmid they have incorporated every time they divide. However, many different plasmids will have been formed, each with a different fragment of human DNA. Therefore it will be necessary to identify the bacteria that are carrying the human fragments of interest and isolate them before cloning the bacteria (and their human genes) in large numbers.

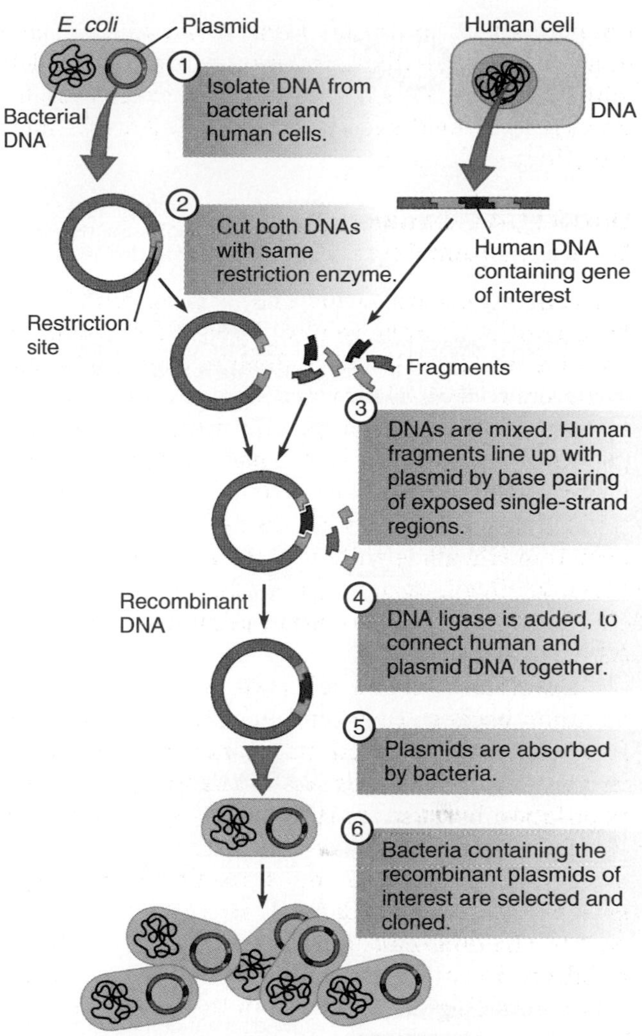

Figure 19.16 Recombinant DNA technique for producing clones of a gene or the protein product of a gene.

This, in essence, is the technique for cloning genes. Taking it one step further, if you can get the bacteria to express (activate) the gene located in the DNA fragment, the bacteria may churn out virtually limitless quantities of a particular protein.

In recent years another technique has emerged for making copies of DNA segments even more rapidly than can be done with live bacteria. The **polymerase chain reaction (PCR)** can make millions of identical copies of DNA in a test tube. This technique is particularly useful for amplifying a small sample of DNA (such as might be found at a crime scene) so it can be identified accurately.

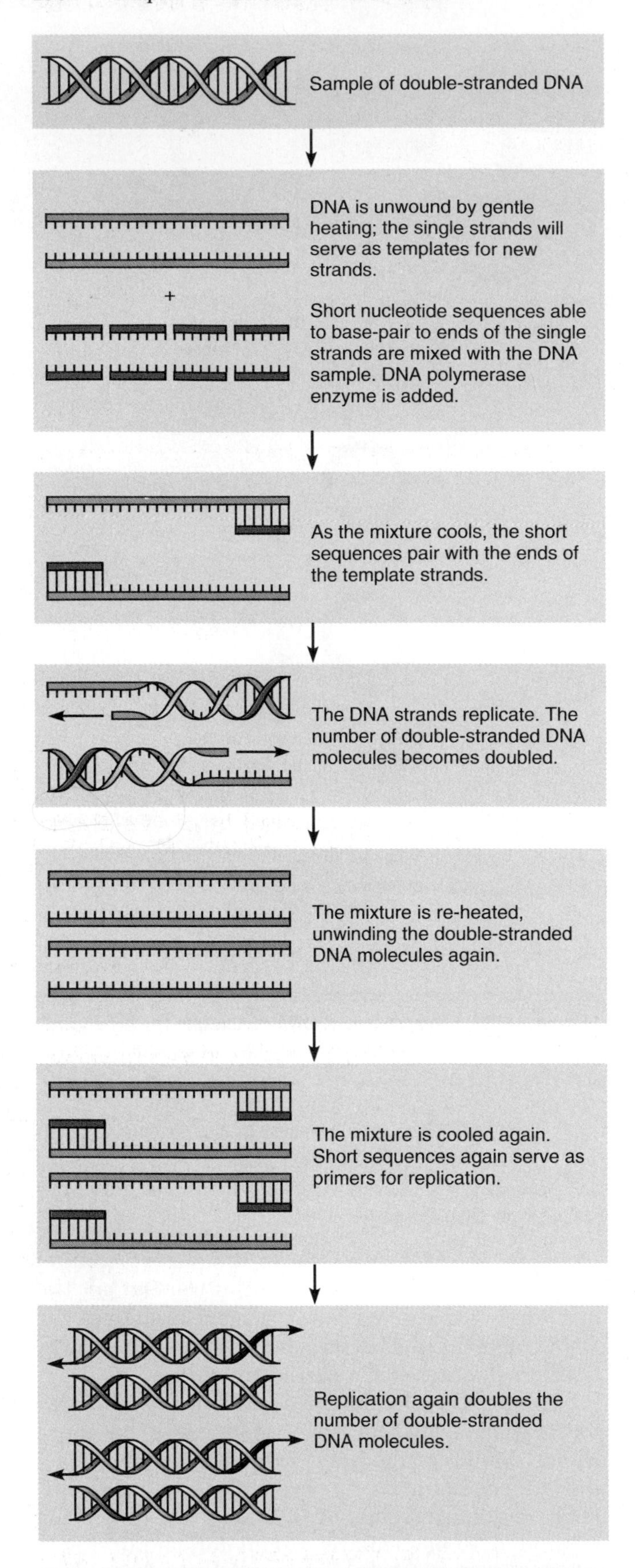

Figure 19.17 The polymerase chain reaction (PCR) for amplifying a small sample of a segment of DNA.

Conceptually, the technique is fairly straightforward (Figure 19.17). The two strands of the DNA molecule are unwound by gentle heating. Then they are mixed with short fragments of DNA and *DNA polymerase,* the enzyme that catalyzes the formation of a complementary new strand for each single strand. As the mixture cools, complementary fragments bind to the ends of the two single strands; these fragments serve as the starting point for replication of each strand. Once both strands of DNA are completely replicated, the sequence is repeated again. Each heating and cooling cycle doubles the amount of DNA. Repeat the reaction just 20 times and you have over a million identical copies of the original DNA.

Recap **Recombinant DNA technology is used to cut and splice DNA into new combinations and make multiple copies of DNA fragments. Copies of DNA fragments can be produced by bacteria or the polymerase chain reaction technique.**

Genetic engineering: Creating new organisms and fixing genes

Scientists can insert foreign genes into animals and plants, creating what are known as *transgenic* organisms. Whether or not you agree that all the feats of genetic engineering are a good idea, here are some of the accomplishments and possibilities.

Genetically engineered microorganisms The creation of modified microorganisms is already a well-advanced field. The bacteria that are used to clone genes or proteins from other organisms are all genetically altered. Genetically engineered microorganisms provide us with human insulin, growth hormone, erythropoietin, blood clotting factors, monoclonal antibodies, vaccines, and several other products as well. Certainly these products have improved the human condition, and so far none of the microorganisms have proven to be dangerous. Genetically engineered bacteria are currently being developed to clean up environmental pollution and to prevent plants from freezing, among other uses.

New plants for agriculture Genetically engineered plants are being created, tested, and put to use around the world. Plants that produce more product per acre; taste better; resist disease, insect pests, and drought; last longer on the shelf; and are not affected by commercial herbicides or pesticides all are in the realm of possibility and in many cases already a reality (Figure 19.18).

Some scientists argue that although genetic engineering permits us to produce highly uniform plants with special features, we cannot predict every possibility and so we may be setting ourselves up for

Figure 19.18 Genetically engineered plants. The cotton plants in the left foreground were genetically engineered for resistance to attack by bollworms. On the right and in the background is a conventional variety of cotton.

Figure 19.19 Gene pharming. These goats have been genetically engineered to produce milk containing antithrombin III, a human protein that may be used to prevent blood clots during surgery.

future failure if some unforeseen disaster strikes such a crop. Others argue that they may disrupt natural ecosystems. Their points are worth considering.

New uses for domestic animals Although large animals can also be genetically engineered, the techniques are somewhat different than for microorganisms. One of the most effective techniques at the moment is to mix the foreign genes of interest with eggs in the presence of tiny silicon needles that make tiny holes in the eggs. Some of the eggs will take up the gene and include it in their genome. The technique is being used to incorporate growth hormone genes into cows, sheep, pigs, and other animals in order to make them larger.

Pharmaceutical companies are inserting genes that code for useful human proteins into goats, sheep, and cows. Large animals are preferred because the proteins they produce are both rare and valuable, and usually it is possible to isolate the protein from the female's milk without having to bleed or sacrifice a valuable animal (Figure 19.19). The process of producing pharmaceuticals in farm animals is called "gene pharming," aptly enough.

Human gene therapy Last but not least, genetic engineering holds out the promise of repairing or replacing damaged genes or alleles that cause human diseases. This is a field that is yet in its infancy. Nevertheless, there have been a few attempts at human gene therapy, starting with diseases caused by genetic defects that involve the function of stem cells, the cells that are the precursors to red blood cells. Most other human cells are just not accessible and don't divide very often.

So far the results have been disappointing. There are as yet no complete defective gene "cures," and no gene therapy technique has advanced beyond the highly experimental stage. Though the hurdles to success seem huge, recall that only a few decades ago we didn't even know how to insert genes into bacteria. Twenty years is a long time in biological research, and a lot will be accomplished in our lifetime.

Recap **Transgenic plants and animals are used to develop new agricultural crops and useful human proteins. The ability to repair human genes is on the horizon.**

CurrentIssue

Consequences of the Human Genome Project

The year is 2030. Your son and his wife are expecting their first child, and the whole family is excited. The one cloud on the horizon is that both of the expectant parents are carriers of the sickle-cell trait. There is one chance in four that your grandchild will have sickle-cell anemia.

The good news is that by the year 2005 scientists had mapped and sequenced the entire set of human genes. Since 2010 there have been diagnostic tests for most genetic disorders and by 2025 there were even genetic treatments for some, including sickle-cell anemia. A simple test of amniotic fluid reveals that the fetus—a boy—has the alleles for sickle-cell anemia. The treatment is fairly straightforward: doctors inject into the fetus a harmless virus that carries a corrected version of the sickle-cell gene. The virus carries the healthy gene into most of your grandson's cells, and his potential disease is cured.

Sound impossible? Perhaps not. At this moment, hundreds of researchers around the world are engaged in one of the most ambitious and far-reaching initiatives ever attempted in biology, the Human Genome Project. Their ultimate goal is to determine the exact sequence of the three billion nucleotide pairs in the entire genome—in essence, to crack our genetic code.

The information obtained from the project will transform our lives in many ways. Some of the information may lead to positive solutions to serious medical problems, such as the one described in the scenario above. But the ability to know one's entire genome raises some unsettling questions regarding how information about a person's genome might be used—or abused. The time to start thinking about these issues is now.

The Biology of Mapping the Entire Human Genome

The Human Genome Project is so big that it can only be tackled by many scientists at once, all working together and sharing information. Sequencing the entire human genome might be considered analogous to mapping the exact location of every single building on earth. In this analogy a DNA base pair would be a building, a gene might be a city block, a chromosome would be a country, and the human genome would be the entire world. To produce such a map, cartographers (map-makers) in every country would probably first survey the land and construct a general world map showing the location of every city and town. For the Human Genome Project the general map, called a *genetic map,* shows the general structure of each chromosome (Figure 1) and pinpoints a number of genes (perhaps 10–100) as located together in a specific region on a specific chromosome.

At the same time the world map would be taking shape, local cartographers in each city would be making a detailed sectional map of the city. Each section might contain 5–10 city blocks. This would be the equivalent of a *physical map* of the human genome. A physical map would be sufficiently detailed

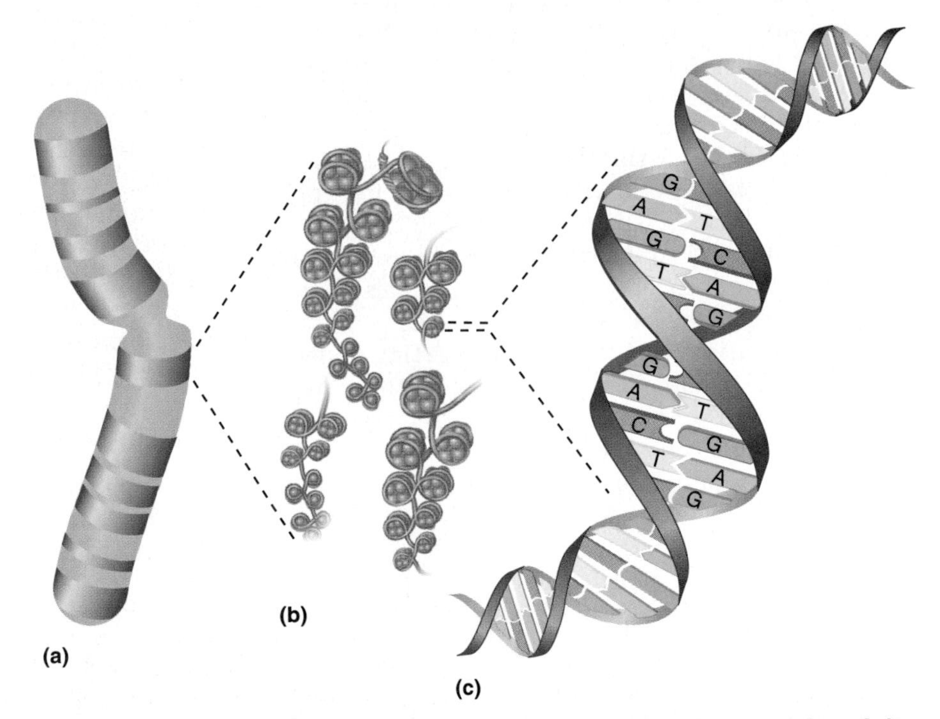

Figure 1. Mapping and sequencing the human genome. (a) A typical genetic map of a chromosome would only be able to locate a particular gene within a specific region of the chromosome. (b) A more detailed physical map is created by breaking sections of the chromosome into small fragments of DNA and then sequencing and piecing together the fragments. (c) Complete decoding of the genome will require determination of the precise sequence of base pairs in all of the DNA.

polygenic inheritance, 475
polymerase chain reaction, 485
Punnett square, 471
recessive, 472
recombinant DNA technology, 485
sex-linked inheritance, 478
sickle-cell anemia, 473
translocation, 483

Test Yourself

Matching Exercise

_____ 1. Variants of a particular gene
_____ 2. A rule of inheritance, proposed by Mendel that states that heritable factors (genes) separate during the formation of sperm or egg
_____ 3. Sickle-cell anemia is a disease caused by this pattern of genetic inheritance
_____ 4. Characteristics such as body height and eye color result from this pattern of genetic inheritance
_____ 5. Huntington disease is caused by this pattern of genetic inheritance
_____ 6. Genes on the same chromosome
_____ 7. The process of cutting, splicing, and copying DNA
_____ 8. Failure of homologous chromosomes or sister chromatids to separate properly
_____ 9. A condition caused by having three copies of chromosome 21
_____10. An X-linked disease

a. polygenic inheritance
b. dominant
c. law of segregation
d. recombinant DNA technology
e. incomplete dominance
f. linked genes
g. alleles
h. hemophilia
i. nondisjunction
j. Down syndrome

Fill in the Blank

11. A(n) __________ occurs when a piece of a chromosome breaks off and is lost.
12. A(n) __________ occurs when a piece of a chromosome breaks off and reattaches at another site.
13. Segments of DNA can be copied in a test tube using the __________.
14. If the alleles at a particular gene locus are identical, the person is __________ for that allele.
15. If the alleles at a particular gene locus are different, the person is __________ for that allele.
16. The physical and functional expression of the genes you inherit is your __________.
17. The complete set of genes you inherit is your __________.
18. If one allele is expressed completely over another allele, we say that the first allele is __________.
19. If the heterozygous genotype results in an intermediate phenotype, this is an example of __________ dominance.
20. Genes that are inherited together comprise a(n) __________.

True or False?

21. The X chromosome determines gender.
22. Cutting and splicing DNA is referred to as the Punnett square technique.
23. In polygenic inheritance, phenotypic traits depend on the interactions of many genes.
24. In incomplete dominance, the heterozygous genotype results in a phenotype that is identical to the mother's phenotype.
25. Huntington disease is caused by inheriting three copies of chromosome 21.

Concept Review

1. Where do we get the two copies we have of every gene?
2. What is an allele?
3. Distinguish between genotype and phenotype.
4. What factors determine phenotype?
5. Explain how alterations of chromosome number and structure can occur.
6. Why is the inheritance of certain genes linked to the inheritance of other genes?
7. What is meant by sex-linked inheritance?
8. What will be the phenotype (male or female) of a person who is XXY? Why?
9. Why are lethal diseases caused by dominant alleles so rare?
10. How do genes affect mood or behavior?

Apply What You Know

1. What fraction of the offspring of two palomino horses would be palomino? Explain using a Punnett square.
2. Why might repeated cloning over many generations have a deleterious effect on the genes of an organism?
3. Why is it that the range of resting blood pressures of humans is best represented by a bell-shaped curve covering a continuous range rather than by two populations, one with high blood pressure and one with normal blood pressure?

Case in Point

The HFE Gene and Hemochromatosis

Hereditary hemochromatosis is a common heritable genetic disorder associated with excessive iron absorption from the small intestine. Eventually the iron builds to dangerous levels in joints and the liver, pancreas, heart, and pituitary gland. Over time, the iron overload damages the liver and leads to cirrhosis, liver failure, and possibly liver cancer. Damage to the pancreas can cause diabetes. Other long-term effects include heart failure, cardiac arrhythmias, and arthritis.

Researchers have linked most cases of hereditary hemochromatosis to a mutation in a single gene on chromosome 6, called the HFE gene. Apparently the protein made by the normal HFE gene is involved in regulating small-intestine iron absorption; in its absence too much iron is absorbed. The initial study that documented the existence of the HFE gene found that 83% of people with hereditary hemochromatosis were homozygous for the mutation, whereas none of the normal persons tested as controls were homozygous for the mutation.

However, this does not mean that everyone who carries a mutated HFE gene gets hemochromatosis or that having the mutated gene is the only cause of the disease. Women are less likely to develop the condition than men, possibly because menstruation removes some of the excess iron from their bodies. Environment and lifestyle factors may also play a role. And in one study, almost 7% of patients with hereditary hemochromatosis showed no HFE mutations at all.

QUESTIONS:

1. What pattern of genetic inheritance is involved in hereditary hemochromatosis? Explain your answer.
2. People with chronic untreated hemochromatosis may eventually develop life-threatening complications, including liver disease and diabetes mellitus. In contrast, people who receive regular treatment before symptoms of hemochromatosis generally can lead normal lives. Would you recommend genetic testing for hereditary hemochromatosis? If you're not certain, what information would you like to have in order to make a decision?

INVESTIGATE:

Katrina Allen and Robert Williamson. Should We Genetically Test Everyone for Haemochromatosis? *Journal of Medical Ethics* (April 1, 1999), p. 209.

David Brandhagen, et al. Update on Hereditary Hemochromatosis and the HFE Gene. *Mayo Clinic Proceedings* (September 1, 1999), p. 917.

References and Additional Reading

Friedmann, Theodore. Overcoming the Obstacles to Gene Therapy. *Scientific American* 276:96–101, 1997. (A good summary of where the science of gene therapy stood in 1997.)

Jessee, Jennifer. Y? *Discover*. Nov. 1997, pp. 88–93. (A short and entertaining account of the evolution and function of the Y chromosome.)

Lyon, Jeff, and Peter Gorner. *Altered Fates: Gene Therapy and the Retooling of Human Life*. New York: W.W. Norton & Co., 1996. 636 pp. (A gripping account of the early scientific and political history of human gene therapy.)

Ridley, Matt. *The Red Queen: Sex and the Evolution of Human Nature*. New York: Penguin Books, 1993. 405 pp. (A readable and well-referenced account of current theories for why we reproduce sexually and how sex has affected human nature.)

Russell, Peter J. *Genetics*. New York: HarperCollins Publishers, 1996. 784 pp. (A standard textbook of genetics: for reference or advanced study only.)

Sapolsky, Robert. A Gene for Nothing. *Discover*. Oct. 1997, pp. 40–46. (A general-interest article about what genes do and don't do.)

CHAPTER 20

DEVELOPMENT AND AGING

How long does it take sperm **to reach the** egg?

What causes puberty?

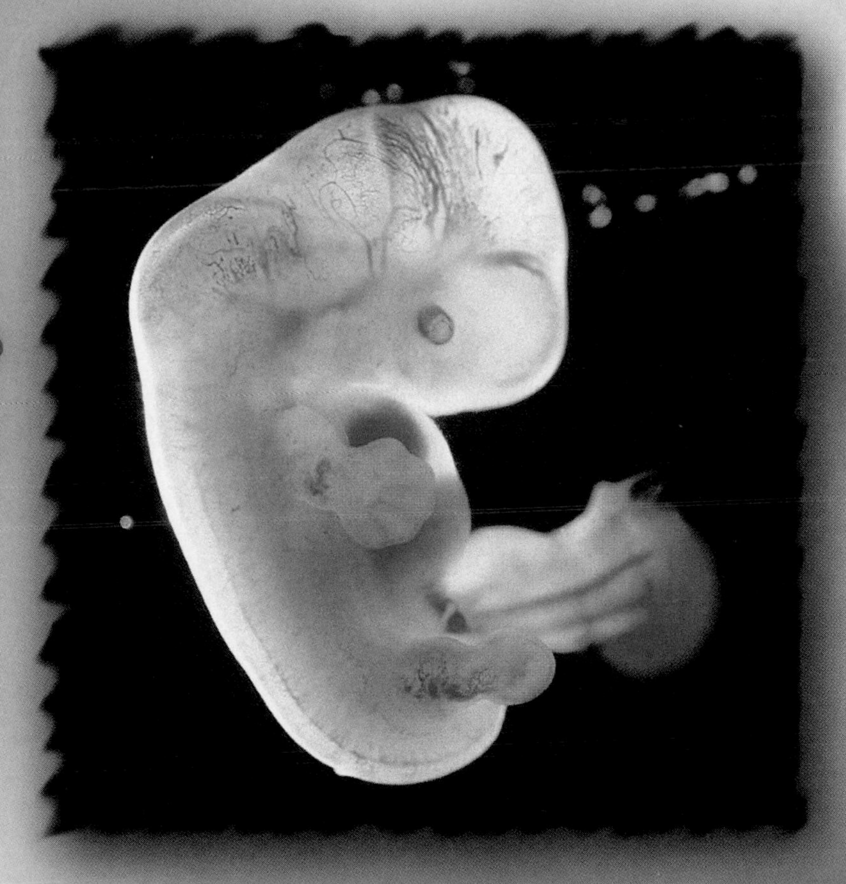

Why do we age?

Why is it important to be able to define death?

What can we do to age well?

LIFE IS A JOURNEY THAT BEGINS AT FERTILIZATION AND ENDS around the time of our last breath. It's an interesting and complicated process in which one developmental stage follows the next like a carefully choreographed dance.

The changes that occur during development are nothing short of amazing: in just nine months a single undifferentiated cell develops into a human baby with complex organ systems comprising more than 10 trillion cells. In less than 20 years, this helpless baby becomes a fully functioning adult capable of producing still more human babies. Then, starting at about age 40, there is a slow decline of function for reasons we don't yet fully understand, ending inevitably in death.

The amazing journey of life is what this chapter is all about. We begin at the beginning when sperm meets egg, and end at the ending when life finally ceases.

20.1 Fertilization occurs when sperm and egg unite

The journeys of egg and sperm

It's remarkable that sperm and egg ever find each other at all, given the journey that each must take. An egg is released from one of the two ovaries at ovulation, and the sperm are deposited in the vagina near the cervix. The egg moves slowly and passively down the oviduct, propelled by cilia that line the tube and sweep it gently downward.

Sperm are not much more than a head containing the all-important DNA, a midpiece containing mitochondria for energy production during the sperm's long journey, and a flagellum for movement (Figure 20.1a). The several hundred million sperm have a hazardous journey if they are to reach the egg. First they must pass through the mucus that blocks the cervical opening and cross the vast (to them) expanse of the uterus. Then they must locate and enter the correct oviduct (half don't). Sperm swim randomly at a rate of one-quarter inch per minute. All the while they must avoid bacteria, the strongly acidic pH of the vagina, and the occasional white blood cells roaming the uterine lining.

A sperm's journey can take hours, and only a few thousand or a few hundred make it successfully to the upper oviduct; the rest are lost along the way and die. Fertilization typically takes place in the upper third of the oviduct approximately 6–24 hours after intercourse, provided that an egg is present (Figure 20.1b). If no egg is present, the sperm may live for a couple of days and then die. If no sperm arrive, the egg dies in about 24 hours.

How long does it take sperm to reach the egg?

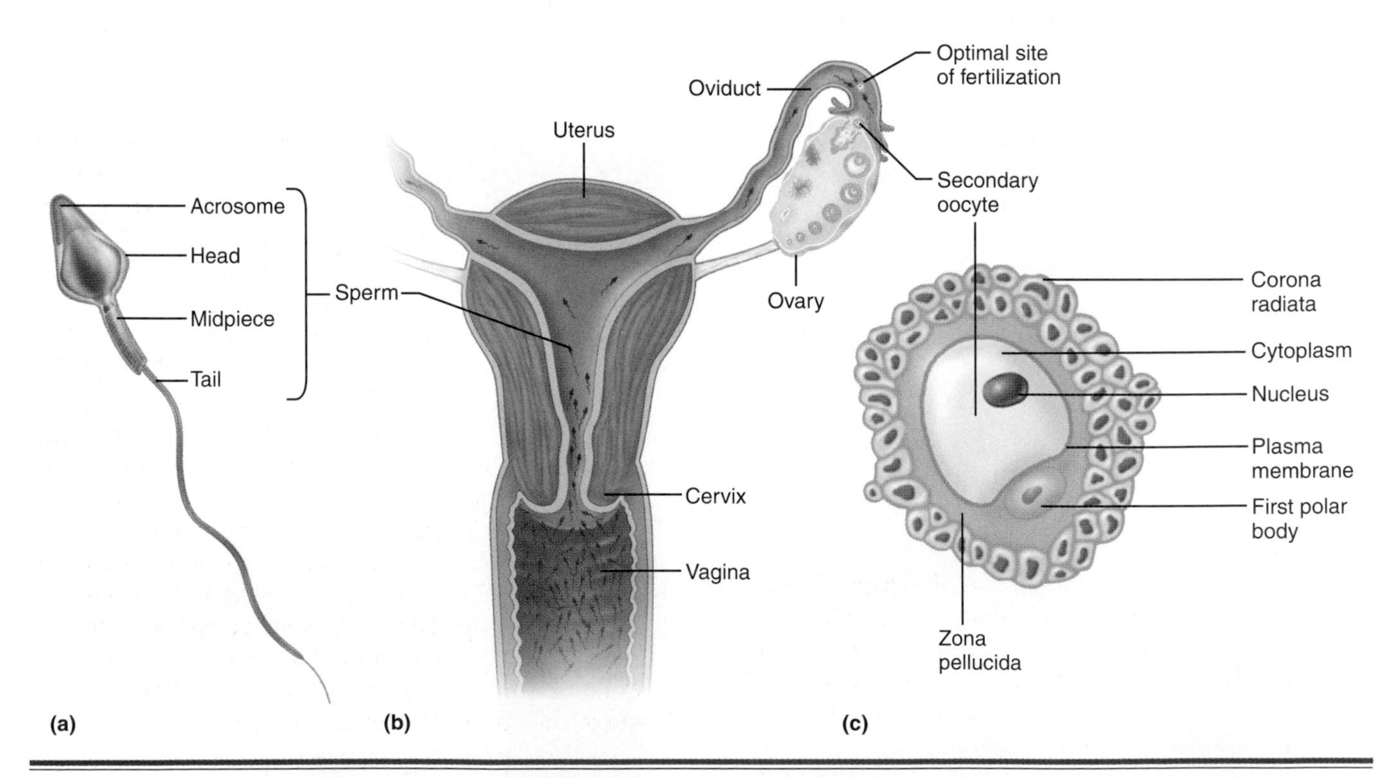

Figure 20.1 Female and male gametes and fertilization. **(a)** The male gamete, or sperm. The size of a sperm relative to the secondary oocyte has been greatly exaggerated. **(b)** Fertilization generally takes place in the upper third of the oviduct. **(c)** The female gamete is a secondary oocyte that is in an arrested state of stage II of meiosis.

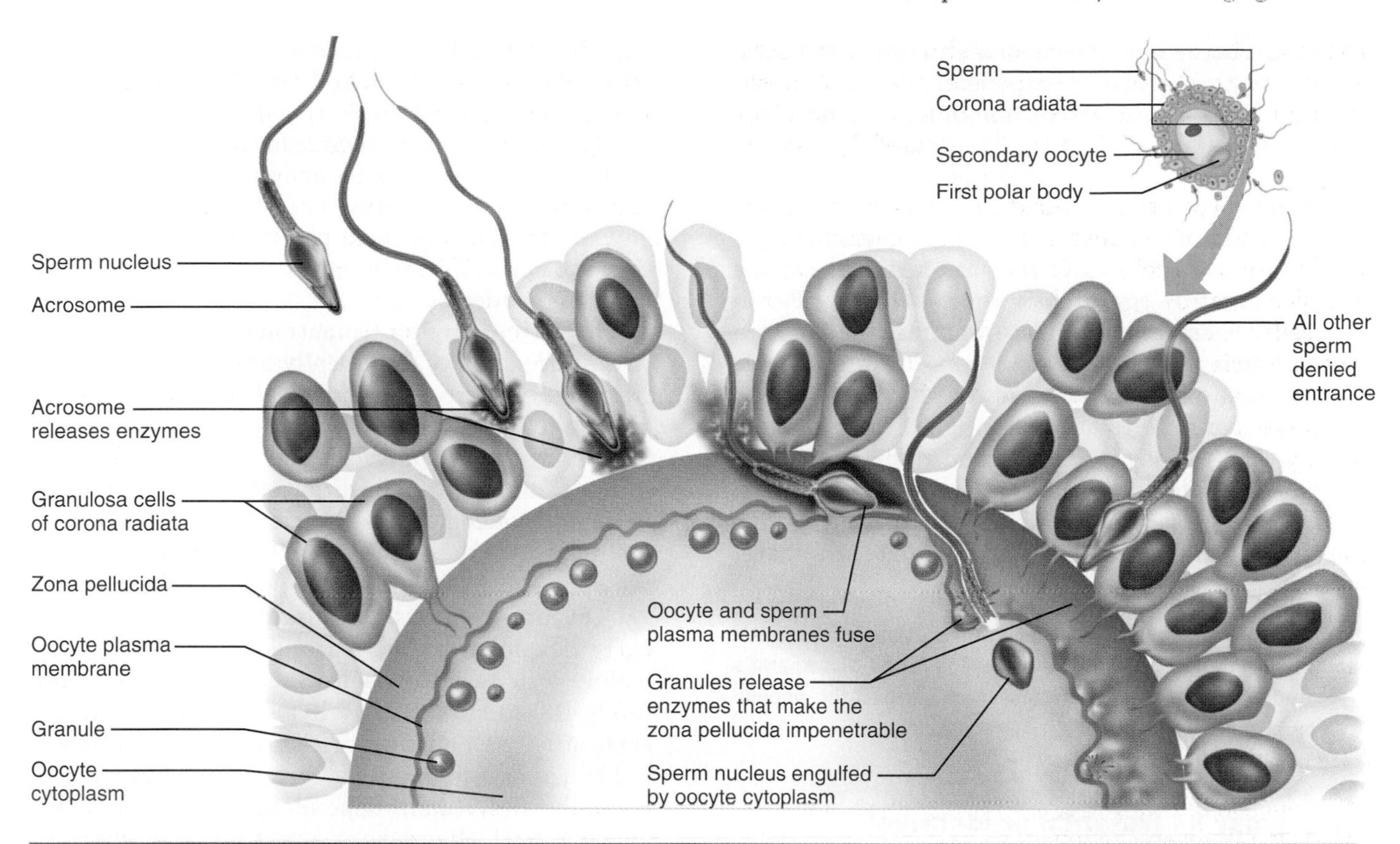

Figure 20.2 Fertilization. Fertilization begins when a sperm makes contact with the secondary oocyte and releases enzymes that digest a path through the zona pellucida. Once a sperm makes contact with the oocyte plasma membrane and enters the oocyte, granules are released from the oocyte that make the zona pellucida impenetrable to other sperm.

Recap **Sperm must swim from the vagina, where they are deposited, through the uterus and up the correct oviduct to meet the egg. Relatively few sperm survive the journey.**

One sperm fertilizes the egg

The odds of success for a single sperm are still not very good, for only one of the several hundred million sperm will fertilize the egg. This is important, for otherwise the zygote would end up with an abnormal number of chromosomes. The process of fertilization has evolved to ensure that there is only one successful sperm.

Spotlight on Structure & Function Recall that before fertilization, the egg is really a secondary oocyte that has started but not finished the second stage of meiosis (Figure 20.1c). The second stage of meiosis will not be completed until fertilization has occurred. At this point the secondary oocyte is surrounded by a protective covering called the *zona pellucida,* and by a layer of cells that were derived from the follicle called the *corona radiata.* The egg is relatively large (nearly 2,000 times the mass of a sperm) because it contains a great deal of cytoplasm. The cytoplasm of the egg must support nearly two weeks' worth of cell divisions until the "pre-embryo," as it is called, makes contact with the uterine lining and begins to receive nutrients from the mother.

When a sperm encounters the egg, the tip of the sperm head, called the *acrosome,* releases powerful enzymes (Figure 20.2). These enzymes digest a path for the sperm between the cells of the corona radiata and through the zona pellucida to the egg cell membrane. Several sperm may be making this journey at the same time. When the first sperm makes contact with the oocyte cell membrane, special recognition protein "keys" of the sperm attach to receptor "locks" in the egg cell membrane in a species-specific manner. The combination of lock and key causes the cell membranes of egg and sperm to fuse so the nucleus of the sperm can enter the egg. ■

Fertilization begins when the sperm's nucleus enters the egg. The entry of one sperm triggers the release of enzymes from granules located just inside the egg. These enzymes produce changes in the zona pellucida that make it impenetrable to all other sperm. One sperm nucleus, and only one, will enter the egg.

The process of fertilization is not yet complete, however, because the "egg" is still a secondary oocyte. Entry of the sperm nucleus triggers the completion of meiosis II and formation of the haploid *ovum,* or mature egg, and another polar body. Fertilization is considered complete when the haploid nuclei of sperm and mature ovum join, forming a single diploid cell with 46 chromosomes (Figure 20.3). Some people refer to the end of fertilization as *conception* and to the product of conception as the *conceptus*. The follicular cells that surrounded the secondary oocyte are shed and the polar bodies die, leaving just the single diploid cell called the **zygote.**

Recap **When sperm contact the egg they release enzymes that clear a path through the corona radiata and zona pellucida. Fertilization begins when one sperm's nucleus enters the oocyte and ends when the haploid nuclei of sperm and egg fuse, creating a new diploid cell.**

Twins may be fraternal or identical

Twins occur once in about every 90 births. Twins are the most common form of multiple births. Occasionally we hear of sextuplets or septuplets (six or seven newborns at once), but these are rare and almost always result from the use of fertility drugs. Even natural triplets are rare (1 in 8,000 births).

Two different processes lead to twins (Figure 20.4). **Fraternal twins** arise from the ovulation of more than one oocyte in a particular cycle. The oocytes are as different genetically as any two oocytes ovulated at different cycles, and each is fertilized by a different sperm. Fraternal twins are as different as any children by the same parents. Indeed, because they are derived from different sperm they can be of different genders.

Identical twins, sometimes called maternal twins, are always genetically identical. Identical twins arise from a single zygote. Recall that up until about the 16-cell stage, all cells of the developing pre-embryo are identical. If the ball of cells breaks into two groups before differentiation has begun, two complete and similar individuals may be formed. Identical twins are always of the same gender and are usually closely alike in phenotypic appearance.

On rare occasions, cells begin to separate during early development but then don't separate completely. When that happens, the embryos and eventually the infants (if they live) remain attached to each other as **conjoined** (Siamese) **twins.** Sometimes it is possible to separate conjoined twins successfully after birth, depending on how they are joined and which organs or structures they share.

Recap **Fraternal twins result from the fertilization of two separate eggs. Identical twins occur when a single fertilized egg divides in two before differentiation has begun.**

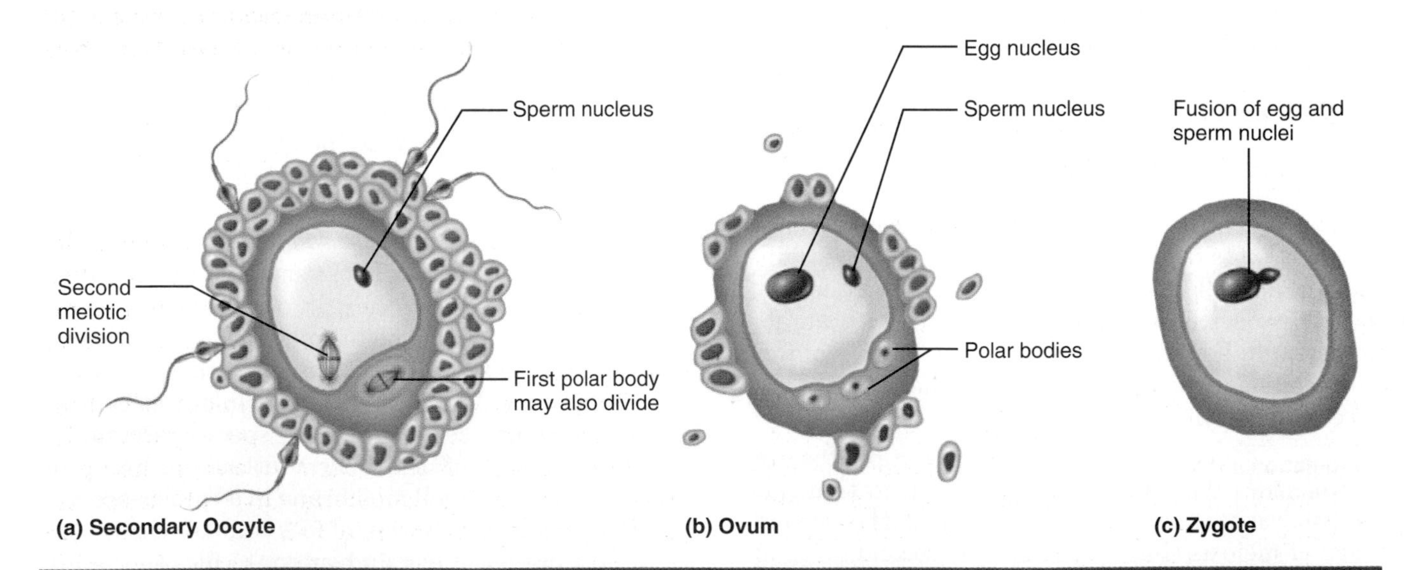

Figure 20.3 Completion of fertilization. (a) A sperm nucleus has just entered the secondary oocyte, triggering completion of meiosis II. **(b)** Completion of meiosis yields an ovum (egg) with a haploid nucleus and another polar body. If it has not already regressed, the original polar body may divide as well. **(c)** The nuclei of the ovum and sperm fuse and the polar bodies die. A single diploid cell called the zygote is formed, and fertilization is complete.

(a)

(b)

(c)

Figure 20.4 Twins. **(a)** Fraternal twins arise when two eggs are ovulated and fertilized in the same monthly cycle. **(b)** Identical twins arise when a zygote divides in two during development. **(c)** Conjoined (Siamese) twins arise when a zygote divides incompletely.

20.2 Development: Cleavage, morphogenesis, differentiation, and growth

During development, rapid and dramatic changes in size and form take place. This involves four processes:

- *Cleavage.* **Cleavage** is a series of cell divisions that occur in the first four days following fertilization that are not accompanied by cell growth or differentiation. Cleavage produces a ball of identical cells that is about the same size as the original zygote. Growth does not occur because the ball of cells is traveling down the oviduct at this time. With no attachment to the mother, the only energy available is that stored in the cytoplasm of the cells plus a little glycogen found within the oviduct itself.
- *Morphogenesis.* Throughout development the organism undergoes dramatic changes in shape and form. Starting as a ball of identical cells at day four, the organism becomes (in succession) several layers of different types of cells, a pre-embryo with a tail and head, an embryo with recognizable human features, and finally a fetus with nearly a complete human form. This process of physical change is called **morphogenesis** (*morpho* means "form; shape" and *genesis* means "origin; production").
- *Differentiation.* Individual cells, too, are beginning to take on specialized forms and functions, a process we know as *differentiation* (review Chapter 17). Differentiation of cells, and the development of organs and organ systems as a result of cell differentiation, is the primary cause of morphogenesis.
- *Growth.* Starting about the time that the developing organism becomes embedded in the endometrial lining of the uterus and begins to receive nutrients from the mother, the organism begins to grow in size. The *growth* of a human infant from fertilization to birth is truly spectacular—from a single cell too small to be seen to over 10 trillion cells with a combined weight of six or seven pounds. Every time the cells divide the two daughter cells double in mass to prepare for the next cell division. The only time cell division is not accompanied by cell growth is the initial period of cleavage.

Pregnancy is considered to comprise three periods called *trimesters*. Each trimester is approximately three months long, and characteristic events take place during each.

> ***Recap*** **The four processes associated with development are: (1) cleavage, a series of cell divisions producing a ball of identical cells; (2) morphogenesis, a sequence of physical changes; (3) differentiation, as cells assume specialized forms and functions; and (4) growth in size. Pregnancy is divided into three trimesters.**

20.3 Pre-embryonic development: The first two weeks

The period of **prenatal** ("before birth") human development can be divided into three stages called pre-embryonic, embryonic, and fetal development. Characteristic processes and changes take place in each stage.

During pre-embryonic development (the first two weeks), the conceptus makes its way down the oviduct, embeds itself in the endometrial lining of the uterus, begins to receive nutrients from the mother, and starts to grow. Cells differentiate into several tissue layers and the first stages of morphogenesis begin.

Throughout the pre-embryonic period, the conceptus is known informally as a **pre-embryo** because many (if not most) of the cells are destined to become part of the placenta, not the embryo. While the conceptus is still in the oviduct, a series of successive cleavages yields a ball of about 32 identical cells called the *morula* (Figure 20.5). The term morula means "little mulberry," aptly describing its spherical, clustered appearance.

On about the fourth day the morula is swept into the uterus, where it undergoes the first stages of differentiation and morphogenesis. Over the next several days it becomes a *blastocyst*, a hollow ball comprised of (1) an outer sphere of cells called a *trophoblast,* (2) a hollow central cavity, and (3) a group of cells called the *inner cell mass.* Only the inner cell mass is destined to become the embryo.

About day six or seven, the trophoblast cells make contact with the endometrial lining and secrete enzymes that digest endometrial cells. The creation of a path for the blastocyst causes it to burrow inward (Figure 20.6). The process in which the blastocyst becomes buried within the endometrium is called **implantation.**

During the second week the inner cell mass begins to separate from the surface of the blastocyst, creating a second hollow cavity. The second cavity will become the amniotic cavity filled with *amniotic fluid.* At this point the cell mass, now called the **embryonic disk,** differentiates into two cell types called *ectoderm* and *endoderm.* The appearance of an amniotic cavity and of ectoderm and endoderm in the embryonic disk marks the end of the pre-embryonic period.

Pre-embryonic development is a hazardous time in which a lot can go wrong. Usually a woman is not even aware that she is pregnant at this time, and so she may continue risky behaviors such as smoking and drinking alcohol.

On rare occasions implantation can occur in the wrong place, resulting in an **ectopic pregnancy.** The most common form of ectopic pregnancy is a *tubal pregnancy*, when the blastocyst implants in an oviduct. It is possible for a tubal pregnancy to result in a healthy baby, but the oviduct is not the ideal site in terms of size or ability to nourish the infant, as is the uterus. If tubal pregnancies are not terminated early for the health of the mother, they must still be watched very closely. Usually the infant will be removed surgically as soon as it is able to survive with intensive care in a hospital.

Recap **During pre-embryonic development, successive cleavages yield a morula. Early stages of differentiation and morphogenesis cause the morula to become a blastocyst, which implants in the lining of the uterus. The embryonic disk is destined to become the embryo.**

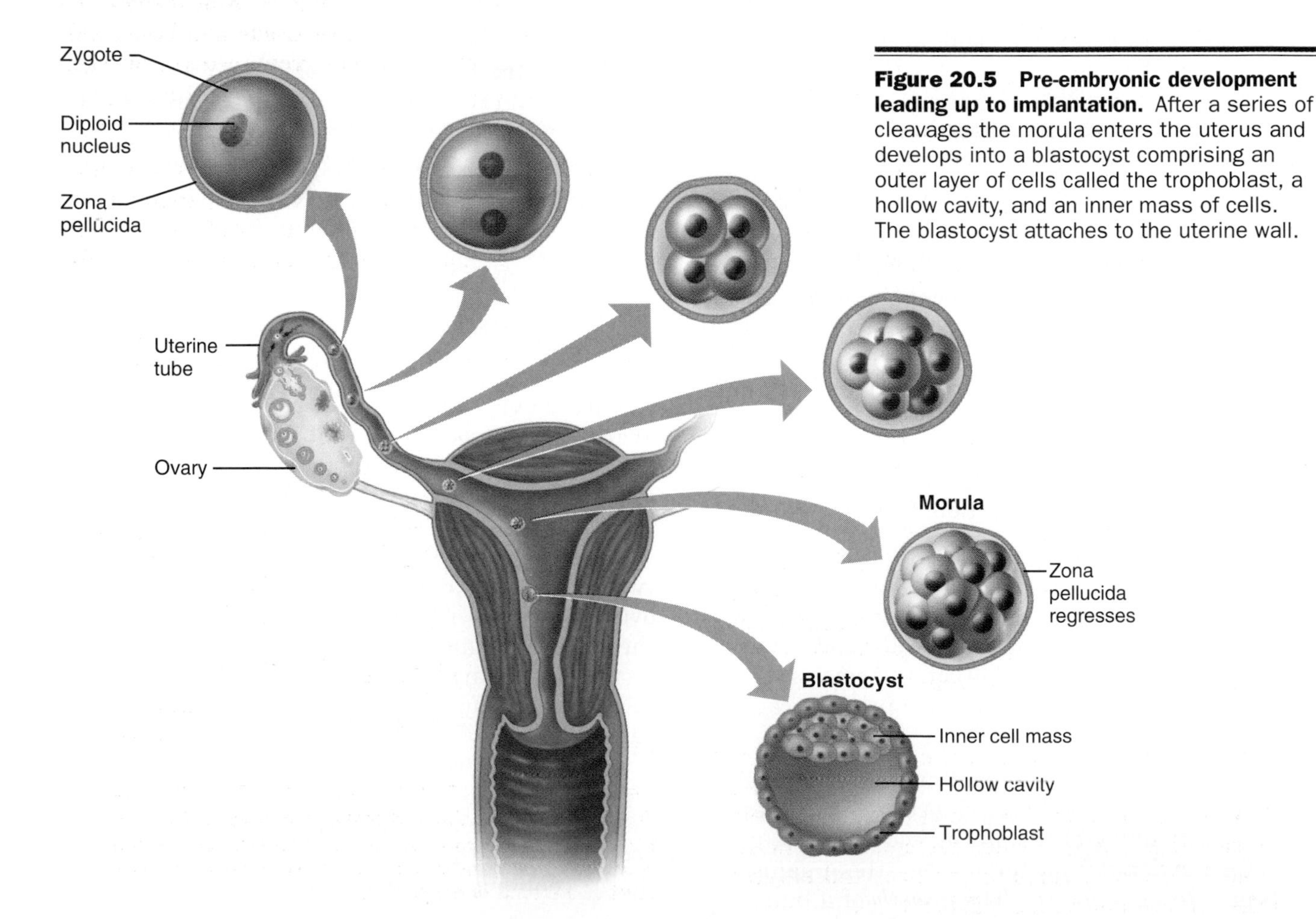

Figure 20.5 Pre-embryonic development leading up to implantation. After a series of cleavages the morula enters the uterus and develops into a blastocyst comprising an outer layer of cells called the trophoblast, a hollow cavity, and an inner mass of cells. The blastocyst attaches to the uterine wall.

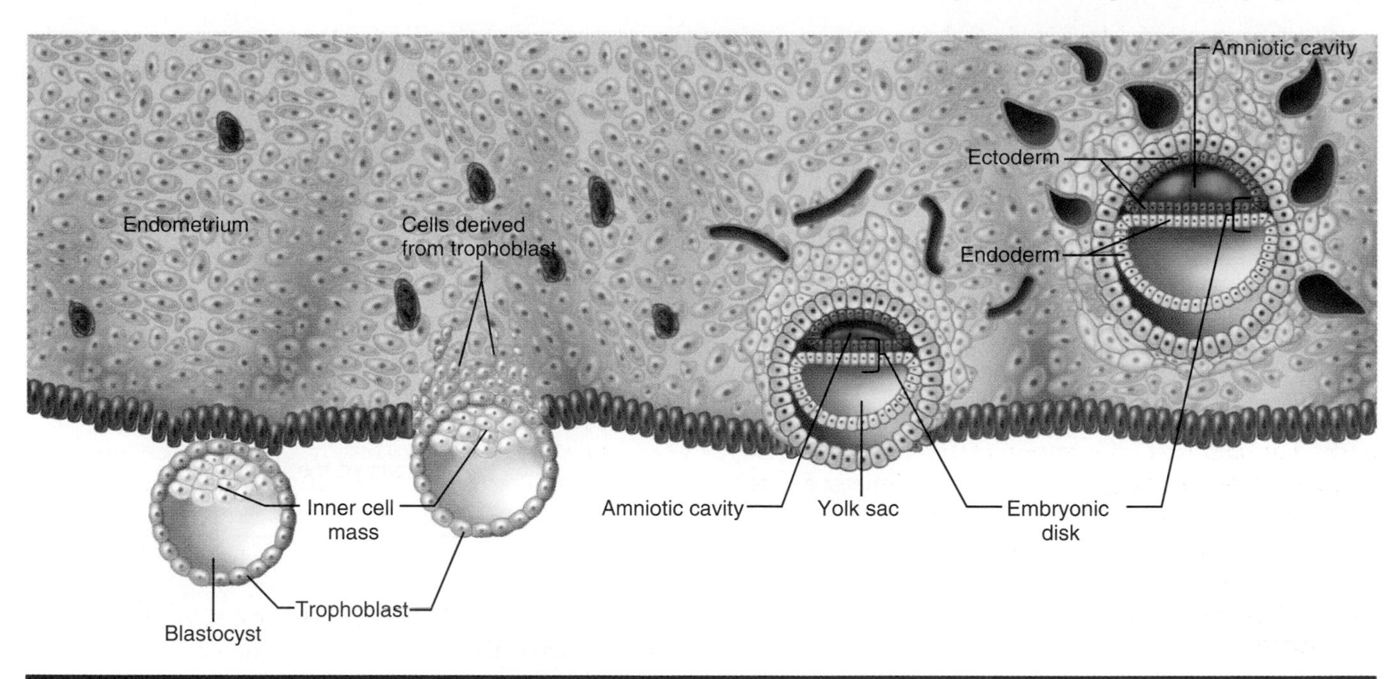

Figure 20.6 Implantation and the end of pre-embryonic development. Implantation takes place during days 7–14, at the same time that the inner cell mass of the blastocyst is developing into the embryonic disk. An inner cell mass forms, and then the inner cell mass separates from the outer ball to become the embryonic disk comprising two cell types. The embryonic disk is destined to become the embryo.

20.4 Embryonic development: Weeks three to eight

Tissues and organs derive from three germ layers

From about the beginning of week three until the end of week eight the developing human is called an **embryo.** During this time growth, differentiation, and morphogenesis are especially rapid. All of the organs and organ systems are established, though most of them are not fully functional. By the end of the embryonic period the embryo has taken on distinctly human features but is still only about an inch long.

The embryonic period begins when a third cell layer (*mesoderm*) appears between the other two in the embryonic disk (Figure 20.7). These three primary tissues, called the *germ layers,* represent the precursor (or germ) cells to the four basic tissue types introduced in Chapter 4 (epithelial, muscle, connective, and nervous) and all organs and organ systems in the body. Differentiation and morphogenesis follow a predictable pattern, so we know which tissues and organs are derived from each germ layer. The three germ layers are:

- *Ectoderm.* The **ectoderm** is the outermost layer, the one exposed to the amniotic cavity. Tissues derived from the ectoderm become the epidermis of the skin, the nervous system, hair, nails, enamel of teeth, parts of the eye, and several other organs and tissues (Table 20.1).
- *Mesoderm.* The middle layer, or **mesoderm,** becomes muscle, connective tissue and bone, kidneys and ureters, bone marrow, testes or ovaries, the lining of the blood vessels, and other organs and tissues.
- *Endoderm.* The innermost layer is the **endoderm.** It gives rise to the liver and pancreas, the alveoli of the lungs, the linings of the urinary bladder, urethra and vagina, and several glands.

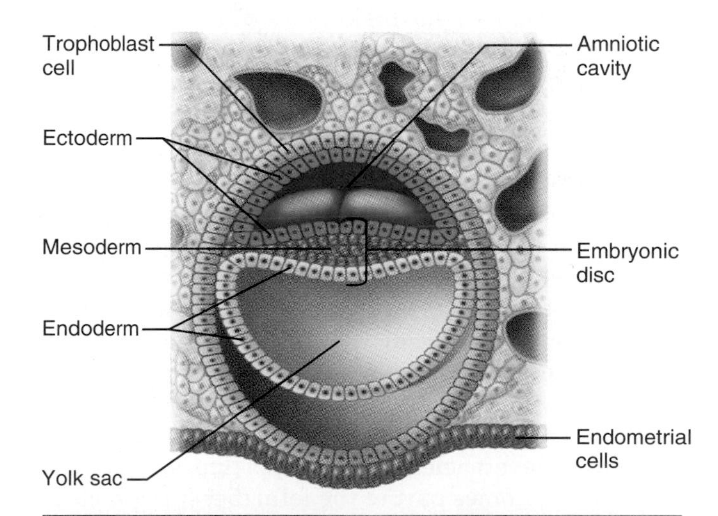

Figure 20.7 The three primary germ layers. The appearance of a third type of cell layer (mesoderm) between the ectoderm and endoderm in the embryonic disk marks the start of the embryonic period.

Table 20.1 Tissues, organs, and organ systems derived from the three primary germ layers

Ectoderm	Mesoderm	Endoderm
Epidermis of skin, including hair, nails, and glands	Dermis of skin	Epithelial tissues lining the digestive tract (except mouth and anus), vagina, bladder, and urethra
Mammary glands	All connective tissue, cartilage, and bone	Alveoli of lungs
The nervous system, including the brain, spinal cord, and all nerves	All muscle tissue	Liver and pancreas
Cornea, retina, and lens of eye	Bone marrow (blood cells)	Thyroid, parathyroid, and thymus glands
Enamel of teeth	Kidneys and ureters	Anterior pituitary gland
Posterior pituitary gland	Testes, ovaries, and reproductive ducts	Tonsils
Adrenal medulla	Lining of blood vessels	Portions of the inner ear
Epithelial lining of nose, mouth, and anus	Lymphatic vessels	

Recap **By the beginning of embryonic development the embryo comprises three primary germ layers, called ectoderm, mesoderm, and endoderm, that ultimately give rise to fetal tissues and organs.**

Extra-embryonic membranes and the placenta

Early in embryonic development, four different *extra-embryonic membranes* form that extend out from or surround the embryo: the amnion, allantois, yolk sac, and chorion. Most of the components of these membranes are either resorbed during later development or discarded at birth.

The innermost layer is the *amnion,* also known as the "bag of waters." The amnion lines the *amniotic cavity,* which is filled with **amniotic fluid.** The amniotic fluid is derived from the mother's interstitial fluid and is in continuous exchange with it. Later in development when the kidneys of the fetus form, the fetus will urinate into the amniotic fluid and the urine will be removed via exchange with the mother's blood. The amnion and amniotic fluid absorb physical shocks, insulate the fetus, and keep it from drying out.

The *allantois* is a temporary membrane that helps form the blood vessels of the umbilical cord. It degenerates during the second month of development.

The *yolk sac* is formed (in part) from endodermal cells of the embryonic disk. It forms a small sac that hangs from the embryo's ventral surface. In many species the yolk sac serves a nutritive function, but in humans that function has been taken over by the placenta. Nevertheless, it is important in humans because it becomes part of the fetal digestive tract. It also produces fetal blood cells until that job is taken over by other tissues, and it is the source of the germ cells that migrate into the gonads and give rise to the gametes (sperm and eggs).

The outermost layer, the *chorion,* is derived primarily from the trophoblast of the early blastocyst. The chorion is the source of the *human chorionic gonadotropin* (hCG) that supports pregnancy for the first three months until the placenta begins producing enough progesterone and estrogen. It also forms structures that will be part of the exchange mechanism in the placenta.

Spotlight on Structure & Function

As the embryo grows and develops, its relationship with the mother becomes more complex. The embryo cannot supply itself with nutrients or get rid of its own wastes, so exchange vessels form between the embryo and the mother for the exchange of nutrients and wastes without direct mixing of their blood. This is the function of the placenta and the umbilical cord. The **placenta** (Figure 20.8) is the entire structure that forms from embryonic tissue (chorion and chorionic villi) and maternal tissue (endometrium). The **umbilical cord** is the two-way lifeline that connects the placenta to the embryo's circulation.

The placenta develops because the cells of the chorion, just like the cells of the trophoblast from which they derive, secrete enzymes that eat away at the endometrial tissues and capillaries in the vicinity of the embryo. These enzymes rupture the capillaries, causing the formation of small blood-filled cavities. The developing chorion extends fingerlike processes called *chorionic villi* into these pools of maternal blood. Each chorionic villus contains small capillaries that are connected to umbilical arteries and veins that are part of the circulation of the fetus. In other words, the chorion damages the endometrium to cause local bleeding, and then it taps that bleeding as a source of nutrients and oxygen for the embryo and as a place to get rid of embryonic wastes, including carbon dioxide. In effect, an infant behaves like a parasite within the mother.

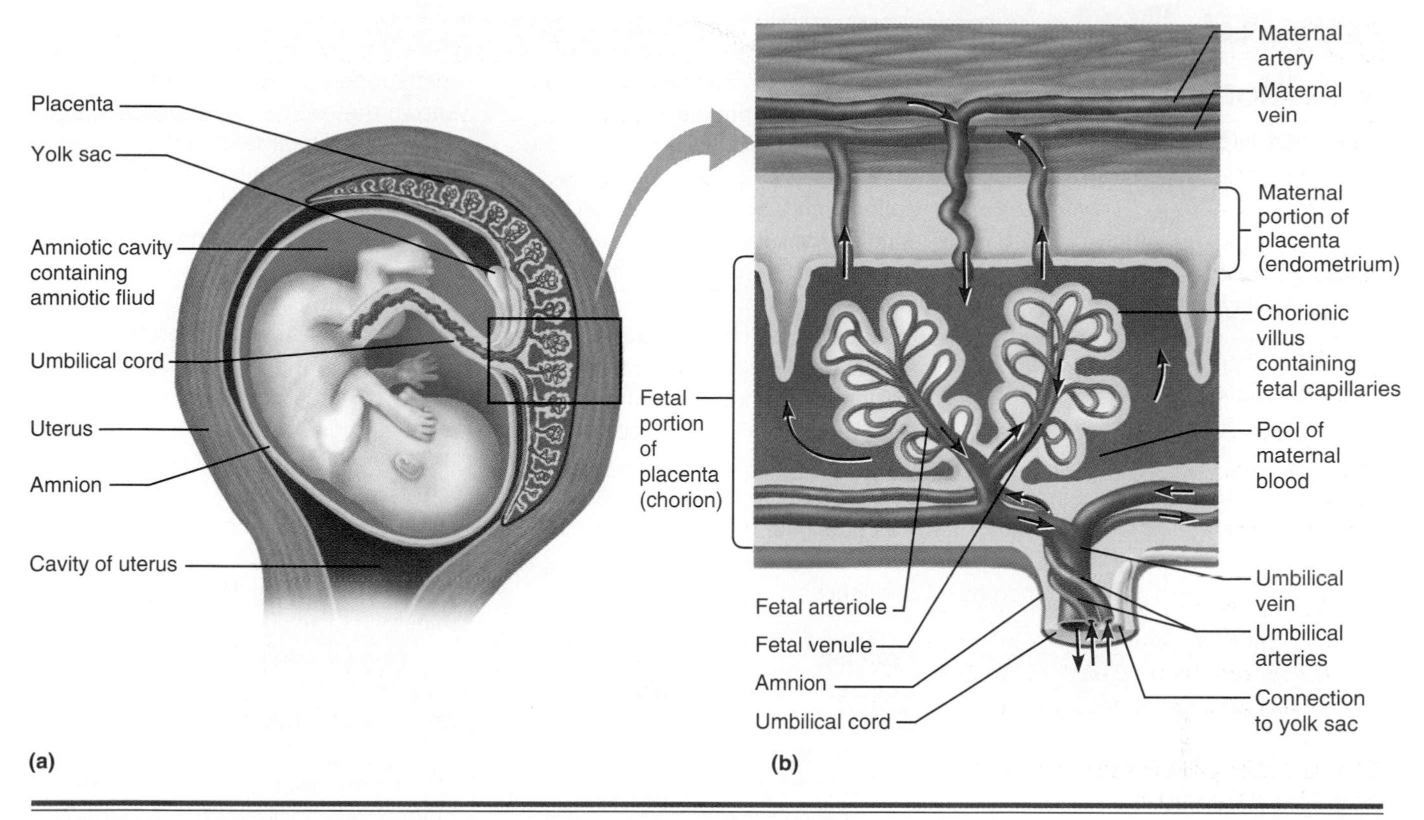

Figure 20.8 The placenta and umbilical cord. (a) The fetus is bathed in amniotic fluid within the amnion. Its only connection to the mother is the umbilical cord. **(b)** A closer view of portions of the placenta and umbilical cord, showing how nutrients and gases are exchanged between maternal and fetal blood without blood mixing.

The placenta is an effective filter. It permits the exchange of nutrients, gases, and antibodies between the maternal and embryonic circulations but not the exchange of large proteins or blood cells. The placenta allows the embryo to take advantage of the mother's organ systems until its own organs develop and become functional.

However, the placenta may permit certain toxic substances and agents of disease to cross over to the fetus as well. Examples include alcohol, cocaine, the HIV virus, and a wide variety of prescription and nonprescription drugs. Many of these substances can do a great deal of harm to the embryo when it is in the earliest stages of differentiation and morphogenesis, even though some of the prescription drugs may be therapeutic for the mother (review the Health Watch box in Chapter 16). Upon learning she is pregnant, a woman should review all her prescription and nonprescription drugs with her physician.

The placenta is also an endocrine organ. Initially the placenta secretes hCG so the corpus luteum continues to secrete the progesterone and estrogen necessary to maintain the pregnancy. Later the placenta secretes its own progesterone and estrogen, and the corpus luteum regresses. Progesterone and estrogen promote the growth of the myometrium in preparation for intense contractions at birth, maintain the endometrial lining so menstruation does not occur, inhibit uterine contractions during pregnancy, and help form a thick mucus plug over the cervix to inhibit uterine infections.

The two umbilical arteries and single umbilical vein are considered part of the embryonic circulation, meaning the umbilical vein carries blood *toward* the embryonic heart and the umbilical arteries carry blood *away* from the embryonic heart and back to the placenta. Because the exchange of nutrients by the placenta requires a functional embryonic heart, the placenta and umbilical vessels do not become fully functional until the embryonic heart develops at about five weeks. Until then the developing chorion supplies the nutrient needs of the embryo by diffusion. ■

Recap **Four extra-embryonic membranes serve varying functions: (1) the amnion cushions the fetus and preserves it from dehydration; (2) the allantois forms blood vessels of the umbilical cord; (3) the yolk sac forms part of the digestive tract and produces fetal blood cells and reproductive germ cells; (4) and the chorion secretes hCG and forms structures that will**

Health Watch

Nutrition and Exercise During Pregnancy

The embryo places quite a burden on the body of the pregnant woman. It's worth knowing how maternal nutrition affects infant development, and whether or not you can (or should) exercise during pregnancy.

Nutrition

The embryo receives all its nutrients from the maternal circulation. Early pregnancy is a time of extremely rapid embryonic differentiation and morphogenesis, and any vitamin or mineral deficiencies can have long-lasting effects on the fetus or may even affect the pregnant woman's health. Protein, iron, and calcium are especially important. If the mother does not consume enough calcium, for example, the fetus may get what it needs from her body, depleting the pregnant woman's bones of much-needed calcium. Some physicians believe that folic acid is important for early nervous system development, and consequently they may recommend folic acid supplements.

During late pregnancy, differentiation and morphogenesis slow somewhat but the fetus grows rapidly. At this time a pregnant woman needs about 300 more calories per day than she normally eats, and it is perfectly normal for a woman to gain 25–30 pounds during pregnancy.

There are plenty of places in the world where all persons, including pregnant women, do not get enough to eat. Studies show that severe caloric restriction increases the infant death rate and causes some infants to be underweight. Interestingly, a study conducted after World War II revealed that underweight female newborns who survived to sexual maturity were also likely to have underweight babies, even though their own nutrition during pregnancy was adequate. In other words, nutritional deficits can have long-lasting consequences.

Exercise

Exercise redistributes blood flow to exercising muscle and skin and away from the uterine area. Exercise also produces heat, and body temperatures several degrees above normal have been recorded in long-distance runners. However, all the evidence so far indicates that moderate exercise is not harmful to pregnant women and may even be beneficial.

Measurements of fetal blood flow by noninvasive ultrasound techniques suggest that moderate exercise during pregnancy does not impede fetal blood flow. Other studies of body temperature during exercise suggest that pregnant women dissipate heat even better than nonpregnant women. In direct support of exercise, it has been shown that women who exercise regularly tend to bear larger infants, and a larger baby generally is healthier. Women who exercise also feel better about themselves and have an easier labor and delivery.

The best types of activity during pregnancy are aerobic exercises involving many muscles. Of course, safety is paramount. Swimming is ideal because the water absorbs body heat. The water also provides support against gravity, preventing injuries to joints. Any exercise during pregnancy should be moderate, with the heart rate not exceeding 160 beats per minute.

Women who already exercise moderately on a regular basis can probably continue during pregnancy. However, it is not advisable to begin a totally unfamiliar exercise program during pregnancy unless advised to do so by a health care provider. After delivery the mother should stop exercising temporarily, usually for about two months.

We have learned a great deal about the nutrition and exercise requirements for a healthy baby. Any woman who learns she is pregnant or is contemplating becoming pregnant should consult her health care provider for guidance.

become part of the placenta. The placenta is the site of nutrient and gas exchange between embryo and mother, and it secretes hormones. The umbilical cord joins the embryo to the placenta.

The embryo develops rapidly

At three weeks the embryo is fully embedded in the endometrium, which provides it with nutrients via the developing chorion. At about this time the embryo begins to take shape as cells of the three primary tissue layers migrate to other locations and begin to form the rudimentary organs and organ systems. First a small groove called the *primitive streak* appears in the flat, round embryonic disk, and the embryonic disk begins to elongate along one axis (Figure 20.9). Then a *neural groove* of ectoderm forms that will later become the brain and spinal cord. Meanwhile, the mesoderm begins to separate into several segments called *somites* that will become most of the bone, muscle, and skin. Prominent bumps called the *pharyngeal arches* appear at one end; they are destined to become part of the face, neck, and mouth. By the end of week four the heart is beginning to develop, the head begins to take shape, the position of the eyes becomes apparent, the neural groove has closed into a *neural tube,* and four *limb buds* appear (Figure 20.10).

The patterns of cell differentiation and organism morphogenesis are so consistent among vertebrates

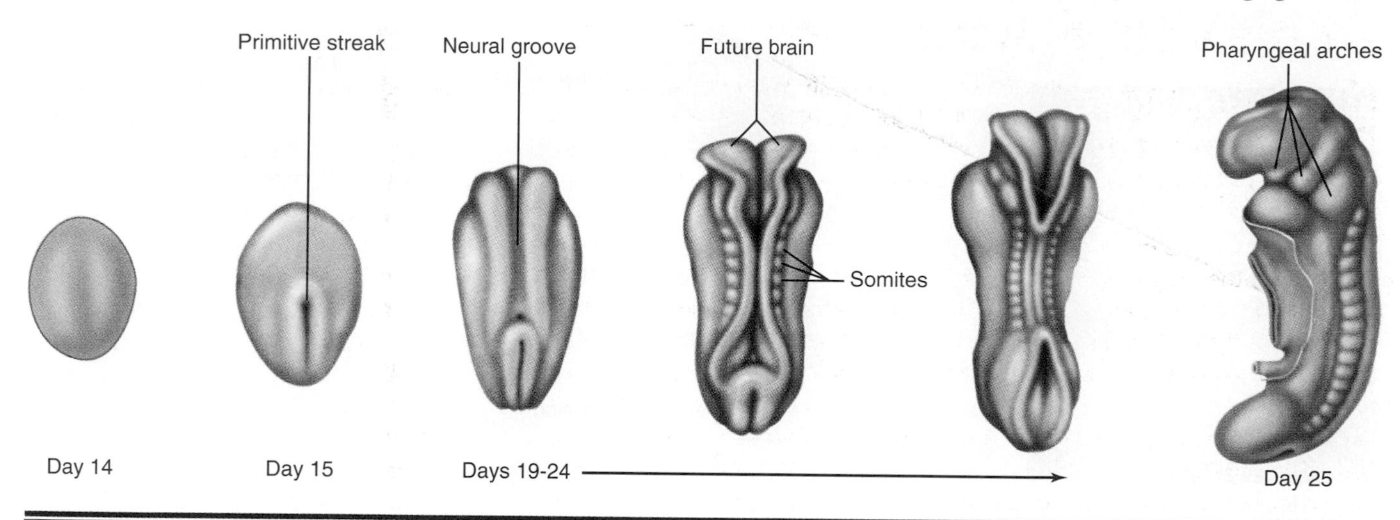

Figure 20.9 Embryonic development during the third and early fourth week. Day 14 represents the flat embryonic disk as it would look from above (from the amniotic cavity). On day 15 the embryo begins to elongate and the primitive streak of ectoderm appears. Days 19–23 mark the appearance of a neural groove of ectoderm. Somites develop in the mesoderm that will become bone, muscle, and skin. By days 24–25, pharyngeal arches that will contribute to structures of the head become visible. The embryo on days 24 and 25 is viewed from the side.

that up until about four weeks of human development all vertebrate embryos are similar. Even an expert would have a hard time telling whether a four-week-old embryo was destined to become a fish, mammal, or bird. Because early differentiation and morphogenesis follow such a common path, it is likely that these patterns developed early in the history of the vertebrates.

Weeks five to eight mark the transition from a general vertebrate form to one that is recognizably human. The head grows in relation to the rest of the body, the eyes and ears are visible, and the four limbs are formed with distinct fingers and toes. A cartilaginous skeleton forms. The heart and circulatory system complete their development and the umbilical cord and placenta become functional, so that now blood circulates throughout the fetus. However, the blood cells for the fetal blood are produced by the yolk sac because the blood-forming tissues of the fetus are not yet mature. Nutrients and wastes are now exchanged more efficiently and in greater quantities. The tail regresses. At the end of the embryonic period, the embryo is an inch long and weighs just about one gram (Figure 20.11).

A major change during this period is the development of male or female gonads. Recall that if the embryo has a Y chromosome the testes develop at about six or seven weeks. A burst of testosterone from the testes influences the subsequent development of male sexual characteristics. Without this burst of testosterone the gonads develop into ovaries.

Although it is difficult to estimate how many embryos fail to survive to the end of the embryonic period, some estimates place the number as high as 20%. Spontaneous termination of pregnancy followed by expulsion of the embryo is called a *miscarriage,* or spontaneous abortion. Sometimes a woman never even realizes she was pregnant. Miscarriages are probably nature's way of weeding out embryos with genetic disorders that might prevent them from developing normally.

> ***Recap*** **Embryonic development is rapid and dramatic. By the fifth week the embryo is becoming distinctly human in form, and by eight weeks it is about an inch long. The natural rate of miscarriages through eight weeks of development may be as high as 20%.**

20.5 Fetal development: Eight weeks to birth

After the eighth week the developing human is called a fetus. During fetal development the fetus grows to about seven pounds and a length of about 20 inches. Fetal organ systems mature and the bones begin to calcify. The mother, too, undergoes obvious changes in physical shape. Each of these stages will now be examined in more detail.

Months three and four

During the third month the kidneys develop sufficiently that the fetus begins eliminating some wastes as urine into the amniotic fluid. The limbs are well developed, the cartilaginous skeleton begins to be replaced with bone, and the teeth form. The spleen participates briefly in red blood cell production, a job that will soon be taken over by the liver and bone marrow. The liver begins to function, and the male and female genitalia are well enough developed that fetal sex can be determined. The end of the third month marks the end of the *first trimester.*

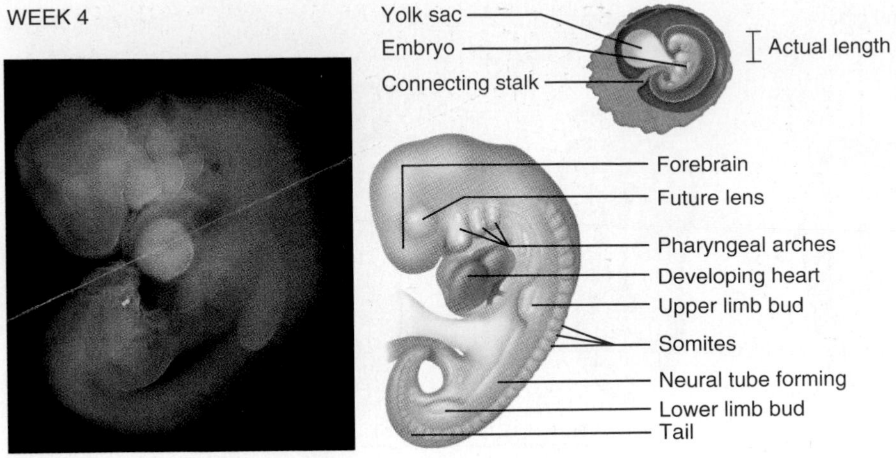

WEEK 4

Yolk sac
Embryo
Connecting stalk
Actual length
Forebrain
Future lens
Pharyngeal arches
Developing heart
Upper limb bud
Somites
Neural tube forming
Lower limb bud
Tail

Figure 20.10 Week four of development. The future head is prominent, the position of the eyes becomes apparent, the heart and neural tube are forming, and limb buds appear.

During the fourth month the liver and bone marrow begin producing blood cells. The face takes on nearly its final form as the eyes and ears become fixed in their permanent locations (Figure 20.12). In the female fetus, follicles are forming in the ovaries.

Overall growth is rapid during this time. By the end of the fourth month the fetus is already about six inches long and weighs about 170 grams (six ounces).

Months five and six

By the fifth month the nervous system and skeletal muscles are sufficiently mature that the fetus begins to move. These movements are called *quickening,* and the mother may feel them for the first time. The skin, which is well formed although red and wrinkled, is covered by soft downy hair. The heartbeat is now loud enough to be heard with a stethoscope. Skeletal hardening continues.

The sixth month marks the first point at which the fetus could, with the best neonatal care available, survive outside the uterus. At this stage the fetus seems to respond to external sounds. Most importantly, the lungs begin to produce surfactant, the substance that reduces surface tension in the lungs and permits the alveoli to fill with air at birth.

The end of the sixth month marks the end of the *second trimester.* At this point the fetus weighs about 700 grams, approximately 1½ pounds.

Months seven through nine

The seventh through ninth months (the *third trimester)* are a period of continued rapid growth and maturation in preparation for birth. The eyes open and close spontaneously and can be conditioned to respond to environmental sounds. Activity increases, as if the fetus is seeking a more comfortable position in the increasingly restrictive space of the uterus. Usually the fetus moves to a position in which the head is positioned downward, near the cervix. The skin begins to lose some of its reddish color and its coat of downy hair. In the male the testes descend into the scrotal sac. Although neither the lungs nor the digestive systems have had a chance to function yet, they are both now ready.

By nine months the fetus is about 20 inches long and weighs approximately 6–7½ pounds. Birth usually occurs at about 38 weeks of development.

Recap **The period of fetal development extends from eight weeks to birth at 38 weeks. Growth is rapid, with the mature fetus weighing approximately 6–7½ pounds at birth. The fetus begins to move at about five months, and life outside the womb is at least possible by about six months when the lungs begin to produce surfactant.**

20.6 Birth and the early postnatal period

Other than perhaps conception and death, there is no period in our lives marked by greater developmental change than the hours during and immediately after birth. Within minutes, the newborn makes the shift from relying solely on the maternal circulation for nutrient and gas exchange to depending on its own circulation and lungs. The digestive tract, too, starts to function with the first swallow of milk. Let's take a quick look at the changes that occur during birth and the **postnatal** ("after birth") period.

Labor ends in delivery

Birth involves a sequence of events that we call *labor* (It's hard work!), which ends in the *delivery* of the newborn into the world.

Spotlight on Structure & Function

Labor begins as a result of a series of events that is triggered by maturation of the fetal pituitary gland, which serves as the timing device to indicate that the fetus is now ready for birth.

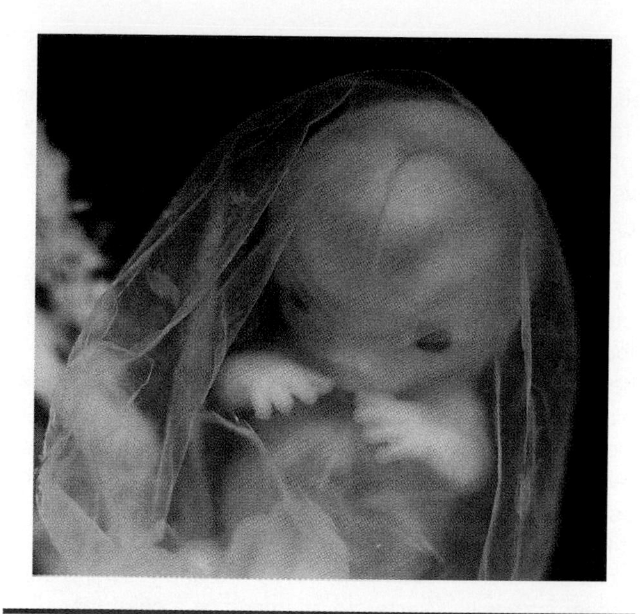

Figure 20.11 Week 8 of development. The embryo is now distinctly human in appearance. The face, limbs, hands, and feet are well formed. The circulatory system is functional and blood now circulates throughout the fetus and to and from the placenta. Actual length is one inch.

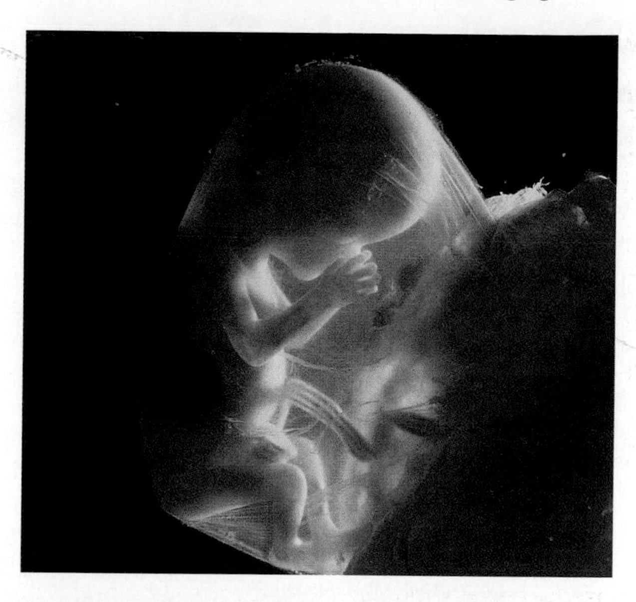

Figure 20.12 The fetus and placenta at four months of development. Note the amnion and the clear amniotic fluid surrounding the fetus. Actual length is 6 inches.

The mature fetal pituitary gland begins to release ACTH, which stimulates the fetal adrenal gland to secrete steroid hormones that cause the placenta to increase its production of estrogen and decrease the production of progesterone. Estrogen, in turn, increases the number of oxytocin receptors and stimulates production of prostaglandins. Together, the increased estrogen, prostaglandins, and oxytocin receptors (along with the mother's oxytocin) stimulate the uterus to contract rhythmically.

A positive feedback cycle begins: rhythmic contractions of the uterus cause the release of still more oxytocin from the maternal pituitary, which increases contractions still further, and so on. As labor progresses the periods of contraction get closer together, last longer, and become stronger, creating enough force to push the fetus toward the cervix and eventually through the vagina. ■

The total period of labor and delivery lasts about 24 hours for the first birth and slightly less time for subsequent births. Labor and delivery are divided into three phases: dilation, expulsion, and afterbirth (Figure 20.13 on page 508).

- *Stage 1—dilation*. The first phase of labor can last 6–12 hours. During this time the rate, duration, and strength of contractions increase over time, pushing the head of the fetus against the cervix. The cervix itself is drawn back toward the uterus, widening the cervical opening and expelling the mucus plug. Continued pressure of the fetus against the cervix widens the cervical opening even further until eventually it is large enough to accommodate the baby's head, or about 10 centimeters. At about this time the pressure of the baby's head ruptures the amnion, releasing the amniotic fluid. This "breaking of the water" is a normal sign that delivery is proceeding. The physician will keep a close watch on the stage of dilation by measuring the degree of cervical dilation at regular intervals.
- *Stage 2—expulsion*. The period of expulsion extends from full cervical dilation through actual delivery. During this time uterine contractions strengthen and the woman experiences an intense urge to assist the expulsion with voluntary contractions as well. In order to make the birth easier, some women opt for a surgical incision called an **episiotomy** to enlarge the vaginal opening. The intense contractions push the baby slowly through the cervix, vagina, and surrounding pelvic girdle. The appearance of the baby's head at the labia is called *crowning*. As soon as the baby's head appears fully, an attendant removes mucus from the baby's nose and mouth to facilitate breathing. The rest of the body emerges rather quickly. When the baby has fully emerged the umbilical cord is clamped to stop blood flow and then cut. By this time the infant should be breathing on its own. The entire expulsion phase lasts less than an hour.
- *Stage 3—afterbirth*. Contractions do not stop with birth. Strong postdelivery contractions serve to detach the placenta from the uterus and expel the umbilical cord and placenta, collectively

Directions in Science

Prenatal Diagnostic Techniques

Fewer than 40 years ago, when a woman became pregnant there was little that either the parents or the physician could do except wait nine months and hope that the baby was born healthy. Late in pregnancy the physician could hear the fetal heartbeat, but that was about it. Given that nearly 7% of all newborns had a minor or major birth defect of some kind, pregnancy was a stressful period of wait-and-see. A healthy baby was cause for a collective sigh of relief.

In the late 1960s a new technique called ***amniocentesis*** became available for the early diagnosis of some types of fetal disease. In amniocentesis, a long hollow needle is inserted through the abdominal and uterine walls, into the amniotic cavity, to take a sample of amniotic fluid. The technique was developed to take advantage of the fact that fetal skin and lung cells are sloughed off into amniotic fluid during fetal development.

Amniocentesis generally is performed during the second trimester, between the 15th and 17th weeks of pregnancy. The harvested fetal cells are then grown in the laboratory for up to four weeks until there are enough of them to examine the chromosomes and perform certain biochemical tests. A drawback to amniocentesis is that test results may not be available until the middle of the second trimester, close to the time that termination of pregnancy will no longer be an option. Still, the technique has gained widespread acceptance because for a long time it was the only technique available.

Amniocentesis is relatively safe. It may be recommended if there is a family history of a disabling or fatal genetic disease such as Huntington disease or cystic fibrosis, or when the risk of Down syndrome is high. Nearly 40 different fetal defects can now be diagnosed using fetal cells obtained by amniocentesis.

The 1970s saw a major advance in fetal imaging called ***ultrasound***. By itself, ultrasound does not test for genetic diseases. Rather it provides a crude but real-time image of the fetus within the womb. The technique is based on the fact that high-frequency (ultrasound) sound waves aimed at the abdomen and uterus rebound when they hit hard tissues in the fetus. These reflecting waves are analyzed and converted to an image by a computer.

Ultrasound is very safe. It can be used to determine whether the fetus is developing properly and to estimate the due date if the couple does not know when fertilization occurred. Ultrasound was originally performed primarily in the third trimester, but with refinements the technique is now

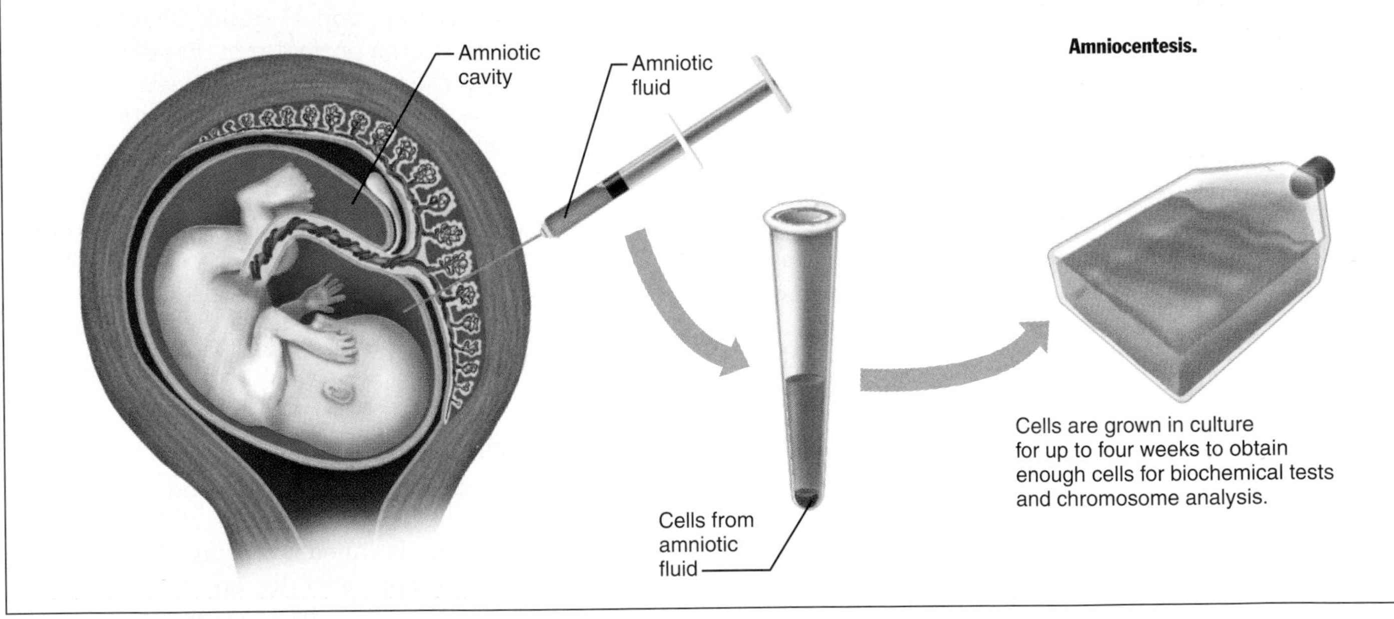

Amniocentesis.

called the afterbirth. The afterbirth stage usually lasts less than half an hour after the birth of the infant.

Recap **Labor and delivery take about 24 hours or less. The dilation phase is marked by widening of the cervix and increasingly frequent and strong uterine contractions. The fetus emerges in the shorter (one hour or less) expulsion phase. The placenta and umbilical cord are expelled during the afterbirth phase.**

Cesarean delivery: Surgical delivery of a baby

Some deliveries are not allowed to proceed as described above. Surgical delivery of a baby is called a used to diagnose fetal structural abnormalities as early as the second trimester. Usually the image is so good that the genitalia are visible and the parents can know the sex of the baby before birth.

At about the same time ultrasound was developed, a few groups of physicians around the world were refining a new technique that allowed the diagnosis of fetal genetic defects much earlier than amniocentesis. ***Chorionic villi sampling*** is a technique for collecting cells of the chorionic villi that originate from embryonic tissue.

At first, chorionic villi sampling was considerably more dangerous and difficult than amniocentesis. The technique did not become widely accepted until the early 1980s, when scientists learned to combine it with ultrasound visualization. As the technique is now performed, a thin flexible tube is inserted through the vagina and into the uterus under the guidance of an ultrasound image. When the tip of the tube is in the chorion as indicated by the ultrasound picture, a small sample of chorionic villi tissue is taken for chromosomal and biochemical analysis.

The great advantage of chorionic villi sampling is that it can be done as early as five weeks of pregnancy. Because there is a slight but real risk of injury to the embryo, the technique is recommended only when a family history indicates that there may be a risk of genetic defects.

Since the 1980s a test for ***alpha feto-protein*** has been used to screen for the presence of a particular fetal protein in maternal blood. Normally this fetal protein is found only in very low concentrations, but when the fetus has certain developmental defects such as ***spina bifida***, large amounts of alpha feto-protein diffuse into the maternal circulation.

The alpha feto-protein screening test is far less invasive than chorionic villi sampling, for only a sample of the pregnant woman's blood is required. The test is usually performed at about 16–18 weeks of pregnancy.

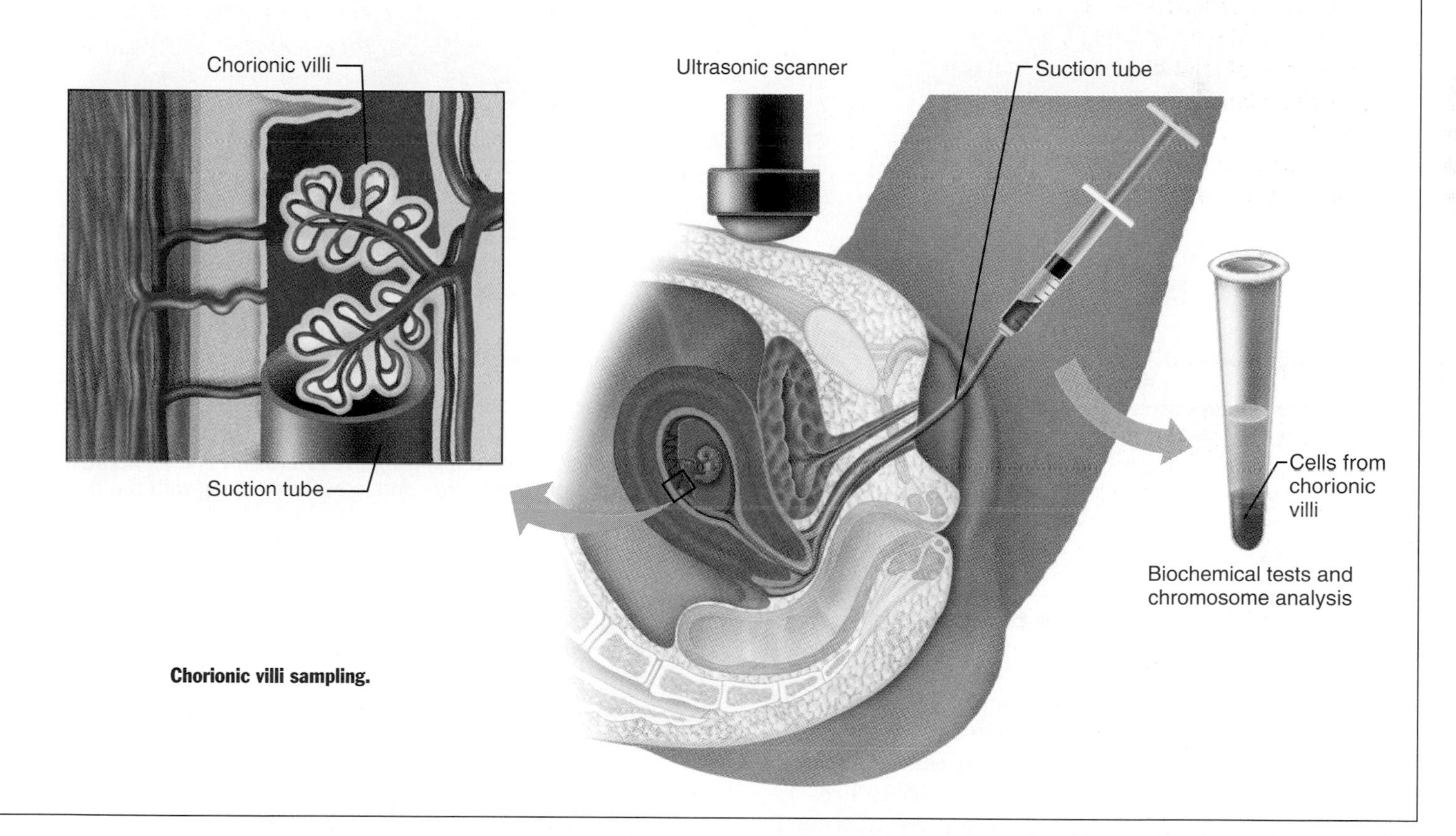

Chorionic villi sampling.

cesarean delivery, or *C-section.* In a C-section, an incision is made through the abdominal wall and uterus so the baby can be removed quickly.

A C-section is often performed when the position or size of the fetus could make a vaginal delivery dangerous. Possible reasons include a fetus that is too large for the birth canal, improper position of the fetus with the legs near the cervix, maternal exhaustion by a long delivery, or signs of fetal distress such as an elevated fetal heart rate.

In the past, if a C-section was performed it was common practice to do all subsequent deliveries by C-section as well. Today most women who have a C-section can still opt for a vaginal delivery with the next child if they wish.

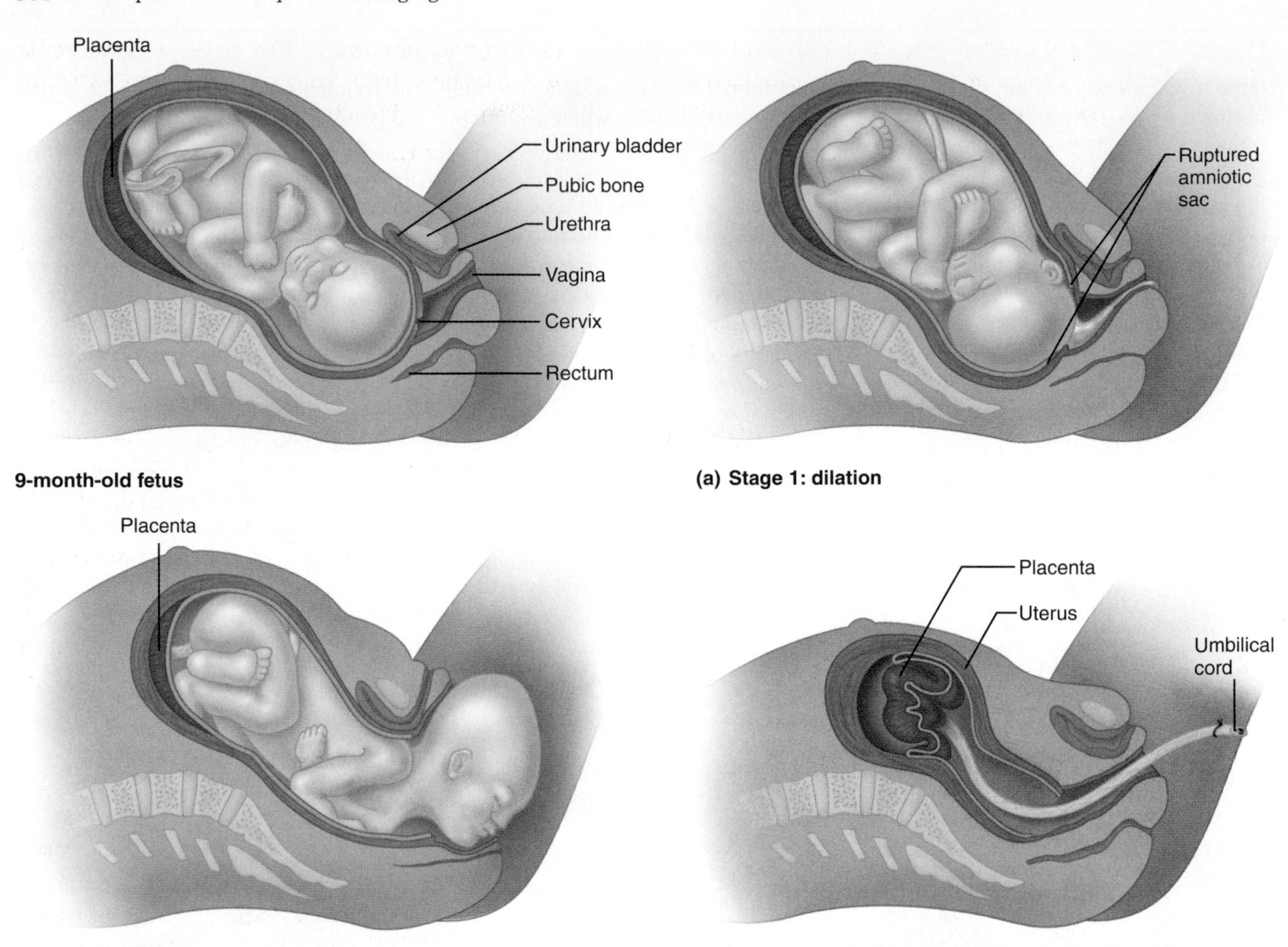

Figure 20.13 The stages of birth. As birth approaches, the fetus usually is positioned with the head down and toward the cervix. **(a)** Stage 1: Dilation. The cervical opening widens. The amnion may break at this stage. **(b)** Stage 2: Expulsion. The fetus passes headfirst through the cervical canal and the vagina. **(c)** Stage 3: Afterbirth. The placenta detaches from the uterus and is expelled along with the remainder of the umbilical cord.

The transition from fetus to newborn

Somewhere between the point when its head emerges and the placenta is clamped, the fetus must become capable of sustaining life on its own. The first necessity is that the newborn (also called the *neonate* for the first 28 days) must start to breathe, preferably even before its umbilical connection to the mother is severed. Second, over the first few days of life, its cardiovascular system undergoes a series of remarkable anatomical changes that reflect the loss of the placental connection for gas and nutrient exchange. The newborn is now entirely on its own.

Taking the first breath When the infant emerges, its lungs have never been inflated. After all, it essentially has been living underwater (more correctly, under amniotic fluid). The first inflation is critical and not all that easy. If you've ever tried to blow up a really stiff balloon, you know that it takes a lot of effort to start the inflation. The reason for this effect is that the surface tension generated by a small sphere is greater than the surface tension generated by a larger one. The first inflation in the infant is facilitated by substances called *pulmonary surfactants* that reduce surface tension in the lungs' alveoli.

But what causes the infant to take that first breath? During labor, the placental connection to the mother begins to separate, reducing gas exchange with the fetus. Clamping the umbilical cord stops gas exchange entirely. This situation is equivalent to someone covering your mouth and nose completely. Within seconds the carbon dioxide concentration in the fetus rises to the point that the respiratory centers in its brain stimulate respiration. With the enormous effort of one who is being smothered, the infant takes its first breath and begins to cry.

Changes in the cardiovascular system The fetal cardiovascular system is different from the cardiovascular system described in Chapter 7. Consider that the fetal lungs are not yet useful for gas exchange and blood enters and leaves the fetus via the umbilical cord. Immediately after birth the umbilical circulation ceases, and pulmonary circulation becomes crucial to survival.

There are four major differences between the fetal cardiovascular system and the cardiovascular system of a newborn several weeks after birth (Figure 20.14):

- Blood enters the fetus via the umbilical vein. A small amount of umbilical vein blood flows through the liver, but most of it bypasses the liver and joins the inferior vena cava of the fetus via the *ductus venosus* (venous duct).
- Blood leaves the fetus via two umbilical arteries that originate from arteries in the legs.
- In the fetus there is a hole between the two atrial chambers of the heart called the *foramen ovale* (oval opening). The hole permits some blood to pass from right atrium to left atrium, bypassing the pulmonary circulation.
- In the fetus there is a shunt, or shortcut, from the pulmonary artery directly to the aorta called the *ductus arteriosus* (arterial duct). Consequently, most of the blood pumped by the right ventricle also bypasses the lungs.

Let us now follow the circulation of blood between fetus and mother and within the fetus. Oxygen-laden and nutrient-rich blood from the placenta enters the fetus via the umbilical vein. Some of the blood is carried to the developing liver either directly or via a branch of the umbilical vein that joins the hepatic portal vein. But the fetal liver is not yet fully developed and cannot handle the entire umbilical blood flow. Therefore, most of the blood bypasses the liver and is carried directly to the inferior vena cava by the ductus venosus. In the inferior vena cava, the nutrient-rich blood mixes with the venous blood of the fetus.

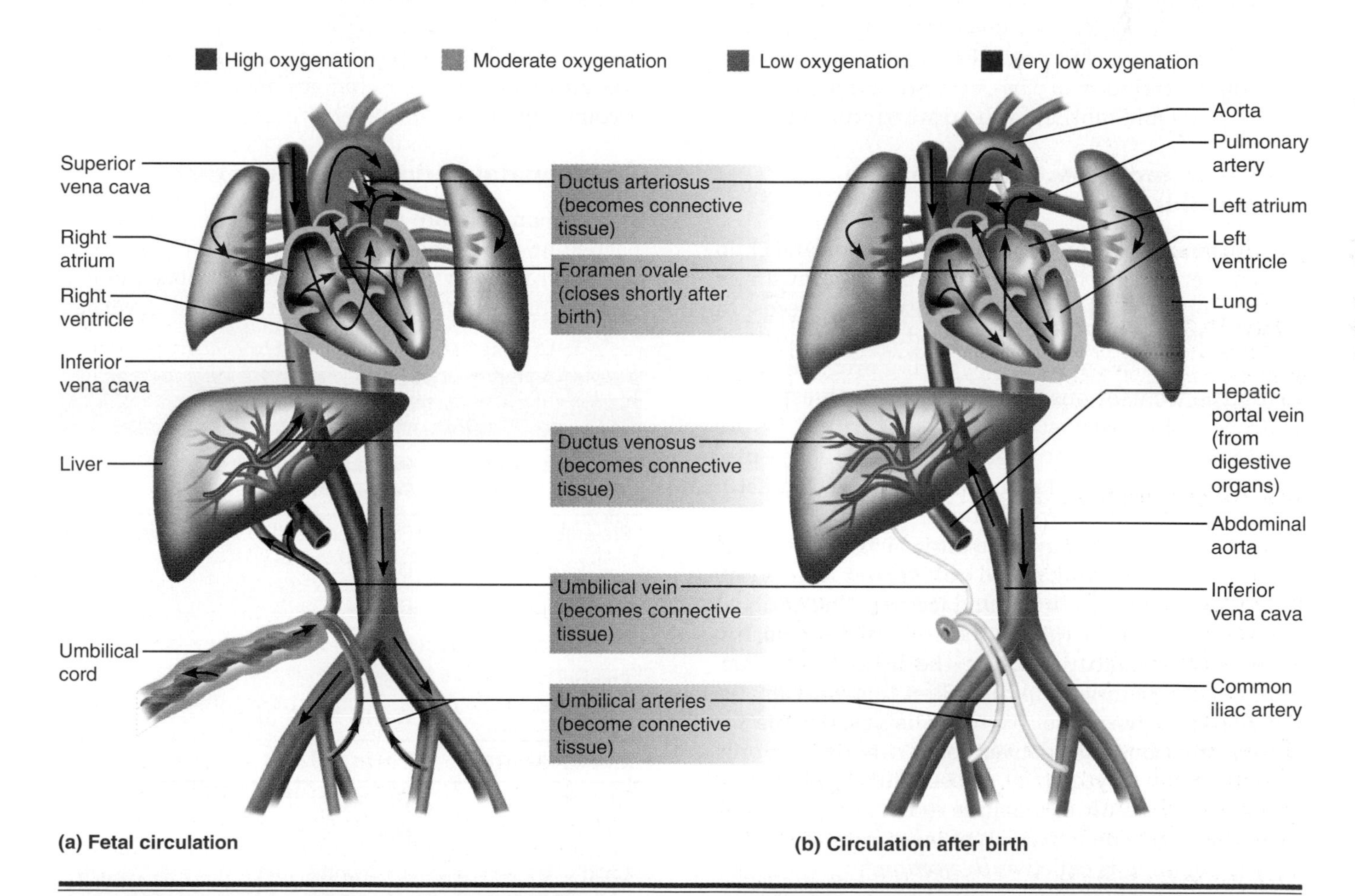

Figure 20.14 The cardiovascular systems of the fetus and of the newborn. In the fetus the immature lungs are not used for gas exchange, so most of the cardiac output bypasses the lungs. Gas exchange occurs via the placenta, so there are special fetal blood vessels that connect the fetus to the placenta. These unique pathways for blood in the fetus close shortly after birth. Starting shortly after birth, the entire cardiac output of the right ventricle passes through the lungs.

Most of the blood that enters the fetal heart must bypass the fetal lungs because they too are not fully developed. Some of the blood passes from the right atrium to the left atrium through the foramen ovale, and some is shunted from the pulmonary artery directly to the aorta via the ductus arteriosus.

Even though the fetal lungs are not yet functional and the digestive tract is not receiving nutrients, the aortic blood still has sufficient oxygen and nutrients (brought to the fetus via the umbilical vein) to supply all the fetal tissues. Some of the fetal arterial blood returns to the placenta via the umbilical arteries. In the placenta the blood again picks up nutrients and oxygen and gets rid of carbon dioxide generated by fetal metabolism.

The situation changes dramatically at birth, for the umbilical circulation is immediately cut off. Over time the umbilical vein, umbilical arteries, and ductus venosus regress to become vestigial connective tissue. The foramen ovale and ductus arteriosus close off in the first few days or weeks after birth as well, so that all the cardiac output passes through the lungs for efficient gas exchange. Nutrients are now absorbed by the newborn's digestive tract and pass through the liver via the hepatic portal vein. The only reminder of the intimate connection between fetus and mother is the **umbilicus** (navel).

Lactation produces milk to nourish the newborn

The intimate relationship between mother and child does not end at birth, for the human infant must rely entirely on its mother for all its nutritional needs. In preparation for lactation (milk production), the pregnant woman's breasts enlarge during pregnancy due to high concentrations of estrogen and progesterone. However, the breasts do not secrete milk before childbirth because estrogen and progesterone prevent the action of prolactin, the hormone that actually stimulates milk production.

During the first day or so after birth, the breasts produce a watery milk called **colostrum** that is rich in antibodies but low in fat and lactose. The antibodies help to protect the newborn until its own immune system matures. Later, the high fat and lactose content of milk help the infant to grow rapidly.

A fourth hormone, oxytocin, is responsible for the contractions that deliver colostrum or milk only when needed. Oxytocin is released during childbirth and also as a result of a neural reflex every time the infant nurses. The expulsion of milk from the breast during suckling is called *milk ejection.*

Recap **At birth a sharp increase in carbon dioxide causes the newborn to take his/her first breath. Shortly after birth the umbilical blood vessels and ductus arteriosus regress, and anatomical changes in the heart reroute all blood through the lungs. Prolactin stimulates lactation, and oxytocin stimulates the release of milk during suckling.**

20.7 From birth to adulthood

The long process of human development continues after birth. Because development is a continuum, experts do not always agree on how to categorize its stages. Some categories may seem somewhat arbitrary, such as the distinction (if there is one) between adulthood and old age. At the risk of oversimplifying, we define the *neonatal* period as the first month, *infancy* as months 2–15, *childhood* as the period from infancy to adolescence, *adolescence* as the transitional period from childhood to adulthood (approximately 12 to 18 or 20 years of age), and *adulthood* as 20 years and beyond.

Table 20.2 summarizes the stages of human development. Bear in mind that these ages are approximate, intended as a guide to orient you with regards to our progression through life. We'll concentrate primarily on biophysical development, leaving human psychosocial and cognitive (learning) development to the fields of psychology and education. Figure 20.15 depicts the changes in body form that accompany human development to adulthood.

The neonatal period: A helpless time

The human neonate (newborn) is fairly helpless. In particular, the nervous system and the muscular system are relatively immature at birth. Consequently, movements are uncoordinated, weak, and governed

Table 20.2 A summary of the stages of human development

Prenatal Stage	Prenatal Age (After Fertilization)
Pre-embryo	From fertilization to two weeks
Zygote	Days 1–3
Morula	Days 4–7
Blastocyst	Days 8–14
Embryo	Weeks 3–8
Fetus	Weeks 8–40 (birth)
Postnatal Stage	**Postnatal Age (After Birth)**
Neonate	The first month
Infant	Months 2–15
Child	16 months to 12 years
Adolescent	12–20 years
Adult	Beyond 20 years
Old age	Aging accelerates after 40 years

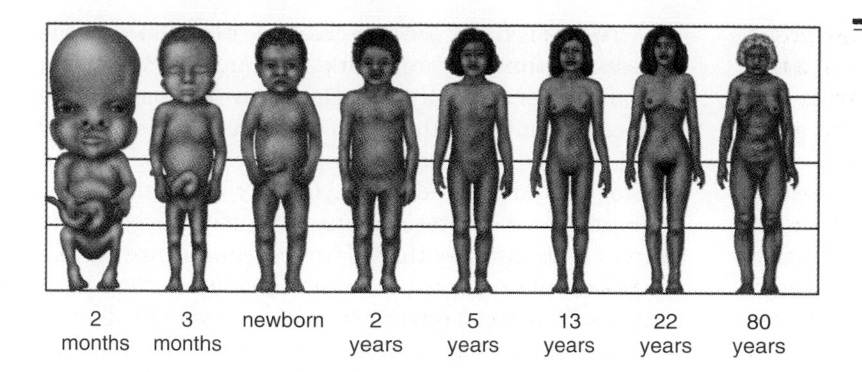

Figure 20.15 Changes in body form and relative proportion throughout prenatal and postnatal growth and development.

primarily by reflexes rather than conscious control. The neonate hardly has enough strength to hold its head up, so its head must be constantly supported whenever it is carried. The eyes are open but often they seem to wander about, unable to focus. Neonates are aware of their environment and can respond to it, but they are unable to retain any long-term memories. (What is *your* first memory?) They hear a narrower range of pitches than adults and are more sensitive to higher-pitched tones.

What neonates do best is suckle, urinate, and defecate. The digestive system can absorb the constituents of milk but is not yet ready for solid foods. The stools are soft, and defecation and urination occur as reflexes.

> ***Recap*** **The neonatal nervous and muscular systems are not well developed. Neonates are physically helpless and cannot form memories.**

Infancy: Rapid development and maturation of organ systems

Infancy is a time of rapid change. In just 15 months the infant more than triples in weight to about 22–24 pounds. The bones harden gradually, and there is a disproportionately large increase in muscle mass and strength. The brain also grows rapidly, with nearly half of all brain growth after birth occurring in infancy. Most of the growth occurs in the cerebral cortex, the area associated with sensory perception, motor function, speech, and learning. Most of the myelination of nerves occurs in infancy as well. The first teeth appear at about six months, and all 20 baby teeth are generally present by about one year. Most infants begin eating solid foods by about six months to a year.

By about 14 months the combined maturation of the musculoskeletal and nervous systems allows the human infant to begin to walk on its own. A rapidly growing body needs lots of rest, and infants sleep a lot. Infants can easily sleep 10 hours a night and still take several naps during the day.

The immune system consistently lags behind most other systems in its development. Most vaccinations are ineffective in infants because their immune systems cannot produce the appropriate antibodies. The schedule of vaccinations in infants and children is designed so that each vaccine is given only after the immune system becomes able to respond to each particular vaccine.

> ***Recap*** **Infancy is marked by rapid growth and continued maturation of organ systems. Infants begin to eat solid foods and to walk. The immune system develops more slowly than other organ systems.**

Childhood: Continued development and growth

The long period of childhood, from about 15 months to 12 years of age, involves continued growth of all systems. The brain grows to 95% of its final size during childhood. The immune system continues its slow process of maturation throughout childhood and even into adolescence. Body weight increases to an average of about 100 pounds in both boys and girls as the result of periods of slow growth interspersed with growth spurts. Both muscle strength and fine motor coordination improve.

In addition, body form alters as the long bones of the arms and legs lengthen. A lumbar curve develops in the small of the back, and abdominal musculature strengthens, so that the childhood "pot belly" disappears.

By the end of childhood the body form is distinctly adultlike, though still sexually immature. Most organ systems are fully functional or nearly so (although not necessarily their full adult size). The lone exceptions are the male and female reproductive systems, which have not yet matured.

> ***Recap*** **Brain growth is nearly complete by the end of childhood. Muscle strength and coordination improve, and body form changes to a more adult form.**

Adolescence: The transition to adulthood

One of the most challenging times of life is the transition from childhood to adulthood. The final growth spurt, one of the largest, occurs in adolescence. The growth spurt begins at different times in different individuals, so there is a wide range of normal body weights and degrees of sexual maturity in the 12–14-year-old age group. Rapid growth can lead to awkwardness and a temporary loss of coordination.

Both the skeletal and muscular systems experience their greatest rates of growth during adolescence. Other organs increase in size as well. Notably, the lungs more than double in mass and the kidneys and stomach increase by more than 50%. The brain, however, gains only about 5%.

Adolescence is marked by maturation of the reproductive systems and the human sexual response, an event known as **puberty.** Puberty is defined as the onset of menstrual periods (menarche) in the female at some time between 9½ and 15½ years and by reflex discharge of semen during sleep (nocturnal emissions) in the male sometime between 11½ and 17½ years.

What causes puberty?

The concentrations of all of the reproductive hormones are low in children, but they begin to rise during the period of sexual maturation (known as the *pubertal period*). The pubertal period appears to be initiated by maturation of GnRH-secreting neurons of the hypothalamus. The timing of maturation of the hypothalamus seems to have a genetic component because there are heritable patterns in the time of onset of puberty. The initiation of GnRH secretion by the hypothalamus stimulates the secretion of luteinizing hormone (LH) and follicle stimulating hormone (FSH) from the anterior pituitary, which in turn stimulate the production of sex hormones and the maturation of sex organs and secondary sexual characteristics.

Recap **Adolescence is accompanied by a growth spurt and by maturation of the reproductive systems. Puberty begins when the hypothalamus matures and starts to secrete GnRH.**

20.8 Aging takes place over time

Physiologically, at least, humans reach their peak in early adulthood at about 20 years of age. Most body systems are fairly stable from 20 to about 40, and then the process that we call aging seems to accelerate. No matter how much we exercise, eat right, and in general take care of ourselves, there still seems to be at least some "natural" aging process, just like the carefully timed sequence of events that occur in human development. To be sure, exercise and a proper diet do help, but in the end we all must face the fact that we will not be the same at age 80 as we were at 20. The big question, of course, is why we age and whether the aging process can be halted or at least slowed.

Aging is the process of change associated with the passage of time (Figure 20.16). Although the word does not necessarily imply deterioration (after all, some wines get better as they age), human aging invariably leads in that direction. Aging as we use the term in biology could be used synonymously with *senescence*, the progressive deterioration of multiple organs and organ systems over time. Human senescence (or aging) has no apparent or obvious cause—that is, it cannot be blamed on a particular disease or event.

A common statistic is that the average life span in the United States has increased from 47 years in 1900 to the upper 70s today. Does that mean we are aging more slowly now? Not at all. **Longevity**—how long a person lives—does not depend solely on the aging process. Most experts believe that our increase in longevity is largely due to a reduction in deaths by accident or disease, not to a slowdown in the aging process itself.

In fact, it appears that nothing we have done (so far!) has decelerated the aging process. If we could solve every known disease and prevent every accidental death, would we still come face to face with a natural endpoint of life? Nobody knows for sure, but so far the answer seems to be a qualified yes. No one has beaten the aging process yet, and there are very few individuals with documented ages of over 100 years.

What causes aging?

Why do we age?

Theories of the causes of aging fall into three major categories. The first is that there is some sort of internal program or counting mechanism, genetic in origin, that determines cell death. The second is that accumulated cell damage or errors eventually limit cells' ability to repair themselves. A third theory is that the decline in function of a critical body system may lead to the parallel senescence of other systems. There is evidence to support each theory.

Figure 20.16 Aging. Human aging is accompanied by progressive changes over time. These photos are of Queen Elizabeth II in 1940 (at age 14) and in 1992 (at age 66).

An internal cellular program counts cell divisions It has long been known that even when cells are grown under laboratory conditions that are ideal for cell growth and division, most cells will not continue to divide indefinitely. Most cells go through approximately 50–90 successive divisions and then just stop. Cancer cells are the exception, for they will continue to divide indefinitely.

Why do normal cells stop dividing? Apparently every strand of DNA ends with a long tail of disposable bits of DNA called *telomeres* that do not code for any genes. The telomeres are like a roll of tickets, and the idea is that every time the DNA is replicated prior to cell division a ticket (telomere) is removed (Figure 20.17). When the cell runs out of telomeres, cell division begins to eat away at vital genetic information. Eventually enough vital genetic information is lost that the cell stops dividing and dies.

However, cancer cells manufacture an enzyme called *telomerase* that builds new telomeres to replace the ones lost during cell division. There has been a lot of excitement about telomerase lately as a result of these findings. Some researchers have even suggested that telomerase may be the magic bullet that could reverse or prevent the aging process, or that an antitelomerase substance might offer the cure for cancer.

These ideas remain far-fetched at the moment; understanding why a single cell stops dividing does not tell us how to arrest the aging process at the level of the whole body. After all, aging is not just a decline in cell division but in cell function as well.

Cells become damaged beyond repair The second theory holds that over time, accumulated damage to DNA and errors in DNA replication become so great that they can no longer be countered by the cell's normal repair mechanisms. The damage is thought to be caused by certain toxic byproducts of metabolism, such as oxygen free radicals. Eventually, like a ticking time bomb, the damage becomes too great for survival.

According to proponents of this theory, how long cells live is a function of the balance between how often damage occurs and how much of it can be repaired. Cells that undergo frequent injury and are slow to repair the damage will live less long than cells that are rarely damaged and repair themselves efficiently.

This theory is supported by the finding that severe caloric restriction in certain experimental animals prolongs life. One possibility is that by lowering body temperature just a little, caloric restriction may slow metabolism and reduce the rate of formation of damaging metabolic byproducts. Some persons believe that taking antioxidant vitamins may slow the aging process and prolong life, but there is not yet enough long-term evidence to know for sure if this is true.

Aging is a whole-body process Why does aging seem to happen to virtually all body systems at once, instead of just one or two systems? The third theory of aging is that because all the organ systems in a complex organism are interdependent; a decline in the function of any one of them eventually impairs the function of the others.

For example, a decline in the secretion of growth hormone by the endocrine system could well tip the balance away from growth and replacement and toward a net loss of cells or organ function. Impaired function of the cardiovascular system would affect nutrient delivery to and waste removal from every cell in the body. A critical system in terms of aging may turn out to be the immune system. Indeed, the elderly have a higher incidence of cancer than the young, which may indicate that the immune system is not recognizing damaged cells as efficiently as it once did.

It is also possible that aging is due to a specific type of tissue change that affects many organ systems at once. One hallmark of aging is that a certain type of protein cross-linking increases with age. This cross-linking could contribute to multiple aging processes, ranging from decreased elasticity of tendons and ligaments to functional changes in lungs, heart, and blood vessels. Of course, this theory still leaves unanswered the question of what initiates the aging process at the cellular level.

Recap **Aging is a complex process that is still poorly understood. It appears that most cells are limited in how many times they can divide and that damage may accumulate in cells until they no longer function properly. At the whole-organism level, declines in function of one organ or system may eventually impair other organ systems.**

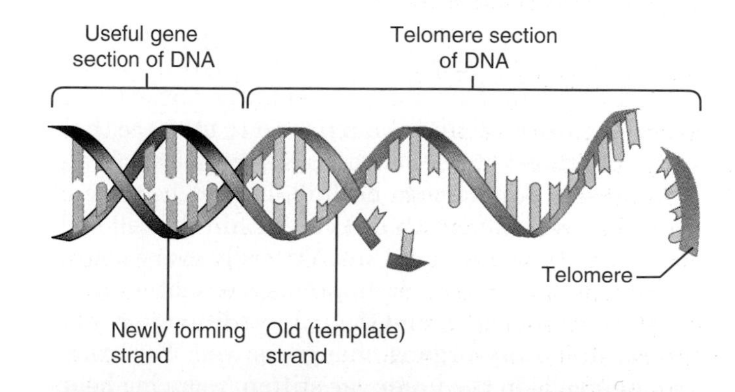

Figure 20.17 Telomeres and aging. Short terminal segments of DNA called telomeres are removed each time the DNA is replicated. As a result, the new strand will be shorter than the original strand. When all telomeres are gone, further cell divisions erode the useful genes.

Body systems age at different rates

What actually happens during aging is fairly well documented, even if we don't know exactly why. Next we describe important developments in each organ system.

Musculoskeletal system and skin Bone and muscle mass begin to decrease in early adulthood. Bone loss is more marked in women than in men, especially after menopause. The decline in muscle mass is a combination of a loss of muscle cells and a decrease in the diameter of remaining cells. By age 80 the average person has lost 30% of his/her muscle mass, and strength declines accordingly. Often the joints become stiff and painful because they secrete less joint fluid. Ligaments and tendons become less elastic.

The skin becomes thinner, less elastic, and wrinkled. There are fewer sweat glands, so it becomes harder to adjust to warm temperatures. Pigmentation may change as well, with the appearance of pigmented blotches called liver spots.

Cardiovascular and respiratory systems
Lung tissue becomes less elastic, decreasing lung capacity. The heart walls become slightly stiffer due to an increase in the ratio of collagen to muscle, with the result that an aging heart generates less force. Blood vessels also lose elasticity, so the systolic pressure and pulse pressure may rise in elderly people. Overall there is a reduced capacity for aerobic exercise as aging progresses.

Immune system The thymus decreases to only 10% of its maximum size by age 60. There are fewer T cells and the activity, but not the number, of B cells declines as well. In addition the immune system fails to recognize "self" as well as it once did, leading to higher rates of autoimmune disorders. Wound healing slows and susceptibility to infections increases.

Nervous and sensory systems Neurons in most areas of the brain do not divide in adults, so neurons that are lost throughout life generally are not replaced. The brain shrinks only slightly, however, because the neurons that remain tend to increase their connections with each other. Nevertheless, brain and sensory function do decline, in part because of degenerative changes that occur within the cell bodies of the remaining neurons. Virtually every neural activity is affected, including balance, motor skills, hearing, sight, smell, and taste. In addition to neural changes, sensory organs change as well. For example, hair cells in the inner ear stiffen, reducing hearing, and the lens of the eye stiffens, making it more difficult to focus on near objects. As a result of these many changes, elderly people become prone to injury. In addition, they may fail to eat properly because food seems less appealing.

Reproductive and endocrine systems Female reproductive capacity ceases when the woman reaches **menopause,** the cessation of menstruation, in her late 40s or early 50s. Menopause occurs because the ovaries lose their responsiveness to LH and FSH and slowly reduce their secretion of estrogen. The onset of menopause may be gradual, with menstrual periods becoming irregular over several years.

The symptoms of menopause are variable. As a result of the decline in estrogen, some women experience "hot flashes" (periods of sweating and uncomfortable warmth). There is a tendency for vaginal lubrication to decline after menopause, making intercourse potentially less comfortable. Mood swings are not uncommon. However, female sexual desire (libido) may be undiminished because the source of female androgens is the adrenal glands, not the ovaries. After menopause, women are at increased risk for cardiovascular disease and osteoporosis.

Low-dose estrogen (Premarin) is sometimes prescribed for postmenopausal women to reduce menopausal symptoms and to decrease the risk of cardiovascular disease and osteoporosis. However, there is some evidence that estrogen alone may increase the risk of breast cancer in postmenopausal women. Combining the estrogen with progesterone may diminish the risk compared to estrogen alone.

Men exhibit a long, slow decline in the number of viable sperm as they age. In addition, after age 50 men may take longer to achieve an erection. Nevertheless, both sexes can and do remain sexually active into old age.

The rate of decline of various components of the endocrine system varies widely. Notably, there is a substantial decrease in the secretion of growth hormone. However, there is not an appreciable decline in renin, TSH, ACTH, or ADH.

Digestion and nutrition The digestive system continues to function well except that some vitamins are not absorbed as efficiently. Nutritional requirements decline somewhat due to slower cellular metabolism, smaller muscle mass, and, often, less physical activity. Adequate nutrition can be a problem if the elderly fail to eat a balanced diet. Tooth loss varies widely among individuals, but people who have lost teeth may find it harder to eat fruit and other healthy but fibrous foods.

The liver's ability to detoxify and remove drugs from the body declines significantly. As a result, elderly people often require smaller doses of medications to achieve the appropriate therapeutic blood concentration.

Urinary system Kidney mass declines, and both blood flow and filtration rate may fall by as much as 50% by age 80. However, there is normally such a tremendous reserve of renal functional capacity that the decline is usually of little consequence. The elderly do lose some capacity to respond to extreme dehydration or overhydration, but otherwise the kidneys carry out their functions rather well.

The primary problem in men is usually prostate hypertrophy. The enlarged (hypertrophic) prostate presses against the ureter, making urination difficult. In women the most common problem is incontinence due to a weakened urethral sphincter.

> ***Recap*** **The actual pattern of aging in various organ systems is well documented even if we don't fully understand the underlying causes. All organs and organ systems decline in function with advancing age, though not necessarily at the same rate.**

Aging well

It would be easy to get depressed after reading the long list of declines in function that accompany aging. Bear in mind that not all of these developments occur to the same degree in everyone, nor do they all occur at the same age. The effects of aging are normal events in our long and exciting journey through life.

What can we do to age well?

Will a healthy lifestyle that includes exercise and a proper diet slow the aging process? The answer is far from certain. However, regular exercise and healthy nutrition can certainly help us age better. Throughout human history, longevity has been determined primarily by the prevalence of disease, not by the "natural" length of a human life span (if indeed there is one). Exercise and a healthy diet improve cardiovascular and skeletomuscular fitness and reduce your risk of major killers such as cancer and cardiovascular disease. A healthy lifestyle also improves the quality of life, providing a heightened sense of well-being and increased energy and vigor (Figure 20.18).

With these benefits it really doesn't matter whether we can extend the "natural" human life span from 100 to, say, 105. If you can't live forever, you can at least age well and gracefully.

> ***Recap*** **Even if the aging process cannot be stopped, it is certainly possible to improve human health and wellness throughout life. Regular exercise and healthy nutrition are important factors in aging well.**

20.9 Death is the final transition

The final transition in the journey of life is death. In terms of human biology, it is as natural as life itself.

Death is defined as the cessation of life. The death of a complex organism such as a human being is not necessarily accompanied by the immediate death of all its organs or cells. Death occurs when an organ system that is essential for life in the short term fails to function, making it impossible to sustain the life of other systems.

The critical organ systems whose failures lead to death very quickly are the brain (because it controls respiration), the respiratory system (without oxygen, life cannot be sustained), and the cardiovascular system (without oxygen and nutrient delivery to all cells, life ends). Failure of other organs with more long-term functions, such as the kidneys or digestive system, does not necessarily kill immediately. You can live without either of these systems for days as long as their loss does not cause one of the critical systems to fail.

The challenge of defining the precise moment of death, then, is that all of the organ systems don't necessarily die at the same time. Death is a process that begins with the failure of certain life-critical organ systems and then proceeds through the death of other organ systems, organs, and cells until the very last of the body's 100 trillion cells is no longer alive. That could take hours or even days.

Even though death may occur gradually, our society requires a practical way to define the moment at which an individual dies. We need a definition of death so the living can begin to grieve and make plans, and so health care workers can stop the medical treatments that sustain life. We need a legal definition so that the police and courts can make a distinction between assault and murder,

Why is it important to be able to define death?

Figure 20.18 Aging well. Although it is not known for certain whether exercise and proper diet actually delay the aging process, they do improve performance and may provide an enhanced sense of wellness throughout life.

and so that estates can be settled. A precise definition of death is especially important because organs that might still be alive can be transplanted into another person and save a life.

Legal and medical criteria for declaring a person dead vary according to state law. Basically, all such criteria attempt to define the moment when, to the best of our knowledge and abilities, the person no longer has even a remote possibility of sustaining life. In general they define death as either (1) *irreversible cessation of circulatory and respiratory functions,* or (2) *irreversible cessation of all functions of the entire brain, including the brain stem.* Note that the definitions focus on critical systems that we cannot live without for more than a few minutes.

The key, of course, is the word "irreversible." Usually it is easy to tell when a person has had no respiration or blood flow for 10 minutes or so, long enough to cause brain death. Brain death alone, especially in a person on cardiopulmonary life support, is harder to define. Medical professionals generally run down a checklist of criteria that must be met in order to declare a person's brain as nonfunctional, and even then they usually repeat the tests 6–12 hours later before they are willing to declare the brain death irreversible. Even for the medical profession, the final transition in life can be difficult to define.

Recap **Death is the termination of life. In a complex organism it is a process that can be gradual or brief. Death starts with the failure of one or more critical organ systems and ends ultimately with the inability to maintain an internal environment consistent with cellular life.**

Chapter Summary

Fertilization occurs when sperm and egg unite

- Sperm and egg meet (and fertilization occurs) in the upper third of the oviduct.
- Enzymes in the head of the sperm create a pathway through the corona radiata and the zona pellucida to the egg cell membrane.
- Entry of the sperm nucleus triggers the completion of meiosis II by the secondary oocyte. Thereafter the nuclei of sperm and egg fuse, creating a single diploid cell, the zygote.

Development: Cleavage, morphogenesis, differentiation, and growth

- Development begins by a process called *cleavage* and then proceeds with *morphogenesis, differentiation,* and *growth.*
- The periods of development prior to birth are known as *pre-embryonic* (the first two weeks), *embryonic* (weeks three through eight), and *fetal* (weeks nine to birth).

Pre-embryonic development: The first two weeks

- During the first two weeks of development, the single cell develops into a hollow ball with an embryonic disk in the center that will eventually become the embryo.
- At about one week the pre-embryo (now called a blastocyst) begins to burrow into the uterine wall, a process called *implantation.*

Embryonic development: Weeks three to eight

- Embryonic development is marked by the presence of three primary germ layers (ectoderm, mesoderm, and endoderm) in the embryonic disk.
- During development the embryo is completely surrounded by two membranes. The *amnion* contains amniotic fluid, and the *chorion* develops into fetal placental tissue. Two other membranes (the allantois and the yolk sac) have temporary functions only.
- By the fifth week the embryo begins to take on distinctly human features.
- By the eighth week the placenta and umbilical cord circulation are fully functional, and sex has been determined.

Fetal development: Eight weeks to birth

- Months three–nine are marked by rapid growth and development of the organ systems. By the fifth month the fetus begins to move.
- By the sixth month, life outside the womb is possible with good medical care.

Birth and the early postnatal period

- Birth occurs at about 38 weeks of development (nine months).
- The three phases of labor are *dilation, expulsion,* and *afterbirth.*
- Shortly after the newborn takes its first breath, the newborn's cardiovascular system undergoes substantial changes. Within hours or a few days the umbilical vessels regress and the ductus ar-

teriosus and foramen ovale close, rerouting all blood through the pulmonary circulation.

From birth to adulthood

- Human neonates are relatively helpless because the nervous and muscular systems are not yet mature.
- There is a disproportionate increase in the development of the musculoskeletal and nervous system during infancy. The immune system remains relatively immature.
- Body shape changes in childhood. The brain achieves 95% of its adult size.
- Adolescence is marked by the last spurt of rapid growth and the attainment of sexual maturity.

Aging takes place over time

- Most cells have an internal mechanism that limits the number of times they can divide.
- Cumulative unrepaired cellular damage as a consequence of metabolic activity may limit the life of cells.
- Aging may affect all organ systems because a decline in one system will affect other systems.

Death is the final transition

- Death is a process. Death begins with the failure of one or more critical organ systems, leading to the failure of other organ systems and eventually to the death of all cells.

Terms You Should Know

aging, 512
amniotic fluid, 500
cesarean delivery, 507
cleavage, 497
colostrum, 510
conjoined twins, 496
death, 515
ectoderm, 499
embryo, 499
embryonic disk, 498
endoderm, 499
episiotomy, 505
fertilization, 495
fraternal twins, 496
identical twins, 496
implantation, 498
longevity, 512
menopause, 514
mesoderm, 499
morphogenesis, 497
placenta, 500
pre-embryo, 497
puberty, 512
umbilical cord, 500
umbilicus, 510
zygote, 496

Test Yourself

Matching Exercise

_____ 1. The outermost sphere of cells in a blastocyst
_____ 2. The germ layer that eventually develops into the epidermis, hair, nails, and teeth enamel
_____ 3. The germ layer that eventually develops into the liver, pancreas, and linings of the urinary bladder and urethra
_____ 4. While the zygote is still in the oviduct, a series of cell divisions produces this ball of about 32 identical cells
_____ 5. This structure enables the exchange of nutrients and wastes between embryo and mother
_____ 6. Surgical delivery of a baby
_____ 7. Surgical incision in the vulva to make delivery easier
_____ 8. A series of cell divisions not accompanied by cell growth or differentiation
_____ 9. The process by which individual cells gradually take on specialized forms and functions
_____10. The process by which a blastocyst becomes embedded within the endometrium

a. ectoderm
b. morula
c. episiotomy
d. cesarean
e. cleavage
f. trophoblast
g. implantation
h. differentiation
i. placenta
j. endoderm

Fill in the Blank

11. The ___________ is the two-way lifeline that connects the embryo's circulation to the placenta.
12. The termination of life is referred to as ___________.
13. ___________ twins derive from a single fertilized ovum.
14. ___________ twins derive from two different oocytes.
15. If implantation occurs in an oviduct, this results in a(n) ___________ pregnancy.
16. Labor is triggered by maturation of the fetal ___________ gland.
17. A watery antibody-rich milk called ___________ nourishes a newborn.
18. A fetus typically begins to move during the ___________ month of pregnancy.
19. ___________ refers to the process of change that we undergo with the passage of time.
20. ___________ is defined as the onset of menstrual periods in females and by nocturnal emissions in males.

True or False?

21. The period of prenatal human development can be divided into three stages called pre-embryonic, post-embryonic, and maturity.
22. The placenta effectively blocks alcohol, drugs, and viruses from entering the embryonic circulation.
23. Until approximately four weeks of development, a human embryo resembles other vertebrate embryos.
24. In human beings, most organ systems die simultaneously.
25. In neonates, the nervous, immune, and muscular systems are relatively immature.

Concept Review

1. Why does only one sperm fertilize the egg, and why is this important?
2. Which extra-embryonic membrane is the source of the reproductive germ cells that become the gametes in the embryo?
3. Which hormones are secreted by the placenta?
4. Why is it important that the foramen ovale close soon after birth?
5. What causes the newborn to take its first breath immediately after birth?
6. Why aren't all of a child's vaccinations given in the first year?
7. What causes the onset of puberty (sexual maturation)?
8. Summarize the three main theories of the causes of aging.
9. What has caused the increase in longevity of the U.S. population during the past 100 years?
10. Why is it sometimes so hard to define the moment when an individual dies?

Apply What You Know

1. Why is it biologically impossible for humans and gorillas to mate and produce offspring?
2. Explain why Siamese twins are always identical twins (not fraternal twins).
3. How is it possible that one's life could be prolonged without slowing the aging process?

Case in Point

Health Benefits of Tai Chi for the Elderly

Tai Chi is a type of moving meditation, originally developed in China, that emphasizes slow, well-coordinated body movements. It has long been popular in Asia as an exercise for elderly citizens.

In 1999, 17 elderly Americans aged 68 to 92 participated in a study to measure the effects of a regular program of Tai Chi exercise. Participants underwent pretest evaluation of their balance, flexibility, and emotional state, then attended hour-long classes three times a week.

After eight weeks, post-test evaluations showed measurable improvement in participants' balance, flexibility, upper-body mobility, and ability to walk. They also scored measurably higher on a standard test for assessing mood. Several with chronic health conditions reported decreased sensations of pain.

QUESTIONS:

1. Why might improved balance and mobility be important for the elderly?
2. From what you have learned here, how long must an exercise program be conducted before one sees measurable benefits?
3. If you could conduct the study, how would you determine whether Tai Chi offered more (or fewer) health benefits for the elderly than some other form of mild exercise (stretching exercises, gentle walking, calisthenics in a swimming pool)?

INVESTIGATE:

M. Candice Ross *et al.,* "The Effects of a Short-Term Exercise Program on Movement, Pain, and Mood in the Elderly." *Journal of Holistic Nursing,* June 1, 1999, p. 139.

References and Additional Reading

Austad, Stephen N. *Why We Age: What Science is Discovering about the Body's Journey through Life*. New York: Wiley & Sons, Inc., 1997. 244 pp. (A colorful and provocative look at aging for both laypersons and professionals.)

Moore, K.L., and T.V.N. Persaud. *Before We are Born: Essentials of Embryology and Birth Defects*. Philadelphia: W.B. Saunders Co., 1993. 354 pp. (A more in-depth treatment of human embryology for the interested student.)

Ricklefs, R.E., and C.E. Finch. *Aging: A Natural History*. New York: Scientific American Library, 1995. 224 pp. (An exploration of patterns of aging among different species.)

CHAPTER 21

EVOLUTION AND THE ORIGINS OF LIFE

What scientific evidence **supports the theory of** evolution?

When did modern humans **first appear?**

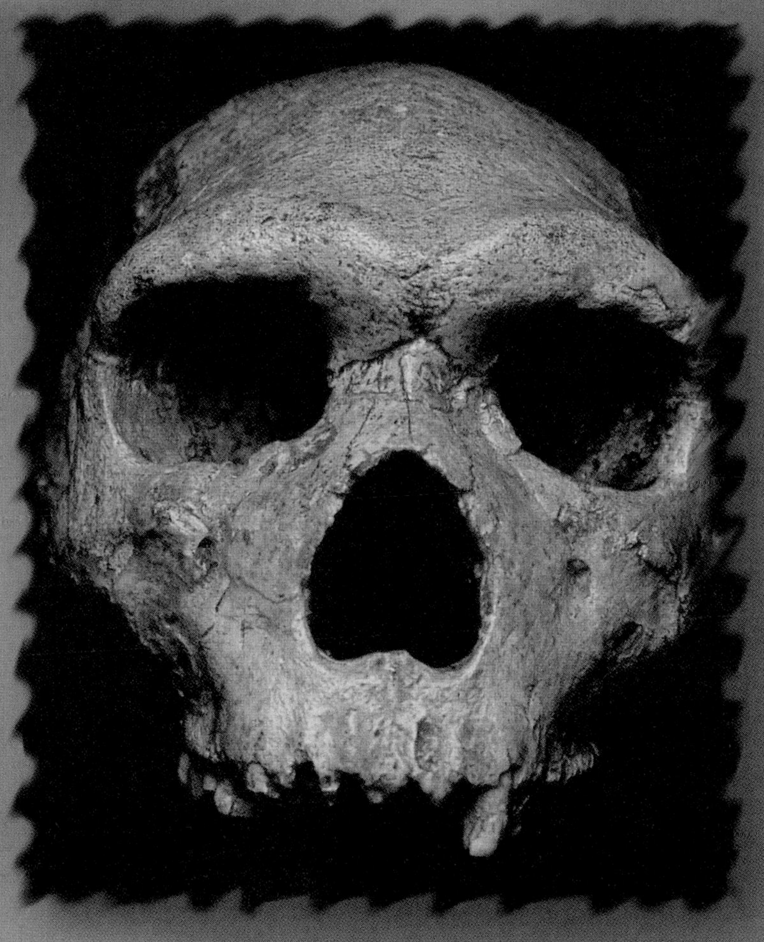

When did our ancestors begin to walk upright?

When did life first appear **on Earth?**

How did racial differences **arise?**

LOOK AROUND AND YOU'LL BE IMPRESSED BY THE ASTOUNDING variety of forms that life takes. From bacteria to beagles, hagfish to humans, living creatures come in a huge range of sizes, shapes, and colors. And yet, the evidence indicates that all living creatures on Earth are related. To the best of our knowledge, all of them descended from a single simple life form that arose from the hot, steamy environment of Earth more than three billion years ago.

How did Earth's inhabitants come to be so different from each other? The diversity of life is thought to result from **evolution,** defined as an unpredictable and natural process of descent over time with genetic modification. The formal definition has three key elements:

- *Descent over time.* As one generation of organisms gives rise to the next, populations of organisms undergo slow change that makes them different from their ancestors. Sometimes a population becomes two populations that are so different from each other that they no longer interbreed, giving rise to new species.
- *Genetic modification.* The process of change depends on changes to the genes of the organisms.
- *Unpredictable and natural.* Evolution is affected by chance, *natural selection,* historical events, and changing environments.

Microevolution refers to evolution as a result of genetic changes that give rise to new species. **Macroevolution,** on the other hand, describes large-scale evolutionary trends or changes that apply to whole groups of species, often as a result of changing environments or major historical events.

The theory that evolutionary processes shaped life on Earth is associated with Charles Darwin, a British naturalist of the mid-1800s. The heart of Darwin's hypothesis was that life arose only once, probably in the primordial sea long ago, and all life as we know it descended from that early life form. Darwin actually used the word evolution only sparingly, preferring instead "descent with change" or "descent with modification." Key to Darwin's hypothesis was the idea that descent with modification resulted from "natural selection," which we will describe later in this chapter.

Scientists reserve the word "theory" for those few concepts that have withstood numerous working hypotheses designed to prove them false and that still offer the best explanation to fit a broad range of established facts. Although rigorous testing has somewhat modified Darwin's original hypothesis, the *theory of evolution* remains the single most unifying concept in all of biology.

In this chapter we review the scientific evidence on which we base our understanding of evolution and discuss processes that affect evolutionary change. Then we pick up the intriguing question of life's origin on Earth. Finally, we trace the relatively recent evolution of human beings and consider the features that make us distinctively human.

Recap **Evolution is a process of descent over time with genetic modification. First proposed by Charles Darwin in the 19th century, the theory of evolution remains the single most unifying concept in biology.**

21.1 Evidence for evolution comes from many sources

What scientific evidence supports the theory of evolution?

The evidence that humans and all other life forms have evolved over time comes from several sources, including the fossil record, comparative anatomy and embryology, biochemistry, and biogeography.

The fossil record: Incomplete but valuable

Fossils, the preserved remnants of organisms, are one of our richest sources of information about life forms that lived in the past. Soft tissues decay quickly, so the fossil record consists of bones, teeth, shells, and occasionally spores and seeds. Even these hard tissues are preserved only if they are covered soon after death with layers of sediment or volcanic ash (Figure 21.1). Over time the remains become mineralized with some of the same minerals that compose rocks, leaving a rocklike impression of the hard tissue of the organism, or a fossil.

Even though we have found fossils from over 200,000 species, there may be millions of species for which we will never have any record because they had no hard tissues. We have many fossils of ancient bony fishes, for example, but no jellyfish from the same time periods. We are also more likely to discover fossils on land (there may be many more, and possibly different, fossils deep under the sea) and to find fossils that were produced in great numbers. For example, we are more apt to find a seashell or spore from 100,000 years ago than to uncover the skeleton of an early human from the same time period. By its nature, then, the fossil record tends to be incomplete and heavily weighted toward fossils that are easy to find.

One of the most important pieces of information about a fossil is its age, which places it at a certain point in evolutionary history and shows how body structures have changed over time. Scientists know that sediment and volcanic ash are laid down layer upon layer, a process known as *stratification.* Where

(a) A body will decompose unless it is covered by sediment or volcanic ash.

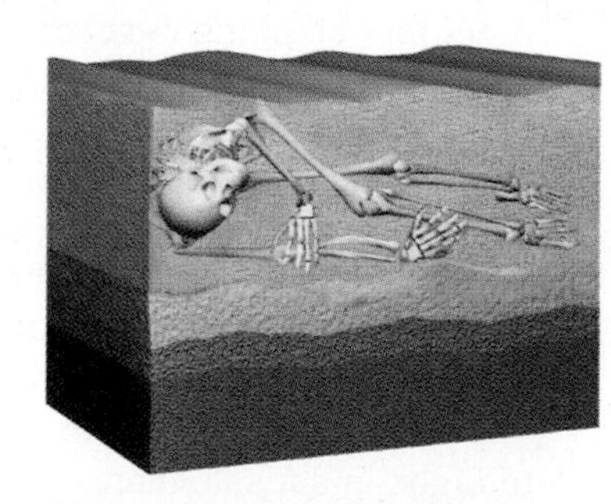

(b) Hard elements of a body may be preserved from decomposition if they are covered quickly.

(c) Over time more layers of sediment, ash, or soil are deposited. The hard elements of the organism become mineralized by the minerals that compose rocks.

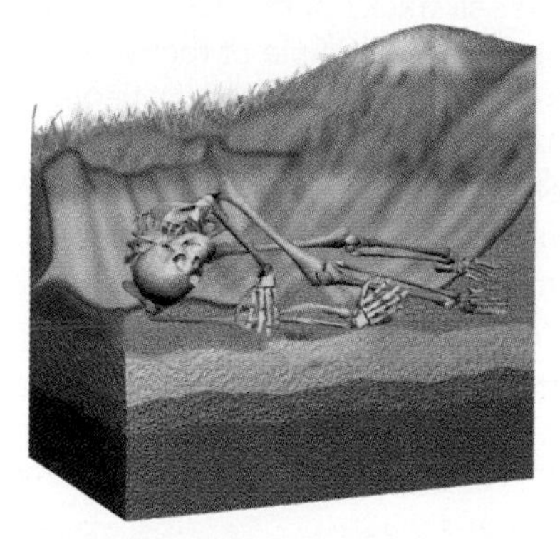

(d) Occasionally erosion, uplifting of the earth's crust, or human excavation may expose fossils to the surface.

(e) A fossilized human skeleton.

Figure 21.1 How fossils are formed.

fossils exist in the same location in different layers, it is usually possible to tell which one came first. This helps scientists establish relative relationships between fossils over time (Figure 21.2). In addition, techniques are now available for dating rocks and fossils (see Directions in Science, on the next page).

Recap **Fossils are the mineralized remains of the hard tissues of past life forms. Although the fossil record is incomplete, pieced together it provides compelling evidence for evolution.**

Comparative anatomy and embryology provide more evidence

Additional evidence for evolution comes from comparing the anatomy of animals and how their embryos develop. When examining them in the context of evolution, scientists describe anatomical structures as *homologous, analogous,* or *vestigial.*

When different organisms share similar anatomical features, often it is because they evolved from a common ancestor. For example, the forelimbs of all vertebrates (animals with backbones) seem to share roughly the same set of bones arranged in similar patterns. This is true even though vertebrate forelimbs have undergone numerous modifications so animals can use them in a variety of specialized ways, such as flying, swimming, running, or grasping objects (Figure 21.3). **Homologous structures** are body structures that share a common origin. The degree to which homologous structures resemble each other can be used to infer the closeness of evolutionary relatedness.

However, structures that share a common function may look alike even if they do not share a common origin. Good examples are the wings of birds and the wings of insects, both of which evolved for flight even though birds and insects are not closely related. Structures that share a common function but not necessarily a common ancestry are called **analogous structures** (Figure 21.4).

Directions in Science

Dating Rocks and Fossils

The distant past is like deep water—murky, dark, and uncertain, with a million years a mere drop in the bucket. How can we even guess at dates that measure in billions of years?

The most common technique for dating the time of formation of rocks and fossils is called *radiometric dating*. Radiometric dating takes advantage of the fact that some atoms, called radioactive isotopes, are inherently unstable. Radioactive isotopes radiate or give off particles until they reach a more stable state, sometimes becoming a different, more stable atom.

The isotope most often used to date very old fossils is radioactive potassium, which slowly converts to argon, a gas. The half-life of radioactive potassium is 1.3 billion years, meaning that it would take 1.3 billion years for half of the radioactive potassium in a fossil to become argon. It would take *another* 1.3 billion years for half of the *remaining* radioactive potassium to become argon, and so on. The argon formed from the decay of radioactive potassium remains trapped in the solid rock or fossil.

The ratio of radioactive potassium to argon in a fossil, then, can be used to estimate the fossil's actual age in billions of years. However the method is only accurate to plus-or-minus 10%, meaning that an age estimate of 4.5 billion years could be any time between 4 and 5 billion years. And because radioactive potassium decays so slowly, it can take several years to perform a single measurement with even this accuracy.

For fossils newer than 50,000 years old, the radioactive isotope of carbon (C^{14}) is more commonly used because it decays with a half-life of only 5,700 years. But 50,000 years represents only about one thousandth of one percent of life's history.

Expect some of the more important milestone numbers to remain contentious for awhile. As an example, in 1998 a debate raged over whether the best estimate of the date of the first appearance of animals was 600 million years ago, or a new unconfirmed estimate of 1.1 billion years ago. The debate centered on alleged fossil tracks of multicellular animals found in a sample of rocks with an estimated age of 1.1 billion years. The findings await confirmation of the age of the rocks and an analysis of the tracks themselves to determine whether they were indeed made by animals. It may be years before we know.

Figure 21.2 Stratification of sedimentary deposits. The rock layers of the Grand Canyon were formed by sequential deposits of sediments. The oldest layers at the bottom of the canyon are over 1.2 billion years old. The fossil trilobite was found in a layer of Grand Canyon rock that is over 500 million years old.

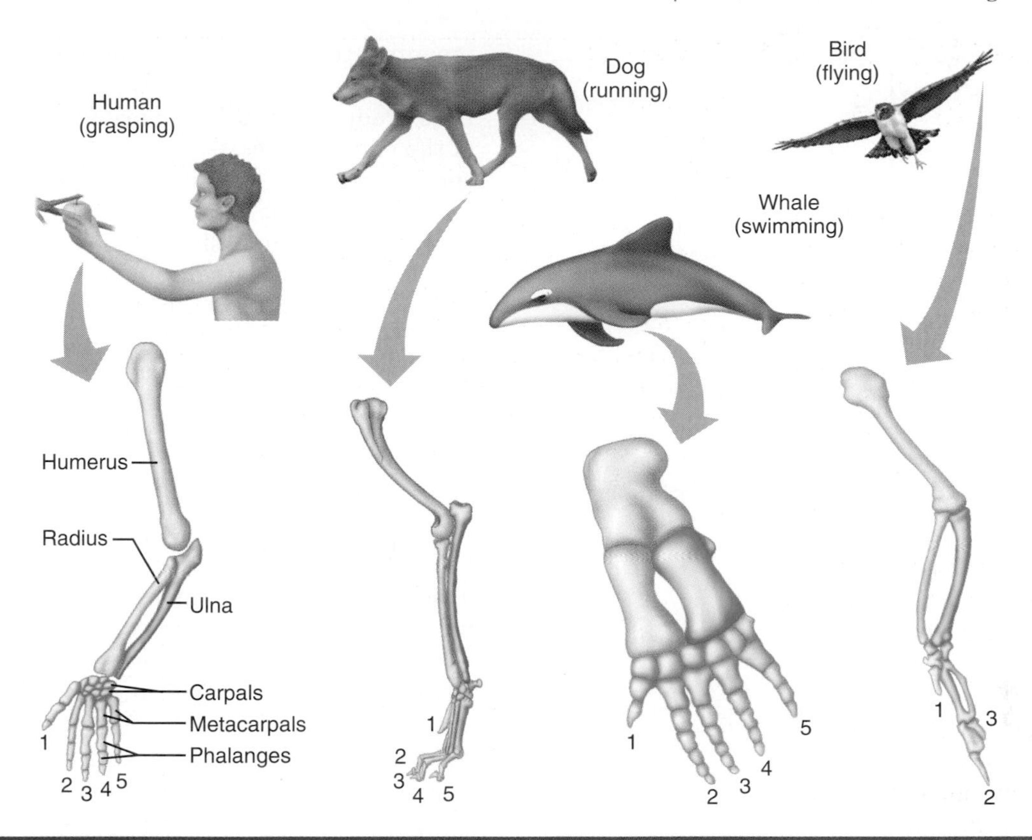

Figure 21.3 Homologous structures. Homologous structures evolve from a common structure. Numbers indicate digits. Notice how homologous bones (indicated by similar colors) have undergone modification in order to perform different functions in different vertebrates.

Finally, **vestigial structures** serve little or no function at all. Vestigial structures may be homologous to body parts in other organisms, where they do still have an important function. The human coccyx (tailbone) is the vestigial remnant of a tail, suggesting that we share a common ancestry with other vertebrates with tails. The muscles that enable you to wiggle your ears are vestigial in humans. Other animals, such as dogs and deer, still use these muscles to rotate their ears toward a sound so that they can hear it better.

Compelling evidence for evolution also comes from comparisons of the embryos of animals, especially vertebrates. The vertebrates are a varied group: fishes, amphibians, reptiles, birds, and mammals. Yet despite their diversity, their early embryonic development follows the same general pathway (Figure 21.5). For example, all vertebrate embryos develop a primitive support structure called a *notocord* and a series of folds called *somites* that will become bone, muscle, and skin. But in addition, all vertebrate embryos develop a series of arches just below the head. In fishes and amphibians these are called *gill arches* because they develop into gills. In humans the homologous

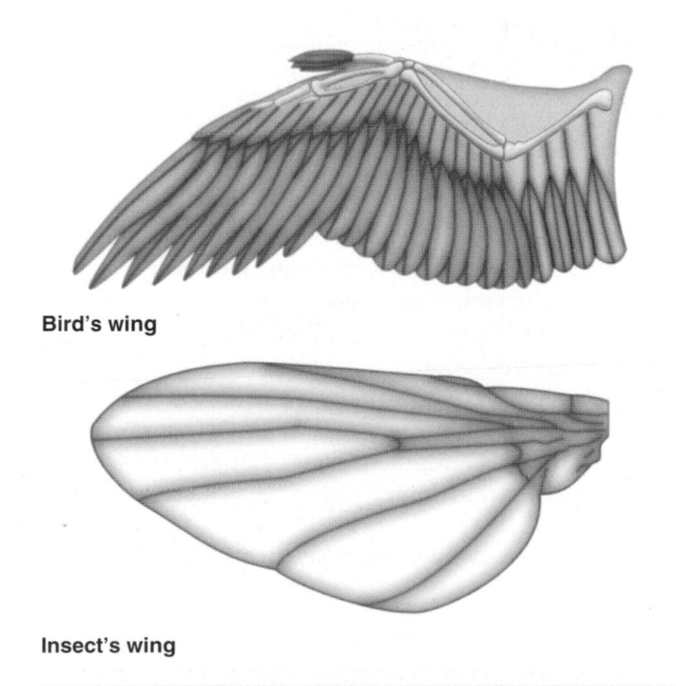

Figure 21.4 Analogous structures. The bird's wing and the insect's wing are both used for the same function, but they evolved from entirely different structures.

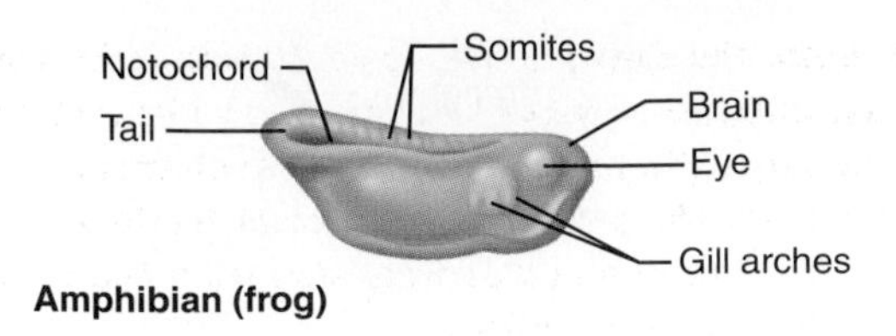

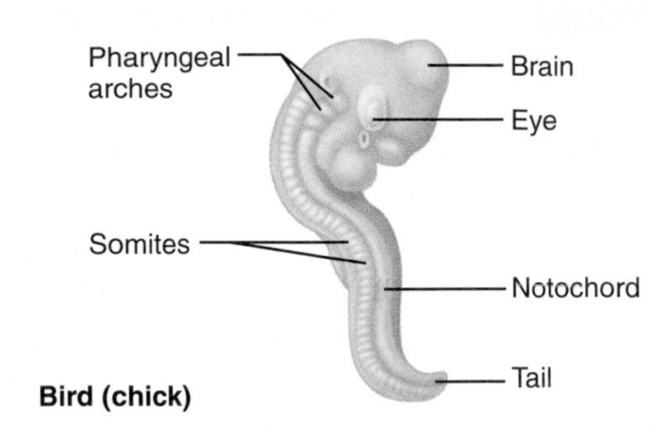

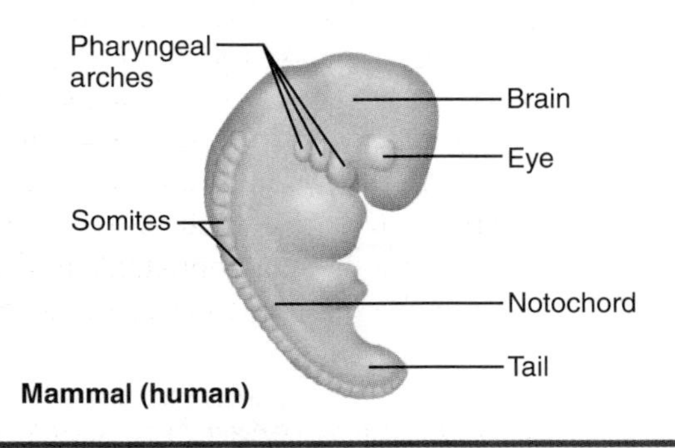

Figure 21.5 Early developmental similarity among vertebrate embryos. The embryos of all vertebrates tend to share certain developmental features, suggesting a common ancestry. In particular, all vertebrate embryos have either gill or pharyngeal arches. These arches ultimately have very different functions.

pharyngeal arches become parts of the face, middle ear, and mouth. The most likely explanation for these common embryonic structures with completely different functions is common ancestry. Later in embryonic and fetal development the paths of morphogenesis and differentiation diverge, giving us the variety of vertebrates we know today. It is almost as if embryonic and fetal development represent an acceleration of the entire evolutionary history of the vertebrates.

Comparative biochemistry examines similarity between molecules

Proteins and genes also provide clues to the relationships among species. When two species possess identical or nearly identical biochemical molecules, common ancestry may be indicated. In principle, the two similar molecules are akin to homologous structures—you might call them "homologous molecules." They arise when evolutionary paths *diverge,* meaning that one species becomes two different species. Mutations occurring over long periods of time gradually modify the original molecule in each of the two new species. The more that two molecules serving the same function in two species become different, the earlier the two species probably diverged from a common ancestor.

For example, cytochrome c is an important protein in the metabolic pathways for energy exchange, and it is found in organisms ranging from yeast to humans. In humans, cytochrome c consists of 104 amino acids. A chimpanzee's cytochrome c is identical to the human form, but it differs by one amino acid in rhesus monkeys, 16 amino acids in chickens, and over 50 amino acids in yeast. From this data you could hypothesize that humans are more closely related to chimpanzees than to rhesus monkeys, and also that humans are more closely related to other primates than to chickens or yeast.

Similarities in other protein sequences and in DNA are also used as indicators of common ancestry. The degree to which DNA base pair sequences resemble each other helps scientists to infer ancestry and relationships between humans and other species.

Recap **Comparisons of homologous structures and the course of embryologic development provide clues to how life forms are related and give us additional evidence for the process of evolution. The degree of structural similarity between the DNA and proteins of different species can be used to estimate the closeness of evolutionary relationships.**

Biogeography: The impact of geographic barriers and continental drift on evolutionary processes

Biogeography is the study of the distribution of plants and animals around the world. Why do we find certain animals and plants in some locations on Earth and not in others? One explanation is that migrations are inhibited in regions isolated by barriers such as water, high mountain ranges, or deserts (Figure 21.6).

Where migrations of plants and animals were possible, the evidence indicates that closely related life forms often evolved in one location and then spread to regions that were accessible to them at the time. Eventually their spread may be blocked by physical barriers or by environments in which they could not survive, such as deserts or oceans. There are no snakes on the Hawaiian Islands, for example, because snakes evolved long before the volcanic islands existed and the Pacific Ocean represents an effective barrier to snake migration. When we see how easily some plants and animals *adapt* (adjust to new conditions) when placed in new locations, we can ap-

Figure 21.6 Geographic barriers. Animal and plant migrations are inhibited by barriers such as these mountains and desert sands in Death Valley, California.

preciate the impact of geography on evolutionary processes and the distribution of life forms.

Geologists tell us that the Earth's continents are located on huge plates whose locations on the Earth's surface are not fixed. The plates slowly move over time, a process called **continental drift.** About 200 million years ago, however, all the continents were joined in one interconnected land mass called *Pangaea* (Figure 21.7). Because the separate continents formed *after* the first animals and plants appeared, the effect of continental drift has been to isolate related groups of organisms from each other. Once separated, organisms tend to evolve along separate but similar paths.

For instance, Australia separated from the other continents about 65 million years ago. Nevertheless, there are striking similarities between the predominant *placental* mammals (mammals that nourish their young with a true placenta) found on all other continents, and the unique *marsupial* (pouched) mammals of Australia. These similarities came about because the two types of mammals evolved over the same time period in the same kinds of environments, even though they evolved separately.

Recap **The slow process of continental drift over the past 200 million years has had a major impact on evolutionary patterns and the worldwide distribution of life forms. Within a continent, physical barriers such as mountain ranges and deserts also influence evolution and life-form distribution.**

21.2 Natural selection contributes to evolution

Random mutations underlie evolution

Recall that an organism's DNA represents the set of instructions for all its life processes. If all creatures had an identical set of DNA they would all *be* identical in form and function. Differences in DNA account for the differences between species and even between individuals within a species.

DNA changes as a result of mutations that produce a slightly different form of a gene, called an *allele*. Mutations are rare accidental events, and many (if not most) are lethal. Those that are not lethal will remain in the gene pool and alter subsequent generations; sometimes noticeably, sometimes imperceptibly.

Without random mutations, there could be no evolution. Many mutations over thousands and thousands of years might be enough to cause one species to diverge into two.

Natural selection encourages "survival of the fittest"

Mutations alone do not amount to evolution. Living organisms interact with their local environments, and as a consequence of that interaction, some live long enough to reproduce and some do not. The frequency with which any allele is found in the gene pool of a species depends critically on whether that allele confers survival value on its holder.

Darwin understood this principle even though he did not fully comprehend the genetic basis for it. He referred to it as "survival of the fittest" by means of natural selection. By **natural selection** he

(a) 200 million years ago

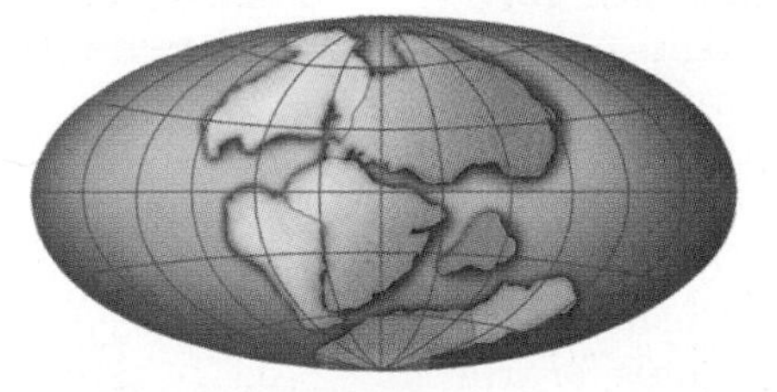

(b) 90 million years ago

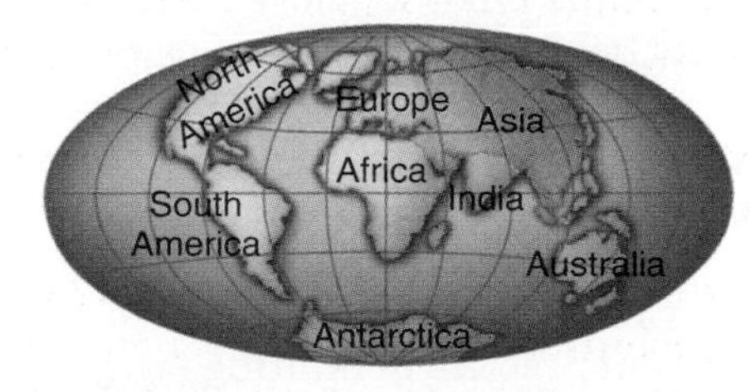

(c) Present

Figure 21.7 The formation of the Earth's continents from a common land mass. The continents were once connected in one giant landmass called Pangaea.

meant that individuals with certain traits that made them more fit for their local environments were naturally more likely to survive and reproduce. Together, mutations coupled with natural selection produce changes in the gene pool of a population.

Recap **Without mutations there would be no variation among life forms. Mutations coupled with natural selection provide much of the basis for evolution.**

Genetic drift and gene flow alter populations

A **population** is a group of individuals of the same species that occupy the same geographical area. Darwin was convinced that the fittest members of a population were always the ones to survive. We now know that there are several other contributors to evolution besides natural selection. Other factors that affect the gene pool of a population include genetic drift, gene flow, and selective hunting.

Genetic drift refers to random changes in allele frequency because of chance events. One cause of genetic drift is the *bottleneck effect,* which occurs when a major catastrophe—such as a fire, a change in climate, or a new predator—wipes out most of a population without regard to any previous measure of fitness. When this happens the few genes left in the gene pool may no longer represent the original population, nor do they necessarily represent the most fit genes. An example of a species suffering from a severe bottleneck effect may be the cheetah. The 15,000 surviving cheetahs are all genetically similar, and their genomes include a few harmful alleles that affect fertility. It is not clear whether cheetahs will survive as a species.

Another cause of genetic drift, the *founder effect,* occurs when a few individuals leave the original group and begin a new population in a different location, or when some environmental change isolates a small population from a larger one. Again, the new gene pool may not be a representative sample of the original population.

Differences in the gene pool of a particular population are also affected by movement of individuals into (immigration) or out of (emigration) the population. This physical movement of alleles, called **gene flow,** tends to mix pools of genes that might not otherwise mingle. The prevalence of international travel today has markedly increased the possibility of gene flow in the human population.

A few events that have an impact on evolution may not be random at all. For example, the distinctly human practice of hunting big game animals for trophies specifically selects for larger, more fit males (Figure 21.8). How this practice might affect evolution is worth considering.

Figure 21.8 Nonrandom impacts on evolution. Big-game hunters specifically select the larger, more fit males in a population. Prince Philip (on the left) shot the tiger in 1961. Queen Elizabeth II is in the center, directly behind the tiger. The skin of this tiger was destined for Windsor Castle.

Mass extinctions eliminate many species

An **extinction** occurs when a life form dies out completely. Throughout evolutionary history, major catastrophic events have sometimes wiped out whole groups of species regardless of fitness. In the last 530 million years (the only period for which we have a good fossil record) there have been at least five *mass extinctions,* each of which destroyed more than 50% of the species existing at that time.

The largest mass extinction, called the Triassic, occurred around 200–250 million years ago. It was accompanied by a period of high carbon dioxide concentration and substantial global warming, and it killed 70% of all land-dwelling species and 90% of all ocean-dwelling species. The most recent and best-known mass extinction, called the Cretaceous, claimed all of the dinosaurs about 65 million years ago. Most scientists attribute the Cretaceous mass extinction to a dramatic change in climate after a comet or asteroid collided with Earth near what is now Cancun, Mexico.

Evolutionary trees trace relationships between species

An **evolutionary tree** (also called a phylogenetic tree) is a map that illustrates evolutionary change and relationships among species. In an evolutionary tree, each branch represents a point of divergence between two species (the creation of a new species) and the length of the branches represent time (Figure 21.9). If a species becomes extinct, the branch for that species ends short of the present time.

When conditions are right, many new species may develop in a relatively short time from a single ancestor. Such short bursts of evolutionary activity are called **adaptive radiation** and are shown as numerous branches from a single point on an evolutionary tree. Periods of adaptive radiation most commonly follow changes in the environment that create new habitats.

Populations of living creatures are constantly undergoing evolutionary change. Sometimes it is necessary to further subdivide species into *subspecies* that look slightly different but can still interbreed. At any given time some species are becoming extinct while others are in the process of diverging to become several new species. Evolutionary trees can be general, covering the entire history of all life forms, or they can be detailed and focus on one species, such as humans.

> ***Recap*** **In addition to natural selection, several events can influence the evolutionary process. The pool of alleles in a population can be affected by genetic drift, gene flow, and selective hunting practices. There have also been at least five mass extinctions in Earth's history. An evolutionary tree visually depicts relationships between species.**

21.3 The young Earth was too hot for life

To the best of our knowledge, the Earth formed about 4.6 billion years ago. At one time our sun was surrounded by a swirling cloud of gases, dust, and debris. As the outer region of the cloud cooled, it condensed into solid masses that today we know as the planets, one of which was Earth.

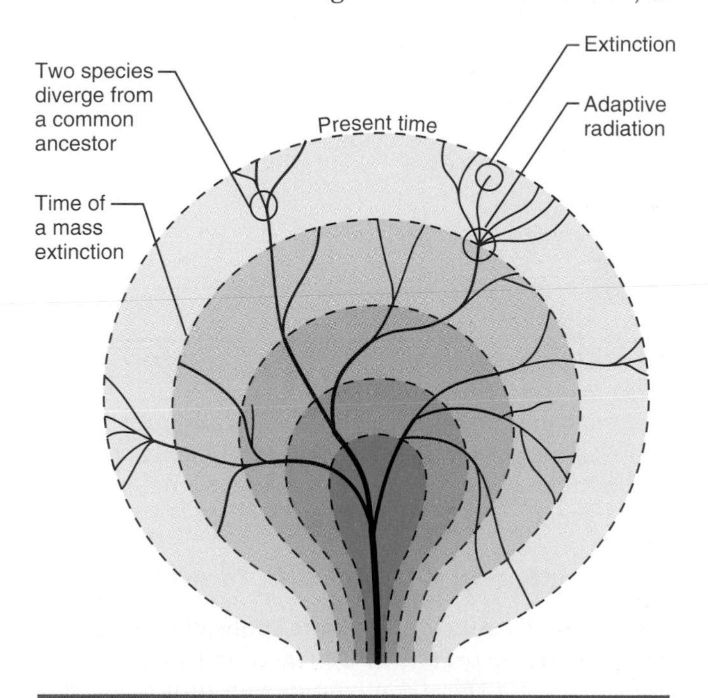

Figure 21.9 An evolutionary (phylogenetic) tree.

Four and one half billion years ago the Earth was an inhospitable place. Its tremendous heat melted the interior, which condensed as a core of liquid nickel, iron, and other metals. The solid but thin outer crust was constantly changing as frequent cracks in the surface and numerous volcanoes released molten rock and hot gases. The early atmosphere consisted primarily of carbon dioxide (CO_2), water vapor (H_2O), hydrogen (H_2), nitrogen (N_2), methane (CH_4), and ammonia (NH_3), but no oxygen (O_2). There were no oceans yet because the tremendous heat vaporized all the water. There was no ozone layer to protect the Earth from ultraviolet rays, so the Earth was constantly bombarded by ultraviolet radiation. The atmosphere crackled with electrical storms.

Over time the Earth cooled enough that water vapor at higher elevations began to condense as rain. Most of the rain immediately vaporized again as it struck hot rock. The hot, steamy cycle of condensation and vaporization continued for millions of years until the Earth's surface cooled enough to permit water to remain liquid. Oceans began to form, but they were warm, small, and not very salty. The land mass remained steamy, hot, and subject to frequent volcanoes and electrical storms. There was still no oxygen in the atmosphere.

Somehow, despite this harsh environment, living organisms appeared about 3.8 billion years ago, less than a billion years after the Earth was formed. Now we turn to the question of how life began, certainly one of the most intriguing questions in all of biology.

> ***Recap*** **Earth was formed about 4.6 billion years ago and life appeared about 3.8 billion years ago. The environment at the time was hot, steamy, and devoid of oxygen.**

21.4 The first cells were able to live without oxygen

We know from the fossil record that by 3.5 billion years ago single-celled creatures resembling bacteria were present. Once life appeared, evolutionary processes over billions of years produced the varied life forms that we know today, as well as many more that have long since become extinct.

When did life first appear on Earth?

Organic molecules formed from atmospheric gases

Current scientific evidence indicates that the first step on the path to life was the creation of simple organic molecules from atmospheric gases (Figure 21.10). In today's living organisms the formation of organic mol-

(a) The primitive Earth was hot and steamy with small, warm oceans.

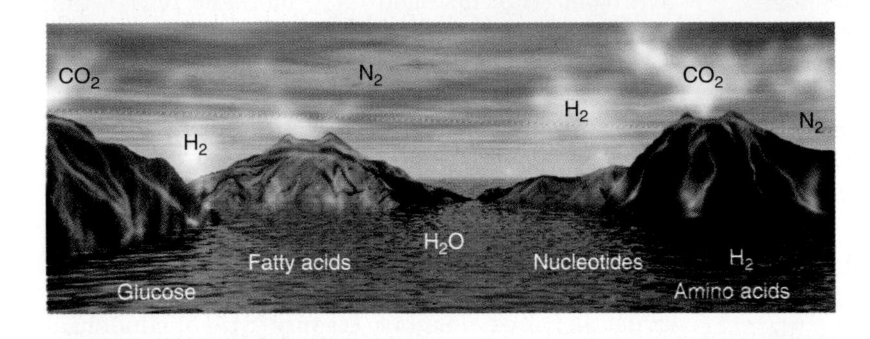

(b) Simple organic molecules formed from atmospheric gases and dissolved in the oceans. The energy for their formation was provided by electrical storms, heat, and intense ultraviolet radiation.

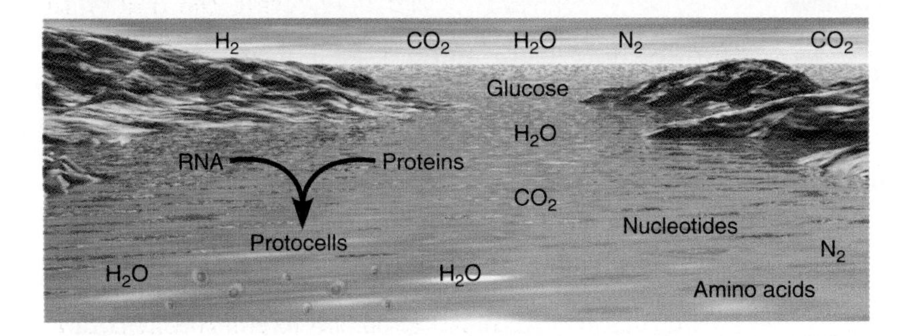

(c) Self-replicating RNA probably formed on clay templates in shallow waters along the shorelines of oceans. RNA and other organic molecules became enclosed in a cell membrane and the first self-replicating cells were formed.

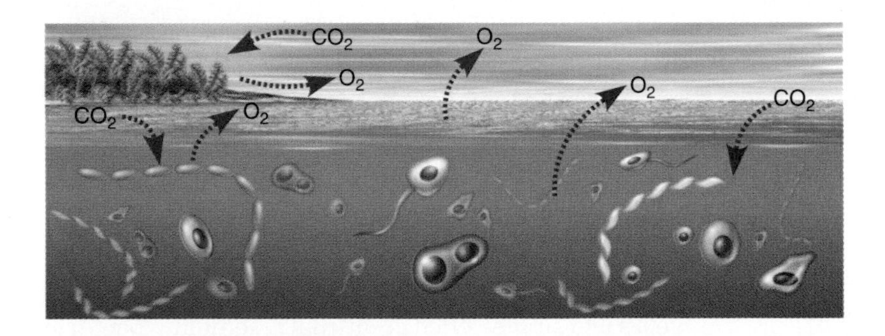

(d) The development of photosynthesis created oxygen gas. The presence of oxygen permitted the later development of aerobic forms of metabolism.

Figure 21.10 The origin and early evolution of life.

ecules requires enzymes, but of course enzymes had not yet evolved. In fact, it appears that it was only *because* the early environment was so inhospitable (by our standards) that organic molecules were able to form. Apparently the intense heat, ultraviolet radiation, and electrical discharges produced an extreme amount of energy, enough to combine molecules in the atmosphere into simple organic compounds, even without enzymes. These compounds (amino acids, simple sugars, fatty acids) dissolved in the sea, so that over time the oceans became a warm soup containing organic molecules and a little salt. The absence of oxygen was important, because oxygen is highly reactive and would have broken down the organic molecules as they formed.

Self-replicating RNA and DNA formed

The mere presence of organic molecules does not create a living organism that can grow and reproduce. How did these compounds join into molecules that could reliably reproduce themselves?

The problem we face in our understanding of this process is this: the only *self-replicating* molecules we know of are primitive single-stranded RNA and our modern double-stranded DNA, neither of which is likely to have formed spontaneously in the sea. Our best guess is that single-stranded self-replicating RNA first formed on templates of clay in mudflats along the ocean edge. Indeed, experiments have shown that, under the right conditions of moisture and heat, thin layers of clay can promote the formation of fairly complex molecules. Alternatively, the dry conditions and intense heat near volcanoes may have provided just the right environment for RNA formation.

If RNA was the first self-replicating molecule, the more stable DNA molecule must have developed from RNA very early in evolution. Modern cells use DNA as their self-replicating molecule and RNA to direct the synthesis of proteins, including enzymes. Only viruses use single-stranded RNA as their genetic material, and consequently viruses do not replicate on their own.

The first living cells were anaerobic

At some point, self-replicating molecules and small organic molecules became enclosed within a lipid-protein membrane. We do not know exactly how this happened. However, experiments have demonstrated that, under the right conditions, certain mixtures of amino acids and lipids will spontaneously form hollow, water-filled spheres.

The first simple cells relied on anaerobic metabolism, meaning metabolism without oxygen. Their ability to synthesize compounds was limited, and so they relied on their immediate environment for the energy and raw materials they needed. When energy and raw materials were not available they died, or at least they did not replicate.

> ***Recap*** **As best we can surmise, the sequence of events leading to self-replicating living cells was: (1) the formation of organic molecules from atmospheric gases, (2) formation of self-replicating RNA on templates of clay, and (3) enclosure of RNA and organic molecules inside a cell membrane. All early cells used an anaerobic form of metabolism.**

21.5 Photosynthetic organisms altered the course of evolution

Photosynthesis increased oxygen in the atmosphere

By the time another billion years had passed (bringing us to about 2.5 billion years ago), certain cells had mutated and developed the ability to produce, internally, some of the molecules they needed. They drew on the abundant CO_2 and water in their environment and the energy available in sunlight to create complex organic molecules containing carbon. This process is called *photosynthesis,* and its byproduct is oxygen. As a result, the oxygen concentration in the atmosphere began to rise. Note that oxygen appeared in Earth's atmosphere only after life began and as a direct consequence of life itself.

Aerobic organisms evolved

The availability of oxygen changed everything. Oxygen was toxic to the anaerobic cells. In addition, because it is highly reactive, the oxygen began to break down the energy-containing molecules in the sea that the anaerobic organisms relied on for energy. Consequently most of the anaerobic organisms that could not create their own organic molecules by photosynthesis died out.

In their place new cells evolved that could harness the reactive nature of oxygen. These cells extracted energy from the abundant organic molecules stored within the cell itself, using the now readily available oxygen. As you may recall, the ability to use oxygen to extract energy from organic molecules is called *aerobic metabolism.*

The rise of animals and our human ancestors

Even from the brief description above, you can see that evolutionary processes are remarkably efficient at shaping life forms to fit their environments. At the risk of condensing several billion years into a few sentences so that we can move to the recent evolution of human beings, here are some highlights:

- DNA became enclosed within a nucleus about 1.7 billion years ago. Cells with nuclei, called eukaryotes, became the dominant cell type.
- The first multicellular organism, a type of seaweed, appeared approximately 1.3 billion years ago.
- Animals appeared about 600 million years ago.
- Dinosaurs became extinct about 65 million years ago.
- Our first distinctly human ancestors appeared only about 5 million years ago.

Figure 21.11 presents a time line of biological milestones in the history of Earth as they relate to humans.

> ***Recap*** **Over time photosynthesis evolved, resulting in increased atmospheric concentrations of oxygen. Some organisms subsequently developed an aerobic form of metabolism. A cell nucleus formed to enclose the DNA in most cells, multicellular organisms arose, and animals appeared. Our first recognizably human ancestors emerged about 5 million years ago.**

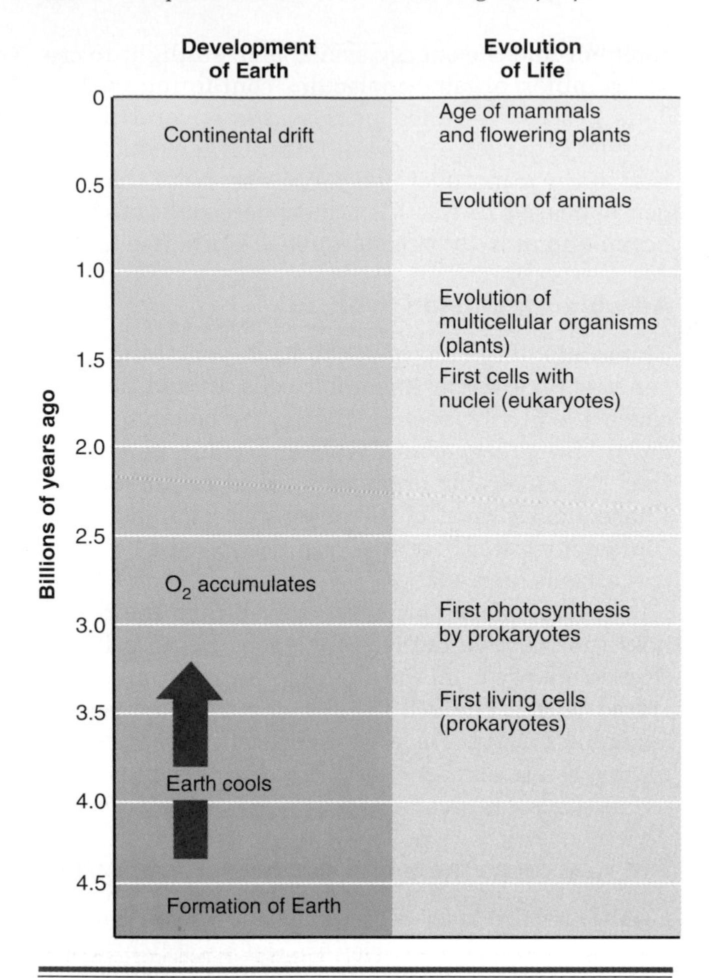

Figure 21.11 A time line of the evolution of life since Earth's formation. The time of appearance of our first distinctly human ancestors would be represented by less than the thickness of the horizontal zero time line at the top of the figure.

21.6 Modern humans came from Africa

Humans are primates

Taxonomy, the branch of science that focuses on classifying and naming life forms, seeks a precise way to describe the relationships between all life forms. The seven levels of the taxonomic classification system are: *kingdom, phylum, class, order, family, genus,* and *species.* A **species** is defined as a group of organisms that under natural conditions tend to interbreed and produce fertile offspring.

Life forms are generally referred to by their genus and species names. (These names are italicized, with the genus name capitalized and listed first.) Table 21.1 shows the seven taxonomic categories as they apply to humans, or *Homo sapiens.*

Spotlight on Structure & Function

Humans belong to the class *Mammalia.* **Mammals** are vertebrates that have hair during all or part of their life, and mammary glands that produce milk. Within the class Mammalia we belong to the order *Primata,* more commonly called **primates.** Primates are mammals with five digits on their hands, fairly flat fingernails and toenails rather than hooves or claws, and forward-facing eyes adapted for stereoscopic vision. They include the lemurs (sometimes called premonkeys), monkeys, apes, and humans. The primate hand has a thumb that opposes four fingers. All of these features apparently evolved as adaptations for an *arboreal lifestyle,* that is, life in trees. The fossil record indicates that all present-day primates share a common ancestor that lived about 60 million years ago (Figure 21.12).

Table 21.1 The Taxonomic Classification System as it applies to modern humans

Taxonomic Category	Scientific Name	Characteristics and Examples of Life Forms in Each Category
Kingdom	Animalia	Many-celled organisms with eukaryotic cells and a complex anatomy. Includes all animals: sponges, worms, insects, fish, birds, mammals, and many more.
Phylum	Chordata	Animals with a nerve cord and a backbone. Fish, birds, and mammals are examples.
Class	Mammalia	Members of the phylum Chordata with hair during part of their life-cycle and mammary glands. Includes mice, dogs, cows, whales, kangaroos, primates, and humans.
Order	Primata	Members of the class Mammalia with five digits, flat fingernails, and stereoscopic vision. Lemurs, monkeys, apes, and humans are examples.
Family	Hominidae	Primates that walk upright on two hindlimbs and that have slightly enlarged brains. Includes the genus *Australopithecus* and our *Homo* ancestors (now all extinct), and modern humans.
Genus	*Homo*	Hominidae with an enlarged brain and a recognizable human skull and body form. Includes our extinct immediate *Homo* ancestors and modern humans.
Species	*Sapiens*	*Homo* with large brains and a complex cultural and social structure that includes spoken language. Modern humans.

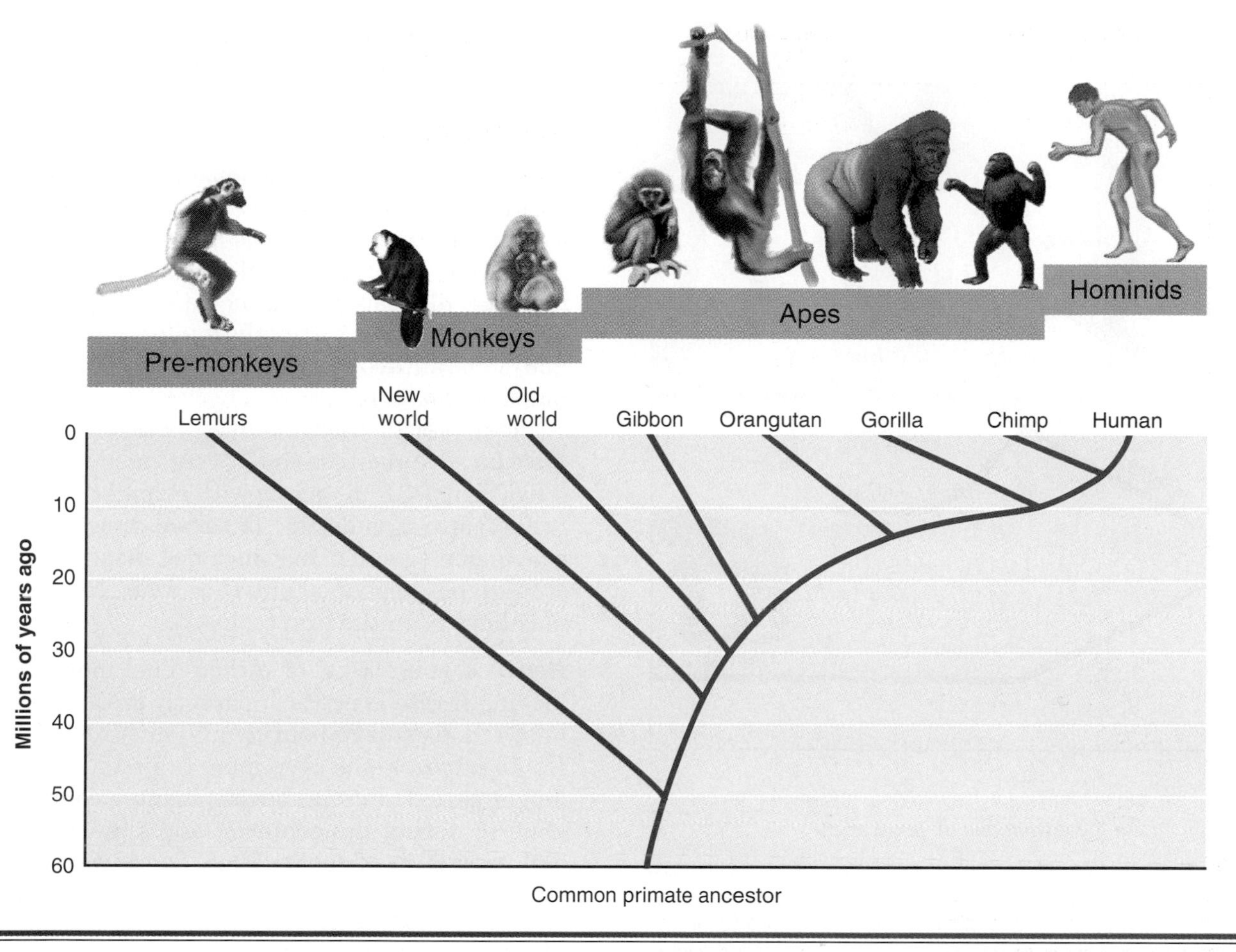

Figure 21.12 Evolution of modern primates from a common primate ancestor.

About 25 million years ago, a subgroup of primates diverged from the old-world monkeys to become the present-day apes and humans. Apes and humans differ from the rest of the primates in that they are larger and have bigger brains. They also lack tails and their behaviors are more complex. Based on the number of differences in DNA, it appears that gibbons diverged first, followed by orangutans, gorillas, chimpanzees, and humans. In other words, chimpanzees are our closest living animal relatives. However, this does not mean that we descended from chimpanzees. It means that chimpanzees and humans apparently evolved from a common ape ancestor (Figure 21.13). ■

Recap **A taxonomic system of seven categories classifies the relationship of any species to all others. Primates, an order of the class Mammalia, have flat fingernails and toenails, hands with five digits and an opposable thumb, and forward-facing eyes. A subset of the primates became the present-day apes and humans. Chimpanzees appear to be our closest living animal relatives.**

Evolution of *Homo sapiens*

Although scientists disagree regarding specific dates and the classification of certain fossils, the general framework of human evolution is fairly well established. Most of the evidence indicates that our story

Figure 21.13 Chimpanzees are our closest living relatives.

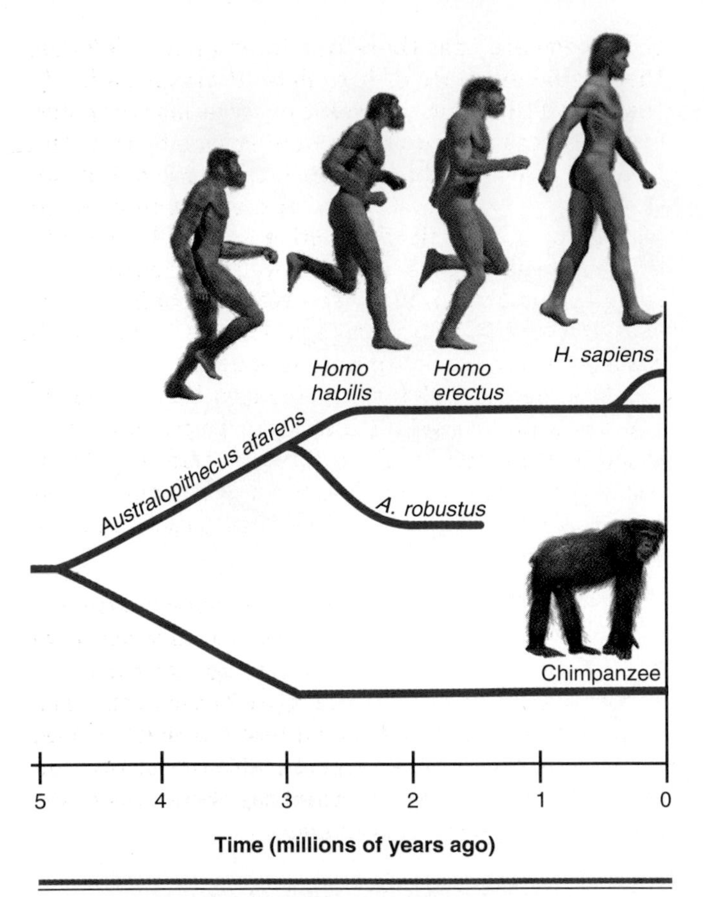

Figure 21.14 A proposed evolutionary pathway for *Homo sapiens.* The ancestors of humans and chimpanzees diverged about 5 million years ago. Although the chimpanzee has undergone evolutionary change as well, the intermediate steps are not shown here. The time represented by this figure is represented by less than the thickness of the zero time line at the top of Figure 21.11.

began in Africa with a primate that began to walk upright.

Australopithecus afarensis: An upright primate The fossil record indicates that the chimpanzee and human lines diverged from a common ancestor about 4–5 million years ago (Figure 21.14). The first of the lineage that eventually led to humans was a member of the family *Hominidae* called ***Australopithecus afarensis*** (Figure 21.15). **Hominids** are defined by the fact that they walk upright and show the beginnings of brain enlargement that is a key feature of the human lineage. *Australopithecus afarensis* was a vegetarian, judging by its teeth, and it was probably distinctly apelike in its anatomy and social structure. The males were considerably larger than the females (called sexual dimorphism), which suggests that *Australopithecus* had a social structure in which large, dominant males exerted control over small bands of females and less-dominant males. The important trait of *Australopithecus afarensis* in terms of human evolution was that it walked upright, freeing the forelimbs for other tasks. The oldest known fossil of *Australopithecus afarensis* is about 3.5 million years old.

When did our ancestors begin to walk upright?

***Homo habilis:* The first toolmaker** About 2.6 million years ago, the *Australopithecus* line split in two. One side of the *Australopithecus* line was a large upright vegetarian hominid, called *Australopithecus robustus,* that ultimately became extinct. The other line began to assume distinctly human characteristics, and thus anthropologists have given it the genus name *Homo.*

Our first distinctly human ancestor was ***Homo habilis.*** Brain enlargement continued in *Homo habilis* (Figure 21.16), along with changes in teeth and facial features, a decline in sexual dimorphism, and conversion to a diet that included meat. As the first stone tools appear about this time, *Homo habilis* may have been the first toolmaker.

Homo erectus:* Out of Africa** The more human-looking ***Homo erectus apparently arose from *Homo habilis* ancestors about 2 million years ago. With *Homo erectus* came even more brain enlargement, a longer period of infant development, a continued decline in sexual dimorphism, and the continuation and expansion of toolmaking. There were probably changes in social structure as well, including male-female pair bonds, shared responsibilities for the offspring, and the formation of hunting and gathering groups that shared food.

Homo erectus was the first of the human line to spread from Africa to Europe and Asia. The last European descendents of *Homo erectus* may have be-

Figure 21.15 Fossilized footprints of *Australopithecus afarensis.* Found by Mary Leakey in Tanzania in 1978, these 3.6 million-year-old footprints provided the first evidence of patterns of early hominid upright posture and locomotion. The prints appear to be of two adults and a child.

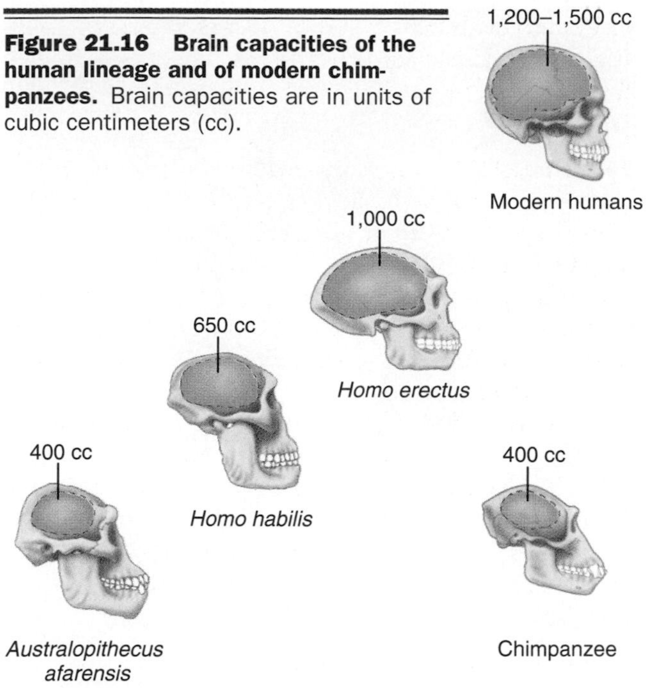

Figure 21.16 Brain capacities of the human lineage and of modern chimpanzees. Brain capacities are in units of cubic centimeters (cc).

come the Neanderthals, a controversial group of humans whose fossils were first discovered in the Neander Valley of Germany. Some researchers assign the Neanderthals to a separate species (*Homo neanderthalensis*), but this assignment remains controversial.

Scientists have long debated whether the Neanderthals, who became extinct about 35,000 years ago, are related to modern humans or were replaced by modern humans and simply became an evolutionary dead end. In 1997, biochemical analysis of DNA led one group of researchers to conclude that the Neanderthals apparently were not our direct ancestors, but this conclusion is not yet universally accepted.

***Homo sapiens:* Out of Africa again** The final step to modern humans was heralded by a substantial increase in brain size, the development of spoken language, and the development of a physical structure that we would call thoroughly human.

When did modern humans first appear?

Although it is by no means certain, a current hypothesis receiving considerable support is that modern humans, or ***Homo sapiens,*** evolved from a single colony of *Homo erectus* that were living in Africa sometime between 140,000 and 100,000 years ago. From there they spread around the world.

A key point is that wherever they went, *Homo sapiens* replaced the existing descendants of *Homo erectus* (including the Neanderthals). Ultimately *Homo erectus* and the Neanderthals vanished completely. Why? We don't know. Although some scientists speculate that there was interspecies violence, there is no evidence of it to date. It is just as likely that they fell victim to disease or were out-competed for resources. Some researchers speculate that the Neanderthals interbred with *Homo sapiens,* but because they were so few in number compared to *Homo sapiens* their distinctive physical features eventually disappeared. Whatever the reason, the last premodern humans (the Neanderthals) disappeared about 35,000 years ago, leaving *Homo sapiens* as the sole surviving human species.

At a mere 140,000 years, modern *Homo sapiens* has not been around all that long. To put our existence in perspective, if the history of life on Earth were compressed into a 24-hour day, the entire history of *Homo sapiens* would represent less than the last 3 seconds.

Recap **The lineage that was to become modern humans arose from a primate ancestor that began to walk upright nearly 5 million years ago. Over time the brain enlarged, and tool-making skills and spoken language developed. The current best hypothesis is that modern humans first appeared in Africa about 140,000 years ago and that all present-day *Homo sapiens* descended from that African group.**

Differences within the human race

All humans belong to a single species. We are so similar that sometimes the sum of all humankind is called the *human race*. Nevertheless, there are certain distinct heritable physical differences between subpopulations of the human species that the dictionary defines as *racial differences*. How did these differences come about?

How did racial differences arise?

Throughout most of human history, the total human population existed mainly as smaller subpopulations that were geographically isolated from each other. Minor random mutations coupled with sexual reproduction within isolated subpopulations apparently led to slightly different human group phenotypes (referred to as *races*) that are still observable today. Human racial differences are a testament to the effectiveness of evolution over tens of thousands of years.

Spotlight on Structure & Function

Some racial differences may have helped humans adapt more effectively to their environment. For example, in a climate with intense sunlight a dark skin color would have offered protection against excessive ultraviolet radiation. But in a Northern climate where sunlight was less intense a light skin color would have been helpful because small amounts of ultraviolet radiation are required to activate vitamin D. Facial features and even blood types may also have had their special advantages at the time. Other differences may simply have been neutral, meaning that they were the result of mutations that did no harm and so were retained in the population in which they arose. ■

In the final analysis, racial differences are nothing more than differences in phenotypes between subgroups of our common species. But then, individual phenotypes are all very different, too (Figure 21.17). In fact, DNA analysis reveals that there are as many differences between individuals of the *same* race as there are between individuals of *different* races. Given that there are no biological barriers to reproduction and that global travel and communication have recently become commonplace, human racial differences may disappear over the next 10,000 to 100,000 years, assuming we survive that long as a species.

Recap **All human beings belong to the same species. Racial differences most likely arose as adaptations to different environments in geographically isolated human subpopulations.**

Figure 21.17 The human race. Genotypic and phenotypic differences between individuals of the same race are as great as the differences between races.

Chapter Summary

Evidence for evolution comes from many sources

- The theory of evolution is the single most unifying concept in biology today.
- The fossil record provides a physical record of life forms that lived in the past.
- Structures with common evolutionary origins and patterns of embryonic development can be used to determine relationships between life forms.
- Structural similarities between proteins and DNA can be used to define closeness of evolutionary relationships.
- Geographical separations can affect the distributions of evolving life forms.

Natural selection contributes to evolution

- Mutations coupled with natural selection are the cause of most microevolution.
- The pool of alleles in a population is affected by chance events such as genetic drift and gene flow.
- Mass extinctions can alter the course of macroevolution by causing the wholesale extinction of many species at once.
- An evolutionary tree visually depicts relationships between species.

The young Earth was too hot for life

- Earth was formed about 4.6 billion years ago. At first, it was a harsh environment with constant cycles of condensation and vaporization.
- Life began about 3.8 billion years ago.

The first cells were able to live without oxygen

- Organic molecules formed from gases in the Earth's atmosphere.
- The first self-replicating molecule was probably single-stranded RNA.
- The first cell was formed when RNA and organic molecules became enclosed in a membrane.
- The early atmosphere was devoid of oxygen, so the first living cells survived by anaerobic metabolism.

Photosynthetic organisms altered the course of evolution

- Oxygen appeared only after certain life forms developed the process of photosynthesis as a way to produce organic molecules inside the cell.

- With oxygen available, other cells developed the process of aerobic metabolism as a way to extract energy from organic molecules.
- Eukaryotes became the dominant cell type and multicellular organisms evolved.

Modern humans came from Africa

- Humans are a relatively recent development on Earth. Humans and modern chimpanzees evolved from a common ancestor that lived about 5 million years ago.
- Hominids are primates that walk upright and are characterized by larger brains. The hominid *Australopithecus afarenus* is one of the earliest hominids in human evolution.
- Modern humans apparently first appeared in Africa about 140,000 years ago. From there they spread around the world.
- Heritable, observable, physical phenotype differences in human subpopulations are called racial differences. Racial differences are no more significant than differences between individuals. We all belong to the same human species.

Terms You Should Know

adaptive radiation, 527
analogous structures, 521
Australopithecus afarensis, 532
continental drift, 525
evolution, 520
evolutionary tree, 526
extinction, 526
fossil, 520
gene flow, 526
genetic drift, 526
hominid, 532
Homo erectus, 532
Homo habilis, 532
Homo sapiens, 533
homologous structures, 521
macroevolution, 520
mammal, 530
microevolution, 520
natural selection, 525
population, 526
primate, 530
species, 530
vestigial structures, 523

Test Yourself

Matching Exercise

_____ 1. The seven taxonomic categories of classification are kingdom, phylum, class, order, _____, genus, and species.
_____ 2. Large-scale trends or changes that apply to whole groups of species.
_____ 3. A group of organisms that under natural conditions tend to breed within that group.
_____ 4. The term for a body structure that no longer has any function.
_____ 5. The term for a lifestyle that is lived in trees.
_____ 6. These body structures serve similar functions but do not have the same evolutionary origin.
_____ 7. The first of our human ancestors to use tools.
_____ 8. The first of our human ancestors to walk upright.
_____ 9. A unifying concept that fits a broad range of established facts.
_____10. Random changes in allele frequency because of chance events.

a. macroevolution
b. vestigial
c. *Homo habilis*
d. arboreal
e. analogous
f. family
g. species
h. genetic drift
i. theory
j. *Australopithecus afarensis*

Fill in the Blank

11. Body structures that share a common origin are said to be _____________.
12. The preserved remnants of organisms are called _____________.
13. _____________ is the study of the distribution of animals and plants around the world.
14. A(n) _____________ is a group of individuals of the same species that occupies the same geographical area.
15. A(n) _____________ destroys more than 50 percent of all species existing at that time.
16. The first self-replicating molecule on Earth appears to have been _____________.
17. _____________ refers to a natural and unpredictable process of descent over time with genetic modification.
18. _____________ are a class of vertebrates that have hair during part or all of their life and mammary glands that produce milk.
19. Identify your genus and species: _____________.
20. Your tailbone is an example of a(n) _____________ structure.

True or False?

21. Due to the nature of the fossilization process, the fossil record consists largely of soft tissues such as skin and internal organs.
22. All humans living today belong to the same species.
23. Continental drift may isolate groups of organisms from each other.

24. When he coined the term *genetic drift,* Darwin meant that individuals with traits that made them more fit for their local environments were more likely to survive and reproduce.
25. Modern humans belong to the phylum called Animalia.

Concept Review

1. What is a fossil and how is a fossil created?
2. What are some of the main sources of evidence for evolution?
3. What process in addition to mutation is required for microevolution of a species?
4. Why were there no oceans on the Earth when it was first formed?
5. Summarize the scientific theory of how life began on Earth.
6. Identify and describe the process that is responsible for the presence of oxygen in the Earth's atmosphere.
7. Explain how rising atmospheric concentrations of oxygen affected life on Earth.
8. What features distinguish the taxonomic family *Hominidae* from other families in the order *primates*?
9. How long ago and on which continent did *Homo sapiens* probably arise?
10. Explain why racial differences may disappear in the next 10,000 to 100,000 years.

Apply What You Know

1. Because dogs are all one species, different breeds of dogs can mate and produce viable offspring. How, then, did different breeds of dogs with distinct characteristics evolve?
2. Explain how more recent knowledge of Earth's history of mass extinctions required modification of Darwin's original explanation for evolution.
3. How would you expect an increase in gene flow in the human population to affect human phenotypes?

Case in Point

Why Was Walking Upright Important in Human Evolution?

The chapter notes that *Australopithecus afarensis* seems to have been the first primate to walk upright. Skeletons discovered in Africa show that this species had a broad pelvis and thigh bones with a dense top that could withstand vertical pressure. As in modern human feet, its big toe was aligned next to its other four toes.

At the same time, *Australopithecus afarensis* had upper arms that were longer than those of modern humans, and markings on its upper arm bones indicate that it had powerful muscles. Shoulder and hip joints were probably more mobile than ours.

QUESTIONS:

1. Explain how the skeletal structure of *Australopithecus afarensis* might have enabled it to walk on its hind limbs.
2. What conclusions can you draw from the differences between *Australopithecus*'s upper body and ours?
3. Scientists hypothesize that the climate in which *Australopithecus* lived was fluctuating dramatically, causing its environment to vary from dense forest to treeless grassland. Explain how the hominid's body structure might have helped it adapt to this environment.

INVESTIGATE:

Rick Potts, "Why Are We Human? Maybe Evolution Favored an All-Terrain Animal," *The Washington Post,* April 14, 1999, p. H01.

References and Additional Reading

Amigoni, D. and J. Wallace, eds. *Charles Darwin's Origin of Species—New Interdisciplinary Essays.* New York: Manchester University Press, 1995. (Essays in history, literature, sociology, and anthropology related to the work of Charles Darwin.)

Cartmill, M. Oppressed by Evolution. *Discover,* p. 78, Mar. 1998. (A look at why so many people don't accept the theory of evolution.)

Leakey, R. and R. Lewin. *Origins Reconsidered: In Search of What Makes Us Human.* New York: Anchor Books, 1992. 375 pp. (An entertaining look at human evolution by one of the leading authorities in the field.)

Ward, P. The Greenhouse Extinction. *Discover,* p. 54, Aug. 1998. (An examination of the possible cause of the largest mass extinction on Earth that occurred 250 million years ago.)

Wong, K. Who Were the Neandertals? *Scientific American,* April 2000, p. 98.

CHAPTER 22

ECOSYSTEMS AND HUMAN POPULATIONS

How can we ensure our future **on Earth?**

Why do all populations eventually stop growing?

What is the source of energy **for living organisms?**

When **did the human population begin** growing rapidly?

How might we be contributing to global warming?

THROUGHOUT THIS BOOK, WE HAVE FOCUSED ON UNDERSTANDing enough basic biology to appreciate *human* biology. As you know by now, all living organisms share certain features in common. We all consist of one or more living cells, and we have the capacity to reproduce. We share a common origin and have evolved over time. And, perhaps most importantly, we share the same planet. Our futures are inextricably linked.

Ecology is the study of the relationships between organisms and their physical environment. In this chapter we begin by describing general principles of ecology. Then we turn to the impact of the human population on Earth's ecosystems. Humans have proven themselves capable of altering the environment, sometimes dramatically. Fortunately, we are also capable of making thoughtful and informed choices about our actions. The choices we make may determine or change the course of life on Earth far into the future.

22.1 Ecosystems: Communities interact with their environment

The basic unit of study in ecology is an *ecosystem*. *Eco-* implies a connection to the environment, and *-system* indicates that an ecosystem is a functional unit. An **ecosystem** consists of a community of organisms *and* the physical environment in which they live. The ecosystem comprises all the living things, all the matter, and all the energy. In an ecosystem we consider not only who eats whom, but how nonliving matter is recycled and how energy flows into and out of the system. The components of an ecosystem form a hierarchy that includes populations, communities, and the physical environment in which they live.

As discussed in the last chapter, a **population** is a group of individuals of the same species that occupy the same geographic area and interact with each other. A population of elk inhabits a mountain valley, and a population of frogs inhabits a small pond.

A **community** consists of the populations of all the species that occupy the same geographic area and therefore interact. Some organisms serve as food or shelter for others. Insects feed on decaying wood. Woodpeckers feed on the insects and make nests in those same trees. Deer browse on shrubs, cougars hunt the deer, and so on. A community is the living part of an ecosystem.

Together, all of the ecosystems on Earth compose the **biosphere,** or planetary ecosystem. The biosphere is so complex that it is difficult to study it as one whole entity. Studies of simpler ecosystems have, however, led to some interesting conclusions that may apply to the entire biosphere as well. Most importantly, it appears that ecosystems are delicately balanced self-sustaining units in which organisms interact with each other and their environment in complex ways. In the process, they can modify or control the physical and chemical nature of the ecosystem itself. Some ecologists suggest that disturbing the delicate balance of the entire biosphere could endanger the survival of the human species—and of all other species as well.

Recap **An ecosystem consists of a community of organisms and their physical environment, and a community consists of all populations that interact within the same area. The biosphere comprises all ecosystems on earth.**

22.2 Populations: The dynamics of one species in an ecosystem

Characteristics of populations include habitat, range, size, actual growth rate, and potential capacity for growth.

Where a species lives: Habitat and range

Spotlight on Structure & Function

A species' **habitat** is the type of location where it chooses to live (Figure 22.1). Typically this is determined by its tolerance for certain environmental conditions. For example, the habitat of the now rare northern spotted owl is mature evergreen forests, the habitat of the bison is grassland, and beavers prefer a habitat of mountain valleys with freshwater streams and small trees.

Each organism's habitat has certain chemical and physical characteristics that favor the organism's comfort and survival. Animals and plants that occupy the same or overlapping habitats form a community, competing with each other for resources in terms of matter or energy and using each other for food and/or shelter.

Because the availability of ideal habitats varies, each species also has a **geographic range** of distribution representing the area over which it may be found. An organism's range is limited by several factors: (1) competition for resources such as sunlight or nutrients, (2) intolerable conditions such as extreme temperatures or altitude, or (3) physical obstacles including mountain ranges, deserts, and bodies of water. ■

In the last few centuries the development of worldwide transportation has enabled some animals to expand their ranges beyond physical barriers. A good example is the ring-necked pheasant, a Chinese bird that has thrived in the United States since its introduction from Europe.

(a) These red-eyed tree frogs inhabit the tree canopy of the Costa Rican rain forest.

(b) The habitat of these musk oxen is the tundra of Northwest Territories, Canada.

(c) The habitat of these students is a university campus in Texas.

Figure 22.1 Habitats. An organism's habitat is the place where it chooses to live.

Population growth rate tends toward biotic potential

One of the most important topics in ecology is how populations change size over time. The maximum rate of growth of any population under ideal conditions is called its **biotic potential.** Biotic potential is a function of certain characteristics of the species itself, including:

- The number of offspring produced by each reproducing member (female, if reproduction is sexual)
- How long it takes the offspring to reach reproductive maturity
- The sex distribution if reproduction is sexual (ratio of males to females)
- The number of reproductive-age members of the population

Typically, the biotic potential of any organism follows an **exponential growth** curve with a characteristic J-shape. Figure 22.2 compares the growth curves of two species of bacteria. For any species, the steepness of the growth curve, representing the biotic potential, will be determined by the four factors above. The J-shape reflects the fact that under ideal conditions, populations would double again and again over similar time periods.

Environmental resistance limits biotic potential

Why do all populations eventually stop growing?

In the real world of finite resources and competition, an opposing force called **environmental resistance** limits any species' ability to consistently realize its biotic potential. Environmental resistance consists of factors that kill organisms and/or prevent them from reproducing. Examples include limitations on nutrients, energy,

Try It Yourself

The Rule of 72

A human population growing at a rate of just 1.5% per year will double in just 48 years. How do we know that? You can calculate it yourself by applying a simple mathematical rule known as the **Rule of 72,** which states that dividing the number 72 by the percent increase per unit of time yields how long it will take for any number to double. In the human population example above, divide 72 by 1.5% per year and you get 48 years. The rule works for any time period and for any percent increase, not just one relating to population. Try it yourself:

1. How long would it take the human population to double if it were increasing at a rate of 2% per year? (_____ years.)
2. How long would it take a bacterial population to double if it were increasing at a rate of 24% per hour? (_____ hours.)
3. How long would it take you to double your money if it were earning interest at a rate of 4% per year? (_____ years.) How about at 7.2% per year? (_____ years.)

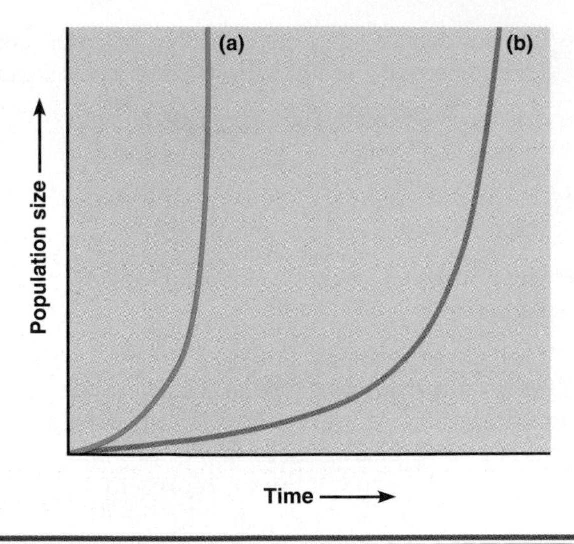

Figure 22.2 Exponential growth curves. J-shaped or exponential growth curves are typical of organisms whose populations are increasing according to their biotic potential. The bacteria represented by curve (a) have a greater biotic potential than those represented by curve (b).

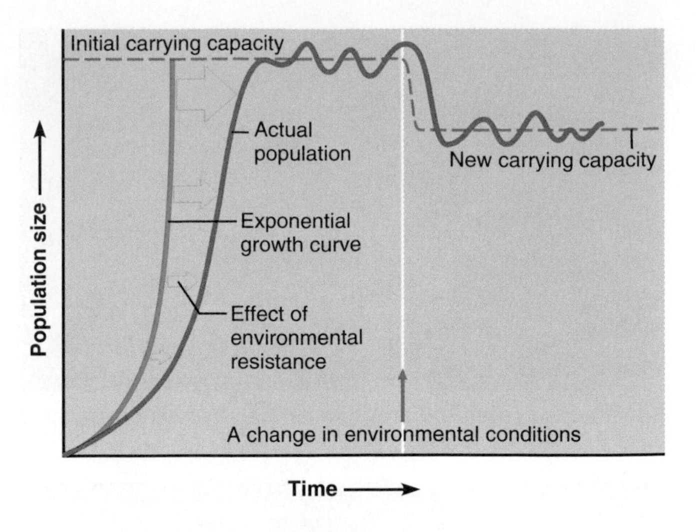

Figure 22.3 Effects of environmental resistance and carrying capacity on population size. As the population increases the environmental resistance to continued ideal exponential growth increases. Eventually the population stabilizes near the carrying capacity. Changing environmental conditions can alter the carrying capacity.

and space; disease; predation by other organisms; and environmental toxins.

Even when conditions are ideal to begin with, eventually every growing population reaches a point where environmental resistance begins to increase. For bacteria in a petri dish it may be the food supply. For a population of mice it may be disease brought about by overcrowding. For a woodpecker it may be a lack of suitable nesting sites. Although we cannot always measure the specific environmental resistance, no population grows at its full biotic potential indefinitely.

Ultimately a population achieves an approximate balance between biotic potential and environmental resistance. At this point it stabilizes, reaching a steady state. The population that the ecosystem can support indefinitely is called the **carrying capacity** (Figure 22.3). Population sizes may vary around their carrying capacity from time to time. In addition, the carrying capacity can change if environmental conditions change. A wet spring might increase the carrying capacity of a prairie ecosystem for certain grasses and the animals that feed on those grasses, and three years of drought might reduce the carrying capacity to below its average value. But in the end, every population tends to stabilize near its carrying capacity.

Recap **Every species has a preferred habitat and a geographic range over which it is distributed. Under ideal conditions a population will grow exponentially according to its biotic potential. Under normal circumstances, however, environmental resistance limits growth. The population of a particular species that an ecosystem can sustain indefinitely is called its carrying capacity.**

22.3 Communities: Different species living together

Overlapping niches foster competition

An organism's **niche** is its role in the community—its functional relationship with all living and nonliving resources of its habitat (Figure 22.4). The woodpecker's habitat is woodland forests. Its niche is to rid trees of their insect infestations and serve as a food supply for hawks or raccoons, or ultimately for bacteria. A well-balanced ecosystem supports a wide variety of species, each with a different niche.

Although two species do not occupy exactly the same niche, niches can overlap enough that **competition** may occur between species for limited resources (Figure 22.5). Plants may compete for available sunlight or mineral nutrients, and animals often compete for similar food supplies (both foxes and hawks prey on mice). Depending on their particular niches, one species may be able to out-compete another in a given location, a phenomenon known as *competitive exclusion.*

Succession leads toward a climax community

Most natural communities undergo constant change. **Succession** is a natural sequence of change in terms of which organisms dominate in a community. Succession is determined by population growth rates, the

Figure 22.4 An organism's role in its community is its niche. A woodpecker's habitat is woodland forest, and its niche involves pest control. The woodpecker shown here is feeding on insects that damage trees.

niches occupied by various species, and the kinds of competition that exist between them.

If you have ever seen an abandoned field grow first to weeds, then to shrubs, and finally to trees, you have seen succession in action. Organisms that can reproduce and grow quickly under the prevailing conditions are the first to arrive in any area. These first organisms provide food and shelter for the next arrivals, which grow more slowly but that occupy more specific niches. The number of species present tends to increase as the succession advances and interrelationships between populations become ever more complex.

If the area suffers no major shocks, such as a fire, succession ends with the establishment of a stable *climax community.* The type of climax community found in any region is largely determined by the physical environment but can also be influenced by the organisms living in surrounding areas. In the Northeast United States, the typical climax community is dominated by large but slow-growing *deciduous* trees (broad-leaved trees such as oaks that lose their leaves each winter, then grow new leaves each spring). In the Midwest the climax community is prairie, in the Southwest it is semi-arid desert, and in the mountainous West it is towering evergreen forests.

Climax communities are the most efficient communities in terms of energy and nutrient utilization and also the most varied in terms of numbers of species present. They may take hundreds of years to develop, and when disturbed they do not recover easily. Their very complexity makes them vulnerable.

Figure 22.5 Competition. African zebras and wildebeest compete for the same food and water supply.

Ecosystems: Communities and their physical environment

An *ecosystem,* as mentioned earlier, has both living and nonliving components. The total living component of an ecosystem is called that ecosystem's **biomass.** The nonliving components consist of the *chemical elements* (the physical matter that makes up the earth) and the *energy* that drives all chemical reactions.

A constant supply of energy is essential for continued life because the transfer of nutrients and energy from one organism to the next is always an energy-absorbing process. Nearly all of earth's energy comes from the sun, with only a tiny amount supplied by the residual heat of the earth itself (thermal vents and volcanoes). In contrast to energy, the chemical matter of the earth is recycled over and over again between the earth and the biomass (Figure 22.6). The next two sections discuss these two key concepts in more detail.

What is the source of energy for living organisms?

Recap **Species with similar niches may be in competition with each other. Communities go through progressive changes over time, called succession, that increase overall efficiency and species variety. Left undisturbed, eventually they achieve a climax community. Ecosystems consist of biomass, energy, and nonliving matter.**

22.4 Energy flows through ecosystems

Producers capture and convert energy, consumers rely on stored energy

The flow of energy in an ecosystem demonstrates two key principles of physics, called the *laws of thermodynamics:*

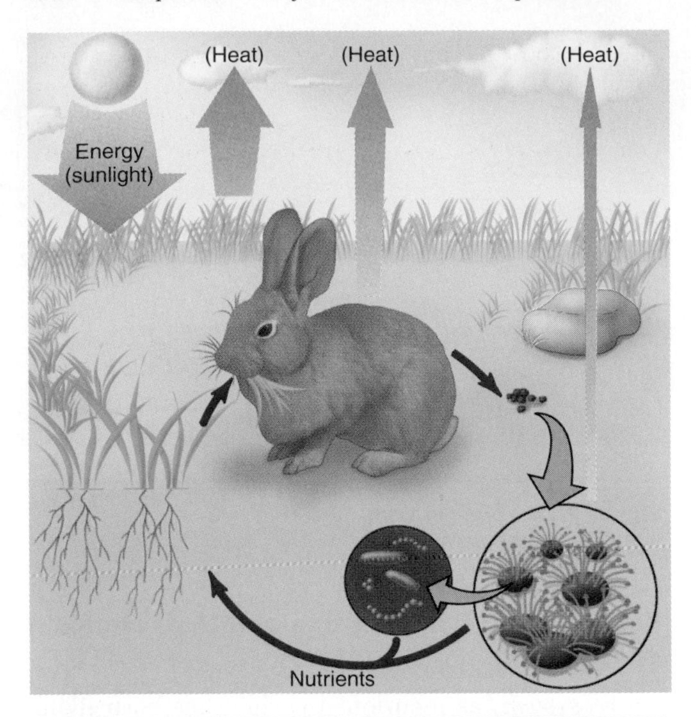

Figure 22.6 The flow of energy and nutrients in ecosystems. Energy provided by the sun passes through living organisms and then back into the atmosphere as heat. Nutrients are continually recycled between the earth and the biomass.

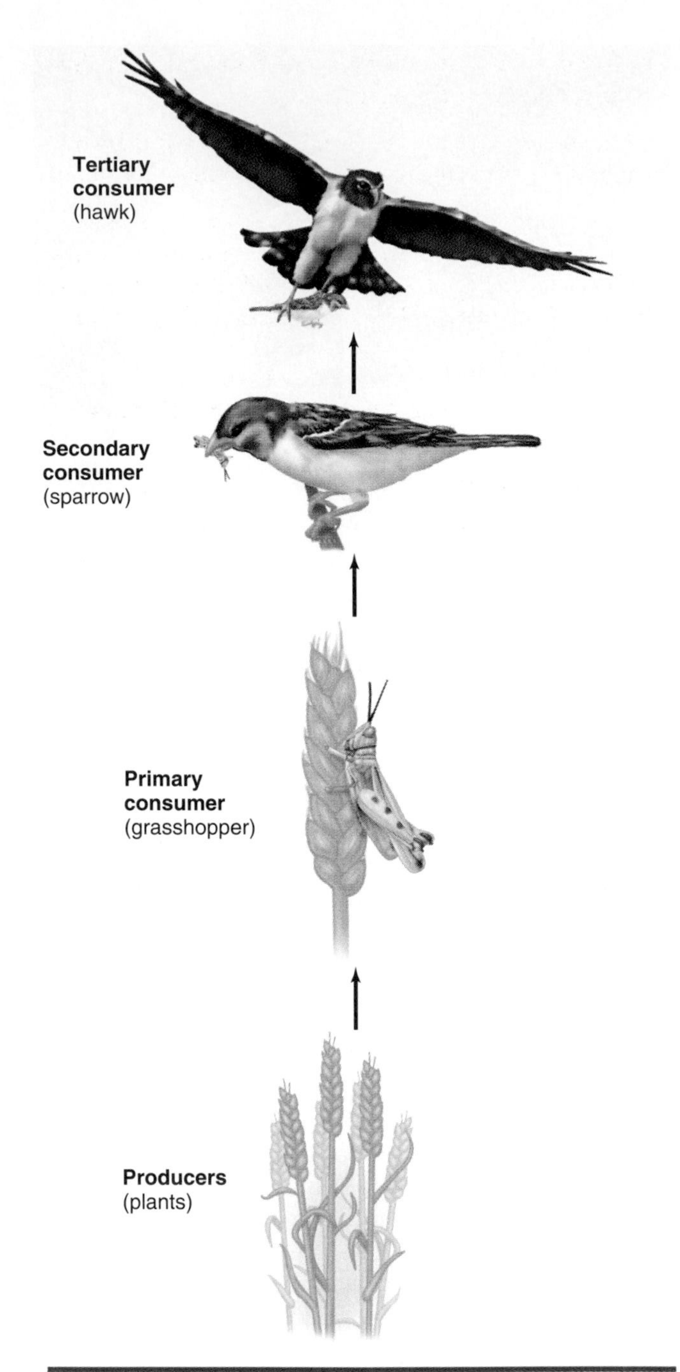

Figure 22.7 A food chain. Food chains show the hierarchical nature of feeding relationships in ecosystems.

- The *first law of thermodynamics* states that energy is neither created nor destroyed. Energy can, however, change form and it can be stored.
- The *second law of thermodynamics* states that whenever energy changes form or is transferred, some energy may be transformed into nonuseful forms. Generally this nonuseful component is heat energy. The point is that energy transfer is an inefficient process.

The biomass consists of *producer* and *consumer* organisms. Energy flows in one direction through an ecosystem, from the sun to producers to consumers.

Spotlight on Structure & Function

Producers are organisms that are capable of photosynthesis. They use the energy of the sun to make food for themselves from the chemical elements of the atmosphere, soil, or water. Producers are also sometimes called *autotrophs,* meaning "self-nutritive" or "self-growing." Only producers are able to capture the abundant energy provided by the sun and convert it to a form usable for themselves and other organisms. On land the producers are mainly green plants, whereas in aquatic ecosystems they are primarily species of algae. Producers store some energy in the form of complex carbohydrates such as starch and cellulose.

Unlike producers, **consumers** (also called *heterotrophs*) cannot utilize the sun's energy to synthesize the molecules they need. Instead they must consume foods that already contain stored forms of energy. Consumers fall into four types depending on what they use as a food source:

- **Herbivores** utilize the energy stored in green plants, and for that reason they are also called *primary* consumers.
- **Carnivores** feed on other animals, and thus they are *secondary* and sometimes *tertiary* (third-level) consumers. The primary forms of

stored energy in animals are fats and carbohydrates such as glycogen.

- **Omnivores**, such as humans, pigs, and bears, can derive energy from either plants or animals.
- **Decomposers** feed on *detritus,* or dead organisms. Bacteria and fungi are decomposers, as are earthworms and certain small arthropods (a phylum of organisms that includes insects and spiders). ■

A food web: Interactions among producers and consumers

Relationships between producers and consumers are sometimes depicted by simple food chains (Figure 22.7). Although such diagrams are useful for presenting the relationships between types of consumers, they understate the degree of complexity in most ecosystems. Most organisms rely on more than one species for food and in turn may be eaten by more than one species.

A **food web** consisting of numerous interacting food chains more accurately depicts the complex balanced nature of an ecosystem (Figure 22.8). As the population of a particular primary consumer species rises, for example, it provides a ready food supply for secondary species. These secondary consumers then rely less on other primary food species whose populations are low at the moment. Consequently, no single population gets completely out of hand, and rarely does any species become extinct.

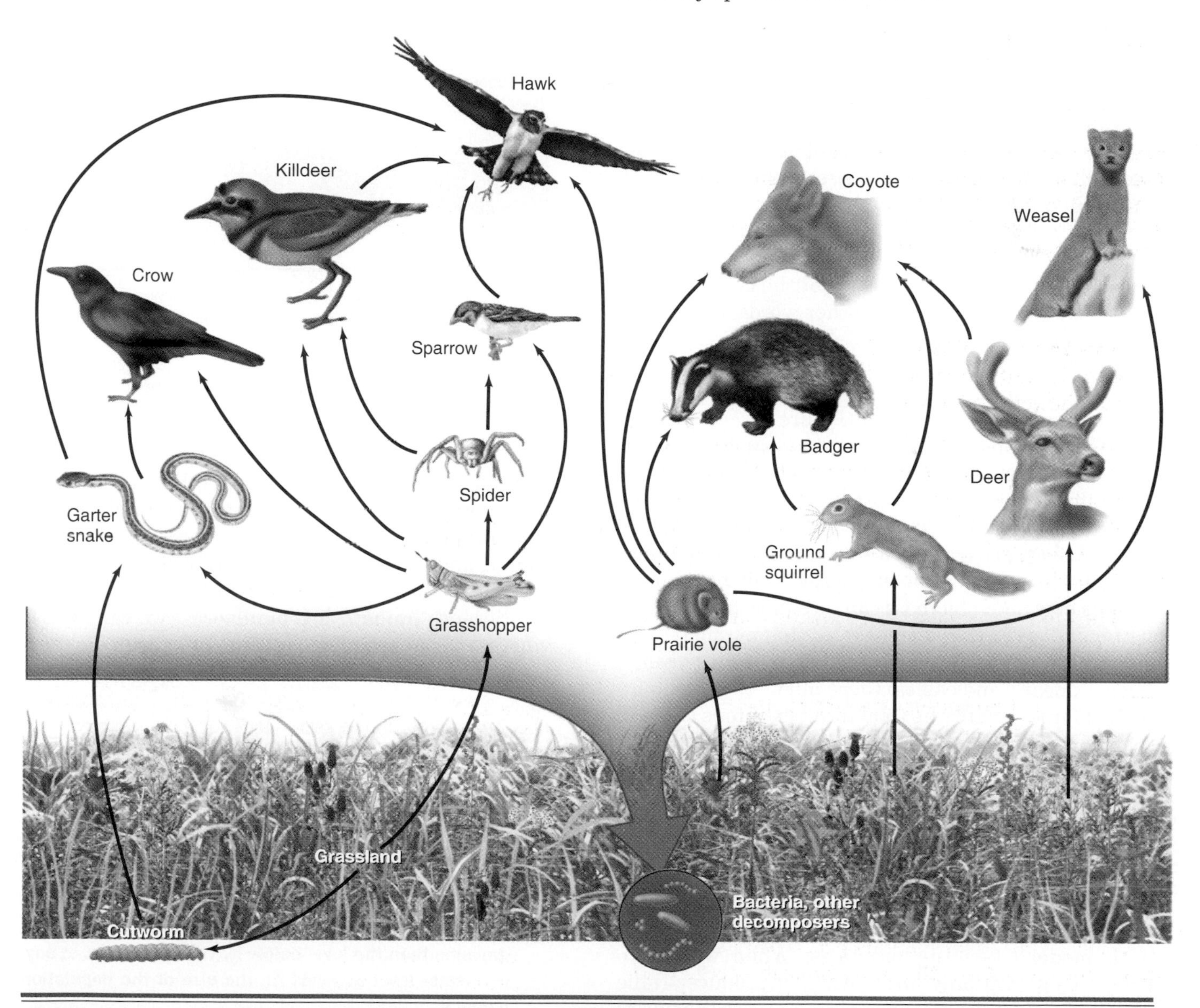

Figure 22.8 A prairie food web. This food web includes the food chain shown in Figure 22.7 and several other food chains as well. Note that bacteria and other decomposers also play a role in the food web by feeding on dead organisms at all levels.

In a stable ecosystem every population rises and falls periodically as part of the normal ebb and flow of life, as shown previously in Figure 22.3. These variations in a species population are always near the carrying capacity of the ecosystem for that particular species.

Recap **Producers are photosynthesizing organisms that use the sun's energy, available nutrients, and atmospheric carbon dioxide to produce energy-containing molecules. All other organisms, called consumers, obtain their energy from plants or other organisms. All organisms in an ecosystem form an interconnecting food web, and energy flows in one direction through the ecosystem. Populations may rise and fall, but generally they hover near the ecosystem's carrying capacity for each species.**

Lower levels of an ecological pyramid support consumer populations

The processes of energy conversion, transfer, and storage are remarkably inefficient at every level. For example, photosynthetic processes utilize and store in the biomass only 2% of the total energy in sunlight. The rest is reflected as light or heat energy. Only about 10% of the energy at one level of a food web can be found in the tissues of the consumers at the next level in the web. In other words, only about 10% of the stored energy available in plants is stored in herbivores, and only about 10% of that (or 1%) is stored in secondary consumers.

An **ecological pyramid** (Figure 22.9) depicts either the total amount of energy stored at each level of an ecosystem or the total biomass at each level. The total amount of energy represented by tertiary consumers at the top of the food chain is very small indeed. Ecological pyramids also generally indicate total biomass, as the amount of energy available usually defines how much biomass can be supported. To get a better image of the steepness of this pyramid, consider how many total pounds of large carnivores (wolves, bears, and eagles) there might be in a square mile of forest versus the total number of tons of trees.

Ecological pyramids illustrate that the population of consumers at any level depends critically on the populations of consumers directly below it. Tertiary consumers are especially vulnerable to ecosystem disruptions because the small amount of energy available to them depends on efficient transfer at all levels below them.

Humans can disrupt this balance rather easily with modern farming practices. When we create ecosystems with only one species of producer (mile after mile of wheat or corn), then harvest the entire crop for our own use, we exclude other species from their natural food web and their place on the ecological pyramid. We do this to feed more people, of course, and it works. But we must recognize that our actions do affect other species on Earth.

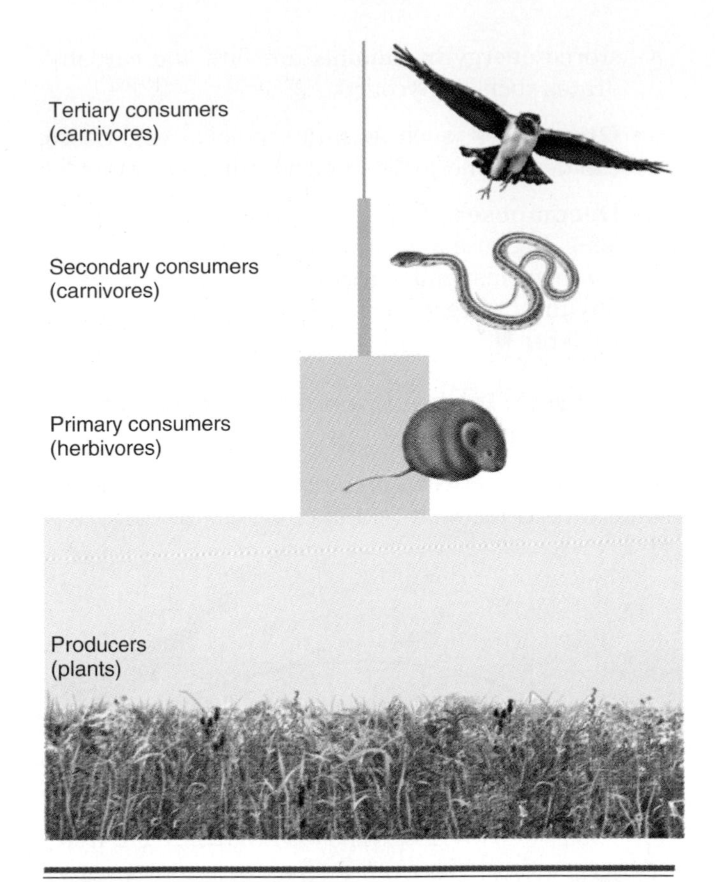

Figure 22.9 An ecological pyramid. This ecological pyramid depicts total energy stored in the biomass at each producer/consumer level. Because energy transfer is inefficient, only about 10% of the energy at one level can be found in the tissues of the consumer at the next highest level of the pyramid.

Because humans are omnivores, we can choose whether to eat producers (plants) or consumers (herbivores or omnivores). Eating meat is energetically expensive. When we feed grain to cattle and then eat the beef, we utilize only 10% of the energy that would otherwise be available to us in the grain. At some point, if we face hard choices about feeding the growing world population, scientists will surely point out that we can be more energy-efficient by eating grain rather than meat.

Recap **Energy transfer is inefficient, so each producer/consumer level can support fewer and fewer organisms than the level below it. The population at any consumer level depends on the size of the population below it.**

22.5 Chemical cycles recycle molecules in ecosystems

In contrast to the continuous input of energy from the sun, the total mass of nutrient matter that makes up the Earth and its biomass remains completely fixed. With only a finite amount of each element available, the chemicals that compose living organisms are recycled over and over again between organisms and the Earth itself. Such **biogeochemical cycles** include living organisms, geologic events, and even weather events.

As shown in Figure 22.10, every molecule and element in a living organism cycles between three different pools: the exchange pool, biomass, and reservoir. Some of the elements or molecules are found in the *exchange pool* (water, soil, and atmosphere), from which the primary producers draw their nutrients. Using the energy of the sun, the primary producers incorporate nutrients into the *biomass*. Organisms in the biomass exchange molecules rapidly as they consume each other. When an organism dies and decomposes, its nutrients return to the exchange pool, where they again become available to primary producers.

Coupled to the exchange pool is a large but hard-to-access pool of nutrients called the *reservoir*. Nutrients in reservoirs include minerals in solid rock and in sediments deposited at the bottom of the oceans. Other important reservoirs are carbon-containing compounds in **fossil fuels** (coal, oil, and gas) that were created millions of years ago from the remains of living organisms, and were then covered by sediments.

We turn now to the biogeochemical cycles for some of the most important molecules and elements of life—water, carbon, nitrogen, and phosphorus.

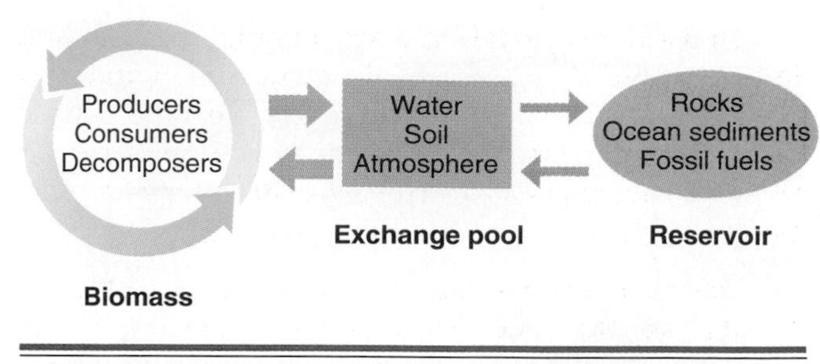

Figure 22.10 The flow of nutrients in biogeochemical cycles. The three components of a biogeochemical cycle are the biomass, the exchange pool, and the reservoir. The chemicals that compose living organisms are constantly recycled among the three components. The relative magnitudes of nutrient flows between nutrient pools and within the biomass are represented by differences in arrow thickness.

Recap **A biogeochemical cycle describes how a chemical cycles between the biomass, exchange pool, and reservoir pool. Most nutrients cycle rapidly within the biomass pool, less rapidly with the exchange pool, and rarely with the reservoir pool.**

The water cycle is essential to other biogeochemical cycles

Water is probably more essential to life than any other molecule. It makes up about 70% of your body and covers approximately 75% of the Earth's surface. Water is the universal solvent for many other elements—indeed, without the water cycle, all other biogeochemical cycles would cease.

Figure 22.11 illustrates the water cycle. Both the water cycle and the biomass are important to the biogeochemical cycles for minerals. Rain and surface water runoff slowly erode rocks and leach the minerals from them. Plants take up these minerals and incorporate them into the biomass. If there are no

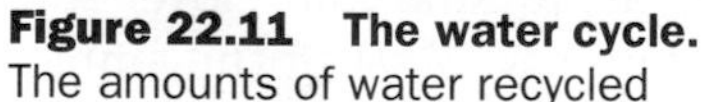

Figure 22.11 The water cycle. The amounts of water recycled relative to the total rates of evaporation and precipitation are indicated as percentages. Heat energy from the sun causes water to evaporate, primarily from oceans but also from surface water on land. Most of the atmospheric water vapor precipitates as rain over the oceans, but approximately a quarter of it precipitates over land, replenishing surface water that was lost by evaporation and providing the extra water that becomes ponds, lakes, streams, and rivers.

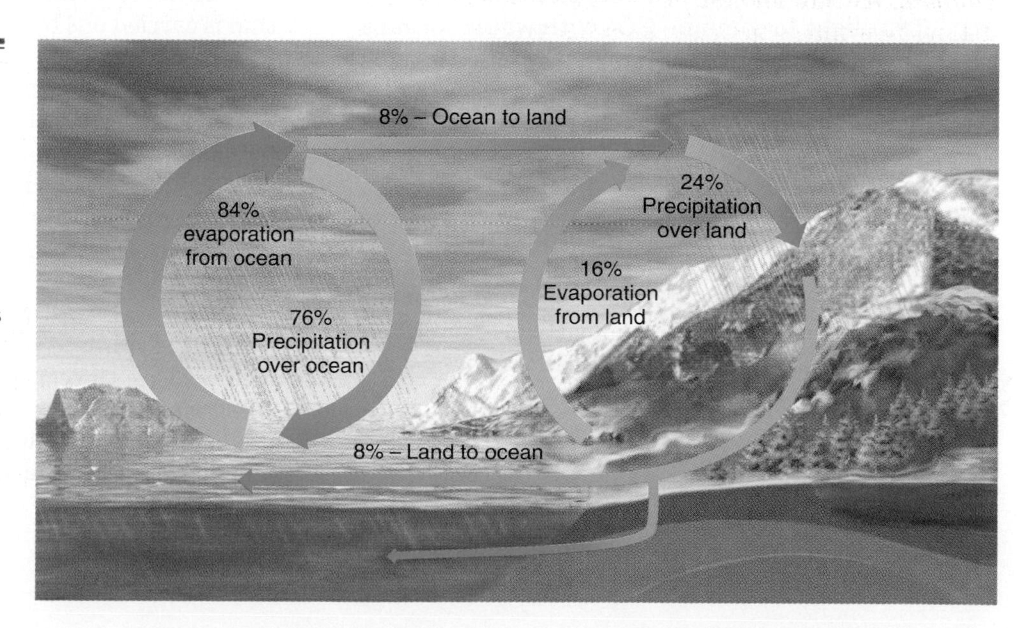

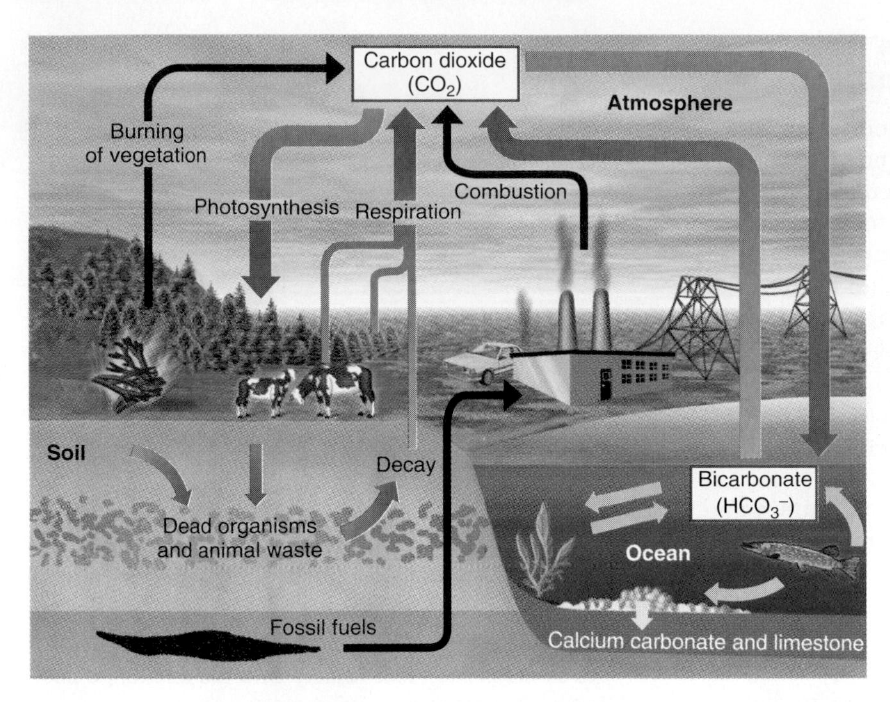

Figure 22.12 The carbon cycle. Carbon dioxide in the atmosphere and bicarbonate in the oceans constitute the exchange pool. The reservoir pool of carbon includes limestone and calcium carbonate from the shells of marine organisms and also fossil fuels formed several hundred million years ago from decomposed plant material. Fossil fuels represent a reservoir for carbon that would not normally be accessible to the exchange pool, but over the last several hundred years humans have tapped them for energy. The effect of human activities is indicated by the darker-colored arrows.

plants, the minerals wash into the sea instead. When forests or grasslands are destroyed, nutrients lost from the soil may not be replenished from the reservoir pool for hundreds of years.

The carbon cycle: Organisms exchange CO_2 with the atmosphere

Carbon forms the backbone of organic molecules and the crystalline structure of bones and shells. The biogeochemical cycle for carbon (Figure 22.12) is called a *gaseous cycle* because carbon in living organisms is exchanged with atmospheric CO_2 gas.

The carbon cycle is closely tied to photosynthesis by plants and aerobic respiration by both plants and animals. During photosynthesis, plants use the energy of sunlight to combine CO_2 with water, forming carbohydrates and releasing oxygen as a by-product. Both plants and animals then utilize aerobic respiration (metabolism in the presence of oxygen) to break down the carbohydrates, releasing the energy to make other complex molecules. In the process they produce CO_2 again.

Recap **The amount of matter that makes up the Earth and its biomass is fixed. The molecules that constitute living and nonliving things are constantly recycled. In the water cycle, water evaporates from the land and oceans. About a quarter of all water vapor precipitates over land, where it becomes surface water that either evaporates again or returns to the oceans. In the carbon cycle, plants use atmospheric CO_2 to make organic molecules, which cycle through the biomass of plants and animals.**

Nitrogen: An essential component of nucleic acids and proteins

Nitrogen is an essential component of proteins and nucleic acids. By far the largest reservoir of nitrogen is the nearly 79% of the air that is nitrogen gas (N_2). However, the nitrogen in air is unavailable to living organisms because the two nitrogen atoms are held tightly together by a triple covalent bond. Consequently plant growth is more often limited by a shortage of useful nitrogen than of any other nutrient.

The nitrogen cycle is illustrated in Figure 22.13. The process of converting nitrogen gas to ammonium (NH_4^+) is called **nitrogen fixation** because the nitrogen becomes trapped, or "fixed," in a form that ultimately can be used by plants. Nitrogen fixation is carried out by certain bacteria in the root nodules of plants such as peas, alfalfa, and clover and also by modern fertilizer plants. Most of the ammonium is converted to nitrate (NO_3^-) before it is used by plants. The formation of nitrate is called **nitrification.** Modern fertilizer plants use fossil fuels to manufacture nitrate from air, and lightning provides the energy for a small amount of nitrate formation as well.

Plants take up ammonium and nitrates from the soil and assimilate them into their proteins and nucleic acids in the course of their cellular metabolism. All other organisms must rely on plants or other animals for their usable nitrogen. When organisms die and decompose, the nitrogen compounds in their proteins and nucleic acids are converted again to ammonia, aided by bacteria. Most of the ammonia un-

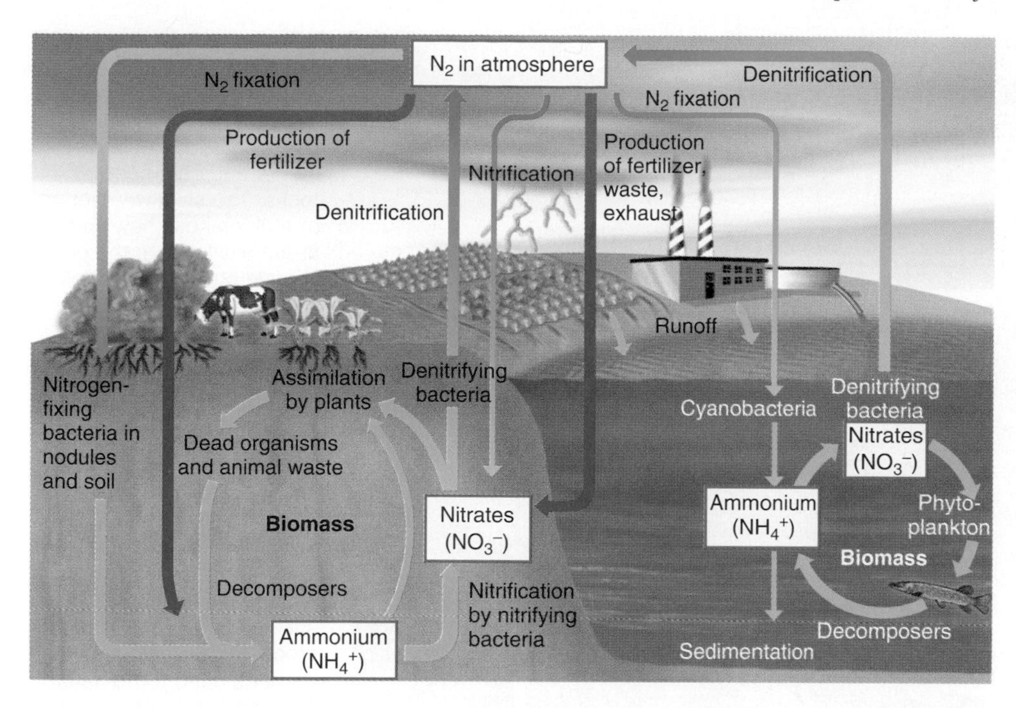

Figure 22.13 The nitrogen cycle. The nitrogen in nitrogen gas is not readily available to living organisms. Once it is "fixed" as ammonium or nitrified to nitrate, the nitrogen can be recycled over and over through the biomass. Nitrogen fixation is carried out by bacteria in the root nodules of certain plants and also by fertilizer factories. Cycles of nitrogen utilization occur both on land and in the sea. Under normal circumstances, the amount of nitrogen gas returned to the atmosphere by denitrification approximately balances nitrogen fixation. Humans upset this balance through the production of fertilizers from air, using fossil fuels as a source of energy.

dergoes nitrification again and recycles through the biomass of living organisms. Similar nitrogen cycles occur on land and in the sea.

Finally, **denitrification** by certain denitrifying bacteria converts some nitrate back to atmospheric nitrogen gas. In a balanced ecosystem, denitrification should equal nitrogen fixation. The production of fertilizers by humans has tipped the balance in favor of nitrogen fixation. Excessive fertilizer runoff can pollute freshwater because the nitrates stimulate excessive growth of plants and algae.

Phosphorus: A sedimentary cycle

Phosphorus is an essential element of life that most commonly exists as phosphate ions ($PO_4^{=}$ and HPO_4^{-}). Producer organisms use phosphate to make ATP for energy, phospholipids for cell membranes, and nucleotides for RNA and DNA. Consumer organisms that feed on the producers use phosphate for these purposes too, but they also incorporate phosphate into bones, teeth, and shells.

Figure 22.14 shows the phosphorus cycle. It is called a *sedimentary cycle* because, unlike carbon or nitrogen, phosphorus never enters the atmosphere. Phosphate ions are deposited in sediments that over the course of millions of years are brought to the earth's surface. The phosphates in sedimentary rocks are mined by humans and weathered by the forces of nature. Phosphate from fertilizers or from human or animal wastes may enter aquatic ecosystems, causing excessive algae growth. The decomposition of the algae when they die uses so much oxygen that other organisms may suffocate.

In our discussion, we have focused on major nutrients of life. There are similar biogeochemical cycles for all other nutrients, including oxygen, calcium, potassium, and sulfur. Pollutants and toxic substances such as heavy metals (mercury, cadmium, chromium, and lead), pesticides (DDT), and industrial chemicals (polychlorinated biphenyls; PCBs) also cycle between living organisms and the Earth. Environmentalists need to understand these cycles to advise us on how to minimize the effects of pollution.

Recap **Nitrogen gas in the atmosphere is not available to most living organisms until it undergoes nitrogen fixation to ammonia or to nitrates. Most phosphorus takes the form of phosphate that recycles within the biomass. The reservoir of phosphorus is in sedimentary rock. Similar biogeochemical cycles exist for all other nutrients as well as for toxic substances.**

22.6 The issue of human population growth

The human population is growing rapidly. Should we be concerned? How will this affect the world of the future? To understand the implications, we return to the concept of carrying capacity. The dynamics of our population on Earth are no different, conceptually, from those of a bacterial population in a laboratory dish—it's just that our world is slightly bigger.

Human population growth is relatively recent

A graph of the human population over the past several hundred thousand years (Figure 22.15) reveals why there is reason for concern. For most of human

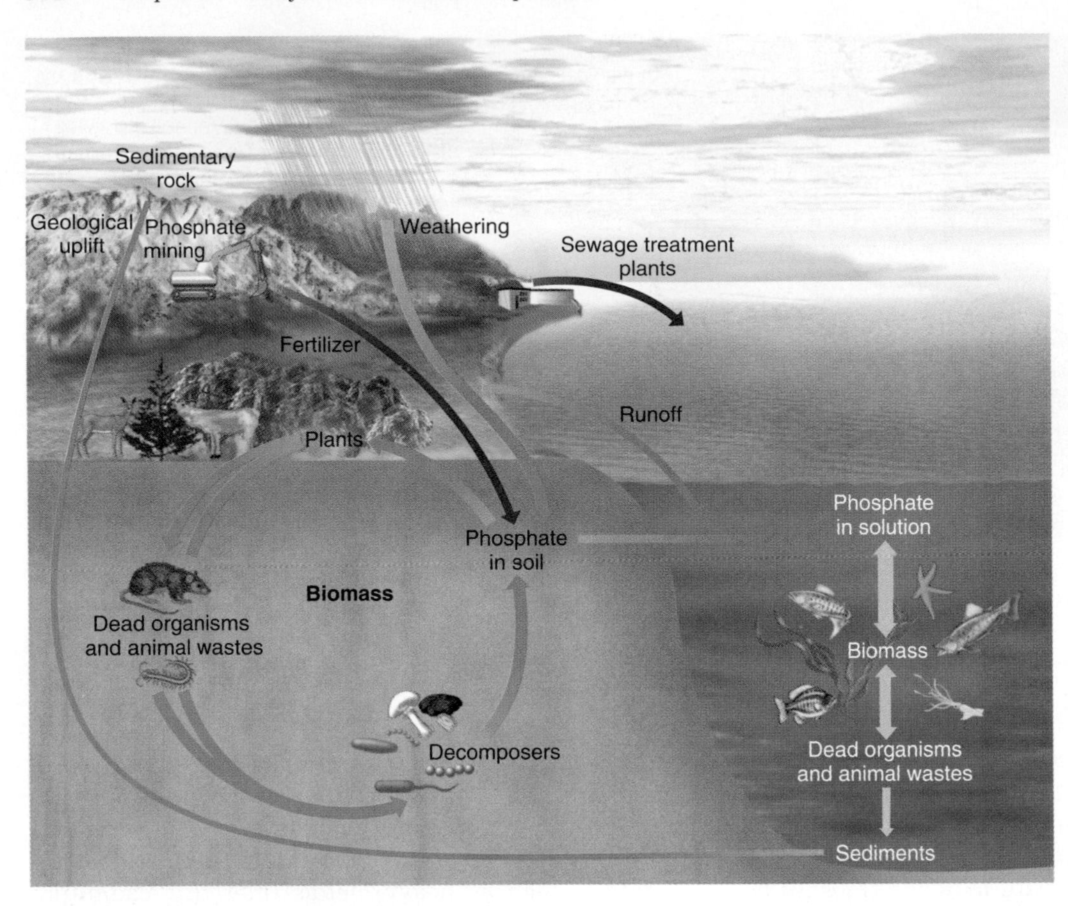

Figure 22.14 The phosphorus cycle. The reservoir pool for phosphorus is phosphate ions in sediments and sedimentary rocks. The slow weathering of rock makes "new" phosphate available to the exchange pool of phosphate in the soil. Some of this phosphate is taken up by plants and some runs off with the surface water to become the phosphorus supply for aquatic ecosystems. Eventually the phosphate ends up in sediments, where it may remain for millions of years until geologic upheaval brings the sedimentary rocks to the surface and a weathering process can begin again. Humans disturb the natural cycle by mining phosphate and applying it as fertilizer where it is in short supply. When organisms die the store of phosphate becomes available to producer organisms again.

history the global population remained relatively low and apparently stable, never exceeding 10 million people. People lived in widely dispersed small groups whose numbers were held in check by disease, famine, and a lack of tools and technology with which to control the environment or even to grow food. Apparently the environmental resistance was so high that our population was already at its carrying capacity.

About 5,000 years ago, the global human population began to rise slowly. The development of agriculture and the subsequent domestication of animals and plants for our own uses led to better conditions (a reduction in environmental resistance) and an increase in carrying capacity. Rapid growth began in the 1700s (barely 300 years ago) with the Industrial Revolution. Advances in communication, transportation, technologies for feeding and housing more people, and better medical care, including the discovery of antibiotics and vaccines, have all favored human reproduction and survival. As a result, the human population has entered a phase of explosive growth. Note that the human population curve resembles a J-shaped curve, typical of a population that is increasing according to its biotic potential, without any apparent environmental resistance.

When **did the human population begin** growing rapidly?

Can we continue on this trend? In fact, we do not know how high the human carrying capacity might be or how much farther it could be raised by future discoveries that we can't even imagine. Nevertheless, at some time our population must stabilize at a point short of having a person standing on every square foot of Earth. There is even the possibility that we will inadvertently *lower* the carrying capacity of the Earth through environmental degradation or some other human action, forcing a reduction in overall human population.

Fertility rates differ around the world

The **growth rate** of a population is calculated as the number of births per year minus the number of deaths per year, divided by the population size. Usually the growth rate is expressed as a percentage by multiplying by 100. The total human growth rate peaked at 2% per year in 1965 and has been declining since that time. Currently it is near 1.5% per year.

Zero population growth would be the point at which births equal deaths. In order to reach that point, of course, we would have to either decrease the birth rate or increase the death rate. If we let nature take its course, eventually environmental resistance will rise to accomplish one or the other. Given

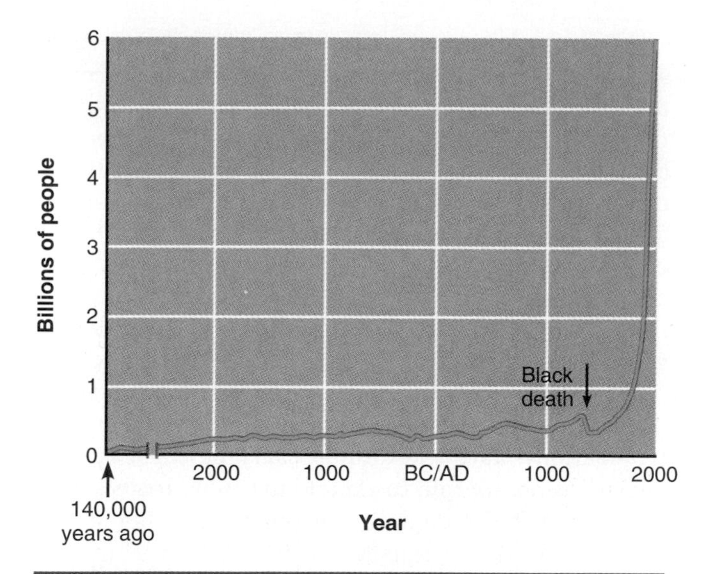

Figure 22.15 Growth of the human population. The dip in the population in the 1300s (arrow) was caused by the Black Death, a bacterial plague that devastated the human population in Europe and Asia. Notice that the horizontal axis is a split axis; most of the graph represents only the last 5,000 years of the nearly 140,000-year history of *Homo sapiens.*

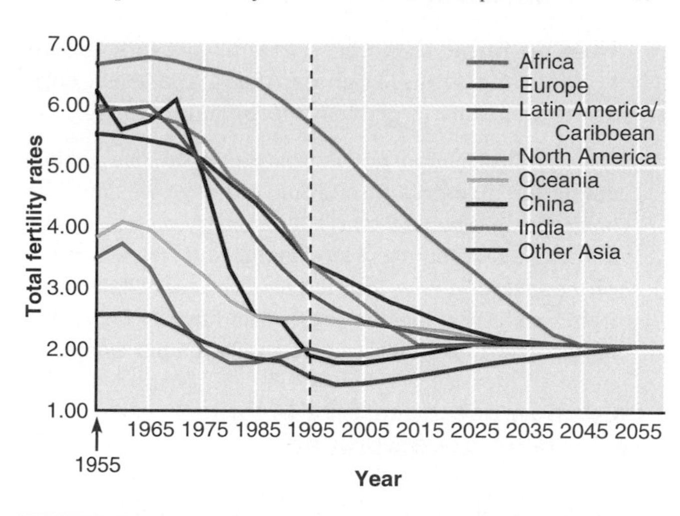

Figure 22.16 Past and projected fertility rates in different regions of the world. Projections (to the right of the vertical line) are based on the assumption that all regions will achieve replacement fertility rate by the year 2050.

SOURCE: Population Division of the Department of Economic and Social Affairs at the United Nations Secretariat, *World Population Projections to 2150,* United Nations 1998.

that no one would advocate increasing the death rate, let's look at how we might reach a zero population growth through changing the birth rate.

The human **fertility rate** is the number of children born to each woman during her lifetime. Currently the human fertility rate averages about 3.5 children per woman. Fertility experts agree that the average fertility rate required to achieve long-term zero population growth, called the **replacement fertility rate,** is only 2.1 children per woman. Those 2.1 children would exactly replace the equivalent of the two people (a woman and a man) who produced them. The replacement fertility rate is slightly higher than 2.0 to compensate for children who die before reaching their reproductive years.

The good news is that the trend toward a more stable human population is already well underway. We've already mentioned that the overall population growth rate has declined from 2.0% to 1.5% per year since 1965. Even more telling, however, is the trend over the past 40 years toward declining fertility rates around the world (Figure 22.16). Note that fertility rates are not consistent for all regions. They are already *below* the replacement fertility rates in China and Europe, and they are close to the replacement rate in North America as well. Figure 22.16 also shows how the lines might continue into the future *if* we assume that all regions will reach the replacement fertility rate by 2050.

The United Nations Population Division estimates that if the replacement fertility rate is achieved by 2050, the world population will reach 9.4 billion by 2050 and stabilize at just under 11 billion by 2150 (Figure 22.17). However, if fertility rates stabilize at higher or lower levels, the outcomes could range from a declining population of less than five billion to a rapidly-increasing population of over 25 billion in less than 150 years. Note that the difference between the low and high projections is only about one child per couple.

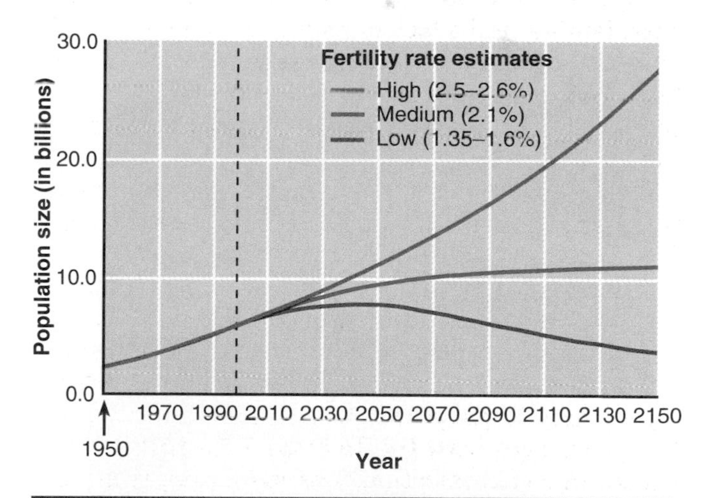

Figure 22.17 World population projections to 2150 based on three different fertility rates. The medium projection is based on the assumption that all world regions achieve replacement fertility rate by 2050. The high and low projections assume that the fertility rates vary slightly by region, within the ranges shown, by 2050.

SOURCE: Population Division of the Department of Economic and Social Affairs at the United Nations Secretariat, *World Population Projections to 2150,* United Nations 1998.

Keep in mind that these graphs present global averages. Some areas, such as Africa and Asia, contribute to population growth more than others.

Recap **The human population began to rise rapidly during the Industrial Revolution. Although we don't yet know Earth's human carrying capacity, eventually the human population must stop increasing. We will reach zero population growth when deaths equal births. This would mean the fertility rate would have to fall to the replacement fertility rate of 2.1 children per woman.**

Population age structure is linked to economic development

Many people assume that as soon as a population has reached the replacement fertility rate of just over two children per couple, the population must be at zero population growth. In fact, a population can continue to grow for decades even after reaching its replacement fertility rate, depending on its *age structure.*

In most populations, the age structure pattern follows a predictable pattern known as a **demographic transition** (*demography* is the study of human populations). This progression is tied to the region's industrial development and the economic well-being of its citizens. The first stage of demographic transition sees a shift from a society dominated by poor living conditions and a high death rate to one of rapidly improving economic conditions, declining death rate, and higher birth rate. In the final stages of transition the birth rate slowly declines until it equals the death rate, achieving zero population growth.

Demographic transitions have been underway around the world since the 1700s, with some countries farther along in the transition than others. Demographers categorize nations with established industry-based economies as **more industrialized countries (MICs)** and those that are only just beginning to industrialize as **less industrialized countries (LICs)**. The MICs, which include most countries of Europe, North America, northeastern Asia, and Australia, are in later stages of demographic transition than the LICs, which include most nations of Africa, Latin America, and most of Asia. The demographic patterns suggest that industrialization raises the human carrying capacity and allows the population to expand to a new steady state.

Figure 22.18 depicts the population age structures of the LICs versus the MICs. Males within various age groups are shown on the left of the figure and females are shown on the right. The population of the less industrialized countries has a pyramidlike age structure, with a large proportion of young people still below reproductive age. This population may continue to grow even though each woman may only have two children, because there are more women entering their reproductive years than there are leaving. In a stable population the age structure is more bullet-shaped, with roughly the same number of people in each age group through the reproductive years.

The problem the world faces, therefore, is that there will be large increases in population over the next 50 years in the very nations that are least able to provide for their citizens. To prevent poverty and starvation in these countries, we must seek solutions that promote economic development while at the

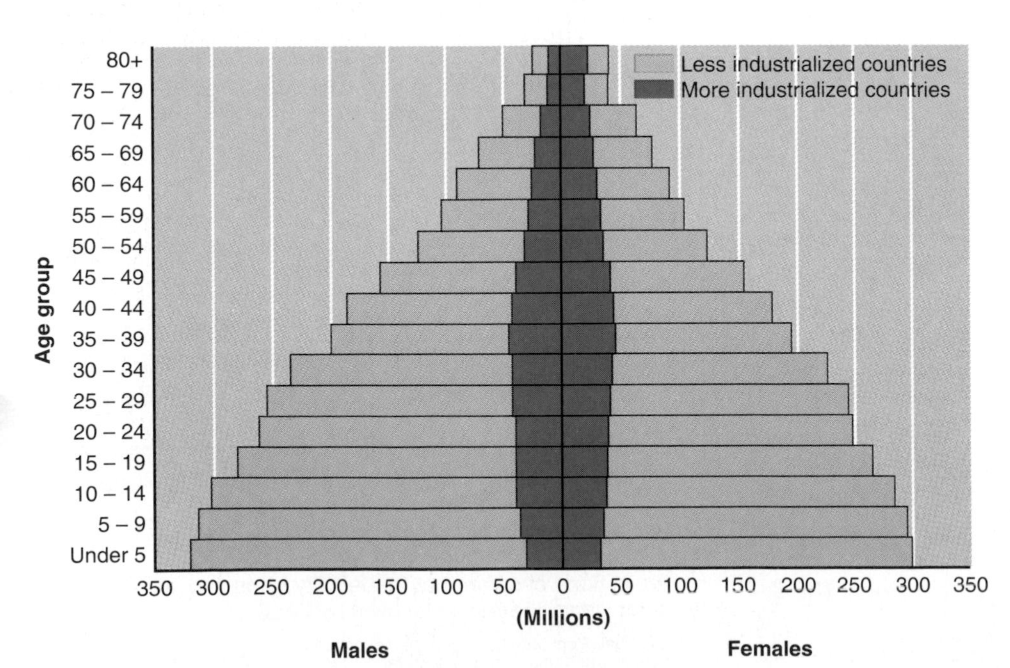

Figure 22.18 Population age structures in the LICs and MICs. The pyramid shape of the age structure of the LIC indicates that much of the population is younger than reproductive age. Therefore it is likely that their populations will continue to expand. In the MIC there is roughly the same number of people in each age group in and below the reproductive years, indicating a stable population in the future.

SOURCE: U.S. Bureau of the Census, International Data Base.

same time slowing their population growth rate. Family planning and other social programs that encourage delaying childbirth or reducing the number of children per couple would certainly help. Even government action has worked in some societies. In China, for example, it is against the law to have more than two children. In addition, substantial evidence supports the hypothesis that one of the most effective ways to reduce the birth rate is to improve the economic, social, and political standing of women in a society. Social and religious beliefs may come in conflict with this approach in some cultures, however, and those beliefs need to be respected.

> ***Recap*** **A population can continue to grow for decades after reaching replacement fertility rate if it has a pyramid-shaped age structure with a large proportion of children under reproductive age. Such an age structure is typical of less industrialized countries as they become more industrialized. Consequently, most of the increase in human population during the next century will occur in LICs.**

22.7 How humans affect ecosystems

Humans have a remarkable capacity to shape our environment, which is why we produce such an impact on our ecosystems. Only humans live in houses that can be thousands of square feet per person. Only humans irrigate lands that otherwise could not grow crops, level entire mountains in search of minerals and fossil fuels, and spread herbicides, insecticides, and fertilizers over thousands of square miles. Our presence has altered air, water, and land quality not only in local ecosystems but globally as well.

Pollutants impair air quality

The air we breathe is a mixture of primarily nitrogen (nearly 79%) and oxygen (nearly 21%), with trace amounts of carbon dioxide (0.03%). Air also contains trace amounts of thousands of **pollutants,** chemicals in the environment that have adverse effects on living organisms. The major concerns regarding air pollution fall into four areas:

- Global warming
- Destruction of the ozone layer
- Acid production and acid rain
- Production of smog

Figure 22.19 shows that for each of these issues there may be more than one contributing pollutant. Air pollution is difficult to remedy because it can be hard to determine who is responsible. Global problems require global solutions.

Excessive greenhouse gases could cause global warming If you've ever been in a glass greenhouse on a sunny day you know that it permits sunlight to penetrate, but as sunlight is converted to heat energy, the greenhouse becomes warmer because the glass prevents the heat from escaping. Certain gases in the upper layer of the atmosphere (the stratosphere) act

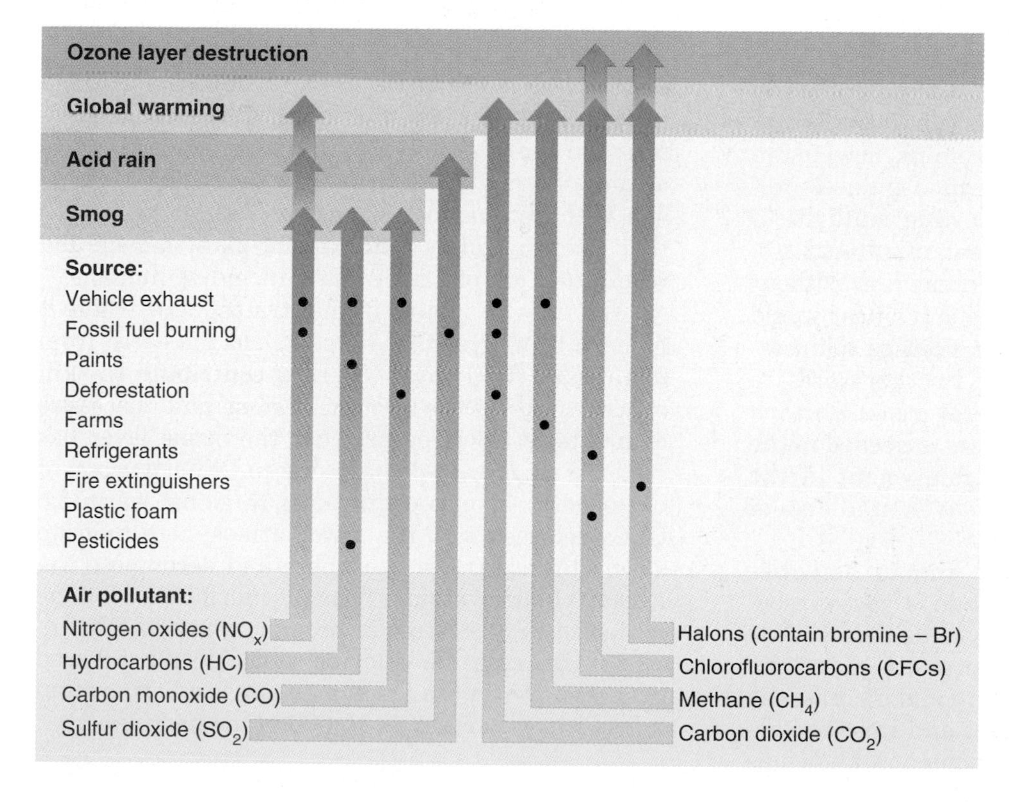

Figure 22.19 Major sources of air pollutants. A single human activity may create more than one type of air pollutant. Many types of pollution are the result of many different human activities.

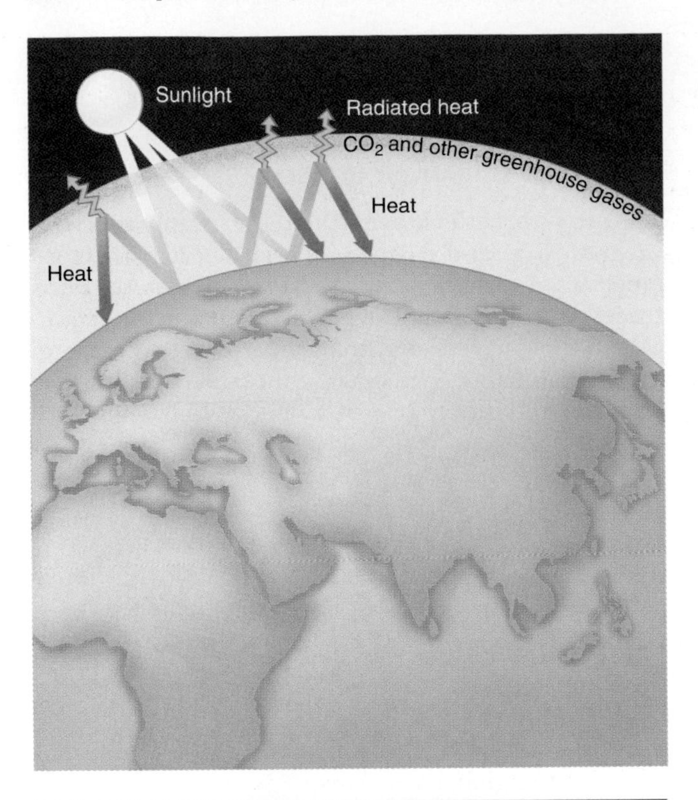

Figure 22.20 The greenhouse effect. A layer of stratospheric greenhouse gases consisting primarily of carbon dioxide (CO_2) allows sunlight to pass but traps most of the heat. An overabundance of greenhouse gases may contribute to global warming.

in much the same way for Earth, therefore they are called *greenhouse gases.*

The primary greenhouse gases are carbon dioxide (CO_2), methane, and small amounts of two air pollutants. The air pollutants are the chlorofluorocarbons (CFCs), used primarily as refrigerants, and gases called halons that contain bromine, used in fire extinguishers. Together the greenhouse gases produce a **greenhouse effect,** allowing sunlight to pass but trapping most of the heat (Figure 22.20). This is a natural and normal phenomenon. Without the greenhouse effect, most of the sun's heat would radiate away from Earth and the average temperature would be well below freezing. In other words, to maintain a normal surface temperature on Earth, a certain amount of greenhouse gases is needed in the stratosphere. Most of the CO_2 component of the greenhouse gases accumulated over millions of years, as a result of respiration.

In modern times, however, human activities have produced additional greenhouse gases, especially CO_2. Many scientists believe that excessive production of the greenhouse gases could increase the greenhouse effect and raise the average global temperature, a phenomenon known as **global warming.** The two main human activities that have raised atmospheric levels of CO_2 are **deforestation** (removing trees from large areas of land) and the burning of fossil fuels. Deforestation by burning is doubly damaging—not only is CO_2 liberated as wood is burned, but with fewer plants, CO_2 uptake is reduced as well. The burning of fossil fuels has returned to the atmosphere CO_2 that has been stored over millions of years.

How might we be contributing to global warming?

Some people believe we can already feel the effects of global warming, but we can't. In fact, the average global temperature has risen by only about 0.5°C in the past 100 years. Most scientists predict an average global temperature rise over the next 20 years of only another 0.5°C, hardly enough to be felt compared to normal seasonal fluctuations. Although the change is small, the rate of temperature change has accelerated to the point that it is now faster than at any time in the past million years.

Experts still debate the effects of global warming, but most agree that if global temperatures continue to increase at this rate, we could expect the sea level to rise by several feet within several centuries as a result of melting of polar ice. This would be enough to destroy coastal ecosystems and even flood major cities. Global warming could also cause regional changes in climate, resulting in crop failures and changing the suitability of certain farm areas for specific crops. There may even be droughts in areas not used to drought conditions, accompanied by a loss of biomass. As a result of our understanding of the potential for global warming, 166 nations signed agreements in 1992 and 1997 that call for reductions of CO_2 and CFC emissions.

Destruction of the ozone layer **Ozone** (O_3) is found in two places in the atmosphere. In the atmosphere near the Earth's surface (the troposphere), ozone is an air pollutant that is formed by the reaction of oxygen with automobile exhaust and industrial pollution. Ozone is mildly toxic, causing plant damage and respiratory distress in animals, including humans.

Ozone is also found in the stratosphere, where it forms a thin layer that helps shield the earth from ultraviolet (UV) rays. UV rays contribute to skin cancer, suppress the immune system, and may cause cataracts. It now appears that the ozone layer has suffered significant damage due to CFCs, a group of chemicals that also contributes to global warming. CFCs released into the lower atmosphere migrate slowly toward the stratosphere and decompose, releasing chlorine atoms. The chlorine atoms combine with ozone and destroy it, producing oxygen (Figure 22.21). Because the chlorine molecule can be used over and over in the reaction, a single chlorine atom may destroy as many as 10,000 ozone molecules.

Directions in Science

Measuring Global Warming

Perhaps no subject has generated so much heated debate these past few years as the question of global warming (no pun intended). It's a sensitive political and economic issue with important implications for the long-range future of the Earth. So just how certain are we that global warming is taking place?

The possibility that the Earth might be getting warmer first arose when scientists became aware that CO_2 and methane trap heat within the stratospheric layer. In the 1950s scientists began to monitor atmospheric levels of CO_2 on a mountaintop in Hawaii. They chose the high and relatively isolated site because they thought the levels there would reflect average values for the Northern hemisphere. They found that there were seasonal cycles of greenhouse gases that correlated inversely with the growing season for plants, as might be expected. They also found that between 1958 and 1992, the average CO_2 concentration rose over 15%. These measurements are still going on, but it seems fairly certain that the level of greenhouse gases really is rising.

But is it actually getting warmer? Reexamination of surface temperatures recorded over the past 100 years for other purposes indicate that mean global temperatures have risen by about 0.5°C over that time. More recently, sensitive measurements going back to 1958 from Earth's surface, from the ocean, and from weather balloons show that the mean temperature has increased by several tenths of a degree over just the past 20 years.

However, some researchers remained perplexed by what were supposed to be the most convincing data of all, the 20-year-long record of temperature measurements taken by satellites circling the globe. Unexpectedly, the satellite data showed a slight *cooling* trend of the troposphere, the atmospheric layer that extends from the earth's surface to about eight miles above sea level. This gave skeptics a powerful tool to resist taking action against what might not be a problem at all. And it left scientists figuratively scratching their heads and pondering which set of data might be wrong and why.

A breakthrough came in 1998 with a series of reports showing the satellite data were flawed. Although not designed for monitoring global temperatures, the nine satellites were pressed into service for that purpose after they were launched for the National Oceanic and Atmospheric Administration (NOAA) starting in 1979. Atmospheric drag has slowly pulled the satellites into lower orbits, causing an error in the reported temperatures. The satellite measurements also include several short-term global weather events such as El Niño and volcanic eruptions that skew the data. When corrected for atmospheric drag and weather disturbances, the satellite data now seem to agree with surface measurements that a slight warming trend has indeed occurred.

Most scientists now accept global warming as probable, and politicians seem sufficiently convinced that they are beginning to act. There are still skeptics, but given the short period of data collection such skepticism is healthy for science. Temperature measurements and the debate over them will undoubtedly continue until we can be more certain.

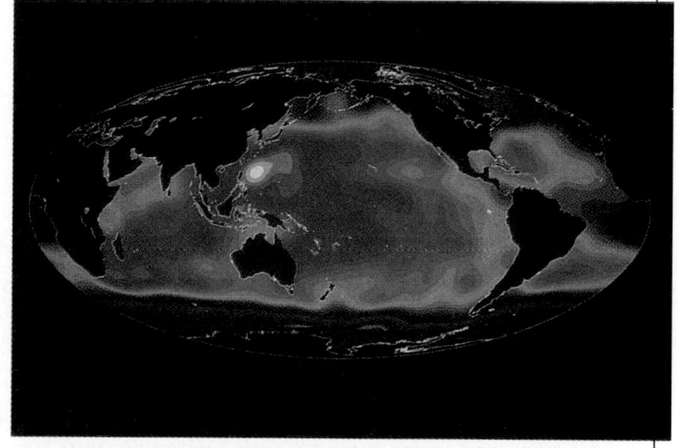

Satellite photos, such as this one of ocean surface temperatures, are used to access global temperature trends.

Already the stratospheric ozone layer has thinned noticeably, and "holes" have even developed at certain times of the year. As a result, researchers at the United Nations expect the incidence of cataracts and skin cancer to rise worldwide over the next several decades.

Fortunately, this is one area where action has been taken. In an international agreement signed in 1987, most industrialized nations are phasing out the production of CFCs. There will be a delay before we experience the full benefits of the phase-out, however, as it takes 10 to 20 years for CFCs to reach the stratosphere.

Acid production and acid rain The major source of acid is sulfur dioxide, which is released into the air as a result of burning high-sulfur coal and oil for power. A secondary source is nitrogen oxide in automobile exhaust. Sulfur dioxide and nitrogen oxides combine with water vapor in the air, become sulfuric acid and nitric acid, and dissolve in raindrops which fall as **acid rain.** Acid rain corrodes metal and stone and damages forests and aquatic ecosystems, particularly in the northeastern United States, southeastern Canada, and Europe (Figure 22.22). Some acid products also precipitate as solid particles of sulfate and nitrate salts.

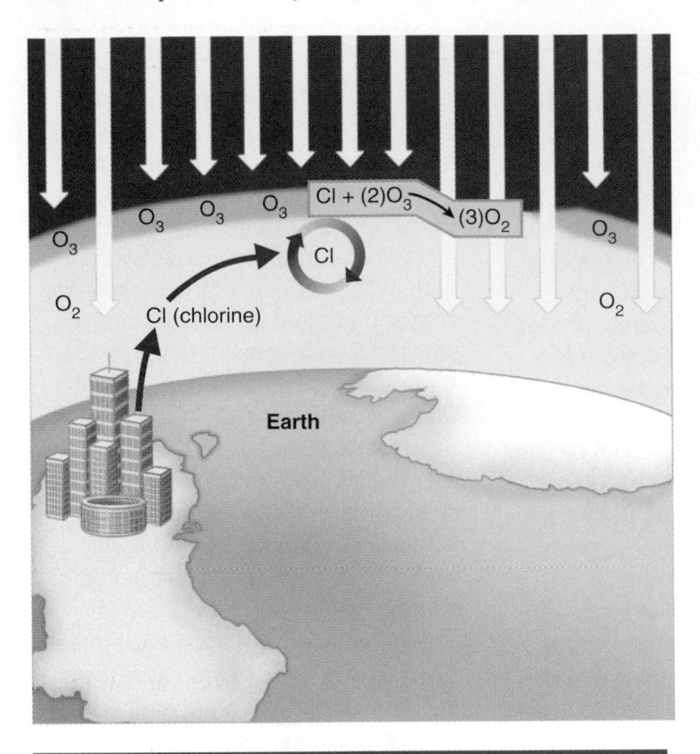

Figure 22.21 Destruction of the ozone layer. Chlorine released into the atmosphere reacts with ozone (O_3) to produce oxygen (O_2). Destruction of the ozone layer exposes the surface of the Earth to more ultraviolet radiation.

Power plants are now reducing their emissions of sulfur dioxide, either by burning low-sulfur fuels or by removing ("scrubbing") the sulfur dioxide from power plant emissions before they reach the atmosphere. Automobile manufacturers are designing vehicles that produce fewer nitrogen oxides. These technologies cost money, of course. Whether to spend money on reducing acid rain or on something else is another of those choices that we make as a society.

Pollutants create smog Several pollutants in air, most notably nitrogen oxides and hydrocarbons (chains of carbons linked to hydrogens), react with each other in the presence of sunlight and water vapor. They form a hazy brown or gray layer of **smog** that tends to hover over the region where it is produced (the term comes from "smoke" + "fog"). The burning of fossil fuels (coal and oil) and automobile exhausts are the primary culprits.

Smog contains ozone and PANs (peroxyacyl nitrates), both of which irritate the eyes and lungs and may lead to chronic respiratory illnesses such as asthma and emphysema. Depending on the source, it may also contain small oil droplets and particles of wood or coal ash, asbestos, lead, dust, and animal waste. Smog can become especially troublesome during a **thermal inversion,** when a warm stagnant upper layer of air traps a cooler air mass containing smog beneath it.

Figure 22.22 Effects of acid rain. Acid rain killed these Fraser firs in the Appalachian mountains of North Carolina.

Smog generally appears after an area has undergone rapid industrialization and growth. Then there is a period during which the problem is realized and acknowledged, followed by cleanup efforts. Smog was once a significant health problem in London and Pittsburgh, but both of these cities have succeeded in major efforts to reduce smog (Figure 22.23). The state of California has passed strict automobile anti-pollution laws. Many regions of the United States now require the use of cleaner-burning oxygenated automobile fuels containing ethanol or methanol during certain winter months, and the use of lead as a fuel additive has been significantly reduced as well. Other areas have not yet solved their smog problems.

Recap **Air quality and air pollution are a worldwide concern. Human activities are destroying the protective layer of ozone around the Earth, producing acid rain and smog, and may be causing a slow global warming trend. Remedial efforts are effective, as shown by the reduction of smog in some cities and application of new technologies for reducing acid rain. Global initiatives to slow global warming and reduce ozone layer destruction are already underway.**

Pollution jeopardizes scarce water supplies

The problem of the human impact on water is twofold. First, fresh water is relatively scarce in some regions of the world. Second, human activities can pollute what water there is.

Water is scarce and unequally distributed In one sense, water is a renewable resource because it evaporates continuously from the oceans and falls on land as rain or snow. However, all of the freshwater

(a)

(b)

Figure 22.23 Smog. (a) Pittsburgh at the height of the steel-making era, before smog cleanup efforts began. **(b)** Pittsburgh after programs were instituted to reduce smog.

on the surface of the land and in **aquifers** (deep underground reservoirs) forms less than 1% of the Earth's total amount of water. Over 97% is salty ocean water and 2% is frozen in glaciers and polar ice caps.

In addition to general scarcity, a second issue is the unequal distribution of water among human populations. Residents of the more industrialized countries may use 10 to 100 times as much water as people in the less industrialized countries. Some desert and semiarid regions of the world have already reached their human carrying capacity as a result of water shortage. When we divert water for our purposes, we take it away from other species or limit their normal migration patterns. Water rights are a controversial subject in the western United States, where water is limited. Already, choices have to be made between irrigating crops and encouraging the reproduction of Pacific salmon.

Human activities pollute freshwater Human activities tend to pollute what freshwater there is. Untreated sewage, chemicals from factories, the runoff of pesticides, fertilizers, and rubber and oil from city streets, all must go somewhere. Either they degrade chemically according to their natural cycle of decomposition or they end up in water or soil.

Some water pollutants are *organic nutrients* that arise from sewage treatment plants, food-packing plants, and paper mills. When these nutrients are degraded by bacteria, the rapid growth rate of bacteria can deplete the water of oxygen, threatening aquatic animals. Others are *inorganic nutrients* such as nitrate and phosphate fertilizers and sulfates in laundry detergents. These cause prolific growth of algae, which die and are also decomposed by bacteria.

Eutrophication refers to the rapid growth of plant life and the death of animal life in a shallow body of water as a result of excessive organic or inorganic nutrients (Figure 22.24). Eutrophication is part of the normal process that converts freshwater into marsh and then dry land, but human activities can accelerate the process.

A special problem is *toxic pollutants,* including polychlorinated biphenyls (PCBs), oil and gasoline, pesticides, herbicides, and heavy metals. Most of these cannot be degraded by biological decomposition, so they remain in the environment for a long time. In addition, they tend to become more concentrated in the tissues of organisms higher up the food chain, a phenomenon known as **biological magnification.** Biological magnification occurs because each animal in a food chain consumes many times its

Figure 22.24 Eutrophication. Water pollution accelerates the normal process of eutrophication, encouraging prolific growth of algae and plants at the expense of animal life.

own weight in food in a lifetime. A classic example of biological magnification is DDT, a pesticide that was once used extensively. Eventually DDT reached such a high concentration in secondary consumers such as red-tailed hawks and peregrine falcons that it seriously threatened the populations of these birds before its use was curtailed.

Other important water pollutants worldwide include *disease-causing organisms* like those that cause typhoid fever and hepatitis, *sediments* from soil erosion that clog waterways and fill in lakes and shipping canals, and even *heat pollution* from power plants. Heat pollution reduces the amount of oxygen that water can carry. It also increases the oxygen demand by aquatic organisms whose activity level is temperature-dependent. The resulting oxygen deprivation may suffocate them.

Groundwater pollution may impair human health Many of the same pollutants that taint surface water also pollute groundwater. However, groundwater pollution poses two additional concerns. First, groundwater is often used as drinking water, so its pollutants can quickly affect human health. Second, groundwater is a slowly exchanging pool. Once it becomes polluted it may stay polluted for a long time.

The Environmental Protection Agency estimates that as many as 50% of all water systems and rural wells contain some type of pollutant. The most common are organic solvents such as carbon tetrachloride, pesticides, and fertilizers such as nitrates. We do not know the full effect of groundwater pollution because it is hard to separate from other possible causes of human disease. Public health officials suspect that some pollutants contribute to miscarriages, skin rashes, nervous disorders, and birth defects.

An issue of special concern regarding possible groundwater contamination is the disposal of radioactive wastes. Radioactive materials are used in nuclear power plants and to diagnose and treat human disease. Radioactive wastes are usually stored in some of the driest locations in the world because they remain radioactive for up to thousands of years. Some are so long-lived that they are dropped into molten glass, which cools and hardens before being buried.

Oil pollution damages oceans and shorelines Pollution of freshwater rivers eventually ends up in the oceans. Human activities can also pollute oceans directly, and the most common and visible way is by spilling oil at sea.

Several million tons of oil enter the oceans each year. About half of this is from natural seepage; therefore, some oil pollution is a natural phenomenon. However, 30% of total oceanic oil pollution is caused by oil disposal on land and its eventual washing to the sea in streams and rivers. The remaining 20% results from accidents at sea.

Figure 22.25 An oil spill. Oil spilled near a shoreline may wash up on shore. This oil spill was in Galveston, Texas.

When oil is spilled at sea about a quarter of the oil evaporates, nearly half is degraded by bacteria, and the remaining quarter eventually settles to the ocean bottom. Nevertheless, an oil spill can cause significant short-term damage to marine and shoreline ecosystems. Before the oil dissipates it coats living organisms, disrupting their ability to function and even choking smaller organisms (Figure 22.25). Shoreline ecosystems may show signs of damage for years. Although cleaning up an oil spill can help an ecosystem recover more quickly, it also shifts some of the pollution to a site on land.

Recap **Freshwater is scarce and often distributed unevenly. Freshwater can become polluted by surface water runoff from urban areas and by fertilizers, pesticides, and herbicides in rural areas. Pollution of rivers eventually reaches the oceans. Oil spills at sea are an example of how human activities directly pollute oceans.**

Pollution and overuse damage the land

Here the issue may not be how we pollute the land (though we do), but how much of it we use. Humans consume a lot, and we tend to alter the landscape to suit our own purposes. We dam river valleys to produce hydroelectric power, strip mountaintops to find coal, and cut down forests for lumber or to clear space for crops.

The economy of the United States alone accounts for the direct consumption of 44,000 pounds (20 metric tons) of fuels, metals, minerals, and biomass (food and forest products) for every person every year. Add to this number the amount of earth moved to build roads and waterways and to find energy, and add the erosion of soil by agricultural and forestry activities, and the total use of natural resources amounts to nearly 80 metric tons per U.S. citizen each year. It has been estimated that human activities have already altered a third of the Earth's land mass, including removing nearly half of its original forest cover.

As the human population grows, a significant trend has been our migration toward cities. Cities often expand into productive farmland because it is nearby and because it is more economical to build where it is flat, even though only a fraction of the earth's surface will ever be suitable as farmland. Cities also require large quantities of water and power and generate waste and pollution in a relatively small area.

The problems of land use in rural areas are quite different but just as significant. More than half the people of the world live in rural poverty. Many rely almost entirely on their local environment to survive. It is not uncommon for impoverished rural communities to cut down all trees for fuel and shelter and overgraze communal lands with livestock. Stripping the biomass from fragile ecosystems leads to erosion and **desertification**—the transformation of marginal lands into near-desert conditions unsuitable for future agriculture. Every year an estimated 15 million more acres of once-productive land become desert. Because the very survival of the rural poor is linked so closely to the use of the available resources, it is unlikely that we can halt land degradation in these regions without first doing something about rural poverty (Figure 22.26).

Wars, too, destroy habitable land. During the Vietnam war, millions of acres were altered by defoliants (chemicals that cause the leaves to fall off plants) and by bombs and bulldozers. More recently, the fires set in Kuwait during the Persian Gulf War destroyed fragile desert habitats.

And finally, there is the issue of how we should dispose of our garbage. Landfills, one solution, deserve our attention, perhaps because they are so clearly intentional and we are questioning our choices. It is wise that we *do* continually examine our choices, but it may be worth considering that landfills are not our biggest environmental problem—as long as they are well-designed and do not contribute to ground or water pollution. Nevertheless, we could all seek ways to generate less garbage and recycle as much as possible.

Recap **Land use and land pollution problems differ by region. In some areas humans alter the landscape in search of fossil fuels and minerals. Cities expand into productive land or place a burden on resources. In rural regions, deforestation and desertification damage ecosystems and limit their future productivity.**

Figure 22.26 Desertification. Many poor residents of rural areas must depend on local resources to survive. Stripping the biomass from the ecosystem causes desertification, as once-productive land becomes eroded and barren. This photo was taken in northern Africa.

Energy: Options and choices

Energy—we just can't seem to get enough of it. We want inexpensive fuel for automobiles weighing thousands of pounds. We heat our homes and even the water in our swimming pools. We need energy to refine raw materials, manufacture the products we use, and process and cook our food.

Energy use is tied to consumption, so we do have some choice in determining how much energy to use. We can also make choices about where to obtain our energy.

First, there are *fossil fuels*—coal, oil, and gas. We've already seen how retrieving them from their reservoir pool and transporting them has an environmental cost, and how using them may contribute to the greenhouse gases by releasing CO_2. In addition, coal in particular generates pollution in the form of acid rain because it contains sulfur. Finally, we need to consider that fossil fuels are nonrenewable resources. Their formation took millions of years, but we could consume them all in a couple of centuries.

For a time it was thought that *nuclear energy* would provide all of our energy needs in the future. In practice nuclear energy has proven to be expensive, in part because of all the safeguards necessary to ensure that the core of the energy plant does not become overheated, suffer a meltdown, and release radioactivity into the environment. One such accident occurred in 1986 at Chernobyl in the Ukraine. In addition, nuclear power plants generate wastes that remain radioactive for thousands of years. How to store those wastes properly is a subject of current debate.

CurrentIssue

Can the Rain Forests and Their Residents be Saved?

Recently, forestry agents from the West African nation of Ivory Coast raided the farming village of Kadoudougou, burning down residents' homes and destroying the fields they had painstakingly carved into the hillside. Farmer Sep Djekoule and his neighbors fled as the conservation officers demolished their mud-and-thatched-roof huts—only to return a few weeks later, rebuild their homes, and plant new crops.

Ivory Coast once possessed 70 million acres of pristine tropical rain forest. Today only 5 to 7 million acres remain, and even these may be gone in 10 years. The government destroyed Djekoule's village to protect the country's dwindling state-protected wilderness. But, as the father of 10 children, Djekoule feels he has no choice but to continue farming.

Like Djekoule, millions of people around the world depend on subsistence farming to survive and feed their families. Unfortunately, their farmlands often encroach on endangered wilderness. How can we preserve both the rain forests and their residents?

Why Should We Care About Rain Forests?

One of the primary reasons we might care is that growing forests use carbon dioxide during photosynthesis, thereby removing carbon dioxide from the atmosphere that might otherwise contribute to the greenhouse gases. Deforestation by burning is considered especially harmful because it liberates CO_2 from the burned plant material while at the same time destroying the trees that utilize it for photosynthesis. Deforestation could lead to an increase in greenhouse gases and contribute to global warming.

Another reason we might care is that the rain forests contain most of the globe's animal and plant species, perhaps more than all other types of ecosystems combined. Their biological diversity offers vast potential for developing important new medicines for human use. A single hectare (about 2 acres) in Brazil's Bahia wilderness, for instance, has been found to contain over 500 tree species—more than any other forest in the world. Consider that 25 percent of all prescription medicines sold in the U.S. are based on substances derived from fewer than 50 species of plants. It's estimated that there are 265,000 flowering plants in the world. Fewer than one percent have been tested for medicinal uses.

In addition, properly farmed tropical rain forests produce such popular items as coffee, cocoa, fruits, and nuts. Ecologists warn that if these unique ecosystems vanish, it would devastate countries whose economies depend on these products.

Rain forests are also a valuable source of clean drinking water. In Brazil, rain forests near Rio de Janeiro provide water to 18 million people. As the forest shrinks, the quality and quantity of Rio's water supply are suffering.

Who Is Destroying the Rain Forests, and Why?

Deforestation is actually nothing new. Much of the eastern half of the United States underwent considerable deforestation in the late 1800s, both for timber and to clear land for farming. The current destruction of the world's rain forests cannot be blamed on just one people, country, continent, or industry. The reasons are often as varied as the geographies of the forests themselves and the people who inhabit them.

One threat is population growth. Rural residents move to urban areas in search of better jobs and living conditions. As cities grow, land is cleared to make way for new homes and businesses.

A small subsistence farm plot in a burned-out area of rain forest, Ivory Coast.

Another threat is farming. For farmers, the rain forest environment offers a unique mix of humid climate and shade—ideal growing conditions for profitable crops such as cocoa and coffee. Despite their lush vegetation, however, rain forests are fragile ecosystems with very little topsoil and scant nutrients. Several years of intensive farming leach the nutrients from the soil and leave it depleted. To continue raising enough crops to survive, farmers move to a fresh section of rain forest and start another destructive cycle. Some types of farming require clear-cutting the forest, which often results in nutrient depletion and soil erosion.

Still another threat is the exploitation of scarce forest resources. In Brazil, poor farmers sneak into Intervales State Park and cut down 20-year-old palm trees for the tender heart of palm, found only at the top of the tree. Hearts of palm are a popular delicacy, so restaurants are willing to pay well for them.

One of the most significant threats is logging. German, French, and Middle Eastern logging companies harvest 800-year-old hardwood trees in Africa and ship them to other continents, where they are processed into furniture, floors, and buildings. In recent years the pace of logging has accelerated. As economies in Africa collapse and local currencies are devalued, it becomes more economically feasible to ship logs great distances to sawmills. The timber industry is simply responding to the demand from industrialized nations for paper, cardboard, and exotic woods.

Should governments set limits on logging? Perhaps, but for cash-starved nations, logging is more likely to be viewed as an important source of income. Cameroon and Gabon turned increasingly to logging after oil exports slumped in the early 1990s. During the next five years, logging in Cameroon soared by 50 percent. Even when governments oppose it, logging may continue: Indonesia sets an annual quota of 29 million hectares, but many millions more are cleared illegally.

Logging affects animals as well as plants in the rain forests, because logging requires roads. As once-remote areas become accessible, poachers follow. Across Africa, illegal hunting of elephants, monkeys, and antelopes has soared.

Current and Future Solutions

Just as the reasons for the loss of rain forests are complex, so too will be the solutions. A number of different approaches are being tried in various places:

Conservation by threat of economic sanction. To force Cameroon to crack down on poaching, the World Bank and European Union froze $52 million in funds for road maintenance. The irony was not lost on African journalists, who pointed out that foreign logging companies built the roads that make poaching possible in the first place.

Selling the rights to bioprospect in rain forests. Scientists and governments are just beginning to appreciate the wealth of potential new products that might be discovered in the vast riches of the rain forests, including drugs, flavors, fragrances, and natural pesticides. In order to save and eventually tap its rain forest resources, Costa Rica has entered into agreements with several major drug, chemical, and food companies for the right to "bioprospect" in the rain forest in exchange for a percentage of the profits. Some of the profits would be used for preservation of the rain forests themselves.

Finding sustainable sources of income from rain forests. A source of income can give local residents an incentive to preserve rain forests. In Guatemala's Peten region, for example, about 6,000 families harvest nontimber items such as chicle (a tree sap used in chewing gum), allspice (a dried fruit that is made into a cooking spice), and xate (a palm popular in floral arrangements). In a typical year, Peten inhabitants earn up to $6 million from these products, far more than they could make by farming.

Encouraging ecotourism. Countries are increasingly finding that people are willing to travel to remote locations to enjoy nature. One of the fastest-growing segments of the travel industry, ecotourism, is a boon to Costa Rica, where it generates $700 million each year—a powerful incentive to preserve natural beauty.

Government or international regulation. CO_2 emission standards can convince companies to take voluntary action. In order to comply with a requirement that the they compensate for the CO_2 output of a new power plant, for example, a U.S power company agreed to pay $500,000 toward the purchase of 5,000 acres of private forest land in Costa Rica. By protecting the forest, the company would be preventing the release of the same amount of CO_2 that the new power plant will release. For the company, preservation of the forest was a cheaper way to comply with CO_2 emission requirements than installing expensive pollution-control equipment in the power plant.

Despite these hopeful signs, rain forests around the world remain endangered. Poachers have removed all large animals from thousands of square miles in Africa. At their present rates of deforestation, the rain forests of Ivory Coast will vanish within a decade, and Brazil's Atlantic wilderness rain forest will be gone within 50 years. The challenge is to preserve at least some of the remaining rain forests before they all disappear.

Where water is readily available, *hydroelectric* power plants generate electricity. Hydroelectric power plants use the kinetic energy of water to turn turbines that drive electric generators. The environmental cost is the need to build dams that disrupt valley and river ecosystems. Of minor significance are "wind farms" that harness wind to generate electricity, and geothermal power that uses heat from underground sources deep within the earth.

Many rural poor use *biomass fuels* such as wood and the dung of herbivores. These are renewable if they are not depleted too fast, but they pollute the air and are not always available. Some countries are experimenting with processes that burn waste material, including municipal garbage, to extract energy.

Ultimately we may have to rely on the one original and renewable source, *solar power* (Figure 22.27), just as plants do. At the moment, however, solar power technologies are not capable of meeting all our energy demands.

Recap **Humans consume a disproportionate amount of energy compared to other species. Most of the energy we use comes from nonrenewable fossil fuels that contribute to pollution. Other possible sources include nuclear energy, hydroelectric power, biomass fuels, and solar energy.**

Figure 22.27 A solar power plant. Although solar energy holds great potential for energy production, the technology is not yet capable of generating sufficient power at low cost. This power plant is in Australia.

22.8 Toward a sustainable future

How can we ensure our future on Earth?

In a **sustainable world,** humans would be an integral part of sustainable ecosystems so that those ecosystems (and we humans) could exist on the planet far into the future. Finding solutions for environmental and human population issues may require more information than we have at the moment. It may also require making difficult economic and political choices once we have the information. Given what we know today, let's focus on several strategies that might help us reach a sustainable world:

- *Consume less.* This approach probably won't be too popular, but by consuming less you automatically use fewer resources, generate less garbage, and cause less pollution. If you use less, this leaves more for another person or even another species. *What* to consume less of may be a very personal choice. In other cases these may be decisions that we make as a society (we have already chosen to use fewer chlorofluorocarbons, for example).

- *Recycle more.* This one is always popular. Recycling reusable materials such as paper, metal, and plastic reduces our consumption of raw materials. It also keeps the recycled substances from becoming either garbage or pollutants. We also need to consider not only how we recycle substances back into our manufacturing processes, but also how we recycle them back into the ecosystem. Some cities collect, compost, and then recycle leaves and organic material. Even landfills might be considered a slow form of recycling back into the ecosystem.

- *Lower the worldwide fertility rate.* Translation: have fewer children. Fewer children means fewer cloth and disposable diapers. It means fewer cars and houses in North America, less deforestation in Brazil, fewer polluted rivers in India, less desertification in Africa. If the current fertility rate of the human species does not decline, eventually environmental degradation will reduce the human carrying capacity and the death rate will rise to equal the birth rate. We can already see this happening in certain areas of Africa where life expectancy is short and infant mortality is high.

- *Reduce rural world poverty.* People who are starving and without shelter must think of their immediate needs first. It is virtually impossible to convey the long-term impact of desertification to someone who relies on a few goats and trees for day-to-day survival. More and more of the world will become uninhabitable due to deforestation and desertification unless the poor people in those areas are given a viable economic alternative to destruction of their own habitat.

Will we be able to achieve a sustainable world? Given our history so far there is every reason for optimism, for we humans have a remarkable capacity for making intelligent decisions and shaping our own environment. Already we can see cooperation on a global scale with the problem of global warming. There will be more choices made in the future and you will have a part in them. Your decisions will help to shape humankind's future.

Recap **A sustainable world will require stabilizing the human population growth rate and the global environment. Achieving this goal will test our capacity to make intelligent decisions on a worldwide scale.**

Chapter Summary

Ecosystems: Communities interact with their environment

- The study of the relationships between organisms and their environment is called ecology.
- An ecosystem consists of a community of organisms and the physical environment in which they live.

Populations: The dynamics of one species in an ecosystem

- When there are no restrictions to population growth, a population will grow exponentially according to its biotic potential.
- Within an ecosystem, the stable size of the population of a particular species ultimately is determined by carrying capacity.

Communities: Different species living together

- Species that occupy overlapping niches may be in competition for limited resources.
- Communities go through a natural sequence of changes called succession that ends with the establishment of a stable climax community.

Energy flows through ecosystems

- Because energy flows only one way through an ecosystem, a constant supply of energy (from the sun) is required.
- Producer organisms are capable of utilizing the energy of the sun in a process called photosynthesis. Consumer organisms must rely on other organisms for energy.
- Less energy is available at each higher consumer level of a food web.

Chemical cycles recycle molecules in ecosystems

- Chemicals (nutrients) recycle in an ecosystem, sometimes quickly, sometimes slowly.
- Water is constantly being recycled from the oceans to land and back to the oceans.
- In the carbon cycle, plants use CO_2 available in the atmosphere and animals obtain organic carbon molecules by consuming plants or other animals.
- Nitrogen gas in the atmosphere must be converted to NH_4^+ or NO_3^- ("fixed") before it is useful to living organisms.
- The phosphorus cycle is a sedimentary cycle. Phosphorus from decaying organisms is available to be used again.

The issue of human population growth

- The human population has been growing rapidly only since the Industrial Revolution beginning in the 1700s.
- Currently there are wide differences in fertility rates between more industrialized countries and the less industrialized countries.
- Even when (or if) the replacement fertility rate of 2.1 children per woman is reached, the world population will continue to grow for several more decades.

How humans affect ecosystems

- Air pollution contributes to global warming, destruction of the ozone layer, acid rain, and smog.
- Water is scarce and often distributed unequally. Pollution of surface freshwater can contaminate groundwater and ultimately pollute the oceans.
- Humans alter the landscape for their own purposes. Deforestation and desertification are transforming productive land into nonproductive land.
- Humans use a great deal of energy, much of it in the form of nonrenewable fossil fuels that contribute to pollution.

Toward a sustainable future

- Actions that would enable us to achieve a sustainable future include consuming less, recycling more, lowering the worldwide fertility rate to the replacement fertility rate, and eliminating rural poverty.
- Humans can shape the environment. It's up to us to do it wisely.

Terms You Should Know

acid rain, 553
aquifer, 555
biogeochemical cycle, 545
biological magnification, 555
biomass, 541
biosphere, 538
biotic potential, 539
carnivore, 542
carrying capacity, 540
community, 538
competition, 540
consumer, 542
decomposer, 543
deforestation, 552
demographic transition, 550
denitrification, 547
desertification, 557
ecological pyramid, 544
ecology, 538
ecosystem, 538
environmental resistance, 539
eutrophication, 555
exponential growth, 539
fertility rate, 549
food web, 543
fossil fuels, 545
geographic range, 538
global warming, 552
greenhouse effect, 552
growth rate, 548
habitat, 538
herbivore, 542
less industrialized countries (LIC), 550
more industrialized countries (MIC), 550
niche, 540
nitrification, 546
nitrogen fixation, 546
omnivore, 543
ozone, 552
pollutant, 551
population, 538
producer, 542
replacement fertility rate, 549
smog, 554
succession, 540
sustainable world, 560
thermal inversion, 554
zero population growth, 548

Test Yourself

Matching Exercise

_____ 1. The process of converting NO_3^- to N_2.
_____ 2. A consumer organism that feeds only on plants.
_____ 3. The production of NO_3^- by lightning or certain bacteria.
_____ 4. The process in which a shallow body of water becomes overgrown with plant life and eventually becomes land.
_____ 5. The factors that oppose unrestricted population growth.
_____ 6. A consumer organism that feeds on plants and animals.
_____ 7. The maximum rate of growth a population can achieve under ideal conditions.
_____ 8. The process in which marginal land is converted into a barren condition unsuitable for future agriculture.
_____ 9. Important reservoirs for carbon.
_____10. Organisms that are capable of photosynthesis.

a. herbivore
b. eutrophication
c. environmental resistance
d. omnivore
e. producers
f. fossil fuels
g. desertification
h. biotic potential
i. denitrification
j. nitrification

Fill in the Blank

11. The type of location where a species chooses to live is its __________.
12. The role in a community that a species fills is its __________.
13. Organisms that can utilize the sun's energy to synthesize the molecules they need are called __________.
14. Organisms that must eat foods that already contain stored forms of energy are called __________.
15. The equation:

$$\frac{\text{(births per year)} - \text{(deaths per year)}}{\text{(Population size)}}$$

is an equation for __________.
16. In the biogeochemical cycle for any nutrient, the three types of nutrient pools are __________, __________, and __________.
17. __________ are chemicals in the environment that have adverse effects on living organisms.
18. The two gases that are the main contributors to the greenhouse effect are __________ and __________.
19. The gas called __________ is found in two places: in the stratosphere and near the Earth's surface.
20. Sulfur dioxide, nitrogen oxides, and atmospheric water vapor combine to form __________.

True or False?

21. Plants use gaseous N_2 as their source of nitrogen.
22. A community consists of the populations of all the species that live in the same geographic area.
23. About 20% of all the water on Earth is freshwater.
24. Smog generally appears over rural areas and disappears after they have undergone industrialization.
25. About half of the ocean's oil pollution is due to natural seepage.

Concept Review

1. What are the components of an ecosystem?
2. What factors limit the size of any population?
3. Why is a constant supply of energy needed to sustain life?
4. What is a fossil fuel, and how were fossil fuels formed?
5. Summarize trends in human population growth over the past 10,000 years. What caused these trends?
6. Describe what happens to a country as it undergoes demographic transition. Give examples of nations in early and late stages of transition.
7. What could cause a population to continue to grow even after reaching its replacement fertility rate? Explain.
8. Why is it important to maintain the ozone layer that surrounds the Earth?
9. Tertiary consumers often are affected adversely by toxic pollutants before primary consumers. Explain.
10. What can we do to improve our chances of achieving a sustainable world?

Apply What You Know

1. Given absolutely no environmental resistance, how long would it take a population that was increasing by 8% a year to double?
2. Your uncle in Iowa builds a farm pond and stocks it with bass and bluegill sunfish. At first the pond water is fairly clear and the bass grow to large size, but within five years the pond is covered with a thick green scum and all the fish are dead. Explain to your uncle what might have happened to the pond.
3. Speculate as to what you think might eventually become the most important environmental resistance factor (or factors) that determines the world's human carrying capacity.

Case in Point

Report Cards for Polluters?

To curb rising pollution in Indonesia, the country's Environmental Impact Management Agency color-codes the compliance of industrial facilities with pollution regulations. A black rating indicates serious violations that are damaging to the environment or community, red indicates facilities that have some pollution controls but not enough, blue means they meet national standards for pollution control, and green shows they are cleaner than required by law. (The top rating, gold, indicates world-class compliance and has not yet been awarded to any factory in Indonesia.)

When the program began, two-thirds of Indonesia's companies failed the test; only 35.3 percent complied with minimum national standards for pollution control. Government officials staged a highly publicized ceremony to recognize the five green-class companies, and privately notified polluters of their failing "grade." Executives at several black- and red-rated firms soon called to ask how they could improve their ratings. Within 18 months, water pollution from all companies had fallen by 40 percent. After two years, more than 49% of the companies met government standards.

QUESTIONS:

1. The World Bank praises Indonesia's program as an example of a success story in fighting pollution. Why do you think it has been successful?

2. What else might the Indonesian government do to encourage companies to achieve a gold ranking?

3. How might public awareness be increased in your community in order to improve the pollution control programs of companies in your area?

INVESTIGATE:

"World Bank: Greening Industry—Communities, Stock Markets and Governments Join Hands to Cut Pollution," *M2 Presswire,* November 23, 1999.

Nivedita Prabhu, "IBRD Shows the Way to Fight Pollution," *The Economic Times,* November 28, 1999.

References and Additional Reading

Diamond, Jared. Easter's End. *Discover,* August 1995. (A fascinating and compelling look at an isolated human society that drove itself to extinction by resource depletion.)

Hammond, Allen. *Which World? Scenarios for the 21st Century*. Washington, D.C.: Island Press / Shearwater Books, 1998. 306 pp. (A readable and well-referenced look at how the choices we make might affect world political, social, economic, and environmental conditions in the next 50 years.)

Smith, Robert L. and Thomas M. Smith. *Elements of Ecology*. Menlo Park, CA: Benjamin/Cummings, 1998. 555 pp. (A more complete treatment of the subject of ecology for the interested student.)

World Population Projections to 2150. New York: United Nations, 1998. 41 pp. (The latest figures, by region, for world population trends.)

GLOSSARY

ABO blood typing Technique for classifying blood according to the presence or absence of two protein markers, A and B.

absorption Process by which nutrients, water, and ions pass through the digestive tract mucosa and into the blood or lymph.

accommodation The process of adjusting the refractive power of the lens of the eye so that light rays are focused on the retina.

acetylcholine (ACh) (AS-uh-teel-KOL-een) Neurotransmitter released by some nerve endings in both the peripheral and central nervous systems.

acid A substance that releases hydrogen ions when in solution. A proton donor.

acromegaly (AK-roh-MEG-uh-lee) A condition of enlargement of the bones and muscles of the head, hands, face, and feet caused by an excess of growth hormone in an adult.

acrosome (AK-roh-sohm) A cap containing enzymes that covers most of the head of a sperm. Helps sperm penetrate an egg.

ACTH (adrenocorticotropic hormone) (uh-DREE-noh-kort-ih-koh-TRO-pic) Anterior pituitary hormone that influences the activity of the adrenal cortex.

actin (AK-tin) A contractile protein of muscle. Comprises the thin filaments in the myofibrils.

action potential A brief change in membrane potential of a nerve or muscle cell due to the movement of ions across the plasma membrane. Also called a nerve impulse.

active transport Process that transfers a substance into or out of a cell, usually against its concentration gradient. Requires a carrier and the expenditure of energy.

adaptation General term for the act or process of adjusting. In environmental biology, the process of changing to better suit a particular set of environmental conditions. Of sensory neurons, a decline in the frequency of nerve impulses even when a receptor is stimulated continuously and without change in stimulus strength.

adaptive radiation A short burst of evolutionary activity in which many new lifeforms develop in a brief span of time from a single ancestor.

addiction The need to continue obtaining and using a chemical substance despite one's better judgment and good intentions.

Addison's disease A condition caused by a failure of the adrenal cortex to secrete sufficient cortisol and aldosterone.

adenosine diphosphate (ADP) (uh-DEN-oh-seen dy-FOSS-fate) A nucleotide composed of adenosine with two phosphate groups. When ADP accepts another phosphate group it becomes the energy-storage molecule ATP.

adenosine triphosphate (ATP) (uh-DEN-oh-seen try-FOSS-fate) A nucleotide composed of adenosine and three phosphate groups. ATP is an important carrier of energy for the cell because energy is released when one of the phosphate groups is removed.

ADH (antidiuretic hormone) (AN-tih-dy-yu-RET-ik) Hormone produced by the hypothalamus and released by the posterior pituitary. Stimulates the kidneys to reabsorb more water, reducing urine volume.

adipocyte (AH-di-po-syt) An adipose (fat) cell.

adipose tissue (AH-di-pohs) A connective tissue consisting chiefly of fat cells.

ADP (adenosine diphosphate) (uh-DEN-oh-seen dy-FOSS-fate) A nucleotide composed of adenosine with two phosphate groups. When ADP accepts another phosphate group it becomes the energy-storage molecule ATP.

adrenal cortex (ah-DREE-nul KOR-teks) External portion of the adrenal gland. Its primary hormones, aldosterone and corticosterone, influence inflammation, metabolism, extracellular fluid volume, and other functions.

adrenal glands (ah-DREE-nul) Hormone-producing glands located atop the kidneys. Each consists of medulla and cortex areas.

adrenal medulla (ah-DREE-nul meh-DUL-uh) Interior portion of the adrenal gland. Its hormones, epinephrine and norepinephrine, influence carbohydrate metabolism and blood circulation.

adrenocorticotropic hormone (ACTH) (uh-DREE-noh-kort-ih-koh-TRO-pic) Anterior pituitary hormone that influences the activity of the adrenal cortex.

aerobic metabolism (air-OH-bik) Metabolic reactions within a cell that utilize oxygen in the process of producing ATP.

agglutination (uh-glue-tin-AY-shun) The clumping together of (foreign) cells induced by cross-linking of antigen-antibody complexes.

aging Progressive, cumulative changes in the body's structure and function that occur throughout a person's life span.

agranular leukocyte A white blood cell that does not contain distinctive granules. Includes lymphocytes and monocytes.

AIDS (acquired immune deficiency syndrome) Disease caused by human immunodeficiency virus (HIV). Symptoms include severe weight loss, night sweats, swollen lymph nodes, opportunistic infections.

aldosterone (al-DOS-ter-ohn or AL-do-STEER-ohn) Hormone produced by the adrenal cortex that regulates sodium reabsorption.

all-or-none principle Principle stating that muscle cells always contract completely each time they are stimulated by their motor neuron, and that they do not contract at all if they are not stimulated by their motor neuron.

allantois (uh-LAN-toh-is) Embryonic membrane. Its blood vessels develop into blood vessels of the umbilical cord.

alleles (uh-LEELS) Genes coding for the same trait and found at the same locus on homologous chromosomes.

allergen (AL-ur-jen) Any substance that triggers an allergic response.

allergy (AL-ur-jee) Overzealous immune response to an otherwise harmless substance.

alveolus (pl. alveoli) (al-VEE-oh-lus) One of the microscopic air sacs of the lungs.

Alzheimer's disease (AHLTZ-hy-merz) Degenerative brain disease resulting in progressive loss of motor control, memory, and intellectual functions.

amino acid (uh-MEE-no) Organic compound containing nitrogen, carbon, hydrogen, and oxygen. The 20 amino acids are the building blocks of protein.

amniocentesis A form of fetal testing in which a small sample of amniotic fluid is removed so that cells of the fetus can be examined for certain known diseases and genetic abnormalities.

amnion (AM-nee-ahn) Fetal membrane that forms a fluid-filled sac around the embryo.

ampulla (am-PYU-la) Base of a semicircular canal in the inner ear.

anabolic steroid Any one of over 100 synthetic and natural compounds, all of which are related to the male sex steroid hormone testosterone. Anabolic steroids increase muscle mass and contribute to the development and maintenance of masculine characteristics.

anabolism (ah-NAB-o-lizm) Energy-requiring building phase of metabolism, during which simpler substances are combined to form more complex substances.

anaphase Third stage of mitosis in which a full set of daughter chromosomes moves toward the poles of a cell.

androgen (AN-dro-jen) A hormone that controls male secondary sex characteristics, such as testosterone.

anemia (uh-NEE-mee-uh) A condition in which the blood's ability to carry oxygen is reduced due to a shortage of normal hemoglobin or too few red blood cells.

aneurysm (AN-yur-iz-um) A ballooning or bulging of the wall of an artery caused by dilation or weakening of the wall.

angina pectoris (AN-jin-uh PEK-tor-is) Sensation of severe pain and tightness in the chest caused by insufficient blood supply to the heart.

angiogram (AN-jee-oh-gram) An x-ray image of blood vessels.

angiotensin II (AN-jee-oh-TEN-sin TOO) A potent vasoconstrictor activated by renin. Also triggers the release of aldosterone from the adrenal gland.

ANH (atrial natriuretic hormone) (AY-tree-ul NAH-tree-u-RET-ik) A hormone secreted by the atria of the heart. Helps regulate blood pressure by causing the kidneys to excrete more salt.

anorexia nervosa (ah-nuh-REK-see-uh nur-VOH-suh) An eating disorder characterized by abnormally low body weight and unrealistic fear of becoming obese.

antibiotic (AN-tih-by-AH-tik) A substance that interferes with bacterial metabolism. Administered to cure bacterial diseases.

antibody A protein molecule released by a B cell or a plasma cell that binds to a specific antigen.

antibody-mediated immunity Mechanism of defense in which B cells and plasma cells produce antibodies that bind to specific antigens.

anticodon The three-base sequence in transfer RNA that pairs with a complementary sequence in messenger RNA.

antidiuretic hormone (ADH) (AN-tih-dy-yu-RET-ik) Hormone produced by the hypothalamus and released by the posterior pituitary. Stimulates the kidneys to reabsorb more water, reducing urine volume.

antigen (AN-tih-jen) A substance or part of a substance (living or nonliving) that the immune system recognizes as foreign. It activates the immune system and reacts with immune cells or their products, such as antibodies.

antigen-presenting cell (APC) A cell that displays antigen-MHC complexes on its surface. Triggers an immune response by lymphocytes.

antioxidant An agent that inhibits oxidation; specifically a substance that can neutralize the damaging effects of free radicals and other substances.

anus The outlet of the digestive tract.

aorta (ay-OR-tah) The main systemic artery. Arises from the left ventricle of the heart.

aortic bodies Receptors in the aorta that are sensitive primarily to the oxygen concentration of the blood. These receptors can also respond to H^+ and CO_2.

appendicular skeleton (ap-un-DIK-yuh-lur) The portion of the skeleton that forms the pectoral and pelvic girdles and the four extremities.

appendix (uh-PEN-diks) A small finger-like pouch with no known digestive function that extends from the cecum of the large intestine.

aqueous humor (AY-kwee-us) Watery fluid between the cornea and the lens of the eye.

aquifer A deep underground reservoir of water.

arrhythmia Irregular heart rhythm. Arrythmias may be caused by defects in the cardiac conduction system or by drugs, stress, or cardiac disease.

arteriole (ar-TEER-ee-ohl) The smallest of the arterial blood vessels. Arterioles supply the capillaries, where nutrient and gas exchange take place.

artery A blood vessel that conducts blood away from the heart and toward the capillaries.

arthritis Inflammation of the joints.

artificial insemination A technique to enhance fertility in which sperm are placed into the vagina or uterus with a syringe.

asthma (AZ-muh) A recurrent, chronic lung disorder characterized by spasmodic contraction of the bronchi, making breathing difficult.

astigmatism (uh-STIG-muh-tiz-um) An eye condition in which unequal curvatures in different parts of the lens (or cornea) lead to blurred vision.

atherosclerosis (ath-er-o-skler-OS-is) Buildup of fatty deposits on and within the inner walls of arteries. Atherosclerosis is a risk factor for cardiovascular disease.

atom Smallest particle of an element that exhibits the properties of that element. Atoms are composed of protons, neutrons, and electrons.

ATP (adenosine triphosphate) (uh-DEN-oh-seen try-FOSS-fate) A nucleotide composed of adenosine and three phosphate groups. ATP is an important carrier of energy for the cell because energy is released when one of the phosphate groups is removed.

atrial natriuretic hormone (ANH) (AY-tree-ul NAH-tree-u-RET-ik) A hormone secreted by the atria of the heart. Helps regulate blood pressure by causing the kidneys to excrete more salt.

atrium (pl. atria) (AY-tree-um) One of the two chambers of the heart that receive blood from veins and deliver it to the ventricles.

auditory canal Part of the outer ear. The auditory canal channels sounds to the eardrum.

Australopithecus (oss-TRAL-o-pith-uh-kus) An early hominid (ancestor of modern humans). *Australopithecus* walked upright and showed the beginnings of brain enlargement.

autoimmune disorder A condition in which the immune system attacks the body's own tissues.

autonomic nervous system (ANS) Division of the peripheral nervous system composed of the sympathetic and parasympathetic divisions. The ANS carries signals from the central nervous system that control automatic functions of the body's internal organs.

autosomes (AW-tow-sohms) Chromosomes 1 to 22. The term autosomes applies to all chromosomes except the sex chromosomes.

autotroph An organism that uses carbon dioxide and energy from its physical environment to build its own large organic molecules.

AV (atrioventricular) bundle Bundle of specialized fibers that conduct impulses from the AV node to the right and left ventricles.

AV (atrioventricular) node Specialized mass of conducting cells located at the atrioventricular junction in the heart.

AV (atrioventricular) valve One of two valves located between the atria and the ventricles that prevents backflow into the atria when the ventricles are contracting.

axial skeleton (AK-see-ul) The portion of the skeleton that forms the main axis of the body, consisting of the skull, ribs, sternum, and backbone.

axon (AK-sahn) An extension of a neuron that carries nerve impulses away from the nerve cell body. The conducting portion of a nerve cell.

axon bulb One of many rounded tips at the end of an axon.

B cell (B lymphocyte) (LIM-fo-syt) A white blood cell that matures in bone marrow and gives rise to antibody-producing plasma cells.

bacteria Prokaryotic microorganisms responsible for many human diseases.

balloon angioplasty Treatment to open a partially blocked blood vessel. The technique involves threading a small balloon into the blood vessel and then inflating the balloon, pressing the blocking material against the sides of the vessel.

baroreceptor Sensory receptor that is stimulated by increases in blood pressure.

basal metabolic rate (BMR) Rate at which energy is expended (heat produced) by the body per unit of time under resting conditions.

base A substance capable of binding with hydrogen ions. A proton acceptor.

basement membrane Layer of nonliving extracellular material that anchors epithelial tissue to the underlying connective tissue.

basophil (BAY-so-fil) White blood cell whose granules stain well with a basic dye. Basophils release histamine and other substances during inflammation.

bile Greenish-yellow or brownish fluid that is secreted by the liver, stored in the gallbladder, and released into the small intestine. Bile is important in emulsifying fat.

biogeochemical cycle A cycle in which the chemicals that comprise living organisms are recycled between organisms and the Earth.

biogeography Study of the distribution of plants and animals around the world.

biological magnification Process by which nonexcreted substances become more concentrated in organisms that are higher in the food chain.

biomass Total living component of an ecosystem.

biosphere (BY-oh-sfeer) The portion of the surface of the Earth (land, water, and air) inhabited by living organisms.

biotic potential (by-AHT-ic) Maximum growth rate of a population under ideal conditions.

birth control pill Oral contraceptive containing hormones that inhibit ovulation.

blastocyst (BLAS-toh-sist) Stage of early embryonic development. The product of cleavage.

blind spot Region of the eye where the optic nerve passes through the retina. There are no photoreceptor cells in the blind spot.

blood A fluid connective tissue consisting of water, solutes, blood cells, and platelets that carries nutrients and waste products to and from all cells.

blood pressure The force exerted by blood against blood vessel walls, generated by the pumping action of the heart.

blood type Classification of blood based on the presence of surface antigens on red blood cells, and the presence of antibodies to surface antigens other than one's own.

blood-brain barrier A functional barrier between the blood and the brain that inhibits passage of certain substances from the blood into brain tissues.

BMR (basal metabolic rate) Rate at which energy is expended (heat produced) by the body per unit of time under resting (basal) conditions.

bone A connective tissue that forms the bony skeleton. Bone consists of a few living cells encased in a hard extracellular matrix of mineral salts.

bottleneck effect Evolutionary effect that occurs when a major catastrophe wipes out most of a population without regard to any previous measure of fitness.

botulism A form of poisoning caused by a bacterium, *clostridium botulinum*, occasionally found in inadequately cooked or preserved foods.

brain Organ of the central nervous system that receives, stores, and integrates information from throughout the body and coordinates the body's responses.

bronchiole (BRAHNG-kee-ohl) Small airways within the lungs that carry air to the alveoli. The walls of bronchioles are devoid of cartilage.

bronchitis (brahng-KI-tis) Inflammation of the bronchi.

bronchus (pl. bronchi) (BRAHNG-kus) Any one of the larger branching airways of the lungs. The walls of bronchi are reinforced with cartilage.

buffer Chemical substance that minimizes changes in pH by releasing or binding hydrogen ions.

bulbourethral gland (bul-boh-yoo-REE-thral) One of a pair of male reproductive glands located below the prostate gland that secretes mucus into the urethra during sexual arousal.

bulimia (byoo-LEE-mee-uh) Eating disorder involving episodes of binging and purging.

C-section (cesarean section) Surgical delivery of a baby.

calcitonin (kal-suh-TOW-nin) Hormone released by the thyroid glands that promotes a decrease in blood calcium levels.

Calorie Amount of energy needed to raise the temperature of 1 kilogram of water 1° Celsius.

cancer A malignant, invasive disease in which cells become abnormal and divide uncontrollably. May spread throughout the body.

capillary (KAP-uh-lair-ee) The smallest type of blood vessel. Capillaries are the site of exchange between the blood and tissue cells.

carbaminohemoglobin (kar-buh-MEE-noh-HEE-muh-gloh-bin) Hemoglobin molecule carrying carbon dioxide.

carbohydrate (kar-boh-HY-drayt) Organic compound composed primarily of CH_2O groups. Includes starches, sugars, and cellulose.

carbonic anhydrase (kar-BAHN-ik an-HY-drase) Enzyme that facilitates the combination of carbon dioxide with water to form carbonic acid.

carcinogen (kar-SIN-o-jen) Cancer-causing agent.

carcinogenesis (kar-sin-o-JEN-uh-sis) The process by which cancer develops.

cardiac cycle (KAR-dee-ak) Sequence of events encompassing one complete contraction and relaxation cycle of the atria and ventricles of the heart.

cardiac muscle Specialized muscle tissue of the heart.

cardiac output Amount of blood pumped out of a ventricle in one minute.

carnivore (KAR-nih-vor) Organism that feeds only on animals.

carotid bodies (kuh-RAHT-id) Receptors in the carotid artery that are sensitive primarily to the oxygen concentration of the blood. These receptors can also respond to H^+ and CO_2. Carotid and aortic bodies are identical in function.

carrier In genetics, an individual who appears normal but who carries an allele for a genetic disorder. In cell physiology, a protein that binds with a specific substance and transports it across the plasma membrane.

carrying capacity The maximum number of individuals in a population that a given environment can sustain indefinitely.

cartilage (KAR-til-ij) White, semiopaque, flexible connective tissue.

cataracts (KAT-uh-rakts) Clouding of the eye's lens. Cataracts are often congenital or age-related.

CCK (cholecystokinin) (kohl-uh-sis-tuh-KYN-in) An intestinal hormone that stimulates gallbladder contraction and the release of pancreatic juice.

cecum (SEE-kum) The blind-end pouch at the beginning of the large intestine.

cell The smallest structure that shows all the characteristics of life. The cell is the fundamental structural and functional unit of all living organisms.

cell body Region of a nerve cell that includes the nucleus and most of the cytoplasmic mass, and from which the dendrites and axons extend.

cell cycle A repeating series of events in which a eukaryotic cell grows, duplicates its DNA, and then undergoes nuclear and cytoplasmic division to become two cells. Includes interphase (growth and DNA synthesis) and mitosis (cell division).

cell doctrine Theory consisting of three basic principles: (1) All living things are composed of cells and cell products. (2) A single cell is the smallest unit that exhibits all the characteristics of life. (3) All cells come only from preexisting cells.

cell-mediated immunity Immunity conferred by activated T cells, which directly lyse infected or cancerous body cells or foreign cells and release chemicals that regulate the immune response.

central nervous system (CNS) The brain and spinal cord.

centriole (SEN-tree-ohl) Small organelle found near the nucleus of the cell, important in cell division.

centromere (SEN-troh-meer) Constricted region of a chromosome where sister chromatids are attached to each other and where they attach to the mitotic spindle during cell division.

cerebellum (ser-ah-BELL-um) Brain region most involved in producing smooth, coordinated skeletal muscle activity.

cerebral cortex (suh-REE-brul KOR-teks) The outer gray matter region of the cerebral hemispheres. Some regions of the cerebral cortex receive sensory information; others control motor responses.

cerebral hemispheres The large paired structures that together form most of the cerebrum of the brain.

cerebrospinal fluid (suh-REE-broh-SPY-nul) Plasmalike fluid that fills the cavities of the CNS and surrounds the CNS externally. Protects the brain and spinal cord.

cerebrum (suh-REE-brum) The cerebral hemispheres and nerve tracts that join them together. Involves higher mental functions.

cesarean section (C-section) Surgical delivery of a baby.

chemoreceptor (KEE-moh-rih-sep-tur) Type of receptor sensitive to various chemicals.

chemotherapy The use of therapeutic drugs to selectively kill cancer cells.

chlamydia (kla-MID-ee-uh) Sexually transmitted disease involving infection by the bacterium *Chlamydia trachomatis*.

cholecystokinin (CCK) (kohl-uh-sis-tuh-KYN-in) An intestinal hormone that stimulates contraction of the gallbladder and the release of pancreatic juice.

cholesterol (ko-LES-ter-ahl) Steroid found in animal fats as well as in most body tissues. Made by the liver.

chorion (KOR-ee-ahn) Outermost fetal membrane. Helps form the placenta.

chorionic villi (kor-ee-AHN-ik VIL-eye) Extensions of the fetal chorion that project into maternal tissues.

choroid (KOR-oyd) The vascular middle layer of the eye.

chromatin (KROH-muh-tin) Threadlike material in the nucleus composed of DNA and proteins.

chromosome (KROH-muh-som) Rodlike structure of tightly coiled chromatin. Visible in the nucleus during cell division.

chyme (KYM) Semifluid mass consisting of partially digested food and gastric juice that is delivered from the stomach into the small intestine.

cilium (pl. cilia) (SYL-ee-um) A tiny hairlike projection on cell surfaces that moves in a wavelike manner.

circadian cycle (sur-KAY-dee-un) In many organisms, a repeating cycle of physiological events that occurs approximately once every 24 hours.

cirrhosis (sir-ROH-sis) Chronic disease of the liver, characterized by an overgrowth of connective tissue (fibrosis).

cleavage An early phase of embryonic development consisting of rapid mitotic cell divisions without cell growth.

climax community The stable result of ecological succession in a given environment.

clitoris (KLIT-uh-ris) In the female, a sensitive erectile organ located in the vulva.

cloning Production of identical copies of a gene, a cell, or an organism.

cochlea (KOK-lee-uh) Snail-shaped chamber in the inner ear that houses the organ of Corti.

codominance Pattern of inheritance in which both alleles of a gene are equally expressed even though the phenotypes they specify are different.

codon (KOH-dahn) The three-base sequence on a messenger RNA molecule that provides the code for a specific amino acid in protein synthesis.

coitus (KOH-ee-tus; koh-EE-tus) Genital sexual intercourse.

colon (KOH-lun) The region of the large intestine between the cecum and the rectum. The colon includes ascending, transverse, descending, and sigmoid portions.

colostrum (koh-LAHS-trum) Milky fluid secreted by the mammary glands shortly before and after delivery. Contains proteins and antibodies.

community Array of several different populations that coexist and interact within the same environment.

compact bone Type of dense bone tissue found on the outer surfaces and shafts of bones.

complement system A group of blood-borne proteins which, when activated, enhance the inflammatory and immune responses and may lead to lysis of pathogens.

complete dominance Pattern of genetic inheritance in which one allele completely masks or suppresses the expression of its complementary allele.

conception The end of fertilization, when a single diploid cell is produced from the union of sperm and egg nuclei.

concussion An injury to the brain resulting from a physical blow that may result in a brief loss of consciousness.

condom A contraceptive sheath, made of latex or animal membrane, that can be used to enclose the penis or line the vagina during sexual intercourse.

cone cell One of the two types of photoreceptor cells in the retina of the eye. Cone cells provide for color vision.

congestive heart failure Progressive condition in which the pumping efficiency of the heart becomes impaired. The result is often an inability to adequately pump the venous blood being returned to the heart, venous congestion, and body fluid imbalances.

connective tissue A primary tissue; form and function vary extensively. Functions include support, energy storage, and protection.

constipation (kahn-sti-PAY-shun) A condition in which chyme remains in the colon too long and the feces become dry and hard, making defecation difficult.

consumer In ecosystems, an organism that cannot produce its own food and must feed on other organisms. Also called a heterotroph.

control group In scientific experiments, a group of subjects that undergoes all the steps in the experiment except the one being tested. The control group is used to evaluate all possible factors that might influence the experiment other than the experimental treatment. Compare experimental group.

controlled variable Any physical or chemical property that might *vary* from time to time and that must be *controlled* in order to maintain homeostasis. Examples of controlled variables are blood pressure, body temperature, and the concentration of glucose in blood.

convergence A coming together. In the nervous system, the phenomenon in which several presynaptic neurons form synapses with one postsynaptic neuron.

core temperature The body's internal temperature. In humans, normal core temperature is about 98.6° F (37° C).

cornea (KOR-nee-uh) The transparent anterior portion of the eyeball.

coronary artery (KOR-uh-nair-ee) One of the two arteries of the heart leading to capillaries that supply blood to cardiac muscle.

coronary artery bypass graft (CABG) Surgery to improve blood flow. A piece of blood vessel is removed from elsewhere in the body and grafted onto a blocked coronary artery to bypass the damaged region.

corpus luteum (KOR-pus LOOT-ee-um) Structure that develops from cells of a ruptured ovarian follicle. The corpus luteum secretes progesterone and estrogen.

cortex (KOR-teks) General term for the outer surface layer of an organ.

cortisol (KOR-tih-sahl) Glucocorticoid produced by the adrenal cortex, also called hydrocortisone.

covalent bond (koh-VAY-lent) Chemical bond created by electron sharing between atoms.

cranial nerve (KRAY-nee-ul) One of the 12 pairs of peripheral nerves that originate in the brain.

creatinine (kree-AT-uhn-een) A nitrogenous waste molecule excreted by the kidneys.

crossing-over Exchange of DNA segments between homologous chromosomes during Prophase I of meiosis.

cystic fibrosis (SYS-tik fy-BRO-sys) Genetic disorder in which oversecretion of mucus clogs the respiratory passages. Cystic fibrosis can lead to fatal respiratory infections.

cytokinesis (sy-tow-kih-NEE-sis) The division of cytoplasm that occurs after a cell nucleus has divided.

cytoplasm (SY-tow-plaz-um) The cellular material surrounding a cell nucleus and enclosed by the plasma membrane.

cytoskeleton A cell's internal "skeleton." The cytoplasm is a system of microtubules and other components that support cellular structures and provide the machinery to generate various cell movements.

cytotoxic T cell (sy-toh-TAHK-sik) A "killer T cell" that directly lyses foreign cells, cancer cells, or virus-infected body cells.

data Facts derived from experiments and observations.

decomposer An organism that obtains energy by chemically breaking down the products, wastes, or remains of other organisms.

defecation (def-ih-KAY-shun) The movement of feces through the rectum and the anus, with their subsequent elimination from the body.

deforestation The loss of wooded areas due to harvesting trees and/or converting forestlands to other uses.

dehydration synthesis Process by which a larger molecule is synthesized by covalently bonding two smaller molecules together, with the subsequent removal of a molecule of water.

demographic transition A progression of changes followed by many nations, in which the society gradually moves from poor living conditions and a high death rate to improved economic conditions and declining death rate.

denaturation (dee-nay-chur-AY-shun) Loss of a protein's normal shape and function produced by disruption of hydrogen bonds and other weak bonds.

dendrite (DEN-dryt) Branching neuron extension that serves as a receptive or input region. Dendrites conduct graded potentials toward the cell body.

denitrification (dee-nyt-ruh-fih-KAY-shun) Process in which nitrate is converted to nitrogen gas (N_2). Denitrification is part of the nitrogen cycle.

dental caries (KAIR-eez) Tooth decay characterized by cavities.

deoxyhemoglobinvHemoglobin that has given up its oxygen.

deoxyribonucleic acid (DNA) (dee-OX-ee-RY-bo-new-CLAY-ik) A nucleic acid found in all living cells. Carries the organism's hereditary information.

depolarization Loss of a state of polarity. In nerve and muscle cells, the loss or reduction of the inside-negative membrane potential.

dermis Layer of skin underlying the epidermis. Consists mostly of dense connective tissue.

desertification (deh-ZERT-uh-fuh-KAY-shun) Process by which marginal lands are converted to desertlike conditions.

detritus (dih-TRY-tus) Nonliving organic matter.

diabetes mellitus (dy-uh-BEE-teez MEL-ih-tus) Disease characterized by a high blood sugar concentration and glucose in the urine, caused by either deficient insulin production or insufficient glucose uptake by cells.

diaphragm (DY-uh-fram) A dome-shaped sheet of muscle that separates the thoracic cavity from the abdominal cavity. Also, a contraceptive device inserted into the vagina to cover the cervical opening.

diarrhea (dy-uh-REE-uh) Watery feces. Can result from a number of factors.

diastole (dy-AS-toh-lee) Period of the cardiac cycle when a heart chamber is relaxed.

diastolic pressure (DY-uh-STAHL-ik) Lowest point of arterial blood pressure during a cardiac cycle. Diastolic pressure is the second of two numbers recorded by the health care provider, the first being systolic pressure.

differentiation Process by which a cell changes in form or function.

diffusion (dih-FYOO-shun) The movement of molecules from one region to another as the result of random motion. Net diffusion proceeds from a region of higher concentration to a region of lower concentration.

digestion Chemical or mechanical process of breaking down foodstuffs to substances that can be absorbed.

diploid The number of chromosomes in a body cell ($2n$), twice the chromosomal number (n) of a gamete. In humans, $2n = 46$.

disaccharide (dy-SAK-uh-ryd) Literally, a double sugar. A disaccharide consists of two monosaccharides linked together. Sucrose and lactose are disaccharides.

diuretic (dy-yoo-RET-ik) Any substance that enhances urinary output.

divergence Going in different directions from a common point. In the nervous system, the phenomenon in which one presynaptic neuron forms synapses with several postsynaptic neurons.

DNA (deoxyribonucleic acid) (dee-OX-ee-RY-bo-new-CLAY-ik) A nucleic acid found in all living cells. Carries the organism's hereditary information.

DNA polymerase (po-LIM-er-ase) Enzyme that links separate nucleotides together as a new strand of DNA is formed.

dominant allele An allele that masks or suppresses the expression of its complementary allele.

Down syndrome A condition caused by inheriting three copies of chromosome 21.

ductus deferens (pl. ductus deferentia) (DUK-tus DEF-er-enz) Tube extending from the epididymis to the urethra through which sperm pass during a man's sexual climax and ejaculation. Also called the vas deferens.

duodenum (doo-oh-DEE-num) First part of the small intestine.

dysplasia (dis-PLAY-zhee-uh) Abnormal changes in the shape and/or organization of cells. Dysplasia may precede the development of cancer.

ECG (electrocardiogram) Graphic record of the electrical activity of the heart.

ecological pyramid Graph representing the biomass, energy content, or number of organisms of each level in a food web. An ecological pyramid includes both producer and consumer populations.

ecology (ee-KAHL-uh-jee) Study of the relationships between organisms and their physical environment.

ecosystem (EEK-oh-sis-tem) All living organisms, all matter, and all energy in a given environment.

ectoderm Embryonic germ layer. Ectoderm forms the epidermis of the skin and its derivatives, and nerve tissues.

ectopic pregnancy (ek-TOP-ik) A pregnancy that results from implantation outside the uterus. The most common form of ectopic pregnancy is a tubal pregnancy.

egg An ovum; a mature female gamete.

elastic fiber Fiber formed from the protein elastin, which gives a rubbery and resilient quality to the matrix of connective tissue.

electrocardiogram (ECG) Graphic record of the electrical activity of the heart.

electron Negatively charged subatomic particle with almost no weight. Electrons orbit the atom's nucleus.

electron transport system An series of electron- and energy-transfer molecules in the inner membrane of mitochondria that provide the energy for the active transport of hydrogen ions across the membrane. The electron transport system is essential to the ability of the mitochondria to produce ATP for the cell.

element One of a limited number of unique varieties of matter that composes substances of all kinds. Elements each comprise just one kind of atom.

embolus (EM-bo-lus) Material, such as a blood clot, that floats in the bloodstream and may block a blood vessel.

embryo (EM-bree-oh) An organism in an early stage of development. In humans, the embryonic period extends from the beginning of week 3 to the end of week 8.

embryonic disk A flattened disk of cells that develops in the blastocyst shortly after implantation. The embryonic disk will develop into the embryo.

emphysema (em-fih-SEE-muh) Chronic respiratory disorder involving damage to the bronchioles and alveoli. In emphysema, loss of elastic tissue in the bronchioles causes them to collapse during expiration, trapping air in the alveoli and eventually damaging them as well.

endocrine gland (EN-duh-krin) Ductless gland that secretes one or more hormones into the bloodstream.

endocrine system Body system that includes all of the hormone-secreting organs and glands. Along with the nervous system, the endocrine system is involved in coordination and control of body activities.

endocytosis (en-do-sy-TO-sis) Process by which fluids, extracellular particles, or even whole bacteria are taken into cells. In endocytosis, the materials become enclosed by a vesicle composed of cell membrane material and then internalized within the cell. Phagocytosis is an example of endocytosis.

endoderm (EN-do-derm) Embryonic germ layer. Endoderm forms the lining of the digestive tube and its associated structures.

endometriosis (en-do-mee-tree-OH-sis) Condition in which endometrial tissue migrates up the oviduct during a menstrual period and implants on other organs such as ovaries, walls of the bladder, or colon.

endometrium (en-do-MEE-tree-um) Inner lining of the uterus. The endometrium becomes thickened and more vascular during the uterine cycle in preparation for pregnancy.

endoplasmic reticulum (en-do-PLAS-mik reh-TIK-yuh-lum) Membranous network of tubular or saclike channels in the cytoplasm of a cell. The endoplasmic reticulum is the site of most of the cell's production of proteins and other cell compounds.

energy The capacity to do work. Energy may be stored (potential energy) or in action (kinetic energy).

environmental resistance Factors in the environment that limit population growth in a particular geographic area.

enzyme (EN-zym) A protein that acts as a biological catalyst to speed up a chemical reaction.

eosinophil (ee-oh-SIN-oh-fil) Granular white blood cell whose granules stain readily with a red stain called eosin. Eosinophils attack parasites and function in allergic responses.

epidermis (ep-ih-DUR-mus) Outermost layer of the skin. The epidermis is composed of keratinized stratified squamous epithelium.

epididymis (ep-ih-DID-ih-mis) Portion of the male reproductive system in which sperm mature. The epididymis empties into the vas deferens.

epiglottis (ep-ih-GLAHT-is) Flaplike structure of elastic cartilage at the back of the throat that covers the opening of the larynx during swallowing.

epilepsy (EP-ih-lep-see) Condition involving abnormal electrical discharges of groups of brain neurons. Epilepsy may cause seizures.

epinephrine (ep-ih-NEF-rin) Primary hormone produced by the adrenal medulla, also called adrenalin.

episiotomy (ih-pee-zee-AHT-uh-mee) Surgical widening of the opening to the vagina during childbirth, performed to prevent tearing of tissues.

epithelial tissue (ep-ih-THEE-lee-ul) A primary tissue that covers the body's surface, lines its internal cavities, and forms glands.

erythropoietin (ih-rith-roh-PO-ih-tin) Hormone that stimulates production of red blood cells.

esophagus (ih-SOF-uh-gus) Muscular tube extending from the throat to the stomach. The esophagus is collapsed when it does not contain food.

essential amino acids Group of eight amino acids that the body cannot synthesize. Essential amino acids must be obtained from the diet.

estrogen (ES-tro-jen) Female sex hormone that stimulates the development of female secondary sex characteristics, helps oocytes mature, and affects the uterine lining during the menstrual cycle and pregnancy.

eukaryote (yoo-KAIR-ih-ot) A organism composed of cells with nuclei and internal membrane-bound organelles. Animals, plants, protists, and fungi are all eukaryotes.

eutrophication (yoo-troh-fih-KAY-shun) Process by which mineral and organic nutrients accumulate in a body of water, leading to plant overgrowth and loss of oxygen. Eutrophication may be accelerated by pollution.

evolution An unpredictable and natural process of descent over time with genetic modification. Evolution is influenced by natural selection, chance, historical events, and changing environments.

evolutionary tree Map that illustrates evolutionary change and relationships between life-forms.

excretion Elimination of waste products from the body.

exocytosis (ex-oh-sy-TO-sis) Mechanism by which substances are moved from the cell interior to the extracellular space. In exocytosis, a membrane-bound secretory vesicle fuses with the plasma membrane and releases its contents to the exterior.

exon A nucleotide sequence of DNA that specifies a useful informational sequence. Compare intron.

experiment A procedure designed to test a working hypothesis. In a controlled experiment, all conditions of the environment are controlled except the specific factor dictated by the working hypothesis.

experimental group In a controlled experiment, the group of subjects that receives the experimental treatment. Compare control group.

expiration Act of exhaling or expelling air from the lungs.

exponential growth Growth pattern of a population that follows a characteristic J-curve. A population undergoing exponential growth will double repeatedly over similar time periods.

extension Movement that increases the angle of a joint, such as straightening a flexed knee.

extinction The loss of a particular life-form completely.

extra-embryonic membranes Membranes that are not part of an embryo's body but are essential for its growth and development (amnion, allantois, yolk sac, and chorion).

facilitated transport Passive transport of a substance into or out of a cell along a concentration gradient. Facilitated transport requires a carrier protein.

FAD A transport molecule within mitochondria that can accept hydrogen ions and electrons liberated by the Krebs cycle, forming high-energy $FADH_2$.

farsightedness Condition in which the eyeball is too short, causing nearby objects to be focused behind the retina; also called hyperopia.

fatigue Condition of feeling tired.

fatty acid Linear chain of carbon and hydrogen atoms (hydrocarbon chains) with an organic acid group at one end. Fatty acids are constituents of fats.

fertility rate Number of children that each female has during her lifetime.

fertilization Fusion of the nuclei of sperm and egg, thereby creating a zygote.

fever Body temperature that has risen above the normal set point (in humans, approximately 98.6° F). A fever often indicates illness.

fiber General term for a thin strand. A muscle fiber is a single muscle cell. A connective tissue fiber is a thin strand of extracellular material. In the digestive system, fiber refers to material that is indigestible but beneficial in the diet.

fibrin (FY-brin) Fibrous insoluble protein formed during blood clotting.

fibrinogen (fy-BRIN-oh-jen) A blood protein that is converted to fibrin during blood clotting.

fibroblast (FY-broh-blast) Young, actively mitotic cell that forms the fibers of connective tissue.

fimbriae (FYM-bree-ay) Fingerlike projections at the open end of an oviduct that help move an oocyte into the oviduct.

flagellum (pl. flagella) (flah-JEL-um) Long, whiplike extension of the plasma membrane used by sperm and some bacteria for propulsion.

flexion (FLEK-shun) Movement that decreases the angle of a joint, such as bending the knee from a straight to an angled position.

flexor (withdrawal) reflex Reflex initiated by a painful stimulus that causes automatic withdrawal of the threatened body part from the stimulus.

follicle (FAHL-ih-kul) Ovarian structure consisting of a developing egg surrounded by one or more layers of granulosa cells.

follicle stimulating hormone (FSH) Hormone produced by the anterior pituitary that stimulates ovarian follicle production in females and sperm production in males.

food web Hierarchy consisting of numerous food chains. A food web depicts interactions between producers and consumers in an ecosystem.

forebrain Anterior portion of the brain that includes the cerebrum, the thalamus, and the hypothalamus.

formed element Portion of the blood consisting of cells or cell components. Formed elements include red blood cells, white blood cells, and platelets.

fossil Preserved remnant of an organism. The fossil record is one of our richest sources of information about life-forms that lived in the past.

fossil fuel A carbon-containing compound (coal, oil, or gas) that was formed millions of years ago by living organisms, then covered by sediments.

fovea centralis (FOH-vee-uh sen-TRA-lis) A small region of the retina that is responsible for the highest visual acuity. The fovea centralis consists of densely packed cone cells.

fraternal twins Twins resulting from the ovulation and fertilization of more than one oocyte in a particular cycle. Fraternal twins are as different as any two children by the same parents.

free radicals Highly reactive chemicals with unpaired electrons that can disrupt the structure of proteins, lipids, and nucleic acids.

FSH (follicle stimulating hormone) Hormone produced by the anterior pituitary that stimulates ovarian follicle production in females and sperm production in males.

fungi (FUN-jy) One of the five kingdoms of life. The fungi include yeasts, molds, and mushrooms.

gallbladder Organ located beneath the right lobe of the liver. The gallbladder stores and concentrates bile.

gallstones Hard crystals containing cholesterol, calcium, and bile salts that can obstruct the flow of bile from the gallbladder.

gamete (GAM-eet) Haploid sex or germ cell. Sperm and eggs are gametes.

ganglion (pl. ganglia) (GANG-lee-on) Collection of nerve cell bodies outside the central nervous system.

gastric glands Glands located beneath the stomach lining. Gastric glands secrete gastric juice.

gastric juice Collective term for the hydrochloric acid, pepsinogen, and fluid secreted by gastric glands.

gastrin (GAS-trin) Hormone secreted into the blood by cells located in the gastric glands of the stomach. Gastrin regulates the secretion of gastric juice by stimulating production of hydrochloric acid.

gastrointestinal (GI) tract A hollow tube that extends from the mouth to the anus. The GI tract includes mouth, pharynx, esophagus, stomach, small intestine, large intestine, rectum, and anus.

gene (JEEN) The unit of heredity. Most genes encode for specific polypeptides, and each gene has a specific location on a particular chromosome.

gene therapy Technique in which defective genes are repaired or replaced with their normal counterparts.

genetic drift Random changes in allele frequency because of chance events.

genetic engineering The planned alteration of the genetic makeup of an organism by modifying, inserting, or deleting genes or groups of genes.

genital herpes (JEN-ih-tul HUR-peez) A sexually transmitted disease that results from infection by the herpes simplex type 2 virus.

genital warts Sexually transmitted disease caused by the human papillomavirus (HPV). Genital warts are a potential risk factor for cancers of the cervix or penis.

genome (JEE-nohm) The complete set of genes in the chromosomes of a particular species of organism.

genotype An individual's particular set of genes.

genus (JEE-nus) Level of taxonomic classification. Modern humans belong to the genus *Homo*.

geographic range The area of the earth over which a particular species is distributed.

GIFT (gamete intrafallopian transfer) Technique to enhance fertility in which unfertilized eggs and sperm are placed directly into an oviduct (Fallopian tube) through a small incision in a woman's abdomen.

gigantism Pituitary disorder caused by hypersecretion of growth hormone during childhood and adolescence.

gingivitis (jin-jih-VY-tis) Condition of inflamed gum tissue (gingiva), often caused by tooth decay and poor oral hygiene.

gland One or more epithelial cells that are specialized to secrete or excrete substances.

glaucoma (glaw-KO-muh) Eye condition caused by blockage of the ducts that drain the aqueous humor in the eye, resulting in high intraocular pressures and compression of the optic nerve. Glaucoma can result in blindness unless detected early.

global warming Increase in average global temperature.

glomerular capsule (glow-MAIR-yoo-lar) Double-walled cup at the beginning of a renal tubule. The glomerular capsule, also called Bowman's capsule, encloses the glomerulus.

glomerular filtrate The fluid filtered from the glomerulus into the glomerular capsule.

glomerulus (glow-MAIR-yoo-lus) Cluster or tuft of capillaries inside the glomerular capsule in a kidney. The glomerulus is involved in the formation of the glomerular filtrate.

glottis (GLAHT-tis) Opening between the vocal cords in the larynx.

glucagon (GLOO-kah-gon) Hormone formed by alpha cells in the islets of Langerhans in the pancreas. Glucagon raises the glucose level of blood by stimulating the liver to break down glycogen.

glucocorticoids (gloo-kah-KOR-tih-koyds) Adrenal cortex hormones that increase blood glucose levels and help the body resist long-term stressors. The principal glucocorticoid is cortisol.

glycogen (GLY-ko-jen) Main carbohydrate stored in animal cells. A polysaccharide.

glycolysis (gly-KOL-ih-sis) Breakdown of glucose to pyruvic acid. Glycolysis is an anaerobic process.

goiter (GOY-tur) An enlarged thyroid, caused by iodine deficiency or other factors.

Golgi apparatus (GOHL-jee) Membranous system within a cell that packages proteins and lipids destined for export, packages enzymes into lysosomes for cellular use, and modifies proteins destined to become part of cellular membranes.

gonad (GOH-nad) An organ that produces sex cells. The gonads are the testes of the male and the ovaries of the female.

gonadotropins (goh-nad-o-TRO-pinz) Gonad-stimulating hormones produced by the anterior pituitary. The gonadotropins are FSH and LH.

gonorrhea (gahn-o-REE-uh) Sexually transmitted disease caused by the bacterium *Neisseria gonorrhoeae*.

graded potential A local change in membrane potential that varies directly with the strength of the stimulus. Graded potentials decline with distance from the stimulus.

granular leukocyte (GRAN-yoo-lur LOO-ko-syt) A type of white blood cell with granules that stain readily.

Graves' disease A disorder in which the thyroid gland produces too much thyroxine.

gray matter Area of the central nervous system that contains cell bodies and unmyelinated fibers of neurons. The gray matter is gray because of the absence of myelin.

greenhouse effect A natural phenomenon in which atmospheric gases allow sunlight to pass through but trap most of the heat that radiates outward from the Earth's surface.

greenhouse gases Atmospheric gases, notably CO_2 and methane, responsible for the greenhouse effect.

growth hormone Hormone produced by the anterior pituitary that stimulates growth in general.

growth plate Cartilage plate located near the ends of bone. Bones become longer during childhood and adolescence because new cartilage is continually being added to the outer surface of the growth plates.

habitat The type of location where an organism chooses to live. An organism's habitat typically is determined by its tolerance for certain environmental conditions.

hair cells The mechanoreceptor cells of the ear that lie between the tectorial and basilar membranes.

haploid (HAP-loyd) Half the diploid number of chromosomes; the number of chromosomes in a gamete.

hCG (human chorionic gonadotropin) (kor-ee-AHN-ik goh-nad-o-TRO-pin) A hormone secreted by the chorion that supports pregnancy for the first three months, until the placenta begins producing enough progesterone and estrogen.

heart attack Condition characterized by the death of heart tissue caused by inadequate oxygen supply. The medical term for a heart attack is myocardial infarction.

helper T cell Type of T lymphocyte that orchestrates cell-mediated immunity by direct contact with other immune cells and by releasing chemicals called lymphokines.

hematocrit (hee-MAHT-oh-krit) The percentage of blood that consists of red blood cells.

hemodialysis (hee-moh-dy-AL-ih-sis) Technique to duplicate the functions of healthy kidneys. A patient's blood is circulated through an artificial kidney machine consisting of semipermeable membranes in contact with a large volume of plasmalike fluid.

hemoglobin (HEE-moh-gloh-bin) Oxygen-transporting protein in red blood cells that gives the cells their characteristic red color.

hemolytic disease of the newborn (HDN) (hee-moh-LIT-ik) Potentially life-threatening disease in newborns characterized by reduced numbers of mature red blood cells and elevated blood levels of bilirubin.

hemophilia (hee-moh-FIL-ee-uh) Collective term for several different hereditary bleeding disorders that show similar symptoms. Hemophilia is caused by a deficiency of one or more clotting factors.

hemorrhoids (HEM-oh-royds) Sometimes painful condition caused by swollen veins in the lining of the anus.

hemostasis (HEE-mo-stay-sis) Stoppage of bleeding or of the circulation of blood to a part.

hepatic portal system (heh-PAT-ik) The system of blood vessels located between the organs of the digestive tract and the liver.

hepatitis (hep-uh-TY-tis) General term for inflammation of the liver. Hepatitis is sometimes caused by viral infections.

herbivore (HUR-buh-vor) An organism that feeds on plants, utilizing the energy stored in them. Also called a primary consumer.

heterotroph (HET-ur-oh-trof) An organism that cannot utilize the sun's energy to synthesize the molecules it needs. Instead, it must consume foods that already contain stored forms of energy. Also called a consumer.

heterozygous (HET-ur-oh-ZY-gus) Having different alleles at the same location (on a pair of homologous chromosomes).

hindbrain Region of brain connected to the spinal cord. The hindbrain comprises the medulla oblongata, cerebellum, pons, and part of the reticular formation.

hippocampus (hip-po-KAM-pus) In the brain, a region of the limbic system that plays a role in converting new information into long-term memories.

histamine (HIS-ta-meen) Substance produced by basophils and mast cells that causes vasodilation and increases vascular permeability.

HIV (human immunodeficiency virus) Virus that destroys helper T cells, thus depressing cell-mediated immunity. Symptomatic AIDS gradually appears when lymph nodes can no longer contain the virus.

homeostasis (ho-mee-oh-STAY-sis) State of body equilibrium characterized by a relatively constant and stable internal environment.

hominids (HAHM-ih-nids) Early ancestors of modern humans. Hominids walked upright and showed the beginnings of brain enlargement.

Homo erectus (HOH-moh ih-REK-tus) Ancestor of modern humans. *Homo erectus* arose from *Homo habilis* and was the first of the human line to spread from Africa to Europe and Asia.

Homo habilis (HOH-moh HAB-ih-lus) First distinctly human ancestor. *Homo habilis* showed brain enlargement, changes in teeth and facial features, a decline in sexual dimorphism, and conversion to a diet that included meat. *Homo habilis* may also have used tools.

Homo sapiens (HOH-moh SAY-pee-enz) Genus and species of modern humans, thought to have evolved about 140,000 to 100,000 years ago.

homologous chromosomes (ho-MAHL-uh-gus) Chromosomes that look identical under the microscope. Homologous chromosomes have the same genes in the same locations.

homologous structures Body structures that share a common origin.

homozygous (hoh-moh-ZY-gus) Having identical alleles at the same location (on a pair of homologous chromosomes).

hormone (HOR-mohn) A chemical messenger molecule secreted by an endocrine gland or cell into the bloodstream that has effects on specific target cells throughout the body.

human chorionic gonadotropin (hCG) (kor-ee-AHN-ik goh-nad-o-TRO-pin) A hormone secreted by the chorion that supports pregnancy for the first three months, until the placenta begins producing enough progesterone and estrogen.

human immunodeficiency virus (HIV) Virus that destroys helper T cells, thus depressing cell-mediated immunity. Symptomatic AIDS gradually appears when lymph nodes can no longer contain the virus.

human papillomavirus (HPV) (pap-ih-LOH-ma-vy-rus) Virus that causes genital warts. HPV is a risk factor for cancers of the cervix or penis.

hydrogen bond (HY-droh-jen) Weak bond that forms between a hydrogen atom with a partial positive charge and a nearby atom with a partial negative charge.

hydrolysis (hy-DRAHL-ih-sis) Process in which water is used to split a molecule into two smaller molecules.

hyperopia (hy-per-OH-pee-uh) Condition in which the eyeball is too short, causing nearby objects to be focused behind the retina; also called farsightedness.

hyperplasia (hy-per-PLAY-zhee-uh) An increase in the number of cells in a tissue caused by an increase in the rate of cell division.

hyperpolarization Transient local change in the resting potential of a neuron that makes it even more negative than usual.

hypertension High blood pressure.

hypertonic General term for above-normal tone or tension. In biology, a hypertonic solution is one with a higher solute concentration than plasma. Cells shrink when placed in a hypertonic solution.

hypotension Low blood pressure.

hypothalamus Region of the brain forming the floor of the third ventricle. The hypothalamus helps regulate the body's internal environment by secreting releasing factors that affect the secretion of the hormones of the anterior pituitary.

hypothesis An explanation or description of a proposed truth. An hypothesis is expressed as a testable statement about the natural world.

hypotonic General term for below-normal tone or tension. In biology, a hypotonic solution is one with a lower solute concentration than plasma. Cells swell when placed in a hypotonic solution.

identical twins Twins that arise when a single fertilized ovum splits in two shortly after fertilization. Identical twins have identical genotypes.

immune response Collective term for the defense activities of the immune system.

immune system A complex group of cells, proteins, and structures of the lymphatic system that work together to provide the immune response.

immunity Resistance to disease and pathogens.

immunization (im-yoo-nuh-ZAY-shun) A strategy for causing the body to develop immunity to a specific pathogen.

immunoglobulin (Ig) (im-yoo-noh-GLAHB-yoo-lin) One of five classes of blood plasma proteins (gamma globulins) that play a crucial role in immunity.

immunosuppressive drugs Medications that block the immune response. Examples are corticosteroid drugs to suppress inflammation and cytotoxic medications to block activated lymphocytes.

immunotherapy Treatments that promote the general responsiveness of the immune system so it can fight cancer more effectively.

incomplete dominance A pattern of genetic inheritance in which the heterozygous genotype results in a phenotype that is intermediate between the two homozygous phenotypes.

independent assortment Mendelian principle stating that genetic factors separate completely independently of each other during the formation of sperm and egg. Independent assortment applies fully only to genes located on different chromosomes.

inflammation (in-fluh-MAY-shun) A nonspecific defensive response of the body to tissue injury characterized by dilation of blood vessels and an increase in vessel permeability. The presence of inflammation is indicated by redness, heat, swelling, and pain.

inheritance Characteristics or traits that are transmitted from parents to offspring via genes.

inner cell mass Accumulation of cells in the blastocyst from which the embryo develops.

inner ear Region of the ear consisting of the cochlea and the vestibular apparatus. The cochlea sorts sounds by tone and converts them into nerve impulses. The vestibular apparatus helps maintain balance.

insertion The end of a muscle attached to the bone that moves when the muscle contracts.

inspiration Inhalation; act of taking a breath.

insulin (IN-suh-lin) A hormone secreted by the pancreas that enhances the uptake of glucose by cells, thus lowering blood glucose levels.

integration General term for the process of combining into a complete and harmonious whole. In the nervous system, the process by which neurons receive multiple inputs and produce a coordinated output.

interferon (in-tur-FEER-ahn) Chemical that is able to provide some protection against virus invasion of the body. Interferon inhibits viral reproduction.

interleukin (in-tur-LOO-kin) A substance that promotes development of immune cells.

internal environment In a multicellular organism, the environment that surrounds the cells; the interstitial fluid.

internal respiration Exchange of gases between blood, interstitial fluid, and cells.

interneuron A neuron within the central nervous system located between two other neurons. Interneurons receive input from one or more neurons and influence the functioning of other neurons.

interphase One of two major periods in the cell life cycle. Includes the period from cell formation to cell division.

interstitial fluid (in-tur-STISH-ul) Fluid between body cells.

intervertebral disk (in-tur-vur-TEE-brul) A disk of fibrocartilage between vertebrae.

intron A noncoding nucleotide sequence of DNA. Compare exon.

ion (EYE-ahn) An atom or molecule with a positive or negative electric charge.

ionic bond (eye-AHN-ik) Chemical bond formed by the attractive force between oppositely charged ions.

iris (EYE-ris) A colored disk-shaped muscle that determines how much light enters the eye.

isotonic General term for normal tone or tension. In biology, an isotonic solution is one with the same solute concentration as plasma. Cells maintain their normal cell volume in isotonic solutions.

isotopes Different atomic forms of the same element, varying only in the number of neutrons they contain. The heavier forms tend to be radioactive.

IUD (intrauterine device) Contraceptive device that is inserted into the uterus by a health care provider.

IVF (*in vitro* fertilization) Fertilization outside the body. Eggs are harvested from a woman and fertilized, and the embryo is inserted into the woman's uterus through her cervix.

jaundice (JAWN-dis) Yellowish color of the skin, mucous membranes, and sometimes the whites of the eyes. Jaundice is often caused by liver malfunction and high circulating levels of bilirubin.

jejunum (jeh-JOO-num) The part of the small intestine between the duodenum and the ileum.

joint The junction or area of contact between two or more bones; also called an articulation.

karyotype (KAIR-ee-oh-typ) The diploid chromosomal complement in any species. The human karyotype typically is shown as a composite display of the 22 pairs of autosomes arranged from longest to shortest, plus the sex chromosomes X and Y.

keratinocyte (kair-uh-TIN-oh-syt) Type of cell in the epidermis that produces a tough, waterproof protein called keratin.

killer T cell T lymphocyte that directly lyses foreign cells, cancer cells, or virus-infected body cells; also called a cytotoxic T cell.

kinetic energy Energy in motion; energy actually doing work.

Klinefelter syndrome A condition caused by an XXY genotype. People with Klinefelter syndrome show a tall male phenotype.

Krebs cycle A cyclic aerobic metabolic pathway within mitochondria in which oxygen is used to completely dissemble acetyl groups into CO_2 and high-energy compounds.

labor Collective term for the series of events that expel the infant from the uterus.

lactation Production and secretion of milk.

lacteal (LAK-tee-ul) Small lymphatic vessel in a villus of the small intestine that takes up lipids.

lactose intolerance (LAK-tohs) A common disorder of digestion and absorption caused by insufficient quantities of the enzyme lactase, responsible for digesting the lactose in milk and dairy products.

large intestine Portion of the digestive tract extending from the small intestine to the anus. Consists of the cecum, appendix, colon, rectum, and anal canal.

larynx (LAIR-inks) Cartilaginous organ containing the vocal cords, located between the trachea and pharynx. The larynx is also called the voice box.

lens A transparent flexible curved structure in the eye responsible for focusing incoming light on the retina.

less industrialized countries (LICs) Nations that are just beginning to industrialize.

leukemia (loo-KEE-mee-uh) A group of cancerous conditions involving overproduction of abnormal white blood cells.

Leydig cell (LAY-dig) Cells located between the seminiferous tubules of the testes. Leydig cells produce testosterone.

LH (luteinizing hormone) (LOO-tee-in-eye-zing) Anterior pituitary hormone that aids maturation of cells in the ovary and triggers ovulation in females. In males, LH causes the interstitial cells of the testis to produce testosterone.

limbic system Functional brain system involved in emotional responses.

lipase (LY-pace) Any lipid-digesting enzyme.

lipid (LIP-id) Organic compounds formed of carbon, hydrogen, and oxygen. Fats, oils, and cholesterol are lipids. Lipids are not very soluble in water.

lipoprotein (ly-poh-PRO-teen) A compound containing both lipid and protein.

Two medically important lipoproteins are the low-density lipoproteins (LDLs) and high-density lipoproteins (HDLs) that transport cholesterol.

liver Large-lobed organ in the abdominal cavity that overlies the stomach. The liver produces bile to help digest fat, and serves other metabolic and regulatory functions.

longevity How long a person lives.

loop of Henle (HEN-lee) Hairpin-shaped tubular structure that extends into the medulla of a nephron. The loop of Henle (also called the loop of the nephron) is important in the ability of the kidney to form either concentrated or dilute urine.

lumen (LOO-men) Cavity inside a tube, blood vessel, or hollow organ.

lupus erythematosis (LOO-pus air-uh-them-uh-TOW-sis) An autoimmune disorder in which the body attacks its own connective tissue.

luteinizing hormone (LH) (LOO-tee-in-eye-zing) Anterior pituitary hormone that aids maturation of cells in the ovary and triggers ovulation. In males, LH causes the interstitial cells of the testis to produce testosterone.

lymph (LIMF) Fluid derived from interstitial fluid transported by lymphatic vessels.

lymph nodes Small masses of tissue and lymphatic vessels that contain macrophages and lymphocytes. Macrophages and lymphocytes remove microorganisms, cellular debris, and abnormal cells from the lymph before it is returned to the cardiovascular system.

lymphatic system System consisting of lymphatic vessels, lymph nodes, and other lymphoid organs and tissues. The lymphatic system returns excess interstitial fluid to the cardiovascular system and provides a site for immune surveillance.

lymphatic vessels Lymphatic capillaries and larger conduits that transport lymph.

lymphocyte (LIM-foh-syt) One of several types of white blood cells (T cells and B cells) that participate in nonspecific and specific (immune) defense responses.

lysosome (LY-soh-sohm) Organelle originating from the Golgi apparatus that contains strong digestive enzymes.

lysozyme (LY-soh-zym) Enzyme in saliva and tears that is capable of destroying certain kinds of bacteria.

macroevolution Refers to large-scale evolutionary trends or changes that apply to whole groups of species, often as a result of changing environments or major historical events.

macrophage (MAK-roh-fayj) Large phagocytic white blood cell derived from a monocyte that engulfs damaged tissue cells, bacteria, and other foreign debris. Macrophages are important as antigen-presenters to T cells and B cells in the immune response.

malignant tumor A mass of cells that has metastasized to become invasive cancer.

malnutrition A nutritional condition in which human development and function are compromised by an unbalanced or insufficient diet. Malnutrition can be caused by overnutrition or undernutrition.

mammal Vertebrates that have hair during all or part of their lives, and mammary glands in the female that produce milk.

mammary gland Milk-producing gland of the breast.

mass extinction Catastrophic event that destroys more than 50% of the species existing at the time. The fossil record indicates the Earth has experienced at least five mass extinctions.

mast cell Immune cell that detects foreign substances in the tissue spaces and releases histamine to initiate a local inflammatory response against them. Mast cells typically are clustered deep in an epithelium or along blood vessels.

matter The material substance, composed of elements, that makes up the natural world. Matter takes up space and has weight.

mechanoreceptor (meh-KAN-oh-ree-sep-tur) Receptor sensitive to mechanical pressure or distortion, such as that produced by touch, sound, or muscle contractions.

medulla (meh-DUL-uh) General term for the inner zone of an organ. In the kidney, an inner pyramid-shaped zone of dense tissue that contains the loops of Henle. The term is also used for an area of the brain (medulla oblongata) and for the inner zone of the adrenal glands.

medulla oblongata (meh-DUL-uh ahb-long-GAH-tah) Portion of the brain stem. Contains reflex centers for vital functions such as respiration and blood circulation.

meiosis (my-OH-sis) Nuclear division process that reduces the chromosomal number by half. Meiosis results in the formation of four haploid (n) cells. It occurs only in certain reproductive organs.

melanocyte (meh-LAN-oh-syt) Type of cell located near the base of the epidermis. Melanocytes produce a dark-brown pigment called melanin that is largely responsible for skin color.

melanoma (mel-ah-NO-mah) Least common but most dangerous form of skin cancer.

melatonin (mel-ah-TON-in) A hormone secreted by the pineal gland that inhibits secretion of gonadotropin releasing hormone by the hypothalamus.

memory Storage of information for later retrieval as needed.

memory cell Members of T cell and B cell clones that provide for immunologic memory.

meninges (meh-NIN-jeez) Protective coverings of the central nervous system.

meningitis (meh-nin-JY-tis) Inflammation of the meninges.

menopause (MEN-oh-pawz) Period of a woman's life when, prompted by hormonal changes, ovulation and menstruation cease.

menstrual cycle (MEN-stroo-ul) Pattern of changes in the ovaries and uterus. The menstrual cycle lasts about 28 days and is controlled by hormones.

menstruation (men-stroo-AY-shun) Process in which the endometrial lining disintegrates, its small blood vessels rupture, and the tissue and blood are shed through the vagina.

mesoderm (MEEZ-oh-durm) Primary germ layer that forms the skeleton and muscles of the body.

messenger RNA (mRNA) Long nucleotide strand that complements the exact nucleotide sequence of genetically active DNA. mRNA carries the genetic message to the cytoplasm.

metabolism Sum total of the chemical reactions occurring in body cells.

metaphase Second stage of mitosis, during which the chromosomes align themselves on one plane at the center of the cell.

metastasize (meh-TAS-tuh-syz) Process by which cancer spreads from one body part or organ to another.

MHC (major histocompatibility complex) proteins A group of proteins on the cell surface that identify an individual's cells as "self" and normally signal the immune system to bypass the individual's cells.

microevolution Refers to evolution as a result of genetic changes that give rise to new species.

microvilli (sing. microvillus) Tiny projections on the surface of some epithelial cells that function to increase a cell's surface area. Microvilli are often found on the luminal surface of cells involved in absorption of fluids, nutrients, and ions.

midbrain Region of the brain stem. The midbrain is a coordination center for reflex responses to visual and auditory stimuli.

middle ear Region of the ear consisting of an air-filled chamber within the temporal bone of the skull, bridged by the malleus, incus, and stapes. The middle ear amplifies sound.

milk ejection Expulsion of milk from the breast during suckling.

mineral An inorganic chemical compound found in nature. Some minerals must be in the diet because they are necessary for normal metabolic functioning.

mineralocorticoid (min-er-al-oh-KOR-tih-koyd) Steroid hormone of the adrenal cortex that regulates salt and fluid balance.

miscarriage Spontaneous termination of pregnancy followed by expulsion of the embryo.

mitochondrion (pl. mitochondria) (my-toh-KAHN-dree-ahn) A cytoplasmic organelle responsible for ATP generation for cellular activities.

mitosis (my-TOH-sis) Process of nuclear division during which duplicated chromosomes are distributed to two daughter nuclei, each with the same number of chromosomes as the parent cell. Mitosis consists of prophase, metaphase, anaphase, and telophase.

mitral valve (MY-trul) Left atrioventricular valve in the heart, also called the bicuspid valve.

molecule Particle consisting of two or more atoms joined together by chemical bonds.

monoclonal antibodies Pure preparations of identical antibodies produced in the laboratory by a colony of genetically identical cells.

monocyte (MAHN-oh-syt) Large agranular leukocyte that can differentiate into a macrophage that functions as a phagocyte.

mononucleosis (mahn-oh-nu-klee-OH-sis) A contagious infection of lymphocytes in blood and lymph tissues, caused by the Epstein-Barr virus.

monosaccharide (mahn-oh-SAK-uh-ryd) Simplest type of carbohydrate with only one sugar unit. Glucose is a monosaccharide.

more industrialized countries (MIC) Nations with established industry-based economies.

morphogenesis (mor-foh-JEN-ih-sis) Process involving dramatic changes in shape and form that an organism goes through during development.

motor neuron A neuron in the peripheral nervous system that conducts nerve impulses from the central nervous system to body tissues and organs.

motor unit A somatic motor neuron and all the muscle cells it stimulates.

MRI (magnetic resonance imaging) Technique that uses short bursts of a powerful magnetic field to produce cross-sectional images of body structures.

mRNA (messenger RNA) Long nucleotide strand that complements the exact nucleotide sequence of genetically active DNA. mRNA carries the genetic message to the cytoplasm.

mucosa (myoo-KO-suh) The innermost tissue layer in the wall of the gastrointestinal tract.

mucus (MYOO-kus) A sticky, thick fluid secreted by mucous glands and mucous membranes. Mucus keeps the surfaces of certain membranes moist.

multiple sclerosis (skler-OH-sis) Progressive disorder of the central nervous system in which the myelin sheath around neuron axons disappears and is replaced by fibrous connective tissue.

muscle spindle Encapsulated receptor found in skeletal muscle that is sensitive to stretch.

muscle tension The force exerted by a contracting muscle.

muscle tissue Tissue consisting of cells that are specialized to shorten or contract, resulting in movement of some kind. There are three types of muscle tissue: skeletal, cardiac, and smooth.

muscular dystrophy (DIS-trof-ee) A group of inherited muscle-destroying diseases.

mutation A change in the DNA base pair sequence of a cell.

myelin sheath (MY-ih-lin) Fatty insulating sheath that surrounds many neuron axons.

myocardial infarction (my-oh-KAR-dee-ul in-FAHRK-shun) Condition characterized by death of heart tissue caused by inadequate oxygen supply; also called a heart attack.

myocardium (my-oh-KAR-dee-um) Middle and largest layer of the heart wall, composed of cardiac muscle.

myopia (my-OH-pee-uh) A condition in which the eyeball is too long, causing distant visual objects to be focused in front of the retina. Also called nearsightedness.

myosin (MY-oh-sin) One of the principal contractile proteins found in muscle. Myosin is composed of thick filaments with cross-bridges.

NAD^+ A transport molecule within mitochondria that can accept hydrogen ions and electrons liberated by the Krebs cycle, forming high-energy FADH.

nasal cavity An internal space directly behind the external nose.

natural killer (NK) cells A type of lymphocyte that can lyse and kill cancer cells and virus-infected body cells before the immune system is activated.

natural selection Darwinian principle stating that individuals with certain traits that make them more fit for their local environments are naturally more likely to survive and reproduce. Most changes in allele frequency in a population are the result of mutations coupled with natural selection.

nearsightedness A condition in which the eyeball is too long, causing distant visual objects to be focused in front of the retina; also called myopia.

negative feedback A homeostatic control mechanism in which a change in a controlled variable triggers a series of events that ultimately opposes ("negates") the initial change, returning the controlled variable to its normal value, or set point.

neonate (NEE-oh-nayt) A newborn baby during its first 28 days after birth.

neoplasm (NEE-oh-plazm) Abnormal mass of proliferating cells; also called a tumor. Neoplasms may be benign or malignant.

nephron (NEF-rahn) Structural and functional unit of the kidney. A nephron consists of the renal tubule and the blood vessels that supply it.

nerve Cable-like bundle of many neuron axons, all wrapped together in a protective connective tissue sheath. Nerves carry nerve impulses to and from the central nervous system.

nerve impulse A self-propagating wave of depolarization, also called an action potential.

nerve tract A bundle of nerve axons in the central nervous system.

nervous tissue Tissue consisting of cells that are specialized for generating and transmitting nerve impulses throughout the body. Nervous tissue forms a rapid communication network for the body.

neuroendocrine cell A type of cell in the hypothalamus that essentially functions as both a nerve cell and an endocrine cell because it generates nerve impulses and releases a hormone into blood vessels.

neuroglial cells (noo-RAHG-lee-ul) Cells that provide physical support and protection to neurons and help maintain healthy concentrations of important chemicals in the fluid surrounding them.

neuromuscular junction Region where a motor neuron comes into close contact with a muscle cell.

neuron (NYOOR-ahn, NOOR-) Cell of the nervous system specialized to generate and transmit nerve impulses.

neurotransmitter A chemical released by a neuron that may stimulate or inhibit other neurons or effector cells.

neutron Uncharged subatomic particle found in the atomic nucleus.

neutrophil (NOO-truh-fil) Most abundant type of white blood cell. Neutrophils respond rapidly during an infection, surrounding and engulfing bacteria.

niche The role of an organism in its community.

nitrification The formation of nitrate (NO_3) by any of several means, including nitrifying bacteria, lightning, and nitrate-producing fertilizer factories.

nitrogen fixation The process whereby certain bacteria convert N_2 gas to ammonia (NH_3). Nitrogen fixation is an important part of the nitrogen cycle.

node of Ranvier (RAHN-vee-ay) A short unmyelinated gap between adjacent Schwann cells where the surface of a neuron axon is exposed.

nondisjunction Failure of sister chromatids to separate during mitosis or failure of homologous pairs to separate during meiosis. Nondisjunction causes abnormal numbers of chromosomes in the resulting daughter cells.

nonsteroid hormone A hormone that consists of, or at least is partly derived from, the amino acid building blocks of proteins. Nonsteroid hormones generally are lipid-insoluble, so they tend to act by binding to a receptor on the cell surface rather than by entering the cell cytoplasm.

norepinephrine (NOR-ep-ih-NEF-rin) A neurotransmitter and adrenal medullary hormone, associated with sympathetic nervous system activation.

nucleolus (pl. nucleoli) (noo-KLEE-oh-lus) A dense spherical body in the cell nucleus involved with ribosomal subunit synthesis and storage.

nucleotide A structural unit consisting of a nitrogen base, a five-carbon sugar, and one or more phosphate groups. Nucleotides are the structural units of DNA and RNA. ATP and ADP are also nucleotides.

nucleus General term for a central or essential part. The nucleus of an atom consists of neutrons and protons and contains most of the atomic mass. The nucleus of a eukaryotic cell contains the cell's DNA.

nutrient A chemical substance taken in via the diet that is used for energy and for cell growth and reproduction.

obesity The condition of weighing more than 20% above ideal body weight.

olfactory receptor A modified sensory neuron, located in the upper part of the nasal passages, that detects odors.

oligosaccharide A short string of monosaccharides (single sugars) linked together by dehydration synthesis.

omnivore (AHM-nih-vor) An organism that can derive energy from either plants or animals. Humans, pigs, and bears are omnivores.

oncogene (AHNG-koh-jeen) A mutated or damaged gene that contributes to cancer.

optic disk The area of the retina where the axons of the optic nerve exit the eye. The optic disk has no photoreceptors, so its presence creates a "blind spot" in each eye.

optic nerve One of the two cranial nerves that transmit nerve impulses from the retina to the brain.

organ A part of the body formed of two or more tissues and adapted to carry out a specific function. The stomach and the heart are examples of organs.

organ of Corti Collective term for the hair cells and tectorial membrane of the inner ear. The organ of Corti converts pressure waves to nerve impulses.

organelle (or-guh-NEL) One of many small cellular structures that perform specific functions for the cell as a whole.

organic molecule A molecule that contains carbon and other elements, held together by covalent bonds.

organ system A group of organs that work together to perform a vital body function. The digestive system is an example of an organ system.

orgasm (OR-gazm) A brief, intensely pleasurable reflex event consisting of rhythmic, involuntary muscular contractions.

origin The end of a muscle that is attached to the bone that does not move during muscular contraction.

osmosis (ahz-MO-sis) The net diffusion of water across a selectively permeable membrane, such as a cell membrane.

osteoblast (AHS-tee-oh-blast) A bone-forming cell.

osteoclast (AHS-tee-oh-klast) A cell that resorbs or breaks down bone.

osteocyte (AHS-tee-oh-syt) A mature bone cell.

osteodystrophy (AHS-tee-oh-DIS-troh-fee) Progressive disorder in which the bones become weak because of an abnormal mineral composition of bone.

osteoporosis (AHS-tee-oh-poh-ROH-sis) Progressive disorder involving increased softening and thinning of the bone. Osteoporosis results from an imbalance in the rates of bone resorption and bone formation.

otoliths (OH-toh-lith) Hard crystals of bone-like material embedded in gel in the inner ear. Otoliths contribute to sensations of head position and movement.

outer ear Region of the ear consisting of the pinna (visible portion) and the auditory canal. The outer ear channels sound waves to the eardrum.

ovarian cycle (oh-VAIR-ee-un) Monthly cycle of follicle development, ovulation, and corpus luteum formation in an ovary.

ovaries (OH-vair-eez) The two female sex organs (gonads) in which ova (eggs) are produced.

oviduct (OH-vih-dukt) Tube that leads from an ovary to the uterus, also called the Fallopian tube or uterine tube.

ovulation (AHV-yoo-LAY-shun) Ejection of an immature egg (oocyte) from the ovary.

ovum (pl. ova) (OH-vum) Female gamete, also called an egg.

oxygen debt The amount of oxygen required after exercise to oxidize the lactic acid formed by anaerobic metabolism during exercise.

oxyhemoglobin (ahk-see-HEE-moh-gloh-bin) Oxygen-bound form of hemoglobin.

oxytocin (ahk-sih-TOH-sin) Hormone secreted from the posterior pituitary by neuroendocrine cells of the hypothalamus. Oxytocin stimulates contraction of the uterus during childbirth and the ejection of milk during nursing.

Pacinian corpuscle (puh-SIN-ee-un) Type of encapsulated mechanoreceptor located in the dermis that responds to deep pressure or high-frequency vibration.

pancreas (PANG-kree-us) Organ located behind the stomach that secretes digestive enzymes and bicarbonate into the small intestine and the hormones insulin and glucagon into the bloodstream.

pancreatic amylase (pang-kree-AT-ik AM-ih-lays) Pancreatic enzyme that continues the digestion of carbohydrates begun by salivary amylase.

parasympathetic division (par-uh-sym-puh-THET-ik) The division of the autonomic nervous system that generally promotes activities associated with the resting state, such as digestion and elimination.

parathyroid glands (par-uh-THY-royd) Small endocrine glands located on the thyroid gland that produce parathyroid hormone.

parathyroid hormone (PTH) Hormone released by the parathyroid glands that regulates blood calcium levels.

Parkinson's disease Progressive disorder of the nervous system involving death and degeneration of dopamine-containing nerve cells in the midbrain. Symptoms of Parkinson's disease include persistent tremor and rigid movements.

passive transport Membrane transport processes that do not require cellular energy. Diffusion is a form of passive transport.

pathogen (PATH-uh-jen) Disease-causing microorganism, such as a bacterium or virus.

pectoral girdle (PEK-tur-ul) Portion of the skeleton that attaches the upper limbs to the axial skeleton, composed of the clavicle and scapula.

pelvic girdle Portion of the skeleton that supports the weight of the upper body and attaches the lower limbs to the axial skeleton, composed of the two coxal bones and the sacrum and coccyx of the vertebral column.

pelvic inflammatory disease (PID) Infection of the internal female reproductive organs. PID can result in scarring that blocks the oviducts.

penis Male organ of sexual intercourse and urination.

pepsin (PEP-sin) Enzyme secreted by the gastric glands of the stomach that digests proteins.

peptic ulcer An open, sometimes bleeding sore that forms in the inner lining of the stomach, the esophagus, or the upper part of the small intestine.

perception The process or act of becoming aware of a sensation and attaching meaning to it.

perforin (PUR-for-in) A chemical released by secretory vesicles of a cytotoxic T cell that perforates the membrane of an abnormal or foreign cell and kills it.

pericardium (pair-ih-KAR-dee-um) Tough fibrous sac that encloses and protects the heart and prevents it from overfilling.

periodontitis (pair-ee-oh-dahn-TY-tus) Inflammation of the periodontal membrane lining the tooth socket.

peripheral nervous system Portion of the nervous system that lies outside of the brain and spinal cord.

peristalsis (pair-ih-STAHL-sis) Progressive wavelike contractions of muscle in a tubular structure. Peristalsis moves food through the digestive tract and urine through the ureters.

peritubular capillariesv(pair-ih-TOO-byoo-lur) Capillaries that form a network surrounding the proximal and distal tubules in the cortex of a nephron.

peroxisome (per-OX-ih-sohm) Membrane-bound vesicle in the cell cytoplasm containing powerful enzymes that detoxify harmful or toxic substances.

PET (positron emission tomography) Technique that employs radioactive substances to create three-dimensional images showing the metabolic activity of body structures.

pH The measure of the relative acidity or alkalinity of a solution. Any pH below 7 is acidic and any pH above 7 is basic.

phagocyte (FAG-uh-syt) A white blood cell that destroys foreign cells through the process of phagocytosis.

phagocytosis (fag-uh-sy-TOH-sis) Process by which phagocytes surround, engulf, and destroy foreign cells.

pharynx (FAIR-inks) The region of the digestive and respiratory systems that extends from behind the nasal cavities to the esophagus, also called the throat. Both air and food pass through the pharynx.

phenotype (FEE-noh-typ) The observable physical and functional traits of an organism. Phenotype is determined by an organism's genotype and by environmental influences.

phospholipid A modified lipid that has a polar (water-soluble) region containing a phosphate group. Phospholipids comprise the main structural components of the cell membrane.

photopigment Any one of several proteins located in the rods and cones of the eye that changes its shape when exposed to energy in the form of light. The change in shape causes the rod or cone to alter the amount of neurotransmitter it releases.

photoreceptor A specialized receptor cell that responds to light energy.

photosynthesis The process by which plants capture the energy in sunlight and convert it into chemical energy for their own use.

pineal gland (PY-nee-ul) Endocrine gland thought to be involved in setting the biological clock and influencing reproductive function. The pineal gland secretes melatonin.

pituitary gland (pih-TOO-ih-tair-ee) Neuroendocrine gland located just below the hypothalamus of the brain. Together, the anterior and posterior lobes of the pituitary gland secrete eight hormones.

pituitary portal system A system of small blood vessels between the hypothalamus and anterior lobe of the pituitary gland.

placebo (pluh-SEE-bow) A false treatment in a controlled experiment, given or performed to minimize the possibility of bias by suggestion.

placenta (plah-SEN-tuh) Temporary organ formed from both fetal and maternal tissues that provides nutrients and oxygen to the developing fetus, removes fetal metabolic wastes, and produces the hormones of pregnancy.

plasma (PLAZ-muh) The fluid component of blood in which the formed elements are suspended.

plasma cell A cell derived from a B cell lymphocyte specifically to mass-produce and release antibodies.

plasma membrane Membrane surrounding the cell, consisting of a phospholipid bilayer with embedded cholesterol and proteins. The plasma membrane regulates the passage of substances into and out of the cell.

plasmid (PLAZ-mid) A small circular molecule of DNA found in the cytoplasm of some bacteria. Plasmids are used in recombinant DNA technology.

platelet (PLAYT-let) Small cell fragments that are derived from certain cells in the bone marrow. Platelets are important in blood clotting.

pleural membranes (PLUR-ul) Membranes that line the thoracic cavity and cover the external surface of the lungs.

pneumonia (noo-MOHN-yuh) Respiratory condition involving inflammation and infection of the lungs. Pneumonia is usually caused by viruses or bacteria.

polar body Small nonfunctional cell with almost no cytoplasm that is formed when the primary oocyte completes stage I of meiosis.

pollutant A chemical in the environment that has adverse effects on living organisms.

polygenic inheritance (pahl-ee-JEN-ik) Pattern of inheritance in which a phenotypic trait depends on many genes. An example is eye color, controlled by three or more genes.

polymerase chain reaction (PCR) (poh-LIM-ur-ays) A technique for making multiple identical copies of DNA in a test tube.

polypeptide A chain of 3–100 amino acids (*poly* means "many"). Chains of more than 100 amino acids are generally referred to as proteins.

polysaccharide A chain of monosaccharides. Starch and glycogen are polysaccharides.

population A group of individuals of the same species that occupy the same geographic area and interact with each other.

positive feedback A control system in which a change in the controlled variable sets in motion a series of events that amplifies the original change. The process of childbirth once labor has started is an example of positive feedback.

potential energy Stored or inactive energy.

precapillary sphincter (SFINK-ter) Band of smooth muscle that controls blood flow into individual capillaries.

pre-embryo An organism during the first two weeks after fertilization. The term pre-embryo is used because many of the cells that result from cleavage and morphogenesis in the first two weeks will eventually not be part of the embryo at all, but will instead constitute extra-embryonic membranes and parts of the placenta.

premenstrual syndrome (PMS) Recurring episodes of discomfort during the menstrual cycle, generally beginning after ovulation and lasting until menstruation. PMS can include symptoms such as food cravings, mood swings, anxiety, back and joint pain, water retention, and headaches.

presbyopia (prez-bee-OH-pee-uh) Eye condition in which an increase in stiffness of the lens results in an inability to focus on near objects. Presbyopia typically begins after age 40.

primary motor area Region of the frontal lobe of the brain that initiates motor activity.

primary somatosensory area Region in the parietal lobe of the brain that receives sensory input from the skin.

primate (PRY-mayt) An order of mammals with five digits on their hands, fairly flat fingernails and toenails rather than hooves or claws, and forward-facing eyes adapted for stereoscopic vision. Primates include lemurs, monkeys, apes, and humans.

producer An organism that is capable of photosynthesis. Plants and algae are producers.

progesterone (proh-JES-tur-ohn) Hormone partly responsible for preparing the uterus for the fertilized ovum. Progesterone is secreted by the corpus luteum of the ovary and by the placenta.

prokaryote (proh-KAIR-ee-oht) A single-celled organism that lacks the nucleus and the membrane-bound organelles characteristic of eukaryotes. Bacteria are prokaryotes.

prolactin (proh-LAK-tin) Hormone secreted by the anterior pituitary gland that stimulates the mammary glands to produce milk.

promoter A unique base sequence that marks the beginning of a gene.

prophase The first stage of mitosis. During prophase the chromosomes condense and thicken, the pairs of centrioles migrate to opposite sides of the cell, and the mitotic spindle forms.

prostaglandins (prahs-tuh-GLAN-dins) A group of lipid-based chemicals synthesized by most tissue cells that act as local messengers. Some prostaglandins dilate blood vessels and others constrict them.

prostate gland (PRAHS-tayt) Accessory reproductive gland in males that produces approximately one-third of the semen volume, including fluids that activate sperm.

protein One or more polypeptide chains of more than 100 amino acids. The three-dimensional structure of a protein is determined by the sequence of amino acids and by hydrogen bonds between different regions of the protein.

proteinuria (proh-tee-NYUR-ee-uh) Appearance of protein in urine. Persistent proteinuria may indicate glomerular damage.

prothrombin (proh-THRAHM-bin) A plasma protein that is converted to thrombin as part of the process of blood clot formation.

prothrombin activator Substance released by blood vessels and nearby platelets when blood vessels are damaged. Prothrombin activator activates the conversion of prothrombin into thrombin, facilitating the process of blood clotting.

proton Positively charged subatomic particle in the nucleus of an atom.

proto-oncogene (proh-toh-AHNG-koh-jeen) Regulatory gene that promotes cell growth, differentiation, division, or adhesion. A proto-oncogene may contribute to cancer if it becomes mutated or damaged in such a way that it is turned on all the time.

proximal tubule Segment of a nephron that starts at the glomerular capsule and extends to the renal medulla.

PTH (parathyroid hormone) Hormone released by the parathyroid gland that regulates blood calcium levels.

puberty (PYOO-bur-tee) Period of life when reproductive maturity is achieved.

pulmonary circuit That part of the vascular system that takes deoxygenated blood to the lungs and returns oxygenated blood to the heart.

pulmonary embolism (EM-buh-liz-um) Condition that occurs when an embolus (blood clot) blocks an artery supplying blood to the lungs. A pulmonary embolism can cause sudden chest pain, shortness of breath, and even sudden death.

pulse Rhythmic expansion and recoil of arteries resulting from heart contraction that can be felt from the surface of the body.

Punnett square (PUN-et) A grid used for predicting patterns of inheritance and the probability that a particular genotype will be inherited.

pupil (PYOO-pul) Opening in the center of the iris through which light enters the eye.

Purkinje fibers (pur-KIN-jee) Modified cardiac muscle fibers that are part of the electrical conduction system of the heart.

pyloric sphincter (py-LOHR-ik SFINK-tur) A thickening of the circular layer of muscle at the distal end of the stomach that controls the rate at which the stomach empties.

pyrogen (PY-roh-jen) A fever-inducing agent. Macrophages and certain bacteria and viruses release pyrogens.

radiation High-energy waves or particles emitted by radioactive isotopes.

recessive allele An allele that does not manifest itself in the presence of a more dominant allele. Two recessive alleles must be present in order for the recessive phenotypic trait to be expressed.

recombinant DNA technology Field of applied science that explores applications of cutting, splicing, and creating DNA.

red blood cell (erythrocyte) Blood cell that transports oxygen from the lungs to the tissues and carbon dioxide from the tissues to the lungs.

reflex An involuntary, automatic response to a stimulus.

refractory period (rih-FRAK-tuh-ree) Period immediately following a nerve impulse during which a neuron is unable to conduct another nerve impulse.

regulatory gene A gene that encodes for a repressor or promoter regulatory protein.

REM (rapid eye movement) sleep Stage of sleep in which rapid eye movements occur. REM sleep is accompanied by an alert EEG pattern and dreaming.

renin (REE-nin) An enzyme secreted into the bloodstream by the kidneys that leads to an increase in blood pressure and the secretion of aldosterone.

replacement fertility rate The fertility rate required to achieve long-term zero population growth. Replacement fertility rate is estimated to be an average of 2.1 children per woman.

replication The process of copying DNA prior to cell division.

repolarization In a nerve or muscle cell, return of the membrane potential to the initial resting (polarized) state after depolarization.

reproductive system Organ system that functions to produce offspring.

residual volume The amount of air remaining in the lungs even after a forceful, maximal expiration. Normal residual volume is approximately 1200 ml.

respiratory system Organ system responsible for gas exchange. The respiratory system includes the nose, pharynx, larynx, trachea, bronchi, and lungs.

resting potential The slight difference in voltage (electrical potential) between the inside and outside of a cell.

restriction enzyme An enzyme found in bacteria that can be used to cut DNA at specific nucleotide sequences called restriction sites.

reticular activating system (RAS) (reh-TIK-yoo-lur) A group of neurons in the reticular formation that control levels of sleep and wakefulness.

reticular fibers Thin, interconnecting fibers of collagen found in fibrous connective tissue. Reticular fibers provide an internal structural framework for soft organs such as the liver, spleen, and lymph nodes.

reticular formation Collective term for neurons in the pons that work with the cerebellum to coordinate the skeletal muscle activity required to maintain posture, balance, and muscle tone. The reticular formation is also responsible for maintaining the level of wakefulness.

retina (RET-ih-nuh) Layers of tissue at the back and sides of the eye. The retina is composed primarily of photoreceptor cells, neurons, and a few blood vessels.

retrovirus (REHT-roh-vy-rus) A class of viruses that replicate by using their RNA to produce double-stranded DNA that is inserted into the host cell's DNA. The host cell then produces RNA for new viruses. HIV is a retrovirus.

Rh factor A red blood cell surface antigen, first discovered in Rhesus monkeys, that is a crucial consideration in blood transfusions.

rheumatoid arthritis (ROO-muh-toyd) A type of chronic arthritis involving inflammation of the synovial membrane that lines certain joints.

rhodopsin (rho-DOP-sin) The photopigment found in the rods of the eye.

rhythm method Contraceptive technique based on the limited time available for fertilization in each menstrual cycle. The rhythm method requires avoiding intercourse from about five days before ovulation to three days after ovulation.

ribonucleic acid (RNA) Nucleic acid that contains ribose and the bases A, G, C, and U. RNA carries out DNA's instructions for protein synthesis. Types of RNA include mRNA, rRNA, and tRNA.

ribosomal RNA (rRNA) The RNA component of a ribosome. Ribosomal RNA assists in protein synthesis.

ribosome (RY-boh-sohm) A cellular structure consisting of rRNA and protein at which amino acids are assembled into proteins. Some ribosomes float freely within the cytoplasm; others are attached to the endoplasmic reticulum.

rigor mortis (RIG-ur MOR-tis) Period of time after death during which a body becomes stiff because it lacks the ATP required for muscle relaxation. Rigor mortis usually begins a few hours after death and ends several days later, when decomposition begins.

RNA (ribonucleic acid) Nucleic acid that contains ribose and the bases A, G, C, and U. RNA carries out DNA's instructions for protein synthesis. Types of RNA include mRNA, rRNA, and tRNA.

RNA polymerase (poh-LIM-ur-ays) One of several enzymes that recognizes a promoter (the starting point of transcription). RNA polymerase attaches to the promoter, starts the DNA unwinding process, and assists in attaching the appropriate RNA segments to the growing chain.

rod cell One of the two types of photosensitive cells in the retina.

rRNA (ribosomal RNA) The RNA component of a ribosome. Ribosomal RNA assists in protein synthesis.

SA (sinoatrial) node (sy-noh-AY-tree-ul) Group of specialized myocardial cells in the wall of the right atrium that initiates the heartbeat. The SA node is called the pacemaker of the heart.

salivary amylase (SAL-ih-vair-ee AM-uh-layz) Enzyme in saliva that begins the process of digesting starch.

salivary glands Three pairs of glands that produce saliva to begin the process of digestion. The three pairs of salivary glands are the parotid, submandibular, and sublingual glands.

saltatory conduction Form of rapid transmission of a nerve impulse along a myelinated fiber, in which the nerve impulse leaps from node to node.

sarcomere (SAR-koh-mir) The smallest contractile unit of a muscle myofibril. A sarcomere extends from one Z-line to the next.

sarcoplasmic reticulum (sar-koh-PLAZ-mik (reh-TIK-yuh-lum) Specialized endoplasmic reticulum of muscle cells that surrounds the myofibrils and stores the calcium needed for the initiation of muscle contraction.

saturated fat Type of fat with two hydrogen atoms for every carbon atom in the fatty acid tails. Saturated fats are found in meat and dairy products and in a few plant sources such as coconut and palm kernel oil.

scanning electron microscope (SEM) Microscope that bombards an object with beams of electrons to reveal what appears to be a three-dimensional view of the object's surface.

Schwann cell A type of supporting cell in the peripheral nervous system. Schwann cells form the myelin sheaths around myelinated neurons and are vital to peripheral nerve fiber regeneration.

science The study of the natural world.

scientific method The process of science; the way scientific knowledge is acquired.

sclera (SKLEAR-uh) White, opaque outer layer of the eyeball.

scrotum (SKROH-tum) External sac of skin enclosing the testes.

second messenger Any intracellular messenger molecule generated by the binding of a chemical messenger (hormone or neurotransmitter) to a plasma membrane receptor on a cell's outer surface. A second messenger mediates intracellular responses to an extracellular (first) messenger.

secondary oocyte The haploid female reproductive cell that is released from an ovary and begins traveling down the oviduct during the ovarian cycle. A secondary oocyte has completed stage I of meiosis but not stage II.

secretin (seh-KREE-tin) Hormone secreted by the duodenum into the bloodstream that stimulates the pancreas to secrete water and bicarbonate to neutralize acid.

secretion General term for the movement of a substance out of a cell of an endocrine or exocrine gland. Also, the substance secreted. The term is also used to denote the movement of any substance into the lumen of a tubule in the kidney.

segmentation Type of gastrointestinal motility in which short sections of smooth muscle contract and relax in seemingly random fashion, mixing the contents of the intestinal lumen.

segregation Mendelian principle stating that diploid organisms inherit two genes for each trait (on a pair of homologous chromosomes), and that during meiosis the two genes are distributed to different gametes.

selective gene expression The principle that, in any cell at any time, only a few of the genes are actually being expressed. Selective gene expression is responsible for differences in structure and function among different cells and for changes in cell function at different times.

selectively permeable The quality of allowing certain substances to pass while restricting the movement of others. The plasma membrane of a cell is selectively permeable.

semen (SEE-men) Thick, whitish fluid mixture containing sperm and the secretions of the male accessory reproductive glands.

semicircular canals Three fluid-filled canals in the vestibular apparatus of the inner ear that are important in sensing rotational movements of the head.

semilunar valves (sem-ee-LOO-nur) Heart valves located between the ventricles and the two major arteries of the body (pulmonary and aortic) that prevent blood from returning to the ventricles after contraction.

seminal vesicles (SEM-ih-nul) Male accessory reproductive structures that add fructose and prostaglandins to semen. Fructose serves as an energy source for sperm, and prostaglandins may induce muscle contractions in the female reproductive tract to help sperm travel more effectively.

seminiferous tubules (sem-ih-NIF-ur-us) Highly convoluted tubes within the testes in which sperm are formed.

senescence (seh-NES-ens) The progressive deterioration of multiple organs and organ systems over time; old age.

sensation The conscious awareness of a stimulus.

sensory neuron A neuron of the peripheral nervous system that is specialized to respond to a certain type of sensory stimulus, such as pressure or light. Sensory neurons conduct information about a stimulus to the central nervous system in the form of nerve impulses.

septicemia (sep-tih-SEE-mee-uh) Systemic disease caused by the spread of microorganisms or their toxins in blood. Also called blood poisoning or toxemia.

septum (SEP-tum) The muscular partition that separates the right and left sides of the heart.

sex chromosomes The chromosomes, X and Y, that determine gender (XX = female, XY = male). The 23rd pair of chromosomes.

sex-influenced trait A trait not inherited with a sex gene but influenced by the actions of the sex genes. An example is the phenotypic trait of male baldness.

sex-linked trait A trait that depends on genes located on the sex chromosomes. Sex-linked inheritance is X-linked if the gene is located only on the X-chromosome, and Y-linked if it is located only on the Y-chromosome.

sickle-cell anemia An inherited disorder in which the red blood cells assume a sickle shape when the oxygen concentration is low.

sickle-cell trait A heterozygous condition for the sickle-cell allele, designated Hb^A-Hb^S. Unlike people with sickle-cell anemia, those with sickle-cell trait can generally lead normal lives because their condition is intermediate in terms of severity.

sinoatrial (SA) node (sy-noh-AY-tree-ul) Group of specialized myocardial cells in the wall of the right atrium that initiates the heartbeat. The SA node is called the pacemaker of the heart.

sinus Mucous-membrane–lined, air-filled cavity in certain cranial bones.

sister chromatids The two identical chromosomes that remain attached at the centromere after replication. The centromere holds the sister chromatids together until they are physically pulled apart during the mitotic phase.

skeletal muscle Muscle composed of cylindrical multinucleate cells with obvious striations. Skeletal muscles attach to the body's skeleton.

skepticism A questioning attitude. Good scientists combine creativity and imagination with skepticism.

sliding filament mechanism The mechanism of muscle contraction. Muscles contract when the thick and thin filaments slide past each other and sarcomeres shorten.

smog Hazy brown or gray layer of atmospheric pollution that tends to hover over the region where it is produced. The burning of fossil fuels (coal and oil) and automobile exhaust are the primary culprits.

smooth muscle Spindle-shaped muscle cells with one centrally located nucleus and no externally visible striations. Smooth muscle is found mainly in the walls of hollow organs.

sodium-potassium pump An active transport protein of the plasma membrane that simultaneously transports three Na^+ out of the cell and two K^+ in. The sodium-potassium pump is important for maintaining cell volume and for generating the resting membrane potential.

somatostatin (soh-mah-toh-STAHT-in) Hormone secreted from several locations in the body. In the pancreas, somatostatin inhibits the secretion of glucagon and insulin.

special senses The five senses (taste, smell, hearing, balance, and vision) originating from receptors that are restricted to special areas of the body. The special senses deliver information about the external world to the central nervous system.

species The smallest classification category of life. A species is a group of organisms that under natural conditions tend to breed within that group.

sperm Male reproductive cell (gamete).

sphincter (SFINK-tur) General term for a ring of muscle around a hollow tube or duct that functions to restrict the passage of materials.

sphygmomanometer (SFIG-moh-mah-NOM-uh-tur) Device used to measure blood pressure.

spinal cord The portion of the central nervous system that lies outside the brain, extending from the base of the brain to the third lumbar vertebra. The spinal cord provides a conduction pathway to and from the brain.

spinal nerve Any one of the 31 paired nerves that arise from the spinal cord.

spleen Largest lymphoid organ. The spleen removes old and damaged red blood cells and helps fight infections.

spongy bone Type of bone tissue characterized by thin, hard interconnecting bony elements enclosing hollow spaces. Red blood cells are produced in the spaces between the bony elements.

starch The storage polysaccharide utilized by plants.

stem cells General term for any cells that have not yet differentiated. Stem cells in bone marrow are the source of all blood cells and platelets.

steroids (STEER-oydz) Group of lipids having four interconnected rings that includes cholesterol and several hormones. Steroids are fat soluble molecules.

stimulus An excitant or irritant. A chemical or physical form of energy such as heat, physical pressure, or sound waves that evokes a response in a nerve, muscle, gland, or receptor.

stomach Organ of the gastrointestinal tract where food is initially stored and where chemical breakdown of proteins begins.

stretch reflex Type of spinal reflex, important in maintaining upright posture and coordinating movement. Stretch receptors in a skeletal muscle stimulate sensory nerves when stretched, causing the sensory nerves to transmit nerve impulses to the spinal cord.

stroke Condition in which brain tissue is deprived of a blood supply, most often caused by blockage of a cerebral blood vessel. Also called a cerebrovascular accident.

stroke volume Volume of blood pumped out of a ventricle during one ventricular contraction.

structural gene A gene that encodes for an enzyme or a structural protein.

submucosa The middle layer of the gastrointestinal tract wall. The submucosa is a layer of connective tissue that contains blood vessels, lymph vessels, and nerves.

succession A natural sequence of change in terms of which organisms dominate in a community.

summation Accumulation of effects, especially those of muscular or neural activity.

suppressor T cell Regulatory T lymphocyte that suppresses the immune response.

surfactant (sur-FAK-tant) Secretion produced by certain cells of the alveoli that reduces the surface tension of water molecules, preventing the collapse of the alveoli after each expiration.

sympathetic division The division of the autonomic nervous system that helps the body cope with stressors (danger, excitement, etc.) and with situations requiring high mental or physical activity.

synapse (SIN-aps) Functional junction or point of close contact between two neurons or between a neuron and an effector cell.

synaptic cleft (sih-NAP-tik) Fluid-filled space at a synapse.

synaptic transmission The process of transmitting information from a neuron to its target across a synapse.

syndrome Medical term for a group of symptoms that occur together.

synovial joint (sih-NO-vee-ul) A movable joint having a thin fluid-filled cavity between the bones.

syphilis (SIF-uh-lis) Sexually transmitted disease caused by infection with the bacterium *Treponema pallidum.*

systemic circuit The system of blood vessels that transport blood to all of the cells of the body except those served by the pulmonary circuit.

systole (SIS-toh-lee) Period when either the ventricles or the atria of the heart are contracting.

systolic pressure (sis-TOL-ik) The highest pressure reached in the arterial blood vessels during the cardiac cycle, and the first of two blood pressure numbers recorded by the health care provider. Systolic pressure is achieved during ventricular systole.

T cells (T lymphocytes) Lymphocytes responsible for cell-mediated immunity, which depends on the actions of several types of T cells.

taste bud One of many sensory receptor organs located on the tongue that contains receptor cells responsive to dissolved food chemicals.

taxonomy (tak-SAHN-uh-mee) The science of classifying and naming life-forms.

technology The application of science.

tectorial membrane (tek-TOR-ee-ul) The gelatinous noncellular material in which the mechanoreceptor hair cells of the inner ear are embedded. With the hair cells, the tectorial membrane forms the organ of Corti, which converts pressure waves to nerve impulses.

telomerase (tel-OH-mer-ays) An enzyme produced primarily by cancer cells and reproductive cells that builds new telomeres to replace telomeres lost during cell division.

telomeres (TEEL-oh-meer) Short segments of DNA at the end of each DNA molecule that do not encode for a protein. In most cells, telomeres are removed every time a cell divides until there are no telomeres left.

TEM (transmission electron microscope) Microscope that bombards a very thin sample with a beam of electrons, some of which pass through the sample. A TEM provides images that are two-dimensional but highly magnified.

tendon A cord of dense fibrous connective tissue attaching muscle to bone.

teratogenic (TAIR-uh-toh-JEN-ik) Capable of producing abnormal development of a fetus. Teratogenic substances are one cause of birth defects.

testis (pl. testes) The male primary sex organ (gonad); produces sperm.

testosterone (tes-TAHS-teh-rohn) Male sex hormone produced primarily by the testes. (The adrenal cortex produces a small amount of testosterone in both sexes.) Testosterone promotes the development of secondary sexual characteristics and is necessary for normal sperm production.

tetanus (TET-uh-nus) In a muscle, a sustained maximal muscle contraction resulting from high-frequency stimulation. In medicine, an infectious disease also called lockjaw that is caused by an anaerobic bacterium.

thalamus (THAL-uh-mus) A region of the forebrain that receives and processes sensory information and relays it to the cerebrum.

theory A main hypothesis that has been extensively tested over time and that explains a broad range of scientific facts with a high degree of reliability. An example is the theory of evolution.

thermal inversion A situation in which a warm stagnant upper layer of air traps a cooler air mass beneath it. Thermal inversions can cause air pollutants to accumulate close to the ground by preventing them from dissipating.

thermoreceptor Receptor sensitive to temperature changes.

threshold In an excitable cell such as a nerve or muscle, the membrane voltage that must be reached in order to trigger an action potential, or nerve impulse.

thrombin Enzyme that facilitates the conversion of fibrinogen into long threads of fibrin. Thrombin promotes the process of blood clotting.

thymus gland Endocrine gland that contributes to immune responsiveness. T cells mature in the thymus.

thyroid gland An endocrine gland located in the neck that produces the hormones thyroxine, triiodothyronine, and calcitonin.

thyroid-stimulating hormone (TSH) Hormone produced by the anterior pituitary that regulates secretion of thyroxine and triiodothyronine.

thyroxine (thy-RAHK-sin) A hormone containing four iodine molecules that is secreted by the thyroid gland. Thyroxine and a closely related three-iodine hormone (triiodothyronine) accelerate cellular metabolism in most body tissues.

tidal volume Volume of air inhaled and exhaled in a single breath. Normal tidal volume is approximately 500 milliliters, or about a pint.

tissue A group of similar cells (and their intercellular substance) specialized to perform a specific function. Primary tissue types of the body are epithelial, connective, muscle, and nervous tissue.

tissue rejection Situation in which the body's immune system attacks foreign tissues. Examples of tissue rejection are transfusion reactions and reactions to organ transplants.

tonsillitis Inflammation of the tonsils at the back of the throat.

tonsils Masses of lymphatic tissue near the entrance to the throat. Lymphocytes in the tonsils remove microorganisms that enter in food or in the air.

trachea (TRAY-kee-uh) Tube extending from the larynx to the bronchi that is the passageway for air to the lungs. Also called the windpipe.

transcription The production of a single strand of RNA from a segment (representing a gene) of one of the two strands of DNA. The base sequence of the RNA is complementary to that of the single strand of DNA.

transfer RNA (tRNA) Short-chain RNA molecule that transfers amino acids to the ribosome.

transfusion reaction The adverse reaction that occurs in a recipient when a donor's red blood cells are attacked by the recipient's antibodies. During a transfusion reaction the donated cells may clump together (agglutinate), blocking blood flow in blood vessels.

transgenic organism A living organism that has had foreign genes inserted into it. Transgenic organisms are sometimes created to produce substances useful to humans, including human proteins.

translation The process by which the genetic code of mRNA is used to string together the appropriate amino acids to produce a specific protein.

translocation A change in chromosome location that occurs when a piece of a chromosome breaks off but reattaches at another site, either on the same chromosome or another chromosome. Translocations can result in subtle changes in gene expression and ability to function.

tricuspid valve (try-KUS-pid) The right atrioventricular valve of the heart.

triglycerides (try-GLIS-uh-rydz) Fats and oils composed of fatty acids and glycerol. Triglycerides are the body's most concentrated source of energy fuel.

triiodothyronine (try-i-oh-doh-THY-roh-neen) Hormone secreted by the thyroid gland, similar to thyroxine except that it contains three atoms of iodine instead of four. Like thyroxine, triiodothyronine accelerates cellular metabolism in most body tissues.

triplet code The genetic code of mRNA and tRNA in which three successive bases encode for one of the 20 amino acids.

Trisomy-X syndrome (TRY-soh-mee) A condition caused by having three X chromosomes. Trisomy-X syndrome individuals are female phenotype, sometimes showing a tendency toward mild mental retardation.

tRNA (transfer RNA) Short-chain RNA molecule that transfers amino acids to the ribosome.

trophoblast Outer sphere of cells of the blastocyst.

TSH (thyroid-stimulating hormone) Hormone produced by the anterior pituitary that regulates secretion of thyroxine and triiodothyronine.

tubal ligation Procedure for female sterilization in which each oviduct is cut and tied.

tuberculosis Infectious disease of the lungs caused by the bacterium *Mycobacterium tuberculosis*.

tubular reabsorption The movement of fluid and solutes (primarily nutrients) from the renal tubules into the blood.

tubular secretion The movement of solutes (primarily undesirable substances such as drugs, metabolic wastes, and excess ions) from the blood into the renal tubules.

tumor A mass of cells derived from a single cell that began to divide at an abnormally high rate. The cells of a benign tumor remain at the site of origin, whereas those of a malignant tumor travel to distant sites (metastasize).

twitch A complete cycle of contraction and relaxation in a muscle cell.

ultrasound Fetal imaging technique in which high-frequency (ultrasound) sound waves are aimed at the mother's uterus. The sound waves rebound when they hit hard tissues in the fetus. These reflecting waves are analyzed and converted to an image by a computer.

umbilical cord (um-BIL-uh-kul) Cable-like structure connecting the fetus to the mother, through which arteries and veins pass.

unsaturated fat Type of fat with fewer than two hydrogen atoms for every carbon atom in their fatty acid tails. Examples include plant oils—such as olive, safflower, canola, and corn oil—and fish oil.

urea (yoo-REE-uh) Main nitrogen-containing waste excreted in urine.

ureter (YUR-uh-tur) One of two tubes that transport urine from the kidneys to the urinary bladder.

urethra (yoo-REE-thruh) Tube through which urine passes from the bladder to the outside of the body.

uterine cycle (YOO-tur-in) A series of changes that occur in the uterus in preparation for the arrival of a fertilized egg. A complete cycle generally takes about 28 days.

uterus (YOO-tur-us) Hollow, thick-walled organ that receives, retains, and nourishes the fertilized egg. The uterus is the site of embryonic/fetal development.

vaccine Antigens prepared in such a way that when injected or taken orally they induce active immunity without causing disease.

vagina (vah-JY-nuh) Hollow muscular female organ extending from the cervix to the body exterior. The vagina functions as an organ of sexual intercourse and as the birth canal.

varicose veins (VAIR-ih-kohs) Permanently dilated veins, often caused by improperly functioning venous valves. Varicose veins can appear anywhere but are most common in the legs and feet.

vasectomy (vuh-SEK-tuh-mee) Procedure for male surgical sterilization in which the vas deferens (ductus deferens) is cut and tied in two places.

vasoconstriction A narrowing of blood vessel diameter caused by contraction of the smooth muscle in the vessel wall.

vasodilation A widening of blood vessel diameter caused by relaxation of the smooth muscle in the vessel wall.

vein A thin-walled blood vessel that returns blood toward the heart from the venules.

ventilation The movement of air into and out of the lungs. Also called breathing.

ventricle (VEN-trih-kul) A general term for a cavity. In the heart, one of the two chambers that receives blood from an atrium and pumps it into an artery. In the central nervous system, one of four hollow cavities in the brain.

ventricular fibrillation (ven-TRIK-yoo-lur fib-rih-LAY-shun) A type of abnormal heart rhythm involving rapid irregular ventricular contractions.

venule (VEEN-yool, VEN-yool) A small blood vessel that transports blood from capillaries to a vein.

vertebral column (VUR-tuh-brul) Structure of the axial skeleton formed of a number of individual bones (vertebrae) and two composite bones (sacrum and coccyx). Also called the spine or backbone.

vertebrate (VUR-tuh-brayt) Any animal with a backbone composed of vertebrae.

vesicle (VES-ih-kul) A small membrane-bound fluid-filled sac that encloses and contains certain substances within a cell.

vestibular apparatus (ves-TIB-yoo-lur) A system of fluid-filled canals and chambers in the inner ear. The vestibular apparatus consists of three semicircular canals for sensing rotational movement of the head, and the vestibule for sensing static (nonmoving) position and linear acceleration and deceleration.

vestigial structure (ves-TIJ-ee-ul) A structure that may have had a function in some ancestor but which no longer serves any function. The human coccyx (tailbone) is a vestigial structure.

villus (pl. villi) (VIL-us) Fingerlike projection of the small intestinal mucosa that functions to increase the surface area for absorption.

virus A noncellular infectious agent that can only replicate within a host cell. Viruses consist only of one or more pieces of genetic material (DNA or RNA) surrounded by a protein coat.

vital capacity The maximum volume of air that can be expelled from the lungs by forcible expiration after the deepest inspiration.

vitamin Any one of more than a dozen organic compounds that are required by the body in minute amounts but generally are not synthesized in the body.

vitreous humor (VIT-ree-us) The clear fluid within the main chamber of the eye, between the lens and the retina.

vocal cords Two folds of connective tissue that extend across the airway. Audible sounds are heard when the vocal cords vibrate as air passes by.

vulva Female external genitalia that surround the opening of the vagina.

white blood cell (WBC or leukocyte) Large blood cells that are part of the body's defense system. WBCs are diverse in structure and specific function, but comprise only about 1% of the volume of blood.

withdrawal method Birth control method in which the man withdraws his penis from the vagina immediately before ejaculation. In practice it may be difficult to time withdrawal correctly, and so the method is considered less effective than other contraceptive techniques.

X chromosome One of the two human sex chromosomes. An embryo that inherits two X chromosomes will develop into a female.

Y chromosome One of the two human sex chromosomes. An embryo that inherits a Y chromosome will develop into a male.

yolk sac Extraembryonic membrane that serves as the first source for red blood cells for the fetus and as the source for primordial germ cells.

zero population growth The condition that exists in a population when the birth rate equals the death rate.

ZIFT (zygote intrafallopian transfer) (ZY-goht IN-trah-fuh-LO-pee-un) A technique for enhancing fertility in which an egg is fertilized outside the body, then inserted into an oviduct through a small incision in the abdomen.

zona pellucida (ZO-nuh peh-LOOS-ih-duh) A noncellular coating around an oocyte.

zygote (ZY-goht) The diploid cell formed by the union of the ovum and a sperm. The zygote is the product of fertilization.

ANSWERS TO TEST YOURSELF QUESTIONS

Chapter 1—Human Biology, Science and Society

1. d **2.** f **3.** a **4.** b **5.** c
6. g **7.** e **8.** biology **9.** matter; energy
10. bipedalism **11.** generalize
12. experiments **13.** experimental; control
14. statistics **15.** abscissa **16.** True
17. True **18.** True **19.** False **20.** False

Chapter 2—The Chemistry of Living Things

1. g **2.** d **3.** a **4.** b **5.** j
6. f **7.** c **8.** e **9.** h **10.** i
11. alkaline **12.** polypeptide **13.** shells (or energy levels) **14.** ATP **15.** glycogen
16. enzymes **17.** covalent **18.** kinetic
19. fatty acids **20.** oils **21.** False
22. True **23.** True **24.** False **25.** True

Chapter 3—Structure and Function of Cells

1. e **2.** h **3.** a **4.** j **5.** b **6.** g
7. c **8.** f **9.** d **10.** i **11.** prokaryotes
12. cholesterol **13.** channels **14.** energy
15. hypertonic **16.** endoplasmic reticulum
17. lysosomes **18.** substrate **19.** Krebs cycle **20.** lactic acid **21.** False
22. True **23.** True **24.** True **25.** False

Chapter 4—From Cells to Organ Systems

1. d **2.** f **3.** a **4.** j **5.** b **6.** i
7. c **8.** h **9.** g **10.** e **11.** internal environment **12.** tissues **13.** endocrine
14. fibroblasts **15.** epidermis **16.** cranial
17. sagittal **18.** keratinocytes **19.** control center (or integrating center) **20.** set point
21. True **22.** False **23.** True
24. False **25.** True

Chapter 5—The Skeletal System

1. e **2.** b **3.** g **4.** j **5.** a **6.** c
7. i **8.** d **9.** f **10.** h **11.** epiphysis
12. central canal (or Haversian canal)
13. spongy **14.** flat **15.** chondroblasts
16. axial **17.** pelvic **18.** pectoral
19. hinge **20.** rheumatoid **21.** False
22. True **23.** True **24.** False **25.** True

Chapter 6—The Muscular System

1. g **2.** d **3.** f **4.** a **5.** j **6.** b
7. i **8.** c **9.** e **10.** h
11. involuntary **12.** synergistic
13. sarcomere **14.** acetylcholine
15. troponin **16.** isotonic **17.** recruitment
18. pacemaker **19.** gap **20.** lockjaw
21. True **22.** False **23.** True **24.** False
25. False

Chapter 7—The Cardiovascular System–Blood

1. g **2.** d **3.** b **4.** j **5.** a **6.** e
7. i **8.** c **9.** h **10.** f **11.** staining
12. fibrinogen **13.** hemophilia **14.** antigens
15. eosinophils **16.** bilirubin **17.** erythroblast
18. stem **19.** spasm **20.** O **21.** False
22. True **23.** False **24.** False **25.** True

Chapter 8—The Cardiovascular System–Heart and Blood Vessels

1. g **2.** c **3.** j **4.** a **5.** b **6.** d
7. i **8.** e **9.** f **10.** h **11.** aorta
12. pulmonary semilunar valve **13.** right atrium
14. right AV valve **15.** right ventricle **16.** left atrium **17.** left AV valve **18.** aortic semilunar valve **19.** left ventricle **20.** septum
21. aorta **22.** pulmonary embolus
23. sphygmomanometer **24.** cardiac output
25. stroke

Chapter 9—The Immune System and Mechanisms of Defense

1. e **2.** j **3.** h **4.** f **5.** c **6.** g
7. a **8.** d **9.** b **10.** i **11.** pathogen
12. interferons **13.** antigen-presenting cells (APCs) **14.** thymus gland **15.** antibiotics
16. tonsils and adenoids **17.** allergen
18. antibody **19.** autoimmune
20. monoclonal **21.** True **22.** True
23. False **24.** False **25.** True

Chapter 10—The Respiratory System: Exchange of Gases

1. g **2.** d **3.** i **4.** b **5.** a **6.** e
7. c **8.** f **9.** j **10.** h **11.** epiglottis
12. trachea **13.** pleural **14.** 21%
15. oxyhemoglobin **16.** carbaminohemoglobin
17. respiratory center **18.** diaphragm
19. carotid; aortic **20.** pharynx **21.** False
22. True **23.** True **24.** True **25.** False

Chapter 11—The Nervous System–Integration and Control

1. f **2.** c **3.** a **4.** i **5.** h **6.** j
7. b **8.** d **9.** e **10.** g **11.** dendrites
12. axon terminals **13.** neuroglial
14. refractory period **15.** synapse

16. Central **17.** sympathetic; parasympathetic **18.** frontal **19.** withdrawal **20.** Rabies **21.** False **22.** False **23.** True **24.** True **25.** True

Chapter 12—Sensory Mechanisms

1. g **2.** c **3.** j **4.** a **5.** d **6.** b **7.** e **8.** i **9.** f **10.** h **11.** adaptation **12.** core **13.** chemoreceptors **14.** sweet, sour, salty, bitter **15.** olfactory **16.** tympanic membrane **17.** rods, lenses **18.** lens **19.** astigmatism **20.** cataracts **21.** True **22.** False **23.** True **24.** False **25.** True

Chapter 13—The Endocrine System

1. e **2.** h **3.** a **4.** b **5.** j **6.** c **7.** d **8.** f **9.** i **10.** g **11.** steroid **12.** nonsteroid **13.** ADH; oxytocin **14.** neuroendocrine **15.** FSH; LH **16.** iodine **17.** aldosterone **18.** calcium **19.** adrenal **20.** epinephrine **21.** False **22.** True **23.** True **24.** True **25.** True

Chapter 14—The Digestive System and Nutrition

1. e **2.** g **3.** a **4.** j **5.** b **6.** c **7.** f **8.** d **9.** i **10.** h **11.** liver **12.** mucosa **13.** canines **14.** heartburn **15.** mucus **16.** constipation **17.** carbohydrates **18.** minerals **19.** intake **20.** bulemia **21.** False **22.** True **23.** True **24.** False **25.** True

Chapter 15—The Urinary System

1. g **2.** e **3.** j **4.** c **5.** h **6.** f **7.** b **8.** d **9.** a **10.** i **11.** erythropoietin **12.** hypothalamus **13.** renin **14.** urea **15.** pelvis **16.** proximal **17.** excretion **18.** acute renal failure **19.** sodium **20.** water **21.** False **22.** True **23.** False **24.** True **25.** True

Chapter 16—Reproductive Systems

1. f **2.** b **3.** j **4.** g **5.** c **6.** a **7.** e **8.** h **9.** d **10.** i **11.** epididymis **12.** prostate **13.** fructose **14.** endometrium **15.** corpus luteum **16.** oviduct **17.** clitoris **18.** FSH **19.** miscarriage **20.** rhythm **21.** True **22.** False **23.** True **24.** True **25.** False

Chapter 17—Cell Reproduction and Differentiation

1. b **2.** g **3.** e **4.** h **5.** c **6.** a **7.** d **8.** f **9.** interphase **10.** gene **11.** mutation **12.** transfer RNA ($_t$RNA) **13.** mitosis **14.** cytokinesis **15.** homologous **16.** False **17.** True **18.** True **19.** False **20.** True

Chapter 18—Cancer

1. b **2.** i **3.** d **4.** j **5.** e **6.** g **7.** a **8.** h **9.** c **10.** f **11.** melanoma **12.** surgerg, radiation, chemotherapy **13.** lung **14.** carcinogen **15.** Prostate **16.** leukemia **17.** tumor **18.** proto-oncogenes **19.** immunotherapy **20.** smoking **21.** True **22.** False **23.** False **24.** True **25.** True

Chapter 19—Genetics and Inheritance

1. g **2.** c **3.** e **4.** a **5.** b **6.** f **7.** d **8.** i **9.** j **10.** h **11.** deletion **12.** translocation **13.** polymerase chain reaction **14.** homozygous **15.** heterozygous **16.** phenotype **17.** genotype **18.** dominant **19.** incomplete **20.** linkage group **21.** False **22.** False **23.** True **24.** False **25.** False

Chapter 20—Development and Aging

1. f **2.** a **3.** j **4.** b **5.** i **6.** d **7.** c **8.** e **9.** h **10.** g **11.** umbilical cord **12.** death **13.** identical **14.** fraternal **15.** ectopic (tubal) **16.** pituitary **17.** colostrum **18.** fifth **19.** aging **20.** puberty **21.** False **22.** False **23.** True **24.** False **25.** True

Chapter 21—Evolution and the Origins of Life

1. f **2.** a **3.** g **4.** b **5.** d **6.** e **7.** c **8.** j **9.** i **10.** h **11.** homologous **12.** fossils **13.** biogeography **14.** population **15.** mass extinction **16.** RNA **17.** evolution **18.** mammals **19.** *Homo sapiens* **20.** vestigial **21.** False **22.** True **23.** True **24.** False **25.** False

Chapter 22—Ecosystems and Human Populations

1. i **2.** a **3.** j **4.** b **5.** c **6.** d **7.** h **8.** g **9.** f **10.** e **11.** habitat **12.** niche **13.** producers **14.** consumers **15.** growth rage **16.** exchange pool; biomass; reservoir **17.** pollutants **18.** carbon dioxide; methane **19.** ozone **20.** acid rain **21.** False **22.** True **23.** False **24.** False **25.** True

ANSWERS to **Try It Yourself—The Rule of 72:**
1) 36 years **2)** 3 hours **3)** 18 and 10 years

CREDITS

Chapter 1

Opener: SuperStock
1.1a: NASA / Liason Agency
1.1b: Frank Siteman / PhotoEdit
1.1c: Norbert Wu / Stone Images
1.1d: A. Gunter / Peter Arnold, Inc.
1.2a Oliver Meckes / Photo Researchers, Inc.
1.2b: Juergen Berger, Max-Planck Institute / Science Photo Library / Custom Medical Stock Photo
1.3 left: T. Dickenson / The Image Works
1.3 middle: Terry Beveridge / BPS / Stone Images
1.3 right: SuperStock
1.4–1.5: Imagineering
1.6: William B. Folsom 1994
1.7–1.11: Imagineering
1.12a: G. Ryann & S. Beyer / Stone Images
1.12b: Gerald Davis / Phototake
1.12c: NASA / Science Source / Custom Medical Stock Photo
Current Issue: Maggie Leonard / Rainbow
1.13: Michael Rosenfeld / Stone Images

Chapter 2

Opener: (An artist's depiction of the structure of an atom) Mike Agliolo / Photo Researchers, Inc.
2.1a: Custom Medical Stock Photo
2.1b: Richard Megna / Fundamental Photographs
2.2: Imagineering
2.3a–b: AllSport Photography, USA
2.4–2.8: Imagineering
2.9: Michael Newman / PhotoEdit
2.10: Imagineering
2.11a: Vaughan Fleming / Science Photo Library / Custom Medical Stock Photo
2.11b: Rudi Van Briel / PhotoEdit
2.11–2.15: Imagineering
2.15c: Don Fawcett / Visuals Unlimited
2.15: Imagineering
2.16: Adapted from Starr / McMillan, *Human Biology* 3/e, Wadsworth Publishing Company, figure 1.20, page 28.
2.17: Adapted from Starr / McMillan, *Human Biology* 3/e, Wadsworth Publishing Company, figure 1.21, page 29.
2.18–2.25: Imagineering

Chapter 3

Opener: (False-color SEM of a lymphocyte, a type of blood cell.) Quest / Science Photo Library / Custom Medical Stock Photo
3.1a: David M. Phillips / Photo Researchers, Inc.
3.1b: Dr. Dennis Kunkel / Phototake
3.1c: Ed Reschke, Peter Arnold, Inc.
3.2 left: M. Abbey / Photo Researchers, Inc
3.2 middle: Biophoto Associates / Photo Researchers, Inc.
3.2 right: Oliver Meckes / Photo Researchers, Inc.
3.2–3.26: Imagineering
3.5 top: Scott Foresman
3.9c: M. Schliwa / Visuals Unlimited
3.12b left: Stanley Flagler / Visuals Unlimited
3.12b middle: David M. Phillips / Visuals Unlimited
3.12b right: David M. Phillips / Visuals Unlimited
3.14: Don W. Fawcett / Visuals Unlimited
3.16: Biophoto Associates / Photo Researchers, Inc.
3.18b: Bill Loncore / Photo Researchers, Inc.

Chapter 4

Opener: (False-color SEM of fat cells) Quest / Science Photo Library / Custom Medical Stock Photo
4.1: Steve Oh and Myriam Kirkman-Oh, artists; adapted from Benjamin / Garman / Funston, *Human Biology,* McGraw-Hill, 1997, page 82.
4.2–4.2: Imagineering
4.3: Adapted from Marieb, *Human Anatomy and Physiology*, 5/e, Addison Wesley Longman, 2000, figure 4.7, page 126.
4.4a: Ed Reschke, Peter Arnold, Inc.
4.4b: Ed Reschke, Peter Arnold, Inc.
4.5a: Ed Reschke, Peter Arnold, Inc.
4.5b: Benjamin / Cummings Publishing Company. Photo by Allen Bell, University of New England
4.6a: Eric Graves / Photo Researchers, Inc.
4.6b: Ed Reschke, Peter Arnold, Inc.
4.6c: Benjamin / Cummings Publishing Company. Photo by Allen Bell, University of New England
4.5–4.7: Adapted from Marieb, *Human Anatomy and Physiology*, 5/e, Addison Wesley Longman, 2000.
Current Issue: Michael Tighe / La Moine Agency
4.7: Ed Reschke, Peter Arnold, Inc.
4.8: Steve Oh and Myriam Kirkman-Oh
Un. spread: Steve Oh and Myriam Kirkman-Oh
4.9–4.13: Imagineering
Health Watch left: Sondra Davis / The Image Works
Health Watch right: Dr. P. Marazzi / Science Photo Library / Custom Medical Stock Photo
Case in Point: CNRI / Phototake

Chapter 5

Opener: (A colorized x-ray of a human wrist) Science Visuals Unlimited
5.1c: Biophoto Associates / Science Source / Photo Researchers, Inc.
5.1–5.2: Imagineering
5.3: Adapted from Benjamin / Garman / Funston, *Human Biology,* McGraw-Hill, 1997, page 139.
5.4: Imagineering
Health Watch, left & middle: Dr. D.W. Dempster / Journal of Bone and Mineral Research, Volume 1, November 1, 1986, Mary Ann Liebert
right: Larry Mulvehill / Science Source / Photo Researchers, Inc.
5.5a–c: Dr. David Hubbard, Department of Orthopedics, West Virginia School of Medicine
5.6–5.8: Imagineering
5.9: Adapted from Benjamin, Garman, Funston, *Human Biology,* McGraw-Hill, page 143.
5.10–5.14: Imagineering

Chapter 6

Opener: Ron Chapple / FPG International
6.1: Imagineering
6.2: Steve Oh and Myriam Kirkman-Oh
6.3–6.10: Imagineering
6.4a: R. Calentine / Visuals Unlimited
6.5c: Biophoto Associates / Science Source / Photo Researchers, Inc.
6.5d: K.G Murtil / Visuals Unlimited
6.9b: Fred Hossler / Visuals Unlimited
6.11a: Vandystadt Angence De Presse / Photo Researchers, Inc.
6.11b: B. Daemmrich / The Image Works
Current Issue, left: Dieter Endicher / Associated Press / AP
Current Issue, right: Imagineering
6.12–6.13: Imagineering

Chapter 7

Opener: (False-color SEM of several red blood cells) Professor P. Motta / Department of Anatomy / University "La Sapienza", Rome / Science Photo Library / Custom Medical Stock Photo
7.1–7.2: Imagineering
7.3: Dr. Dennis Kunkel / CNRI / Phototake
7.4–7.7: Imagineering
Health Watch: David M. Grossman / Phototake
7.8: Volker Steger / Peter Arnold, Inc.
7.9–7.10: Imagineering
7.11a: David M. Phillips / Visuals Unlimited
7.11b: Martin Rotker / Phototake
7.12: Imagineering

Chapter 8

Opener: (False-color angiogram of blood vessels in the heart) CNRI / Science Photo Library / Custom Medical Stock Photo
8.1: Adapted from Benjamin, Garman, Funston, *Human Biology,* McGraw-Hill, page 241.
8.2: Imagineering
8.3a: Courtesy of Mathew A. Boegehold, Ph.D., West Virginia University Department of Physiology
8.3b: Lennart Nilsson
8.4–8.9: Imagineering
8.7: Visuals Unlimited
8.10: Steve Oh and Myriam Kirkman-Oh
8.11–8.12: Imagineering
8.13a: Jean Claude Revy / Phototake
8.13b: Visuals Unlimited
Health Watch a & b: Science Photo Library / Custom Medical Stock Photo
8.14–8.18: Imagineering
8.19: ISM / Phototake
8.20: Adapted from Benjamin, Garman, Funston, *Human Biology,* McGraw-Hill, page 259.
8.21: Robert E. Daemmrich / Stone Images
Chapter Summary Art: Imagineering

Chapter 9

Opener: (False-color SEM of a macrophage ensnaring bacteria) Boehringer Ingelheim International, GMBH, Lennart Nilsson
9.1a: S. Lowry / University of Ulster / Stone Images
9.1b: Dr. Dennis Kunkel / Phototake
9.1c: S. Lowry / University of Ulster / Stone Images
9.2: Imagineering
9.3–9.4: Steve Oh and Myriam Kirkman-Oh
9.5: Matt Meadows / Peter Arnold, Inc.
9.6–9.17: Imagineering
9.8 lower left: Lennart Nilsson / "The Body Victorious", Delacorte Publishing
9.8 lower right: Lennart Nilsson / "The Body Victorious", Delacorte Publishing
9.14a: Jean Claude Revy / Phototake
9.18 left: Visuals Unlimited
9.18 right: CNRI / Science Photo Library / Custom Medical Stock Photo
9.19–9.23: Imagineering
Current Issue: Eric Bouvet / Matrix

Chapter 10

Opener: (False-color chest x-ray of a normal 11-year-old boy) Science Photo Library / Custom Medical Stock Photo
10.1: Steve Oh and Myriam Kirkman-Oh
10.2–10.5: Imagineering
10.4a: CNRI / Phototake
10.4b: CNRI / Phototake
Health Watch: Imagineering
10.6a: CNRI / Science Photo Library / Custom Medical Stock Photo
10.6b: Dr. Andrew Evan, Indiana University Medical Center, Department of Anatomy
10.7–10.13: Imagineering
10.8b: R. Kessel / Visuals Unlimited
10.10b: 1994 L. Steinmark / Custom Medical Stock Photo
10.14a: John D. Cunningham / Visuals Unlimited
10.14b: John Burbridge / Science Photo Library / Custom Medical Stock Photo

Chapter 11

Opener: (Positron emission tomography (PET) scans of human brains) Dr. Monty Buchsbaum / Peter Arnold, Inc.
11.1: Steve Oh and Myriam Kirkman-Oh
11.2–11.9: Imagineering
11.3b: C. Raines / Visuals Unlimited
11.7 top: E. Lewis / T. Everhart / Y. Zaevi / Visuals Unlimited
11.7 bottom: CRNI / Science Photo Library / Custom Medical Stock Photo
11.10: Jim Selby / Science Photo Library / Custom Medical Stock Photo
11.11: Steve Oh and Myrima Kirkman-Oh; adapted from Benjamin, Garman, Funston, *Human Biology,* McGraw-Hill, page 376.

11.12–11.13: Imagineering
11.13c: Manfred Kage / Peter Arnold, Inc.
11.14: Martin Rotker / Phototake
11.15–11.18: Imagineering
Directions in Science left: Jeff Hunter / The Image Bank
Directions in Science right: Damasio H, Grabowski T, Frank R, Galaburba AM, Damasio AR: "The Return of Phineas Gage: Clues about the Brain from a Famous Patient". *Science,* 264: 11:02-1105, 1994. Department of Neurology and Image Analysis Facility, University of Iowa,
Health Watch: Sygma / Brooks Kraft / Sygma
11.19a&b: Dr. Mony de Leon / Peter Arnold, Inc.

Chapter 12
Opener: (The retina of an eye) Howard Sochurek / The Stock Market
12.1–12.3: Imagineering
12.4: Steve Oh and Myriam Kirkman-Oh
12.5a: Imagineering
12.5b: Omikron / Science / Photo Researchers, Inc.
12.5c: Astrid & Hanns-Frieder Michler / Science Photo Library / Custom Medical Stock Photo
12.6: Imagineering
12.7: Jenny Thomas Photography
12.8–12.17: Imagineering
Health Watch a&b: J.E. Hawkins, Jr. Ph.D./ Krsge Hearing Research Institute, Department of Otorhinolaryngology
12.11c: Richard J. Mount, Auditory Science Library, The Hospital for Sick Children, Toronto, Canada
12.12c: Lennart Nilsson "Behold Man"
12.17b: Ralph C. Eagle / Photo Researchers, Inc.
12.18: Reproduced from Ishihara's *Tests for Colour Blindness* published by Kanehara and Company, Ltd., Tokyo, Japan.

Chapter 13
Opener: (False-color SEM of a section through a hormone-secreting cell) Science Photo Library / Custom Medical Stock Photo
13.1: Steve Oh and Myriam Kirkman-Oh
13.2–13.4: Imagineering
13.5: Steve Oh and Myriam Kirkman-Oh
13.6–13.7: Adapted from Benjamin, Garman, Funston, *Human Biology,* McGraw-Hill, page 426.
Directions in Science: Gentech
13.8: Science Photo Library / Custom Medical Stock Photo
13.9: Courtesy of Dr. William H. Daughaday, Washington University School of Medicine, from A. Mendelhoff and D.E. Smith, eds. American Journal of Medicine 20:133, 1956
13.10–13.14: Imagineering
13.15: Bob Daemmrich / Stock Boston
13.16: Imagineering
13.17: National Medical Slide / Custom Medical Stock Photo

Chapter 14
Opener: (False-color SEM of the surface of the stomach) Professor P. M. Motta, K. R. Porter, and P. M. Andrews / Science Photo Library / Custom Medical Stock Photo
14.1: Steve Oh and Myriam Kirkman-Oh
14.2–14.17: Imagineering
14.4c: David Scarf / Peter Arnold, Inc.
14.7c: Fred Hossler / Visuals Unlimited
14.18a: Rapho Agency / Photo Researchers, Inc.
14.18b: A. Ramel / PhotoEdit

Chapter 15
Opener: (False-color SEM of cells in the glomular capsule which surround the glomular capillaries in the kidney) Manfred Kage / Peter Arnold, Inc.
15.1: Precision Graphics Imagineering
15.7: Dr. William Kriz, Ruprecgt-Karls-UniversitatHeidelberg, Institut fur Anatomie and Zellbiologie
15.8 left: Fred Hossler / Visuals Unlimited
15.8–15.15: Imagineering
Current Issue: Imagineering
15.16: Hank Morgan / Science Source / Photo Researchers, Inc.

Chapter 16
Opener: (False-color SEM of a human egg surrounded by sperm) Yorgos Nikas / Stone Images
16.1–16.8: Imagineering
16.9: David Phillips / Science Source / Photo Researchers, Inc.
16.10–16.12: Imagineering
16.11a: Mauritius / GMBH / Phototake
16.11b: Scott Camazine / Photo Researchers, Inc.
16.11c: SIU / Photo Researchers, Inc.
16.11d: J. Griffin / The Image Works
16.11e: M. Long / Visuals Unlimited
16.13: Richard Rawlins, Ph.D. / Custom Medical Stock Photo
16.14: Precision Graphics
16.15: E. Gray / Science Photo Library / Custom Medical Stock Photo

Chapter 17
Opener: (False-color TEM of a cell in anaphase of mitosis) 1992 SPL / Custom Medical Stock Photo
17.1–17.2: Imagineering
17.2 bottom: Biophoto Associates / Photo Researchers, Inc.
17.3: Precision Graphics
17.4–17.6: Imagineering
Directions in Science left: Precision Graphics
Directions in Science right: Courtesy of Cellmark Diagnostics
17.7: Precision Graphics
17.8–17.9: Imagineering
17.9: Michael Abbey / Photo Researchers, Inc.
17.10: David M. Phillips / Visuals Unlimited
17.11–17.12: Imagineering
17.13: Precision Graphics
17.14: Imagineering
Current Issue: Imagineering
Health Watch: Paul Fievez / Hulton Getty Picture Library / Liason Agency

Chapter 18
Opener: (False-color chest x-ray showing a cancerous tumor) Simon Fraser / Science Photo Library / Custom Medical Stock Photo
18.1–18.2: Imagineering
18.3: Myrleen Cate / PhotoEdit
18.4–18.5: Precision Graphics
18.6: Chris Bjornberg / Photo Researchers, Inc.
18.7: Simon Fraser / Science Photo Library / Newcastle-upon-Tyne / Photo Researchers, Inc.
18.8a: Ken Greer / Visuals Unlimited
18.8b: 1995 Caliendo / Custom Medical Stock Photo
18.8c: Ken Greer / Visuals Unlimited
Health Watch: Precision Graphics

Chapter 19
Opener: (Color-enhanced SEM of a set of male human chromosomes) Adrian T. Sumner / Stone Images
19.1 top left: Joseph Nettis / Stock Boston
19.1 top right: Esbin Anderson / The Image Works
19.1 bottom left: Skjold / The Image Works
19.1 bottom right: Bill Backmann / The Image Works
19.2: Imagineering
19.3: Precision Graphics
19.3 left: W. Bokelberg / The Image Bank
19.3 right: Roy Morsch / The Stock Market
19.4 left: John Turner / Stone Images
19.4 right: Tony Craddock / Science Photo Library / Custom Medical Stock Photo
19.5: Precision Graphics
19.5 White: International American Albino Association, Inc.
19.5 Chestnut: Larry Lefever / Grant Heilman Photography
19.5 Palamino: Fritz Prenzel / Animals Animals / Earth Scenes
19.6: Meckes / Ottawa / Photo Researchers, Inc.
Health Watch: Simon Fraser / RVI, Newcastle-upon-Tyne/Science Photo Library / Custom Medical Stock Photo
19.7 left–right: George Hausman / The Stock Market
Peter Saloutos / The Stock Market
Jon Feingersh / The Stock Market
David A. Wagner / Phototake
George Hausman / The Stock Market
Steve Dunwell / The Image Bank,
19.8: Precision Graphics
19.9: Imagineering
19.10: CNRI / Phototake
19.11: Imagineering
19.12: Precision Graphics
Directions in Science: Imagineering
19.13: Imagineering
19.14: Richard Hutching / Photo Researchers, Inc.
19.15a–b: Courtesy of Dr. Nadler / Children's Memorial Hospital, Chicago, Illinois
19.16–19.17: Precision Graphics
19.17: Adapted from Starr, *Human Biology* 2/e, Wadsworth Publishing Company, page 446.
19.18: Courtesy Monsanto Company
19.19: Courtesy of Genzyme Transgenics
Current Issue: Imagineering

Chapter 20
Opener: (A human fetus) Lennart Nilsson "A Child is Born"
20.1–20.3: Imagineering
20.4a: Anthony B. Wood / Stock Boston
20.4b: Roy Morsch / The Stock Market
20.4c: Bettman / Corbis
20.5–20.9: Imagineering
20.10–20.12: Lennart Nilsson, "A Child is Born"
Directions in Science: Imagineering
20.13: Adapted from Mader, *Human Biology* 5/e, McGraw-Hill, page 370.
20.14–20.14: Imagineering
20.15: Precision Graphics
20.16 left–right: Bettmann / Corbis
20.17: Imagineering
20.18: Ariel Skelley / The Stock Market

Chapter 21
Opener: (The oldest known skull of a human in Europe, thought to be over 400,000 years old) Liason Agency
21.1: Adapted from Benjamin, Garman, Funston, *Human Biology,* McGraw-Hill, page 534.
21.1e: Kenneth Garrett
21.2 left: Harvey Lloyd / The Stock Market
21.2 right: Tom Bean Photography
21.3–21.5: Imagineering
21.6: Geoffrey Clifford / Woodfin Camp & Associates
21.7: Imagineering
21.8: Hulton Getty Picture Collection / Archive Photos
21.9–21.9: Imagineering
21.10: Adapted from Mader, *Human Biology* 5/e, McGraw-Hill, page 461.
21.11–21.12: Imagineering
21.13: Tom Brakeield / The Image Works
21.14: Imagineering
21.15: John Reader / Science Photo Library / Custom Medical Stock Photo
21.16: Imagineering
21.17: Robert Mort / Stone Images

Chapter 22
Opener: (Deforested hillsides of a Costa Rican rain forest, showing erosion of topsoil) John Mitchell / The Image Works, Inc.
22.1 top: R. Andrew Odum / Peter Arnold, Inc.
22.1 middle: Thomas Kitchin / Tom Stack & Associates
22.1 bottom: B. Daemmrich / The Image Works
22.2–22.3: Imagineering
22.4: Dr. Harvey Barnett / Peter Arnold, Inc.
22.5: Len Rue / Stock Boston
22.6–22.21: Imagineering
Directions in Science: Tom Stack
22.22: Tom & Theresa Stack
22.23a: Bettmann / Corbis
22.23b: Mark E. Gibson / Visuals Unlimited
22.24: David Hoffman / Stone Images
22.25: Jim Olive / Peter Arnold, Inc.
22.26: R. Arndt / Visuals Unlimited
Current Issue: Charles O. Cecil / Visuals Unlimited
22.27: BIOS / Peter Arnold

INDEX

NOTE: A *t* following a page number indicates tabular material, an *f* indicates a figure, a *b* indicates a boxed or highlighted feature, and a *c* indicates a "Case in Point" feature.

Dear Student:

My goal for writing ***Human Biology: Concepts and Current Issues*** was to not only help you succeed in the course, but also to make the learning process exciting and rewarding. I would appreciate hearing about your experiences with this textbook and its multimedia, and I invite your suggestions for improvements.

Sincerely,

Michael D. Johnson

Michael D. Johnson

1. Did you use the **Interactive Physiology for *Human Biology*** CD-ROM? ☐ Yes ☐ No

If so, which features were most useful to you? Which module(s) or animation(s) did you find most helpful? ____________

____

____

____

Did you use the quizzes after each module? ☐ Yes ☐ No

Did you use the CD-ROM on your own, or did your instructor require you to use it? ____________

____

2. Did you visit **The Human Biology Place**? ☐ Yes ☐ No

If so, approximately how frequently did you visit the web site? ____________

Which area(s) of the web site did you find most useful (e.g., Concept Review, Test Yourself, Issues Forum, etc.)?

____

____

____

Did you visit the web site on your own, or did your instructor require it? ____________

____

How can we improve **The Human Biology Place**? What type of activities do you think are most valuable? What additional topics should we add? ____________

____

____

____

3. Which features of the text helped you to learn the material? Please choose only **three** from the following checklist.

☐ Text Headings ☐ Recaps ☐ Spotlights on Structure and Function ☐ Chapter Summaries

☐ Test Yourself questions ☐ Apply What You Know Questions ☐ Case in Point stories

☐ Health Watch ☐ Directions in Science ☐ Try It Yourself ☐ Current Issues

Did you read these on your own, or did your instructor assign them? ____________

____

4. What did you like most about ***Human Biology: Concepts and Current Issues***? Please provide three examples, in order of priority. ____________________

____

____

____

5. Do you have any suggestions for improving this text? Please list them, citing page numbers as appropriate. ____________________

____

____

____

BENJAMIN CUMMINGS
PEARSON EDUCATION
1301 SANSOME STREET
SAN FRANCISCO CA 94111-9328

School: ____________________

Instructor's Name: ____________________

Optional:

Your name: ____________________

Email: ____________________

Date: ____________________

May Benjamin Cummings have permission to quote your comments in promotions for ***Human Biology: Concepts and Current Issues***? ☐ Yes ☐ No